VALIDATION COMPLIANCE ANNUAL

1995

VALIDATION COMPLIANCE ANNUAL

1995

INTERNATIONAL VALIDATION FORUM, INC.

Key West, Florida

CRC Press is an imprint of the
Taylor & Francis Group, an **informa** business

CRC Press
Taylor & Francis Group
6000 Broken Sound Parkway NW, Suite 300
Boca Raton, FL 33487-2742

ISBN-13: 9780824794590 (hbk)

Visit the Taylor & Francis Web site at
http://www.taylorandfrancis.com

and the CRC Press Web site at
http://www.crcpress.com

PREFACE

Computer system validation issues are not the most important problems facing the pharmaceutical, medical device, biologics, food, and cosmetic industries. Consumers are more concerned with the accuracy of blood tests, the cleanliness of food products, the purity and effectiveness of drugs, the functionality of medical devices, and the stability of biologics than with the computer systems that underlie the manufacturing, testing, and analyzing of these products.

Scientists, however, are well aware of the critical importance of computers conducting those blood tests, controlling that food inventory, researching those pharmaceutical properties, manufacturing those medical devices, tracking those biologics, and analyzing data in environmental laboratories. The industry awareness of the importance and vulnerability of those systems has been created largely by effective emphasis that agencies like the Food and Drug Administration and the Environmental Protection Agency are placing on the validation of those systems.

Why should validation be on the regulatory agenda? Two important facts can re-emphasize what those in the regulated industries already know. Fact one: In the past five years, we have completed in-depth studies of more than 500 regulated computer systems. We have yet to see any system that has met all appropriate standards for control, documentation, and performance prior to the initial audit. Our experience suggests that validation does not merely provide assurances; it concretely and demonstrably improves the quality of data and of the processes which systems control.

Fact two: The dependence of the industries upon computers is overwhelming. We all find computers dominating the laboratory testing facility, the Quality Assurance laboratory, the manufacturing floor, the inventory control process, the clinical design and study process, and the statistical analysis of data. In fact, every aspect of researching, testing, manufacturing, and distribution is computer-dominated. The managers who are responsible for the safe and effective control of those processes are more and more computer-dependent.

There is a side issue. In the last two years, a new issue has captured our national health agenda -- cost effectiveness. Every manufacturer is forced now to focus on lowering the costs of discovering, perfecting, and delivering products. Their other goals have not changed, however. Manufacturers must still focus on the quality and safety of those products.

Are these goals conflicting? Does quality decline when cost effectiveness is maximized? Probably not, as long as controls are maintained. With proper quality controls built in, cost effectiveness can be maximized through high quality and safe products.

In the modern company, that control is a function of computerized systems -- watching inventory, improving testing, maximizing manufacturing efficiency, and tracking distribution. Systems that are proven to be effective assure safety and quality in a cost-effective world.

This brings us to the reason for this volume. The *Validation Compliance Annual* is intended to give those individuals involved in computer system validation a compendium of the regulations, guidance documents, issues, compliance tools, terminology, and literature involved in computer system validation. We intend to update this volume annually. The field of system validation is still relatively young and rapidly growing. Regulations change worldwide, new procedures are devised, and new issues emerge regularly, and we want to keep up with this changing world.

The preparation of this book required the assistance of many individuals. In particular, Dr. Sandy Weinberg, Janie Spelton, and Philip Sax inaugurated the project. Additional assistance and contributions were received from Les Banks, Stan Schulman, Debbie Cenci, and especially John English. Special thanks also go to Christine Stoner, without whom many of the documents in this volume would not be here.

Finally, our grateful appreciation to Dr. Gary C. Stein, Director of Regulatory Affairs for Weinberg, Spelton & Sax, Inc. Although he inherited this project late in its evolution, his skills in coordinating, compiling, writing, re-writing and editing brought it to successful fruition.

This volume is a tool to be used for those who are both familiar and unfamiliar with the changing world of system validation. Like any other tool, it is useful only if you use it.

CONTENTS

VALIDATION COMPLIANCE ANNUAL

1995

Chapter 1

BEHIND VALIDATION

REGULATORY CONCERNS

Computers affect almost every aspect of our daily lives. It is difficult to imagine any modern industry that is not automated to some extent. This holds true for today's pharmaceutical, medical device, and biologics industries and environmental laboratories. The proliferation of computerized systems used in these industries -- to manufacture drugs and medical devices (and as components of medical devices), or to process blood for transfusion, or to analyze laboratory data -- has led to greater regulatory scrutiny of these systems. Investigators from the U.S. Food and Drug Administration (FDA) and the U.S. Environmental Protection Agency (EPA) are finding more and more computer-attributable errors resulting from erroneous programming, hardware flaws, calibration errors, or electronically corrupted data.

The findings of these investigators have led to a great deal of skepticism surrounding the accuracy of computer operations, and those agencies that regulate some of these industries -- the FDA for medical devices, pharmaceuticals, and biologics and the EPA for environmental laboratories -- continually demand more and more proof that a wide range of controls are present in any computerized operation.

This skepticism which underlies the demand for proof of systems control is not alien to either the scientist or the regulatory professional, yet it often emerges as a personal affront when representatives from the two camps interact. Perhaps this resentment springs from the history of interaction between the two groups. Pharmaceutical and medical device manufacturers and laboratory scientists have seen regulatory demands grow beyond levels that seem reasonable. Regulators, on the other hand, have seen too many dubious data submissions claiming to be solid evidence.

In the computer automation field, that skepticism may graduate into full-scale cynicism. The technical complexities involved in computerization may exceed the expertise of both scientists and regulators, who have grown increasing uncomfortable with the jargon-filled, non-explanations of computer professionals.

In the earliest days of computer systems, highly inflated estimates of the power, potential, and accuracy of systems created a strong faith in their reliability. "The computer says so" became the rallying cry and defense of billing agents, government clerks, and bureaucrats the world over. But as stories of enormous computer errors -- both humorous and tragic -- flooded the media in later years, a "computer error" became as common a punchline as "the check's in the mail." Computer professionals fell from god-like status to a reputation probably far below the reasonable norm of accurate and reliable system function.

Those computer professionals contribute to the atmosphere of cynicism, too, with their resentments. Their world has never previously had to surrender the shroud of authority for the ego-reducing discipline of double-check and confirmation.

Finally, experience has created the need for evidence supporting computer accuracy. Too many systems have failed in the past, despite all the best promises of controls and safeguards.

The result of this combination of history, reality, and attitude is a general regulatory dismissal of a presumption of system control. The "default situation," the unproven norm expectation, is that a system is not adequately controlled. Until firm evidence of control is actually provided, an automated manufacturing process or laboratory system is considered to be without appropriate controls, and both industry management and the data that is presented to the regulatory agencies are suspect.

Validation defines the proofs of systems control that are necessary to establish regulatory compliance.

IMPORTANCE OF VALIDATION

A computer validation study translates to a directed total quality effort on a computerized system (the computer and the business function it is controlling). Validation guides a manufacturer towards the total quality goal of moving away from defect detection and towards DEFECT PREVENTION. Computer System Validation can be defined as:

> The independent and systematic review of a computer system and its related activities to insure management that the system is operating as they intend and that there is reasonable assurance the system will continue to operate in such a manner.

This definition is taken right from the definition of "validation" in federal regulations. We interpret this definition as the accumulation of evidence -- what we see as a series of proofs -- in support of expert testimony demonstrating present and continuing control of a computer system utilized in manufacturing, research, quality assurance, or in the function of a medical device. These proofs can be described as:

Evidence of Current Accuracy and Reliability

The validation study must present clear and compelling evidence that the system as configured can be considered to provide consistent and faithful processing of entered data and activated programs.

A system user has a responsibility to design or adopt an appropriate function test, to conduct that test periodically, and to use normal scientific procedures of establishing acceptance criteria and performance expectations in advance of the testing. That advance definition of acceptability is generally established as the "needs statement" or "system goal," and is the first documentary step in the effective utilization of a system life cycle approach for system management, the industry standard for control of system development and implementation.

Evidence of Continued Accuracy and Reliability

The validation study must present clear and compelling evidence that the system as configured will continue to provide consistent and faithful processing of entered data and activated programs until such time as an internal auditing or monitoring system provides warning of aberration.

System users or reviewers have a responsibility to look into the software and confirm the appropriateness of the formulae and algorithms used. Not to do so is to defer decision responsibility to an unknown -- and unaccountable -- programmer. The results may be serious: We have seen LIMS products with inaccurate statistical formulae, laboratory equipment with erroneous calibration algorithms, data acquisition systems that misassign data fields, and decision systems that simply make erroneous decisions. Unless the source code is examined, these problems are undetectable, and unless that code review is included, the validation study has simply failed to assume proper responsibility and control of the system.

Evidence of Management Awareness and Control

The validation study must present clear and compelling evidence that the organization and/or system management maintains appropriate control mechanisms to assure proper operations, use, maintenance, and procedural supervision of the system.

This is such a common industry requirement as to be considered self-obvious, but there are environments in which, despite the most rigorous attention to testing and reporting Standard Operating Procedure (SOP) documentation, instructions for using, controlling, and recovering the computer system is casual and unwritten. SOPs for system control are as important as SOPs in any other area, and they are a necessary demonstration of the awareness and control of effective management.

Evidence of Auditability

The validation study must present clear and compelling evidence that the system contains within its infrastructure the constructs, tools, and controls necessary to allow full and complete audit of functions and activities.

A manual system of notebooks and laboratory records provides an audit trail of changes and deletions. Unless a system is carefully constructed, that audit trail may not be maintained in an electronic system. Yet a record of all changes, and the rationale for those changes, is an important control issue. Audit trail evidence assures that control and assures that a data change would not be inadvertently or fraudulently unrecorded.

Evidence of Data Integrity

The validation study must present clear and compelling evidence that the system receives, stores, transmits, operates upon, processes, recalls, sorts, and otherwise manipulates user-provided or -generated data without system imposed inappropriate modification, distortion, addition, or deletion.

Confidence that a system is not inappropriately or spontaneously changing data points is critical to reliance upon that system. Data integrity evidence provides that confidence, demonstrating that a system has not truncated, misassigned, or otherwise modified a key entry.

Evidence of Reviewer Credentials

The validation study must present clear and compelling evidence that the expert judgements upon which its conclusions are based will pass Expert Witness credibility review for technical and theoretical expertise, appropriate experience, and complete independence.

This is a critical issue, as will be discussed in further detail below. It is beyond the scope of regulatory agencies to have investigators examine source code, test systems, or analyze complex protocols. A user must assume that responsibility. The investigator's role is to determine that an effective, independent study has been conducted. Not surprisingly, the majority of all validation 483 observations issued by the FDA last year were issued for failure to provide a validation study or summary certificate. To be useful, that study must be examined as expert testimony, with a review of the testifying person or agency to assure independence and qualification. Well-trained, internal Quality Assurance staff, outside auditors, or a combination of persons may be involved, but their joint credentials demonstrating independence of reporting responsibility, system hardware and software expertise, and knowledge of appropriate applications and regulations, are a critical part of the investigator's review.

MANAGERIAL COMMITMENT

Regulatory pressures to validate computer systems are now clear and widely accepted. The U.S. Food and Drug Administration (FDA), the Bundesgesundheitsamtes (BGA) in Germany, the UK Department of Health, the Canadian Health Protection Bureau, the U.S. Environmental Protection Agency (EPA), and other international bodies have issued unambiguous statements establishing that the development of evidence in support of system accuracy and reliability -- *VALIDATION* -- is a top priority.

But how does one most effectively manage the process of validating systems? Once a commitment has been made to the testing and review of computer control and quality, what is the most effective method of implementing system validation?

Experience with more than four hundred studies in almost three hundred companies has led to the evolution of a successful six-step approach to the development of a validation management process:

1. securing timely ***information*** concerning regulatory intent;
2. developing internal ***guidelines*** or standards;
3. establishing an appropriate validation team ***organizational structure***;
4. determining the appropriate ***priority*** for validation studies;
5. completing the ***follow-through*** efforts validation reviews may determine to be appropriate; and
6. ***auditing and certifying*** the final results.

Information

Validation is an extremely diverse and complex area of regulatory concern, impacting all areas of pharmaceutical, medical device, and biologic research, manufacturing, and clinical testing. In the United States, validation clarification and evolutionary conceptualization is continuously released by more than a dozen different FDA spokespersons and offices and interpreted in twenty-two field offices. Simultaneously, nine different EPA offices are addressing laboratory validation issues. In Europe, two European Union (EU) offices and national agencies in the United Kingdom, Germany, France, the Netherlands, Switzerland, Sweden, Belgium, Spain, Italy, and Ireland have all issued statements on validation policy and requirements. In other parts of the world, regulatory agencies in Canada, Japan, Israel, Australia, Brazil, and India have addressed validation concerns.

This constant flow of important information is critical to the validation process, since it serves to dynamically define the two limiting dimensions of validation activity: How much evidence is required? What elements affect the credibility of that evidence? Without answers to these two questions, validation can be an all but endless process, cost inefficient and hopelessly convoluted. Sorting and digesting the data that provides these limited dimensions is the most significant first step in managing the validation process.

Newsletters can serve an important function in summarizing that informational morass. *The International Pharmaceutical Regulatory Monitor*, and its sister publication, *The International Medical Device Regulatory Monitor*, provide monthly access to the key documents and policy statements that address validation (and other issues). A European publication, *The Regulatory Affairs Journal*, provides similar international information. There are also newsletters published by industry organizations, such as the Drug Information Association (DIA), the Parenteral Drug Association (PDA), and the Regulatory Affairs Professional Society (RAPS). Similarly, a number of books have recently been released to summarize publication date state-of-the-art policies.

Guidelines

Using the information identified in step 1, and updating that information regularly in response to an evolving environment, an organization addressing validation issues must develop a set of clear guidelines for evaluation and control. One source of these guidelines, of course, is the formal regulations themselves, found in the CFR, the Code of Federal Regulations (see chart, below). But the regulations of the FDA and other regulatory groups only indirectly address validation concerns. The situation is not unlike traffic control: the law requires "safe conduct." A transportation agency is empowered to establish clear definitions of safe practices and posts speed limits, warning signs, and other devices. Ultimately, though, it is the field police officers, in deciding the speeds to enforce, the deviation from norms to accept, and the regulations to cite who define through their activities the actual traffic rules.

Rather than relying strictly upon the vague and general principles contained in the CFR, guidelines should be formulated upon the self-regulating question: "What kinds of evidence are appropriate in this environment of application, risk, software, hardware, and product to provide assurances of the continuing control of this system?" Generally, the answer will include a combination of system testing (in both normal and stress modes), source code review or analysis, documentation review (including the development of appropriate user Standard Operating Procedures), and an audit of the data or results produced by the system.

For the greatest efficiency in developing appropriate guidelines, many organizations begin with an established set of standards, such as the System Validation Standards (Weinberg, 1990), the EPA's Good Automated Laboratory Practices Implementation Guide, or System Validation Automated Checklist (Weinberg, 1993). By adapting one or more of these documents to the operating conditions and risk levels of a specific environment, a clear set of widely-accepted standards can be very rapidly and cost effectively developed.

FDA AND EPA REGULATIONS

Requirement	Register	GMP	Validation
Human Medical Devices	21 CFR 807	21 CFR 820	21 CFR 820 812
Human Pharmaceuticals	21 CFR 207/314	21 CFR 210/211	21 CFR 211/312
Animal Pharmaceuticals	21 CFR 207/514	21 CFR 210/211	21 CFR 211/511
Human Biologics	21 CFR 601	21 CFR 606/640	21 CFR 606/312
Animal Biologics	21 CFR 102	21 CFR 108	21 CFR 108/103
Human In-Vitro Diagnostics	21 CFR 807	21 CFR 820	21 CFR 820/812
Animal In-Vitro Diagnostics	9 CFR 102	9 CFR 108	9 CFR 103/108
Pesticides (EPA)	40 CFR 167	NONE	40 CFR 160
Toxic Substances Control (EPA)	40 CFR 720	NONE	40 CFR 792

Those standards, developed "from scratch" or through adoption and adaptation, serve as the basis for validation studies throughout the organization.

Organization

The organization of validation teams is complicated by a single factor: A level of independence, similar to the Quality Assurance role in manufacturing, is fundamental to the credibility of a validation study. Vendors should not validate their own products; users should not be providing assurances that they are developing and following appropriate Standard Operating Procedures. The unconfirmed statement by a programmer that proper design and testing procedures have been followed does not meet validation criteria. The audit of such claims is critical to the credibility of the validation process.

Most organizations have evolved a team model for validation. Representatives from the user department, along with representatives from Information Systems, provide the important background, application, and technical expertise for a validation study. But the team is not complete, and its conclusions not credible, until an independent element is added.

In some organizations, the team is completed or enhanced with the addition of an outside consultant who can bring experience, a methodology, regulatory information, valuable review tools, and highly credible credentials (including a Certified Validation Professional or equivalent credential review). In other organizations, a representative of the Quality Assurance department brings this level of independence, experience, and expertise to the validation team. In either variation, the joint team model seems an effective approach to managing the validation process.

Priority

A team has been formed; it has been equipped with a continuing source of up-to-date information of regulatory guidelines and requirements; it carries with it a set of validation standards or a Standard Operating Procedure on Validation that coincides with those regulatory positions. One step remains before the actual testing, review, and analysis that constitutes the validation process.

A major pharmaceutical company may have one hundred or more systems that require validation: Laboratory Management Information Systems, computerized manufacturing systems, computerized laboratory and manufacturing equipment, labeling systems, inventory and MRP systems, and so forth. To effectively manage validation it is necessary to establish some sort of logical system for prioritizing these projects.

Having a logical priority list not only tells the team where to begin and how to proceed. It also shows regulators that, while not all systems may yet have been completely validated, the organization is in clear control of the process. With a good prioritization model, it is possible to tell a regulator exactly when (to within a month) every system in house will be validated; it will, as well, demonstrate the progress already made on the most critical or vulnerable systems.

Follow Through

While actually conducting the validation study is, of course, the most important part of the process, only a few managerial suggestions will be offered here.

First, focus is a key issue. Organizing and conducting a validation study requires a concentrated, uninterrupted period of time. Tests can not be halted in mid data set. If weeks pass between one test and the next, a true picture of the system at a single point in time is not provided. A review of the source code should be concentrated to avoid system down time. If a clear priority focus is not a practicality for the validation team, consideration should be given to the cost effective use of outside validation specialists.

Second, do not waste time in "reinventing the wheel." Have the team use pre-developed validation tools, guidelines, and procedures wherever possible. These templates will need modification and customization, of course, but they can provide an excellent and efficient starting point.

Finally, maintain careful records. Documentation is the key to any regulatory process, including validation. Signed and dated notes can document code walk-through analyses. Test result printouts will prove useful evidence. SOPs should be accompanied by inspection reports documenting confirmatory observations. As far as possible, document everything.

Auditing and Certifying

Certification, or confirmatory auditing, is not a required step in system validation under the regulatory requirements of most agencies. It may, however, be a cost effective method of quality control and "insurance" in high risk situations.

In the certifying process, a recognized and credible outside party reviews the validation study and audits the conclusions. In confirming the findings, a certificate under the name and responsibility of the auditing group is issued. This process serves to review the efforts of the validation team, provides a recognized (and demonstrable) assurance of validation, and provides a level of assurance of regulatory acceptance of the result.

Under the guidelines of the ISO 9000 program, for example, one group prepares for the audit (in parallel to the validation team role), while an outside group (BSI, Underwriters Laboratories, etc.) certifies the result. This outside group can serve as the validation team itself (or as a member of the team), or it can serve as the certifying group, adding the outside independence or credibility of compliance with FDA or EPA requirements; equivalent requirements from other regulatory agencies; or with international guidelines of groups like ISO, World Health Organization, and CITECH.

When is it cost effective to have a system certified? Most companies conclude that the risk of regulatory rejection or questioning of a validation study is too great with a system critical to a New Drug Application, or fundamental to the manufacture or control of a high risk product (controlled substances, blood products, etc.). Other organizations have used the certification process as an important managerial step in controlling the costs, time, and quality of validation teams, using outside auditors in much the same role as a CPA is used to audit the effectiveness and efficiency of an in-house accounting department. Still other companies have added the certification process in order to have an advertisable standard of quality: Hewlett-Packard, Gettinge, Pharmacia, and other companies, for example, advertise the certification of their laboratory instruments.

Summary

These six steps of validation management, identified above as information, guidelines, organization, prioritization, follow through, and certifying, provide a sound structure for controlling the process of gathering evidence in support of a system. Together, this framework can serve as the fundamental principles for developing and implementing a validation plan.

The key to developing such a plan, and the key to the implementation of that plan, is commitment. Validation is an efficient process, but it still represents an expenditure. That investment produces significant savings in lost data, erroneous conclusions, regulatory delays, and potential fines. It produces a positive cost/benefit ratio, but it still represents an out-of-pocket corporate expenditure.

In an era of cost consciousness, such expenditures are still necessary but must be carefully analyzed. The managerial controls suggested above can provide the necessary efficiencies to minimize expenditures, eliminate waste, and improve the effective use of automated systems in regulated industries.

BENEFITS OF VALIDATION

A pharmaceutical, medical device, or biologics product manufacturer is in business to obtain satisfactory market acceptance of existing and new products in a cost-effective and profitable manner. Computer-system validation can meet this goal.

The FDA has stated its intention to consider all data derived directly or indirectly from systems not validated through a policy meeting the above guidelines to be "suspect data," presumably delaying all applications and approvals and triggering additional inspection visits. Furthermore, the FDA has stated its intention in the future to reject automatically as "un-calibrated" all data derived directly or indirectly from systems not validated through a conforming policy.

The cost of **NOT** validating can be immense. For example, if an FDA inspector wants to see evidence of validation of a computer system utilized in a pharmaceutical manufacturing facility, and that evidence is not available, that manufacturer will receive a warning letter from the FDA. In the United States, a company's stock price drops three to six points when a warning letter becomes public information. But what happens if the inspector chooses to halt operations until the manufacturer takes corrective action? The Pharmaceutical Research and Manufacturers of America (PhRMA formerly the PMA) has estimated that a company can lose between $50,000 and $100,000 for every **DAY** of delay during the pharmaceutical submission process.

Compliance with regulatory guidelines is a major justification for performing computer validation, but there are other reasons why validation is simply good business. The Quality Assurance specialist of one major pharmaceutical company recently stated that validation is good for business in an economic sense and in a quality sense. In fact, he believes that if the regulatory agencies entirely did away with their regulations requiring validation, 99% of the manufacturing firms would still validate.

From a business point of view, regulatory and industry spokespersons have recently expressed their conviction that computer validation will also save a company money. Validation makes a positive contribution in several areas, such as:

- Higher reliability of systems, resulting in higher output quality:
  -- A decrease in the number of errors, which are sometimes expensive to correct
  -- Better acceptance among customers and users
- Improved control over process efficiency and timing, leading to faster output
- Defense against potential adverse events and timely and comprehensive recovery when an adverse event occurs.

For example, have you ever come in contact with a new employee at your company trying to use a computer system to perform a task, only to be frustrated because of inadequate documentation on how to use the system? It is very common for systems to lack appropriate documentation, or the software version is wrong, or it is hard to understand, or controlling Standard Operating Procedures (SOPs) are out of date. What does it cost a company in lost productivity and decreased learning curve because of inadequate documentation and no planned training? Validation addresses these issues.

Have you seen someone who is your company's computer expert and has all of the SOPs and testing documentation ***MENTALLY*** recorded? What happens when this person wins the lottery, gets a major inheritance or gets run over by a bus -- in all cases leaving you with no documentation. How much will it cost you to recreate this documentation and how many best practices will you lose in the process? This is not a fanciful situation; a recent (10/28/93) FDA Warning Letter to a medical device manufacturer cited the following violation: "The validation protocol is inadequate.... The data is maintained in an employee's laboratory notebook and only he can explain it."

The above information is thought-provoking, but **HOW** do we really calculate if the cost of validating a particular system (internal and/or consulting costs) is really worth it? One way is to calculate the cost of **NOT** validating the system. A method to do this involves analyzing the adverse events that can occur with an un-validated system and then calculating the "Expected Cost" of the adverse event (where *Expected Cost* = the *Actual Costs* to the company when the adverse event happens multiplied by the *Probability* that the particular adverse event will occur). This method is based on a cost of quality model, and it may illustrate that there are many obvious and less obvious costs to certain adverse events.

To expand on this "cost of not validating" further, what follows are actual adverse events that can occur to regulated systems with recommendations on how to calculate the actual costs to the company when this event occurs:

1. The probability that the FDA will delay an NDA/ PMA/ 510(k) submission because clinical data was generated on a non-validated system; where the Event Cost = # of planned submissions X # of delay days X $50,000-$100,000 per delay day.

"A computer validation report will be requested before any statistical data in a submission is reviewed" -- Steve Wilson, who is in charge of Statistical Analysis and Review at the FDA, at SQA conference, October 1992.

2. The probability that the FDA will only issue a 483/Warning Letter for lack of computer validation; where Event Cost = cost of corrective action + cost of senior management effort to bring stock price back to pre-483 level (+ cost of career-limiting adverse visibility?).

   "When FDA inspectors encounter vendor-supplied software, they expect that the user firm has adequate documented evidence to demonstrate that the system is capable of meeting predetermined specifications and quality attributes. When the FDA finds the user firm's documentation to be lacking or incomplete, there may be adverse consequences." -- Former FDA Investigator, *Pharmaceutical Technology*, April, 1992.

3. The probability that lack of computer validation will cause the FDA to order a production line or plant-wide shutdown; where the Event Cost = the daily cost of halted production X # of days it takes for FDA review of corrective action + cost of corrective action. (Note: This happens to many foreign firms!)

4. The probability that erroneous data from a computer system could cause a product recall; where the Event Cost = cost of a recall + cost of halted production + cost of internal investigation + cost of corrective action + cost of regaining market share + cost of senior management effort to bring stock price back to pre-recall level + cost of increased FDA scrutiny.

5. The probability that erroneous data from a computer system could cause a product-related injury or death; where the Event Cost = Event cost in #4 above X 150% + the cost of lawsuits/settlements + the incremental increase in product liability insurance.

6. The probability that a system disaster will occur (flood, fire, power outage, disk crash, etc.); where Event Cost = daily cost of halted production X # of incremental days lost because of inadequate disaster recovery plan + the cost of any lost data.

7. The probability that a software vendor will make a software change that effects data integrity; where Event Cost = cost of rework (redo analyses, etc.) because of inadequate version and change control procedures + cost of bad decisions made on faulty data. (Note: you may want to consider adding the event costs from #4 or #5, above, if appropriate.

8. The probability that a company will hire a new employee who interacts with an un-validated system; where Event Cost = cost of decreased learning curve

because of inadequate or no documentation exists + the incremental motivation techniques needed by management to counter the new employee's frustration.

9. The probability that a company employs a highly skilled employee, with many SOPs mentally recorded, and that employee wins the lottery, gets run over by a bus, or quits to join a competitor; where Event Cost = production loss + cost of trying to recreate SOPs + cost of losing best practices + cost of #8, above.

10. For Vendors: The probability that lost sales will occur because a competitor offers a system validation package with the software product; where Event Cost = cost of lost market share.

11. For Vendors: The probability that a non-validated software product will fail in the field; where Event Cost = Cost of rework + cost of customer dissatisfaction.

COST/BENEFITS ANALYSIS

DATA:

Total *Expected* Costs for 1 though 11, above $________

Total External Validation Costs $________

Total Internal Validation Costs $________

==========

Total Validation Costs ($________)

===========

Adverse Event Expected Costs -- Validation Costs $________
(COST SAVINGS!)

QUALIFICATIONS OF VALIDATORS

The validation of computer systems, like the validation of any product or procedure, is essentially a process of collecting credible quality-related evidence of the control and functionality of that system. That evidence may include test results, analyses of Standard Operating Procedures, inspections, and a variety of other inputs that provide a comprehensive picture of the system in question and the way in which it is being utilized.

The credibility of that evidentiary picture resides in four key factors:

1. the methodology of standards employed;
2. the breadth and depth of the test and analytical data generated under that methodology;

3. the degree of documentation available for review of that methodology and data; and,
4. the least tangible factor in the quadrad: the credibility of the validator or validating group itself.

It is this fourth critical, but ephemeral, foundation of system validation that forms a major focus in validation.

The credibility of a validation study should not be viewed from a law enforcement perspective: the antithesis of a credible study is not fraud but rather sloppy or incomplete work. It is possible, no doubt, to fabricate data, to forge review documents, or to otherwise distort analyses to provide convincing but spurious evidence. To some degree, the credible reputation that the validators provide proves convincing in the rebuttal of fraud. But the appropriate and realized regulatory position is not a prior assumption of dishonesty, but rather skepticism that a validation study meets two critical tests for credibility: first, that the certifying or signing validator has the knowledge and skills to conduct an appropriate quality assurance review of the system; second, that the validator had the appropriate level of independence to assure the non-contamination of that review. These two criteria of expertise and independence together establish the basic credibility of the validation study. A deficiency in either area may throw the value of that study into question. If the validators lacked the knowledge to determine and conduct the appropriate reviewing, testing and analyses; or if the validators failed the test of independence designed to assure that their conclusions are unbiased, the value of the study is reduced below acceptable levels.

Expertise Required

The expertise required to give a validation study the weight of credibility may reside in either the validation team (or person) itself, or in an audit group (or person) responsible for the review and certification or approval of the validation study. If an audit group is the primary expert source, its review should focus upon the methodologies utilized as well as the results of the validation study. Examining only the results may lead to erroneous acceptance of conclusions based upon inadequate or inappropriate sampling, limiting testing approaches, or other methodological inadequacies.

Expertise can be further divided into six competency areas: system hardware, system software, applicational environment, validation methods, statistical techniques, and experience. While these six competencies need not be represented by any one member of the validation (or audit) team, all six should be incorporated in the union of skills of that team. Strength in one or more areas is not sufficient to compensate for absence or significant weakness of other skill areas.

Although there have been attempts made in recent years to identify academic or training programs that represent proficiency in these six areas, to date there is no single certification or training program that will assure, upon completion, expertise in all six skill sets. A combination of training courses, however, coupled with academic or experiential education, may together provide appropriate confidence in the mastery of validation competencies.

We suggest the following combined training and skills competency definition for validation experts:

a) completion of a recognized and accredited validation program of twenty or more hours of classroom work;
b) participation as an observer in three or more validations;
c) completion of an advanced degree in computer system hardware and software (or experiential equivalent);
d) completion of an advanced degree or training in statistical analysis, methods, and test design; and
e) continuing education training in regulatory affairs, systems, and testing.

In addition, validation studies should be subject to a final internal audit by a senior quality control professional with the above credentials plus:

a) doctoral-level technical hardware and software skills;
b) certification in electrical, industrial, or computer engineering; and
c) doctoral-level testing and statistical analysis skills.

Hardware

The design of appropriate testing, and the identification of key elements to be tested, often requires an engineering-level understanding of the computer equipment (hardware) that runs a particular system. This requirement is maximized in the cases of robotics, medical device, or manufacturing control systems and is probably minimized in centralized data oriented analytical and reporting systems. In distributed data systems, engineering concerns generally focus upon the networking and communication elements in the systems.

To the degree that system validation may overlap process validation (for diagnostic or production equipment, for example), the hardware expertise may be reflected most strongly in the design and implementation of Installation Qualification, Operations Qualification, and Performance Qualification testing.

Hardware expertise may be demonstrated experientially, or it may be "certified" through a variety of channels. Most vendors offer training programs for customers that may include hardware issues in their curricula. Academic programs, generally in engineering, provide strong credibility. Engineering credentials, through licensure and/ or membership in IEEE, AIIE, and equivalent professional organizations, provide even stronger evidence of hardware competence. Particular care must be taken in the review of qualifications of American academically prepared persons in "Computer Science," "Information Systems," or "Communication Science," whose skills may be limited to software or mathematical application areas without a grounding in system hardware.

Software

Software skills parallel hardware skills in specificality, demonstration, and concerns. Unfortunately, the software arena lacks even the limited hardware "certification" represented by an engineering society's membership; ACM, SIM, DPMA, and the other software associations are generally "open membership," demonstrating little more than an interest and a willingness to pay an annual fee. Many software professionals are self-trained, either in reality or claim, and may have skills tightly focused upon a single programming language, technique, or sub-method.

Generally, software experience should include three key elements. First, the validator must have the ability to read the software language(s) utilized. This literacy must be of sufficient depth to permit a trace or walkthrough of the code, including mastery of all symbols, positioning codes, and syntax features. Second, the software expert must be qualified to diagnose, develop, and evaluate algorithms. This skill includes the ability to applicationally understand the various software approaches to a given problem, and to selectively evaluate those approaches. Finally, the software expertise required for validation includes an understanding of the innate characteristics of various computer languages, incorporating syntax checking, interpretation and compilation characteristics, and internal diagnostic capabilities and limitations.

U.S. academic programs in computer science, information systems, communication science, data management, and programming generally provide strong software skills, although care must be taken to evaluate the match between mastered languages and code languages in use for a specific validation project. Again, care must be taken in assuming that a computer engineering curriculum includes applicational software topics. Many do not.

Applicational Environment

Expertise in just computer hardware and software is not sufficient; a validator must also understand the applicational environment in which the system operates. Consider a blood testing system, for example: without at least a rudimentary grasp of the blood factors under analysis it is not possible to assure that the system in question is properly interpreting the results received. Identification of the critical inputs that may appropriately require double entry, or the life-threatening impact of power fluctuation in a pacemaker, or of any other situationally-specific conditions that may identify significant control issues, is fundamental to the validation process.

Generally, system validation environments can be placed in one of three categories: data, control, and device. The data systems, generally seen in GLP environments, receive, analyze, interpret, report, and potentially evaluate based upon information analysis. Control systems (usually in GMP environments), often incorporating A- to D-converters, add a feedback mechanism to the operation of fermentation tanks, mixers, stamping machines, and other manufacturing equipment. Computerized devices, such as laboratory diagnostic equipment, pacemakers, infusions pumps, and similar types of equipment may incorporate both kinds of systems, sensing and analyzing data while controlling or operating processes. All three environments, and the myriad subcategories possible, have unique issues that require at least a fundamental understanding of functionality.

Expertise in the applicational area is generally experiential, supplemented where appropriate by academic training (bioengineering, chemistry, medicine/nursing, biology, etc.).

Validation Methods

Validation of a system provides confidence to Quality Assurance, management, and the regulatory community. It is not intended to simply satisfy a regulator, but rather as an integral part of the entire quality picture of an operation. As a relatively new element in that picture, however, validation is still open to interpretation. While the essence has been distilled, the ramifications of that essence are still evolving. An important part of the validation process, then, centers around "why" questions. Why is this testing sufficient? Why have you chosen this approach? Why have you decided to validate this system while another is assigned a lower priority? In effect, these questions ask for a validation methodology, and the value placed upon the responses is largely a matter of the credibility of the validator providing answers. If those answers are "method free," that credibility is immediately suspect; in effect, the lack of a rationale and an a priori definition of the validation methods is as glaring a weakness as the absence of a system design document. Both problems suggest that no one thought through the issues and goals before starting.

Expertise in validation methods is generally demonstrated through the completion of a validation training program built around a specific methodology. Currently, such system validation courses are available from a number of organizations including the Center for Professional Advancement (New Jersey and Amsterdam), and the International Validation Forum (Boothwyn, PA).

Statistics and Testing

Advanced academic education in most fields provides a methodological emphasis: doctoral degrees in the sciences, social sciences, and some humanities disciplines include testing and statistical analysis in their disciplines. Other sources of testing and analysis credentials include undergraduate academic programs in statistics, research methods courses and practicums, and special training programs in statistics methods (often sponsored by vendors of statistical packages such as SAS, SPSS, and BMD). A Quality Assurance or Quality Control background is likely to incorporate and demonstrate experiential training in testing, particularly if coupled with training in sampling, test design, or statistics.

All of these credential sources are designed to provide a clear demonstration of expertise in the development and implementation of testing methodologies and in the analysis and interpretation of the results of those methodologies. Most validation testing utilizes unique stratified sampling scenarios. Statistical interpretation is generally limited to descriptive tools such as standard deviations, ranges, and chi-square analyses (except, of course, when the statistical programs form the subject rather than the method of the study, as in validating statistical analysis computer systems).

Test design is perhaps the most complex of the methodology expertises. Determining the appropriate test strategy, and the logical testing construct, is based upon a complicated ability to adapt theory to practice. Many organizations have attempted to standardize testing for all systems (often with a resulting "overkill" approach) to minimize the testing expertise necessary. This approach is possible, although it may lead to billing abuses with consulting firms operating on an hourly billing basis.

Experience

The "experience" criterion is the clearest to demonstrate, the easiest to define, and the hardest to theoretically justify. Every non-parent psychologist is a master of child-rearing theory, but he or she may not be so successful when faced with a real tantrum-throwing two-year-old. If a validator can demonstrate all five other areas of expertise, experience may not be apparently critical. But the gap between the hypothetical and the real is huge, and a validator with experience in that real environment carries more credible weight than one who has studied but never applied.

The degree of experience necessary is not a constant. Since there are three distinct kinds of systems (data, control, and device), a validator should have a minimum of three experiences in order to provide credible testimony in all three areas. Some jobs, however, may limit responsibility to a single environment. In the same manner, a validator providing guidance for a GLP environment LIMS may extrapolate from a database system of another type, but will be more credible with more direct applicability. In short, more is better, and some experience is critical. The validator who has never applied his or her skills before is to be avoided (or, better yet, paired with an experienced validator); the value of the testimony increases with the similarity and extent of the experience, until some reasonable asymptote is reached.

Required Independence

The basic quality assurance precept of "independence of review" is appropriately carried to the qualification of a validation expert. If a coder checks his or her own work, that check is suspect for three fundamental reasons: a) there is a natural bias toward NOT finding fault internally; b) there is the human tendency, perhaps most evident in proofreading, not to see errors that one is "too close to;" and c) since the role of a validation (or any other quality review) is in part to see if rules and procedures have been followed, self-examination falls to a logical fallacy. That is, if validation is necessary (there is a possibility that procedures are not followed), self-validation is impossible (the validation procedures are likely not to be followed either); if self-validation is possible, it is not necessary. For all three reasons, validation can only carry the "second check" conformational credibility of its intention if the validators meet an independence criterion.

Validation studies may be conducted by either internal or external sources, each with different independence criteria. In both cases, however, independence of source provides an aura for a complete team. It is perfectly possible, and often appropriate, to include in a validation team representatives of the users (who may have a dependency upon the design and functionality of the system) and of the developers, whose independence may be suspect because of financial gain or work limitations flowing from acceptance. Their expertise is important, even if their independence is restricted. The validation team, however, may meet the criteria of independence through the inclusion within it (in a veto-potential role) of an independent member who may review and evaluate the contributions of the rest of the team. This individual must have sufficient expertise to understand, if of lesser depth than that defined above.

In-House Independence

The general rule of thumb for in-house independence is chain-of-command branching. As with any other quality assurance officer, the career path of the validator should not be influenced by the results of a validation study. That is, the validator should not directly report to the "owner" or developer of the system, nor to the manager to whom that owner or developer reports. In small shops this independence may by necessity be compromised, but overlapping reporting relations may affect the credibility of the validation report.

Incidentally, commercial vendors of systems used in high risk or high hazard environments may be subject to even more stringent independence requirements. Because of sales pressures, all members of the vendor organization may be suspect, effectively requiring outside validation.

The recently-released, draft Good Automated Laboratory Practices (GALPs) recommendations of the U.S. EPA provide a very clear guideline for internal independence. The GALPs introduce the concept of "Responsible Person" (RP), an individual of any title or position who ultimately is charged with the effective use of the system. The GALPs recommend a clear distinction in person and reporting structure between the RP and the quality assurance or validation role, defining the validation as the recheck or review of the RP system control role.

External Independence

When external validators are used, independence can be assumed as long as three criteria are met:

a) *Conflict of Interest.* The validators should not be drawn from a division or company that also markets services of software development, system analysis, or other conflicting activities. A software company offering validation services does not pass the presumption of independence. Similarly, the company that consulted in the design or development of the system should not also be charged with the validation of that design.
b) *Conflict of Incentive.* The audit and testing of a system may be suspect if the outside validators operate on a time billing basis. In such an environment a conflict of incentive (to find or avoid problems based upon financial considerations) may subjectively compromise independence.
c) *Conflict of Objectivity.* A bias toward a problem solution, computer brand, or software language may bring into question the independence of a validator. The independent validator is consistently committed to a methodology, not to an environment, computer, program, or device. In effect, this criterion of independence is best demonstrated through extensive experience.

Of course, it may be practical for an outside validator to demonstrate independence despite a failure in one or more of these three areas, but the presumption is carried only with full conformity. Of particular suspicion are outside validators who combine their services with a wide range of non-validation offerings and the validation groups too small in numbers to offer depth of expertise or experience.

Summary

The value of validation reports rests largely with the credibility of the validator; the credentials of that validator (or validation team) should form an integral part of that report. In fact, many field investigators have confidentially reported that the credentials of the validators are the single most important part of that report; two separately confessed that the credentials were the only part of the report they reviewed.

That credibility rests upon two foundation stones: expertise and independence. The most credible validation, providing the largest degree of managerial, quality assurance, and regulatory confidence and acceptability, will be signed and certified by a person or group with demonstrable expertise in the specific hardware and software involved, in an appropriate validation methodology, in statistical analysis and testing, and in the specific applicational environment. That expertise will include extensive prior experience, as closely related as possible. And, finally, that experience and expertise will be supported by internal or external independence, assuring the credibility of the conclusions and appropriately maximizing that credibility.

VALIDATION RESPONSIBILITY

The current regulatory requirement is that all computer systems utilized in the support of data submission, in the manufacture of regulated products, or in the monitoring of regulated processes be regularly validated and tested to assure accuracy, reliability, and appropriateness of control. The question of who should assume responsibility for that validation process, however, has been left largely unaddressed.

Should a pharmaceutical, blood processing, or medical device company follow the current Quality Assurance Model, establishing an oversight group to perform retrospective and periodic validations? Can the internal data processing or management information system staff perform the validation function as a part of the development and installation process? Will FDA or EPA inspectors in effect validate a suspect system through their review processes? Or is another model appropriate to insure the objectivity and accuracy of system validations?

There are five possible models corresponding to these options: the Quality Assurance approach, the MIS model, the FDA Review alternative, a Consultant model, and a proposed "SEC" model. Each approach must be evaluated on criteria of appropriateness, performance capability, and -- perhaps most importantly -- objectivity.

In-House Alternatives

Most internal checks on data submitted to the regulatory agencies are provided by an in-house department or staff, generally called Quality Assurance, Quality Control, or Compliance. This internal group acts as a second check on experimental design, data collection, statistical analysis, and related functions, working closely with appropriate departments while simultaneously remaining aloof from the operations of those departments. Quality Assurance staffers are often (inappropriately) perceived as "big brother" inspectors. If appropriately utilized, however, they confirm the quality control measures employed on all levels and serve as an internal resource in the design of those measures.

The addition of system validation responsibility places a bi-directional strain upon Quality Assurance operations, however. First, appropriate validation procedures require a high level of technical expertise, including the ability to read and evaluate source code, experience in the evaluation of security control and disaster recovery procedures, and a hardware background that includes reliability testing and performance standards. Few, if any, existing Quality Assurance departments possess such expertise, which is acquired only through lengthy professional training. It is, of course, possible to add experienced system validation personnel to the Quality Assurance operation, although such individuals are likely to prove scarce and expensive.

One of the key organizational constructs of Quality Assurance is an independent reporting relationship. Quality Assurance personnel should (by regulation) report through a separate channel of responsibility, independent of the chain of supervision controlling the departments they audit. This independence, assuring a freedom from undue influence, is critical to their objectivity and, in turn, their credibility.

The Quality Assurance Model, then, meets the test of objectivity despite the lack of pure independence inherent in an in-house operation. Unfortunately, however, it is unlikely that the Quality Assurance staff will cost-effectively possess the depth of expertise and experience necessary for effective system testing and validation.

That expertise is likely to be present, however, on at least the top levels of the management information systems staff. The system management personnel are presumably fluent in the coding languages in use; they know at least the rudimentary details of the hardware configuration and functionality; and they have been exposed to or can master the techniques of effective system testing and review. In terms of the ability to perform validations, they stand temptingly as the ideal group for the assignment.

But the MIS model fails to meet the critical test of objectivity. Every effective programmer, supervisor of programmers, and user accepting a new program performs routine tests of the functionality of the code. In fact, the most commonly taught and used programming procedure relies upon a process of coding, testing, "debugging," and retesting. However, to rely upon the assurances of that programmer, both psychologically and politically biased in the evaluation of that program, is a chancy proposition at best. While the reliance upon a supervisor to double check work represents a slight improvement, it too fails to meet the independence criterion. Within a single department, reporting to a single line of responsibility, no true objectivity of validation is assumable.

With the dilemma of expertise generally concentrated in the MIS department, while the objectivity criterion is met only through the traditional Quality Assurance formulation, these two tempting in-house models are jointly doomed. It has been argued that Quality Assurance people do not require great depth of expertise in their other areas of responsibility (statistics, for example), but both the sad history of computer validation (or its lack) and the depth of knowledge inherent in current regulatory guidelines suggests that the analogy does not hold. Both expertise and objectivity are required, but the combination is unlikely to be found in-house.

Outside Alternatives

In the past, the validation function has tacitly been performed by the regulatory agencies, and at least in the short term this model can be continued. For example, when FDA reviewers find data discrepancies, they track down the cause, placing new drug applications on hold in the interim (at an estimated cost, remember, of $50,000 to $100,000 per day). Should the cause prove computer related, an increased schedule of inspections will be instituted, presumably including the assignment of an FDA computer specialist (a half-dozen or so such individuals have been so qualified). While this model may cause increased discomfort for the organization -- and result in costly delays as problems are discovered -- it can not be disqualified as a current validation alternative. The FDA does have inspectors with credentials indicating sufficient expertise to perform at least a preliminary validation, and this clearly meets the criterion of objectivity. The model, however, has a very short-term future.

There are an estimated 1,200 pharmaceutical companies with computer systems and an additional 240 computerized blood processing organizations in the United States, with approximately another one-third that number under FDA regulation but operating in Europe and Japan. Each of those 1,880 systems requires annual validation, testing, or review, and that number is growing at a rate of more than 12% per year. It is clear that the number of FDA computer-qualified inspectors will prove vastly inadequate to the task, forcing the FDA to require adoption of another model. One FDA computer inspector, pointing out that he now has a workload allowing less than one day per system review (a process generally requiring forty-five to fifty person-days), suggested that the process of removing the FDA from direct validation has already begun and that within two years the FDA will require that all data submitted be generated by pre-validated systems. Clearly, the FDA model has little or no life expectancy, and does not represent a viable option.

The remaining model available today is the Consultant alternative. In effect, outside consultants specializing in system validation and testing, using a methodology pre-approved by the regulatory agencies, act as adjuncts to the Quality Assurance operation. A consulting team conducts a validation, providing an in-house report to Quality Assurance, and presumably increasing the expertise in the Quality Assurance department until it is capable of assuming increasing shares of the validation responsibility. While the consultants may be expensive, their services are generally cost effective if the validation is performed on a fixed cost basis (allowing the consultant to use cost saving division of labor team assignments) and if the consultants bring both a high level of experience in validation and a highly focused commitment to validation procedures in the regulatory environment. The Consultant Model clearly meets the joint effectiveness and objectivity criteria, although a further improvement may be possible.

The "SEC" Model

The one potential weakness in the Consultant Model lies in the area of responsibility. Unless the consulting group brings high levels of credibility and a previous reputation with the regulatory agencies, the regulated industry is left in the situation of having its Quality Assurance department assume responsibility for an area in which they admittedly lack expertise, relying wholly on the advice of outsiders.

The Security and Exchange Commission has evolved a model designed to cope with the same issues of expertise, objectivity, and responsibility in its own area of regulatory responsibility. This SEC paradigm provides a very powerful model that the FDA and EPA should consider carefully. By all inside accounts, in fact, that consideration is already underway.

A public company uses its in-house controller to accumulate regulatory data and relies on its corporate treasurer to act as a quality control check on that process. But more than the in-house reports are submitted to the SEC. The company is required to employ an outside, independent group -- in this case, a public accounting firm -- to audit and verify the submitted data. The accounting firm applies clear standards, draws a conclusion, and assumes responsibility and liability for its conclusions (with reasonable constraints).

While the public accounting firms are not appropriate to the pharmaceutical, medical device, biologics, and environmental laboratory situations, both due to a lack of system validation expertise and a conflict of interest as financial auditors, the model has high applicability. If the regulatory agencies can establish guidelines for validation consulting firms, including credentialing of validators, review of methodology, evaluation of previous experience, and similar criteria, the system validation equivalent of CPAs can be identified. These validators would then presumably be invited into a client company to conduct a highly objective, independent evaluation of a system and to certify, with assumption of (perhaps limited) legal responsibility, the conformity of that system to the established criteria.

In the interim, as the regulators and industry are jointly designing the institutional mechanisms to establish an "SEC" Model for System Validation and Testing, the executive responsible for system control can best serve the organization by following the Consultant Model strategy, taking care to select an outside consulting firm with the credentials necessary to move into the certification role. Briefly, those credentials include a strongly defined validation focus, experience with regulatory guidelines of validation, and utilization of a regulatory-reviewed methodology.

As the "SEC" Model evolves for the industry, true validation control will be achieved. Executives and managers will have the assurance that systems have been tested and approved according to established standards, in a process conducted by experts meeting the joint criteria of objectivity and knowledge. That assurance will be reinforced by the assignment of responsibility (and, presumably, liability) to the independent group providing the review and certification. Regulators, in turn, will be able to fulfill their appropriate mission within their resource constraints, in effect transferring their focus from the system to the certifying system inspectors, much as the SEC concentrates upon the auditors rather than attempting to randomly re-audit the vast number of organizations under their regulatory umbrellas.

Chapter 2

THE REGULATORY WORLD

PHILOSOPHY: AUDITED SELF-REGULATION

If you drive along a major highway in any part of the world, you are likely to encounter a road sign telling you the maximum legal speed limit: 100 kilometers per hour, or 55 miles per hour, or whatever the local limit may be. Those signs serve a dual purpose: They give the driver a general idea of the safe speed for that road, and they serve as an enforcement warning for police action in restricting that limit. Of course, the posted limit is only a rough approximation. Bad weather, a poorly tuned automobile, or an inexperienced driver may all argue for a lower safe speed. And in many areas, the upper limit is really much in excess of the posted maximum. Traffic may routinely move much faster than the posted speed, and traffic officers may well enforce a limit well above the maximum allowed.

Despite these variations of condition and of reasonableness, the posted limit does provide the driver with some general guidelines, and with a defensible position in the event of capricious or inappropriate enforcement. While the prudent driver will strive to understand the "real" enforced top speed, the posted limit will help in arriving at a reasoned judgement of the proper rate for unique combination of enforcement, weather, and road conditions.

Effective regulation works much as a speed limit sign. It provides a general guideline, around which experts ought to be able to make reasoned decisions. It provides a framework for those decisions. And it serves as a definition for the limitations of regulatory authority. But effective regulation can not and should not replace that reasoned and professional judgement. High allowable speeds may not be safe in some conditions of snow or ice, and unusual circumstances may justify the risk of a still safe limit in some reasoned excess of posted warnings.

The effective road sign, and effective regulation, serve the primary purposes of establishing an agenda for decision; evaluative criteria for that decision; a financial justification for assessing the risks and benefits of that decision; and a framework for the reasoned defense of that decision.

As increased industry reliance upon computerized systems necessitates regulation of data and the software systems that generate that data, so effective road signs are appropriate to define the guidelines for system validation. Today, those signs are increasingly being posted, and the fuzzy limits of interpretation and application of those guidelines are increasingly being brought into sharp focus.

The United States Food and Drug Administration (FDA) has taken the leading role in setting the international agenda for system validation, and industry awareness of the importance and vulnerability of computer systems has been created largely by the effective agenda-setting function of the FDA. Working closely with colleagues throughout Europe, North America, and Japan, the FDA has used its public speeches and forums to establish an agenda for proving control of systems used throughout the industry. Private companies, as well as associations like the Pharmaceutical Research and Manufacturers of America (PhRMA), the Health Industry Manufacturers Association (HIMA), the Organization for Economic Cooperation and Development (OECD), the Drug Information Association (DIA) and others have responded. And international regulatory agencies worldwide have provided the reasoned argument, the analytical data, and the legal incentive to make system validation a high priority.

We now have road signs in many areas, defining appropriate standards of testing, control, documentation, and planning for laboratory systems, medical devices, manufacturing systems, and controllers. As field visits examine the validation studies that confirm those controls, the fuzzy limits and interpretations of validation are coming into focus. Here are a few examples of the clear consensus issues that have emerged:

Application of Validation Guidelines

System Validation guidelines clearly apply to all computerized systems in a regulated environment, with the isolated exceptions of financial and personnel systems. There is not any persuasive rationale for exclusion of any system operating in a regulated environment. The depth of controlling evidence, however, can be reasonably limited by criteria of system hazard, increasing test samples when human life is at risk; by the universality of the system, limiting review efforts in the case of operating systems or standardized packages; and by the credibility of the system developer, established through independent validation, ISO certification, and audit of the development and support organization. These three criteria may call for pre-established but varying levels of evidence of control, all encompassing the totality of regulated systems in an environment. All systems equally need to be validated, but all validation studies need not be equal in extent or rigor.

Credibility

All validations must be equal, however, in the credibility of the effort. That credibility is supported by two factors: the expertise of the validation team, and the independence of that team. Defining the fuzzy limits for a particular environment, and reasonably justifying any variation from the posted guideline, requires an expertise intersecting computer system knowledge, experience in the application area, and an understanding of the detail and philosophy of regulatory-legal requirements. Appropriately, the credentials of the validation team should be scrutinized for evidence of these three areas of expertise before accepting their collective professional opinion of the validation of a system.

But that opinion is suspect if it is biased. A regulator, or an industry executive, has a right to know that the opinion on the validation of a system is not tainted by vested interests. Vendors who self-validate will always face difficulty in justifying their independence. The use of outside, third-party experts to validate systems has emerged as a common sense necessity, and commonplace reality. Within regulated organizations, a validation team should be assembled with all the care inherent in any quality assurance role. We would not ask a bench scientist to validate the accuracy of his own research, nor can we ask programmers or information systems professionals to validate the computer system that their own department has constructed, selected, or manages. Similarly, the users of a system can not self-assess their own use and operating procedures without compromising their credibility and their capability. Independence, either through the use of an outside auditing group or an expert in-house validation department, is a practical, theoretical, and regulatory necessity.

Software Code Review

Yet another consensus issue surrounds the software code itself. Functional testing can determine current performance of a system. But there exists to our knowledge no way of assessing the future performance, the supportability, reliability, and formulae level of accuracy, of any system without reviewing the source code of that system. A review need not include a character-by-character assessment, but should examine the code structure, the internal organization and documentation, the appropriateness of the algorithms employed, and the general professionalism of the construction. Inability to obtain access to the code of a system immediately raises suspicions. Failure to review that code once access is obtained appropriately leads to serious questions about the expertise and professionalism of the validation team.

Classification of Software Code

The last point of consensus is a logical extension of the previous discussion. Clearly, computer software is utilized in the laboratory for the interpretation of results, in the manufacturing site for the production of biologics and pharmaceuticals, and in the clinical setting for the analysis of findings. The classification of such software as "medical devices" is an extension of reason without any unnecessary stretching. Even HIMA has chosen not to disagree with such a designation. This was emphasized most forcefully by the FDA's Center for Biologics Evaluation and Research. On March 31, 1994, CDER's director, Dr. Kathryn Zoon, wrote the following letter to 23 computer software manufacturers in the United States and Canada:

To: Blood Establishment Computer Software Manufacturers

Dear Sir/Madam:

The purpose of this letter is to advise you that the Food and Drug Administration (FDA) considers software products intended for use in the manufacture of blood and blood components or for the maintenance of data that personnel use in making decisions regarding the suitability of donors and the release of blood or blood components for transfusion or further manufacture to be devices under section 201(h) of the Federal Food, Drug and Cosmetic Act (the Act) [21 U.S.C. §321(h)].

These software products are designed to receive and store data used by blood establishments during the manufacturing process, from determining donor suitability through component processing, testing, and labeling to product release. They are designed to receive and store data regarding blood donor status, including donors' answers to health history questions and the results of laboratory tests, including blood grouping and typing, hepatitis, and antibody to the human immunodeficiency virus (anti-HIV). Blood establishment personnel later access and use the data to determine whether donors are suitable and whether blood or blood components are free from disease-causing agents transmissible by blood, such as hepatitis and HIV. In addition, the data are used to label blood and blood components prior to release for use in hospitals and other health care facilities or for further manufacturing. Because they aid in the prevention of disease (e.g., hepatitis, HIV, etc.) by identifying unsuitable donors and preventing release of unsuitable blood and blood components for transfusion or for further manufacturing use, these software products meet the definition of device under the Act.

Facilities that manufacture and distribute these software products are subject to the device provisions of the Act and FDA's device regulations, including establishment registration, product listing, premarket notification or approval, Current Good Manufacturing Practices (CGMP), and adverse event reporting. FDA's CGMP regulations for devices appear at Title 21, Code of Federal Regulations (CFR), Part 820 and the MDR regulations at 21 CFR, Part 803.

According to FDA's information, your facility manufactures software intended for use in the manufacture of blood and blood components. Consequently, you are required under the Act to register your establishment and list your devices. In addition, your manufacturing operations are required to be in compliance with CGMP for devices, and you must report adverse events and other problems as required by FDA's Medical Device Reporting (MDR) regulations.

We are enclosing a registration package for your convenience. We will forward a device listing package to you in the near future. When completing the device listing form (Form FDA 2892), identify your product as Software, Blood-Bank, Stand Alone Products, Product Code 75MMH. The registration form should be submitted within 60 days of receipt of this letter if you intend to continue to distribute software products to blood establishments for use in manufacturing.

In addition, you are required to submit a premarket notification or application for premarket approval for each of your devices unless you can demonstrate that you commercially distributed the devices in interstate commerce prior to May 28, 1976 (the date of enactment of the Medical Device Amendments) and have continued to do so without any significant changes to these devices. If you claim such preamendment status for any product(s), please complete only the registration form, and send it to: Center for Devices and Radiological Health (HFZ-300), 2098 Gaither Road, Rockville, MD, 20850.

If you do not intend to submit a premarket submission, and intend to submit proof of the claimed preamendment distribution, this information should be sent to: Center for Biologics Evaluation and Research (CBER), Division of Blood Applications (HFM-370), 1401 Rockville Pike, Rockville, MD, 20852-1448. Finally, if you do not currently manufacture software products for blood establishments, please advise CBER promptly.

When we forward the product listing information to you, we will include guidance on how to prepare your premarket submissions. If you have questions about the content or format of a premarket submission once you have reviewed our guidance, CBER staff are available to help answer such questions. Premarket submissions should be submitted to CBER no later than March 31, 1995.

In the interim, FDA will continue to conduct inspections of blood establishment software vendors. These inspections will include, among other things, a review of standards for software development, testing, validation, and quality assurance. The primary focus of these inspections will be to assess compliance with the CGMP regulations for devices (21 CFR, Part 820).

The agency will also review and assess your procedures for investigating reports of product problems and defects and for implementing and evaluating corrective actions. We will also review and assess your procedures for notifying your customers and the agency when you take corrective actions.

Please be advised that during this interim period if a manufacturer of software products for blood establishments is not making good faith efforts to comply with the Act and FDA's regulations as stated above, the agency will not hesitate to take appropriate steps to bring the firm into compliance.

If you have questions concerning: (1) the preparation of the establishment registration and device listing notification, contact Center for Devices and Radiological Health, Division of Small Manufacturers Assistance (HFZ-220), at 301-443-6597, or (2) guidance for premarket submissions, contact Center for Biologics Evaluation and Research, Division of Blood Applications (HFM-370), at 301-594-2012.

Sincerely yours,
Kathryn C. Zoon, Ph.D.
Director
Center for Biologics Evaluation
and Research

Clearly, consensus has been reached: In many applications, the software utilized in testing and production falls under the regulatory guidelines applying to medical devices. While an educational period will be necessary, and time will be required to organize the inspection and enforcement mechanisms, there is no doubt that within the next five years the international Good Manufacturing Practices will be extended to the developers and vendors of application software. Already a number of responsible vendors, including Hewlett-Packard, Johnson-Yokagawa, Pharmacia LKB, and others are arranging the preparation of independent validation certification for their software products.

BENEFITS OF REGULATION

In addition to setting an agenda, there are four good reasons why regulation really works. There are probably some other good reasons why they do not work, and there are probably some bad reasons why they work, too.

Defining Responsibility

The first good reason for regulation in the medical device, pharmaceutical, biologics and blood processing industries and in environmental laboratories is that regulation -- that clear guidance -- has the effect of defining and limiting responsibility. In effect, regulation limits the liability of any manufacturer or laboratory manager by defining good common practice.

Without that definition, a group of non-peers, often a lay jury, has to make determinations of whether proper protections have been made, whether proper analytical procedures have been followed, and whether proper kinds of standards have been enforced. With some kind of regulation, the fact that a medical device manufacturer, for example, is compliant provides a clear definition, usually decided by the right level of professionals, defining what responsibility and liability is. A manufacturer of a LIMS system, for example, has almost infinite liability if that particular system malfunctions and gives a false reading that ultimately hurts the public. However, with regulation restricting liability, that particular device no longer has unlimited liability. For example, if a product is registered as a medical device under the 510(k) submission process, the federal courts have ruled that such registration limits a manufacturer's liability to pure negligence. So the first real value of regulatory compliance is that it limits and defines responsibility and liability.

This is not only a legal definition, but probably a moral one as well. For a manufacturer or laboratory manager to know, in their altruistic goals, that they have done everything practical and possible to assure quality is a very positive gain. With regulation, we can define what kind of quality is practical and possible.

Balancing Competing Requirements

A second advantage of the regulatory process is that it balances competing public requirements. At the same time that we have a public requirement for certain kinds of safety guidelines or health standards, there are competing public requirements for price controls and for reasonable cost effectiveness.

As nations throughout the world debate health-care reform, those particular interests are being thrown out of balance. Suddenly, the importance of cost effectiveness is overshadowing the importance of ascertaining that health care is actually being delivered. Pharmaceutical companies, medical device manufacturers, and environmental and biological laboratories are aware of those competing areas, and they are being pressured to find ways of providing less expensive products, or conducting fewer and less expensive tests, even if it means that the products or tests are not quite as effective. But at what point do we say that we will take a greater risk to save greater dollars?

The need to balance these competing priorities can be defined best by clear regulations. With them, we can know when we need to be cost effective, and when we need to put safety forward as our predominant consideration.

Leveling the Playing Field

A third reason why regulation works is that regulation has the effect of leveling the playing field. If we are placed in a position in which some companies put a great deal of money and effort into quality control and safety maintenance, while other companies make a decision to perhaps cut corners and skimp in that part of their operation, we end up with an unfair competitive advantage for low-quality production.

By having the quality of production clearly defined, we then have a level playing field throughout the United States and throughout the world.

Rationale for Quality

The fourth reason why regulation works is that compliance really has the effect of making quality cost effective, and that provides a clear rationale for quality. By having requirements for validating computer systems, it becomes cost-effective to go ahead and do that kind validation. It becomes cost effective very specifically because doing that kind of validation reduces the economic risks of potential fines, potential seizures of property, potential recalls, or potential legal action that will add greater cost to companies that are not compliant.

SUMMARY

The problem, of course, is that this discussion is based on an idealized kind of regulation. Reality, perhaps, is failing in all four areas. The reality is that we do not have clear definitions, or clear regulations on liability. We do not have a very tight balance between competing public concerns, because we often have different agencies, and sometimes different portions of agencies, emphasizing different kinds of regulatory concerns. Because we have those differences, our playing field is far from level; it is filled with valleys and peaks so that regulations often affect one company, while another is left to go ahead and use a lower level of standards. Or we are left in a situation where we do not have good, clear cost-effectiveness guidelines, because we have not had a clear definition of what is and what is not really required. Often, some companies decide to be pro-active, and they go ahead and invest more money and more effort in quality issues, while companies without those financial resources are forced to wait and see whether they will be cited by the regulatory agencies. And there is no certainty that that citation will come.

Finally, despite the record of the FDA and other agencies in setting the agenda for system validation, we are often left without a very precise agenda and, therefore, without a very real incentive to go ahead and take some kind of action. It is like driving down the turnpike where we do not know what the real speed limit is. We only know what the posted limit is, and that is clearly not what is being enforced. We do not know how often or how diligently the police might enforce whatever speed limit there is. If we are stopped, we do not know whether we will get a fine. If we do get a fine, we are almost certain that that fine is not proportionate to our particular income or ability to pay; while a $20.00 fine may not be too significant for some, it may break the bank of some other driver.

In effect, that lack of standardization, that lack of certainty, that lack of clear definition, leaves us with some very serious problems in this whole regulatory area.

In response to this kind of need, as well as in response to detailed problems in the regulated industries, there are some regulatory incentives and initiatives being undertaken that are intended to try and solve those very problems. The recognition that we need clearer standards comes not just from industry, but from the regulators themselves. If they want to emphasize cost effectiveness, and be cost effective themselves, they realize that they, too, need to establish some standard that gives that kind of uniformity. To achieve this, we are seeing two actions -- two kinds of standardization -- coming out, one from the EPA and one from the FDA.

The EPA has established the Good Automated Laboratory Practices (GALPs), a combination of a variety of standards that are intended to provide one, concrete definition for an automated laboratory doing EPA work. The FDA, on its own initiative, is rewriting and reinterpreting its Good Manufacturing Practice (GMP) and Good Laboratory Practice (GLP) regulations to try and get some broader picture in terms of what is happening in terms of automation in a broad spectrum of its regulated industries.

This trend will no doubt call for the placement of more road signs. Similarly, the speeds posted will be subject to adjustment as experience provides greater warning of dangerous curves and a comparative relaxation in less threatening areas. No doubt, too, the pharmaceutical, biologics, and medical devices industries' sophistication in calculating the risks of exceeding posted limits will grow, and the range considered tolerable will be more clearly defined.

There is no doubt, however, that regulatory system validation "speed limits" will continue to be set; that those limits will provide expert guidelines with tight and parsimonious standards of exactness; and that we will all continue to test, expand, and learn from the practical and reasonable limits under which the proper control of regulated computer systems can be demonstrated.

Chapter 3

FDA REGULATIONS

THE FDA APPROACH

The U.S. Food and Drug Administration, an agency within the Department of Health and Human Services, Public Health Service, regulates the manufacture and distribution of foods, human and veterinary drugs, cosmetics, medical devices, and biological products by authority of the Food, Drug and Cosmetic Act (21 U.S.C.) and the Public Health Service Act (42 U.S.C.), as amended.

FDA requirements for computer system validation generally fall under the following categories of FDA regulations, all of which can be found in Title 21 of the *Code of Federal Regulations* (CFR):

- *Good Manufacturing Practices* (GMPs) for drugs, biologics, medical devices, and veterinary products.
- *Good Laboratory Practices* (GLPs) for nonclinical laboratory studies.
- *Good Clinical Practices* (GCPs) for clinical investigations supporting research applications or marketing permits.

The FDA's Center for Drug Evaluation and Research (CDER) supervises regulatory compliance for drug (both prescription and non-prescription) manufacturers. The Center for Devices and Radiological Health (CDRH) supervises regulatory compliance for manufacturers of medical devices used in the diagnosis, treatment, prevention, and cure of disease, as well as *in vitro* diagnostic products used to test tissue specimens. The Center for Biologics Evaluation and Research (CBER) supervises regulatory compliance for manufacturers of biological products, including serums, vaccines, blood, and blood components. The Center for Veterinary Medicine supervises regulatory compliance for manufacturers of animal drugs and feeds.

GOOD MANUFACTURING PRACTICES

FDA's Current Good Manufacturing Practice (CGMP) regulations for the manufacturing, processing, packing, or holding of drugs (21 CFR 210) and for finished pharmaceuticals (21 CFR 211) were first proposed in February 1976 to update the drug GMPs in light of new technology and to adopt specific requirements to assure the quality of drug products. They were finalized in September 1978 and last revised in March 1990. Requirements in the regulations relating to system validation have been highlighted in boldface.

21 CFR PART 210--CURRENT GOOD MANUFACTURING PRACTICE IN MANUFACTURING, PROCESSING, PACKING, OR HOLDING OF DRUGS; GENERAL

§*210.1 Status of current good manufacturing practice regulations.*

(a) The regulations set forth in this part and in parts 211 through 226 of this chapter contain the minimum current good manufacturing practice for methods to be used in, and the facilities or controls to be used for, the manufacture, processing, packing, or holding of a drug to assure that such drug meets the requirements of the act as to safety, and has the identity and strength and meets the quality and purity characteristics that it purports or is represented to possess.

(b) The failure to comply with any regulation set forth in this part and in parts 211 through 226 of this chapter in the manufacture, processing, packing, or holding of a drug shall render such drug to be adulterated under section 501(a) (2)(B) of the act and such drug, as well as the person who is responsible for the failure to comply, shall be subject to regulatory action.

§*210.2 Applicability of current good manufacturing practice regulations.*

(a) The regulations in this part and in parts 211 through 226 of this chapter as they may pertain to a drug and in parts 600 through 680 of this chapter as they may pertain to a biological product for human use, shall be considered to supplement, not supersede, each other, unless the regulations explicitly provide otherwise. In the event that it is impossible to comply with all applicable regulations in these parts, the regulations specifically applicable to the drug in question shall supersede the more general.

(b) If a person engages in only some operations subject to the regulations in this part and in parts 211 through 226 and parts 600 through 680 of this chapter, and not in others, that person need only comply with those regulations applicable to the operations in which he or she is engaged.

§*210.3 Definitions.*

(a) The definitions and interpretations contained in section 201 of the act shall be applicable to such terms when used in this part and in parts 211 through 226 of this chapter.

(b) The following definitions of terms apply to this part and to
parts 211 through 226 of this chapter.

(1) *Act* means the Federal Food, Drug, and Cosmetic Act, as amended (21 U.S.C. 301 et seq.).

(2) *Batch* means a specific quantity of a drug or other material that is intended to have uniform character and quality, within specified limits, and is produced according to a single manufacturing order during the same cycle of manufacture.

(3) *Component* means any ingredient intended for use in the manufacture of a drug product, including those that may not appear in such drug product.

(4) *Drug product* means a finished dosage form, for example, tablet, capsule, solution, etc., that contains an active drug ingredient

etc., that contains an active drug ingredient generally, but not necessarily, in association with inactive ingredients. The term also includes a finished dosage form that does not contain an active ingredient but is intended to be used as a placebo.

(5) *Fiber* means any particulate contaminant with a length at least three times greater than its width.

(6) *Non-fiber-releasing filter* means any filter, which after any appropriate pretreatment such as washing or flushing, will not release fibers into the component or drug product that is being filtered. All filters composed of asbestos are deemed to be fiber-releasing filters.

(7) *Active ingredient* means any component that is intended to furnish pharmacological activity or other direct effect in the diagnosis, cure, mitigation, treatment, or prevention of disease, or to affect the structure or any function of the body of man or other animals. The term includes those components that may undergo chemical change in the manufacture of the drug product and be present in the drug product in a modified form intended to furnish the specified activity or effect.

(8) *Inactive ingredient* means any component other than an active ingredient.

(9) *In-process material* means any material fabricated, compounded, blended, or derived by chemical reaction that is produced for, and used in, the preparation of the drug product.

(10) *Lot* means a batch, or a specific identified portion of a batch, having uniform character and quality within specified limits; or, in the case of a drug product produced by continuous process, it is a specific identified amount produced in a unit of time or quantity in a manner that assures its having uniform character and quality within specified limits.

(11) *Lot number, control number, or batch number* means any distinctive combination of letters, numbers, or symbols, or any combination of them, from which the complete history of the manufacture, processing, packing, holding, and distribution of a batch or lot of drug product or other material can be determined.

(12) *Manufacture, processing, packing, or holding of a drug product* includes packaging and labeling operations, testing, and quality control of drug products.

(13) The term *medicated feed* means any Type B or Type C medicated feed as defined in §558.3 of this chapter. The feed contains one or more drugs as defined in section 201 (g) of the act. The manufacture of medicated feeds is subject to the requirements of part 225 of this chapter.

(14) The term *medicated premix* means a Type A medicated article as defined in §558.3 of this chapter. The article contains one or more drugs as defined in section 201(g) of the act. The manufacture of medicated premixes is subject to the requirements of part 226 of this chapter.

(15) *Quality control unit* means any person or organizational element designated by the firm to be responsible for the duties relating to quality control.

(16) *Strength* means:

(i) The concentration of the drug substance (for example, weight/weight, weight/volume, or unit dose/volume basis), and/or

(ii) The potency, that is, the therapeutic activity of the drug product as indicated by appropriate laboratory tests or by adequately developed and controlled clinical data (expressed, for example, in terms of units by reference to a standard).

(17) *Theoretical yield* means the quantity that would be produced at any appropriate phase of manufacture, processing, or packing of a particular drug product, based upon the quantity of components to be used, in the absence of any loss or error in actual production.

(18) *Actual yield* means the quantity that is actually produced at any appropriate phase of manufacture, processing, or packing of a particular drug product.

(19) *Percentage of theoretical yield* means the ratio of the actual yield (at any appropriate phase of manufacture, processing, or packing of a particular drug product) to the theoretical yield (at the same phase), stated as a percentage.

(20) *Acceptance criteria* means the product

specifications and acceptance/rejection criteria, such as acceptable quality level and unacceptable quality level, with an associated sampling plan, that are necessary for making a decision to accept or reject a lot or batch (or any other convenient subgroups of manufactured units).

(21) *Representative sample* means a sample that consists of a number of units that are drawn based on rational criteria such as random sampling and intended to assure that the sample accurately portrays the material being sampled.

21 CFR PART 211--CURRENT GOOD MANUFACTURING PRACTICE FOR FINISHED PHARMACEUTICALS

Subpart A--General Provisions

§211.1 Scope.

(a) The regulations in this part contain the minimum current good manufacturing practice for preparation of drug products for administration to humans or animals.

(b) The current good manufacturing practice regulations in this chapter, as they pertain to drug products, and in parts 600 through 680 of this chapter, as they pertain to biological products for human use, shall be considered to supplement, not supersede, the regulations in this part unless the regulations explicitly provide otherwise. In the event it is impossible to comply with applicable regulations both in this part and in other parts of this chapter or in parts 600 through 680 of this chapter, the regulation specifically applicable to the drug product in question shall supersede the regulation in this part.

(c) Pending consideration of a proposed exemption, published in the FEDERAL REGISTER of September 29, 1978, the requirements in this part shall not be enforced for OTC drug products if the products and all their ingredients are ordinarily marketed and consumed as human foods, and which products may also fall within the legal definition of drugs by virtue of their intended use. Therefore, until further notice, regulations under part 110 of this chapter, and where applicable, parts 113 to 129 of this chapter, shall be applied in determining whether these OTC drug products that are also foods are manufactured, processed, packed, or held under current good manufacturing practice.

§211.3 Definitions.

The definitions set forth in §210.3 of this chapter apply in this part.

Subpart B--Organization and Personnel

§211.22 Responsibilities of quality control unit.

(a) There shall be a quality control unit that shall have the responsibility and authority to approve or reject all components, drug product containers, closures, in-process materials, packaging material, labeling, and drug products, and the authority to review production records to assure that no errors have occurred or, if errors have occurred, that they have been fully investigated. The quality control unit shall be responsible for approving or rejecting drug products manufactured, processed, packed, or held under contract by another company.

(b) Adequate laboratory facilities for the testing and approval (or rejection) of components, drug product containers, closures, packaging materials, in-process materials, and drug products shall be available to the quality control unit.

(c) The quality control unit shall have the responsibility for approving or rejecting all procedures or specifications impacting on the identity, strength, quality, and purity of the drug product.

(d) The responsibilities and procedures applicable to the quality control unit shall be in writing; such written procedures

shall be followed.

§211.25 Personnel qualifications.

(a) Each person engaged in the manufacture, processing, packing, or holding of a drug product shall have education, training, and experience, or any combination thereof, to enable that person to perform the assigned functions. Training shall be in the particular operations that the employee performs and in current good manufacturing practice (including the current good manufacturing practice regulations in this chapter and written procedures required by these regulations) as they relate to the employee's functions. Training in current good manufacturing practice shall be conducted by qualified individuals on a continuing basis and with sufficient frequency to assure that employees remain familiar with CGMP requirements applicable to them.

(b) Each person responsible for supervising the manufacture, processing, packing, or holding of a drug product shall have the education, training, and experience, or any combination thereof, to perform assigned functions in such a manner as to provide assurance that the drug product has the safety, identity, strength, quality, and purity that it purports or is represented to possess.

(c) There shall be an adequate number of qualified personnel to perform and supervise the manufacture, processing, packing, or holding of each drug product.

§211.28 Personnel responsibilities.

(a) Personnel engaged in the manufacture, processing, packing, or holding of a drug product shall wear clean clothing appropriate for the duties they perform. Protective apparel, such as head, face, hand, and arm coverings, shall be worn as necessary to protect drug products from contamination.

(b) Personnel shall practice good sanitation and health habits.

(c) Only personnel authorized by supervisory personnel shall enter those areas of the buildings and facilities designated as limited-access areas.

(d) Any person shown at any time (either by medical examination or supervisory observation) to have an apparent illness or open lesions that may adversely affect the safety or quality of drug products shall be excluded from direct contact with components, drug product containers, closures, in-process materials, and drug products until the condition is corrected or determined by competent medical personnel not to jeopardize the safety or quality of drug products. All personnel shall be instructed to report to supervisory personnel any health conditions that may have an adverse effect on drug products.

§211.34 Consultants.

Consultants advising on the manufacture, processing, packing, or holding of drug products shall have sufficient education, training, and experience, or any combination thereof, to advise on the subject for which they are retained. Records shall be maintained stating the name, address, and qualifications of any consultants and the type of service they provide.

Subpart C--Buildings and Facilities

§211.42 Design and construction features.

(a) Any building or buildings used in the manufacture, processing, packing, or holding of a drug product shall be of suitable size, construction and location to facilitate cleaning, maintenance, and proper operations.

(b) Any such building shall have adequate space for the orderly placement of equipment and materials to prevent mixups between different components, drug product containers, closures, labeling, in-process materials, or drug products, and to prevent contamination. The flow of components, drug product containers, closures, labeling, in-process materials, and drug products through the building or buildings shall be designed to prevent contamination.

(c) Operations shall be performed within specifically defined areas of adequate size. There shall be separate or defined areas for the firm's operations to prevent contamination or mixups as follows:

(1) Receipt, identification, storage, and withholding from use of components, drug product containers, closures, and labeling, pending the appropriate sampling, testing, or examination by the quality control unit before release for manufacturing or packaging;

(2) Holding rejected components, drug product containers, closures, and labeling before disposition;

(3) Storage of released components, drug product containers, closures, and labeling;

(4) Storage of in-process materials;

(5) Manufacturing and processing operations;

(6) Packaging and labeling operations;

(7) Quarantine storage before release of drug products;

(8) Storage of drug products after release;

(9) Control and laboratory operations;

(10) Aseptic processing, which includes as appropriate:

(i) Floors, walls, and ceilings of smooth, hard surfaces that are easily cleanable;

(ii) Temperature and humidity controls;

(iii) An air supply filtered through high-efficiency particulate air filters under positive pressure, regardless of whether flow is laminar or nonlaminar;

(iv) A system for monitoring environmental conditions;

(v) A system for cleaning and disinfecting the room and equipment to produce aseptic conditions;

(vi) A system for maintaining any equipment used to control the aseptic conditions.

(d) Operations relating to the manufacture, processing, and packing of penicillin shall be performed in facilities separate from those used for other drug products for human use.

§211.44 Lighting.

Adequate lighting shall be provided in all areas.

§211.46 Ventilation, air filtration, air heating and cooling.

(a) Adequate ventilation shall be provided.

(b) Equipment for adequate control over air pressure, micro-organisms, dust, humidity, and temperature shall be provided when appropriate for the manufacture, processing, packing, or holding of a drug product.

(c) Air filtration systems, including prefilters and particulate matter air filters, shall be used when appropriate on air supplies to production areas. If air is recirculated to production areas, measures shall be taken to control recirculation of dust from production. In areas where air contamination occurs during production, there shall be adequate exhaust systems or other systems adequate to control contaminants.

(d) Air-handling systems for the manufacture, processing, and packing of penicillin shall be completely separate from those for other drug products for human use.

§211.48 Plumbing.

(a) Potable water shall be supplied under continuous positive pressure in a plumbing system free of defects that could contribute contamination to any drug product. Potable water shall meet the standards prescribed in the Environmental Protection Agency's Primary Drinking Water Regulations set forth in 40 CFR part 141. Water not meeting such standards shall not be permitted in the potable water system.

(b) Drains shall be of adequate size and, where connected directly to a sewer, shall be provided with an air break or other mechanical device to prevent back-siphonage.

§211.50 Sewage and refuse.

Sewage, trash, and other refuse in and from the building and immediate premises shall be disposed of in a safe and sanitary manner.

§*211.52 Washing and toilet facilities.*

Adequate washing facilities shall be provided, including hot and cold water, soap or detergent, air driers or single-service towels, and clean toilet facilities easily accessible to working areas.

§*211.56 Sanitation.*

(a) Any building used in the manufacture, processing, packing, or holding of a drug product shall be maintained in a clean and sanitary condition, Any such building shall be free of infestation by rodents, birds, insects, and other vermin (other than laboratory animals). Trash and organic waste matter shall be held and disposed of in a timely and sanitary manner.

(b) There shall be written procedures assigning responsibility for sanitation and describing in sufficient detail the cleaning schedules, methods, equipment, and materials to be used in cleaning the buildings and facilities; such written procedures shall be followed.

(c) There shall be written procedures for use of suitable rodenticides, insecticides, fungicides, fumigating agents, and cleaning and sanitizing agents. Such written procedures shall be designed to prevent the contamination of equipment, components, drug product containers, closures, packaging, labeling materials, or drug products and shall be followed. Rodenticides, insecticides, and fungicides shall not be used unless registered and used in accordance with the Federal Insecticide, Fungicide, and Rodenticide Act (7 U.S.C. 135).

(d) Sanitation procedures shall apply to work performed by contractors or temporary employees as well as work performed by full-time employees during the ordinary course of operations.

§*211.58 Maintenance.*

Any building used in the manufacture, processing, packing, or holding of a drug product shall be maintained in a good state of repair.

Subpart D--Equipment

§*211.63 Equipment design, size, and location.*

Equipment used in the manufacture, processing, packing, or holding of a drug product shall be of appropriate design, adequate size, and suitably located to facilitate operations for its intended use and for its cleaning and maintenance.

§*211.65 Equipment construction.*

(a) Equipment shall be constructed so that surfaces that contact components, in-process materials, or drug products shall not be reactive, additive, or absorptive so as to alter the safety, identity, strength, quality, or purity of the drug product beyond the official or other established requirements.

(b) Any substances required for operation, such as lubricants or coolants, shall not come into contact with components, drug product containers, closures, in-process materials, or drug products so as to alter the safety, identity, strength, quality, or purity of the drug product beyond the official or other established requirements.

§*211.67 Equipment cleaning and maintenance.*

(a) Equipment and utensils shall be cleaned, maintained, and sanitized at appropriate intervals to prevent malfunctions or contamination that would alter the safety, identity, strength, quality, or purity of the drug product beyond the official or other established requirements.

(b) Written procedures shall be established and followed for cleaning and maintenance of equipment, including utensils, used in the manufacture, processing, packing, or holding of a drug product. These procedures shall include,

but are not necessarily limited to, the following:

(1) Assignment of responsibility for cleaning and maintaining equipment;

(2) Maintenance and cleaning schedules, including, where appropriate, sanitizing schedules;

(3) A description in sufficient detail of the methods, equipment, and materials used in cleaning and maintenance operations, and the methods of disassembling and reassembling equipment as necessary to assure proper cleaning and maintenance;

(4) Removal or obliteration of previous batch identification;

(5) Protection of clean equipment from contamination prior to use;

(6) Inspection of equipment for cleanliness immediately before use.

(c) Records shall be kept of maintenance, cleaning, sanitizing, and inspection as specified in §§211.180 and 211. 182.

§211.68 Automatic, mechanical, and electronic equipment.

(a) Automatic, mechanical, or electronic equipment or other types of equipment, including computers, or related systems that will perform a function satisfactorily, may be used in the manufacture, processing, packing, and holding of a drug product. If such equipment is so used, it shall be routinely calibrated, inspected, or checked according to a written program designed to assure proper performance. Written records of those calibration checks and inspections shall be maintained.

(b) Appropriate controls shall be exercised over computer or related systems to assure that changes in master production and control records or other records are instituted only by authorized personnel. Input to and output from the computer or related system of formulas or other records or data shall be checked for accuracy. A backup file of data entered into the computer or related system shall be maintained except where certain data, such as calculations performed in connection with laboratory analysis, are eliminated by computerization or other automated processes. In such instances a written record of the program shall be maintained along with appropriate validation data. Hard copy or alternative systems, such as duplicates, tapes, or microfilm, designed to assure that backup data are exact and complete and that it is secure from alteration, inadvertent erasures, or loss shall be maintained.

§211.72 Filters.

Filters for liquid filtration used in the manufacture, processing, or packing of injectable drug products intended for human use shall not release fibers into such products. Fiber-releasing filters may not be used in the manufacture, processing, or packing of these injectable drug products unless it is not possible to manufacture such drug products without the use of such filters. If use of a fiber-releasing filter is necessary, an additional non-fiber-releasing filter of 0.22 micron maximum mean porosity (0.45 micron if the manufacturing conditions so dictate) shall subsequently be used to reduce the content of particles in the injectable drug product. Use of an asbestos-containing filter, with or without subsequent use of a specific non-fiber-releasing filter, is permissible only upon submission of proof to the appropriate bureau of the Food and Drug Administration that use of a non-fiber-releasing filter will, or is likely to, compromise the safety or effectiveness of the injectable drug product.

Subpart E--Control of Components and Drug Product Containers and Closures

§211.80 General requirements.

(a) There shall be written procedures describing in sufficient detail the receipt, identification, storage, handling, sampling, testing, and approval or rejection of components and drug product containers and closures; such written procedures shall be followed.

(b) Components and drug product containers and closures shall at all times be handled and stored in a manner to prevent contamination.

(c) Bagged or boxed components of drug product containers, or closures shall be stored off the floor and suitably spaced to permit cleaning and inspection.

(d) Each container or grouping of containers for components or drug product containers, or closures shall be identified with a distinctive code for each lot in each shipment received. This code shall be used in recording the disposition of each lot. Each lot shall be appropriately identified as to its status (i.e., quarantined, approved, or rejected).

§211.82 Receipt and storage of untested components, drug product containers, and closures.

(a) Upon receipt and before acceptance, each container or grouping of containers of components, drug product containers, and closures shall be examined visually for appropriate labeling as to contents, container damage or broken seals, and contamination.

(b) Components, drug product containers, and closures shall be stored under quarantine until they have been tested or examined, as appropriate, and released. Storage within the area shall conform to the requirements of §211.80.

§211.84 Testing and approval or rejection of components, drug product containers, and closures.

(a) Each lot of components, drug product containers, and closures shall be withheld from use until the lot has been sampled, tested, or examined, as appropriate, and released for use by the quality control unit.

(b) Representative samples of each shipment of each lot shall be collected for testing or examination. The number of containers to be sampled, and the amount of material to be taken from each container, shall be based upon appropriate criteria such as statistical criteria for component variability, confidence levels, and degree of precision desired, the past quality history of the supplier, and the quantity needed for analysis and reserve where required by §211.170.

(c) Samples shall be collected in accordance with the following procedures:

(1) The containers of components selected shall be cleaned where necessary, by appropriate means.

(2) The containers shall be opened, sampled, and resealed in a manner designed to prevent contamination of their contents and contamination of other components, drug product containers, or closures.

(3) Sterile equipment and aseptic sampling techniques shall be used when necessary.

(4) If it is necessary to sample a component from the top, middle, and bottom of its container, such sample subdivisions shall not be composited for testing.

(5) Sample containers shall be identified so that the following information can be determined: name of the material sampled, the lot number, the container from which the sample was taken, the data on which the sample was taken, and the name of the person who collected the sample.

(6) Containers from which samples have been taken shall be marked to show that samples have been removed from them.

(d) Samples shall be examined and tested as follows:

(1) At least one test shall be conducted to verify the identity of each component of a drug product. Specific identity tests, if they exist, shall be used.

(2) Each component shall be tested for conformity with all appropriate written specifications for purity, strength, and quality. In lieu of such testing by the

manufacturer, a report of analysis may be accepted from the supplier of a component, provided that at least one specific identity test is conducted on such component by the manufacturer, and provided that the manufacturer establishes the reliability of the supplier's analyses through appropriate validation of the supplier's test results at appropriate intervals.

(3) Containers and closures shall be tested for conformance with all appropriate written procedures. In lieu of such testing by the manufacturer, a certificate of testing may be accepted from the supplier, provided that at least a visual identification is conducted on such containers/closures by the manufacturer and provided that the manufacturer establishes the reliability of the supplier's test results through appropriate validation of the supplier's test results at appropriate intervals.

(4) When appropriate, components shall be microscopically examined.

(5) Each lot of a component, drug product container, or closure that is liable to contamination with filth, insect infestation, or other extraneous adulterant shall be examined against established specifications for such contamination.

(6) Each lot of a component, drug product container, or closure that is liable to microbiological contamination that is objectionable in view of its intended use shall be subjected to microbiological tests before use.

(e) Any lot of components, drug product containers, or closures that meets the appropriate written specifications of identity, strength, quality, and purity and related tests under paragraph (d) of this section may be approved and released for use. Any lot of such material that does not meet such specifications shall be rejected.

§211.86 Use of approved components, drug product containers, and closures.

Components, drug product containers, and closures approved for use shall be rotated so that the oldest approved stock is used first. Deviation from this requirement is permitted if such deviation is temporary and appropriate.

§211.87 Retesting of approved components, drug product containers, and closures.

Components, drug product containers, and closures shall be retested or reexamined, as appropriate, for identity, strength, quality, and purity and approved or rejected by the quality control unit in accordance with §211.84 as necessary, e.g., after storage for long periods or after exposure to air, heat or other conditions that might adversely affect the component, drug product container, or closure.

§211.89 Rejected components, drug product containers, and closures.

Rejected components, drug product containers, and closures shall be identified and controlled under a quarantine system designed to prevent their use in manufacturing or processing operations for which they are unsuitable.

§211.94 Drug product containers and closures.

(a) Drug product containers and closures shall not be reactive, additive, or absorptive so as to alter the safety, identity, strength, quality, or purity of the drug beyond the official or established requirements.

(b) Container closure systems shall provide adequate protection against foreseeable external factors in storage and use that can cause deterioration or contamination of the drug product.

(c) Drug product containers and closures shall be clean and, where indicated by the nature of the drug, sterilized and processed to remove pyrogenic properties to assure that they are suitable for their intended use.

(d) Standards or specifications, methods of testing, and, where indicated, methods of cleaning, sterilizing, and processing to remove pyrogenic properties shall be written and followed for drug product containers and closures.

Subpart F--Production and Process Controls

§*211.100 Written procedures; deviations.*

(a) There shall be written procedures for production and process control designed to assure that the drug products have the identity, strength, quality, and purity they purport or are represented to possess. Such procedures shall include all requirements in this subpart. These written procedures, including any changes, shall be drafted, reviewed, and approved by the appropriate organizational units and reviewed and approved by the quality control unit.

(b) Written production and process control procedures shall be followed in the execution of the various production and process control functions and shall be documented at the time of performance. Any deviation from the written procedures shall be recorded and justified.

§*211.101 Charge-in of components.*

Written production and control procedures shall include the following, which are designed to assure that the drug products produced have the identity, strength, quality, and purity they purport or are represented to possess:

(a) The batch shall be formulated with the intent to provide not less than 100 percent of the labeled or established amount of active ingredient.

(b) Components for drug product manufacturing shall be weighed, measured, or subdivided as appropriate. If a component is removed from the original container to another, the new container shall be identified with the following information:

(1) Component name or item code;

(2) Receiving or control number;

(3) Weight or measure in new container;

(4) Batch for which component was dispensed, including its product name, strength, and lot number.

(c) Weighing, measuring, or subdividing operations for components shall be adequately supervised. Each container of component dispensed to manufacturing shall be examined by a second person to assure that:

(1) The component was released by the quality control unit;

(2) The weight or measure is correct as stated in the batch production records;

(3) The containers are properly identified.

(d) Each component shall be added to the batch by one person and verified by a second person.

§*211.103 Calculation of yield.*

Actual yields and percentages of theoretical yield shall be determined at the conclusion of each appropriate phase of manufacturing, processing, packaging, or holding of the drug product. Such calculations shall be performed by one person and independently verified by a second person.

§*211.105 Equipment identification.*

(a) All compounding and storage containers, processing lines, and major equipment used during the production of a batch of a drug product shall be properly identified at all times to indicate their contents and, when necessary, the phase of processing of the batch.

(b) Major equipment shall be identified by a distinctive identification number or code that shall be recorded in the batch production record to show the specific equipment used in the manufacture of each batch of a drug product. In cases where only one of a particular type of equipment exists in a manufacturing facility, the name of the equipment may be used in lieu of a distinctive identification number or code.

§*211.110 Sampling and testing of in-process materials and drug products.*

(a) To assure batch uniformity and integrity of drug products, **written procedures shall be established and followed that describe the inprocess controls, and tests, or examinations to be conducted on appropriate samples of in-process materials of each batch. Such control procedures shall be established to monitor the output and to validate the performance of those manufacturing processes that may be responsible for causing variability in the characteristics of in-process material and the drug product.** Such control procedures shall include, but are not limited to, the following, where appropriate:

(1) Tablet or capsule weight variation;

(2) Disintegration time;

(3) Adequacy of mixing to assure uniformity and homogeneity;

(4) Dissolution time and rate;

(5) Clarity, completeness, or pH of solutions.

(b) Valid in-process specifications for such characteristics shall be consistent with drug product final specifications and shall be derived from previous acceptable process average and process variability estimates where possible and determined by the application of suitable statistical procedures where appropriate. Examination and testing of samples shall assure that the drug product and in-process material conform to specifications.

(c) In-process materials shall be tested for identity, strength, quality, and purity as appropriate, and approved or rejected by the quality control unit, during the production process, e.g., at commencement or completion of significant phases or after storage for long periods.

(d) Rejected in-process materials shall be identified and controlled under a quarantine system designed to prevent their use in manufacturing or processing operations for which they are unsuitable.

§211.111 *Time limitations on production.*

When appropriate, time limits for the completion of each phase of production shall be established to assure the quality of the drug product. Deviation from established time limits may be acceptable if such deviation does not compromise the quality of the drug product. Such deviation shall be justified and documented.

§211.113 *Control of microbiological contamination.*

(a) Appropriate written procedures, designed to prevent objectionable microorganisms in drug products not required to be sterile, shall be established and followed.

(b) Appropriate written procedures, designed to prevent microbiological contamination of drug products purporting to be sterile, shall be established and followed. Such procedures shall include validation of any sterilization process.

§211.115 *Reprocessing.*

(a) Written procedures shall be established and followed prescribing a system for reprocessing batches that do not conform to standards or specifications and the steps to be taken to insure that the reprocessed batches will conform with all established standards, specifications, and characteristics.

(b) Reprocessing shall not be performed without the review and approval of the quality control unit.

Subpart G--Packaging and Labeling Control

§211.122 *Materials examination and usage criteria.*

(a) There shall be written procedures describing in sufficient detail the receipt, identification, storage, handling, sampling, examination, and/or testing of labeling and packaging materials; such written procedures shall be followed. Labeling and packaging materials shall be representatively sampled, and examined or tested upon receipt and before use in packaging or labeling of a drug product.

(b) Any labeling or packaging materials meeting appropriate written specifications may be approved and released for use. Any labeling or packaging materials that do not meet such specifications shall be rejected to prevent their use in operations for which they are unsuitable.

(c) Records shall be maintained for each shipment received of each different labeling and packaging material indicating receipt, examination or testing, and whether accepted or rejected.

(d) Labels and other labeling materials for each different drug product, strength, dosage form, or quantity of contents shall be stored separately with suitable identification. Access to the storage area shall be limited to authorized personnel.

(e) Obsolete and outdated labels, labeling, and other packaging materials shall be destroyed.

(f) Gang printing of labeling to be used for different drug products or different strengths of the same drug product (or labeling of the same size and identical or similar format and/or color schemes) shall be minimized. If gang printing is employed, packaging and labeling operations shall provide for special control procedures, taking into consideration sheet layout, stacking, cutting, and handling during and after printing.

(g) Printing devices on, or associated with, manufacturing lines used to imprint labeling upon the drug product unit label or case shall be monitored to assure that all imprinting conforms to the print specified in the batch production record.

§211.125 Labeling issuance.

(a) Strict control shall be exercised over labeling issued for use in drug product labeling operations.

(b) Labeling materials issued for a batch shall be carefully examined for identity and conformity to the labeling specified in the master or batch production records.

(c) Procedures shall be utilized to reconcile the quantities of labeling issued, used, and returned, and shall require evaluation of discrepancies found between the quantity of drug product finished and the quantity of labeling issued when such discrepancies are outside narrow preset limits based on historical operating data. Such discrepancies shall be investigated in accordance with §211.192.

(d) All excess labeling bearing lot or control numbers shall be destroyed.

(e) Returned labeling shall be maintained and stored in a manner to prevent mixups and provide proper identification.

(f) Procedures shall be written describing in sufficient detail the control procedures employed for the issuance of labeling; such written procedures shall be followed.

§211.130 Packaging and labeling operations.

There shall be written procedures designed to assure that correct labels, labeling, and packaging materials are used for drug products; such written procedures shall be followed. These procedures shall incorporate the following features:

(a) Prevention of mixups and cross-contamination by physical or spatial separation from operations on other drug products.

(b) Identification of the drug product with a lot or control number that permits determination of the history of the manufacture and control of the batch.

(c) Examination of packaging and labeling materials for suitability and correctness before packaging operations, and documentation of such examination in the batch production record.

(d) Inspection of the packaging and labeling facilities immediately before use to assure that all drug products have been removed from previous operations. Inspection shall also be made to assure that packaging and labeling materials not suitable for subsequent operations have been removed. Results of inspection shall be documented in the batch production records.

§211.132 Tamper-resistant packaging requirements for over-the-counter (OTC) human drug products.

(a) General. The Food and Drug Administration has the authority under the Federal Food, Drug, and Cosmetic Act (the act) to establish a uniform national requirement for tamper-resistant packaging of OTC drug products that will improve the security of OTC drug packaging and help assure the safety and effectiveness of OTC drug products. An OTC drug product (except a dermatological, dentifrice, insulin, or throat lozenge product) for retail sale that is not packaged in a tamper-resistant package or that is not properly labeled under this section is adulterated under section 501 of the act or misbranded under section 502 of the act, or both.

(b) Requirement for tamper-resistant package. Each manufacturer and packer who packages an OTC drug product (except a dermatological, dentifrice, insulin, or throat lozenge product) for retail sale shall package the product in a tamper-resistant package, if this product is accessible to the public while held for sale. A tamper-resistant package is one having one or more indicators or barriers to entry which, if breached or missing, can reasonably be expected to provide visible evidence to consumers that tampering has occurred. To reduce the likelihood of successful tampering and to increase the likelihood that consumers will discover if a product has been tampered with, the package is required to be distinctive by design (e.g., an aerosol product container) or by the use of one or more indicators or barriers to entry that employ an identifying characteristic (e.g., a pattern, name, registered trademark, logo, or picture). For purposes of this section, the term "distinctive by design" means the packaging cannot be duplicated with commonly available materials or through commonly available processes. For purposes of this section, the term "aerosol product" means a product which depends upon the power of a liquified or compressed gas to expel the contents from the container. A tamper-resistant package may involve an immediate-container and closure system or secondary-container or carton system or any combination of systems intended to provide a visual indication of package integrity. The tamper-resistant feature shall be designed to and shall remain intact when handled in a reasonable manner during manufacture, distribution, and retail display.

(1) For two-piece, hard gelatin capsule products subject to this requirement, a minimum of two tamper-resistant packaging features is required, unless the capsules are sealed by a tamper-resistant technology.

(2) For all other products subject to this requirement, including two-piece, hard gelatin capsules that are sealed by a tamper-resistant technology, a minimum of one tamper-resistant feature is required.

(c) Labeling. Each retail package of an OTC drug product covered by this section, except ammonia inhalant in crushable glass ampules, aerosol products as defined in paragraph (b) of this section, or containers of compressed medical oxygen, is required to bear a statement that is prominently placed so that consumers are alerted to the specific tamper-resistant feature of the package. The labeling statement is also required to be so placed that it will be unaffected if the tamper-resistant feature of the package is breached or missing. If the tamper-resistant feature chosen to meet the requirement in paragraph (b) of this section is one that uses an identifying characteristic, that characteristic is required to be referred to in the labeling statement. For example, the labeling statement on a bottle with a shrink band could say "For your protection, this bottle has an imprinted seal around the neck."

(d) Request for exemptions from packaging and labeling requirements. A manufacturer or packer may request an exemption from the packaging and labeling requirements of this section. A request for an exemption is required to be submitted in the form of a citizen petition under §10.30 of this chapter and should be clearly identified on the envelope as a "Request for Exemption from Tamper-Resistant Rule." The petition is required to contain the following:

(1) The name of the drug product or, if the petition seeks an exemption for a drug class, the name of the drug class, and a list of products within that class.

(2) The reasons that the drug product's compliance with the tamper-resistant packaging or labeling requirements of this section is unnecessary or cannot be achieved.

(3) A description of alternative steps that are available, or that the petitioner has already

taken, to reduce the likelihood that the product or drug class will be the subject of malicious adulteration.

(4) Other information justifying an exemption.

(e) OTC drug products subject to approved new drug applications. Holders of approved new drug applications for OTC drug products are required under §314.70 of this chapter to provide the agency with notification of changes in packaging and labeling to comply with the requirements of this section. Changes in packaging and labeling required by this regulation may be made before FDA approval, as provided under §314. 70(c) of this chapter. Manufacturing changes by which capsules are to be sealed require prior FDA approval under §314.70(b) of this chapter.

(f) Poison Prevention Packaging Act of 1970. This section does not affect any requirements for "special packaging" as defined under §310.3(l) of this chapter and required under the Poison Prevention Packaging Act of 1970.

§*211.134 Drug product inspection.*

(a) Packaged and labeled products shall be examined during finishing operations to provide assurance that containers and packages in the lot have the correct label.

(b) A representative sample of units shall be collected at the completion of finishing operations and shall be visually examined for correct labeling.

(c) Results of these examinations shall be recorded in the batch production or control records.

§*211.137 Expiration dating.*

(a) To assure that a drug product meets applicable standards of identity, strength, quality, and purity at the time of use, it shall bear an expiration date determined by appropriate stability testing described in §211.166.

(b) Expiration dates shall be related to any storage conditions stated on the labeling, as determined by stability studies described in §211.166.

(c) If the drug product is to be reconstituted at the time of dispensing, its labeling shall bear expiration information for both the reconstituted and unreconstituted drug products.

(d) Expiration dates shall appear on labeling in accordance with the requirements of §201. 17 of this chapter.

(e) Homeopathic drug products shall be exempt from the requirements of this section.

(f) Allergenic extracts that are labeled "No U.S. Standard of Potency" are exempt from the requirements of this section.

(g) Pending consideration of a proposed exemption, published in the FEDERAL REGISTER of September 29, 1978, the requirements in this section shall not be enforced for human OTC drug products if their labeling does not bear dosage limitations and they are stable for at least 3 years as supported by appropriate stability data.

Subpart H--Holding and Distribution

§*211.142 Warehousing procedures.*

Written procedures describing the warehousing of drug products shall be established and followed. They shall include:

(a) Quarantine of drug products before release by the quality control unit.

(b) Storage of drug products under appropriate conditions of temperature, humidity, and light so that the identity, strength, quality, and purity of the drug products are not affected.

§*211.150 Distribution procedures.*

Written procedures shall be established, and followed, describing the distribution of drug products. They shall include:

(a) A procedure whereby the oldest approved stock of a drug product is distributed first. Deviation from this requirement is permitted if such deviation is temporary and appropriate.

(b) A system by which the distribution of each lot of drug product can be readily determined to facilitate its recall if necessary.

Subpart I--Laboratory Controls

§211.160 General requirements.

(a) The establishment of any specifications, standards, sampling plans, test procedures, or other laboratory control mechanisms required by this subpart, including any change in such specifications, standards, sampling plans, test procedures, or other laboratory control mechanisms, shall be drafted by the appropriate organizational unit and reviewed and approved by the quality control unit. The requirements in this subpart shall be followed and shall be documented at the time of performance. Any deviation from the written specifications, standards, sampling plans, test procedures, or other laboratory control mechanisms shall be recorded and justified.

(b) Laboratory controls shall include the establishment of scientifically sound and appropriate specifications, standards, sampling plans, and test procedures designed to assure that components, drug product containers, closures, in-process materials, labeling, and drug products conform to appropriate standards of identity, strength, quality, and purity. Laboratory controls shall include:

(1) Determination of conformance to appropriate written specifications for the acceptance of each lot within each shipment of components, drug product containers, closures, and labeling used in the manufacture, processing, packing, or holding of drug products. The specifications shall include a description of the sampling and testing procedures used. Samples shall be representative and adequately identified. Such procedures shall also require appropriate retesting of any component, drug product container, or closure that is subject to deterioration.

(2) Determination of conformance to written specifications and a description of sampling and testing procedures for in-process materials. Such samples shall be representative and properly identified.

(3) Determination of conformance to written descriptions of sampling procedures and appropriate specifications for drug products. Such samples shall be representative and properly identified.

(4) The calibration of instruments, apparatus, gauges, and recording devices at suitable intervals in accordance with an established written program containing specific directions, schedules, limits for accuracy and precision, and provisions for remedial action in the event accuracy and/or precision limits are not met. Instruments, apparatus, gauges, and recording devices not meeting established specifications shall not be used.

§211.165 Testing and release for distribution.

(a) For each batch of drug product, there shall be appropriate laboratory determination of satisfactory conformance to final specifications for the drug product, including the identity and strength of each active ingredient, prior to release. Where sterility and/or pyrogen testing are conducted on specific batches of shortlived radiopharmaceuticals, such batches may be released prior to completion of sterility and/or pyrogen testing, provided such testing is completed as soon as possible.

(b) There shall be appropriate laboratory testing, as necessary, of each batch of drug product required to be free of objectionable microorganisms.

(c) Any sampling and testing plans shall be described in written procedures that shall include the method of sampling and the number of units per batch to be tested; such written procedure shall be followed.

(d) Acceptance criteria for the sampling and testing conducted by the quality control unit shall be adequate to assure that batches of drug products meet each appropriate specification and appropriate statistical quality control criteria as a condition for their approval and release. The statistical quality control criteria shall include appropriate acceptance levels and/or appropriate rejection levels.

(e) The accuracy, sensitivity, specificity, and reproducibility of test methods employed by the firm shall be established and documented. Such validation and documentation may be accomplished in accordance with §211.194(a) (2).

(f) Drug products failing to meet established standards or specifications and any other relevant quality control criteria shall be rejected. Reprocessing may be performed. Prior to acceptance and use, reprocessed material must meet appropriate standards, specifications, and any other relevant critieria.

§*211.166 Stability testing.*

(a) There shall be a written testing program designed to assess the stability characteristics of drug products. The results of such stability testing shall be used in determining appropriate storage conditions and expiration dates. The written program shall be followed and shall include:

(1) Sample size and test intervals based on statistical criteria for each attribute examined to assure valid estimates of stability;

(2) Storage conditions for samples retained for testing;

(3) Reliable, meaningful, and specific test methods;

(4) Testing of the drug product in the same container-closure system as that in which the drug product is marketed;

(5) Testing of drug products for reconstitution at the time of dispensing (as directed in the labeling) as well as after they are reconstituted.

(b) An adequate number of batches of each drug product shall be tested to determine an appropriate expiration date and a record of such data shall be maintained. Accelerated studies, combined with basic stability information on the components, drug products, and container/closure system, may be used to support tentative expiration dates provided full shelf life studies are not available and are being conducted. Where data from accelerated studies are used to project a tentative expiration date that is beyond a date supported by actual shelf life studies, there must be stability studies conducted, including drug product testing at appropriate intervals, until the tentative expiration date is verified or the appropriate expiration date determined.

(c) For homeopathic drug products, the requirements of this section are as follows:

(1) There shall be a written assessment of stability based at least on testing or examination of the drug product for compatibility of the ingredients, and based on marketing experience with the drug product to indicate that there is no degradation of the product for the normal or expected period of use.

(2) Evaluation of stability shall be based on the same container-closure system in which the drug product is being marketed.

(d) Allergenic extracts that are labeled "No U.S. Standard of Potency" are exempt from the requirements of this section.

§*211.167 Special testing requirements.*

(a) For each batch of drug product purporting to be sterile and/or pyrogen-free, there shall be appropriate laboratory testing to determine conformance to such requirements. The test procedures shall be in writing and shall be followed.

(b) For each batch of ophthalmic ointment, there shall be appropriate testing to determine conformance to specifications regarding the presence of foreign particles and harsh or abrasive substances. The test procedures shall be in writing and shall be followed.

(c) For each batch of controlled-release dosage form, there shall be appropriate laboratory testing to determine conformance to the specifications for the rate of release of each active ingredient. The test procedures shall be in writing and shall be followed.

§*211.170 Reserve samples.*

(a) An appropriately identified reserve sample that is representative of each lot in each shipment of each active ingredient shall be retained. The reserve sample consists of at least twice the quantity necessary for all tests required to determine whether the active ingredient meets its established specifications, except for sterility and pyrogen testing. The retention time is as follows:

(1) For an active ingredient in a drug product other than those described in paragraphs (a)(2) and (3) of this section, the

reserve sample shall be retained for 1 year after the expiration date of the last lot of the drug product containing the active ingredient.

(2) For an active ingredient in a radioactive drug product, except for nonradioactive reagent kits, the reserve sample shall be retained for:

(i) Three months after the expiration date of the last lot of the drug product containing the active ingredient if the expiration dating period of the drug product is 30 days or less; or

(ii) Six months after the expiration date of the last lot of the drug product containing the active ingredient if the expiration dating period of the drug product is more than 30 days.

(3) For an active ingredient in an OTC drug product that is exempt from bearing an expiration date under §211.137, the reserve sample shall be retained for 3 years after distribution of the last lot of the drug product containing the active ingredient.

(b) An appropriately identified reserve sample that is representative of each lot or batch of drug product shall be retained and stored under conditions consistent with product labeling. The reserve sample shall be stored in the same immediate container-closure system in which the drug product is marketed or in one that has essentially the same characteristics. The reserve sample consists of at least twice the quantity necessary to perform all the required tests, except those for sterility and pyrogens. Reserve samples, except those drugs products described in paragraph (b)(2), shall be examined visually at least once a year for evidence of deterioration unless visual examination would affect the integrity of the reserve samples. Any evidence of reserve sample deterioration shall be investigated in accordance with §211. 192. The results of the examination shall be recorded and maintained with other stability data on the drug product. Reserve samples of compressed medical gases need not be retained. The retention time is as follows:

(1) For a drug product other than those described in paragraphs (b) (2) and (3) of this section, the reserve sample shall be retained for 1 year after the expiration date of the drug product.

(2) For a radioactive drug product, except for nonradioactive reagent kits, the reserve sample shall be retained for:

(i) Three months after the expiration date of the drug product if the expiration dating period of the drug product is 30 days or less; or

(ii) Six months after the expiration date of the drug product if the expiration dating period of the drug product is more than 30 days.

(3) For an OTC drug product that is exempt for bearing an expiration date under §211.137, the reserve sample must be retained for 3 years after the lot or batch of drug product is distributed.

§211.173 Laboratory animals.

Animals used in testing components, in-process materials, or drug products for compliance with established specifications shall be maintained and controlled in a manner that assures their suitability for their intended use. They shall be identified, and adequate records shall be maintained showing the history of their use.

§211.176 Penicillin contamination.

If a reasonable possibility exists that a non-penicillin drug product has been exposed to cross-contamination with penicillin, the non- penicillin drug product shall be tested for the presence of penicillin. Such drug product shall not be marketed if detectable levels are found when tested according to procedures specified in 'Procedures for Detecting and Measuring Penicillin Contamination in Drugs,' which is incorporated by reference. Copies are available from the Division of Research and Testing (HFD-470), Center for Drug Evaluation and Research, Food and Drug Administration, 200 C St. SW., Washington, DC 20204, or available for inspection at the Office of the Federal Register, 800

North Capitol Street, NW., suite 700, Washington, DC 20408.

Subpart J--Records and Reports

§*211.180 General requirements.*

(a) Any production, control, or distribution record that is required to be maintained in compliance with this part and is specifically associated with a batch of a drug product shall be retained for at least 1 year after the expiration date of the batch or, in the case of certain OTC drug products lacking expiration dating because they meet the criteria for exemption under §211.137, 3 years after distribution of the batch.

(b) Records shall be maintained for all components, drug product containers, closures, and labeling for at least 1 year after the expiration date or, in the case of certain OTC drug products lacking expiration dating because they meet the criteria for exemption under §211.137, 3 years after distribution of the last lot of drug product incorporating the component or using the container, closure, or labeling.

(c) All records required under this part, or copies of such records, shall be readily available for authorized inspection during the retention period at the establishment where the activities described in such records occurred. These records or copies thereof shall be subject to photocopying or other means of reproduction as part of such inspection. Records that can be immediately retrieved from another location by computer or other electronic means shall be considered as meeting the requirements of this paragraph.

(d) Records required under this part may be retained either as original records or as true copies such as photocopies, microfilm, microfiche, or other accurate reproductions of the original records. Where reduction techniques, such as microfilming, are used, suitable reader and photocopying equipment shall be readily available.

(e) Written records required by this part shall be maintained so that data therein can be used for evaluating, at least annually, the quality standards of each drug product to determine the need for changes in drug product specifications or manufacturing or control procedures. Written procedures shall be established and followed for such evaluations and shall include provisions for:

(1) A review of every batch, whether approved or rejected, and, where applicable, records associated with the batch.

(2) A review of complaints, recalls, returned or salvaged drug products, and investigations conducted under §211.192 for each drug product.

(f) Procedures shall be established to assure that the responsible officials of the firm, if they are not personally involved in or immediately aware of such actions, are notified in writing of any investigations conducted under §§211.198, 211.204, or 211.208 of these regulations, any recalls, reports of inspectional observations issued by the Food and Drug Administration, or any regulatory actions relating to good manufacturing practices brought by the Food and Drug Administration.

§ *211.182 Equipment cleaning and use log.*

A written record of major equipment cleaning, maintenance (except routine maintenance such as lubrication and adjustments), and use shall be included in individual equipment logs that show the date, time, product, and lot number of each batch processed. If equipment is dedicated to manufacture of one product, then individual equipment logs are not

required, provided that lots or batches of such product follow in numerical order and are manufactured in numerical sequence. In cases where dedicated equipment is employed, the records of cleaning, maintenance, and use shall be part of the batch record. The persons performing and double-checking the cleaning and maintenance shall date and sign or initial the log indicating that the work was performed. Entries in the log shall be in chronological order.

§211.184 Component, drug product container, closure, and labeling records.

These records shall include the following:

(a) The identity and quantity of each shipment of each lot of components, drug product containers, closures, and labeling; the name of the supplier; the supplier's lot number(s) if known; the receiving code as specified in §211. 80; and the date of receipt. The name and location of the prime manufacturer, if different from the supplier, shall be listed if known.

(b) The results of any test or examination performed (including those performed as required by §211.82(a), §211.84(d), or §211. 122(a)) and the conclusions derived therefrom.

(c) An individual inventory record of each component, drug product container, and closure and, for each component, a reconciliation of the use of each lot of such component. The inventory record shall contain sufficient information to allow determination of any batch or lot of drug product associated with the use of each component, drug product container, and closure.

(d) Documentation of the examination and review of labels and labeling for conformity with established specifications in accord with §§211.122(c) and 211.130(c).

(e) The disposition of rejected components, drug product containers, closure, and labeling.

§211.186 Master production and control records.

(a) To assure uniformity from batch to batch, master production and control records for each drug product, including each batch size thereof, shall be prepared, dated, and signed (full signature, handwritten) by one person and independently checked, dated, and signed by a second person. The preparation of master production and control records shall be described in a written procedure and such written procedure shall be followed.

(b) Master production and control records shall include:

(1) The name and strength of the product and a description of the dosage form;

(2) The name and weight or measure of each active ingredient per dosage unit or per unit of weight or measure of the drug product, and a statement of the total weight or measure of any dosage unit;

(3) A complete list of components designated by names or codes sufficiently specific to indicate any special quality characteristic;

(4) An accurate statement of the weight or measure of each component, using the same weight system (metric, avoirdupois, or apothecary) for each component. Reasonable variations may be permitted, however, in the amount of components necessary for the preparation in the dosage form, provided they are justified in the master production and control records;

(5) A statement concerning any calculated excess of component;

(6) A statement of theoretical weight or measure at appropriate phases of processing;

(7) A statement of theoretical yield, including the maximum and minimum percentages of theoretical yield beyond which investigation according to §211.192 is required;

(8) A description of the drug product containers, closures, and packaging materials, including a specimen or copy of each label and all other labeling signed and dated by the person or persons responsible for approval of such labeling;

(9) Complete manufacturing and control instructions, sampling and testing procedures, specifications, special notations, and precautions to be followed.

§211.188 Batch production and control records.

Batch production and control records shall be prepared for each batch of drug product produced and shall include complete information relating to the production and control of each batch. These records shall include:

(a) An accurate reproduction of the appropriate master production or control record, checked for accuracy, dated, and signed;

(b) Documentation that each significant step in the manufacture, processing, packing, or holding of the batch was accomplished, including:

(1) Dates;

(2) Identity of individual major equipment and lines used;

(3) Specific identification of each batch of component or in-process material used;

(4) Weights and measures of components used in the course of processing;

(5) In-process and laboratory control results;

(6) Inspection of the packaging and labeling area before and after use;

(7) A statement of the actual yield and a statement of the percentage of theoretical yield at appropriate phases of processing;

(8) Complete labeling control records, including specimens or copies of all labeling used;

(9) Description of drug product containers and closures;

(10) Any sampling performed;

(11) Identification of the persons performing and directly supervising or checking each significant step in the operation;

(12) Any investigation made according to §211.192.

(13) Results of examinations made in accordance with §211.134.

§211.192 Production record review.

All drug product production and control records, including those for packaging and labeling, shall be reviewed and approved by the quality control unit to determine compliance with all established, approved written procedures before a batch is released or distributed. Any unexplained discrepancy (including a percentage of theoretical yield exceeding the maximum or minimum percentages established in master production and control records) or the failure of a batch or any of its components to meet any of its specifications shall be thoroughly investigated, whether or not the batch has already been distributed. The investigation shall extend to other batches of the same drug product and other drug products that may have been associated with the specific failure or discrepancy. A written record of the investigation shall be made and shall include the conclusions and followup.

§211.194 Laboratory records.

(a) Laboratory records shall include complete data derived from all tests necessary to assure compliance with established specifications and standards, including examinations and assays, as follows:

(1) A description of the sample received for testing with identification of source (that is, location from where sample was obtained), quantity, lot number or other distinctive code, date sample was taken, and date sample was received for testing.

(2) A statement of each method used in the testing of the sample. The statement shall indicate the location of data that establish that the methods used in the testing of the sample meet proper standards of accuracy and reliability as applied to the product tested. (If the method employed is in the current revision of the United States Pharmacopeia, National Formulary, Association of Official Analytical Chemists, Book of Methods, or in other recognized standard references, or is detailed

in an approved new drug application and the referenced method is not modified, a statement indicating the method and reference will suffice). The suitability of all testing methods used shall be verified under actual conditions of use.

(3) A statement of the weight or measure of sample used for each test, where appropriate.

(4) A complete record of all data secured in the course of each test, including all graphs, charts, and spectra from laboratory instrumentation, properly identified to show the specific component, drug product container, closure, in-process material, or drug product, and lot tested.

(5) A record of all calculations performed in connection with the test, including units of measure, conversion factors, and equivalency factors.

(6) A statement of the results of tests and how the results compare with established standards of identity, strength, quality, and purity for the component, drug product container, closure, in-process material, or drug product tested.

(7) The initials or signature of the person who performs each test and the date(s) the tests were performed.

(8) The initials or signature of a second person showing that the original records have been reviewed for accuracy, completeness, and compliance with established standards.

(b) Complete records shall be maintained of any modification of an established method employed in testing. Such records shall include the reason for the modification and data to verify that the modification produced results that are at least as accurate and reliable for the material being tested as the established method.

(c) Complete records shall be maintained of any testing and standardization of laboratory reference standards, reagents, and standard solutions.

(d) Complete records shall be maintained of the periodic calibration of laboratory instruments, apparatus, gauges, and recording devices required by §211.160(b)(4).

(e) Complete records shall be maintained of all stability testing performed in accordance with §211.166.

§211.196 Distribution records.

Distribution records shall contain the name and strength of the product and description of the dosage form, name and address of the consignee, date and quantity shipped, and lot or control number of the drug product. For compressed medical gas products, distribution records are not required to contain lot or control numbers.

§211.198 Complaint files.

(a) Written procedures describing the handling of all written and oral complaints regarding a drug product shall be established and followed. Such procedures shall include provisions for review by the quality control unit, of any complaint involving the possible failure of a drug product to meet any of its specifications and, for such drug products, a determination as to the need for an investigation in accordance with §211.192. Such procedures shall include provisions for review to determine whether the complaint represents a serious and unexpected adverse drug experience which is required to be reported to the Food and Drug Administration in accordance with §310.305 of this chapter.

(b) A written record of each complaint shall be maintained in a file designated for drug product complaints. The file regarding such drug product complaints shall be maintained at the establishment where the drug product involved was manufactured, processed, or packed, or such file may be maintained at another facility if the written records in such files are readily available for inspection at that other facility. Written records involving a drug product shall be maintained until at least 1 year after the expiration date of the drug product, or 1 year after the date that the complaint was received, whichever is longer. In the case of certain OTC drug products lacking expiration dating because they meet the criteria for exemption under §211.137, such written records shall be maintained for 3 years after distribution of the drug product.

(1) The written record shall include the following information, where known: the name and strength of the drug product, lot number, name of complainant, nature of complaint, and reply to complainant.

(2) Where an investigation under §211.192 is conducted, the written record shall include the findings of the investigation and followup. The record or copy of the record of the investigation shall be maintained at the establishment where the investigation occurred in accordance with §211.180(c).

(3) Where an investigation under §211.192 is not conducted, the written record shall include the reason that an investigation was found not to be necessary and the name of the responsible person making such a determination.

Subpart K--Returned and Salvaged Drug Products

§211.204 Returned drug products.

Returned drug products shall be identified as such and held. If the conditions under which returned drug products have been held, stored, or shipped before or during their return, or if the condition of the drug product, its container, carton, or labeling, as a result of storage or shipping, casts doubt on the safety, identity, strength, quality or purity of the drug product, the returned drug product shall be destroyed unless examination, testing, or other investigations prove the drug product meets appropriate standards of safety, identity, strength, quality, or purity. A drug product may be reprocessed provided the subsequent drug product meets appropriate standards, specifications, and characteristics. Records of returned drug products shall be maintained and shall include the name and label potency of the drug product dosage form, lot number (or control number or batch number), reason for the return, quantity returned, date of disposition, and ultimate disposition of the returned drug product. If the reason for a drug product being returned implicates associated batches, an appropriate investigation shall be conducted in accordance with the requirements of §211.192. Procedures for the holding, testing, and reprocessing of returned drug products shall be in writing and shall be followed.

§211.208 Drug product salvaging.

Drug products that have been subjected to improper storage conditions including extremes in temperature, humidity, smoke, fumes, pressure, age, or radiation due to natural disasters, fires, accidents, or equipment failures shall not be salvaged and returned to the marketplace. Whenever there is a question whether drug products have been subjected to such conditions, salvaging operations may be conducted only if there is (a) evidence from laboratory tests and assays (including animal feeding studies where applicable) that the drug products meet all applicable standards of identity, strength, quality, and purity and (b) evidence from inspection of the premises that the drug products and their associated packaging were not subjected to improper storage conditions as a result of the disaster or accident. Organoleptic examinations shall be acceptable only as supplemental evidence that the drug products meet appropriate standards of identity, strength, quality, and purity. Records including name, lot number, and disposition shall be maintained for drug products subject to this section.

FDA's CGMPs for blood and blood products (21 CFR 606) were proposed in May 1974 to assure uniform adherence to standards used in collecting, processing, storing, and distributing blood and blood products. They were finalized in November 1975 and last revised in April 1992. Requirements in the regulation relating to system validation are highlighted in boldface.

21 CFR PART 606--CURRENT GOOD MANUFACTURING PRACTICE FOR BLOOD AND BLOOD COMPONENTS

Subpart A--General Provisions

§*606.3 Definitions.*

As used in this part:

(a) *Blood* means whole blood collected from a single donor and processed either for transfusion or further manufacturing.

(b) *Unit* means the volume of blood or one of its components in a suitable volume of anticoagulant obtained from a single collection of blood from one donor.

(c) *Component* means that part of a single-donor unit of blood separated by physical or mechanical means.

(d) *Plasma for further manufacturing* means that liquid portion of blood separated and used as material to prepare another product.

(e) *Plasmapheresis* means the procedure in which blood is removed from the donor, the plasma is separated from the formed elements and at least the red blood cells are returned to the donor. This process may be immediately repeated, once.

(f) *Plateletpheresis* means the procedure in which blood is removed from the donor, a platelet concentrate is separated, and the remaining formed elements and residual plasma are returned to the donor.

(g) *Leukapheresis* means the procedure in which blood is removed from the donor, a leukocyte concentrate is separated, and the remaining formed elements and residual plasma are returned to the donor.

(h) *Facilities* means any area used for the collection, processing, compatibility testing, storage or distribution of blood and blood components.

(i) *Processing* means any procedure employed after collection and before compatibility testing of blood and includes the identification of a unit of donor blood, the preparation of components from such unit of donor blood, serological testing, labeling and associated recordkeeping.

(j) *Compatibility testing* means the in vitro serological tests performed on donor and recipient blood samples to establish the serological matching of a donor's blood or blood components with that of a potential recipient.

Subpart B--Organization and Personnel

§*606.20 Personnel.*

(a) A blood establishment shall be under the direction of a designated, qualified person who shall exercise control of the establishment in all matters relating to compliance with the provisions of this subchapter. This person shall also have the authority to represent the establishment in all pertinent matters with the Center for Biologics Evaluation and Research and to enforce, or direct the enforcement of, discipline and the performance of assigned functions by employees engaged in the collection, processing, compatibility testing, storage and distribution of blood and blood components. The designated director shall have an understanding of the scientific principles and techniques involved in the manufacture of blood products and shall have the responsibility for ensuring that employees are adequately trained in standard operating procedures and that they are aware of the application of the pertinent provisions of this chapter to their respective functions.

(b) The personnel responsible for the collection, processing, compatibility testing, storage or distribution of blood or blood components shall be adequate in number, educational background, training and experience, including professional training as necessary, or combination

thereof, to assure competent performance of their assigned functions, and to ensure that the final product has the safety, purity, potency, identity and effectiveness it purports or is represented to possess. All personnel shall have capabilities commensurate with their assigned functions, a thorough understanding of the procedures or control operations they perform, the necessary training or experience, and adequate information concerning the application of pertinent provisions of this part to their respective functions.

(c) Persons whose presence can adversely affect the safety and purity of the products shall be excluded from areas where the collection, processing, compatibility testing, storage or distribution of blood or blood components is conducted.

Subpart C--Plant and Facilities

§606.40 Facilities

Facilities shall be maintained in a clean and orderly manner, and shall be of suitable size, construction and location to facilitate adequate cleaning, maintenance and proper operations. The facilities shall:

(a) Provide adequate space for the following when applicable:

(1) Private and accurate examinations of individuals to determine their suitability as blood donors.

(2) The withdrawal of blood from donors with minimal risk of contamination, or exposure to activities and equipment unrelated to blood collection.

(3) The storage of blood or blood components pending completion of tests.

(4) The quarantine storage of blood or blood components in a designated location pending repetition of those tests that initially gave questionable serological results.

(5) The storage of finished products prior to distribution.

(6) The quarantine storage, handling and dispostion of products and reagents not suitable for use.

(7) The orderly collection, processing compatibility testing, storage and distribution of blood and blood components to prevent contamination.

(8) The adequate and proper performance of all steps in plasmapheresis, plateletpheresis and leukapheresis procedures.

(9) The orderly conduction of all packaging, labeling and other finishing oeprations.

(b) Provide adequate lighting, ventilation and screening of open windows and doors.

(c) Provide adequate, clean, and convenient handwashing facilities for personnel, and adequate, clean, and convenient toilet facilities for donors and personnel. Drains shall be of adequate size and, where connected directly to a sewer, shall be equipped with traps to prevent back-siphonage.

(d) Provide for safe and sanitary disposal for the following:

(1) Trash and items used during the collection, processing and compatibility testing of blood and blood components.

(2) Blood and blood components not suitable for use or distribution.

Subpart D--Equipment

§606.60 Equipment.

(a) Equipment used in the collection, processing, compatibility testing, storage and distribution of blood and blood components shall be maintained in a clean and orderly manner and located so as to facilitate cleaning and maintenance. The equipment shall be observed, standardized and calibrated on a regularly scheduled basis as prescribed in the Standard Operating Procedures Manual and shall perform in the manner for which it was designed so as to assure compliance with the official requirements prescribed in this chapter for blood and blood products.

(b) Equipment that shall be observed, standardized and calibrated with at least the following frequency, include but are not limited to:

Equipment	Performance check	Frequency	Frequency of Calibration
Temperature recorder	Compare against thermometer	Daily	As necessary
Refrigerated centrifuge	Observe speed and temperature	Each day of use	Do.
Hematocrit centrifuge			Standardize before initial use, after repairs or adjustments, and annually. Timer every 3 mo.
General lab centrifuge			Tachometer every 6 mo.
Automated blood-typing machine	Observe controls for correct results	Each day of use	
Hemoglobinometer	Standardize against cyanmethemoglobin standard	do.	
Refractometer	Standardize against distilled water	do.	
Blood container scale	Standardize against container of known weight	do.	As necessary
Water bath	Observe temperature	do.	Do.
Rh view box	do.	do.	Do.
Autoclave	do.	Each time of use	Do.
Serologic rotators	Observe controls for correct results	Each day of use	Speed as necessary
Laboratory thermometers			Before initial use
Electronic thermometers			Monthly
Vacuum blood agitator	Observe weight of the first container of blood filled for correct results	Each day of use	Standardize with container of known mass or volume before initial use, and after repairs or adjustments

(c) Equipment employed in the sterilization of materials used in blood collection or for disposition of contaminated products shall be designed, maintained and utilized to ensure the destruction of contaminating microorganisms. The effectiveness of the sterilization procedure shall be no less than that achieved by an attained temperature of 121.5° C (251° F) maintained for 20 minutes by saturated steam or by an attained temperature of 170° C (338° F) maintained for 2 hours with dry heat.

§606.65 Supplies and reagents.

All supplies and reagents used in the collection, processing, compatibility testing, storage and distribution of blood and blood components shall be stored in a safe, sanitary and orderly manner.

(a) All surfaces coming in contact with blood and blood components intended for transfusion shall be sterile, pyrogen- free, and shall not interact with the product in such a manner as to have an adverse effect upon the safety, purity, potency or effectiveness of the product. All final containers and closures for blood and blood components not intended for transfusion shall be clean and free of surface solids and other contaminants.

(b) Each blood collecting container and its satellite container(s), if any, shall be examined visually for damage or evidence of contamination prior to its use and immediately after filling. Such examination shall include inspection for breakage of seals, when indicated, and abnormal discoloration. Where any defect is observed, the container shall not be used, or, if detected after filling, shall be properly discarded.

(c) Representative samples of each lot of the following reagents or solutions shall be tested on a regularly scheduled basis by methods described in the Standard Operating Procedures Manual to determine their capacity to perform as required:

Reagent or solution	Frequency of testing
Anti-human serum	Each day of use.
Blood grouping serums	Do.
Lectins	Do.
Antibody screening and reverse grouping cells	Do.
Hepatitis test reagents	Each run.
Syphilis serology reagents	Do.
Enzymes	Each day of use.

(d) Supplies and reagents that do not bear an expiration date shall be stored in such a manner that the oldest is used first.

(e) Supplies and reagents shall be used in a manner consistent with instructions provided by the manufacturer.

(f) Items that are required to be sterile and come into contact with blood should be disposable whenever possible.

Subpart E--[Reserved]

Subpart F--Production and Process Controls

§606.100 Standard operating procedures.

(a) In all instances, except clinical investigations, standard operating procedures shall comply with published additional standards in part 640 of this chapter for the products being processed; except that, references in part 640 relating to licenses, licensed establishments and submission of material or data to or approval by the Director, Center for Biologics Evaluation and Research, are not applicable to establishments not subject to licensure under section 351 of the Public Health Service Act.

(b) Written standard operating procedures shall be maintained and shall include all steps to be followed in the collection, processing, compatibility testing, storage and distribution of blood and blood components for homologous transfusion, autologous transfusion and further manufacturing purposes. Such procedures shall be available to the personnel for use in the areas where the procedures are performed, unless this is impractical. The written standard operating procedures shall include, but are not limited to, descriptions of the following, when applicable:

(1) Criteria used to determine donor suitability, including acceptable medical history criteria.

(2) Methods of performing donor qualifying tests and measurements, including minimum and maximum values for a test or procedure when a factor in determining acceptability.

(3) Solutions and methods used to prepare the site of phlebotomy to give maximum assurance of a sterile container of blood.

(4) Method of accurately relating the product(s) to the donor.

(5) Blood collection procedure, including in-process precautions taken to measure accurately the quantity of blood removed from the donor.

(6) Methods of component preparation, including any time restrictions for specific steps in processing.

(7) All tests and repeat tests performed on blood and blood components during processing, including testing for hepatitis B surface antigen as prescribed in §610.40 of this chapter.

(8) Pretransfusion testing, where applicable, including precautions to be taken to identify accurately the recipient blood samples and crossmatched donor units.

(9) Procedures for investigating adverse donor and recipient reactions.

(10) Storage temperatures and methods of controlling storage temperatures for all blood products and reagents as prescribed in §§600. 15 and 610.53 of this chapter.

(11) Length of expiration dates, if any, assigned for all final products as prescribed in §610.53 of this chapter.

(12) Criteria for determining whether returned blood is suitable for reissue.

(13) Procedures used for relating a unit of blood or bogous transfusion and further manufacturing purposes. Such procedures shall be available to the personnel for use in the areas where the procedures are performed, unless this is impractical. The written standard operating procedures shall include, but are not limited to, descriptions of the following, when applicable:

(1) Criteria used to determine donor suitability, including acceptable medical history criteria.

(2) Methods of performing donor qualifying tests and measurements, including minimum and maximum values for a test or procedure when a factor in determining acceptability.

(3) Solutions and methods used to prepare the site of phlebotomy to give maximum assurance of a sterile container of blood.

(4) Method of accurately relating the product(s) to the donor.

(5) Blood collection procedure, including in-process precautions taken to measure accurately the quantity of blood removed from the donor.

(6) Methods of component preparation, including any time restrictions for specific steps in processing.

(7) All tests and repeat tests performed on blood and blood components during processing, including testing for hepatitis B surface antigen as prescribed in §610.40 of this chapter.

(8) Pretransfusion testing, where applicable, including precautions to be taken to identify accurately the recipient blood samples and crossmatched donor units.

(9) Procedures for investigating adverse donor and recipient reactions.

(10) Storage temperatures and methods of controlling storage temperatures for all blood products and reagents as prescribed in §§600. 15 and 610.53 of this chapter.

(11) Length of expiration dates, if any, assigned for all final products as prescribed in §610.53 of this chapter.

(12) Criteria for determining whether returned blood is suitable for reissue.

(13) Procedures used for relating a unit of blood or blood component from the donor to its final disposition.

(14) Quality control procedures for supplies and reagents employed in blood collection, processing and pretransfusion testing.

(15) Schedules and procedures for equipment maintenance and calibration.

(16) Labeling procedures, including safeguards to avoid labeling mixups.

(17) Procedures of plasmapheresis, plateletpheresis, and leukapheresis, if performed, including precautions to be taken to ensure reinfusion of a donor's own cells.

(18) Procedure for preparing recovered (salvaged) plasma, if performed, including details of separation, pooling, labeling, storage and distribution.

(c) All records pertinent to the lot or unit maintained pursuant to these regulations shall be reviewed before the release or distribution of a lot or unit of final product. The review or portions of the review may be performed at appropriate periods during or after blood collecting, processing, compatibility testing and storing. A thorough investigation, including the conclusions and followup, of any unexplained discrepancy or the failure of a lot or unit to meet any of its specifications shall be made and recorded.

(d) In addition to the requirements of this subpart and in conformity with this section, any facility may utilize current standard operating procedures such as the manuals of the following organizations, as long as such specific procedures are consistent with, and at least as stringent as, the requirements contained in this part.

(1) American Association of Blood Banks.

(2) American National Red Cross.

(3) Other organizations or individual blood banks, subject to approval by the Director, Center for Biologics Evaluation and Research.

§*606.110 Plateletpheresis, leukapheresis, and plasmapheresis.*

(a) The use of plateletpheresis and leukapheresis procedures to obtain a product for a specific recipient may be at variance with the additional standards for specific products prescribed in this part provided that: (1) A physician has determined that the recipient must be transfused with the leukocytes or platelets from a specific donor, and (2) the procedure is performed under the supervision of a qualified licensed physician who is aware of the health status of the donor, and the physician has certified in writing that the donor's health permits plateletpheresis or leukapheresis.

(b) Plasmapheresis of donors who do not meet the donor requirements of §§640.63, 640.64 and 640.65 of this chapter for the collection of plasma containing rare antibodies shall be permitted only with the prior approval of the Director, Center for Biologics Evaluation and Research.

Subpart G--Finished Product Control

§*606.120 Labeling, general requirements.*

(a) Labeling operations shall be separated physically or spatially from other operations in a manner adequate to prevent mixups.

(b) The labeling operation shall include the following labeling controls:

(1) Labels shall be held upon receipt, pending review and proofing against an approved final copy, to ensure accuracy regarding identity, content, and conformity with the approved copy.

(2) Each type of label representing different products shall be stored and maintained in a manner to prevent mixups, and stocks of obsolete labels shall be destroyed.

(3) All necessary checks in labeling procedures shall be utilized to prevent errors in translating test results to container labels.

(c) All labeling shall be clear and legible.

§*606.121 Container label.*

(a) The container label requirements are designed to facilitate the use of a uniform container label for blood and blood components (except Source Plasma) by all blood establishments. Single copies of an FDA guideline entitled "Guideline for the Uniform Labeling of Blood and Blood Components" are available upon request (under Docket No. 80N-0120) from the Dockets Management Branch (HFA-305), Food and Drug Administration, Rm. 1-23, 12420 Parklawn Dr., Rockville, MD 20857 (copies of the guideline are available also from the American Blood Commission, 1901 North Ft. Myer Drive, Suite 300, Arlington, VA 22209).

(b) The label provided by the collecting facility and the initial processing facility shall not be removed, altered, or obscured, except that the label may be altered to indicate the proper name and other information required to identify accurately the contents of a container after blood components have been prepared.

(c) The container label shall include the following information, as well as other specialized information as required in this section for specific products:

(1) The proper name of the product in a prominent position, and modifier(s), if appropriate.

(2) The name, address, registration number, and, if a licensed product, the license number of each manufacturer.

(3) The donor, pool, or lot number relating the unit to the donor.

(4) The expiration date, including the day, month, and year, and, if the dating period for the product is 72 hours or less, the hour of expiration.

(5) If the product is intended for transfusion, the appropriate donor classification statement, i.e., "paid donor" or "volunteer donor", in no less prominence than the proper name of the product.

(i) A paid donor is a person who receives monetary payment for a blood donation.

(ii) A volunteer donor is a person who does not receive monetary payment for a blood donation.

(iii) Benefits, such as time off from work, membership in blood assurance programs, and cancellation of nonreplacement fees that are not readily convertible to cash, do not constitute monetary payment within the meaning of this paragraph.

(6) For Whole Blood, Plasma, Platelets, and partial units of Red Blood Cells, the volume of the product, accurate to within 10 percent; or optionally for Platelets, the volume range within reasonable limits.

(7) The recommended storage temperature (in degrees Celsius).

(8) If the product is intended for transfusion, the statements:

(i) "Caution: Federal law prohibits dispensing without prescription."

(ii) "See circular of information for indications, contraindications, cautions, and methods of infusion."

(iii) "Properly identify intended recipient."

(9) The statement: "This product may transmit infectious agents."

(10) Where applicable, the name and volume of source material.

(11) The statement: "Caution: For Manufacturing Use Only", when applicable.

(12) If the product is intended for transfusion, the ABO and Rh groups of the donor shall be designated conspicuously. For Cryoprecipitated AHF, the Rh group may be omitted. The Rh group shall be designated as follows:

(i) If the test using Anti-D Blood Grouping Serum is positive, the product shall be labeled: "Rh positive."

(ii) If the test using Anti-D Blood Grouping Serum is negative but the test for Du is positive, the product shall be labeled: "Rh positive."

(iii) If the test using Anti-D Blood Grouping Serum is negative and the test for Du is negative, the product shall be labeled: "Rh negative."

(13) The container label may bear encoded information in the form of machine-readable symbols approved for use by the Director, Center for Biologics Evaluation and Research (HFB-1).

(d) Except for recovered plasma intended for manufacturing use or as otherwise approved by the Director, Center for Biologics Evaluation and Research (HFB-1), the paper of the container label shall be white and print shall be solid black, with the following additional exceptions:

(1) The Rh blood group shall be printed as follows:

(i) Rh positive: Use black print on white background.

(ii) Rh negative: Use white print on black background.

(2) The proper name of the product, any appropriate modifier(s), the donor classification statement, and the statement "properly identify intended recipient" shall be printed in solid red.

(3) The following color scheme may be used optionally for differentiating ABO Blood groups:

Blood group	Color of label paper
O	Blue.
A	Yellow.
B	Pink.
AB	White.

(4) Ink colors used for the optional color coding system described in paragraph (d)(3) of this section shall be a visual match to specific color samples designated by the Director, Center for Biologics Evaluation and Research (HFB-1).

(5) Special labels, such as those described in paragraphs (h) and (i) of this section, may be color coded using the colors recommended in the guideline (see paragraph (a) of this section), or colors otherwise approved for use by the Director, Center for Biologics Evaluation and Research (HFB-1).

(e) Container label requirements for particular products or groups of products.

(1) Whole Blood labels shall include:

(i) The volume of anticoagulant.

(ii) The name of the applicable anticoagulant immediately preceding and of no less prominence than the proper name and expressd as follows: (a) ACD, (b) CPD, (c) Heparin, (d) CPDA-1, (e) CP2D, or by other nomenclature approved for use by the Director, Office of Biologics Research and Review (HFN-800), Center for Drugs and Biologics.

(iii) If tests for unexpected antibodies are positive, blood intended for transfusion shall be labeled: "Contains (name of antibody)."

(2) Except for frozen, deglycerolized, or washed Red Blood Cell products, red blood cell labels shall include:

(i) The volume and kind of Whole Blood, including the type of anticoagulant, from which the product was prepared.

(ii) If tests for unexpected antibodies are positive and the product is intended for transfusion, the statement: "Contains (name of antibody)."

(3) Labels for products with a dating period of 72 hours or less, including any product prepared in a system that may compromise sterility, shall bear the hour of expiration.

(4) If tests for unexpected antibodies are positive, Plasma intended for transfusion shall be labeled: "Contains (name of antibody)."

(5) Recovered plasma labels shall include:

(i) In lieu of an expiration date, the date of collection of the oldest material in the container.

(ii) The statement: "Caution: For Manufacturing Use Only"; or "Caution: For Use in Manufacturing Noninjectable Products Only", as applicable.

(iii) For recovered plasma not meeting the requirements for manufacture into licensable products, the statement: "Not for Use in Products Subject to License Under Section 351 of the Public Health Service Act."

(f) Blood and blood components determined to be unsuitable for transfusion shall be prominently labeled: "NOT FOR TRANSFUSION", and the label shall state the reason the unit is considered unsuitable. The provision does not apply to recovered plasma labeled according to paragraph (e)(5) of this section.

(g) As required under §610.40 of this chapter, labels for blood and blood components that are reactive for Hepatitis B Surface Antigen, but that are intended for further manufacturing, shall state conspicuously that the material is reactive when tested for hepatitis B surface antigen and may transmit viral hepatitis or, as applicable, that blood was collected from a donor known to be reactive for hepatitis B surface antigen and is presumed to be infectious, although confirmatory hepatitis testing has not been done.

(h) The following additional information shall appear on the label for blood or blood components shipped in an emergency, prior to completion of required tests, in accordance with §640.2(f) of this chapter:

(1) The statement: "FOR EMERGENCY USE ONLY BY ____."

(2) Results of any tests prescribed under §§610.40, 610.45, and 640.5 (a), (b), or (c) of this chapter completed before shipment.

(3) Indication of any tests prescribed under §§610.40, 610.45, and 640.5 (a), (b), or (c)

of this chapter and not completed before shipment.

(i) The following additional information shall appear on the label for Whole Blood or Red Blood Cells intended for autologous infusion:

(1) Information adequately identifying the patient, e.g., name, blood group, hospital, and identification number.

(2) Date of donation.

(3) The statement: "FOR AUTOLOGOUS USE ONLY."

(4) In place of the blood group label, each container of blood intended for autologous use and obtained from a donor who fails to meet any of the donor suitability requirements under §640.3 of this chapter or who is reactive in the hepatitis tests prescribed under §610.40 of this chapter shall be prominently and permanently labeled: "FOR AUTOLOGOUS USE ONLY."

(5) Units of blood originally intended for autologous use, except those labeled as prescribed under paragraph (i)(4) of this section, may be issued for homologous transfusion provided the container label complies with all applicable provisions of paragraphs (b) through (e) of this section. In such case, the special label required under paragraph (i) (1), (2), and (3) of this section shall be removed or otherwise obscured.

(j) A tie-tag attached to the container may be used for providing the information required by paragraph (e) (1)(iii), (2)(ii), and (4), (h), or (i)(1), (2), and (3) of this section.

§606.122 Instruction circular.

An instruction circular shall be available for distribution if the product is intended for transfusion. The instruction circular shall provide adequate directions for use, including the following information:

(a) Instructions to mix the product before use.

(b) Instructions to use a filter in the administration equipment.

(c) The statement "Do Not Add Medications" or an explanation concerning allowable additives.

(d) A description of the product, its source, and preparation, including the name and proportion of the anticoagulant used in collecting the Whole Blood from each product is prepared.

(e) Statements that the product was prepared from blood that was negative when tested for antibody to Human Immunodeficiency Virus (HIV) and nonreactive for hepatitis B surface antigen by FDA required tests and nonreactive when tested for syphilis by a serologic test for syphilis (STS).

(f) The statements: "Warning. The risk of transmitting hepatitis is present. Careful donor selection and available laboratory tests do not eliminate the hazard."

(g) The names of cryoprotective agents and other additives that may still be present in the product.

(h) The names and results of all tests performed when necessary for safe and effective use.

(i) The use of the product, indications, contradications, side effects and hazards, dosage and administration recommendations.

(j) [Reserved]

(k) For Red Blood Cells, the instruction circular shall contain:

(1) Instructions to administer a suitable plasma volume expander if Red Blood Cells are substituted when Whole Blood is the indicated product.

(2) A warning not to add Lactated Ringer's Injection U.S.P. solution to Red Blood Cell products.

(l) For Platelets, the instruction circular shall contain:

(1) The approximate volume of plasma from which a sample unit of Platelets is prepared.

(2) Instructions to begin administration as soon as possible, but not more than 4 hours after entering the container.

(m) For Plasma, the instruction circular shall contain:

(1) A warning against further processing of the frozen product if there is evidence of breakage or thawing.

(2) Instructions to thaw the frozen product at a temperature between 30 and 37°C.

(3) When applicable, instructions to begin administration of the product within 6 hours

after thawing.

(4) Instructions to administer to ABO-group-compatible recipients.

(5) A statement that this product has the same hepatitis risk as Whole Blood; other plasma volume expanders without this risk are available for treating hypovolemia.

(n) For Cryoprecipitated AHF, the instruction circular shall contain:

(1) A statement that the average potency is 80 or more International Units of antihemophilic factor.

(2) The statement: "Usually contains at least 150 milligrams of fibrinogen"; or, alternatively, the average fibrinogen level determined by assay of representative units.

(3) A warning against further processing of the product if there is evidence of breakage or thawing.

(4) Instructions to thaw the product for no more than 15 minutes at a temperature of 37°C.

(5) Instructions to store at room temperature after thawing and to begin administration as soon as possible but no more than 4 hours after entering the container or after pooling and within 6 hours after thawing.

(6) A statement that 0.9 percent Sodium Chloride Injection U.S.P. is the preferred diluent.

(7) Adequate instructions for pooling to ensure complete removal of all concentrated material from each container.

(8) The statement: "Good patient management requires monitoring treatment responses to Cryoprecipitated AHF transfusions with periodic plasma factor VIII or fibrinogen assays in hemophilia A and hypofibrinogenemic recipients, respectively."

Subpart H--Laboratory Controls

§606.140 Laboratory controls.

Laboratory control procedures shall include:

(a) The establishment of scientifically sound and appropriate specifications, standards and test procedures to assure that blood and blood components are safe, pure, potent and effective.

(b) Adequate provisions for monitoring the reliability, accuracy, precision and performance of laboratory test procedures and instruments.

(c) Adequate identification and handling of all test samples so that they are accurately related to the specific unit of product being tested, or to its donor, or to the specific recipient, where applicable.

§606.151 Compatibility testing.

Standard operating procedures for compatibility testing shall include the following:

(a) A method of collecting and identifying the blood samples of recipients to ensure positive identification.

(b) The use of fresh recipient serum samples less than 48 hours old for all pretransfusion testing.

(c) The testing of the donor's cells with the recipient's serum (major crossmatch) by a method that will demonstrate agglutinating, coating and hemolytic antibodies, which shall include the antiglobulin method.

(d) A provision that, if the unit of donor's blood has not been screened by a method that will demonstrate agglutinating, coating and hemolytic antibodies, the recipient's cells shall be tested with the donor's serum (minor crossmatch) by a method that will so demonstrate.

(e) Procedures to expedite transfusions in life-threatening emergencies. Records of all such incidents shall be maintained, including complete documentation justifying the emergency action, which shall be signed by the physician requesting the procedure.

Subpart I--Records and Reports

§606.160 Records

(a)(1) Records shall be maintained concurrently with the performance of each significant step in the collection, processing, compatibility testing, storage and distribution of each unit of blood and blood components so that all steps can be

clearly traced. All records shall be legible and indelible, and shall identify the person performing the work, include dates of the various entries, show test results as well as the interpretation of the results, show the expiration date assigned to specific products, and be as detailed as necessary to provide a complete history of the work performed.

(2) Appropriate records shall be available from which to determine lot numbers of supplies and reagents used for specific lots or units of the final product.

(b) Records shall be maintained that include, but are not limited to, the following when applicable:

(1) Donor records:

(i) Donor selection, including medical interview and examination and where applicable, informed consent.

(ii) Permanent and temporary deferrals for health reasons including reason(s) for deferral.

(iii) Donor adverse reaction complaints and reports, including results of all investigations and followup.

(iv) Therapeutic bleedings, including signed requests from attending physicians, the donor's disease and disposition of units.

(v) Immunization, including informed consent, identification of the antigen, dosage and route of administration.

(vi) Blood collection, including identification of the phlebotomist.

(2) Processing records:

(i) Blood processing, including results and interpretation of all tests and retests.

(ii) Component preparation, including all relevant dates and times.

(iii) Separation and pooling of recovered plasma.

(iv) Centrifugation and pooling of source plasma.

(v) Labeling, including initials of person(s) responsible.

(3) Storage and distribution records:

(i) Distribution and disposition, as appropriate, of blood and blood products.

(ii) Visual inspection of whole blood and red blood cells during storage and immediately before distribution.

(iii) Storage temperature, including initialed temperature recorder charts.

(iv) Reissue, including records of proper temperature maintenance.

(v) Emergency release of blood, including signature of requesting physician obtained before or after release.

(4) Compatibility test records:

(i) Results of all compatibility tests, including crossmatching, testing of patient samples, antibody screening and identification.

(ii) Results of confirmatory testing.

(5) Quality control records:

(i) Calibration and standardization of equipment.

(ii) Performance checks of equipment and reagents.

(iii) Periodic check on sterile technique.

(iv) Periodic tests of capacity of shipping containers to maintain proper temperature in transit.

(v) Proficiency test results.

(6) Transfusion reaction reports and complaints, including records of investigations and followup.

(7) General records:

(i) Sterilization of supplies and reagents prepared within the facility, including date, time interval, temperature and mode.

(ii) Responsible personnel.

(iii) Errors and accidents.

(iv) Maintenance records for equipment and general physical plant.

(v) Supplies and reagents, including name of manufacturer or supplier, lot numbers, expiration date and date of receipt.

(vi) Disposition of rejected supplies and reagents used in the collection, processing and compatibility testing of blood and blood components.

(c) A donor number shall be assigned to each accepted donor, which relates the unit of blood collected to that donor, to his medical record, to any component or blood product from that donor's unit of blood, and to all records describing the history and ultimate disposition of these products.

(d) Records shall be retained for such interval beyond the expiration date for the blood or blood component as necessary to facilitate the reporting of any unfavorable clinical reactions. The retention period shall be no less than 5 years after the records of processing have been completed or 6 months after the latest expiration date for the individual product, whichever is a later date. When there is no expiration date, records shall be retained indefinitely.

(e) A record shall be available from which unsuitable donors may be identified so that products from such individuals will not be distributed.

§606.165 Distribution and receipt; procedures and records.

(a) Distribution and receipt procedures shall include a system by which the distribution or receipt of each unit can be readily determined to facilitate its recall, if necessary.

(b) Distribution records shall contain information to readily facilitate the identification of the name and address of the consignee, the date and quantity delivered, the lot number of the unit(s), the date of expiration or the date of collection, whichever is applicable, or for crossmatched blood and blood components, the name of the recipient.

(c) Receipt records shall contain the name and address of the collecting facility, date received, donor or lot number assigned by the collecting facility and the date of expiration or the date of collection, whichever is applicable.

§606.170 Adverse reaction file.

(a) Records shall be maintained of any reports of complaints of adverse reactions regarding each unit of blood or blood product arising as a result of blood collection or transfusion. A thorough investigation of each reported adverse reaction shall be made. A written report of the investigation of adverse reactions, including conclusions and followup, shall be prepared and maintained as part of the record for that lot or unit of final product by the collecting or transfusing facility. When it is determined that the product was at fault in causing a transfusion reaction, copies of all such written reports shall be forwarded to and maintained by the manufacturer or collecting facility.

(b) When a complication of blood collection or transfusion is confirmed to be fatal, the Director, Office of Compliance, Center for Biologics Evaluation and Research, shall be notified by telephone or telegraph as soon as possible; a written report of the investigation shall be submitted to the Director, Office of Compliance, Center for Biologics Evaluation and Research, within 7 days after the fatality by the collecting facility in the event of a donor reaction, or by the facility that performed the compatibility tests in the event of a transfusion reaction.

FDA's CGMPs for Medical Devices (21 CFR 820) were proposed in March 1977 to assure that medical devices are safe and effective by defining the methods used in, and the facilities and controls used for, the manufacture, packing, storage, aand installation of medical devices. They were finalized in July 1978. A major revision to the medical device GMPs was proposed in November 1993. This revision would replace current quality assurance *PROGRAM* requirements with quality *SYSTEM* requirements and ensure that the GMP requirements for medical devices are compatible with specifications for quality systems that are contained in the international ISO 9001 standards. Because of the emphasis on system requirements, the entire regulation is highlighted in boldface.

21 CFR PART 820--GOOD MANUFACTURING PRACTICE FOR MEDICAL DEVICES (PROPOSED)

Subpart A--General Provisions

§820.1 Scope.

(a) Applicability. (1) The regulations set forth in this part describe current good manufacturing practices (CGMP's) for methods used in, and the facilities and controls used for, the design, purchasing, manufacture, packaging, labeling, storage, installation, and servicing of all finished devices intended for human use. The regulations in this part are intended to ensure that finished devices will be safe and effective and otherwise in compliance with the Federal Food, Drug, and Cosmetic Act (the act). This part establishes minimum requirements applicable to manufacturers of finished devices, including additional requirements for critical devices. With respect to class I devices, design controls apply only to those devices listed in §820. 30(a)(2). The regulations in this part do not apply to manufacturers of components or parts of finished devices when such components or parts are not intended specifically for use as part of a medical device, but such manufacturers are encouraged to use appropriate provisions of this regulation as guidelines. Manufacturers of human blood and blood components are not subject to this part, but are subject to part 606 of this chapter.

(2) The provisions of this part shall be applicable to any finished device, as defined in this part, intended for human use, that is manufactured, imported, or offered for import in any State or Territory of the United States.

(b) Limitations. The CGMP regulation in this part supplements regulations in other parts of this chapter except where explicitly stated otherwise. In the event it is impossible to comply with all applicable regulations, both in this part and in other parts of this chapter, the regulations specifically applicable to the device in question shall supersede any other regulations.

(c) Consequences of failure to comply with the regulations. (1) The failure to comply with any applicable provision in this part in the design, purchasing, manufacture, packaging, labeling, storage, installation, or servicing of a device renders the device adulterated under section 501(h) of the act. Such a device, as well as any person responsible for the failure to comply, is subject to regulatory action under sections 301, 302, 303, 304, and 801 of the act.

(2) If a manufacturer who imports devices into the United States refuses to schedule an FDA inspection of a foreign facility for compliance with this part or refuses to permit FDA to conduct or complete a scheduled inspection at a foreign facility, it shall appear, for purposes of 801(a) of the act, that the methods used in, and the facilities and controls used for, the design, purchasing, manufacture, packaging, labeling, storage, installation, or servicing of any devices produced at such facility that are offered for import into the United States do not conform to the requirements of section 520(f) of the act and this part and that the devices manufactured at that facility are adulterated under section 501(h) of the act. Foreign CGMP inspections will be scheduled in advance by FDA in writing.

(d) Exemptions or variances. Any person who wishes to petition for an exemption or variance from any device

good manufacturing practice requirement is subject to the requirements of section 520(f)(2) of the act. Petitions for an exemption or variance shall be submitted according to the procedures set forth in §10.30 of this chapter, the Food and Drug Administration's administrative procedures. Guidance is available from the Center for Devices and Radiological Health, Division of Small Manufacturers Assistance, Regulatory Assistance Branch (HFZ-220), 1901 Chapman Ave., Rockville, MD 20857, telephone 1-800-638-2041. Maryland and foreign residents, 1-301-443-6597, FAX 301-443-8818.

§820.3 Definitions.

(a) *Act* means the Federal Food, Drug, and Cosmetic Act, as amended (secs. 201-903, 52 Stat. 1040 et seq., as amended (21 U.S.C. 321-394)). All definitions in section 201 of the act shall apply to these regulations.

(b) *Complaint* means any written, electronic, or oral communication that relates to or concerns the unacceptability of the identity, quality, durability, reliability, safety, effectiveness, or performance of a device.

(c) *Component* means any raw material, substance, piece, part, software, firmware, packaging, labeling, or assembly used during device manufacture which is intended to be included as part of the finished, packaged, and labeled device.

(d) *Control number* means any distinctive combination of letters or numbers, or both, from which the complete history of the purchasing, manufacturing, packaging, labeling, and distribution of a lot or batch of finished devices can be determined.

(e) *Critical device* means a device that is intended to be surgically implanted into the body or to support or sustain life the failure of which to perform when properly used in accordance with instructions for use provided in the labeling can be reasonably expected to result in a serious injury to the user. Examples of critical devices are identified by the Commissioner of Food and Drugs after consultation with the Device Good Manufacturing Practice Advisory Committee authorized under section 520(f) of the act, and an illustrative list of critical devices is available from the Center for Devices and Radiological Health, Food and Drug Administration, at the addresses given in §820.1(d).

(f) *Design history record* means a compilation of records containing the complete design history of a finished device.

(g) *Design input* means the physical and performance requirements of a device that are used as a basis for device design.

(h) *Design output* means the results of a design effort at each design phase and at the end of the total design effort. The total finished design output consists of the device, its packaging and labeling, and the associated specifications and drawings and the production and quality system specifications and procedures which are included in the device master record (DMR).

(i) *Design review* means a comprehensive, systematic examination of a design to evaluate the adequacy of the device requirements, to evaluate the capability of the design to meet these requirements, and to identify problems with the design and design requirements and to propose solutions to all such problems.

(j) *Device history record* means a compilation of records containing the complete production history of a finished device.

(k) *Device master record (DMR)* means a compilation of records containing a device's complete design, formulation, and specifications, the purchasing and manufacturing procedures and specifications, the quality system requirements and procedures, and the packaging, labeling, servicing, maintenance, and installation procedures of a finished device.

(l) *Establish* means define, document, and implement.

(m) *Executive management* means those senior employees of a manufacturer who have the authority to establish or make changes to the manufacturer's quality policy, quality system requirements, or to a device's design specifications or its production, distribution, servicing, maintenance, or installation procedures.

(n) *Finished device* means any device or accessory to any device that is suitable for use, whether or not it is packaged or labeled for commercial distribution. A finished device includes a device that is intended to be sterile that is not yet sterilized.

(o) *Lot or batch* means a unit of components or finished devices that consists of a single type, model, class, size, composition, and software version that are manufactured under essentially the same conditions and that are intended to have uniform character and quality within specified limits.

(p) *Manufacturer* means any person who designs, manufactures, fabricates, assembles, or processes a finished device, including contract sterilizers, specification developers, repackers, relabelers, and initial distributors of imported devices.

(q) *Manufacturing material* means any material or substance used in, or to facilitate, a manufacturing process that is not intended by the manufacturer to be included in the finished device, including cleaning agents, mold-release agents, lubricating oils, ethylene oxide or other sterilant residues, or other byproducts of the manufacturing process.

(r) *Nonconforming* means the failure of a component, manufacturing material, or finished device to meet its specifications, either before or after distribution of the finished device.

(s) *Production* means all activities subsequent to design transfer and to the point of distribution.

(t) *Quality* means the totality of features and characteristics that bear on the ability of a device to satisfy fitness-for-use, including safety and performance.

(u) *Quality audit* means an established systematic, independent, examination of a manufacturer's entire quality system that is performed at defined intervals and at sufficient frequency to ensure that both quality system activities and the results of such activities comply with specified quality system procedures, that these procedures are implemented effectively, and that these procedures are suitable to achieve quality system objectives. "Quality audit" is different from, and in addition to, the other quality system activities required by or under this part.

(v) *Quality policy* means the overall quality intentions and direction of an organization with respect to quality, as formally expressed by executive management.

(w) *Quality system* means the organizational structure, responsibilities, procedures, specifications, processes, and resources for implementing quality management.

(x) *Record* means any written or automated document, including specifications, procedures, protocols, standards, methods, instructions, plans, files, forms, notes, reviews, analyses, and reports.

(y) *Reprocessing* means all or part of a manufacturing operation which is intended to correct nonconformance in a component or finished device.

(z) *Servicing* means maintenance or repair of a finished device for purposes of returning a device to its specifications.

(aa) *Special process* means any process the results of which cannot be completely verified by subsequent inspection and testing.

(bb) *Specifications* means the documents that prescribe the requirements with which a device, component, production or servicing activity, or quality system must conform.

(cc) *Validation* means, with respect to a device, establishing and documenting evidence that the device is fit for its intended use. With respect to a process, "validation" means establishing and documenting evidence that the process will consistently produce a result or product meeting its predetermined specifications and quality attributes.

(dd) *Verification* means confirming and documenting, with valid, objective evidence, that specified requirements have been met. Verification includes the process of examining the results of an activity to determine conformity with the stated specifications for that activity and ensuring that the device is adequate for its intended use.

§820.5 Quality system.

Each manufacturer shall establish and maintain a quality system that ensures that the requirements of this part are met, and that devices produced are safe, effective, and otherwise fit for their intended uses. As part of its quality system activities, each manufacturer shall:

(a) Establish effective quality system instructions and procedures in accordance with the requirements of this part; and

(b) Maintain the established quality system instructions and procedures effectively.

Subpart B--Quality System Requirements

§820.20 Management responsibility.

(a) Quality policy. Each manufacturer's executive management shall establish its policy and objectives for, and commitment to, quality. Executive management shall maintain the policy at all levels in the organization. Executive management shall ensure that this policy is understood by all employees who may affect or influence the quality of a device.

(b) Organization. Each manufacturer shall establish and maintain an adequate organizational structure with sufficient personnel to ensure that devices are produced in accordance with the requirements of this part.

(1) Responsibility and authority. With respect to each section in this part, each manufacturer shall establish the responsibility, authority, and interrelation of all personnel who manage, perform, and verify work affecting quality, particularly for personnel who need the organizational freedom and authority to:

(i) Initiate or implement action to prevent the occurrence or use of nonconforming components, manufacturing materials, or finished devices;

(ii) Identify or document quality problems with devices, production, or the quality system;

(iii) Initiate, recommend, provide, or implement solutions or corrective actions to quality problems;

(iv) Verify the adequacy or implementation of solutions or corrective actions to quality problems; and

(v) Direct or control further processing, distribution, or installation of nonconforming components, manufacturing materials, or finished devices.

(2) Verification resources and personnel. Each manufacturer shall establish verification functions and shall provide adequate resources and assign adequately trained personnel to perform verification activities.

(3) Management representative. Each manufacturer's executive management shall appoint an individual in executive

management, who irrespective of other responsibilities, shall have established authority over and responsibility for:

(i) Ensuring that quality system requirements are established and maintained in accordance with this part; and

(ii) Reporting on the performance of the quality system to executive management for review and to provide information for improvement of the quality system; and the appointment shall be documented.

(c) Management review. Each manufacturer's executive management shall review the suitability and effectiveness of the quality system at defined intervals and at sufficient frequency to ensure that the quality system satisfies the requirements of this part and the manufacturer's established quality policy objectives. The management review shall be conducted in accordance with established review procedures, and the results of each quality system review shall be documented.

§820.22 Quality audit.

(a) Each manufacturer shall conduct quality audits to verify that the quality system is in compliance with the established quality system requirements. Quality audits shall be conducted in accordance with established audit procedures by appropriately trained individuals who do not have direct responsibilities for the matters being audited. A report of the results of each quality audit shall be made and the audit reports shall be reviewed by management having responsibility for the matters audited. Followup corrective action, including reaudit of deficient matters, shall be taken when necessary and shall be documented in the audit report.

(b) Section 820.180 does not apply to quality audit reports required under this section, except reports written to satisfy §820.50(a), but does apply to established quality audit procedures. Audit reports written as part of the assessment of suppliers or contractors (§820.50(a)) are subject to review and copying by FDA. Upon request of a designated employee of the Food and Drug Administration, an employee in executive management shall certify in writing that the audits of the quality system required under this section have been performed and documented and that any required corrective action has been taken.

§820.25 Personnel.

(a) General. Each manufacturer shall employ sufficient personnel ith the necessary education, background, training, and experience to ensure that all activities required by this part are correctly performed.

(b) Training. Each manufacturer shall ensure that all personnel are trained to adequately perform their assigned responsibilities. Training shall be conducted in accordance with established procedures by qualified individuals to ensure that employees have a thorough understanding of their current job functions and with the CGMP requirements applicable to their job functions. As part of their training, all employees shall be made aware of device defects which may occur from the improper performance of their specific jobs. Personnel who perform verification activities shall be made aware of defects and errors that may be encountered as part of their verification functions. Employee training shall be documented.

(c) Consultants. (1) Each manufacturer shall ensure that any consultant advising on the methods used in, or facilities or controls used for, the design, purchasing, manufacture, packaging, labeling, storage, installation, or servicing of devices has sufficient qualifications (education, training, and experience) to advise on the subjects about which the consultant will advise.

(2) Each manufacturer shall maintain records pertaining to each consultant. Such records shall include the consultant's name and address, the consultant's qualifications, including a copy of the curriculum vitae and a list of previous jobs, and a specific description of the subjects on which the consultant advised.

Subpart C--Design Controls

§*820.30 Design controls.*

(a) General. (1) Each manufacturer of any class III, or class II device, and the class I devices listed in paragraph (a)(2) of this section shall establish and maintain procedures to control and verify the design of the device in order to ensure that specified design requirements are met.

(2) The following class I devices are subject to design controls:

Section	Device
862.2050 through 862.2920	Instruments, Clinical Laboratory.
868.6810...	Catheter, Tracheobronchial Suction.
878.4460...	Glove, Surgeon's.
880.4680...	Apparatus, Single Patient, Portable Suction.
880.6760...	Restraint, Protection.
892.1100...	Camera, Scintillation (gamma).
892.1110...	Camera, Positron.
892.1130...	Counter, Whole Body, Nuclear.
892.1300...	Scanner, Rectilinear, Nuclear.
892.1320...	Probe, Uptake, Nuclear.
892.1330...	Scanner, Rectilinear, Nuclear.
892.1410...	Synchornizer, Electrocardiograph, Nuclear.
892.1970...	Synchronizer, Radiographic, ECG/Respirator.
892.5650...	System, Applicator, Radionuclido, Manual.
892.5740...	Source, Radionuclide, Teletherapy.

(b) Design and development planning. Each manufacturer shall establish and maintain plans that identify each design and development activity and the persons responsible for each activity. The plans shall describe or reference the description of these design and development activities, including any interaction between or among different organizational and technical groups. The plans shall be updated as design and development evolves.

(c) Design input. Each manufacturer shall establish design input requirements relating to a device. The design input requirements shall completely address the intended use of the device, including the needs of the user and patient, and shall be reviewed and approved by a designated qualified individual. The approval of design input requirements, including the date and the person(s) approving the requirements, shall be documented.

(d) Design verification. Each manufacturer shall establish and maintain procedures for verification of the device design and assign such functions to competent personnel. Design verification shall be performed in a timely manner and shall confirm that design output meets the design input requirements and that the design is adequate for its intended use. The results of the design verification, including identification of the design verified, verification method(s), the date, and the person(s) performing the verification shall be documented in the design history record. Where applicable, design verification shall include software validation and hazard analysis.

(e) Design review. Each manufacturer shall conduct a formal design review of the design output according to established procedures. Each manufacturer shall assign design review responsibility to qualified individuals who do not have direct responsibility for the design development. The assignments shall be documented. The results of a design review shall be documented in the design history record.

(f) Design output. Each manufacturer shall define and document design output in terms that allow an adequate evaluation of conformance to design input requirements. Design output shall meet the design input requirements and shall include those design characteristics that are essential for the intended use of the device.

(g) Design transfer. Each manufacturer shall establish and maintain procedures to ensure that the design basis for a device and its components are correctly translated into production specifications. The production specifications shall be approved by an individual designated by the manufacturer. The approval, including identification of the design, the date, and the person(s) approving the specifications, shall be documented. Each manufacturer shall select a representative sample of a device from the first three production lots or batches and test such sample under actual or simulated use conditions. Each manufacturer shall conduct such testing according to established procedures and shall maintain records of all results of the testing. Each manufacturer shall also conduct such testing when changes are made in the device or manufacturing process.

(h) Design release. Each manufacturer shall ensure that a design is not released for production until the design is approved by individuals designated by the manufacturer. The designated individuals shall review all records required for the design history record to ensure that the design history file is complete and that the final design is consistent with the approved design plan before releasing the design. The release, including the date and signature of the individual(s) approving release, shall be documented.

(i) Design changes. Each manufacturer shall establish and maintain procedures for the identification, documentation, validation, review, and approval of design changes.

(j) Design history record. Each manufacturer shall establish and maintain a design history record for each device. Each design history record shall contain or reference all records necessary to demonstrate that the design was developed in accordance with the approved design plan and the requirements of this part.

Subpart D--Document and Record Controls

§820.40 *Document controls.*

Each manufacturer shall establish and maintain document control procedures to ensure that all documents that must be established and maintained under this part meet the requirements of this part and are accurate and adequate for their intended use.

(a) Document approval and issue. Each manufacturer shall designate individuals to review and approve all documents established under this part for adequacy prior to issuance. The approval, including the date and signature of the individual(s) approving the document, shall be documented.

(b) Document distribution. Each manufacturer shall ensure that all documents are current and available at all locations for which they are designed, and that all unneeded or obsolete documents are removed from all points of use in a timely manner.

(c) Documentation changes. Changes to specifications, methods, or procedures for components, finished devices, manufacturing materials, production, installation, servicing, or the quality system shall be documented, reviewed, and approved by individuals in the same functions/organizations that performed the original review and approval unless specifically designated otherwise. In addition, any change to a specification, method, or procedure that may affect quality shall be validated as adequate for their intended use before approval and issuance. Validation results shall be recorded. Approved changes shall be communicated to the appropriate personnel in a timely manner. When changes are made to a specification, method, or procedure, each manufacturer shall evaluate the change in accordance with an established procedure to determine if the submission of a premarket notification (510(k)) under §807. 81(a)(3) of this chapter, or the submission of a supplement to a premarket approval application (PMA) under §814. 39(a) of this chapter is required, as applicable. Records of this evaluation and its results shall be maintained.

(d) Documentation change records. Each manufacturer shall maintain records of changes to documents. Documentation change records shall include a description of the change, identification of the affected documents, the signature of the approving individuals, the approval date, and the date the change becomes effective. A list, index, or equivalent document control procedure shall be established and maintained to identify the current revision of documents in order to ensure that only current, approved documents are in use.

Subpart E--Purchasing Controls

§820.50 Purchasing controls.

Each manufacturer shall establish and maintain procedures to ensure that all components, manufacturing materials, and finished devices that are manufactured, processed, labeled, or packaged by other persons or held by other persons under contract conform to specifications. Each manufacturer shall also ensure that services provided by other persons conform to specifications.

(a) Assessment of suppliers and contractors. Each manufacturer shall establish and maintain assessment criteria for suppliers and contractors that specify the requirements, including quality requirements that suppliers and contractors must meet. Each manufacturer shall assess and select potential suppliers and contractors on the basis of their ability to meet requirements, including quality requirements and shall establish and maintain a list of suppliers and contractors that meet the manufacturer's documented assessment criteria. Records of the assessment, and assessment results shall be maintained.

(b) Purchasing forms. Each manufacturer shall establish and maintain purchasing forms that clearly describe or reference the specifications, including quality requirements, for the components, manufacturing materials, finished devices, or services ordered or contracted for. Purchasing forms shall include an agreement that the suppliers agree to notify the manufacturer of any changes in the product or service so that manufacturers may determine whether the change may affect the quality of a finished device. Each manufacturer shall review and approve purchasing documents prior to release. The approval, including the date and signature of the individual(s) approving the form, shall be documented.

Subpart F--Identification and Traceability

§820.60 Identification and traceability.

Each manufacturer shall establish and maintain procedures for identifying components, manufacturing materials, and finished devices during all stages of production, distribution, and installation to prevent mixups and to ensure orderly handlings. For certain devices, additional traceability requirements apply under section 519(e) of the act and part §§820. 65 and 820.165 of this chapter.

§820.65 Critical devices, traceability.

Each manufacturer shall identify each unit, batch, or lot of critical devices with a control number. Such identification shall be recorded in the device history record.

Subpart G--Production and Process Controls

§820.70 Production and process controls.

(a) General. Each manufacturer shall design, conduct, and control all production processes to ensure that a device conforms to its specifications. Where any deviation from device specifications could occur as a result of the manufacturing process, the manufacturer shall establish and maintain process control procedures that describe all process controls necessary to ensure conformance to specifications. Process controls shall include:

(1) Documented instructions, standard operating procedures (SOPs), and methods that define and control the manner of production, installation, and servicing;

(2) Monitoring and control of process parameters and component and device characteristics during production, installation, and servicing;

(3) Compliance with applied reference standards or codes and process control procedures;

(4) The approval of processes and process equipment; and

(5) Criteria for workmanship which shall be expressed in documented standards or by means of representative samples.

(b) Environmental control. Each manufacturer shall establish and maintain a control system to prevent contamination or other adverse effects on the device and to provide proper conditions for all operations. Conditions to be considered for control include: Lighting, ventilation, space, temperature, humidity, air pressure, filtration, airborne contamination, static electricity, and other environmental conditions. Each manufacturer shall periodically inspect its facilities and review its control system to verify that the system is adequate and functioning properly. Records of the results of such inspections shall be made and reviewed.

(c) Cleaning and sanitation. Each manufacturer shall establish and maintain adequate cleaning procedures and schedules to meet manufacturing process specifications. Each manufacturer shall ensure that the appropriate personnel understand such procedures.

(d) Personnel health and cleanliness. Each manufacturer shall ensure that personnel in contact with a device or its environment are clean, healthy, and suitably attired where lack of cleanliness, good health, or suitable attire could adversely affect the device. Any person who appears to be unclean or inappropriately attired shall be excluded from operations until he or she is clean and suitably attired. Any person who, by medical examination or supervisory observation, appears to have a condition which could adversely affect the device shall be excluded from operations until the condition is corrected. Each manufacturer shall instruct personnel to report such conditions to their supervisors.

(1) Clothing. When special clothing requirements are necessary to ensure that a device is fit for its intended use, each manufacturer shall provide clean dressing for personnel.

(2) Hygiene. Each manufacturer shall provide clean and adequate washing and toilet facilities.

(3) Personnel practices. When eating, drinking, smoking, and other activities by personnel may have an adverse effect on a device, each manufacturer shall limit such practices to designated areas. Each manufacturer shall ensure that its personnel understand any such limits. Each manufacturer shall designate selected areas to avoid any adverse effects on a device.

(e) Contamination control. Each manufacturer shall establish and maintain procedures to prevent contamination of equipment, components, manufacturing materials, and in-process and finished devices by rodenticides, insecticides, fungicides, fumigants, cleaning and sanitizing substances, and hazardous substances, including hazardous substances or contaminants generated by the manufacturing process.

(f) Sewage and refuse disposal. Each manufacturer shall dispose of sewage, trash, byproducts, chemical effluents, and other refuse in a safe, timely, and sanitary manner.

(g) Equipment. Each manufacturer shall ensure that all equipment used in the manufacturing process is adequate for its intended use and is appropriately designed, constructed, placed, and installed to facilitate maintenance, adjustment, cleaning, and use.

(1) Maintenance schedule. Each manufacturer shall establish and maintain schedules for the maintenance, adjustment, and, where applicable, cleaning of equipment to ensure that manufacturing specifications are met. The maintenance schedule shall be visibly posted on or near each piece of equipment or shall be readily available to personnel performing maintenance activities. A written record shall be maintained documenting the date when scheduled maintenance activities were performed and the individual(s) performing the maintenance activity.

(2) Inspection. Each manufacturer shall conduct periodic inspections in accordance with established procedures to ensure adherence to applicable equipment maintenance schedules. The inspections, including the date and individual conducting the inspections, shall be documented.

(3) Adjustment. Each manufacturer shall ensure that any inherent limitations or allowable tolerances are visibly posted on or near equipment requiring periodic adjustments or are readily available to personnel performing these adjustments.

(4) Manufacturing material. Each manufacturer shall establish and maintain procedures for the use and removal of manufacturing material to ensure that such material is removed from the device or limited to a specified amount that does not adversely affect the device's quality. The removal of such manufacturing material shall be documented.

(h) Automated processes. When computers are used as part of production, the quality system, or automated data processing systems, individuals designated by the manufacturer shall validate the computer software according to an established protocol. The results shall be documented. All software changes shall be made by a designated individual(s) through an established validation and approval procedure in accordance with §820.40(c) document changes.

§820.75 Special processes.

(a) Each manufacturer shall ensure that special processes are:

(1) Validated according to an established protocol and records shall be made of the results of validation, including the date of and individual responsible for the validation;

(2) Conducted according to established procedures that describe all processing controls necessary to ensure conformance to specifications;

(3) Monitored according to established procedures to ensure process parameters are met; and

(4) Performed by qualified, designated individuals.

(b) The individual(s) responsible for the performance of a special process shall record the completion of the process in the device history record. The record shall include identification of the process, the date performed, each individual that performed the special process, and the equipment used.

Subpart H--Inspection and Testing

§820.80 Inspection and testing.

(a) General. Each manufacturer shall establish and maintain the inspection and testing activities necessary to ensure that specified requirements are met. The results of all inspection and testing shall be documented.

(b) Receiving inspection and testing. Each manufacturer shall establish and maintain procedures for acceptance of incoming components, manufacturing materials, and finished devices. Incoming components, manufacturing materials, and finished devices shall not be used or processed until they have been verified as conforming to specified requirements. Individual(s) designated by the manufacturer shall accept or reject incoming components, finished devices, and manufacturing materials. Acceptance and rejection shall be documented.

(c) In-process inspection and testing. Each manufacturer shall establish and maintain procedures for inspecting and testing in-process components, finished devices, and manufacturing materials. Each manufacturer shall establish and maintain procedures for holding in-process components, finished devices, and manufacturing materials until the required inspection and tests have been completed or necessary reports have been received and verified.

(d) Final inspection and testing. Each manufacturer shall establish and maintain procedures for finished device inspection to ensure each lot or batch meets device specifications. Finished devices shall be held in quarantine or otherwise adequately controlled until released by an individual designated by the manufacturer. Finished devices shall not be released until all the required activities specified in the DMR have been completed and the associated data and documentation are reviewed to ensure all acceptance criteria have been met. Release, including the date and signature of the designated individual(s) responsible for release, shall be documented.

(e) Inspection and test records. Each manufacturer shall maintain records of the results of all inspections and tests required by this part. These records shall include the acceptance criteria, inspection checks performed; results;

equipment used; and the date and signature of the individual(s) conducting the inspection and testing. These records shall be part of the device history record.

§820.84 Inspection, measuring, and test equipment.

Each manufacturer shall ensure that all measurement and test equipment, including mechanical, automated, or electronic inspection and test equipment, is suitable for its intended purposes and is capable of producing valid results. Each manufacturer shall establish and maintain procedures to ensure that equipment is routinely calibrated, inspected, and checked. Records documenting these activities shall be maintained.

(a) Calibration. Each manufacturer shall establish and maintain calibration procedures that include specific directions and limits for accuracy and precision and provisions for remedial action when accuracy and precision limits are not met. Calibration shall be performed by personnel who have the necessary education, training, background, and experience.

(b) Calibration standards. Each manufacturer shall establish and maintain calibration standards for measurement equipment that are traceable to the national standards of the National Institute for Standards and Technology, Department of Commerce. If national standards are not practical or available, the manufacturer shall use an independent reproducible standard. If no applicable standard exists, the manufacturer shall establish and maintain an in-house standard.

(c) Calibration records. Each manufacturer shall ensure that records of calibration dates, the individual performing each calibration, and the next calibration date are maintained. These records shall be maintained by individuals designated by the manufacturer and displayed on or near each piece of equipment or shall be readily available to the personnel using such equipment and the individuals responsible for calibrating the equipment.

(d) Maintenance. Each manufacturer shall establish and maintain procedures to ensure that the handling, preservation, and storage of inspection, measuring, and test equipment is such that their accuracy and fitness-for-use are maintained.

(e) Facilities. Each manufacturer shall protect inspection, measuring, and test facilities and equipment, including both test hardware and test software, from adjustments that would invalidate the calibration.

§820.86 Inspection and test status.

(a) Each manufacturer shall identify the inspection and test status of all components, manufacturing materials, and finished devices. The identification shall be visible, shall indicate the conformance or nonconformance of these items with respect to acceptance criteria, and shall be maintained, as necessary, throughout component acceptance, manufacturing, packaging, labeling, installation, and servicing of the device to ensure that only components, finished devices, and manufacturing materials which have passed the required inspections and tests are distributed, used, or installed.

(b) Each manufacturer shall ensure that records shall identify the individual(s) responsible for the release of components, of manufacturing materials, and of finished devices.

Subpart I--Nonconforming Components and Devices

§820.90 Nonconforming components and devices.

(a) Control of nonconforming components and devices. Each manufacturer shall establish and maintain procedures to ensure that components, manufacturing materials, finished devices, and returned devices that do not conform to specified requirements are not inadvertently used or installed. The procedures shall provide for the identification, documentation, investigation, segregation, and disposition of nonconforming components, manufacturing materials, finished devices, and returned devices, and for notification of the persons or organizations responsible for the nonconformance.

(b) Nonconformity review and disposition. (1) The responsibility for review and the authority for the disposition of nonconforming components, manufacturing materials, finished devices, and returned devices shall be defined.

(2) Each manufacturer shall establish and maintain procedures for the reprocessing, retesting, and reinspection of nonconforming components and finished devices, to ensure that they meet their original, or subsequently modified and approved, specifications. The procedures shall be contained or referenced in the device master record. Reprocessed devices or components shall be clearly identified as reprocessed, and the reprocessing and reinspection results shall be recorded in the device history record. Reprocessed devices or components shall be subject to another complete reinspection for any characteristic of the device which may be adversely affected by such reprocessing. When there is repeated reprocessing of a device or component, a determination of the effect of the reprocessing upon the device or component shall be made and documented.

Subpart J--Corrective Action

§820.100 Corrective action.

(a) Each manufacturer shall establish and maintain procedures for:

(1) Analyzing all processes, work operations, concessions, quality audit reports, quality records, service records, complaints, returned product, and other sources of quality data to identify existing and potential causes of nonconforming components, finished devices, or other quality problems (analysis shall include trend analysis to detect recurring quality problems);

(2) Investigating the failure of any distributed device to meet specifications;

(3) Identifying action needed to correct the cause and prevent recurrence of nonconforming components or finished devices and other quality problems;

(4) Verifying or validating the adequacy of the corrective action to ensure that the corrective action does not adversely affect the finished device and that the corrective action is effective;

(5) Implementing and recording changes in methods and procedures needed as a result of the identification of quality problems and corrective action; and

(6) Ensuring that quality problem information is disseminated to those directly responsible for ensuring quality and is reviewed by management.

(b) All activities required under this section, and their results, shall be documented.

Subpart K--Handling, Storage, Distribution, and Installation

§820.120 Handling.

Each manufacturer shall establish and maintain procedures to ensure that mixups, damage, deterioration, or other adverse effects to components, finished devices, and manufacturing materials do not occur during any stage of handling.

§820.122 Storage.

(a) Each manufacturer shall establish and maintain procedures for the control of storage areas or stock rooms for components, manufacturing materials, and finished devices to prevent mixups, damage, deterioration, or other adverse effects pending use or distribution.

(b) Each manufacturer shall establish and maintain procedures for authorizing receipt from and dispatch to such designated areas. Any control number or other identification used shall be legible and clearly visible. When the quality of components or finished devices deteriorates over time, such devices shall be stored in a manner to facilitate proper stock rotation and their condition shall be assessed at appropriate intervals. Each manufacturer shall establish and maintain procedures to ensure that all obsolete, rejected, or deteriorated manufacturing materials, components, and devices located in storage are not inadvertently used or distributed.

§820.124 Distribution.

(a) Each manufacturer shall establish and maintain procedures for control and distribution of finished devices to ensure that only those devices approved for release are distributed. Where a device's fitness-for-use or quality deteriorates over time, the procedures shall ensure that the oldest approved devices are distributed first and that expired devices are not distributed.

(b) Each manufacturer shall maintain distribution records which include or make reference to the location of:

(1) The name and address of the consignee;

(2) The identification and quantity of devices shipped, the date shipped; and

(3) Any control number used for traceability.

§820.126 Installation.

Each manufacturer shall establish and maintain adequate instructions and procedures for proper device installation. Instructions and procedures shall include directions for verifying proper performance of the installation. When a manufacturer or its authorized representative installs a device, the manufacturer or representative shall verify that the device(s) will perform as intended after installation. The results of verification shall be recorded. When a person other than the manufacturer or its authorized representative installs a device, the manufacturer shall ensure that the installation instructions and procedures are distributed with the device or otherwise available to the person installing the device.

Subpart L--Packaging and Labeling Control

§820.160 Device packaging.

Each manufacturer shall design and construct device packaging and shipping containers to protect the device from alteration or damage during the customary conditions of processing, storage, handling, and distribution.

§820.162 Device labeling.

Each manufacturer shall establish and maintain procedures to maintain labeling integrity and to prevent labeling mixups.

(a) Labeling integrity. Each manufacturer shall ensure that labels are designed, printed, and, where applicable, applied so as to remain legible and affixed to the device during the customary conditions of processing, storage, handling, distribution, and use.

(b) Labeling inspection. Labels shall not be released for storage or use until a designated individual(s) has examined the labeling for accuracy including, where applicable, the correct expiration date, control number, storage instructions, handling instructions, and additional processing instructions. The release, including the date, name and signature of the individuals performing the examination, shall be documented in the device history record.

(c) Labeling storage. Each manufacturer shall store and maintain labeling in a manner that provides proper identification and is designed to prevent mixups.

(d) Labeling control. Each manufacturer shall control labeling and packaging operations to prevent labeling mixups.

§820.165 Critical devices, labeling.

Labeling for critical devices shall contain a control number.

Subpart M--Records

§820.180 General requirements.

All records shall be legible and shall be stored to minimize deterioration, prevent loss, and allow rapid retrieval. All records stored in automated data processing systems shall be backed up. All records required by this part shall be maintained at the manufacturing establishment or other location that is reasonably accessible to responsible officials of the manufacturer and to employees of the Food and Drug Administration designated to perform inspections. Such records shall be available for review and copying by such employee. Except as specifically provided elsewhere, the following general provisions shall apply to all records required by this part.

(a) Confidentiality. Those records deemed confidential by the manufacturer may be marked to aid the Food and Drug Administration in determining whether information may be disclosed under the public information regulation in part 20 of this chapter.

(b) Record retention period. All required records pertaining to a device shall be retained for a period of time equivalent to the design and expected life of the device, but in no case less than 2 years from the date of release for commercial distribution by the manufacturer. Photostatic or other reproductions of records required by this part may be used. Where reduction techniques such as microfilming are used, suitable reading and photocopying equipment shall be available for use with the records.

§820.181 Device master record (DMR).

Each manufacturer shall maintain device master records (DMR's). Each manufacturer shall ensure that each DMR is prepared, dated, and signed by qualified individual(s) designated by the manufacturer. Any changes in a DMR shall meet the applicable requirements of §820.40. The DMR for each type of device shall include, or refer to the location of, the following information:

(a) Device specifications including appropriate drawings, composition, formulation, component specifications, software design specifications, and software source code;

(b) Production process specifications including the appropriate equipment specifications, production methods, production procedures, and production environment specifications;

(c) Quality system documents, including verification checks used, the verification apparatus used, and validation protocols and results;

(d) Packaging and labeling specifications, including methods and processes used; and

(e) Installation, maintenance, and servicing procedures and methods.

§820.184 Device history record.

Each manufacturer shall maintain device history records. Each manufacturer shall establish and maintain procedures to ensure that device history records are maintained for each batch lot, or unit to demonstrate that the device(s) was manufactured in accordance with the device master record and the requirements of this part. Device history records shall be readily accessible and maintained by a designated individual(s). The device history record shall include, or refer to the location of, the following information:

(a) The dates of manufacture;

(b) The quantity manufactured;

(c) The quantity released for distribution;

(d) The labeling; and

(e) Any control number(s) used.

§820.198 Complaint files.

(a) Each manufacturer shall maintain complaint files. Each manufacturer shall establish and maintain procedures for receiving, reviewing, evaluating, and maintaining complaints. Such procedures shall ensure that:

(1) Complaints are received, reviewed, evaluated, investigated, and maintained by a formally designated unit;

(2) Oral complaints are documented upon receipt; and

(3) The complaint is reviewed to determine whether an investigation is necessary. When no investigation is made, the unit shall maintain a record that includes the reason no investigation was made and the name of the individual responsible for the decision not to investigate.

(b) Each manufacturer shall review, evaluate, and investigate all complaints involving the possible failure of a device, labeling, or packaging to meet any of its specifications. Any complaint pertaining to death, injury, or any hazard to safety shall be immediately reviewed, evaluated, and investigated by a designated individual(s) and shall be maintained in a separate portion of the complaint files. Investigations shall include a determination of whether there was an actual device failure to perform pursuant to specifications; whether the device was being used to treat or diagnose a patient; whether a death, injury, or serious illness was involved; and the relationship, if any, of the device to the reported incident or adverse event.

(c) When an investigation is made, a written record of each investigation shall be maintained by the formally designated unit identified in paragraph (a) of this section. The record of investigation shall include:

(1) The name of the device;

(2) The date the complaint was received;

(3) Any control number used;

(4) The name, address, and phone number of the complainant;

(5) The nature of the complaint; and

(6) The results of the investigation.

(d) The investigation results shall include:

(1) The corrective action taken;

(2) The dates of the investigation;

(3) The details of the complaint; and

(4) The reply to the complainant.

(e) When no reply is made to the complainant, the reason shall be recorded.

(f) Records of investigations of events that are determined to be reportable under medical device reporting (MDR) requirements of part 803 of this chapter shall include the information required by part 803 of this chapter. When such information cannot be obtained, a record of the reason shall be made and retained in the record of investigation.

(g) When the formally designated complaint unit is located at a site separate from the actual manufacturing establishment and a complaint involves the manufacturing site, a duplicate copy of the complaint and the record of investigation of the complaint shall be transmitted to and maintained at the actual manufacturing establishment in a file designated for device complaints.

(h) If a manufacturer's formally designated complaint unit is located outside of the United States, a copy of all of each records required under this section shall be maintained in the United States. If a manufacturer has a location in the United States where records are regularly kept, the copies required under this paragraph may be maintained at such location. Otherwise, the copies required under this paragraph shall be provided to and kept by the agent designated under §803.26(g)(3) of this chapter.

(i) Each manufacturer shall establish and maintain procedures for processing complaints to ensure that all complaints are processed in a uniform and timely manner. Such procedures shall include provisions for determining whether the complaint represents an event which is required to be reported to the Food and Drug Administration under part 803 of this chapter.

(j) Any written or oral complaint that is also a reportable event under part 803 of this chapter shall be identified in the complaint file as such.

Subpart N--Servicing

§820.200 Servicing.

Each manufacturer shall establish and maintain procedures to ensure that finished devices that are serviced by the manufacturer or its representatives meet specifications. Procedures for servicing shall include provisions for determining if service requests represent an event which must be reported to the Food and Drug Administration under the requirements of part 803 of this chapter.

(a) Service records. Each manufacturer shall establish and maintain procedures to ensure that service records are maintained that identify the device serviced, including any control number used, the date of service, the service performed, and individual(s) servicing the device.

(b) Service record evaluation. Each manufacturer shall analyze servicing records in accordance with §820.100; except that when a service report involves a death, serious injury, or safety hazard, the report shall be considered a complaint and shall be investigated in accordance with the requirements of §820.198.

Subpart O--Statistical Techniques

§820.250 Statistical techniques.

(a) Where appropriate, each manufacturer shall establish and maintain procedures for identifying valid statistical techniques required for verifying the

acceptability of process capability and product characteristics.

(b) Sampling plans shall be written and based on a valid statistical rationale. Each manufacturer shall establish and maintain procedures to ensure that sampling methods are adequate for their intended use and are regularly reviewed, especially for events such as nonconforming devices, adverse quality audit reports, or complaints.

GOOD LABORATORY PRACTICES

In the mid-1970s, the FDA identified significant deficiencies in the quality and integrity of data submitted to the agency from nonclinical laboratory studies on the safety and effectiveness of drugs and color additives for foods. Responding to this problem, the FDA issued proposed Good Laboratory Practice (GLP) regulations for nonclinical laboratory studies in November 1976 to ensure that such studies are conducted according to scientifically sound protocols and proper procedures. The GLP regulations (21 CFR 58) were finalized in December 1978, and the last revision was made in July 1991.

Requirements in the regulation relating to system validation are highlighted in boldface.

21 CFR PART 58--GOOD LABORATORY PRACTICE FOR NONCLINICAL LABORATORY STUDIES

Subpart A--General Provisions

§58.1 *Scope.*

(a) This part prescribes good laboratory practices for conducting nonclinical laboratory studies that support or are intended to support applications for research or marketing permits for products regulated by the Food and Drug Administration, including food and color additives, animal food additives, human and animal drugs, medical devices for human use, biological products, and electronic products. Compliance with this part is intended to assure the quality and integrity of the safety data filed pursuant to sections 406, 408, 409, 502, 503, 505, 506, 507, 510, 512-516, 518-520, 706, and 801 of the Federal Food, Drug, and Cosmetic Act and sections 351 and 354-360F of the Public Health Service Act.

(b) References in this part to regulatory sections of the Code of Federal Regulations are to chapter I of title 21, unless otherwise noted.

§58.3 *Definitions.*

As used in this part, the following terms shall have the meanings specified:

(a) *Act* means the Federal Food, Drug, and Cosmetic Act, as amended (secs. 201-902, 52 Stat. 1040 et seq., as amended (21 U.S.C. 321392)).

(b) *Test article* means any food additive, color additive, drug, biological product, electronic product, medical device for human use, or any other article subject to regulation under the act or under sections 351 and 354-360F of the Public Health Service Act.

(c) *Control article* means any food additive, color additive, drug, biological product, electronic product, medical device for human use, or any article other than a test article, feed, or water that is administered to the test system in the course of a nonclinical laboratory study for the purpose of establishing a basis for comparison with the test article.

(d) *Nonclinical laboratory study* means in vivo or in vitro experiments in which test articles are studied prospectively in test systems under laboratory conditions to determine their safety. The term does not include studies utilizing human subjects or clinical studies or field trials in animals. The term does not include basic exploratory studies carried out to determine whether a test article has any potential utility or to determine physical or chemical characteristics of a test article.

(e) *Application for research or marketing permit* includes:

(1) A color additive petition, described in part 71.

(2) A food additive petition, described in parts 171 and 571.

(3) Data and information regarding a substance submitted as part of the procedures for establishing that a substance is generally recognized as safe for use, which use results or may reasonably be expected to result, directly or indirectly, in its becoming a component or otherwise affecting the characteristics of any food, described in §§170.35 and 570.35.

(4) Data and information regarding a food additive submitted as part of the procedures regarding food additives permitted to be used on an interim basis pending additional study, described in §180.1.

(5) An investigational new drug application, described in part 312 of this chapter.

(6) A new drug application, described in part 314.

(7) Data and information regarding an over-the-counter drug for human use, submitted as part of the procedures for classifying such drugs as generally recognized as safe and effective and not misbranded, described in part 330.

(8) Data and information about a substance submitted as part of the procedures for establishing a tolerance for unavoidable contaminants in food and food-packaging materials, described in parts 109 and 509.

(9) Data and information regarding an antibiotic drug submitted as part of the procedures for issuing, amending, or repealing regulations for such drugs, described in §314. 300 of this chapter.

(10) A Notice of Claimed Investigational Exemption for a New Animal Drug, described in part 511.

(11) A new animal drug application, described in part 514.

(12) [Reserved]

(13) An application for a biological product license, described in part 601.

(14) An application for an investigational device exemption, described in part 812.

(15) An Application for Premarket Approval of a Medical Device, described in section 515 of the act.

(16) A Product Development Protocol for a Medical Device, described in section 515 of the act.

(17) Data and information regarding a medical device submitted as part of the procedures for classifying such devices, described in part 860.

(18) Data and information regarding a medical device submitted as part of the procedures for establishing, amending, or repealing a performance standard for such devices, described in part 861.

(19) Data and information regarding an electronic product submitted as part of the procedures for obtaining an exemption from notification of a radiation safety defect or failure of compliance with a radiation safety performance standard, described in subpart D of part 1003.

(20) Data and information regarding an electronic product submitted as part of the procedures for establishing, amending, or repealing a standard for such product, described in section 358 of the Public Health Service Act.

(21) Data and information regarding an electronic product submitted as part of the procedures for obtaining a variance from any electronic product performance standard as described in §1010.4.

(22) Data and information regarding an electronic product submitted as part of the procedures for granting, amending, or extending an exemption from any electronic product performance standard, as described in §1010. 5.

(f) *Sponsor* means:

(1) A person who initiates and supports, by provision of financial or other resources, a nonclinical laboratory study;

(2) A person who submits a nonclinical study to the Food and Drug Administration in support of an application for a research or marketing permit; or

(3) A testing facility, if it both initiates and actually conducts the study.

(g) *Testing facility* means a person who actually conducts a nonclinical laboratory study, i.e., actually uses the test article in a test system. Testing facility includes any establishment required to register under section 510 of the act that conducts nonclinical laboratory studies and any consulting laboratory described in section 704 of the act that conducts such studies. Testing facility encompasses only those operational units that are being or have been used to conduct nonclinical laboratory studies.

(h) *Person* includes an individual, partnership, corporation, association, scientific or academic establishment, government agency, or organizational unit thereof, and any other legal entity.

(i) *Test system* means any animal, plant, microorganism, or subparts thereof to which the test or control article is administered or added for study. Test system also includes appropriate groups or components of the system not treated with the test or control articles.

(j) *Specimen* means any material derived from a test system for examination or analysis.

(k) *Raw data* means any laboratory worksheets, records, memoranda, notes, or exact copies thereof, that are the result of original observations and activities of a nonclinical laboratory study and are necessary for the reconstruction and evaluation of the report of that study. In the event that exact transcripts of raw data have been prepared (e.g., tapes which have been transcribed verbatim, dated, and verified accurate by signature), the exact copy or exact transcript may be substituted for the original source as raw data. Raw data may include photographs, microfilm or microfiche copies, computer printouts, magnetic media, including dictated observations, and recorded data from automated instruments.

(l) *Quality assurance unit* means any person or organizational element, except the study director, designated by testing facility management to perform the duties relating to quality assurance of nonclinical laboratory studies.

(m) *Study director* means the individual responsible for the overall conduct of a nonclinical laboratory study.

(n) *Batch* means a specific quantity or lot of a test or control article that has been characterized according to §58.105(a).

(o) *Study initiation date* means the date the protocol is signed by the study director.

(p) *Study completion date* means the date the final report is signed by the study director.

§58.10 Applicability to studies performed under grants and contracts.

When a sponsor conducting a nonclinical laboratory study intended to be submitted to or reviewed by the Food and Drug Administration utilizes the services of a consulting laboratory, contractor, or grantee to perform an analysis or other service, it shall notify the consulting laboratory, contractor, or grantee that the service is part of a nonclinical laboratory study that must be conducted in compliance with the provisions of this part.

§58.15 Inspection of a testing facility.

(a) A testing facility shall permit an authorized employee of the Food and Drug Administration, at reasonable times and in a reasonable manner, to inspect the facility and to inspect (and in the case of records also to copy) all records and specimens required to be maintained regarding studies within the scope of this part. The records inspection and copying requirements shall not apply to quality assurance unit records of findings and problems, or to actions recommended and taken.

(b) The Food and Drug Administration will not consider a nonclinical laboratory study in support of an application for a research or marketing permit if the testing facility refuses to permit inspection. The determination that a nonclinical laboratory study will not be considered in support of an application for a research or marketing permit does not, however, relieve the applicant for such a permit of any obligation under any applicable statute or regulation to submit the results of the study to the Food and Drug Administration.

Subpart B--Organization and Personnel

§58.29 *Personnel.*

(a) Each individual engaged in the conduct of or responsible for the supervision of a nonclinical laboratory study shall have education, training, and experience, or combination thereof, to enable that individual to perform the assigned functions.

(b) Each testing facility shall maintain a current summary of training and experience and job description for each individual engaged in or supervising the conduct of a nonclinical laboratory study.

(c) There shall be a sufficient number of personnel for the timely and proper conduct of the study according to the protocol.

(d) Personnel shall take necessary personal sanitation and health precautions designed to avoid contamination of test and control articles and test systems.

(e) Personnel engaged in a nonclinical laboratory study shall wear clothing appropriate for the duties they perform. Such clothing shall be changed as often as necessary to prevent microbiological, radiological, or chemical contamination of test systems and test and control articles.

(f) Any individual found at any time to have an illness that may adversely affect the quality and integrity of the nonclinical laboratory study shall be excluded from direct contact with test systems, test and control articles and any other operation or function that may adversely affect the study until the condition is corrected. All personnel shall be instructed to report to their immediate supervisors any health or medical conditions that may reasonably be considered to have an adverse effect on a nonclinical laboratory study.

§58.31 *Testing facility management.*

For each nonclinical laboratory study, testing facility management shall:

(a) Designate a study director as described in §58.33, before the study is initiated.

(b) Replace the study director promptly if it becomes necessary to do so during the conduct of a study.

(c) Assure that there is a quality assurance unit as described in §58.35.

(d) Assure that test and control articles or mixtures have been appropriately tested for identity, strength, purity, stability, and uniformity, as applicable.

(e) Assure that personnel, resources, facilities, equipment, materials, and methodologies are available as scheduled.

(f) Assure that personnel clearly understand the functions they are to perform.

(g) Assure that any deviations from these regulations reported by the quality assurance unit are communicated to the study director and corrective actions are taken and documented.

§58.33 *Study director.*

For each nonclinical laboratory study, a scientist or other professional of appropriate education, training, and experience, or combination thereof, shall be identified as the study director. The study director has overall responsibility for the technical conduct of the study, as well as for the interpretation, analysis, documentation and reporting of results, and represents the single point of study control. The study director shall assure that:

(a) The protocol, including any change, is approved as provided by §58.120 and is followed.

(b) All experimental data, including observations of unanticipated responses of the test system are accurately recorded and verified.

(c) Unforeseen circumstances that may affect the quality and integrity of the nonclinical laboratory study are noted when they occur, and corrective action is taken and documented.

(d) Test systems are as specified in the protocol.

(e) All applicable good laboratory practice regulations are followed.

(f) All raw data, documentation, proto-

cols, specimens, and final reports are transferred to the archives during or at the close of the study.

§58.35 *Quality assurance unit.*

(a) A testing facility shall have a quality assurance unit which shall be responsible for monitoring each study to assure management that the facilities, equipment, personnel, methods, practices, records, and controls are in conformance with the regulations in this part. For any given study, the quality assurance unit shall be entirely separate from and independent of the personnel engaged in the direction and conduct of that study.

(b) The quality assurance unit shall:

(1) Maintain a copy of a master schedule sheet of all nonclinical laboratory studies conducted at the testing facility indexed by test article and containing the test system, nature of study, date study was initiated, current status of each study, identity of the sponsor, and name of the study director.

(2) Maintain copies of all protocols pertaining to all nonclinical laboratory studies for which the unit is responsible.

(3) Inspect each nonclinical laboratory study at intervals adequate to assure the integrity of the study and maintain written and properly signed records of each periodic inspection showing the date of the inspection, the study inspected, the phase or segment of the study inspected, the person performing the inspection, findings and problems, action recommended and taken to resolve existing problems, and any scheduled date for reinspection. Any problems found during the course of an inspection which are likely to affect study integrity shall be brought to the attention of the study director and management immediately.

(4) Periodically submit to management and the study director written status reports on each study, noting any problems and the corrective actions taken.

(5) Determine that no deviations from approved protocols or standard operating procedures were made without proper authorization and documentation.

(6) Review the final study report to assure that such report accurately describes the methods and standard operating procedures, and that the reported results accurately reflect the raw data of the nonclinical laboratory study.

(7) Prepare and sign a statement to be included with the final study report which shall specify the dates inspections were made and findings reported to management and to the study director.

(c) The responsibilities and procedures applicable to the quality assurance unit, the records maintained by the quality assurance unit, and the method of indexing such records shall be in writing and shall be maintained. These items including inspection dates, the study inspected, the phase or segment of the study inspected, and the name of the individual performing the inspection shall be made available for inspection to authorized employees of the Food and Drug Administration.

(d) A designated representative of the Food and Drug Administration shall have access to the written procedures established for the inspection and may request testing facility management to certify that inspections are being implemented, performed, documented, and followed-up in accordance with this paragraph.

Subpart C--Facilities

§58.41 *General.*

Each testing facility shall be of suitable size and construction to facilitate the proper conduct of nonclinical laboratory studies. It shall be designed so that there is a degree of separation that will prevent any function or activity from having an adverse effect on the study.

§58.43 *Animal care facilities.*

(a) A testing facility shall have a sufficient number of animal rooms or areas, as needed, to assure proper: (1) Separation of species or test systems, (2) isolation of individual projects, (3) quarantine of animals, and (4) routine or specialized housing of animals.

(b) A testing facility shall have a number of animal rooms or areas separate from those described in paragraph (a) of this section to ensure isolation of studies being done with test systems or test and control articles known to be biohazardous, including volatile substances, aerosols, radioactive materials, and infectious agents.

(c) Separate areas shall be provided, as appropriate, for the diagnosis, treatment, and control of laboratory animal diseases. These areas shall provide effective isolation for the housing of animals either known or suspected of being diseased, or of being carriers of disease, from other animals.

(d) When animals are housed, facilities shall exist for the collection and disposal of all animal waste and refuse or for safe sanitary storage of waste before removal from the testing facility. Disposal facilities shall be so provided and operated as to minimize vermin infestation, odors, disease hazards, and environmental contamination.

§58.45 *Animal supply facilities.*

There shall be storage areas, as needed, for feed, bedding, supplies, and equipment. Storage areas for feed and bedding shall be separated from areas housing the test systems and shall be protected against infestation or contamination. Perishable supplies shall be preserved by appropriate means.

§58.47 *Facilities for handling test and control articles.*

(a) As necessary to prevent contamination or mixups, there shall be separate areas for:

(1) Receipt and storage of the test and control articles.

(2) Mixing of the test and control articles with a carrier, e.g., feed.

(3) Storage of the test and control article mixtures.

(b) Storage areas for the test and/or control article and test and control mixtures shall be separate from areas housing the test systems and shall be adequate to preserve the identity, strength, purity, and stability of the articles and mixtures.

§58.49 *Laboratory operation areas.*

Separate laboratory space shall be provided, as needed, for the performance of the routine and specialized procedures required by nonclinical laboratory studies.

§58.51 *Specimen and data storage facilities.*

Space shall be provided for archives, limited to access by authorized personnel only, for the storage and retrieval of all raw data and specimens from completed studies.

Subpart D--Equipment

§58.61 *Equipment design.*

Equipment used in the generation, measurement, or assessment of data and equipment used for facility environmental control shall be of appropriate design and adequate capacity to function according to the protocol and shall be suitably located for operation, inspection, cleaning, and maintenance.

§58.63 *Maintenance and calibration of equipment.*

(a) Equipment shall be adequately inspected, cleaned, and maintained. Equipment used for the generation, measurement, or assessment of data shall be adequately tested, calibrated and/or standardized.

(b) The written standard operating procedures required under §58.81(b)(11) shall set forth in sufficient detail the methods, materials, and schedules to be used in the routine inspection, cleaning, maintenance, testing, calibration, and/or standardization of equipment, and shall specify, when appropriate, remedial action to be taken in the event of failure

or malfunction of equipment. The written standard operating procedures shall designate the person responsible for the performance of each operation.

(c) Written records shall be maintained of all inspection, maintenance, testing, calibrating and/or standardizing operations. These records, containing the date of the operation, shall describe whether the maintenance operations were routine and followed the written standard operating procedures. Written records shall be kept of nonroutine repairs performed on equipment as a result of failure and malfunction. Such records shall document the nature of the defect, how and when the defect was discovered, and any remedial action taken in response to the defect.

Subpart E--Testing Facilities Operation

§58.81 *Standard operating procedures.*

(a) A testing facility shall have standard operating procedures in writing setting forth nonclinical laboratory study methods that management is satisfied are adequate to insure the quality and integrity of the data generated in the course of a study. All deviations in a study from standard operating procedures shall be authorized by the study director and shall be documented in the raw data. Significant changes in established standard operating procedures shall be properly authorized in writing by management.

(b) Standard operating procedures shall be established for, but not limited to, the following:

(1) Animal room preparation.

(2) Animal care.

(3) Receipt, identification, storage, handling, mixing, and method of sampling of the test and control articles.

(4) Test system observations.

(5) Laboratory tests.

(6) Handling of animals found moribund or dead during study.

(7) Necropsy of animals or postmortem examination of animals.

(8) Collection and identification of specimens.

(9) Histopathology.

(10) Data handling, storage, and retrieval.

(11) Maintenance and calibration of equipment.

(12) Transfer, proper placement, and identification of animals.

(c) Each laboratory area shall have immediately available laboratory manuals and standard operating procedures relative to the laboratory procedures being performed. Published literature may be used as a supplement to standard operating procedures.

(d) A historical file of standard operating procedures, and all revisions thereof, including the dates of such revisions, shall be maintained.

§58.83 *Reagents and solutions.*

All reagents and solutions in the laboratory areas shall be labeled to indicate identity, titer or concentration, storage requirements, and expiration date. Deteriorated or outdated reagents and solutions shall not be used.

§58.90 *Animal care.*

(a) There shall be standard operating procedures for the housing, feeding, handling, and care of animals.

(b) All newly received animals from outside sources shall be isolated and their health status shall be evaluated in accordance with acceptable veterinary medical practice.

(c) At the initiation of a nonclinical laboratory study, animals shall be free of any disease or condition that might interfere with the purpose or conduct of the study. If, during the course of the study, the animals contract such a disease or condition, the diseased animals shall be isolated, if necessary. These animals may be treated for disease or signs of disease provided that such treatment does not interfere with the study. The diagnosis, authorizations of treatment, description of treatment, and each date of treatment shall be documented and shall be retained.

(d) Warm-blooded animals, excluding suckling rodents, used in laboratory procedures that require manipulations and observations over an extended period of time or in studies that require the animals to be removed from and returned to their home cages for any reason (e.g., cage cleaning, treatment, etc.), shall receive appropriate identification. All information needed to specifically identify each animal within an animal-housing unit shall appear on the outside of that unit.

(e) Animals of different species shall be housed in separate rooms when necessary. Animals of the same species, but used in different studies, should not ordinarily be housed in the same room when inadvertent exposure to control or test articles or animal mixup could affect the outcome of either study. If such mixed housing is necessary, adequate differentiation by space and identification shall be made.

(f) Animal cages, racks and accessory equipment shall be cleaned and sanitized at appropriate intervals.

(g) Feed and water used for the animals shall be analyzed periodically to ensure that contaminants known to be capable of interfering with the study and reasonably expected to be present in such feed or water are not present at levels above those specified in the protocol. Documentation of such analyses shall be maintained as raw data.

(h) Bedding used in animal cages or pens shall not interfere with the purpose or conduct of the study and shall be changed as often as necessary to keep the animals dry and clean.

(i) If any pest control materials are used, the use shall be documented. Cleaning and pest control materials that interfere with the study shall not be used.

Subpart F--Test and Control Articles

§58.105 Test and control article characterization.

(a) The identity, strength, purity, and composition or other characteristics which will appropriately define the test or control article shall be determined for each batch and shall be documented. Methods of synthesis, fabrication, or derivation of the test and control articles shall be documented by the sponsor or the testing facility. In those cases where marketed products are used as control articles, such products will be characterized by their labeling.

(b) The stability of each test or control article shall be determined by the testing facility or by the sponsor either: (1) Before study initiation, or (2) concomitantly according to written standard operating procedures, which provide for periodic analysis of each batch.

(c) Each storage container for a test or control article shall be labeled by name, chemical abstract number or code number, batch number, expiration date, if any, and, where appropriate, storage conditions necessary to maintain the identity, strength, purity, and composition of the test or control article. Storage containers shall be assigned to a particular test article for the duration of the study.

(d) For studies of more than 4 weeks' duration, reserve samples from each batch of test and control articles shall be retained for the period of time provided by §58.195.

§58.107 Test and control article handling.

Procedures shall be established for a system for the handling of the test and control articles to ensure that:

(a) There is proper storage.

(b) Distribution is made in a manner designed to preclude the possibility of contamination, deterioration, or damage.

(c) Proper identification is maintained

throughout the distribution process.

(d) The receipt and distribution of each batch is documented. Such documentation shall include the date and quantity of each batch distributed or returned.

§58.113 Mixtures of articles with carriers.

(a) For each test or control article that is mixed with a carrier, tests by appropriate analytical methods shall be conducted:

(1) To determine the uniformity of the mixture and to determine, periodically, the concentration of the test or control article in the mixture.

(2) To determine the stability of the test and control articles in the mixture as required by the conditions of the study either:

(i) Before study initiation, or

(ii) Concomitantly according to written standard operating procedures which provide for periodic analysis of the test and control articles in the mixture.

(b) [Reserved]

(c) Where any of the components of the test or control article carrier mixture has an expiration date, that date shall be clearly shown on the container. If more than one component has an expiration date, the earliest date shall be shown.

Subpart G--Protocol for and Conduct of a Nonclinical Laboratory Study

§58.120 Protocol.

(a) Each study shall have an approved written protocol that clearly indicates the objectives and all methods for the conduct of the study. The protocol shall contain, as applicable, the following information:

(1) A descriptive title and statement of the purpose of the study.

(2) Identification of the test and control articles by name, chemical abstract number, or code number.

(3) The name of the sponsor and the name and address of the testing facility at which the study is being conducted.

(4) The number, body weight range, sex, source of supply, species, strain, substrain, and age of the test system.

(5) The procedure for identification of the test system.

(6) A description of the experimental design, including the methods for the control of bias.

(7) A description and/or identification of the diet used in the study as well as solvents, emulsifiers, and/or other materials used to solubilize or suspend the test or control articles before mixing with the carrier. The description shall include specifications for acceptable levels of contaminants that are reasonably expected to be present in the dietary materials and are known to be capable of interfering with the purpose or conduct of the study if present at levels greater than established by the specifications.

(8) Each dosage level, expressed in milligrams per kilogram of body weight or other appropriate units, of the test or control article to be administered and the method and frequency of administration.

(9) The type and frequency of tests, analyses, and measurements to be made.

(10) The records to be maintained.

(11) The date of approval of the protocol by the sponsor and the dated signature of the study director.

(12) A statement of the proposed statistical methods to be used.

(b) All changes in or revisions of an approved protocol and the reasons therefor shall be documented, signed by the study director, dated, and maintained with the protocol.

§58.130 Conduct of a nonclinical laboratory study.

(a) The nonclinical laboratory study shall be conducted in accordance with the protocol.

(b) The test systems shall be monitored in conformity with the protocol.

(c) Specimens shall be identified by test system, study, nature, and date of collection. This information shall be located on the specimen container or shall accompany the specimen in a manner that precludes error in the recording and storage of data.

(d) Records of gross findings for a specimen from postmortem observations should be available to a pathologist when examining that specimen histopathologically.

(e) All data generated during the conduct of a nonclinical laboratory study, except those that are generated by automated data collection systems, shall be recorded directly, promptly, and legibly in ink. All data entries shall be dated on the date of entry and signed or initialed by the person entering the data. Any change in entries shall be made so as not to obscure the original entry, shall indicate the reason for such change, and shall be dated and signed or identified at the time of the change. In automated data collection systems, the individual responsible for direct data input shall be identified at the time of data input. Any change in automated data entries shall be made so as not to obscure the original entry, shall indicate the reason for change, shall be dated, and the responsible individual shall be identified.

Subparts H-I--[Reserved]

Subpart J--Records and Reports

§58.185 Reporting of nonclinical laboratory study results.

(a) A final report shall be prepared for each nonclinical laboratory study and shall include, but not necessarily be limited to, the following:

(1) Name and address of the facility performing the study and the dates on which the study was initiated and completed.

(2) Objectives and procedures stated in the approved protocol, including any changes in the original protocol.

(3) Statistical methods employed for analyzing the data.

(4) The test and control articles identified by name, chemical abstracts number or code number, strength, purity, and composition or other appropriate characteristics.

(5) Stability of the test and control articles under the conditions of administration.

(6) A description of the methods used.

(7) A description of the test system used. Where applicable, the final report shall include the number of animals used, sex, body weight range, source of supply, species, strain and substrain, age, and procedure used for identification.

(8) A description of the dosage, dosage regimen, route of administration, and duration.

(9) A description of all circumstances that may have affected the quality or integrity of the data.

(10) The name of the study director, the names of other scientists or professionals, and the names of all supervisory personnel, involved in the study.

(11) A description of the transformations, calculations, or operations performed on the data, a summary and analysis of the data, and a statement of the conclusions drawn from the analysis.

(12) The signed and dated reports of each of the individual scientists or other professionals involved in the study.

(13) The locations where all specimens, raw data, and the final report are to be stored.

(14) The statement prepared and signed by the quality assurance unit as described in §58.35 (b)(7).

(b) The final report shall be signed and dated by the study director.

(c) Corrections or additions to a final report shall be in the form of an amendment by the study director. The amendment shall clearly identify that part of the final report that is being added to or corrected and the reasons for the correction or addition, and shall be signed and dated by the person responsible.

§58.190 Storage and retrieval of records and data.

(a) All raw data, documentation, protocols, final reports, and specimens (except those specimens obtained from mutagenicity tests and wet specimens of blood, urine, feces, and biological fluids) generated as a result of a nonclinical laborato-

ry study shall be retained.

(b) There shall be archives for orderly storage and expedient retrieval of all raw data, documentation, protocols, specimens, and interim and final reports. Conditions of storage shall minimize deterioration of the documents or specimens in accordance with the requirements for the time period of their retention and the nature of the documents or specimens. A testing facility may contract with commercial archives to provide a repository for all material to be retained. Raw data and specimens may be retained elsewhere provided that the archives have specific reference to those other locations.

(c) An individual shall be identified as responsible for the archives.

(d) Only authorized personnel shall enter the archives.

(e) Material retained or referred to in the archives shall be indexed to permit expedient retrieval.

§58.195 Retention of records.

(a) Record retention requirements set forth in this section do not supersede the record retention requirements of any other regulations in this chapter.

(b) Except as provided in paragraph (c) of this section, documentation records, raw data and specimens pertaining to a nonclinical laboratory study and required to be made by this part shall be retained in the archive(s) for whichever of the following periods is shortest:

(1) A period of at least 2 years following the date on which an application for a research or marketing permit, in support of which the results of the nonclinical laboratory study were submitted, is approved by the Food and Drug Administration. This requirement does not apply to studies supporting investigational new drug applications (IND's) or applications for investigational device exemptions (IDE's), records of which shall be governed by the provisions of paragraph (b)(2) of this section.

(2) A period of at least 5 years following the date on which the results of the nonclinical laboratory study are submitted to the Food and Drug Administration in support of an application for a research or marketing permit.

(3) In other situations (e.g., where the nonclinical laboratory study does not result in the submission of the study in support of an application for a research or marketing permit), a period of at least 2 years following the date on which the study is completed, terminated, or discontinued.

(c) Wet specimens (except those specimens obtained from mutagenicity tests and wet specimens of blood, urine, feces, and biological fluids), samples of test or control articles, and specially prepared material, which are relatively fragile and differ markedly in stability and quality during storage, shall be retained only as long as the quality of the preparation affords evaluation. In no case shall retention be required for longer periods than those set forth in paragraphs (a) and (b) of this section.

(d) The master schedule sheet, copies of protocols, and records of quality assurance inspections, as required by §58.35(c) shall be maintained by the quality assurance unit as an easily accessible system of records for the period of time specified in paragraphs (a) and (b) of this section.

(e) Summaries of training and experience and job descriptions required to be maintained by §58.29(b) may be retained along with all other testing facility employment records for the length of time specified in paragraphs (a) and (b) of this section.

(f) Records and reports of the maintenance and calibration and inspection of equipment, as required by §58.63(b) and (c), shall be retained for the length of time specified in paragraph (b) of this section.

(g) Records required by this part may be retained either as original records or as true copies such as photocopies, microfilm, microfiche, or other accurate reproductions of the original records.

(h) If a facility conducting nonclinical testing goes out of business, all raw data, documentation, and other material specified in this section shall be transferred to the archives of the sponsor of the study. The Food and Drug Administration shall be notified in writing of such a transfer.

Subpart K--Disqualification of Testing Facilities

§58.200 *Purpose.*

(a) The purposes of disqualification are: (1) To permit the exclusion from consideration of completed studies that were conducted by a testing facility which has failed to comply with the requirements of the good laboratory practice regulations until it can be adequately demonstrated that such noncompliance did not occur during, or did not affect the validity or acceptability of data generated by, a particular study; and (2) to exclude from consideration all studies completed after the date of disqualification until the facility can satisfy the Commissioner that it will conduct studies in compliance with such regulations.

(b) The determination that a nonclinical laboratory study may not be considered in support of an application for a research or marketing permit does not, however, relieve the applicant for such a permit of any obligation under any other applicable regulation to submit the results of the study to the Food and Drug Administration.

§58.202 *Grounds for disqualification.*

The Commissioner may disqualify a testing facility upon finding all of the following:

(a) The testing facility failed to comply with one or more of the regulations set forth in this part (or any other regulations regarding such facilities in this chapter);

(b) The noncompliance adversely affected the validity of the nonclinical laboratory studies; and

(c) Other lesser regulatory actions (e.g., warnings or rejection of individual studies) have not been or will probably not be adequate to achieve compliance with the good laboratory practice regulations.

§58.204 *Notice of and opportunity for hearing on proposed disqualification.*

(a) Whenever the Commissioner has information indicating that grounds exist under §58. 202 which in his opinion justify disqualification of a testing facility, he may issue to the testing facility a written notice proposing that the facility be disqualified.

(b) A hearing on the disqualification shall be conducted in accordance with the requirements for a regulatory hearing set forth in part 16 of this chapter.

§58.206 *Final order on disqualification.*

(a) If the Commissioner, after the regulatory hearing, or after the time for requesting a hearing expires without a request being made, upon an evaluation of the administrative record of the disqualification proceeding, makes the findings required in §58.202, he shall issue a final order disqualifying the facility. Such order shall include a statement of the basis for that determination. Upon issuing a final order, the Commissioner shall notify (with a copy of the order) the testing facility of the action.

(b) If the Commissioner, after a regulatory hearing or after the time for requesting a hearing expires without a request being made, upon an evaluation of the administrative record of the disqualification proceeding, does not make the findings required in §58.202, he shall issue a final order terminating the disqualification proceeding. Such order shall include a statement of the basis for that determination. Upon issuing a final order the Commissioner shall notify the testing facility and provide a copy of the order.

§58.210 *Actions upon disqualification.*

(a) Once a testing facility has been disqualified, each application for a research or marketing permit, whether approved or not, containing or relying upon any nonclinical laboratory study conducted by the disqualified testing facility may be examined to determine whether such study was or would be essential to a decision. If it is determined that a study was or would be essential, the Food and Drug Administration shall also determine whether the study is acceptable, notwithstanding the disqualification of the facility. **Any study done by a testing facility before or after disqualification may be presumed to be unacceptable, and the person relying on the study may be required to establish that the study was not affected by the circumstances that led to the disqualification, e.g., by submitting validating information.** If the study is then determined to be unacceptable, such data such be eliminated from consideration in support of the application; and such elimination may serve as new information justifying the termination or withdrawal of approval of the application.

(b) No nonclinical laboratory study begun by a testing facility after the date of the facility's disqualification shall be considered in support of any application for a research or marketing permit, unless the facility has been reinstated under §58.219. The determination that a study may not be considered in support of an application for a research or marketing permit does not, however, relieve the applicant for such a permit of any obligation under any other applicable regulation to submit the results of the study to the Food and Drug Administration.

§58.213 Public disclosure of information regarding disqualification.

(a) Upon issuance of a final order disqualifying a testing facility under §58.206(a), the Commissioner may notify all or any interested persons. Such notice may be given at the discretion of the Commissioner whenever he believes that such disclosure would further the public interest or would promote compliance with the good laboratory practice regulations set forth in this part. Such notice, if given, shall include a copy of the final order issued under §58.206(a) and shall state that the disqualification constitutes a determination by the Food and Drug Administration that nonclinical laboratory studies performed by the facility will not be considered by the Food and Drug Administration in support of any application for a research or marketing permit. If such notice is sent to another Federal Government agency, the Food and Drug Administration will recommend that the agency also consider whether or not it should accept nonclinical laboratory studies performed by the testing facility. If such notice is sent to any other person, it shall state that it is given because of the relationship between the testing facility and the person being notified and that the Food and Drug Administration is not advising or recommending that any action be taken by the person notified.

(b) A determination that a testing facility has been disqualified and the administrative record regarding such determination are disclosable to the public under part 20 of this chapter.

§58.215 Alternative or additional actions to disqualification.

(a) Disqualification of a testing facility under this subpart is independent of, and neither in lieu of nor a precondition to, other proceedings or actions authorized by the act. The Food and Drug Administration may, at any time, institute against a testing facility and/or against the sponsor of a nonclinical laboratory study that has been submitted to the Food and Drug Administration any appropriate judicial proceedings (civil or criminal) and any other appropriate regulatory action, in addition to or in lieu of, and prior to, simultaneously with, or subsequent to, disqualification. The Food and Drug Administration may also refer the matter to another Federal, State, or local government law enforcement or regulatory agency for such action as that agency deems appropriate.

(b) The Food and Drug Administration may refuse to consider any particular nonclinical laboratory study in support of an application for a research or marketing permit, if it finds that the study was not conducted in accordance with the good laboratory practice regulations set forth in this part, without disqualifying the testing facility that conducted the study or undertaking other regulatory action.

§58.217 *Suspension or termination of a testing facility by a sponsor.*

Termination of a testing facility by a sponsor is independent of, and neither in lieu of nor a precondition to, proceedings or actions authorized by this subpart. If a sponsor terminates or suspends a testing facility from further participation in a nonclinical laboratory study that is being conducted as part of any application for a research or marketing permit that has been submitted to any Center of the Food and Drug Administration (whether approved or not), it shall notify that Center in writing within 15 working days of the action; the notice shall include a statement of the reasons for such action. Suspension or termination of a testing facility by a sponsor does not relieve it of any obligation under any other applicable regulation to submit the results of the study to the Food and Drug Administration.

§58.219 *Reinstatement of a disqualified testing facility.*

A testing facility that has been disqualified may be reinstated as an acceptable source of nonclinical laboratory studies to be submitted to the Food and Drug Administration if the Commissioner determines, upon an evaluation of the submission of the testing facility, that the facility can adequately assure that it will conduct future nonclinical laboratory studies in compliance with the good laboratory practice regulations set forth in this part and, if any studies are currently being conducted, that the quality and integrity of such studies have not been seriously compromised. A disqualified testing facility that wishes to be so reinstated shall present in writing to the Commissioner reasons why it believes it should be reinstated and a detailed description of the corrective actions it has taken or intends to take to assure that the acts or omissions which led to its disqualification will not recur. The Commissioner may condition reinstatement upon the testing facility being found in compliance with the good laboratory practice regulations upon an inspection. If a testing facility is reinstated, the Commissioner shall so notify the testing facility and all organizations and persons who were notified, under §58.213 of the disqualification of the testing facility. A determination that a testing facility has been reinstated is disclosable to the public under part 20 of this chapter.

GOOD CLINICAL PRACTICES

FDA's regulations for Good Clinical Practices (GCPs) are found in four parts of the *Code of Federal Regulations*:

- 21 CFR 312 -- Regulations for Investigational New Drug (IND) applications which were proposed in June 1983, finalized in March 1987, and last revised in April 1992, define the FDA's requirements for reviewing IND applications and monitoring the progress of investigational drug use to facilitate the development, evaluation, and approval of new drugs without compromising safety and effectiveness.
- 21 CFR 314 -- Regulations for New Drug Applications (NDA), which were proposed in October 1982, finalized in February 1985, and last revised in December 1992, describe the format of NDAs, define how the FDA reviews NDAs, and the

responsibilities of drug manufacturers after NDAs are approved.

- 21 CFR 50 -- Regualtions for the Protection of Human Subjects, which were proposed in January 1978, finalized in January 1981, and last revised in June 1991, set forth informed consent requirements for use of human subjects in clinical investigations.
- 21 CFR 56 -- Regulations for Institutional Review Boards (IRB), which were proposed in August 1978, finalized in February 1981, and last revised in June 1991, establish standards for the activities of IRBs that review clinical investigations involving human subjects to assure the quality and integrity of data submitted to the FDA in support of research or marketing applications.

Requirements in the regulation relating to system validation have been highlighted in boldface.

21 CFR PART 50--PROTECTION OF HUMAN SUBJECTS

Subpart A--General Provisions

§*50.1 Scope.*

(a) This part applies to all clinical investigations regulated by the Food and Drug Administration under sections 505(i), 507(d), and 520(g) of the Federal Food, Drug, and Cosmetic Act, as well as clinical investigations that support applications for research or marketing permits for products regulated by the Food and Drug Administration, including food and color additives, drugs for human use, medical devices for human use, biological products for human use, and electronic products. Additional specific obligations and commitments of, and standards of conduct for, persons who sponsor or monitor clinical investigations involving particular test articles may also be found in other parts (e.g., parts 312 and 812). Compliance with these parts is intended to protect the rights and safety of subjects involved in investigations filed with the Food and Drug Administration pursuant to sections 406, 409, 502, 503, 505, 506, 507, 510, 513-516, 518-520, 706, and 801 of the Federal Food, Drug, and Cosmetic Act and sections 351 and 354-360F of the Public Health Service Act.

(b) References in this part to regulatory sections of the Code of Federal Regulations are to chapter I of title 21, unless otherwise noted.

§*50.3 Definitions.*

As used in this part:

(a) *Act* means the Federal Food, Drug, and Cosmetic Act, as amended (secs. 201-902, 52 Stat. 1040 et seq. as amended (21 U.S.C. 321-392)).

(b) *Application for research or marketing permit* includes:

(1) A color additive petition, described in part 71.

(2) A food additive petition, described in parts 171 and 571.

(3) Data and information about a substance submitted as part of the procedures for establishing that the substance is generally recognized as safe for use that results or may reasonably be expected to result, directly or indirectly, in its becoming a component or otherwise affecting the characteristics of any food, described in §§170.30 and 570.30.

(4) Data and information about a food additive submitted as part of the procedures for food additives permitted to be used on an interim basis pending additional study, described in §180.1.

(5) Data and information about a substance submitted as part of the procedures for establishing a tolerance for unavoidable contaminants in food and food-packaging materials, described in section 406 of the act.

(6) An investigational new drug application, described in part 312 of this chapter.

(7) A new drug application, described in part 314.

(8) Data and information about the bioavailability or bioequivalence of drugs for human use submitted as part of the procedures for issuing, amending, or repealing a bioequivalence requirement, described in part 320.

(9) Data and information about an over-the-counter drug for human use submitted as part of the procedures for classifying these drugs as generally recognized as safe and effective and not misbranded, described in part 330.

(10) Data and information about a prescription drug for human use submitted as part of the procedures for classifying these drugs as generally recognized as safe and effective and not misbranded, described in this chapter.

(11) Data and information about an antibiotic drug submitted as part of the procedures for issuing, amending, or repealing regulations for these drugs, described in §314.300 of this chapter.

(12) An application for a biological product license, described in part 601.

(13) Data and information about a biological product submitted as part of the procedures for determining that licensed biological products are safe and effective and not misbranded, described in part 601.

(14) Data and information about an in vitro diagnostic product submitted as part of the procedures for establishing, amending, or repealing a standard for these products, described in part 809.

(15) An Application for an Investigational Device Exemption, described in part 812.

(16) Data and information about a medical device submitted as part of the procedures for classifying these devices, described in section 513.

(17) Data and information about a medical device submitted as part of the procedures for establishing, amending, or repealing a standard for these devices, described in section 514.

(18) An application for premarket approval of a medical device, described in section 515.

(19) A product development protocol for a medical device, described in section 515.

(20) Data and information about an electronic product submitted as part of the procedures for establishing, amending, or repealing a standard for these products, described in section 358 of the Public Health Service Act.

(21) Data and information about an electronic product submitted as part of the procedures for obtaining a variance from any electronic product performance standard, as described in §1010.4.

(22) Data and information about an electronic product submitted as part of the procedures for granting, amending, or extending an exemption from a radiation safety performance standard, as described in §1010.5.

(c) *Clinical investigation* means any experiment that involves a test article and one or more human subjects and that either is subject to requirements for prior submission to the Food and Drug Administration under section 505(i), 507(d), or 520(g) of the act, or is not subject to requirements for prior submission to the Food and Drug Administration under these sections of the act, but the results of which are intended to be submitted later to, or held for inspection by, the Food and Drug Administration as part of an application for a research or marketing permit. The term does not include experiments that are subject to the provisions of part 58 of this chapter, regarding nonclinical laboratory studies.

(d) *Investigator* means an individual who actually conducts a clinical investigation, i.e., under whose immediate direction the test article is administered or dispensed to, or used involving, a subject, or, in the event of an investigation conducted by a team of individuals, is the responsible leader of that team.

(e) *Sponsor* means a person who initiates a clinical investigation, but who does not actually conduct the investigation, i.e., the test article is administered or dispensed to or used involving, a subject under the immediate direction of another individual. A person other than an individual (e.g., corporation or agency) that uses one or more of its own employees to conduct a clinical investigation it has initiated is considered to be a sponsor (not a sponsor-investigator), and the employees are considered to be investigators.

(f) *Sponsor-investigator* means an individual

who both initiates and actually conducts, alone or with others, a clinical investigation, i.e., under whose immediate direction the test article is administered or dispensed to, or used involving, a subject. The term does not include any person other than an individual, e.g., corporation or agency.

(g) *Human subject* means an individual who is or becomes a participant in research, either as a recipient of the test article or as a control. A subject may be either a healthy human or a patient.

(h) *Institution* means any public or private entity or agency (including Federal, State, and other agencies). The word facility as used in section 520(g) of the act is deemed to be synonymous with the term institution for purposes of this part.

(i) *Institutional review board (IRB)* means any board, committee, or other group formally designated by an institution to review biomedical research involving humans as subjects, to approve the initiation of and conduct periodic review of such research. The term has the same meaning as the phrase institutional review committee as used in section 520(g) of the act.

(j) *Prisoner* means any individual involuntarily confined or detained in a penal institution. The term is intended to encompass individuals sentenced to such an institution under a criminal or civil statute, individuals detained in other facilities by virtue of statutes or commitment procedures that provide alternatives to criminal prosecution or incarceration in a penal institution, and individuals detained pending arraignment, trial, or sentencing.

(k) *Test article* means any drug (including a biological product for human use), medical device for human use, human food additive, color additive, electronic product, or any other article subject to regulation under the act or under sections 351 and 354-360F of the Public Health Service Act (42 U.S.C. 262 and 263b-263n).

(l) *Minimal risk* means that the probability and magnitude of harm or discomfort anticipated in the research are not greater in and of themselves than those ordinarily encountered in daily life or during the performance of routine physical or psychological examinations or tests.

(m) *Legally authorized representative* means an individual or judicial or other body authorized under applicable law to consent on behalf of a prospective subject to the subject's participation in the procedure(s) involved in the research.

Subpart B--Informed Consent of Human Subjects

§50.20 *General requirements for informed consent.*

Except as provided in §50.23, no investigator may involve a human being as a subject in research covered by these regulations unless the investigator has obtained the legally effective informed consent of the subject or the subject's legally authorized representative. An investigator shall seek such consent only under circumstances that provide the prospective subject or the representative sufficient opportunity to consider whether or not to participate and that minimize the possibility of coercion or undue influence. The information that is given to the subject or the representative shall be in language understandable to the subject or the representative. No informed consent, whether oral or written, may include any exculpatory language through which the subject or the representative is made to waive or appear to waive any of the subject's legal rights, or releases or appears to release the investigator, the sponsor, the institution, or its agents from liability for negligence.

§50.21 *Effective date.*

The requirements for informed consent set out in this part apply to all human subjects entering a clinical investigation that commences on or after July 27, 1981.

§50.23 *Exception from general requirements.*

(a) The obtaining of informed consent shall be deemed feasible unless, before use of the test article (except as provided in paragraph (b) of this section), both the investigator and a physician who is not otherwise participating in the clinical investigation certify in writing all of the following:

(1) The human subject is confronted by a life-threatening situation necessitating the use of the test article.

(2) Informed consent cannot be obtained from the subject because of an inability to communicate with, or obtain legally effective consent from, the subject.

(3) Time is not sufficient to obtain consent from the subject's legal representative.

(4) There is available no alternative method of approved or generally recognized therapy that provides an equal or greater likelihood of saving the life of the subject.

(b) If immediate use of the test article is, in the investigator's opinion, required to preserve the life of the subject, and time is not sufficient to obtain the independent determination required in paragraph (a) of this section in advance of using the test article, the determinations of the clinical investigator shall be made and, within 5 working days after the use of the article, be reviewed and evaluated in writing by a physician who is not participating in the clinical investigation.

(c) The documentation required in paragraph (a) or (b) of this section shall be submitted to the IRB within 5 working days after the use of the test article.

(d)(1) The Commissioner may also determine that obtaining informed consent is not feasible when the Assistant Secretary of Defense (Health Affairs) requests such a determination in connection with the use of an investigational drug (including an antibiotic or biological product) in a specific protocol under an investigational new drug application (IND) sponsored by the Department of Defense (DOD). DOD's request for a determination that obtaining informed consent from military personnel is not feasible must be limited to a specific military operation involving combat or the immediate threat of combat. The request must also include a written justification supporting the conclusions of the physician(s) responsible for the medical care of the military personnel involved and the investigator(s) identified in the IND that a military combat exigency exists because of special military combat (actual or threatened) circumstances in which, in order to facilitate the accomplishment of the military mission, preservation of the health of the individual and the safety of other personnel require that a particular treatment be provided to a specified group of military personnel, without regard to what might be any individual's personal preference for no treatment or for some alternative treatment. The written request must also include a statement that a duly constituted institutional review board has reviewed and approved the use of the investigational drug without informed consent. The Commissioner may find that informed consent is not feasible only when withholding treatment would be contrary to the best interests of military personnel and there is no available satisfactory alternative therapy.

(2) In reaching a determination under paragraph (d)(1) of this section that obtaining informed consent is not feasible and withholding treatment would be contrary to the best interests of military personnel, the Commissioner will review the request submitted under paragraph (d)(1) of this section and take into account all pertinent factors, including, but not limited to:

(i) The extent and strength of the evidence of the safety and effectiveness of the investigational drug for the intended use;

(ii) The context in which the drug will be administered, e.g., whether it is intended for use in a battlefield or hospital setting or whether it will be self-administered or will be administered by a health professional;

(iii) The nature of the disease or condition for which the preventive or therapeutic treatment is intended; and

(iv) The nature of the information to be provided to the recipients of the drug concerning the potential benefits and risks of taking or not taking the drug.

(3) The Commissioner may request a recommendation from appropriate experts before reaching a determination on a request submitted under paragraph (d)(1) of this section.

(4) A determination by the Commissioner that obtaining informed consent is not feasible

and withholding treatment would be contrary to the best interests of military personnel will expire at the end of 1 year, unless renewed at DOD's request, or when DOD informs the Commissioner that the specific military operation creating the need for the use of the investigational drug has ended, whichever is earlier. The Commissioner may also revoke this determination based on changed circumstances.

§*50.25 Elements of informed consent.*

(a) Basic elements of informed consent. In seeking informed consent, the following information shall be provided to each subject:

(1) A statement that the study involves research, an explanation of the purposes of the research and the expected duration of the subject's participation, a description of the procedures to be followed, and identification of any procedures which are experimental.

(2) A description of any reasonably foreseeable risks or discomforts to the subject.

(3) A description of any benefits to the subject or to others which may reasonably be expected from the research.

(4) A disclosure of appropriate alternative procedures or courses of treatment, if any, that might be advantageous to the subject.

(5) A statement describing the extent, if any, to which confidentiality of records identifying the subject will be maintained and that notes the possibility that the Food and Drug Administration may inspect the records.

(6) For research involving more than minimal risk, an explanation as to whether any compensation and an explanation as to whether any medical treatments are available if injury occurs and, if so, what they consist of, or where further information may be obtained.

(7) An explanation of whom to contact for answers to pertinent questions about the research and research subjects' rights, and whom to contact in the event of a research-related injury to the subject.

(8) A statement that participation is voluntary, that refusal to participate will involve no penalty or loss of benefits to which the subject is otherwise entitled, and that the subject may discontinue participation at any time without penalty or loss of benefits to which the subject is otherwise entitled.

(b) Additional elements of informed consent. When appropriate, one or more of the following elements of information shall also be provided to each subject:

(1) A statement that the particular treatment or procedure may involve risks to the subject (or to the embryo or fetus, if the subject is or may become pregnant) which are currently unforeseeable.

(2) Anticipated circumstances under which the subject's participation may be terminated by the investigator without regard to the subject's consent.

(3) Any additional costs to the subject that may result from participation in the research.

(4) The consequences of a subject's decision to withdraw from the research and procedures for orderly termination of participation by the subject.

(5) A statement that significant new findings developed during the course of the research which may relate to the subject's willingness to continue participation will be provided to the subject.

(6) The approximate number of subjects involved in the study.

(c) The informed consent requirements in these regulations are not intended to preempt any applicable Federal, State, or local laws which require additional information to be disclosed for informed consent to be legally effective.

(d) Nothing in these regulations is intended to limit the authority of a physician to provide emergency medical care to the extent the physician is permitted to do so under applicable Federal, State, or local law.

§*50.27 Documentation of informed consent.*

(a) Except as provided in §56.109(c), informed consent shall be documented by the use of a written consent form approved by the IRB and signed by the subject or the subject's legally authorized representative. A copy shall be given to the person signing the form.

(b) Except as provided in §56.109(c), the consent form may be either of the following:

(1) A written consent document that embod-

ies the elements of informed consent required by §50.25. This form may be read to the subject or the subject's legally authorized representative, but, in any event, the investigator shall give either the subject or the representative adequate opportunity to read it before it is signed.

(2) A short form written consent document stating that the elements of informed consent required by §50.25 have been presented orally to the subject or the subject's legally authorized representative. When this method is used, there shall be a witness to the oral presentation. Also, the IRB shall approve a written summary of what is to be said to the subject or the representative. Only the short form itself is to be signed by the subject or the representative. However, the witness shall sign both the short form and a copy of the summary, and the person actually obtaining the consent shall sign a copy of the summary. A copy of the summary shall be given to the subject or the representative in addition to a copy of the short form.

Subpart C--Protections Pertaining to Clinical Investigations Involving Prisoners as Subjects

§50.40 Applicability.

(a) The regulations in this subpart apply to all clinical investigations involving prisoners as subjects that are regulated by the Food and Drug Administration under sections 505(i), 507(d), or 520(g) of the Federal Food, Drug, and Cosmetic Act, as well as clinical investigations involving prisoners that support applications for research or marketing permits for products regulated by the Food and Drug Administration.

(b) Nothing in this subpart shall be construed as indicating that compliance with the procedures set forth herein will authorize research involving prisoners as subjects to the extent such research is limited or barred by applicable State or local law.

§50.42 Purpose.

Inasmuch as prisoners may be under constraints because of their incarceration which could affect their ability to make a truly voluntary uncoerced decision whether or not to participate as subjects in research, it is the purpose of this subpart to provide additional safeguards for the protection of prisoners involved in activities to which this subpart is applicable.

§50.44 Restrictions on clinical investigations involving prisoners.

(a) Except as provided in §50.44(b), clinical investigations regulated by the Food and Drug Administration under sections 505(i), 507(d), and 505(g) of the Federal Food, Drug, and Cosmetic Act, as well as clinical investigations that support applications for research or marketing permits for products regulated by the Food and Drug Administration may not involve prisoners as subjects.

(b) Clinical investigations that are regulated by the Food and Drug Administration under sections 505(i), 507(d), or 520(g) of the Federal Food, Drug, and Cosmetic Act, as well as clinical investigations that support applications for research or marketing permits for products regulated by the Food and Drug Administration, may involve prisoners as subjects only if the institution responsible for the conduct of the clinical investigation has certified to the Food and Drug Administration that the institutional review board has approved the clinical investigation under §50.48; and

(1)(i) In the judgment of the Food and Drug Administration, the proposed clinical investigation involves solely research on practices both innovative and accepted, which have the intent and reasonable probability of improving, the health and well-being of the subjects;

(ii) In cases in which these studies require the assignment of prisoners in a manner consistent with protocols approved by the institutional review board to control groups that may not benefit from the research, the study may proceed only after the Food and Drug Administration has consulted with appropriate experts, including experts in penology, medicine, and ethics, and has published notice in the FEDERAL REGISTER of its intent to approve such research; or

(2) Research on conditions particularly affecting prisoners as a class (for example, vaccine trials and other research on hepatitis, which is much more prevalent in prisons than elsewhere) provided that the Food and Drug Administration has consulted with appropriate experts, including experts in penology, medicine, and ethics, and has published notice in the FEDERAL REGISTER of its intent to approve such research; subject to the approval of the Food and Drug Administration, prisoners may participate in the research even though they are assigned, in a manner consistent with protocols approved by the institutional review board, to control groups that may not benefit from the research.

§50.46 Composition of institutional review boards where prisoners are involved.

In addition to satisfying any other requirements governing institutional review boards set forth in this chapter, an institutional review board, in carrying out responsibilities under this part with respect to research covered by this subpart, shall also meet the following specific requirements:

(a) A majority of the institutional review board (exclusive of prisoner members) shall have no association with the prison(s) involved, apart from their membership on the institutional review board.

(b) At least one member of the institutional review board shall be a prisoner, or a prisoner advocate with appropriate background and experience to serve in that capacity, except that if a particular research project is reviewed by more than one institutional review board, only one institutional review board need satisfy this requirement.

§50.48 Additional duties of the institutional review boards where prisoners are involved.

(a) In addition to all other responsibilities prescribed for institutional review boards under this chapter, the institutional review board shall review clinical investigations covered by this subpart and approve such clinical investigations only if it finds that:

(1) The research under review represents one of the categories of research permitted under §50.44(b) (1) and (2);

(2) Any possible advantages accruing to the prisoner through his or her participation in the clinical investigation, when compared to the general living conditions, medical care, quality of food, amenities, and opportunity for earnings in prison, are not of such a magnitude that his or her ability to weigh the risks of the clinical investigation against the value of such advantages in the limited-choice environment of the prison is impaired;

(3) The risks involved in the clinical investigation are commensurate with risks that would be accepted by nonprisoner volunteers;

(4) Procedures for the selection of subjects within the prison are fair to all prisoners and immune from arbitrary intervention by prison authorities or prisoners; unless the principal investigator provides to the institutional review board justification in writing for following some other procedures, control subjects must be selected randomly from the group of available prisoners who meet the characteristics needed for that research project;

(5) Any information given to subjects is presented in language which is appropriate for the subject population;

(6) Adequate assurance exists that parole boards will not take into account a prisoner's participation in the clinical investigation in making decisions regarding parole, and each prisoner is clearly informed in advance that participation in the clinical investigation will have no effect on his or her parole; and

(7) Where the institutional review board finds there may be need for followup examination or care of participants after the end of their participation, adequate provision has been made for such examination or care, taking into account the varying lengths of individual prisoners' sentences, and for informing participants of this fact.

(b) The institutional review board shall carry out such other duties as may be assigned by the Food and Drug Administration.

(c) The institution shall certify to the Food and Drug Administration, in such form and manner as the Food and Drug Administration may require, that the duties of the institutional review board under this section have been fulfilled.

21 CFR PART 56--INSTITUTIONAL REVIEW BOARDS

Subpart A--General Provisions

§56.101 Scope.

(a) This part contains the general standards for the composition, operation, and responsibility of an Institutional Review Board (IRB) that reviews clinical investigations regulated by the Food and Drug Administration under sections 505(i), 507(d), and 520(g) of the act, as well as clinical investigations that support applications for research or marketing permits for products regulated by the Food and Drug Administration, including food and color additives, drugs for human use, medical devices for human use, biological products for human use, and electronic products. Compliance with this part is intended to protect the rights and welfare of human subjects involved in such investigations.

(b) References in this part to regulatory sections of the Code of Federal Regulations are to chapter I of title 21, unless otherwise noted.

§56.102 Definitions.

As used in this part:

(a) *Act* means the Federal Food, Drug, and Cosmetic Act, as amended (secs. 201-902, 52 Stat. 1040 et seq., as amended (21 U.S.C. 321-392)).

(b) *Application for research or marketing permit* includes:

(1) A color additive petition, described in part 71.

(2) Data and information regarding a substance submitted as part of the procedures for establishing that a substance is generally recognized as safe for a use which results or may reasonably be expected to result, directly or indirectly, in its becoming a component or otherwise affecting the characteristics of any food, described in §170.35.

(3) A food additive petition, described in part 171.

(4) Data and information regarding a food additive submitted as part of the procedures regarding food additives permitted to be used on an interim basis pending additional study, described in §180.1.

(5) Data and information regarding a substance submitted as part of the procedures for establishing a tolerance for unavoidable contaminants in food and food-packaging materials, described in section 406 of the act.

(6) An investigational new drug application, described in part 312 of this chapter.

(7) A new drug application, described in part 314.

(8) Data and information regarding the bioavailability or bioequivalence of drugs for human use submitted as part of the procedures for issuing, amending, or repealing a bioequivalence requirement, described in part 320.

(9) Data and information regarding an over-the-counter drug for human use submitted as part of the procedures for classifying such drugs as generally recognized as safe and effective and not misbranded, described in part 330.

(10) Data and information regarding an antibiotic drug submitted as part of the procedures for issuing, amending, or repealing regulations for such drugs, described in §314.300 of this chapter.

(11) An application for a biological product license, described in part 601.

(12) Data and information regarding a biological product submitted as part of the procedures for determining that licensed biological products are safe and effective and not misbranded, as described in part 601.

(13) An Application for an Investigational Device Exemption, described in parts 812 and 813.

(14) Data and information regarding a medical device for human use submitted as part of the procedures for classifying such devices, described in part 860.

(15) Data and information regarding a medical device for human use submitted as part of the procedures for establishing, amending, or repealing a standard for such device, described in part 861.

(16) An application for premarket approval of a medical device for human use, described in section 515 of the act.

(17) A product development protocol for a medical device for human use, described in section 515 of the act.

(18) Data and information regarding an electronic product submitted as part of the procedures for establishing, amending, or repealing a standard for such products, described in section 358 of the Public Health Service Act.

(19) Data and information regarding an electronic product submitted as part of the procedures for obtaining a variance from any electronic product performance standard, as described in §1010.4.

(20) Data and information regarding an electronic product submitted as part of the procedures for granting, amending, or extending an exemption from a radiation safety performance standard, as described in §1010. 5.

(21) Data and information regarding an electronic product submitted as part of the procedures for obtaining an exemption from notification of a radiation safety defect or failure of compliance with a radiation safety performance standard, described in subpart D of part 1003.

(c) *Clinical investigation* means any experiment that involves a test article and one or more human subjects, and that either must meet the requirements for prior submission to the Food and Drug Administration under section 505(i), 507(d), or 520(g) of the act, or need not meet the requirements for prior submission to the Food and Drug Administration under these sections of the act, but the results of which are intended to be later submitted to, or held for inspection by, the Food and Drug Administration as part of an application for a research or marketing permit. The term does not include experiments that must meet the provisions of part 58, regarding nonclinical laboratory studies. The terms research, clinical research, clinical study, study, and clinical investigation are deemed to be synonymous for purposes of this part.

(d) *Emergency use* means the use of a test article on a human subject in a life-threatening situation in which no standard acceptable treatment is available, and in which there is not sufficient time to obtain IRB approval.

(e) *Human subject* means an individual who is or becomes a participant in research, either as a recipient of the test article or as a control. A subject may be either a healthy individual or a patient.

(f) *Institution* means any public or private entity or agency (including Federal, State, and other agencies). The term facility as used in section 520(g) of the act is deemed to be synonymous with the term institution for purposes of this part.

(g) *Institutional Review Board (IRB)* means any board, committee, or other group formally designated by an institution to review, to approve the initiation of, and to conduct periodic review of, biomedical research involving human subjects. The primary purpose of such review is to assure the protection of the rights and welfare of the human subjects. The term has the same meaning as the phrase institutional review committee as used in section 520(g) of the act.

(h) *Investigator* means an individual who actually conducts a clinical investigation (i.e., under whose immediate direction the test article is administered or dispensed to, or used involving, a subject) or, in the event of an investigation conducted by a team of individuals, is the responsible leader of that team.

(i) *Minimal risk* means that the probability and magnitude of harm or discomfort anticipated in the research are not greater in and of themselves than those ordinarily encountered in daily life or during the performance of routine physical or psychological examinations or tests.

(j) *Sponsor* means a person or other entity that initiates a clinical investigation, but that does not actually conduct the investigation, i.e., the test article is administered or dispensed to, or used involving, a subject under the immediate direction of another individual. A person other than an individual (e.g., a corporation or agency) that uses one or more of its own employees to conduct an investigation that it has initiated is considered to be a sponsor (not a sponsor-investigator), and the employees are considered to be investigators.

(k) *Sponsor-investigator* means an individual who both initiates and actually conducts, alone or with others, a clinical investigation, i.e., under whose immediate direction the test article is administered or dispensed to, or used involving, a subject. The term does not include any person other than an individual, e.g., it does not include a corporation or agency. The obligations of a sponsor-investigator under this part include both those of a sponsor and those of an investigator.

(l) *Test article* means any drug for human use, biological product for human use, medical device for human use, human food additive, color additive, electronic product, or any other article subject to regulation under the act or under sections 351 or 354-360F of the Public Health Service Act.

(m) *IRB approval* means the determination of the IRB that the clinical investigation has been reviewed and may be conducted at an institution within the constraints set forth by the IRB and by other institutional and Federal requirements.

§56.103 *Circumstances in which IRB review is required.*

(a) Except as provided in §§56.104 and 56.105, any clinical investigation which must meet the requirements for prior submission (as required in parts 312, 812, and 813) to the Food and Drug Administration shall not be initiated unless that investigation has been reviewed and approved by, and remains subject to continuing review by, an IRB meeting the requirements of this part.

(b) Except as provided in §§56.104 and 56.105, the Food and Drug Administration may decide not to consider in support of an application for a research or marketing permit any data or information that has been derived from a clinical investigation that has not been approved by, and that was not subject to initial and continuing review by, an IRB meeting the requirements of this part. The determination that a clinical investigation may not be considered in support of an application for a research or marketing permit does not, however, relieve the applicant for such a permit of any obligation under any other applicable regulations to submit the results of the investigation to the Food and Drug Administration.

(c) Compliance with these regulations will in no way render inapplicable pertinent Federal, State, or local laws or regulations.

§56.104 *Exemptions from IRB requirement.*

The following categories of clinical investigations are exempt from the requirements of this part for IRB review:

(a) Any investigation which commenced before July 27, 1981 and was subject to requirements for IRB review under FDA regulations before that date, provided that the investigation remains subject to review of an IRB which meets the FDA requirements in effect before July 27, 1981.

(b) Any investigation commenced before July 27, 1981 and was not otherwise subject to requirements for IRB review under Food and Drug Administration regulations before that date.

(c) Emergency use of a test article, provided that such emergency use is reported to the IRB within 5 working days. Any subsequent use of the test article at the institution is subject to IRB review.

(d) Taste and food quality evaluations and consumer acceptance studies, if wholesome foods without additives are consumed or if a food is consumed that contains a food ingredient at or below the level and for a use found to be safe, or agricultural, chemical, or environmental contaminant at or below the level found to be safe, by the Food and Drug Administration or approved by the Environmental Protection Agency or the Food Safety and Inspection Service of the U.S. Department of Agriculture.

§56.105 *Waiver of IRB requirement.*

On the application of a sponsor or sponsor-investigator, the Food and Drug Administration may waive any of the requirements contained in these regulations, including the requirements for IRB review, for specific research activities or for classes of research activities, otherwise covered by these regulations.

Subpart B--Organization and Personnel

§56.107 *IRB membership.*

(a) Each IRB shall have at least five members, with varying backgrounds to promote complete and adequate review of research activities commonly conducted by the institution. The IRB shall be sufficiently qualified through the experience and expertise of its members, and the diversity of the members, including consideration of race, gender, cultural backgrounds, and sensitivity to such issues as community attitudes, to promote respect for its advice and counsel in safeguarding the rights and welfare of human subjects. In addition to possessing the professional competence necessary to review the specific research activities, the IRB shall be able to ascertain the acceptability of proposed research in terms of institutional commitments and regulations, applicable law, and standards or professional conduct and practice. The IRB shall therefore include persons knowledgeable in these areas. If an IRB regularly reviews research that involves a vulnerable category of subjects, such as children, prisoners, pregnant women, or handicapped or mentally disabled persons, consideration shall be given to the inclusion of one or more individuals who are knowledgeable about and experienced in working with those subjects.

(b) Every nondiscriminatory effort will be made to ensure that no IRB consists entirely of men or entirely of women, including the institution's consideration of qualified persons of both sexes, so long as no selection is made to the IRB on the basis of gender. No IRB may consist entirely of members of one profession.

(c) Each IRB shall include at least one member whose primary concerns are in the scientific area and at least one member whose primary concerns are in nonscientific areas.

(d) Each IRB shall include at least one member who is not otherwise affiliated with the institution and who is not part of the immediate family of a person who is affiliated with the institution.

(e) No IRB may have a member participate in the IRB's initial or continuing review of any project in which the member has a conflicting interest, except to provide information requested by the IRB.

(f) An IRB may, in its discretion, invite individuals with competence in special areas to assist in the review of complex issues which require expertise beyond or in addition to that available on the IRB. These individuals may not vote with the IRB.

Subpart C--IRB Functions and Operations

§56.108 *IRB functions and operations.*

In order to fulfill the requirements of these regulations, each IRB shall:

(a) Follow written procedures: (1) For conducting its initial and continuing review of research and for reporting its findings and actions to the investigator and the institution; (2) for determining which projects require review more often than annually and which projects need verification from sources other than the investigator that no material changes have occurred since previous IRB review; (3) for ensuring prompt reporting to the IRB of changes in research activity; and (4) for ensuring that changes in approved research, during the period for which IRB approval has already been given, may not be initiated without IRB review and approval except where necessary to eliminate apparent immediate hazards to the human subjects.

(b) Follow written procedures for ensuring prompt reporting to the IRB, appropriate institutional officials, and the Food and Drug Administration of: (1) Any unanticipated problems involving risks to human subjects or others; (2) any instance of serious or continuing noncompliance with these regulations or the requirements or determinations of the IRB; or (3) any suspension or termination of IRB approval.

(c) Except when an expedited review procedure is used (see §56.110), review proposed research at convened meetings at which a majority of the members of the IRB are present, including at least one member whose primary concerns are in nonscientific areas. In order for the research to be approved, it shall receive the approval of a majority of those members present at the meeting.

§56.109 IRB review of research.

(a) An IRB shall review and have authority to approve, require modifications in (to secure approval), or disapprove all research activities covered by these regulations.

(b) An IRB shall require that information given to subjects as part of informed consent is in accordance with §50.25. The IRB may require that information, in addition to that specifically mentioned in §50.25, be given to the subjects when in the IRB's judgment the information would meaningfully add to the protection of the rights and welfare of subjects.

(c) An IRB shall require documentation of informed consent in accordance with §50.27, except that the IRB may, for some or all subjects, waive the requirement that the subject or the subject's legally authorized representative sign a written consent form if it finds that the research presents no more than minimal risk of harm to subjects and involves no procedures for which written consent is normally required outside the research context. In cases where the documentation requirement is waived, the IRB may require the investigator to provide subjects with a written statement regarding the research.

(d) An IRB shall notify investigators and the institution in writing of its decision to approve or disapprove the proposed research activity, or of modifications required to secure IRB approval of the research activity. If the IRB decides to disapprove a research activity, it shall include in its written notification a statement of the reasons for its decision and give the investigator an opportunity to respond in person or in writing.

(e) An IRB shall conduct continuing review of research covered by these regulations at intervals appropriate to the degree of risk, but not less than once per year, and shall have authority to observe or have a third party observe the consent process and the research.

§56.110 Expedited review procedures for certain kinds of research involving no more than minimal risk, and for minor changes in approved research.

(a) The Food and Drug Administration has established, and published in the FEDERAL REGISTER, a list of categories of research that may be reviewed by the IRB through an expedited review procedure. The list will be amended, as appropriate, through periodic republication in the FEDERAL REGISTER.

(b) An IRB may use the expedited review procedure to review either or both of the following: (1) Some or all of the research appearing on the list and found by the reviewer(s) to involve no more than minimal risk, (2) minor changes in previously approved research during the period (of 1 year or less) for which approval is authorized. Under an expedited review procedure, the review may be carried out by the IRB chairperson or by one or more experienced reviewers designated by the IRB chairperson from among the members of the IRB. In reviewing the research, the reviewers may exercise all of the authorities of the IRB except that the reviewers may not disapprove the research. A research activity may be disapproved only after review in accordance with the nonexpedited review procedure set forth in §56.108(c).

(c) Each IRB which uses an expedited review procedure shall adopt a method for keeping all members advised of research proposals which have been approved under the procedure.

(d) The Food and Drug Administration may restrict, suspend, or terminate an institution's or IRB's use of the expedited review procedure when necessary to protect the rights or welfare of subjects.

§56.111 Criteria for IRB approval of research.

(a) In order to approve research covered by these regulations the IRB shall determine that all of the following requirements are satisfied:

(1) Risks to subjects are minimized: (i) By using procedures which are consistent with sound research design and which do not

unnecessarily expose subjects to risk, and (ii) whenever appropriate, by using procedures already being performed on the subjects for diagnostic or treatment purposes.

(2) Risks to subjects are reasonable in relation to anticipated benefits, if any, to subjects, and the importance of the knowledge that may be expected to result. In evaluating risks and benefits, the IRB should consider only those risks and benefits that may result from the research (as distinguished from risks and benefits of therapies that subjects would receive even if not participating in the research). The IRB should not consider possible long-range effects of applying knowledge gained in the research (for example, the possible effects of the research on public policy) as among those research risks that fall within the purview of its responsibility.

(3) Selection of subjects is equitable. In making this assessment the IRB should take into account the purposes of the research and the setting in which the research will be conducted and should be particularly cognizant of the special problems of research involving vulnerable populations, such as children, prisoners, pregnant women, handicapped, or mentally disabled persons, or economically or educationally disadvantaged persons.

(4) Informed consent will be sought from each prospective subject or the subject's legally authorized representative, in accordance with and to the extent required by part 50.

(5) Informed consent will be appropriately documented, in accordance with and to the extent required by §50.27.

(6) Where appropriate, the research plan makes adequate provision for monitoring the data collected to ensure the safety of subjects.

(7) Where appropriate, there are adequate provisions to protect the privacy of subjects and to maintain the confidentiality of data.

(b) When some or all of the subjects, such as children, prisoners, pregnant women, handicapped, or mentally disabled persons, or economically or educationally disadvantaged persons, are likely to be vulnerable to coercion or undue influence additional safeguards have been included in the study to protect the rights and welfare of these subjects.

§56.112 *Review by institution.*

Research covered by these regulations that has been approved by an IRB may be subject to further appropriate review and approval or disapproval by officials of the institution. However, those officials may not approve the research if it has not been approved by an IRB.

§56.113 *Suspension or termination of IRB approval of research.*

An IRB shall have authority to suspend or terminate approval of research that is not being conducted in accordance with the IRB's requirements or that has been associated with unexpected serious harm to subjects. Any suspension or termination of approval shall include a statement of the reasons for the IRB's action and shall be reported promptly to the investigator, appropriate institutional officials, and the Food and Drug Administration.

§56.114 *Cooperative research.*

In complying with these regulations, institutions involved in multi-institutional studies may use joint review, reliance upon the review of another qualified IRB, or similar arrangements aimed at avoidance of duplication of effort.

Subpart D--Records and Reports

§56.115 *IRB records.*

(a) An institution, or where appropriate an IRB, shall prepare and maintain adequate documentation of IRB activities, including the following:

(1) Copies of all research proposals reviewed, scientific evaluations, if any, that accompany the proposals, approved sample consent documents, progress reports submitted by investigators, and reports of injuries to subjects.

(2) Minutes of IRB meetings which shall be in sufficient detail to show attendance at the meetings; actions taken by the IRB; the vote

on these actions including the number of members voting for, against, and abstaining; the basis for requiring changes in or disapproving research; and a written summary of the discussion of controverted issues and their resolution.

(3) Records of continuing review activities.

(4) Copies of all correspondence between the IRB and the investigators.

(5) A list of IRB members identified by name; earned degrees; representative capacity; indications of experience such as board certifications, licenses, etc., sufficient to describe each member's chief anticipated contributions to IRB deliberations; and any employment or other relationship between each member and the institution; for example: full-time employee, part-time employee, a member of governing panel or board, stockholder, paid or unpaid consultant.

(6) Written procedures for the IRB as required by §56.108 (a) and (b).

(7) Statements of significant new findings provided to subjects, as required by §50.25.

(b) The records required by this regulation shall be retained for at least 3 years after completion of the research, and the records shall be accessible for inspection and copying by authorized representatives of the Food and Drug Administration at reasonable times and in a reasonable manner.

(c) The Food and Drug Administration may refuse to consider a clinical investigation in support of an application for a research or marketing permit if the institution or the IRB that reviewed the investigation refuses to allow an inspection under this section.

Subpart E--Administrative Actions for Noncompliance

§56.120 *Lesser administrative actions.* determined to be responsible for formal designation of the IRB.

§56.121 *Disqualification of an IRB or an institution.*

(a) Whenever the IRB or the institution has failed to take adequate steps to correct the noncompliance stated in the letter sent by the agency under §56.120(a), and the Commissioner of Food and Drugs determines that this noncompliance may justify the disqualification of the IRB or of the parent institution, the Commissioner will institute proceedings in accordance with the requirements for a regulatory hearing set forth in part 16.

(b) The Commissioner may disqualify an IRB or the parent institution if the Commissioner determines that:

(1) The IRB has refused or repeatedly failed to comply with any of the regulations set forth in this part, and

(2) The noncompliance adversely affects the rights or welfare of the human subjects in a clinical investigation.

(c) If the Commissioner determines that disqualification is appropriate, the Commissioner will issue an order that explains the basis for the determination and that prescribes any actions to be taken with regard to ongoing clinical research conducted under the review of the IRB. The Food and Drug Administration will send notice of the disqualification to the IRB and the parent institution. Other parties with a direct interest, such as sponsors and clinical investigators, may also be sent a notice of the disqualification. In addition, the agency may elect to publish a notice of its action in the FEDERAL REGISTER.

(d) The Food and Drug Administration will not approve an application for a research permit for a clinical investigation that is to be under the review of a disqualified IRB or that is to be conducted at a disqualified institution, and it may refuse to consider in support of a marketing permit the data from a clinical investigation that was reviewed by a disqualified IRB as conducted at a disqualified institution, unless the IRB or the parent institution is reinstated as provided in §56.123.

§56.122 *Public disclosure of information regarding revocation.*

A determination that the Food and Drug Administration has disqualified an institution and the administrative record regarding that determination are disclosable to the public under part 20.

§56.123 Reinstatement of an IRB or an institution.

An IRB or an institution may be reinstated if the Commissioner determines, upon an evaluation of a written submission from the IRB or institution that explains the corrective action that the institution or IRB plans to take, that the IRB or institution has provided adequate assurance that it will operate in compliance with the standards set forth in this part. Notification of reinstatement shall be provided to all persons notified under §56.121(c).

§56.124 Actions alternative or additional to disqualification.

Disqualification of an IRB or of an institution is independent of, and neither in lieu of nor a precondition to, other proceedings or actions authorized by the act. The Food and Drug Administration may, at any time, through the Department of Justice institute any appropriate judicial proceedings (civil or criminal) and any other appropriate regulatory action, in addition to or in lieu of, and before, at the time of, or after, disqualification. The agency may also refer pertinent matters to another Federal, State, or local government agency for any action that that agency determines to be appropriate.

21 CFR PART 312--INVESTIGATIONAL NEW DRUG APPLICATION

Subpart A--General Provisions

§312.1 Scope.

(a) This part contains procedures and requirements governing the use of investigational new drugs, including procedures and requirements for the submission to, and review by, the Food and Drug Administration of investigational new drug applications (IND's). An investigational new drug for which an IND is in effect in accordance with this part is exempt from the premarketing approval requirements that are otherwise applicable and may be shipped lawfully for the purpose of conducting clinical investigations of that drug.

(b) References in this part to regulations in the Code of Federal Regulations are to chapter I of title 21, unless otherwise noted.

§312.2 Applicability.

(a) Applicability. Except as provided in this section, this part applies to all clinical investigations of products that are subject to section 505 or 507 of the Federal Food, Drug, and Cosmetic Act or to the licensing provisions of the Public Health Service Act (58 Stat. 632, as amended (42 U.S.C. 201 et seq.)).

(b) Exemptions. (1) The clinical investigation of a drug product that is lawfully marketed in the United States is exempt from the requirements of this part if all the following apply:

(i) The investigation is not intended to be reported to FDA as a well-controlled study in support of a new indication for use nor intended to be used to support any other significant change in the labeling for the drug;

(ii) If the drug that is undergoing investigation is lawfully marketed as a prescription drug product, the investigation is not intended to support a significant change in the advertising for the product;

(iii) The investigation does not involve a route of administration or dosage level or use in a patient population or other factor that significantly increases the risks (or decreases the acceptability of the risks) associated with the use of the drug product;

(iv) The investigation is conducted in compliance with the requirements for institutional review set forth in part 56 and with the requirements for informed consent set forth in part 50; and

(v) The investigation is conducted in compliance with the requirements of §312.7.

(2)(i) A clinical investigation involving an in vitro diagnostic biological product listed in paragraph (b)(2)(ii) of this section is exempt from the requirements of this part if (a) it is intended to be used in a diagnostic procedure that confirms the diagnosis made by another, medically established, diagnostic product or procedure and (b) it is shipped in compliance with §312.160.

(ii) In accordance with paragraph (b)(2)(i) of this section, the following products are exempt from the requirements of this part: (a) blood grouping serum; (b) reagent red blood cells; and (c) anti-human globulin.

(3) A drug intended solely for tests in vitro or in laboratory research animals is exempt from the requirements of this part if shipped in accordance with §312.160.

(4) FDA will not accept an application for an investigation that is exempt under the provisions of paragraph (b)(1) of this section.

(5) A clinical investigation involving use of a placebo is exempt from the requirements of this part if the investigation does not otherwise require submission of an IND.

(c) Bioavailability studies. The applicability of this part to in vivo bioavailability studies in humans is subject to the provisions of §320. 31.

(d) Unlabeled indication. This part does not apply to the use in the practice of medicine for an unlabeled indication of a new drug or antibiotic drug product approved under part 314 or of a licensed biological product.

(e) Guidance. FDA may, on its own initiative, issue guidance on the applicability of this part to particular investigational uses of drugs. On request, FDA will advise on the applicability of this part to a planned clinical investigation.

§312.3 Definitions and interpretations.

(a) The definitions and interpretations of terms contained in section 201 of the Act apply to those terms when used in this part:

(b) The following definitions of terms also apply to this part:

Act means the Federal Food, Drug, and Cosmetic Act (secs. 201- 902, 52 Stat. 1040 et seq., as amended (21 U.S.C. 301-392)).

Clinical investigation means any experiment in which a drug is administered or dispensed to, or used involving, one or more human subjects. For the purposes of this part, an experiment is any use of a drug except for the use of a marketed drug in the course of medical practice.

Contract research organization means a person that assumes, as an independent contractor with the sponsor, one or more of the obligations of a sponsor, e.g., design of a protocol, selection or monitoring of investigations, evaluation of reports, and preparation of materials to be submitted to the Food and Drug Administration.

FDA means the Food and Drug Administration.

IND means an investigational new drug application. For purposes of this part, "IND" is synonymous with "Notice of Claimed Investigational Exemption for a New Drug."

Investigational new drug means a new drug, antibiotic drug, or biological drug that is used in a clinical investigation. The term also includes a biological product that is used in vitro for diagnostic purposes. The terms "investigational drug" and "investigational new drug" are deemed to be synonymous for purposes of this part.

Investigator means an individual who actually conducts a clinical investigation (i.e., under whose immediate direction the drug is administered or dispensed to a subject). In the event an investigation is conducted by a team of individuals, the investigator is the responsible leader of the team. "Subinvestigator" includes any other individual member of that team.

Marketing application means an application for a new drug submitted under section 505(b) of the Act, a request to provide for certification of an antibiotic submitted under section 507 of the Act, or a product license application for a biological product submitted under the Public Health Service Act.

Sponsor means a person who takes responsibility for and initiates a clinical investigation. The sponsor may be an individual or pharmaceutical company, governmental agency, academic institution, private organization, or other organization. The sponsor does not actually conduct the investigation unless the sponsor is a sponsor-investigator. A person other than an individual that uses one or more of its own employees to conduct an investigation that it has initiated is a sponsor, not a sponsor-investigator, and the employees are investigators.

Sponsor-Investigator means an individual who both initiates and conducts an investigation, and under whose immediate direction the investigational drug is administered or dispensed. The term does not include any person other than an individual. The requirements applicable to a sponsor-investigator under this part include both those applicable to an investigator and a sponsor.

Subject means a human who participates in an investigation, either as a recipient of the investigational new drug or as a control. A subject may be a healthy human or a patient with a disease.

§*312.6 Labeling of an investigational new drug.*

(a) The immediate package of an investigational new drug intended for human use shall bear a label with the statement "Caution: New Drug-Limited by Federal (or United States) law to investigational use."

(b) The label or labeling of an investigational new drug shall not bear any statement that is false or misleading in any particular and shall not represent that the investigational new drug is safe or effective for the purposes for which it is being investigated.

§*312.7 Promotion and charging for investigational drugs.*

(a) Promotion of an investigational new drug. A sponsor or investigator, or any person acting on behalf of a sponsor or investigator, shall not represent in a promotional context that an investigational new drug is safe or effective for the purposes for which it is under investigation or otherwise promote the drug. This provision is not intended to restrict the full exchange of scientific information concerning the drug, including dissemination of scientific findings in scientific or lay media. Rather, its intent is to restrict promotional claims of safety or effectiveness of the drug for a use for which it is under investigation and to preclude commercialization of the drug before it is approved for commercial distribution.

(b) Commercial distribution of an investigational new drug. A sponsor or investigator shall not commercially distribute or test market an investigational new drug.

(c) Prolonging an investigation. A sponsor shall not unduly prolong an investigation after finding that the results of the investigation appear to establish sufficient data to support a marketing application.

(d) Charging for and commercialization of investigational drugs (1) Clinical trials under an IND. Charging for an investigational drug in a clinical trial under an IND is not permitted without the prior written approval of FDA. In requesting such approval, the sponsor shall provide a full written explanation of why charging is necessary in order for the sponsor to undertake or continue the clinical trial, e.g., why distribution of the drug to test subjects should not be considered part of the normal cost of doing business.

(2) Treatment protocol or treatment IND. A sponsor or investigator may charge for an investigational drug for a treatment use under a treatment protocol or treatment IND provided: (i) There is adequate enrollment in the ongoing clinical investigations under the authorized IND; (ii) charging does not constitute commercial marketing of a new drug for which a marketing application has not been approved; (iii) the drug is not being commercially promoted or advertised; and (iv) the sponsor of the drug is actively pursuing marketing approval with due diligence. FDA must be notified in writing in advance of commencing any such charges, in an information amendment submitted under §312.31. Authorization for charging goes into effect automatically 30 days after receipt by FDA of the information amendment, unless the sponsor is notified to the contrary.

(3) Noncommercialization of investigational drug. Under this section, the sponsor may not commercialize an investigational drug by charging a price larger than that necessary to recover costs of manufacture, research, development, and handling of the investigational drug.

(4) Withdrawal of authorization. Autho-

rization to charge for an investigational drug under this section may be withdrawn by FDA if the agency finds that the conditions underlying the authorization are no longer satisfied.

§*312.10 Waivers.*

(a) A sponsor may request FDA to waive applicable requirement under this part. A waiver request may be submitted either in an IND or in an information amendment to an IND. In an emergency, a request may be made by telephone or other rapid communication means. A waiver request is required to contain at least one of the following:

(1) An explanation why the sponsor's compliance with the requirement is unnecessary or cannot be achieved;

(2) A description of an alternative submission or course of action that satisfies the purpose of the requirement; or

(3) Other information justifying a waiver.

(b) FDA may grant a waiver if it finds that the sponsor's noncompliance would not pose a significant and unreasonable risk to human subjects of the investigation and that one of the following is met:

(1) The sponsor's compliance with the requirement is unnecessary for the agency to evaluate the application, or compliance cannot be achieved;

(2) The sponsor's proposed alternative satisfies the requirement; or

(3) The applicant's submission otherwise justifies a waiver.

Subpart B--Investigational New Drug Application (IND)

§*312.20 Requirement for an IND.*

(a) A sponsor shall submit an IND to FDA if the sponsor intends to conduct a clinical investigation with an investigational new drug that is subject to §312.2(a).

(b) A sponsor shall not begin a clinical investigation subject to §312.2(a) until the investigation is subject to an IND which is in effect in accordance with §312.40.

§*312.21 Phases of an investigation.*

An IND may be submitted for one or more phases of an investigation. The clinical investigation of a previously untested drug is generally divided into three phases. Although in general the phases are conducted sequentially, they may overlap. These three phases of an investigation are a follows:

(a) Phase 1. (1) Phase 1 includes the initial introduction of an investigational new drug into humans. Phase 1 studies are typically closely monitored and may be conducted in patients or normal volunteer subjects. These studies are designed to determine the metabolism and pharmacologic actions of the drug in humans, the side effects associated with increasing doses, and, if possible, to gain early evidence on effectiveness. During Phase 1, sufficient information about the drug's pharmacokinetics and pharmacological effects should be obtained to permit the design of well-controlled, scientifically valid, Phase 2 studies. The total number of subjects and patients included in Phase 1 studies varies with the drug, but is generally in the range of 20 to 80.

(2) Phase 1 studies also include studies of drug metabolism, structure-activity relationships, and mechanism of action in humans, as well as studies in which investigational drugs are used as research tools to explore biological phenomena or disease processes.

(b) Phase 2. Phase 2 includes the controlled clinical studies conducted to evaluate the effectiveness of the drug for a particular indication or indications in patients with the disease or condition under study and to determine the common short-term side effects and risks associated with the drug. Phase 2 studies are typically well controlled, closely monitored, and conducted in a relatively small number of patients, usually involving no more than several hundred subjects.

(c) Phase 3. Phase 3 studies are expanded controlled and uncontrolled trials. They are performed after preliminary evidence suggesting effectiveness of the drug has been obtained, and are intended to gather the additional information about effectiveness and safety that is needed to evaluate the overall benefit-risk relationship of the drug and to provide an adequate basis for physician labeling. Phase 3 studies usually include from several hundred to several thousand subjects.

§312.22 General principles of the IND submission.

(a) FDA's primary objectives in reviewing an IND are, in all phases of the investigation, to assure the safety and rights of subjects, and, in Phase 2 and 3, to help assure that the quality of the scientific evaluation of drugs is adequate to permit an evaluation of the drug's effectiveness and safety. Therefore, although FDA's review of Phase 1 submissions will focus on assessing the safety of Phase 1 investigations, FDA's review of Phases 2 and 3 submissions will also include an assessment of the scientific quality of the clinical investigations and the likelihood that the investigations will yield data capable of meeting statutory standards for marketing approval.

(b) The amount of information on a particular drug that must be submitted in an IND to assure the accomplishment of the objectives described in paragraph (a) of this section depends upon such factors as the novelty of the drug, the extent to which it has been studied previously, the known or suspected risks, and the developmental phase of the drug.

(c) The central focus of the initial IND submission should be on the general investigational plan and the protocols for specific human studies. Subsequent amendments to the IND that contain new or revised protocols should build logically on previous submissions and should be supported by additional information, including the results of animal toxicology studies or other human studies as appropriate. Annual reports to the IND should serve as the focus for reporting the status of studies being conducted under the IND and should update the general investigational plan for the coming year.

(d) The IND format set forth in §312.23 should be followed routinely by sponsors in the interest of fostering an efficient review of applications. Sponsors are expected to exercise considerable discretion, however, regarding the content of information submitted in each section, depending upon the kind of drug being studied and the nature of the available information. Section 312.23 outlines the information needed for a commercially sponsored IND for a new molecular entity. A sponsor-investigator who uses, as a research tool, an investigational new drug that is already subject to a manufacturer's IND or marketing application should follow the same general format, but ordinarily may, if authorized by the manufacturer, refer to the manufacturer's IND or marketing application in providing the technical information supporting the proposed clinical investigation. A sponsor-investigator who uses an investigational drug not subject to a manufacturer's IND or marketing application is ordinarily required to submit all technical information supporting the IND, unless such information may be referenced from the scientific literature.

§312.23 IND content and format.

(a) A sponsor who intends to conduct a clinical investigation subject to this part shall submit an "Investigational New Drug Application" (IND) including, in the following order:

(1) Cover sheet (Form FDA-1571). A cover sheet for the application containing the following:

(i) The name, address, and telephone number of the sponsor, the date of the application, and the name of the investigational new drug.

(ii) Identification of the phase or phases of the clinical investigation to be conducted.

(iii) A commitment not to begin clinical investigations until an IND covering the investigations is in effect.

(iv) A commitment that an Institutional Review Board (IRB) that complies with the requirements set forth in part 56 will be responsible for the initial and continuing review and approval of each of the studies in the proposed clinical investigation and that the investigator will report to the IRB proposed changes in the research activity in accordance with the requirements of part 56.

(v) A commitment to conduct the investigation in accordance with all other applicable regulatory requirements.

(vi) The name and title of the person responsible for monitoring the conduct and progress of the clinical investigations.

(vii) The name(s) and title(s) of the person(s) responsible under §312.32 for review and evaluation of information relevant to the safety of the drug.

(viii) If a sponsor has transferred any obligations for the conduct of any clinical study to a contract research organization, a statement containing the name and address of the contract research organization, identification of the clinical study, and a listing of the obligations transferred. If all obligations governing the conduct of the study have been transferred, a general statement of this transfer -- in lieu of a listing of the specific obligations transferred -- may be submitted.

(ix) The signature of the sponsor or the sponsor's authorized representative. If the person signing the application does not reside or have a place of business within the United States, the IND is required to contain the name and address of, and be countersigned by, an attorney, agent, or other authorized official who resides or maintains a place of business within the United States.

(2) A table of contents.

(3) Introductory statement and general investigational plan. (i) A brief introductory statement giving the name of the drug and all active ingredients, the drug's pharmacological class, the structural formula of the drug (if known), the formulation of the dosage form(s) to be used, the route of administration, and the broad objectives and planned duration of the proposed clinical investigation(s).

(ii) A brief summary of previous human experience with the drug, with reference to other IND's if pertinent, and to investigational or marketing experience in other countries that may be relevant to the safety of the proposed clinical investigation(s).

(iii) If the drug has been withdrawn from investigation or marketing in any country for any reason related to safety or effectiveness, identification of the country(ies) where the drug was withdrawn and the reasons for the withdrawal.

(iv) A brief description of the overall plan for investigating the drug product for the following year. The plan should include the following: (a) The rationale for the drug or the research study; (b) the indication(s) to be studied; (c) the general approach to be followed in evaluating the drug; (d) the kinds of clinical trials to be conducted in the first year following the submission (if plans are not developed for the entire year, the sponsor should so indicate); (e) the estimated number of patients to be given the drug in those studies; and (f) any risks of particular severity or seriousness anticipated on the basis of the toxicological data in animals or prior studies in humans with the drug or related drugs.

(4) [Reserved]

(5) Investigator's brochure. If required under §312.55, a copy of the investigator's brochure, containing the following information:

(i) A brief description of the drug substance and the formulation, including the structural formula, if known.

(ii) A summary of the pharmacological and toxicological effects of the drug in animals and, to the extent known, in humans.

(iii) A summary of the pharmacokinetics and biological disposition of the drug in animals and, if known, in humans.

(iv) A summary of information relating to safety and effectiveness in humans obtained from prior clinical studies. (Reprints of published articles on such studies may be appended when useful.)

(v) A description of possible risks and side effects to be anticipated on the basis of prior experience with the drug under investigation or with related drugs, and of precautions or special monitoring to be done as part of the investigational use of the drug.

(6) Protocols. (i) A protocol for each planned study. (Protocols for studies not submitted initially in the IND should be submitted in accordance with §312.30(a).) In general, protocols for Phase 1 studies may be less detailed and more flexible than protocols for Phase 2 and 3 studies. Phase 1 protocols should be directed primarily at providing an outline of the investigation -- an estimate of the number of patients to be involved, a description of safety exclusions, and a descrip-

tion of the dosing plan including duration, dose, or method to be used in determining dose -- and should specify in detail only those elements of the study that are critical to safety, such as necessary monitoring of vital signs and blood chemistries. Modifications of the experimental design of Phase 1 studies that do not affect critical safety assessments are required to be reported to FDA only in the annual report.

(ii) In Phases 2 and 3, detailed protocols describing all aspects of the study should be submitted. A protocol for a Phase 2 or 3 investigation should be designed in such a way that, if the sponsor anticipates that some deviation from the study design may become necessary as the investigation progresses, alternatives or contingencies to provide for such deviation are built into the protocols at the outset. For example, a protocol for a controlled short-term study might include a plan for an early crossover of nonresponders to an alternative therapy.

(iii) A protocol is required to contain the following, with the specific elements and detail of the protocol reflecting the above distinctions depending on the phase of study:

(a) A statement of the objectives and purpose of the study.

(b) The name and address and a statement of the qualifications (curriculum vitae or other statement of qualifications) of each investigator, and the name of each subinvestigator (e.g., research fellow, resident) working under the supervision of the investigator; the name and address of the research facilities to be used; and the name and address of each reviewing Institutional Review Board.

(c) The criteria for patient selection and for exclusion of patients and an estimate of the number of patients to be studied.

(d) A description of the design of the study, including the kind of control group to be used, if any, and a description of methods to be used to minimize bias on the part of subjects, investigators, and analysts.

(e) The method for determining the dose(s) to be administered, the planned maximum dosage, and the duration of individual patient exposure to the drug.

(f) A description of the observations and measurements to be made to fulfill the objectives of the study.

(g) A description of clinical procedures, laboratory tests, or other measures to be taken to monitor the effects of the drug in human subjects and to minimize risk.

(7) Chemistry, manufacturing, and control information. (i) As appropriate for the particular investigations covered by the IND, a section describing the composition, manufacture, and control of the drug substance and the drug product. Although in each phase of the investigation sufficient information is required to be submitted to assure the proper identification, quality, purity, and strength of the investigational drug, the amount of information needed to make that assurance will vary with the phase of the investigation, the proposed duration of the investigation, the dosage form, and the amount of information otherwise available. FDA recognizes that modifications to the method of preparation of the new drug substance and dosage form and changes in the dosage form itself are likely as the investigation progresses. Therefore, the emphasis in an initial Phase 1 submission should generally be placed on the identification and control of the raw materials and the new drug substance. Final specifications for the drug substance and drug product are not expected until the end of the investigational process.

(ii) It should be emphasized that the amount of information to be submitted depends upon the scope of the proposed clinical investigation. For example, although stability data are required in all phases of the IND to demonstrate that the new drug substance and drug product are within acceptable chemical and physical limits for the planned duration of the proposed clinical investigation, if very short-term tests are proposed, the supporting stability data can be correspondingly limited.

(iii) As drug development proceeds and as the scale or production is changed from the pilot-scale production appropriate for the limited initial clinical investigations to the larger-scale production needed for expanded clinical trials, the sponsor should submit information amendments to supplement the initial information submitted on the chemistry, manufacturing, and control processes with information appropriate to the expanded scope of the investigation.

(iv) Reflecting the distinctions described in this paragraph (a)(7), and based on the phase(s) to be studied, the submission is required to contain the following:

(a) Drug substance. A description of the drug substance, including its physical, chemical, or biological characteristics; the name and address of its manufacturer; the general method of preparation of the drug substance; the acceptable limits and analytical methods used to assure the identity, strength, quality, and purity of the drug substance; and information sufficient to support stability of the drug substance during the toxicological studies and the planned clinical studies. Reference to the current edition of the United States Pharmacopeia-National Formulary may satisfy relevant requirements in this paragraph.

(b) Drug product. A list of all components, which may include reasonable alternatives for inactive compounds, used in the manufacture of the investigational drug product, including both those components intended to appear in the drug product and those which may not appear but which are used in the manufacturing process, and, where applicable, the quantitative composition of the investigational drug product, including any reasonable variations that may be expected during the investigational stage; the name and address of the drug product manufacturer; a brief general description of the manufacturing and packaging procedure as appropriate for the product; the acceptable limits and analytical methods used to assure the identity, strength, quality, and purity of the drug product; and information sufficient to assure the product's stability during the planned clinical studies. Reference to the current edition of the United States Pharmacopeia-National Formulary may satisfy certain requirements in this paragraph.

(c) A brief general description of the composition, manufacture, and control of any placebo used in a controlled clinical trial.

(d) Labeling. A copy of all labels and labeling to be provided to each investigator.

(e) Environmental analysis requirements. A claim for categorical exclusion under §25.24 or an environmental assessment under §25.31.

(8) Pharmacology and toxicology information. Adequate information about pharmacological and toxicological studies of the drug involving laboratory animals or in vitro, on the basis of which the sponsor has concluded that it is reasonably safe to conduct the proposed clinical investigations. The kind, duration, and scope of animal and other tests required varies with the duration and nature of the proposed clinical investigations. Guidelines are available from FDA that describe ways in which these requirements may be met. Such information is required to include the identification and qualifications of the individuals who evaluated the results of such studies and concluded that it is reasonably safe to begin the proposed investigations and a statement of where the investigations were conducted and where the records are available for inspection. As drug development proceeds, the sponsor is required to submit informational amendments, as appropriate, with additional information pertinent to safety.

(i) Pharmacology and drug disposition. A section describing the pharmacological effects and mechanism(s) of action of the drug in animals, and information on the absorption, distribution, metabolism, and excretion of the drug, if known.

(ii) Toxicology.

(a) An integrated summary of the toxicological effects of the drug in animals and in vitro. Depending on the nature of the drug and the phase of the investigation, the description is to include the results of acute, subacute, and chronic toxicity tests; tests of the drug's effects on reproduction and the developing fetus; any special toxicity test related to the drug's particular mode of administration or conditions of use (e.g., inhalation, dermal, or ocular toxicology); and any in vitro studies intended to evaluate drug toxicity.

(b) For each toxicology study that is intended primarily to support the safety of the proposed clinical investigation, a full tabulation of data suitable for detailed review.

(iii) For each nonclinical laboratory study subject to the good laboratory practice regulations under part 58, a statement that the study was conducted in compliance with the good laboratory practice regulations in part 58, or, if the study was not conducted in compliance with those regulations, a brief statement of the reason for the noncompliance.

(9) Previous human experience with the investigational drug. A summary of previous human experience known to the applicant, if any, with the investigational drug. The information is required to include the following:

(i) If the investigational drug has been investigated or marketed previously, either in the United States or other countries, detailed information about such experience that is relevant to the safety of the proposed investigation or to the investigation's rationale. If the drug has been the subject of controlled trials, detailed information on such trials that is relevant to an assessment of the drug's effectiveness for the proposed investigational use(s) should also be provided. Any published material that is relevant to the safety of the proposed investigation or to an assessment of the drug's effectiveness for its proposed investigational use should be provided in full. Published material that is less directly relevant may be supplied by a bibliography.

(ii) If the drug is a combination of drugs previously investigated or marketed, the information required under paragraph (a)(9)(i) of this section should be provided for each active drug component. However, if any component in such combination is subject to an approved marketing application or is otherwise lawfully marketed in the United States, the sponsor is not required to submit published material concerning that active drug component unless such material relates directly to the proposed investigational use (including publications relevant to component- component interaction).

(iii) If the drug has been marketed outside the United States, a list of the countries in which the drug has been marketed and a list of the countries in which the drug has been withdrawn from marketing for reasons potentially related to safety or effectiveness.

(10) Additional information. In certain applications, as described below, information on special topics may be needed. Such information shall be submitted in this section as follows:

(i) Drug dependence and abuse potential. If the drug is a psychotropic substance or otherwise has abuse potential, a section describing relevant clinical studies and experience and studies in test animals.

(ii) Radioactive drugs. If the drug is a radioactive drug, sufficient data from animal or human studies to allow a reasonable calculation of radiation-absorbed dose to the whole body and critical organs upon administration to a human subject. Phase 1 studies of radioactive drugs must include studies which will obtain sufficient data for dosimetry calculations.

(iii) Other information. A brief statement of any other information that would aid evaluation of the proposed clinical investigations with respect to their safety or their design and potential as controlled clinical trials to support marketing of the drug.

(11) Relevant information. If requested by FDA, any other relevant information needed for review of the application.

(b) Information previously submitted. The sponsor ordinarily is not required to resubmit information previously submitted, but may incorporate the information by reference. A reference to information submitted previously must identify the file by name, reference number, volume, and page number where the information can be found. A reference to information submitted to the agency by a person other than the sponsor is required to contain a written statement that authorizes the reference and that is signed by the person who submitted the information.

(c) Material in a foreign language. The sponsor shall submit an accurate and complete English translation of each part of the IND that is not in English. The sponsor shall also submit a copy of each original literature publication for which an English translation is submitted.

(d) Number of copies. The sponsor shall submit an original and two copies of all submissions to the IND file, including the original submission and all amendments and reports.

(e) Numbering of IND submissions. Each submission relating to an IND is required to be numbered serially using a single, three-digit serial number. The initial IND is required to be numbered 000; each subsequent submission (e.g., amendment, report, or correspondence) is required to be numbered chronologically in sequence.

§312.30 Protocol amendments.

Once an IND is in effect, a sponsor shall amend it as needed to ensure that the clinical investigations are conducted according to protocols included in the application. This section sets forth the provisions under which new protocols may be submitted and changes in previously submitted protocols may be made.

(a) New protocol. Whenever a sponsor intends to conduct a study that is not covered by a protocol already contained in the IND, the sponsor shall submit to FDA a protocol amendment containing the protocol for the study. Such study may begin provided two conditions are met: (1) The sponsor has submitted the protocol to FDA for its review; and (2) the protocol has been approved by the Institutional Review Board (IRB) with responsibility for review and approval of the study in accordance with the requirements of part 56. The sponsor may comply with these two conditions in either order.

(b) Changes in a protocol. (1) A sponsor shall submit a protocol amendment describing any change in a Phase 1 protocol that significantly affects the safety of subjects or any change in a Phase 2 or 3 protocol that significantly affects the safety of subjects, the scope of the investigation, or the scientific quality of the study. Examples of changes requiring an amendment under this paragraph include:

(i) Any increase in drug dosage or duration of exposure of individual subjects to the drug beyond that in the current protocol, or any significant increase in the number of subjects under study.

(ii) Any significant change in the design of a protocol (such as the addition or dropping of a control group).

(iii) The addition of a new test or procedure that is intended to improve monitoring for, or reduce the risk of, a side effect or adverse event; or the dropping of a test intended to monitor safety.

(2)(i) A protocol change under paragraph (b)(1) of this section may be made provided two conditions are met:

(a) The sponsor has submitted the change to FDA for its review; and

(b) The change has been approved by the IRB with responsibility for review and approval of the study. The sponsor may comply with these two conditions in either order.

(ii) Notwithstanding paragraph (b)(2)(i) of this section, a protocol change intended to eliminate an apparent immediate hazard to subjects may be implemented immediately provided FDA is subsequently notified by protocol amendment and the reviewing IRB is notified in accordance with §56.104(c).

(c) New investigator. A sponsor shall submit a protocol amendment when a new investigator is added to carry out a previously submitted protocol, except that a protocol amendment is not required when a licensed practitioner is added in the case of a treatment protocol under §312. 34. Once the investigator is added to the study, the investigational drug may be shipped to the investigator and the investigator may begin participating in the study. The sponsor shall notify FDA of the new investigator within 30 days of the investigator being added.

(d) Content and format. A protocol amendment is required to be prominently identified as such (i.e., "Protocol Amendment: New Protocol", "Protocol Amendment: Change in Protocol", or "Protocol Amendment: New Investigator"), and to contain the following:

(1)(i) In the case of a new protocol, a copy of the new protocol and a brief description of the most clinically significant differences between it and previous protocols.

(ii) In the case of a change in protocol, a brief description of the change and reference (date and number) to the submission that contained the protocol.

(iii) In the case of a new investigator, the investigator's name, the qualifications to conduct the investigation, reference to the previously submitted protocol, and all additional information about the investigator's study as is required under §312.23(a)(6)(iii)(b).

(2) Reference, if necessary, to specific

technical information in the IND or in a concurrently submitted information amendment to the IND that the sponsor relies on to support any clinically significant change in the new or amended protocol. If the reference is made to supporting information already in the IND, the sponsor shall identify by name, reference number, volume, and page number the location of the information.

(3) If the sponsor desires FDA to comment on the submission, a request for such comment and the specific questions FDA's response should address.

(e) When submitted. A sponsor shall submit a protocol amendment for a new protocol or a change in protocol before its implementation. Protocol amendments to add a new investigator or to provide additional information about investigators may be grouped and submitted at 30-day intervals. When several submissions of new protocols or protocol changes are anticipated during a short period, the sponsor is encouraged, to the extent feasible, to include these all in a single submission.

§312.31 Information amendments.

(a) Requirement for information amendment. A sponsor shall report in an information amendment essential information on the IND that is not within the scope of a protocol amendment, IND safety reports, or annual report. Examples of information requiring an information amendment include:

(1) New toxicology, chemistry, or other technical information; or

(2) A report regarding the discontinuance of a clinical investigation.

(b) Content and format of an information amendment. An information amendment is required to bear prominent identification of its contents (e.g., "Information Amendment: Chemistry, Manufacturing, and Control", "Information Amendment: Pharmacology-Toxicology", "Information Amendment: Clinical"), and to contain the following:

(1) A statement of the nature and purpose of the amendment.

(2) An organized submission of the data in a format appropriate for scientific review.

(3) If the sponsor desires FDA to comment on an information amendment, a request for such comment.

(c) When submitted. Information amendments to the IND should be submitted as necessary but, to the extent feasible, not more than every 30 days.

§312.32 IND safety reports.

(a) Definitions. The following definitions of terms apply to this section:

Associated with the use of the drug means that there is a reasonable possibility that the experience may have been caused by the drug.

Serious adverse experience means any experience that suggests a significant hazard, contraindication, side effect, or precaution. With respect to human clinical experience, a serious adverse drug experience includes any experience that is fatal or life-threatening, is permanently disabling, requires inpatient hospitalization, or is a congenital anomaly, cancer, or overdose. With respect to results obtained from tests in laboratory animals, a serious adverse drug experience includes any experience suggesting a significant risk for human subjects, including any finding of mutagenicity, teratogenicity, or carcinogenicity.

Unexpected adverse experience means any adverse experience that is not identified in nature, severity, or frequency in the current investigator brochure; or, if an investigator brochure is not required, that is not identified in nature, severity, or frequency in the risk information described in the general investigational plan or elsewhere in the current application, as amended.

(b) Review of safety information. The sponsor shall promptly review all information relevant to the safety of the drug obtained or otherwise received by the sponsor from any source, foreign or domestic, including information derived from clinical investigations, animal investigations, commercial marketing experience, reports in the scientific literature, and unpublished scientific papers.

(c) IND safety reports. (1) Written reports. (i) The sponsor shall notify FDA and all participating investigators in a written IND safety report of any adverse experience associated with use of the drug that is both serious and unexpected. Such notification shall be made as soon as possible and in no event later than 10 working days after the sponsor's initial receipt of the information. Each written notification shall bear prominent identification of its contents, i.e., "IND Safety Report." Each written notification to FDA shall be transmitted to the FDA division of the Center for Drug Evaluation and Research or the Center for Biologics Evaluation and Research which has responsibility for review of the IND.

(ii) In each written IND safety report, the sponsor shall identify all safety reports previously filed with the IND concerning a similar adverse experience, and shall analyze the significance of the adverse experience in light of the previous, similar reports.

(2) Telephone report. The sponsor shall also notify FDA by telephone of any unexpected fatal or life-threatening experience associated with use of the drug in the clinical studies conducted under the IND no later than 3 working days after receipt of the information. Each telephone call to FDA shall be transmitted to the FDA division of the Center for Drug Evaluation and Research or the Center for Biologics Evaluation and Research which has responsibility for review of the IND. For purposes of this section, life-threatening means that the patient was, in the view of the investigator, at immediate (emphasis added) risk of death from the reaction as it occurred, i.e., it does not include a reaction that, had it occurred in a more serious form, might have caused death. For example, drug- induced hepatitis that resolved without evidence of hepatic failure would not be considered life-threatening even though drug-induced hepatitis can be fatal.

(3) Reporting format or frequency. FDA may request a sponsor to submit IND safety reports in a format or at a frequency different than that required under this paragraph. The sponsor may also propose and adopt a different reporting format or frequency if the change is agreed to in advance by the director of the division in the Center for Drug Evaluation and Research or the Center for Biologics Evaluation and Research which is responsible for review of the IND.

(4) A sponsor of a clinical study of a marketed drug is not required to make a safety report for any adverse experience associated with use of the drug that is not from the clinical study itself.

(d) Followup. (1) The sponsor shall promptly investigate all safety information received by it.

(2) Followup information to a safety report shall be submitted as soon as the relevant information is available.

(3) If the results of a sponsor's investigation show that an adverse experience not initially determined to be reportable under paragraph (c) of this section is so reportable, the sponsor shall report such experience in a safety report as soon as possible after the determination is made, but in no event longer than 10-working days.

(4) Results of a sponsor's investigation of other safety information shall be submitted, as appropriate, in an information amendment or annual report.

(e) Disclaimer. A safety report or other information submitted by a sponsor under this section (and any release by FDA of that report or information) does not necessarily reflect a conclusion by the sponsor or FDA that the report or information constitutes an admission that the drug caused or contributed to an adverse experience. A sponsor need not admit, and may deny, that the report or information submitted by the sponsor constitutes an admission that the drug caused or contributed to an adverse experience.

§312.33 Annual reports.

A sponsor shall within 60 days of the anniversary date that the IND went into effect, submit a brief report of the progress of the investigation that includes:

(a) Individual study information. A brief summary of the status of each study in progress and each study completed during the previous year. The summary is required to include the following information for each study:

(1) The title of the study (with any appro-

priate study identifiers such as protocol number), its purpose, a brief statement identifying the patient population, and a statement as to whether the study is completed.

(2) The total number of subjects initially planned for inclusion in the study, the number entered into the study to date, the number whose participation in the study was completed as planned, and the number who dropped out of the study for any reason.

(3) If the study has been completed, or if interim results are known, a brief description of any available study results.

(b) Summary information. Information obtained during the previous year's clinical and nonclinical investigations, including:

(1) A narrative or tabular summary showing the most frequent and most serious adverse experiences by body system.

(2) A summary of all IND safety reports submitted during the past year.

(3) A list of subjects who died during participation in the investigation, with the cause of death for each subject.

(4) A list of subjects who dropped out during the course of the investigation in association with any adverse experience, whether or not thought to be drug related.

(5) A brief description of what, if anything, was obtained that is pertinent to an understanding of the drug's actions, including, for example, information about dose response, information from controlled trails, and information about bioavailability.

(6) A list of the preclinical studies (including animal studies) completed or in progress during the past year and a summary of the major preclinical findings.

(7) A summary of any significant manufacturing or microbiological changes made during the past year.

(c) A description of the general investigational plan for the coming year to replace that submitted 1 year earlier. The general investigational plan shall contain the information required under §312.23(a)(3)(iv).

(d) If the investigator brochure has been revised, a description of the revision and a copy of the new brochure.

(e) A description of any significant Phase 1 protocol modifications made during the previous year and not previously reported to the IND in a protocol amendment.

(f) A brief summary of significant foreign marketing developments with the drug during the past year, such as approval of marketing in any country or withdrawal or suspension from marketing in any country.

(g) If desired by the sponsor, a log of any outstanding business with respect to the IND for which the sponsor requests or expects a reply, comment, or meeting.

§312.34 Treatment use of an investigational new drug.

(a) General. A drug that is not approved for marketing may be under clinical investigation for a serious or immediately life-threatening disease condition in patients for whom no comparable or satisfactory alternative drug or other therapy is available. During the clinical investigation of the drug, it may be appropriate to use the drug in the treatment of patients not in the clinical trials, in accordance with a treatment protocol or treatment IND. The purpose of this section is to facilitate the availability of promising new drugs to desperately ill patients as early in the drug development process as possible, before general marketing begins, and to obtain additional data on the drug's safety and effectiveness. In the case of a serious disease, a drug ordinarily may be made available for treatment use under this section during Phase 3 investigations or after all clinical trials have been completed; however, in appropriate circumstances, a drug may be made available for treatment use during Phase 2. In the case of an immediately life-threatening disease, a drug may be made available for treatment use under this section earlier than Phase 3, but ordinarily not earlier than Phase 2. For purposes of this section, the "treatment use" of a drug includes the use of a drug for diagnostic purposes. If a protocol for an investigational drug meets the criteria of this section, the protocol is to be submitted as a treatment protocol under the provisions of this section.

(b) Criteria. (1) FDA shall permit an investigational drug to be used for a treatment use under a treatment protocol or treatment IND if:

(i) The drug is intended to treat a serious or immediately life-threatening disease;

(ii) There is no comparable or satisfactory alternative drug or other therapy available to treat that stage of the disease in the intended patient population;

(iii) The drug is under investigation in a controlled clinical trial under an IND in effect for the trial, or all clinical trials have been completed; and

(iv) The sponsor of the controlled clinical trial is actively pursuing marketing approval of the investigational drug with due diligence.

(2) Serious disease. For a drug intended to treat a serious disease, the Commissioner may deny a request for treatment use under a treatment protocol or treatment IND if there is insufficient evidence of safety and effectiveness to support such use.

(3) Immediately life-threatening disease. (i) For a drug intended to treat an immediately life-threatening disease, the Commissioner may deny a request for treatment use of an investigational drug under a treatment protocol or treatment IND if the available scientific evidence, taken as a whole, fails to provide a reasonable basis for concluding that the drug:

(A) May be effective for its intended use in its intended patient population; or

(B) Would not expose the patients to whom the drug is to be administered to an unreasonable and significant additional risk of illness or injury.

(ii) For the purpose of this section, an "immediately life-threatening" disease means a stage of a disease in which there is a reasonable likelihood that death will occur within a matter of months or in which premature death is likely without early treatment.

(c) Safeguards. Treatment use of an investigational drug is conditioned on the sponsor and investigators complying with the safeguards of the IND process, including the regulations governing informed consent (21 CFR part 50) and institutional review boards (21 CFR part 56) and the applicable provisions of part 312, including distribution of the drug through qualified experts, maintenance of adequate manufacturing facilities, and submission of IND safety reports.

(d) Clinical hold. FDA may place on clinical hold a proposed or ongoing treatment protocol or treatment IND in accordance with §312.42.

§312.35 Submissions for treatment use.

(a) Treatment protocol submitted by IND sponsor. Any sponsor of a clinical investigation of a drug who intends to sponsor a treatment use for the drug shall submit to FDA a treatment protocol under §312.34 if the sponsor believes the criteria of §312.34 are satisfied. If a protocol is not submitted under §312.34, but FDA believes that the protocol should have been submitted under this section, FDA may deem the protocol to be submitted under §312.34. A treatment use under a treatment protocol may begin 30 days after FDA receives the protocol or on earlier notification by FDA that the treatment use described in the protocol may begin.

(1) A treatment protocol is required to contain the following:

(i) The intended use of the drug.

(ii) An explanation of the rationale for use of the drug, including, as appropriate, either a list of what available regimens ordinarily should be tried before using the investigational drug or an explanation of why the use of the investigational drug is preferable to the use of available marketed treatments.

(iii) A brief description of the criteria for patient selection.

(iv) The method of administration of the drug and the dosages.

(v) A description of clinical procedures, laboratory tests, or other measures to monitor the effects of the drug and to minimize risk.

(2) A treatment protocol is to be supported by the following:

(i) Informational brochure for supplying to each treating physician.

(ii) The technical information that is relevant to safety and effectiveness of the drug for the intended treatment purpose. Information contained in the sponsor's IND may be incorporated by reference.

(iii) A commitment by the sponsor to assure compliance of all participating investigators with the informed consent requirements of 21

CFR part 50.

(3) A licensed practitioner who receives an investigational drug for treatment use under a treatment protocol is an "investigator" under the protocol and is responsible for meeting all applicable investigator responsibilities under this part and 21 CFR parts 50 and 56.

(b) Treatment IND submitted by licensed practitioner. (1) If a licensed medical practitioner wants to obtain an investigational drug subject to a controlled clinical trial for a treatment use, the practitioner should first attempt to obtain the drug from the sponsor of the controlled trial under a treatment protocol. If the sponsor of the controlled clinical investigation of the drug will not establish a treatment protocol for the drug under paragraph (a) of this section, the licensed medical practitioner may seek to obtain the drug from the sponsor and submit a treatment IND to FDA requesting authorization to use the investigational drug for treatment use. A treatment use under a treatment IND may begin 30 days after FDA receives the IND or on earlier notification by FDA that the treatment use under the IND may begin. A treatment IND is required to contain the following:

(i) A cover sheet (Form FDA 1571) meeting §312.23(g)(1).

(ii) Information (when not provided by the sponsor) on the drug's chemistry, manufacturing, and controls, and prior clinical and nonclinical experience with the drug submitted in accordance with §312.23. A sponsor of a clinical investigation subject to an IND who supplies an investigational drug to a licensed medical practitioner for purposes of a separate treatment clinical investigation shall be deemed to authorize the incorporation-by-reference of the technical information contained in the sponsor's IND into the medical practitioner's treatment IND.

(iii) A statement of the steps taken by the practitioner to obtain the drug under a treatment protocol from the drug sponsor.

(iv) A treatment protocol containing the same information listed in paragraph (a)(1) of this section.

(v) A statement of the practitioner's qualifications to use the investigational drug for the intended treatment use.

(vi) The practitioner's statement of familiarity with information on the drug's safety and effectiveness derived from previous clinical and nonclinical experience with the drug.

(vii) Agreement to report to FDA safety information in accordance with §312.32.

(2) A licensed practitioner who submits a treatment IND under this section is the sponsor-investigator for such IND and is responsible for meeting all applicable sponsor and investigator responsibilities under this part and 21 CFR parts 50 and 56.

§312.36 Emergency use of an investigational new drug.

Need for an investigational drug may arise in an emergency situation that does not allow time for submission of an IND in accordance with §312. 23 or §312.34. In such a case, FDA may authorize shipment of the drug for a specified use in advance of submission of an IND. A request for such authorization may be transmitted to FDA by telephone or other rapid communication means. For investigational biological drugs, the request should be directed to the Division of Biological Investigational New Drugs (HFB-230), Center for Biologics Evaluation and Research, 8800 Rockville Pike, Bethesda, MD 20892, 301-443-4864. For all other investigational drugs, the request for authorization should be directed to the Document Management and Reporting Branch (HFD-53), Center for Drug Evaluation and Research, 5600 Fishers Lane, Rockville, MD 20857, 301-443-4320. After normal working hours, eastern standard time, the request should be directed to the FDA Division of Emergency and Epidemiological Operations, 202-857-8400. Except in extraordinary circumstances, such authorization will be conditioned on the sponsor making an appropriate IND submission as soon as practicable after receiving the authorization.

§312.38 Withdrawal of an IND.

(a) At any time a sponsor may withdraw an effective IND without prejudice.

(b) If an IND is withdrawn, FDA shall be so notified, all clinical investigations conducted under the IND shall be ended, all current investigators notified, and all stocks of the drug returned to the sponsor or otherwise disposed of at the request of the sponsor in accordance with §312.59.

(c) If an IND is withdrawn because of a safety reason, the sponsor shall promptly so inform FDA, all participating investigators, and all reviewing Institutional Review Boards, together with the reasons for such withdrawal.

Subpart C--Administrative Actions

§312.40 General requirements for use of an investigational new drug in a clinical investigation.

(a) An investigational new drug may be used in a clinical investigation if the following conditions are met:

(1) The sponsor of the investigation submits an IND for the drug to FDA; the IND is in effect under paragraph (b) of this section; and the sponsor complies with all applicable requirements in this part and parts 50 and 56 with respect to the conduct of the clinical investigations; and

(2) Each participating investigator conducts his or her investigation in compliance with the requirements of this part and parts 50 and 56.

(b) An IND goes into effect:

(1) Thirty days after FDA receives the IND, unless FDA notifies the sponsor that the investigations described in the IND are subject to a clinical hold under §312.42; or

(2) On earlier notification by FDA that the clinical investigations in the IND may begin. FDA will notify the sponsor in writing of the date it receives the IND.

(c) A sponsor may ship an investigational new drug to investigators named in the IND:

(1) Thirty days after FDA receives the IND; or

(2) On earlier FDA authorization to ship the drug.

(d) An investigator may not administer an investigational new drug to human subjects until the IND goes into effect under paragraph (b) of this section.

§312.41 Comment and advice on an IND.

(a) FDA may at any time during the course of the investigation communicate with the sponsor orally or in writing about deficiencies in the IND or about FDA's need for more data or information.

(b) On the sponsor's request, FDA will provide advice on specific matters relating to an IND. Examples of such advice may include advice on the adequacy of technical data to support an investigational plan, on the design of a clinical trial, and on whether proposed investigations are likely to produce the data and information that is needed to meet requirements for a marketing application.

(c) Unless the communication is accompanied by a clinical hold order under §312.42, FDA communications with a sponsor under this section are solely advisory and do not require any modification in the planned or ongoing clinical investigations or response to the agency.

§312.42 Clinical holds and requests for modification.

(a) General. A clinical hold is an order issued by FDA to the sponsor to delay a proposed clinical investigation or to suspend an ongoing investigation. The clinical hold order may apply to one or more of the investigations covered by an IND. When a proposed study is placed on clinical hold, subjects may not be given the investigational drug. When an ongoing study is placed on clinical hold, no new subjects may be recruited to the study and placed on the investigational drug; patients already in the study should be taken off therapy involving the investigational drug unless specifically permitted by FDA in the interest of patient safety.

(b) Grounds for imposition of clinical hold.

(1) Clinical hold of a Phase 1 study under an IND. FDA may place a proposed or ongoing Phase 1 investigation on clinical hold if it finds that:

(i) Human subjects are or would be exposed to an unreasonable and significant risk of illness or injury;

(ii) The clinical investigators named in the IND are not qualified by reason of their scientific training and experience to conduct the investigation described in the IND;

(iii) The investigator brochure is misleading, erroneous, or materially incomplete; or

(iv) The IND does not contain sufficient information required under §312.23 to assess the risks to subjects of the proposed studies.

(2) Clinical hold of a Phase 2 or 3 study under an IND. FDA may place a proposed or ongoing Phase 2 or 3 investigation on clinical hold if it finds that:

(i) Any of the conditions in paragraph (b)(1)(i) through (iv) of this section apply; or

(ii) The plan or protocol for the investigation is clearly deficient in design to meet its stated objectives.

(3) Clinical hold of a treatment IND or treatment protocol.

(i) Proposed use. FDA may place a proposed treatment IND or treatment protocol on clinical hold if it is determined that:

(A) The pertinent criteria in §312.34(b) for permitting the treatment use to begin are not satisfied; or

(B) The treatment protocol or treatment IND does not contain the information required under §312.35 (a) or (b) to make the specified determination under §312.34(b).

(ii) Ongoing use. FDA may place an ongoing treatment protocol or treatment IND on clinical hold if it is determined that:

(A) There becomes available a comparable or satisfactory alternative drug or other therapy to treat that stage of the disease in the intended patient population for which the investigational drug is being used;

(B) The investigational drug is not under investigation in a controlled clinical trial under an IND in effect for the trial and not all controlled clinical trials necessary to support a marketing application have been completed, or a clinical study under the IND has been placed on clinical hold;

(C) The sponsor of the controlled clinical trial is not pursuing marketing approval with due diligence;

(D) If the treatment IND or treatment protocol is intended for a serious disease, there is insufficient evidence of safety and effectiveness to support such use; or

(E) If the treatment protocol or treatment IND was based on an immediately life-threatening disease, the available scientific evidence, taken as a whole, fails to provide a reasonable basis for concluding that the drug:

(1) May be effective for its intended use in its intended population; or

(2) Would not expose the patients to whom the drug is to be administered to an unreasonable and significant additional risk of illness or injury.

(iii) FDA may place a proposed or ongoing treatment IND or treatment protocol on clinical hold if it finds that any of the conditions in paragraph (b)(4)(i) through (b)(4)(viii) of this section apply.

(4) Clinical hold of any study that is not designed to be adequate and well-controlled. FDA may place a proposed or ongoing investigation that is not designed to be adequate and well-controlled on clinical hold if it finds that:

(i) Any of the conditions in paragraph (b)(1) or (b)(2) of this section apply; or

(ii) There is reasonable evidence the investigation that is not designed to be adequate and well-controlled is impeding enrollment in, or otherwise interfering with the conduct or completion of, a study that is designed to be an adequate and well-controlled investigation of the same or another investigational drug; or

(iii) Insufficient quantities of the investigational drug exist to adequately conduct both the investigation that is not designed to be adequate and well-controlled and the investigations that are designed to be adequate and well-controlled; or

(iv) The drug has been studied in one or more adequate and well-controlled investigations that strongly suggest lack of effectiveness; or

(v) Another drug under investigation or approved for the same indication and available to the same patient population has demonstrated a better potential benefit/risk balance; or

(vi) The drug has received marketing approval for the same indication in the same patient population; or

(vii) The sponsor of the study that is designed to be an adequate and well-controlled investigation is not actively pursuing marketing approval of the investigational drug with due diligence; or

(viii) The Commissioner determines that it would not be in the public interest for the study to be conducted or continued. FDA ordinarily intends that clinical holds under paragraphs (b)(4)(ii), (b)(4)(iii) and (b)(4)(v) of this section would only apply to additional enrollment in nonconcurrently controlled trials rather than eliminating continued access to individuals already receiving the investigational drug.

(c) Discussion of deficiency. Whenever FDA concludes that a deficiency exists in a clinical investigation that may be grounds for the imposition of clinical hold FDA will, unless patients are exposed to immediate and serious risk, attempt to discuss and satisfactorily resolve the matter with the sponsor before issuing the clinical hold order.

(d) Imposition of clinical hold. The clinical hold order may be made by telephone or other means of rapid communication or in writing. The clinical hold order will identify the studies under the IND to which the hold applies, and will briefly explain the basis for the action. The clinical hold order will be made by or on behalf of the Division Director with responsibility for review of the IND. As soon as possible, and no more than 30 days after imposition of the clinical hold, the Division Director will provide the sponsor a written explanation of the basis for the hold.

(e) Resumption of clinical investigations. If, by the terms of the clinical hold order, resumption of the affected investigation is permitted without prior notification by FDA once a stated correction or modification is made, the investigation may proceed as soon as the correction or modification is made. In all other cases, an investigation may only resume after the Division Director (or the Director's designee) with responsibility for review of the IND has notified the sponsor that the investigation may proceed. In these cases resumption of the affected investigation(s) will be authorized when the sponsor corrects the deficiency(ies) previously cited or otherwise satisfied the agency that the investigation(s) can proceed. Resumption of a study may be authorized by telephone or other means of rapid communication.

(f) Appeal. If the sponsor disagrees with the reasons cited for the clinical hold, the sponsor may request reconsideration of the decision in accordance with §312.48.

(g) Conversion of IND on clinical hold to inactive status. If all investigations covered by an IND remain on clinical hold for 1 year or more, the IND may be placed on inactive status by FDA under §312.45.

§312.44 Termination.

(a) General. This section describes the procedures under which FDA may terminate an IND. If an IND is terminated, the sponsor shall end all clinical investigations conducted under the IND and recall or otherwise provide for the disposition of all unused supplies of the drug. A termination action may be based on deficiencies in the IND or in the conduct of an investigation under an IND. Except as provided in paragraph (d) of this section, a termination shall be preceded by a proposal to terminate by FDA and an opportunity for the sponsor to respond. FDA will, in general, only initiate an action under this section after first attempting to resolve differences informally or, when appropriate, through the clinical hold procedures described in §312.42.

(b) Grounds for termination-(1) Phase 1. FDA may propose to terminate an IND during Phase 1 if it finds that:

(i) Human subjects would be exposed to an unreasonable and significant risk of illness or injury.

(ii) The IND does not contain sufficient information required under §312.23 to assess the safety to subjects of the clinical investigations.

(iii) The methods, facilities, and controls used for the manufacturing, processing, and packing of the investigational drug are inadequate to establish and maintain appropriate standards of identity, strength, quality, and purity as need-

ed for subject safety.

(iv) The clinical investigations are being conducted in a manner substantially different than that described in the protocols submitted in the IND.

(v) The drug is being promoted or distributed for commercial purposes not justified by the requirements of the investigation or permitted by §312.7.

(vi) The IND, or any amendment or report to the IND, contains an untrue statement of a material fact or omits material information required by this part.

(vii) The sponsor fails promptly to investigate and inform the Food and Drug Administration and all investigators of serious and unexpected adverse experiences in accordance with §312.32 or fails to make any other report required under this part.

(viii) The sponsor fails to submit an accurate annual report of the investigations in accordance with §312.33.

(ix) The sponsor fails to comply with any other applicable requirement of this part, part 50, or part 56.

(x) The IND has remained on inactive status for 5 years or more.

(xi) The sponsor fails to delay a proposed investigation under the IND or to suspend an ongoing investigation that has been placed on clinical hold under §312.42(b)(4).

(2) Phase 2 or 3. FDA may propose to terminate an IND during Phase 2 or Phase 3 if FDA finds that:

(i) Any of the conditions in paragraphs (b)(1)(i) through (b)(1)(xi) of this section apply; or

(ii) The investigational plan or protocol(s) is not reasonable as a bona fide scientific plan to determine whether or not the drug is safe and effective for use; or

(iii) There is convincing evidence that the drug is not effective for the purpose for which it is being investigated.

(3) FDA may propose to terminate a treatment IND if it finds that:

(i) Any of the conditions in paragraphs (b)(1)(i) through (x) of this section apply; or

(ii) Any of the conditions in §312.42 (b)(3) apply.

(c) Opportunity for sponsor response. (1) If FDA proposes to terminate an IND, FDA will notify the sponsor in writing, and invite correction or explanation within a period of 30 days.

(2) On such notification, the sponsor may provide a written explanation or correction or may request a conference with FDA to provide the requested explanation or correction. If the sponsor does not respond to the notification within the allocated time, the IND shall be terminated.

(3) If the sponsor responds but FDA does not accept the explanation or correction submitted, FDA shall inform the sponsor in writing of the reason for the nonacceptance and provide the sponsor with an opportunity for a regulatory hearing before FDA under Part 16 on the question of whether the IND should be terminated. The sponsor's request for a regulatory hearing must be made within 10 days of the sponsor's receipt of FDA's notification of nonacceptance.

(d) Immediate termination of IND. Notwithstanding paragraphs (a) through (c) of this section, if at any time FDA concludes that continuation of the investigation presents an immediate and substantial danger to the health of individuals, the agency shall immediately, by written notice to the sponsor from the Director of the Center for Drug Evaluation and Research or the Director of the Center for Biologics Evaluation and Research, terminate the IND. An IND so terminated is subject to reinstatement by the Director on the basis of additional submissions that eliminate such danger. If an IND is terminated under this paragraph, the agency will afford the sponsor an opportunity for a regulatory hearing under part 16 on the question of whether the IND should be reinstated.

§312.45 Inactive status.

(a) If no subjects are entered into clinical studies for a period of 2 years or more under an IND, or if all investigations under an IND remain on clinical hold for 1 year or more, the IND may be placed by FDA on inactive status. This action may be taken by FDA either on request of the sponsor or on FDA's own initiative. If FDA seeks to act on its own initiative under this section, it shall first notify the sponsor in writing of the proposed inactive status. Upon receipt of such notification, the sponsor shall have 30 days to respond as to why the IND should continue to remain active.

(b) If an IND is placed on inactive status, all investigators shall be so notified and all stocks of the drug shall be returned or otherwise disposed of in accordance with §312.59.

(c) A sponsor is not required to submit annual reports to an IND on inactive status. An inactive IND is, however, still in effect for purposes of the public disclosure of data and information under §312.130.

(d) A sponsor who intends to resume clinical investigation under an IND placed on inactive status shall submit a protocol amendment under §312.30 containing the proposed general investigational plan for the coming year and appropriate protocols. If the protocol amendment relies on information previously submitted, the plan shall reference such information. Additional information supporting the proposed investigation, if any, shall be submitted in an information amendment. Notwithstanding the provisions of §312.30, clinical investigations under an IND on inactive status may only resume (1) 30 days after FDA receives the protocol amendment, unless FDA notifies the sponsor that the investigations described in the amendment are subject to a clinical hold under §312.42, or (2) on earlier notification by FDA that the clinical investigations described in the protocol amendment may begin.

(e) An IND that remains on inactive status for 5 years or more may be terminated under §312.44.

§312.47 Meetings.

(a) General. Meetings between a sponsor and the agency are frequently useful in resolving questions and issues raised during the course of a clinical investigation. FDA encourages such meetings to the extent that they aid in the evaluation of the drug and in the solution of scientific problems concerning the drug, to the extent that FDA's resources permit. The general principle underlying the conduct of such meetings is that there should be free, full, and open communication about any scientific or medical question that may arise during the clinical investigation. These meetings shall be conducted and documented in accordance with part 10.

(b) "End-of-Phase 2" meetings and meetings held before submission of a marketing application. At specific times during the drug investigation process, meetings between FDA and a sponsor can be especially helpful in minimizing wasteful expenditures of time and money and thus in speeding the drug development and evaluation process. In particular, FDA has found that meetings at the end of Phase 2 of an investigation (end-of-Phase 2 meetings) are of considerable assistance in planning later studies and that meetings held near completion of Phase 3 and before submission of a marketing application ("pre-NDA" meetings) are helpful in developing methods of presentation and submission of data in the marketing application that facilitate review and allow timely FDA response.

(1) End-of-Phase 2 meetings-(i) Purpose. The purpose of an end-of-Phase 2 meeting is to determine the safety of proceeding to Phase 3, to evaluate the Phase 3 plan and protocols, and to identify any additional information necessary to support a marketing application for the uses under investigation.

(ii) Eligibility for meeting. While the end-of-Phase 2 meeting is designed primarily for IND's involving new molecular entities or major new uses of marketed drugs, a sponsor of any IND may request and obtain an end-of-Phase 2 meeting.

(iii) Timing. To be most useful to the sponsor, end-of-Phase 2 meetings should be held before major commitments of effort and resources to specific Phase 3 tests are made. The scheduling of an end-of-Phase 2 meeting is not, however, intended to delay the transition of an investigation from Phase 2 to Phase 3.

(iv) Advance information. At least 1 month in advance of an end-of-Phase 2 meeting, the sponsor should submit background information on the sponsor's plan for Phase 3, including summaries of the Phase 1 and 2 investigations, the specific protocols for Phase 3 clinical studies, plans for any additional nonclinical studies, and, if available, tentative labeling for the drug. The recommended contents of such a submission are described more fully in FDA Staff Manual Guide 4850.7 that is publicly available under FDA's public information regulations in Part 20.

(v) Conduct of meeting. Arrangements for an end-of-Phase 2 meeting are to be made with the division in FDA's Center for Drug Evaluation and Research or the Center for Biologics Evaluation and Research which is responsible for review of the IND. The meeting will be scheduled by FDA at a time convenient to both FDA and the sponsor. Both the sponsor and FDA may bring consultants to the meeting. The meeting should be directed primarily at establishing agreement between FDA and the sponsor of the overall plan for Phase 3 and the objectives and design of particular studies. The adequacy of technical information to support Phase 3 studies and/or a marketing application may also be discussed. Agreements reached at the meeting on these matters will be recorded in minutes of the conference that will be taken by FDA in accordance with §10.65 and provided to the sponsor. The minutes along with any other written material provided to the sponsor will serve as a permanent record of any agreements reached. Barring a significant scientific development that requires otherwise, studies conducted in accordance with the agreement shall be presumed to be sufficient in objective and design for the purpose of obtaining marketing approval for the drug.

(2) "Pre-NDA" meetings. FDA has found that delays associated with the initial review of a marketing application may be reduced by exchanges of information about a proposed marketing application. The primary purpose of this kind of exchange is to uncover any major unresolved problems, to identify those studies that the sponsor is relying on as adequate and well-controlled to establish the drug's effectiveness, to acquaint FDA reviewers with the general information to be submitted in the marketing application (including technical information), to discuss appropriate methods for statistical analysis of the data, and to discuss the best approach to the presentation and formatting of data in the marketing application. Arrangements for such a meeting are to be initiated by the sponsor with the division responsible for review of the IND. To permit FDA to provide the sponsor with the most useful advice on preparing a marketing application, the sponsor should submit to FDA's reviewing division at least 1 month in advance of the meeting the following information:

(i) A brief summary of the clinical studies to be submitted in the application.

(ii) A proposed format for organizing the submission, including methods for presenting the data.

(iii) Any other information for discussion at the meeting.

§312.48 Dispute resolution.

(a) General. The Food and Drug Administration is committed to resolving differences between sponsors and FDA reviewing divisions with respect to requirements for IND's as quickly and amicably as possible through the cooperative exchange of information and views.

(b) Administrative and procedural issues. When administrative or procedural disputes arise, the sponsor should first attempt to resolve the matter with the division in FDA's Center for Drug Evaluation and Research or Center for Biologics Evaluation and Research which is responsible for review of the IND, beginning with the consumer safety officer assigned to the application. If the dispute is not resolved, the sponsor may raise the matter with the person designated as ombudsman, whose function shall be to investigate what has happened and to facilitate a timely and equitable resolution. Appropriate issues to raise with the ombudsman include resolving difficulties in scheduling meetings and obtaining timely replies to inquiries. Further details on this procedure are contained in FDA Staff Manual Guide 4820.7 that is publicly available under FDA's public information regulations in part 20.

(c) Scientific and medical disputes. (1) When scientific or medical disputes arise during the drug investigation process, sponsors should discuss the matter directly with the responsible reviewing officials. If necessary, sponsors may request a meeting with the appropriate reviewing officials and management representatives in order to seek a resolution. Requests for such meetings shall be directed to the director of the division in FDA's Center for Drug Evaluation and Research or Center for Biologics Evaluation and Research which is responsible for review of the IND. FDA will make every attempt to grant requests for meetings that involve important issues and that can be scheduled at mutually convenient times.

(2) The "end-of-Phase 2" and "pre-NDA" meetings described in §312.47(b) will also provide a timely forum for discussing and resolving scientific and medical issues on which the sponsor disagrees with the agency.

(3) In requesting a meeting designed to resolve a scientific or medical dispute, applicants may suggest that FDA seek the advice of outside experts, in which case FDA may, in its discretion, invite to the meeting one or more of its advisory committee members or other consultants, as designated by the agency. Applicants may rely on, and may bring to any meeting, their own consultants. For major scientific and medical policy issues not resolved by informal meetings, FDA may refer the matter to one of its standing advisory committees for its consideration and recommendations.

Subpart D--Responsibilities of Sponsors and Investigators

§312.50 General responsibilities of sponsors.

Sponsors are responsible for selecting qualified investigators, providing them with the information they need to conduct an investigation properly, ensuring proper monitoring of the investigation(s), ensuring that the investigation(s) is conducted in accordance with the general investigational plan and protocols contained in the IND, maintaining an effective IND with respect to the investigations, and ensuring that FDA and all participating investigators are promptly informed of significant new adverse effects or risks with respect to the drug. Additional specific responsibilities of sponsors are described elsewhere in this part.

§312.52 Transfer of obligations to a contract research organization.

(a) A sponsor may transfer responsibility for any or all of the obligations set forth in this part to a contract research organization. Any such transfer shall be described in writing. If not all obligations are transferred, the writing is required to describe each of the obligations being assumed by the contract research organization. If all obligations are transferred, a general statement that all obligations have been transferred is acceptable. Any obligation not covered by the written description shall be deemed not to have been transferred.

(b) A contract research organization that assumes any obligation of a sponsor shall comply with the specific regulations in this chapter applicable to this obligation and shall be subject to the same regulatory action as a sponsor for failure to comply with any obligation assumed under these regulations. Thus, all references to "sponsor" in this part apply to a contract research organization to the extent that it assumes one or more obligations of the sponsor.

§312.53 Selecting investigators and monitors.

(a) Selecting investigators. A sponsor shall select only investigators qualified by training and experience as appropriate experts to investigate the drug.

(b) Control of drug. A sponsor shall ship investigational new drugs only to investigators participating in the investigation.

(c) Obtaining information from the investigator. Before permitting an investigator to begin participation in an investigation, the sponsor shall obtain the following:

(1) A signed investigator statement (Form FDA-1572) containing:

(i) The name and address of the investigator;

(ii) The name and code number, if any, of

the protocol(s) in the IND identifying the study(ies) to be conducted by the investigator;

(iii) The name and address of any medical school, hospital, or other research facility where the clinical investigation(s) will be conducted;

(iv) The name and address of any clinical laboratory facilities to be used in the study;

(v) The name and address of the IRB that is responsible for review and approval of the study(ies);

(vi) A commitment by the investigator that he or she:

(a) Will conduct the study(ies) in accordance with the relevant, current protocol(s) and will only make changes in a protocol after notifying the sponsor, except when necessary to protect the safety, the rights, or welfare of subjects;

(b) Will comply with all requirements regarding the obligations of clinical investigators and all other pertinent requirements in this part;

(c) Will personally conduct or supervise the described investigation(s);

(d) Will inform any patients, or any persons used as controls, that the drugs are being used for investigational purposes and will ensure that the requirements relating to obtaining informed consent and institutional review board review and approval are met;

(e) Will report to the sponsor adverse experiences that occur in the course of the investigation(s) in accordance with §312.64;

(f) Has read and understands the information in the investigator's brochure, including the potential risks and side effects of the drug; and

(g) Will ensure that all associates, colleagues, and employees assisting in the conduct of the study(ies) are informed about their obligations in meeting the above commitments.

(vii) A commitment by the investigator that, for an investigation subject to an institutional review requirement under part 56, an IRB that complies with the requirements of that part will be responsible for the initial and continuing review and approval of the clinical investigation and that the investigator will promptly report to the IRB all changes in the research activity and all unanticipated problems involving risks to human subjects or others, and will not make any changes in the research without IRB approval, except where necessary to eliminate apparent immediate hazards to the human subjects.

(viii) A list of the names of the subinvestigators (e.g., research fellows, residents) who will be assisting the investigator in the conduct of the investigation(s).

(2) Curriculum vitae. A curriculum vitae or other statement of qualifications of the investigator showing the education, training, and experience that qualifies the investigator as an expert in the clinical investigation of the drug for the use under investigation.

(3) Clinical protocol. (i) For Phase 1 investigations, a general outline of the planned investigation including the estimated duration of the study and the maximum number of subjects that will be involved.

(ii) For Phase 2 or 3 investigations, an outline of the study protocol including an approximation of the number of subjects to be treated with the drug and the number to be employed as controls, if any; the clinical uses to be investigated; characteristics of subjects by age, sex, and condition; the kind of clinical observations and laboratory tests to be conducted; the estimated duration of the study; and copies or a description of case report forms to be used.

(d) Selecting monitors. A sponsor shall select a monitor qualified by training and experience to monitor the progress of the investigation.

§312.55 Informing investigators.

(a) Before the investigation begins, a sponsor (other than a sponsor-investigator) shall give each participating clinical investigator an investigator brochure containing the information described in §312.23(a)(5).

(b) The sponsor shall, as the overall investigation proceeds, keep each participating investigator informed of new observations discovered by or reported to the sponsor on the drug, particularly with respect to adverse effects and safe use. Such information may be distributed to investigators by means of periodically revised investigator brochures, reprints or published studies, reports or letters to clinical investigators, or other appropriate means. Important safety information is required to be relayed to investigators in accordance with §312.32.

§312.56 Review of ongoing investigations.

(a) The sponsor shall monitor the progress of all clinical investigations being conducted under its IND.

(b) A sponsor who discovers that an investigator is not complying with the signed agreement (Form FDA-1572), the general investigational plan, or the requirements of this part or other applicable parts shall promptly either secure compliance or discontinue shipments of the investigational new drug to the investigator and end the investigator's participation in the investigation. If the investigator's participation in the investigation is ended, the sponsor shall require that the investigator dispose of or return the investigational drug in accordance with the requirements of §312.59 and shall notify FDA.

(c) The sponsor shall review and evaluate the evidence relating to the safety and effectiveness of the drug as it is obtained from the investigator. The sponsors shall make such reports to FDA regarding information relevant to the safety of the drug as are required under §312.32. The sponsor shall make annual reports on the progress of the investigation in accordance with §312.33.

(d) A sponsor who determines that its investigational drug presents an unreasonable and significant risk to subjects shall discontinue those investigations that present the risk, notify FDA, all institutional review boards, and all investigators who have at any time participated in the investigation of the discontinuance, assure the disposition of all stocks of the drug outstanding as required by §312.59, and furnish FDA with a full report of the sponsor's actions. The sponsor shall discontinue the investigation as soon as possible, and in no event later than 5 working days after making the determination that the investigation should be discontinued. Upon request, FDA will confer with a sponsor on the need to discontinue an investigation.

§312.57 Recordkeeping and record retention.

(a) A sponsor shall maintain adequate records showing the receipt, shipment, or other disposition of the investigational drug. These records are required to include, as appropriate, the name of the investigator to whom the drug is shipped, and the date, quantity, and batch or code mark of each such shipment.

(b) A sponsor shall retain the records and reports required by this part for 2 years after a marketing application is approved for the drug; or, if an application is not approved for the drug, until 2 years after shipment and delivery of the drug for investigational use is discontinued and FDA has been so notified.

(c) A sponsor shall retain reserve samples of any test article and reference standard used in a bioequivalence or bioavailability study and release the reserve samples to FDA upon request, in accordance with, and for the period specified in, §320.32 of this chapter.

§312.58 Inspection of sponsor's records and reports.

(a) FDA inspection. A sponsor shall upon request from any properly authorized officer or employee of the Food and Drug Administration, at reasonable times, permit such officer or employee to have access to and copy and verify any records and reports relating to a clinical investigation conducted under this part. Upon written request by FDA, the sponsor shall submit the records or reports (or copies of them) to FDA. The sponsor shall discontinue shipments of the drug to any investigator who has failed to maintain or make available records or reports

of the investigation as required by this part.

(b) Controlled substances. If an investigational new drug is a substance listed in any schedule of the Controlled Substances Act (21 U.S.C. 801; 21 CFR part 1308), records concerning shipment, delivery, receipt, and disposition of the drug, which are required to be kept under this part or other applicable parts of this chapter shall, upon the request of a properly authorized employee of the Drug Enforcement Administration of the U.S. Department of Justice, be made available by the investigator or sponsor to whom the request is made, for inspection and copying. In addition, the sponsor shall assure that adequate precautions are taken, including storage of the investigational drug in a securely locked, substantially constructed cabinet, or other securely locked, substantially constructed enclosure, access to which is limited, to prevent theft or diversion of the substance into illegal channels of distribution.

§312.59 Disposition of unused supply of investigational drug.

The sponsor shall assure the return of all unused supplies of the investigational drug from each individual investigator whose participation in the investigation is discontinued or terminated. The sponsor may authorize alternative disposition of unused supplies of the investigational drug provided this alternative disposition does not expose humans to risks from the drug. **The sponsor shall maintain written records of any disposition of the drug in accordance with §312.57.**

§312.60 General responsibilities of investigators.

An investigator is responsible for ensuring that an investigation is conducted according to the signed investigator statement, the investigational plan, and applicable regulations; for protecting the rights, safety, and welfare of subjects under the investigator's care; and for the control of drugs under investigation. An investigator shall, in accordance with the provisions of part 50, obtain the informed consent of each human subject to whom the drug is administered, except as provided in §50.23. Additional specific responsibilities of clinical investigators are set forth in this part and in parts 50 and 56.

§312.61 Control of the investigational drug.

An investigator shall administer the drug only to subjects under the investigator's personal supervision or under the supervision of a subinvestigator responsible to the investigator. The investigator shall not supply the investigational drug to any person not authorized under this part to receive it.

§312.62 Investigator recordkeeping and record retention.

(a) Disposition of drug. An investigator is required to maintain adequate records of the disposition of the drug, including dates, quantity, and use by subjects. If the investigation is terminated, suspended, discontinued, or completed, the investigator shall return the unused supplies of the drug to the sponsor, or otherwise provide for disposition of the unused supplies of the drug under §312.59.

(b) Case histories. An investigator is required to prepare and maintain adequate and accurate case histories designed to record all observations and other data pertinent to the investigation on each individual treated with the investigational drug or employed as a control in the investigation.

(c) Record retention. An investigator shall retain records required to be maintained under this part for a period of 2 years following the date a marketing application is approved for the drug for the indication for which it is being investigated; or, if no application is to be filed or if the application is not approved for such indication, until 2 years after the investigation is discontinued and FDA is notified.

§312.64 Investigator reports.

(a) Progress reports. The investigator shall furnish all reports to the sponsor of the drug who is responsible for collecting and evaluating the results obtained. The sponsor is required under §312.33 to submit annual reports to FDA on the progress of the clinical investigations.

(b) Safety reports. An investigator shall promptly report to the sponsor any adverse effect that may reasonably be regarded as caused by, or probably caused by, the drug. If the adverse effect is alarming, the investigator shall report the adverse effect immediately.

(c) Final report. An investigator shall provide the sponsor with an adequate report shortly after completion of the investigator's participation in the investigation.

§312.66 Assurance of IRB review.

An investigator shall assure that an IRB that complies with the requirements set forth in Part 56 will be responsible for the initial and continuing review and approval of the proposed clinical study. The investigator shall also assure that he or she will promptly report to the IRB all changes in the research activity and all unanticipated problems involving risk to human subjects or others, and that he or she will not make any changes in the research without IRB approval, except where necessary to eliminate apparent immediate hazards to human subjects.

§312.68 Inspection of investigator's records and reports.

An investigator shall upon request from any properly authorized officer or employee of FDA, at reasonable times, permit such officer or employee to have access to, and copy and verify any records or reports made by the investigator pursuant to §312.62. The investigator is not required to divulge subject names unless the records of particular individuals require a more detailed study of the cases, or unless there is reason to believe that the records do not represent actual case studies, or do not represent actual results obtained.

§312.69 Handling of controlled substances.

If the investigational drug is subject to the Controlled Substances Act, the investigator shall take adequate precautions, including storage of the investigational drug in a securely locked, substantially constructed cabinet, or other securely locked, substantially constructed enclosure, access to which is limited, to prevent theft or diversion of the substance into illegal channels of distribution.

§312.70 Disqualification of a clinical investigator.

(a) If FDA has information indicating that an investigator has repeatedly or deliberately failed to comply with the requirements of this part, Part 50, or part 56, or has submitted to the sponsor false information in any required report, the Center for Drug Evaluation and Research or the Center for Biologics Evaluation and Research will furnish the investigator written notice of the matter complained of and offer the investigator an opportunity to explain the matter in writing, or, at the option of the investigator, in an informal conference. If an explanation is offered but not accepted by the Center for Drug Evaluation and Research or the Center for Biologics Evaluation and Research, the investigator will be given an opportunity for a regulatory hearing under part 16 on the question of whether the investigator is entitled to receive investigational new drugs.

(b) After evaluating all available information, including any explanation presented by the investigator, if the Commissioner determines that the investigator has repeatedly or deliberately failed to comply with the requirements of this part, part 50, or part 56, or has deliberately or repeatedly submitted false information to the sponsor in any required report, the Commissioner will notify the investigator and the sponsor of any investigation in which the investigator has been named as a participant that the investigator is not entitled to receive investigational drugs. The notification will provide a statement of basis for such determination.

(c) Each IND and each approved application submitted under part 314 containing data reported by an investigator who has been determined to be ineligible to receive investigational drugs will be examined to determine whether the investigator has submitted unreliable data that are essential to the continuation of the investigation or essential to the approval of any marketing application.

(d) If the Commissioner determines, after the unreliable data submitted by the investigator are eliminated from consideration, that the data remaining are inadequate to support a conclusion that it is reasonably safe to continue the investigation, the Commissioner will notify the sponsor who shall have an opportunity for a regulatory hearing under part 16. If a danger to the public health exists, however, the Commissioner shall terminate the IND immediately and notify the sponsor of the determination. In such case, the sponsor shall have an opportunity for a regulatory hearing before FDA under part 16 on the question of whether the IND should be reinstated.

(e) If the Commissioner determines, after the unreliable data submitted by the investigator are eliminated from consideration, that the continued approval of the drug product for which the data were submitted cannot be justified, the Commissioner will proceed to withdraw approval of the drug product in accordance with the applicable provisions of the act.

(f) An investigator who has been determined to be ineligible to receive investigational drugs may be reinstated as eligible when the Commissioner determines that the investigator has presented adequate assurances that the investigator will employ investigational drugs solely in compliance with the provisions of this part and of parts 50 and 56.

Subpart E--Drugs Intended to Treat Life Threatening and Severely-debilitating Illnesses

§*312.80 Purpose.*

The purpose of this section is to establish procedures designed to expedite the development, evaluation, and marketing of new therapies intended to treat persons with life-threatening and severely-debilitating illnesses, especially where no satisfactory alternative therapy exists. As stated §314.105(c) of this chapter, while the statutory standards of safety and effectiveness apply to all drugs, the many kinds of drugs that are subject to them, and the wide range of uses for those drugs, demand flexibility in applying the standards. The Food and Drug Administration (FDA) has determined that it is appropriate to exercise the broadest flexibility in applying the statutory standards, while preserving appropriate guarantees for safety and effectiveness. These procedures reflect the recognition that physicians and patients are generally willing to accept greater risks or side effects from products that treat life-threatening and severely-debilitating illnesses, than they would accept from products that treat less serious illnesses. These procedures also reflect the recognition that the benefits of the drug need to be evaluated in light of the severity of the disease being treated. The procedure outlined in this section should be interpreted consistent with that purpose.

§*312.81 Scope.*

This section applies to new drug, antibiotic, and biological products that are being studied for their safety and effectiveness in treating life-threatening or severely-debilitating diseases.

(a) For purposes of this section, the term "life-threatening" means:

(1) Diseases or conditions where the likelihood of death is high unless the course of the disease is interrupted; and

(2) Diseases or conditions with potentially fatal outcomes, where the end point of clinical trial analysis is survival.

(b) For purposes of this section, the term "severely debilitating" means diseases or conditions that cause major irreversible morbidity.

(c) Sponsors are encouraged to consult with FDA on the applicability of these procedures to specific products.

§312.82 Early consultation.

For products intended to treat life-threatening or severely-debilitating illnesses, sponsors may request to meet with FDA-reviewing officials early in the drug development process to review and reach agreement on the design of necessary preclinical and clinical studies. Where appropriate, FDA will invite to such meetings one or more outside expert scientific consultants or advisory committee members. To the extent FDA resources permit, agency reviewing officials will honor requests for such meetings

(a) Pre-investigational new drug (IND) meetings. Prior to the submission of the initial IND, the sponsor may request a meeting with FDA-reviewing officials. The primary purpose of this meeting is to review and reach agreement on the design of animal studies needed to initiate human testing. The meeting may also provide an opportunity for discussing the scope and design of phase 1 testing, and the best approach for presentation and formatting of data in the IND.

(b) End-of-phase 1 meetings. When data from phase 1 clinical testing are available, the sponsor may again request a meeting with FDA-reviewing officials. The primary purpose of this meeting is to review and reach agreement on the design of phase 2 controlled clinical trials, with the goal that such testing will be adequate to provide sufficient data on the drug's safety and effectiveness to support a decision on its approvability for marketing. The procedures outlined in §312.47(b)(1) with respect to end-of-phase 2 conferences, including documentation of agreements reached, would also be used for end-of- phase 1 meetings.

§312.83 Treatment protocols.

If the preliminary analysis of phase 2 test results appears promising, FDA may ask the sponsor to submit a treatment protocol to be reviewed under the procedures and criteria listed in §§312.34 and 312.35. Such a treatment protocol, if requested and granted, would normally remain in effect while the complete data necessary for a marketing application are being assembled by the sponsor and reviewed by FDA (unless grounds exist for clinical hold of ongoing protocols, as provided in §312.42(b)(3)(ii)).

§312.84 Risk-benefit analysis in review of marketing applications for drugs to treat life-threatening and severely-debilitating illnesses.

(a) FDA's application of the statutory standards for marketing approval shall recognize the need for a medical risk-benefit judgment in making the final decision on approvability. As part of this evaluation, consistent with the statement of purpose in §312.80, FDA will consider whether the benefits of the drug outweigh the known and potential risks of the drug and the need to answer remaining questions about risks and benefits of the drug, taking into consideration the severity of the disease and the absence of satisfactory alternative therapy.

(b) In making decisions on whether to grant marketing approval for products that have been the subject of an end-of-phase 1 meeting under §312.82, FDA will usually seek the advice of outside expert scientific consultants or advisory committees. Upon the filing of such a marketing application under §314.101 or part 601 of this chapter, FDA will notify the members of the relevant standing advisory committee of the application's filing and its availability for review.

(c) If FDA concludes that the data presented are not sufficient for marketing approval, FDA will issue (for a drug) a not approvable letter pursuant to §314.120 of this chapter, or (for a biologic) a deficiencies letter consistent with the biological product licensing procedures. Such letter, in describing the deficiencies in the application, will address why the results of the research design agreed to under §312.82, or in subsequent meetings, have not provided sufficient evidence for marketing approval. Such letter will also describe any recommendations made by the advisory committee regarding the application.

(d) Marketing applications submitted under the procedures contained in this section will be subject to the requirements and procedures contained in part 314 or part 600 of this chapter, as well as those in this subpart.

§312.85 Phase 4 studies.

Concurrent with marketing approval, FDA may seek agreement from the sponsor to conduct certain postmarketing (phase 4) studies to delineate additional information about the drug's risks, benefits, and optimal use. These studies could include, but would not be limited to, studying different doses or schedules of administration than were used in phase 2 studies, use of the drug in other patient populations or other stages of the disease, or use of the drug over a longer period of time.

§312.86 Focused FDA regulatory research.

At the discretion of the agency, FDA may undertake focused regulatory research on critical rate-limiting aspects of the preclinical, chemical/manufacturing, and clinical phases of drug development and evaluation. When initiated, FDA will undertake such research efforts as a means for meeting a public health need in facilitating the development of therapies to treat life-threatening or severely debilitating illnesses.

§312.87 Active monitoring of conduct and evaluation of clinical trials.

For drugs covered under this section, the Commissioner and other agency officials will monitor the progress of the conduct and evaluation of clinical trials and be involved in facilitating their appropriate progress.

§312.88 Safeguards for patient safety.

All of the safeguards incorporated within parts 50, 56, 312, 314, and 600 of this chapter designed to ensure the safety of clinical testing and the safety of products following marketing approval apply to drugs covered by this section. This includes the requirements for informed consent (part 50 of this chapter) and institutional review boards (part 56 of this chapter). These safeguards further include the review of animal studies prior to initial human testing (§312.23), and the monitoring of adverse drug experiences through the requirements of IND safety reports (§312.32), safety update reports during agency review of a marketing application (§314.50 of this chapter), and postmarketing adverse reaction reporting (§314.80 of this chapter).

Subpart F--Miscellaneous

§312.110 Import and export requirements.

(a) Imports. An investigational new drug offered for import into the United States complies with the requirements of this part if it is subject to an IND that is in effect for it under §312.40 and: (1) The consignee in the United States is the sponsor of the IND; (2) the consignee is a qualified investigator named in the IND; or (3) the consignee is the domestic agent of a foreign sponsor, is responsible for the control and distribution of the investigational drug, and the IND identifies the consignee and describes what, if any, actions the consignee will take with respect to the investigational drug.

(b) Exports. An investigational new drug intended for export from the United States complies with the requirements of this part as follows:

(1) If an IND is in effect for the drug under §312.40 and each person who receives the drug is an investigator named in the application; or

(2) If FDA authorizes shipment of the drug for use in a clinical investigation. Authorization may be obtained as follows:

(i) Through submission to the International Affairs Staff (HFY-50), Associate Commissioner for Health Affairs, Food and Drug Administration, 5600 Fishers Lane, Rockville, MD 20857, of a written request from the person that seeks to export the drug. A request must provide adequate information about the drug to satisfy FDA that the drug is appropriate for the proposed investigational use in humans, that the drug will be used for investigational purposes only, and that the drug may be legally used by that consignee in the importing country for the proposed investigational use. The request shall specify the quantity of the drug to be shipped per shipment and the frequency of expected shipments. If FDA authorizes exportation under this paragraph, the agency shall concurrently notify the government of the importing country of such authorization.

(ii) Through submission to the International Affairs Staff (HFY-50), Associate Commissioner for Health Affairs, Food and Drug Administration, 5600 Fishers Lane, Rockville, MD 20857, of a formal request from an authorized official of the government of the country to which the drug is proposed to be shipped. A request must specify that the foreign government has adequate information about the drug and the proposed investigational use, that the drug will be used for investigational purposes only, and that the foreign government is satisfied that the drug may legally be used by the intended consignee in that country. Such a request shall specify the quantity of drug to be shipped per shipment and the frequency of expected shipments.

(iii) Authorization to export an investigational drug under paragraph (b)(2)(i) or (ii) of this section may be revoked by FDA if the agency finds that the conditions underlying its authorization are not longer met.

(3) This paragraph applies only where the drug is to be used for the purpose of clinical investigation.

(4) This paragraph does not apply to the export of an antibiotic drug product shipped in accordance with the provisions of section 801(d) of the act.

(5) This paragraph does not apply to the export of new drugs (including biological products) approved for export under section 802 of the act or section 351(h)(1)(A) of the Public Health Service Act.

§312.120 Foreign clinical studies not conducted under an IND.

(a) Introduction. This section describes the criteria for acceptance by FDA of foreign clinical studies not conducted under an IND. In general, FDA accepts such studies provided they are well designed, well conducted, performed by qualified investigators, and conducted in accordance with ethical principles acceptable to the world community. Studies meeting these criteria may be utilized to support clinical investigations in the United States and/or marketing approval. Marketing approval of a new drug or antibiotic drug based solely on foreign clinical data is governed by §314.106.

(b) Data submissions. A sponsor who wishes to rely on a foreign clinical study to support an IND or to support an application for marketing approval shall submit to FDA the following information:

(1) A description of the investigator's qualifications;

(2) A description of the research facilities;

(3) A detailed summary of the protocol and results of the study, and, should FDA request, case records maintained by the investigator or additional background data such as hospital or other institutional records;

(4) A description of the drug substance and drug product used in the study, including a description of components, formulation, specifications, and bioavailability of the specific drug product used in the clinical study, if available; and

(5) If the study is intended to support the effectiveness of a drug product, information showing that the study is adequate and well controlled under §314.126.

(c) Conformance with ethical principles. (1) Foreign clinical research is required to have been conducted in accordance with the ethical principles stated in the "Declaration of Helsinki" (see paragraph (c)(4) of this section) or the laws and regulations of the country in which the research was conducted, whichever represents the greater protection of the individual.

(2) For each foreign clinical study submitted under this section, the sponsor shall explain how the research conformed to the ethical principles contained in the "Declaration of Helsinki" or the foreign country's standards, whichever were used. If the foreign country's standards were used, the sponsor shall explain in detail how those standards differ from the

"Declaration of Helsinki" and how they offer greater protection.

(3) When the research has been approved by an independent review committee, the sponsor shall submit to FDA documentation of such review and approval, including the names and qualifications of the members of the committee. In this regard, a "review committee" means a committee composed of scientists and, where practicable, individuals who are otherwise qualified (e.g., other health professionals or laymen). The investigator may not vote on any aspect of the review of his or her protocol by a review committee.

(4) The "Declaration of Helsinki" states as follows:

RECOMMENDATIONS GUIDING PHYSICIANS IN BIOMEDICAL RESEARCH INVOLVING HUMAN SUBJECTS

Introduction

It is the mission of the physician to safeguard the health of the people. His or her knowledge and conscience are dedicated to the fulfillment of this mission.

The Declaration of Geneva of the World Medical Association binds the physician with the words, "The health of my patient will be my first consideration," and the International Code of Medical Ethics declares that, "A physician shall act only in the patient's interest when providing medical care which might have the effect of weakening the physical and mental condition of the patient."

The purpose of biomedical research involving human subjects must be to improve diagnostic, therapeutic and prophylactic procedures and the understanding of the aetiology and pathogenesis of disease.

In current medical practice most diagnostic, therapeutic or prophylactic procedures involve hazards. This applies especially to biomedical research.

Medical progress is based on research which ultimately must rest in part on experimentation involving human subjects.

In the field of biomedical research a fundamental distinction must be recognized between medical research in which the aim is essentially diagnostic or therapeutic for a patient, and medical research, the essential object of which is purely scientific and without implying direct diagnostic or therapeutic value to the person subjected to the research.

Special caution must be exercised in the conduct of research which may affect the environment, and the welfare of animals used for research must be respected.

Because it is essential that the results of laboratory experiments be applied to human beings to further scientific knowledge and to help suffering humanity, the World Medical Association has prepared the following recommendations as a guide to every physician in biomedical research involving human subjects. They should be kept under review in the future. It must be stressed that the standards as drafted are only a guide to physicians all over the world. Physicians are not relieved from criminal, civil and ethical responsibilities under the laws of their own countries.

I. Basic Principles

1. Biomedical research involving human subjects must conform to generally accepted scientific principles and should be based on adequately performed laboratory and animal experimentation and on a thorough knowledge of the scientific literature.

2. The design and performance of each experimental procedure involving human subjects should be clearly formulated in an experimental protocol which should be transmitted for consideration, comment and guidance to a specially appointed committee independent of the investigator and the sponsor provided that this independent committee is in conformity with the laws and regulations of the country in which the research experiment is performed.

3. Biomedical research involving human subjects should be conducted only by scientifically qualified persons and under the supervision of a clinically competent medical person. The responsibility for the human subject must always rest with a medically qualified person and never rest on the subject of the research, even though the subject has given his or her consent.

4. Biomedical research involving human

subjects cannot legitimately be carried out unless the importance of the objective is in proportion to the inherent risk to the subject.

5. Every biomedical research project involving human subjects should be preceded by careful assessment of predictable risks in comparison with foreseeable benefits to the subject or to others. Concern for the interests of the subject must always prevail over the interests of science and society.

6. The right of the research subject to safeguard his or her integrity must always be respected. Every precaution should be taken to respect the privacy of the subject and to minimize the impact of the study on the subject's physical and mental integrity and on the personality of the subject.

7. Physicians should abstain from engaging in research projects involving human subjects unless they are satisfied that the hazards involved are believed to be predictable. Physicians should cease any investigation if the hazards are found to outweigh the potential benefits.

8. In publication of the results of his or her research, the physician is obliged to preserve the accuracy of the results. Reports of experimentation not in accordance with the principles laid down in this Declaration should not be accepted for publication.

9. In any research on human beings, each potential subject must be adequately informed of the aims, methods, anticipated benefits and potential hazards of the study and the discomfort it may entail. He or she should be informed that he or she is at liberty to abstain from participation in the study and that he or she is free to withdraw his or her consent to participation at any time. The physician should then obtain the subject's freely-given informed consent, preferably in writing.

10. When obtaining informed consent for the research project the physician should be particularly cautious if the subject is in a dependent relationship to him or her or may consent under duress. In that case the informed consent should be obtained by a physician who is not engaged in the investigation and who is completely independent of this official relationship.

11. In case of legal incompetence, informed consent should be obtained from the legal guardian in accordance with national legislation. Where physical or mental incapacity makes it impossible to obtain informed consent, or when the subject is a minor, permission from the responsible relative replaces that of the subject in accordance with national legislation.

Whenever the minor child is in fact able to give a consent, the minor's consent must be obtained in addition to the consent of the minor's legal guardian.

12. The research protocol should always contain a statement of the ethical considerations involved and should indicate that the principles enunciated in the present Declaration are complied with.

II. Medical Research Combined with Professional Care (Clinical Research)

1. In the treatment of the sick person, the physician must be free to use a new diagnostic and therapeutic measure, if in his or her judgment it offers hope of saving life, reestablishing health or alleviating suffering.

2. The potential benefits, hazards and discomfort of a new method should be weighed against the advantages of the best current diagnostic and therapeutic methods.

3. In any medical study, every patient -- including those of a control group, if any -- should be assured of the best proven diagnostic and therapeutic method.

4. The refusal of the patient to participate in a study must never interfere with the physician-patient relationship.

5. If the physician considers it essential not to obtain informed consent, the specific reasons for this proposal should be stated in the experimental protocol for transmission to the independent committee (I, 2).

6. The physician can combine medical research with professional care, the objective being the acquisition of new medical knowledge, only to the extent that medical research is justified by its potential diagnostic or therapeutic value for the patient.

III. Non-Therapeutic Biomedical Research Involving Human Subjects (Non-Clinical Biomedical Research)

1. In the purely scientific application of medical research carried out on a human

being, it is the duty of the physician to remain the protector of the life and health of that person on whom biomedical research is being carried out.

2. The subjects should be volunteers--either healthy persons or patients for whom the experimental design is not related to the patient's illness.

3. The investigator or the investigating team should discontinue the research if in his/her or their judgment it may, if continued, be harmful to the individual.

4. In research on man, the interest of science and society should never take precedence over considerations related to the well- being of the subject.

§312.130 Availability for public disclosure of data and information in an IND.

(a) The existence of an investigational new drug application will not be disclosed by FDA unless it has previously been publicly disclosed or acknowledged.

(b) The availability for public disclosure of all data and information in an investigational new drug application for a new drug or antibiotic drug will be handled in accordance with the provisions established in §314.430 for the confidentiality of data and information in applications submitted in part 314. The availability for public disclosure of all data and information in an investigational new drug application for a biological product will be governed by the provisions of §§601.50 and 601.51.

(c) Notwithstanding the provisions of §314. 430, FDA shall disclose upon request to an individual to whom an investigational new drug has been given a copy of any IND safety report relating to the use in the individual.

§312.140 Address for correspondence.

(a) Except as provided in paragraph (b) of this section, a sponsor shall send an initial IND submission to the Central Document Room, Center for Drug Evaluation and Research, Food and Drug Administration, Park Bldg., Rm. 214, 12420 Parklawn Dr., Rockville, MD 20852. On receiving the IND, FDA will inform the sponsor which one of the divisions in the Center for Drug Evaluation and Research or the Center for Biologics Evaluation and Research is responsible for the IND. Amendments, reports, and other correspondence relating to matters covered by the IND should be directed to the appropriate division. The outside wrapper of each submission shall state what is contained in the submission, for example, "IND Application", "Protocol Amendment", etc.

(b) Applications for the products listed below should be submitted to the Division of Biological Investigational New Drugs (HFB-230), Center for Biologics Evaluation and Research, Food and Drug Administration, 8800 Rockville Pike, Bethesda, MD 20892. (1) Products subject to the licensing provisions of the Public Health Service Act of July 1, 1944 (58 Stat. 682, as amended (42 U.S.C. 201 et seq.)) or subject to part 600; (2) ingredients packaged together with containers intended for the collection, processing, or storage of blood or blood components; (3) urokinase products; (4) plasma volume expanders and hydroxyethyl starch for leukapheresis; and (5) coupled antibodies, i.e., products that consist of an antibody component coupled with a drug or radionuclide component in which both components provide a pharmacological effect but the biological component determines the site of action.

(c) All correspondence relating to biological products for human use which are also radioactive drugs shall be submitted to the Division of Oncology and Radiopharmaceutical Drug Products (HFD-150), Center for Drug Evaluation and Research, Food and Drug Administration, 5600 Fishers Lane, Rockville, MD 20857, except that applications for coupled antibodies shall be submitted in accordance with paragraph (b) of this section.

(d) All correspondence relating to export of an investigational drug under §312.110 (b)(2) shall be submitted to the International Affairs Staff (HFY-50), Office of Health Affairs, Food and Drug Administration, 5600 Fishers Lane, Rockville, MD 20857.

§312.145 Guidelines.

(a) FDA has made available guidelines under §10.90(b) to help persons to comply with certain requirements of this part.

(b) The Center for Drug Evaluation and Research and the Center for Biologics Evaluation and Research maintain lists of guidelines that apply to the Centers' regulations. The lists state how a person can obtain a copy of each guideline. A request for a copy of the lists should be directed to the CDER Executive Secretariat Staff (HFD-8), Center for Drug Evaluation and Research, Food and Drug Administration, 5600 Fishers Lane, Rockville, MD 20857, for drug products, and the Congressional, Consumer, and International Affairs Staff (HFB-142), Center for Biologics Evaluation and Research, Food and Drug Administration, 8800 Rockville Pike, Bethesda, MD 20892, for biological products.

Subpart G--Drugs for Investigational Use in Laboratory Research Animals or In Vitro Tests

§312.160 Drugs for investigational use in laboratory research animals or in vitro tests.

(a) Authorization to ship. (1)(i) A person may ship a drug intended solely for tests in vitro or in animals used only for laboratory research purposes if it is labeled as follows:

CAUTION: Contains a new drug for investigational use only in laboratory research animals, or for tests in vitro. Not for use in humans.

(ii) A person may ship a biological product for investigational in vitro diagnostic use that is listed in §312.2(b)(2)(ii) if it is labeled as follows:

CAUTION: Contains a biological product for investigational in vitro diagnostic tests only.

(2) A person shipping a drug under paragraph (a) of this section shall use due diligence to assure that the consignee is regularly engaged in conducting such tests and that the shipment of the new drug will actually be used for tests in vitro or in animals used only for laboratory research.

(3) A person who ships a drug under paragraph (a) of this section shall maintain adequate records showing the name and post office address of the expert to whom the drug is shipped and the date, quantity, and batch or code mark of each shipment and delivery. Records of shipments under paragraph (a)(1)(i) of this section are to be maintained for a period of 2 years after the shipment. Records and reports of data and shipments under paragraph (a)(1)(ii) of this section are to be maintained in accordance with §312.57(b). The person who ships the drug shall upon request from any properly authorized officer or employee of the Food and Drug Administration, at reasonable times, permit such officer or employee to have access to and copy and verify records required to be maintained under this section.

(b) Termination of authorization to ship. FDA may terminate authorization to ship a drug under this section if it finds that:

(1) The sponsor of the investigation has failed to comply with any of the conditions for shipment established under this section; or

(2) The continuance of the investigation is unsafe or otherwise contrary to the public interest or the drug is used for purposes other than bona fide scientific investigation. FDA will notify the person shipping the drug of its finding and invite immediate correction. If correction is not immediately made, the person shall have an opportunity for a regulatory hearing before FDA pursuant to part 16.

(c) Disposition of unused drug. The person who ships the drug under paragraph (a) of this section shall assure the return of all unused supplies of the drug from individual investigators whenever the investigation discontinues or the investigation is terminated. The person who ships the drug may authorize in writing alternative disposition of unused supplies of the drug provided this alternative disposition does not expose humans to risks from the drug, either directly or indirectly (e.g., through food-producing animals). The shipper shall maintain records of any alternative disposition.

21 CFR PART 314--APPLICATIONS FOR FDA APPROVAL TO MARKET A NEW DRUG OR AN ANTIBIOTIC DRUG

Subpart A--General Provisions

§314.1 Scope of this part.

(a) This part sets forth procedures and requirements for the submission to, and the review by, the Food and Drug Administration of applications and abbreviated applications, as well as amendments, supplements, and postmarketing reports to them, by persons seeking or holding approval from FDA of the following:

(1) An application or abbreviated application under section 505 of the Federal Food, Drug, and Cosmetic Act to market a new drug.

(2) An application or abbreviated application under section 507 of the Federal Food, Drug, and Cosmetic Act to market an antibiotic drug.

(b) This part does not apply to drug products subject to licensing by FDA under the Public Health Service Act (58 Stat. 632 as amended (42 U.S.C. 201 et seq.)) and subchapter F of chapter I of title 21 of the Code of Federal Regulations.

(c) References in this part to regulations in the Code of Federal Regulations are to chapter I of title 21, unless otherwise noted.

§314.2 Purpose.

The purpose of this part is to establish an efficient and thorough drug review process in order to: (a) Facilitate the approval of drugs shown to be safe and effective; and (b) ensure the disapproval of drugs not shown to be safe and effective. These regulations are also intended to establish an effective system for FDA's surveillance of marketed drugs. These regulations shall be construed in light of these objectives.

§314.3 Definitions.

(a) The definitions and interpretations contained in section 201 of the act apply to those terms when used in this part.

(b) The following definitions of terms apply to this part:

Abbreviated application means the application described under §314.94, including all amendments and supplements to the application. "Abbreviated application" applies to both an abbreviated new drug application and an abbreviated antibiotic application.

Act means the Federal Food, Drug, and Cosmetic Act (sections 201- 901 (21 U.S.C. 301-392)).

Applicant means any person who submits an application or abbreviated application or an amendment or supplement to them under this part to obtain FDA approval of a new drug or an antibiotic drug and any person who owns an approved application or abbreviated application.

Application means the application described under §314.50, including all amendments and supplements to the application.

505(b)(2) Application means an application submitted under section 505(b)(1) of the act for a drug for which the investigations described in section 505(b)(1)(A) of the act and relied upon by the applicant for approval of the application were not conducted by or for the applicant and for which the applicant has not obtained a right of reference or use from the person by or for whom the investigations were conducted.

Approvable letter means a written communication to an applicant from FDA stating that the agency will approve the application or abbreviated application if specific additional information or material is submitted or specific conditions are met. An approvable letter does not constitute approval of any part of an application or abbreviated application and does not permit marketing of the drug that is the subject of the application or abbreviated application.

Approval letter means a written communication to an applicant from FDA approving an application or an abbreviated application.

Drug product means a finished dosage form, for example, tablet, capsule, or solution, that contains a drug substance, generally, but not necessarily, in association with one or more other ingredients.

Drug substance means an active ingredient that is intended to furnish pharmacological activity or other direct effect in the diagnosis, cure, mitigation, treatment, or prevention of disease or to affect the structure or any function of the human body, but does not include intermediates use in the synthesis of such ingredient.

FDA means the Food and Drug Administration.

Listed drug means a new drug product that has an effective approval under section 505(c) of the act for safety and effectiveness or under section 505(j) of the act, which has not been withdrawn or suspended under section 505 (e)(1) through (e)(5) or (j)(5) of the act, and which has not been withdrawn from sale for what FDA has determined are reasons of safety or effectiveness. Listed drug status is evidenced by the drug product's identification as a drug with an effective approval in the current edition of FDA's "Approved Drug Products with Therapeutic Equivalence Evaluations" (the list) or any current supplement thereto, as a drug with an effective approval. A drug product is deemed to be a listed drug on the date of effective approval of the application or abbreviated application for that drug product.

Not approvable letter means a written communication to an applicant from FDA stating that the agency does not consider the application or abbreviated application approvable because one or more deficiencies in the application or abbreviated application preclude the agency from approving it.

Reference listed drug means the listed drug identified by FDA as the drug product upon which an applicant relies in seeking approval of its abbreviated application.

Right of reference or use means the authority to rely upon, and otherwise use, an investigation for the purpose of obtaining approval of an application, including the ability to make available the underlying raw data from the investigation for FDA audit, if necessary.

The list means the list of drug products with effective approvals published in the current edition of FDA's publication "Approved Drug Products with Therapeutic Equivalence Evaluations" and any current supplement to the publication.

Subpart B--Applications

§314.50 Content and format of an application.

Applications and supplements to approved applications are required to be submitted in the form and contain the information, as appropriate for the particular submission, required under this section. Two copies of the application are required, an archival copy and a review copy. An application for a new chemical entity will generally contain an application form, an index, a summary, five or six technical sections, case report tabulations of patient data, case report forms, drug samples, and labeling. Other applications will generally contain only some of those items, and information will be limited to that needed to support the particular submission. These include an application of the type described in section 505(b)(2) of the act, an amendment, and a supplement. The application is required to contain reports of all investigations of the drug product sponsored by the applicant, and all other information about the drug pertinent to an evaluation of the application that is received or otherwise obtained by the applicant from any source. The Food and Drug Administration will maintain guidelines on the format and content of applications to assist applicants in their preparation.

(a) Application form. The applicant shall submit a completed and signed application form that contains the following:

(1) The name and address of the applicant; the date of the application; the application number if previously issued (for example, if the application is a resubmission, an amendment, or a supplement); the name of the drug product, including its established, proprietary, code, and chemical names; the dosage form and strength; the route of administration; the identification numbers of all investigational

new drug applications that are referenced in the application; the identification numbers of all drug master files and other applications under this part that are referenced in the application; and the drug product's proposed indications for use.

(2) A statement whether the submission is an original submission, a 505(b)(2) application, a resubmission, or a supplement to an application under §314.70.

(3) A statement whether the applicant proposes to market the drug product as a prescription or an over-the-counter product.

(4) A check-list identifying what enclosures required under this section the applicant is submitting.

(5) The applicant, or the applicant's attorney, agent, or other authorized official shall sign the application. If the person signing the application does not reside or have a place of business within the United States, the application is required to contain the name and address of, and be countersigned by, an attorney, agent, or other authorized official who resides or maintains a place of business within the United States.

(b) Index. The archival copy of the application is required to contain a comprehensive index by volume number and page number to the summary under paragraph (c) of this section, the technical sections under paragraph (d) of this section, and the supporting information under paragraph (f) of this section.

(c) Summary. (1) An application is required to contain a summary of the application in enough detail that the reader may gain a good general understanding of the data and information in the application, including an understanding of the quantitative aspects of the data. The summary is not required for supplements under §314.70. Resubmissions of an application should contain an updated summary, as appropriate. The summary should discuss all aspects of the application, and synthesize the information into a well-structured and unified document. The summary should be written at approximately the level of detail required for publication in, and meet the editorial standards generally applied by, refereed scientific and medical journals. In addition to the agency personnel reviewing the summary in the context of their review of the application, FDA may furnish the summary to FDA advisory committee members and agency officials whose duties require an understanding of the application. **To the extent possible, data in the summary should be presented in tabular and graphic forms. FDA has prepared a guideline under §10.90(b) that provides information about how to prepare a summary.** The summary required under this paragraph may be used by FDA or the applicant to prepare the Summary Basis of Approval document for public disclosure (under §314.430(e)(2)(ii)) when the application is approved.

(2) The summary is required to contain the following information:

(i) The proposed text of the labeling for the drug, with annotations to the information in the summary and technical sections of the application that support the inclusion of each statement in the labeling, and, if the application is for a prescription drug, statements describing the reasons for omitting a section or subsection of the labeling format in §201. 57.

(ii) A statement identifying the pharmacologic class of the drug and a discussion of the scientific rationale for the drug, its intended use, and the potential clinical benefits of the drug product.

(iii) A brief description of the marketing history, if any, of the drug outside the United States, including a list of the countries in which the drug has been marketed, a list of any countries in which the drug has been withdrawn from marketing for any reason related to safety or effectiveness, and a list of countries in which applications for marketing are pending. The description is required to describe both marketing by the applicant and, if known, the marketing history of other persons.

(iv) A summary of the chemistry, manufacturing, and controls section of the application.

(v) A summary of the nonclinical pharmacology and toxicology section of the application.

(vi) A summary of the human pharmacokinetics and bioavailability section of the application.

(vii) A summary of the microbiology section of the application (for anti-infective drugs only).

(viii) A summary of the clinical data section of the application, including the results of statistical analyses of the clinical trials.

(ix) A concluding discussion that presents the benefit and risk considerations related to the drug, including a discussion of any proposed additional studies or surveillance the applicant intends to conduct postmarketing.

(d) Technical sections. The application is required to contain the technical sections described below. Each technical section is required to contain data and information in sufficient detail to permit the agency to make a knowledgeable judgement about whether to approve the application or whether grounds exist under section 505(d) or 507 of the act to refuse to approve the application. The required technical sections are as follows:

(1) Chemistry, manufacturing, and controls section. A section describing the composition, manufacture, and specification of the drug substance and the drug product, including the following:

(i) Drug substance. A full description of the drug substance including its physical and chemical characteristics and stability; the name and address of its manufacturer; the method of synthesis (or isolation) and purification of the drug substance; **the process controls used during manufacture and packaging;** and such specifications and analytical methods as are necessary to assure the identity, strength, quality, and purity of the drug substance and the bioavailability of the drug products made from the substance, including, for example, specifications relating to stability, sterility, particle size, and crystalline form. **The application may provide additionally for use of alternatives to meet any of these requirements, including alternative sources, process controls, methods, and specifications.** Reference to the current edition of the U.S. Pharmacopeia and the National Formulary may satisfy relevant requirements in this paragraph.

(ii) Drug product. A list of all components used in the manufacture of the drug product (regardless of whether they appear in the drug product); and a statement of the composition of the drug product; a statement of the specifications and analytical methods for each component; the name and address of each manufacturer the drug product; **a description of the manufacturing and packaging procedures and in-process controls for the drug product;** such specifications and analytical methods as are necessary to assure the identity, strength, quality, purity, and bioavailability of the drug product, including, for example, specifications relating to sterility, dissolution rate, containers and closure systems; and stability data with proposed expiration dating. **The application may provide additionally for the use of alternatives to meet any of these requirements, including alternative components, manufacturing and packaging procedures, in-process controls, methods, and specifications.** Reference to the current edition of the U.S. Pharmacopeia and the National Formulary may satisfy relevant requirements in this paragraph.

(iii) Environmental impact. The application is required to contain either a claim for categorical exclusion under §25.24 of this chapter or an environmental assessment under §25.31 of this chapter.

(iv) The applicant may, at its option, submit a complete chemistry, manufacturing, and controls section 90 to 120 days before the anticipated submission of the remainder of the application. FDA will review such early submissions as resources permit.

(2) Nonclinical pharmacology and toxicology section. A section describing, with the aid of graphs and tables, animal and in vitro studies with drug, including the following:

(i) Studies of the pharmacological actions of the drug in relation to its proposed therapeutic indication and studies that otherwise define the pharmacologic properties of the drug or are pertinent to possible adverse effects.

(ii) Studies of the toxicological effects of the drug as they relate to the drug's intended clinical uses, including, as appropriate, studies assessing the drug's acute, subacute, and chronic toxicity; carcinogenicity; and studies of toxicities related to the drug's particular mode of administration or conditions of use.

(iii) Studies, as appropriate, of the effects of the drug on reproduction and on the developing fetus.

(iv) Any studies of the absorption, distribution, metabolism, and excretion of the drug in animals.

(v) For each nonclinical laboratory study subject to the good laboratory practice regulations under part 58 a statement that it was conducted in compliance with the good laboratory practice regulations in part 58, or, if the study was not conducted in compliance with those regulations, a brief statement of the reason for the noncompliance.

(3) Human pharmacokinetics and bioavailability section. A section describing the human pharmacokinetic data and human bioavailability data, or information supporting a waiver of the submission of in vivo bioavailability data under subpart B of part 320, including the following:

(i) A description of each of the bioavailability and pharmacokinetic studies of the drug in humans performed by or on behalf of the applicant that includes a description of the analytical and statistical methods used in each study and a statement with respect to each study that it either was conducted in compliance with the institutional review board regulations in part 56, or was not subject to the regulations under §56.104 or §56.105, and that it was conducted in compliance with the informed consent regulations in part 50.

(ii) If the application describes in the chemistry, manufacturing, and controls section specifications or analytical methods needed to assure the bioavailability of the drug product or drug substance, or both, a statement in this section of the rationale for establishing the specification or analytical methods, including data and information supporting the rationale.

(iii) A summarizing discussion and analysis of the pharmacokinetics and metabolism of the active ingredients and the bioavailability or bioequivalence, or both, of the drug product.

(4) Microbiology section. If the drug is an anti-infective drug, a section describing the microbiology data, including the following:

(i) A description of the biochemical basis of the drug's action on microbial physiology.

(ii) A description of the antimicrobial spectra of the drug, including results of in vitro preclinical studies to demonstrate concentrations of the drug required for effective use.

(iii) A description of any known mechanisms of resistance to the drug, including results of any known epidemiologic studies to demonstrate prevalence of resistance factors.

(iv) A description of clinical microbiology laboratory methods (for example, in vitro sensitivity discs) needed for effective use of the drug.

(5) Clinical data section. A section describing the clinical investigations of the drug, including the following:

(i) A description and analysis of each clinical pharmacology study of the drug, including a brief comparison of the results of the human studies with the animal pharmacology and toxicology data.

(ii) A description and analysis of each controlled clinical study pertinent to a proposed use of the drug, including the protocol and a description of the statistical analyses used to evaluate the study. If the study report is an interim analysis, this is to be noted and a projected completion date provided. Controlled clinical studies that have not been analyzed in detail for any reason (e.g., because they have been discontinued or are incomplete) are to be included in this section, including a copy of the protocol and a brief description of the results and status of the study.

(iii) A description of each uncontrolled clinical study, a summary of the results, and a brief statement explaining why the study is classified as uncontrolled.

(iv) A description and analysis of any other data or information relevant to an evaluation of the safety and effectiveness of the drug product obtained or otherwise received by the applicant from any source, foreign or domestic, including information derived from clinical investigations, including controlled and uncontrolled studies of uses of the drug other than those proposed in the application, commercial marketing experience, reports in the scientific literature, and unpublished scientific papers.

(v) An integrated summary of the data demonstrating substantial evidence of effectiveness for the claimed indications. Evidence is also required to support the dosage and administration section of the labeling, including support for the dosage and dose interval recommended, and modifications for specific subgroups (for example, pediatrics, geriatrics, patients with renal failure).

(vi) A summary and updates of safety information, as follows:

(a) The applicant shall submit an integrated summary of all available information about the safety of the drug product, including pertinent animal data, demonstrated or potential adverse effects of the drug, clinically significant drug/drug interactions, and other safety considerations, such as data from epidemiological studies of related drugs. A description of any statistical analyses performed in analyzing safety data should also be included, unless already included under paragraph (a)(5)(ii) of this section.

(b) The applicant shall, under section 505(i) of the act, update periodically its pending application with new safety information learned about the drug that may reasonably affect the statement of contraindications, warnings, precautions, and adverse reactions in the draft labeling. These "safety update reports" are required to include the same kinds of information (from clinical studies, animal studies, and other sources) and are required to be submitted in the same format as the integrated summary in paragraph (d)(5)(vi)(a) of this section. In addition, the reports are required to include the case report forms for each patient who died during a clinical study or who did not complete the study because of an adverse event (unless this requirement is waived). The applicant shall submit these reports (1) 4 months after the initial submission; (2) following receipt of an approvable letter; and (3) at other times as requested by FDA. Prior to the submission of the first such report, applicants are encouraged to consult with FDA regarding further details on its form and content.

(vii) If the drug has a potential for abuse, a description and analysis of studies or information related to abuse of the drug, including a proposal for scheduling under the Controlled Substances Act. A description of any studies related to overdosage is also required, including information on dialysis, antidotes, or other treatments, if known.

(viii) An integrated summary of the benefits and risks of the drug, including a discussion of why the benefits exceed the risks under the conditions stated in the labeling.

(ix) A statement with respect to each clinical study involving human subjects that it either was conducted in compliance with the institutional review board regulations in part 56, or was not subject to the regulations under §56.104 or §56.105, and that it was conducted in compliance with the informed consent regulations in part 50.

(x) If a sponsor has transferred any obligations for the conduct of any clinical study to a contract research organization, a statement containing the name and address of the contract research organization, identification of the clinical study, and a listing of the obligations transferred. If all obligations governing the conduct of the study have been transferred, a general statement of this transfer -- in lieu of a listing of the specific obligations transferred -- may be submitted.

(xi) If original subject records were audited or reviewed by the sponsor in the course of monitoring any clinical study to verify the accuracy of the case reports submitted to the sponsor, a list identifying each clinical study so audited or reviewed.

(6) Statistical section. A section describing the statistical evaluation of clinical data, including the following:

(i) A copy of the information submitted under paragraph (d)(5)(ii) of this section concerning the description and analysis of each controlled clinical study, and the documentation and supporting statistical analyses used in evaluating the controlled clinical studies.

(ii) A copy of the information submitted under paragraph (d)(5)(vi)(a) of this section concerning a summary of information about the safety of the drug product, and the documentation and supporting statistical analyses used in evaluating the safety information.

(e) Samples and labeling. (1) Upon request from FDA, the applicant shall submit the samples described below to the places identified in the agency's request. FDA will generally ask applicants to submit samples directly to two or more agency laboratories that will perform all necessary tests on the samples and validate the applicant's analytical methods.

(i) Four representative samples of the following, each sample in sufficient quantity to permit FDA to perform three times each test described in the application to determine whether the drug substance and the drug product meet the specifications given in the application:

(a) The drug product proposed for marketing;

(b) The drug substance used in the drug product from which the samples of the drug product were taken; and

(c) Reference standards and blanks (except that reference standards recognized in an official compendium need not be submitted).

(ii) Samples of the finished market package, if requested by FDA.

(2) The applicant shall submit the following in the archival copy of the application:

(i) Three copies of the analytical methods and related descriptive information contained in the chemistry, manufacturing, and controls section under paragraph (d)(1) of this section for the drug substance and the drug product that are necessary for FDA's laboratories to perform all necessary tests on the samples and to validate the applicant's analytical methods. The related descriptive information includes a description of each sample; the proposed regulatory specifications for the drug; a detailed description of the methods of analysis; supporting data for accuracy, specificity, precision and ruggedness; and complete results of the applicant's tests on each sample.

(ii) Copies of the label and all labeling for the drug product (4 copies of draft labeling or 12 copies of final printed labeling).

(f) Case report forms and tabulations. The archival copy of the application is required to contain the following case report tabulations and case report forms:

(1) Case report tabulations. The application is required to contain tabulations of the data from each adequate and well-controlled study under §314.126 (Phase 2 and Phase 3 studies as described in §§312.21 (b) and (c) of this chapter), tabulations of the data from the earliest clinical pharmacology studies (Phase 1 studies as described in §312.21(a) of this chapter), and tabulations of the safety data from other clinical studies. Routine submission of other patient data from uncontrolled studies is not required. The tabulations are required to include the data on each patient in each study, except that the applicant may delete those tabulations which the agency agrees, in advance, are not pertinent to a review of the drug's safety or effectiveness. Upon request, FDA will discuss with the applicant in a "pre-NDA" conference those tabulations that may be appropriate for such deletion. Barring unforeseen circumstances, tabulations agreed to be deleted at such a conference will not be requested during the conduct of FDA's review of the application. If such unforeseen circumstances do occur, any request for deleted tabulations will be made by the director of the FDA division responsible for reviewing the application, in accordance with paragraph (f)(3) of this section.

(2) Case report forms. The application is required to contain copies of individual case report forms for each patient who died during a clinical study or who did not complete the study because of an adverse event, whether believed to be drug related or not, including patients receiving reference drugs or placebo. This requirement may be waived by FDA for specific studies if the case report forms are unnecessary for a proper review of the study.

(3) Additional data. The applicant shall submit to FDA additional case report forms and tabulations needed to conduct a proper review of the application, as requested by the director of the FDA division responsible for reviewing the application. The applicant's failure to submit information requested by FDA within 30 days after receipt of the request may result in the agency viewing any eventual submission as a major amendment

under §314.60 and extending the review period as necessary. If desired by the applicant, the FDA division director will verify in writing any request for additional data that was made orally.

(4) Applicants are invited to meet with FDA before submitting an application to discuss the presentation and format of supporting information. If the applicant and FDA agree, the applicant may submit tabulations of patient data and case report forms in a form other than hard copy, for example, on microfiche or computer tapes.

(g) Other. The following general requirements apply to the submission of information within the summary under paragraph (c) of this section and within the technical sections under paragraph (d) of this section.

(1) The applicant ordinarily is not required to resubmit information previously submitted, but may incorporate the information by reference. A reference to information submitted previously is required to identify the file by name, reference number, volume, and page number in the agency's records where the information can be found. A reference to information submitted to the agency by a person other than the applicant is required to contain a written statement that authorizes the reference and that is signed by the person who submitted the information.

(2) The applicant shall submit an accurate and complete English translation of each part of the application that is not in English. The applicant shall submit a copy of each original literature publication for which an English translation is submitted.

(3) If an applicant who submits a new drug application under section 505(b) of the act obtains a "right of reference or use," as defined under §314.3(b), to an investigation described in clause (A) of section 505(b)(1) of the act, the applicant shall include in its application a written statement signed by the owner of the data from each such investigation that the applicant may rely on in support of the approval of its application, and provide FDA access to, the underlying raw data that provide the basis for the report of the investigation submitted in its application.

(h) Format of an original application. (1) The applicant shall submit a complete archival copy of the application that contains the information required under paragraphs (a) through (f) of this section. FDA will maintain the archival copy during the review of the application to permit individual reviewers to refer to information that is not contained in their particular technical sections of the application, to give other agency personnel access to the application for official business, and to maintain in one place a complete copy of the application. An applicant may submit on microfiche the portions of the archival copy of the application described in paragraphs (b) through (d) of this section. Information relating to samples and labeling, described in paragraph (e) of this section, is required to be submitted in hard copy. Tabulations of patient data and case report forms, described in paragraph (f) of this section, may be submitted on microfiche only if the applicant and FDA agree. If FDA agrees, the applicant may use another suitable microform system.

(2) The applicant shall submit a review copy of the application. Each of the technical sections (described in paragraphs (d) (1) through (6) of this section) in the review copy is required to be separately bound with a copy of the application form required under paragraph (a) of this section and a copy of the summary required under paragraph (c) of this section. The applicant may obtain from FDA sufficient folders to bind the archival and review copies of the application.

§314.54 Procedure for submission of an application requiring investigations for approval of a new indication for, or other change from, a listed drug.

(a) The act does not permit approval of an abbreviated new drug application for a new indication, nor does it permit approval of other changes in a listed drug if investigations, other than bioavailability or bioequivalence studies, are essential to the approval of the change. Any person seeking approval of a drug product that represents a modification of a listed drug (e.g., a new indication or new dosage form) and for which investigations, other than bioavailability or bioequivalence studies, are essential to the approval of the changes may, except as provided in paragraph (b) of this section, submit a 505(b)(2) application. This application need contain only that information needed to support the modification(s) of the listed drug.

(1) The applicant shall submit a complete archival copy of the application that contains the following:

(i) The information required under §314.50 (a), (b), (c), (d)(1) and (d)(3), (e), and (g).

(ii) The information required under §314.50 (d)(2), (d)(4) (if an anti-infective drug), (d) (5), (d)(6), and (f) as needed to support the safety and effectiveness of the drug product.

(iii) Identification of the listed drug for which FDA has made a finding of safety and effectiveness and on which finding the applicant relies in seeking approval of its proposed drug product by established name, if any, proprietary name, dosage form, strength, route of administration, name of listed drug's application holder, and listed drug's approved application number.

(iv) If the applicant is seeking approval only for a new indication and not for the indications approved for the listed drug on which the applicant relies, a certification so stating.

(v) Any patent information required under section 505(b)(1) of the act with respect to any patent which claims the drug for which approval is sought or a method of using such drug and to which a claim of patent infringement could reasonably be asserted if a person not licensed by the owner of the patent engaged in the manufacture, use, or sale of the drug product.

(vi) Any patent certification or statement required under section 505(b)(2) of the act with respect to any relevant patents that claim the listed drug or that claim any other drugs on which investigations relied on by the applicant for approval of the application were conducted, or that claim a use for the listed or other drug.

(2) The applicant shall submit a review copy that contains the technical sections described in §314.50(d)(1) and (d)(3), and the technical sections described in §314.50(d)(2), (d)(4), (d)(5), (d)(6), and (f) when needed to support the modification. Each of the technical sections in the review copy is required to be separately bound with a copy of the information required under §314.50 (a), (b), and (c) and a copy of the proposed labeling.

(3) The information required by §314.50 (d)(2), (d)(4) (if an anti-infective drug), (d) (5), (d)(6), and (f) for the listed drug on which the applicant relies shall be satisfied by reference to the listed drug under paragraph (a)(1)(iii) of this section.

(b) An application may not be submitted under this section for a drug product whose only difference from the reference listed drug is that:

(1) The extent to which its active ingredient(s) is absorbed or otherwise made available to the site of action is less than that of the reference listed drug; or

(2) The rate at which its active ingredient(s) is absorbed or otherwise made available to the site of action is unintentionally less than that of the reference listed drug.

§314.60 Amendments to an unapproved application.

(a) Except as provided in paragraph (b) of this section, the applicant may submit an amendment to an application that is filed under §314.100, but not yet approved. The submission of a major amendment (for example, an amendment that contains significant new data from a previously unreported study or detailed new analyses of previously submitted data), whether on the applicant's own initiative or at the invitation of the agency, constitutes an agreement by the applicant under section 505(c) of the act to extend the date by which the agency is required to reach a decision on the application. Ordinarily, the agency will extend the review period for a major amendment but only for the time necessary to review the new information. However, the agency may not extend the review period more than 180 days. If the agency extends the review period for the application, the director of the division responsible for reviewing the application will notify the applicant of the length of the extension. The submission of an amendment that is not a major amendment will not extend the review period.

(b)(1) An unapproved application may not be amended if all of the following conditions apply:

(i) The unapproved application is for a drug for which a previous application has been approved and granted a period of exclusivity in accordance with section 505(c)(3)(D)(ii) of the act that has not expired;

(ii) The applicant seeks to amend the unapproved application to include a published report of an investigation that was conducted or sponsored by the applicant entitled to exclusivity for the drug;

(iii) The applicant has not obtained a right of reference to the investigation described in paragraph (b)(1)(ii) of this section; and

(iv) The report of the investigation described in paragraph (b)(1)(ii) of this section would be essential to the approval of the unapproved application.

(2) The submission of an amendment described in paragraph (b)(1) of this section will cause the unapproved application to be deemed to be withdrawn by the applicant under §314.65 on the date of receipt by FDA of the amendment. The amendment will be considered a resubmission of the application, which may not be accepted except as provided in accordance with section 505(c)(3)(D)(ii) of the act.

§314.65 Withdrawal by the applicant of an unapproved application.

An applicant may at any time withdraw an application that is not yet approved by notifying the Food and Drug Administration in writing. The agency will consider an applicant's failure to respond within 10 days to an approvable letter under §314.110 or a not approvable letter under §314.120 to be a request by the applicant to withdraw the application. A decision to withdraw the application is without prejudice to refiling. The agency will retain the application and will provide a copy to the applicant on request under the fee schedule in §20.42 of FDA's public information regulations.

§314.70 Supplements and other changes to an approved application.

(a) Changes to an approved application. The applicant shall notify the Food and Drug Administration about each change in each condition established in an approved application beyond the variations already provided for in the application. The notice is required to describe the change fully. Depending on the type of change, the applicant shall notify FDA about it in a supplemental application under paragraph (b) or (c) of this section or by inclusion of the information in the annual report to the application under paragraph (d) of this section. Notwithstanding the requirements of paragraphs (b) and (c) of this section, an applicant shall make a change provided for in those paragraphs (for example, the deletion of an ingredient common to many drug products) in accordance with a guideline, notice, or regulation published in the FEDERAL REGISTER that provides for a less burdensome notification of the change (for example, by notification at the time a supplement is submitted or in the next annual report).

(b) Supplements requiring FDA approval before the change is made. An applicant shall submit a supplement, and obtain FDA approval of it, before making the changes listed below in the conditions in an approved application, unless the change is made to comply with an official compendium. An applicant may ask FDA to expedite its review of a supplement if a delay in making the change described in it would impose an extraordinary hardship on the applicant. Such a supplement and its mailing cover should be plainly marked: "Supplement-Expedited Review Requested."

(1) Drug substance. A change affecting the drug substance to accomplish any of the following:

(i) To relax the limits for a specification;

(ii) To establish a new regulatory analytical method;

(iii) To delete a specification or regulatory analytical method;

(iv) To change the synthesis of the drug substance, including a change in solvents and a change in the route of synthesis.

(v) To use a different facility or establishment to manufacture the drug substance, where: (a) the manufacturing process in the new facility or establishment differs materially from that in the former facility or establishment, or (b) the new facility or establishment has not received a satisfactory current good manufacturing practice (CGMP) inspection within the previous 2 years covering that manufacturing process.

(2) Drug product. A change affecting the drug product to accomplish any of the following:

(i) To add or delete an ingredient, or otherwise to change the composition of the drug product, other than deletion of an ingredient intended only to affect the color of the drug product;

(ii) To relax the limits for a specification;

(iii) To establish a new regulatory analytical method;

(iv) To delete a specification or regulatory analytical method;

(v) To change the method of manufacture of the drug product, including changing or relaxing an in-process control;

(vi) To use a different facility or establishment, including a different contract laboratory or labeler, to manufacture, process, or pack the drug product;

(vii) To change the container and closure system for the drug product (for example, glass to high density polyethylene (HDPE), or HDPE to polyvinyl chloride) or change a specification or regulatory analytical method for the container and closure system;

(viii) To change the size of the container, except for solid dosage forms, without a change in the container and closure system.

(ix) To extend the expiration date of the drug product based on data obtained under a new or revised stability testing protocol that has not been approved in the application.

(x) To establish a new procedure for reprocessing a batch of the drug product that fails to meet specifications.

(3) Labeling. Any change in labeling, except one described in paragraph (c)(2) or (d) of this section.

(c) Supplements for changes that may be made before FDA approval. An applicant shall submit a supplement at the time the applicant makes any kind of change listed below in the conditions in an approved application, unless the change is made to comply with an official compendium. A supplement under this paragraph is required to give a full explanation of the basis for the change, identify the date on which the change is made, and, if the change concerns labeling, include 12 copies of final printed labeling. The applicant shall promptly revise all promotional labeling and drug advertising to make it consistent with any change in the labeling. The supplement and its mailing cover should be plainly marked: "Special Supplement-Changes Being Effected."

(1) Adds a new specification or test method or changes in the methods, facilities (except a change to a new facility), or controls to provide increased assurance that the drug will have the characteristics of identity, strength, quality, and purity which it purports or is represented to possess;

(2) Changes labeling to accomplish any of the following:

(i) To add or strengthen a contraindication, warning, precaution, or adverse reaction;

(ii) To add or strengthen a statement about drug abuse, dependence, or overdosage; or

(iii) To add or strengthen an instruction about dosage and administration that is intended to increase the safe use of the product.

(iv) To delete false, misleading, or unsupported indications for use or claims for effectiveness.losure system.

(ix) To extend the expiration date of the drug product based on data obtained under a new or revised stability testing protocol that has not been approved in the application.

(x) To establish a new procedure for reprocessing a batch of the drug product that fails to meet specifications.

(3) Labeling. Any change in labeling, except one described in paragraph (c)(2) or (d) of this section.

(c) Supplements for changes that may be made before FDA approval. An applicant shall submit a supplement at the time the applicant makes any kind of change listed below in the conditions in an approved application, unless the change is made to comply with an official compendium. A supplement under this paragraph is required to give a full explanation of the basis for the change, identify the date on which the change is made, and, if the change concerns labeling, include 12 copies of final printed labeling. The applicant shall promptly revise all promotional labeling and drug advertising to make it consistent with any change in the labeling. The supplement and its mailing cover should be plainly marked: "Special Supplement-Changes Being Effected."

(1) Adds a new specification or test method or changes in the methods, facilities (except a change to a new facility), or controls to provide increased assurance that the drug will have the characteristics of identity, strength, quality, and purity which it purports or is represented to possess;

(2) Changes labeling to accomplish any of the following:

(i) To add or strengthen a contraindication, warning, precaution, or adverse reaction;

(ii) To add or strengthen a statement about drug abuse, dependence, or overdosage; or

(iii) To add or strengthen an instruction about dosage and administration that is intended to increase the safe use of the product.

(iv) To delete false, misleading, or unsupported indications for use or claims for effectiveness.

(3) To use a different facility or establishment to manufacture the drug substance, where: (i) The manufacturing process in the new facility or establishment does not differ materially from that in the former facility or establishment, and (ii) the new facility or establishment has received a satisfactory current good manufacturing practice (CGMP) inspection within the previous 2 years covering that manufacturing process.

(d) Changes described in the annual report. An applicant shall not submit a supplement to make any change in the conditions in an approved application, unless otherwise required under paragraph (b) or (c) of this section, but shall describe the change in the next annual report required under §314.81. Some examples of changes that can be described in the annual report are the following:

(1) Any change made to comply with an official compendium.

(2) A change in the labeling concerning the description of the drug product or in the information about how the drug product is supplied, that does not involve a change in the dosage strength or dosage form.

(3) An editorial or similar minor change in labeling.

(4) The deletion of an ingredient intended only to affect the color of the drug product.

(5) An extension of the expiration date based upon full shelf-life data obtained from a protocol approved in the application.

(6) A change within the container and closure system for the drug product (for example, a change from one high density polyethylene (HDPE) to another HDPE), except a change in container size for nonsolid dosage forms, based upon a showing of equivalency to the approved system under a protocol approved in the application or published in an official compendium.

(7) The addition or deletion of an alternate analytical method.

(8) A change in the size of a container for a solid dosage form, without a change from one container and closure system to another.

(e) Patent information. The applicant shall comply with the patent information requirements under section 505(c)(2) of the act.

§314.71 Procedures for submission of a supplement to an approved application.

(a) Only the applicant may submit a supplement to an application.

(b) All procedures and actions that apply to an application under §314.50 also apply to supplements, except that the information required in the supplement is limited to that needed to support the change. A supplement is required to contain an archival copy and a review copy that include an application form and appropriate technical sections, samples, and labeling.

(c) All procedures and actions that apply to applications under this part, including actions by applicants and the Food and Drug Administration, also apply to supplements.

§314.72 Change in ownership of an application.

(a) An applicant may transfer ownership of

its application. At the time of transfer the new and former owners are required to submit information to the Food and Drug Administration as follows:

(1) The former owner shall submit a letter or other document that states that all rights to the application have been transferred to the new owner.

(2) The new owner shall submit an application form signed by the new owner and a letter or other document containing the following:

(i) The new owner's commitment to agreements, promises, and conditions made by the former owner and contained in the application;

(iii) The date that the change in ownership is effective; and

(iii) Either a statement that the new owner has a complete copy of the approved application, including supplements and records that are required to be kept under §314.81, or a request for a copy of the application from FDA's files. FDA will provide a copy of the application to the new owner under the fee schedule in §20.42 of FDA's public information regulations.

(b) The new owner shall advise FDA about any change in the conditions in the approved application under §314.70, except the new owner may advise FDA in the next annual report about a change in the drug product's label or labeling to change the product's brand or the name of its manufacturer, packer, or distributor.

§314.80 Postmarketing reporting of adverse drug experiences.

(a) Definitions. The following definitions of terms apply to this section:

Adverse drug experience means any adverse event associated with the use of a drug in humans, whether or not considered drug related, including the following: an adverse event occurring in the course of the use of a drug product in professional practice; an adverse event occurring from drug overdose, whether accidental or intentional; an adverse event occurring from drug abuse; an adverse event occurring from drug withdrawal; and any failure of expected pharmacological action.

Increased frequency means an increase in the rate of occurrence of a particular adverse drug experience, e.g., an increased number of reports of a particular adverse drug experience after appropriate adjustment for drug exposure.

Serious means an adverse drug experience that is fatal or life-threatening, is permanently disabling, requires inpatient hospitalization, or is a congenital anomaly, cancer, or overdose.

Unexpected means an adverse drug experience that is not listed in the current labeling for the drug and includes an event that may be symptomatically and pathophysiologically related to an event listed in the labeling, but differs from the event because of greater severity or specificity. For example, under this definition, hepatic necrosis would be unexpected (by virtue of greater severity) if the labeling only referred to elevated hepatic enzymes or hepatitis. Similarly, cerebral thromboembolism and cerebral vasculitis would be unexpected (by virtue of greater specificity) if the labeling only listed cerebral vascular accidents.

(b) Review of adverse drug experiences. Each applicant having an approved application under §314.50 or, in the case of a 505(b)(2) application, an effective approved application, shall promptly review all adverse drug experience information obtained or otherwise received by the applicant from any source, foreign or domestic, including information derived from commercial marketing experience, postmarketing clinical investigations, postmarketing epidemiological/surveillance studies, reports in the scientific literature, and unpublished scientific papers.

(c) Reporting requirements. The applicant shall report to FDA adverse drug experience information, as described in this section. The applicant shall submit two copies of each report described in this section to the Central Document Room, Park Bldg., Rm. 214, 12420 Parklawn Dr., Rockville, MD 20852. FDA may waive the requirement for the second copy in appropriate instances.

(1) Fifteen-day "Alert reports." (i) The applicant shall report each adverse drug expe-

rience that is both serious and unexpected, regardless of source, as soon as possible but in any case within 15 working days of initial receipt of the information. These reports are required to be submitted on Form FDA-1639 (Adverse Reaction Report). The applicant shall promptly investigate all adverse drug experiences that are the subject of these 15day Alert reports and shall submit followup reports within 15 working days of receipt of new information or as requested by FDA. If additional information is not obtainable, a followup report may be required that describes briefly the steps taken to seek additional information and the reasons why it could not be obtained. These 15-day Alert reports and followups to them are required to be submitted under separate cover and may not be included, except for summary or tabular purposes, in a periodic report.

(ii) The applicant shall review periodically (at least as often as the periodic reporting cycle) the frequency of reports of adverse drug experiences that are both serious and expected and reports of therapeutic failure (lack of effect), regardless of source, and report any significant increase in frequency as soon as possible but in any case within 15 working days of determining that a significant increase in frequency exists. Upon written notice, FDA may require that applicants review the frequency of reports of serious, expected adverse drug experiences at intervals different than the periodic reporting cycle. Reports of a significant increase in frequency are required to be submitted in narrative form (including the time period on which the increased frequency is based, the method of analysis, and the interpretation of the results), rather than using Form FDA-1639. Fifteen-day Alert reports based on increased frequency are required to be submitted under separate cover and may not be included, except for summary purposes, in a periodic report.

(iii) The requirements of paragraphs (c)(1) (i) and (ii) of this section, concerning the submission of 15-day alert reports, shall also apply to any person (other than the applicant) whose name appears on the label of an approved drug product as a manufacturer, packer, or distributor. However, in order to avoid unnecessary duplication in the submission to FDA, and followup to, reports required by paragraph (c)(1) (i) and (ii) of this section, obligations of a nonapplicant may be met by submission of all reports of serious adverse drug experiences to the applicant. If a nonapplicant elects to submit adverse drug experience reports to the applicant rather than to FDA, it shall submit each report to the applicant within 3 working days of its receipt by the nonapplicant, and the applicant shall then comply with the requirements of this section. Under this circumstance, the nonapplicant shall maintain a record of this action which shall include:

(a) A copy of the drug experience report.

(b) Date the report was received by the nonapplicant.

(c) Date the report was submitted to the applicant.

(d) Name and address of the applicant.

(iv) Each report submitted under this paragraph shall bear prominent identification as to its contents, i.e., "15-day Alert report" or "15-day Alert report-followup."

(2) Periodic adverse drug experience reports. (i) The applicant shall report each adverse drug experience not reported under paragraph (c)(1)(i) of this section at quarterly intervals, for 3 years from the date of approval of the application, and then at annual intervals. The applicant shall submit each quarterly report within 30 days of the close of the quarter (the first quarter beginning on the date of approval of the application) and each annual report within 60 days of the anniversary date of approval of the application. Upon written notice, FDA may extend or reestablish the requirement that an applicant submit quarterly reports, or require that the applicant submit reports under this section at different times than those stated. For example, the agency may reestablish a quarterly reporting requirement following the approval of a major supplement. Followup information to adverse drug experiences submitted in a periodic report may be submitted in the next periodic report.

(ii) Each periodic report is required to contain: (a) a narrative summary and analysis of the information in the report and an analysis of the 15-day Alert reports submitted during the reporting interval (all 15-day Alert reports being appropriately referenced by the applicant's patient identification number, adverse reaction term(s), and date of submission to FDA); (b) a Form FDA-1639 (Adverse Reaction Report) for each adverse drug experience not reported under paragraph (c)(1)(i) of this section (with an index consisting of a line listing of the applicant's patient identification number and adverse reaction term(s)); and (c) a history of actions taken since the last report because of adverse drug experiences (for example, labeling changes or studies initiated).

(iii) Periodic reporting, except for information regarding 15-day Alert reports, does not apply to adverse drug experience information obtained from postmarketing studies (whether or not conducted under an investigational new drug application), from reports in the scientific literature, and from foreign marketing experience.

(d) Scientific literature. (1) A 15-day Alert report based on information from the scientific literature is required to be accompanied by a copy of the published article. The 15-day reporting requirements in paragraph (c)(1)(i) of this section (i.e., serious, unexpected adverse drug experiences) apply only to reports found in scientific and medical journals either as case reports or as the result of a formal clinical trial. The 15-day reporting requirements in paragraph (c)(1)(ii) of this section (i.e., a significant increase in frequency of a serious, expected adverse drug experience or of a therapeutic failure) apply only to reports found in scientific and medical journals either as the result of a formal clinical trial, or from epidemiological studies or analyses of experience in a monitored series of patients.

(2) As with all reports submitted under paragraph (c)(1)(i) of this section, reports based on the scientific literature shall be submitted on Form FDA-1639 or comparable format as prescribed by paragraph (f) of this section. In cases where the applicant believes that preparing the Form FDA-1639 constitutes an undue hardship, the applicant may arrange with the Division of Epidemiology and Surveillance for an acceptable alternative reporting format.

(e) Postmarketing studies. (1) An applicant is not required to submit a 15-day Alert report under paragraph (c) of this section for an adverse drug experience obtained from a postmarketing study (whether or not conducted under an investigational new drug application) unless the applicant concludes that there is a reasonable possibility that the drug caused the adverse experience.

(2) The applicant shall separate and clearly mark reports of adverse drug experiences that occur during a postmarketing study as being distinct from those experiences that are being reported spontaneously to the applicant.

(f) Reporting Form FDA-1639. (1) Except as provided in paragraphs (c)(1)(ii) and (f)(3) of this section, the applicant shall complete a Form FDA-1639 (Adverse Reaction Report) for each report of an adverse drug experience.

(2) Each completed Form FDA-1639 should refer only to an individual patient or a single attached publication.

(3) Instead of using Form FDA-1639, an applicant may use a computer- generated FDA-1639 or other alternative format (e.g., a computer-generated tape or tabular listing) provided that: (i) The content of the alternative format is equivalent in all elements of information to those specified in Form FDA-1639; and (ii) the format is agreed to in advance by the Division of Epidemiology and Surveillance (HFD-730).

(4) Single copies of Form FDA-1639 may be obtained from the Division of Epidemiology and Surveillance (HFD-730), Center for Drug Evaluation and Research, Food and Drug Administration, 5600 Fishers Lane, Rockville, MD 20857. Supplies of Form FDA-1639 may be obtained from the PHS Forms and Publications Distribution Center, 12100 Parklawn Dr., Rockville, MD 20857.

(g) Multiple reports. An applicant should not include in reports under this section any adverse drug experiences that occurred in clinical trials if they were previously submitted as part of the approved application. If a report applies to a drug for which an applicant holds more than one approved application, the applicant should submit the report to the application that was first approved. If a report refers to more than one drug marketed by an applicant, the applicant should submit the report to the application for the drug listed first in the report.

(h) Patient privacy. An applicant should not include in reports under this section the names and addresses of individual patients; instead, the applicant should assign a unique code number to each report, preferably not more than eight characters in length. The applicant should include the name of the reporter from whom the information was received. Names of patients, health care professionals, hospitals, and geographical identifiers in adverse drug experience reports are not releasable to the public under FDA's public information regulations in Part 20.

(i) Recordkeeping. The applicant shall maintain for a period of 10 years records of all adverse drug experiences known to the applicant, including raw data and any correspondence relating to adverse drug experiences.

(j) Guideline. FDA has prepared under §10.90(b) a guideline for the submission of reports of adverse drug experiences and suggested followup investigation of reports.

(k) Withdrawal of approval. If an applicant fails to establish and maintain records and make reports required under this section, FDA may withdraw approval of the application and, thus, prohibit continued marketing of the drug product that is the subject of the application.

(l) Disclaimer. A report or information submitted by an applicant under this section (and any release by FDA of that report or information) does not necessarily reflect a conclusion by the applicant or FDA that the report or information constitutes an admission that the drug caused or contributed to an adverse effect. An applicant need not admit, and may deny, that the report or information submitted under this section constitutes an admission that the drug caused or contributed to an adverse effect. For purposes of this provision, the term "applicant" also includes any person reporting under paragraph (c)(1)(iii) of this section.

§314.81 Other postmarketing reports.

(a) Applicability. Each applicant shall make the reports for each of its approved applications and abbreviated applications required under this section and sections 505(k) and 507(g) of the act.

(b) Reporting requirements. The applicant shall submit to the Food and Drug Administration at the specified times two copies of the following reports:

(1) NDA-Field alert report. The applicant shall submit information of the following kinds about distributed drug products and articles to the FDA district office that is responsible for the facility involved within 3 working days of receipt by the applicant. The information may be provided by telephone or other rapid communication means, with prompt written followup. The report and its mailing cover should be plainly marked: "NDA-Field Alert Report."

(i) Information concerning any incident that causes the drug product or its labeling to be mistaken for, or applied to, another article.

(ii) Information concerning any bacteriological contamination, or any significant chemical, physical, or other change or deterioration in the distributed drug product, or any failure of one or more distributed batches of the drug product to meet the specifications established for it in the application.

(2) Annual report. The applicant shall submit the following information in the order listed each year within 60 days of the anniversary date of approval of the application. The applicant shall submit the report to the FDA division responsible for reviewing the application. Each annual report is required to be accompanied by a completed transmittal Form FDA-2252 (Transmittal of Periodic Reports for Drugs for Human Use) which may be obtained from the PHS Forms and Publications Distribution Center, 12100 Parklawn Dr., Rockville, MD 20857, and is required to include all the information required under this

section that the applicant received or otherwise obtained during the annual reporting interval which ends on the anniversary date. The report is required to contain the following:

(i) Summary. A brief summary of significant new information from the previous year that might affect the safety, effectiveness, or labeling of the drug product. The report is also required to contain a brief description of actions the applicant has taken or intends to take as a result of this new information, for example, submit a labeling supplement, add a warning to the labeling, or initiate a new study.

(ii) Distribution data. Information about the quantity of the drug product distributed under the approved application, including that distributed to distributors. The information is required to include the National Drug Code (NDC) number, the total number of dosage units of each strength or potency distributed (e.g., 100,000/5 milligram tablets, 50,000/10 milliliter vials), and the quantities distributed for domestic use and the quantities distributed for foreign use. Disclosure of financial or pricing data is not required.

(iii) Labeling. Currently used professional labeling, patient brochures or package inserts (if any), a representative sample of the package labels, and a summary of any changes in labeling that have been made since the last report listed by date in the order in which they were implemented, or if no changes, a statement of that fact.

(iv) Chemistry, manufacturing, and controls changes. (a) Reports of experiences, investigations, studies, or tests involving chemical or physical properties, or any other properties of the drug (such as the drug's behavior or properties in relation to microorganisms, including both the effects of the drug on microorganisms and the effects of microorganisms on the drug). These reports are only required for new information that may affect FDA's previous conclusions about the safety or effectiveness of the drug product.

(b) A full description of the manufacturing and controls changes not requiring a supplemental application under §314.70 (b) and (c), listed by date in the order in which they were implemented.

(v) Nonclinical laboratory studies. Copies of unpublished reports and summaries of published reports of new toxicological findings in animal studies and in vitro studies (e.g., mutagenicity) conducted by, or otherwise obtained by, the applicant concerning the ingredients in the drug product. The applicant shall submit a copy of a published report if requested by FDA.

(vi) Clinical data. (a) Published clinical trials of the drug (or abstracts of them), including clinical trials on safety and effectiveness; clinical trials on new uses; biopharmaceutic, pharmacokinetic, and clinical pharmacology studies; and reports of clinical experience pertinent to safety (for example, epidemiologic studies or analyses of experience in a monitored series of patients) conducted by or otherwise obtained by the applicant. Review articles, papers describing the use of the drug product in medical practice, papers and abstracts in which the drug is used as a research tool, promotional articles, press clippings, and papers that do not contain tabulations or summaries of original data should not be reported.

(b) Summaries of completed unpublished clinical trials, or prepublication manuscripts if available, conducted by, or otherwise obtained by, the applicant. Supporting information should not be reported. (A study is considered completed 1 year after it is concluded.)

(vii) Status reports. A statement on the current status of any postmarketing studies performed by, or on behalf of, the applicant. To facilitate communications between FDA and the applicant, the report may, at the applicant's discretion, also contain a list of any open regulatory business with FDA concerning the drug product subject to the application.

(3) Other reporting-(i) Advertisements and promotional labeling. The applicant shall submit specimens of mailing pieces and any other labeling or advertising devised for promotion of the drug product at the time of initial dissemination of the labeling and at the time of initial publication of the advertisement

for a prescription drug product. Mailing pieces and labeling that are designed to contain samples of a drug product are required to be complete, except the sample of the drug product may be omitted. Each submission is required to be accompanied by a completed transmittal Form FDA-2253 (Transmittal of Advertisements and Promotional Labeling for Drugs for Human Use) and is required to include a copy of the product's current professional labeling. Form FDA-2253 may be obtained from the PHS Forms and Publications Distribution Center, 12100 Parklawn Dr., Rockville, MD 20857.

(ii) Special reports. Upon written request the agency may require that the applicant submit the reports under this section at different times than those stated.

(iii) Withdrawal of approved drug product from sale. (a) The applicant shall submit on Form FDA 2657 (Drug Product Listing), within 15 working days of the withdrawal from sale of a drug product, the following information:

(1) The National Drug Code (NDC) number.

(2) The identity of the drug product by established name and by proprietary name.

(3) The new drug application or abbreviated application number.

(4) The date of withdrawal from sale. It is requested but not required that the reason for withdrawal of the drug product from sale be included with the information.

(b) The applicant shall submit each Form FDA- 2657 to the Drug Listing Branch (HFD-334), Center for Drug Evaluation and Research, Food and Drug Administration, 5600 Fishers Lane, Rockville, MD 20857.0857.

(c) Reporting under paragraph (b)(3)(iii) of this section constitutes compliance with the requirements under §207.30(a) of this chapter to report "at the discretion of the registrant when the change occurs."

(c) General requirements-(1) Multiple applications. For all reports required by this section, the applicant shall submit the information common to more than one application only to the application first approved, and shall not report separately on each application. The submission is required to identify all the applications to which the report applies.

(2) Patient identification. Applicants should not include in reports under this section the names and addresses of individual patients; instead, the applicant should code the patient names whenever possible and retain the code in the applicant's files. The applicant shall maintain sufficient patient identification information to permit FDA, by using that information alone or along with records maintained by the investigator of a study, to identify the name and address of individual patients; this will ordinarily occur only when the agency needs to investigate the reports further or when there is reason to believe that the reports do not represent actual results obtained.

(d) Withdrawal of approval. If an applicant fails to make reports required under this section, FDA may withdraw approval of the application and, thus, prohibit continued marketing of the drug product that is the subject of the application.

§314.90 Waivers.

(a) An applicant may ask the Food and Drug Administration to waive under this section any requirement that applies to the applicant under §§314.50 through 314.81. An applicant may ask FDA to waive under §314. 126(c) any criteria of an adequate and well- controlled study described in §314.126(b). A waiver request under this section is required to be submitted with supporting documentation in an application, or in an amendment or supplement to an application. The waiver request is required to contain one of the following:

(1) An explanation why the applicant's compliance with the requirement is unnecessary or cannot be achieved;

(2) A description of an alternative submission that satisfies the purpose of the requirement; or

(3) Other information justifying a waiver.

(b) FDA may grant a waiver if it finds one of the following:

(1) The applicant's compliance with the requirement is unnecessary for the agency to evaluate the application or compliance cannot be achieved;

(2) The applicant's alternative submission

satisfies the requirement; or

(3) The applicant's submission otherwise justifies a waiver.

Subpart C--Abbreviated Applications

§314.92 Drug products for which abbreviated applications may be submitted.

(a) Abbreviated applications are suitable for the following drug products within the limits set forth under §314.93:

(1) Drug products that are the same as a listed drug. A "listed drug" is defined in §314. 3. For determining the suitability of an abbreviated new drug application, the term "same as" means identical in active ingredient(s), dosage form, strength, route of administration, and conditions of use, except that conditions of use for which approval cannot be granted because of exclusivity or an existing patent may be omitted. If a listed drug has been voluntarily withdrawn from or not offered for sale by its manufacturer, a person who wishes to submit an abbreviated new drug application for the drug shall comply with §314.122.

(2) Drug products that are duplicates of, or that meet the monograph for, an antibiotic drug for which FDA has approved an application.

(3) Drug products that have been declared suitable for an abbreviated new drug application submission by FDA through the petition procedures set forth under §10.30 of this chapter and §314.93.

(b) FDA will publish in the list listed drugs for which abbreviated applications may be submitted. The list is available from the Superintendent of Documents, U.S. Government Printing Office, Washington, DC 20402, 202-783-3238.

§314.93 Petition to request a change from a listed drug.

(a) The only changes from a listed drug for which the agency will accept a petition under this section are those changes described in paragraph (b) of this section. Petitions to submit abbreviated new drug applications for other changes from a listed drug will not be approved.

(b) A person who wants to submit an abbreviated new drug application for a drug product which is not identical to a listed drug in route of administration, dosage form, and strength, or in which one active ingredient is substituted for one of the active ingredients in a listed combination drug, must first obtain permission from FDA to submit such an abbreviated application.

(c) To obtain permission to submit an abbreviated new drug application for a change described in paragraph (b) of this section, a person must submit and obtain approval of a petition requesting the change. A person seeking permission to request such a change from a reference listed drug shall submit a petition in accordance with §10.20 of this chapter and in the format specified in §10.30 of this chapter. The petition shall contain the information specified in §10.30 of this chapter and any additional information required by this section. If any provision of §10.20 or §10.30 of this chapter is inconsistent with any provision of this section, the provisions of this section apply.

(d) The petitioner shall identify a listed drug and include a copy of the proposed labeling for the drug product that is the subject of the petition and a copy of the approved labeling for the listed drug. The petitioner may, under limited circumstances, identify more than one listed drug, for example, when the proposed drug product is a combination product that differs from the combination reference listed drug with regard to an active ingredient, and the different active ingredient is an active ingredient of a listed drug. The petitioner shall also include information to show that:

(1) The active ingredients of the proposed drug product are of the same pharmacological or therapeutic class as those of the reference listed drug.

(2) The drug product can be expected to have the same therapeutic effect as the reference listed drug when administered to patients for each condition of use in the reference listed drug's labeling for which the applicant seeks approval.

(3) If the proposed drug product is a com-

bination product with one different active ingredient, including a different ester or salt, from the reference listed drug, that the different active ingredient has previously been approved in a listed drug or is a drug that does not meet the definition of "new drug" in section 201(b) of the act.

(e) No later than 90 days after the date a petition that is permitted under paragraph (a) of this section is submitted, FDA will approve or disapprove the petition.

(1) FDA will approve a petition properly submitted under this section unless it finds that:

(i) Investigations must be conducted to show the safety and effectiveness of the drug product or of any of its active ingredients, its route of administration, dosage form, or strength which differs from the reference listed drug; or

(ii) For a petition that seeks to change an active ingredient, the drug product that is the subject of the petition is not a combination drug; or

(iii) For a combination drug product that is the subject of the petition and has an active ingredient different from the reference listed drug:

(A) The drug product may not be adequately evaluated for approval as safe and effective on the basis of the information required to be submitted under §314.94; or

(B) The petition does not contain information to show that the different active ingredient of the drug product is of the same pharmacological or therapeutic class as the ingredient of the reference listed drug that is to be changed and that the drug product can be expected to have the same therapeutic effect as the reference listed drug when administered to patients for each condition of use in the listed drug's labeling for which the applicant seeks approval; or

(C) The different active ingredient is not an active ingredient in a listed drug or a drug that meets the requirements of section 201(p) of the act; or

(D) The remaining active ingredients are not identical to those of the listed combination drug; or

(iv) Any of the proposed changes from the listed drug would jeopardize the safe or effective use of the product so as to necessitate significant labeling changes to address the newly introduced safety or effectiveness problem; or

(v) FDA has determined that the reference listed drug has been withdrawn from sale for safety or effectiveness reasons under §314.161, or the reference listed drug has been voluntarily withdrawn from sale and the agency has not determined whether the withdrawal is for safety or effectiveness reasons.

(2) For purposes of this paragraph, "investigations must be conducted" means that information derived from animal or clinical studies is necessary to show that the drug product is safe or effective. Such information may be contained in published or unpublished reports.

(3) If FDA approves a petition submitted under this section, the agency's response may describe what additional information, if any, will be required to support an abbreviated new drug application for the drug product. FDA may, at any time during the course of its review of an abbreviated new drug application, request additional information required to evaluate the change approved under the petition.

(f) FDA may withdraw approval of a petition if the agency receives any information demonstrating that the petition no longer satisfies the conditions under paragraph (e) of this section.

§314.94 Content and format of an abbreviated application.

Abbreviated applications are required to be submitted in the form and contain the information required under this section. Two copies of the application are required, an archival copy and a review copy. FDA will maintain guidelines on the format and content of applications to assist applicants in their preparation.

(a) Abbreviated new drug applications. Except as provided in paragraph (b) of this section, the applicant shall submit a complete archival copy of the abbreviated new drug application that includes the following:

(1) Application form. The applicant shall submit a completed and signed application form that contains the information described under §314.50(a)(1), (a)(3), (a)(4), and (a)(5). The applicant shall state whether the submission is an abbreviated application under this section or a supplement to an abbreviated application under §314.97.

(2) Table of contents. the archival copy of the abbreviated new drug application is required to contain a table of contents that shows the volume number and page number of the contents of the submission.

(3) Basis for abbreviated new drug application submission. An abbreviated new drug application must refer to a listed drug. Ordinarily, that listed drug will be the drug product selected by the agency as the reference standard for conducting bioequivalence testing. The application shall contain:

(i) The name of the reference listed drug, including its dosage form and strength. For an abbreviated new drug application based on an approved petition under §10.30 of this chapter or §314.93, the reference listed drug must be the same as the listed drug approved in the petition.

(ii) A statement as to whether, according to the information published in the list, the reference listed drug is entitled to a period of marketing exclusivity under section 505(j)(4)(D) of the act.

(iii) For an abbreviated new drug application based on an approved petition under §10.30 of this chapter or §314.93, a reference to FDA-assigned docket number for the petition and a copy of FDA's correspondence approving the petition.

(4) Conditions of use. (i) A statement that the conditions of use prescribed, recommended, or suggested in the labeling proposed for the drug product have been previously approved for the reference listed drug.

(ii) A reference to the applicant's annotated proposed labeling and to the currently approved labeling for the reference listed drug provided under paragraph (a)(8) of this section.

(5) Active ingredients. (i) For a single-active-ingredient drug product, information to show that the active ingredient is the same as that of the reference single-active-ingredient listed drug, as follows:

(A) A statement that the active ingredient of the proposed drug product is the same as that of the reference listed drug.

(B) A reference to the applicant's annotated proposed labeling and to the currently approved labeling for the reference listed drug provided under paragraph (a)(8) of this section.

(ii) For a combination drug product, information to show that the active ingredients are the same as those of the reference listed drug except for any different active ingredient that has been the subject of an approved petition, as follows:

(A) A statement that the active ingredients of the proposed drug product are the same as those of the reference listed drug, or if one of the active ingredients differs from one of the active ingredients of the reference listed drug and the abbreviated application is submitted under the approval of a petition under §314.93 to vary such active ingredient, information to show that the other active ingredients of the drug product are the same as the other active ingredients of the reference listed drug, information to show that the different active ingredient is an active ingredient of another listed drug or of a drug that does not meet the definition of "new drug" in section 201(p) of the act, and such other information about the different active ingredient that FDA may require.

(B) A reference to the applicant's annotated proposed labeling and to the currently approved labeling for the reference listed drug provided under paragraph (a)(8) of this section.

(6) Route of administration, dosage form, and strength. (i) Information to show that the route of administration, dosage form, and strength of the drug product are the same as those of the reference listed drug except for any differences that have been the subject of an approved petition, as follows:

(A) A statement that the route of administration, dosage form, and strength of the proposed drug product are the same as those of the reference listed drug.

(B) A reference to the applicant's annotated

proposed labeling and to the currently approved labeling for the reference listed drug provided under paragraph (a)(8) of this section.

(ii) If the route of administration, dosage form, or strength of the drug product differs from the reference listed drug and the abbreviated application is submitted under an approved petition under §314.93, such information about the different route of administration, dosage form, or strength that FDA may require.

(7) Bioequivalence. (i) Information that shows that the drug product is bioequivalent to the reference listed drug upon which the applicant relies; or

(ii) If the abbreviated new drug application is submitted under a petition approved under §314.93, the results of any bioavailability or bioequivalence testing required by the agency, or any other information required by the agency to show that the active ingredients of the proposed drug product are of the same pharmacological or therapeutic class as those in the reference listed drug and that the proposed drug product can be expected to have the same therapeutic effect as the reference listed drug. If the proposed drug product contains a different active ingredient than the reference listed drug, FDA will consider the proposed drug product to have the same therapeutic effect as the reference listed drug if the applicant provides information demonstrating that:

(A) There is an adequate scientific basis for determining that substitution of the specific proposed dose of the different active ingredient for the dose of the member of the same pharmacological or therapeutic class in the reference listed drug will yield a resulting drug product whose safety and effectiveness have not been adversely affected.

(B) The unchanged active ingredients in the proposed drug product are bioequivalent to those in the reference listed drug.

(C) The different active ingredient in the proposed drug product is bioequivalent to an approved dosage form containing that ingredient and approved for the same indication as the proposed drug product or is bioequivalent to a drug product offered for that indication which does not meet the definition of "new drug" under section 201(p) of the act.

(iii) For each in vivo bioequivalence study contained in the abbreviated new drug application, a description of the analytical and statistical methods used in each study and a statement with respect to each study that it either was conducted in compliance with the institutional review board regulations in part 56 of this chapter, or was not subject to the regulations under §56.104 or §56.105 of this chapter and that each study was conducted in compliance with the informed consent regulations in part 50 of this chapter.

(8) Labeling-(i) Listed drug labeling. A copy of the currently approved labeling for the listed drug referred to in the abbreviated new drug application, if the abbreviated new drug application relies on a reference listed drug.

(ii) Proposed labeling. Copies of the label and all labeling for the drug product (4 copies of draft labeling or 12 copies of final printed labeling).

(iii) A statement that the applicant's proposed labeling is the same as the labeling of the reference listed drug except for differences annotated and explained under paragraph (a)(8)(iv) of this section.

(iv) A side-by-side comparison of the applicant's proposed labeling with the approved labeling for the reference listed drug with all differences annotated and explained. Labeling (including the container label and package insert) proposed for the drug product must be the same as the labeling approved for the reference listed drug, except for changes required because of differences approved under a petition filed under §314.93 or because the drug product and the reference listed drug are produced or distributed by different manufacturers. Such differences between the applicant's proposed labeling and labeling approved for the reference listed drug may include differences in expiration date, formulation, bioavailability, or pharmacokinetics, labeling revisions made to comply with current FDA labeling guidelines or other guidance, or omission of an indication or other aspect of labeling protected by patent or accorded exclusivity under section 505(j)(4)(D) of the act.

(9) Chemistry, manufacturing, and controls. (i) The information required under §314.50(d)(1).

(ii) Inactive ingredients. Unless otherwise stated in paragraphs (a)(9)(iii) through (a)(9)(v) of this section, an applicant shall identify and characterize the inactive ingredients in the proposed drug product and provide information demonstrating that such inactive ingredients do not affect the safety of the proposed drug product.

(iii) Inactive ingredient changes permitted in drug products intended for parenteral use. Generally, a drug product intended for parenteral use shall contain the same inactive ingredients and in the same concentration as the reference listed drug identified by the applicant under paragraph (a)(3) of this section. However, an applicant may seek approval of a drug product that differs from the reference listed drug in preservative, buffer, or antioxidant provided that the applicant identifies and characterizes the differences and provides information demonstrating that the differences do not affect the safety of the proposed drug product.

(iv) Inactive ingredient changes permitted in drug products intended for ophthalmic or otic use. Generally, a drug product intended for ophthalmic or otic use shall contain the same inactive ingredients and in the same concentration as the reference listed drug identified by the applicant under paragraph (a)(3) of this section. However, an applicant may seek approval of a drug product that differs from the reference listed drug in preservative, buffer, substance to adjust tonicity, or thickening agent provided that the applicant identifies and characterizes the differences and provides information demonstrating that the differences do not affect the safety of the proposed drug product, except that, in a product intended for ophthalmic use, an applicant may not change a buffer or substance to adjust tonicity for the purpose of claiming a therapeutic advantage over or difference from the listed drug, e.g., by using a balanced salt solution as a diluent as opposed to an isotonic saline solution, or by making a significant change in the pH or other change that may raise questions of irritability.

(v) Inactive ingredient changes permitted in drug products intended for topical use. Generally, a drug product intended for topical use shall contain the same inactive ingredients as the reference listed drug identified by the applicant under paragraph (a)(3) of this section. However, an applicant may seek approval of a drug product that differs from the reference listed drug provided that the applicant identifies and characterizes the differences and provides information demonstrating that the differences do not affect the safety of the proposed drug product.

(10) Samples. The information required under §314.50(e)(1) and (e)(2)(i). Samples need not be submitted until requested by FDA.

(11) Other. The information required under §314.50(g).

(b) Drug products subject to the Drug Efficacy Study Implementation (DESI) review. If the abbreviated new drug application is for a duplicate of a drug product that is subject to FDA's DESI review (a review of drug products approved as safe between 1938 and 1962) or other DESI-like review and the drug product evaluated in the review is a listed drug, the applicant shall comply with the provisions of paragraph (a) of this section.

(c) Abbreviated antibiotic application. For applications submitted under section 507 of the act, the applicant shall submit a complete archival copy of the abbreviated application that contains the information described under §314.50 (a)(1), (a)(3), (a)(4), and (a)(5), (b), (d)(1) and (d)(3), (e), and (g). The applicant shall state whether the submission is an abbreviated application under this section or a supplement to an abbreviated application under §314.97.

(d) Format of an abbreviated application. (1) The applicant shall submit a complete archival copy of the abbreviated application as required under paragraphs (a) and (c) of this section. FDA will maintain the archival copy during the review of the application to permit individual reviewers to refer to information that is not contained in their particular technical sections of the application, to give other agency personnel access to the application for official business, and to maintain in one place a complete copy of the application. **An applicant may submit all or portions of the archival copy of the abbreviated application in any form (e.g., microfiche, optical disc, and magnetic tape) that the applicant and FDA agree is acceptable.**

(2) For abbreviated new drug applications, the applicant shall submit a review copy of the abbreviated application that contains two separate sections. One section shall contain the information described under paragraphs (a)(2) through (a)(6), (a)(8), and (a)(9) of this section 505(j)(2)(A)(vii) of the act and one copy of the analytical methods and descriptive information needed by FDA's laboratories to perform tests on samples of the proposed drug product and to validate the applicant's analytical methods. The other section shall contain the information described under paragraphs (a)(3), (a)(7), and (a)(8) of this section. Each of the sections in the review copy is required to contain a copy of the application form described under §314.50(a).

(3) For abbreviated antibiotic applications, the applicant shall submit a review copy that contains the technical sections described in §314.50 (d)(1) and (d)(3). Each of the technical sections in the review copy is required to be separate with a copy of the application form required under §314.50(a).

(4) The applicant may obtain from FDA sufficient folders to bind the archival and the review copies of the abbreviated application.

§314.96 Amendments to an unapproved abbreviated application.

(a) Abbreviated new drug application. (1) An applicant may amend an abbreviated new drug application that is submitted under §314. 94, but not yet approved, to revise existing information or provide additional information.

(2) Submission of an amendment containing significant data or information constitutes an agreement between FDA and the applicant to extend the review period only for the time necessary to review the significant data or information and for no more than 180 days.

(3) Submission of an amendment containing significant data or information to resolve deficiencies in the application as set forth in a not approvable letter issued under §314.120 constitutes an agreement between FDA and the applicant under section 505(j)(4)(A) of the act to extend the date by which the agency is required to reach a decision on the abbreviated new drug application only for the time necessary to review the significant data or information and for no more than 180 days.

(b) Abbreviated antibiotic application. The applicant shall comply with the provisions of §314.60.

§314.97 Supplements and other changes to an approved abbreviated application.

The applicant shall comply with the requirements of §§314.70 and 314.71 regarding the submission of supplemental applications and other changes to an approved abbreviated application.

§314.98 Postmarketing reports.

(a) Except as provided in paragraph (b) of this section, each applicant having an approved abbreviated antibiotic application under §314.94 or approved abbreviated new drug application under §314.94 that is effective shall comply with the requirements of §314.80 regarding the reporting and recordkeeping of adverse drug experiences.

(b) Each applicant shall submit one copy of each report required under §314.80 to the Division of Epidemiology and Surveillance (HFD-730), Center for Drug Evaluation and Research, Food and Drug Administration, 5600 Fishers Lane, Rockville, MD 20857.

(c) Each applicant shall make the reports required under §314.81 and sections 505(k) and 507(g) of the act for each of its approved abbreviated applications.

§314.99 Other responsibilities of an applicant of an abbreviated application.

(a) An applicant shall comply with the requirements of §314.65 regarding withdrawal by the applicant of an unapproved abbreviated application and §314.72 regarding a change in ownership of an abbreviated application.

(b) An applicant may ask FDA to waive under this section any requirement that applies to the applicant under §§314.92 through 314.99. The applicant shall comply with the requirements for a waiver under §314.90.

Subpart D--FDA Action on Applications and Abbreviated Applications

§314.100 Time frames for reviewing applications and abbreviated applications.

(a) Within 180 days of receipt of an application for a new drug under section 505(b) of the act, or of an abbreviated application for a new drug under section 505(j) of the act, or of an application or abbreviated application for an antibiotic drug under section 507 of the act, FDA will review it and send the applicant either an approval letter under §314.105, or an approvable letter under §314.110, or a not approvable letter under §314.120. This 180-day period is called the "review clock."

(b) During the review period, an applicant may withdraw an application under §314.65 or an abbreviated application under §314.99 and later resubmit it. FDA will treat the resubmission as a new application or abbreviated application.

(c) The review clock may be extended by mutual agreement between FDA and an applicant or as provided in §§314.60 and 314.96, as the result of a major amendment.

§314.101 Filing an application and an abbreviated antibiotic application and receiving an abbreviated new drug application.

(a)(1) Within 60 days after FDA receives an application or abbreviated antibiotic application, the agency will determine whether the application or abbreviated antibiotic application may be filed. The filing of an application or abbreviated antibiotic application means that FDA has made a threshold determination that the application or abbreviated antibiotic application is sufficiently complete to permit a substantive review.

(2) If FDA finds that none of the reasons in paragraphs (d) and (e) of this section for refusing to file the application or abbreviated antibiotic apply, the agency will file the application or abbreviated antibiotic application and notify the applicant in writing. The date of filing will be the date 60 days after the date FDA received the application or abbreviated antibiotic application. The date of filing begins the 180-day period described in section 505(c) of the act. This 180-day period is called the "filing clock."

(3) If FDA refuses to file the application or abbreviated antibiotic application, the agency will notify the applicant in writing and state the reason under paragraph (d) or (e) of this section for the refusal. If FDA refuses to file the application or abbreviated antibiotic application under paragraph (d) of this section, the applicant may request in writing within 30 days of the date of the agency's notification an informal conference with the agency about whether the agency should file the application or abbreviated antibiotic application. If, following the informal conference, the applicant requests that FDA file the application or abbreviated antibiotic application (with or without amendments to correct the deficiencies), the agency will file the application or abbreviated antibiotic application over protest under paragraph (a)(2) of this section, notify the applicant in writing, and review it as filed. If the application or abbreviated antibiotic application is filed over protest, the date of filing will be the date 60 days after the date the applicant requested the informal conference. The applicant need not resubmit a copy of an application or abbreviated antibiotic application that is filed over protest. If FDA refuses to file the application or abbreviated antibiotic application under paragraph (e) of this section, the applicant may amend the application or abbreviated antibiotic application and resubmit it, and the agency will make a determination under this section whether it may be filed.

(b)(1) An abbreviated new drug application will be reviewed after it is submitted to determine whether the abbreviated application may be received. Receipt of an abbreviated new drug application means that FDA has made a threshold determination that the abbreviated application is sufficiently complete to permit a substantive review.

(2) If FDA finds that none of the reasons in paragraphs (d) and (e) of this section for considering the abbreviated new drug application not to have been received applies, the agency will receive the abbreviated new drug application and notify the applicant in writing.

(3) If FDA considers the abbreviated new drug application not to have been received under paragraph (d) or (e) of this section, FDA will notify the applicant, ordinarily by telephone. The applicant may then:

(i) Withdraw the abbreviated new drug application under §314.99; or

(ii) Amend the abbreviated new drug application to correct the deficiencies; or

(iii) Take no action, in which case FDA will refuse to receive the abbreviated new drug application.

(c) [Reserved]

(d) FDA may refuse to file an application or abbreviated antibiotic application or may not consider an abbreviated new drug application to be received if any of the following applies:

(1) The application or abbreviated application does not contain a completed application form.

(2) The application or abbreviated application is not submitted in the form required under §314.50 or §314.94.

(3) The application or abbreviated application is incomplete because it does not on its face contain information required under section 505(b), section 505(j), or section 507 of the act and §314.50 or §314.94.

(4) The applicant fails to submit a complete environmental assessment, which addresses each of the items specified in the applicable format under §25.31 of this chapter or fails to provide sufficient information to establish that the requested action is subject to categorical exclusion under §25.24 of this chapter.

(5) The application or abbreviated application does not contain an accurate and complete English translation of each part of the application that is not in English.

(6) The application does not contain a statement for each nonclinical laboratory study that it was conducted in compliance with the requirements set forth in part 58 of this chapter, or, for each study not conducted in compliance with part 58 of this chapter, a brief statement of the reason for the noncompliance.

(7) The application does not contain a statement for each clinical study that it was conducted in compliance with the institutional review board regulations in part 56 of this chapter, or was not subject to those regulations, and that it was conducted in compliance with the informed consent regulations in part 50 of this chapter, or, if the study was subject to but was not conducted in compliance with those regulations, the application does not contain a brief statement of the reason for the noncompliance.

(8) The drug product that is the subject of the submission is already covered by an approved application or abbreviated application and the applicant of the submission:

(i) Has an approved application or abbreviated application for the same drug product; or

(ii) Is merely a distributor and/or repackager of the already approved drug product.

(9) The application is submitted as a 505 (b)(2) application for a drug that is a duplicate of a listed drug and is eligible for approval under section 505(j) of the act.

(e) The agency will refuse to file an application or an abbreviated antibiotic application or will consider an abbreviated new drug application not to have been received if the drug product is subject to licensing by FDA under the Public Health Service Act (42 U.S.C. 201 et seq.) and subchapter F of this chapter.

(f)(1) Within 180 days after the date of filing, plus the period of time the review period was extended (if any), FDA will either:

(i) Approve the application or abbreviated antibiotic application; or

(ii) Issue a notice of opportunity for hearing if the applicant asked FDA to provide it an opportunity for a hearing on an application or abbreviated antibiotic application in response to an approvable letter or a not approvable letter.

(2) Within 180 days after the date of receipt, plus the period of time the review clock was extended (if any), FDA will either ap-

prove or disapprove the abbreviated new drug application. If FDA disapproves the abbreviated new drug application, FDA will issue a notice of opportunity for hearing if the applicant asked FDA to provide it an opportunity for a hearing on an abbreviated new drug application in response to a not approvable letter.

(3) This paragraph does not apply to applications or abbreviated applications that have been withdrawn from FDA review by the applicant.

§314.102 Communications between FDA and applicants.

(a) General principles. During the course of reviewing an application or an abbreviated application, FDA shall communicate with applicants about scientific, medical, and procedural issues that arise during the review process. Such communication may take the form of telephone conversations, letters, or meetings, whichever is most appropriate to discuss the particular issue at hand. Communications shall be appropriately documented in the application in accordance with §10.65 of this chapter. Further details on the procedures for communication between FDA and applicants are contained in a staff manual guide that is publicly available.

(b) Notification of easily correctable deficiencies. FDA reviewers shall make every reasonable effort to communicate promptly to applicants easily correctable deficiencies found in an application or an abbreviated application when those deficiencies are discovered, particularly deficiencies concerning chemistry, manufacturing, and controls issues. The agency will also inform applicants promptly of its need for more data or information or for technical changes in the application or the abbreviated application needed to facilitate the agency's review. This early communication is intended to permit applicants to correct such readily identified deficiencies relatively early in the review process and to submit an amendment before the review period has elapsed. Such early communication would not ordinarily apply to major scientific issues, which require consideration of the entire pending application or abbreviated application by agency managers as well as reviewing staff. Instead, major scientific issues will ordinarily be addressed in an action letter.

(c) Ninety-day conference. Approximately 90 days after the agency receives the application, FDA will provide applicants with an opportunity to meet with agency reviewing officials. The purpose of the meeting will be to inform applicants of the general progress and status of their applications, and to advise applicants of deficiencies that have been identified by that time and that have not already been communicated. This meeting will be available on applications for all new chemical entities and major new indications of marketed drugs. Such meetings will be held at the applicant's option, and may be held by telephone if mutually agreed upon. Such meetings would not ordinarily be held on abbreviated applications because they are not submitted for new chemical entities or new indications.

(d) End-of-review conference. At the conclusion of FDA's review of an application or an abbreviated application as designated by the issuance of an approvable or not approvable letter, FDA will provide applicants with an opportunity to meet with agency reviewing officials. The purpose of the meeting will be to discuss what further steps need to be taken by the applicant before the application or abbreviated application can be approved. This meeting will be available on all applications or abbreviated applications, with priority given to applications for new chemical entities and major new indications for marketed drugs and for the first duplicates for such drugs. Requests for such meetings shall be directed to the director of the division responsible for reviewing the application or abbreviated application.

(e) Other meetings. Other meetings between FDA and applicants may be held, with advance notice, to discuss scientific, medical, and other issues that arise during the review process. Requests for meetings shall be directed to the director of the division responsible for reviewing the application or abbreviated application. FDA will make every attempt to grant requests for meetings that involve important issues and that can be scheduled at mutually convenient times. However, "dropin" visits (i.e., an unannounced and unscheduled visit by a company representative) are discouraged except for urgent matters, such as to discuss an important new safety issue.

§314.103 Dispute resolution.

(a) General. FDA is committed to resolving differences between applicants and FDA reviewing divisions with respect to technical requirements for applications or abbreviated applications as quickly and amicably as possible through the cooperative exchange of information and views.

(b) Administrative and procedural issues. When administrative or procedural disputes arise, the applicant should first attempt to resolve the matter with the division responsible for reviewing the application or abbreviated application, beginning with the consumer safety officer assigned to the application or abbreviated application. If resolution is not achieved, the applicant may raise the matter with the person designated as ombudsman, whose function shall be to investigate what has happened and to facilitate a timely and equitable resolution. Appropriate issues to raise with the ombudsman include resolving difficulties in scheduling meetings, obtaining timely replies to inquiries, and obtaining timely completion of pending reviews. Further details on this procedure are contained in a staff manual guide that is publicly available under FDA's public information regulations in part 20.

(c) Scientific and medical disputes. (1) Because major scientific issues are ordinarily communicated to applicants in an approvable or not approvable letter pursuant to §314.110 or §314.120, respectively, the "end-of-review conference" described in §314.102(d) will provide a timely forum for discussing and resolving, if possible, scientific and medical issues on which the applicant disagrees with the agency. In addition, the "ninety-day conference" described in §314.102(c) will provide a timely forum for discussing and resolving, if possible, issues identified by that date.

(2) When scientific or medical disputes arise at other times during the review process, applicants should discuss the matter directly with the responsible reviewing officials. If necessary, applicants may request a meeting with the appropriate reviewing officials and management representatives in order to seek a resolution. Ordinarily, such meetings would be held first with the Division Director, then with the Office Director, and finally with the Center Director if the matter is still unresolved. Requests for such meetings shall be directed to the director of the division responsible for reviewing the application or abbreviated application. FDA will make every attempt to grant requests for meetings that involve important issues and that can be scheduled at mutually convenient times.

(3) In requesting a meeting designed to resolve a scientific or medical dispute, applicants may suggest that FDA seek the advice of outside experts, in which case FDA may, in its discretion, invite to the meeting one or more of its advisory committee members or other consultants, as designated by the agency. Applicants may also bring their own consultants. For major scientific and medical policy issues not resolved by informal meetings, FDA may refer the matter to one of its standing advisory committees for its consideration and recommendations.

§314.104 Drugs with potential for abuse.

The Food and Drug Administration will inform the Drug Enforcement Administration under section 201(f) of the Controlled Substances Act (21 U.S.C. 801) when an application or abbreviated application is submitted for a drug that appears to have an abuse potential.

§314.105 Approval of an application and an abbreviated application.

(a) The Food and Drug Administration will approve an application or an abbreviated antibiotic application and send the applicant an approval letter if none of the reasons in §314.

125 for refusing to approve the application or abbreviated antibiotic application applies. An approval becomes effective on the date of the issuance of the approval letter, except with regard to an approval under section 505(b)(2) of the act with a delayed effective date. An approval with a delayed effective date is tentative and does not become final until the effective date. When FDA sends an applicant an approval letter for an antibiotic, it will promulgate a regulation under §314.300 providing for certification of the drug, if necessary. A new drug product or antibiotic approved under this paragraph may not be marketed until an approval is effective. Marketing of an antibiotic need not await the promulgation of a regulation under §314.300.

(b) FDA will approve an application or abbreviated antibiotic application and issue the applicant an approval letter (rather than an approvable letter under §314.110) on the basis of draft labeling if the only deficiencies in the application or abbreviated antibiotic application concern editorial or similar minor deficiencies in the draft labeling. Such approval will be conditioned upon the applicant incorporating the specified labeling changes exactly as directed, and upon the applicant submitting to FDA a copy of the final printed labeling prior to marketing.

(c) FDA will approve an application after it determines that the drug meets the statutory standards for safety and effectiveness, manufacturing and controls, and labeling, and an abbreviated application after it determines that the drug meets the statutory standards for manufacturing and controls, labeling, and, where applicable, bioequivalence. While the statutory standards apply to all drugs, the many kinds of drugs that are subject to the statutory standards and the wide range of uses for those drugs demand flexibility in applying the standards. Thus FDA is required to exercise its scientific judgment to determine the kind and quantity of data and information an applicant is required to provide for a particular drug to meet the statutory standards. FDA makes its views on drug products and classes of drugs available through guidelines, recommendations, and other statements of policy.

(d) FDA will approve an abbreviated new drug application and send the applicant an approval letter if none of the reasons in §314. 127 for refusing to approve the abbreviated new drug application applies. The approval becomes effective on the date of the issuance of the agency's approval letter unless the approval letter provides for a delayed effective date. An approval with a delayed effective date is tentative and does not become final until the effective date. A new drug product approved under this paragraph may not be introduced or delivered for introduction into interstate commerce until approval of the abbreviated new drug application is effective. Ordinarily, the effective date of approval will be stated in the approval letter.

§314.106 Foreign data.

(a) General. The acceptance of foreign data in an application generally is governed by §312.120 of this chapter.

(b) As sole basis for marketing approval. An application based solely on foreign clinical data meeting U.S. criteria for marketing approval may be approved if: (1) The foreign data are applicable to the U.S. population and U.S. medical practice; (2) the studies have been performed by clinical investigators of recognized competence; and (3) the data may be considered valid without the need for an on-site inspection by FDA or, if FDA considers such an inspection to be necessary, FDA is able to validate the data through an on-site inspection or other appropriate means. Failure of an application to meet any of these criteria will result in the application not being approvable based on the foreign data alone. FDA will apply this policy in a flexible manner according to the nature of the drug and the data being considered.

(c) Consultation between FDA and applicants. Applicants are encouraged to meet with agency officials in a "presubmission" meeting when approval based solely on foreign data will be sought.

§314.110 Approvable letter to the applicant.

(a) In selected circumstances, it is useful at the end of the review period for the Food and Drug Administration to indicate to the applicant that the application or abbreviated application is basically approvable providing certain issues are resolved. An approvable letter may be issued in such circumstances. FDA will send the applicant an approvable letter if the application or abbreviated application substantially meets the requirements of this part and the agency believes that it can approve the application or abbreviated application if specific additional information or material is submitted or specific conditions (for example, certain changes in labeling) are agreed to by the applicant. The approvable letter will describe the information or material FDA requires or the conditions the applicant is asked to meet. As a practical matter, the approvable letter will serve in most instances as a mechanism for resolving outstanding issues on drugs that are about to be approved and marketed. For an application or an abbreviated antibiotic application, the applicant shall, within 10 days after the date of the approvable letter:

(1) Amend the application or abbreviated antibiotic application or notify FDA of an intent to file an amendment. The filing of an amendment or notice of intent to file an amendment constitutes an agreement by the applicant to extend the review period for 45 days after the date FDA receives the amendment. The extension is to permit the agency to review the amendment;

(2) Withdraw the application or abbreviated antibiotic application. FDA will consider the applicant's failure to respond within 10 days to an approvable letter to be a request by the applicant to withdraw the application under §314.65 or the abbreviated antibiotic application under §314.99. A decision to withdraw an application or abbreviated antibiotic application is without prejudice to a refiling;

(3) For a new drug application or abbreviated antibiotic application, ask the agency to provide the applicant an opportunity for a hearing on the question of whether there are grounds for denying approval of the application under section 505(d) of the act. The applicant shall submit the request to the Division of Regulatory Affairs (HFD-360), Center for Drug Evaluation and Research, Food and Drug Administration, 5600 Fishers Lane, Rockville, MD 20857. Within 60 days of the date of the approvable letter, or within a different time period to which FDA and the applicant agree, the agency will either approve the application or abbreviated antibiotic application under §314.105 or refuse to approve the application or abbreviated antibiotic application under §314.125 and give the applicant written notice of an opportunity for a hearing under §314.200 and section 505(c)(2) of the act on the question of whether there are grounds for denying approval of the application under section 505(d) of the act;

(4) For an antibiotic, file a petition or notify FDA of an intent to file a petition proposing the issuance, amendment, or repeal of a regulation under §314.300 and section 507(f) of the act; or

(5) Notify FDA that the applicant agrees to an extension of the review period under section 505(c) of the act, so that the applicant can determine whether to respond further under paragraph (a)(1), (a)(2), (a)(3), or (a)(4) of this section. The applicant's notice is required to state the length of the extension. FDA will honor any reasonable request for such an extension. FDA will consider the applicant's failure to respond further within the extended review period to be a request to withdraw the application under §314. 65 or the abbreviated antibiotic application under §314.99. A decision to withdraw an application or abbreviated antibiotic application is without prejudice to a refiling.

(b) FDA will send the applicant of an abbreviated new drug application an approvable letter only if the application substantially meets the requirements of this part and the agency believes that it can approve the abbreviated application if minor deficiencies (e.g., labeling deficiencies) are corrected. The approvable letter will describe the deficiencies and state a time period within which the applicant must respond. Unless the applicant corrects the deficiencies by amendment within the specified time period, FDA will refuse to approve the abbreviated application under §314.127. Within 10 days after the date of the approvable letter, the applicant may also ask the agency to provide the applicant an opportunity for a hearing on the question of whether there are grounds for denying approval of the abbreviated new drug application. Applicants who request a hearing shall submit the request to the Division of Regulatory Affairs (HFD-360), Center for Drug Evaluation and Research, Food and Drug Administration, 5600 Fishers Lane, Rockville, MD 20857.

§314.120 Not approvable letter to the applicant.

(a) The Food and Drug Administration will send the applicant a not approvable letter if the agency believes that the application or abbreviated antibiotic application may not be approved for one of the reasons given in §314.125 or the abbreviated new drug application may not be approved for one of the reasons given in §314.127. The not approvable letter will describe the deficiencies in the application or abbreviated application. Except as provided in paragraph (b) of this section, within 10 days after the date of the not approvable letter, the applicant shall:

(1) Amend the application or abbreviated application or notify FDA of an intent to file an amendment. The filing of an amendment or a notice of intent to file an amendment constitutes an agreement by the applicant to extend the review period under §314.60 or §314.96;

(2) Withdraw the application or abbreviated application. Except as provided in paragraph (b) of this section, FDA will consider the applicant's failure to respond within 10 days to a not approvable letter to be a request by the applicant to withdraw the application under §314.65 or abbreviated application under §314. 99. A decision to withdraw the application or abbreviated application is without prejudice to refiling;

(3) For a new drug application or an abbreviated application, ask the agency to provide the applicant an opportunity for a hearing on the question of whether there are grounds for denying approval of the application under section 505(d) or (j)(3) of the act. The applicant shall submit the request to the Division of Regulatory Affairs (HFD-360), Center for Drug Evaluation and Research, Food and Drug Administration, 5600 Fishers Lane, Rockville, MD 20857. Within 60 days of the date of the not approvable letter, or within a different time period to which FDA and the applicant agree, the agency will either approve the application or abbreviated application under §314.105 or refuse to approve the application or abbreviated antibiotic application under §314.125 or abbreviated new drug application under §314.127 and give the applicant written notice of an opportunity for a hearing under §314.200 and section 505(c)(1)(B) or (j)(4)(C) of the act on the question of whether there are grounds for denying approval of the application under section 505(d) or (j)(3) of the act;

(4) For an antibiotic application, file a petition or notify FDA of an intent to file a petition proposing the issuance, amendment, or repeal of a regulation under §314.300 and section 507(f) of the act; or

(5) Notify FDA that the applicant agrees to an extension of the review period under section 505(c)(1) or (j)(4)(A) of the act, so that the applicant can determine whether to respond further under paragraph (a)(1), (a)(2), (a)(3), or (a)(4) of this section. The applicant's notice is required to state the length of the extension. FDA will honor any reasonable request for such an extension. FDA will consider the applicant's failure to respond further within the extended review period to be a request to withdraw the application under §314.65 or abbreviated application under §314.99. A decision to withdraw an application or abbreviated application is without prejudice to a refiling.

(b) With the exception of a request for an opportunity for a hearing under paragraph (a)(3) of this section, the 10-day time period in this section for responding to a not approvable letter does not apply to abbreviated new drug applications. FDA may consider the applicant's failure to respond within 180 days to a not approvable letter to be a request by the applicant to withdraw the abbreviated new drug application under §314.99.

§314.122 Submitting an abbreviated application for, or a 505(j)(2)(C) petition that relies on, a listed drug that is no longer marketed.

(a) An abbreviated new drug application that refers to, or a petition under section 505(j)(2)(C) of the act and §314.93 that relies on, a listed drug that has been voluntarily withdrawn from sale in the United States must be accompanied by a petition seeking a determination whether the listed drug was withdrawn for safety or effectiveness reasons. The petition must be submitted under §§10.25(a) and 10.30 of this chapter and must contain all evidence available to the petitioner concerning the reasons for the withdrawal from sale.

(b) When a petition described in paragraph (a) of this section is submitted, the agency will consider the evidence in the petition and any other evidence before the agency, and determine whether the listed drug is withdrawn from sale for safety or effectiveness reasons, in accordance with the procedures in §314.161.

(c) An abbreviated new drug application described in paragraph (a) of this section will be disapproved, under §314.127(a)(11), and a 505(j)(2)(C) petition described in paragraph (a) of this section will be disapproved, under §314.93(e)(1)(iv), unless the agency determines that the withdrawal of the listed drug was not for safety or effectiveness reasons.

(d) Certain drug products approved for safety and effectiveness that were no longer marketed on September 24, 1984, are not included in the list. Any person who wishes to obtain marketing approval for such a drug product under an abbreviated new drug application must petition FDA for a determination whether the drug product was withdrawn from the market for safety or effectiveness reasons and request that the list be amended to include the drug product. A person seeking such a determination shall use the petition procedures established in §10.30 of this chapter. The petitioner shall include in the petition information to show that the drug product was approved for safety and effectiveness and all evidence available to the petitioner concerning the reason that marketing of the drug product ceased.

§314.125 Refusal to approve an application or abbreviated antibiotic application.

(a) The Food and Drug Administration will refuse to approve the application or abbreviated antibiotic application and for a new drug give the applicant written notice of an opportunity for a hearing under §314.200 on the question of whether there are grounds for denying approval of the application under section 505(d) of the act, or for an antibiotic publish a proposed regulation based on an acceptable petition under §314.300, if:

(1) FDA sends the applicant an approvable or a not approvable letter under §314.110 or §314.120;

(2) The applicant requests an opportunity for hearing for a new drug on the question of whether the application is approvable or files a petition for an antibiotic proposing the issuance, amendment, or repeal of a regulation; and

(3) FDA finds that any of the reasons given in paragraph (b) of this section apply.

(b) FDA may refuse to approve an application or abbreviated antibiotic application for any of the following reasons:

(1) The methods to be used in, and the facilities and controls used for, the manufacture, processing, packing, or holding of the drug substance or the drug product are inadequate to preserve its identity, strength, quality, purity, stability, and bioavailability.

(2) The investigations required under section 505(b) or 507 of the act do not include adequate tests by all methods reasonably applicable to show whether or not the drug is safe for use under the conditions prescribed, recommended, or suggested in its proposed labeling.

(3) The results of the tests show that the drug is unsafe for use under the conditions prescribed, recommended, or suggested in its proposed labeling or the results do not show that the drug product is safe for use under those conditions.

(4) There is insufficient information about the drug to determine whether the product is safe for use under the conditions prescribed, recommended, or suggested in its proposed labeling.

(5) There is a lack of substantial evidence consisting of adequate and well-controlled investigations, as defined in §314.126, that the drug product will have the effect it purports or is represented to have under the conditions of use prescribed, recommended, or suggested in its proposed labeling.

(6) The proposed labeling is false or misleading in any particular.

(7) The application contains an untrue statement of a material fact.

(8) The drug product's proposed labeling does not comply with the requirements for labels and labeling in Part 201.

(9) The application does not contain bioavailability or bioequivalence data required under Part 320.

(10) A reason given in a letter refusing to file the application under §314.101(d), if the deficiency is not corrected.

(11) The drug will be manufactured or processed in whole or in part in an establishment that is not registered and not exempt from registration under section 510 of the act and Part 207.

(12) The applicant does not permit a properly authorized officer or employee of the Department of Health and Human Services an adequate opportunity to inspect the facilities, controls, and any records relevant to the application or abbreviated antibiotic application.

(13) The methods to be used in, and the facilities and controls used for, the manufacture, processing, packing, or holding of the drug substance or the drug product do not comply with the current good manufacturing practice regulations in parts 210 and 211.

(14) The application or abbreviated antibiotic application does not contain an explanation of the omission of a report of any investigation of the drug product sponsored by the applicant, or an explanation of the omission of other information about the drug pertinent to an evaluation of the application or abbreviated antibiotic application that is received or otherwise obtained by the applicant from any source.

(15) A nonclinical laboratory study that is described in the application or abbreviated antibiotic application and that is essential to show that the drug is safe for use under the conditions prescribed, recommended, or suggested in its proposed labeling was not conducted in compliance with the good laboratory practice regulations in part 58 of this chapter and no reason for the noncompliance is provided or, if it is, the differences between the practices used in conducting the study and the good laboratory practice regulations do not support the validity of the study.

(16) Any clinical investigation involving human subjects described in the application or abbreviated antibiotic application, subject to the institutional review board regulations in part 58 of this chapter or informed consent regulations in part 50 of this chapter, was not conducted in compliance with those regulations such that the rights or safety of human subjects were not adequately protected.

(17) The applicant or contract research organization that conducted a bioavailability or bioequivalence study contained in the application or abbreviated antibiotic application refuses to permit an inspection of facilities or records relevant to the study by a properly authorized officer or employee of the Department of Health and Human Services or refuses to submit reserve samples of the drug products used in the study when requested by FDA.

(18) For a new drug, the application failed to contain the patent information required by section 505(b)(1) of the act.

(c) For drugs intended to treat life-threatening or severely-debilitating illnesses that are developed in accordance with §§312.80 through 312.88 of this chapter, the criteria contained in paragraphs (b)(3), (4), and (5) of this section shall be applied according to the considerations contained in §312.84 of this chapter.

§314.126 Adequate and well-controlled studies.

(a) The purpose of conducting clinical investigations of a drug is to distinguish the effect of a drug from other influences, such as spontaneous change in the course of the disease, placebo effect, or biased observation. The characteristics described in paragraph (b) of this section have been developed over a period of years and are recognized by the scientific community as the essentials of an adequate and well-controlled clinical investigation. The Food and Drug Administration considers these characteristics in determining whether an investigation is adequate and well-controlled for purposes of sections 505 and 507 of the act. Reports of adequate and well-controlled investigations provide the primary basis for determining whether there is "substantial evidence" to support the claims of effectiveness for new drugs and antibiotics. Therefore, the study report should provide sufficient details of study design, conduct, and analysis to allow critical evaluation and a determination of whether the characteristics of an adequate and well-controlled study are present.

(b) An adequate and well-controlled study has the following characteristics:

(1) There is a clear statement of the objectives of the investigation and a summary of the proposed or actual methods of analysis in the protocol for the study and in the report of its results. In addition, the protocol should contain a description of the proposed methods of analysis, and the study report should contain a description of the methods of analysis ultimately used. If the protocol does not contain a description of the proposed methods of analysis, the study report should describe how the methods used were selected.

(2) The study uses a design that permits a valid comparison with a control to provide a quantitative assessment of drug effect. The protocol for the study and report of results should describe the study design precisely; for example, duration of treatment periods, whether treatments are parallel, sequential, or crossover, and whether the sample size is predetermined or based upon some interim analysis. Generally, the following types of control are recognized:

(i) Placebo concurrent control. The test drug is compared with an inactive preparation designed to resemble the test drug as far as possible. A placebo-controlled study may include additional treatment groups, such as an active treatment control or a dose-comparison control, and usually includes randomization and blinding of patients or investigators, or both.

(ii) Dose-comparison concurrent control. At least two doses of the drug are compared. A dose-comparison study may include additional treatment groups, such as placebo control or active control. Dose comparison trials usually include randomization and blinding of patients or investigators, or both.

(iii) No treatment concurrent control. Where objective measurements of effectiveness are available and placebo effect is negligible, the test drug is compared with no treatment. No treatment concurrent control trials usually include randomization.

(iv) Active treatment concurrent control. The test drug is compared with known effective therapy; for example, where the condition treated is such that administration of placebo or no treatment would be contrary to the interest of the patient. An active treatment study may include additional treatment groups, however, such as a placebo control or a dose-comparison control. Active treatment trials usually include randomization and blinding of patients or investigators, or both. If the intent of the trial is to show similarity of the test and control drugs, the report of the study should assess the ability of the study to have detected a difference between treatments. Similarity of test drug and active control can mean either that both drugs were effective or that neither was effective. The analysis of the study should explain why the drugs should be considered effective in the study, for example, by reference to results in previous placebo-controlled studies of the active control drug.

(v) Historical control. The results of treatment with the test drug are compared with experience historically derived from the adequately documented natural history of the disease or condition, or from the results of active treatment, in comparable patients or populations. Because historical control populations usually cannot be as well assessed with respect to pertinent variables as can concurrent control populations, historical control designs are usually reserved for special circumstances. Examples include studies of diseases with high and predictable mortality (for example, certain malignancies) and studies in which the effect of the drug is self-evident (general anesthetics, drug metabolism).

(3) The method of selection of subjects provides adequate assurance that they have the disease or condition being studied, or evidence of susceptibility and exposure to the condition against which prophylaxis is directed.

(4) The method of assigning patients to treatment and control groups minimizes bias and is intended to assure comparability of the groups with respect to pertinent variables such as age, sex, severity of disease, duration of disease, and use of drugs or therapy other than the test drug. The protocol for the study and the report of its results should describe how subjects were assigned to groups. Ordinarily, in a concurrently controlled study, assignment is by randomization, with or without stratification.

(5) Adequate measures are taken to minimize bias on the part of the subjects, observers, and analysts of the data. The protocol and report of the study should describe the procedures used to accomplish this, such as blinding.

(6) The methods of assessment of subjects' response are well-defined and reliable. The protocol for the study and the report of results should explain the variables measured, the methods of observation, and criteria used to assess response.

(7) There is an analysis of the results of the study adequate to assess the effects of the drug. The report of the study should describe the results and the analytic methods used to evaluate them, including any appropriate statistical methods. The analysis should assess, among other things, the comparability of test and control groups with respect to pertinent variables, and the effects of any interim data analyses performed.

(c) The Director of the Center for Drug Evaluation and Research may, on the Director's own initiative or on the petition of an interested person, waive in whole or in part any of the criteria in paragraph (b) of this section with respect to a specific clinical investigation, either prior to the investigation or in the evaluation of a completed study. A petition for a waiver is required to set forth clearly and concisely the specific criteria from which waiver is sought, why the criteria are not reasonably applicable to the particular clinical investigation, what alternative procedures, if any, are to be, or have been employed, and what results have been obtained. The petition is also required to state why the clinical investigations so conducted will yield, or have yielded, substantial evidence of effectiveness, notwithstanding nonconformance with the criteria for which waiver is requested.

(d) For an investigation to be considered adequate for approval of a new drug, it is required that the test drug be standardized as to identity, strength, quality, purity, and dosage form to give significance to the results of the investigation.

(e) Uncontrolled studies or partially controlled studies are not acceptable as the sole basis for the approval of claims of effectiveness. Such studies carefully conducted and documented, may provide corroborative support of well-controlled studies regarding efficacy and may yield valuable data regarding safety of the test drug. Such studies will be considered on their merits in the light of the principles listed here, with the exception of the requirement for the comparison of the treated subjects with controls. Isolated case reports, random experience, and reports lacking the details which permit scientific evaluation will not be considered.

§314.127 Refusal to approve an abbreviated new drug application.

(a) FDA will refuse to approve an abbreviated application for a new drug under section 505(j) of the act for any of the following reasons:

(1) The methods used in, or the facilities and controls used for, the manufacture, processing, and packing of the drug product are inadequate to ensure and preserve its identity, strength, quality, and purity.

(2) Information submitted with the abbreviated new drug application is insufficient to show that each of the proposed conditions of use has been previously approved for the listed drug referred to in the application.

(3)(i) If the reference listed drug has only one active ingredient, information submitted with the abbreviated new drug application is insufficient to show that the active ingredient is the same as that of the reference listed drug;

(ii) If the reference listed drug has more than one active ingredient, information submitted with the abbreviated new drug application is insufficient to show that the active ingredients are the same as the active ingredients of the reference listed drug; or

(iii) If the reference listed drug has more than one active ingredient and if the abbreviated new drug application is for a drug product that has an active ingredient different from the reference listed drug:

(A) Information submitted with the abbreviated new drug application is insufficient to show:

(1) That the other active ingredients are the same as the active ingredients of the reference listed drug; or

(2) That the different active ingredient is an active ingredient of a listed drug or a drug that does not meet the requirements of section 201(p) of the act; or

(B) No petition to submit an abbreviated application for the drug product with the different active ingredient was approved under §314.93.

(4)(i) If the abbreviated new drug application is for a drug product whose route of administration, dosage form, or strength purports to be the same as that of the listed drug referred to in the abbreviated new drug application, information submitted in the abbreviated new drug application is insufficient to show that the route of administration, dosage form, or strength is the same as that of the reference listed drug; or

(ii) If the abbreviated new drug application is for a drug product whose route of administration, dosage form, or strength is different from that of the listed drug referred to in the application, no petition to submit an abbreviated new drug application for the drug product with the different route of administration, dosage form, or strength was approved under §314.93.

(5) If the abbreviated new drug application was submitted under the approval of a petition under §314.93, the abbreviated new drug application did not contain the information required by FDA with respect to the active ingredient, route of administration, dosage form, or strength that is not the same as that of the reference listed drug.

(6)(i) Information submitted in the abbreviated new drug application is insufficient to show that the drug product is bioequivalent to the listed drug referred to in the abbreviated new drug application; or

(ii) If the abbreviated new drug application was submitted under a petition approved under §314.93, information submitted in the abbreviated new drug application is insufficient to show that the active ingredients of the drug product are of the same pharmacological or therapeutic class as those of the reference listed drug and that the drug product can be expected to have the same therapeutic effect as the reference listed drug when administered to patients for each condition of use approved for the reference listed drug.

(7) Information submitted in the abbreviated new drug application is insufficient to show that the labeling proposed for the drug is the same as the labeling approved for the listed drug referred to in the abbreviated new drug application except for changes required because of differences approved in a petition

under §314.93 or because the drug product and the reference listed drug are produced or distributed by different manufacturers or because aspects of the listed drug's labeling are protected by patent, or by exclusivity, and such differences do not render the proposed drug product less safe or effective than the listed drug for all remaining, nonprotected conditions of use.

(8)(i) Information submitted in the abbreviated new drug application or any other information available to FDA shows that:

(A) The inactive ingredients of the drug product are unsafe for use, as described in paragraph (a)(8)(ii) of this section, under the conditions prescribed, recommended, or suggested in the labeling proposed for the drug product; or

(B) The composition of the drug product is unsafe, as described in paragraph (a)(8)(ii) of this section, under the conditions prescribed, recommended, or suggested in the proposed labeling because of the type or quantity of inactive ingredients included or the manner in which the inactive ingredients are included.

(ii)(A) FDA will consider the inactive ingredients or composition of a drug product unsafe and refuse to approve an abbreviated new drug application under paragraph (a)(8)(i) of this section if, on the basis of information available to the agency, there is a reasonable basis to conclude that one or more of the inactive ingredients of the proposed drug or its composition raises serious questions of safety. From its experience with reviewing inactive ingredients, and from other information available to it, FDA may identify changes in inactive ingredients or composition that may adversely affect a drug product's safety. The inactive ingredients or composition of a proposed drug product will be considered to raise serious questions of safety if the product incorporates one or more of these changes. Examples of the changes that may raise serious questions of safety include, but are not limited to, the following:

(1) A change in an inactive ingredient so that the product does not comply with an official compendium.

(2) A change in composition to include an inactive ingredient that has not been previously approved in a drug product for human use by the same route of administration.

(3) A change in the composition of a parenteral drug product to include an inactive ingredient that has not been previously approved in a parenteral drug product.

(4) A change in composition of a drug product for ophthalmic use to include an inactive ingredient that has not been previously approved in a drug for ophthalmic use.

(5) The use of a delivery or a modified release mechanism never before approved for the drug.

(6) A change in composition to include a significantly greater content of one or more inactive ingredients than previously used in the drug product.

(7) If the drug product is intended for topical administration, a change in the properties of the vehicle or base that might increase absorption of certain potentially toxic active ingredients thereby affecting the safety of the drug product, or a change in the lipophilic properties of a vehicle or base, e.g., a change from an oleaginous to a water soluble vehicle or base.

(B) FDA will consider an inactive ingredient in, or the composition of, a drug product intended for parenteral use to be unsafe and will refuse to approve the abbreviated new drug application unless it contains the same inactive ingredients, other than preservatives, buffers, and antioxidants, in the same concentration as the listed drug, and, if it differs from the listed drug in a preservative, buffer, or antioxidant, the application contains sufficient information to demonstrate that the difference does not affect the safety of the drug product.

(C) FDA will consider an inactive ingredient in, or the composition of, a drug product intended for ophthalmic or otic use unsafe and will refuse to approve the abbreviated new drug application unless it contains the same inactive ingredients, other than preservatives, buffers, substances to adjust tonicity, or thickening agents, in the same concentration as the listed drug, and if it differs from the listed drug in a preservative, buffer, substance to adjust tonicity, or thickening agent, the application contains sufficient information to demonstrate that the difference does not affect the safety of the drug product and the labeling does not claim any therapeutic advantage over or difference from the listed drug.

(9) Approval of the listed drug referred to in the abbreviated new drug application has been withdrawn or suspended for grounds described in §314.150(a) or FDA has published a notice of opportunity for hearing to withdraw approval of the reference listed drug under §314.150(a).

(10) Approval of the listed drug referred to in the abbreviated new drug application has been withdrawn under §314.151 or FDA has proposed to withdraw approval of the reference listed drug under §314.151(a).

(11) FDA has determined that the reference listed drug has been withdrawn from sale for safety or effectiveness reasons under §314.161, or the reference listed drug has been voluntarily withdrawn from sale and the agency has not determined whether the withdrawal is for safety or effectiveness reasons, or approval of the reference listed drug has been suspended under §314.153, or the agency has issued an initial decision proposing to suspend the reference listed drug under §314.153(a)(1).

(12) The abbreviated new drug application does not meet any other requirement under section 505(j)(2)(A) of the act.

(13) The abbreviated new drug application contains an untrue statement of material fact.

(b) FDA may refuse to approve an abbreviated application for a new drug if the applicant or contract research organization that conducted a bioavailability or bioequivalence study contained in the abbreviated new drug application refuses to permit an inspection of facilities or records relevant to the study by a properly authorized officer or employee of the Department of Health and Human Services or refuses to submit reserve samples of the drug products used in the study when requested by FDA.

§314.150 Withdrawal of approval of an application or abbreviated application.

(a) The Food and Drug Administration will notify the applicant, and, if appropriate, all other persons who manufacture or distribute identical, related, or similar drug products as defined in §§310.6 and 314.151(a) of this chapter and for a new drug afford an opportunity for a hearing on a proposal to withdraw approval of the application or abbreviated new drug application under section 505(e) of the act and under the procedure in §314.200, or, for an antibiotic, rescind a certification or release, or amend or repeal a regulation providing for certification under section 507 of the act and under the procedure in §314.300, if any of the following apply:

(1) The Secretary of Health and Human Services has suspended the approval of the application or abbreviated application for a new drug on a finding that there is an imminent hazard to the public health. FDA will promptly afford the applicant an expedited hearing following summary suspension on a finding of imminent hazard to health.

(2) FDA finds:

(i) That clinical or other experience, tests, or other scientific data show that the drug is unsafe for use under the conditions of use upon the basis of which the application or abbreviated application was approved; or

(ii) That new evidence of clinical experience, not contained in the application or not available to FDA until after the application or abbreviated application was approved, or tests by new methods, or tests by methods not deemed reasonably applicable when the application or abbreviated application was approved, evaluated together with the evidence available when the application or abbreviated application was approved, reveal that the drug is not shown to be safe for use under the conditions of use upon the basis of which the application or abbreviated application was approved; or

(iii) Upon the basis of new information before FDA with respect to the drug, evaluated together with the evidence available when the application or abbreviated application was approved, that there is a lack of substantial evidence from adequate and well-controlled investigations as defined in §314.126, that the drug will have the effect it is purported or represented to have under the conditions of use prescribed, recommended, or suggested in its labeling; or

(iv) That the application or abbreviated application contains any untrue statement of a material fact; or

(v) That the patent information prescribed by section 505(c) of the act was not submitted within 30 days after the receipt of written notice from FDA specifying the failure to submit such information; or

(b) FDA may notify the applicant, and, if appropriate, all other persons who manufacture or distribute identical, related, or similar drug products as defined in §310.6, and for a new drug afford an opportunity for a hearing on a proposal to withdraw approval of the application or abbreviated new drug application under section 505(e) of the act and under the procedure in §314.200, or, for an antibiotic, rescind a certification or release, or amend or repeal a regulation providing for certification under section 507 of the act and the procedure in §314.300, if the agency finds:

(1) That the applicant has failed to establish a system for maintaining required records, or has repeatedly or deliberately failed to maintain required records or to make required reports under section 505(k) or 507(g) of the act and §314.80, §314.81, or §314.98, or that the applicant has refused to permit access to, or copying or verification of, its records.

(2) That on the basis of new information before FDA, evaluated together with the evidence available when the application or abbreviated application was approved, the methods used in, or the facilities and controls used for, the manufacture, processing, and packing of the drug are inadequate to ensure and preserve its identity, strength, quality, and purity and were not made adequate within a reasonable time after receipt of written notice from the agency.

(3) That on the basis of new information before FDA, evaluated together with the evidence available when the application or abbreviated application was approved, the labeling of the drug, based on a fair evaluation of all material facts, is false or misleading in any particular, and the labeling was not corrected by the applicant within a reasonable time after receipt of written notice from the agency.

(4) That the applicant has failed to comply with the notice requirements of section 510 (j)(2) of the act.

(5) That the applicant has failed to submit bioavailability or bioequivalence data required under part 320 of this chapter.

(6) The application or abbreviated application does not contain an explanation of the omission of a report of any investigation of the drug product sponsored by the applicant, or an explanation of the omission of other information about the drug pertinent to an evaluation of the application or abbreviated application that is received or otherwise obtained by the applicant from any source.

(7) That any nonclinical laboratory study that is described in the application or abbreviated application and that is essential to show that the drug is safe for use under the conditions prescribed, recommended, or suggested in its labeling was not conducted in compliance with the good laboratory practice regulations in part 58 of this chapter and no reason for the noncompliance was provided or, if it was, the differences between the practices used in conducting the study and the good laboratory practice regulations do not support the validity of the study.

(8) Any clinical investigation involving human subjects described in the application or abbreviated application, subject to the institutional review board regulations in part 56 of this chapter or informed consent regulations in part 50 of this chapter, was not conducted in compliance with those regulations such that the rights or safety of human subjects were not adequately protected.

(9) That the applicant or contract research organization that conducted a bioavailability or bioequivalence study contained in the application or abbreviated application refuses to permit an inspection of facilities or records relevant to the study by a properly authorized officer or employee of the Department of Health and Human Services or refuses to submit reserve samples of the drug products used in the study when requested by FDA.

(10) That the labeling for the drug product

that is the subject of the abbreviated new drug application is no longer consistent with that for the listed drug referred to in the abbreviated new drug application, except for differences approved in the abbreviated new drug application or those differences resulting from:

(i) A patent on the listed drug issued after approval of the abbreviated new drug application; or

(ii) Exclusivity accorded to the listed drug after approval of the abbreviated new drug application that do not render the drug product less safe or effective than the listed drug for any remaining, nonprotected condition(s) of use.

(c) FDA will withdraw approval of an application or abbreviated application if the applicant requests its withdrawal because the drug subject to the application or abbreviated application is no longer being marketed, provided none of the conditions listed in paragraphs (a) and (b) of this section applies to the drug. FDA will consider a written request for a withdrawal under this paragraph to be a waiver of an opportunity for hearing otherwise provided for in this section. Withdrawal of approval of an application or abbreviated application under this paragraph is without prejudice to refiling.

(d) FDA may notify an applicant that it believes a potential problem associated with a drug is sufficiently serious that the drug should be removed from the market and may ask the applicant to waive the opportunity for hearing otherwise provided for under this section, to permit FDA to withdraw approval of the application or abbreviated application for the product, and to remove voluntarily the product from the market. If the applicant agrees, the agency will not make a finding under paragraph (b) of this section, but will withdraw approval of the application or abbreviated application in a notice published in the FEDERAL REGISTER that contains a brief summary of the agency's and the applicant's views of the reasons for withdrawal.

§314.151 Withdrawal of approval of an abbreviated new drug application under section 505(j) (5) of the act.

(a) Approval of an abbreviated new drug application approved under §314.105(d) may be withdrawn when the agency withdraws approval, under §314.150(a) or under this section, of the approved drug referred to in the abbreviated new drug application. If the agency proposed to withdraw approval of a listed drug under §314.150(a), the holder of an approved application for the listed drug has a right to notice and opportunity for hearing. The published notice of opportunity for hearing will identify all drug products approved under §314.105(d) whose applications are subject to withdrawal under this section if the listed drug is withdrawn, and will propose to withdraw such drugs. Holders of approved applications for the identified drug products will be provided notice and an opportunity to respond to the proposed withdrawal of their applications as described in paragraphs (b) and (c) of this section.

(b)(1) The published notice of opportunity for hearing on the withdrawal of the listed drug will serve as notice to holders of identified abbreviated new drug applications of the grounds for the proposed withdrawal.

(2) Holders of applications for drug products identified in the notice of opportunity for hearing may submit written comments on the notice of opportunity for hearing issued on the proposed withdrawal of the listed drug. If an abbreviated new drug application holder submits comments on the notice of opportunity for hearing and a hearing is granted, the abbreviated new drug application holder may participate in the hearing as a nonparty participant as provided for in §12.89 of this chapter.

(3) Except as provided in paragraphs (c) and (d) of this section, the approval of an abbreviated new drug application for a drug product identified in the notice of opportunity for hearing on the withdrawal of a listed drug will be withdrawn when the agency has completed the withdrawal of approval of the listed drug.

(c)(1) If the holder of an application for a drug identified in the notice of opportunity for hearing has submitted timely comments but does not have an opportunity to participate in a hearing because a hearing is not requested or is settled, the submitted comments will be considered by the agency, which will issue an initial decision. The initial decision will respond to the comments, and contain the agency's decision whether there are grounds to withdraw approval of the listed drug and of the abbreviated new drug applications on which timely comments were submitted. The initial decision will be sent to each abbreviated new drug application holder that has submitted comments.

(2) Abbreviated new drug application holders to whom the initial decision was sent may, within 30 days of the issuance of the initial decision, submit written objections.

(3) The agency may, at its discretion, hold a limited oral hearing to resolve dispositive factual issues that cannot be resolved on the basis of written submissions.

(4) If there are no timely objections to the initial decision, it will become final at the expiration of 30 days.

(5) If timely objections are submitted, they will be reviewed and responded to in a final decision.

(6) The written comments received, the initial decision, the evidence relied on in the comments and in the initial decision, the objections to the initial decision, and, if a limited oral hearing has been held, the transcript of that hearing and any documents submitted therein, shall form the record upon which the agency shall make a final decision.

(7) Except as provided in paragraph (d) of this section, any abbreviated new drug application whose holder submitted comments on the notice of opportunity for hearing shall be withdrawn upon the issuance of a final decision concluding that the listed drug should be withdrawn for grounds as described in §314.150(a). The final decision shall be in writing and shall constitute final agency action, reviewable in a judicial proceeding.

(8) Documents in the record will be publicly available in accordance with §10.20(j) of this chapter. Documents available for examination or copying will be placed on public display in the Dockets Management Branch (HFA-305), Food and Drug Administration, room. 1-23, 12420 Parklawn Dr., Rockville, MD 20857, promptly upon receipt in that office.

(d) If the agency determines, based upon information submitted by the holder of an abbreviated new drug application, that the grounds for withdrawal of the listed drug are not applicable to a drug identified in the notice of opportunity for hearing, the final decision will state that the approval of the abbreviated new drug application for such drug is not withdrawn.

§314.152 Notice of withdrawal of approval of an application or abbreviated application for a new drug.

If the Food and Drug Administration withdraws approval of an application or abbreviated application for a new drug, FDA will publish a notice in the FEDERAL REGISTER announcing the withdrawal of approval. If the application or abbreviated application was withdrawn for grounds described in §314.150(a) or §314.151, the notice will announce the removal of the drug from the list of approved drugs published under section 505(j)(6) of the act and shall satisfy the requirement of §314.162(b).

§314.153 Suspension of approval of an abbreviated new drug application.

(a) Suspension of approval. The approval of an abbreviated new drug application approved under §314.105(d) shall be suspended for the period stated when:

(1) The Secretary of the Department of Health and Human Services, under the imminent hazard authority of section 505(e) of the act or the authority of this paragraph, suspends approval of a listed drug referred to in the abbreviated new drug application, for the period of the suspension;

(2) The agency, in the notice described in paragraph (b) of this section, or in any subsequent written notice given an abbreviated new drug application holder by the agency, concludes that the risk of continued marketing and use of the drug is inappropriate, pending completion of proceedings to withdraw or suspend approval under §314.151 or paragraph (b) of this section; or

(3) The agency, under the procedures set forth in paragraph (b) of this section, issues a final decision stating the determination that the

abbreviated application is suspended because the listed drug on which the approval of the abbreviated new drug application depends has been withdrawn from sale for reasons of safety or effectiveness or has been suspended under paragraph (b) of this section. The suspension will take effect on the date stated in the decision and will remain in effect until the agency determines that the marketing of the drug has resumed or that the withdrawal is not for safety or effectiveness reasons.

(b) Procedures for suspension of abbreviated new drug applications when a listed drug is voluntarily withdrawn for safety or effectiveness reasons. (1) If a listed drug is voluntarily withdrawn from sale, and the agency determines that the withdrawal from sale was for reasons of safety or effectiveness, the agency will send each holder of an approved abbreviated new drug application that is subject to suspension as a result of this determination a copy of the agency's initial decision setting forth the reasons for the determination. The initial decision will also be placed on file with the Dockets Management Branch (HFA-305), Food and Drug Administration, room 1-23, 12420 Parklawn Dr., Rockville, MD 20857.

(2) Each abbreviated new drug application holder will have 30 days from the issuance of the initial decision to present, in writing, comments and information bearing on the initial decision. If no comments or information is received, the initial decision will become final at the expiration of 30 days.

(3) Comments and information received within 30 days of the issuance of the initial decision will be considered by the agency and responded to in a final decision.

(4) The agency may, in its discretion, hold a limited oral hearing to resolve dispositive factual issues that cannot be resolved on the basis of written submissions.

(5) If the final decision affirms the agency's initial decision that the listed drug was withdrawn for reasons of safety or effectiveness, the decision will be published in the FEDERAL REGISTER in compliance with §314.152, and will, except as provided in paragraph (b)(6) of this section, suspend approval of all abbreviated new drug applications identified under paragraph (b)(1) of this section and remove from the list the listed drug and any drug whose approval was suspended under this paragraph. The notice will satisfy the requirement of §314.162(b). The agency's final decision and copies of materials on which it relies will also be filed with the Dockets Management Branch (address in paragraph (b)(1) of this section).

(6) If the agency determines in its final decision that the listed drug was withdrawn for reasons of safety or effectiveness but, based upon information submitted by the holder of an abbreviated new drug application, also determines that the reasons for the withdrawal of the listed drug are not relevant to the safety and effectiveness of the drug subject to such abbreviated new drug application, the final decision will state that the approval of such abbreviated new drug application is not suspended.

(7) Documents in the record will be publicly available in accordance with §10.20(j) of this chapter. Documents available for examination or copying will be placed on public display in the Dockets Management Branch (address in paragraph (b)(1) of this section) promptly upon receipt in that office.

§314.160 Approval of an application or abbreviated application for which approval was previously refused, suspended, or withdrawn.

Upon the Food and Drug Administration's own initiative or upon request of an applicant, FDA may, on the basis of new data, approve an application or abbreviated application which it had previously refused, suspended, or withdrawn approval. FDA will publish a notice in the FEDERAL REGISTER announcing the approval.

§314.161 Determination of reasons for voluntary withdrawal of a listed drug.

(a) A determination whether a listed drug that has been voluntarily withdrawn from sale was withdrawn for safety or effectiveness reasons may be made by the agency at any time after the drug has been voluntarily withdrawn from sale, but must be made:

(1) Prior to approving an abbreviated new drug application that refers to the listed drug;

(2) Whenever a listed drug is voluntarily withdrawn from sale and abbreviated new drug applications that referred to the listed drug have been approved; and

(3) When a person petitions for such a determination under §§10.25(a) and 10.30 of this chapter.

(b) Any person may petition under §§10. 25(a) and 10.30 of this chapter for a determination whether a listed drug has been voluntarily withdrawn for safety or effectiveness reasons. Any such petition must contain all evidence available to the petitioner concerning the reason that the drug is withdrawn from sale.

(c) If the agency determines that a listed drug is withdrawn from sale for safety or effectiveness reasons, the agency will, except as provided in paragraph (d) of this section, publish a notice of the determination in the FEDERAL REGISTER.

(d) If the agency determines under paragraph (a) of this section that a listed drug is withdrawn from sale for safety and effectiveness reasons and there are approved abbreviated new drug applications that are subject to suspension under section 505(j)(5) of the act, FDA will initiate a proceeding in accordance with §314. 153(b).

(e) A drug that the agency determines is withdrawn for safety or effectiveness reasons will be removed from the list, under §314. 162. The drug may be relisted if the agency has evidence that marketing of the drug has resumed or that the withdrawal is not for safety or effectiveness reasons. A determination that the drug is not withdrawn for safety or effectiveness reasons may be made at any time after its removal from the list, upon the agency's initiative, or upon the submission of a petition under §§10.25(a) and 10.30 of this chapter. If the agency determines that the drug is not withdrawn for safety or effectiveness reasons, the agency shall publish a notice of this determination in the FEDERAL REGISTER. The notice will also announce that the drug is relisted, under §314.162(c). The notice will also serve to reinstate approval of all suspended abbreviated new drug applications that referred to the listed drug.

§314.162 Removal of a drug product from the list.

(a) FDA will remove a previously approved new drug product from the list for the period stated when:

(1) The agency withdraws or suspends approval of a new drug application or an abbreviated new drug application under §314. 150(a) or §314.151 or under the imminent hazard authority of section 505(e) of the act, for the same period as the withdrawal or suspension of the application; or

(2) The agency, in accordance with the procedures in §314.153(b) or §314.161, issues a final decision stating that the listed drug was withdrawn from sale for safety or effectiveness reasons, or suspended under §314.153 (b), until the agency determines that the withdrawal from the market has ceased or is not for safety or effectiveness reasons.

(b) FDA will publish in the FEDERAL REGISTER a notice announcing the removal of a drug from the list.

(c) At the end of the period specified in paragraph (a)(1) or (a)(2) of this section, FDA will relist a drug that has been removed from the list. The agency will publish in the FEDERAL REGISTER a notice announcing the relisting of the drug.

§314.170 Adulteration and misbranding of an approved drug.

All drugs, including those the Food and Drug Administration approves, or provides for certification of, under sections 505, 506, and 507 of the act and this part, are subject to the adulteration and misbranding provisions in sections 501, 502, and 503 of the act. FDA is authorized to regulate approved new drugs and approved antibiotic drugs by regulations issued through informal rulemaking under sections 501, 502, and 503 of the act.

Subpart E--Hearing Procedures for New Drugs

§314.200 Notice of opportunity for hearing; notice of participation and request for hearing; grant or denial of hearing.

(a) Notice of opportunity for hearing. The Director of the Center for Drug Evaluation and Research, Food and Drug Administration, will give the applicant, and all other persons who manufacture or distribute identical, related, or similar drug products as defined in §310.6 of this chapter, notice and an opportunity for a hearing on the Center's proposal to refuse to approve an application or to withdraw the approval of an application or abbreviated application under section 505(e) of the act. The notice will state the reasons for the action and the proposed grounds for the order.

(1) The notice may be general (that is, simply summarizing in a general way the information resulting in the notice) or specific (that is, either referring to specific requirements in the statute and regulations with which there is a lack of compliance, or providing a detailed description and analysis of the specific facts resulting in the notice).

(2) FDA will publish the notice in the FEDERAL REGISTER and will state that the applicant, and other persons subject to the notice under §310.6, who wishes to participate in a hearing, has 30 days after the date of publication of the notice to file a written notice of participation and request for hearing. The applicant, or other persons subject to the notice under §310.6, who fails to file a written notice of participation and request for hearing within 30 days, waives the opportunity for a hearing.

(3) It is the responsibility of every manufacturer and distributor of a drug product to review every notice of opportunity for a hearing published in the FEDERAL REGISTER to determine whether it covers any drug product that person manufactures or distributes. Any person may request an opinion of the applicability of a notice to a specific product that may be identical, related, or similar to a product listed in a notice by writing to the Division of Drug Labeling Compliance (HFD-310), Center for Drug Evaluation and Research, Food and Drug Administration, 5600 Fishers Lane, Rockville, MD 20857. A person shall request an opinion within 30 days of the date of publication of the notice to be eligible for an opportunity for a hearing under the notice. If a person requests an opinion, that person's time for filing an appearance and request for a hearing and supporting studies and analyses begins on the date the person receives the opinion from FDA.

(b) FDA will provide the notice of opportunity for a hearing to applicants and to other persons subject to the notice under §310.6, as follows:

(1) To any person who has submitted an application or abbreviated application, by delivering the notice in person or by sending it by registered or certified mail to the last address shown in the application or abbreviated application.

(2) To any person who has not submitted an application or abbreviated application but who is subject to the notice under §310.6 of this chapter, by publication of the notice in the FEDERAL REGISTER.

(c)(1) Notice of participation and request for a hearing, and submission of studies and comments. The applicant, or any other person subject to the notice under §310.6, who wishes to participate in a hearing, shall file with the Dockets Management Branch (HFA-305), Food and Drug Administration, Rm. 4-62, Rockville, MD 20857, (i) within 30 days after the date of the publication of the notice (or of the date of receipt of an opinion requested under paragraph (a)(3) of this section) a written notice of participation and request for a hearing and (ii) within 60 days after the date of publication of the notice, unless a different period of time is specified in the notice of opportunity for a hearing, the studies on which the person relies to justify a hearing as specified in paragraph (d) of this section. The applicant, or other person, may incorporate by reference the raw data underlying a study if the data were previously submitted to FDA as part of an application, abbreviated application, or other report.

(2) FDA will not consider data or analyses

submitted after 60 days in determining whether a hearing is warranted unless they are derived from well-controlled studies begun before the date of the notice of opportunity for hearing and the results of the studies were not available within 60 days after the date of publication of the notice. Nevertheless, FDA may consider other studies on the basis of a showing by the person requesting a hearing of inadvertent omission and hardship. The person requesting a hearing shall list in the request for hearing all studies in progress, the results of which the person intends later to submit in support of the request for a hearing. The person shall submit under paragraph (c)(1) (ii) of this section a copy of the complete protocol, a list of the participating investigators, and a brief status report of the studies.

(3) Any other interested person who is not subject to the notice of opportunity for a hearing may also submit comments on the proposal to withdraw approval of the application or abbreviated application. The comments are requested to be submitted within the time and under the conditions specified in this section.

(d) The person requesting a hearing is required to submit under paragraph (c)(1)(ii) of this section the studies (including all protocols and underlying raw data) on which the person relies to justify a hearing with respect to the drug product. Except, a person who requests a hearing on the refusal to approve an application is not required to submit additional studies and analyses if the studies upon which the person relies have been submitted in the application and in the format and containing the summaries required under §314.50.

(1) If the grounds for FDA's proposed action concern the effectiveness of the drug, each request for hearing is required to be supported only by adequate and well-controlled clinical studies meeting all of the precise requirements of §314.126 and, for combination drug products, §300.50, or by other studies not meeting those requirements for which a waiver has been previously granted by FDA under §314.126. Each person requesting a hearing shall submit all adequate and well-controlled clinical studies on the drug product, including any unfavorable analyses, views, or judgments with respect to the studies. No other data, information, or studies may be submitted.

(2) The submission is required to include a factual analysis of all the studies submitted. If the grounds for FDA's proposed action concern the effectiveness of the drug, the analysis is required to specify how each study accords, on a point-by-point basis, with each criterion required for an adequate well-controlled clinical investigation established under §314. 126 and, if the product is a combination drug product, with each of the requirements for a combination drug established in §300.50, or the study is required to be accompanied by an appropriate waiver previously granted by FDA. If a study concerns a drug or dosage form or condition of use or mode of administration other than the one in question, that fact is required to be clearly stated. Any study conducted on the final marketed form of the drug product is required to be clearly identified.

(3) Each person requesting a hearing shall submit an analysis of the data upon which the person relies, except that the required information relating either to safety or to effectiveness may be omitted if the notice of opportunity for hearing does not raise any issue with respect to that aspect of the drug; information on compliance with §300.50 may be omitted if the drug product is not a combination drug product. FDA can most efficiently consider submissions made in the following format.

I. Safety data.

A. Animal safety data.

1. Individual active components.

a. Controlled studies.

b. Partially controlled or uncontrolled studies.

2. Combinations of the individual active components.

a. Controlled studies.

b. Partially controlled or uncontrolled studies.

B. Human safety data.

1. Individual active components.

a. Controlled studies.

b. Partially controlled or uncontrolled studies.

c. Documented case reports.

d. Pertinent marketing experiences that may influence a determination about the safety of each individual active component.

2. Combinations of the individual active components.

a. Controlled studies.

b. Partially controlled or uncontrolled studies.

c. Documented case reports.

d. Pertinent marketing experiences that may influence a determination about the safety of each individual active component.

II. Effectiveness data.

A. Individual active components: Controlled studies, with an analysis showing clearly how each study satisfies, on a pointby-point basis, each of the criteria required by §314.126.

B. Combinations of individual active components.

1. Controlled studies with an analysis showing clearly how each study satisfies on a point-by-point basis, each of the criteria required by §314.126.

2. An analysis showing clearly how each requirement of §300.50 has been satisfied.

III. A summary of the data and views setting forth the medical rationale and purpose for the drug and its ingredients and the scientific basis for the conclusion that the drug and its ingredients have been proven safe and/or effective for the intended use. If there is an absence of controlled studies in the material submitted or the requirements of any element of §300.50 or §314.126 have not been fully met, that fact is required to be stated clearly and a waiver obtained under §314.126 is required to be submitted.

IV. A statement signed by the person responsible for such submission that it includes in full (or incorporates by reference as permitted in §314.200(c)(2)) all studies and information specified in §314.200(d).

(WARNING: A willfully false statement is a criminal offense, 18 U.S.C. 1001.)

(e) Contentions that a drug product is not subject to the new drug requirements. A notice of opportunity for a hearing encompasses all issues relating to the legal status of each drug product subject to it, including identical, related, and similar drug products as defined in §310.6. A notice of appearance and request for a hearing under paragraph (c)(1)(i) of this section is required to contain any contention that the product is not a new drug because it is generally recognized as safe and effective within the meaning of section 201(p) of the act, or because it is exempt from part or all of the new drug provisions of the act under the exemption for products marketed before June 25, 1938, contained in section 201(p) of the act or under section 107(c) of the Drug Amendments of 1962, or for any other reason. Each contention is required to be supported by a submission under paragraph (c)(1)(ii) of this section and the Commissioner of Food and Drugs will make an administrative determination on each contention. The failure of any person subject to a notice of opportunity for a hearing, including any person who manufactures or distributes an identical, related, or similar drug product as defined in §310.6, to submit a notice of participation and request for hearing or to raise all such contentions constitutes a waiver of any contentions not raised.

(1) A contention that a drug product is generally recognized as safe and effective within the meaning of section 201(p) of the act is required to be supported by submission of the same quantity and quality of scientific evidence that is required to obtain approval of an application for the product, unless FDA has waived a requirement for effectiveness (under §314.126) or safety, or both. The submission should be in the format and with the analyses required under paragraph (d) of this section. A person who fails to submit the required scientific evidence required under paragraph (d) waives the contention. General recognition of safety and effectiveness shall ordinarily be based upon published studies which may be corroborated by unpublished studies and other data and information.

(2) A contention that a drug product is exempt from part or all of the new drug provisions of the act under the exemption for products marketed before June 25, 1938,

contained in section 201(p) of the act, or under section 107(c) of the Drug Amendments of 1962, is required to be supported by evidence of past and present quantitative formulas, labeling, and evidence of marketing. A person who makes such a contention should submit the formulas, labeling, and evidence of marketing in the following format.

I. Formulation.

A. A copy of each pertinent document or record to establish the exact quantitative formulation of the drug (both active and inactive ingredients) on the date of initial marketing of the drug.

B. A statement whether such formulation has at any subsequent time been changed in any manner. If any such change has been made, the exact date, nature, and rationale for each change in formulation, including any deletion or change in the concentration of any active ingredient and/or inactive ingredient, should be stated, together with a copy of each pertinent document or record to establish the date and nature of each such change, including, but not limited to, the formula which resulted from each such change. If no such change has been made, a copy of representative documents or records showing the formula at representative points in time should be submitted to support the statement.

II. Labeling.

A. A copy of each pertinent document or record to establish the identity of each item of written, printed, or graphic matter used as labeling on the date the drug was initially marketed.

B. A statement whether such labeling has at any subsequent time been discontinued or changed in any manner. If such discontinuance or change has been made, the exact date, nature, and rationale for each discontinuance or change and a copy of each pertinent document or record to establish each such discontinuance or change should be submitted, including, but not limited to, the labeling which resulted from each such discontinuance or change. If no such discontinuance or change has been made, a copy of representative documents or records showing labeling at representative points in time should be submitted to support the statement.

III. Marketing.

A. A copy of each pertinent document or record to establish the exact date the drug was initially marketed.

B. A statement whether such marketing has at any subsequent time been discontinued. If such marketing has been discontinued, the exact date of each such discontinuance should be submitted, together with a copy of each pertinent document or record to establish each such date.

IV. Verification.

A statement signed by the person responsible for such submission, that all appropriate records have been searched and to the best of that person's knowledge and belief it includes a true and accurate presentation of the facts.

(WARNING: A willfully false statement is a criminal offense, 18 U.S.C. 1001.)

(3) The Food and Drug Administration will not find a drug product, including any active ingredient, which is identical, related, or similar, as described in §310.6, to a drug product, including any active ingredient for which an application is or at any time has been effective or deemed approved, or approved under section 505 of the act, to be exempt from part or all of the new drug provisions of the act.

(4) A contention that a drug product is not a new drug for any other reason is required to be supported by submission of the factual records, data, and information that are necessary and appropriate to support the contention.

(5) It is the responsibility of every person who manufactures or distributes a drug product in reliance upon a "grandfather" provision of the act to maintain files that contain the data and information necessary fully to document and support that status.

(f) Separation of functions. Separation of functions commences upon receipt of a request for hearing. The Director of the Center for Drug Evaluation and Research, Food and Drug Administration, will prepare an analysis of the request and a proposed order ruling on the matter. The analysis and proposed order, the request for hearing, and any proposed order denying a hearing and response under paragraph (g) (2) or (3) of this section will be submitted to the Office of the Commissioner of Food and Drugs for review and decision. When the Center for Drug Evaluation and Research recommends denial of a hearing on all issues on which a hearing is requested, no representative of the Center will participate or advise in the review and decision by the Commissioner. When the Center for Drug Evaluation and Research recommends that a hearing be granted on one or more issues on which a hearing is requested, separation of functions terminates as to those issues, and representatives of the Center may participate or advise in the review and decision by the Commissioner on those issues. The Commissioner may modify the text of the issues, but may not deny a hearing on those issues. Separation of functions continues with respect to issues on which the Center for Drug Evaluation and Research has recommended denial of a hearing. The Commissioner will neither evaluate nor rule on the Center's recommendation on such issues and such issues will not be included in the notice of hearing. Participants in the hearing may make a motion to the presiding officer for the inclusion of any such issue in the hearing. The ruling on such a motion is subject to review in accordance with §312.35(b). Failure to so move constitutes a waiver of the right to a hearing on such an issue. Separation of functions on all issues resumes upon issuance of a notice of hearing. The Office of the General Counsel, Department of Health and Human Services, will observe the same separation of functions.

(g) Summary judgment. A person who requests a hearing may not rely upon allegations or denials but is required to set forth specific facts showing that there is a genuine and substantial issue of fact that requires a hearing with respect to a particular drug product specified in the request for hearing.

(1) Where a specific notice of opportunity for hearing (as defined in paragraph (a)(1) of this section) is used, the Commissioner will enter summary judgment against a person who requests a hearing, making findings and conclusions, denying a hearing, if it conclusively appears from the face of the data, information, and factual analyses in the request for the hearing that there is no genuine and substantial issue of fact which precludes the refusal to approve the application or abbreviated application or the withdrawal of approval of the application or abbreviated application; for example, no adequate and well-controlled clinical investigations meeting each of the precise elements of §314.126 and, for a combination drug product, §300.50 of this chapter, showing effectiveness have been identified. Any order entering summary judgment is required to set forth the Commissioner's findings and conclusions in detail and is required to specify why each study submitted fails to meet the requirements of the statute and regulations or why the request for hearing does not raise a genuine and substantial issue of fact.

(2) When following a general notice of opportunity for a hearing (as defined in paragraph (a)(1) of this section) the Director of the Center for Drug Evaluation and Research concludes that summary judgment against a person requesting a hearing should be considered, the Director will serve upon the person requesting a hearing by registered mail a proposed order denying a hearing. This person has 60 days after receipt of the proposed order to respond with sufficient data, information, and analyses to demonstrate that there is a genuine and substantial issue of fact which justifies a hearing.

(3) When following a general or specific notice of opportunity for a hearing a person requesting a hearing submits data or information of a type required by the statute and regulations, and the Director of the Center for Drug Evaluation and Research concludes that summary judgment against the person should be considered, the Director will serve upon the person by registered mail a proposed order denying a hearing. The person has 60 days after receipt of the proposed order to respond with sufficient data, information, and analyses to demonstrate that there is a genuine and substantial issue of fact which justifies a hearing.

(4) If review of the data, information, and analyses submitted show that the grounds cited in the notice are not valid, for example, that substantial evidence of effectiveness exists, the

Commissioner will enter summary judgment for the person requesting the hearing, and rescind the notice of opportunity for hearing.

(5) If the Commissioner grants a hearing, it will begin within 90 days after the expiration of the time for requesting the hearing unless the parties otherwise agree in the case of denial of approval, and as soon as practicable in the case of withdrawal of approval.

(6) The Commissioner will grant a hearing if there exists a genuine and substantial issue of fact or if the Commissioner concludes that a hearing would otherwise be in the public interest.

(7) If the manufacturer or distributor of an identical, related, or similar drug product requests and is granted a hearing, the hearing may consider whether the product is in fact identical, related, or similar to the drug product named in the notice of opportunity for a hearing.

(8) A request for a hearing, and any subsequent grant or denial of a hearing, applies only to the drug products named in such documents.

(h) FDA will issue a notice withdrawing approval and declaring all products unlawful for drug products subject to a notice of opportunity for a hearing, including any identical, related, or similar drug product under §310.6, for which an opportunity for a hearing is waived or for which a hearing is denied. The Commissioner may defer or stay the action pending a ruling on any related request for a hearing or pending any related hearing or other administrative or judicial proceeding.

§314.201 Procedure for hearings.

Parts 10 through 16 apply to hearings relating to new drugs under section 505 (d) and (e) of the act.

§314.235 Judicial review.

(a) The Commissioner of Food and Drugs will certify the transcript and record. In any case in which the Commissioner enters an order without a hearing under §314.200(g), the record certified by the Commissioner is required to include the requests for hearing together with the data and information submitted and the Commissioner's findings and conclusion.

(b) A manufacturer or distributor of an identical, related, or similar drug product under §310.6 may seek judicial review of an order withdrawing approval of a new drug application, whether or not a hearing has been held, in a United States court of appeals under section 505(h) of the act.

Subpart F--Administrative Procedures for Antibiotics

§314.300 Procedure for the issuance, amendment, or repeal of regulations.

(a) The procedures in Part 10 apply to the issuance, amendment, or repeal of regulations under section 507 of the act.

(b)(1) The Commissioner of Food and Drugs, on his or her own initiative or on the application or request of any interested person, may publish in the FEDERAL REGISTER a notice of proposed rulemaking and order to issue, amend, or repeal any regulation contemplated by section 507 of the act. The notice and order may be general (that is, simply summarizing in a general way the information resulting in the notice and order) or specific (that is, either referring to specific requirements in the statute and regulations with which there is a lack of compliance, or providing a detailed description and analysis of the specific facts resulting in the notice and order).

(2) The Food and Drug Administration will give interested persons an opportunity to submit written comments and to request an informal conference on the proposal, unless the notice and opportunity for comment and informal conference have already been provid ed in connection with the announcement of the reports of the National Academy of Sciences/National Research Council, Drug Efficacy Study Group, to persons who will be adversely affected, or as provided in §§10.40(e) and 12.20(c)(2). A person is required to request an informal conference within 30 days of the notice of proposed rulemaking unless otherwise specified in the notice. If an informal

conference is requested and granted, those persons participating in the conference may submit comments, within 30 days of the conference, unless otherwise specified in the proposal.

(3) It is the responsibility of every manufacturer and distributor of an antibiotic drug product to review every proposal published in the FEDERAL REGISTER to determine whether it covers any drug product that person manufactures or distributes.

(4) After considering the written comments, the results of any conference, and the data available, the Commissioner will publish an order in the FEDERAL REGISTER acting on the proposal, with an opportunity for any person who will be adversely affected to file objections, to request a hearing, and to show reasonable grounds for the hearing. Any person who wishes to participate in a hearing, shall file with the Dockets Management Branch (HFA-305), Food and Drug Administration, Rm. 4-62, 5600 Fishers Lane, Rockville, MD 20857,

(i) within 30 days after the date of the publication of the order a written notice of participation and request for a hearing and

(ii) within 60 days after the date of publication of the order, unless a different period of time is specified in the order, the studies on which the person relies to justify a hearing as specified in paragraph (b)(6) of this section. The person may incorporate by reference the raw data underlying a study if the data were previously submitted to FDA as part of an application or other report.

(5) FDA will not consider data or analysis submitted after 60 days in determining whether a hearing is warranted unless they are derived from well-controlled studies begun before the date of the order and the results of the studies were not available within 60 days after the date of publication of the order. Nevertheless, FDA may consider other studies on the basis of a showing by the person requesting a hearing of inadvertent omission and hardship. The person requesting a hearing shall list in the request for hearing all studies in progress, the results of which the person intends later to submit in support of the request for hearing. The person shall submit under paragraph (b)(4)(ii) of this section a copy of the complete protocol, a list of the participating investigators, and a brief status report of the studies.

(6) The person requesting a hearing is required to submit as required under §314.200 (c)(1)(ii) the studies (including all protocols and underlying raw data) on which the person relies to justify a hearing with respect to the drug product. Except, a person who requests a hearing on a proposal is not required to submit additional studies and analyses if the studies upon which the person relies have been submitted in an application and in the format and containing the summaries required under §314.50.

(i) If the grounds for FDA proposed action concern the effectiveness of the drug, each request for hearing is required to be supported only by adequate and well-controlled clinical studies meeting all of the precise requirements of §314.126 and, for combination drug products, §300.50, or by other studies not meeting those requirements for which a waiver has been previously granted by FDA under §314. 126. Each person requesting a hearing shall submit all adequate and well-controlled clinical studies on the drug product, any unfavorable analyses, views, or judgments with respect to the studies. No other data, information, or studies may be submitted.

(ii) The submission is required to include a factual analysis of all the studies submitted. If the grounds for FDA proposed action concern the effectiveness of the drug, the analysis is required to specify how each study accords, on a point-by-point basis, with each criterion required for an adequate well-controlled clinical investigation established under §314. 126 and, if the product is a combination drug product, with each of the requirements for a combination drug established in §300.50, or the study is required to be accompanied by an appropriate waiver previously granted by FDA. If a study concerns a drug entity or dosage form or condition of use or mode of administration other than the one in question, that fact is required to be clearly stated. Any study conducted on the final marketed form of the drug product is required to be clearly identified.

(iii) Each person requesting a hearing shall submit an analysis of the data upon which the person relies, except that the required information relating either to safety or to effectiveness may be omitted if the notice of opportunity for hearing does not raise any issue with respect to that aspect of the drug; information on compliance with §300.50 may be omitted if the drug product is not a combination drug product. FDA can most efficiently consider submissions made in the following format.

I. Safety data.

A. Animal safety data.

1. Individual active components.

a. Controlled studies.

b. Partially controlled or uncontrolled studies.

2. Combinations of the individual active components.

a. Controlled studies.

b. Partially controlled or uncontrolled studies.

B. Human safety data.

1. Individual active components.

a. Controlled studies.

b. Partially controlled or uncontrolled studies.

c. Documented case reports.

d. Pertinent marketing experiences that may influence a determination about the safety of each individual active component.

2. Combinations of the individual active components.

a. Controlled studies.

b. Partially controlled or uncontrolled studies.

c. Documented case reports.

d. Pertinent marketing experiences that may influence a determination about the safety of each individual active component.

II. Effectiveness data.

A. Individual active components: Controlled studies, with an analysis showing clearly how each study satisfies, on a pointby-point basis, each of the criteria required by §314.126.

B. Combinations of individual active components.

1. Controlled studies with an analysis showing clearly how each study satisfies on a point-by-point basis, each of the criteria required by §314.126.

2. An analysis showing clearly how each requirement of §300.50 has been satisfied.

III. A summary of the data and views setting forth the medical rationale and purpose for the drug and its ingredients and the scientific basis for the conclusion that the drug and its ingredients have been proven safe and/or effective for the intended use. If there is an absence of controlled studies in the material submitted or the requirements of any element of §300.50 or §314.126 have not been fully met, that fact is required to be stated clearly and a waiver obtained under §314.126 is required to be submitted.

IV. A statement signed by the person responsible for such submission that it includes in full (or incorporates by reference as permitted in §314.200(c)(2)) all studies and information specified in §314.200(d).

(WARNING: A willfully false statement is a criminal offense, 18 U.S.C. 1001.)

(7) Separation of functions. Separation of functions commences upon receipt of a request for hearing. The Director of the Center for Drug Evaluation and Research will prepare an analysis of the request and a proposed order ruling on the matter. The analysis and proposed order, the request for hearing, and any proposed order denying a hearing and response under paragraph (b)(8) (ii) or (iii) of this section will be submitted to the Office of the Commissioner for review and decision. When the Center for Drug Evaluation and Research recommends denial of a hearing on all issues on which a hearing is requested, no representative of the Center will participate or advise in the review and decision by the Commissioner. When the Center for Drug Evaluation and Research recommends that a hearing be granted on one or more issues on which a hearing is requested, separation of functions terminates as to those issues, and representatives of the Center may participate or advise in the review and decision by the Commissioner on those issues. The Commissioner may modify the text of the issues, but may not deny a hearing on those issues. Separation of functions continues with respect to issues on which the Center for Drug Evaluation and Research has recommended denial

of a hearing. The Commissioner will neither evaluate nor rule on the Center's recommendation on such issues and such issues will not be included in the notice of hearing. Participants in the hearing may make a motion to the presiding officer for the inclusion of any such issue in the hearing. The ruling on such a motion is subject to review in accordance with §12.35(b). Failure to so move constitutes a waiver of the right to a hearing on such an issue. Separation of functions on all issues resumes upon issuance of a notice of hearing. The Office of the General Counsel, Department of Health and Human Services, will observe the same separation of functions.

(8) Summary judgment. A person who requests a hearing may not rely upon allegations or denials but is required to set forth specific facts showing that there is a genuine and substantial issue of fact that requires a hearing with respect to a particular drug product specified in the request for hearing.

(i) Where a specific notice of opportunity for hearing (as defined in paragraph (b)(1) of this section) is used, the Commissioner will enter summary judgment against a person who requests a hearing, making findings and conclusions, denying a hearing, if it conclusively appears from the face of the data, information, and factual analyses in the request for the hearing that there is no genuine and substantial issue of fact which precludes the refusal to approve the application or the withdrawal of approval of the application; for example, no adequate and well-controlled clinical investigations meeting each of the precise elements of §314.126 and, for a combination drug product, §300.50, showing effectiveness have been identified. Any order entering summary judgment is required to set forth the Commissioner's findings and conclusions in detail and is required to specify why each study submitted fails to meet the requirements of the statute and regulations or why the request for hearing does not raise a genuine and substantial issue of fact.

(ii) When following a general notice of opportunity for a hearing (as defined in paragraph (b)(1) of this section) the Director of the Center for Drug Evaluation and Research concludes that summary judgment against a person requesting a hearing should be considered, the Director will serve upon the person requesting a hearing by registered mail a proposed order denying a hearing. This person has 60 days after receipt of the proposed order to respond with sufficient data, information, and analyses to demonstrate that there is a genuine and substantial issue of fact which justifies a hearing.

(iii) When following a general or specific notice of opportunity for a hearing a person requesting a hearing submits data or information of a type required by the statute and regulations, and the Director of the Center for Drug Evaluation and Research concludes that summary judgment against the person should be considered, the Director will serve upon the person by registered mail a proposed order denying a hearing. The person has 60 days after receipt of the proposed order to respond with sufficient data, information, and analyses to demonstrate that there is a genuine and substantial issue of fact which justifies a hearing.

(iv) If review of the data, information, and analyses submitted show that the basis for the order is not valid, for example, that substantial evidence of effectiveness exists, the Commissioner will enter summary judgment for the person requesting the hearing, and revoke the order. If a hearing is not requested, the order will become effective as published. (v) If the Commissioner grants a hearing, it will be conducted under part 12.

(vi) The Commissioner will grant a hearing if there exists a genuine and substantial issue of fact or if the Commissioner concludes that a hearing would otherwise be in the public interest.

(9) The repeal of any regulation constitutes a revocation of all outstanding certificates based upon such regulation. However, the Commissioner may, in his or her discretion, defer or stay such action pending a ruling on any related request for a hearing or pending any related hearing or other administrative or judicial proceeding.

(c) Whenever any interested person submits an application or request under section 507 of the act and part 314 and FDA sends the person an approvable letter under §314.110 or a not approvable letter under §314.120, the person may file a petition proposing the issuance, amendment, or repeal of the regulation under the provisions of section 507(f) of the act and part 10. The Commissioner shall cause the regulation proposed in the petition to be published in the FEDERAL REGISTER within 60 days of the receipt of an acceptable petition and further proceedings shall be in accord with the provisions of sections 507(f) and 701 (f) and (g) of the act and part 10.

(d)(1) FDA will not promulgate a regulation providing for the certification of any batch of any drug composed wholly or in part of any kind of penicillin, streptomycin, chlortetracycline, chloramphenicol, bacitracin, or any other antibiotic drug, or any derivative thereof, intended for human use and no existing regulation will be continued in effect unless it is established by substantial evidence that the drug will have such characteristics of identity, strength, quality, and purity necessary to adequately ensure safety and efficacy of use. "Substantial evidence" has been defined by Congress to mean "evidence consisting of adequate and well-controlled investigations, including clinical investigations, by experts qualified by scientific training and experience to evaluate the effectiveness of the drug involved, on the basis of which it could fairly and responsibly be concluded by such experts that the drug will have the effect it purports or is represented to have under the conditions prescribed, recommended, or suggested in the labeling or proposed labeling thereof." This definition is made applicable to a number of antibiotic drugs by section 507(h) of the act and it is the test of efficacy that FDA will apply in promulgating, amending, or repealing regulations for all antibiotics under section 507(a) of the act as well.

(2) The scientific essentials of an adequate and well-controlled clinical investigation are described in §314.126.

Subpart G--Miscellaneous Provisions

§*314.410 Imports and exports of new drugs and antibiotics.*

(a) Imports.

(1) A new drug or an antibiotic may be imported into the United States if:

(i) It is the subject of an approved application under this part or, in the case of an antibiotic not exempt from certification under part 433, it is also certified or released; or

(ii) it complies with the regulations pertaining to investigational new drugs under part 312; and it complies with the general regulations pertaining to imports under subpart E of part 1.

(2) A drug substance intended for use in the manufacture, processing, or repacking of a new drug may be imported into the United States if it complies with the labeling exemption in §201.122 pertaining to shipments of drug substances in domestic commerce.

(b) Exports.

(1) A new drug or an antibiotic may be exported if it is the subject of an approved application under this part, and, in the case of an antibiotic, it is certified or released, or it complies with the regulations pertaining to investigational new drugs under part 312.

(2) A new drug substance that is covered by an application approved under this part for use in the manufacture of an approved drug product may be exported by the applicant or any person listed as a supplier in the approved application, provided the drug substance intended for export meets the specifications of, and is shipped with a copy of the labeling required for, the approved drug product.

(3) An antibiotic drug product or drug substance that is subject to certification under section 507 of the act, but which has not been certified or released, may be exported under section 801(d) of the act if it meets the following conditions:

(i) It meets the specifications of the foreign purchaser;

(ii) It is not in conflict with the laws of the country to which it is intended for export;

(iii) It is labeled on the outside of the shipping package that it is intended for export; and

(iv) It is not sold or offered for sale in the United States.

§314.420 Drug master files.

(a) A drug master file is a submission of information to the Food and Drug Administration by a person (the drug master file holder) who intends it to be used for one of the following purposes: To permit the holder to incorporate the information by reference when the holder submits an investigational new drug application under part 312 or submits an application or an abbreviated application or an amendment or supplement to them under this part, or to permit the holder to authorize other persons to rely on the information to support a submission to FDA without the holder having to disclose the information to the person. FDA ordinarily neither independently reviews drug master files nor approves or disapproves submissions to a drug master file. Instead, the agency customarily reviews the information only in the context of an application under part 312 or this part. **A drug master file may contain information of the kind required for any submission to the agency, including information about the following:**

(1) Manufacturing site, facilities, operating procedures, and personnel (because an FDA on-site inspection of a foreign drug manufacturing facility presents unique problems of planning and travel not presented by an inspection of a domestic manufacturing facility, this information is only recommended for foreign manufacturing establishments);

(2) Drug substance, drug substance intermediate, and materials used in their preparation, or drug product;

(3) Packaging materials;

(4) Excipient, colorant, flavor, essence, or materials used in their preparation;

(5) FDA-accepted reference information. (A person wishing to submit information and supporting data in a drug master file (DMF) that is not covered by Types I through IV DMF's must first submit a letter of intent to the Drug Master File Staff, Food and Drug Administration, 12420 Parklawn Dr., Rm. 2-14, Rockville, MD 20852. FDA will then contact the person to discuss the proposed submission.)

(b) An investigational new drug application or an application, abbreviated application, amendment, or supplement may incorporate by reference all or part of the contents of any drug master file in support of the submission if the holder authorizes the incorporation in writing. Each incorporation by reference is required to describe the incorporated material by name, reference number, volume, and page number of the drug master file.

(c) A drug master file is required to be submitted in two copies. The agency has prepared under §10.90(b) a guideline that provides information about how to prepare a well-organized drug master file. If the drug master file holder adds, changes, or deletes any information in the file, the holder shall notify in writing, each person authorized to reference that information. Any addition, change, or deletion of information in a drug master file (except the list required under paragraph (d) of this section) is required to be submitted in two copies and to describe by name, reference number, volume, and page number the information affected in the drug master file.

(d) The drug master file is required to contain a complete list of each person currently authorized to incorporate by reference any information in the file, identifying by name, reference number, volume, and page number the information that each person is authorized to incorporate. If the holder restricts the authorization to particular drug products, the list is required to include the name of each drug product and the application number, if known, to which the authorization applies.

(e) The public availability of data and information in a drug master file, including the availability of data and information in the file to a person authorized to reference the file, is determined under part 20 and §314. 430.

§314.430 Availability for public disclosure of data and information in an application or abbreviated application.

(a) The Food and Drug Administration will determine the public availability of any part of an application or abbreviated application under this section and part 20 of this chapter. For purposes of this section, the application or abbreviated application includes all data and information submitted with or incorporated by reference in the application or abbreviated application, including investigational new drug applications, drug master files under §314. 420, supplements submitted under §314.70 or §314.97, reports under §314.80 or §314.98, and other submissions. For purposes of this section, safety and effectiveness data include all studies and tests of a drug on animals and humans and all studies and tests of the drug for identity, stability, purity, potency, and bioavailability.

(b) FDA will not publicly disclose the existence of an application or abbreviated application before an approvable letter is sent to the applicant under §314.110, unless the existence of the application or abbreviated application has been previously publicly disclosed or acknowledged. The Center for Drug Evaluation and Research will maintain and make available for public disclosure a list of applications or abbreviated applications for which the agency has sent an approvable letter to the applicant.

(c) If the existence of an unapproved application or abbreviated application has not been publicly disclosed or acknowledged, no data or information in the application or abbreviated application is available for public disclosure.

(d) If the existence of an application or abbreviated application has been publicly disclosed or acknowledged before the agency sends an approval letter to the applicant, no data or information contained in the application or abbreviated application is available for public disclosure before the agency sends an approval letter, but the Commissioner may, in his or her discretion, disclose a summary of selected portions of the safety and effectiveness data that are appropriate for public consideration of a specific pending issue; for example, for consideration of an open session of an FDA advisory committee.

(e) After FDA sends an approval letter to the applicant, the following data and information in the application or abbreviated application are immediately available for public disclosure, unless the applicant shows that extraordinary circumstances exist. A list of approved applications and abbreviated applications, entitled "Approved Drug Products with Therapeutic Equivalence Evaluations," is available from the Government Printing Office, Washington, DC 20402. This list is updated monthly.

(1) [Reserved]

(2) If the application applies to a new drug, all safety and effectiveness data previously disclosed to the public as set forth in §20.81 and a summary or summaries of the safety and effectiveness data and information submitted with or incorporated by reference in the application. The summaries do not constitute the full reports of investigations under section 505(b)(1) of the act (21 U.S.C. 355(b)(1)) on which the safety or effectiveness of the drug may be approved. The summaries consist of the following:

(i) For an application approved before July 1, 1975, internal agency records that describe safety and effectiveness data and information, for example, a summary of the basis for approval or internal reviews of the data and information, after deletion of the following:

(a) Names and any information that would identify patients or test subjects or investigators.

(b) Any inappropriate gratuitous comments unnecessary to an objective analysis of the data and information.

(ii) For an application approved on or after July 1, 1975, a Summary Basis of Approval (SBA) document that contains a summary of the safety and effectiveness data and information evaluated by FDA during the drug approval process. The SBA is prepared in one of the following ways:

(a) Before approval of the application, the applicant may prepare a draft SBA which the Center for Drug Evaluation and Research will review and may revise. The draft may be submitted with the application or as an amendment.

(b) The Center for Drug Evaluation and Research may prepare the SBA.

(3) A protocol for a test or study, unless it is shown to fall within the exemption estab-

lished for trade secrets and confidential commercial information in §20.61.

(4) Adverse reaction reports, product experience reports, consumer complaints, and other similar data and information after deletion of the following:

(i) Names and any information that would identify the person using the product.

(ii) Names and any information that would identify any third party involved with the report, such as a physician or hospital or other institution.

(5) A list of all active ingredients and any inactive ingredients previously disclosed to the public as set forth in §20.81.

(6) An assay method or other analytical method, unless it serves no regulatory or compliance purpose and is shown to fall within the exemption established for trade secrets and confidential commercial information in §20.61.

(7) All correspondence and written summaries of oral discussions between FDA and the applicant relating to the application, under the provisions of Part 20.

(8) All records showing the testing of an action on a particular lot of a certifiable antibiotic by FDA.

(f) All safety and effectiveness data and information which have been submitted in an application and which have not previously been disclosed to the public are available to the public, upon request, at the time any one of the following events occurs unless extraordinary circumstances are shown:

(1) No work is being or will be undertaken to have the application approved.

(2) A final determination is made that the application is not approvable and all legal appeals have been exhausted.

(3) Approval of the application is withdrawn and all legal appeals have been exhausted.

(4) A final determination has been made that the drug is not a new drug.

(5) For applications submitted under section 505(b) of the act, the effective date of the approval of the first abbreviated application submitted under section 505(j) of the act which refers to such drug, or the date on which the approval of an abbreviated application under section 505(j) of the act which refers to such drug could be made effective if such an abbreviated application had been submitted.

(6) For applications or abbreviated applications submitted under sections 505(j), 506, and 507 of the act, when FDA sends an approval letter to the applicant.

(g) The following data and information in an application or abbreviated application are not available for public disclosure unless they have been previously disclosed to the public as set forth in §20.81 of this chapter or they relate to a product or ingredient that has been abandoned and they do not represent a trade secret or confidential commercial or financial information under §20.61 of this chapter:

(1) Manufacturing methods or processes, including quality control procedures.

(2) Production, sales distribution, and similar data and information, except that any compilation of that data and information aggregated and prepared in a way that does not reveal data or information which is not available for public disclosure under this provision is available for public disclosure.

(3) Quantitative or semiquantitative formulas.

(h) The compilations of information specified in §20.117 are available for public disclosure.

§314.440 Addresses for applications and abbreviated applications.

(a) Applicants shall send applications, abbreviated applications, and other correspondence relating to matters covered by this part, except for products listed in paragraph (b) of this section, to the Center for Drug Evaluation and Research, Food and Drug Administration, 5600 Fishers Lane, Rockville, MD 20857, and directed to the appropriate office identified below:

(1) An application under §314.50 or §314.54 submitted for filing should be directed to the Document and Records Section, 12420 Parklawn Dr., Rockville, MD 20852. Applicants may obtain folders for binding applications from the Forms and Publications Warehouse, 12100 Parklawn Dr., Rockville, MD 20852. After FDA has filed the application,

the agency will inform the applicant which division is responsible for the application. Amendments, supplements, resubmissions, requests for waivers, and other correspondence about an application that has been filed should be directed to the appropriate division.

(2) An abbreviated application under §314. 94, and amendments, supplements, and resubmissions should be directed to the Office of Generic Drugs (HFD-600), Center for Drug Evaluation and Research, Food and Drug Administration, 5600 Fishers Lane, Rockville, MD 20857. Items sent by parcel post or overnight courier service should be directed to the Office of Generic Drugs (HFD-600), Center for Drug Evaluation and Research, Food and Drug Administration, Metro Park North II, 7500 Standish Place, rm. 150, Rockville, MD 20855. Correspondence not associated with an application should be addressed specifically to the intended office or division and to the person as follows: Center for Drug Evaluation and Research, Food and Drug Administration, Attn: [insert name of person], MPN II, HFD- [insert mail code of office or division], 5600 Fishers Lane, Rockville, MD 20857. The mail code for the Office of Generic Drugs is HFD-600, the mail code for the Division of Chemistry is HFD-630, and the mail code for the Division of Bioequivalence is HFD-650.

(3) A request for an opportunity for a hearing under §314.110 or §314.120 on the question of whether there are grounds for denying approval of an application, except an application under paragraph (b) of this section, should be directed to the Division of Regulatory Affairs (HFD-360).

(b) Applicants shall send applications and other correspondence relating to matters covered by this part for the drug products listed below to the Division of Product Certification (HFB-240), Center for Biologics Evaluation and Research, Food and Drug Administration, 8800 Rockville Pike, Bethesda, MD 20892, except applicants shall send a request for an opportunity for a hearing under §314. 110 or §314.120 on the question of whether there are grounds for denying approval of an application to the Director, Center for Biologics Evaluation and Research (HFB-1), at the same address.

(1) Ingredients packaged together with containers intended for the collection, processing, or storage of blood and blood components.

(2) Urokinase products.

(3) Plasma volume expanders and hydroxyethyl starch for leukapheresis.

§314.445 Guidelines.

(a) The Food and Drug Administration prepares guidelines under §10.90(b) to help persons comply with requirements in this part.

(b) The Center for Drug Evaluation and Research will maintain and make publicly available a list of guidelines that apply to the Center's regulations. The list states how a person can obtain a copy of each guideline. A request for a copy of the list should be directed to the CDER Executive Secretariat Staff (HFD-8), Center for Drug Evaluation and Research, Food and Drug Administration, 5600 Fishers Lane, Rockville, MD 20857.

Subpart H--Accelerated Approval of New Drugs for Serious or Life-Threatening Illnesses

§314.500 Scope.

This subpart applies to certain new drug and antibiotic products that have been studied for their safety and effectiveness in treating serious or life-threatening illnesses and that provide meaningful therapeutic benefit to patients over existing treatments (e.g., ability to treat patients unresponsive to, or intolerant of, available therapy, or improved patient response over available therapy).

§314.510 Approval based on a surrogate endpoint or on an effect on a clinical endpoint other than survival or irreversible morbidity.

FDA may grant marketing approval for a new drug product on the basis of adequate and well-controlled clinical trials establishing that the drug product has an effect on a surrogate endpoint that is reasonably likely, based on epidemiologic, therapeutic, pathophysiologic, or other evidence, to predict clinical benefit or on the basis of an effect on a clinical endpoint other than survival or irreversible morbidity. Approval under this section will be subject to the requirement that the applicant study the drug further, to verify and describe its clinical benefit, where there is uncertainty as to the relation of the surrogate endpoint to clinical benefit, or of the observed clinical benefit to ultimate outcome. Postmarketing studies would usually be studies already underway. When required to be conducted, such studies must also be adequate and well-controlled. The applicant shall carry out any such studies with due diligence.

§314.520 Approval with restrictions to assure safe use.

(a) If FDA concludes that a drug product shown to be effective can be safely used only if distribution or use is restricted, FDA will require such postmarketing restrictions as are needed to assure safe use of the drug product, such as:

(1) Distribution restricted to certain facilities or physicians with special training or experience; or

(2) Distribution conditioned on the performance of specified medical procedures.

(b) The limitations imposed will be commensurate with the specific safety concerns presented by the drug product.

§314.530 Withdrawal procedures.

(a) For new drugs and antibiotics approved under §§314.510 and 314.520, FDA may withdraw approval, following a hearing as provided in part 15 of this chapter, as modified by this section, if:

(1) A postmarketing clinical study fails to verify clinical benefit;

(2) The applicant fails to perform the required postmarketing study with due diligence;

(3) Use after marketing demonstrates that postmarketing restrictions are inadequate to assure safe use of the drug product;

(4) The applicant fails to adhere to the postmarketing restrictions agreed upon;

(5) The promotional materials are false or misleading; or

(6) Other evidence demonstrates that the drug product is not shown to be safe or effective under its conditions of use.

(b) Notice of opportunity for a hearing. The Director of the Center for Drug Evaluation and Research will give the applicant notice of an opportunity for a hearing on the Center's proposal to withdraw the approval of an application approved under §314.510 or §314.520. The notice, which will ordinarily be a letter, will state generally the reasons for the action and the proposed grounds for the order.

(c) Submission of data and information.

(1) If the applicant fails to file a written request for a hearing within 15 days of receipt of the notice, the applicant waives the opportunity for a hearing.

(2) If the applicant files a timely request for a hearing, the agency will publish a notice of hearing in the FEDERAL REGISTER in accordance with §§12.32(e) and 15.20 of this chapter.

(3) An applicant who requests a hearing under this section must, within 30 days of receipt of the notice of opportunity for a hearing, submit the data and information upon which the applicant intends to rely at the hearing.

(d) Separation of functions. Separation of functions (as specified in §10.55 of this chapter) will not apply at any point in withdrawal proceedings under this section.

(e) Procedures for hearings. Hearings held under this section will be conducted in accordance with the provisions of part 15 of this chapter, with the following modifications:

(1) An advisory committee duly constituted under part 14 of this chapter will be present at the hearing. The committee will be asked to review the issues involved and to provide advice and recommendations to the Commissioner of Food and Drugs.

(2) The presiding officer, the advisory committee members, up to three representatives of the applicant, and up to three representatives of the Center may question any person during or at the conclusion of the person's presentation. No other person attending the hearing may question a person making

a presentation. The presiding officer may, as a matter of discretion, permit questions to be submitted to the presiding officer for response by a person making a presentation.

(f) Judicial review. The Commissioner's decision constitutes final agency action from which the applicant may petition for judicial review. Before requesting an order from a court for a stay of action pending review, an applicant must first submit a petition for a stay of action under §10.35 of this chapter.

§314.540 Postmarketing safety reporting.

Drug products approved under this program are subject to the postmarketing recordkeeping and safety reporting applicable to all approved drug products, as provided in §§314.80 and 314.81.

§314.550 Promotional materials.

For drug products being considered for approval under this subpart, unless otherwise informed by the agency, applicants must submit to the agency for consideration during the preapproval review period copies of all promotional materials, including promotional labeling as well as advertisements, intended for dissemination or publication within 120 days following marketing approval. After 120 days following marketing approval, unless otherwise informed by the agency, the applicant must submit promotional materials at least 30 days prior to the intended time of initial dissemination of the labeling or initial publication of the advertisement.

§314.560 Termination of requirements.

If FDA determines after approval that the requirements established in §314.520, §314.530, or §314.550 are no longer necessary for the safe and effective use of a drug product, it will so notify the applicant. Ordinarily, for drug products approved under §314.510, these requirements will no longer apply when FDA determines that the required postmarketing study verifies and describes the drug product's clinical benefit and the drug product would be appropriate for approval under traditional procedures. For drug products approved under §314.520, the restrictions would no longer apply when FDA determines that safe use of the drug product can be assured through appropriate labeling. FDA also retains the discretion to remove specific postapproval requirements upon review of a petition submitted by the sponsor in accordance with §10.30.

Chapter 4

EPA REGULATIONS

THE EPA APPROACH

Under the authority of section 4 of the Toxic Substances Control Act (TSCA), the EPA evaluates laboratory data submitted to the agency regarding tests of the health effects of chemical substances and mixtures. The EPA also, under authority of the Federal Insecticide, Fungicide and Rodenticide Act (FIFRA), evaluates laboratory test data relating to hazards to humans arising from the use of a pesticide product when the agency evaluates pesticide registration applications.

The EPA was aware of problems, uncovered by the FDA in the mid-1970s, relating to unacceptable laboratory practices. The EPA responded by forming the Toxicology Auditing Program in the agency's Office of Pesticide Programs. The EPA also held public hearings to solicit comments on how appropriate the agency's approach was to data quality assurance for pesticide testing.

GOOD LABORATORY PRACTICES

Like the FDA, the EPA determined that the promulgation of Good Laboratory Practice regulations would most effectively handle the problem of compliance with adequate control standards, and the agency published proposed health effects standards for testing under TSCA on May 9, 1979. Proposed GLP regulations applicable to laboratory studies submitted to the EPA in compliance with FIFRA were published on April 18, 1980. Supplemental GLP standards for the development of data on physical, chemical, persistence and ecological effects of chemical substances for which the EPA requires testing under section 4 of TSCA were published November 21, 1980. The EPA's FIFRA and TSCA GLP regulations were both issued in final form on November 29, 1983.

Requirements in the regulations relating to system validation have been highlighted in boldface.

40 CFR 160 GOOD LABORATORY PRACTICE STANDARDS [Pursuant to the Federal Insecticide, Fungicide, and Rodenticide Act (FIFRA)]

Subpart A--General Provisions

§160.1 Scope

(a) This part prescribes good laboratory practices for conducting studies that support or are intended to support applications for research or marketing permits for pesticide products regulated by the EPA. This part is intended to assure the quality and integrity of data submitted pursuant to sections 3, 4, 5, 8, 18 and 24(c) of the Federal Insecticide, Fungicide, and Rodenticide Act (FIFRA), as amended (7 U.S.C. 136a, 136c, 136f, 136q and 136v(c)) and sections 408 and 409 of the Federal Food, Drug and Cosmetic Act (FFDCA) (21 U.S.C. 346a, 348).

(b) This part applies to any study described by paragraph (a) of this section which any person conducts, initiates, or supports on or after October 16, 1989.

§160.3 Definitions.

As used in this part the following terms shall have the meanings specified:

Application for research or marketing permits includes:

(1) An application for registration, amended registration, or reregistration of a pesticide product under FIFRA sections 3, 4 or 24(c).

(2) An application for an experimental use permit under FIFRA section 5.

(3) An application for an exemption under FIFRA section 18.

(4) A petition or other request for establishment or modification of a tolerance, for an exemption for the need for a tolerance, or for other clearance under FFDCA section 408.

(5) A petition or other request for establishment or modification of a food additive regulation or other clearance by EPA under FFDCA section 409.

(6) A submission of data in response to a notice issued by EPA under FIFRA section 3(c)(2)(B).

(7) Any other application, petition, or submission sent to EPA intended to persuade EPA to grant, modify, or leave unmodified a registration or other approval required as a condition of sale or distribution of a pesticide.

***Batch* means a specific quantity or lot of a test, control, or reference substance that has been characterized according to §160.105(a).**

Carrier means any material, including but not limited to feed, water, soil, nutrient media, with which the test substance is combined for administration to a test system.

Control substance means any chemical substance or mixture, or any other material other than a test substance, feed, or water, that is administered to the test system in the course of a study for the purpose of establishing a basis for comparison with the test substance for known chemical or biological measurements.

EPA means the U.S. Environmental Protection Agency.

Experimental start date means the first date the test substance is applied to the test system.

Experimental termination date means the last date on which data are collected directly from the study.

FDA means the U.S. Food and Drug Administration.

FFDCA means the Federal Food, Drug and Cosmetic Act, as amended (21 U.S.C. 321 et seq).

FIFRA means the Federal Insecticide, Fungicide and Rodenticide Act as amended (7 U.S.C. 136 et seq).

Person includes an individual, partnership, corporation, association, scientific or academic establishment, government agency, or organizational unit thereof, and any other legal entity.

Quality assurance unit means any person or organizational element, except the study director, designated by testing facility management to perform the duties relating to quality assurance of the studies.

***Raw data* means any laboratory worksheets, records, memoranda, notes, or exact copies thereof, that are the result of original observations and activities of a**

study and are necessary for the reconstruction and evaluation of the report of that study. In the event that exact transcripts of raw data have been prepared (e.g., tapes which have been transcribed verbatim, dated, and verified accurate by signature), the exact copy or exact transcript may be substituted for the original source as raw data. "Raw data" may include photographs, microfilm or microfiche copies, computer printouts, magnetic media, including dictated observations, and recorded data from automated instruments.

Reference substance means any chemical substance or mixture, or analytical standard, or material other than a test substance, feed, or water, that is administered to or used in analyzing the test system in the course of a study for the purposes of establishing a basis for comparison with the test substance for known chemical or biological measurements.

Specimens means any material derived from a test system for examination or analysis.

Sponsor means:

(1) A person who initiates and supports, by provision of financial or other resources, a study;

(2) A person who submits a study to the EPA in support of an application for a research or marketing permit; or

(3) A testing facility, if it both initiates and actually conducts the study.

Study means any experiment at one or more test sites, in which a test substance is studied in a test system under laboratory conditions or in the environment to determine or help predict its effects, metabolism, product performance (efficacy studies only as required by 40 CFR 158.640), environmental and chemical fate, persistence and residue, or other characteristics in humans, other living organisms, or media. The term "study" does not include basic exploratory studies carried out to determine whether a test substance or a test method has any potential utility.

Study completion date means the date the final report is signed by the study director.

Study director means the individual responsible for the overall conduct of a study.

Study initiation date means the date the protocol is signed by the study director.

Test substance means a substance or mixture administered or added to a test system in a study, which substance or mixture:

(1) Is the subject of an application for a research or marketing permit supported by the study, or is the contemplated subject of such an application; or

(2) Is an ingredient, impurity, degradation product, metabolite, or radioactive isotope of a substance described by paragraph (1) of this definition, or some other substance related to a substance described by that paragraph, which is used in the study to assist in characterizing the toxicity, metabolism, or other characteristics of a substance described by that paragraph.

Test system means any animal, plant, microorganism, chemical or physical matrix, including but not limited to soil or water, or subparts thereof, to which the test, control, or reference substance is administered or added for study. "Test system" also includes appropriate groups or components of the system not treated with the test, control, or reference substance.

Testing facility means a person who actually conducts a study, i.e., actually uses the test substance in a test system. "Testing facility" encompasses only those operational units that are being or have been used to conduct studies.

Vehicle means any agent which facilitates the mixture, dispersion, or solubilization of a test substance with a carrier.

§160.10 Applicability to studies performed under grants and contracts.

When a sponsor or other person utilizes the services of a consulting laboratory, contractor, or grantee to perform all or a part of a study to which this part applies, it shall notify the consulting laboratory, contractor, or grantee that the service is, or is part of, a study that must be conducted in compliance with the provisions of this part.

§160.12 Statement of compliance or non-compliance.

Any person who submits to EPA an application for a research or marketing permit and who, in connection with the application, submits data from a study to which this part applies shall include in the application a true and correct statement, signed by the applicant, the sponsor, and the study director, of one of the following types:

(a) A statement that the study was conducted in accordance with this part; or

(b) A statement describing in detail all differences between the practices used in the study and those required by this part; or

(c) A statement that the person was not a sponsor of the study, did not conduct the study, and does not know whether the study was conducted in accordance with this part.

§160.15 Inspection of a testing facility.

(a) A testing facility shall permit an authorized employee or duly designated representative of EPA or FDA, at reasonable times and in a reasonable manner, to inspect the facility and to inspect (and in the case of records also to copy) all records and specimens required to be maintained regarding studies to which this part applies. The records inspection and copying requirements should not apply to quality assurance unit records of findings and problems, or to actions recommended and taken, except that EPA may seek production of these records in litigation or formal adjudicatory hearings.

(b) EPA will not consider reliable for purposes of supporting an application for a research or marketing permit any data developed by a testing facility or sponsor that refuses to permit inspection in accordance with this part. The determination that a study will not be considered in support of an application for a research or marketing permit does not, however, relieve the applicant for such a permit of any obligation under any applicable statute or regulation to submit the results of the study to EPA.

§160.17 Effects of non-compliance.

(a) EPA may refuse to consider reliable for purposes of supporting an application for a research or marketing permit any data from a study which was not conducted in accordance with this part.

(b) Submission of a statement required by §160.12 which is false may form the basis for cancellation, suspension, or modification of the research or marketing permit, or denial or disapproval of an application for such a permit, under FIFRA section 3, 5, 6, 18, or 24 or FFDCA section 406 or 409, or for criminal prosecution under 18 U.S.C. 2 or 1001 or FIFRA section 14, or for imposition of civil penalties under FIFRA section 14.

Subpart B--Organization and Personnel

§160.29 Personnel.

(a) Each individual engaged in the conduct of or responsible for the supervision of a study shall have education, training, and experience, or combination thereof, to enable that individual to perform the assigned functions.

(b) Each testing facility shall maintain a current summary of training and experience and job description for each individual engaged in or supervising the conduct of a study.

(c) There shall be a sufficient number of personnel for the timely and proper conduct of the study according to the protocol.

(d) Personnel shall take necessary personal sanitation and health precautions designed to avoid contamination of test, control, and reference substances and test systems.

(e) Personnel engaged in a study shall wear clothing appropriate for the duties they perform. Such clothing shall be changed as often as necessary to prevent microbiological, radiological, or chemical contamination of test systems and test, control, and reference substances.

(f) Any individual found at any time to have an illness that may adversely affect the quality and integrity of the study shall be excluded from direct contact with test systems, and test, control, and reference substances, and any other operation or function that may adversely affect the study until the condition is corrected. All personnel shall be instructed to report to their immediate supervisors any health or medical conditions that may reasonably be considered to have an adverse effect on a study.

§160.31 Testing facility management.

For each study, testing facility management shall:

(a) Designate a study director as described in §160.33 before the study is initiated.

(b) Replace the study director promptly if it becomes necessary to do so during the conduct of a study.

(c) Assure that there is a quality assurance unit as described in §160.35.

(d) Assure that test, control, and reference substances or mixtures have been appropriately tested for identity, strength, purity, stability, and uniformity, as applicable.

(e) Assure that personnel, resources, facilities, equipment, materials and methodologies are available as scheduled.

(f) Assure that personnel clearly understand the functions they are to perform.

(g) Assure that any deviations from these regulations reported by the quality assurance unit are communicated to the study director and corrective actions are taken and documented.

§160.33 Study director.

For each study, a scientist or other professional of appropriate education, training, and experience, or combination thereof, shall be identified as the study director. The study director has overall responsibility for the technical conduct of the study, as well as for the interpretation, analysis, documentation, and reporting of results, and represents the single point of study control. The study director shall assure that:

(a) The protocol, including any change, is approved as provided by §160.120 and is followed.

(b) All experimental data, including observations of unanticipated responses of the test system are accurately recorded and verified.

(c) Unforseen circumstances that may affect the quality and integrity of the study are noted when they occur, and corrective action is taken and documented.

(d) Test systems are as specified in the protocol.

(e) All applicable good laboratory practice regulations are followed.

(f) All raw data, documentation, protocols, specimens, and final reports are transferred to the archives during or at the close of the study.

§160.35 Quality assurance unit.

(a) A testing facility shall have a quality assurance unit which shall be responsible for monitoring each study to assure management that the facilities, equipment, personnel, methods, practices, records, and controls are in conformance with the regulations in this part. For any given study, the quality assurance unit shall be entirely separate from and independent of the personnel engaged in the direction and conduct of that study. The quality assurance unit shall conduct inspections and maintain records appropriate to the study.

(b) The quality assurance unit shall:

(1) Maintain a copy of a master schedule sheet of all studies conducted at the testing facility indexed by test substance, and containing the test system, nature of study, date study was initiated, current status of each study, identity of the sponsor, and name of the study director.

(2) Maintain copies of all protocols pertaining to all studies for which the unit is

responsible.

(3) Inspect each study at intervals adequate to ensure the integrity of the study and maintain written and properly signed records of each periodic inspection showing the date of the inspection, the study inspected, the phase or segment of the study inspected, the person performing the inspection, findings and problems, action recommended and taken to resolve existing problems, and any scheduled date for reinspection. Any problems which are likely to affect study integrity found during the course of an inspection shall be brought to the attention of the study director and management immediately.

(4) Periodically submit to management and the study director written status reports on each study, noting any problems and the corrective actions taken.

(5) Determine that no deviations from approved protocols or standard operating procedures were made without proper authorization and documentation.

(6) Review the final study report to assure that such report accurately describes the methods and standard operating procedures, and that the reported results accurately reflect the raw data of the study.

(7) Prepare and sign a statement to be included with the final study report which shall specify the dates inspections were made and findings reported to management and to the study director.

(c) The responsibilities and procedures applicable to the quality assurance unit, the records maintained by the quality assurance unit, and the method of indexing such records shall be in writing and shall be maintained. These items including inspection dates, the study inspected, the phase or segment of the study inspected, and the name of the individual performing the inspection shall be made available for inspection to authorized employees or duly designated representatives of EPA or FDA.

(d) An authorized employee or a duly designated representative of EPA or FDA shall have access to the written procedures established for the inspection and may request testing facility management to certify that inspections are being implemented, performed, documented, and followed up in accordance with this paragraph.

Subpart C--Facilities

§*160.41 General.*

Each testing facility shall be of suitable size and construction to facilitate the proper conduct of studies. Testing facilities which are not located within an indoor controlled environment shall be of suitable location to facilitate the proper conduct of studies. Testing facilities shall be designed so that there is a degree of separation that will prevent any function or activity from having an adverse effect on the study.

§*160.43 Test system care facilities.*

(a) A testing facility shall have a sufficient number of animal rooms or other test system areas, as needed, to ensure: proper separation of species or test systems, isolation of individual projects, quarantine or isolation of animals or other test systems, and routine or specialized housing of animals or other test systems.

(1) In tests with plants or aquatic animals, proper separation of species can be accomplished within a room or area by housing them separately in different chambers or aquaria. Separation of species is unnecessary where the protocol specifies the simultaneous exposure of two or more species in the same chamber, aquarium, or housing unit.

(2) Aquatic toxicity tests for individual projects shall be isolated to the extent necessary to prevent cross-contamination of different chemicals used in different tests.

(b) A testing facility shall have a number of

animal rooms or other test system areas separate from those described in paragraph (a) of this section to ensure isolation of studies being done with test systems or test, control, and reference substances known to be biohazardous, including volatile substances, aerosols, radioactive materials, and infectious agents.

(c) Separate areas shall be provided, as appropriate, for the diagnosis, treatment, and control of laboratory test system diseases. These areas shall provide effective isolation for the housing of test systems either known or suspected of being diseased, or of being carriers of disease, from other test systems.

(d) Facilities shall have proper provisions for collection and disposal of contaminated water, soil, or other spent materials. When animals are housed, facilities shall exist for the collection and disposal of all animal waste and refuse or for safe sanitary storage of waste before removal from the testing facility. Disposal facilities shall be so provided and operated as to minimize vermin infestation, odors, disease hazards, and environmental contamination.

(e) Facilities shall have provisions to regulate environmental conditions (e.g., temperature, humidity, photoperiod) as specified in the protocol.

(f) For marine test organisms, an adequate supply of clean sea water or artificial sea water (prepared from deionized or distilled water and sea salt mixture) shall be available. The ranges of composition shall be as specified in the protocol.

(g) For freshwater organisms, an adequate supply of clean water of the appropriate hardness, pH, and temperature, and which is free of contaminants capable of interfering with the study, shall be available as specified in the protocol.

(h) For plants, an adequate supply of soil of the appropriate composition, as specified in the protocol, shall be available as needed.

§160.45 Test system supply facilities.

(a) There shall be storage areas, as needed, for feed, nutrients, soils, bedding, supplies, and equipment. Storage areas for feed nutrients, soils, and bedding shall be separated from areas where the test systems are located and shall be protected against infestation or contamination. Perishable supplies shall be preserved by appropriate means.

(b) When appropriate, plant supply facilities shall be provided. As specified in the protocol, these include:

(1) Facilities for holding, culturing, and maintaining algae and aquatic plants.

(2) Facilities for plant growth, including, but not limited to greenhouses, growth chambers, light banks, and fields.

(c) When appropriate, facilities for aquatic animal tests shall be provided. These include, but are not limited to, aquaria, holding tanks, ponds, and ancillary equipment, as specified in the protocol.

§160.47 Facilities for handling test, control, and reference substances.

(a) As necessary to prevent contamination or mixups, there shall be separate areas for:

(1) Receipt and storage of the test, control, and reference substances.

(2) Mixing of the test, control, and reference substances with a carrier, e.g., feed.

(3) Storage of the test, control, and reference substance mixtures.

(b) Storage areas for test, control, and/or reference substance and for test, control, and/or reference mixtures shall be separate from areas housing the test systems and shall be adequate to preserve the identity, strength, purity, and stability of the substances and mixtures.

§160.49 Laboratory operation areas.

Separate laboratory space and other space shall be provided, as needed, for the performance of the routine and specialized procedures required by studies.

§160.51 Specimen and data storage facilities.

Space shall be provided for archives, limited to access by authorized personnel only, for the storage and retrieval of all

raw data and specimens from completed studies.

Subpart D--Equipment

§160.61 Equipment design.

Equipment used in the generation, measurement, or assessment of data and equipment used for facility environmental control shall be of appropriate design and adequate capacity to function according to the protocol and shall be suitably located for operation, inspection, cleaning and maintenance.

§160.63 Maintenance and calibration of equipment.

(a) Equipment shall be adequately inspected, cleaned, and maintained. Equipment used for the generation, measurement, or assessment of data shall be adequately tested, calibrated, and/or standardized.

(b) The written standard operating procedures required under §160.81(b)(11) shall set forth in sufficient detail the methods, materials, and schedules to be used in the routine inspection, cleaning, maintenance, testing, calibration, and/or standardization of equipment, and shall specify, when appropriate, remedial action to be taken in the event of failure or malfunction of equipment. The written standard operating procedures shall designate the person responsible for the performance of each operation.

(c) Written records shall be maintained of all inspection, maintenance, testing, calibrating, and/or standardizing operations. These records, containing the dates of the operations, shall describe whether the maintenance operations were routine and followed the written standard operating procedures. Written records shall be kept of nonroutine repairs performed on equipment as a result of failure and malfunction. Such records shall document the nature of the defect, how and when the defect was discovered, and any remedial action taken in response to the defect.

Subpart E--Testing Facilities Operation

§160.81 Standard operating procedures.

(a) A testing facility shall have standard operating procedures in writing setting forth study methods that management is satisfied are adequate to insure the quality and integrity of the data generated in the course of a study. All deviations in a study from standard operating procedures shall be authorized by the study director and shall be documented in the raw data. Significant changes in established standard operating procedures shall be properly authorized in writing by management.

(b) Standard operating procedures shall be established for, but not limited to, the following:

(1) Test system area preparation.

(2) Test system care.

(3) Receipt, identification, storage, handling, mixing, and method of sampling of the test, control, and reference substances.

(4) Test system observations.

(5) Laboratory or other tests.

(6) Handling of test systems found moribund or dead during study.

(7) Necropsy of test systems or post-mortem examination of test systems.

(8) Collection and identification of specimens.

(9) Histopathology.

(10) Data handling, storage and retrieval.

(11) Maintenance and calibration of equipment.

(12) Transfer, proper placement, and identification of test systems.

(c) Each laboratory or other study area shall have immediately available manuals and standard operating procedures relative to the laboratory or field procedures being performed. Published literature may be used as a supplement to standard operating procedures.

(d) A historical file of standard operating procedures, and all revisions thereof, including the dates of such revisions, shall be maintained.

§160.83 *Reagents and solutions.*

All reagents and solutions in the laboratory areas shall be labeled to indicate identity, titer or concentration, storage requirements, and expiration date. Deteriorated or outdated reagents and solutions shall not be used.

§160.90 *Animal and other test system care.*

(a) There shall be standard operating procedures for the housing, feeding, handling, and care of animals and other test systems.

(b) All newly received test systems from outside sources shall be isolated and their health status or appropriateness for the study shall be evaluated. This evaluation shall be in accordance with acceptable veterinary medical practice or scientific methods.

(c) At the initiation of a study, test systems shall be free of any disease or condition that might interfere with the purpose or conduct of the study. If during the course of the study, the test systems contract such a disease or condition, the diseased test systems should be isolated, if necessary. These test systems may be treated for disease or signs of disease provided that such treatment does not interfere with the study. The diagnosis, authorization of treatment, description of treatment, and each date of treatment shall be documented and shall be retained.

(d) Warm-blooded animals, adult reptiles, and adult terrestrial amphibians used in laboratory procedures that require manipulations and observations over an extended period of time or in studies that require these test systems to be removed from and returned to their test system-housing units for any reason (e.g., cage cleaning, treatment, etc.), shall receive appropriate identification (e.g., tattoo, color code, ear tag, ear punch, etc.). All information needed to specifically identify each test system within the test system-housing unit shall appear on the outside of that unit. Suckling mammals and juvenile birds are excluded from the requirement of individual identification unless otherwise specified in the protocol.

(e) Except as specified in paragraph (e)(1) of this section, test systems of different species shall be housed in separate rooms when necessary. Test systems of the same species, but used in different studies, should not ordinarily be housed in the same room when inadvertent exposure to test, control, or reference substances or test system mixup could affect the outcome of either study. If such mixed housing is necessary, adequate differentiation by space and identification shall be made.

(1) Plants, invertebrate animals, aquatic vertebrate animals, and organisms that may be used in multispecies tests need not be housed in separate rooms, provided that they are adequately segregated to avoid mixup and cross contamination.

(2) [Reserved]

(f) Cages, racks, pens, enclosures, aquaria, holding tanks, ponds, growth chambers, and other holding, rearing and breeding areas, and accessory equipment, shall be cleaned and sanitized at appropriate intervals.

(g) Feed, soil, and water used for the test systems shall be analyzed periodically to ensure that contaminants known to be capable of interfering with the study and reasonably expected to be present in such feed, soil, or water are not present at levels above those specified in the protocol. Documentation of such analyses shall be maintained as raw data.

(h) Bedding used in animal cages or pens shall not interfere with the purpose or conduct of the study and shall be changed as often as necessary to keep the animals dry and clean.

(i) If any pest control materials are used, the use shall be documented. Cleaning and pest control materials that interfere with the study shall not be used.

(j) All plant and animal test systems shall be

acclimatized to the environmental conditions of the test, prior to their use in a study.

Subpart F--Test, Control, and Reference Substances

§106.105 Test, control, and reference substance characterization.

(a) The identity, strength, purity, and composition, or other characteristics which will appropriately define the test, control, or reference substance shall be determined for each batch and shall be documented before its use in a study. Methods of synthesis, fabrication, or derivation of the test, control, or reference substance shall be documented by the sponsor or the testing facility, and the location of such documentation shall be specified.

(b) When relevant to the conduct of the study the solubility of each test, control, or reference substance shall be determined by the testing facility or the sponsor before the experimental start date. The stability of the test, control, or reference substance shall be determined before the experimental start date or concomitantly according to written standard operating procedures, which provide for periodic analysis of each batch.

(c) Each storage container for a test, control, or reference substance shall be labeled by name, chemical abstracts service number (CAS) or code number, batch number, expiration date, if any, and, where appropriate, storage conditions necessary to maintain the identity, strength, purity, and composition of the test, control, or reference substance. Storage containers shall be assigned to a particular test substance for the duration of the study.

(d) For studies of more than 4 weeks experimental duration, reserve samples from each batch of test, control, and reference substances shall be retained for the period of time provided by §160.195.

(e) The stability of test, control, and reference substances under storage conditions at the test site shall be known for all studies.

§160.107 Test, control, and reference substance handling.

Procedures shall be established for a system for the handling of the test, control, and reference substances to ensure that:

(a) There is proper storage.

(b) Distribution is made in a manner designed to preclude the possibility of contamination, deterioration, or damage.

(c) Proper identification is maintained throughout the distribution process.

(d) The receipt and distribution of each batch is documented. Such documentation shall include the date and quantity of each batch distributed or returned.

§160.113 Mixtures of substances with carriers.

(a) For each test, control, or reference substance that is mixed with a carrier, tests by appropriate analytical methods shall be conducted:

(1) To determine the uniformity of the mixture and to determine, periodically, the concentration of the test, control, or reference substance in the mixture.

(2) When relevant to the conduct of the study, to determine the solubility of each test, control, or reference substance in the mixture by the testing facility or the sponsor before the experimental start date.

(3) To determine the stability of the test, control, or reference substance in the mixture before the experimental start date or concomitantly according to written standard operating procedures, which provide for periodic analysis of each batch.

(b) Where any of the components of the test, control, or reference substance carrier mixture has an expiration date, that date shall be clearly shown on the container. If more than one component has an expiration date, the earliest date shall be shown.

(c) If a vehicle is used to facilitate the mixing of a test substance with a carrier, assurance shall be provided that the vehicle does not interfere with the integrity of the test.

Subpart G--Protocol for and Conduct of a Study

§160.120 Protocol.

(a) Each study shall have an approved written protocol that clearly indicates the objectives and all methods for the conduct of the study. The protocol shall contain but shall not necessarily be limited to the following information:

(1) A descriptive title and statement of the purpose of the study.

(2) Identification of the test, control, and reference substance by name, chemical abstracts service (CAS) number or code number.

(3) The name and address of the sponsor and the name and address of the testing facility at which the study is being conducted.

(4) The proposed experimental start and termination dates.

(5) Justification for selection of the test system.

(6) Where applicable, the number, body weight range, sex, source of supply, species, strain, substrain, and age of the test system.

(7) The procedure for identification of the test system.

(8) A description of the experimental design, including methods for the control of bias.

(9) Where applicable, a description and/or identification of the diet used in the study as well as solvents, emulsifiers and/or other materials used to solubilize or suspend the test, control, or reference substances before mixing with the carrier. The description shall include specifications for acceptable levels of contaminants that are reasonably expected to be present in the dietary materials and are known to be capable of interfering with the purpose or conduct of the study if present at levels greater than established by the specifications.

(10) The route of administration and the reason for its choice.

(11) Each dosage level, expressed in milligrams per kilogram of body or test system weight or other appropriate units, of the test, control, or reference substance to be administered and the method and frequency of administration.

(12) The type and frequency of tests, analyses, and measurements to be made.

(13) The records to be maintained.

(14) The date of approval of the protocol by the sponsor and the dated signature of the study director.

(15) A statement of the proposed statistical method to be used.

(b) All changes in or revisions of an approved protocol and the reasons therefore shall be documented, signed by the study director, dated, and maintained with the protocol.

§160.130 Conduct of a study.

(a) The study shall be conducted in accordance with the protocol.

(b) The test systems shall be monitored in conformity with the protocol.

(c) Specimens shall be identified by test system, study, nature, and date of collection. This information shall be located on the specimen container or shall accompany the specimen in a manner that precludes error in the recording and storage of data.

(d) In animal studies where histopathology is required, records of gross findings for a specimen from postmortem observations shall be available to a pathologist when examining that specimen histopathologically.

(e) All data generated during the conduct of a study, except those that are generated by automated data collection systems, shall be recorded directly, promptly, and legibly in ink. All data entries shall be dated on the day of entry and signed or initialed by the person entering the data. Any change in entries shall be made so as not to obscure the original entry, shall indicate the reason for such change, and shall be dated and signed or identified at the time of the change. In automated data collection systems, the individual responsible for direct data input shall be identified at the time of data input. Any change in automated data entries shall be made so as not to obscure the original entry, shall indicate the reason for the change, shall be dated, and the responsible individual shall be identified.

§160.135 Physical and chemical characterization studies.

(a) All provisions of the GLP standards shall apply to physical and chemical characterization studies designed to determine stability, solubility, octanol water partition coefficient, volatility, and persistence (such as biodegradation, photodegradation, and chemical degradation studies) of test, control, or reference substances.

(b) The following GLP standards shall not apply to studies, other than those designated in paragraph (a) of this section, designed to determine physical and chemical characteristics of a test, control, or reference substance:

§160.31 (c), (d), and (g)
§160.35 (b) and (c)
§160.43
§160.45
§160.47
§160.49
§160.81(b)(1), (2), (6) through (9), and (12)
§160.90
§160.105(a) through (d)
§160.113
§160.120(a)(5) through (12), and 15
§160.185(a)(5) through (8), (10), (12), and (14)
§160.195(c) and (d)

Subparts H-I--[Reserved]

Subpart J--Records and Reports

§160.185 Reporting of study results.

(a) A final report shall be prepared for each study and shall include, but not necessarily be limited to, the following:

(1) Name and address of the facility performing the study and the dates on which the study was initiated and was completed, terminated, or discontinued.

(2) Objectives and procedures stated in the approved protocol, including any changes in the original protocol.

(3) Statistical methods employed for analyzing the data.

(4) The test, control, and reference substances identified by name, chemical abstracts service (CAS) number or code number, strength, purity, and composition, or other appropriate characteristics.

(5) Stability and, when relevant to the conduct of the study the solubility of the test, control, and reference substances under the conditions of administration

(6) A description of the methods used.

(7) A description of the test system used. Where applicable, the final report shall include the number of animals used, sex, body weight range, source of supply, species, strain and substrain, age, and procedure used for identification.

(8) A description of the dosage, dosage regimen, route of administration, and duration.

(9) A description of all circumstances that may have affected the quality or integrity of the data.

(10) The name of the study director, the names of other scientists or professionals and the names of all supervisory personnel, involved in the study.

(11) A description of the transformations, calculations, or operations performed on the data, a summary and analysis of the data, and a statement of the conclusions drawn from the analysis.

(12) The signed and dated reports of each of the individual scientists or other professionals involved in the study, including each person who, at the request or direction of the testing facility or sponsor, conducted an analysis or evaluation of data or specimens from the study after data generation was completed

(13) The locations where all specimens, raw data, and the final report are to be stored.

(14) The statement prepared and signed by the quality assurance unit as described in §160. 35(b)(7).

(b) The final report shall be signed and dated by the study director.

(c) Corrections or additions to a final report

shall be in the form of an amendment by the study director. The amendment shall clearly identify that part of the final report that is being added to or corrected and the reasons for the correction or addition, and shall be signed and dated by the person responsible. Modification of a final report to comply with the submission requirements of EPA does not constitute a correction, addition, or amendment to a final report.

(d) A copy of the final report and of any amendment to it shall be maintained by the sponsor and the test facility.

§160.190 Storage and retrieval of records and data.

(a) All raw data, documentation, records, protocols, specimens, and final reports generated as a result of a study shall be retained. Specimens obtained from mutagenicity tests, specimens of soil, water, and plants, and wet specimens of blood, urine, feces, and biological fluids, do not need to be retained after quality assurance verification. Correspondence and other documents relating to interpretation and evaluation of data, other than those documents contained in the final report, also shall be retained.

(b) There shall be archives for orderly storage and expedient retrieval of all raw data, documentation, protocols, specimens, and interim and final reports. Conditions of storage shall minimize deterioration of the documents or specimens in accordance with the requirements for the time period of their retention and the nature of the documents of [*sic*] specimens. A testing facility may contract with commercial archives to provide a repository for all material to be retained. Raw data and specimens may be retained elsewhere provided that the archives have specific reference to those other locations.

(c) An individual shall be identified as responsible for the archives.

(d) Only authorized personnel shall enter the archives.

(e) Material retained or referred to in the archives shall be indexed to permit expedient retrieval.

§160.195 Retention of records.

(a) Record retention requirements set forth in this section do not supersede the record retention requirements of any other regulations in this subchapter.

(b) Except as provided in paragraph (c) of this section, documentation records, raw data, and specimens pertaining to a study and required to be retained by this part shall be retained in the archive(s) for whichever of the following periods is longest:

(1) In the case of any study used to support an application for a research or marketing permit approved by EPA, the period during which the sponsor holds any research or marketing permit to which the study is pertinent.

(2) A period of at least 5 years following the date on which the results of the study are submitted to the EPA in support of an application for a research or marketing permit.

(3) In other situations (e.g., where the study does not result in the submission of the study in support of an application for a research or marketing permit), a period of at lest 2 years following the date on which the study is completed, terminated, or discontinued.

(c) Wet specimens, samples of test, control, or reference substances, and specially prepared material which are relatively fragile and differ markedly in stability and quality during storage, shall be retained only as long as the quality of the preparation affords evaluation. Specimens obtained from mutagenicity tests, specimens of soil, water, and plants, and wet specimens of blood, urine, feces, and biological fluids, do not need to be retained after quality assurance verification. In no case shall retention be required for longer periods than those set forth in paragraph (b) of this section.

(d) The master schedule sheet, copies of

protocols, and records of quality assurance inspections, as required by §160. 35(c) shall be maintained by the quality assurance unit as an easily accessible system or records for the period of time specified in paragraph (b) of this section.

(e) Summaries of training and experience and job descriptions required to be maintained by §160.29(b) may be retained along with all other testing facility employment records for the length of time specified in paragraph (b) of this section.

(f) Records and reports of the maintenance and calibration and inspection of equipment, as required by §160.63(b) and (c), shall be retained for the length of time specified in paragraph (b) of this section.

(g) If a facility conducting testing or an archive contracting facility goes out of business, all raw data, documentation, and other material specified in this section shall be transferred to the archives of the sponsor of the study. The EPA shall be notified in writing of such a transfer.

(h) Specimens, samples, or other non-documentary materials need not be retained after EPA has notified in writing the sponsor or testing facility holding the materials that retention is no longer required by EPA. Such notification normally will be furnished upon request after EPA or FDA has completed an audit of the particular study to which the materials relate and EPA has concluded that the study was conducted in accordance with this part.

(i) Records required by this part may be retained either as original records or as true copies such as photocopies, microfilm, microfiche, or other accurate reproductions of the original records.

40 CFR 792--GOOD LABORATORY PRACTICE STANDARDS [Pursuant to the Toxic Substances Control Act (TSCA)]

Subpart A--General Provisions

§792.1 Scope.

(a) This part prescribes good laboratory practices for conducting studies relating to health effects, environmental effects, and chemical fate testing. This part is intended to ensure the quality and integrity of data submitted pursuant to testing consent agreements and test rules issued under section 4 of the Toxic Substances Control Act (TSCA) (Pub. L. 94-469, 90 Stat. 2006, 15 U.S.C. 2603 *et seq.*).

(b) This part applies to any study described by paragraph (a) of this section which any person conducts, initiates, or supports on or after September 18, 1989.

(c) It is EPA's policy that all data developed under section 5 of TSCA be in accordance with provisions of this part. If data are not developed in accordance with the provisions of this part, EPA will consider such data insufficient to evaluate the health and environmental effects of the chemical substances unless the submitter provides additional information demonstrating that the data are reliable and adequate.

§792.3 Definitions.

As used in this part the following terms shall have the meanings specified:

***Batch* means a specific quantity or lot of a test, control, or reference substance that has been characterized according to §792.105(a).**

Carrier means any material, including but not limited to, feed, water, soil, nutrient media, with which the test substance is combined for administration to a test system.

Control substance means any chemical substance or mixture, or any other material other than a test substance, feed, or water, that is administered to the test system in the course of a study for the purpose of establishing a basis for comparison with the test substance for chemical or biological measurements.

EPA means the U.S. Environmental Protection Agency.

Experimental start date means the first date the test substance is applied to the test system.

Experimental termination date means the last date on which data are collected directly from the study.

FDA means the U.S. Food and Drug Administration.

Person includes an individual, partnership, corporation, association, scientific or academic establishment, government agency, or organizational unit thereof, and any other legal entity.

Quality assurance unit means any person or organizational element, except the study director, designated by testing facility management to perform the duties relating to quality assurance of the studies.

***Raw data* means any laboratory worksheets, records, memoranda, notes, or exact copies thereof, that are the result of original observations and activities of a study and are necessary for the reconstruction and evaluation of the report of that study. In the event that exact transcripts of raw data have been prepared (e.g., tapes which have been transcribed verbatim, dated, and verified accurate by signature), the exact copy or exact transcript may be substituted for the original source as raw data. "Raw data" may include photographs, microfilm or microfiche copies, computer printouts, magnetic media, including dictated observations, and recorded data from automated instruments.**

Reference substance means any chemical substance or mixture, or analytical standard, or material other than a test substance, feed, or water, that is administered to or used in analyzing the test system in the course of a study for the purposes of establishing a basis for comparison with the test substance for known chemical or biological measurements.

Specimen means any material derived from a test system for examination or analysis.

Sponsor means:

(1) A person who initiates and supports, by provision of financial or other resources, a study;

(2) A person who submits a study to the EPA in response to a TSCA section 4(a) test rule and/or a person who submits a study under a TSCA section 4 testing consent agreement or a TSCA section 5 rule or order to the extent the agreement, rule or order references this part; or

(3) A testing facility, if it both initiates and actually conducts the study.

Study means any experiment at one or more test sites, in which a test substance is studied in a test system under laboratory conditions or in the environment to determine or help predict its effects, metabolism, environmental and chemical fate, persistence, or other characteristics in humans, other living organisms, or media. The term "study" does not include basic exploratory studies carried out to determine whether a test substance or a test method has any potential utility.

Study completion date means the date the final report is signed by the study director.

Study director means the individual responsible for the overall conduct of a study.

Study initiation date means the date the protocol is signed by the study director.

Test substance means a substance or mixture administered or added to a test system in a study, which substance or mixture is used to develop data to meet the requirements of a TSCA section 4(a) test rule and/or is developed under a TSCA section 4 testing consent agreement or section 5 rule or order to the extent the agreement, rule or order references this part.

Test system means any animal, plant, microorganism, chemical or physical matrix, including but not limited to, soil or water, or components thereof, to which the test, control, or reference substance is administered or added for study. "Test system" also includes appropriate groups or components of the system not treated with the test, control, or reference substance.

Testing facility means a person who actually conducts a study, i.e., actually uses the test substance in a test system. "Testing facility" encompasses only those operational units that are being or have been used to conduct studies.

TSCA means the Toxic Substances Control Act (15 U.S.C. 2601 *et seq.*)

Vehicle means any agent which facilitates the mixture, dispersion, or solubilization of a test substance with a carrier.

§792.10 *Applicability to studies performed under grants and contracts.*

When a sponsor or other person utilizes the services of a consulting laboratory, contractor, or grantee to perform all or a part of a study to which this part applies, it shall notify the consulting laboratory, contractor, or grantee that the service is, or is part of, a study that must be conducted in compliance with the provisions of this part.

§792.12 *Statement of compliance or non-compliance.*

Any person who submits to EPA a test required by a testing consent agreement or a test rule issued under section 4 of TSCA shall include in the submission a true and correct statement, signed by the sponsor and the study director, of one of the following types:

(a) A statement that the study was conducted in accordance with this part; or

(b) A statement describing in detail all differences between the practices used in the study and those required by this part; or

(c) A statement that the person was not a sponsor of the study, did not conduct the study, and does not know whether the study was conducted in accordance with this part.

§792.15 *Inspection of a testing facility.*

(a) A testing facility shall permit an authorized employee or duly designated representative of EPA or FDA, at reasonable times and in a reasonable manner, to inspect the facility and to inspect (and in the case of records also to copy) all records and specimens required to be maintained regarding studies to which this part applies. The records inspection and copying requirements shall not apply to quality assurance unit records of findings and problems, or to actions recommended and taken, except the EPA may seek production of these records in litigation or formal adjudicatory hearings.

(b) EPA will not consider reliable for purposes of showing that a chemical substance or mixture does not present a risk of injury to health or the environment any data developed by a testing facility or sponsor that refuses to permit inspection in accordance with this part. The determination that a study will not be considered reliable does not, however, relieve the sponsor for such a permit of any obligation under any applicable statute or regulation to submit the results of the study to EPA.

(c) Since a testing facility is a place where chemicals are stored or held, it is subject to inspection under section 11 of TSCA.

§792.17 *Effects of non-compliance.*

(a) The sponsor or any other person who is conducting or has conducted a test to fulfill the requirements of a testing consent agreement or a test rule issued under section 4 of TSCA will be in violation of section 15 of TSCA if:

(1) The test is not being or was not conducted in accordance with any requirement of this part;

(2) Data or information submitted to EPA under this part (including the statement required by §792.12) include information or data that are false or misleading, contain significant omissions, or otherwise do not fulfill the requirements of this part; or

(3) Entry in accordance with §792.15 for the purpose of auditing test data or inspecting test facilities is denied. Persons who violate the provisions of this part may be subject to civil or criminal penalties under section 16 of

TSCA, legal action in United States district court under section 17 of TSCA, or criminal prosecution under 18 U.S.C. 2 or 1001.

(b) EPA, at its discretion, may not consider reliable for purposes of showing that a chemical substance or mixture does not present a risk of injury to health or the environment any study which was not conducted in accordance with this part. EPA, at its discretion, may rely upon such studies for purposes of showing adverse effects. The determination that a study will not be considered reliable does not, however, relieve the sponsor of a required test of the obligation under any applicable statute or regulation to submit the results of the study to EPA.

(c) If data submitted to fulfill a requirement of a testing consent agreement or a test rule issued under section 4 of TSCA are not developed in accordance with this part, EPA may determine that the sponsor has not fulfilled its obligations under section 4 of TSCA and may require the sponsor to develop data in accordance with the requirements of this part in order to satisfy such obligations.

Subpart B--Organization and Personnel

§*792.29 Personnel.*

(a) Each individual engaged in the conduct of or responsible for the supervision of a study shall have education, training, and experience, or combination thereof, to enable that individual to perform the assigned functions.

(b) Each testing facility shall maintain a current summary of training and experience and job description for each individual engaged in or supervising the conduct of a study.

(c) There shall be a sufficient number of personnel for the timely and proper conduct of the study according to the protocol.

(d) Personnel shall take necessary personal sanitation and health precautions designed to avoid contamination of test, control, and reference substances and test systems.

(e) Personnel engaged in a study shall wear clothing appropriate for the duties they perform. Such clothing shall be changed as often as necessary to prevent microbiological, radiological, or chemical contamination of test systems and test, control, and reference substances.

(f) Any individual found at any time to have an illness that may adversely affect the quality and integrity of the study shall be excluded from direct contact with test systems, test, control, and reference substances and any other operation or function that may adversely affect the study until the condition is corrected. All personnel shall be instructed to report to their immediate supervisors any health or medical conditions that may reasonably be considered to have an adverse effect on a study.

§*792.31 Testing facility management.*

For each study, testing facility management shall:

(a) Designate a study director as described in §792.33 before the study is initiated.

(b) Replace the study director promptly if it becomes necessary to do so during the conduct of a study.

(c) Assure that there is a quality assurance unit as described in §792.35.

(d) Assure that test, control, and reference substances or mixtures have been appropriately tested for identity, strength, purity, stability, and uniformity, as applicable.

(e) Assure that personnel, resources, facilities, equipment, materials and methodologies are available as scheduled.

(f) Assure that personnel clearly understand the functions they are to perform.

(g) Assure that any deviations form these regulations reported by the quality assurance unit are communicated to the study director and corrective actions are taken and documented.

§*792.33 Study director.*

For each study, a scientist or other professional of appropriate education, training, and experience, or combination thereof, shall be identified as the study director. The study director has overall responsibility for the technical conduct of the study, as well as for the interpretation, analysis, documentation, and reporting of results, and represents the single point of study control. The study director shall assure that:

(a) The protocol, including any change, is approved as provided by §792.120 and is followed.

(b) All experimental data, including observations of unanticipated responses of the test system are accurately recorded and verified.

(c) Unforseen circumstances that may affect the quality and integrity of the study are noted when they occur, and corrective action is taken and documented.

(d) Test systems are as specified in the protocol.

(e) All applicable good laboratory practice regulations are followed.

(f) All raw data, documentation, protocols, specimens, and final reports are transferred to the archives during or at the close of the study.

§792.35 Quality assurance unit.

(a) A testing facility shall have a quality assurance unit which shall be responsible for monitoring each study to assure management that the facilities, equipment, personnel, methods, practices, records, and controls are in conformance with the regulations in this part. For any given study, the quality assurance unit shall be entirely separate from and independent of the personnel engaged in the direction and conduct of that study. The quality assurance unit shall conduct inspections and maintain records appropriate to the study.

(b) The quality assurance unit shall:

(1) Maintain a copy of a master schedule sheet of all studies conducted at the testing facility indexed by test substance and containing the test system, nature of study, date study was initiated, current status of each study, identity of the sponsor, and name of the study director.

(2) Maintain copies of all protocols pertaining to all studies for which the unit is responsible.

(3) Inspect each study at intervals adequate to ensure the integrity of the study and maintain written and properly signed records of each periodic inspection showing the date of the inspection, the study inspected, the phase or segment of the study inspected, the person performing the inspection, findings and problems, action recommended and taken to resolve existing problems, and any scheduled date for re-inspection. Any problems which are likely to affect study integrity found during the course of an inspection shall be brought to the attention of the study director and management immediately.

(4) Periodically submit to management and the study director written status reports on each study, noting any problems and the corrective actions taken.

(5) Determine that no deviations from approved protocols or standard operating procedures were made without proper authorization and documentation.

(6) Review the final study report to assure that such report accurately describes the methods and standard operating procedures, and that the reported results accurately reflect the raw data of the study.

(7) Prepare and sign a statement to be included with the final study report which shall specify the dates inspections were made and findings reported to management and to the study director.

(c) The responsibilities and procedures applicable to the quality assurance unit, the records maintained by the quality assurance unit, and the method of indexing such records shall be in writing and shall be maintained. These items including inspection dates, the study inspected, the phase or segment of the study in-

spected, and the name of the individual performing the inspection shall be made available for inspection to authorized employees or duly designated representatives of EPA or FDA.

(d) An authorized employee or a duly designated representative of EPA or FDA shall have access to the written procedures established for the inspection and may request testing facility management to certify that inspections are being implemented, performed, documented, and followed up in accordance with this paragraph.

Subpart C--Facilities

§*792.41 General.*

Each testing facility shall be of suitable size and construction to facilitate the proper conduct of studies. Testing facilities which are not located within an indoor controlled environment shall be of suitable location to facilitate the proper conduct of studies. Testing facilities shall be designed so that there is a degree of separation that will prevent any function or activity from having an adverse effect on the study.

§*792.43 Test system care facilities.*

(a) A testing facility shall have a sufficient number of animal rooms or other test system areas, as needed, to ensure: proper separation of species or test systems, isolation of individual projects, quarantine or isolation of animals or other test systems, and routine or specialized housing of animals or other test systems.

(1) In tests with plants or aquatic animals, proper separation of species can be accomplished within a room or area by housing them separately in different chambers or aquaria. Separation of species is unnecessary where the protocol specifies the simultaneous exposure of two or more species in the same chamber, aquarium, or housing unit.

(2) Aquatic toxicity tests for individual projects shall be isolated to the extent necessary to prevent cross-contamination of different chemicals used in different tests.

(b) A testing facility shall have a number of animal rooms or other test system areas separate from those described in paragraph (a) of this section to ensure isolation of studies being done with test systems or test, control, and reference substances known to be biohazardous, including volatile substances, aerosols, radioactive materials, and infectious agents.

(c) Separate areas shall be provided, as appropriate, for the diagnosis, treatment, and control of laboratory test system diseases. These areas shall provide effective isolation for the housing of test systems either known or suspected of being diseased, or of being carriers of disease, from other test systems.

(d) Facilities shall have proper provisions for collection and disposal of contaminated water, soil, or other spent materials. When animals are housed, facilities shall exist for the collection and disposal of all animal waste and refuse or for safe sanitary storage of waste before removal from the testing facility. Disposal facilities shall be so provided and operated as to minimize vermin infestation, odors, disease hazards, and environmental contamination.

(e) Facilities shall have provisions to regulate environmental conditions (e.g., temperature, humidity, photoperiod) as specified in the protocol.

(f) For marine test organisms, an adequate supply of clean sea water or artificial sea water (prepared from deionized or distilled water and sea salt mixture) shall be available. The ranges of composition shall be as specified in the protocol.

(g) For freshwater organisms, an adequate supply of clean water of the appropriate hardness, pH, and temperature, and which is free of contaminants capable of interfering with the study, shall be available as specified in the protocol.

(h) For plants, an adequate supply of soil of

the appropriate composition, as specified in the protocol, shall be available as needed.

§792.45 *Test system supply facilities.*

(a) There shall be storage areas, as needed, for feed, nutrients, soils, bedding, supplies, and equipment. Storage areas for feed nutrients, soils, and bedding shall be separated from areas where the test systems are located and shall be protected against infestation or contamination. Perishable supplies shall be preserved by appropriate means.

(b) When appropriate, plant supply facilities shall be provided. These include:

(1) Facilities, as specified in the protocol, for holding, culturing, and maintaining algae and aquatic plants.

(2) Facilities, as specified in the protocol, for plant growth, including but not limited to, greenhouses, growth chambers, light banks, and fields.

(c) When appropriate, facilities for aquatic animal tests shall be provided. These include but are not limited to aquaria, holding tanks, ponds, and ancillary equipment, as specified in the protocol.

§792.47 *Facilities for handling test, control, and reference substances.*

(a) As necessary to prevent contamination or mixups, there shall be separate areas for:

(1) Receipt and storage of the test, control, and reference substances.

(2) Mixing of the test, control, and reference substances with a carrier, e.g., feed.

(3) Storage of the test, control, and reference substance mixtures.

(b) Storage areas for test, control, and/or reference substance and for test, control, and/or reference mixtures shall be separate from areas housing the test systems and shall be adequate to preserve the identity, strength, purity, and stability of the substances and mixtures.

§792.49 *Laboratory operation areas.*

Separate laboratory space and other space shall be provided, as needed, for the performance of the routine and specialized procedures required by studies.

§792.51 *Specimen and data storage facilities.*

Space shall be provided for archives, limited to access by authorized personnel only, for the storage and retrieval of all raw data and specimens from completed studies.

Subpart D--Equipment

§792.61 *Equipment design.*

Equipment used in the generation, measurement, or assessment of data and equipment used for facility environmental control shall be of appropriate design and adequate capacity to function according to the protocol and shall be suitably located for operation, inspection, cleaning and maintenance.

§792.63 *Maintenance and calibration of equipment.*

(a) Equipment shall be adequately inspected, cleaned, and maintained. Equipment used for the generation, measurement, or assessment of data shall be adequately tested, calibrated, and/or standardized.

(b) The written standard operating procedures required under §792.81(b)(11) shall set forth in sufficient detail the methods, materials, and schedules to be used in the routine inspection, cleaning, maintenance, testing, calibration, and/or standardization of equipment, and shall specify, when appropriate, remedial action to be taken in the event of failure or malfunction of equipment. The writ-

ten standard operating procedures shall designate the person responsible for the performance of each operation.

(c) Written records shall be maintained of all inspection, maintenance, testing, calibrating, and/or standardizing operations. These records, containing the date of the operation, shall describe whether the maintenance operations were routine and followed the written standard operating procedures. Written records shall be kept of nonroutine repairs performed on equipment as a result of failure and malfunction. Such records shall document the nature of the defect, how and when the defect was discovered, and any remedial action taken in response to the defect.

Subpart E--Testing Facilities Operation

§792.81 *Standard operating procedures.*

(a) A testing facility shall have standard operating procedures in writing, setting forth study methods that management is satisfied are adequate to insure the quality and integrity of the data generated in the course of a study. All deviations in a study from standard operating procedures shall be authorized by the study director and shall be documented in the raw data. Significant changes in established standard operating procedures shall be properly authorized in writing by management.

(b) Standard operating procedures shall be established for, but not limited to, the following:

(1) Test system room preparation.

(2) Test system care.

(3) Receipt, identification, storage, handling, mixing, and method of sampling of the test, control, and reference substances.

(4) Test system observations.

(5) Laboratory or other tests.

(6) Handling of test systems found moribund or dead during study.

(7) Necropsy of test systems or post-mortem examination of test systems.

(8) Collection and identification of specimens.

(9) Histopathology.

(10) Data handling, storage and retrieval.

(11) Maintenance and calibration of equipment.

(12) Transfer, proper placement, and identification of test systems.

(c) Each laboratory or other study area shall have immediately available manuals and standard operating procedures relative to the laboratory or field procedures being performed. Published literature may be used as a supplement to standard operating procedures.

(d) A historical file of standard operating procedures, and all revisions thereof, including the dates of such revisions, shall be maintained.

§792.83 *Reagents and solutions.*

All reagents and solutions in the laboratory areas shall be labeled to indicate identity, titer or concentration, storage requirements, and expiration date. Deteriorated or outdated reagents and solutions shall not be used.

§792.90 *Animal and other test system care.*

(a) There shall be standard operating procedures for the housing, feeding, handling, and care of animals and other test systems.

(b) All newly received test systems from outside sources shall be isolated and their health status or appropriateness for the study shall be evaluated. This evaluation shall be in accordance with acceptable veterinary medical practice or scientific methods.

(c) At the initiation of a study, test systems shall be free of any disease or condition that might interfere with the purpose or conduct of the study. If during the course of the study, the test systems contract such a disease or condition, the diseased test systems should be isolated, if necessary. These test systems may be treated for disease or signs of disease provided that such treatment does not interfere with the study. The diagnosis, authorization of treatment, description of treatment, and each date of treatment shall be documented and shall be retained.

(d) Warm-blooded animals, adult reptiles, and adult terrestrial amphibians used in laboratory procedures that require manipulations and observations over an extended period of time, or in studies that require these test systems to be removed from and returned to their test system-housing units for any reason (e.g., cage cleaning, treatment, etc.), shall receive appropriate identification (e.g., tattoo, color code, ear tag, ear punch, etc.). All information needed to specifically identify each test system within the test system-housing unit shall appear on the outside of that unit. Suckling mammals and juvenile birds are excluded from the requirement of individual identification unless otherwise specified in the protocol.

(e) Except as specified in paragraph (e)(1) of this section, test systems of different species shall be housed in separate rooms when necessary. Test systems of the same species, but used in different studies, should not ordinarily be housed in the same room when inadvertent exposure to test, control, or reference substances or test system mixup could affect the outcome of either study. If such mixed housing is necessary, adequate differentiation by space and identification shall be made.

(1) Plants, invertebrate animals, aquatic vertebrate animals, and organisms that may be used in multispecies tests need not be housed in separate rooms, provided that they are adequately segregated to avoid mixup and cross contamination.

(2) [Reserved]

(f) Cages, racks, pens, enclosures, aquaria, holding tanks, ponds, growth chambers, and other holding, rearing and breeding areas, and accessory equipment, shall be cleaned and sanitized at appropriate intervals.

(g) Feed, soil, and water used for the test systems shall be analyzed periodically to ensure that contaminants known to be capable of interfering with the study and reasonably expected to be present in such feed, soil, or water are not present at levels above those specified in the protocol. Documentation of such analyses shall be maintained as raw data.

(h) Bedding used in animal cages or pens shall not interfere with the purpose or conduct of the study and shall be changed as often as necessary to keep the animals dry and clean.

(i) If any pest control materials are used, the use shall be documented. Cleaning and pest control materials that interfere with the study shall not be used.

(j) All plant and animal test systems shall be acclimatized to the environmental conditions of the test, prior to their use in a study.

Subpart F--Test, Control, and Reference Substances

§792.105. Test, control, and reference substance characterization.

(a) The identity, strength, purity, and composition, or other characteristics which will appropriately define the test, control, or reference substance shall be determined for each batch and shall be documented before its use in a study. Methods of synthesis, fabrication, or derivation of the test, control, or reference substance shall be documented by the sponsor or the testing facility, and such location of documentation shall be specified.

(b) When relevant to the conduct of the study the solubility of each test, control, or reference substance shall be determined by the testing facility or the sponsor before the experimental start date. The stability of the test, control or reference substance shall be determined before the experimental start date or concomitantly according to written standard operating procedures, which provide for periodic analysis of each batch.

(c) Each storage container for a test, control, or reference substance shall be labeled by name, chemical abstracts service number (CAS) or code number, batch number, expiration date, if any, and, where appropriate, storage conditions necessary to maintain the identity, strength, purity, and composition of the test, control, or reference substance. Storage containers shall be assigned to a particular test substance for the duration of the study.

(d) For studies of more than 4 weeks experimental duration, reserve samples from each batch of test, control, and reference substances shall be retained for the period of time provided by §792.195.

(e) The stability of test, control, and reference substances under storage conditions at the test site shall be known for all studies.

§792.107 Test, control, and reference substance handling.

Procedures shall be established for a system for the handling of the test, control, and reference substances to ensure that:

(a) There is proper storage.

(b) Distribution is made in a manner designed to preclude the possibility of contamination, deterioration, or damage.

(c) Proper identification is maintained throughout the distribution process.

(d) The receipt and distribution of each batch is documented. Such documentation shall include the date and quantity of each batch distributed or returned.

§792.113 Mixtures of substances with carriers.

(a) For each test, control, or reference substance that is mixed with a carrier, tests by appropriate analytical methods shall be conducted:

(1) To determine the uniformity of the mixture and to determine, periodically, the concentration of the test, control, or reference substance in the mixture.

(2) When relevant to the conduct of the experiment, to determine the solubility of each test, control, or reference substance in the mixture by the testing facility or the sponsor before the experimental start date.

(3) To determine the stability of the test, control, or reference substance in the mixture before the experimental start date or concomitantly according to written standard operating procedures, which provide for periodic analysis of each batch.

(b) Where any of the components of the test, control, or reference substance carrier mixture has an expiration date, that date shall be clearly shown on the container. If more than one component has an expiration date, the earliest date shall be shown.

(c) If a vehicle is used to facilitate the mixing of a test substance with a carrier, assurance shall be provided that the vehicle does not interfere with the integrity of the test.

Subpart G--Protocol for and Conduct of a Study

§792.120 Protocol.

(a) Each study shall have an approved written protocol that clearly indicates the objectives and all methods for the conduct of the study. The protocol shall contain but shall not necessarily be limited to the following information:

(1) A descriptive title and statement of the purpose of the study.

(2) Identification of the test, control, and reference substance by name, chemical abstracts service (CAS) number or code number.

(3) The name and address of the sponsor and the name and address of the testing facility at which the study is being conducted.

(4) The proposed experimental start and termination dates.

(5) Justification for selection of the test system.

(6) Where applicable, the number, body weight, sex, source of supply, species, strain, substrain, and age of the test system.

(7) The procedure for identification of the test system.

(8) A description of the experimental design, including methods for the control of bias.

(9) Where applicable, a description and/or identification of the diet used in the study as well as solvents, emulsifiers and/or other materials used to solubilize or suspend the test, control, or reference substances before mixing with the carrier. The description shall include specifications for acceptable levels of contaminants that are reasonably expected to be present in the dietary materials and are known to be capable of interfering with the purpose or conduct of the study if present at levels greater than established by the specifications.

(10) The route of administration and the reason for its choice.

(11) Each dosage level, expressed in milli-

grams per kilogram of body or test system weight or other appropriate units, of the test, control, or reference substance to be administered and the method and frequency of administration.

(12) The type and frequency of tests, analyses, and measurements to be made.

(13) The records to be maintained.

(14) The date of approval of the protocol by the sponsor and the dated signature of the study director.

(15) A statement of the proposed statistical method.

(b) All changes in or revisions of an approved protocol and the reasons therefore shall be documented, signed by the study director, dated, and maintained with the protocol.

§*792.130 Conduct of a study.*

(a) The study shall be conducted in accordance with the protocol.

(b) The test systems shall be monitored in conformity with the protocol.

(c) Specimens shall be identified by test system, study, nature, and date of collection. This information shall be located on the specimen container or shall accompany the specimen in a manner that precludes error in the recording and storage of data.

(d) In animal studies where histopathology is required, records of gross findings for a specimen from postmortem observations shall be available to a pathologist when examining that specimen histopathologically.

(e) All data generated during the conduct of a study, except those that are generated by automated data collection systems, shall be recorded directly, promptly, and legibly in ink. All data entries shall be dated on the day of entry and signed or initialed by the person entering the data. Any change in entries shall be made so as not to obscure the original entry, shall indicate the reason for such change, and shall be dated and signed or identified at the time of the change. In automated data collection systems, the individual responsible for direct data input shall be identified at the time of data input. Any change in automated data entries shall be made so as not to obscure the original entry, shall indicate the reason for the change, shall be dated, and the responsible individual shall be identified.

§*792.135 Physical and chemical characterization studies.*

(a) All provisions of the GLP's shall apply to physical and chemical characterization studies designed to determine stability, solubility, octanol water partition coefficient, volatility, and persistence (such as biodegradation, photodegradation, and chemical degradation studies) of test, control, or reference substances.

(b) The following GLP standards shall not apply to studies designed to determine physical and chemical characteristics of a test, control, or reference substance:

§792.31 (c), (d), and (g)
§792.35 (b) and (c)
§792.43
§792.45
§792.47
§792.49
§792.81(b)(1), (2), (6) through (9), and (12)
§792.90
§792.105(a) through (d)
§792.113
§792.120(a)(5) through (12), and 15
§792.185(a)(5) through (8), (10), (12), and (14)
§792.195 (c) and (d)

Subparts H-I--[Reserved]

Subpart J--Records and Reports

§*792.185 Reporting of study results.*

(a) A final report shall be prepared for each study and shall include, but not necessarily be limited to, the following:

(1) Name and address of the facility performing the study and the dates on which the study was initiated and was completed, terminated, or discontinued.

(2) Objectives and procedures stated in the approved protocol, including any changes in

the original protocol.

(3) Statistical methods employed for analyzing the data.

(4) The test, control, and reference substances identified by name, chemical abstracts service (CAS) number or code number, strength, purity, and composition, or other appropriate characteristics.

(5) Stability and, when relevant to the conduct of the study the solubility of the test, control, and reference substances under the conditions of administration

(6) A description of the methods used.

(7) A description of the test system used. Where applicable, the final report shall include the number of animals or other test organisms used, sex, body weight range, source of supply, species, strain and substrain, age, and procedure used for identification.

(8) A description of the dosage, dosage regimen, route of administration, and duration.

(9) A description of all circumstances that may have affected the quality or integrity of the data.

(10) The name of the study director, the names of other scientists or professionals and the names of all supervisory personnel, involved in the study.

(11) A description of the transformations, calculations, or operations performed on the data, a summary and analysis of the data, and a statement of the conclusions drawn from the analysis.

(12) The signed and dated reports of each of the individual scientists or other professionals involved in the study, including each person who, at the request or direction of the testing facility or sponsor, conducted an analysis or evaluation of data or specimens from the study after data generation was completed.

(13) The locations where all specimens, raw data, and the final report are to be stored.

(14) The statement prepared and signed by the quality assurance unit as described in §792. 35(b)(7).

(b) The final report shall be signed and dated by the study director.

(c) Corrections or additions to a final report shall be in the form of an amendment by the study director. The amendment shall clearly identify that part of the final report that is being added to or corrected and the reasons for the correction or addition, and shall be signed and dated by the person responsible. Modification of a final report to comply with the submission requirements of EPA does not constitute a correction, addition, or amendment to a final report.

(d) A copy of the final report and of any amendment to it shall be maintained by the sponsor and the test facility.

§792.190 Storage and retrieval of records and data.

(a) All raw data, documentation, records, protocols, specimens, and final reports generated as a result of a study shall be retained. Specimens obtained from mutagenicity tests, specimens of soil, water, and plants, and wet specimens of blood, urine, feces, and biological fluids, do not need to be retained after quality assurance verification. Correspondence and other documents relating to interpretation and evaluation of data, other than those documents contained in the final report, also shall be retained.

(b) There shall be archives for orderly storage and expedient retrieval of all raw data, documentation, protocols, specimens, and interim and final reports. Conditions of storage shall minimize deterioration of the documents or specimens in accordance with the requirements for the time period of their retention and the nature of the documents of [*sic*] specimens. A testing facility may contract with commercial archives to provide a repository for all material to be retained. Raw data and specimens may be retained elsewhere provided that the archives have specific reference to those other locations.

(c) An individual shall be identified as responsible for the archives.

(d) Only authorized personnel shall enter the archives.

(e) Material retained or referred to in the archives shall be indexed to permit expedient retrieval.

§792.195 Retention of records.

(a) Record retention requirements set forth in this section do not supersede the record retention requirements of any other regulations in this subchapter.

(b)(1) Except as provided in paragraph (c) of this section, documentation records, raw data, and specimens pertaining to a study and required to be retained by this part shall be retained in the archive(s) for a period of at least ten years following the effective date of the applicable final test rule.

(2) In the case of negotiated testing agreements, each agreement will contain a provision that, except as provided in paragraph (c) of this section, documentation records, raw data, and specimens pertaining to a study and required to be retained by this part shall be retained in the archive(s) for a period of at least ten years following the publication date of the acceptance of a negotiated test agreement.

(3) In the case of testing submitted under section 5, except for those items listed in paragraph (c) of this section, documentation records, raw data, and specimens pertaining to a study and required to be retained by this part shall be retained in the archive(s) for a period of at least five years following the date on which the results of the study are submitted to the agency.

(c) Wet specimens, samples of test, control, or reference substances, and specially prepared material which are relatively fragile and differ markedly in stability and quality during storage, shall be retained only as long as the quality of the preparation affords evaluation. Specimens obtained from mutagenicity tests, specimens of soil, water, and plants, and wet specimens of blood, urine, feces, biological fluids, do not need to be retained after quality assurance verification. In no case shall retention be required for longer periods than those set forth in paragraph (b) of this section.

(d) The master schedule sheet, copies of protocols, and records of quality assurance inspections, as required by §792.35(c) shall be maintained by the quality assurance unit as an easily accessible system or records for the period of time specified in paragraph (b) of this section.

(e) Summaries of training and experience and job descriptions required to be maintained by §792.29(b) may be retained along with all other testing facility employment records for the length of time specified in paragraph (b) of this section.

(f) Records and reports of the maintenance and calibration and inspection of equipment, as required by §792.63(b) and (c), shall be retained for the length of time specified in paragraph (b) of this section.

(g) If a facility conducting testing or an archive contracting facility goes out of business, all raw data, documentation, and other material specified in this section shall be transferred to the archives of the sponsor of the study. The EPA shall be notified in writing of such a transfer.

(h) Specimens, samples, or other non-documentary materials need not be retained after EPA has notified in writing the sponsor or testing facility holding the materials that retention is no longer required by EPA. Such notification normally will be furnished upon request after EPA or FDA has completed an audit of the particular study to which the materials relate and EPA has concluded

that the study was conducted in accordance with this part.

(i) Records required by this part may be retained either as original records or as true copies such as photocopies, microfilm, microfiche, or other accurate reproductions of the original records.

GOOD AUTOMATED LABORATORY PRACTICES

In many laboratories, computers are replacing many manual operations: they manage operations, interface with equipment, and generate reports, for example. The proliferation of automated data collection, however, has resulted in new problems of corruption, loss, and inappropriate modification in data provided to the EPA. The GALPs are EPA's answer to the need for standardized laboratory data management principles. The lack of such principles was underscored after an investigation by EPA's Office of Inspector General, and an EPA survey of automated laboratory practices disclosed that the integrity of computer-resident data was at risk in many laboratories providing data to EPA because of gaps in system security, data validation, and basic documentation. At the same time, laboratory staffs expressed the need for -- and were frustrated with their inability to obtain -- adequate EPA guidance to protect integrity of computer-resident data. These laboratories wanted a single source of guidance for automated operations.

The EPA determined that data management procedures should be standardized in laboratories supporting EPA programs, and that the Agency itself should assume the responsibility to establish those standards. The GALPs and the GALP *Implementation Guidance* were published in draft form on December 28, 1990. The intent of these documents is to enable laboratories that provide data to have a clear understanding of what EPA considers to be adequate controls to assure data integrity.

EPA'S GOOD AUTOMATED LABORATORY PRACTICES [Draft, December 28, 1990]

7.0 *Policy*

It is EPA policy that data collected, analyzed, processed or maintained on automated data collection system(s) in support of health and environmental effects studies be accurate and of sufficient integrity to support effective environmental management.

The Good Automated Laboratory Practices (GALPs) ensure the integrity of computer-resident data. They recommend minimum practices and procedures for laboratories that provide data to EPA in support of its health and environmental programs to follow when automating their operations.

7.1 *Personnel*

When an automated data collection system is used in the conduct of a laboratory study, all personnel involved in the design or operation of the automated system shall:

1) have adequate education, training, and experience to enable those individuals to perform the assigned system functions.

2) have a current summary of their training, experience, and job description, including information relevant to system design and operation maintained at the

facility.

3) be of sufficient number for timely and proper conduct of the study, including timely and proper operation of the automated data collection system(s).

7.2 *Laboratory Management*

When an automated data collection system is used in the conduct of a study, the laboratory management shall:

1) designate an individual primarily responsible for the automated data collection system(s), as described in Section 7.3.

2) assure that there is a quality assurance unit that oversees the automated data collections system(s), as described in Section 7.4.

3) assure that the personnel, resources, facilities, computer and other equipment, materials, and methodologies are available as scheduled.

4) receive reports of quality assurance inspections or audits of computers and/or computer-resident data and promptly take corrective actions in response to any deficiencies.

5) assure that personnel clearly understand the functions they are to perform on automated data collection system(s).

6) assure that any deviations from this guide for automated data collection system(s) are reported to the designated Responsible Person and that corrective actions are taken and documented.

7.3 *Responsible Person*

The laboratory shall designate a computer scientist or other professional of appropriate education, training, and experience or combination thereof as the individual primarily responsible for the automated data collection system(s) (the Responsible Person). This individual shall ensure that:

1) there are sufficient personnel with adequate training and experience to supervise and/or conduct, design, and operate the automated data collection system(s).

2) the continuing competency of staff who design or use the automated data collection system is maintained by documentation of their training, review of work performance, and verification of required skills.

3) a security risk assessment has been made, points of vulnerability of the system have been determined, and all necessary security measures have been implemented.

4) the automated data collection system(s) have written operating procedures and appropriate software documentation that are complete, current, and available to all staff.

5) all significant changes to operating procedures and/or software are approved by review and signature.

6) there are adequate acceptance procedures for software and software changes.

7) there are procedures to assure that data are accurately recorded in the automated data collection system.

8) problems with the automated data collection system that could affect data quality are documented when they occur, are subject to corrective action, and the corrective action is documented.

9) all applicable good laboratory practices are followed.

7.4 *Quality Assurance Unit*

The laboratory shall have a quality assurance unit that shall be responsible for monitoring those aspects of a study where an automated data collection system is used. The quality assurance unit shall be entirely separate from and independent of the personnel engaged in the direction and conduct of a study or contract. The quality assurance unit shall inspect and audit the automated data collection system(s) at intervals adequate to ensure the integrity of the study. The quality assurance unit shall:

1) maintain a copy of the written procedures that include operation of the automated data collection system.

2) perform periodic inspections of the laboratory operations that utilize automated data collection system(s) and submit properly signed records of each inspection, the study inspected, the person performing the inspection, findings and problems, action recommended and taken to resolve existing problems, and any scheduled dates for reinspection. Any problems noted in the automated data collection system that are likely to affect study integrity found during the course of an inspection shall be brought to the immediate attention of the designated Responsible Person.

3) determine that no deviations from approved written operating instructions and software were made without prior proper authorization and sufficient documentation.

4) periodically review final data reports to ensure that results reported by the automated data collection system accurately represent the raw data.

5) ensure that the responsibilities and procedures applicable to the quality assurance unit, the records maintained by the quality assurance unit, and the method of indexing such records shall be in writing and shall be maintained. These items include inspection dates of automated data collections systems, name of the individual performing each inspection, and results of the inspection.

7.5 *Facilities*

When an automated data collection system is used in the conduct of a study, the laboratory shall:

1) ensure that the facility used to house the automated data collection system(s) has provisions to regulate the environmental conditions (e.g., temperature, humidity, adequacy of electrical requirements) adequate to protect the system(s) against data loss due to environment problems.

2) provide adequate storage capability of the automated data collection system(s) or of the facility itself to provide retention of raw data, including archives of computer-resident data.

7.6 *Equipment*

1) Automated data collection equipment used in the generation, measurement, or assessment of data shall be of appropriate design and adequate capacity to function according to specifications and shall be suitably located for operation, inspection, cleaning, and maintenance. There shall be a written description of the computer system(s) hardware. Automated data collection equipment shall be installed in accordance with manufacturer's recommendations and undergo appropriate acceptance testing following written acceptance criteria at installation. Significant changes to automated data collection system(s) shall be made only by approved review, testing, and signature of the designated Responsible Person and the Quality Assurance Unit.

2) Automated data collection system(s) shall be adequately tested, inspected, cleaned, and maintained. The laboratory shall:

2.1) have written operating procedures for routine maintenance operations.

2.2) designate in writing an individual responsible for performance of each operation.

2.3) maintain written records of all maintenance testing containing the dates of the operation, describing whether the operation was routine and followed the written procedure.

2.4) maintain records of non-routine repairs performed on the equipment as a result of a failure and/or malfunction. Such records shall document the problem, how and when the problem occurred, and describe the remedial action taken in response to the problem along with acceptance criteria to ensure the return of function of the repaired system.

3) The laboratory shall institute backup and recovery procedures to ensure that operating instructions (i.e., software) for the automated data collection system(s) can be recovered after a system failure.

7.7 *Security*

1) When an automated data collection system is used in the conduct of a study, the laboratory shall evaluate the need for system security. The laboratory shall have procedures that assure that the automated data collection system is secured if that system:

1.1) contains confidential information that requires protection from unauthorized disclosure.

1.2) contains data whose integrity must be protected against unintentional error or intentional fraud.

1.3) performs time-critical functions that require that data be available to sample tracking critical to prompt data analysis, monitors quality control criteria critical to timely release of data, or generates reports which are critical to the timely submission of data.

2) When the automated data collection system contains data that must be secured, the laboratory shall ensure that the system is physically secured, that physical and functional access to the system is limited to only authorized personnel, and that introduction of unauthorized external programs/software is prohibited.

2.1) Only personnel with specifically documented authorization shall be allowed physical access to areas where automated data collection systems are maintained.

2.2) Log-ons, restricted passwords, call-backs on modems, voiceprints, fingerprints, etc., shall be used to ensure that only personnel with documented authorization can access automated data collection systems.

2.3) Procedures shall be in place to ensure that only personnel with documented authorization to access automated data collection system functions shall be able to access those functions.

2.4) In order to protect the operational integrity of the automated data collection system, the laboratory shall have procedures for protecting the system from introduction of external programs/software (e.g., to prevent introduction of viruses, worms, etc.).

7.8 *Standard Operating Procedures*

1) In laboratories where automated data collection systems are used in the conduct of a study, the laboratory shall have written standard operating procedures (SOPs). Standard operating procedures shall be established for, but not limited to:

1.1) maintaining the security of the automated data collection system(s) (i.e., physical security, securing access to the system and its functions, and restricting installation of external programs/software).

1.2) defining raw data for the laboratory operation and providing a working definition of raw data.

1.3) entry of data and proper identification of the individual entering the data.

1.4) verification of manually or electronically input data.

1.5) interpretation of error codes or flags and the corrective action to follow when these occur.

1.6) changing data and proper methods

for execution of data changes to include the original data element, the changed data element, identification of the date of change, the individual responsible for the change, and the reason for the change.

1.7) data analysis, processing, storage, and retrieval.

1.8) backup and recovery of data.

1.9) maintaining automated data collection system(s) hardware.

1.10) electronic reporting, if applicable.

2) In laboratories where automated data collection systems are used in the conduct of a study, the laboratory shall have written standard operating procedures (SOPs). Each laboratory or other study area shall have readily available manuals and standard operating procedures that document the procedures being performed. Published literature or vendor documentation may be used as a supplement to the standard operating procedures if properly referenced therein.

3) In laboratories where automated data collection systems are used in the conduct of a study, the laboratory shall have written standard operating procedures (SOPs). A historical file of standard operating procedures shall be maintained. All revisions, including the dates of such revisions, shall be maintained within the historical file.

7.9 *Software*

1) The laboratory shall consider software to be the operational instructions for automated data collection systems and shall, therefore, have written standard operating procedures setting forth methods that management is satisfied are adequate to ensure that the software is accurately performing the intended functions. All deviations from the operational instructions for automated data collection systems shall be authorized by the designated Responsible Person. Changes in the established operational instructions shall be properly authorized, reviewed, and accepted in writing by the designated Responsible Person.

2) The laboratory shall have documentation to demonstrate the validity of software used in the conduct of a study as outlined in Section 7.9.3.

2.1) For new systems the laboratory shall have documentation throughout the life cycle of the system (i.e., beginning with identification of user requirements and continuing through design, integration, qualification, validations, control, and maintenance, until use of the system is terminated).

2.2) Automated data collection system(s) currently in existence or purchased from a vendor shall be, to the greatest extent possible, similarly documented to demonstrate validity.

3) Documentation of operational instructions (i.e., software) shall be established and maintained for, but not be limited to:

3.1) detailed written description of the software in use and what the software is expected to do or the functional requirements that the system is designed to fulfill.

3.2) identification of software development standards used, including coding standards and requirements for adding comments to the code to identify its functions.

3.3) listing of all algorithms or formulas used for data analysis, processing, conversion, or other manipulations.

3.4) acceptance testing that outlines acceptance criteria, identifies when the tests were done and the individual(s) responsible for the testing, summarizes the results of the tests, and documents review and written approval of tests performed.

3.5) change control procedures that include instructions for requesting, testing, approving, and issuing software changes.

3.6) procedures that document the version of software used to generate data sets.

3.7) procedures for reporting software problems, evaluation of problems, and documentation of corrective actions.

4) Manuals or written procedures for documentation of operational instructions shall be readily available in the areas where these procedures are performed. Published literature or vendor documentation may be used as a supplement to software documentation if properly referenced therein.

5) A historical file of operating instructions, changes, or version numbers shall be maintained. All software revisions, including the dates of such revisions, shall be maintained within the historical file. The laboratory shall have appropriate historical documentation to determine the software version used for the collection, analysis, processing or maintenance of all data sets on automated data collection systems.

7.10 *Data Entry*

When a laboratory uses an automated data collection system in the conduct of a study, the laboratory shall ensure the integrity of the computer-resident data collected, analyzed, processed, or maintained on the system. The laboratory shall ensure that in automated data collection systems:

1) The individual responsible for direct data input shall be identified at the time of data input.

2) The instruments transmitting data to the automated data collection system shall be identified, and the time and date of transmittal shall be documented.

3) Any change in automated data entries shall not obscure the original entry, shall indicate the reason for the change, shall be dated, and shall identify the individual making the change.

4) Data integrity in an automated data collection system is most vulnerable during data entry whether done via manual input or by electronic transfer from automated instruments. The laboratory shall have written procedures and practices in place to verify the accuracy of manually entered and electronically transferred data collected on automated system(s).

7.11 *Raw Data*

Raw data collected, analyzed, processed, or maintained on automated data collection system(s) are subject to the procedures outlined below for storage and retention of records. Raw data may include microfilm, microfiche, computer printouts, magnetic media, and recorded data from automated data collection systems. Raw data is defined as data that cannot be easily derived or recalculated from other information. The laboratory shall:

1) define raw data for its own laboratory operation.

2) include this definition in the laboratory's standard operating procedures.

7.12 *Records and Archives*

1) All raw data, documentation, and records generated in the design and operation of automated data collection system(s) shall be retained. Correspondence and other documents relating to interpretation and evaluation of data collected, analyzed, processed, or maintained on the automated data collection system(s) also shall be retained. Records to be maintained include, but are not limited to:

1.1) A written definition of computer-resident "raw data" (see Section 7.11 of this document).

1.2) A written description of the hardware and software used in the collection, analysis, processing, or maintenance of data on automated data collection system(s). This description shall identify expectations of computer system performance and shall list the hardware and software used for data handling. Where multiple automated data collection systems are used, the written description shall include how the systems interact with one another.

1.3) Software and/or hardware acceptance test records which identify the item tested, the method of testing, the date(s) the tests were performed, and the individuals who conducted and reviewed the tests.

1.4) Summaries of training and experience and job descriptions of staff as required by Section 7.1 of this document.

1.5) Records and reports of maintenance of automated data collection system(s).

1.6) Records of problems reported with software and corrective actions taken.

1.7) Records of quality assurance inspections (but not the findings of the inspections) of computer hardware, software, and computer-resident data.

1.8) Records of backups and recoveries, including backup schedules or logs, type and storage location of backup media used, and logs of system failures and recoveries.

2) There shall be archives for orderly storage and expedient retrieval of all raw data, documentation, and records generated in the design and operation of the automated data collection system. conditions of storage shall minimize potential deterioration of documents or magnetic media in accordance with the requirements for the retention period and the nature of the document or magnetic media.

3) An individual shall be designated in writing as a records custodian for the archives.

4) Only personnel with documented authorization to access the archives shall be permitted this access.

5) Raw data collected, analyzed, processed, or maintained on automated data collection system(s) and documentation and records for the automated data collection system(s) shall be retained for the period specified by EPA contract or EPA statute.

7.13 *Reporting*

A laboratory may choose to report or may be required to report data electronically. If the laboratory reports data electronically, the laboratory shall:

1) Ensure that electronic reporting of data from analytical instruments is reported in accordance with the EPA's standards for electronic transmission of laboratory measurements. Electronic reporting of laboratory measurements must be provided on standard magnetic media (i.e., magnetic tapes and/or floppy disks) and shall adhere to standard requirements for record identification, sequence, length, and content as specified in EPA Order 2180.2 -- Data Standards for Electronic Transmission of Laboratory Measurement Results.

2) Ensure that electronically reported data are transmitted in accordance with the recommendations of the Electronic Reporting Standards Workgroup (to be identified when the recommendations are finalized).

7.14 *Comprehensive Ongoing Testing*

Laboratories using automated data collection systems must conduct comprehensive tests of overall system performance, including document review, at least once every 24 months. These tests must be documented, and the documentation must be retained and available for inspection or audit.

Chapter 5

EUROPEAN UNION REGULATIONS

OVERVIEW

The European Union (EU), formerly the European Community (EC), provides guidelines for medical devices and pharmaceuticals through directives issued by the Commission of the European Communities. Three directives are major factors in complying with systems validation requirements in the EU: the directive on good manufacturing practices for medicinal products, the active implantable medical device directive, and the general medical device directive. The Commission's guideline on Good Clinical Practices provides additional guidance on system validation.

Good Manufacturing Practice Directive

The European Union's directive establishing principles and guidelines of good manufacturing practices for medicinal products was adopted on June 13, 1991. It covers generally the same good manufacturing practice guidelines as the U.S. FDA's Good Manufacturing Practice regulations, i.e., principles for management control, personnel activities and training, equipment design and operation, documentation, production procedures, quality control, and self-audits.

Annex 11 of the directive, however, sets forth express requirements for computerized systems utilized in pharmaceutical manufacture. A guideline of 18 points relating to effective testing, use, and documentation of system accuracy is provided in this annex, including a point on validation:

> The extent of validation necessary will depend on a number of factors including the use to which the system is to be put, whether the validation is to be prospective or retrospective and whether or not novel elements are incorporated. Validation should be considered as part of the complete life cycle of a computer system. This cycle includes the stages of planning, specification, programming, testing, commissioning, documentation, operation, monitoring and modifying.

Medical Device Directives

The Commission issued its Active Implantable Medical Device (AIMD) Directive on June 20, 1990. Active implantable medical devices are defined in the directive as those "active" devices either partially or totally introduced into the human body by medical or surgical procedures. Medical devices are "active " if they depend upon electrical or other outside source of energy for their power.

The EU recognizes the importance of increasing computerization of medical devices. The definitions section of the directive includes software in the definition of the term "medical device," classifying software as an accessory for the "proper functioning" of a medical device.

Annex 1 of the AIMD Directive lists the "essential requirements" which must be met in the manufacture of active implantable medical devices. Paragraph 9 of this annex requires manufacturers to pay particular attention to the "proper functioning of the programming and control systems, including software," of any active implantable medical device. Paragraph 15 of the directive requires that when active implantable medical devices are placed on the EU market, they must be accompanied by instructions that include "information allowing the physician to select a suitable device and the corresponding software and accessories," and "information constituting the instructions for use allowing the physician and, where appropriate, the patient to use the device, its accessories and software correctly, as well as information on the nature, scope and times for operating controls and trials and, where appropriate, maintenance measures."

The Commission's Medical Device (MD) Directive, issued on June 14, defines "medical devices" as any device -- including the software necessary for its proper application -- used to diagnose, prevent, monitor, treat, or alleviate a disease or injury. In this directive, software is considered specifically to be a medical device, rather than classed with accessories as in the AIMD Directive.

Among the essential requirements in Annex I of the MD Directive, there are requirements to ensure the reliable performance of devices with electronic programmable systems. In addition, a recent examination of the essential requirements of this directive determined that in more than half of the 54 essential requirements listed in Annex I, software requirements must be considered.[1] Annex IX, which contains criteria for classification of medical devices according to risk analysis, states that software used in a device falls into the same class as the device in which it is used.

Good Clinical Practices

On July 13, 1990, the European Commission's Committee for Proprietary Medicinal Products (CPMP) approved a guideline developed by the CPMP Working Party on Efficacy of Medicinal Products on good clinical practice standards. The portions of this guideline dealing with system validation requirements in data handling, statistics, and

---

[1]Oliver P. Christ, "Software Validation: Requirements for Programmable Electronic Medical Systems, Part I, How to Comply with the Essential Requirements of the EC Directives. *Medical Device Technology*, 5(March 1994):28-32.

quality assurance are, to date, more explicit than the system validation requirements in the U.S. FDA's Good Clinical Practice regulations.

The complete texts of the European Commission's directives on Good Manufacturing Practice, Active Implantable Medical Devices, and Medical Devices are reprinted here. Only the portions of the GCP guidelines relating to system validation standards is reprinted. Requirements relating to system validation in these documents have been highlighted in boldface.

COMMISSION DIRECTIVE
of 13 June 1991
laying down the principles and guidelines of good manufacturing practice for medicinal products for human use
(91/356/EEC)

THE COMMISSION OF THE EUROPEAN COMMUNITIES,

Having regard to the Treaty establishing the European Economic Community,

Having regard to Council Directive 75/31/EEC of 20 May 1975 on the approximation of provisions laid down by law, regulation or administrative action relating to proprietary medicinal products, as last amended by Directive 89/381/EEC, and in particular Article 19a thereof,

Whereas all medicinal products for human use manufactured or imported into the Community, including medicinal products intended for export, should be manufactured in accordance with the principles and guidelines of good manufacturing practice;

Whereas, in accordance with national legislation, Member States may require compliance with these principles of good manufacturing practice during the manufacture of products intended for use in clinical trials;

Whereas the detailed guidelines mentioned in Article 19a of Directive 75/319/EEC have been published by the Commission after consultation with the pharmaceutical inspection services of the Member States in the form of a *'Guide to good manufacturing practice for medicinal products';*

Whereas it is necessary that all manufacturers should operate an effective quality management of their manufacturing operations, and that this requires the implementation of a pharmaceutical quality assurance system;

Whereas officials representing the competent authorities should report on whether the manufacturer complies with good manufacturing practice and that these reports should be communicated upon reasoned request to the competent authorities of another Member State;

Whereas the principles and guidelines of good manufacturing practice should primarily concern personnel, premises and equipment, documentation, production, quality control, contracting out, complaints and product recall, and self inspection;

Whereas the principles and guidelines envisaged by this Directive are in accordance with the opinion of the Committee for the Adaptation to Technical Progress of the Directives on the Removal of Technical Barriers to Trade in the Proprietary Medicinal Products Sector set up by Article 2b of Council Directive 75/318/EEC of 20 May 1975 on the approximation of the laws of Member States relating to analytical, pharmaco-toxicological and clinical standards and protocols in respect of the testing of proprietary medicinal products, as last amended by Directive 89/341/EEC,

HAS ADOPTED THIS DIRECTIVE:

CHAPTER I

GENERAL PROVISIONS

Article 1

This Directive lays down the principles and guidelines of good manufacturing practice for medicinal products for human use whose manufacture requires the authorization referred to in Article 16 of Directive 75/319/EEC.

Article 2

For the purposes of this Directive, the definition of medicinal products set out in Article 1(2) of Council Directive 65/65/EEC, shall apply.

In addition,

-- 'manufacturer' shall mean any holder of the authorization referred to in Article 16 of Directive 75/319/EEC,
-- 'qualified person' shall mean the person referred to in Article 21 of Directive 75/319/EEC,
-- 'pharmaceutical quality assurance' shall mean the sum total of the organized arrangements made with the object of ensuring that medicinal products are of the quality required for their intended use,
-- 'good manufacturing practice' shall mean the part of quality assurance which ensures that products are consistently produced and controlled to the quality standards appropriate to their intended use.

Article 3

By means of the repeated inspections referred to in Article 26 of Directive 75/319/EEC, the Member States shall ensure that manufacturers respect the principles and guidelines of good manufacturing practice laid down by this Directive.

For the interpretation of these principles and guidelines of good manufacturing practice, the manufacturers and the agents of the competent authorities shall refer to the detailed guidelines referred to in Article 19a of Directive 75/319/ EEC. These detailed guidelines are published by the Commission in the *'Guide to good manufacturing practice for medicinal products'* and in its Annexes (Office for Official Publications of the European Communities, *The rules governing medicinal products in the European Community*, Volume IV).

Article 4

The manufacturer shall ensure that the manufacturing operations are carried out in accordance with good manufacturing practice and with the manufacturing authorization.

For medicinal products imported from third countries, the importer shall ensure that the medicinal products have been manufactured by manufacturers duly authorized and conforming to good manufacturing practice standards, at least equivalent to those laid down by the Community.

Article 5

The manufacturer shall ensure that all manufacturing operations subject to an authorization for marketing are carried out in accordance with the information given in the application for marketing authorization as accepted by the competent authorities.

The manufacturer shall regularly review their manufacturing methods in the light of scientific and technical progress. When a modification to the marketing authorization dossier is necessary, the application for modification must be submitted to the competent authorities.

CHAPTER II

PRINCIPLES AND GUIDELINES OF GOOD MANUFACTURING PRACTICE

Article 6

Quality Management

The manufacturer shall establish and implement an effective pharmaceutical quality assurance system, involving the active participation of the management and personnel of the different services involved.

Article 7

Personnel

1. At each manufacturing site, the manufacturer shall have competent and appropriately qualified personnel at his disposal in sufficient number to achieve the pharmaceutical quality assurance objective.
2. The duties of managerial and supervisory staff, including the qualified person(s), responsible for implementing and operating good manufacturing practice shall be defined in job descriptions. Their hierarchical relationships shall be defined in an organization chart. Organization charts and job descriptions shall be approved in accordance with the manufacturer's internal procedures.
3. Staff referred to in paragraph 2 shall be given sufficient authority to discharge their responsibilities correctly.
4. Personnel shall receive initial and continuing training including the theory and application of the concept of quality assurance and good manufacturing practice.
5. Hygiene programmes adapted to the activities to be carried out shall be established and observed. These programmes include procedures relating to health, hygiene and clothing of personnel.

Article 8

Premises and equipment

1. Premises and manufacturing equipment shall be located, designed, constructed, adapted and maintained to suit the intended operations.
2. Lay out, design and operation must aim to minimize the risk of errors and permit effective cleaning and maintenance in order to avoid contamination, cross contamination and, in general, any adverse effect on the quality of the product.
3. Premises and equipment intended to be used for manufacturing operations which are critical for the quality of the products shall be subjected to appropriate qualification.

Article 9

Documentation

1. The manufacturer shall have a system of documentation based upon specifications, manufacturing formulae and processing and packaging instructions, procedures and records covering the various manufacturing operations that they perform. Documents shall be clear, free from errors and kept up to date. Pre-established procedures for general manufacturing operations and conditions shall be available, together with specific documents for the manufacture of each batch. This set of documents shall make it possible to trace the history of the manufacture of each batch. The batch documentation shall be retained for at least one year after the expiry date of the batches to which it relates or at least five years after the certification referred to in Article 22(2) of Directive 75/319/EEC whichever is the longer.
2. When electronic, photographic or other data processing systems are used instead of written documents, the manufacturer shall have validated the systems by proving that the data will be appropriately stored during the anticipated period of storage. Data stored by these systems shall be made readily available in legible form. The electronically stored data shall be protected against loss or damage of data (e.g. by duplication or back-up and transfer onto another storage system).

Article 10

Production

The different production operations shall be carried out according to pre-established instructions and procedures and in accordance with good manufacturing practice. Adequate and sufficient resources shall be made available for the in-process controls.

Appropriate technical and/or organizational

Any new manufacture or important modification of a manufacturing process shall be validated. Critical phases of manufacturing processes shall be regularly revalidated.

Article 11

Quality control

1. The manufacturer shall establish and maintain a quality control department. This department shall be placed under the authority of a person having the required qualifications and shall be independent of the other departments.
2. The quality control department shall have at its disposal one or more quality control laboratories appropriately staffed and equipped to carry out the necessary examination and testing of starting materials, packaging materials and intermediate and finished products testing. Resorting to outside laboratories may be authorized in accordance with Article 12 of this Directive after the authorization referred to in Article 5b of Directive 75/319/EEC has been granted.
3. During the final control of finished products before their release for sale or distribution, in addition to analytical results, the quality control department shall take into account essential information such as the production conditions, the results of in-process controls, the examination of the manufacturing documents and the conformity of the products to their specifications (including the final finished pack).
4. Samples of each batch of finished products shall be retained for at least one year after the expiry date. Unless in the Member States of manufacture a longer period is required, samples of starting materials (other than solvents, gases and water) used shall be retained for at least two years after the release of the product. This period may be shortened if their stability, as mentioned in the relevant specification, is shorter. All these samples shall be maintained at the disposal of the competent authorities.

For certain medicinal products manufactured individually or in small quantities, or when their storage could raise special problems, other sampling and retaining conditions may be defined in agreement with the competent authority.

Article 12

Work contracted out

1. Any manufacturing operation or operation linked with the manufacture which is carried out under contract, shall be the subject of a written contract between the contract giver and the contract acceptor.
2. The contract shall clearly define the responsibilities of each party and in particular the observance of good manufacturing practice by the contract acceptor and the manner in which the qualified person responsible for releasing each batch shall undertake his full responsibilities.
3. The contract acceptor shall not subcontract any of the work entrusted to him by the contract giver without the written authorization of the contract giver.
4. The contract acceptor shall respect the principles and guidelines of good manufacturing practice and shall submit to inspections carried out by the competent authorities as provided for by Article 26 of Directive 75/319/EEC.

Article 13

Complaints and product recall

The manufacturer shall implement a system for recording and reviewing complaints together with an effective system for recalling promptly and at any time the medicinal products in the distribution network. Any complaint concerning a defect shall be recorded and investigated by the manufacturer. The competent authority shall be informed by the manufacturer of any defect that could result in a recall or abnormal restriction on the supply. In so far as possible, the countries of destination shall also be indicated. Any recall shall be made in accordance with the requirements referred to tin Article 33 of Directive 75/319/EEC.

be made in accordance with the requirements referred to in Article 33 of Directive 75/319/EEC.

Article 14

Self-inspection

The manufacturer shall conduct repeated self-inspections as part of the quality assurance system in order to monitor the implementation and respect of good manufacturing practice and to propose any necessary corrective measures. Records of such self-inspections and any subsequent corrective action shall be maintained.

CHAPTER III

FINAL PROVISIONS

Article 15

Member States shall bring into force the laws, regulations and administrative provisions necessary to comply with this Directive not later than 1 January 1992. They shall forthwith inform the Commission thereof.

When Member States adopt these provisions, these shall contain a reference to this Directive or shall be accompanied by such reference at the time of their official publication. The procedure for such reference shall be adopted by Member States.

Article 16

This Directive is addressed to the Member States.

Done at Brussels, 13 June 1991.

For the Commission
Martin BANGEMANN
Vice-President

ANNEX 11

COMPUTERIZED SYSTEMS

Principle

The introduction of computerized systems into systems of manufacturing, including storage, distribution and quality control does not alter the need to observe the relevant principles given elsewhere in the Guide. Where a computerized system replaces a manual operation, there should be no resultant decrease in product quality or quality assurance. Consideration should be given to the risk of losing aspects of the previous system which could result from reducing the involvement of operators.

Personnel

1. It is essential that there is the closest cooperation between key personnel and those involved with computer systems. Persons in responsible positions should have the appropriate training for the management and use of systems within their field of responsibility which utilizes computers. This should include ensuring that appropriate expertise is available and used to provide advice on aspects of design, validation, installation and operation of computerized systems.

Validation

2. The extent of validation necessary will depend on a number of factors including the use to which the system is to be put, whether the validation is to be prospective or retrospective and whether or not novel elements are incorporated. Validation should be considered as part of the

complete life cycle of a computer system. This cycle includes the stages of planning, specification, programming, testing, commissioning, documentation, operation, monitoring and modifying.

System

3. Attention should be paid to the siting of equipment in suitable conditions where extraneous factors cannot interfere with the system.
4. A written detailed description of the system should be produced (including diagrams as appropriate) and kept up to date. It should describe the principles, objectives, security measures and scope of the system and the main features of the way in which the computer is used and how it interacts with other systems and procedures.
5. The software is a critical component of a computerized system. The user of such software should take all reasonable steps to ensure that it has been produced in accordance with a system of Quality Assurance.
6. The system should include, where appropriate, built-in checks of the correct entry and processing of data.
7. Before a system using a computer is brought into use, it should be thoroughly tested and confirmed as being capable of achieving the desired results. If a manual system is being replaced, the two should be run in parallel for a time, as a part of this testing and validation.
8. Data should only be entered or amended by persons authorized to do so. Suitable methods of deterring unauthorized entry of data include the use of keys, pass cards, personal codes and restricted access to computer terminals. There should be a defined procedure for the issue, cancellation, and alteration of authorization to enter and amend data, including the changing of personal passwords. Consideration should be given to systems allowing for recording of attempts to access by unauthorized persons.
9. When critical data are being entered manually (for example the weight and batch number of an ingredient during dispensing), there should be an additional check on the accuracy of the record which is made. This check may be done by a second operator or by validated electronic means.
10. The system should record the identity of operators entering or confirming critical data. Authority to amend entered data should be restricted to nominated persons. Any alteration to an entry of critical data should be authorized and recorded with the reason for the change. Consideration should be given to building into the system the creation of a complete record of all entries and amendments (an "audit trail").
11. Alterations to a system or to a computer program should only be made in accordance with a defined procedure which should include provision for validating, checking, approving and implementing the change. Such an alteration should only be implemented with the agreement of the person responsible for the part of the system concerned, and the alteration should be recorded. Every significant modification should be validated.
12. For quality auditing purposes, it should be possible to obtain clear printed copies of electronically stored data.
13. Data should be secured by physical or electronic means against willful or accidental damage, in accordance with item 4.9 of the Guide. Stored data should be checked for accessibility, durability and accuracy. If changes are proposed to the computer equipment or its programs, the above mentioned checks should be performed at a frequency appropriate to the storage medium being used.
14. Data should be protected by backing-up at regular intervals. Back-up data

should be stored as long as necessary at a separate and secure location.
15. There should be available adequate alternative arrangements for systems which need to be operated in the event of a breakdown. The time required to bring the alternative arrangements into use should be related to the possible urgency of the need to use them. For example, information required to effect a recall must be available at short notice.
16. The procedures to be followed if the system fails or breaks down should be defined and validated. Any failures and remedial action taken should be recorded.
17. A procedure should be established to record and analyze errors and to enable corrective action to be taken.
18. When outside agencies are used to provide a computer service, there should be a formal agreement including a clear statement of the responsibilities of that outside agency.
19. When the release of batches for sale or supply is carried out using a computerized system, the system should allow for only a Qualified Person to release the batches and it should clearly identify and record the person releasing the batches.

COUNCIL DIRECTIVE
of 20 June 1990
on the approximation of the laws of the Member States relating to active implantable medical devices

(90/385/EEC)

THE COUNCIL OF THE EUROPEAN COMMUNITIES,

Having regard to the Treaty establishing the European Economic Community, and in particular Article 100a thereof,

Having regard to the proposal from the Commission,

In cooperation with the European Parliament,

Having regard to the opinion of the Economic and Social Committee,

Whereas in each Member State active implantable medical devices must give patients, users and other persons a high level of protection and achieve the intended level of performance when implanted in human beings;

Whereas several Member States have sought to ensure that level of safety by mandatory specifications relating both to the technical safety features and the inspection procedures for such devices; whereas those specifications differ from one Member State to another;

Whereas national provisions ensuring that safety level should be harmonized in order to guarantee the free movement of active implantable medical devices without lowering existing and justified levels of safety in the member states;

Whereas harmonized measures must be distinguished from measures taken by Member States to manage the financing of public health and sickness insurance schemes directly or indirectly concerning such devices; whereas, therefore, such provisions do not affect the right of Member States to implement the abovementioned measures in compliance with Community law;

Whereas maintaining or improving the level of protection achieved in Member States constitutes one of this Directive's essential objectives as defined by the essential requirements;

Whereas rules governing active implantable medical devices can be confined to those provisions needed to satisfy the essential requirements; whereas, because they are essential, these requirements must replace corresponding national provisions;

Whereas, in order to facilitate proof of conformity with these essential requirements and to permit monitoring of that conformity, it is desirable to have Europe-wide harmonized standards in respect of the prevention of risks in connection with the design, manufacture and packaging of active implantable medical devices; whereas such standards harmonized at European level are drawn up by private-law bodies and must retain their status as non-mandatory texts; whereas, to that end, the European Committee for Standardization (CEN) and the European Committee for Electrotechnical Standardization (Cenelec) are recognized as being the competent bodies to adopt harmonized standards in accordance with the general guidelines for cooperation between the Commission and these two bodies, signed on 13 November 1984; whereas, for the purposes of this Directive, a harmonized standard is a technical specification (European standard or harmonization document) adopted by either or both of these bodies, as instructed by the Commission pursuant to the provisions of Council Directive 83/189/EEC of 28 March 1983 laying down a procedure for the provision of information in the field of technical standards and regulations, as last amended by Directive 88/182/EEC, and under the abovementioned general guidelines;

Whereas evaluation procedures have to be established and accepted by common accord between the Member States in accordance with Community criteria;

Whereas the specific nature of the medical sector makes it advisable to make provision for the notified body and the manufacturer or his agent established in the Community to fix, by common accord, the time limits for completion of the evaluation and verification operations for the conformity of devices,

HAS ADOPTED THIS DIRECTIVE:

Article 1

1. This Directive shall apply to active implantable medical devices.
2. For the purposes of this Directive, the following definitions shall apply:

(a) "medical device" means any instrument, apparatus, appliance, material or other article, whether used alone or in combination, together with any accessories or software for its proper functioning, intended by the manufacturer to be used for human beings in the:

-- diagnosis, prevention, monitoring, treatment or alleviation of disease or injury,

-- investigation, replacement or modification of the anatomy or of a physiological process,

-- control of conception,

and which does not achieve its principal intended action by pharmacological, chemical, immunological or metabolic means, but which may be assisted in its function by such means;

(b) "active medical device" means any medical device relying for its functioning on a source of electrical energy or any source of power other than that directly generated by the human body or gravity;

(c) 'active implantable medical device" means any active medical device which is intended to be totally or partially introduced, surgically or medically, into the human body or by medical intervention into a natural orifice, and which is intended to remain after the procedure;

(d) "custom-made device" means any active implantable medical device specifically made in accordance with a medical specialist's written prescription which gives, under his responsibility, specific design characteristics and is intended to be used only for an individual named patient;

(e) "device intended for clinical investigation" means any active implantable medical device intended for use by a specialist doctor when conducting investigations in an adequate human clinical environment;

(f) "intended purpose" means the use for which the medical device is intended and for which it is suited according to the data supplied by the manufacturer in the instructions;

(g) "putting into service" means making available to the medical profession for implantation.

3. Where an active implantable medical device is intended to administer a substance defined as a medicinal product within the

meaning of Council Directive 65/65/EEC of 26 January 1965 on the approximation of provisions laid down by law, regulation or administrative action relating to proprietary medicinal products, as last amended by Directive 87/21/EEC, that substance shall be subject to the system of marketing authorization provided for in that Directive.
4. Where an active implantable medical device incorporates, as an integral part, a substance which, if used separately, may be considered to be a medicinal product within the meaning of Article 1 of Directive 65/65/EEC, that device must be evaluated and authorized in accordance with the provisions of this Directive.
5. This Directive constitutes a specific Directive within the meaning of Article 2 (2) of Council Directive 89/336/EEC of 3 May 1989 on the approximation of the laws of the Member States relating to electromagnetic compatibility.

Article 2

Member States shall take all necessary steps to ensure that the devices referred to in Article 1 (2) (c) and (d) may be placed on the market and put into service only if they do not compromise the safety and health of patients, users and, where applicable, other persons when properly implanted, maintained and used in accordance with their intended purposes.

Article 3

The active implantable medical devices referred to in Article 1 (2) (c) (d) and (e), hereinafter referred to as "devices," must satisfy the essential requirements set out in Annex 1, which shall apply to them account being taken of the intended purpose of the devices concerned.

Article 4

1. Member States shall not impede the placing on the market or the putting into service within their territory of devices bearing the CE mark.
2. Member States shall not create any obstacles to:
-- devices intended for clinical investigations being made available to specialist doctors for that purpose if they satisfy the conditions laid down in Article 10 and in Annex 6,
-- custom-made devices being placed on the market and put into service if they satisfy the conditions laid down in Annex 6 and are accompanied by the statement referred to in that Annex.

These devices shall not bear the CE mark.
3. At trade fairs, exhibitions, demonstrations, etc., Member States shall not prevent the showing of devices which do not conform to this Directive, provided that a visible sign clearly indicates that such devices do not conform and cannot be put into service until they have been made to comply by the manufacturer or his authorized representative established within the Community.
4. When a device is put into service, Member States may require the information described in sections 13, 14 and 15 of Annex 1 to be in their national language.

Article 5

Member States shall presume compliance with the essential requirements referred to in Article 3 in respect of devices which are in conformity with the relevant national standards adopted pursuant to the harmonized standards the references of which have been published in the *Official Journal of the European Communities;* Member States shall publish the references of such national standards.

Article 6

1. Where a Member State or the Commission considers that the harmonized standards referred to in Article 5 do not entirely meet the essential requirements referred to in Article 3, the Commission or the Member State concerned shall bring the matter before the Standing Committee set up under Directive 83/189/EEC, giving the reasons therefor. The Committee shall deliver an opinion without delay.

In the light of the opinion of the Committee, the Commission shall inform Member States of the measures to be taken with regard to the standards and the publication referred to in Article 5.

2. A Standing Committee, hereinafter referred to as the "Committee," shall be set up, composed of the representatives of the Member States and chaired by the representative of the Commission.

The Committee shall draw up its rules of procedure.

Any matter relating to the implementation and practical application of this Directive may be brought before the Committee, in accordance with the procedure set out below.

The representative of the Commission shall submit to the Committee a draft of the measures to be taken. The Committee shall deliver its opinion according to the urgency of the matter, if necessary by taking a vote.

The opinion shall be recorded in the minutes; in addition, each Member State shall have the right to ask to have its position recorded in the minutes.

The Commission shall take the utmost account of the opinion delivered by the Committee. It shall inform the Committee of the manner in which its opinion has been taken into account.

Article 7

1. Where a Member State finds that the devices referred to in Article 1 (2) (c) and (d), correctly put into service and used in accordance with their intended purpose, may compromise the health and/or safety of patients, users or, where applicable, other persons, it shall take all appropriate measures to withdraw such devices from the market or prohibit or restrict their being placed on the market or their being put into service.

The Member State shall immediately inform the Commission of any such measure, indicating the reasons for its decision and, in particular, whether non-compliance with this Directive is due to:

(a) failure to meet the essential requirements referred to in Article 3, where the device does not meet in full or in part the standards referred to in Article 5;
(b) incorrect application of those standards;
(c) shortcomings in the standards themselves.

2. The Commission shall enter into consultation with the parties concerned as soon as possible. Where, after such consultation, the Commission finds that:

-- the measures are justified, it shall immediately so inform the Member State which took the initiative and the other Member States; where the decision referred to in paragraph 1 is attributed to shortcomings in the standards, the Commission shall, after consulting the parties concerned, bring the matter before the Committee referred to in Article 6 (1) within two months if the Member State which has taken the decision intends to maintain it and shall initiate the procedures referred to in Article 6 (1),
-- the measures are unjustified, it shall immediately so inform the Member State which took the initiative and the manufacturer or his authorized representative established within the Community.

3. Where a device which does not comply bears the CE mark, the competent Member State shall take appropriate action against whomsoever has affixed the mark and shall inform the Commission and the other Member States thereof.

4. The Commission shall ensure that the Member States are kept informed of the progress and outcome of this procedure.

Article 8

1. Member States shall take the necessary steps to ensure that information brought to their knowledge regarding the incidents mentioned below involving a device is recorded and evaluated in a centralized manner:

(a) any deterioration in the characteristics and performances of a device, as well as any inaccuracies in the instruction leaflet which might lead to or might have led to the death of a patient or to a deterioration in his state of health;
(b) any technical or medical reason resulting in withdrawal of a device from the market by the manufacturer.

2. Member States shall, without prejudice to Article 7, forthwith inform the Commission and the other Member States of the incidents referred to in paragraph 1 and of the relevant measures taken or contemplated.

Article 9

1. In the case of devices other than those which are custom-made or intended for clinical investigations, the manufacturer must, or order to affix the CE mark, at this own choice:

(a) follow the procedure relating to the EC declaration of conformity set out in Annex 2; or
(b) follow the procedure relating to EC type-examination set out in Annex 3, coupled with:
(i) the procedure relating to EC verification set out in Annex 4, or
(ii) the procedure relating to the EC declaration of conformity to type set out in Annex 5.

2. In the case of custom-made devices, the manufacturer must draw up the declaration provided for in Annex 6 before placing each device on the market.
3. Where appropriate, the procedures provided for in Annexes 3, 4 and 6 may be discharged by the manufacturer's authorized representative established in the Community.
4. The records and correspondence relating to the procedures referred to in paragraphs 1, 2 and 3 shall be in an official language of the Member State in which the said procedures will be carried out and/or in a language acceptable to the notified body defined in article 11.

Article 10

1. In the case of devices intended for clinical investigations, the manufacturer or his authorized representative established in the Community shall, at least 60 days before the commencement of the investigations, submit the statement referred to in Annex 6 to the competent authorities of the Member State in which the investigations are to be conducted.
2. The manufacturer may commence the relevant clinical investigations at the end of a period of 60 days after notification, unless the competent authorities have notified him within that period of a decision to the contrary, based on considerations of public health or public order.
3. The Member States shall, if necessary, take the appropriate steps to ensure public health and order.

Article 11

1. Each Member State shall notify the other Member States and the Commission of the bodies which they have designated for carrying out the tasks pertaining to the procedures referred to in Articles 9 and 13, the specific tasks for which each body has been designated and the identifying logo of these bodies, hereinafter referred to as "notified bodies."

The Commission shall publish a list of these notified bodies, together with the tasks for which they have been notified, in the *Official Journal of the European Communities* and shall ensure that the list is kept up to date.

2. Member States shall apply the minimum criteria, set out in Annex 8, for the designation of bodies. Bodies that satisfy the criteria fixed by the relevant harmonized standards shall be presumed to satisfy the relevant minimum criteria.
3. A Member State that has notified a body shall withdraw that notification if its finds that the body no longer meets the criteria referred to in paragraph 2. It shall immediately inform the other Member States and the Commission thereof.
4. The notified body and the manufacturer or his agent established in the Community shall fix, by common accord, the time limits for completion of the evaluation and verification operations referred to in Annexes 2 to 5.

Article 12

1. Devices other than those which are custom made or intended for clinical investigations considered to meet the essential requirements

referred to in Article 3 must bear the EC mark of conformity.
2. The EC mark of conformity, as shown in Annex 9, must appear in a visible, legible and indelible form on the sterile pack and, where appropriate, on the sales packaging, if any, and on the instruction leaflet.

It must be accompanied by the logo of the notified body responsible for implementation of the procedures set out in Annexes 2, 4 and 5.
3. The affixing of marks likely to be confused with the EC mark of conformity shall be prohibited.

Article 13

Where it is established that the EC mark has been wrongly affixed, in particular, in respect of devices:
-- that do not conform to the relevant standards referred to in Article 5, should the manufacturer have opted for conformity therewith,
-- that do not conform to an approved type,
-- that conform to an approved type which does not meet the relevant essential requirements,
-- regarding which the manufacturer has failed to fulfil his obligations under the relevant EC declaration of conformity,

the notified body shall take appropriate measures and forthwith inform the competent Member State thereof.

Article 14

Any decision taken pursuant to this Directive and resulting in the refusal of or restrictions on the placing on the market and/or putting into service of a device shall state the exact grounds on which it is based. Such decision shall be notified without delay to the party concerned, who shall at the same time be informed of the remedies available to him under the laws in force in the Member State in question and of the time limits to which such remedies are subject.

Article 15

Member States shall ensure that all the parties involved in the application of this Directive are bound to observe confidentiality with regard to all information obtained in carrying out their tasks. This does not affect the obligations of Member States and notified bodies with regard to mutual information and the dissemination of warnings.

Article 16

1. before 1 July 1992, Member States shall adopt and publish the laws, regulations and administrative provisions necessary in order to comply with this Directive. They shall forthwith inform the Commission thereof.

They shall apply such provisions from 1 January 1993.
2. Member States shall communicate to the Commission the texts of the provisions of national law which they adopt in the field covered by this Directive.
3. Member States shall, for the period up to 31 December 1994, permit the placing on the market and putting into service of devices complying with national rules in force in their territory on 31 December 1992.

Article 17

This Directive is addressed to the Member States.

Done at Luxembourg, 10 June 1990.

For the Council
The President
D.J. O'MALLEY

ANNEX 1

ESSENTIAL REQUIREMENTS

I. GENERAL REQUIREMENTS

1. The devices must be designed and manufactured in such a way that, when implanted

under the conditions and for the purposes laid down, their use does not compromise the clinical condition or the safety of patients. They must not present any risk to the persons implanting them or, where applicable, to other persons.

2. The device must achieve the performances intended by the manufacturer, viz. be designed and manufactured in such a way that they are suitable for one or more of the functions referred to in Article 1 (2)(a) as specified by him.

3. The characteristics and performances referred to in sections 1 and 2 must not be adversely affected to such a degree that the clinical condition and safety of the patients or, as appropriate, of other persons are compromised during the lifetime of the device anticipated by the manufacturer, where the device is subjected to stresses which may occur during normal conditions of use.

4. The devices must be designed, manufactured and packed in such a way that their characteristics and performances are not adversely affected in the storage and transport conditions laid down by the manufacturer (temperature, humidity, etc.).

5. Any side effects or undesirable conditions must constitute acceptable risks when weighed against the performances intended.

II. REQUIREMENTS REGARDING DESIGN AND CONSTRUCTION

6. the solutions adopted by the manufacturer for the design and construction of the devices must comply with safety principles taking account of the generally acknowledged state of the art.

7. Implantable devices must be designed, manufactured and packed in a non-reusable pack according to appropriate procedures to ensure they are sterile when placed on the market and, in the storage and transport conditions stipulated by the manufacturer, remain so until the packaging is removed and they are implanted.

8. Devices must be designed and manufactured in such a way as to remove or minimize as far as possible:

-- the risk of physical injury in connection with their physical, including dimensional, features,

-- risks connected with the use of energy sources with particular reference, where electricity is used, to insulation, leakage currents and overheating of the devices,

-- risks connected with reasonably foreseeable environmental conditions such as magnetic fields, external electrical influences, electrostatic discharge, pressure or variations in pressure and acceleration.

-- risks connected with medical treatment, in particular those resulting from the use of defibrillators or high-frequency surgical equipment.

-- risks connected with ionizing radiation from radioactive substances included in the device, in compliance with the protection requirements laid down in Directive 80/836/Euratom, as amended by Directives 84/467/Euratom (2) and 84/466/Euratom,

-- risks which may arise where maintenance and calibration are impossible, including:
  -- excessive increase of leakage currents,
  -- ageing of the materials used,
  -- excess heat generated by the device,
  -- decreased accuracy of any measuring or control mechanism

9. The device must be designed and manufactured in such a way as to guarantee the characteristics and performances referred to in I. "General requirements," with particular attention being paid to:

-- the choice of materials used, particularly as regards toxicity aspects,

-- mutual compatibility between the materials used and biological tissues, cells and body fluids, account being taken of the anticipated use of the device,

-- compatibility of the devices with the substances they are intended to administer,

-- the quality of the connections, particularly in respect of safety,
-- the reliability of the source of energy,
-- if appropriate, that they are leakproof,
-- **proper functioning of the programming and control systems, including software.**

10. Where a device incorporates, as an integral part, a substance which, when used separately, is likely to be considered to be a medicinal product as defined in Article 1 of Directive 65/65/EEC, and whose action in combination with the device may result in its bioavailability, the safety, quality and usefulness of the substance, account being taken of the purpose of the device, must be verified by analogy with the appropriate methods specified in Directive 75/318/EEC, as last amended by Directive 89/341/EEC.

11. The devices and, if appropriate, their component parts must be identified to allow any necessary measure to be taken following the discovery of a potential risk in connection with the devices and their component parts.

12. Devices must bear a code by which they are their manufacturer can be unequivocably identified (particularly with regard o the type of device and year of manufacture); it must be possible to read this code, if necessary, without the need for a surgical operation.

13. When a device or its accessories bear instructions required for the operation of the device or indicate operating or adjustment parameters, by means of a visual system, such information must be understandable to the user and, as appropriate, the patient.

14. Every device must bear, legibly and indelibly, the following particulars, where appropriate in the form of generally recognized symbols:

14.1 On the sterile pack:
-- the method of sterilization,
-- an indication permitting this packaging to be recognized as such,
-- the name and address of the manufacturer,
-- a description of the device,
-- if the device is intended for clinical investigations, the words: "exclusively for clinical investigations,"
-- if the device is custom-made, the words "custom-made device,"
-- a declaration that the implantable device is in a sterile condition,
-- the month and year of manufacture,
-- an indication of the time limit for implanting a device safely.

14.2 On the sales packaging:
-- the name and address of the manufacturer,
-- a description of the device,
-- the purpose of the device,
-- the relevant characteristics for its use,
-- if the device is intended for clinical investigations, the words: "exclusively for clinical investigations,"
-- if the device is custom-made, the words "custom-made device,"
-- a declaration that the implantable device is in a sterile condition,
-- the month and year of manufacture,
-- an indication of the time limit for implanting a device safely.
-- the conditions for transporting and storing the device.

15. When placed on the market, each device must be accompanied by instructions for use giving the following particulars:
-- the year of authorization to affix the CE mark,
-- the details referred to in 14.1 and 14.2, with the exception of those referred to in the eighth and ninth indents,
-- the performances referred to in section 2 and any undesirable side effects,
-- **information allowing the physician to select a suitable device and the corresponding software and accessories,**
-- **information constituting the instructions for use allowing the physician and, where appropriate, the patient to use the device, its accessories and software correctly, as well as information on the nature, scope and times for operating controls and trials and, where appropriate, maintenance measures.**
-- information allowing, if appropriate, certain risks in connection with implantation of the device to be avoided,

-- information regarding the risks of reciprocal interference in connection with the presence of the device during specific investigations or treatment,
-- the necessary instructions in the event of the sterile pack being damaged and, where appropriate, details of appropriate methods of resterilization,
-- an indication, if appropriate, that a device can be reused only if it is reconditioned under the responsibility of the manufacturer to comply with the essential requirements.

The instruction leaflet must also include details allowing the physician to brief the patient on the contra-indications and the precautions to be taken. These details should cover in particular:

-- information allowing the lifetime of the energy source to be established.
-- precautions to be taken should change occur in the device's performance,
-- precautions to be taken as regards exposure, in reasonably foreseeable environmental conditions, to magnetic fields, external electrical influences, electrostatic discharge, pressure or variations in pressure, acceleration, etc.,
-- adequate information regarding the medicinal products which the device in question is designed to administer.

16. Confirmation that the device satisfies the requirements in respect of characteristics and performances, as referred to in I. "General requirements," in normal conditions of use, and the evaluation of the side effects or undesirable effects must be based on clinical data established in accordance with Annex 7.

ANNEX 2

EC DECLARATION OF CONFORMITY
(Complete quality assurance system)

1. The manufacturer shall apply the quality system approved for the design, manufacture and final inspection of the products concerned as specified in sections 3 and 4 and shall be subject to EC surveillance as specified in section 5.

2. The declaration of conformity is the procedure by means of which the manufacturer who satisfies the obligations of section 1 ensures and declares that the products concerned meet the provisions of this Directive which apply to them

The manufacturer shall apply the CE mark in accordance with Article 12 and draw up a written declaration of conformity. This declaration shall cover one or more identified specimens of the product and shall be kept by the manufacturer. The CE mark shall be accompanied by the identifying logo of the notified body responsible.

3. Quality System

3.1 The manufacturer shall make an application for evaluation of his quality system to a notified body.

The application shall include:

-- all the appropriate items of information for the category of products manufacture of which is envisaged,
-- the quality-system documentation,
-- an undertaking to fulfil the obligations arising from the quality system as approved.
-- an undertaking to maintain the approved quality system in such a way that it remains adequate and efficacious.
-- an undertaking by the manufacturer to institute and keep up-dated a post-marketing surveillance system. The undertaking shall include an obligation for the manufacturer to notify the competent authorities of the following incidents immediately on learning of them:

(i) any deterioration in the characteristics or performances, and any inaccuracies in the instruction leaflet for a device which might lead to or have led to the death of a patient or a deterioration in his state of health.

(ii) any technical or medical reason resulting in withdrawal of a device from the market by the manufacturer.

3.2 The application of the quality system must ensure that the products conform to the provisions of this Directive which apply to them at every stage, from design to final controls.

All the elements, requirements and provisions adopted by the manufacturer for his quality system shall be documented in a systematic and orderly manner in the form of written policies and procedures. This quality-system documentation must make possible a uniform interpretation of the quality policies and procedures such as quality programmes, quality plans, quality manuals and quality records.

It shall include in particular an adequate description of:

(a) the manufacturer's quality objectives;

(b) the organization of the business and in particular:

-- the organizational structures, the responsibilities of the managerial staff and their organizational authority where quality of design and manufacture of the products is concerned,

-- the methods of monitoring the efficient operation of the quality system and in particular its ability to achieve the desired quality of the design and of the products, including control of products which do not conform;

(c) the procedures for monitoring and verifying the design of the products and in particular:

-- the design specifications, including the standards which will be applied and a description of the solutions adopted to fulfil the essential requirements which apply to the products when the standards referred to in Article 5 are not applied in full,

-- the techniques of control and verification of the design, the processes and systematic actions which will be used when the products are being designed;

(d) the techniques of control and of quality assurance at the manufacturing stage and in particular:

-- the processes and procedures which will be used, particularly as regards sterilization, purchasing and the relevant documents,

-- product-identification procedures drawn up and kept up to date from drawings, specifications or other relevant documents at every stage of manufacture;

(e) the appropriate tests and trials which will be effected before, during and after production, the frequency with which they will take place, and the test equipment used.

3.3 Without prejudice to Article 13 of this Directive, the notified body shall effect an audit of the quality system to determine whether it meets the requirements referred to in 3.2. It shall presume conformity with these requirements for the quality systems which use the corresponding harmonized standards.

The team entrusted with the evaluation shall include at least one member who has already had experience of evaluations of the technology concerned. The evaluation procedure shall include an inspection on the manufacturer's premises.

The decision shall be notified to the manufacturer after the final inspection. It shall contain the conclusions of the control and a reasoned evaluation.

3.4 The manufacturer shall inform the notified body which has approved the quality system of any plan to alter the quality system.

The notified body shall evaluate the proposed modifications and shall verify whether the quality system so modified would meet the requirements referred to in 3.2; it shall notify the manufacturer of its decision. This decision shall contain the conclusions of the control and a reasoned evaluation.

4. Examination of the design of the product

4.1 In addition to the obligations incumbent on him under section 3, the manufacturer shall make an application for examination of the design dossier relating to the product which he plans to manufacture and which falls into the category referred to in 3.1.

4.2 The application shall describe the design, manufacture, and performances of the product in question and shall include the necessary particulars which make it possible to evaluate whether it complies with the requirements of this Directive.

It shall include *inter alia:*

-- the design specifications, including the standards which have been applied.

-- the necessary proof of their appropriations, in particular where the standards referred to in Article 5 have not been applied in full. This proof must include the results of the appropriate tests carried out by the manufacturer or carried out under his responsibility,

-- a statement as to whether or not the device incorporates, as an integral part, a substance as referred to in section 10 of Annex 1, whose action in combination with the device may result in its bioavailability, together with data on the relevant trials conducted,

-- the clinical data referred to in Annex 7,

-- the draft instruction leaflet.

4.3 The notified body shall examine the application and, where the product complies with the relevant provisions of this Directive, shall issue the applicant with an EC design examination certificate. The notified body may require the application to be supplemented by further tests or proof so that compliance with the requirements of the Directive may be evaluated. The certificate shall contain the conclusions of the examination, the conditions of its validity, the data needed for identification of the approved design and, where appropriate, a description of the intended use of the product.

4.4 The applicant shall inform the notified body which issued the EC design examination certificate of any modifications made to the approved design. Modifications made to the approved design must obtain supplementary approval from the notified body which issued the EC design examination certificate where such modifications may affect conformity with the essential requirements of this Directive or the conditions prescribed for the use of the product. This supplementary approval shall be given in the form of an addendum to the EC design examination certificate.

5. Surveillance

5.1 The aim of surveillance is to ensure that the manufacturer duly fulfils the obligations arising from the approved quality system.

5.2 The manufacturer shall authorize the notified body to carry out all necessary inspections and shall supply it with all appropriate information, in particular:

-- the quality-system documentation,

-- the data stipulated in the part of the quality system relating to design, such as the results of analyses, calculations, texts, etc.,

-- the data stipulated in the part of the quality system relating to manufacture, such as reports concerning inspections, tests, standardizations/calibrations and the qualifications of the staff concerned, etc.

5.3 The notified body shall periodically carry out appropriate inspections and evaluations in order to ascertain that the manufacturer is applying the approved quality system, and shall supply the manufacturer with an evaluation report.

5.4 In addition, the notified body may make unannounced visits to the manufacturer, and shall supply him with an inspection report.

6. The notified body shall communicate to the other notified bodies all relevant information concerning approvals of quality systems issued, refused and withdrawn.

ANNEX 3

EC TYPE-EXAMINATION

1. EC type-examination is the procedure whereby a notified body observes and certifies that a representative sample of the production envisaged satisfies the relevant provisions of this Directive.

2. The application for EC type-examination shall be made by the manufacturer, or by his authorized representative established in the Community, to a notified body.

The application shall include:

-- the name and address of the manufacturer and the name and address of the authorized representative if the application is made by the latter,

-- a written declaration specifying that an application has not been made to any other notified body,

-- the documentation described in section 3 needed to allow an evaluation to be made of the conformity of a representative sample of the production in question, hereinafter referred to as "type," with the requirements of this Directive.

The applicant shall make a "type" available to the notified body. The notified body may request other samples as necessary.

3. The documentation must make it possible to understand the design, the manufacture and the performances of the product. The documentation shall contain the following items in particular:

-- a general description of the type,
-- design drawings, methods of manufacture envisaged, in particular as regards sterilization, and diagrams of parts, sub-assemblies, circuits, etc.,
-- the descriptions and explanations necessary for the understanding of the above-mentioned drawings and diagrams and of the operation of the product,
-- a list of the standards referred to in Article 5, applied in full or in part, and a description of the solutions adopted to satisfy the essential requirements where the standards referred to in Article 5 have not been applied,
-- the results of design calculations, investigations and technical tests carried out, etc.,
-- a statement as to whether or not the device incorporates, as an integral part, a substance as referred to in section 10 of Annex 1, whose action in combination with the device may result in its bioavailability, together with data on the relevant trials conducted,
-- the clinical data referred to in Annex 7.
-- the draft instruction leaflet.

4. The notified body shall:

4.1 examine and evaluate the documentation, verify that the type has been manufactured in accordance with that documentation; it shall also record the items which have been designed in accordance with the applicable provisions of the standards referred to in Article 5, as well as the items for which the design is not based on the relevant provisions of the said standards;

4.2 carry out or have carried out the appropriate inspections and the tests necessary to verify whether the solutions adopted by the manufacturer satisfy the essential requirements of this Directive where the standards referred to in Article 5 have not been applied;

4.3 carry out or have carried out the appropriate inspections and the tests necessary to verify whether, where the manufacturer has chosen to apply the relevant standards, these have actually been applied;

4.4 agree with the applicant on the place where the necessary inspections and tests will be carried out.

5. Where the type meets the provisions of this Directive, the notified body shall issue an EC type-examination certificate to the applicant. The certificate shall contain the name and address of the manufacturer, the conclusions of the control, the conditions under which the certificate is valid and the information necessary for identification of the type approved.

The significant parts of the documentation shall be attached to the certificate and a copy shall be kept by the notified body.

6. The applicant shall inform the notified body which issued the EC type-examination certificate of any modification made to the approved product.

Modifications to the approved product must receive further approval from the notified body which issued the EC type-examination certificate where such modifications may affect conformity with the essential requirements or with the conditions of use specified for the product. This new approval shall be issued, where appropriate, in the form of a supplement to the initial EC type-examination certificate.

7. Each notified body shall communicate to the other notified bodies all relevant information on EC type-examination certificates and supplements issued, refused or withdrawn.

8. Other notified bodies may obtain a copy of the EC type-examination certificates and/or the supplements to them. The annexes to the certificates shall be made available to the

other notified bodies when a reasoned application is made and after first informing the manufacturer.

ANNEX 4

EC VERIFICATION

1. EC verification is the act by which a notified body verifies and certifies that products conform to the type described in the EC type-examination certificate and satisfy the relevant requirements of this Directive.
2. The manufacturer shall, before the start of manufacture, prepare documents defining the manufacturing process, in particular as regards sterilization, together with all the routine, pre-established provisions to be implemented to ensure homogeneity of production and conformity of the products with the type described in the EC type-examination certificate as well as with the relevant requirements of the Directive.
3. The manufacture shall undertake to institute and keep up-dated a post-marketing surveillance system. The undertaking shall include an obligation for the manufacture to notify the competent authorities of the following events immediately on learning of them:
(i) any deterioration in the characteristics or performances, and any inaccuracies in the instruction leaflet for a device which might lead to or have led to the death of a patient or a deterioration in his state of health;
(ii) any technical or medical reason resulting in withdrawal of a device from the market by the manufacturer.
4. The notified body shall carry out EC verification by controls and tests on the products on a statistical basis as specified in 5. The manufacturer must authorize the notified body to evaluate the efficiency of the measures taken pursuant to section 2, by audit where appropriate.
5. Statistical verification
5.1 The manufacturer shall present the manufactured products in the form of homogeneous batches.
5.2 A random sample shall be taken from each batch. The products which make up the sample shall be examined individually and appropriate tests, defined in the relevant standard(s) referred to in Article 5, or equivalent tests, shall be carried out to verify the conformity of the products with the type described in the EC type-examination certificate, in order to determine whether the batch is to be accepted or rejected.
5.3 Statistical control of products will be based on attributes, entailing a sampling system with the following characteristics:
-- a level of quality corresponding to a probability of acceptance of 95%, with a non-conformity percentage of between 0.29 and 1%.
-- a limit quality corresponding to a probability of acceptance of 5%, with a non-conformity percentage of between 3 and 7%.
5.4 If a batch is accepted, the notified body shall draw up a written certificate of conformity. All the products in the batch may be placed on the market, with the exception of those products in the sample which were found not to conform.

If a batch is rejected, the notified body which is responsible shall take the appropriate measures to prevent the batch from being placed on the market.

If justified on practical grounds, the manufacturer may affix the CE mark during manufacture, under the responsibility of the notified body, in accordance with Article 12, accompanied by the identifying logo of the notified body responsible for statistical verification.

ANNEX 5

EC DECLARATION OF CONFORMITY TO TYPE

(Assurance of production quality)

1. The manufacturer shall apply the quality system approved for the manufacture and shall conduct the final inspection of the products concerned as specified in 3; he shall be subject to the surveillance referred to in section 4.
2. This declaration of conformity is the procedural element whereby the manufacturer who satisfies the obligations of section 1 guarantees and declares that the products

concerned conform to the type described in the EC type-examination certificate and meet the provisions of the Directive which apply to them.

The manufacturer shall affix the CE mark in accordance with Article 12 and draw up a written declaration of conformity. This declaration shall cover one or more identified specimens of the product and shall be kept by the manufacturer. The CE mark shall be accompanied by the identifying logo of the notified body responsible.

3. Quality system

3.1 The manufacturer shall make an application for evaluation of his quality system to a notified body.

The application shall include:

-- all appropriate information concerning the products which it is intended to manufac ture,
-- the quality-system documentation,
-- an undertaking to fulfil the obligations arising from the quality system as approved,
-- an undertaking to maintain the approved quality system in such a way that it remains adequate and efficacious,
-- where appropriate, the technical documentation relating to the approved type and a copy of the EC type examination certificate,
-- an undertaking by the manufacturer to institute and keep up-dated a post-marketing surveillance system. The undertaking shall include an obligation for the manufacturer to notify the competent authorities of the following incidents immediately on learning of them:

  (i) any deterioration in the characteristics or performances, and any inaccuracies in the instruction leaflet for a device which might lead to or have led to the death of a patient or a deterioration in his state of health;

  (ii) any technical or medical reason resulting in withdrawal of a device from the market by the manufacturer.

3.2. Application of the quality system must ensure that the products conform to the type described in the EC type-examination certificate.

All the elements, requirements and provisions adopted by the manufacturer for his quality system shall be documented in a systematic and orderly manner in the form of written policies and procedures. This quality-system documentation must make possible a uniform interpretation of the quality policies and procedures such as quality programmes, quality plans, quality manuals and quality records.

It shall include in particular an adequate description of:

(a) the manufacturer's quality objectives;

(b) the organization of the business and in particular:

-- the organizational structures, the responsibilities of the managerial staff and their organizational authority where manufacture of the products is concerned,
-- the methods of monitoring the efficient operation of the quality system and in particular its ability to achieve the desired quality of the design and of the products, including control of products which do not conform;

(c) the techniques of control and of quality assurance at the manufacturing state and in particular:

-- the processes and procedures which will be used, particularly as regards sterilization, purchasing and the relevant documents,
-- product identification procedures drawn up and kept up-to-date from drawings, specifications or other relevant documents at every stage of manufacture;

(d) the appropriate tests and trials which will be effected before, during and after production, the frequency with which they will take place, and the test equipment used.

3.3 Without prejudice to Article 13, the notified body shall effect an audit of the quality system to determine whether it meets the requirements referred to in 3.2. It shall presume conformity with these requirements for the quality systems which use the corresponding harmonized standards.

The team entrusted with the evaluation shall include at least one member who has already had experience of evaluations of the technology concerned. The evaluation procedure shall include an inspection on the manufacturer's premises.

The decision shall be notified to the manufacturer after the final inspection. It shall contain the conclusions of the control and a reasoned evaluation.

3.4 The manufacturer shall inform the notified body which has approved the quality system of any plan to alter the quality system.

The notified body shall evaluate the proposed modifications and shall verify whether the quality system so modified would meet the requirements referred to in 3.2; it shall notify the manufacturer of its decision. This decision shall contain the conclusions of the control and a reasoned evaluation.

4. Surveillance

4.1 The aim of surveillance is to ensure that the manufacturer duly fulfils the obligations which arise from the approved quality system.

4.2 The manufacturer shall authorize the notified body to carry out all necessary inspections and shall supply it with all appropriate information, in particular:

-- the quality-system documentation,
-- the data stipulated in the part of the quality system relating to manufacture, such as reports concerning inspections, tests, standardizations/calibrations and the qualifications of the staff concerned, etc.

4.3 The notified body shall periodically carry out appropriate inspections and evaluations in order to ascertain that the manufacturer is applying the approved quality system, and shall supply the manufacturer with an evaluation report.

4.4 In addition, the notified body may make unannounced visits to the manufacturer, and shall supply him with an inspection report.

5. The notified body shall communicate to the other notified bodies all relevant information concerning approvals of quality systems issued, refused or withdrawn.

ANNEX 6

STATEMENT CONCERNING DEVICES INTENDED FOR SPECIAL PURPOSES

1. The manufacturer or his authorized representative established within the Community shall draw up for custom-made devices or for devices intended for clinical investigations the statement comprising the elements stipulated in section 2.

2. The statement shall comprise the following information:

2.1 For custom-made devices:

-- data allowing the device in question to be identified,
-- a statement affirming that the device is intended for exclusive use by a particular patient, together with his name,
-- the name of the doctor who drew up the prescription and, if applicable, the name of the clinic concerned.
-- the particular features of the device as described by the medical prescription concerned.
-- a statement affirming that the device in question complies with the essential requirements apart from the aspects constituting the object of the investigations and that, with regard to these aspects, every precaution has been taken to protect the health and safety of the patient.

3. The manufacturer shall undertake to keep available for the competent national authorities:

3.1 For custom-made devices, documentation enabling the design, manufacture and performances of the product, including the expected performances, to be understood, so as to allow conformity with the requirement of this Directive to be assessed.

The manufacturer shall take all necessary measures to see that the manufacturing process ensures that the products manufactured conform to the documentation referred to in the first paragraph.

3.2 For devices intended for clinical investigations, the documentation shall also contain:

-- a general description of the product,
-- design drawings, manufacturing methods, in particular as regards sterilization, and diagrams of parts, sub-assemblies, circuits, etc.,
-- the descriptions and explanations necessary for the understanding of the said drawings

and diagrams and of the operation of the product,

- a list of the standards laid down in Article 5, applied in full or in part, and a description of the solutions adopted to satisfy the essential requirements of the Directive where the standards in Article 5 have not been applied,
- the results of the design calculations, checks and technical tests carried out, etc.

The manufacturer shall take all necessary measures to see that the manufacturing process ensures that the products manufactured conform to the documentation referred to in 3.1 and in the first paragraph of this section.

The manufacturer may authorize the evaluation, by audit where necessary, of the effectiveness of these measures.

ANNEX 7

CLINICAL EVALUATION

1. General provisions

1.1. Adequacy of the clinical data presented, as referred to in section 4.2 of Annex 2, and in section 3 of Annex 3, shall be based, account being taken as appropriate of the relevant harmonized standards, on either:

1.1.1 a collation of currently available relevant scientific literature covering the intended use of the device and the techniques therefor, as well as, if appropriate, a written report making a critical assessment of this collation; or

1.1.2 the results of all clinical investigations made, including those carried out in accordance with section 2.

1.2 All data must remain confidential unless it is deemed essential that they be divulged.

2. Clinical investigation

2.1 *Purpose*

The purpose of clinical investigation is to:

- verify that, under normal conditions of use, the performances of the device comply with those indicated in section 2 of Annex 1,
- determine any undesirable side effects, under normal conditions of use, and assess whether they are acceptable risks having regard to the intended performance of the device.

2.2 *Ethical consideration*

Clinical investigations shall be made in accordance with the Declaration of Helsinki approved by the 18th World Medical Assembly in Helsinki, Finland, in 1964, and amended by the 29th World Medical Assembly in Tokyo, Japan, in 1975 and the 35th World Medical Assembly in Venice, Italy, in 1983. It is mandatory that all measures relating to the protection of human subjects are carried out in the spirit of the Declaration of Helsinki. This includes every step in the clinical investigation from first consideration of need and justification of the study to publication of results.

2.3 *Methods*

2.3.1. Clinical investigations shall be performed according to an appropriate state of the art plan of investigation defined in such a way as to confirm or refute the manufacturer's claims for the device; the investigations shall include an adequate number of observations to guarantee the scientific validity of the conclusions.

2.3.2. The procedures utilized to perform the investigations shall be appropriate to the device under examination.

2.3.3. Clinical investigations shall be performed in circumstances equivalent to those which would be found in normal conditions of use of the device.

2.3.4. All appropriate features, including those involving the safety and performances of the device, and its effects on the patients, shall be examined.

2.3.5. All adverse events shall be fully recorded.

2.3.6. The investigations shall be performed under the responsibility of an appropriately qualified medical specialist, in an appropriate environment.

The medical specialist shall have access to the technical data regarding the device.

2.3.7. The written report, signed by the responsible medical specialist, shall comprise a critical evaluation of all the data collected during the clinical investigation.

ANNEX 8

MINIMUM CRITERIA TO BE MET WHEN DESIGNATING INSPECTION BODIES TO BE NOTIFIED

1. The body, its director and the staff responsible for carrying out the evaluation and verification operations shall not be the designer, manufacturer, supplier or installer of devices which they control, nor the authorized representative of any of those parties. They may not become directly involved in the design, construction, marketing or maintenance of the devices, nor represent the parties engaged in these activities. This does not preclude the possibility of exchanges of technical information between the manufacturer and the body.
2. The body and its staff must carry out the evaluation and verification operations with the highest degree of professional integrity and technical competence and must be free from all pressures and inducements, particularly financial, which might influence their judgement or the results of the inspection, especially from persons or groups of persons with an interest in the results of verifications.
3. The body must be able to carry out all the tasks in one of Annexes 2 to 5 assigned to such a body and for which it has been notified, whether those tasks are carried out by the body itself or under its responsibility. In particular, it must have at its disposal the necessary staff and possess the necessary facilities to enable it to perform properly the technical and administrative tasks connected with evaluation and verification; it must also have access to the equipment necessary for the verifications required.
4. The staff responsible for control operations must have:
-- sound vocational training covering all the evaluation and verification operations for which the body has been designated,
-- satisfactory knowledge of the requirements of the controls they carry out and adequate experience of such operations,
-- the ability required to draw up the certificates, records and reports to demonstrate that the controls have been carried out.
5. The impartiality of inspection staff must be guaranteed. Their remuneration must not depend on the number of controls carried out, nor on the results of such controls.
6. The body must take out liability insurance unless liability is assumed by the State in accordance with national law, or the Member State itself is directly responsible for controls.
7. The staff of the body are bound to observe professional secrecy with regard to all information gained in carrying out their tasks (except *vis-a-vis* the competent administrative authorities of the State in which their activities are carried out) under this Directive or any provision of national law giving effect to it.

COUNCIL DIRECTIVE 93/42/EEC
of 14 June 1993
concerning medical devices

THE COUNCIL OF THE EUROPEAN COMMUNITIES,

Having regard to the Treaty establishing the European Economic Community, and in particular Article 100a thereof,
Having regard to the proposal from the Commission,

In cooperation with the European Parliament,

Having regard to the opinion of the Economic and Social Committee,

Whereas measures should be adopted in the context of the internal market; whereas the internal market is an area without internal frontiers in which the free movement of goods, persons, services and capital is ensured;

Whereas the content and scope of the laws, regulations and administrative provisions in force in the Member States with regard to the safety, health protection and performance characteristics of medical devices are different; whereas the certification and inspection procedures for such devices differ from one Member State to another; whereas such disparities constitute barriers to trade within the Community;

Whereas the national provisions for the safety and health protection of patients, users and, where appropriate, other persons, with regard to the use of medical devices should be harmonized in order to guarantee the free movement of such devices within the internal market;

Whereas the harmonized provisions must be distinguished from the measures adopted by the Member States to manage the funding of public health and sickness insurance schemes relating directly or indirectly to such devices; whereas, therefore, the provisions do not affect the ability of the Member States to implement the abovementioned measures provided Community law is complied with;

Whereas medical devices should provide patients, users and third parties with a high level of protection and attain the performance levels attributed to them by the manufacturer; whereas, therefore, the maintenance or improvement of the level of protection attained in the Member States is one of the essential objectives of this Directive.

Whereas certain medical devices are intended to administer medicinal products within the meaning of Council Directive 65/65/EEC of 26 January 1965 on the approximation of provisions laid down by law, regulation or administrative action relating to proprietary medicinal products; whereas, in such cases, the placing on the market of the medical device as a general rule is governed by the present Directive and the placing on the market of the medicinal product is governed by Directive 65/65/ EEC; whereas if, however, such a device is placed on the market in such a way that the device and the medicinal product form a single integral unit which is intended exclusively for use in the given combination and which is not reusable, that single-unit product shall be governed by Directive 65/65/ EEC; whereas a distinction must be drawn between the abovementioned devices and medical devices incorporating, ***inter alia***, substances which, if used separately, may be considered to be a medicinal substance within the meaning of Directive 65/65/EEC; whereas in such cases, if the substances incorporated in the medical devices are liable to act upon the body with action ancillary to that of the device, the placing of the devices on the market is governed by this Directive; whereas, in this context, the safety, quality and usefulness of the substances must be verified by analogy with the appropriate methods specified in Council Directive 75/318/EEC of 20 May 1975 on the approximation of the laws of the Member States relating to analytical, pharmaco-toxicological and clinical standards and protocols in respect of the testing of proprietary medicinal products;

Whereas the essential requirements and other requirements set out in the Annexes to this Directive, including any reference to "minimizing" or "reducing" risk must be interpreted and applied in such a way as to take account of technology and practice existing at the time of design and of technical and economical considerations compatible with a high level of protection of health and safety;

Whereas in accordance with the principles set out in the Council resolution of 7 May 1985 concerning a new approach to technical harmonization and standardization, rules regarding the design and manufacture of medical devices must be confined to the provisions required to meet the essential requirements; whereas, because they are essential, such requirements should replace the corresponding national provisions; whereas the essential requirements should be applied with discretion to take account of the technological level existing at the time of design and of technical and economic considerations compatible with a high level of protection of health and safety;

Whereas Council Directive 90/385/EEC of 20 June 1990 on the approximation of the laws of the Member States relating to active implantable medical devices is the first case of application of the new approach to the field of medical devices; whereas in the interest of uniform Community rules applicable to all medical devices, this Directive is based largely on the provisions of Directive 90/385/EEC; whereas for the same reasons Directive 90/3845/EEC must be amended to insert the general provisions laid down in this Directive;

Whereas the electromagnetic compatibility aspects form an integral part of the safety of medical devices; whereas this Directive should contain specific rules on this subject with regard to Council Directive 89/336/EEC of 3 May 1989 on the approximation of the laws of the Member States relating to electromagnetic compatibility;

Whereas this Directive should include requirements regarding the design and manufacture of devices emitting ionizing radiation; whereas this Directive does not affect the authorization required by Council Directive 80/836/ Euratom of 15 July 1980 amending the Directives laying down the basic safety standards for the health protection of the general public and workers against the dangers of ionizing radiation, nor application of Council Directive 84/466/Euratom of 3 September 1984 laying down basic measures for the radiation protection of persons undergoing medical examination or treatment; whereas Council Directive 89/391/EEC of 17 June 1989 on the introduction of measures to encourage improvements in the safety and health of workers at work and the specific directives on the same subject should continue to apply;

Whereas, in order to demonstrate conformity with the essential requirements and to enable conformity to be verified, it is desirable to have harmonized European standards to protect against the risks associated with the design, manufacture and packaging of medical devices; whereas such harmonized European standards are drawn up by private-law bodies and should retain their status as non-mandatory texts; whereas, to this end, the European Committee for Standardization (CEN) and the European Committee for Electrotechnical Standardization (Cenelec) are recognized as the competent bodies for the adoption of harmonized standards in accordance with the general guidelines on cooperation between the Commission and these two bodies signed on 13 November 1984;

Whereas, for the purpose of this Directive, a harmonized standard is a technical specification (European standard or harmonization document) adopted, on a mandate from the Commission, by either or both of these bodies in accordance with Council Directive 83/189/EEC of 28 March 1983 laying down a procedures for the provision of information in the field of technical standards and regulations, and pursuant to the abovementioned general guidelines; whereas with regard to possible amendment of the harmonized standards, the Commission should be assisted by the Committee set up pursuant to Directive 83/189/EEC; whereas the measures to be taken must be defined in line with procedure I, as laid down in Council Decision 87/373/EEC; whereas, for specific fields, what already exists in the form of European *Pharmacopoeia* monographs should be incorporated within the framework of this Directive; whereas, therefore, several European *Pharmacopoeia* monographs may be considered equal to the abovementioned harmonized standards;

Whereas, in Decision 90/683/EEC of 13 December 1990 concerning the modules for the various phases of the conformity assessment procedures which are intended to be used in the technical harmonization directives, the Council has laid down harmonized conformity assessment procedures; whereas the application of these modules to medical devices enables the responsibility of manufacturers and notified bodies to be determined during conformity assessment procedures on the basis of the type of devices concerned; whereas the details added to these modules are justified by the nature of the verification required for medical devices;

Whereas it is necessary, essentially for the purpose of the conformity assessment procedures, to group the devices into four product classes; whereas the classification rules are based on the vulnerability of the human body taking account of the potential risks associated with the technical design and manufacture of the devices; whereas the conformity assessment procedures for Class I devices can be carried out, as a general rule, under the sole responsibility of the manufacturers in view of the low level of vulnerability associated with these products; whereas, for Class IIa devices, the intervention of a notified body should be compulsory at the production stage; whereas, for devices falling within Classes IIb and III which constitute a high risk potential, inspection by a notified body is required with regard to the design and manufacture of the devices; whereas Class III is set aside for the most critical devices for which explicit prior authorization with regard to conformity is required for them to be placed on the market;

Whereas in cases where the conformity of the devices can be assessed under the responsibility of the manufacturer the competent authorities must be able, particularly in emergencies, to contact a person responsible for placing the device on the market and established in the Community, whether the manufacturer or another person established in the Community and designated by the manufacturer for the purpose;

Whereas medical devices should, as a general rule, bear the CE mark to indicate their conformity with the provisions of this Directive to enable them to move freely within the Community and to be put into service in accordance with their intended purpose;

Whereas, in the fight against AIDS and in the light of the conclusions of the Council adopted on 16 May 1989 regarding future activities on AIDS prevention and control at Community level, medical devices used for protection against the HIV virus must afford a high level of protection; whereas the design and manufacture of such products should be verified by a notified body;

Whereas the classification rules generally enable medical devices to be appropriately classified; whereas, in view of the diverse nature of the devices and technological progress in this field, steps must be taken to include amongst the implementing powers conferred on the Commission the decisions to be taken with regard to the proper classification or reclassification of the devices or, where appropriate, the adjustment of the classification rules themselves; whereas since these issues are closely connected with the protection of health, it is appropriate that these decisions should come under procedure IIIa, as provided for in Directive 87/373/EEC;

Whereas the confirmation of compliance with the essential requirements may mean that clinical investigations have to be carried out under the responsibility of the manufacturer; whereas, for the purpose of carrying out the clinical investigations, appropriate means have to be specified for the protection of public health and public order;

Whereas the protection of health and the associated controls may be made more effective by means of medical device vigilance systems which are integrated at Community level;

Whereas this Directive covers the medical devices referred to in Council Directive 76/764/EEC of 27 July 1976 on the approximation of the laws of the Member States on clinical mercury-in-glass, maximum reading thermometers; whereas the abovementioned Directive must therefore be repealed; whereas for the same reasons Council Directive 84/539/EEC on 17 September 1984 on the approximation of the laws of the Member States relating to electro-medical equipment used in human or veterinary medicine must be amended,

HAS ADOPTED THIS DIRECTIVE:

Article 1

Definitions, scope

1. This Directive shall apply to medical devices and their accessories. For the purposes of this Directive, accessories shall be treated as medical devices in their own right. Both medical devices and accessories shall hereinafter be termed devices.

2. For the purposes of this Directive, the following definitions shall apply:

(a) "medical device" means any instrument, apparatus, appliance, material or other article, whether used alone or in combination, including the software necessary for its proper application in-

tended by the manufacturer to be used for human beings for the purpose of:

-- diagnosis, prevention, monitoring, treatment or alleviation of disease,

-- diagnosis, monitoring, treatment, alleviation of or compensation for an injury or handicap,

-- investigation, replacement or modification of the anatomy or of a physiological process,

-- control of conception,

and which does not achieve its principal intended action in or on the human body by pharmacological, immunological or metabolic means, but which may be assisted in its function by such means;

(b) "accessory" means an article which whilst not being a device is intended specifically by its manufacturer to be used together with a device to enable it to be used in accordance with the use of the device intended by the manufacturer of the device;

(c) "device used for *in vitro* diagnosis" means any device which is a reagent, reagent product, kit, instrument, equipment or system, whether used alone or in combination, intended by the manufacturer to be used *in vitro* for the examination of samples derived from the human body with a view to providing information on the physiological state, state of health or disease, or congenital abnormality thereof;

(d) "custom-made device" means any device specifically made in accordance with a duly qualified medical practitioner's written prescription which gives, under his responsibility, specific design characteristics and is intended for the sole use of a particular patient;

The abovementioned prescription may also be made out by any other person authorized by virtue of his professional qualifications to do so.

Mass-produced devices which need to be adapted to meet the specific requirements of the medical practitioner or any other professional user are not considered to be custom-made devices;

(e) "device intended for clinical investigation" means any device intended for use by a duly qualified medical practitioner when conducting investigations as referred to in Section 2.1 of Annex X in an adequate human clinical environment.

For the purpose of conducting clinical investigation, any other person who, by virtue of his professional qualifications, is authorized to carry out such investigation shall be accepted as equivalent to a duly qualified medical practitioner;

(f) "manufacturer" means the natural or legal person with responsibility for the design, manufacture, packaging and labelling of a device before it is placed on the market under his own name, regardless of whether these operations are carried out by that person himself or on his behalf by a third party.

The obligations of this Directive to be met by manufacturers also apply to the natural or legal person who assembles, packages, processes, fully refurbishes and/or labels one or more ready-made products and/or assigns to them their intended purpose as a device with a view to their being placed on the market under his own name. This subparagraph does not apply to the person who, while not a manufacturer within the meaning of the first subparagraph, assembles or adapts devices already on the market to their intended purpose for an individual patient.

(g) "intended purpose" means the use for which the device is intended according to the data supplied by the manufacturer on the labelling, in the instructions and/or in promotional materials.

(h) "placing on the market" means the first making available in return for payment or free of charge of a device other than a device intended for clinical investigation, with a view to distribution and/or use on the Community market, regardless of whether it is new or fully refurbished.

(i) "putting into service" means the stage at which a device is ready for use on the Community market for the first time for its intended purpose.

3. Where a device is intended to administer a medicinal product within the meaning of Article 1 of Directive 65/65/EEC, that device

shall be governed by the present Directive, without prejudice to the provisions of Directive 65/65/EEC with regard to the medicinal product.

If, however, such a device is placed on the market in such a way that the device and the medicinal product form a single integral product which is intended exclusively for use in the given combination and which is not reusable, that single product shall be governed by Directive 65/65/EEC. The relevant essential requirements of Annex I to the present Directive shall apply as far as safety and performance related device features are concerned.

5. This Directive does not apply to:

(a) *in vitro* diagnostic devices;

(b) active implantable devices covered by Directive 90/385/EEC;

(c) medicinal products covered by Directive 65/65/EEC;

(d) cosmetic products covered by Directive 76/768/EEC;

(e) human blood, human blood products, human plasma or blood cells of human origin or to devices which incorporate at the time of placing on the market such blood products, plasma or cells;

(f) transplants or tissues or cells of human origin nor to products incorporating or derived from tissues or cells of human origin;

(g) transplants or tissues or cells of animal origin, unless a device is manufactured utilizing animal tissue which is rendered non-viable or non-viable products derived from animal tissue.

6. This Directive does not apply to personal protective equipment covered by Directive 89/686/EEC. In deciding whether a product falls under that Directive or the present Directive, particular account shall be taken of the principal intended purpose of the product.

7. This Directive is a specific Directive within the meaning of Article 2(2) of Directive 89/336/EEC.

8. This Directive does not affect the application of Directive 80/836/Euratom, nor of Directive 84/466/Euratom.

Article 2

Placing on the market and putting into service

Member States shall take all necessary steps to ensure that devices may be placed on the market and put into service only if they do not compromise the safety and health of patients, users and, where applicable, other persons when properly installed, maintained and used in accordance with their intended purpose.

Article 3

Essential requirements

The devices must meet the essential requirements set out in Annex I which apply to them, taking account of the intended purpose of the devices concerned.

Article 4

Free movement, devices intended for special purposes

1. Member States shall not create any obstacle to the placing on the market or the putting into service within their territory of devices bearing the CE marking provided for in Article 17 which indicate that they have been the subject of an assessment of their conformity in accordance with the provisions of Article 11.

2. Member States shall not create any obstacle to:

- devices intended for clinical investigation being made available to medical practitioners or authorized persons for that purpose if they meet the conditions laid down in Article 15 and Annex VIII,
- custom-made devices being placed on the market and put into service if they meet the conditions laid down in Article 11 in combination with Annex VIII; Class IIa, IIb and III devices shall be accompanied by the statement referred to in Annex VIII.

These devices shall not bear the CE marking.

3. At trade fairs, exhibitions, demonstrations, etc. Member States shall not create any obstacle to the showing of devices which do not conform to this Directive, provided that a visible sign clearly indicates that such devices cannot be marketed or put into service until they have been made to comply.
4. Member States may require the information, which must be made available to the user and the patient in accordance with Annex I, point 13, to be in their national language(s) or in another Community language, when a device reaches the final user, regardless of whether it is for professional or other use.
5. Where the devices are subject to other Directives concerning other aspects and which also provide for the affixing of the CE marking, the latter shall indicate that the devices also fulfil the provisions of the other Directives.

However, should one or more of these directives allow the manufacturer, during a transitional period, to choose which arrangements to apply, the CE marking shall indicate that the devices fulfil the provisions only of those directives applied by the manufacturer. In this case, the particulars of these directives, as published in the *Official Journal of the European Communities*, must be given in the documents, notices or instructions required by the directive and accompanying such devices.

Article 5

1. Member States shall presume compliance with the essential requirements referred to in Article 3 in respect of devices which are in conformity with the relevant national standards adopted pursuant to the harmonized standards the references of which have been published in the *Official Journal of the European Communities;* Member States shall publish the references of such national standards.
2. For the purposes of this Directive, reference to harmonized standards also includes the monographs of the European *Pharmacopoeia* notably on surgical sutures and on interaction between medicinal products and materials used in devices containing such medicinal products, the references of which have been published in the *Official Journal of the European Communities*.
3. If a Member State or the Commission considers that the harmonized standards do not entirely meet the essential requirements referred to in Article 3, the measures to be taken by the Member States with regard to these standards and the publication referred to in paragraph 1 of this Article shall be adopted by the procedure defined in Article 6(2).

Article 6

Committee on Standards and Technical Regulations

1. The Commission shall be assisted by the Committee set up by Article 5 of Directive 83/189/EEC.
2. The representative of the Commission shall submit to the Committee a draft of the measures to be taken. The Committee shall deliver its opinion on the draft within a time limit which the chairman may lay down according to the urgency of the matter, if necessary by taking a vote.

The opinion shall be recorded in the minutes; in addition, each Member State shall have the right to ask to have its position recorded in the minutes.

The Commission shall take the utmost account of the opinion delivered by the Committee. It shall inform the Committee of the manner in which its opinion has been taken into account.

Article 7

Committee on Medical Devices

1. The Commission shall be assisted by the Committee set up by Article 6(2) of Directive 90/385/EEC.
2. The representative of the Commission shall submit to the Committee a draft of the measures to be taken. The Committee shall deliver its opinion on the draft within a time limit which the chairman may lay down

according to the urgency of the matter. The opinion shall be delivered by the majority laid down in Article 148(2) of the Treaty in the case of decisions which the Council is required to adopt on a proposal from the Commission. The votes of the representatives of the Member States within the Committee shall be weighted in the manner set out in that Article. The chairman shall not vote.

The Commission shall adopt the measures envisaged if they are in accordance with the opinion of the Committee.

If the measures envisaged are not in accordance with the opinion of the Committee, or if no opinion is delivered, the Commission shall, without delay, submit to the Council a proposal relating to the measures to be taken. The Council shall act by a qualified majority.

If, on the expiry of a period of three months from the date of referral to the Council, the Council has not acted, the proposed measures shall be adopted by the Commission.

4. The Committee may examine any question connected with implementation of this Directive.

Article 8

Safeguard clause

1. Where a Member State ascertains that the devices referred to in Article 4(1) and (2) second indent, when correctly installed, maintained and used for their intended purpose, may compromise the health and/or safety of patients, users or, where applicable, other persons, it shall take all appropriate interim measures to withdraw such devices from the market or prohibit or restrict their being placed on the market or put into service. The Member States shall immediately inform the Commission of any such measures, indicating the reasons for its decision and, in particular, whether non-compliance with this Directive is due to:

(a) failure to meet the essential requirements referred to in Article 3;

(b) incorrect application of the standards referred to in Article 5, in so far as it is claimed that the standards have been applied;

(c) shortcomings in the standards themselves.

2. The Commission shall enter into consultation with the parties concerned as soon as possible. Where, after such consultation, the Commission finds that:

-- the measures are justified, it shall immediately so inform the Member State which took the initiative and the other Member States; where the decision referred to in paragraph 1 is attributed to shortcomings in the standards, the Commission shall, after consulting the parties concerned, bring the matter before the Committee referred to in Article 6 (1) within two months if the Member State which has taken the decision intends to maintain it and shall initiate the procedures referred to in Article 6,

-- the measures are unjustified, it shall immediately so inform the Member State which took the initiative and the manufacturer or his authorized representative established within the Community.

3. Where a non-complying device bears the CE mark, the competent Member State shall take appropriate action against whomsoever has affixed the mark and shall inform the Commission and the other Member States thereof.

4. The Commission shall ensure that the Member States are kept informed of the progress and outcome of this procedure.

Article 9

Classification

1. Devices shall be divided in Classes I, IIa, IIb and III. Classification shall be carried out in accordance with Annex IX.

2. In the event of a dispute between the manufacturer and the notified body concerned, resulting from the application of the classification rules, the matter shall be referred for decision to the competent authority to which the notified body is subject.

3. The classification rules set out in Annex IX may be adapted in accordance with the procedure referred to in Article 7(2) in the light of technical progress and any information which becomes available under the information system provided for in Article 10.

Article 10

Information on incidents occuring following placing of devices on the market

1. Member States shall take the necessary steps to ensure that any information brought to their knowledge, in accordance with the provisions of this Directive, regarding the incidents mentioned below involving a Class I, IIa, IIb or III device is recorded and evaluated centrally:
(a) any malfunction or deterioration in the characteristics and/or performance of a device, as well as any inadequacy in the labelling or the instructions for use which might lead to or might have led to the death of a patient or user or to a serious deterioration in his state of health;
(b) any technical or medical reason in relation to the characteristics or performance of a device for the reasons referred to in subparagraph (a), leading to systematic recall of devices of the same type by the manufacturer.
2. Where a Member State requires medical practitioners or the medical institutions to inform the competent authorities of any incidents referred to in paragraph 1, it shall take the necessary steps to ensure that the manufacturer of the device concerned, or his authorized representative established in the Community, is also informed of the incident.
3. After carrying out an assessment, if possible together with the manufacturer, Member States shall, without prejudice to Article 8, immediately inform the Commission and the other Member States of the incidents referred to in paragraph 1 for which relevant measures have been taken or are contemplated.

Article 11

Conformity assessment procedures

1. In the case of devices falling within Class III, other than devices which are custom-made or intended for clinical investigations, the manufacturer shall, in order to affix the CE marking, either:
(a) follow the procedure relating to the EC declaration of conformity set out in Annex II (full quality assurance); or
(b) follow the procedure relating to the EC type-examination set out in Annex III, coupled with:

(i) the procedure relating to the EC verification set out in Annex IV;

or

(ii) the procedure relating to the EC declaration of conformity set out in Annex V (production quality assurance).

2. In the case of devices falling within Class IIa, other than devices which are custom-made or intended for clinical investigations, the manufacturer shall, in order to affix the CE marking, follow the procedure relating to the EC declaration of conformity set out in Annex VII, coupled with either:
(a) the procedure relating to the EC verification set out in Annex IV;

or

(b) the procedure relating to the EC declaration of conformity set out in Annex V (production quality assurance);

or

(c) the procedure relating to the EC declaration of conformity set out in Annex VI (product quality assurance).

Instead of applying these procedures, the manufacturer may also follow the procedure referred to in paragraph 3(a).

3. In the case of devices falling within Class IIb, other than devices which are custom-made or intended for clinical investigations, the manufacturer shall, in order to affix the CE marking, either:
(a) follow the procedure relating to the EC declaration of conformity set out in Annex II (full quality assurance); in this case, point 4 of Annex II is not applicable; or
(b) follow the procedure relating to the EC type-examination set out in Annex III, coupled with:

(i) the procedure relating to the EC verification set out in Annex IV;

or

(ii) the procedure relating to the EC declaration of conformity set out in Annex V (production quality assurance).

or

(iii) the procedure relating to the EC declaration of conformity set out in Annex VI (product quality assurance).

4. The Commission shall, no later than five years from the date of implementation of this Directive, submit a report to the Council on the operation of the provisions referred to in Article 10(1), Article 15(1), in particular in respect of Class I and Class IIa devices, and on the operation of the provisions referred to in Annex II, Section 4.3 second and third subparagraphs and in Annex III, Section 5 second and third subparagraphs to this Directive, accompanied, if necessary, by appropriate proposals.

5. In the case of devices falling within Class I, other than devices which are custom-made or intended for clinical investigations, the manufacturer shall, in order to affix the CE marking, follow the procedure referred to the Annex VII and draw up the EC declaration of conformity required before placing the device on the market.

6. In the case of custom-made devices, the manufacturer shall follow the procedure referred to in Annex VIII and draw up the statement set out in that Annex before placing each device on the market.

Member States may require that the manufacturer shall submit to the competent authority a list of such devices which have been put into service in their territory.

7. During the conformity assessment procedure for a device, the manufacturer and/or the notified body shall take account of the results of any assessment and verification operations which, where appropriate, have been carried out in accordance with this Directive at an intermediate stage of manufacture.

8. The manufacturer may instruct his authorized representative established in the Community to initiate the procedures provided for in Annexes III, IV, VII and VIII.

9. Where the conformity assessment procedure involves the intervention of a notified body, the manufacturer, or his authorized representative established in the Community, may apply to a body of his choice within the framework of the tasks for which the body has been notified.

10. The notified body may require, where duly justified, any information or data, which is necessary for establishing and maintaining the attestation of conformity in view of the chosen procedure.

11. Decisions taken by the notified bodies in accordance with Annexes II and III shall be valid for a maximum of five years and may be extended on application, made at a time agreed in the contract signed by both parties, for further periods of five years.

12. The records and correspondence relating to the procedures referred to in paragraphs 1 to 6 shall be in an official language of the Member State in which the procedures are carried out and/or in another Community language acceptable to the notified body.

13. By derogation from paragraphs 1 to 6, the competent authorities may authorize, on duly justified request, the placing on the market and putting into service, within the territory of the Member State concerned, of individual devices for which the procedures referred to in paragraphs 1 to 6 have not been carried out and the use of which is in the interest of protection of health.

Article 12

Particular procedure for systems and procedure packs

1. By way of derogation from Article 11 this Article shall apply to systems and procedure packs.

2. Any natural or legal person who puts devices bearing the CE marking together within their intended purpose and within the limits of use specified by their manufacturers, in order to place them on the market as a system or procedure pack, shall draw up a declaration by which he states that:

(a) he has verified the mutual compatibility of the devices in accordance with the manufacturers' instructions and has carried out his operations in accordance with these instructions; and

(b) he has packaged the system or procedure

pack and supplied relevant information to users incorporating relevant instructions from the manufacturers; and

(c) the whole activity is subjected to appropriate methods of internal control and inspection.

Where the conditions above are not met, as in cases where the system or procedure pack incorporate devices which do not bear a CE marking or where the chosen combination of devices is not compatible in view of their original intended use, the system or procedure pack shall be treated as a device in its own right and as such be subjected to the relevant procedure pursuant to Article 11.

3. Any natural or legal person who sterilized, for the purpose of placing on the market, systems or procedure packs referred to in paragraph 2 or other CE-marked medical devices designed by their manufacturers to be sterilized before use, shall, at his choice, follow one of the procedures referred to in Annex IV, V or VI. The application of the abovementioned Annexes and the intervention of the notified body are limited to the aspects of the procedure relating to the obtaining of sterility. The person shall draw up a declaration stating that sterilization has been carried out in accordance with the manufacturer's instructions.

4. The products referred to in paragraphs 2 and 3 themselves shall not bear an additional CE marking. They shall be accompanied by the information referred to in point 13 of Annex I which includes, where appropriate, the information supplied by the manufacturers of the devices which have been put together. The declaration referred to in paragraphs 2 and 3 above shall be kept at the disposal of competent authorities for a period of five years.

Article 13

Decisions with regard to classification, derogation clause

1. Where a Member State considers that:

(a) application of the classification rules set out in Annex IX requires a decision with regard to the classification of a given device or category of devices;

or

(b) a given device or family of devices should be classified by way of derogation from the provisions of Annex IX, in another class;

or

(c) the conformity of a device or family of devices should be established, by way of derogation from the provisions of Article 11, by applying solely one of the given procedures chosen from among those referred to in Article 11, it shall submit a duly substantiated request to the Commission and ask it to take the necessary measures. These measures shall be adopted in accordance with the procedure referred to in Article 7(2).

2. The Commission shall inform the Member States of the measures taken and, where appropriate, publish the relevant parts of these measures in the *Official Journal of the European Communities*.

Article 14

Registration of persons responsible for placing devices on the market

1. Any manufacturer who, under his own name, places devices on the market in accordance with the procedures referred to in Article 11(5) and (6) and any other natural or legal person engaged in the activities referred to in Article 12 shall inform the competent authorities of the Member State in which he has his registered place of business of the address of the registered place of business and the description of the devices concerned.

2. Where a manufacturer who places devices referred to in paragraph 1 on the market under his own name does not have a registered place of business in a Member State, he shall designate the person(s) responsible for marketing them who is (are) established in the Community. These persons shall inform the competent authorities of the Member State in which they have their registered place of business of the address of the registered place of business and the category of devices concerned.

3. The Member States shall on request inform the other Member States and the Commission of the details referred to in paragraphs 1 and 2.

Article 15

Clinical investigation

1. In the case of devices intended for clinical investigations, the manufacturer, or his authorized representative established in the Community, shall follow the procedure referred to in Annex VIII and notify the competent authorities of the Member States in which the investigations are to be conducted.
2. In the case of devices falling within Class III and implantable and long-term invasive devices falling within Class IIa or IIb, the manufacturer may commence the relevant clinical investigation at the end of a period of 60 days after notification, unless the competent authorities have notified him within that period of a decision to the contrary based on considerations of public health or public policy.

Member States may however authorize manufacturers to commence the relevant clinical investigations before the expiry of the period of 60 days, in so far as the relevant ethics committee has issued a favourable opinion on the programme of investigation in question.
3. In the case of devices other than those referred to in the second paragraph, Member States may authorize manufacturers to commence clinical investigations, immediately after the date of notification, provided that the ethics committee concerned has delivered a favourable opinion with regard to the investigational plan.
4. The authorization referred to in paragraph 2 second subparagraph and paragraph 3, may be made subject to authorization from the competent authority.
5. The clinical investigations must be conducted in accordance with the provisions of Annex X. The provisions of Annex X may be adjusted in accordance with the procedure laid down in Article 7(2).
6. The Member States shall, if necessary, take the appropriate steps to ensure public health and public policy.
7. The manufacturer or his authorized representative established in the Community shall keep the report referred to in point 2.3.7 of Annex X at the disposal of the competent authorities.
8. The provisions of paragraphs 1 and 2 do not apply where the clinical investigations are conducted using devices which are authorized in accordance with Article 11 to bear the CE marking unless the aim of these investigations is to use the devices for a purpose other than that referred to in the relevant conformity assessment procedure. The relevant provisions of Annex X remain applicable.

Article 16

Notified bodies

1. The Member States shall notify the Commission and other Member States of the codes which they have designated for carrying out the tasks pertaining to the procedures referred to in Article 11 and the specific tasks for which the bodies have been designated. The Commission shall assign identification numbers to these bodies, hereinafter referred to as "notified bodies."

The Commission shall publish a list of the notified bodies, together with the identification numbers it has allocated to them and the tasks for which they have been notified, in the *Official Journal of the European Communities*. It shall ensure that the list is kept up to date.
2. Member States shall apply the criteria set out in Annex XI for the designation of bodies. Bodies that meet the criteria laid down in the national standards which transpose the relevant harmonized standards shall be presumed to meet the relevant criteria.
3. A Member State that has notified a body shall withdraw that notification if it finds that the body no longer meets the criteria referred to in paragraph 2. It shall immediately inform the other Member States and the Commission thereof.

4. The notified body and the manufacturer, or his authorized representative established in the Community, shall lay down, by common accord, the time limits for completion of the assessment and verification operations referred to in Annexes II to VI.

Article 17

CE marking

1. Devices, other than devices which are custom-made or intended for clinical investigations, considered to meet the essential requirements referred to in Article 3 must bear the CE marking of conformity when they are placed on the market.
2 The CE marking of conformity, as shown in Annex XII, must appear in a visible, legible and indelible form on the device or its sterile pack, where practicable and appropriate, and on the instructions for use. Where applicable, the CE marking must also appear on the sale packaging.

It shall be accompanied by the identification number of the notified body responsible for implementation of the procedures set out in Annexes II, IV, V and VI.
3. It is prohibited to affix marks or inscriptions which are likely to mislead third parties with regard to the meaning or the graphics of the CE marking. Any other mark may be affixed to the device, to the packaging or to the instruction leaflet accompanying the device provided that the visibility and legibility of the CE marking is not thereby reduced.

Article 18

Wrongly affixed CE marking

Without prejudice to Article 8:
(a) where a Member State establishes the CE marking has been affixed unduly, the manufacturer or his authorized representative established in the Community shall be obliged to end the infringement under conditions imposed by the Member State;
(b) where non-compliance continues, the Member State must take all appropriate measures to restrict or prohibit the placing on the market of the product in question or to ensure that it is withdrawn from the market, in accordance with the procedure in Article 8.

Article 19

Decision in respect of refusal or restriction
1. Any decision taken pursuant to this Directive:
(a) to refuse or restrict the placing on the market or the putting into service of a device or the carrying out of clinical investigations;
or
(b) to withdraw devices from the market, shall state the exact grounds on which it is based. Such decisions shall be notified without delay to the party concerned, who shall at the same time be informed of the remedies available to him under the national law in force in the Member State in question and of the time limits to which such remedies are subject.
2. In the event of a decision as referred to in paragraph 1, the manufacturer, or his authorized representative established in the Community, shall have an opportunity to put forward his viewpoint in advance unless such consultation is not possible because of the urgency of the measure to be taken.

Article 20

Confidentiality

Without prejudice to the existing national provisions and practices on medical secrets, Member States shall ensure that all the parties involved in the application of this Directive are bound to observe confidentiality with regard to all information obtained in carrying out their tasks. This does not affect the obligation of Member Stats and notified bodies with regard to mutual information and the dissemination of warnings, nor the obligations of the persons concerned to provide information under criminal law.

Article 21

Repeal and amendment of Directives

1. Directive 76/764/EEC is hereby repealed with effect from 1 January 1995.
2. In the title and Article 1 of Directive 84/539/EEC, "human or" is deleted.

In Article 2 of Directive 84/539/EEC, the following subparagraph is added to paragraph 1:

"If the appliance is at the same time a medical device within the meaning of Directive 93/42/EEC and if it satisfies the essential requirements laid down therein for that device, the device shall be deemed to be in conformity with the requirements of this Directive."

3. Directive 90/385/EEC is hereby amended as follows:

1. In Article 1(2) the following two subparagraphs are added:

"(h) 'placing on the market' means the first making available in return for payment or free of charge of a device other than a device intended for clinical investigation, with a view to distribution and/or use on the Community market, regardless of whether it is new or fully refurbished;

(i) 'manufacturer' means the natural or legal person with responsibility for the design, manufacture, packaging and labelling of a device before it is placed on the market under his own name, regardless of whether these operations are carried out by that person himself or on his behalf by a third party.

The obligations of this Directive to be met by manufacturers also apply to the natural or legal person who assembles, packages, processes, fully refurbishes and/or labels one or more ready-made products and/or assigns to them their intended purpose as a device with a view to their being placed on the market under his own name. This subparagraph does not apply to the person who, while not a manufacturer within the meaning of the first subparagraph, assembles or adapts devices already on the market to their intended purpose for an individual patient;"

2. in Article 9 the following paragraphs are added:

"5. During the conformity assessment procedure for a device, the manufacturer and/or the notified body shall take account of the results of any assessment and verification operations which, where appropriate, have been carried out in accordance with this Directive at an intermediate state of manufacture.

6. Where the conformity assessment procedure involves the intervention of a notified body, the manufacturer, or his authorized representative established in the Community, may apply to a body of his choice within the framework of the tasks for which the body has been notified.

7. The notified body may require, where duly justified, any information or data which is necessary for establishing and maintaining the attestation of conformity in view of the chosen procedure.

8. Decisions taken by the notified bodies in accordance with Annexes II and III shall be valid for a maximum of five years and may be extended on application, made at a time agreed in the contract signed by both parties, for further periods of five years.

9. By derogation from paragraphs 1 and 2 the competent authorities may authorize, on duly justified request, the placing on the market and putting into service, within the territory of the Member State concerned, of individual devices for which the procedures referred to in paragraphs 1 and 2 have not been carried out and the use of which is in the interest of protection of health;"

3. the following Article 9a is inserted after Article 9:

"*Article 9a*

1. Where a Member State considers that the conformity of a device or family of

devices should be established, by way of derogation from the provisions of Article 9, by applying solely one of the given procedures chosen from among those referred to in Article 9, it shall submit a duly substantiated request to the Commission and ask it to take the necessary measures. These measures shall be adopted in accordance with the procedure referred to in Article 7(2) of Directive 93/42/EEC.

2. The Commission shall inform the Member States of the measures taken and, where appropriate, publish the relevant parts of these measures in the *Official Journal of the European Communities.*"

4. Article 10 shall be amended as follows:

-- the following subparagraph shall be added to paragraph 2:

"Member States may however authorize manufacturers to start the clinical investigations in question before the expiry of the 60-day period, provided that the Ethical Committee concerned has delivered a favourable opinion with respect to the investigation programme in question.",

-- the following paragraph shall be inserted:

"2a. The authorization referred to in the second subparagraph of paragraph 2 may be subject to approval by the competent authority.";

5. the following is added to Article 14:

"In the event of a decision as referred to in the previous paragraph the manufacturer, or his authorized representative established in the Community, shall have an opportunity to put forward his viewpoint in advance, unless such consultation is not possible because of the urgency of the measures to be taken."

Article 22

Implementation, transitional provisions

1. Member States shall adopt and publish the laws, regulations and administrative provisions necessary to comply with this Directive not later than 1 July 1994. They shall immediately inform the Commission thereof.

The Standing Committee referred to in Article 7 may assume its tasks from the date of notification of this Directive [29 June 1993]. The Member States may take the measures referred to in Article 16 on notification of this Directive.

When Member States adopt these provisions, these shall contain a reference to this Directive or shall be accompanied by such a reference at the time of their official publication. The procedure for such reference shall be adopted by Member States.

Member States shall apply these provisions with effect from 1 January 1995.

2. Member States shall communicate to the Commission the texts of the provisions of national law which they adopt in the field covered by this Directive.

3. Member States shall take the necessary action to ensure that the notified bodies which are responsible pursuant to Article 11(1) to (5) for conformity assessment take account of any relevant information regarding the characteristics and performance of such devices, including in particular the results of any relevant tests and verification already carried out under pre-existing national law, regulations or administrative provisions in respect of such devices.

4. Member States shall accept the placing on the market and putting into service of devices which conform to the rules in force in their territory on 31 December 1994 during a period of five years following adoption of this Directive.

In the case of devices which have been subjected to EEC pattern approval in accordance with Directive 76/764/EEC, Member States shall accept their being placed on the market and put into service during the period up to 30 June 2004.

Article 23

This Directive is addressed to the Member States.

Done at Luxembourg, 14 June 1993.

For the Council
The President
J. TRØJBORG

ANNEX I

ESSENTIAL REQUIREMENTS

I. GENERAL REQUIREMENTS

1. The devices must be designed and manufactured in such a way that, when used under the conditions and for the purposes intended, they will not compromise the clinical condition or the safety of patients, or the safety and health of users or, where applicable, other persons, provided that any risks which may be associated with their use constitute acceptable risks when weighed against the benefits to the patient and are compatible with a high level of protection of health and safety.
2. The solutions adopted by the manufacturer for the design and construction of the devices must conform to safety principles, taking account of the generally acknowledged state of the art.

In selecting the most appropriate solutions, the manufacturer must apply the following principles in the following order:
-- eliminate or reduce risks as far as possible (inherently safe design and construction),
-- where appropriate take adequate protection measures including alarms if necessary, in relation to risks that cannot be eliminated.
-- inform users of the residual risks due to any shortcomings of the protection measures adopted.

3. The devices must achieve the performances intended by the manufacturer and be designed, manufactured and packaged in such a way that they are suitable for one or more of the functions referred to in Article 1(2)(a), as specified by the manufacturer.
4. The characteristics and performances referred to in Sections 1, 2 and 3 must not be adversely affected to such a degree that the clinical conditions and safety of the patients and, where applicable, of other persons are compromised during the lifetime of the device as indicated by the manufacturer, when the device is subjected to the stresses which can occur during normal conditions of use.
5. The devices must be designed, manufactured and packed in such a way that their characteristics and performances during their intended use will not be adversely affected during transport and storage taking account of the instructions and information provided by the manufacturer.
6. Any undesirable side-effect must constitute an acceptable risk when weighed against the performances intended.

II. REQUIREMENTS REGARDING DESIGN AND CONSTRUCTION

7. Chemical, physical and biological properties
7.1. The devices must be designed and manufactured in such a way as to guarantee the characteristics and performances referred to in Section I on the "General requirements." Particular attention must be paid to:
-- the choice of materials used, particularly as regards toxicity and, where appropriate, flammability,
-- the compatibility between the materials used and biological tissues, cells and body fluids, taking account of the intended purpose of the device.

7.2. The devices must be designed, manufactured and packed in such a way as to minimize the risk posed by contaminants and residues to the persons involved in the transport, storage and use of the devices and to the patients, taking account of the intended purpose of the product. Particular attention must be paid to the tissues exposed and to the duration and frequency of exposure.
7.3. The devices must be designed and manufactured in such a way that they can be used safely with the materials, substances and gases with which they enter into contact during their normal use or during the routine procedures; if the devices are intended to administer medicinal products they must be designed and manufactured in such a way as to be compatible with the medicinal products concerned according to the provisions and restrictions governing these products and that their performance is maintained in accordance with the intended use.
7.4. Where a device incorporates, as an integral part, a substance which, if used separately, may be considered to be a medici-

nal product as defined in Article 1 of Directive 65/65/EEC and which is liable to act upon the body with action ancillary to that of the device, the safety, quality and usefulness of the substance must be verified, taking account of the intended purpose of the device, by analogy with the appropriate methods specified in Directive 75/318/EEC.

7.5. The devices must be designed and manufactured in such a way as to reduce to a minimum the risks posed by substances leaking from the device.

7.6. Devices must be designed and manufactured in such a way as to reduce, as much as possible, the risks posed by the unintentional ingress of substances into the device taking into account the device and the nature of the environment in which it is intended to be used.

8. Infection and microbial contamination

8.1. The devices and manufacturing processes must be designed in such a way as to eliminate or reduce as far as possible the risk of infection to the patient, user and third parties. The design must allow easy handling and, where necessary, minimize contamination of the device by the patient or vice versa during use.

8.2. Tissues of animal origin must originate from animals that have been subjected to veterinary controls and surveillance adapted to the intended use of the tissues.

Notified bodies shall retain information on the geographical origin of the animals.

Processing, preservation, testing and handling of tissues, cells and substances of animal origin must be carried out so as to provide optimal security. In particular safety with regard to viruses and other transferable agents must be addressed by implementation of validated methods of elimination or viral inactivation in the course of the manufacturing process.

8.3. Devices delivered in a sterile state must be designed, manufactured and packed in a non-reusable pack and/or according to appropriate procedures to ensure that they are sterile when placed on the market and remain sterile, under the storage and transport conditions laid down, until the protective packaging is damaged or opened.

8.4. Devices delivered in a sterile state must have been manufactured and sterilized by an appropriate, validated method.

8.5. Devices intended to be sterilized must be manufactured in appropriately controlled (e.g. environmental) conditions.

8.6. Packaging systems for non-sterile devices must keep the product without deterioration at the level of cleanliness stipulated and, if the devices are to be sterilized prior to use, minimize the risk of microbial contamination; the packaging system must be suitable taking account of the method of sterilization indicated by the manufacturer.

8.7. The packaging and/or label of the device must distinguish between identical or similar products sold in both sterile and non-sterile condition.

9. Construction and environmental properties

9.1. If the device is intended for use in combination with other devices or equipment, the whole combination, including the connection system must be safe and must not impair the specified performances of the devices. Any restrictions on use must be indicated on the label or in the instructions for use.

9.2. Devices must be designed and manufactured in such a way as to remove or minimize as far as it is possible:

- the risk of injury, in connection with their physical features, including the volume/pressure ratio, dimensional and where appropriate ergonomic features,
- risks connected with reasonably foreseeable environmental conditions, such as magnetic fields, external electrical influences, electrostatic discharge, pressure, temperature or variations in pressure and acceleration,
- the risks of reciprocal interference with other devices normally used in the investigations or for the treatment given,
- risks arising where maintenance or calibration are not possible (as with implants), from ageing of materials used or loss of accuracy of any measuring or control mechanism.

9.3. Devices must be designed and manufactured in such a way as to minimize the risks of fire or explosion during normal use and in single fault condition. Particular attention must be paid to devices whose intended use includes exposure to flammable substances or to substances which could cause combustion.

10. Devices with a measuring function

10.1. Devices with a measuring function must be designed and manufactured in such a way as to provide sufficient accuracy and stability within appropriate limits of accuracy and taking account of the intended purpose of the device. The limits of accuracy must be indicated by the manufacturer.

10.2. The measurement, monitoring and display scale must be designed in line with ergonomic principles, taking account of the intended purpose of the device.

10.3. The measurements made by devices with a measuring function must be expressed in legal units conforming to the provisions of Council Directive 80/181/EEC.

11. Protection against radiation

11.1. *General*

11.1.1. Devices shall be designed and manufactured in such a way that exposure of patients, users and other persons to radiation shall be reduced as far as possible compatible with the intended purpose, whilst not restricting the application of appropriate specified levels for therapeutic and diagnostic purposes.

11.2 *Intended radiation*

11.2.1. Where devices are designed to emit hazardous levels of radiation necessary for a specific medical purpose the benefit of which is considered to outweigh the risks inherent in the emission, it must be possible for the user to control the emissions. Such devices shall be designed and manufactured to ensure reproducibility and tolerance of relevant variable parameters.

11.2.2. Where devices are intended to emit potentially hazardous, visible and/or invisible radiation, they must be fitted, where practicable, with visual displays and/or audible warnings of such emissions.

11.3. *Unintended radiation*

11.3.1. devices shall be designed and manufactured in such a way that exposure of patients, users and other persons to the emission of unintended, stray or scattered radiation is reduced as far as possible.

11.4. *Instructions*

11.4.1. The operating instructions for devices emitting radiation must give detailed information as to the nature of the emitted radiation, means of protecting the patient and the user and on ways of avoiding misuse and of eliminating the risks inherent in installation.

11.5. *Ionizing radiation*

11.5.1. Devices intended to emit ionizing radiation must be designed and manufactured in such a way as to ensure that, where practicable, the quantity, geometry and quality of radiation emitted can be varied and controlled taking into account the intended use.

11.5.2. Devices emitting ionizing radiation intended for diagnostic radiology shall be designed and manufactured in such a way as to achieve appropriate image and/or output quality for the intended medical purpose whilst minimizing radiation exposure of the patient and user.

11.5.3. Devices emitting ionizing radiation, intended for therapeutic radiology shall be designed and manufactured in such a way as to enable reliable monitoring and control of the delivered dose, the beam type and energy and where appropriate the quality of radiation.

12. Requirements for medical devices connected to or equipped with an energy source

12.1. Devices incorporating electronic programmable systems must be designed to ensure the repeatability, reliability and performance of these systems according to the intended use. In the event of a single fault condition (in the system) appropriate means should be adopted to eliminate or reduce as far as possible consequent risks.

12.2. Devices where the safety of the patients depends on an internal power supply must be equipped with a means of determining the state of the power supply.

12.3. Devices where the safety of the patients depends on an external power supply must include an alarm system to signal any power failure.

12.4. Devices intended to monitor one or more clinical parameters of a patient must be equipped with appropriate alarm systems to alert the user of situations which could lead to death or severe deterioration of the patient's state of health.

12.5. Devices must be designed and manufactured in such a way as to minimize the risks of creating electromagnetic fields which could impair the operation of other devices or equipment in the usual environment.

12.6 *Protection against electrical risks*

Devices must be designed and manufactured in such a way as to avoid, as far as possible, the risk of accidental electric shocks during normal use and in single fault condition, provided the devices are installed correctly.

12.7. *Protection against mechanical and thermal risks*

12.7.1. Devices must be designed and manufactured in such a way as to protect the patient and user against mechanical risks connected with, for example, resistance, stability and moving parts.

12.7.2. Devices must be designed and manufactured in such a way as to reduce to the lowest possible level the risks arising from vibration generated by the devices, taking account of technical progress and of the means available for limiting vibrations, particularly at source, unless the vibrations are part of the specified performance.

12.7.3. Devices must be designed and manufactured in such a way as to reduce to the lowest possible level the risks arising from the noise emitted, taking account of technical progress and of the means available to reduce noise, particularly at source, unless the noise emitted is part of the specified performance.

12.7.4. Terminals and connectors to the electricity, gas or hydraulic and pneumatic energy supplies which the user has to handle must be designed and constructed in such a way as to minimize all possible risks.

12.7.5. Accessible parts of the devices (excluding the parts or areas intended to supply heat or reach given temperatures) and their surroundings must not attain potentially dangerous temperatures under normal use.

12.8. *Protection against the risks posed to the patient by energy supplies or substances*

12.8.1. Devices for supplying the patient with energy or substances must be designed and constructed in such a way that the flow-rate can be set and maintained accurately enough to guarantee the safety of the patient and of the user.

12.8.2. Devices must be fitted with the means of preventing and/or indicating any inadequacies in the flow-rate which could pose a danger.

Devices must incorporate suitable means to prevent, as far as possible, the accidental release of dangerous levels of energy from an energy and/or substance source.

12.9. The function of the controls and indicators must be clearly specified on the devices. Where a device bears instructions required for its operation or indicates operating or adjustment parameters by means of a visual system, such information must be understandable to the user and, as appropriate, the patient.

13. Information supplied by the manufacturer

13.1. Each device must be accompanied by the information needed to use it safely and to identify the manufacturer, taking account of the training and knowledge of the potential users.

This information comprises the details on the label and the data in the instructions for use.

As far as practicable and appropriate, the information needed to use the device safely must be set out on the device itself and/or on the packaging for each unit or, where appropriate, on the sales packaging. If individual packaging of each unit is not practicable, the information must be set out in the leaflet supplied with one or more devices.

Instructions for use must be included in the packaging for every device. By way of exception, no such instructions for use are needed for devices in Class I or IIa if they can be used safely without any such instructions.

13.2. Where appropriate, this information should take the form of symbols. Any symbol or identification colour used must conform to the harmonized standards. In areas for which no standards exist, the symbols and colours must be described in the documentation supplied with the device.

13.3. *The label* must bear the following particulars:
(a) the name or trade name and address of the manufacturer. For devices imported into the Community, in view of their distribution in the Community, the label, or the outer packaging, or instructions for use, shall contain in addition the name and address of either the person responsible referred to in Article 14(2) or of the authorized representative of the manufacturer established within the Community or of the importer established within the Community, as appropriate;
(b) the details strictly necessary for the user to identify the device and the contents of the packaging.
(c) where appropriate, the word "STERILE;"
(d) where appropriate, the batch code, preceded by the word "LOT," or the serial number;
(e) where appropriate, an indication of the date by which the device should be used, in safety, expressed as the year and month;
(f) where appropriate, an indication that the device is for single use;
(g) if the device is custom-made, the words "custom-made device;"
(h) if the device is intended for clinical investigations, the words "exclusively for clinical investigations;"
(i) any special storage and/or handling conditions;
(j) any special operating instructions;
(k) any warnings and/or precautions to take;
(l) year of manufacture for active devices other than those covered by (e). This indication may be included in the batch or serial number;
(m) where applicable, method of sterilization.
13.4. If the intended purpose of the device is not obvious to the user, the manufacturer must clearly state it on the label and in the instructions for use.
13.5. Wherever reasonable and practicable, the devices and detachable components must be identified, where appropriate in terms of batches, to allow all appropriate action to detect any potential risk posed by the devices and detachable components.

13.6. Where appropriate, the instructions for use must contain the following particulars:
(a) the details referred to in Section 13.3, with the exception of (d) and (e);
(b) the performances referred to in Section 3 and any undesirable side-effects;
(c) if the device must be installed with or connected to other medical devices or equipment in order to operate as required for its intended purpose, sufficient details of its characteristics to identify the correct devices or equipment to use in order to obtain a safe combination;
(d) all the information needed to verify whether the device is properly installed and can operate correctly and safely, plus details of the nature and frequency of the maintenance and calibration needed to ensure that the devices operate properly and safely at all times;
(e) where appropriate, information to avoid certain risks in connection with implantation of the device;
(f) information regarding the risks of reciprocal interference posed by the presence of the device during specific investigations or treatment;
(g) the necessary instructions in the event of damage to the sterile packaging and, where appropriate, details of appropriate methods of resterilization.
(h) if the device is reusable, information on the appropriate processes to allow reuse, including cleaning, disinfection, packaging and, where appropriate, the method of sterilization of the device to be resterilized, and any restriction on the number of reuses.

Where devices are supplied with the intention that they be sterilized before use, the instructions for cleaning and sterilization must be such that, if correctly followed, the device will still comply with the requirements in Section I;
(i) details of any further treatment or handling needed before the device can be used (for example, sterilization, final assembly, etc.);
(j) in the case of devices emitting radiation for medical purposes, details of the nature, type, intensity and distribution of the radiation.

The instructions for use must also include details allowing the medical staff to brief the patient on any contra-indications and any precautions to be taken. These details should cover in particular:

(k) precautions to be taken in the event of changes in the performance of the device;

(l) precautions to be taken as regards exposure, in reasonably foreseeable environmental conditions, to magnetic fields, external electrical influences, electrostatic discharge, pressure or variation in pressure, acceleration, thermal ignition sources, etc.;

(m) adequate information regarding the medicinal product or products which the device in question is designed to administer, including any limitations in the choice of substances to be delivered.

(n) precautions to be taken against any special, unusual risks related to the disposal of the device;

(o) medicinal substances incorporated into the device as an integral part in accordance with Section 7.4;

(p) degree of accuracy claimed for devices with a measuring function.

14. Where conformity with the essential requirements must be based on clinical data, as in Section I(6), such data must be established in accordance with Annex X.

ANNEX II

EC DECLARATION OF CONFORMITY
(Full quality assurance system)

1. The manufacturer must ensure application of the quality system approved for the design, manufacture and final inspection of the products concerned, as specified in Section 3 and is subject to audit as laid down in Sections 3.3 and 4 and to Community surveillance as specified in Section 5.

2. The declaration of conformity is the procedure whereby the manufacturer who fulfils the obligations imposed by Section 1 ensures and declares that the products concerned meet the provisions of this Directive which apply to them.

The manufacturer must affix the CE marking in accordance with Article 17 and draw up a written declaration of conformity. This declaration must cover a given number of the products manufactured and be kept by the manufacturer.

3. Quality system

3.1. The manufacturer must lodge an application for assessment of his quality system with a notified body.

The application must include:

- the name and address of the manufacturer and any additional manufacturing site covered by the quality system,
- all the relevant information on the product or product category covered by the procedure,
- a written declaration that no application has been lodged with any other notified body for the same product-related quality system,
- the documentation on the quality system,
- an undertaking by the manufacturer to fulfil the obligations imposed by the quality system approved,
- an undertaking by the manufacturer to keep the approved quality system adequate and efficacious,
- an undertaking by the manufacturer to institute and keep up to date a systematic procedure to review experience gained from devices in the post-production phase and to implement appropriate means to apply any necessary corrective action. This undertaking must include an obligation for the manufacturer to notify the competent authorities of the following incidents immediately on learning of them:

  (i) any malfunction or deterioration in the characteristics and/or performance of a device, as well as any inadequacy in the instructions for use which might lead to or might have led to the death of a patient or user or to a serious deterioration in his state of health;

  (ii) any technical or medical reason connected with the characteristics or performance of a device leading for the reasons referred to in subparagraph (i) to systematic recall of devices of the same type by the manufacturer.

3.2 Application of the quality system must ensure that the products conform to the provisions of this Directive which apply to them at every stage, from design to final inspection. All the elements, requirements and provisions adopted by the manufacturer for his quality system must be documented in a systematic and orderly manner in the form of written policies and procedures such as quality programmes, quality plans, quality manuals and quality records.

It shall include in particular an adequate description of:

(a) the manufacturer's quality objectives;

(b) the organization of the business and in particular:

-- the organizational structures, the responsibilities of the managerial staff and their organizational authority where quality of design and manufacture of the products is concerned,

-- the methods of monitoring the efficient operation of the quality system and in particular its ability to achieve the desired quality of design and of product, including control of products which fail to conform;

(c) the procedures for monitoring and verifying the design of the products and in particular:

-- a general description of the product, including any variants planned,

-- the design specifications, including the standards which will be applied and the results of the risk analysis, and also a description of the solutions adopted to fulfil the essential requirements which apply to the products if the standards referred to in Article 5 are not applied in full,

-- the techniques used to control and verify the design and the processes and systematic actions which will be used when the products are being designed,

-- if the device is to be connected to other device(s) in order to operate as intended, proof must be provided that it conforms to the essential requirements when connected to any such device(s) having the characteristics specified by the manufacturer,

-- a statement indicating whether or not the device incorporates, as an integral part, a substance as referred to in Section 7.4 of Annex I and data on the tests conducted in this connection,

-- the clinical data referred to in Annex X,

-- the draft label and, where appropriate, instructions for use;

(d) the inspection and quality assurance techniques at the manufacturing stage and in particular:

-- the processes and procedures which will be used, particularly as regards sterilization, purchasing and the relevant documents,

-- the product identification procedures drawn up and kept up to date from drawings, specifications or other relevant documents at every stage of manufacture;

(e) the appropriate tests and trials which will be carried out before, during and after manufacture, the frequency with which they will take place, and the test equipment used; it must be possible to trace back the calibration of the test equipment adequately.

3.3. The notified body must audit the quality system to determine whether it meets the requirements referred to in Section 3.2. It must presume that quality systems which implement the relevant harmonized standards conform to these requirements.

The assessment team must include at least one member with past experience of assessments of the technology concerned. The assessment procedure must include an inspection on the manufacturer's premises and, in duly substantiated cases, on the premises of the manufacturer's suppliers and/or subcontractors to inspect the manufacturing process.

The decision is notified to the manufacturer. It must contain the conclusions of the inspection and a reasoned assessment.

3.4. The manufacturer must inform the notified body which approved the quality system of any plan for substantial change to the quality system or the product-range covered. The notified body must assess the changes proposed and verify whether after these changes the quality system still meets the requirements referred to in Section 3.2. It must notify the manufacturer of its decision. This decision must contain the conclusions of the inspection and a reasoned assessment.

4. Examination of the design of the product

4.1 In addition to the obligations imposed by Section 3, the manufacturer must lodge with the notified body an application for examination of the design dossier relating to the product which he plans to manufacture and which falls into the category referred to in Section 3.1.

4.2 The application must describe the design, manufacture and performances of the product in question. It must include the documents needed to assess whether the product conforms with the requirements of this Directive, as referred to in Section 3.2(c).

4.3. The notified body must examine the application and, if the product conforms to the relevant provisions os this Directive, issue the application with an EC design-examination certificate. The notified body may require the application to be completed by further tests or proof to allow assessment of conformity with the requirements of the Directive. The certificate must contain the conclusions of the examination, the conditions of validity, the data needed for identification of the approved design, where appropriate, a description of the intended purpose of the product.

In the case of devices referred to in Annex I, paragraph 7.4, the notified body shall, in view of the aspects addressed in that paragraph, consult one of the competent bodies established by the Member States in accordance with Directive 65/65/EEC before taking a decision.

The notified body will give due consideration to the views expressed in this consultation when making a decision. It will convey its final decision to the competent body concerned.

4.4. Changes to the approved design must receive further approval from the notified body which issued the EC design-examination certificate wherever the changes could affect conformity with the essential requirements of the Directive or with the conditions prescribed for use of the product. The applicant shall inform the notified body which issued the EC design-examination certificate of any such changes made to the approved design. This additional approval must take the form of a supplement to the EC design-examination certificate.

5. Surveillance

5.1 The aim of surveillance is to ensure that the manufacturer duly fulfils the obligations imposed by the approved quality system.

5.2 The manufacturer must authorize the notified body to carry out all the necessary inspections and supply it with all relevant information, in particular:

-- the documentation on the quality system,
-- the data stipulated in the part of the quality system relating to design, such as the results of analyses, calculation texts, etc.,
-- the data stipulated in the part of the quality system relating to manufacture, such as inspection reports and test data, calibration data, qualification reports of the personnel concerned, etc.

5.3. The notified body must periodically carry out appropriate inspections and assessments to make sure that the manufacturer applies the approved quality system and must supply the manufacturer with an assessment report.

5.4. In addition, the notified body may pay unannounced visits to the manufacturer. At the time of such visits, the notified body may, where necessary, carry out or ask for tests in order to check that the quality system is working properly. It must provide the manufacturer with an inspection report and, if a test has been carried out, with a test report.

6. Administrative provisions

6.1. The manufacturer must, for a period ending at least five years after the last product has been manufactured, keep at the disposal of the national authorities:

-- the declaration of conformity,
-- the documentation referred to in the fourth indent of Section 3.1,
-- the changes referred to in Section 3.4,
-- the documentation referred to in Section 4.2, and
-- the decisions and reports from the notified body as referred to in Sections 3.3, 4.3, 4.4, 5.3 and 5.4.

6.2. The notified body must make available to the other notified bodies and the competent authority, on request, all relevant information concerning quality system approvals issued, refused or withdrawn.

6.3. In respect of devices subject to the procedure in Section 4, when neither the manufacturer nor his authorized representative established in the Community, the obligation to keep available the technical documentation shall fall to the person responsible for placing the device on the Community market or the importer referred to the Annex I, Section 13.3(a).

7. Application to devices in Classes IIa and IIb

In line with Article 11(2) and (3), this Annex may apply to products in Classes IIa and IIb. Section 4, however, does not apply.

ANNEX III

EC TYPE-EXAMINATION

1. EC type-examination is the procedure whereby a notified body ascertains and certifies that a representative sample of the production covered fulfils the relevant provisions of this Directive.

2. The application includes:

-- the name and address of the manufacturer and the name and address of the authorized representative if the application is lodged by the representative,

-- the documentation described in Section 3 needed to assess the conformity of the representative sample of the production in question, hereinafter referred to as the "type," with the requirements of this Directive. The applicant must make a "type" available to the notified body. The notified body may request other samples as necessary,

-- a written declaration that no application has been lodged with any other notified body for the same type.

3. The documentation must allow an under standing of the design, the manufacture and the performances of the product and must contain the following items in particular:

-- a general description of the type, including any variants planned,

-- design drawings, methods of manufacture envisaged, in particular as regards sterilization, and diagrams of components, subassemblies, circuits, etc.,

-- the descriptions and explanations necessary to understand the abovementioned drawings and diagrams and the operation of the product,

-- a list of the standards referred to in Article 5, applied in full or in part, and descriptions of the solutions adopted to meet the essential requirements if the standards referred to in Article 5 have not been applied in full,

-- the results of the design calculations, risk analysis, investigations, technical tests, etc. carried out,

-- a statement indicating whether or not the device incorporates, as an integral part, a substance as referred to in Section 7.4 of Annex I and data on the tests conducted in this connection,

-- the clinical data referred to in Annex X,

-- the draft label and, where appropriate, instructions for use.

4. The notified body must:

4.1 examine and assess the documentation and verify that the type has been manufactured in conformity with that documentation; it must also record the items designed in conformity with the applicable provisions of the standards referred to in Article 5, as well as the items not designed on the basis of the relevant provisions of the abovementioned standards;

4.2 carry out or arrange for the appropriate inspections and the tests necessary to verify whether the solutions adopted by the manufacturer meet the essential requirements of this Directive if the standards referred to in Article 5 have not been applied; if the device is to be connected to other device(s) in order to operate as intended, proof must be provided that it conforms to the essential requirements when connected to any such device(s) having the characteristics specified by the manufacturer;

4.3 carry out or arrange for the appropriate inspections and the tests necessary to verify whether, if the manufacturer has chosen to

apply the relevant standards, these have actually been applied;
4.4 agree with the applicant on the place where the necessary inspections and tests will be carried out.
5. If the type conforms to the provisions of this Directive, the notified body issues the applicant with an EC type-examination certificate. The certificate must contain the name and address of the manufacturer, the conclusions of the inspection, the conditions of validity and the data needed for identification of the type approved. The relevant parts of the documentation must be annexed to the certificate and a copy kept by the notified body.

In the case of devices referred to in Annex I, paragraph 7.4, the notified body shall, in view of the aspects addressed in that paragraph, consult one of the competent bodies established by the Member States in accordance with Directive 65/65/EEC before taking a decision.

The notified body will give due consideration to the views expressed in this consultation when making a decision. It will convey its final decision to the competent body concerned.
6. The applicant must inform the notified body which issued the EC type-examination certificate of any significant change made to the approved product.

Changes to the approved product must receive further approval from the notified body which issued the EC type-examination certificate wherever the changes may affect conformity with the essential requirements or with the conditions prescribed for use of the product. This new approval must, where appropriate, take the form of a supplement to the initial EC type-examination certificate.
7. Administrative provisions
7.1. The notified body must make available to the other notified bodies on request, all relevant information on EC type-examination certificates and supplements issued, refused or withdrawn.
7.2. Other notified bodies may obtain a copy of the EC type-examination certificates and/or the supplements thereto. The Annexes to the certificates must be made available to other notified bodies on reasoned application, after the manufacturer has been informed.
7.3. The manufacturer or his authorized representative must keep with the technical documentation copies of EC type-examination certificates and their additions for a period ending at least five years after the last device has been manufactured.
7.4. When neither the manufacturer nor his authorized representative is established in the Community, the obligation to keep available the technical documentation shall fall to the person responsible for placing the device on the Community market or the importer referred to in Annex I, Section 13.3(a).

ANNEX IV

EC VERIFICATION

1. EC verification is the procedure whereby the manufacturer or his authorized representative established in the Community ensures and declares that the products which have been subject to the procedure set out in Section 4 conform to the type described in the EC type-examination certificate and meet the requirements of this Directive which apply to them.
2. The manufacturer must take all the measures necessary to ensure that the manufacturing process produces products which conform to the type described in the EC type-examination certificate and to the requirements of the Directive which apply to them. Before the start of manufacture, the manufacturer must prepare documents defining the manufacturing process, in particular as regards sterilization where necessary, together with all the routine, pre-established provisions to be implemented to ensure homogeneous production and, where appropriate, conformity of the products with the type described in the EC type-examination certificate and with the requirements of this Directive which apply to them. The manufacturer must affix the CE marking in accordance with Article 17 and draw up a declaration of conformity.

In addition, for products placed on the market in sterile condition, and only for those aspects of the manufacturing process designed to secure and maintain sterility, the manufacturer must apply the provisions of Annex V, Sections 3 and 4.

3. The manufacturer must undertake to institute and keep up to date a systematic procedure to review experience gained from devices in the post-production phase and to implement appropriate means to apply any necessary corrective action. This undertaking must include an obligation for the manufacturer to notify the competent authorities of the following incidents immediately on learning of them:

(i) any malfunction or deterioration in the characteristics and/or performance of a device, as well as any inadequacy in the labelling or the instructions for use which might lead to or might have led to the death of a patient or user or to a serious deterioration in his state of health;

(ii) any technical or medical reason connected with the characteristics or performance of a device leading for the reasons referred to in subparagraph (i) to systematic recall of devices of the same type by the manufacturer.

4. The notified body must carry out the appropriate examinations and tests in order to verify the conformity of the product with the requirements of the Directive either by examining and testing every product as specified in Section 5 or by examining and testing products on a statistical basis as specified in Section 6, as the manufacturer decides.

The aforementioned checks do not apply to those aspects of the manufacturing process designed to secure sterility.

5. Verification by examination and testing of every product

5.1. Every product is examined individually and the appropriate tests defined in the relevant standard(s) referred to in Article 5 or equivalent tests must be carried out in order to verify, where appropriate, the conformity of the products with the EC type described in the type-examination certificate and with the requirements of the Directive which apply to them.

5.2. The notified body must affix, or have affixed its identification to each approved product and must draw up a written certificate of conformity relating to the tests carried out.

6. Statistical verification

6.1. The manufacturer must present the manufactured products in the form of homogeneous batches.

6.2. A random sample is taken from each batch. The products which make up the sample are examined individually and the appropriate tests defined in the relevant standard(s) referred to in Article 5 or equivalent tests must be carried out to verify, where appropriate, the conformity of the products with the EC type described in the type-examination certificate and with the requirements of the Directive which apply to them in order to determine whether to accept or reject the batch.

6.3. Statistical control of products will be based on attributes, entailing a sampling system ensuring a limit quality corresponding to a probability of acceptance of 5%, with a non-conformity percentage of between 3 and 7%. The sampling method will be established by the harmonized standards referred to in Article 5, taking account of the specific nature of the product categories in question.

6.4. If the batch is accepted, the notified body affixes or has affixed its identification number to each product and draws up a written certificate of conformity relating to the tests carried out. All products in the batch may be put on the market except any in the sample which failed to conform.

If a batch is rejected, the competent notified body must take appropriate measures to prevent the batch from being placed on the market. In the event of frequent rejection of batches, the notified body may suspend the statistical verification.

The manufacturer may, on the responsibility of the notified body, affix the notified body's identification number during the manufacturing process.

7. Administrative provisions

The manufacturer or his authorized representative must, for a period ending at least five years after the last product has been manufactured, make available to the national authorities:

-- the declaration of conformity,

-- the documentation referred to in Section 2,
-- the certificates referred to in Sections 5.2 and 6.4,
-- where appropriate, the type-examination certificate referred to in Annex III.

8. Application to devices in Class IIa
In line with Article 11(2), this Annex may apply to products in Class IIa, subject to the following exemptions:
8.1. in derogation from Sections 1 and 2, by virtue of the declaration of conformity the manufacturer ensures and declares that the products in Class IIa are manufactured in conformity with the technical documentation referred to in Section 3 of Annex VII and meet the requirements of this Directive which apply to them;
8.2. in derogation from Sections 1, 2, 5 and 6, the verifications conducted by the notified body are intended to confirm the conformity of the products in Class IIa with the technical documentation referred to in Section 3 of Annex VII.

ANNEX V

EC DECLARATION OF CONFORMITY
(Production quality assurance)

1. The manufacturer must ensure application of the quality system approved for the manufacture of the products concerned and carry out the final inspection, as specified in Section 3, and is subject to the Community surveillance referred to in Section 4.
2. The declaration of conformity is the part of the procedure whereby the manufacturer who fulfils the obligations imposed by Section 1 ensures and declares that the products concerned conform to the type described in the EC type-examination certificate and meets the provisions of this Directive which apply to them.

The manufacturer must affix the CE marking in accordance with Article 17 and draw up a written declaration of conformity. This declaration must cover a given number of identified specimens of the products manufactured and must be kept by the manufacturer.
3. Quality system
3.1. The manufacturer must lodge an application for assessment of his quality system with a notified body.

The application must include:
-- the name and address of the manufacturer and any additional manufacturing site covered by the quality system,
-- all the relevant information on the product or product category covered by the procedure,
-- a written declaration that no application has been lodged with any other notified body for the same products,
-- the documentation on the quality system,
-- an undertaking to fulfil the obligations imposed by the quality system is approved,
-- an undertaking to maintain the practicability and effectiveness of the approved quality system,
-- where appropriate, the technical documentation on the types approved and a copy of the EC type-examination certificates,
-- an undertaking by the manufacturer to institute and keep up to date a systematic procedure to review experience gained from devices in the post-production phase and to implement appropriate means to apply any necessary corrective action. This undertaking must include an obligation for the manufacturer to notify the competent authorities of the following incidents immediately on learning of them:
(i) any malfunction or deterioration in the characteristics and/or performance of a device, as well as any inadequacy in the labelling or the instructions for use which might lead to or might have led to the death of a patient or user or to a serious deterioration in his state of health;
(ii) any technical or medical reason connected with the characteristics or performance of a device leading for the reasons referred to in subparagraph (i) to systematic recall of devices of the same type by the manufacturer.

3.2 Application of the quality system must ensure that the products conform to the type described in the EC type-examination certificate.

All the elements, requirements and provisions adopted by the manufacturer for his quality system must be documented in a systematic and orderly manner in the form of written policy statements and procedures. This quality system documentation must permit uniform interpretation of the quality policy and procedures such as quality programmes, plans, manuals and records.

It must include in particular an adequate description of:

(a) the manufacturer's quality objectives;

(b) the organization of the business and in particular:

-- the organizational structures, the responsibilities of the managerial staff and their organizational authority where manufacture of the products is concerned,
-- the methods of monitoring the efficient operation of the quality system and in particular its ability to achieve the desired quality of product, including control of products which fail to conform;

(c) the inspection and quality assurance techniques at the manufacturing state and in particular:

-- the processes and procedures which will be used, particularly as regards sterilization, purchasing and the relevant documents,
-- the product identification procedures drawn up and kept up to date from drawings, specifications or other relevant documents at every stage of manufacture;

(d) the appropriate tests and trials to be carried out before, during and after manufacture, the frequency with which they will take place, and the test equipment used; it must be possible adequately to trace back the calibration of the test equipment.

3.3. The notified body must audit the quality system to determine whether it meets the requirements referred to in Section 3.2. It must presume that quality systems which implement the relevant harmonized standards conform to these requirements.

The assessment team must include at least one member with past experience of assessments of the technology concerned. The assessment procedure must include an inspection on the manufacturer's premises and, in duly substantiated cases, on the premises of the manufacturer's suppliers to inspect the manufacturing process.

The decision must be notified to the manufacturer after the final inspection and contain the conclusions of the inspection and a reasoned assessment.

3.4. The manufacturer must inform the notified body which approved the quality system of any plan for substantial changes to the quality system.

The notified body must assess the changes proposed and verify whether after these changes the quality system still meets the requirements referred to in Section 3.2.

After the abovementioned information has been received the decision is notified to the manufacturer. It must contain the conclusions of the inspection and a reasoned assessment.

4. Surveillance

4.1 The aim of surveillance is to ensure that the manufacturer duly fulfils the obligations imposed by the approved quality system.

4.2 The manufacturer authorizes the notified body to carry out all the necessary inspections and must supply it with all relevant information, in particular:

-- the documentation on the quality system,
-- the data stipulated in the part of the quality system relating to manufacture, such as inspection reports and test data, calibration data, qualification reports of the personnel concerned, etc.

4.3. The notified body must periodically carry out appropriate inspections and assessments to make sure that the manufacturer applies the approved quality system and supply the manufacturer with an assessment report.

4.4. In addition, the notified body may pay unannounced visits to the manufacturer. At the time of such visits, the notified body may, where necessary, carry out or ask for tests in order to check that the quality system is working properly. It must provide the manufacturer with an inspection report and, if a test has been carried out, with a test report.

5. Administrative provisions

5.1. The manufacturer must, for a period ending at least five years after the last product has been manufactured, make available to the national authorities:

-- the declaration of conformity,
-- the documentation referred to in the fourth indent of Section 3.1,
-- the changes referred to in Section 3.4,
-- the documentation referred to in the seventh indent of Section 3.1,
-- the decisions and reports from the notified body as referred to in Sections 4.3 and 4.4,
-- where appropriate, the type-examination certificate referred to in Annex III.

5.2. The notified body must make available to the other notified bodies, on request, all relevant information concerning the quality system approvals issued, refused or withdrawn.

6. Application to devices in Class IIa

In line with Article 11(2), this Annex may apply to products in Class IIa, subject to the following exemption:

6.1. in derogation from Sections 2, 3.1 and 3.2, by virtue of the declaration of conformity the manufacturer ensures and declares that the products in Class IIa are manufactured in conformity with the technical documentation referred to in Section 3 of Annex VII and meet the requirements of this Directive which apply to them.

ANNEX VI

EC DECLARATION OF CONFORMITY
(Product quality assurance)

1. The manufacturer must ensure application of the quality system approved for the final inspection and testing of the product, as specified in Section 3 and must be subject to the surveillance referred to in Section 4.

In addition, for products placed on the market in sterile condition, and only for those aspects of the manufacturing process designed to secure and maintain sterility, the manufacturer must apply the provisions of Annex V, Sections 3 and 4.

2. The declaration of conformity is the part of the procedure whereby the manufacturer who fulfils the obligations imposed by Section 1 ensures and declares that the products concerned conform to the type described in the EC type-examination certificate and meet the provisions of this Directive which apply to them.

The manufacturer affixes the CE marking in accordance with Article 17 and draws up a written declaration of conformity. This declaration must cover a given number of identified specimens of the products manufactured and be kept by the manufacturer. The CE marking must be accompanied by the identification number of the notified body which performs the tasks referred to in this Annex.

3. Quality system

3.1. The manufacturer lodges an application for assessment of his quality system with a notified body.

The application must include:
-- the name and address of the manufacturer,
-- all the relevant information on the product or product category covered by the procedure,
-- a written declaration specifying that no application has been lodged with any other notified body for the same products,
-- the documentation on the quality system,
-- an undertaking by the manufacturer to fulfil the obligations imposed by the quality system approved,
-- an undertaking by the manufacturer to keep the approved quality system adequate and efficacious,
-- where appropriate, the technical documentation on the types approved and a copy of the EC type-examination certificates,
-- an undertaking by the manufacturer to institute and keep up to date a systematic procedure to review experience gained from devices in the post-production phase and to implement appropriate means to apply any necessary corrective action. This undertaking must include an obligation for the manufacturer to notify the competent authorities of the following incidents immediately on learning of them:

(i) any malfunction or deterioration in the characteristics and/or performance of a device, as well as any inadequacy in the labelling or the instructions for use which might lead to or might have led to the death of a patient or user or to a serious deterioration in his state of health;

(ii) any technical or medical reason connected with the characteristics or performance of a device leading for the reasons referred to in subparagraph (i) to systematic recall of devices of the same type by the manufacturer.

3.2. Under the quality system, each product or a representative sample of each batch is examined and the appropriate tests defined in the relevant standard(s) referred to in Article 5 or equivalent tests are carried out to ensure that the products conform to the type described in the EC type-examination certificate and fulfil the provisions of this Directive which apply to them. All the elements, requirements and provisions adopted by the manufacturer must be documented in a systematic and orderly manner in the form of written measures, procedures and instructions. This quality system documentation must permit uniform interpretation of the quality programmes, quality plans, quality manuals and quality records.

It must include in particular an adequate description of:

-- the quality objectives and the organizational structure, responsibilities and powers of the managerial staff with regard to product quality,
-- the examinations and tests that will be carried out after manufacture; it must be possible to trace back the calibration of the test equipment adequately,
-- the methods of monitoring the efficient operation of the quality system,
-- the quality records, such as reports concerning inspections, tests, calibration and the qualifications of the staff concerned, etc.

The aforementioned checks do not apply to those aspects of the manufacturing process designed to secure sterility.

3.3. The notified body audits the quality system to determine whether it meets the requirements referred to in Section 3.2. It must presume that quality systems which implement the relevant harmonized standards conform to these requirements.

The assessment team must include at least one member with past experience of assessments of the technology concerned. The assessment procedure must include an inspection on the manufacturer's premises and, in duly substantiated cases, on the premises of the manufacturer's suppliers to inspect the manufacturing process.

The decision must be notified to the manufacturer. It must contain the conclusions of the inspection and a reasoned assessment.

3.4. The manufacturer must inform the notified body which approved the quality system of any plan for substantial changes to the quality system.

The notified body must assess the changes proposed and verify whether after these changes the quality system will still meet the requirements referred to in Section 3.2.

After receiving the abovementioned information it must notify the manufacturer of its decision. This decision must contain the conclusions of the inspection and a reasoned assessment.

4. Surveillance

4.1 The aim of surveillance is to ensure that the manufacturer duly fulfils the obligations imposed by the approved quality system.

4.2 The manufacturer must allow the notified body access for inspection purposes to the inspection, testing and storage locations and supply it with all relevant information, in particular:

-- the documentation on the quality system,
-- the technical documentation,
-- the quality records, such as inspection reports, test data, calibration data, qualification reports of the staff concerned, etc.

4.3. The notified body must periodically carry out appropriate inspections and assessments to make sure that the manufacturer applies the quality system and must supply the manufacturer with an assessment report.

4.4. In addition, the notified body may pay unannounced visits to the manufacturer. At the time of such visits, the notified body may, where necessary, carry out or ask for tests in order to check that the quality system is working properly and that the production conforms to the requirements of the Directive which apply to them. To this end, an adequate sample of the final products, taken on site by the notified body, must be examined and the appropriate tests defined in the relevant standard(s) referred to in Article 5 or equivalent tests must be carried out. Where one or more of the samples fails to conform, the notified body must take the appropriate measures.

It must provide the manufacturer with an inspection report and, if a test has been carried out, with a test report.

5. Administrative provisions

5.1. The manufacturer must, for a period ending at least five years after the last product has been manufactured, make available to the national authorities:

-- the declaration of conformity,
-- the documentation referred to in the seventh indent of Section 3.1,
-- the changes referred to in Section 3.4,
-- the decisions and reports from the notified body as referred to in the final indent of Section 3.4 and in Sections 4.3 and 4.4,
-- where appropriate, the certificate of conformity referred to in Annex III.

5.2. The notified body must make available to the other notified bodies, on request, all relevant information concerning the quality system approvals issued, refused or withdrawn.

6. Application to devices in Class IIa

In line with Article 11(2), this Annex may apply to products in Class IIa, subject to this derogation:

6.1. in derogation from Sections 2, 3.1 and 3.2 by virtue of the declaration of conformity the manufacturer ensures and declares that the products in Class IIa are manufactured in conformity with the technical documentation referred to in Section 3 of Annex VII and meet the requirements of this Directive which apply to them.

ANNEX VII

EC DECLARATION OF CONFORMITY

1. The EC declaration of conformity is the procedure whereby the manufacturer or his authorized representative established in the Community who fulfils the obligations imposed by Section 2 and, in the case of products placed on the market in a sterile condition and devices with a measuring function, the obligations imposed by Section 5 ensures and declares that the products concerned meet the provisions of the Directive which apply to them.

2. The manufacturer must prepare the technical documentation described in Section 3. The manufacturer or his authorized representative established in the Community must make this documentation, including the declaration of conformity, available to the national authorities for inspection purposes for a period ending at least five years after the last product has been manufactured.

Where neither the manufacturer nor his authorized representative are established in the Community, this obligation to keep the technical documentation available must fall to the person(s) who place(s) the product on the Community market.

3. The technical documentation must allow assessment of the conformity of the product with the requirements of the Directive. It must include in particular:

-- a general description of the product, including any variants planned,
-- design drawings, methods of manufacture envisaged and diagrams of components, sub-assemblies, circuits, etc.,
-- the descriptions and explanations necessary to understand the abovementioned drawings and diagrams and the operations of the product,
-- the results of the risk analysis and a list of the standards referred to in Article 5, applied in full or in part, and descriptions of the solutions adopted to meet the essential requirements of the Directive if the standards referred to in Article 5 have not been applied in full,
-- in the case of products placed on the market in a sterile condition, description of the methods used,
-- the results of the design calculations and of the inspections carried out, etc.; if the device is to be connected to other device(s) in order to operate as intended, proof must be provided that it conforms to the essential requirements when connected to any such device(s) having the characteristics

specified by the manufacturer,
- the test reports and, where appropriate, clinical data in accordance with Annex X,
- the label and instructions for use.

4. The manufacturer shall institute and keep up to date a systematic procedure to review experience gained from devices in the post-production phase and to implement appropriate means to apply any necessary corrective actions, taking account of the nature and risks in relation to the product. He shall notify the competent authorities of the following incidents immediately on learning of them:

(i) any malfunction or deterioration in the characteristics and/or performance of a device, as well as any inadequacy in the labelling or the instructions for use which might lead to or might have led to the death of a patient or user or to a serious deterioration in his state of health;

(ii) any technical or medical reason connected with the characteristics or performance of a device leading for the reasons referred to in subparagraph (i) to systematic recall of devices of the same type by the manufacturer.

5. With products placed on the market in sterile condition and Class I devices with a measuring function, the manufacturer must observe not only the provisions laid down in this Annex but also one of the procedures referred to in Annex IV, V or VI. Application of the abovementioned Annexes and the intervention by the notified body is limited to:
- in the case of products placed on the market in sterile condition, only the aspects of manufacture concerned with securing and maintaining sterile conditions,
- in the case of devices with a measuring function, only the aspects of manufacture concerned with the conformity of the products with the metrological requirements,

Section 6.1 of this Annex is applicable.

6. Application to devices in Class IIa

In line with Article 11(2), this Annex may apply to products in Class IIa, subject to the following derogation:

6.1. Where this Annex is applied in conjunction with the procedure referred to in Annex IV, V or VI, the declaration of conformity referred to in the abovementioned Annexes forms a single declaration. As regards the declaration based on this Annex, the manufacturer must ensure and declare that the product design meets the provisions of this Directive which apply to it.

ANNEX VIII

STATEMENT CONCERNING DEVICES FOR SPECIAL PURPOSES

1. For custom-made devices or for devices intended for clinical investigations the manufacturer or his authorized representative established in the Community must draw up the statement containing the information stipulated in Section 2.

2. The statement must contain the following information:

2.1. for custom-made devices:
- data allowing identification of the device in question,
- a statement that the device is intended for exclusive use by a particular patient, together with the name of the patient,
- the name of the medical practitioner or other authorized person who made out the prescription and, where applicable, the name of the clinic concerned,
- the particular features of the device as specified in the relevant medical prescription,
- a statement that the device in question conforms to the essential requirements set out in Annex I and, where applicable, indicating which essential requirements have not been fully met, together with the grounds;

2.2. for devices intended for the clinical investigations covered by Annex X:
- data allowing identification of the device in question,
- an investigation plan stating in particular the purpose, scientific, technical or medical grounds, scope and number of devices concerned,

-- the opinion of the ethics committee concerned and details of the aspects covered by its opinion,
-- the name of the medical practitioner or other authorized person and of the institution responsible for the investigations,
-- the place, starting date and scheduled duration for the investigations,
-- a statement that the device in question conforms to the essential requirements apart from the aspects covered by the investigations and that, with regard to these aspects, every precaution has been taken to protect the health and safety of the patient.

3. The manufacturer must also undertake to keep available for the competent national authorities:

3.1. for custom-made devices, documentation allowing an understanding of the design, manufacture and performances of the product, including the expected performances, so as to allow assessment of conformity with the requirements of this Directive.

The manufacturer must take all the measures necessary to ensure that the manufacturing process produces products which are manufactured in accordance with the documentation mentioned in the first paragraph;

3.2. for devices intended for clinical investigations, the documentation must contain:
-- a general description of the product,
-- design drawings, methods of manufacture envisaged, in particular as regards sterilization, and diagrams of components, subassemblies, circuits, etc.,
-- the descriptions and explanations necessary to understand the abovementioned drawings and diagrams and the operation of the product,
-- the results of the risk analysis and a list of the standards referred to in Article 5, applied in full or in part, and descriptions of the solutions adopted to meet the essential requirements of this Directive if the standards referred to the Article 5 have not been applied,
-- the results of the design calculations, and of the inspections and technical tests carried out, etc.

The manufacturer must take all the measures necessary to ensure that the manufacturing process produces products which are manufactured in accordance with the documentation referred to in the first paragraph of this Section.

The manufacturer must authorize the assessment, or audit where necessary, of the effectiveness of these measures.

The information contained in the declarations concerned by this Annex should be kept for a period of time of at least five years.

ANNEX IX

CLASSIFICATION CRITERIA

I. DEFINITIONS

1. Definitions for the classification rules

1.1. *Duration*

Transient

Normally intended for continuous use for less than 60 minutes.

Short term

Normally intended for continuous use for not more than 30 days.

Long term

Normally intended for continuous use for more than 30 days.

1.2 *Invasive devices*

Invasive device

A device which, in whole or in part, penetrates inside the body, either through a body orifice or through the surface of the body.

Body orifice

Any natural opening in the body, as well as the external surface of the eyeball, or any permanent artificial opening, such as a stoma.

Surgically invasive device

An invasive device which penetrates inside the body through the surface of the body, with the aid or in the context of a surgical operation.

For the purposes of this Directive devices other than those referred to in the previous subparagraph and which product penetration other than through an established body orifice, shall be treated as surgically invasive devices.

Implantable device

Any device which is intended:
-- to be totally introduced into the human body or,

-- to replace an epithelial surface or the surface of the eye,

by surgical intervention which is intended to be partially introduced into the human body through surgical intervention and intended to remain in place after the procedure for at least 30 days is also considered an implantable device.

1.3. *Reusable surgical instrument*
Instrument intended for surgical use by cutting, drilling, sawing, scratching, scraping, clamping, retracting, clipping or similar procedures, without connection to any active medical device and which can be reused after appropriate procedures have been carried out.

1.4. *Active medical device*
Any medical device operation of which depends on a source of electrical energy or any source of power other than that directly generated by the human body or gravity and which acts by converting this energy. Medical devices intended to transmit energy, substances or other elements between an active medical device and the patient, without any significant change, are not considered to be active medical devices.

1.5. *Active therapeutical device*
Any active medical device, whether used alone or in combination with other medical devices, to support, modify, replace or restore biological functions or structures with a view to treatment or alleviation of an illness, injury or handicap.

1.6. *Active device for diagnosis*
Any active medical device, whether used alone or in combination with other medical devices, to supply information for detecting, diagnosing, monitoring or treating physiological conditions, states of health, illnesses or congenital deformities.

1.7. *Central circulatory system*
For the purposes of this Directive, "central circulatory system" means the following vessels:
arteriae pulmonales, aorta ascendens, arteriae coronariae, arteria carotis communis, arteria carotis externa, arteria carotis interna, arteriae cerebrales, truncus brachicephalicus, venae cordis, venae pulmonales, vena cava superior, vena cava inferior.

1.8. *Central nervous system* For the purposes of this Directive, "central nervous system" means brain, meninges and spinal cord.

II. IMPLEMENTING RULES

2. Implementing rules

2.1. Application of the classification rules shall be governed by the intended purpose of the devices.

2.2. If the device is intended to be used in combination with another device, the classification rules shall apply separately to each of the devices. Accessories are classified in their own right separately from the device with which they are used.

2.3. Software, which drives a device or influences the use of a device, falls automatically in the same class.

2.4. if the device is not intended to be used solely or principally in a specific part of the body, it must be considered and classified on the basis of the most critical specified use.

2.5. If several rules apply to the same device, based on the performance specified for the device by the manufacturer, the strictest rules resulting in the higher classification shall apply.

III. CLASSIFICATION

1. Non-invasive devices

1.1. *Rule 1*
All non-invasive devices are in Class I, unless one of the rules set out hereinafter applies.

1.2. *Rule 2*
All non-invasive devices intended for channelling or storing blood, body liquids or tissues, liquids or gases for the purpose of eventual infusion, administration or introduction into the body are in Class IIa:

-- if they may be connected to an active medical device in Class IIa or a higher class,
-- if they are intended for use for storing or channelling blood or other body liquids or for storing organs, parts of organs or body tissues,

in all other cases they are in Class I.

1.3. *Rule 3*

All non-invasive devices intended for modifying the biological or chemical composition of blood, other body liquids or other liquids intended for infusion into the body are in Class IIb, unless the treatment consists of filtration, centrifugation or exchanges of gas, heat, in which case they are in Class IIa.

1.4. *Rule 4*

All non-invasive devices which come into contact with injured skin:

-- are in Class I if they are intended to be used as a mechanical barrier, for compression or for absorption of exudates,

-- are in Class IIb if they are intended to be used principally with wounds which have breached the dermis and can only heal by secondary intent,

-- are in Class IIa in all other cases, including devices principally intended to manage the micro-environment of a wound.

2. Invasive devices

2.1. *Rule 5*

All invasive devices with respect to body orifices, other than surgically invasive devices and which are not intended for connection to an active medical device:

-- are in Class I if they are intended for transient use,

-- are in Class IIa if they are intended for short-term use, except if they are used in the oral cavity as far as the pharynx, in an ear canal up to the ear drum or in a nasal cavity, in which case they are in Class I,

-- are in Class IIb if they are intended for long-term use, except if they are used in the oral cavity as far as the pharynx, in an ear canal up to the ear drum or in a nasal cavity and are not liable to be absorbed by the mucous membrane, in which case they are in Class IIa.

All invasive devices with respect to body orifices, other than surgically invasive devices, intended for connection to an active medical device in Class IIa or a higher class, are in Class IIa.

2.2. *Rule 6*

All surgically invasive devices intended for transient use are in Class IIa unless they are:

-- intended specifically to diagnose, monitor or correct a defect of the heart or of the central circulatory system through direct contact with these parts of the body, in which case they are in Class III,

-- reusable surgical instruments, in which case they are in Class I,

-- intended to supply energy in the form of ionizing radiation in which case they are in Class IIb,

-- intended to have a biological effect or to be wholly or mainly absorbed in which case they are in Class IIb,

-- intended to administer medicines by means of a delivery system, if this is done in a manner that is potentially hazardous taking account of the mode of application, in which they are in Class IIb.

2.3. *Rule 7*

All surgically invasive devices intended for short-term use are in Class IIa unless they are intended:

-- either specifically to diagnose, monitor or correct a defect of the heart or of the central circulatory system through direct contact with these parts of the body, in which case they are in Class III,

-- or specifically for use in direct contact with the central nervous system, in which case they are in Class III,

-- or to supply energy in the form of ionizing radiation in which case they are in Class IIb,

-- or to have a biological effect or to be wholly or mainly absorbed in which case they are in Class III,

-- or to undergo chemical change in the body, except if the devices are placed in the teeth, or to administer medicines, in which case they are in Class IIb.

2.4. *Rule 8*

All implantable devices and long-term surgically invasive devices are in Class IIb unless they are intended:

-- to be placed in the teeth, in which case they are in Class IIa,

-- to be used in direct contact with the heart, the central circulatory system or the central nervous system, in which case they are in Class III,

-- to have a biological effect or to be wholly or mainly absorbed, in which case they are in Class III,

-- or to undergo chemical change in the body, except if the devices are placed in the teeth, or to administer medicines, in which case they are in Class III.

3. Additional rules applicable to active devices

3.1. *Rule 9*

All active therapeutic devices intended to administer or exchange energy are in Class IIa unless their characteristics are such that they may administer or exchange energy to or from the human body in a potentially hazardous way, taking account of the nature, the density and site of application of the energy, in which case they are in Class IIb.

All active devices intended to control or monitor the performance of active therapeutic devices in Class IIb, or intended directly to influence the performance of such devices are in Class IIb.

3.2. *Rule 10*

Active devices intended for diagnosis are in Class IIa:

-- if they are intended to supply energy which will be absorbed by the human body, except for devices used to illuminate the patient's body, in the visible spectrum,

-- if the are intended to image *in vivo* distribution of radiopharmaceuticals,

-- if they are intended to allow direct diagnosis or monitoring of vital physiological processes, unless they are specifically intended for monitoring of vital physiological parameters, where the nature of variations is such that it could result in immediate danger to the patient, for instance variations in cardiac performance, respiration, activity of CNS in which case they are in Class IIb.

Active devices intended to emit ionizing radiation and intended for diagnostic and therapeutic interventional radiology including devices which control or monitor such devices, or which directly influence their performance, are in Class IIb.

Rule 11

All active devices intended to administer and/or remove medicines, body liquids or other substances to or from the body are in Class IIa, unless this is done in a manner:

-- that is potentially hazardous, taking account of the nature of the substances involved, of the part of the body concerned and of the mode of application in which case they are in Class IIb.

3.3. *Rule 12*

All other active devices are in Class I.

4. Special Rules

4.1. *Rule 13*

All devices incorporating, as an integral part, a substance which, if used separately, can be considered to be a medicinal product, as defined in Article 1 of Directive 65/65/EEC, and which is liable to act on the human body with action ancillary to that of the devices, are in Class III.

4.2. *Rule 14*

All devices used for contraception or the prevention of the transmission of sexually transmitted diseases are in Class IIb, unless they are implantable or long term invasive devices, in which case they are in Class III.

4.3. *Rule 15*

All devices intended specifically to be used for disinfecting, cleaning, rinsing or, when appropriate, hydrating contact lenses are in Class IIb.

All devices intended specifically to be used for disinfecting medical devices are in Class IIa.

This rule does not apply to products that are intended to clean medical devices other than contact lenses by means of physical action.

4.4. *Rule 16*

Non-active devices specifically intended for recording of X-ray diagnostic images are in Class IIa.

4.5. *Rule 17*

All devices manufactured utilizing animal tissues or derivatives rendered non-viable are Class III except where such devices are intended to come into contact with intact skin only.

5. *Rule 18*

By derogation from other rules, blood bags are in Class IIb.

ANNEX X

CLINICAL EVALUATION

1. General provisions

1.1. As a general rule, conformation of conformity with the requirements concerning the characteristics and performances referred to in Sections 1 and 3 of Annex I under the normal conditions of use of the device and the evaluation of the undesirable side-effects must be based on clinical data in particular in the case of implantable devices and devices in Class III. Taking account of any relevant harmonized standards, where appropriate, the adequacy of the clinical data must be based on:

1.1.1. either a compilation of the relevant scientific literature currently available on the intended purpose of the device and the techniques employed as well as, if appropriate, a written report containing a critical evaluation of this compilation;

1.1.2. or the results of all the clinical investigations made, including those carried out in conformity with section 2.

1.2. All the data must remain confidential, in accordance with the provisions of Article 20.

2. Clinical investigations

2.1. *Objectives*

The objectives of clinical investigation are:

-- to verify that, under normal conditions of use, the performance of the devices conform to those referred to in Section 3 of Annex I, and

-- to determine any undesirable side-effects, under normal conditions of use, and assess whether they constitute risks when weighed against the intended performance of the device.

2.2. *Ethical considerations*

Clinical investigations must be carried out in accordance with the Helsinki Declaration adopted by the 18th World Medical Assembly in Helsinki, Finland, in 1964, as last amended by the 41st World Medical Assembly in Hong Kong in 1989. It is mandatory that all measures relating to the protection of human subjects are carried out in the spirit of the Helsinki Declaration. This includes every step in the clinical investigation from first consideration of the need and justification of the study to publication of the results.

2.3. *Methods*

2.3.1. Clinical investigations must be performed on the basis of an appropriate plan of investigation reflecting the latest scientific and technical knowledge and defined in such a way as to confirm or refute the manufacturer's claims for the device; these investigations must include an adequate number of observations to guarantee the scientific validity of the conclusions.

2.3.2. The procedures used to perform the investigations must be appropriate to the device under examination.

2.3.3. Clinical investigations must be performed in circumstances similar to the normal conditions of use of the device.

2.3.4. All the appropriate features, including those involving the safety and performances of the device, and its effect on patients must be examined.

2.3.5. All adverse incidents such as those specified in Article 10 must be fully recorded and notified to the competent authority.

2.3.6. The investigations must be performed under the responsibility of a medical practitioner or another authorized qualified person in an appropriate environment.

The medical practitioner or other authorized person must have access to the technical and clinical data regarding the device.

2.3.7. The written report, signed by the medical practitioner or other authorized person responsible, must contain a critical evaluation of all the data collected during the clinical investigation.

ANNEX XI

CRITERIA TO BE MET FOR THE DESIGNATION OF NOTIFIED BODIES

1. The notified body, its Director and the assessment and verification staff shall not be the designer, manufacturer, supplier, installer or user of the devices which they inspect, nor the authorized representative of any of these persons. They may not be directly involved in

the design, construction, marketing or maintenance of the devices, nor represent the parties engaged in these activities. This in no way precludes the possibility of exchanges of technical information between the manufacturer and the body.

2. The notified body and its staff must carry out the assessment and verification operations with the highest degree of professional integrity and the requisite competence in the field of medical devices and must be free from all pressures and inducements, particularly financial, which might influence their judgement or the results of the inspection, especially from persons or groups of persons with an interest in the results of the verifications.

Should the notified body subcontract specific tasks connected with the establishment and verification of the facts, it must first ensure that the subcontractor meets the provisions of the Directive and, in particular, of this Annex. The notified body shall keep at the disposal of the national authorities the relevant documents assessing the subcontractor's qualifications and the work carried out by the subcontractor under this Directive.

3. The notified body must be able to carry out all the tasks assigned to such bodies by one of Annexes II to VI and for which it has been notified, whether these tasks are carried out by the body itself or on its responsibility. In particular, it must have the necessary staff and possess the facilities needed to perform properly the technical and administrative tasks entailed in assessment and verification. It must also have access to the equipment necessary for the verifications required.

4. The notified body must have:

-- sound vocational training covering all the assessment and verification operations for which the body has been designated,
-- satisfactory knowledge of the rules on the inspections which they carry out and adequate experience of such inspections,
-- the ability required to draw up the certificates, records and reports to demonstrate that the inspections have been carried out.

5. The impartiality of the notified body must be guaranteed. Their remuneration must not depend on the number of inspections carried out, nor on the results of the inspections.

6. The body must take out civil liability insurance, unless liability is assumed by the State under domestic legislation or the Member State itself carries out the inspections directly.

7. The staff of the notified body are bound to observe professional secrecy with regard to all information gained in the course of their duties (except *vis-à-vis* the competent administrative authorities of the State in which their activities are carried out) pursuant to this Directive or any provision of national law putting it into effect.

COMMISSION OF THE EUROPEAN COMMUNITIES
CPMP Working Party on Efficacy of Medicinal Products
Note For Guidance

Good Clinical Practice for Trials on Medicinal Products in the European Community
July 13, 1990

FOREWORD

This document should be read and interpreted in the light of Directive 65/65/EEC and 75/318/EEC.

The objective of this guideline is to establish the principles for the standard of Good Clinical Practice for trials on medicinal products in human beings within the EEC. It is directed primarily towards the pharmaceutical industry, but also to all who are involved in the generation of clinical data for inclusion in regulatory submissions for medicinal products. These principles are pertinent to all four phases of clinical investigation of medicinal products, including bioavailability and bioequivalence studies and can be applied more widely by those who undertake experimental investigation in human subjects.

All parties involved in the evaluation of medicinal products share the responsibility of accepting and working according to such standards in mutual trust and confidence. **Pre-established, systematic written procedures for the organization, conduct, data collection, documentation and verification of clinical trials are necessary to ensure that the rights and integrity of the trial subjects are thoroughly protected and to establish the credibility of data and to improve the ethical, scientific and technical quality of trials. These procedures also include good statistical design as an essential prerequisite for credibility of data and moreover, it is unethical to enlist the cooperation of human subjects in trials which are not adequately designed. By these means all data, information and documents may be confirmed as being properly generated, recorded and reported.**

GLOSSARY

Audit (of a trial)
A comparison of Raw Data and associated records with the Interim or Final Report in order to determine whether the Raw Data have been accurately reported, to determine whether testing was carried out in accordance with the Protocol and Standard Operating Procedures (SOP), to obtain additional information not provided in the Final Report, and to establish whether practices were employed in the development of data that would impair their validity.

Audits must be conducted either through an internal facility at the sponsor but independent of the units responsible for clinical research, or through an external contractor.

Case Report Form (CRF)
A record of the data and other information on each subject in a trial as defined by the protocol. **The data may be recorded on any medium, including magnetic and optical carriers, provided that there is assurance of accurate input and presentation, and allows verification.**

Documentation
All records in any form (including documents, magnetic and optical records) describing methods and conduct of the trail, factors affecting the trial and the action taken. These include protocol, copies of submissions and approvals from the authorities and the Ethics Committee, investigator(s)' curriculum vitae, consent forms, monitor reports, audit certificates, relevant letters, reference ranges, raw data, completed CRF and the Final Report.

Good Clinical Practice (GCP)
A standard by which clinical trials are designed, implemented and reported so that there is public assurance that the data are credible, and that the rights, integrity and confidentiality of subjects are protected.

Good Manufacturing Practice (GMP)
The part of the pharmaceutical quality assurance which ensures that products are consistently produced and controlled to the quality standards appropriate for their intended use and as required by the product specification.

Quality Assurance
Systems and processes established to ensure that the trial is performed and the data are generated in compliance with Good Clinical Practice including procedures for ethical conduct, SOPs, reporting, personal qualifications, etc.

This is validated through in-process quality control and in- and post-process auditing, both being applied to the clinical trial process as well as to the data.

Personnel involved in Quality Assurance audit activities must be independent

of those involved in or managing a particular trial.

Quality Control

The operational techniques and activities undertaken within the system of Quality Assurance to verify that the requirements for quality of the trial have been fulfilled.

Quality Control activities concern all members of the investigational team, including the staff of the sponsor or CRO involved with planning, conducting, monitoring, evaluating and reporting a trial including data processing, with the objective of avoiding trial subjects being exposed to unnecessary risk, or false conclusions being drawn from unreliable data.

Raw Data

Patient files, original recordings from automated instruments, tracings (ECG, EEG), x-ray films, laboratory notes, etc.

Verification/Validation of Data

The procedures carried out to ensure that the data contained in the final clinical trial report (Final Report) match original observations. These procedures may apply to raw data, hard copy, or electronic CRFs, computer print-outs and statistical analyses and tables.

CHAPTER 3

DATA HANDLING

INVESTIGATOR

3.1 The investigator undertakes to ensure that the observations and findings are recorded correctly and completely in the CRFs and signed.

3.2 Entry to a computerized system is acceptable when controlled as recommended in the EEC guide to GMP.

3.3 If trial data are entered directly into a computer there must always be adequate safeguard to ensure validation including a signed and dated print-out and back-up records. Computerized systems should be validated and a detailed description for their use be produced and kept up-to-date.

3.4 All corrections on a CRF and elsewhere in the hard copy raw data must be made in a way which does not obscure the original entry. The correct data must be inserted with the reason for the correction, dated and initialled by the investigator. For electronic data processing only authorized persons should be able to enter or modify data in the computer and there should be a record of changes and deletions.

3.5 If data are altered during processing, the alteration must be documented and the system validated.

3.6 Laboratory values with normal reference ranges should always be recorded on CRF or attached to it. Values outside a clinically accepted reference range or values that differ importantly from previous values must be evaluated and commented upon by the investigator.

3.7 Data other than those requested by the protocol may appear on the CRF clearly marked as additional findings, and their significance should be described by the investigator.

3.8 Units of measurement must always be stated, and transformation of units must always be indicated and documented.

3.9 The investigator should always make a confidential record to allow the unambiguous identification of each patient.

SPONSOR/MONITOR

3.10 The sponsor must use validated, error-free data processing programmes with adequate user documentation.

3.11 Appropriate measures should be taken by the monitor to avoid overlooking missing data or including logical inconsistencies. If a computer assigns missing values automatically, this should be made clear.

3.12 When electronic data handling systems or remote electronic data entry are employed, SOPs for such systems must be available. Such systems should be designed to allow correction after loading, and the correction must appear in an audit file (cf. 3.4 & 3.16).

3.13 The sponsor must ensure the greatest possible accuracy when transforming data. It should always be possible to compare the data printout with the original observations and findings.

3.14 The sponsor must be able to identify all data entered pertaining to each subject by means of an unambiguous code (cf. 3.9).

3.15 If data are transformed during processing, the transformation must be documented and the method validated.

3.16 The sponsor must maintain a list of persons authorized to make corrections and protect access to the data by appropriate security systems.

ARCHIVING OF DATA

3.17 The investigator must arrange for the retention of the patient identification codes for at least 15 years after the completion or discontinuation of the trial. Patient files and other source data must be kept for the maximum period of time permitted by the hospital, institution or private practice, but not less than 15 years. The sponsor, or subsequent owner, must retain all other documentation pertaining to the trial for the lifetime of the product. Archived data may be held on microfiche or electronic record, provided that a back-up exists and that hard copy can be obtained from it if required.

3.18 The protocol, documentation, approvals and all other documents related to the trial, including certificates that satisfactory audit and inspection procedures have been carried out, must be retained by the sponsor in the Trial Master File.

3.19 Data on AEs must always be included in the Trial Master File.

3.20 The Final Report must be retained by the sponsor, or subsequent owner, for five years beyond the lifetime of his product. Any change of ownership of the data should be documented.

3.21 All data and documents should be made available if requested by relevant authorities.

LANGUAGE

3.22 All written information and other material to be used by patients and paraclinical staff must use language which is clearly understood.

3.23 Competent authorities have agreed to accept CRFs completed in English.

CHAPTER 4

STATISTICS

4.1 Access to biostatistical expertise is necessary before and throughout the entire trial procedure, commencing with designing of the protocol and ending with completion of the Final Report.

4.2 Where and by whom the statistical work shall be carried out should be agreed upon by both the sponsor and the investigator.

EXPERIMENTAL DESIGN

4.3 The scientific integrity of a clinical trial and the credibility of the data produced depend first on the design of the trial. In cases of comparative trials the protocol should, therefore, describe:

a) an "a priori" rationale for the target difference between treatments which the trial is being designed to detect, and the power to detect that difference, taking into account clinical and scientific information and professional judgement on the clinical significance of statistical differences;

b) measures taken to avoid bias, particularly methods of randomisation when relevant.

RANDOMISATION AND BLINDING

4.4 In case of randomisation of subjects the procedure must be documented. Where a sealed code for each individual treatment has been supplied in a blinded, randomised study, it should be kept at the site of the investigation and with the sponsor.

4.5 In case of a blinded trial the protocol must state the conditions for which the code may/must be broken. A system is required enabling access to the treatment of individual subjects in case of an emergency. The system must only permit access to treatment key of one subject at a time. If the code is broken it must be justified in the CRF.

STATISTICAL ANALYSIS

4.6 The type(s) of statistical analyses to be used must be specified in the protocol, and any other subsequent deviations from this plan should be described and justified in the Final Report of the trial. The planning of the analysis and its subsequent execution must be carried out or confirmed by an identified, appropriately qualified and experienced statistician. The possibility and circumstances of interim analyses must also be specified in the protocol.

4.7 The investigator and monitor must ensure that the data are of high quality at the point of collection and the statistician must ensure the integrity of the data during their processing.

4.8 The results of analyses should be presented in a manner likely to facilitate the interpretation of their clinical importance, e.g. by estimates of the magnitude of the treatment effect/difference and confidence intervals, rather than sole reliance on significance testing.

4.9 An account must be made of missing and unused and spurious data during statistical analyses. All omissions of this type must be documented to enable review to be performed.

CHAPTER 5

QUALITY ASSURANCE

5.1 A system of Quality Assurance, including all the elements described in this chapter and the relevant parts of the Glossary, must be employed and implemented by the sponsor.

5.2 All observations and findings should be verifiable. This is particularly important for the credibility of data and to assure that the conclusions presented are correctly derived from the raw data. Verification processes must, therefore, be specified and justified. Statistically controlled sampling may be an acceptable method of data verification in each trial.

5.3 Quality control must be applied to each stage of data handling to ensure that all data are reliable and have been processed correctly.

5.4 Sponsor audit must be conducted by persons/facilities independent of those responsible for the trial.

5.5 Any or all of the recommendations, request or documents addressed in this Guideline may be subject to, and must be available for, an audit through the sponsor or a nominated independent organization and/or competent authorities (inspection).

5.6 Investigational sites, facilities and laboratories, and all data (including source data) and documentation must be available for inspection by competent authorities.

Chapter 6

SITUATIONAL REVIEW BY INDUSTRY

PHARMACEUTICAL INDUSTRY

Background

The Pharmaceutical Industry is almost totally dependent upon computer systems. Computerized laboratory devices test compounds and report to Laboratory Information Management Systems. Clinical studies are computer designed, results are computer recorded, and statistical analyses are computer calculated. Computer Aided Manufacturing Systems use robotic devices to move inventory into computerized manufacturing systems, while other computers test final product for stability and purity.

This reliance is in part a result of the need for complex regulatory documentation of all aspects of the process. But it is also a result of financial pressures to maximize production efficiency in a production process ideally suited to automation. In effect, technology makes it possible; costs make it desirable, and regulation makes it appropriate. The result is the ideal climate for automation.

As production methods grow in complexity and sophistication, and as OSHA and public health safety requirements for workers continue to tighten, pharmaceutical manufacturing will become increasingly reliant on robotic devices for production, testing, and distribution. That reliance will necessitate even greater systems controls, in much the same way that safety concerns have increased regulatory scrutiny of systems in the blood processing industry.

That move toward automation has generated increasing needs to assure control of the computers involved. As a result, the pharmaceutical industry is highly oriented toward validation, the documentation of those controls. All automated aspects of research, laboratory analysis, production, and distribution are subject to validation guidelines. Good Manufacturing Practice (GMP) and Good Laboratory Practice (GLP) inspections focus on a number of areas, but almost always include computer validation as one of their major themes.

History

Demands for evidence of computer validation were first expressed in 1983 and 1984, as an extension of process validation requirements. Early process equipment was generally electric or manual, but as electronic and automated pieces of equipment were introduced most approaches simply "tested around" the computer components. That is, a new mixer now controlled by a computerized programmable logic controller (PLC) was validated by comparing input to output and expectations, ignoring the controller in the middle. By the mid 1980s, though, this "test around" procedure had lost favor in financial auditing, and was being seriously questioned elsewhere. "Black box" testing could tell if a system component was functional at present, but without understanding the inner workings of that system or software there was no certainty of continued performance and no ability to develop appropriate controls to prevent or detect likely future problems.

In 1987, an FDA spokesperson provided the Drug Information Association with a possible timetable, later proven impractical. All systems, it was announced, should be thoroughly tested and validated by 1992, providing a five year window for compliance. Unfortunately, the key parameters and guidelines were not defined until late in 1989, and in 1992 clear definitions of systems to be included were still emerging. It was not until 1994 that the Center for Biologics Evaluation and Research and the Center for Devices and Radiological Health announced a mechanism for registering compliance (using the 510(k) medical device registration, identifying software as a medical device). In 1995, most pharmaceutical companies were still in the process of defining the appropriate validation standards for their operations.

Present

The pharmaceutical industry has been struggling to understand, define, and comply with validation requirements for more than ten years. Currently, acceptance of the need to validate all major computer systems is widespread. Most major pharmaceutical companies have established a policy or Standard Operating Procedure (SOP) for validation that includes a definition of systems to validate; a categorization of those systems with a multi-tier approach to the extend of evidence necessary, based largely upon uniqueness and importance of the systems; and methodology for accomplishing that validation. The standard validation methodology of Dr. Sandy Weinberg, first issued in 1989, has become the general basis for these policies, and for the FDA approach to auditing the results.

The most common validation model that has emerged depends upon the formation of ad hoc task groups with representatives from the user community, the technical support group, and the organization's quality assurance group. These task groups develop SOPs, design and conduct acceptance tests, develop disaster recovery and security plans, and document system and user controls. An outside audit team or internal audit group then periodically reviews the validation effort and certifies compliance to the company and the regulatory authorities.

While all relevant systems have not yet been fully validated, most companies have in place a plan for completing all validations within the next few years. In addition, the validation of new systems as they are introduced is rapidly diminishing the number of non-validated systems currently in use in the pharmaceutical industry.

Future

In the *Federal Register* of June 3, 1994, the FDA announced that as part of a review of regulations for blood establishments and blood products, it would also review other regulations applicable to blood products, 21 CFR 210 and 211 (the Current Good Manufacturing Practice regulations in the manufacture, processing, packing and holding of human and veterinary drugs). Any changes in the GMP would probably follow changes already made to the Medical Device Good Manufacturing Practices, and increase the requirements for system validation.

Whatever the outcome of proposed revisions to the drug GMPs, the future of pharmaceutical validation will include a continuation of the current plans for validation of all relevant systems and the periodic (generally annual) reaudit of validated systems. But this classic approach to validation, critical as it is, is rapidly being supplemented by an important development in the software industry: the inclusion of validation modules in LIMS, MRP, CAM, and other major systems.

An algorithm has been developed, and is currently being integrated into major software, for the industry to perform a continuous statistical co-validation monitoring on all variable data streams. This process involves the detection of unexpected changes in datapoints, in the relationships between data elements, and in the relationships between current and previous elements. The calculations are performed thousands of times per second, and flag discrepancies for investigation and possible diagnostic interpretation. A cybernetic feature allows the system to continuously "learn" from prior functionality, and to feedback that experience to current control situations.

Unlike other automated tools that simply check for consistency in software design, this algorithm identifies problems in the application of the software to specific pharmaceutical regulatory functions. When coupled with a classic validation study to develop appropriate control and use documentation, the validation algorithm provides a high level of confidence in the results of any controlled system.

In the future, this and similar algorithms will provide all software and software driven devices, including medical devices, blood processing systems, and biologics systems, with the ability to supplement classic validation efforts with continuous monitoring of functionality. This combination will meet the original regulatory intent, and will provide the necessary and continuous controls of systems that the high dependency of the pharmaceutical industry requires.

Summary

The Pharmaceutical industry has paid particular attention to the need to develop and document controls for computer systems in use, largely in response to a high dependency upon automation and to the regulatory attention that dependency has generated. Currently, most major companies have developed and are implementing a strategic plan for classic validation of all relevant systems, including internal testing, SOP and user documentation, and security. In the future, an algorithm built into major software systems will supplement these classic validation studies with a continuous monitoring of data streams, using sophisticated statistical analysis to diagnose possible system performance and control problems.

BLOOD PROCESSING INDUSTRY

Background

It is in the Blood Processing industry that regulatory concerns regarding the validation of computer systems have been most clearly focused and most rigorously pursued. This intense scrutiny is a result of an interaction of two vectors. First, the AIDS crisis has heightened public awareness of the potential danger an untested blood supply represents as a carrier of pathogens. Since there is yet to be developed any processing method that effectively sterilizes blood, undetected disease is passed from donor to recipient unimpeded.

Second, since most of the screening tests utilize sophisticated, computerized equipment to detect contaminations, and since virtually all blood processing centers rely upon computer systems to manage the process of evaluating and labeling blood products, the dependency of the blood processing industry upon computer systems is extremely high. In most blood processing centers, the decision to label as safe for transfusion or to destroy as potentially dangerous is placed in the "hands" of a computer system, well beyond conventional human controls.

In most environments, blood donors are initially screened for self-reported disease risks, and a donor profile (health history, donation history, and personal identification) is created. Blood is then "harvested" in bags for processing and in tubes for testing.

The tubes are tested for type and for a series of disease conditions, including hepatitises, syphilis, and AIDS. Simultaneously, the bags are separated into blood products, including packed cells, plasma, and whole blood.

When the testing is completed, the results of the test are married to the created products, which are then labeled for safe use or destroyed.

To accomplish these steps, a computer program with a series of software "switches" is utilized. A donation is assigned a unique number, and all switches are set as "off." The donor profile is evaluated, and the initial switch is left as "off" if a disqualifying factor is identified, or turned "on" if the profile is appropriate.

The donor identification is checked against an extensive donor deferral file, seeking a match with a donor whose blood may have been previously disqualified. If no match is found, the next switch is turned "on;" a match results in a disqualifying "off." Each additional test establishes the disposition profile of the donation, eventually resulting in a file in which all significant "offs" have been changed (safe for use) or in which one or more disqualifications remain, forcing destruction (or, in some circumstances, other appropriate disposition). Finally, the unique donation number is matched to the number on the products, and a label is generated. For all practical purposes, human personnel are not involved in the entire process.

In such an environment, non-conventional secondary human controls, in the form of the computer system controls that validation demands and documents, are critical.

History

In the early 1980s, FDA investigators observed an increased reliance upon computer systems in the processing, testing, and distribution of human blood products. As scattered but highly publicized incidents of inappropriate release of possibly AIDS-contaminated blood reached the public later in the decade, numerous industry conferences attempted to move the industry toward a more extensive and rigorous control over computer systems. The result was a widespread industry shakedown, as small blood processing centers consolidated to reach cost effective levels that would permit the extensive validation controls necessary in such a vulnerable setting.

The remaining centers have been struggling to comply ever since, working to improve systems while developing new alternatives and organizing their operations around new control principles. This reorganization is somewhat complicated by the variety of blood processing operations in place in the United States. Some, like the American Red Cross, are not-for-profit organizations with interests and goals extending well beyond blood processing. These groups generally have highly involved public boards as well as professional staffs, adding to the difficulty of change and of long-term planning for new solutions. Other blood processing centers are for-profit companies, organized much as pharmaceutical companies. These for-profit companies are responsive to regulatory compliance issues and generally can afford the cost of innovation, but they have a historical adversarial relationship with the FDA, and hence await clear and unambiguous guidelines before proceeding. Those guidelines have been long delayed in deference to the planning cycles of the not-for-profits. Finally, a number of small hospitals operate their own blood processing centers. These units fall under joint regulation of the FDA and the voluntary hospital regulatory associations, and hence are often ignored by both groups.

This mix has led to chaos, only now leading to a clear regulatory position and stiff enforcement.

Present

At the Food and Drug Law Institute's annual Education Conference in December 1993, Food and Drug Commissioner Dr. David Kessler delineated his agency's current emphasis on vigilance and scrutiny of the blood processing industry. He has committed the FDA to hold the industry to rigorous Good Manufacturing Practice standards, and announced that the agency would revoke licenses or seek court injunctions against blood centers to ensure compliance with those standards. In addition, he emphasized that CBER would ensure compliance with the "Draft Guideline for the Validation of Blood Establishment Computer Systems" issued in September 1993.

In March of 1994, the FDA sent a letter to all software vendors selling to the blood processing industry. This letter explicitly declared software used in the processing and testing of blood (and, by extension, in other FDA regulatory areas) to be medical devices, subject to the Food, Drug, and Cosmetic Act and the Safe Medical Devices Act of 1990's requirements for establishment registration, product listing, adverse event reporting, premarket approval or premarket notification, and the documentation of software validation. The letter also announced that the FDA would conduct inspections of vendors supplying software to the blood processing industry, and that these inspections "will include, among other things, a review of your standards for software development, testing, validation, and quality assurance. The primary focus of these inspections will be to assess compliance with the CGMP regulations for devices.... [I]f a manufacturer of software products for blood establishments is not making good faith efforts to comply with the Act and FDA's regulations ... the agency will not hesitate to take appropriate steps to bring the firm into compliance."[1] Sixty days after the release of that letter, the FDA began visiting software vendors to enforce those validation guidelines.

Concurrently, the FDA increased the frequency and extent of its visitations to blood processing centers, asking for clear evidence of system controls and validation documentation. Centers unable to comply have been cited for violations of the Good Manufacturing Practice (GMP) regulations, regardless of their ownership or organizational status.

This two-prong thrust, pressuring both the vendors and users of systems, has provided a clear message to the blood processing industry. While full compliance will no doubt require some time as well as expense and effort, there is now industry certainty that only systems that can be documented as validated can be sold, supported, or used. After almost ten years of less-than-subtle FDA signals, the message has finally been received.

Future

In the *Federal Register* of June 3, 1994, the FDA announced that as part of a review of regulations for blood establishments and blood products, it would also review other regulations applicable to blood products, 21 CFR 606 (the Current Good Manufacturing Practice regulations for blood and blood components). Any changes in the GMP would probably follow changes already made to the Medical Device Good Manufacturing Practices, and increase the requirements for system validation.

The increased cost of well designed, controlled, and validated systems will probably result in a further consolidation in the blood processing industry. Small centers, processing ten thousand or fewer bags of blood a year, will have great difficulty demonstrating cost efficiency. Even the largest facilities are likely to have management problems as they try and balance community service priorities, volunteer involvement, and strict federal requirements.

Vendors, too, are likely to decrease in number. Several vendors have already re-evaluated their opportunities, unwilling to invest in one market segment when their same software can be sold in other segments without the overhead and liability that blood regulation implies. Those vendors whose products were equally suited to blood processing and other non-regulated applications are simply withdrawing from the market.

In the not-for-profit environment, discussions are still proceeding. The American Red Cross has publicly announced a plan to adapt a single system for all of its processing sites, while simultaneously reducing the number of those sites. Interim plans, during the development and phase-in of that adapted system, are still hazy.

---

[1]Letter from Dr. Kathryn C. Zoon, Director of the Center for Biologics Evaluation and Research, to Blood Establishment Computer Software Manufacturers, March 31, 1994. See Chapter 2 for the complete text of Dr. Zoon's letter.

Several large, independent blood processors have encouraged the formation of a remote teleconnected central blood systems organization that is offering validated and controlled computerized services to blood facilities all over the country. This "time share" concept may well prove the ideal trend of the future, providing the tight controls necessary in a regulated and critical environment but placing those controls in the hands of experienced and focused computer professionals rather than medical personnel.

Summary

The Blood Processing industry is clearly in transition. It is perhaps poetically appropriate that the validation "leading edge - bleeding edge" is in blood; regardless of the pun potential, the high reliance upon systems and high vulnerability of this industry has caused a sharp regulatory focus. Organizational complexities related to the mixed status of blood processing centers have delayed the implementation and enforcement of the guidelines resulting from that focus, but official patience has been exhausted. Today, and tomorrow, blood processing centers will utilize compliance-validated software selected from 510(k) (medical device Premarket Notification Submission) approved systems, or will be cited for GMP violations.

BIOLOGICS MANUFACTURING

Background

In the regulated pharmaceutical industry, biologics are commonly defined as any substances which are obtained from animal products or other biological sources for use in the treatment or prevention of disease. They are governed by the Good Manufacturing Practice (GMP) Regulations which have been in existence, inclusive of various modifications, since 1963. The current GMP regulations were initially published in 1978 and include subsequent revisions.

As the use of computers dramatically increased to support the demand for automation in the manufacturing of biologic products, those recognized principles and standards used in conjunction with process validation formed the basis for demonstrating evidence of control in the use of computer systems.

History

The regulation of biologics in the U.S. is mandated in the *Code of Federal Regulations* (CFR), Title 21 as follows: Part 210, Current Good Manufacturing Practice in Manufacturing, Processing, Packing, or Holding of Drugs, Subpart A -- General Provisions and Part 211 Current Good Manufacturing Practice for Finished Pharmaceuticals, Subpart A -- General Provisions.

It is apparent that the discipline of process validation became the basis for providing evidence of control with respect to computer systems involved in the manufacture of biologic products. In many cases, the procedures to confirm processing integrity and to provide assurance of computer system functionality and inherent resident data were applied from previously acknowledged standards, practices, and techniques. These methods were used to regulate the use of process control equipment in production operations.

The Division of Field Investigations, Office of Regional Operations, Office of Regulatory Affairs and the Division of Manufacturing and Product Quality, Office of Compliance, Center for Drug Evaluation and Research originally issued the *Guide to Inspection of Bulk Pharmaceutical Chemicals* in April 1984. These reference materials and training aids were intended for use by field investigators; they were revised in February 1987 to include changes submitted by the Pharmaceutical Manufacturers Association (PMA). In May 1987, the FDA published the *General Principles of Process Validation* to cover drugs and medical devices. These principles provided those practices generally accepted by the FDA, along with guidance to assist manufacturers in fulfilling process validation requirements contained in 21 CFR Parts 210 and 211.

Before being marketed in the United States, new biological products must be approved by the FDA. Based on the Safe Medical Devices Act of 1990, the FDA implemented a mechanism for extending regulation to the registration of products that incorporate drug, device, and device/biologic combinations. As part of the approval process, manufacturers of biological products are required to submit to the Center for Biologics Evaluation and Research (CBER) adequate data concerning the safety and efficacy of its products. These submissions often include reports, analyses, tabulations, and case reports.

In July 1990, CBER published a number of considerations with respect to computer-assisted submissions for license applications. These materials provided biologic product and establishment license manufacturers with information regarding computer-assisted submissions. In order to evaluate such computer assisted submissions, CBER instituted a program to acquire practical experience. Procedures were developed leading to generalized acceptance of portions or complete license application submitted to the center on computer-readable media. By using computer technology, it was envisioned that efficiency improvements in the transmission, storage, retrieval, and analysis of data submitted to CBER as part of the development, licensing, and marketing of a biological products could be realized.

CBER issued an updated *Guide to Inspection of Bulk Pharmaceutical Chemicals* in September 1991. Revisions were included that were sought by a number of federal regulatory and industry association sources. The relevant stipulation regarding validation process and control procedures in this document is the following:

- An important factor in the assurance of product quality includes the adequate design and control of manufacturing process. Each step of the manufacturing process must be controlled to the extent necessary to assure that the product meets established specifications. The concept of validation is a key element in assuring that these quality assurance goals are met.

Many guidelines on process validation were made applicable to bulk pharmaceutical chemicals (BPCs) to assure that they were manufactured in accordance with GMPs. Sections in the *Guide to Inspection of Bulk Pharmaceutical Chemicals* became an integral standard practice for computer systems supporting biologics manufacturing. The areas of validation consideration included:

- A formal process change system.
- Standard operating procedures.
- Independence of quality assurance to have responsibility and authority for final approval of process changes.
- Performance and recording of tests and results.
- Calibration and maintenance of equipment.
- Independence of an operational function having the responsibility and authority to reject quality assurance activities including procedure approvals, investigation of product failures, process change approvals, and product record reviews.
- Procedures to assure that the correct label is applied.
- Formalization of a written testing program.
- Determination of a test sampling.
- Documentation of the manufacturing process, including a written description of production records.

As a practical matter, when computers and other sophisticated equipment are employed in the biologic manufacturing process emphasis focuses on:

- Systems and procedures that demonstrate that equipment is in fact performing as intended.
- Checking the calibration of the equipment at appropriate intervals.
- Retention of suitable backup systems.
- Assurance that changes in the program are clearly documented and are made only by authorized personnel.

The objective of the *Biotechnology Inspection Guide, Reference Materials and Training Aids*, issued by the FDA in November 1991, was to provide a basis to determine whether a biologic manufacturer was operating in a state of control and was in compliance with federal regulations and requirements. One important aspect of an inspection was identification of system defect or failure. The inspection approach of reviewing standard operating procedures to assure adequacy of controls became the criterion for performing such regulatory review and audit of biologic manufacturers.

Present

In December 1993, Dr. David Kessler, Commissioner of the FDA, stated that CBER would emphasize compliance with computer system validation guidelines issued earlier in the year. In March 1994, CBER Director Dr. Kathryn Zoon stated in a letter to software vendors that computer system software would be treated as a medical device, a significant statement of policy.

The first draft of the *Application of the Medical Device GMPs to Computerized Devices and Manufacturing Processes, Medical Device GMP Guidance For FDA Investigators* was issued in November 1990 by the FDA Office of Compliance and Surveillance, Division of Compliance Programs. This document outlines requirements which apply to the control of computerized manufacturing and quality assurance systems.

It relates to manufacturers who utilize automated systems for manufacturing, quality assurance, and/or record keeping purposes. It also refers to manufacturers of medical devices that are driven or controlled by software. This document is intended to provide assistance to FDA investigators and to supplement FDA document 84-4191, *Medical Device GMP Guidance for FDA Investigators*. As such, these materials were designed to supplement FDA-issued compliance policy statements and references.

Future

The use of computer systems has integrated into every area of the manufacture of biologics. The reliance on computer systems for automation will continue to increase as computer integrated manufacturing (CIM), electronic batch record (EBR), and automated document management systems are used to their full potential and capability by the pharmaceutical and medical device industries. Predictions can be made that even the notorious Title 21 CFR Part 211 Subpart J Section §211.186 calling for "full signature, handwritten" will be modified to enable acceptance of electronic signatures.

As mandated by Section 351 of the Public Health Service Act; the Federal Food, Drug and Cosmetic Act, Section 502; and 21 CFR Part 600 (licensing of the manufacture of biological products) can only be processed when a completed "Application For Establishment License For Manufacture of Biological Products" form is received. Section nine of the updated Department of Health and Human Services, Public Health Service, Food and Drug Administration, FDA Form 3210 (5/94 Form Approved: OMB No. 0910-0124, Expiration Date: April 30, 1997) specifically requires information relating to computer systems used in the biologic manufacturing process. Questions requiring response are:

A. Describe all processing steps which are computer controlled. Describe each system used in these steps. Include a schematic diagram of the system and the functional statement.
B. Describe any computer system used to control the tracking and/or status of raw materials, in-process materials, or final product.
C. Describe the validation and security of the hardware and software of each system.
D. Indicate where the software was developed and describe procedures for program update.

Section fourteen of the same form provides calibration and validation requirements:

A. For all major systems (except those where validation has been previously described in another section) such as waste disposal, clean-in-place (CIP), steam-in-place (SIP), compressed air, etc., please provide a description of the validation studies performed including the results. Include a table of all validations done and the related protocols and a summary of the results for each major system. Include:

    1. Installation Qualification
    2. Operation Qualification
    3. Performance Qualification
    4. Revalidation

B. For major pieces of equipment (autoclaves, dry heat ovens, lyophilizers, hoods, etc; reference chromatographic columns to Product License Application (PLA) if appropriate) please provide a description of the validation studies performed including the results. Include a table of all validations done and the related protocols and a summary of the results for each piece of equipment. Include:

   1. Installation Qualification
   2. Operation Qualification
   3. Performance Qualification
   4. Revalidation

C. For critical processes such as cleaning of product contact parts and major equipment in product contact, disinfection, product changeover, etc., please describe the validation studies that were performed and the results of each.

Section fifteen describes the records requirements:

A. Describe the relationship between the master production record and each batch production record.
B. Explain preparation of records and sign off authority within the organization.
C. How are records assembled, stored and accessed?
D. Who reviews the records for each lot prior to release of the lot? Describe the method of documenting such reviews.
E. How long are records kept?
F. Describe the records used for distribution. How do these records allow for efficient recall?
G. Describe the complaint file and adverse reaction reporting procedures.
H. Describe audit procedures and when and how trend analyses are performed.

In addition, applicants must provide an overview of the GMP training program for biologics establishment employees.

Summary

Certainly, there is increasing pressure to provide evidence of control for the development, installation, and utilization of computer systems. This pressure has produced an appropriate response in the pharmaceutical, medical device, and biologics industries regulated by the FDA with the result being regulation, field inspection, and impending threat of regulatory response. Since computer systems are increasingly used by manufacturers of biologic products in their production operations, the connectivity between computer systems, regulation, and economic realities becomes inseparable. Clearly, validation becomes a solution to protect the best interests of all parties concerned with biologic manufacture.

Validation reviews of computer systems correlate directly to providing evidence of control, with the ultimate goal of producing a total quality product. The strategic validation protocol, tactical validation plan, and the entire validation process serve as guideposts leading biologic manufacturers away from defect detection and towards defect prevention.

MEDICAL DEVICE INDUSTRY

Background

The Medical Device Industry encompasses a wide range of apparatuses, from the most sophisticated computerized imaging systems through the lowest, technologically mundane bandages. In this universe of devices, validation has had a significant impact in three ways:

- in the approval and control of computerized devices;
- in the control of computerized systems involved in the manufacture and distribution of all medical devices; and
- in the classification of software itself as a medical device.

Medical devices that utilize computers as either control or data systems must submit evidence of the validation of the computer elements of the product as a part of the Pre-Market Approval (PMA) or Premarket Notification (510(k) submission) process. The validation of software, and evidence of the controls built into the code, use, or design is an important review element, and it seems that the scrutiny of that validation evidence is most rigorous when the device is a direct (no professional human intervention) delivery system of medical treatment, when the computer elements are utilized in a critical decision-making capacity, or when the computer serves as the safety or protective mechanism for the device.

Systems involved in the testing, manufacture, labeling, recall, or distribution of medical devices are required to be supported with appropriate levels of validation evidence. The depth of that evidence -- and the rigor with which the requirement is enforced -- is dependent largely upon the risks or hazards associated with the final product and with the contribution made to the safety of that product by the computer system. A system that counts and boxes band-aids may be subject to only minimal controls; a system that checks the sterility of those band-aids would be, appropriately, subject to much more rigorous validation testing.

In recent years, federal regulators have begun to discuss the classification of computer software as a medical device when that software is utilized in the production or testing of regulated products, including biologics, other medical devices, and pharmaceuticals. The theory underlying this classification derives from the broad definition of a medical device as anything utilized in providing medical care, triggering FDA Commissioner David Kessler's famous, if melodramatic, comment that a pencil, used by a physician to record data on a patient's chart, is a medical device subject to FDA regulations. While such a statement is clearly an exaggeration intended for effect, the principle is sound. Current regulation does give the FDA broad interpretative powers, and the extension through those powers to the regulation of software is not absurd. In situations in which that software is deeply ingrained in decision-making processes involving human health and safety, either through diagnostic testing, labeling, manufacture, or clinical analysis, the FDA's regulatory responsibility is clear and appropriate.

History

Validation has been a criterion for medical device registration since the early 1960s. In fact, registration of a medical device is ultimately nothing more than a validation process -- a third-party review of the design, control, and performance aspects of a (medical) system. In the early 1970s, the concept of validation was formalized as an element of process review in the production of medical devices (and pharmaceuticals). Subsequent decades have expanded the concept in depth to the review of software used to control devices and their manufacture.

The key element in the validation of medical device software is the hazard analysis, first introduced in the early 1990s. Evaluating the potential danger (in degree and likelihood) of a system failure, varying levels of control may be acceptable, and hence acceptably tested and documented. The same manufacturing process might be used to test batteries in internal pacemakers and in external electronic thermometers. But a failure of the testing software in the pacemaker case may prove fatal, or may require emergency surgery, while a failure of that software in the thermometer may result in an inoperative diagnostic instrument. Under the hazard analysis umbrella, proof of control is more critical, and would require higher standards of evidence, in the first of these two cases.

With a reorganization of the Center for Devices and Radiological Health (CDRH) at FDA in 1993, and subsequent higher reliance upon PMAs and the 510(k) registration, validation has taken on a much greater importance than was previously assigned. Field investigators are checking the manufacturing of devices, including the validation evidence in support of systems involved in that manufacturing process, while registration officers are examining and reviewing validation evidence with a high level of scrutiny. In many ways, validation concerns, particularly in the review of systems, have emerged as the main issues in the regulation of medical devices.

Present

Currently, the FDA is informing medical device manufacturers of the need to file validation evidence as a part of the 510(k) submission process. Spot inspections of medical devices production facilities utilize the Good Manufacturing Practices (GMPs) as a guideline, and include the requirements for validation of software used in the manufacturing and tracking process as an emphasis of those inspections. In situations in which a computer is the main vehicle for potential product recalls, validation of that computer system is given special scrutiny.

Since other regulatory divisions of FDA are now calling for registration of software used by their regulatees (see the section on the Blood Processing Industry, above), CDRH is expecting an increase in the number of software registrations over the next five years. The Center for Biologics Evaluation and Research (CBER), which regulates the blood processing industry, has been particularly active in this area. GMP inspections of software vendors have been held, and citations of GMP violations have been issued. A notification program informing all software vendors of their registration responsibilities has been introduced, and those registrations are now under receipt and review.

Future

There is always a lag period between an announced FDA interpretation of a regulation and the widespread enforcement of that interpretation. During that period, the FDA and industry publicize the interpretation and the methods of documenting compliance.

When the new interpretation impacts upon an industry segment not previously regulated, that lag time is necessarily lengthened, since affected organizations are not generally monitoring the communication channels normally used by FDA to publicize new requirements. Because the software industry has not been subject to any previous regulation, extension of the GMPs and registration requirements to software used in the manufacture or testing of regulated products will be a slow and gradual process. By 1996 virtually all blood processing software will be registered as medical devices; by 1998, most software used by the pharmaceutical and biologics industries will similarly be in compliance.

To achieve that compliance, some software companies will develop and document their internal testing and control procedures, and register that evidence of compliance with FDA. Still other software companies will withdraw from the market, rather than risk operating in a now-regulated environment. These companies will eventually be replaced by companies with FDA regulatory experience, introducing new and compliant products.

All of this shift, re-introduction, and registration will be managed and defined through the definition of software as a medical device, subject to the PMA and 510(k) requirements.

Summary

Medical devices utilizing software will continue to be regulated; software used in the manufacture of medical devices will be subject to the validation requirements of the GMPs. The real change in medical device validation is an outgrowth of the definition of virtually all software used in the pharmaceutical and biologics industries for manufacture, testing, clinical studies, laboratory management, labeling, tracking, and design as medical devices. This "software as a medical device" position is forcing 510(k) registration, which includes a clear definition of, and requirement for, the validation of the registered software.

VETERINARY PRODUCTS

Background

The majority of drug regulation responsibility belongs to the FDA. New drugs -- both prescription and over-the-counter -- must be approved by the FDA before being marketed. The FDA is responsible for ensuring the safety and effectiveness of drugs and overseeing their manufacture. These responsibilities are carried out under the Federal

Food, Drug and Cosmetic Act. Veterinary products -- such as feeds, pet foods, and veterinary drugs and devices -- come under FDA's jurisdiction. However, the U.S. Department of Agriculture's Animal and Plant Health Inspection Service tests and licenses all animal vaccines and serums.

History

Because of the lower standards used in the 1940s and 1950s, and the subsequent need for improving these standards, the 1962 amendments to the Food, Drug and Cosmetic Act required firms to establish that their products were effective as well as safe.

In response to the need for a systematic approach to evaluating the effectiveness and safety of new veterinary drugs, the New Animal Drug Application (NADA) Approval Process was adopted. The NADA assembles data to provide the Center for Veterinary Medicine (CVM) reviewers a complete rationale for approval of veterinary drugs by defining the ingredients, the manufacturing procedures and quality control checks, the environmental impact of the manufacturing process and use of the drug, the safety and effectiveness of the drug, and a summary for publication which relieves the FDA from the burden of defending decisions made behind closed doors.

The process of approval is a dynamic one, affected by scientific and technical developments, personalities of the regulatory staff, social trends and -- very importantly -- by court decisions in cases where disputes between the FDA and the regulated industry or the public have been decided. In reality, the "approval process," such as it is, never ends, because commitments made in the original NADA and subsequent supplemental filings require continued surveillance and evaluation by regulatory personnel.

Present

Eleven major sections comprise the NADA. Each section assists different FDA personnel to analyze the legal and scientific grounds for ultimate acceptance of a given product.

- *Section One* identifies the submission as an original or supplemental application and includes the applicant's name and address, the active ingredient's generic name, and the proposed product brand name.
- *Section Two*, the Table of Contents and Summary, organizes the NADA in order to introduce the active ingredients in the product and provide a summary of the pharmacological rationale upon which to base the drug review.
- *Section Three*, Label Copy, is the sponsor's proposal for product labeling which would permit the intended user (whether lay person or veterinarian) to determine conditions for which the drug is to be used, the proper route of administration and dosage level, and the frequency of use.
- *Section Four*, Components and Composition. The information provided in this section allows the FDA reviewer to check the batch formula quantitatively against the composition of the final dosage unit and facilitates a quick review of all ingredients, active and inactive, including those removed during the manufacturing process itself, such as water removed during a drying process.

- *Section Five*, the Manufacturing Section, provides crucial information concerning the quality and reliability of manufacturing procedures used in drug formulation and packaging. The sponsor must submit a certificate of compliance with the Good Manufacturing Practice (GMP) regulations codified under 21 CFR 210 and 211. This is usually a letter from a responsible company official affirming the sponsor's compliance with the regulations.
- *Section Six*, the Samples Section, states that samples of the drug components and/or the finished drug product may be requested.
- *Section Seven*, the Analytical Methods for Residues Section, requires the firm to specify a method for determining the quantity and identity of the drug, its impurities, and its metabolites present in edible tissue of treated food animals.
- *Section Eight*, the Safety and Effectiveness Section, comprises evidence demonstrating that the drug product is safe and effective when administered according to label directions. The data in this section include information regarding laboratory trials, chronic and acute toxicity tests, and clinical or field observations of the drug.
- *Section Nine*, Good Laboratory Practice Compliance, applies to each non-clinical laboratory study contained in the application. In this regard, the NADA should contain either a statement that each study was conducted in compliance with the Good Laboratory Practice regulations set forth in 21 CFR 58, or, if the study was not conducted in compliance with those regulations, a statement that describes in detail all differences between the practices used in the study and those required in the regulations.
- *Section Ten*, the Environmental Assessment (EA), details the probable and potential environmental impact of all aspects of the manufacture, distribution, administration, metabolism and excretion of the product. The preparation of an EA is an involved process, but in many cases some portion of the requirement can be waived under an exemption.
- *Section Eleven*, the Freedom of Information (FOI) Summary, completes the NADA. The purpose of this section is to provide as much information to the general public concerning the approval of new animal drugs without jeopardizing the applicant's proprietary rights. Failure to submit an adequate FOI Summary constitutes grounds for refusal to approve a pending NADA.

What happens after the separate NADA sections are assembled is even more important to the future of the new animal drug product. Once the sponsoring firm has satisfied itself that the NADA is adequate, the file is forwarded in triplicate to CVM. The three copies are logged into the Document Control Unit, a large storage and dispersal library, where they are assigned an NADA number. This six-digit number permanently identifies the newly-submitted file. The NADA is then routed to the Office of New Animal Drug Evaluation, where it is given to the proper CVM statisticians and others who are involved in various aspects of the review process, depending on the nature of the drug and whether it is intended for use in food-producing species. This process takes about 180 days. If acceptable, the NADA is deemed approved when either a regulation to that effect is officially published in the *Federal Register* or when the sponsor firm receives an approval letter from the FDA.

One or more of the 425-plus discrete chemical entities approved for veterinary use by the CVM is incorporated into each of the 1,350 approved NADAs which exist today. Many others have emerged long enough to be published as approved and then have been withdrawn. In reality, the "approval Process" never ends, because commitments made in the original NADA and subsequent supplemental filings require continued surveillance and evaluation by regulatory personnel.

Future

Currently, any documentation in support of the validation of computer equipment critical to the manufacture or handling of a veterinary drug product is submitted as part of the response in Section Five (the Manufacturing Section) of the NADA. This situation may change in the near future, however. If the proposed changes to 21 CFR 210 and 211 (see discussion above, in the Pharmaceutical Industry section) include significantly increased requirements for validation of computer systems, this will also affect veterinary product manufacturers.

Although there are currently no requirements for medical devices manufactured for veterinary use to comply with medical device GMPs, this situation, too, may be undergoing changes at the FDA. A recent (4/7/94) warning letter to a veterinary device manufacturer recommended "that veterinary devices be manufactured in a state of the art facility using the device GMP as guidance."

ENVIRONMENTAL LABORATORY

Background

Laboratories that provide data to the Environmental Protection Agency (EPA) are governed by the jurisdiction, requirements, and specifications of the EPA program(s) which they support. Consequently, standards for data management procedures concerning automated equipment, system and application software, and associated operating environments which maintain these program(s) are included under the scope of the agency's authority.

History

Certain EPA programs adopted and required laboratories to follow Good Laboratory Practice (GLP) regulations to provide confidence regarding the integrity of laboratory-generated data produced through manual operations, procedures, and practices. Over the years, as the GLP guidelines were implemented, they became accepted by agency and independent contract laboratories as well as by private industry. These groups submitted information and data to the EPA. Based on the results of proven field experience, these standards became, and continue to represent, the expectations and requirements for laboratory management and performance. The EPA's authority to impose controls over laboratories dealing with the agency was initially stipulated in the following federal legislation and published documents:

- Computer Security Act of 1987. Public Law 100-235, January 8, 1988.
- Environmental Protection Agency System Design and Development Guidance. OIRM 87-02, June 1989.
- Environmental Protection Agency Data Standards for Electronic Transmission of Laboratory Measurement Results (EPA Order 2180.2,12/10/87).
- Environmental Protection Agency Security Manual for Personal Computers, December 1989.
- Toxic Substances Control Act (TSCA); Good Laboratory Practices. 40 CFR part 792.
- Federal Fungicide, Insecticide and Rodenticide Act (FIFRA); Good Laboratory Practices. 40 CFR Part 160.
- Findings of EPA's Electronic Reporting Standards Workgroup.

Present

As the use of computer power increased and replaced manual laboratory operations, the EPA had neither the standards to guide laboratories nor the means to provide direction for agency auditors and inspectors. Findings from investigations by the EPA's Office of Inspector General and Office of Information Resources Management (OIRM) raised concerns regarding the integrity of data received by the EPA and highlighted the need for standard data-management practices. The end-result was the *Good Automated Laboratory Practices*, published in draft by the United States Environmental Protection Agency, Office of Administration and Resources Management, on December 28, 1990. This two-part document quickly became known by the acronym GALPs.

The GALPs is a guidance document providing minimum standards, practices, and procedures for Laboratory Information Management (LIMS) and Laboratory Data (LDS) computer systems used to collect, analyze, process, transmit, report, summarize, maintain, and/or manipulate data for EPA health and/or environmental programs. They provide specifications regarding the demonstration of evidence of control in 14 areas of validation consideration: Personnel, Laboratory Management, Responsible Person, Quality Assurance, Facilities, Equipment, Security, Standard Operating Procedures, Software, Data Entry, Raw Data, Records and Archives, Reporting, and Comprehensive Ongoing Testing. Recommendations are presented to ensure data integrity in automated laboratory operations, and an implementation guidance serves as a tool for assuring successful and cost-effective compliance with the GALPs.

The topical areas of coverage of GALP validation acceptance criteria fall into distinct categories. When taken as a composite, they define requirements that would enable laboratories to demonstrate control over activities which develop, maintain, and support computerized environments. Standards in the GALPs are based upon a foundation of the following six concepts for computer systems which should provide:

- A method of assuring the integrity of all data.

- Confidence that the formulae and decision algorithms employed by a system are accurate and appropriate.
- An audit trail, which allows the tracking of data entry and modification to the responsible individual.
- A consistent and appropriate change control procedure capable of tracking system operation and application software.
- A requirement for establishment of and compliance to standard operating procedures.
- For alternate plans of recovery operation in the event of system outage or failure.

Together, these concepts specify the evidence of control defined by the GALPs.

Recognizing that there is an ever-growing availability of a wide variety of computer system design, technology, laboratory purpose, and application functionality, the GALPs were formulated as general guidelines to enable the evolution of alternative control strategies which could not have been anticipated by the original guideline standards. Therefore, the GALPs are intended to serve as guideposts in the interpretation of the standards and as evaluation criteria in considering proposed and warranted forthcoming equivalencies. Until the EPA issues a final version of the *Good Automated Laboratory Practices*, the draft GALPs will continue to serve as the most comprehensively organized set of validation standards for computer systems used by environmental laboratories.

Future

Regarding implementation, adoption of the GALPs is dependent solely on the requirements of the EPA programs under which participating laboratories fall. Furthermore, compliance to the GALPs is also contingent upon consideration of the jurisdiction and by the specifications of the governing EPA program. In those instances where the GALPs are not requirements of a particular EPA program, compliance is voluntary. It is expected that the EPA will complete its plans to:

- Publish official responses to its request for comments on the GALPs.
- Finalize the draft version of the Good Automated Laboratory Practices.
- Develop and issue a standardized checklist to be used for audit and inspection of systems governed by the GALPs.

Summary

Since laboratories are relying increasingly upon computer systems to provide a wide range of operational services, situations have occurred in which the manual GLPs provide little or ambiguous guidance. To avoid confusion and potential problems, and to deal with the newly arising problems associated with computerized systems, the Environmental Protection Agency developed a series of supplemental guidelines called the Good Automated Laboratory Practices (GALPs). In essence, the GALPs provide the elements to be included in a validation strategic protocol and tactical program. These supplemental guidelines are intended to provide clarification of expectations of performance and control for laboratories electing to utilize computer systems.

The "bottom line" issue in expanding GLP concepts to encompass the GALPs in environmental laboratories using computer systems is that of control. Effective management and confidence in these environments cannot be assured unless the design, use, and operation of systems are based on a consistent set of standards.

Chapter 7

SITUATIONAL REVIEW IN THE U.S.A.

OVERVIEW

There are two kinds of industries in the United States -- those that are regulated, and wish they weren't; and those that aren't regulated, and wish they were. Regulation adds costly overhead and removes critical management decisions out of the hands of company officers. But in a highly litigious society, regulation can be a valuable tool in limiting liability and in defining the extent of protective actions that are necessary and appropriate. In effect, regulation can serve as an operationalization of industry standards that provide an umbrella of protection from a variety of lawsuits.

Yet regulated industries complain loudly about the interference and restrictions that regulation represents. In the pharmaceutical industry, for example, the PMA (Pharmaceutical Manufacturers Association), launched a 1993 advertising campaign to counter claims of unfair profits by blaming the Food and Drug Administration for the high cost of pharmaceuticals.

The FDA, on the other hand, considers the pharmaceutical industry to be "self-regulated." This "self-regulation" designation is common in the United States. With very few exceptions (largely in the setting of interstate rail transportation rates), the regulatory agencies have asked their industries to set standards of performance. Rather than directly regulating, the agencies simply evaluate and randomly review that self-regulation.

Self-regulation may be necessary as the complexity and number of regulated companies exceeds the financial capacity of the regulatory agency to conduct appropriate evaluations, but this approach has produced two generic criticism of United States regulatory agencies.

These criticisms are widespread, although stating them explicitly may surprise the critics themselves.

First, the United States regulatory agencies -- the Food and Drug Administration, the Environmental Protection Agency, the Nuclear Regulatory Commission, the Department of Energy, and other agencies -- have been criticized as being insufficiently bureaucratic. The critics, of course, would deny this characterization. Stripped of its pejorative

pejorative meaning, however, bureaucracy refers to a clear, explicit, standardized code of requirements applied uniformly. The Social Security Administration, for example, is highly bureaucratic. When that agency is functioning as designed, an applicant in any part of the United States would receive the same disposition on an application of disability benefits regardless of the state of application, the reviewer, or the date of application. Employees of the SSA have no discretionary power; the rules they follow are explicit, rigid, and uniform.

In direct contrast, the FDA uses vague standards of conduct enforced by investigators with wide latitude. The Good Manufacturing Practices, Good Laboratory Practices, and Good Clinical Practices are frameworks, rather than blueprints. In inspections and application reviews, FDA professionals are expected to use their education, training, and experience to apply these general guidelines to specific situations that vary greatly. There is no assumption that two investigators will come to identical conclusions, or that the same investigator will remain inflexible in a variety of environments.

Such flexibility allows for a variety of industry solutions to common problems, permits variation of interpretation in situations of unique circumstances, and allows a "learning process" for both regulators and the regulated. But this flexibility leads to frustration for managers who want clear directives and explicitly defined requirements.

Whether the current situation is good or bad is a matter for debate; the direct relationship between bureaucratic explicitness and inflexibility of application is established. Where United States regulatory agencies become more specific, they are less able to deal with unusual situations and to respond to changing conditions. In the validation arena, flexibility is critical; the frustrating lack of explicit clarification is inevitable. New technologies emerge much more rapidly than new regulation. A guideline can take five years from drafting to publication, including the legally-required comment processes. A memorandum detailing requirements for testing a software feature or hardware device is likely to be obsolete by the time it is published. The result would likely be the kind of anti-effectiveness requirements that haunt OSHA -- requirements that call for obsolete types of fire detectors, for example.

The second regulatory criticism flows directly from the first: That many of the employees of these agencies are inadequately qualified for their decision-making roles. In the absence of explicit requirements, judgment becomes all the more crucial. Too often, that judgement is based upon too little information.

Eroding federal employee benefits, lucrative opportunities in private industry, and decreased budgets for participation in conventions and training programs have all contributed to this problem. No doubt, most regulatory professionals are well motivated, but access to the latest technologies, and to information about those technologies, may be limited. With possible exceptions in the Department of Defense, government is not the cutting edge of innovation.

In the validation area, this problem has been accelerated. Few regulators have first-class, technical backgrounds. All too often, the field investigator is forced to review complex technologies with inadequate support, training, and background. An investigator hired as a chemist or biologist, for example, may find himself or herself in a computer validation review for FDA, EPA, or DOD, staring at indecipherable code or examining test protocols well beyond understanding.

Increasingly, regulators are demanding third-party evidence in support of validation efforts, relying upon outside experts to evaluate the systems at issue. But the criticism is well founded, and it is acknowledged by regulators who, appropriately, see increased training and support as the appropriate remedy.

Criticism of lack of regulatory support and lack of bureaucratic standardization are widespread and pervasive. The simple solution of firm and inflexible standardization of requirements is, like most simple solutions, inappropriate. Both industry and the public mount significant pressures for standardization and clarity of definition, citing rising costs and lower profits as rationales. Weaknesses in regulatory expertise seem to reinforce the argument. Yet standardized approaches to widely varying and non-standardized problems is a recipe for chaos. Despite pressures to speed approval, lower expenses, and clarify requirements, regulatory agencies must jealously protect their flexibility if they are to be responsive in a technologically innovative world. Bureaucracies, staffed by well-directed but inflexible personnel, handle large case loads rapidly and efficiently. But the well-deserved, negative reputation of bureaucracies is an inevitable outgrowth of a lack of flexibility in a changing environment. In the validation area, bureaucratic inflexibility would be disastrous.

THE FOOD AND DRUG ADMINISTRATION

The FDA is facing a number of major problems. Budget limitations are preventing real growth in the number of investigators, even as the number of regulated companies increases every year. Pressures to accelerate the review process are counterbalanced by increased consumer demands (and their law suits) to assure increasingly far-reaching levels of product safety. Simultaneously, the range of products falling under FDA regulation is in an upward spiral.

Regulation of cosmetics increases as treatment claims are used to enhance products. Dentifrices, for example, are laced with chemical additives, forcing reclassification as Over-the-Counter (OTC) drugs. Shampoos and soaps bear bacteria-fighting claims that require clinical studies, review, and evaluation. Medical devices, vitamins, and so-called New Age treatments require further review within already stretched budgets.

Faced with these mounting pressures, the FDA has subtly adopted four coping tactics:

- User Fees
- Deferred Costs
- Cooperative Agreements
- Selective Targeting

Together, these four tactics are proving successful, even as they rewrite the nature of FDA regulation.

User Fees, the most overt of the new tactics, calls for direct agency billing of regulated industries for identified services. In effect, preferential scheduling is available through a complex fee formula. The user fee concept, first introduced for the pharmaceutical industry in 1992 -- and currently being looked at for medical devices and biologics regulations as well -- carries the promise of accelerated action industry-wide, although that promise has been somewhat subjugated as demand has increased in excess of the services the FDA is legally required to provide.

Deferred Costs are indirectly billed to regulated industries that, under this tactic, are required to provide reports to the FDA which, previously, FDA would have generated itself. For example, in the early 1980s, FDA investigators routinely made validation inspections and cited companies for failures in design, manufacturing, or delivery. Today, companies are required to conduct their own validation studies, generally through the services of outside, independent consulting companies, and to submit those validation reports to the FDA. FDA investigators then review these reports, rather than the system or process. The costs of the investigation have been deferred from the agency to the regulated company.

Cooperative Agreements, still in their infancy, allow the FDA to accept investigations by other international agencies. Some agreements with Swiss, EU, and Swedish agencies have eased the process for companies importing certain kinds of products, although this trend is still slow and cumbersome. While future agreements will permit further cooperation, counter pressures to restrict trade and favor domestic companies may offset any gains this tactic offers.

Finally, *Selective Targeting* opens FDA to the criticism that its technique for randomly investigating companies is, in reality, far from random. Instead, anecdotal evidence seems to indicate that regulators tend to concentrate on a single, high-profile company for an in-depth investigation of a particular issue. Presumably, the publicity generated by that investigation pressures the rest of the industry into compliance. While this tactic may raise some fairness issues, it is an effective method of regulating (albeit by press pressure) in an era of budgetary limitations.

These tactics, and the pressures to which they respond, combine to define an agency under great industry and legislative pressure, yet expanding its power base even as that expansion threatens its effectiveness. Under several current health care reform proposals, the FDA will increase its interest in hospitals and medical treatment. Recent revelations suggest that regulation of tobacco will shift from the Bureau of Alcohol, Tobacco, and Firearms to the FDA. And an increasingly vocal vitamin therapy movement will drive increased FDA activity in that area.

Yet as the FDA increases the width of its focus, it will have to use the diffusion tactics described above to maintain itself within the inevitable budgetary limitations of a United States federal deficit environment. The result will be greater FDA concern, but a shallower depth of regulation and an increased reliance on outside assistance, particularly felt in the validation area.

OTHER U.S. REGULATORY AGENCIES

The Environmental Protection Agency first approached the problems of system validation in the early 1990s, particularly in dealing with its own contract testing laboratories. These private laboratories perform compliance testing and forward results to a variety of EPA databases. Concern for the integrity of that data led to the development of the Good

Automated Laboratory Practices (GALPs), a series of guidelines for controlling computer systems.

Today, the EPA is increasingly relying upon computerized data, and it is in the process of designing a series of common databases that can collect, combine, and analyze water, air, and pesticide data from a variety of sources. These computer databases require careful control, bringing validation issues to the forefront of EPA planning.

Unfortunately, the structure of the EPA makes the establishment of a cohesive validation policy a near impossibility. The agency is actually a collection of semi-autonomous programs, each responsible internally for its own standards and policies. The Superfund Program, for example, has different priorities and procedures from Contract Laboratories, which shares little in common with Pesticides, Air Quality, or other EPA programs. While the GALPs stand as guidelines for all EPA activity areas, enforcement is fragmented.

A recently-introduced legislative initiative, tentatively entitled the Laboratory Accreditation Act, is designed to establish uniform standards for all testing facilities utilized by the agency and by industries reporting results to the EPA. The system validation requirements contained within those standards should represent significant progress in establishing uniform system controls in all EPA establishments.

The Department of Energy is actively involved in a massive effort to clean up nuclear weapons production facilities which were originally established in the early 1950s. That effort is beginning with the creation of measurement laboratories, which will be used to assess progress throughout the multi-year program.

In order to assure uniformity and reliability of measurements at the DOE laboratories, the department has decided to incorporate the GALPs into its design standards. Energy laboratories will be required to meet system validation standards for all LIMS and related data collection and analysis software.

The Nuclear Regulatory Commission regularly inspects nuclear power facilities throughout the United States. While most of the current facilities were originally designed with manual analog guages, the current industry trend is to replace these original systms with computerized measurement and control devices. In this increasingly automated environment, the NRC has requested, and is incorporating, the System Validation Standards developed by Weinberg, Spelton & Sax, Inc. (the basis for the EPA's GALPs) into its internal guidelines. In much the same way, regulators from the department of Agriculture and the Department of Defense are incorporating the standards developed for the pharmaceutical and health-care industries into the validation requirements for laboratory and control systems under their authority.

STATE OF THE ART

A variety of new guidelines, regulations, and standards are currently in release or preparation, further defining and narrowing the responsiblities of system vendors, uders, and developers. These directives are both a result of increasing regulatory emphasis upon validation in field investigations or NDA/IND reviews and have been anticipated by investigators to cybernetically increase that scrutiny.

The new clarifications fall in the upper two sectors of the regulatory pyramid illustrated below, most commonly in the middle, or "guideline" sections. But they are not only potentially valuable as interpretations and clarifications of necessarily vague requirements. These new interpretations are critical for the development of specific, in-house standard operating procedures and policies. The interpretive guidelines serve as

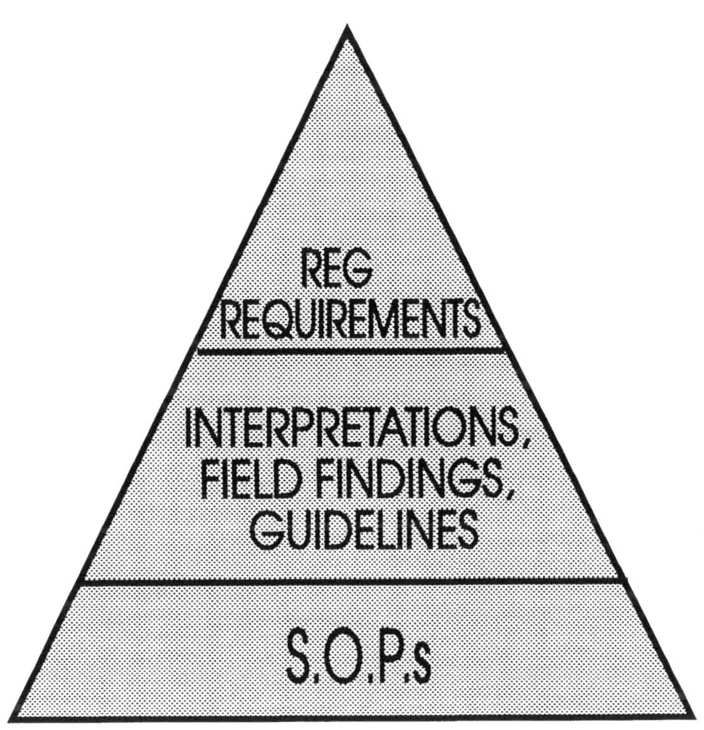

The Regulatory Pyramid

the template for organizational policy review and as the assurance that those internal polcies are compliant with regulatory interpretations.

Four new guidelines are emerging:

- The latest revision of the ISO 9000 series
- The finalized Good Automated Laboratory Practices
- Clarifications of requirements for OTC dental products
- Medical device (diagnostics) requirements

ISO 9000

The ISO 9000 quality guidelines have never had the impact on the pharmaceutical-related industries that they have commanded in other economic sectors. The major U.S. automobile manufacturers seek ISO 9000 compliance from all of their suppliers. Heavy manufacturing industries use ISO 9000 as a baseline for performance assurances. In agriculture, ISO certification lends tacit credibility to seed quality and product purity.

But to heavily regulated industries, ISO 9000 represents a sub-level standard. Because GMP and GLP requirements are more rigorous than ISO 9000 guidelines, certification seems like additional paperwork without additional value or credibility. While this position is sound, there is a subtle shift in the wind that is making ISO 9000 an increasingly valuable tool in the pharmaceutical industries.

FDA has had a long-standing concern with the production environment. While products, not sites, are licensed, the GMPs prescribe site conditions -- cleanliness, organization, control -- as requisites of product approval. Attempts to reconcile differences between the GMPs and ISO 9000 have stumbled over this site vs. product inconsistency, but the realization that the ISO guidelines can help to clarify a significant portion of the GMPs has given a new impetus to harmonization.

Perhaps coincidentally, the fact that the sections of the GMPs taht ISO clarifies are transferable from product to product -- that is, the same quality facility may support a number of licensed products -- makes ISO certification cost effective, where it otherwise would not be. An ISO 9000 review costs upwards of $30,000 to $50,000. That investment would not survive a cost/benefit analysis for a single product line. However, if a facility is used to manufacture 10 products, the $5,000 investment (1/10 the cost of the review) is a realistic and cost-effective method of assuring facility compliance with GMP requriements.

For facilities investing in new equipment, of course, ISO 9000 certification of suppliers provides the same baseline quality assurances that any manufacturer seeks. A coater that comes from a registered company may not be any better than a non-registered piece of equipment, but the assurances of quality are established.

ISO 9000 guidelines are currently in final revision, with new versions promised for June 1994 (and expected, more realistically, in September). A review of the pre-release drafts suggests that this set of revisions is clarificational rather than substantive, at least as the guidelines affect the pharmaceutical industries. In either their current or the soon-to-be-released form, the ISO 9000 guidelines represent a reasonable and cost-effective tool for self-assessing compliance with site controls of GMP facilities.

GALPs

The EPA began developing the GALPs almost five years ago. These guidelines, providing a practical interpretation of Good Laboratory Practices for laboratories relying increasingly upon the use of computerized instruments, data management systems, and reporting systems, have been available in draft form for almost three years. In the interim, the guidelines have been reviewed, field tested, and found to be useful as both a management tool for the control of laboratories and as a regulatory tool for the assessment of that control.

The final GALP guidelines are scheduled for imminent release (planned for late spring, 1995). Galley proofs have been reviewed and returned to the GPO for printing. Full distribution is expected within the 1995 calendar year.

The revision is not significantly different from the original draft, which has been used by investigators from Superfund, DOE, CLP, and other programs. A number of FDA field people have even been seen using the GALPs as a guideline, despite official FDA refusal to acknowledge the EPA document.

Most importantly, however, the GALPs have proven to be useful for laboratories assessing, organizing, and managing their operations. Although a non-regulated environment would, no doubt, dispense with a number of the documentary requirements the GALPs suggest, the Good Automated Laboratory Practices serve an important function in defining areas of concern and recommending methods of control that are fundamental to any automated scientific arena.

One FDA-regulated, pre-clinical animal laboratory used the GALPs as the defining document for developing their own internal Standard Operating Procedures, cutting their start-up time by an estimated 80%. A non-regulated university research laboratory is using the GALPs as an internal template for quality control. And a major energy (oil and natural gas) company has adopted the GALPs as the basis for a corporate-wide, system validation policy.

With final publication and wide distribution, the Good Automated Laboratory Practices will serve as a valuable guideline in a variety of settings, and as an invaluable self-assessment tool for laboratory managers in regulated environments.

OTC -- Dental

The focus of regulatory interest in validation is constantly broadening, partly in response to real or perceived problems with data and systems, and partly in a "what else" philosophy that expands investigations as more mainstream concerns are routinely satisfied. When investigators find good evidence of organization and control, for example, they begin to devote more of their on-site time and energy to the factors taht precede and assure that control.

One area of new examination by the validation microscope involves that most cross-related of product categories -- dental hygiene tools, dentifrices, and washes. While classification of the ethical drug products is clear, the OTC products vary widely, falling under medical device (toothbrushes), OTC drug (most dentifrices), and even cosmetic (some polishes) categories.

Some clarification is now available, through a notice of preliminary 21 CFR 356 regulations published in the *Federal Register* on February 9, 1994 (Vol. 59, No. 27, p. 6084). While dealing with a variety of important dental topics, the notice includes an indication of the widening of validation requirements to the manufacturers of OTC dentifrice products. By extension, the same policy will utlimately include other border-area products, including anti-bacterial soaps and related detergents.

Simultaneously, the major manufacturers of these product lines have reported increased FDA correspondence and visitations. Clearly, the FDA is following through on its commitment to extend validation requirements to all sectors of its regulated industries; the OTC dentifrice areas are just the latest focus group.

There is good news for these companies, incidentally. Validation compliance has been consistently found to be cost effective. While the regulatory agenda is critical to focusing attention, and to leveling the cost playing field for all competitors, validation can be justified purely on a cost analysis basis. Uncontrolled, inadequately documented, poorly designed, or mismanaged computer systems can represent a huge expense sink. Validation requirements are, of course, documentation heavy. But that documentation establishes and assures the proper designs and controls. And while everyone may gripe about the need for carefully-written SOPs and well-organized logs, flow diagrams, and support materials, those gripes turn to sighs of relief when a programmer resigns to pursue a different challenge, a system manager is off on vacation during a critical incident, or a vendor decides to no longer support an important software program.

Medical Devices

While the organizational shakedown in the FDA's medical devices offices is still not complete, an important validation trend is emerging. International consensus has been reached: software, particularly software used in control and decision making with the pharmaceutical industries, out to be regulated as a medical device.

The concept is an expansion of regulation of diagnostic devices. A blood analyzer, for example, is clearly a medical device. It assists in classification and treatment; it is decision based; and its functionality has an associated human health risk factor. All of these elements are important in the definition of medical devices, although the FDA tends to err on the side of a very loose classification. As a top FDA spokesperson half-jokingly stated in a recent address: "If a physician uses a pencil to write a diagnosis on a patient chart, that pencil is a medical device subject to appropriate registration and regulation."

In past years, the FDA field investigators responsible for biologics -- particularly blood -- have been most aggressive in demanding documentation of software as medical devices. Biologics people have visited software companies involved in the design and coding of data management systems and of tracking systems used in blood banks, requesting evidence of compliance with medical device GMPs. By widespread statement and recent -- if thus far, narrow -- action, other FDA groups have expanded the same concept to other critical software systems, including Laboratory Information Management Systems (LIMS), Material Requirements Planning (MRP) systems, lot tracking software, and CIM (manufacturing) systems. In investigator training programs and directives, investigators in at least two of the three largest FDA districts have been instructed to treat all software used in laboratory and manufacturing decision making as medical devices.

This trend is not limited to the United States. At a recent conference in Toronto, a spokesperson for the Canadian Health Protection Bureau (HPB) stated that the Canadian FDA-equivalent would consider all laboratory computer systems to be subject to medical-device regulation. In subsequent conversation, he indicated that his organization's policy had been harmonized to coincide with emergent interpretations by the UK's Department of Health and the Australian Ministry of Health. In a not unrelated announcement, the UK DOH has indicated that they had harmonized their guidelines with BGA in Germany to form the core of the European Union policy.

Should a software company be concerned? Probably not, in the short run. Regulators are unlikely to visit sites of otherwise unregulated industries for many years. Some spot checking, of course, will continue in the interim, but the regulatory concentration will focus on the users rather than developers of software.

On the other hand, anticipating the future regulations, and registering validation evidence for software under the 510(k) provision has three immediate advantages. First, registration can be advertised, providing an inexpensive quality assurance to potential customers. The marketing advantage is as yet unmeasurable, but persuasive. Second, under a ruling by the First Circuit Court of Appeals last year, registration of medical devices provides some kinds of liability immunity. In a litigious society, that advantage can be significant, with saved insurance costs well exceeding registration and validation expenses. Finally, the trend has been to continually tighten the requirements for 510(k) registration, particularly through increasing requests for clinical trials. Arguably, if registration is inevitable, better now rather than later.

SUMMARY

Validation regulation is subject to continual refinement, both in response to emergent clarification of guidelines and to an extension of the environments to which those guidelines apply. While the four areas of change summarized above -- ISO 9000, GALPs, dental OTC products, and medical devices -- are dramatic and significant, they are not unique. Each month new documents are released; every quarter new policies are developed; every year new requirements are announced.

Validation is the foundation of regulation, for it is the evidence upon which all other analyses builds. A toxicology review examines data stored and summarized in an electronic database; a product sampling relies upon the systems that select that sample, the devices that conduct that analysis, and the software that statistically reviews and interprets the results. The validation of the systems underlying the industry provides not just the tools for all other reviews, but the calibration and honing of the edges on those tools.

In such an environment, a clear understanding of validation guidelines, principles, and requirements is critical to maintain a compliant and efficient operation. And that understanding is dependent largely upon a continual updating of the ever-emerging body of validation information.

Chapter 8

INTERNATIONAL SITUATIONAL REVIEW

OVERVIEW

The international situation with regards to validation standards and compliance is in a state of flux. ISO standards are undergoing major revisions. European Union directives are taking effect in only portions of Europe. Asian nations are developing their own standards, or borrowing from everyone else. And former Soviet-influenced portions of the world are both growing markets and suppliers of pharmaceuticals, medical devices, and biological products.

ISO 9000

The ISO 9000 series of voluntary international standards for quality management and quality control were first issued by the International Organization for Standardization in 1987. Since their first publication, these standards have become internationally recognized for their importance in developing quality management practices in a large number of industries.

There are four basic standards in the ISO 9000 series:

- ISO 9001 -- *Quality Systems -- Model for Quality Assurance in Design/Development, Production, Installation and Servicing.* This is the most comprehensive of the ISO 9000 standards. It deals with all the requirements in ISO 9002 and ISO 9003, and also covers design, development, and servicing.

- ISO 9002 -- *Quality Systems -- Model for Quality Assurance in Production and Installation.* This standard addresses installation, procurement and production aspects.
- ISO 9003 -- *Quality Systems -- Model for Quality Assurance in Final Inspection and Test.* Although this is the least comprehensive of the ISO 9000 standards (covering only final inspection and testing in the production process), it is important to system validation because it provides some guidance in applying ISO 9001 standards to software development.
- ISO 9004 -- *Quality Management and Quality System Elements -- Guidelines.* This standard provides guidelines for specific industry application.

The ISO is currently rewriting the ISO 9000 series standards. The revision was planned to be released in the Fall of 1994, but their release has been delayed until early in 1995. Although the revision will supposedly have more specific systems validation guidance, elements of the currently-applied standards are still useful for quality management of automated equipment and software.

Elements of ISO 9001

4.1 Management Responsibility

- Quality policy
  - -- Defined and documented by management
  - -- Includes objectives
  - -- Appears at all levels of the organization
- Worker authority and responsibility
  - -- For all whose work affects quality
  - -- Includes those who verify quality
- Verification of resources and personnel
  - -- Verification requirements identified
  - -- Adequate resources assigned
  - -- Includes those who audit the quality system
  - -- Independent of the line organization
- Management review
  - -- Representative responsible for ISO 9000 requirements assigned
  - -- Quality system and audits reviewed and documented

4.2 Quality System

- Documented procedures
- Effectively implemented
  - -- Quality plan
  - -- Quality manual
  - -- Clarification of standards of acceptance
  - -- Quality records planned

4.3 Contract Review

- Requirements adequately defined
- Differences with tender offer resolved
- Capabilities exist
- Documentation of review

4.4 Design Control

- Planned responsibilities for all design and development activities
- Interfaces between groups documented
- Documented design input requirements
    - -- Meet design input requirements
    - -- Embody acceptance criteria
    - -- Conform to regulatory requirements
    - -- Identify design characteristics crucial to safe and proper functioning of the product
- Documented design verification
- Design reviews
- Qualification tests and demonstrations
- Alternative calculations
- Comparison with any available proven design(s)
- Design changes
    - -- Documented
    - -- Reviewed
    - -- Approved

4.5 Document Control

- Approval and issue
    - -- Procedures exist
    - -- Documents reviewed and approved for adequacy
    - -- Documents available where needed
    - -- Removal of obsolete documents
    - -- Listing maintained of current versions of documents
- Documentation changes
    - -- Reviewed and approved by originators
    - -- Documents reissued whenever changes become confusing

4.6 Purchasing Control

- Subcontractors
    - -- Acceptable ones listed
    - -- Previous performance recorded
- Purchasing data

- -- Clear and complete definition of product
- -- Reviewed and approved before issuance
- Verification of purchased product
  - -- As necessary
  - -- At supplier's site or upon receipt
  - -- Not a substitute for supplier's verification

4.7 Purchaser Supplied Product

- Procedures in place for storage, maintenance, and verification

4.8 Product Identification and Traceability

- Where appropriate, procedures exist
- For traced products
  - -- During all stages of production, deliver, and installation
  - -- Individual items traceable as necessary

4.9 Process Control

- Work instruction documented if lack would impair quality
- Approval of processes and equipment as appropriate
- Workmanship standards are written or in samples
- Special processes are documented
- Appropriate records exist for qualified processes, equipment, and personnel

4.10 Inspection and Testing

- Incoming Product
  - -- Verification procedures exist
  - -- Product held until release
  - -- Product used in emergency will be positively identified
- In-process product
  - -- Procedures exist to inspect, test, and identify
  - -- Product held until release
  - -- Product released in emergency will be positively identified
- Final product
  - -- Procedures exist to inspect and test
  - -- Require passing of all previous tests
  - -- Product held until release
  - -- No emergency product release permitted
- Identification of nonconforming product
- Test records maintained

4.11 Inspection Test Equipment

- Calibration procedures established and checked
- Control, calibrate, and maintain
- Mark equipment to show calibration status
- Calibration records maintained
- Whenever out of calibration, document recent testing
- Safeguard from in-validating adjustments
- Testing hardware and software
  - Proved capable of verifying product acceptability
  - Procedures in place for rechecking
  - Records for evidence of control

4.12 Inspection Test Status

- Identification of conformance or nonconformance to performed tests
- Inspection authority responsible for release identified
- Maintained throughout production, shipping, and installation

4.13 Control of Nonconforming Product

- Procedures to prevent inadvertent use
- Review procedures documented with responsibility defined
- Disposition
  - Authority defined
  - Actions documented
  - Product reinspected when repaired and/or reworked

4.14 Corrective Action

- Procedures documented
  - Investigate nonconforming product for cause and corrective action
  - Analyze all operations, exceptions, and complaints to eliminate
  - Ensure that corrective actions are taken and are effective
  - Implement necessary changes in procedures

4.15 Handling, Storage, Packaging, and Delivery

- Procedures documented
- Secure storage areas and stock rooms
- Periodically reassess stored product for deterioration
- Quality protected after final inspection(s) and test(s)

4.16 Quality Records

- Procedures to control records for their full life cycle
- Subcontractor records included
- Keyed to the product involved

- Retention records recorded
- Safeguarded

4.17 Internal Quality Audits

- Procedures documented
- Comprehensive, planned, documented
- Results given to responsible auditee
- Timely corrective action taken by auditee

4.18 Training

- Procedures for
  - -- Identifying training needs
  - -- Providing training
  - -- Qualifying those performing specific assigned tasks
  - -- Maintaining training records

4.19 Servicing

- Procedures for performance and verifying requirements are met

4.20 Statistical Techniques

- Procedures for identifying adequacy for verifying
  - -- Process capability
  - -- Product characteristics

The chart on the following page compares these elements as they are included ISO 9001, ISO 9002, and ISO 9003:

THE WORLD VIEW

Many nations have not attempted to develop their own regulations regarding systems validation in the manufacturing of drugs and medical devices, but have instead relied on world-wide or regional Good Manufacturing Practice requirements -- primarily those developed by the World Health Organization (WHO) -- or regional requirements such as those established by the by the Pharmaceutical Inspection Convention (PIC) in Europe or the Association of South-East Asian Nations (ASEAN).

The World Health Organization's Good Manufacturing Practice requirements do not specifically address system validation. The PIC's "Guide to Good Manufacturing Practice for Pharmaceutical Products," however, generally defines validation as an assurance that manufacturing procedures, processes, equipment, material, and systems lead to the results expected of them. Under PIC guidelines, validation responsibility is vested in a firm's quality control unit. In relation to documentation, the integrity of data processed by automated equipment must be assured through detailed procedures to assure their accuracy, security, and change control.

Requirement	ISO 9001	ISO 9002	ISO 9003
Management Responsibility	■	●	○
Quality System	■	■	○
Contract Review	■	■	
Design Control	■		
Document Control	■	■	○
Purchasing	■	■	
Purchaser-Supplied Product	■	■	
Product Identification and Traceability	■	■	○
Process Control	■	■	
Inspection and Testing	■	■	○
Inspection Test Equipment	■	■	○
Inspection Test Status	■	■	○
Control of Nonconforming Product	■	■	○
Corrective Action	■	■	
Handling, Storage, Packaging and Delivery	■	■	○
Quality Records	■	■	○
Internal Quality Audits	■	●	
Training	■	●	○
Servicing	■		
Statistical Techniques	■	■	○

● Requirements less stringent than in ISO 9001
○ Requirements less stringent than in ISO 9002

LOCAL REQUIREMENTS

There is a three-tier system at work in the world when it comes to GMP, GLP, and GCP requirements. Locally-issued regulations, if there are any, take precedence. If there are no local regulations, regional requirements, such as those issued by the PIC or ASEAN, are followed. In those areas where there are neither local nor regional requirements, international requirements, such as those of the WHO, are applied.

Many national Good Manufacturing Practice (GMP), Good Laboratory Practice (GLP), and Good Clinical Practice (GCP) standards, guidelines, and regulations do not specifically address systems validation requirements. However, such requirements can generally be considered necessary under other aspects of these requirements. For example, requirements for adequate personnel training and experience in manufacturing or laboratory equipment, equipment operation and maintenance, quality assurance/quality control activities, documentation of procedures, and record retention would all include appropriate actions when automated equipment is used in place of manual operations.

The following discussion will examine some of the more significant, specific national requirements for system validation and their application in GMPs, GLPs, and GCPs.

Japan

The Koseisho is the Japanese government's Ministry of Health and Welfare responsible for drug and medical device registration, with responsibilities similar to those of the FDA in the United States. Japan is in the forefront of establishing guidelines for computer system validation in GLPs and GMPs for pharmaceutical products and medical devices.

Japan's "GLP Inspection of Computer System" covers eight basic points:

1. Computer Hardware and Computer Room

   Computers must be located in an appropriate, controlled environment, where authorized access can be regulated and data adequately archived.

2. Development of Computer Systems

   Computer systems must be developed with appropriate design and with adequate functioning capacity.

3. Software Developed in-house

   Written development and programming standards must be approved and followed. Functional specifications must be met. Testing documents must be reviewed, approved, and retained. Validation testing must be followed and documented.

4. Vendor-Supplied Systems

System specifications must be documented. Vendors must provide confirmation that systems were developed, reviewed, and approved appropriately. Vendors must have established change-control procedures. Vendors must supply development documentation as a validation package in order to evaluate the system's reliability.

5. Existing Systems

The system's historical documentation must be reviewed to accomplish retrospective validation. Testing must be reviewed and approved to confirm the system's reliability. Documentation must be maintained as a retrospective validation package.

6. Operation and Maintenance

Computer systems must be operated and maintained to perform their required functions through supervision by a responsible person and established operation and maintenance procedures.

7. GLP Compliance of the Data Directly Recorded in Computers

Data processing must comply with Japan's general GLP standard. Adequate data change control procedures must be in place and followed.

8. Computer System Inspections by the Quality Assurance Unit

Quality assurance unit must inspect the system to assure that its design complies with the GLP standard, and that the system was developed and is operated to function as designed. QA units must have SOPs for inspecting systems, must inspect all developmental phases of the system, and must periodically inspect the system's operation, maintenance, and data reliability.

The Japanese Ministry of Health and Welfare issued a "Guideline on Control of Computerized Systems in Drug Manufacturing" in 1993 for use by pharmaceutical firms governed by GMP regulations. The guideline applies to systems used for manufacturing process control and management, production control (storage and inventory), manufacturing documentation, and quality control. The Guideline covers three areas:

1. System Development

Responsible individuals must be designated to supervise each step of system development. Development must be documented in a System Development Manual which includes development procedures, evaluation procedures, and approval procedures. A development schedule must be established to describe system development objectives, timetables, staffing, and equipment selection. System engineering and program development must be controlled to ensure development of appropriate system specifications. System development tests, installation tests, and operation tests must be conducted according to established procedures and adequately evaluated and approved.

2. Control of System Operation

   Adequate operation procedures must be established for hardware operation, system inspection and maintenance, disaster recovery, and security control. A "system alteration standard code" must be prepared to describe procedures for alteration or amendment of system operation. System hardware must be operated according to appropriate SOPs. Security control procedures must be established to prevent system use by unauthorized individuals. Continued system reliability is ensured by periodic self-inspection.

3. System Documentation

   Pharmaceutical manufacturers must retain documentation relating to system development and operation.

In October 1993, a Study Group on Policies for Medical Devices submitted a "Proposal for Reclassification and Regulation of Medical Devices." One of the recommendations was to expand the scope of medical devices and include computer software in the category of regulated medical devices because of the necessity to standardize its quality, effectiveness and safety.

United Kingdom

In 1989, the GLP Compliance Programme of the U.K.'s Department of Health (DOH) issued "The Application of GLP Principles to Computer Systems" as an Advisory Leaflet to accompany its 1989 GLP Compliance Programme. The guideline outlined how DOH inspectors would examine computer systems in laboratories conducting health and environmental safety studies. Inspectors would focus on three areas:

1. Identification and Functional Definition of Systems

   Inspectors must require definitions for hardware used in data processing, data input, and data output. Information would be gathered on operation management, system users, and QA system monitoring.

2. Control Procedures

   Inspectors will check procedures for developing, installing, maintaining, and documenting systems and examine data security, change control, archiving, and disaster recovery procedures.

3. Evaluation of Effectiveness of Specific Functions

Inspectors will examine function documentation, methods used by laboratories to test system function, and observe specific system operations.

The DOH's "Guide to Good Pharmaceutical Manufacturing Practices" contains a section on Electronic Data Processing to regulate computer systems used in drug manufacturing, storage, and distribution. Specific requirements relating to such systems include:

1. Personnel must have appropriate expertise and training for system use.
2. System descriptions must be maintained and kept up to date.
3. Before placing a system on line it must be tested and its functions confirmed.
4. Adequate, periodic operational checks must be performed.
5. Change control procedures should be well defined.
6. Only authorized users should enter data into a system, and a procedure for system security should be defined.
7. Independent checks should be made on entry of critical data.
8. Archived data should be accessible and checked for accuracy when equipment or software programs are changed.
9. Disaster recovery procedures should be defined and tested.
10. Formal agreements should be made with system vendors to define their responsibilities.

Australia

Australia's Good Manufacturing Practices for pharmaceutical products also contains a section specifically addressing automated processes and procedures. These requirements include:

1. Computer systems must be designed and installed by personnel with appropriate training and expertise; systems must be tested before being put on line.
2. A specifications document must be prepared detailing system objectives, the types of data entered, testing procedures, operations, and maintenance.
3. System users must be appropriately trained and have access to written SOPs.
4. Change-control procedures must be defined.
5. System operation must be periodically checked and corrective actions taken when necessary.
6. Security procedures must be in place to allow access by authorized system users.
7. When critical data is entered, data entry will be checked by a second individual.
8. Written disaster recovery procedures must be tested, and any recovery actions must be documented.
9. Audit trails must exist to confirm the accurate and reliable transfer of data to or from manufacturing or monitoring equipment.

France

France's Good Pharmaceutical Manufacturing Practices contains four specific requirements listed for electronic data processing:

1. Preliminary Study

   Preliminary studies must be undertaken to determine what kind of automated equipment is used and its location. Personnel from the firm's departments which will be concerned with data processing (such as quality control and production), must undertake the study. These individuals, and personnel who will use the system, must have appropriate training.

2. System Implementation

   Systems descriptions, specifying their functions, must be kept up to date. System validation procedures must be implemented to assure reliability of function.

3. System Operations

   Data entry or modification, and any other access to a system must be authorized. A specific security procedure must be established, as well as a validated disaster recovery procedure. Modifications to the system must be authorized, verified, validated, and documented.

4. Batch Records

   Data relating to the manufacture of batches must be verified by responsible individual(s) and must be maintained.

The Netherlands

The Good Manufacturing Practice guidelines of the Netherlands contain a section on Electronic Data Processing which includes the following requirements:

1. Hardware should be tested for reliability, and the test results recorded.
2. Documentation must be maintained for systems used.
3. Software must be validated, and validated again after any modification.
4. There must be a specific procedure for data entry and modification, and data entry must be verified.
5. Operation of computer/equipment combinations must be validated.
6. To prevent data loss due to equipment malfunction, copies of electronic data must be retained in another medium.

Chapter 9

VALIDATION CHECKLIST

OVERVIEW

All systems in U.S. FDA- and and EPA-regulated industries are subject to validation requirements. The "depth" of that validation effort depends on both the degree of evidence demanded of a particular system and the skepticism of the regulatory agencies toward specific products, processes, and procedures.

Maximum depth of validation (based on regulatory skepticism and a demand for a high standard of evidence) is probably necessary for systems related to the following:

- Blood Processing Systems
- Blood Inventory and Labeling Systems
- Biologics Production Systems
- Quality Assurance/Quality Control Laboratories (LIMS)
- MRP Inventory Systems
- Medical Devices used to treat an illness or injury

A medium depth of validation is usually necessary for the following:

- Pharmaceutical Manufacturing Control Systems
- Diagnostic Medical Devices
- Non-QA/QC LIMS
- Tracking and Recall Systems
- Radiological Systems
- Non-MRP Inventory Systems
- General Label and Shipment Systems
- Laboratory Instrumentation

Although we are seeing increased regulatory concern with -- and therefore attention toward -- systems in the following areas, a cursory depth of validation has been seen as necessary for:

- Veterinary Control Systems
- Cosmetic Control Systems
- Food Control Systems

CHECKLISTS

Different checklists can be constructed to reflect regulatory requirements applied by different agencies to different products and different aspects of the development and manufacturing processes. Since many software and hardware components are themselves increasingly seen as medical devices, checklists can often be keyed to those sections of the *Code of Federal Regulations* (CFR) that refer to automated medical device systems.

Validation checklists should be organized around key requirements found most often in the regulations. The following discussion is based on ten such requirements.

Background

Prior to any validation analysis, background issues -- such as system designation, title, limits, or parameters -- should be described. In addition, any relevant regulatory findings or warnings regarding similar systems should be noted.

Quality Assurance

The validation of a computerized system provides quality assurance of that system. Generally, this QA role can be divided into four phases:

- design review and documentation of the system life cycle prior to implementation;
- user acceptance testing upon implementation;
- confirmatory independent validation of the system upon implementation; and
- periodic validation review of the system in operation.

A quality assurance program should have clear quality criteria, an independent audit to assure compliance with those criteria,and documentation of both the criteria and the audit.

Organization

System validation provides evidence of control. That control includes the ability to respond to software and hardware warnings and recalls and to maintain software functionality. System configuration and design represent evidence of effective organization. This configuration and design should include, but is not limited to: inventory of system components, diagrammatic configuration of system hardware, description of modular-level applicational software design, and data descriptions.

Appropriate control of software structure includes auditing the hardware environment, reviewing the application code (source code review), and, where applicable, review of vendor maintenance capability.

Test Program Requirements

Systems used for production, analysis, storage, sorting, interpretation, modelling, or other purposes related to regulated health care and environmental industries should be subject to rigorous, third-party validation testing. That testing should compare performance characteristics to previously defined system performance requirements.

Software testing involves evaluation of conformance to specifications and the ability of the software to perform as intended in both normal (average) and reasonably abnormal environments. Tests may be continuous, quantitative analyses, or they may be qualitative tests of performance conformity to known results conducted on an appropriate sample of software pathways. Data-related testing should confirm appropriate responses to acceptable data as well as appropriate rejection of improperly structured, coded, designed, or received data points.

Personnel Training

All personnel must have adequate training to perform their assigned responsibilities.This requirement includes system users, system administrators, system developers/designers, system support/maintenance personnel, and system operators. Training should include basic GMP/GLP/GCP orientation as applicable and such other skills and activities as may be defined by specific responsibilities.

In cooperation with appropriate systems personnel, training managers should develop a definition of the necessary job skills for each responsible position. As each skill module is presented, the names, dates, and relevant mastery demonstrations should be recorded in a central archive.

Environmental Control

An analysis should be conducted to determine the appropriate operating environment for all major hardware components. This analysis should include the minimally-acceptable storage environment for all appropriate disks, tapes, optical disks, and other media.

Operating and storage environments of systems must be controlled for factors such as temperature, humidity, power standards, electrostatic discharge (ESD), and magnetic interference. An engineering analysis should determine the relevance of these and other potential threats to systems, as well as identify mitigation or elimination strategies for those threats which are relevant to specific situations.

Hazard Analysis

A "hazard analysis" refers to a determination of the potential risk to health and safety in the event of a mis-performance by the system under review. A hazard analysis begins with considering the intended goals or purposes of the system, assesses the dangers if those goals are not achieved or misachieved, and assesses the seriousness and likelihood of that event in the existing or planned system environment.

An effective hazard analysis must consider four factors:

- the possible adverse events (problem areas) if the system does not perform as intended;
- the reasoned or experience-determined likelihood of those adverse events;
- the seriousness of those adverse events; and
- the appropriate level of prevention or response.

Change and Version Control

While re-validation is a periodic event, change is often a continuous and on-going process. A procedure must be in place to track all changes and to maintain a state of control and validation compliance on an on-going basis.

For any given system -- and for any given time frame -- the hardware and software configuration in use should be controlled. That control should include identification of change-control procedures, documentation of those procedures, and appropriate user instructions and/or SOPs for the operation of those procedures.

Security/Audit Analysis

Unauthorized system access should always be controlled through electronic or mechanical means. The integrity of all audit trails, tracing activity, change, and responsible individual(s) should be protected to assure accuracy and reliability.

Security levels should be designed to balance between user hindrance/intimidation and authorized user access. Audit trails must assure the ability to trace the unique authorizing individual making any adjustment, change, or override procedure.

Disaster Prevention/Recovery Analysis

The design of a system to minimize or adequately respond to disaster is critical to system reliability when computer records serve as the primary method of tracking, controlling, or tracing a product or procedure, when manufacturing is dependent upon computer monitors or controls, when computerized laboratory systems are critical to quality testing, or when device performance is dependent upon computer performance. The response to this criticality depends upon an analysis of the likelihood of disaster events, the development of preventive strategies where justified, and the adoption of corrective or recovery actions when necessary.

Disaster recovery and prevention strategies can vary widely, depending on the type of system, the kind of environment, and the degree of reliance on automation. In situations of redundant, free-standing systems SOPs for disaster recovery should be developed, implemented, distributed, and understood by appropriate personnel.

VENDOR INVESTIGATIONS

One of the clear responsibilities in validating a computer system centers around an investigation of the vendor or developer of that system. An investigation of the vendor is necessary to determine the appropriateness of the development process, the adequacy of the code produced, and the extent of the maintenance support available. But as clear as the need for such an investigation may be, the defining extent of that investigation is obscured by fog.

How detailed an investigation is really necessary? Should a formal audit be scheduled? Is a teleconference interview sufficient? Is a site visit required? Should financial aspects of the vendor's stability be considered as relevant supporting issues? What documentation is appropriate?

The answers to these and other questions, of course, depend upon the circumstances. Consistently evaluating those circumstances, however, requires a firm but flexible set of guidelines. Such guidelines are herein proposed to allow consistent and appropriate determination of the proper extent of vendor investigations. These guidelines are designed to be carefully evaluated, modified as situationally appropriate, and eventually incorporated in a company Validation Standard Operating Procedure or working document.

Most importantly, the development or adoption of formal guidelines such as those proposed here will prevent circumstances in which inappropriate criteria, like cost and bother, will be used to determine when and where audits are required. If a visit is important to the credible completion of a validation, the time, cost, and inconvenience of a cross-country trip should not discount that requirement. Clear guidelines will provide regulatory assurance of a rational policy, as well as provide the internal clout that a policy appropriately commands.

The following guideline is intended to provide validators with specific criteria for determining the appropriate depth and extent of vendor investigations. It does NOT apply to circumstances in which a vendor has decided to provide a comprehensive and certified, supporting, third-party validation report to purchasers of a system or software product. In vendor validation situations, a more thorough audit is required, regardless of other factors.

Criteria

1. **Vendor Interview**: A teleconference interview is appropriate when the following four criteria are ALL fulfilled:
   - 1.1 Detail: An interview is appropriate if the project calls for only a general confirmation of procedures and policies. If there are questions about how policies are followed, a visit will be necessary. Use the teleconference interview to request a copy of and confirm the existence (or absence) of a policy.
   - 1.2 Substantiation: An interview is appropriate if there is no question about the "reality" of the vendor or its facility. This substantiation of reputation may be based upon a history of previous site visits, or upon the size, scope, and reputation of the vendor. This variable is intended to differentiate between a major player (DEC, IBM, Waters, etc.) and an unknown vendor ("Joe's Medical Devices and Hair Salon").
   - 1.3 Absence of Critical Issue: An interview is appropriate if the investigation does not involve a critical vendor-related issue.

1.4 Low Hazard: An interview is appropriate if the system under review is evaluated as low hazard.
1.5 Validation Certificate: An interview is appropriate if the system vendor provides a validation certificate and report issued by an independent, reputable, recognized third party.
1.6 Conservative Default: If in doubt about any of the issues identified above, substitute a **Vendor Visit** for an interview.

2. **Vendor Visit**: A brief (one-day), on-site vendor visit is appropriate when any or all of the criteria defined above ar NOT met, but when ALL of the two additional criteria described below are fulfilled:

2.1 Detail: A visit is appropriate if the project calls for a confirmation that development, support, and/or archive procedures are generally followed.
2.2 Medium Hazard: A visit is appropriate if the system under review is evaluated as a low or medium hazard.
2.3 Conservative Default: If in doubt about any of the issues identified above, substitute a **Vendor Audit** for the visit.

3. **Vendor Audit**: An extensive and significant vendor audit, including observation of standard operating practices, a review of randomly sampled code, and testing of control and maintenance procedures, is appropriate whenever the criteria identified above are NOT met. Note that ALL **high hazard** systems will require a vendor audit.

Procedures

All three levels of investigation require the cooperation of the vendor, including appropriate confidentiality safeguards. Each level described below is described in the context of a carefully arranged interaction with a knowledgeable and authorized spokesperson for the vendor.

1. **Interview**: To conduct a teleconference interview of a vendor, follow these general guidelines:

   - Questions: Supply a list of general questions or topics in advance, to allow the vendor representative ample opportunity to determine accurate responses. Use follow-up questions for clarification and confirmation.
   - Documents: During the interview, request copies of (or samples of) relevant documents such as logs, code, or Standard Operating Procedures. These documents should be placed in the working file for later archiving.
   - Records: Either record in writing the general response to each question, or prepare a memo summarizing the conversation in general. Include in the record the time and date of the interview and the name(s) of the representative(s) involved. FAX (or mail) the record to the vendor representative for confirma-

tion (correction or comment), and place a copy of the original record in the working file along with the vendor representative's comments.

2. **Visit and Audit**: Conduct a visit or audit as you would a normal validation investigation, recording all findings in report format and archiving the report.

Summary

Utilize an interview (by teleconference) when all relevant criteria are met. Substitute a general Vendor Site Visit when confirmation of findings is required or other conditions are not met. Substitute a Vendor Audit when a more detailed analysis is required, or when high hazard situations are involved.

In each case, make a careful record of findings and archive the supporting evidence and notes relevant to the investigation.

GALPs CHECKLIST

The intent of the Good Automated Laboratory Practice (GALP) guidelines published by the EPA is to provide standards which encompass all automated equipment, system and application software, and their associated operating environments. Recognizing that each laboratory environment is unique, provisions of the GALPs present general requirements and direction.

With regard to implementation, adoption of the GALPs is solely dependent on the requirements of the EPA program(s) under which the participating laboratory(s) fall. Furthermore, compliance to the GALPs is also dependent on the jurisdiction and specification of each individual EPA program. In those instances where the GALPs are not requirements of a particular EPA program, compliance is voluntary.

The GALPs Compliance Checklist below provides the acceptance criteria that can be used to monitor compliance to the provisions of EPA's draft Good Automated Laboratory Practices. Any modifications to the finalized text of the GALPs should be incorporated in future versions of this checklist.

Although this checklist is most appropriate for use by automated laboratories subject to EPA GLP regulations, it should also prove useful for laboratories in FDA-regulated industries in complying with that agency's GLP requirements, as well as for QC laboratories required to comply with Good Manufacturing Practice (GMP) regulations.

In the Checklist, every GALP specification has been addressed, affording at least two elements of acceptance criteria. The checklist uses two types of acceptance criteria:

1. Primary (essential) Criteria. These are statements of conditions that an inspector necessarily must establish exist in order to reach a determination that the laboratory is in compliance with the provision(s) of the GALP specification. These primary conditions are identified by the use of "( )" in front of the stated condition.
2. Secondary (highly desirable, but not essential) Criteria. These are statements of conditions that are not necessary for the inspector to establish their existence.

However, if the inspector establishes their compliance, it will provide additional evidence of compliance with the GALP specification over and above the "Primary Criteria." These secondary conditions are identified by the use of " < > " in front of the stated condition.

The checklist will identify documentation that, if established by an inspector to exist, will indicate evidence of adherence to the provision of the GALP specification. These items are identified by the use of "| |" in front of the explanation of the stated available documentation.

Inspector(s) should indicate compliance of laboratory(s) to GALP specification provisions by inserting an "X" within the character enclosure marks which precede each element of acceptance criteria. In those instances where an inspector cannot establish compliance to the designated primary or secondary criteria or to documentation evidence, explanation should be provided in the comment section block. Such explanation(s) will provide the reviewing EPA Agency with the basis to determine whether variance(s) or equivalency(s) are warranted, if applicable.

SECTION 7.1 PERSONNEL

"When an automated data collection system is used in the conduct of a laboratory study, all personnel involved in the design or operation of the automated system shall:"

7.1.1 Background

"Have adequate education, training, and experience to enable individuals to perform the assigned system function."

( ) A standard encompasses all LIMS (laboratory information management systems) and/or LDS (laboratory data systems) computer systems used to collect, transmit, report, analyze, summarize, store, or otherwise manipulate data.

( ) Appropriate professional hiring and assignment criteria, coupled with appropriate training ensure that all users are able to use the system effectively.

< > Users are provided with clear operating instructions, manuals, and SOPs to enable them to perform assigned system functions.

< > Sufficient training to clarify instructions has been provided to users.

< > Users unable to meet the performance criteria are screened out of automated responsibilities prior to hiring or subsequent to a probationary review.

( ) If design of the system has been left to outside vendors, a project leader has been selected whose resume demonstrates some formal computer training, coupled with prior experience in the design or coding of similar systems.

( ) Laboratory maintains a separate education and training file for each employee that documents job description, job requirements, skills, education, and training.

Comment:

7.1.2 Training

"Have a current summary of their training, experience, and job description, including information relevant to system design and operation maintained at the facility."

( ) Documentation of personnel backgrounds including education, training, and experience is available to laboratory management.

( ) Knowledge of personnel pertinent to system design and operations is documented.

( ) Evidence of training and experience indicating knowledge sufficient for job requirements is recorded.

    | | Outside vendors may be presumed to have the required education, training, knowledge and experience.

| | In-house personnel have demonstrated prior success in similar responsibilities.

< > In-house personnel backgrounds show pertinent system design and operations knowledge through the following means filed centrally in the laboratory Personnel Office:

| | Resumes.

| | Reports of completed training.

| | Up-to-date job descriptions.

| | Successful job performance evaluations which demonstrate proper levels of job knowledge and experience.

Comment:

7.1.3 Number of Persons

"Be of sufficient number for timely and proper conduct of the study, including timely and proper operation of the automated data collection system(s)."

( ) The laboratory maintains a staff which is adequate in size to ensure that studies can be performed in an accurate and timely manner, including all system-related tasks.

( ) The person to whom QA is assigned is independent of the laboratory unit.

< > Work plans are designed and are followed for each study so that the Laboratory Manager or designee can anticipate staffing requirements necessary for a particular need.

< > The automated laboratory is staffed with at least two individuals whose qualifications satisfy GALP Section 7.1.1.

< > The Laboratory management is cognizant of any delays in operations due to inadequate staffing and takes proper action.

Comment:

SECTION 7.2 LABORATORY MANAGEMENT

"When an automated data collection system is used in the conduct of a study, the laboratory management shall:"

7.2.1 Designee

"Designate an individual primarily responsible for the automated data collection system(s), as described in Section 7.3."

( ) A single individual has been designated as the Responsible Person, to whom the integrity of the database can be entrusted.
( ) A back-up has been appointed who can manage the automated system if the Responsible Person is not available.
< > An organizational plan has been developed to define lines of communication and reporting within the laboratory structure.
< > One person has been designated as the "owner" ultimately responsible for the automated data collection system and its database.

Comment:

7.2.2 QA

"Assure that there is a quality assurance unit that oversees the automated data collection system(s) as described in Section 7.4."

( ) The Laboratory has designated a group or individual as Quality Assurance. This designation is consistent with the guidelines set forth in Section 7.4.
( ) The responsibilities of the Quality Assurance team are primarily those of system and data inspection, audit, and review.
( ) The QA team or individual maintains a degree of independence, and therefore, does not report to, or is not, the System Responsible Person.
< > An organizational plan has been developed to define lines of communication and reporting within the laboratory structure.
< > Although a single individual may have many managerial responsibilities, the QA individual/head is not the Responsible Person.

Comment:

7.2.3 Resources

"Assure that the personnel, resources, facilities, computer and other equipment, materials, and methodologies are available as scheduled."

( ) The Laboratory Manager guarantees that resources necessary to accurately run a given study in a timely fashion are accessible. These resources include personnel, facilities, computers and other equipment, materials, and related methodologies.

( ) The policy of resource preparedness is clearly stated in written format and adhered to by laboratory management.

< > The experienced Laboratory Manager possesses the acumen and skills necessary to determine that adequate resources for the study are always accessible.

< > The laboratory has provided backup staffing for critical functions.

Comment:

7.2.4 Reporting

"Receive reports of quality assurance inspections or audits of computers and/or computer-resident data and promptly take corrective actions in response to any deficiencies."

( ) The flow of information concerning all laboratory operations including system review and audit are conveyed to upper levels of management.

( ) The Laboratory Manager guarantees that reports generated as a result of Quality Assurance audits are presented for review.

( ) The Laboratory Manager is ultimately responsible for assuring that errors or deficiencies that have been discovered through QA activities are acted upon and rectified in a prompt manner.

< > Laboratory policy or SOPs clearly state that all QA review or audit reports are to be presented to the Laboratory Manager for review.

< > Review documents have a cover sheet (or similar) which the Laboratory Manager can sign and date.

< > An SOP or policy is in place that defines the responsibility of the Laboratory Manager to follow-up on all deficiencies found in the report.

Comment:

7.2.5 Training

"Assure that personnel clearly understand the functions they are to perform on automated data collection system(s)."

( ) The Laboratory Manager guarantees that all laboratory personnel are fully trained in their responsibilities.

( ) Comprehensive employee training programs have been established with appropriate training personnel provided.

( ) Review of training "check-off" sheets is documented.

( ) An annual assessment or evaluation of employee skills and performance has been recorded.

( ) All training procedures undergo periodic review at least yearly, or whenever new or upgraded equipment or methodologies are installed.

< > The comprehensive training of all individuals interacting with the automated data collection system is delineated in a laboratory policy or SOP.

< > The basic operational skills of the system users are clearly defined.

< > Training fully documents all phases of normal system function as they pertain to particular users so that all users clearly understand the functions they perform.

< > Training enables the users to understand enough about normal system function to permit any abnormal system function to be recognized and reported to the appropriate laboratory individual.

< > Problems are routinely reviewed to determine whether their frequency has increased or decreased and how they have been resolved.

Comment:

7.2.6 Deviations

"Assure that any deviations from this guide for automated data collection system(s) are reported to the designated Responsible Person and that corrective actions are taken and documented."

( ) The Laboratory Manager(s) is ultimately responsible for all activity within the confines of the laboratory based on the Guide for Automated Data Collection System.

( ) SOPs or general policy states that any departure from the standards listed within the Guide will be reported to the Responsible Person or designee.

( ) Deviations are properly documented and appropriate corrective actions are taken and documented by the Responsible Person or designee.

< > As part of a comprehensive system policy, there is written assurance that responsible parties are made aware of deficiencies or departures from the standards set forth in the Guide.

< > The policy states that the Responsible Person will handle all deviations and satisfactorily document actions taken.

< > Documentation of deviations includes an indication of the violating party, the date of the violation (if known), and the corrective action and date.

< > A signature block area is included for the Responsible Person or other reviewer.

Comment:

SECTION 7.3 RESPONSIBLE PERSON

"The laboratory shall designate a computer scientist or other professional of appropriate education, training, and experience or combination thereof as the individual primarily responsible for the automated data collection system(s) (the Responsible Person). This individual shall ensure that:"

7.3.1 Personnel

"There are sufficient personnel with adequate training and experience to supervise and/or conduct, design and operate the automated data collection system(s)."

( ) The Responsible Person ensures that the facility is properly staffed with personnel qualified for the systems tasks pertinent to the site and that such personnel are properly managed.

( ) The Responsible Person ensures that staff levels are appropriate and that the staff receives all necessary training including:

- | | Knowledge of SOPs.
- | | Regulatory Requirements.
- | | System-related work flow.
- | | Procedures.
- | | Conventions.

( ) The Responsible Person ensures that staff adequately performs all required system activity.

< > Adequate staffing levels for system supervision, support, and operations are assessed periodically by the proper Operations and Personnel management to determine if established levels need to be changed.

< > The Responsible Person reviews training records to maintain awareness of the current status of training received and needed.

< > Observation of job performance indicates performance levels of current staff and possible need for additional help.

< > Project schedules and work backlogs are examined to determine adequacy of current staff and whether the system is receiving proper staffing support.

Comment:

7.3.2 Training

"The continuing competency of staff who design or use the automated data collection system is maintained by documentation of their training, review of work performance, and verification of required skills."

( ) The Responsible Person ensures that personnel who use or support the system maintain the skills and knowledge necessary for proper performance of their responsibilities.

( ) On-going training and training necessitated by changes in the system are conducted to ensure that skills do not become outdated or forgotten.

( ) The Responsible Person ensures that job performance reviews indicate proper skill levels and that any recommended training is conducted promptly.

< > Written procedures have been established which require that all training needs identified by job performance reviews or observations of job activities are reported to the Responsible Person.

< > SOPs requiring documentation of training and testing have been created.

< > Employees are encouraged to obtain training in use of:

| | System utilities.

| | The operating system.

| | Proper use of available program libraries and databases for testing and production purposes.

| | Sort tools and options.

| | End-user programming languages or report writers.

| | Education they believe is needed.

< > The Responsible Person calls to the attention of staff and users any available in-house or vendor-provided training that might be pertinent.

Comment:

7.3.3 Security

"A security risk assessment has been made and all necessary security measures have been implemented."

( ) The Responsible Person has ensured that an analysis of system vulnerability has been performed and that reasonable measures for preventing unauthorized system access have been taken, as warranted by the degree of exposure that exists.

( ) All aspects of system input, processing, and output requiring security control are identified and measured.

( ) Measures for restricting access to these system functions have been established and are operating in a way that satisfies the stated objectives.

< > An analysis of all entry methods to the system has been conducted to determine possible areas of exposure, such as:

| | Remote modem access by vendors or other users.
| | All persons and methods involved in initiating processing.
| | All persons receiving system output.

< > Precautionary measures to prevent intentional or unintentional data corruption or disruption of system performance have been instituted, consisting of:

| | Password Security.
| | Dial-back procedures for remote access.
| | Procedures for updating security files.
| | Distribution of system output to authorized persons only.

< > Physical access to sensitive records stored magnetically or in hard copy format is appropriately controlled.

< > A system for updating passwords periodically, such as every six months, is used.

< > Automatic system logging of unauthorized access attempts is utilized.

< > Notification procedures have been established for updating security when users resign or change their job responsibilities.

Comment:

7.3.4 SOPs

"The automated data collection system(s) have written operating procedures and appropriate software documentation that are complete, current, and available to all staff."

Documentation on the system:

( ) The Responsible Person ensures that system documentation is comprehensive, current (showing evidence of management review and approval within the last 12 months), and is readily accessible to users.

( ) For purchased systems:

| | Documentation has been provided by the vendor.
| | Vendor-supplied materials have been supplemented and tailored by additionally developed in-house documentation, if required.

( ) Technical documentation has been developed in accordance with in-house standards and is available to Operations and support personnel.

( ) A User's Manual provides all pertinent information for proper system use.

( ) Written procedures for control of the system are available to all persons whose duties involve them with the system.

< > SOPs supporting system activity have been developed covering subjects such as:

| | System Security.
| | Training.
| | Hardware and software change control.
| | Data change procedures and audit trails.
| | Procedures for manual operation during system downtime.
| | Disaster Recovery.
| | Backup and restore procedures.
| | General system safety.

< > Documentation of the software and hardware is available either through on-line help text or manuals.

< > Documentation is numbered and logged out to departments or individuals in order to facilitate the update process.

Comment:

7.3.5 SOP Review

"All significant changes to operating procedures and/or software are approved by review and signature."

( ) System-related SOPs and software changes are subject to a formal approval process that itself is defined in written SOPs.

( ) The Responsible Person ensures that no changes are made to operating procedures or software without proper approval and documentation.

( ) Software changes are made only in accordance with an approved Change Control Procedure.

< > The Responsible Person has established a Change Control Procedure that creates a mechanism for requesting software changes and which defines review and approval measures for changes.

< > The Responsible Person is part of the change control process and can prohibit any software change from moving to the production environment without appropriate signed approval.

< > The Responsible Person is included in the approval process for changes to system-related procedures.

< > Requirements have been established that no changes can be instituted without the approval signature of the Responsible Person.

Comment:

7.3.6 Change Control

"There are adequate acceptance procedures for software and software changes."

( ) Before software changes or new software are put into the production environment, the Responsible Person ensures that:

| | The software is performing in accordance with the needs of the users.

| | Users have had adequate opportunity to evaluate it in a test environment.

< > Documentation of acceptance testing is part of the approval process that precedes putting new or changed software into production.

< > A Software Change Control SOP has been instituted requiring that:

| | Test protocols be created.

| | Tests be conducted in accordance with the protocols.

| | Test data with anticipated and actual results be permanently filed.

< > The change control SOP directs that written approvals from users and MIS are required before changes are put into production.

< > The change control SOP indicates the procedures and conventions to be followed for version control of programs maintained.

< > A test environment has been established for users to test whether new software or software changes meet their needs or requests.

< > User sign-off is obtained to indicate that new program versions are working satisfactorily.

Comment:

7.3.7 Data Recording

"There are procedures to assure that data are accurately recorded in the automated data collection system."

( ) The Responsible Person has instituted practical methods and procedures that control data entry, change, and storage, resulting in ensuring data integrity.

< > Procedures have been established to require that audit trails are produced indicating all new data entered, changed, or deleted, and that these reports are reviewed thoroughly by appropriate personnel.

< > Data changes require appropriate comments or codes.

< > Audit trails indicate;

| | User identification.

| | Date and time stamps.

| | Field names.

| | Old and new values.

| | Authorization codes.

< > Access to data entry/change/delete functions has been restricted to authorized personnel.
< > Double keying can be required where appropriate.
< > Audit trails for data passing through interfaces produce batch control totals of records.
< > Automatic entry of data by test devices is checked by means of audit trail reports.
< > Manual rechecking of data entered against source documents is undertaken, when appropriate.
< > Randomly selected inputs are spot-checked.

Comment:

7.3.8 Problem Reporting

"Problems with the automated collection system that could affect data quality are documented when they occur, are subject to corrective action, and the action is documented."

( ) The Responsible Person has ensured that a problem reporting procedure or method is in effect to:
    | | Log system problems that could impact data integrity.
    | | Record actions taken on those problems.
    | | Denote problem resolutions.
< > A written problem reporting procedure and forms for reporting and describing problems are in place.
< > Actions taken and resolutions are documented on the same forms, which can be retained for later reference and inspection.
< > The Responsible Person monitors compliance with the procedures by periodically reviewing the log and signing it.
< > Summaries are prepared for management review.

Comment:

7.3.9 GLP Compliance

"All applicable good laboratory practices are followed."

( ) The Responsible Person ensures that:

[ ] All laboratory personnel are familiar with current GLPs.
[ ] GLPs are easily accessible.
[ ] Laboratory activities are conducted in accordance with GLPs.

( ) Copies of GLPs are easily accessible to laboratory personnel.
( ) The Responsible Person periodically reviews all pertinent GLPs with laboratory personnel.
( ) The Quality Assurance Unit periodically inspects for compliance with GLPs.
< > Training sessions cover applicable GLPs.
< > Testing is used to confirm knowledge and understanding of GLPs.
< > Copies of relevant GLPs are kept in a designated area for reference by laboratory personnel.

Comment:

SECTION 7.4 QUALITY ASSURANCE UNIT

"The laboratory shall have a quality assurance unit that shall be responsible for monitoring those aspects of a study where an automated data collection system is used. The quality assurance unit shall be entirely separate from and independent of the personnel engaged in the direction and conduct of a study or contract. The quality assurance unit shall inspect and audit the automated data collection system(s) at intervals adequate to ensure the integrity of the study. The quality assurance unit shall:"

7.4.1 SOPs

"Maintain a copy of the written procedures that include operation of the automated data collection system."

( ) The Quality Assurance Unit (QAU) has provided proof that the automated data collection system(s) operate in an accurate and correct manner consistent with its recommended function.
( ) A complete and current set of Standard Operating Procedures is available and accessible to the QAU.
( ) The QAU has access to the most current and version-specific set of system operations technical manuals.
< > A complete and current copy of system SOPs and technical documents exist as part of standard documentation found in the office of the QAU head (or individual).
< > SOPs are written and formalized as standard laboratory (QAU) policy.

< > If SOPs are maintained on line, the QAU keeps a hard copy version and verifies that machine-readable and hard copy versions are identical.

Comment:

7.4.2 Inspections

"Perform periodic inspections of the laboratory operations that utilize automated data collection system(s) and submit properly signed records of each inspection, the study inspected, the person performing the inspection, findings and problems, action recommended and taken to resolve existing problems, and any scheduled dates for reinspection. Any problems noted in the automated data collection system that are likely to affect study integrity found during the course of an inspection shall be brought to the immediate attention of the designated Responsible Person."

( ) The system has been audited and/or validated on a regular basis.
  | | At least once yearly.
  | | Immediately after any change that affects overall system operation or function.

< > As set by SOP, the periodic inspection policy includes provisions for:
  | | Description of the inspection study.
  | | The personnel involved in the inspection activities.
  | | Findings and recommended resolutions to any discovered problems.

< > All documentation of the inspection has been properly signed-off by the inspection unit (QAU).

< > If problems were detected, the Responsible Person was immediately notified and a date for reinspection was/has been established.

Comment:

7.4.3 Deviations

"Determine that no deviations from approved written operating instructions or software were made without proper authorization and sufficient documentation."

( ) The automated data collection system is consistently operated in a manner congruous with its recommended functionality.

( ) Changes are not made to the existing software package that are inconsistent with accepted change authorization procedures.

< > As defined by SOP, the QAU ensures that changes are not made to either software or system operations instructions without prior consent and full documentation of the change.

< > Changes to either software or system operations instructions are permitted as long as the proper change control procedures are followed (refer to Sections 7.3, 7.8, and 7.9; Change Control)

Comment:

7.4.4 Final Data Report Reviews

"Periodically review final data reports to ensure that results reported by the automated data collection system accurately represent the raw data."

( ) Periodic system performance review is conducted by QAU to check that final report data correlates with the raw data for a specific system run.

< > A written SOP directs a weekly review of several final data reports and their corresponding raw data.

< > Problems or deviations arising from QAU review are handled as mentioned in Section 7.4.3.

< > A performance review is a part of the validation study and does not comprise the entire study.

Comment:

7.4.5 Archiving Records

"Ensure that the responsibilities and procedures applicable to the quality assurance unit, the records maintained by the quality assurance unit, and the method of indexing such records shall be in writing and shall be maintained. These items include inspection dates of automated data collection systems, names of the individual performing the inspection, and results of the inspection."

( ) All of the QAU's methods and procedures are fully documented and followed.

( ) The QAU's inspections and results are:

| | Labeled.

| | Identified by date, time, and investigator.
| | Easily accessible.

( ) The filing and/or index system under which the document is stored is fully described and provides easy accessibility.

< > A policy has been established requiring the QAU to maintain all records and documentation pertaining to their activities, methodologies, and investigations (including results).

< > Documentation includes all SOPs that pertain to the unit.

< > The complete set of documents includes an index or description of the indexing method used, to act as a guide for those individuals who need quick access to the information contained within archived files.

Comment:

SECTION 7.5 FACILITIES

"When an automated data collection system is used in the conduct of a study, the laboratory shall:"

7.5.1 Environment

"Ensure that the facility used to house the automated data collection system(s) has provisions to regulate the environmental conditions (e.g., temperature, humidity, adequacy of electrical requirements) adequate to protect the system(s) against data loss due to environmental problems."

( ) The system has been provided with the environment it needs to operate correctly. This applies to Environmental factors that might impact data loss, such as:

| | Proper temperature.
| | Freedom from dust and debris.
| | Adequate power supply and grounding.

( ) System hardware has been installed in accordance with the environmental standards specified by the manufacturer.

< > Climate control systems adequate to provide the proper operating environment have been dedicated to the computer room or other location of the hardware.

< > Backup climate control systems are provided.

< > Hardware has been installed in accordance with the manufacturer's specifications concerning climate and power requirements as specified in the manufacturer's site preparation manual and as installed by the manufacturer.

< > Control devices and alarms have been installed to warn against variances from acceptable temperature ranges.

< > UPS devices are used to protect against the loss of power.

Comment:

7.5.2 Archives

"Provide adequate storage capability of the automated data collection system(s) or of the facility itself to provide for retention of raw data, including archives of computer-resident data."

( ) Adequate storage space is available for raw data to be retained in hard-copy format or on magnetic media.

( ) Storage for system-related records, both electronic and hard-copy, is sufficient to allow orderly conduct of laboratory activities, including complying with reporting and records retention requirements for both on- and off-line storage.

( ) Physical file space requirements (hard copy, microfilm, microfiche) are properly planned and managed to meet laboratory needs and responsibilities.

< > Operations personnel maintain an adequate supply of required tapes or disks and ensure that space to store them is sufficient to meet current and anticipated needs.

< > Storage facilities for retention of raw data in hard-copy or electronic format are planned and available.

< > Procedures defining how raw data is to be retained have been instituted.

< > Offsite storage is available for backup tapes or other media.

< > Backups are recycled through the offsite location by retaining the most recent version in-house for convenience while securing another version offsite for use in the event of a disaster.

Comment:

SECTION 7.6 EQUIPMENT

7.6.1 Design

"Automated data collection equipment used in the generation, measurement, or assessment of data shall be of appropriate design and adequate capacity to function according to specifications and shall be suitably located for operation, inspection, cleaning, and maintenance. There shall be a written description of the computer system(s) hardware. Automated data collection equipment shall be installed in accordance with manufacturer's recommendations and undergo appropriate acceptance testing following written acceptance criteria at installation. Significant changes to automated data collection system(s) shall be made only by approved review, testing, and signature of the designated Responsible Person and the quality assurance unit."

( ) The system's hardware performs in accordance with specifications provided by the manufacturer and is appropriately configured to meet task requirements.
( ) Storage capacity and response times meet user needs.
( ) The installation site has been planned to facilitate use and maintenance.
( ) A current system configuration chart is maintained.
( ) Vendor manuals describing system hardware components, including their installation specification, functions, and usage, are available to proper laboratory personnel and are kept current.
( ) Equipment installation was in accordance to manufacturer's specifications and meets formal, written acceptance test criteria before having been used in production mode.
( ) The Responsible Person ensures that a hardware change control procedure, involving formal approvals and testing, is followed before hardware changes are permitted.
< > Manufacturer's manuals have been obtained for guidance with installation and initial acceptance testing.
< > Diagnostics provided with equipment in accordance with specifications through acceptance testing.
< > Adequacy has been addressed as part of capacity planning.
< > A formal SOP for Hardware Change Control:
| | Is used to require acceptance testing and recommend ways to structure it.
| | Indicates reviews.
| | Specifies authorizations required.

Comment:

7.6.2 Maintenance

"Automated data collection system(s) shall be adequately tested, inspected, cleaned, and maintained. The laboratory shall:"

7.6.2.1 SOPs

"Have written operating procedures for routine maintenance operations."

( ) SOPs have been established to ensure that hardware is maintained, tested, and cleaned on a schedule that will minimize problems and downtime.

( ) The procedures have been reviewed and signed at least every 12 months by the Responsible Person and appropriate management.

< > A Hardware Maintenance SOP addresses:

| | The feasibility of contracting for maintenance through the manufacturer or other outside vendor.

| | What testing, cleaning, and maintenance should be performed in-house by users or Operations personnel

| | The objectives of maintaining equipment performance in accordance with specifications and minimizing downtime and data loss or corruption.

Comment:

7.6.2.2 Responsibility

"Designate, in writing, an individual responsible for performance of each operation."

( ) Specific responsibilities for testing, inspection, cleaning, and maintenance have been assigned in writing and distinguish between the various hardware devices on site.

< > Operations personnel are responsible for:

| | Inspecting and cleaning mainframe and mini-computer equipment.

| | Appropriate maintenance.

< > Contracts with the manufacturer or third-party cover major hardware performance problems and preventive maintenance.

< > Terminal and PC users are required to:

| | Clean their own terminals, keyboards, and personal printers.

| | Test, inspect, and clean their own equipment.

| | Coordinate maintenance activities under a contract with an outside vendor.

| | Coordinate maintenance activities with in-house personnel.

Comment:

7.6.2.3 Records

"Maintain written records of all maintenance testing containing the dates of the operation, describing whether the operation was routine and followed the written procedure."

( ) A log of regularly-scheduled hardware tests is maintained which includes:
  | | Names of persons who conducted them.
  | | Dates.
  | | Indication of results.
( ) Written test procedures with anticipated results are followed, and a log is maintained to document any deviations from them.
( ) The log is reviewed and signed at least annually by management.
( ) The log is reviewed regularly by the Responsible Person.
< > For each type of hardware device utilized on-site, an appropriate test schedule has been developed and on-going testing can be conducted accordingly by the persons assigned.
< > A log of tests, including their schedule and results, is kept centrally by Operations personnel or the Responsible Person.
< > Testing performed by outside vendors as part of preventive maintenance is documented in the log along with results.

Comment:

7.6.2.4 Problems

"Maintain records of non-routine repairs performed on the equipment as a result of a failure and/or malfunction. Such records shall document the problem, how and when the problem occurred, and describe the remedial action taken in response to the problem along with acceptance criteria to ensure the return of function of the repaired system."

( ) All repairs of malfunctioning or inoperable hardware are logged.
( ) This log is retained permanently and is reviewed on a regular basis by management.
( ) All substantive information relevant to problems and their resolutions is recorded.
( ) Formal acceptance testing with documented criteria has been conducted to ensure proper performance prior to returning repaired devices to normal operations.
< > Operations maintains an Equipment Repair log centrally.
< > If repairs are performed by the manufacturer or other vendors, a written report provided by the serviceman is retained to document the problem in addition to the information provided by the user or operator.
< > Responsibility for contracting outside service support is centralized to keep records of such service comprehensive.

< > When repairs are performed in-house by Operations personnel or users, a form has been implemented to obtain the necessary information for the log.

Comment:

7.6.3 Operating Instructions

"The laboratory shall institute backup and recovery procedures to ensure that operating instructions (i.e., software) for the automated data collection system(s) can be recovered after a system failure."

( ) Applications software and systems software currently in use at the laboratory (including operating system) is backed up (i.e., saved on disk or tape) to prevent complete loss due to a system problem.
( ) At least one generation of each software system version is stored off-line.
( ) Procedures for backups and restores have been established and personnel responsible for performing these tasks have been properly trained.
( ) Copyrights pertinent to vendor-supplied software are observed and backups serve only the purpose intended.
< > One generation of each software system used by the laboratory is stored in a usable format and kept in a secure vault or offsite location. Storage is on:
    | | Magnetic disk.
    | | Tape.
< > Written procedures indicate the reasons why backups should be made. Examples include changes to the software.
< > Operations personnel are responsible for backups and restores to the:
    | | Mainframe.
    | | Mini-computer.
    | | Network software.
< > Users of stand-alone PCs are required to perform their own backups and restores of any software they have developed or modified.

Comment:

SECTION 7.7 SECURITY

7.7.1 Security Risk Assessment

"When an automated data collection system is used in the conduct of a study, the laboratory shall evaluate the need for system security. The laboratory shall have procedures that assure that the automated data collection system is secured if that system:"

7.7.1.1 Confidential Information

"Contains confidential information that requires protection from unauthorized disclosure."

( ) Laboratory(s) using automated data collection systems evaluate the needs for system security by determining whether their systems contain confidential data to which access must be restricted.
( ) If it is determined that access needs to be restricted, security procedures have been instituted.
< > Management is familiar with the studies being conducted at its laboratories and is sensitive to issues requiring confidentiality.
< > Management surveys users, when necessary, to assist in determining the need of confidentiality.
< > The Responsible Person ensures that all parties are communicating sufficiently about security needs, and tools are available to meet such needs.
< > Access categories have been established at various levels, and persons are then assigned the appropriate access level according to their needs.

Comment:

7.7.1.2 Data Integrity

"Contains data whose integrity must be protected against unintentional error or intentional fraud."

( ) Security has been instituted on automated data collection systems in laboratory(s) where data integrity has been deemed to be an area of exposure and potential hazard.
< > If data loss or corruption could negate or degrade the value of a laboratory study, security measures have been established on the software systems.
  | | As indicated in Section 3.7.2 below.

| | Restricting the degree of access through use of various levels of password privileges.
| | Using security built into laboratory operations, if adequate.
| | Security is supplemented or replaced by use of software dedicated specifically to security.

< > A double level of protection against intentional security breaches has been implemented.

Comment:

7.7.1.3 Critical Functions

"Performs time-critical functions that require that data be available for sample tracking critical to prompt data analysis, monitors quality control criteria critical to timely release of data, or generates reports which are critical to timely submission of the data."

( ) Security has been instituted on automated data collection systems at laboratories if systems are used for time-critical functions of laboratory studies or reporting study results.

< > If system functions are critical to the performance of laboratory studies, a measure of protection has been added by implementing security procedures that could prevent loss of system use resulting from access by unauthorized persons, such as:

| | User IDs.
| | Passwords.
| | Callback modems.
| | Locked devices.
| | Limited access to computer rooms.
| | Similar restrictions.

Comment:

7.7.2 Security Requirements

"When the automated data collection system contains data that must be secured, the laboratory shall ensure that the system is physically secured, that physical and functional access to the system is limited only to authorized personnel, and that introduction of unauthorized external programs/software is prohibited."

7.7.2.1 Physical Security

"Only personnel with specifically documented authorization shall be allowed physical access to areas where automated data collection systems are maintained."

( ) Physical security of the system is used when it stores data that must be secured.
( ) Access is restricted to hardware devices which physically comprise the system.
( ) Only those persons with documented authorization are allowed access to hardware devices.
 | | Area housing the central processing units (CPUs).
 | | Storage devices.
 | | Terminals.
 | | Printers.
 | | Other user input/output devices.
< > Physical access to systems is restricted to Operations personnel, to the extent possible.
< > CPUs, disk drives, and media on which backups are stored are housed in a locked computer room.
< > Access to computer rooms is restricted by:
 | | Card-controlled entry way(s).
 | | Key-controlled entry way(s).
 | | Alarm systems installed to prevent unauthorized access.
< > Visitor logs are used to log in and log out all personnel accessing the computer room other than those assigned to work in that area.
< > When CPUs or storage media are located in other areas, access to system use is:
 | | Restricted to non-critical functions.
 | | Controlled through measures similar to those used for computer room access.

Comment:

7.7.2.2 System Access Security

"Log-ons, restricted passwords, call-backs on modems, voiceprints, fingerprints, etc., shall be used to ensure that only personnel with documented authorization can access automated data collection systems."

( ) System access security has been implemented when the system stores data that must be secured.

( ) All necessary and reasonable measures of restricting logical access to the system have been instituted to prevent loss or corruption of secured data.

< > Procedures have been established for:

| | Management authorization of system access.

| | Restricting access to persons requiring it for the performance of their jobs.

< > Multiple levels of system access have been established, and users have been assigned to the level appropriate to their work needs.

< > A Security Administrator has been appointed with the responsibility and sole authority to update system security files.

< > If it is not possible to restrict access to personal computers through log-ons or otherwise, then the PCs have been physically secured so that only authorized individuals can gain access.

Comment:

7.7.2.3 Functional Access

"Procedures shall be in place to ensure that only personnel with documented authorization to access automated data collection system functions shall be able to access those functions."

( ) When the system stores data that must be secured, the laboratory has established a hierarchy of passwords which limit access, by function, to those who need to use such functions in the performance of their jobs and are properly authorized.

( ) Security is structured in a way that allows access to needed functions and restricts access to functions not needed or authorized.

< > Security functions of software systems permit establishment of passwords which:

| | Allow limited access to system functions.

| | Screen security.

| | Field level security.

< > The laboratory utilizes security features to limit exposure to system problems and data corruption by restricting users to only those functions or screens they need.

Comment:

7.7.2.4 External Programs/Software

"In order to protect the operational integrity of the automated data collection system, the laboratory shall have procedures for protecting the system from introduction of external programs/software (e.g., to prevent introduction of viruses, worms, etc.)."

( ) If the system stores data that must be secured, the laboratory has established procedures that protect the system against software sabotage in the form of intentionally introduced software bugs that might corrupt or destroy programs, data, or system directories.

( ) External software is not intentionally imported into the system. Measures to ensure that external software is not transferred to the system through the following circumstances have been instituted and are enforced.

| | Telecommunication lines.
| | Modems.
| | Disk packs.
| | Tapes.
| | Other media.

< > SOPs are in place requiring that:

| | Dedicated telecommunication lines be used, where practical, instead of dial-in access.
| | Usage of modems be tightly controlled.
| | Modems be switched off when usage is not required.
| | Call-back systems are used to grant dial-in access.

< > All system access from external sources is documented and confined to persons or organizations on an authorized list maintained by Management.

< > The use of disk packs, diskettes, or tapes from external sources is prohibited or permitted only after all reasonable precautions have been taken to include:

| | Back-ups
| | Identification of source and content of disks.
| | Dumping of the contents of the media on a backup system.
| | Other methods.

Comment:

SECTION 7.8 STANDARD OPERATING PROCEDURES

7.8.1 Scope

"In laboratories where automated data collection system(s) are used in the conduct of a study, the laboratory shall have written standard operating procedures (SOPs). Standard operating procedures shall be established for, but not limited to:"

7.8.1.1 Security

"Maintaining the security of the automated data collection system(s) (i.e., physical security, securing access to the system and its functions, and restricting installation of external programs/software)."

( ) System programs and databases are protected.
( ) SOPs have been installed to maintain security to protect the system(s).
( ) Physical and in-program system security have been implemented.
( ) Management has specified exactly which security measures are to be enacted and maintained.
< > SOPs have been written to establish security of the automated data system(s).
< > System security encompasses three components:
  | | Software and data which must be made secure through program (logical) locks, such as secure levels of password protection.
  | | Hardware which may be protected through passwords or through physical security, such as keyboard or disk drive locks.
  | | Physical protection of the system(s) and/or computer room.
< > Each system user has a unique identification or password.
< > SOPs defining password protection are detailed enough to cover levels of system access and user privileges.
< > An SOP describes the extent of physical protection of system hardware or equipment.

Comment:

7.8.1.2 Raw Data

"Defining raw data for the laboratory operation and providing a working definition of raw data."

( ) Whether entered into the system automatically or manually, the raw data is clearly identified and characterized.
( ) A distinction is made between raw data and processed data (see also Section 7.11 of this checklist.)
< > Analyzer readings of specific samples are considered raw data.
< > The correlation or demography of many samples are regarded as processed data.
< > Hand-written data collections (such as field reading or reports) are raw data.

< > After information is entered into the automated data collection system and is manipulated by calculations and formulas, it becomes processed data.

Comment:

7.8.1.3 Data Entry

"Entry of data and proper identification of the individual entering the data."

( ) SOPs clearly define special requirements pertaining to the entry of data into the automated data entry system(s).

( ) All system users entering data must be identifiable to the system via a unique user identification and/or password.

< > Specific methods for entry of data are required and used.

< > Operators are aware of data entry requirements and have guidelines to follow so that the data is always entered in the same (correct) manner. This procedure contributes to the integrity of the system and the results produced.

< > Methods such as unique user IDs exist whereby the operator actually entering the data can be easily identified.

Comment:

7.8.1.4 Verification

"Verification of manually or electronically input data."

( ) A technique exists that permits an analysis of entered data to confirm that this data is accurate.

( ) Verification defines the correctness of the entered data.

< > Verification methods used include:

- | | The double-blind method of data entry, where two people independently enter the same data.
- | | The simple double entry of data by the same user.
- | | Program edits, whereby input is checked against parameters or system tables.

Comment:

7.8.1.5 Error Codes

"Interpreting of error codes or flags and corrective action to be followed when these occur."

( ) An SOP has been formalized listing possible error messages that appear in printed form or on-screen to let the user know that there is an inconsistency or problem, along with their probable causes.

( ) This SOP documents the methodology by which errors are corrected, and who, if anybody, should be notified.

< > A Chart is used to cross-reference potential error messages, their cause, and the methodology for correction.

Comment:

7.8.1.6 Change Control

"Changing data and proper methods for execution of data changes to include the original data element, the changed data element, identification of the date of change, the individual responsible for the change, and the reason for the change."

( ) Safeguards are in place to protect against unauthorized change of data (either raw or processed).

( ) Audit trails have been installed into automated systems that show:

| | Both changed and original data elements.

| | The date and user making the change.

( ) SOPs have been written to ensure that audit trails for changes are maintained.

( ) Any time data is changed, for whatever reason, the following information is recorded:

| | The date of the change.

| | The reason for the change.

| | The individual making the change.

| | The old and new values of the data elements that have been changed.

< > Separate programs are used for data entry and data maintenance, or separate modules within the same program are used to facilitate capturing the required information for data changes.

< > The system can be programmed to produce audit trails in the form of change logs.

< > An SOP requires that change logs be printed on a regular basis for review by proper supervisors or management.

< > All records added, changed, or deleted can be either flagged, or audit trail records for these updates are written to an audit trail file for printing as follows:

| | A print program provides the option of listing all updates or only selected records, such as deletes.

| | Sort options enable updates to be shown chronologically, or by record type, or both.

< > The responsibility of maintaining the copy of record of audit reports has been assigned.

< > Audit Trail Reports for sensitive records are microfilmed for archive purposes.

< > The audit trail captures the identity of the software module or program making the change.

Comment:

7.8.1.7 Archiving

"Data analysis, processing, storage, and retrieval."

( ) Standard Operating Procedures have been produced that clearly describe consistent methodologies and the techniques used for data processing and analysis encompassing:

| | All manners of manipulating raw data into information that may be easily interpreted.

| | The interpretation of data itself (data analysis).

( ) Methodologies have been formalized that detail how data:

| | Is stored.

| | On what media.

| | Is to be brought back into the automated system for further processing.

( ) Storage encompasses the physical storage of data saved to various magnetic media (such as diskettes, tapes, etc.).

< > SOPs indicate:

| | How formulas used to analyze or process data are to be verified.

| | How standard routines to perform processing or analysis could be utilized.

| | How storage of magnetic media minimizes deterioration.

| | How archived computer records are indexed.

< > SOPs cover and/or indicate:

| | Authorization mechanisms for accessing or retrieving stored data.

| | Responsibilities for maintaining system archives.

Comment:

7.8.1.8 Backup and Recovery

"Backup and recovery of data."

( ) An SOP documenting procedure(s) for system data backup and recovery exists which covers:

| | Proper maintenance of files critical to the system.

| | Assistance to return to operation in the event of corruption or loss of any files critical for system processing.

< > The SOP clearly describes the procedure(s) necessary to create and store a backup copy of system data.

< > Data backup frequency has been established per system or file as specified by an SOP.

| | Daily.

| | Weekly.

| | Monthly.

| | Annually.

< > The SOP clearly delineates where both on-site and off-site backup copies are to be stored, as well as the individual responsible for making the backup copies.

< > A Backup Logbook is used to track the backups if no system utility generates such records automatically.

< > The laboratory has developed procedures for applying "work arounds" in the case of temporary failure or inaccessibility of the automated data collection system which cover:

| | "Rolling back" or "undoing" changes that have not been completed, to a previous, stable documented state of the database.

| | "Rolling forward" the automated system or applying changes to the automated system that were implemented manually during the temporary failure of the automated system.

| | The laboratory has established and implemented procedures that rollback uncommitted transactions or roll the database forward to synchronize it with changes made manually, so that the "current state" of the data base is known and valid at all times.

Comment:

7.8.1.9 Maintenance

"Maintaining automated data collection system(s) hardware."

( ) An SOP has been established that institutes a preventive maintenance plan for all units of the automated data collection hardware to:

| | Generally identify how such maintenance is to be documented.
| | Assure its consistent accurate operation.
| | Assure the proper and vital upkeep of equipment.

< > Vendor-prescribed schedules for preventive maintenance are applied.

< > A Responsible Person has been named to follow up on hardware maintenance to ensure that it is accomplished at the proper time and documented according to the requirements of the SOP(s).

Comment:

7.8.1.10 Electronic Reporting

"Electronic reporting, if applicable."

( ) If electronic reporting is used by the laboratory(s), an SOP exists to establish controls for this process.

( ) Standards, protocols, and procedures used are indicated, and uniformity of such reporting is structured through an SOP.

< > The SOP addresses such issues such as:

| | When electronic reporting is to be done.
| | Which records are involved.
| | How and by whom transmission is to be performed.

< > Guidance in determining the standards to be followed in the process and what audit trails are necessary have been provided (see also Section 7.13 of this checklist).

Comment:

7.8.2 Document Availability

"In laboratories where automated data collection system(s) are used in the conduct of a study, the laboratory shall have written standard operating procedures (SOPs). Each laboratory or other study area shall have readily available manuals and standard operating procedures that document the procedures being performed. Published literature or vendor documentation may be used as a supplement to the standard operating procedures if properly referenced therein."

( ) Written documentation of the procedures being performed are kept available.

( ) If vendor-supplied documentation is used to supplement these written procedures, that documentation is properly referenced in SOPs.

< > Cross references to system documentation supplied by vendors is made in SOPs developed in-house.

Comment:

7.8.3 Historical Files

"In laboratories where automated data collection system(s) are used in the conduct of a study, the laboratory shall have written standard operating procedures (SOPs). A historical file of standard operating procedures shall be maintained. All revisions, including the dates of such revisions, shall be maintained within the historical file."

( ) All versions of SOPs, including expired ones, indicate effective dates and are retained in historical files.

< > A chronological file of SOPs indicating their effective dates is retained in hard copy format.

< > Effective dates are contained in individual SOPs.

Comment:

SECTION 7.9 SOFTWARE

7.9.1 Purpose and Use

"The laboratory shall consider software to be the operational instructions for automated data collection systems and shall, therefore, have written standard operating procedures setting forth methods that satisfy management and are adequate to ensure that the software is accurately performing the intended functions. All deviations from the operational instructions for automated data collection systems shall be authorized by the designated Responsible Person. Changes in the established operational instructions shall be properly authorized, reviewed and accepted in writing by the designated Responsible Person."

( ) Methods for determining that software is performing its functions properly have been documented in SOPs and are followed.

( ) The Responsible Person controls the software change process to prevent any changes which have not been documented, reviewed, authorized, and accepted in writing by the Responsible Person.

( ) Variances from any instructions relevant to the system are first authorized by the Responsible Person before they are instituted.

( ) Formulas are checked and source code is reviewed.

< > A Software Change Control SOP requires that no software changes to the system are implemented unless the proper request, review, authorization, and acceptance procedures are followed.

< > Control of program libraries is restricted to a small number of Operations personnel, where practical, so that no programmers or users are allowed to move changed software into the production environment without following required procedures.

< > User surveys and post-implementation reviews of software performance are required to evaluate whether software is properly performing its functions, as documented.

< > The laboratory distinguishes among different categories of software:

| | Operating systems.

| | "Layered software products" such as programming languages, with which applications are developed.

| | Actual operations of system application(s).

< > Procedures for authorization, review, and acceptance of changes in software may differ across different categories of software.

Comment:

7.9.2 Life Cycle

"The laboratory shall have documentation to demonstrate the validity of software used in the conduct of a study in accordance with the *EPA System Design and Development Guidance* and as outlined in Section 7.9.3."

7.9.2.1 Development

"For new systems the laboratory shall have documentation throughout the life cycle of the system (i.e., beginning with identification of user requirements and continuing through design, integration, qualification, validation, control, and maintenance, until use of the system is terminated)."

( ) For all new systems (systems not in a production mode at the time of the publication of this Checklist) to be used in the conduct of an EPA study, laboratory(s) have established and maintain documentation for all steps of the system's life cycle, in accordance with Section 7.9.3 of this checklist, to include:

| | Documentation of user requirements.
| | Design documents (such as functional specifications, program specifications, file layouts, database design, and hardware configurations).
| | Documentation of unit testing.
| | Qualification.
| | Validation procedures and testing
| | Control of production start-up.
| | Software version update(s) and change(s).
| | Post-implementation reviews.
| | On-going support procedures.

< > SOPs require that each system development life cycle phase of a software project be properly documented before that phase can be regarded as complete.
< > Management reviews development project milestones to ensure that required documentation is available before giving approval for projects to proceed.
< > Where third-party software is used, the Laboratory data sets reference the version of software used.

Comment:

7.9.2.2 Documentation

"Automated data collection system(s) currently in existence or purchased from a vendor shall be, to the greatest extent possible, similarly documented to demonstrate validity."

( ) Systems existing in a production mode prior to publication of this Checklist and purchased systems are documented in the same way as systems developed in accordance with 7.9.2 of this checklist, to the degree possible.
( ) Documentation relevant to certain phases of the system life cycle, such as validation, change control, acceptance testing, and maintenance, for example, are similar for all systems.
< > Reconstruction of documentation for user requirements and design documents is done, when possible, for systems that already exist in a production mode prior to the publication of this Checklist.
< > System descriptions and flow charts have been developed.
< > Evidence of integration and validation testing is maintained for inspection purposes.
< > For vendor-supplied software, user requirements are normally developed prior to software evaluation and selection.
< > Design documentation is provided for vendor-supplied and in-house developed systems to include:

| | File layouts.
| | System descriptions.
| | Functional specifications.
| | Program specifications.
| | Source code.

< > If critical documentation was not provided, attempts to obtain it from the vendor or re-construct it in-house have been undertaken, to the degree possible.

Comment:

7.9.3 Scope

"Documentation of operational instructions (i.e., software) shall be established and maintained for, but not be limited to:"

7.9.3.1 Inventory

"Detailed written description of the software in use and what the software is expected to do or the functional requirements that the system is designed to fulfill."

( ) Functional requirements which document what the system is designed to accomplish are available for use in the system description.

( ) A written system description, which provides detailed information on the software's functionality, has been developed and is maintained for each software application in use at the lab.

< > System flow charts, work flow charts, and data flow charts have been developed by those most knowledgeable user(s) of the system, or by the software vendor.

< > A written system description has been provided by vendors for purchased systems or has been developed in the design phase of in-house software projects.

< > Documentation is made available in a designated area within the laboratory.

Comment:

7.9.3.2 Coding Standards

"Identification of software development standards used, including coding standards and requirements for adding comments to the code to identify its functions."

( ) Written documentation of software development standards exists which includes:

| | Programming conventions.

| | Shop programming standards.

| | Development standards to be followed by design and development staff at the site.

( ) Standards for internal documentation of programs developed or modified have been included.

< > Programming and design standards have been established to ensure that minimum requirements are met and to foster consistency and uniformity in the software.

< > In the area of design, issues such as consistency of file layout formats, screen formats, and report formats are addressed.

< > Other design issues are included:

| | Documentation standards for user requirements definition.

| | Functional specifications.

| | System descriptions.

< > Programs are internally documented based on programming standards to include:

| | Explanatory comments.

| | Section and function labels.

| | Indications of programming language.

| | Programmer name.

| | Dates of original writing and all changes.

| | Use of logical variable names.

Comment:

7.9.3.3 Formulas

"Listing of all algorithms or formulas used for data analysis, processing, conversion, or other manipulations."

( ) All algorithms or formulas used in programs run at the laboratory, including user developed programs and purchased software packages which allow user entry of formulas or algorithms, are:

| | Documented.

| | Retained for reference and inspection.

| | Can be located easily.

( ) Listings of the algorithms and formulas should:

| | Exclude all other information, such as program listing or specifications.

| | Identify programs in which the formulas and algorithms occur.

< > A file or log of all formulas and/or algorithms is maintained centrally in a location designated by the Responsible Person.

< > For purchased software, formulas and algorithms have been:

| | Obtained from vendor-provided documentation.
| | Abstracted from software.

< > Documentation of algorithms and formulas in appropriate listings is made a required part of the design and development process to insure compliance.

Comment:

7.9.3.4 Acceptance Testing

"Acceptance testing that outlines acceptance criteria; identifies when the tests were done and the individual(s) responsible for the testing; summarizes the results of the tests; and documents review and written approval of tests performed."

( ) Acceptance testing, which involves responsible users testing new or changed software to determine that it performs correctly and meets their requirements, has been conducted and is documented.

( ) Written procedures indicate:
| | When such testing is required.
| | How it is to be conducted.

( ) That documentation of testing includes:
| | Acceptance criteria.
| | Summary of results.
| | Names of persons who performed testing.
| | Indication of review and written approval.

< > Acceptance testing procedures, which also apply to implementation of new software, are integral parts of the change control process.

< > Users are given the opportunity to test programs for which they have requested changes in a test environment that will not impact the production system.

< > New software is tested in a similar way by users who will be expected to work with it.

< > Acceptance criteria are documented before testing begins to ensure that testing is predicated on meeting those standards.

< > Quality assurance department or Management review the tests and results to ensure that criteria are appropriate and are met to their satisfaction.

Comment:

7.9.3.5 Change Control

"Change control procedures that include instructions for requesting, testing, approval, and issuance of software changes."

( ) Written documentation of Change Control Procedures exists to provide reference and guidance to MIS and users for management of the on-going software change and maintenance process.

( ) All steps in the change control procedure are explained or clarified, and the procedures are available to all system users and MIS personnel at the laboratory.

( ) Software or software changes that have not been implemented in compliance with the Change Control Procedure are not utilized at the laboratory, except in test mode.

< > Change Control Procedures refer to:

- | | Persons authorized to request software changes.
- | | Forms designed for that purpose.
- | | Requirements to be met before approval of such requests.
- | | Change requests prioritized by defined method.
- | | Program libraries from which to take copies of programs to be amended.
- | | Libraries for program copies undergoing change.
- | | Responsibility for documenting testing.
- | | Approving of changed versions.
- | | Moving changed versions to the production environment.

< > Access to the function of moving changed versions to production is restricted.

Comment:

7.9.3.6 Version Control

"Procedures that document the version of software used to generate data sets."

( ) An audit trail has been established and is retained that permits identification of the software version in use at the time each data set was created.

< > The date and time of generation of all data sets is documented (usually within the data record itself).

< > The software system generating the data set is identifiable.

< > The laboratory ensures that historical files are established and maintained to include:

- | | The current and all previous versions of the software releases and individual programs.

| | Dates and times they were put into and removed from the production system environment.

Comment:

* * *

7.9.3.7 Problem Reporting

"Procedures for reporting software problems, evaluation of problems, and documentation of corrective actions."

( ) A written problem reporting procedure exists to:
  | | Structure the process of documenting software problems encountered by users and MIS staff.
  | | Record the follow-up and resolution of problems.
< > Problem Report forms with written instructions for completion have been developed.
< > Problem Logs are maintained by a person designated by the Responsible Person.
< > Analysis and initial reporting is required within a specific time frame.
< > Periodic follow-up of open problems is done by the Responsible Person until resolution is reached.
< > Documentation of resolved problems is retained in case of recurrences.

Comment:

* * *

7.9.4 Document Availability

"Manuals or written procedures for documentation of operational instructions shall be readily available in the areas where these procedures are performed. Published literature or vendor documentation may be used as a supplement to software documentation if properly referenced therein."

( ) All written SOPs or software documentation mentioned in paragraph 7.3, subparagraphs 7.3.4 to 7.3.9 of this checklist are available, as applicable:
  | | In work areas.
  | | To system users.
  | | To persons involved in software development or maintenance.

( ) For purchased systems, vendor-supplied documentation, if properly referenced, supplements documentation developed in-house.
< > SOP manuals are available to each department or work group within a laboratory.
< > Persons responsible for producing SOP manuals maintain a log of manuals issued, by number, and to whom they were issued in order to ensure that all manual holders receive updates.
< > A distribution key, indicating departments or persons receiving SOPs and the SOPs which were issued to them, has been established.
< > SOPs pertinent only to design, development, and maintenance personnel are available centrally at a specified location in the systems area.
< > User manuals are provided to all user departments or kept in a central documentation area.
< > Technical manuals.
< > Sign-out procedures are followed in all centralized documentation areas to prevent loss or misplacement of these documents.

Comment:

7.9.5 Historical Files

"A historical file of operating instructions, changes, or version numbers shall be maintained. All software revisions including the dates of such revisions, shall be maintained within the historical file. The laboratory shall have appropriate historical documentation to determine the software version used for the collection, analysis, processing, or maintenance of all data sets on automated data collection systems."

( ) Files of all versions of software programs are created and maintained so that the history of each program is evident.
( ) Differences between the various versions and the time of their use are made evident in these files.
( ) An audit trail is established and retained that permits identification of the software version in use at the time each data set was created.
< > The laboratory ensures that historical files are established and maintained to indicate:
  | | The current and all previous versions of the software releases and individual programs.
  | | Dates and times they were put into and removed from the production system environment.
< > Program listings with sufficient internal documentation of changes, dates, and persons making changes are used.
< > Internal references back to a project number or change request form are used.

< > The date and time generation of all data sets within the data record itself are logged, and the software system generating the data set is identifiable.

Comment:

SECTION 7.10 DATA ENTRY

"When a laboratory uses an automated data collection system in the conduct of a study, the laboratory shall ensure integrity of the computer-resident data collected, analyzed, processed, or maintained on the system. The laboratory shall ensure that in automated data collection systems:"

7.10.1 Tracking Person

"The individual responsible for direct data input shall be identified at the time of data input."

( ) Laboratory(s) using automated data collection systems ensure that data input is traceable to the person who entered it (i.e., the person responsible for the data entered can be identified).
< > The system records the user identification code as part of all records entered.
< > User IDs can be referenced back to the associated data entry person to allow identification per each record entered.

Comment:

7.10.2 Tracking Equipment, Time, Date

"The instruments transmitting data to the automated data collection system shall be identified, and the time and date of transmittal shall be documented."

( ) Laboratory(s) using instruments which transmit data to automated data collection systems ensure that an audit trail exists and indicates:
    | | Date and time stamps for each record transmitted.
    | | Which instrument was the source for each entry.
( ) It is possible to trace each record transmitted back to the:

| | Source instrument.
| | Date and time of generation.
< > An instrument identification code along with date and time stamp is entered into each record transmitted to the system and:
| | Is stored as part of the record.
| | Generates an audit trail report with similar information.

Comment:

7.10.3 Data Change

"Any change in automated data entries shall not obscure the original entry, shall indicate the reason for change, shall be dated, and shall identify the individual making the change."

( ) When data in the system is changed after initial entry, an audit trail exists which indicates:
| | The new value entered.
| | The old value.
| | A reason for change.
| | The person who entered the change.
| | Date and time stamp.
< > All values needed so that the history of any data record can always be reconstructed are stored in:
| | The changed record.
| | A permanently kept audit file.
< > Audit Trail reports are maintained.
< > If any electronic data is purged, the reports are kept permanently on microfiche or microfilm.
< > The laboratory has adopted a policy by which only one individual is authorized to change data, rather than implementing a system that records the name of any and all individuals making data changes.

Comment:

7.10.4 Data Verification

"Data integrity in an automated data collection system is most vulnerable during data entry whether done via manual input or by electronic transfer from automated instruments. The laboratory shall have written procedures and practices in place to verify the accuracy of manually entered or automatically transferred data collected on automated system(s)."

( ) Written SOPs exist for validating the data entered manually or automatically to the laboratory's automated data collection systems.
( ) The practice of such procedures is enforced.
< > Data validation methods are practiced to ensure data integrity:
    | | Double-keying of manually entered data.
    | | Blind re-keying of data entered automatically.
    | | Other proven methods.

Comment:

SECTION 7.11 RAW DATA

"Raw data collected, analyzed, processed, or maintained on automated data collection system(s) are subject to the procedures outlined below for storage and retention of records. Raw data may include microfilm, microfiche, computer printouts, magnetic media, and recorded data from automated collection systems. Raw data is defined as data that cannot be easily derived or recalculated from other information. The laboratory shall:"

7.11.1 Definition

"Define raw data for its own laboratory operation."

( ) The operational definition of raw data for the lab, especially as it related to automated data collection systems used, is documented by the laboratory and made known to employees.
( ) Raw data consists of:
    | | Original records of environmental conditions.
    | | Animal weights.
    | | Food consumed by study animals throughout the course of study.
    | | Similar original records.
    | | Documentation necessary for the reconstruction of a study and which cannot be recalculated, as can a statistical value such as a mean or median, given all the original raw data of the study.
( ) Raw data includes data stored on the system or output on various media.
( ) The definition of raw data includes the results of original observations and activities of a study that are necessary for the reconstruction and evaluation of the study and/or which include any statistical manipulation of data.
( ) Raw data source documents used include:

| | Scientist notebooks.
| | Laboratory work sheets.
| | Records.
| | Memoranda.
| | Notes.
| | Exact copies.
| | Photographs.
| | Microfilm.
| | Microfiche copies.
| | Computer printouts and printouts of data bases summarizing the results of testing equipment output.
| | Magnetic media and/or electronic copies of databases.
| | Recorded data from automated instruments.

( ) Data entered into the system directly (not from a source document) by keyboard or automatically by laboratory test devices is considered raw data.

Comment:

7.11.2 Standard Operating Procedures

"Include this definition in the laboratory's standard operating procedures."

( ) The laboratory has included its definition of raw data in the SOPs it publishes.
( ) SOPs are made available to laboratory personnel so that interpretation of what constitutes raw data and retention procedures for such data are uniform for all laboratory studies performed.
< > A policy statement has been issued by the laboratory to make this definition clear to employees.
< > Consideration has been given to a preferred storage media and retention requirements.

Comment:

SECTION 7.12 RECORDS AND ARCHIVES

7.12.1 Records to be Maintained

"All raw data, documentation, and records generated in the design and operation of automated data collection system(s) shall be retained. Correspondence and other documents relating to interpretation and evaluation of data collected, analyzed, processed, or maintained on the automated data collection system(s) also shall be retained. Records to be maintained include, but are not limited to:"

7.12.1.1 Raw Data

"A written definition of computer-resident "raw data" (see Section 7.11 of this checklist)."

( ) Labs retain their written definition of computer resident raw data for inspection or audit.
< > The policy or SOP containing raw data definition, including all prior versions of it, is:
  | | Permanently retained.
  | | Available for inspection or audit.

Comment:

7.12.1.2 Hardware and Software

"A written description of the hardware and software used in the collection, analysis, processing, or maintenance of data on automated data collection system(s). This description shall identify expectations of computer system performance and shall list the hardware and software used for data handling. Where multiple automated data collection systems are used, the written description shall include how the systems interact with one another."

( ) The laboratory retains written descriptions of all hardware and software used in data handling on the system.
( ) Overall descriptions of the purpose and use of the system and specific listing of hardware and software involved in data handling is available.
( ) If more than one system exists, the relationship between them, including what data is passed from one system to another, is documented and retained.
< > Hardware descriptions have been provided by the vendor.

< > System configurations have been documented in-house.

< > General descriptions of software are available from the vendor for purchased software.

< > Vendor supplied software descriptions have been enhanced in-house:

| | If the software has been modified.

| | To describe how important software options are being used.

< > For software developed in-house, the required descriptions have been developed as part of the design documentation

Comment:

7.12.1.3 Acceptance Test Records

"Software and/or hardware acceptance test records identifying the item tested, the method of testing, the date(s) the tests were performed, and the individuals who conducted and reviewed the tests."

( ) Acceptance testing has been performed and documented for new or changed software.

( ) Documentation of testing, including the information mentioned in this standard, are permanently retained.

< > Documentation of acceptance testing by users is made a part of the project file associated with the new or changed software.

< > Documentation of acceptance testing is retained in the MIS department or other designated area for audit purposes.

Comment:

7.12.1.4 Training and Experience

"Summaries of training and experience and job descriptions of staff as required by Section 7.1 of this document."

( ) Laboratories retain summary records for their personnel of their job descriptions, experience, qualifications, and training received.

< > Documentation of personnel backgrounds is retained centrally, and kept available to laboratory management and inspectors or auditors.
< > Documentation of personnel backgrounds includes:
| | Education.
| | Training.
| | Experience.
| | Pertinent systems design and operations knowledge, in accordance with Section 7.1 of this checklist.

Comment:

7.12.1.5 Maintenance

"Records and reports of maintenance of automated data collection system(s)."

( ) All written documentation or logs of repair or preventive maintenance to automated data collection system hardware are retained by laboratory(s) for subsequent reference, inspection, or audit.
( ) Documentation indicates:
| | The devices repaired or maintained (preferably with model and serial numbers).
| | Dates.
| | Nature of the problem for repair.
| | Resolutions.
| | Indications of testing, when appropriate.
| | Authorizations for return of devices to service.
( ) Maintenance documentation includes records pertaining to work performed by in-house personnel as well as that done by vendors or outside service contractors (See Section 7.6 of this checklist).
< > Policies have been implemented to ensure that all required documentation is forwarded to a central archive point, including that for peripheral devices.
< > Documentation of equipment repairs and maintenance is maintained for reference.

Comment:

7.12.1.6 Problem Reporting

"Records of problems reported with software and corrective actions taken."

( ) Laboratory(s) retain all software-related Problem Reports and Logs for subsequent reference and inspection.

( ) Problem Reports and Logs include all information pertinent to the problems and actions taken to resolve the problems (See also Section 7.9.3 of this checklist).

< > Software problems are reported centrally to a system support group or person.

< > Software problems can be reported by both users and Operations personnel.

< > In written procedures, guidelines have been established for documenting, filing, and retention of problem(s) and their resolution(s).

< > Primary responsibility for maintenance and retention of problem records has been specifically delegated to a Responsible Person.

Comment:

7.12.1.7 QA Inspections

"Records of quality assurance inspections (but not the findings of the inspections) of computer hardware, software, and computer-resident data."

( ) In automated laboratories, the Quality Assurance Unit is responsible for conducting periodic inspections of laboratory operations to include that no deviations from the proper design or use, as documented in written procedures or pertinent manuals, is evident for:

  | | Hardware.

  | | Software.

  | | Computer-resident data.

( ) The QAU documents these inspections, and this documentation of inspections is retained.

< > The QAU creates suitable forms or checklists to document inspections.

< > The QAU retains inspection documentation in appropriate files or on microfilm.

< > The QAU staff inspects automated operations for:

  | | Compliance with applicable GLPs and SOPs.

  | | Evidence of proper authorization.

  | | Documentation of deviations.

Comment:

7.12.1.8 Backup and Recovery

"Records of backups and recoveries, including backup schedules or logs, type and storage location of backup media used, and logs of system failures and recoveries."

( ) Laboratory(s) retain all schedules, logs, and reports of:
  | | System backups (data and programs).
  | | System failures.
  | | Recovery(s) and restore(s).
( ) Records indicate:
  | | The type of activity (e.g., normal backup, recovery due to system failure, restore of a particular file due to data corruption).
  | | Location of backup storage media.
< > Suitable files have been established for retention of the forms on which all backup(s), recovery(s), or restore(s) are documented.
< > Documentation is subject to scheduled managerial review.
< > When operations are disrupted, or when PCs are involved, persons responsible for backup, recovery, and for documentation backup and recovery, are subject to frequent managerial review or follow-up to ensure that all necessary records are generated and retained according to SOPs.

Comment:

7.12.2 Conditions of Archives

"There shall be archives for orderly storage and expedient retrieval of all raw data, documentation, and records generated in the design and operation of the automated data collection system. Conditions of storage shall minimize potential deterioration of documents or magnetic media in accordance with the requirements for the retention period and the nature of the documents or magnetic media."

( ) All raw data, documentation, and records generated in the design and operation of the automated data collection system has been archived in a manner that is orderly and facilitates retrieval.
( ) Filing logic and sequences are easily understood.
( ) If stored on the system, data is backed up at intervals appropriate to the importance of the data and potential difficulty of reconstructing it, and the backups are retained.
( ) The storage environments are suitable to accommodate the media involved and prolong the usefulness of the backups or documents in accordance with their retention period requirements.

< > Backup tapes or disks are stored in the computer room, if available, which provides the proper environment to prevent deterioration due to temperature, dust, or other potentially harmful conditions.

< > Documents that must be retained are filed in cabinets that are water and fireproof, which are located in areas appropriately protected from water and fire damage.

< > If retention requirements for data stored on magnetic tapes exceeds two years, procedures for periodically copying such tapes have been established.

< > Filing procedures and sequences have been documented to ensure uniformity.

Comment:

7.12.3 Records Custodian

"An individual shall be designated in writing as a records custodian for the archives."

( ) Laboratory(s) assign responsibility, in writing, for maintenance and security of archives to a designated individual.

< > A job description for the responsibility of the archivist is available.

< > The archivist has a backup person to assume such duties in case of absence.

Comment:

7.12.4 Limited Access

"Only personnel with documented authorization to access the archives shall be permitted this access."

( ) Access to all data and documentation archived in accordance with Section 7.12 of this checklist and related subparagraphs of this checklist is limited to those with documented authorization.

< > Archived data and documentation is accorded the same level of protection as data stored on the system.

< > Procedures defining how access authorization is granted and the proper use of the archived data, including restrictions on how and where it can be used by authorized persons, have been established.

< > Logs are maintained that document access to archives to include:

| | When.

| | By whom.

| | For what reason access was granted to the archives.
| | Identification of the particular records accessed.

< > If removal of records from the archive area is to be permitted, strictly enforced sign-out and return procedures are documented and have been implemented.

Comment:

7.12.5 Retention Periods for Records

"Raw data collected, analyzed, processed, or maintained on and documentation and records for the automated data collection system(s) shall be retained for the period specified by contract or EPA statute."

( ) Raw data and all system-related data or documentation pertaining to laboratory work submitted in support of health or environmental programs is retained by the laboratory(s) for the period specified in the control or by EPA statute.

< > Contract clauses or EPA statutes pertinent to record retention periods are copied and forwarded to the Archivist, who can then ensure compliance and disposal or destruction, as appropriate, when retention periods have expired.

< > The Archivist follows up to determine retention periods for any records lacking such information.

< > The Archivist ensures that the storage media used are adequate to meet retention requirements.

< > The Archivist institutes procedures to periodically copy data stored on magnetic media whose retention capabilities do not meet requirements.

Comment:

SECTION 7.13 REPORTING

"A laboratory may choose to report or may be required to report data electronically. If the laboratory electronically reports data, the laboratory shall:"

7.13.1 Standards

"Ensure that electronic reporting of data from analytical instruments is reported in accordance with the EPA's standards for electronic transmission of laboratory measurements. Electronic reporting of laboratory measurements must be provided on magnetic media (i.e., magnetic tapes and/or floppy disks) and adhere to standard requirements for record identification, sequence, length, and content as specified in EPA Order 2180.2 - Data Standards for Electronic Transmission of Laboratory Measurement Results."

( ) When a laboratory reports data from analytical instruments electronically to the EPA, that data is submitted on standard magnetic media, such as tapes or diskettes, and conforms to all requirements, such as those for record identification, length, and content.

< > Electronically submitted information complies with the content of EPA Order 2180.2 and includes:

| | All character data are in upper case, with two exceptions:

(1) When using the symbols for chemical elements, they are shown as one upper case letter or one upper case letter followed by a lower case letter.

(2) In comment fields, no restrictions are made.

| | Missing or unknown values are left blank.

| | All character fields are left-justified.

| | All numeric fields are right-justified. A decimal point is used with a non-integer if exponential notation is not used. Commas are not used.

| | All temperature fields are in degrees centigrade, and values are presumed non-negative unless preceded by a minus sign (-).

| | Records are 80 bytes in length, ASCII format.

| | Disks or diskettes have a parent directory listing all files present.

| | Tape files are separated by single tape marks with the last file ending with two tape marks.

| | External labels indicate volume ID, number of files, creation date, name, address, and phone number of submitter.

| | Tape labels also contain density, block size, and record length.

< > Data that is electronically submitted to the EPA conforms to:

| | One of the six different record format types.

| | Contains prescribed definition and other important information.

Comment:

7.13.2 Other Data

"Ensure that other electronically reported data are transmitted in accordance with the recommendations of the Electronic Reporting Standards Workgroup (to be identified at such time as the recommendations are finalized)."

( ) If laboratory(s) electronically report data other than that from analytical instruments (covered in subparagraph 1, above), that data is transmitted in accordance with the recommendation made by the ERS Workgroup.

< > The laboratory(s) have adopted the same Federal Information Process Standard (FIPS) proposed by the National Institute of Standards and Technology (NIST) relative to EDI.

Comment:

SECTION 7.14 COMPREHENSIVE ONGOING TESTING

"Laboratories using automated data collection systems must conduct comprehensive tests of overall system performance, including document review, at least once every 24 months. These tests must be documented and the documentation must be retained and available for inspection or audit."

( ) In order to ensure ongoing compliance with EPA requirements for security and integrity of data and continued system reliability and accuracy, a complete test of laboratory system(s) is conducted at least once every 24 months.

( ) This test includes a complete document review for:

- | | SOPs.
- | | Change.
- | | Security.
- | | Training Documentation.
- | | Audit Trails.
- | | Error logs.
- | | Problem reports.
- | | Disaster plans.

< > A test team has been assembled which includes users, QAU, personnel, data processing personnel, and management so that the interests, skills, and backgrounds of individuals from these different areas can best be drawn to the testing process.

< > A system test data set has been developed which significantly exercises all important functions of the system.

< > The test data set can be retained and re-used for future system tests.

< > The test data set is enhanced periodically if new functionality is added to the system.

< > System test protocols and test objectives have been developed and are reused.

< > A checklist has been developed to ensure that all important areas of testing and document review have been addressed.

< > If there have been no changes to the system within the previous 24 months, actual retesting and review of the system is conducted on a limited scope.

< > Documentation is reviewed to determine that it is current and accurate.

Comment:

Chapter 10

THE USE OF AN SDLC IN SYSTEM VALIDATION

OVERVIEW

The U.S. Food and Drug Administration (FDA) is increasing the amount of attention it gives to system validation and related documentation. Many drug manufacturers are using automated systems that have not been validated and do not have complete documentation. When developing a new system, the Software Development Life Cycle (SDLC) approach should be implemented. What can a company do if a system was created without an SDLC approach? What are the company's options?

SDLC is a disciplined method of developing software for computerized systems. This method requires a continuous, well-documented approach to software development and will improve chances for success in the evolution of the system. SDLC is a development technique, divided in a varying number of stages but usually consisting of five general phases. These phases are: REQUIREMENTS, in which the system goals are clearly defined; DESIGN, proving an approach to achieving those goals; CODING in which the computer commands are generated; TESTING, the integration and confirmation of that code; and MAINTENANCE, the ongoing functional condition of the system.

Requirements Phase

This initial phase studies the user's needs, which are developed into a written set of guidelines within which the system is expected to perform.

This phase is a very important step in the SDLC and receives the greatest FDA attention. It is crucial to keep well-documented records when developing the SDLC, because the developer will constantly refer back to them when creating, testing, and maintaining the software. The requirements phase is completed once the guidelines and the objectives of the system have been compiled into requirement specifications and approved by the analysts.

Design Phase

During this phase, the requirements are translated into a design specification, which defines how the system will be coded. The specification is reviewed, modified, and finally approved by the programmers and analysts.

Coding Phase

At this point, the design specifications are utilized by the programming staff in conjunction with formal coding standards to develop the software. After the coding phase, testing phase activities are conducted by persons other than those involved in the development of the code.

Testing Phase

This phase of the SDLC verifies whether the entire system is performing properly and if it complies with the user's requirements.

Once the testing phase is complete and all known errors are corrected, the software is released and the maintenance phase begins.

Maintenance Phase

During this phase, the software is used, errors are corrected, and modifications are made to enhance the system as the requirements may change. If the requirements change, the design, coding, and testing must begin again.

ONGOING ACTIVITIES

The sub-phases or ongoing activities mentioned above are activities that overlap the entire SDLC; they include fact-finding, documentation, presentation, and ongoing testing.

The Good Manufacturing Practice (GMP) regulations for finished pharmaceuticals (21 CFR 211.68) require that regulated systems be inspected on a routine basis, according to written specifications, to assure proper performance of the system. Records of the inspections must be maintained in order to show sufficient performance monitoring. Each company must implement a documentation system to clearly define what the system is supposed to do and to provide documented evidence that processes performed as expected. In order to validate a system, it is important to understand its purpose and goals. It is essential for a system to have a clear maintenance schedule to insure that control of the system will continue. As modifications are made, all the steps in the SDLC should be followed. The FDA urges the use of an SDLC approach to validating regulated systems. By using this approach, the FDA expects firms to be able to prove that processes will consistently work as specified.

The FDA strongly recommends retrospectively developing a system definitions document for systems in place. This document should define the needs the system addresses, the goals, and any additional purposes the system will accomplish. Also, develop and maintain a "from this point forward" SDLC approach policy of development documentation and system maintenance. The emphasis is on precise and organized records and specification maintenance. Keep all previous historical records before implementing a new system.

For all new systems being developed, adopt a formal SDLC and record development policy. Standards should be established and approved which define the individuals or divisions within the company who are to interact in the software development process, the specific responsibilities and the authority of each individual or division, and the procedures to be followed and approvals which must be obtained in the review and approval of a proposal for the development and implementation of a computer system.

Written standards that detail the requirements and restrictions to be followed in developing the computer program should be in place. These standards are to serve as a guide and reference of established programming procedures for those responsible for the actual development of the computer programs. In defining the software requirements, care should be taken to assure that the requirements are as correct and complete as possible. The design specifications should include all of the details and specifics for the program. These specifications should be updated with each software revision and documented so that they can be followed during later phases of the SDLC. Users' documentation should explain the desired functions and performance capabilities for the software and should be documented. Also, users' documentation should explain the procedures to be followed in running the system, with detailed discussions on the meaning and significance of all reports, alarms, and error messages as well as corrective actions which must be taken. This documentation should be developed in conjunction with the coding phase as a sub-phase (implementation) of the SDLC. If this documentation is prepared at the time the program is written, it will be more accurate and clearer than if prepared at the end of the project.

It is important that software is tested following an approved test plan or protocol; that the test results were properly reviewed and compared with the anticipated results; and that all test procedures, records, and documents have been reviewed and approved by the appropriate personnel. The records generated by the system should be routinely reviewed for accuracy and for the presence of discrepancies. In addition, the system should also be placed on a periodic maintenance schedule, and the maintenance activities should include verification that both the hardware and software remain as described in the design specification for the current revision level. Sufficient documentation reflecting the history and revisions to the software must be maintained. A Standard Operating Procedure (SOP) to this end should be developed and implemented.

Succinctly put, the SDLC approach is the documented process of development that is available for use as the system is required to change and adapt. The ability to clearly identify the intended functionality makes verification possible. As the FDA applies pressure on companies for documented evidence that their systems perform as specified, companies are urged to use the SDLC approach. By applying an SDLC approach, systems will have greater chances of passing an FDA inspection.

The FDA's technical report on software development activities follows:

TECHNICAL REPORT -- SOFTWARE DEVELOPMENT ACTIVITIES

REFERENCE MATERIALS AND TRAINING AIDS FOR INVESTIGATORS

FDA

JULY 1987

U.S. DEPARTMENT OF HEALTH AND HUMAN SERVICES
PUBLIC HEALTH SERVICE
FOOD AND DRUG ADMINISTRATION

OFFICE OF REGULATORY AFFAIRS

PREFACE

The use of computerized systems to perform process control and quality assurance activities within the industries regulated by the Food And Drug Administration is becoming more prevalent as the size and cost of this technology decreases. Computer systems are now commonly found controlling activities ranging from the manufacture of animal feeds to the operation of medical device products destined for implant. As these systems become instrumental in assuring the quality, safety, and integrity of FDA regulated products, it becomes extremely important for the Agency to verify that proper controls were employed to assure the correct performance of the computer system prior to its implementation and for the maintenance and monitoring of the system once it has been installed.

This is the second in a series of references designed to assist the investigator in his/her understanding of computerized systems and their controls. The first, "Guide to Inspection of Computerized Systems in Drug Processing" provided an overview of a complete computer processing system. This second publication is intended to focus on methods and techniques for the development and management of software.

It is not the intent of this guide to serve as a medium for the establishment of new software development procedures, standards, or requirements. Ample industry standards and procedures have already been published by such organizations as the American National Standards Institute (ANSI); Institute of Electrical and Electronic Engineers, Inc. (IEEE); International Standards Organization (ISO); National Bureau of Standards (NBS); Department of Defense (DOD); and American Nuclear Society (ANS). This document is intended to provide a synopsis of these requirements in a straight forward manner for use as a technical reference by the field staff of the Food and Drug Administration.

CHAPTER 1 -- WHAT IS SOFTWARE?

There was a time when the components of a computer system were clearly defined. Hardware consisted of those elements which you could see, hold, or physically touch. Software made up the intangible components, the computer programs and data. But computer technology is a dynamic and changing field in which new discoveries, methods, and techniques are encountered each day. With changes in technology it often becomes necessary to change the established definitions to enable a more accurate reflection of the true meaning of a word. Software is one such word.

It is now realized that software encompasses far more than just computer programs and/or data. Today, software is recognized as consisting of not only the computer code, but also the documentation necessary to allow the execution and understanding of the program by other knowledgeable individuals. The International Standards Organization (ISO) supported this concept when it defined SOFTWARE as "programs, procedures, rules and any associated documentation pertaining to the operation of a computer system". The idea that software consists of more than just the computer program is fundamental to a sound software development program and is the basis of this technical guide.

Categories of Software

This manual considers two basic categories of software, operating systems and application programs. The distinction between the two is determined by the function which they perform. An OPERATING SYSTEM is "software that provides services such as resource allocation scheduling, input-output control, and data management."[1] APPLICATION SOFTWARE is "software specifically produced for the functional use of a computer system."[1] By definition then, operating system software controls the activities of the computer system while the application program accomplishes the desired task.

An example of an operating system and an application program can be seen in the review of a computerized process control system. In most instances, the operating system software for the process controller has been written by the hardware manufacturer and purchased as part of the computer system package. Unless specially ordered, the same operating system software being used in our example could also be in use in hundreds of different manufacturing facilities producing many different types of products. The actual processes being controlled by the computer system are generally transparent to the operating system software. The operating system software is concerned only with the function and operation of the computer system.

The application program on the other hand is almost always unique to the specific process being controlled. This program, which may have been written by the supplier of the computer system, a third party software vendor, or the users own programmers, has been developed to meet the specifications of the process. It is the application program that actually monitors and controls the temperature, pressures, times, and all other important elements associated with the process. It is also the applications program that sounds the process deviation alarms and generates the desired reports and records.

From this example it may appear that our quality assurance concerns can be limited to the procedures and controls established for the application software. However, this is not the case. The operating system software and the application system software must work together if the system is to properly function. A revision to the operating system could be as significant to the proper function of the process control activities as a revision to the application program.

In our process control example, we were discussing a system with distinct hardware and software components employing classical operating system and application program software. These distinctions, however, are not always easily discernible. With the desire to reduce the size of microprocessing systems (especially in the area of medical devices), many of the elements of the hardware and software have been combined both physically and functionally.

For example, many of the medical device products being marketed today have combined the software and the hardware together forming a hybrid which is referred to as firmware. FIRMWARE is "hardware that contains a computer program and data that cannot be changed in its user environment" (detailed explanation of firmware is contained within Appendix A). Similarly, there are a number of software products marketed today which do not have a clear cut distinction between the operating system aid the application software. In these devices, the two categories of software have been combined in such a manner that the device is actually driven by the operating software and no true application program exists.

Software Characteristics

There is no mystery or magic to a computer program. Quite simply a computer program consists of nothing more than an extremely detailed set of instructions which are executed by the computer system in a specific predetermined order. The apparent magic results from the intertwining of the instructions and the ability to branch from one set of instructions to another. This would almost appear as a contradiction. How can the order of the execution of the instructions be predetermined while at the same time maintaining the freedom to branch from location to location or instruction to instruction within the program? There is, however, no contradiction. The computer program is so structured that even the branching capabilities are all predetermined. It is the branching ability of a computer program that gives it its versatility. This same branching ability also results in its complexity.

Let us compare for a moment the function of an electromechanical controller in a process and the function of a computerized controller in a process. In the electromechanical system the activities or actions performed are accomplished in exactly the same manner each and every time. If a temperature reaches a predetermined set point, a relay is triggered, an electrical circuit is completed, and a fan motor is turned on to cool the process. The electromechanical system accomplishes this task following exactly the same procedures time after time after time. This is not necessarily the case for a computer driven system.

In the computer driven system, the software monitoring and controlling the process temperature may also be monitoring and controlling other processing functions. As a result we don't know exactly where we will be in the execution of the computer program when a process deviation occurs. Similarly, the program may be written in such a manner that we may have more than one type of corrective action available to us. If the temperature is approaching the set point the program may be written to spend more time monitoring the temperature; if the temperature reaches the set point the program may be written to take some form of corrective action; and if the set point is exceeded the program may be written to divert the process while at the same time sounding an alarm and preparing a process deviation report.

Through the use of the computer driven controller, the options available to us have increased beyond simply turning on the fan. Through the diversity of the program, there is also more than one way (or path) available for reaching the temperature monitoring portion of the program. Consequently, the actions taken by the program may follow one set of instructions one time and a completely different set the next. This is an extremely important difference between electromechanical systems and computerized systems, primarily when we consider the validation of the system. As will be discussed in later chapters of this guide, it is important to realize that although a computerized system may function properly during two, three, four, or more test runs this in itself does not reflect the proper function of the system. It is possible that we followed the same paths or branches through the program during each of our test runs and that the remaining paths or branches contain errors or faults. For this reason, the validation of most computer controlled system must consist of far more than simply testing the computer through the performance of production runs.

There are two additional software characteristics which should also be discussed as they relate directly to the software's ability to follow different instruction paths during execution. It is important for us to realize that unlike the electromechanical systems with which we are familiar, software:

a. software can be improved over time; and
b. failures occur without advance warning.

From what we've discussed to this point about software, the reason for these characteristics should be apparent. It would stand to reason that software can be improved with time as the number of possible branches or paths within the program that are actually executed will normally increase with the use of the program. Consequently, latent or previously undetected program errors are found and changes or corrections to the program can be made. Also contributing to this characteristic is the fact that unlike electromechanical system, software does not wear out from use. A computer program can be executed hundreds of thousands of times with the physical condition of the software remaining unchanged from the first to the last run. (It should be noted that this is not case for computer hardware - disk, tape, module - which serves as the media to hold the software. Hardware again is a physical element that deteriorates with time.)

Similarly, the characteristic that software failures occur without advance warning should also be understood at this point. With mechanical components we realize that they physically deteriorate from use. For this reason we establish maintenance schedules for the replacement of parts that are showing wear and we determine from our experience with the device when or how often these parts will need replacing. We anticipate the wear and subsequent failure. With software, however, there is no physical wear on the program. Consequently, the failure of the software does not result from deterioration but rather from unforeseen or undetected programming errors. These types of errors may be encountered at any time and occur without any warning.

The Software Life Cycle

Software can be very complex. A process control program may consist of thousands of lines of code with <u>numerous</u> paths and branches from instruction to instruction. Nevertheless, the logic of the program must be understood by those responsible for writing, testing, and maintaining it. Because of this complexity, and the fact that computer systems have progressed from performing only basic arithmetic calculations to life dependent and life sustaining systems, computer software development has evolved into a recognized science of established procedures and disciplines. Among the basic teachings of this science are two premises:

1. Quality begins at the design level.
2. You cannot test quality into your software.

In an attempt to assure that these premises are followed, all categories of software should go through an organized development process consisting of a series of distinct phases which are collectively referred to as the SOFTWARE LIFE CYCLE. The existence of these identified phases allows reviews and tests to be conducted at distinct points during the development and installation of software. This in turn helps build quality into the software and provides a measure of that quality.

Although a consensus has not developed as to the naming of the phases, it is generally recognized that the Software Life Cycle is the period of time that starts when the software product is conceived and includes at least:

a. a requirements phase;
b. a design phase;
c. an implementation phase;
d. a test phase;
e. an installation and checkout phase; and
f. an operation and maintenance phase.

Chapters 3 through 8 will focus on each of the phases of the software life cycle. But first, Chapter 2 will discuss the standards and procedures which should be in place prior to entering the Software Life Cycle.

CHAPTER 2 -- STANDARDS AND PROCEDURES

Development Standards

As the development of a computer system generally requires the coordination of talent between a number of different individuals or divisions within the structure of an organization, each division must fully understand its authority and responsibility to the project. Each participant must be aware of what the others are doing and of their own specific responsibilities. For this purpose standards should be established and approved which clearly define:

a. the individuals by position description or the divisions within the organization which are to interact in the development process;
b. the specific responsibilities and the authority of each individual or division; and
c. the sequence of steps to be taken, including the procedures to be followed and the approvals which must be obtained, in the submission, review, and approval of a proposal for the development, implementation, or modification of a computer system.

In many instances the standards and procedures which are established for the development of software address the activities of the engineers, programmers, analysts, and managers. These are the individuals most commonly associated with the software development task. It is important, however, that the quality assurance group not be omitted from consideration at this point. The development process should be periodically audited to assure requirements are actually being adhered to. These audit procedures should describe the involvement of a quality assurance group detailing all reporting requirements and responsibilities for reviews and audits at various stages during the development process.

Programming Standards

Equally important in the development of quality software is the establishment of standards and procedures for the actual writing or coding of the computer program. A computer program is often a lengthy and very complex document which must be understood by those responsible for writing, testing, and maintaining it. To facilitate this understanding every effort should be made to maintain uniformity and clarity from module to module, routine to routine, or program to program regardless of who wrote them. For this purpose, written standards should be in place which detail the requirements and restrictions which are to be followed in writing computer programs. These standards may address:

1. Structural standards (e.g., structured programming with primary top down logic flow, modular program design, etc.).
2. Software documentation procedures including:
   a. description of the supporting documentation which is to be used (e.g., functional and/or process flow charts, data dictionaries, decision trees, pseudocode, etc.); and
   b. a statement as to where the documentation is to be maintained and who is responsible for keeping it current.
3. Coding standards covering:
   a. conventions for naming programs, modules, and variables;
   b. conventions for designating revision levels;
   c. identification of applicable programming manuals;
   d. restrictions and conventions on code format;
   e. methods for measuring the complexity of the program; and
   f. complexity controls (e.g., limiting the number of lines of code per module, control of branching instructions, etc.).
4. Software testing standards and procedures which include:
   a. designation of who will be responsible for the testing of the software;
   b. discussion of how the test data is to be generated;
   c. description of the test records and error logs which must be maintained; and
   d. a discussion of how errors or outstanding problems are to be resolved.

Standards and procedures such as these should be in place prior to beginning the development of the software if a quality product with a reasonable degree of safety and performance confidence is to be produced. Clearly everyone involved in the development project should have a sound understanding of his/her responsibilities and should be able to readily determine the responsibilities of the other members of the team. All expectations should be defined with nothing being left to chance.

CHAPTER 3 -- REQUIREMENTS PHASE

The first phase in the life cycle of any software product should be a definition or REQUIREMENTS PHASE. In defining software requirements, care should be taken to assure that they are as correct and complete as possible. Consider the process control program which we discussed in Chapter 1. If the process temperature exceeded an established set point, an alarm would sound notifying the operator of a process deviation. If the requirements for this program were written to reflect that when a deviation occurs, an alarm sounds, the programmer who knows little about the technical aspects of the process we are controlling would write the code to meet this requirement. The simple sounding of the alarm, however, would probably not meet the needs or requirements of the operator. In our fictitious process control system we stated that the program is monitoring and controlling any number of important factors such as temperature, pressure, time, and product flow. You can imagine the confusion in the production area as the operator tries to determine where the deviation in the process occurred each time the alarm sounded. There is also a good chance that by the time the operator identified the specific deviation, it may be too late to save the product. In addition, this particular manufacturing activity may require that records of process deviations be maintained. If this is not detailed in the program requirements, provisions for a deviation report and a method for alarm identification will more than likely not be included in the software.

Ideally, problems such as this will be realized and resolved while our project is still in the requirements phase. The failure to do so will eventually result in a need to rewrite, add, or delete sections of program. As each revision introduces the possibility for error, the modified code must be accompanied by new documentation and retested to the extent necessary to establish a new degree of confidence in the revised program. This process takes time. Meanwhile, the operator in our example is becoming worn out trying to determine the meaning of each alarm, process deviation reports are not being maintained, and finished product is being discarded.

Identifying the actual requirements for the finished product is one of the most difficult tasks associated with software development. It requires the ability to understand both the process or activity being computerized as well as the needs of the individuals or operators who will be using the system. In many cases, these needs may not even be known by the users of the system at this phase, but they must nevertheless be anticipated to the maximum degree possible by those responsible for planning and coordinating the project.

To this point only the importance of defining the project requirements have been discussed. It is, however, equally important to document the desired functions and performance capabilities for the software product during this phase. Such documentation is essential for the proper conveyance of ideas and needs to those individuals responsible for other phases of the software life cycle.

Prior to proceeding to the next phase, time should be taken for the review and approval of the documents generated during this phase to assure that they are complete, and that the requirements decided upon are correct and compatible with the system being developed. Any unresolved problems at this point should be documented within these records along with a description of the actions being taken to rectify them, the identity of the individual responsible for tracking them, and the points or times at which they will be periodically reviewed.

The requirements phase should not be taken lightly. As stated in Chapter 1, quality begins in the design level and the establishment of the complete requirements is the first step in the design process.

CHAPTER 4 -- DESIGN PHASE

Software design is "the process of defining the software architecture, components, modules, interfaces, test approach, and test data for a software system according to system requirements."[1] The keyword is "defining."

From the Requirements Phase the needs and expectations for the finished software should be known. During the DESIGN PHASE these requirements will be converted into specifications.

For each systems application, a formally developed and approved design specification should be established. This document, which should be written as clearly and precisely as possible, is the guide/instructions to those responsible for writing and maintaining the computer programs. The design specifications should include all of the details and specifics for the program. These may be:

1. A description of the activities, operations, and processes that are being controlled, monitored, or reported by the computer system.
2. A detailed description of the hardware which is to be used.
3. A detailed description of the software which will be used including the algorithms which are to be used in processing the data.
4. A detailed description of the files which will be created/accessed by the system and the reports which are to be generated.
5. A description of all limits and parameters which are to be measured by the system and the points at which the alarms are to sound or the data is to be rejected.
6. A description as to how often or when the limits and specified parameters are to be measured.
7. A description of all error and alarm messages, their cause, and the corrective actions which must be taken should they occur.
8. A description of all interfaces, connections, and communication capabilities between the processor and the external world and between parts of the software.
9. A description of the security measures which are to be followed to verify the accuracy of the program being used and to protect the program from both accidental and intentional abuse.

Prior to the acceptance of these design specifications, a review should be conducted to determine if the design is following accepted standards and is consistent with the software requirements.

The design specifications should be updated with each software revision and documented so that they can be followed during later phases of the life cycle.

CHAPTER 5 -- IMPLEMENTATION PHASE

By now the needs and expectations for the system, as well as the procedures and specifications which are to be followed in meeting these requirements, are fully known. The documentation which has been prepared in the earlier phases of the life cycle is now ready to be provided to the programmer(s) for conversion into a software product. But, there is far more to the implementation phase than the conversion of concepts and ideas into a computer program. It is during this phase that the software developers must also concern themselves with documenting the program which they are writing and begin taking the steps necessary to "debug" or eliminate faults (errors) from it.

In most instances, software is developed by the programmer as a SOURCE CODE. The source code may be in any number of different languages including BASIC, PASCAL, C, ADA, ASSEMBLER, FORTRAN, or COBOL to name a few. Different portions of the source code may even be written in different languages.

Source codes were developed as an aid to the programmer as they allow the program to be written in a mathematical or English based format. As written, however, the source code cannot be understood or executed by the computer. Consequently, prior to being run (executed) the source code must be converted into a machine executable format using another software package which is referred to as an assembler, compiler, or interpreter depending on the language being used. The result of this conversion process is an object code version of our program.

The OBJECT CODE (or machine level binary code) is "a fully compiled or assembled program that is ready to be loaded directly into the computer" (ISO). Where source code is easily readable, object code is represented by a series of 1's and 0's which would require a great deal of knowledge and understanding to interpret their meaning.

It is important to understand that programming even in source code is difficult. Computer programs are difficult for the programmer to write and, in themselves, are even more difficult for someone else to understand. As a result, efforts are made to establish at least some degree of continuity from programmer to programmer. This is the purpose of Standards and Procedures which we discussed in Chapter 2. There are also established rules and procedures, called SYNTAX, which must be followed based on the programming language which we are using. But even with rules, procedures, standards, and syntax requirements, no two programmers will write a computer program in exactly the same way. As a result the documentation of the computer program is extremely important.

Documentation

Software documentation refers to the records, documents, and other information prepared to support the program. This information ranges from the program requirements which were defined in the Requirements Phase of the software life cycle through to the manuals which explain the program to its users. Historical documentation reflecting the history of the software product including records of all changes or revisions is referred to as an AUDIT TRAIL and is further discussed in Chapter 8.

Entire books have been written on the subject of software documentation. To cover this subject in detail is beyond the scope of this guide. Essentially, however, computer software should be supported with sufficient information to allow a knowledgeable individual to understand the program requirements and specifications, as well as the logic and algorithms being used.

To assure this understanding the basic information maintained should include the documents and records generated during the Requirements Phase and the Design Phase of the software life cycle; the records and documents which reflect the test protocols which were used to test the software; test data; test results; and records explaining the rationale behind the tests which were or were not conducted. The information required in support of the software logic, however, will vary from program to program.

The type and degree of documentation needed to support the software logic is dependent upon both the language being used and the complexity of the program. For example, high level languages such as Basic and Pascal may require less support documentation than languages such as Assembler or Fortran. The reason for this is that languages such as Basic and Pascal are more English oriented, making them easier to read and understand. This is not to say that software does not require support in the form of flowchart, data dictionary, comment statements, etc., but rather that the support may not need to be as extensive as it would be with other programming languages.

Regardless of the language used, the deciding factor as to the degree of documentation needed rests in the ability to understand the code by those responsible for maintaining it. It should be possible for these individuals and their supervisors to locate sections of the program responsible for the performance of specific control functions and explain the basic logic being followed without undue difficulty.

Although the focus of this guide is on software, it is important to realize that systems documentation and user documentation are also important to the quality of the finished product.

As computer systems become more complex, it becomes extremely difficult to retain an understanding of how the various peripheral devices, software, and the established records and files interact. SYSTEMS DOCUMENTATION assist in this area by providing pictorial and/or narrative explanations of the interaction of hardware and software. The need for such documentation is again proportionate to the complexity of the system.

USERS DOCUMENTATION consists of operating manuals containing instructions written in lay terminology which clearly define the activities and processes being performed by the computer and those which must be performed manually. This documentation should explain the procedures to be followed in running the system with detailed discussions on the meaning and significance of all advisories, alarms and error messages as well as the corrective action which must be taken.

Documentation, such as that discussed in this section, should be developed in conjunction with the implementation phase of the software life cycle. If prepared at the time that the program is actually written, it is likely to be clearer and more accurate than if prepared at the conclusion of the project.

Debugging

The process of "debugging", or "locating, analyzing, and correcting suspected faults"[1] in the computer program should begin with the actual writing of the code. Control of programming is achieved by keeping the scope of any part of a program small and functional. Breaking a large program into workable pieces called modules allows for this degree of control. It is both more efficient and far easier to review and analyze small modules of code locating and correcting potential errors than it is to try to locate these same faults in the finished complex program.

The concept of testing the software will be discussed in detail within the next chapter. It is important, however, that you realize that it is at THIS point in the software development process that we begin to review the code for suspected errors. You should not wait until the program is completed or, as has been found in several instances, the program is placed into the finished process control system or medical device to begin analyzing the software for possible faults.

CHAPTER 6 -- TEST PHASE

The Test Phase is actually "the period of time in the software life cycle during which the components of a software product are evaluated and integrated, and the software product is evaluated to determine whether or not requirements have been satisfied."[1] This is NOT the point at which we first begin thinking of testing the program. By now, the software should have already undergone some basic debugging by the programmer. In addition, a complete test plan should be based on the program requirements and specifications as detailed in the earlier Requirements and the Design Phases of the project.

Although some discussions of the Software Life Cycle distinguish between the "debugging" of the software during the Implementation Phase and the "testing" of the software during the Test Phase, this document will deal with both the testing and debugging of software as a single entity.

Software Errors

Programming errors can be classified into one of two types, syntax errors or logic errors.

SYNTAX ERRORS are language or format errors. They are caused by typos or oversights in the structure of expressions in a programming language. Examples of syntax errors include the omission of a comma, period, or parenthesis in a program instruction. The failure to properly structure a portion of the program, or on some computer systems, attempting to access a file that does not exist can also result in a syntax error. Syntax errors are relatively straight forward and in most cases are identified by the computer system as it attempts to convert the source code into an object code. Most computer system are programmed to identify syntax errors and will not attempt to execute the program until they have been corrected.

LOGIC ERRORS by contrast are difficult to detect as they reflect an error in the programmers reasoning or method of problem solving. Logic errors are sometimes undetected by the programmer as he/she is unable to see the mistake in his or her own reasoning. Of further concern is the fact that logic errors are totally transparent to the computer. If in the process control example, which we discussed in Chapter 1, we were to program our system to look for a temperature of 250 F with 0 psi, the system would look for exactly that temperature and pressure. The system is totally ignorant to the fact that this temperature and pressure is an impossibility.

Test Plan

The intent of software debugging/testing is to identify and correct as many errors as possible and to the impact of the unforeseen fault when it does occur. The testing of software is an extremely complex task requiring a great degree of planning. The steps taken in testing of the software should include the separation of a test plan which is "a document prescribing the approach to be taken for intended testing activities. The plan typically identifies the items to be tested, the testing to be performed, test schedules, personnel requirements, reporting requirements, evaluation criteria, and any risks requiring contingency planning."[1]

Often, the philosophy used in the development of the test plan is critical to the success or failure of the test exercise. The definition of the word "testing" is worthy of consideration. If the purpose of the test plan is to demonstrate that the software performs its intended function or that the program does what is supposed to do, then the plan will undoubtedly be designed to show that the program works. This concept is not satisfactory for finding errors. A more appropriate approach is to begin the design of the test plan by acknowledging that the program contains errors and then design the plan with the intent of finding as many as possible. From this viewpoint, testing is more correctly defined as "the process of executing a program with the intent of finding errors"[3].

Ideally the test plan and the selection of the test data will be assigned to a team of individuals rather than to the programmer. There is good reason for this approach. As was mentioned in our discussion of logic errors, there is a good chance that the logic errors contained within the program are totally transparent to the programmer. If the programmer was able to see the error, it would already be corrected. In addition the testing approach taken by the programmer will almost always be from the position of trying to show what the program will do. The programmer is undoubtedly proud of his/her work and will have a great deal of difficulty developing an impartial test plan designed to reveal program failures.

This is not to say that programmer should be totally excluded from the development of the test plan. An appropriately designed plan must take into consideration the basic structure of the software when selecting test methods and test data. The programmer will have an intimate knowledge of this structure which he/she will be able to contribute to the project.

Although the team approach is preferred, it's not absolute. But, in situations where the programmer is also responsible for developing the test plan, additional care must be taken in an attempt to arrive at a plan that is as objective as possible.

Regardless of whether the test plan is developed using the team approach or by the individual programmer, consideration should be given to basic testing principles such as the following which were developed by Myers[3]:

1. Do not plan a testing effort under the tacit assumption that no errors will be found.

2. The probability of the existence of more errors in a section of a program is proportional to the number of errors already found in that section.
3. The test plan should include a definition of the expected output or result.
4. The results of each test should be thoroughly reviewed and evaluated.
5. The results of each test should be retained and not discarded upon completion of the test.
6. Test cases must be written for invalid and unexpected, as well as valid and expected, input conditions.
7. The test plan should be designed to challenge the program to not only assure that it does what it is supposed to do, but to also assure that it does not do what it is not supposed to do.

Test Methods

To fully test the capabilities and limitations of software, the software must be tested in simulation. Only through simulation can the necessary test cases be generated to adequately challenge the ability of the software to function under both normal and abnormal conditions. Appropriate testing cannot be accomplished solely through the performance of production or quality assurance runs or through the evaluation of the finished device.

The proper testing of the software includes the challenge of EACH decision path within the program as well as all alarms, error routines, process specifications, etc. This is not to say that all possible combinations of paths must be challenged, but rather that each individual path within the program must be reviewed and evaluated as to its correctness and effect on subsequent program execution.

As this degree of testing is often hampered both the size and complexity of the finished software package, large programs should be divided into smaller units by module or subroutine which can be evaluated individually to assure that the design specifications have been met and that the code structure is challenged according to the defined test criteria. This procedure is referred to as MODULAR TESTING and may be initiated within either the Implementation Phase or the Test Phase of the Software Life Cycle.

Once the modules or routines have been tested individually, they should be integrated and tested to identify design inconsistencies and to remove errors which may result from data transfers and interfaces. This INTEGRATION TESTING should not repeat the module tests but, rather, it should build upon the previous test results.

There are two different methods of integrating the modules during the performance of integration testing. The first, and preferred method, is INCREMENTAL INTEGRATION which is the reformation of the program module by module or function by function with an integration test being preferred following each addition.

The second method, NONINCREMENTAL INTEGRATION, consists of immediately relinking the entire program following the testing of each independent module. Integration testing is then conducted on the program as a whole. This method is not recommended as errors are usually very hard to isolate and correct.

Types of Tests

There are a large number of different tests which can be used for the challenge and evaluation of software. To list and discuss each is beyond the scope of this guide. For identity purposes, several of the more commonly encountered test types are briefly discussed in Appendix C.

In general, the test procedures used for the evaluation of software can be categorized into two types, functional or black-box tests and structural or white box tests. The test plan should consist of tests of both types.

FUNCTIONAL TESTING (black-box testing) challenges the design specifications for the software. The emphasis is on input and output, not processing. It is the testing of the program without using knowledge of program design and implementation. Functional testing can be subdivided into two specific areas, equivalence class partitioning and boundary value analysis.

The theory behind EQUIVALENCE CLASS PARTITIONING is that input and output data can be partitioned into classes with the assumption being that all data in one class should produce an identical result. The equivalent classes cover valid as well as invalid input/output data.

In addition to equivalence class testing, frequently software errors can be related to parameter limits or boundaries (BOUNDARY VALUE ANALYSIS). Therefore, test cases on and close to the class limits are included.

STRUCTURAL TESTING (white-box testing) tests the code. The emphasis is on testing the flow of control within the program. The criteria is to design test cases until each statement is executed at least once. Therefore each true/false decision must be executed with both true and false outcomes.

Depending on the type of software being developed, the data in the performance of these test procedures will normally include:

a. normal cases;
b. limits;
c. exceptions
d. special values (e.g., "-", "0", "1", empty strings);
e. initialization;
f. missing paths;
g. wrong path selection; and
h. wrong action.

Additionally, testing experience and intuition can be combined with knowledge and curiosity about a system under test to add some categorized test cases to the design of the test case set. This may involve the use of special values or a particular combination of values which may be error prone.

It is customary to test various aspects of the software at different times throughout the software development process. As a result, the records of software verification will not necessarily be consolidated in one location or one document. What is of significance is that the software has in fact been tested following an approved test plan or protocol; that the test results have been appropriately reviewed and compared with the anticipated results; and that all test procedures, records and documents, including the test results, have been reviewed and approved by the appropriate personnel.

The testing of the software in the manner described in this chapter is not limited to the initial release of the program. Rather, the software must be retested, as necessary, following each revision. The testing procedures which should be used following the revision of the software will be discussed in Chapter 8.

CHAPTER 7 -- INSTALLATION AND CHECKOUT PHASE

By now it should be apparent that far more goes into assuring the quality and reliability of a computerized system than simply validating its ability to function in an application environment. This premise does not reflect, however, that systems validation is to be omitted. Rather, the validation of the computer system is an integral element, but only one element, in the criteria for system acceptance.

Until now, the tests and evaluations which have been conducted on the software have been performed in simulation and independent of the application environment. We are now ready to begin testing the software package in combination with the actual device or production hardware and within the environment with which it has been designed to function.

The validation of a computer system is accomplished through the performance of either actual or simulated production runs or through the actual or simulated use of the finished medical device product. During the performance of system validation, the computer system must be properly monitored. The physical parameter's, which it is designed to measure, record, and/or control should be measured by an independent method until it is demonstrated that the computer system will function properly in the application environment.

In addition to evaluation the system's ability to properly perform its control functions, the validation of the system should also include an evaluation of the ability of the users or operators of the system to understand and correctly interface with it.

The length of time or the number of production runs necessary for the validation of the system will vary from application to application and program to program depending upon the complexity of the software and the number of faults encountered. The validation efforts should continue for a sufficient period of time to allow the system to encounter a wide spectrum of processing conditions and events in an effort to detect any latent faults which are not apparent during the normal processing activities.

Records must be maintained during the performance of system validation of both the system's capability to properly perform and the system's failures, if any, which are encountered. The revision of the system to compensate for faults detected during the validation process should follow the same procedures and controls as any other software modification or change. These procedures are discussed in Chapter 8.

CHAPTER 8 -- OPERATION AND MAINTENANCE PHASE

The software has been written, tested, installed, and is now operational. Unfortunately, this is the point where some developers close the files and move on to the next project. Actually, the current project is far from complete. By certifying the software as operational we have simply moved it from one phase in its life cycle to another. It has now entered "the period of time .... during which a software product is employed in its operational environment, monitored for satisfactory performance, and modified as necessary to correct problems or to respond to changing requirements."[1]

There are two critical elements expressed within this definition of this phase of the Software Life Cycle: software monitoring and modification. It is these two elements which are the focus of this chapter.

Monitoring Software

As discussed in the previous chapters on software and systems testing, software contains faults. Although the system may have passed the testing and review regimen which we developed for it, there is still no assurance that it is totally fault free. More than likely latent errors remain which have gone by undetected. Because of this, it is important that we do not place the new system into operation with no monitoring plans or procedures.

The degree to which a computerized system should be monitored is dependent upon both the functions which it performs and the length of time that it has been in operation. Consequently, to some extent the degree of confidence which we place in a computer system is proportional to the length of time that the revision level has been in use and whether the process being controlled has remained unchanged.

There are no set or established rules for how often a system should be monitored. It stands to reason, however, that the records generated by the system should be routinely reviewed for accuracy and for the presence of obvious discrepancies. This is especially true when new or unusual circumstances have occurred which affect either the system directly or the process which it is monitoring or controlling. In addition, the system should also be placed on a periodic maintenance schedule with the maintenance activities including the verification that both the hardware and the software remain as described in the design specification for the current revision level.

Modifying Software

One of the major advantages of using a software controlled system is the ease with which the characteristics of the system can be changed. As a result, very few programs stay unchanged for long. New capabilities can be added and existing ones can be revised or deleted and, once the change has been completed, there are no erasures or crosscuts reflecting that it occurred.

In the next chapter we will touch on several security methods which can be used in an effort to protect the system from unauthorized modification or revision. These protection methods, however, are fallible. If someone truly desires to circumvent the system they will find a way to do so. As a result, it is important that everyone concerned be informed of the possible consequences of unapproved revisions and that proper controls be established detailing the procedures which must be followed in revising or modifying software. These procedures should include:

1. Controls which assure that the modification is reviewed and approved by those knowledgeable of the intended function of the system and the programming language being used.
2. Controls which assure that with each modification or revision, the design specifications and support documentation are also revised to the highest level necessary to assure that all documentation accurately reflects the functions and operations of the system.
3. Procedures for determining the degree of testing necessary to assure the proper function of the system following any modification.
4. Procedures for distributing revised software to the operational level.

In addition to these controls and procedures, documents should also be prepared which:

1. Detail the specific changes made to the system. An appropriately identified printout or a copy maintained on a magnetic storage media of each revision level of each application program should be retained within the firm's records. From these records it should be possible to reconstruct the entire history of the software from the initial Requirements Phase to the installation of the current revision level. Collectively, these records are referred to as an AUDIT TRAIL.
2. Record all tests performed following each revision. If the software and total system were not subjected to a complete retesting of all functions and operations following the revision, the test documentation should include a statement explaining the rationale for the performance of only limited testing. The decision to only test portions of the system should be reviewed and approved by the appropriate personnel.

When reviewing software modifications do not overlook the possibility of revisions being made to the operating system software. Any revision to the operating system must be thoroughly reviewed and evaluated by the firm's systems personnel, as even the smallest change to an operating system could have a major impact on the established application software.

CHAPTER 9 -- QUALITY ASSURANCE AND CONTROLS

Throughout the various stages of the Software Life Cycle, documents are being prepared, various forms of data are generated, and control records are being maintained. As we attempt to represent these activities within this reference, all of these materials appear straight forward and manageable.

In reality, the pressures and frustrations of a software development project are such that often there doesn't appear to be time to stop and record an event or document what appears to be a simple and obvious procedure. Experience dictates that as deadlines pass, the amount of support documentation generated becomes less and less. For this reason it is important that QUALITY ASSURANCE reviews be performed of the development process to assure that the process remains in control.

It is recommended within ANSI/IEEE Standard 730-1984[4] that the following reviews be made:

1. Software Requirements Review to ensure the adequacy of the requirements stated in the software requirements specification.
2. Design Review to evaluate the technical adequacy of the preliminary design of the software as depicted in the software design specifications.
3. Software Verification and Validation Review to evaluate the adequacy and completeness of the verification and validation methods.
4. Functional Audit prior to software delivery to verify that all requirements specified in the software requirements specification have been met.
5. Physical Audit to verify that the software and its documentation are internally consistent and are ready for delivery.
6. In-Process Audit of a sample of the design to verify consistency of the design, including:
   a. code versus design document;
   b. interface specifications (hardware and software);
   c. design implementations vs. functional requirements; and
   d. functional requirements vs. test descriptions.
7. Managerial Reviews to assess the execution of this plan. These reviews should be performed by an organizational element independent of the unit being audited or by a qualified third party.

It should be noted that these requirements are those of the ANSI/IEEE, not the Food & Drug Administration. This information is provided for reference only.

APPENDIX A -- FIRMWARE

As briefly mentioned in the first chapter of this guide, firmware has been defined as "hardware that contains a computer program and data that cannot be changed in the user environment. The computer programs and data contained in firmware is classified as software; the circuitry containing the computer program and data is classified as hardware."[1] From this definition it should be apparent that firmware is a hybrid of both hardware and software.

The software prepared for use as firmware should be subjected to the same controls as any other software package. The procedures described throughout our discussion of the software life cycle still apply. Once the software is developed, however, it in itself is not the finished product. Rather, the development project continues with the integration of the hardware and the software. This integration process utilizes memory types, programming devices, and erasing equipment.

Memory Types

What does firmware look like? For the most part we have no problem visualizing the physical characteristics of hardware or software, but what about firmware?

In the finished state, firmware appears as an integrated or silicon chip. The chip used may vary among several different types with their classification being based on their memory capabilities. Those commonly found in use are:

1. READ-ONLY-MEMORY (ROM) -- "a type of memory whose locations can be accessed directly and read, but cannot be written into"[2] The data content is determined by the structure of the memory and is unalterable.
2. PROGRAMMABLE READ-ONLY-MEMORY (PROM) -- a field programmable read-only-memory that can have the data content of all memory cells altered once.
3. ERASABLE PROGRAMMABLE READ-ONLY-MEMORY (EPROM) -- a reprogrammable read-only-memory in which all cells may be simultaneously erased using ultra violet light and in which each cell can be reprogrammed electrically.
4. ELECTRICALLY ERASABLE PROGRAMMABLE READ-ONLY- MEMORY (EEPROM) -- a reprogrammable read-only-memory in which cells may be erased electrically and in which each individual cell may be reprogrammed electrically.
5. VOLATILE MEMORY COMPONENTS -- battery backed random- access memory (RAM) is another type of memory component. This memory requires a power supply but lends itself to modification and reprogramming.

Programming Devices

The integration of our finished software product into one or more of these memory chips is accomplished using a device referred to as a PROM Programmer or PROM Burner. The procedures normally followed in the initial programming of the memory chip consist of connecting the PROM Burner to a computer system and then "downloading" or transferring through the computer the software program from the conventional storage media on which it was developed, to the memory component. (It is possible with some types of PROM Programs to program the memory chip directly using a bit by bit programming method. This procedure, however, is not commonly used as it is extremely time consuming and error prone.)

Once the first memory chip has been successfully programmed, this chip will normally be used as the master with duplicate chips being programmed or cloned on the PROM Programmer directly from it. At this point there is no longer a need for the conventional computer system and the duplication of chips can be accomplished quite rapidly.

Erasing Equipment

The erasing of programmed memory chips is a relatively straight forward process. The programmed chips are removed from the circuitry and are placed in an erasing device. The chip is then subjected to ultraviolet light or an electrical current which resets all of the circuits on the chip back to a non-programmed state. Upon completion of the erasing process, the chip is removed and is ready for reprogramming. (It is important that the state of the memory chip be checked and verified at least prior to programming and preferably also following erasure to assure that all bits on the chip are in a non-programmed state. The failure to do so could result in the erroneous programming of the chip.)

APPENDIX B -- PROM PROGRAMMING

MEMORY PROGRAM DEVICES

These devices perform two key function:

1. Transfer data or software programs from typical storage media (magnetic tape, disks, etc.) to a "master" read-only-memory component; and
2. The cloning of other memory components from the master component.

As with other electronic equipment, operation of these memory programming devices should be controlled and monitored. Documents pertaining to standard operating procedures (SOP's) should be available covering operation of the unit, preventive maintenance, and a schedule and documentation of calibration efforts.

MASTER MEMORY COMPONENTS

The master memory components are the components from which production memory components are cloned or copied. Procedures should be in place covering the following areas:

1. Security and Maintenance -- There should be a SOP for the storage/maintenance of master components which outline the procedures the firm has taken to protect the components against damage.
2. Change Control -- During its life cycle, software is usually modified and specific change control measures should be established. Here it is not intended to repeat software change control but rather that established procedures should be in place to remove old master components from production areas. Additionally, procedures should be established for the archiving of programs so that they can be reconstructed if the need arises.

COMPONENT CONTROL

Acceptance Criteria -- There should be established procedures and acceptance criteria for memory components.

Component History -- A component history record should be maintained which tracks the number of memory components from each lot that were either successfully or unsuccessfully programmed. This record serves a useful purpose in the evaluation of programming device performance, and memory component failure trends.

Component Identification -- A procedure should be established that clearly identifies the methods used to identify programmed components. The identification should include a method of identifying the revision level of the software. The methods have been employed:

1. Stick-On-labels -- Stick-on labels have been used to identify memory components. These labels can be used to identify revision levels and also cover the chip access window which protects the component from UV light and unintentional changes in the program.
2. Ink printing -- Ink printing involves two steps: first the previous identification marking must be covered with black epoxy ink; and second the new identification marking must be applied.

BENEFITS OF A FIRMWARE CONTROL SYSTEM

Destroyed or malfunctioning memory components can be quickly replaced with proper components.

Memory components impacted by changes in software can be immediately identified.

A clear audit trail (a history of the device) to "Burned in" software is established that assures proper marking, testing, and placement.

It is easy to verify that the proper software is on the circuit board.

AREAS OF SPECIAL CONCERN

INTERFACE CONNECTIONS

As with any other peripheral device care must be taken that the memory programmer is compatible with the computer in which software is stored. Additionally, some memory programmers have their own software.

GANG EXPANSIONS

Although some devices must be programmed individually, most of the commercially available programmers are marketed with the capability of expanding the number of components which can be programmed at one time. Expansion boards known as "Gang" expanders are used for this purpose.

CALIBRATION

Programming equipment requires periodic calibration checks. Specifically, the voltage used to program memory locations is important and usually within a narrow range (i.e., 21 V +/- .5 V).

APPENDIX C -- SOFTWARE TEST METHODS

As discussed in Chapter 6, the ability of software to meet its functional requirements and design specifications can best be demonstrated through the execution of a proper test plan. This section describes techniques and methods which are being used to perform this testing function. The methods described in this section are not all inclusive of the test methods available nor are the test methods fully detailed. The intent of this section is to briefly overview some of the common test methods being used.

CODE, MODULE, AND INTEGRATION TESTS

DESK CHECKING

The manual simulation of program execution to detect faults through step by step examination of the source code for errors in logic or syntax. (ANSI/IEEE).

INSPECTION

Inspection is a formal evaluation technique in which software requirements, design, or code are examined in detail by a person or group other than the author to detect faults, violations of development standards, or other problems. (ANSI/IEEE).

WALK-THROUGH

A walk-through is a review process in which a designer or programmer leads one or more other members of the development team through a segment of design or code that he/she has written, while the other members ask questions and make comments about technique, style, possible errors, violation of development standards, and other problems. (ANSI/IEEE).

STATIC ANALYSIS

This technique is one which is used to identify weaknesses in the source code by examination via paper review but without execution. The intent is to find logical errors (program errors dealing with the logical sequence of events) and to pinpoint questionable coding practices which may deviate from established standards.

SYMBOLIC EXECUTION

A verification technique in which program execution is simulated using symbols rather than actual values for input data, and program cutouts are expressed as logical or mathematical expressions involving these symbols. (ANSI/IEEE).

SOFTWARE AUTOMATED TESTING TOOLS

ACCURACY STUDY PROCESSOR

Used to perform calculations or determine accuracy of computer manipulated program variables.

AUTOMATED TEST GENERATOR

A software tool that accepts as input a computer program and test criteria; generates test input data that meet these criteria; and, sometimes, determines the expected results. (ANSI/IEEE).

COMPARATOR

Used to compare two computer programs, files, or sets of data to identify commonalities or differences. Typical objects of comparison are similar versions of source code, object code, data base files, or test results. (ANSI/IEEE).

CONSISTENCY CHECKER

Used to test for requirements in design specifications for both consistency and completeness.

INTERPHASE TESTING

Testing conducted to ensure that program or system components pass information or control correctly to one another. (ANSI/IEEE).

INTERRUPT ANALYZER

Analyzes potential conflicts to a system. As a result of the occurrence of an interrupt.

SIMULATOR

A device, data processing system, or computer program that represents certain features of the behavior of a physical or abstract system. (ANSI).

A simulator provides inputs or responses that resemble anticipated process parameters. Its function is to present data to the system at known speeds and in a proper format.

STATIC ANALYZER

A software tool that aides in the evaluation of a computer program without executing the program. Examples include syntax checkers, compilers, cross-reference generators, standards enforcers, and flowcharters. (ANSI/IEEE).

TEST RESULT ANALYZER

This is used to test output data reduction, formatting, and printing.

CHANGE TRACKER

This program to documents all changes made to a program.

SYSTEMS TESTING

LOAD/STRESS

A method of testing real time process control systems to determine their ability to cope with program interrupts while maintaining adequate process control.

VOLUME

Testing designed to challenge the systems ability to manage the maximum amount of data over a period of time. This type of testing also evaluates the systems ability to handle overload situations in an orderly fashion.

USABILITY

Tests designed to evaluate the machine-user interface. Are the communication device(s) designed in a manner such that the information is displayed in a understandable fashion enabling the operator to correctly interact with the system?

PERFORMANCE TESTING

A test designed to measure the ability of a computer system or subsystem to perform its functions; for example response times, throughput, number of transactions. (ANSI/IEEE).

STORAGE TESTING

This test determines whether or not certain processing conditions use more storage than estimated.

CONFIGURATION AUDIT

The process of verifying that all required configuration items have been produced; that the current version agrees with the specified requirements; that the technical documentation completely and accurately describes the configuration items; and, that all change requests have been resolved. (ANSI/IEEE).

COMPATIBILITY TESTING

The process of determining the ability of two or more systems to exchange information.

In a situation where the developed software replaces an already working program, an investigation should be conducted to assess possible compatibility problems between the new software and other programs or systems.

RELIABILITY ASSESSMENT

The process of determining the achieved level of reliability for an existing system or system component. (ANSI/IEEE).

Theoretical models exist for estimation of mean-time-between-failures based on the number of errors found, and an estimated number of errors left in the software. In general, reliability is also related to documentation and quality assurance efforts performed throughout the software life cycle.

ROBUSTNESS ASSESSMENT

The evaluation of the software to determine the extent to which it can continue to operate correctly despite the introduction of invalid inputs. (ANSI/IEEE).

Test documents should be maintained which demonstrate the performance of the system when exposed to errors including power failures and invalid inputs.

METRICS

In challenging a software or system with test cases, it would be helpful to have a means of evaluating the adequacy of the test being used. It does little good to have a test result which indicates that the software will function according to the requirements and design specification without knowing the accuracy of the test methods used to arrive at this conclusion. As a result, a number of techniques have been developed which are designed to measure the adequacy, or METRICS, of the software test methods being used. Three of the more commonly used methods are:

COVERAGE ANALYSIS

A test procedure in which the program is processed with counters and flags in all different branches. When a branch is executed, the corresponding flag is updated. After test execution with test cases, the flag values are listed, accumulated and compared to the optimum value depending on actual criteria.

FAULT SEEDING

The process of intentionally adding a known number of faults to those already in a computer program for the purpose of estimating the number of indigenous faults in the program. (ANSI/IEEE).

In performing this test, known errors are inserted in a copy of the program being evaluated and the copy is executed with selected test cases. If only some of the seeded errors are found, the test set is not adequate. However, the ratio of found seeded errors to the total number of seeded errors is an estimate of the ratio of real errors to the total number of real errors. This test result is an estimate of the number of errors remaining and thus the amount of future testing required.

PROGRAM MUTATION

A program version purposely altered from the intended version to evaluate the ability of program test cases to detect the alteration. (ANSI/IEEE).

If the test set is adequate, it should be able to identify the mutants through error detection.

The method of seeding is crucial to the success of the technique and consists of modifying single statements of the program in a finite number of "reasonable" ways. The developers of this method conjecture a coupling effect which implies that these "first order mutants" cover the deeper, more subtle errors which might be represented by higher order mutants.

APPENDIX D - REFERENCES

The following references were used in the preparation of this guide.

1. ANSI/IEEE Standard 729-1983. Glossary of Software Engineering Terminology.
2. Harry Helms, The McGraw-Hill Computer Handbook. New York: McGraw-Hill Book Company, 1983.
3. G.J. Myers, The Art of Software Testing. New York: Wiley-Interscience, 1979.
4. ANSI/IEEE Standard 730-1984. Software Quality Assurance Plans.
5. ANSI/IEEE Standard 828-1983. Software Configuration Management Plans.
6. ANSI/IEEE Standard 229-1983. Software Test Documentation.
7. ANSI/IEEE Standard 830-1984. Software Requirements Specifications.
8. G.J. Myers, Software Reliability Principles and Practices. New York: Wiley-Interscience, 1976.
9. R. Dunn and R. Ullman, Quality Assurance for Computer Software. New York: McGraw-Hill, 1982.
10. Susanne Klim, Systematic Software Testing for Micro Computer System, International Planning Information, 1984.

Chapter 11

DATA INTEGRITY AND SECURITY[1]

AUTOMATION PROBLEMS AND DATA INTEGRITY

Recent evidence of corruption, loss, and inappropriate modification of computerized data provided to the FDA and EPA prompted both agencies to begin investigations of automated laboratory and manufacturing practices and procedures.

The primary findings of these investigations was that the integrity of computer-resident data is at risk in many laboratories providing scientific and technical data to the FDA and EPA. The first principle upon which the Good Automated Laboratory Practices (GALPs) are founded addresses this risk in automated data collection systems:

> DATA: The system must provide a method of assuring the integrity of all entered data. Communication, transfer, manipulation, and the storage/recall process all offer potential for data corruption. The demonstration of control necessitates the collection of evidence to prove that the system provides reasonable protection against data corruption.

The first portion of this chapter will address the problems found in laboratories as they move more and more toward automated procedures. The problems found in these laboratories reflect similar problems encountered in the pharmaceutical, medical device, and biologics industries. This first section will be followed by a discussion, based on the GALPs, of systems security.

---

[1]This chapter is based on portions of the *GALP Regulatory Handbook*, by Weinberg, Spelton and Sax, Inc. (Boca Raton: Lewis Publishers, 1994), pp. 59-75. Reprinted by permission of Lewis Publishers, an imprint of CRC Press, Boca Raton, Florida.

THREATS TO DATA

There are three primary threats to data integrity: restricting confidential data to only authorized users, safeguarding the access to automated systems, and maintaining the accuracy of data as it is transferred to the automated system.

Confidentiality

Confidential data is data which must be restricted to authorized personnel. Passwords are an example of confidential data. Blood unit test results are another type of confidential data. When confidential data is made available to individuals who do not have a need to know that information, damage can result in a number of ways. System security can be compromised, a person's personal health history could be adversely documented, or research and development findings could be changed. The ramifications mount in direct proportion to the nature and degree of the confidentiality of the data.

In order to rely on the findings of our automated processes, we must be able to control access to confidential information. In order to comply with this requirement, levels of authorization and linking identifiers must be controlled. A formal access procedure must be implemented in order to assure that only authorized personnel can gain access to the confidential information. In addition, effective control over the linking identifiers would ensure that even if confidential data fell into unauthorized hands, the source of that data would remain unknown, thus limiting the damage that would result if an unauthorized person obtained the confidential information, or if the confidential information had been inadvertently disclosed.

The clear identification of what is confidential information is the first step in compliance. The next step is to limit the availability of that information to employees who have been authorized to have access to it. Restricting access to this confidential information must include the "live" data, meaning that data which is residing on the computer system, as well as the "old" data, or the data which is stored magnetically or otherwise physically residing in storage.

Clearly-defined, authorized procedures must be established which delineate in what cases this information may be disclosed and to whom. Formal review and sign-off mechanisms should be incorporated into these procedures in order to prove that the proper procedures were followed in cases of approved disclosure.

Confidential information is potentially damaging information, and it must be protected to reduce the vulnerability of the institution charged with handling it. The disclosure of confidential information in an inappropriate manner invites civil and regulatory penalties.

Accessibility

Implementing system access security standards is a proven alternative if there is a need to safeguard data input, modification, or retrieval capabilities. The restrictions can consist of, but are not limited to, data edit flags, individual passwords, limited access files, and personalized log-on requirements. To ensure the integrity of the information in any data management system, access security should be implemented across the board.

All data collected and processed in the laboratory must be protected so that only those employees authorized to view and/or edit such data are capable of doing so.

The physical location of the host computer in a centralized processing environment should be such that only trained and authorized employees can gain access. If persons other than those authorized employees must gain access -- auditors and regulators, for example -- they must be accompanied by an authorized employee. This restriction of access includes, but is not limited to: the hardware, the software, the magnetic media, and any hardcopy reports. Physical access must also be restricted in a distributed processing environment.

If regulated data is stored and/or processed in a PC environment, the access to those PC's must be physically and logically restricted to ensure that inappropriate data manipulation does not occur. Restricting physical access is the first line of defense in protecting data in an automated environment. When data discrepancies occur, it will be much simpler to trace the source if the physical access has been restricted to the point where the error must have been caused by an authorized employee. The use of identifying audit trails will then reveal who was logged in at the time the discrepancy took place. Logical system security should have restricted individuals from making inappropriate changes if properly enforced.

The logical access to the system must also be restricted to only those employees trained and authorized to use the system. By restricting levels of access and capability by log-in and password accounts, the likelihood of inappropriate data manipulation is decreased; each transaction will be traced on an audit trail identifying the person who made the change. The users are granted only the capabilities required to perform their jobs, and they should not be capable of accidentally or otherwise gaining access to information they are not authorized to edit.

In this two-tiered restriction configuration, only authorized users can gain physical access. Of those individuals authorized to have physical access, they in turn are restricted to only the level of access defined by their job description.

The other issue of accessibility which combines physical access with logical access is the use of modems. Modems must be controlled to restrict any attempt at unauthorized access. Once such attempts are permitted, someone will succeed. The use of call-back modems is one type of control. This configuration allows incoming calls, but in order to gain access, the system will call back only to an authorized number. This is a type of physical security. If the incoming call does not originate from an authorized phone number, the connection cannot physically be made. In addition, once a connection is made, log-in ID and password are required in order to gain logical access.

The bottom line is to reduce the ability of authorized and unauthorized individuals from manipulating data and/or hardware by restricting their ability to gain system access whether it be physical or logical.

Data Accuracy

The integrity of data is vulnerable in laboratory, manufacturing, and blood processing configurations where data elements are transferred to the automated system from another source. Data integrity is particularly vulnerable when collected information is typed into an automated system during data entry. Input data should therefore be validated. This can be accomplished by implementing the following procedures:

- Verifying that data are entered in the correct format (alphabetic or numeric).
- Verifying the data against predetermined acceptable ranges (pH values between 0 and 14).
- Verifying that data against values in previously entered tables of information (samples must originate from certain locations or from certain client accounts).
- Re-keying data by a second individual.
- Reviewing entered data by an individual not responsible for its entry.

SECURITY ISSUES

Good Automated Laboratory Practices (GALPs) encompass a broad range of security issues that will be examined in this chapter. The GALPs are broken down into two main categories: Risk Assessment and Security Requirements. These main categories are further broken down into sub-categories: Confidential Information, Data Integrity, Critical Factors, Physical Security, System Access Security, Functional Access, and External Programs/Software. These categories will be used to structure the presentation of the rest of this chapter.

The GALPs clarify the scope and depth of measures required to define and implement a complete security system. Security is specified as Principle Six on which the GALPs are formulated:

> Security. Consistent control of a system requires the development of alternative plans for system failure, disaster recovery, and unauthorized access. The principle of control must extend to planning for reasonable unusual events and system stresses.

A viable security system demands identification of a policy, analysis of potential risks, enforcement of appropriate security procedures, and regular review of all security measures. Formulation of a security system begins with a complete risk analysis. Each organization should evaluate their security needs based on potential loss if any portion of the automated data collection system is damaged, tampered with, or destroyed. Areas requiring security evaluation range from the physical facility to the laboratory area. Based on analysis, physical and logical access should be limited in accordance with the potential impact of partial or complete loss of laboratory information.

The type of security necessary to protect data will be determined using proprietary classification and a data control life cycle, which will be discussed later in this chapter. Both physical and logical access should be granted to each authorized individual on a need-to-know basis. Monitoring, control, and coordination among effected areas of the organization should be evaluated during the risk assessment for determination of access authorization procedures for computer and laboratory facilities. The risk assessment is ultimately used to define security requirements in order to provide a balance between security measures and the legitimate need for access to data.

The risk assessment lays the foundation for definition and implementation of security requirements for computer and laboratory facilities. A complete program for access authorization for the entire automated data collection system is devised based on evaluation and regular review of the risk assessment.

Risk Assessment

Laboratory data and analysis are critical to a variety of corporate activities, such as development of new products and quality assurance of raw material samples. Risk assessment of physical sites, computer operations, and laboratories should be completed by an independent group of experts familiar with relevant security monitoring procedures. Written assessment criteria should be used as the initial basis for examination of all areas related to processing laboratory data. Skilled individuals should be selected on the basis of background and experience to assess various areas of security. The areas examined should include, but not be limited to, physical facilities and computer record storage locations (on- and off-site). Risk assessment should also focus on prevention, identification, and resolution of detected security violations.

Proprietary classifications should be defined to more clearly assign risk potential for all laboratory data which interacts with the automated data collection systems. This definition facilitates expert assessment of risk exposure. Analysis of access control risks in accordance with proprietary classifications will help protect data within the automated data collection system from unauthorized review and possible inappropriate disclosure.

Development of a data control life cycle evolves as risk assessment clarifies where the points for potential data corruption exist throughout the automated data collection system. Each reevaluation of risk must include review of this life cycle.

1. DATA INTEGRITY

The test of data integrity is predictability. If, for every stimulus to the system, the response can be predicted both when the system is performing properly and when it is failing, it has integrity. Security procedures must be instituted anywhere in the automated data collection system where data integrity is deemed to be in an area of exposure to potential hazards.

Potential hazards make access control vital to ensuring data integrity. Evaluation of access controls must encompass all stages of the automated data collection system. Assessing access requirements should include access control safeguards, access accountability, usage safeguards, sample handling and disposition, data distribution, off-site access control and destruction of data retained on storage media and/or hard copy. Access controls instituted within the organization must be designed to protect data generation and storage. Protection requirements should be based on proprietary classifications previously discussed.

Multilevel access controls for all personnel are recommended for data with a high proprietary classification. Controls should begin during entry to the operating system software and then to individual software applications. When a LAN and/or WAN are access components, access controls should be designed for each configuration interacting with automated data collection. Access controls for authorized personnel must be designed to limit admittance to segments of the collection system specifically related to performance of currently assigned job responsibilities. Separation of duties and data access will provide an additional level of data integrity assurance. Access must be denied to data and functionality not identified and verified as legitimate for performance of assigned duties.

Access accountability properly implemented will permit identification of authorized and unauthorized attempts to access data and/or functionality. Diligent surveillance is critical to assigning responsibility for activities performed at each level of access for every authorized individual. Accountability tracking initiated upon attempted access through final signoff is essential for effective surveillance necessary to monitor data integrity. Legitimate and erroneous access of the automated data collection system should be traceable throughout the entire access sequence.

Along with access surveillance, capture and review of data modification must be implemented in order to protect data integrity. An access monitoring system should possess the ability to trace modifications back to the original transaction. This extensive tracing record will safeguard data integrity by providing personnel accountability for data modifications. Unauthorized data modification, once detected, must be investigated to a written final resolution. The sequence of unauthorized data access must be carefully reviewed to determine what remedial actions will be necessary to eliminate the threat to data integrity.

Data integrity can also be compromised by changes made to the automated data collection system. Upgrades, changes, fixes, and/or alteration of any portion of the data collection system will require assessment of the scope of testing necessary to verify continued maintenance of data integrity. The type and extent of changes made to the data collection system impact the testing regime, including verification of data access controls and mechanisms for monitoring data modification.

2. CRITICAL FUNCTIONS

Performance of time-critical functions requires that data be available for sample tracking critical to prompt data analysis and monitors quality controls criteria to timely release of data or generated reports, which are critical to timely submission of the data. For example, the drug development and approval process from preclinical testing through FDA approval rely on time-critical laboratory functions relating to automated data collection systems. Identification of such time-critical functions must be an integral part of the risk assessment.

The GALPs recommend devising a plan to prevent loss of critical system use resulting from access by unauthorized personnel. Identifying the elements of this plan must be a component of risk assessment. A data control life cycle should also be developed as part of the risk assessment. The data control life cycle can be represented as a data flow diagram depicting information travel from inception to demise. This diagram will enable security, computer, and laboratory management to identify and evaluate each point of possible vulnerability. In combination with the proprietary classifications, the data control life cycle can add an important dimension to identifying potential weaknesses to be analyzed during the risk assessment process. The data control life cycle includes eleven basic data exposure control points: data gathering, data input movement, data conversion, data communication [input], data receipt, data processing, data preparation [output], data output movement, data communication [output], data usage, and data disposition. System security necessary to protect time-critical functions can be assessed individually and as part of the entire automated data collection system using the data control life cycle.

Security Requirements

Another result of completing the risk assessment is initiation of definition and implementation of risk-appropriate security requirements. Risk assessment observations will enable computer and laboratory personnel to formulate comprehensive security requirements. These requirements range from security for physical equipment and sites to assignment of access limits. Security requirements should be designed to protect the automated data collection system physically and logically from unauthorized access, while limiting authorized access to current job responsibilities.

Evaluation of each potential point of physical and logical access to the automated data collection systems is essential for defining security requirements. Access limitations should encompass computer operation systems, software applications, databases, data, computer room(s), LAN and WAN configurations, workstations, modem communication, data storage media, computer and equipment interfaces, laboratory equipment, and human and/or equipment interaction with the automated data collection systems. Verification of the allocation of system access authorization must be strictly controlled and monitored for all computer and laboratory personnel. Regular review of security policies and procedures by management personnel will confirm adequate security or identify appropriate revision of security requirements. This review process is designed to assure security adequate for deterrence of unauthorized security breaches or identification and correction of intentional errors.

1. PHYSICAL SECURITY

Physical security transcends computer and laboratory operations to encompass the entire facility housing automated data collection systems. Physical security must be applied throughout the facility to achieve a suitable level of protection. The levels of security can be devised as concentric circles with increasing security measures implemented in the inner circles closer to computer and laboratory equipment.

Physical security should be designed as a deterrent to unauthorized access, a monitoring regime to detect security breaches, and protection against environmental variances. Prevention of environmental variances such as electrical supply, temperature control, and fire protection are all physical security considerations. Maintaining current, accurate inventories of computers, laboratory equipment, and all related peripherals can also enhance ability to monitor physical security for all automated data collection system equipment. The degree of physical security required can be determined by identifying the appropriate data classification and data control life cycle components.

The location and configuration of the automated data collection systems are also a physical security consideration. Varying degrees of physical accessibility can be devised using data proprietary classifications to create sufficient controls. Physical security measures must be in place for individual PCs, LANs, and WANs, as well as computers housed in a traditional computer room. Weaknesses and vulnerabilities unique to PC configurations include ability to monitor access to a variety of computers, servers, and peripherals spread over a large area. Restricting physical access to areas housing computer and laboratory equipment can be accomplished using data proprietary classifications and the data control life cycle. Any physical security weaknesses detected during monitoring efforts require swift and complete investigation.

2. SYSTEM ACCESS SECURITY

Evaluation of system access security should include verification of password schemes at various levels of computer access, segregation of software applications, and limitation of access rights to computer software and data, along with controlled access to sensitive areas of the facility. The following list includes a variety of areas subject to access considerations (Enger and Howerton 1981):

- Training all individuals authorized to access secure automated data collections systems.
- Tracing logical requests through the entire written request procedure.
- Comparing access requests to employee user profiles for logical proper access authorization.
- Tracing legitimate computer access transactions through the entire processing sequence.
- Tracing each access route available to programming and computer operations staff.
- Verifying that programmers have generated appropriate internal documentation.
- Verifying proper internal and external documentation for fixes implemented in the source code.
- Tracing data modifications through the entire processing sequence.
- Verifying that the last test staging environment(s) contain the same access restraints as production environments where programs and data will be moved.
- Examining logs and records for installation, operation, and maintenance of hardware, software, telecommunications, and peripheral equipment used to collect, analyze, store, or report laboratory information and activities.
- Verifying compliance to security procedures by comparing written procedures to actual practices in the automated data collection system environment.

The access issues identified above will be examined further to clarify system access security requirement necessary to ensure integrity, availability, and confidentiality of automated data collections systems under the GALP guidelines.

All necessary and reasonable measures for restricting logical access to the system should be instituted and maintained to prevent and/or identify loss or corruption of secured data. Data security requirements should be implemented based on the proprietary data classifications and the data control life cycle issues identified during the risk assessment. In each test and production environment, the ability must exist to track all data input or modification back to the responsible individual. System access security must also restrict unauthorized remote or direct access to automated data collection systems.

Training requirements should be defined for protection of sensitive data. Personnel interacting with the automated data collection system should receive training appropriate to completing their system-related responsibilities. Providing adequate training requires selection of training instructors based on their expertise and security-related credentials. These instructors should maintain a log for each security training session that they conduct. Responsibility for maintenance and retention of these logs should be assigned and all training activities verified by management-level personnel.

Training session attendees should be made aware of the importance of security procedures for protection of the data collection systems. The system(s) in place to permit identification of each unique user for tracing of data input, modification, interpretation, and storage should be included in the session curriculum. The password systems available at access levels throughout automated data collection systems also require in-depth training on their use and function in maintaining data integrity. As the access systems evolve, retraining requirements must be evaluated prior to implementation of changes that impact system access functions.

The request procedure for gaining logical access to the automated data collection systems must include written request, authorization, and review. Approval points throughout the request process will provide a high level of assurance that a bogus access request does not reach implementation in the automated data collection system. Authorization of access requests will include at least one, possibly several, levels of management authorization based on ownership of collection system data and equipment. Tracing a logical access request then requires comparison of the access request with the employee user profile. The user profile can be based on the employee's job description or a combination of job description and automated data collection system responsibilities.

Retention of all the documentation for approved logical access requests will enable authorized security personnel to track computer access transactions through the entire data collection system. Limiting authorized access and preventing or detecting unauthorized attempts to gain entry are integral parts of securing the automated data collection system. The regular review of access tracking documentation, whether on-line or hard copy, will enable security personnel to determine whether access to the data collection system is authorized or an attempt to breach access security. Monitoring the entire data collection system access process enhances all of the levels of security available to deter or detect unauthorized access.

Authorized access granted to programmers usually consists of several ways to access source code, program files, application modules, and laboratory data.

Tracing mechanisms required to monitor programmer activities deserve close scrutiny during the risk assessment. System access security should not be compromised by allowing programmers access paths that cannot be adequately monitored regularly by security personnel. Programmers granted the most comprehensive access to the automated data collection system should be monitored most frequently.

Verification of internal documentation generated by programmers in accordance with predetermined documentation criteria can serve as a method of cross-referencing access to and modification of source code and/or data within the automated data collection system. Change control documentation, internal source code documentation, and access monitoring documentation used together can confirm authorized access or detect unauthorized access to the data collection system available for entry, modification, and storage of programming-related information. Identification of responsible personnel, date of change, and change description will facilitate coordinating examination of all documentation necessary for monitoring programmer access to the automated data collection system and laboratory data.

Procedures should be in place to limit or eliminate access paths available to programmers whose attempts to breach system access security are identified by security personnel. Identified security breaches reported to appropriate supervisory personnel require written documentation of the entire incident and resulting actions. Historical records maintained by security and supervisory personnel will enable both groups to monitor recurrence of unauthorized security access attempts. Response to recurrent security violations can be defined in the security access policy.

Tracing data modifications through the entire processing sequence is intended to provide a significant degree of data integrity. Data modification must be traceable back to a specific individual, regardless of the level of system access authorization. Traceability of modifications completed by any level of system users is an important part of maintaining data integrity. Verification of appropriate access to, and authorized modification of, laboratory data within the automated data collection system enhances the ability of the system to provide accurate data. Other facets of monitoring data modification encompass written documentation and authorization of the change, along with inclusion on-line of a reason for initiating the change. Proving data accuracy will increase reliability of analyses and reports generated by the automated data collection system.

Many organizations complete data modifications and changes to source code in a test environment. When this method of change control is enforced, the final test environment used prior to movement of the changes to production must control logical access in exactly the same manner as the production environment. All levels of logical system access to source code and data require rigorous testing before changes can be transferred to production. Test protocols should address each level of access and authorized system access abilities granted in production. Each combination of authorized access available to programmers, computer operations personnel, and the user community must be exercised to assure that access functionality has not been compromised by the changes being tested for implementation in production. All access problems identified during testing must be documented and resolved. These changes require review and approval by management personnel prior to moving any portion of the change into the production environment.

Prudent system access security procedures should require regular review of logs and records for installation, operation, and maintenance of equipment and software used to collect, analyze, store, or report laboratory information and activities within the automated data collection system. Knowledge of ongoing changes throughout the environment that supports the automated data collection system enhances the ability of security personnel to recognize new functionality or installation of changes impacting system access security. Evaluation of these records and logs must be documented to provide the historical perspective necessary to trace changes in the system access sequence and/or access authorization tracing capability.

Comparison of access security practices to written security procedures should be the final component of system access security monitoring. Practices that deviate from written procedures may compromise the ability of security and management personnel to control system access security. Inability to limit authorized access to automated data collection systems can severely diminished confidence about data integrity and enable excessive availability to the system and to sensitive laboratory data, which might ultimately result in inappropriate release of confidential laboratory information. The potential for data corruption or inappropriate release of information demonstrates the importance of regular review schedules for system access security procedures and documentation of review results. Without a mechanism for assessing need for change, the system access security procedures would be vulnerable to inaccuracy and ineffective control of the automated data control systems.

3. FUNCTIONAL ACCESS

Assessment of functional access requirements should be the next phase of system security evaluation. Functional security embedded in each software system demands examination in an effort to ensure data integrity. The automated data collection system could include several distinct software operating and application systems, each with a potential impact on security. Every system must be scrutinized at each level of access to determine what functions are available for access to, and manipulation of, data. Security access capabilities can then be identified for all of the system functions for each authorized user. Proprietary classifications and the data control life cycle should be a component of each phase of functional access analysis.

Function-driven software capability allows security personnel to control the type of access granted within a system to a specific individual. Functions at both the operating and application system levels require analysis to ensure correct allocation of functions within the automated data collection system. Physical segregation of data within the automated collection system is another method available for ensuring data integrity and confidentiality while granting availability to relevant data.

Data accessibility must be evaluated from all possible automated data collection system access paths. Both menu- and command-driven access sequences require restrictions based on a job performance need-to-know basis. Assignment of operating system functions may have significant impact on the security of the application software. File attributes defining sharing of files and access rights to data collection system resources should be carefully assigned and monitored. At the software application level, menu options and data manipulation capabilities are system functions that should be evaluated for allocation to authorized personnel. Limiting access rights at both the operation and application software access tiers should be designed to support pertinent access, while prohibiting completion of unauthorized functions.

Configuration design for physical segregation of data and software depend on assessment of proprietary classifications along with review of the data control life cycle documentation. Proprietary classifications determine how rigorous the physical segregation partitions must be, based on the sensitivity of the data. The data control life cycle identifies where points for potential data corruption are located throughout the automated data collection system.

Failure to grant appropriate access to system functions can severely compromise the integrity, appropriate availability, and confidentiality of data. Compliance with the GALPs would be jeopardized if control of data integrity, availability, or confidentiality could not be consistently demonstrated.

4. EXTERNAL PROGRAMS/SOFTWARE

In order to protect the operational integrity of the automated data collection system, the laboratory shall have procedures for protecting the system from introducing external programs/software (e.g., to prevent introduction of viruses). Installation and change control must be carefully monitored to prevent introduction of unwanted programs or software into the automated data collection system.

Movement of any programs or software external to a specific software application should be monitored by more than one individual. For instance, once the programmer has completed coding changes, computer operations personnel might assume responsibility for moving the coding into the correct production environment. The installation procedure should be reviewed by more than one programmer prior to completion of the procedure. Devising a backout procedure is also a prudent action. This procedure can be used if problems occur after installation of software that warrant backing the software out of production. Written documentation defining the changes made and the ultimate location of the code within the automated data collection system should be generated for each phase of movement during the installation process. Running debugging and virus detection programs should also be considered prior to moving any portion of the software to production.

Change control procedures should encompass authorization, documentation, testing, notification, and implementation of changes to software in a production environment. This process is intended to ensure control of the change process and afford the ability to trace responsibility for each phase of change through the entire change control process. Authorization requirements should be based on ownership of the software and ownership of the portion of the automated data collection system that houses the software. Documentation standards for a change should include internal reference to the change origin and external documentation of the entire change control process.

Testing of a change is essential for ensuring accurate completion of the intended change and assurance that no other software functionality has been corrupted by the change. A test plan, scripts, result documentation, and problem resolution are integral components of the testing process. Without a clear procedure and appropriate documentation of testing, the chances for introduction of viruses or unintentional corruption of functionality increase significantly.

All departments and personnel involved with the change process should be notified at appropriate stages of the change. Notification duties should be segregated in a way that minimizes the possibility of collusion to sabotage the change process. Notification should be captured in writing to create a trail of responsibility for each phase of change control. Implementation of a change in a production environment is a critical step in the change process. Documentation should be available to prove that from definition to completion of the change, controls were exercised to assure accuracy of all programs and software affected by the change.

CONCLUSION

Automated data collection systems require security adequate to protect the individual components of each system as well as the entire system. Each phase of security definition, from risk assessment to implementation of procedures, will enable a laboratory to demonstrate evidence of control. This evidence is intended to ensure data integrity, availability, and confidentiality throughout the automated data collection system.

Assessment of security should be considered a dynamic process. This process requires monitoring and review based on changes to the automated data collection system, reclassification of proprietary data, and regular appraisal of the data control life cycle.

Chapter 12

DOCUMENTATION: THE SOP[1]

OVERVIEW

The FDA and EPA depend heavily on data submitted to these agencies to reach decisions on public health. Since the accuracy and integrity of these data are fundamental to reaching correct decisions, and since computers are increasingly replacing manual operations in the laboratory and in manufacturing processes, standards and definitive guidelines for automation are needed.

Documentation issues are central to these guidelines. Based upon the survey "Automated Laboratory Standards: Evaluation of the Standards and Procedures Used in Automated Laboratories" completed by the EPA in May 1990, serious deficiencies in documentation were cited as one of the three primary findings which place the integrity of computer data at risk. Similar findings were made by FDA investigators.

In general, six types or categories of documents are necessary for validation compliance. They are **Personnel**, **Equipment**, **Operations**, **Facilities**, **Software**, and **Operational Logs**.

Personnel

In the category of Personnel, Quality Assurance Reports on inspections demonstrate QA oversight. Also in this category, Personnel Records help support the competency of various employees assigned to system responsibilities.

Equipment

In the category of Equipment, a Hardware Description Log records and identifies which hardware is currently in use for a system. A record of Acceptance Testing demonstrates

[1]This chapter is based on portions of the *GALP Regulatory Handbook*, by Weinberg, Spelton and Sax, Inc. (Boca Raton: Lewis Publishers, 1994), pp. 43-57 and Appendix C. Reprinted by permission of Lewis Publishers, an imprint of CRC Press, Boca Raton, Florida.

the initial functioning of the hardware, while Maintenance Records help ensure the continuing operational integrity of the hardware.

Facilities

In the related category of Facilities, written Environmental Specifications guard against data loss or corruption from various environmental threats. This information may be specified by SOP or by the system vendor and verified during system Installation Qualification or Acceptance Testing.

Operations

In the category of Operations, a Security Risk Document identifies likely and possible risks to the security of computer-resident data. Standard Operating Procedures ensure the consistent, controlled use of the system; SOP topics will receive further discussion in the next sections.

Software

In the category of Software, a Software Description records and identifies which software is currently in use for a system. Software Life Cycle documentation helps ensure the operational integrity of the software. These documents may include design specifications, testing and test results, change control procedures, problem resolution procedures, and version control.

Operational Logs

In the category of Operational Logs, Backup and Recovery Logs and Drills help guard against data loss or corruption. A record of Software Acceptance Testing and Software Maintenance or Change Control documents also ensure future software integrity.

In order to simplify, these six categories may be further reduced to SOPs, Logs, and Training documentation. These three major categories are discussed in more detail in the following sections.

GENERAL CRITERIA FOR SOPs

In general, an SOP must establish guidelines for the specific activities, procedures, and records required to demonstrate and maintain control over the system. Certain criteria must be considered when developing and implementing these procedures:

- ACCESSIBILITY: All SOPs should be readily available for reference by appro priate users and by regulators, validators, managers, and Quality Assurance or Quality Control personnel.

- CURRENCY: All SOPs must be current; they should be dated, show a method of version control, and should be revised as procedures change or reviewed at least every year.
- PRACTICE: Unobtrusive observation by management, QA, or another independent group should confirm that SOPs are actually being followed, and are appropriate to the context for which they are written.
- COMPREHENSIVENESS: SOPs should address specific procedures for normal and irregular circumstances and applications. All procedures appropriate to the controlled functioning of the system should be included in the complete set of SOPs.
- CREDIBILITY: The SOPs should be able to withstand regulatory scrutiny by exhibiting proof of management awareness and control, auditability, and reviewer independence and accuracy.

SPECIFIC SOPs

An automated laboratory or pharmaceutical, medical device, or biologics manufacturer requires written Standard Operating Procedures to demonstrate adequate control over automated systems. A minimum set of SOP topics include:

1. Security (System Access Security and Physical Security).
2. Raw Data (Working definition used within a laboratory).
3. Data Entry (Identification of person entering data).
4. Data Verification (Verification of input data).
5. Error Codes (Interpretation of codes and corrective action).
6. Data Change Control (Format and contents of Log also specified).
7. Data Storage and Retrieval (Data analysis may also be included).
8. Backup and Recovery of Data.
9. Hardware Maintenance.
10. Electronic Reporting (where applicable).
11. Problem Reports.
12. Software Change Control.
13. User Training.
14. Testing.
15. Documentation.
16. SOP on SOPs.

In reality, several of these topics may need to be divided into two or more separate procedures in order to address the subject in a clear and concise manner; conversely, two topics can sometimes be condensed into one procedure without sacrificing clarity. Most systems will require additional SOPs to describe and define procedures unique to the specific application of each system. The following descriptions define each of the above topics in more detail.

Security

This SOP describes the Physical Security and Access Security of the computer system. Physical Security focuses primarily on the computer room and any related workstations. The initial security barrier may be a locked door limiting access to authorized personnel. Additional security barriers may include keyboard locks, disk drive locks, and other similar restrictions depending upon the particular hardware layout of the system.

Physical Safety is often included in the SOP on Physical Security, in that it also focuses primarily on the computer room. Protective measures are outlined to guard against fire, flood, prolonged power outage, and other potential threats that apply to the local environment.

Access Security or System Security focuses on access into the computer system. Access is typically limited to authorized users by implementation of a Password system. Various procedures may be outlined to grant, change, or revoke authorized passwords to maintain security. Criteria are specified for creating and maintaining password secrecy, including forbidden words, minimum password length and complexity, frequency and enforcement of password changes, and restriction and review of the password file.

Modem usage presents a separate Access Security issue. Although modem usage facilitates efficient data entry from remote sites, the corresponding increased threat to data integrity is significant. Methods for controlling this increased risk include secure communication lines, call-back arrangements, data check systems, and other emerging techniques.

Raw Data

This SOP provides a working definition of Raw Data for use within the operating environment; a distinction between raw data and processed data is implicit in this definition. Since the definition may be unique to particular applications, it is not possible to specify it more clearly here. However, some guidelines are available. Hand-written or manually-entered data collections, such as field readings or reports, are raw data. Information which has been manipulated by calculations or formulas becomes processed data.

An alternative definition involves determining the point at which given data cannot be derived from subsequent data. By this definition, statistical values such as means or standard deviations are processed data in that the original input data points cannot be derived from the statistical values. Similarly, digitized data from downstream chromatographic sensors is considered raw data by this definition because the original analog signal can indeed be derived from the digitized data.

Given the ultimate nature of raw data, it is appropriate to specify in this SOP the special retention and retrieval procedures used for this information. Restricted access to the raw data archive and the uncorrupted restoration of this data from the archive are prime considerations of this SOP.

Data Entry

For some applications, special or heightened requirements may apply for entering data into the system. Typically, extra or specialized operator training may be required to evaluate, code, and enter the data properly, or special procedures such as double-key entry may be specified. In these situations, the SOP must clearly describe the special procedures in order to ensure that data is entered correctly at all times.

In addition, each user entering data must be identified to the system with a unique password or user ID. The password system outlined in the SECURITY SOP, above, should be adequate to ensure this unique identification. However, in situations of heightened or critical security, a requirement for a second password or the use of a restricted-level password system may be appropriate.

Data Verification

For applications such as those described in the DATA ENTRY SOP, above, i.e. when critical data is input either manually or electronically, there must be a procedure for confirming or verifying the correctness of that data. This procedure must be clearly described in the appropriate SOP.

Three methods are most commonly used for data verification. The double-blind method involves having two users independently enter the same data, and then verifying that the values are identical. This is the most secure but also the most time-consuming and labor-intensive method.

The double-key method involves having the same user enter all data twice, and then checking that the values are identical.

The program-edit method involves checking the single input of data against pre-specified system parameters, ranges, or tables. This method is the least secure, but it is adequate when data entry is less critical.

Error Codes

Error codes are messages that bring the user's attention to problems created by the user or experienced by the system. Encoded, abbreviated, or generic error messages should be listed and more fully explained in an appropriate SOP. Likely causes of these messages should be explained along with error correction methods and proper notification procedures.

In some situations, explanations, causes, correction methods, and notification procedures are all contained within the error message itself. In this case, a dedicated SOP may not be necessary, although extra acceptance testing of the system may be warranted.

Data Change Control

Change control is directed toward minimizing the risk of any unwanted or untested changes taking place within a system. This SOP describes the process of controlling changes made to data once it has been entered into the system. Since credible changes to entered data may be an occasional or frequent necessity, the focus of this SOP is on the safeguards to protect against unauthorized changes and the traceability of authorized changes.

The Change Control SOP also includes documentation of how authorized changes have been tested; proof that the changes do not represent changes which could lead to loss or corruption of data; and cost, scheduling, and impact statements. Hardware and Software Change Control is considered in a separate section.

Two methods of control are commonly used to regulate change. In some systems, the ability to alter data is restricted to users with a higher level of security clearance than is assigned to data entry users. This clearance is typically attributed by password. In addition, and regardless of security clearance, the change is automatically entered into the audit trail system. At a minimum, this system records the date of the change, the reason for the change, the individual making the change, and the old and new values of the data.

The SOP must specify the contents of the Audit Trail and the procedures for printing, reviewing, and archiving the Audit Log. By this process, the Audit Trail may be used to demonstrate management control over change and can provide a key link in any recall situation.

Changes made to the software which potentially result in data modification must also be controlled by this SOP. In this case, the SOPs on Software Change Control and Data Change Control must be written in a coordinated fashion to avoid conflicting requirements.

Data Archiving

A well-controlled system must be able to store data in a clear, logical, repeatable manner, and it must be able to retrieve stored or archived data in a useable, unaltered manner for further processing or analysis. This SOP must specify the detailed methods used to store data, including the frequency of storage, media used, and persons responsible for the storage routine.

Media deterioration must be anticipated and minimized. An indexing system for stored data must be specified in order to provide easy access and record keeping. The procedure for retrieving stored data must also be specified. This includes the authorization process for retrieving archived data and the procedure for loading it back onto the system.

In addition to archiving, this SOP may specify a consistent method used for data processing and analysis. Here, data processing refers to the conversion of raw data into easily interpreted data; data analysis refers to the interpretation of the processed data. For example, the SOP could describe the verification procedure for formulas or decision rules, or the utilization of standard analytical subroutines.

Backup and Recovery

This SOP seeks to ensure the integrity and availability of stored data in the event of a serious breach in security or a system-wide failure. It specifies the procedures for making and storing backup copies of system data and software. Responsibility to complete, deliver, and recycle the backup copies must be assigned to an appropriate individual. The frequency for data backups and the sequence of complete or incremental backups are established. The types of storage media are specified, along with a limit to the number of media recycle times and frequency. Both on-site and off-site backup media storage facilities are specified, again with a delivery and retrieval schedule. Data recovery drills are scheduled, and the dates, people involved, and procedures followed are recorded.

In the event that a manual system is used when the automated system is temporarily unavailable, in lieu of a backup procedure or until the backup copy of the data or software becomes available, then procedures should be written into the SOP for rolling back uncommitted transactions or for rolling the database forward to synchronize it with the manual changes.

Hardware Maintenance

This SOP maximizes the likelihood that hardware will continue to function reliably in the future. A schedule for preventative maintenance of most hardware is generally found in documentation provided by the vendor, and the SOP can formalize this schedule.

If maintenance is to be performed in-house, an SOP must assign a Responsible Person for following this schedule and procedure and for documenting the performance. If the vendor is responsible for maintaining the hardware, the SOP must assign a responsible individual(s) for documenting that the maintenance was performed on schedule. Reference should be made to the current maintenance contract.

In the event that hardware is specified "maintenance-free," a replacement contract with the vendor might be appropriate and should be referenced in the SOP.

Electronic Reporting

This SOP, in specific relation to the Good Automated Laboratory Practice (GALP) guidelines, describes the procedures to be used by laboratories which supply information directly to the EPA by electronic reporting. However, it applies equally well to any system that reports data to remote locations, receives data from remote locations, or generates reports automatically from information residing in a database.

The SOP should specify the standards, protocols, and procedures used in data collection and analysis. It should also identify the format used for reporting data and results to ensure uniformity and comprehensiveness. Detailed issues, such as when the reporting is to be done and how and by whom the report is to be sent, can also be included in this SOP.

Security issues described previously may apply if the transmission is sent by modem. In addition, the electronic reporting function or module of the system may warrant special acceptance testing considerations.

Problem Reports

The Problem Reports SOP provides a record of system errors which require help or input from outside sources, such as the centralized computer center, a telephone hotline, or the software/hardware vendor. The report itself may contain a time section to help evaluate the efficiency of problem resolution and a response section to describe the solution provided for the reported problem.

The Problem Report SOP defines the procedures for initiating reports, recording solutions, and evaluating responses. A standard form is generally provided for this purpose. A responsible individual(s) is assigned to monitor the problem report system and follow up on unresolved issues.

Software Change Control

This SOP describes the process of controlling changes made to system software. Any routine software maintenance must be performed in a manner that creates no changes in the functioning of the system. The SOP may specify the initiation and implementation of software changes and determine the level of testing involved to demonstrate performance. A responsible individual(s) is assigned to monitor and approve software changes. Where software changes may potentially cause data modification, this change procedure must be coordinated with the Data Change Control SOP.

User Training

The purpose of this SOP is to specify the training given to all users of a system in order to ensure the informed, proper use of the system. A record is kept of this training which includes who was trained, the date and outline of training given, the trainer's name, and any test results from the training.

Testing

The purpose of this SOP is to specify the circumstances under which testing must be performed and the degree or depth of testing required to ensure that the system remains functioning in a controlled, reliable manner. This topic may warrant a separate SOP as outlined here, or testing requirements may be included in other appropriate SOPs such as Backup, Security, Disaster Recovery, and Change Control.

Documentation

This SOP describes the various documents related to all current system records, hardware, and software and describes the archiving procedure for these documents. These documents include SOPs, Users Manuals, Technical and Operations Manuals, Error Logs, Problem Reports, Audit Trail records, Training records, Disaster Recovery Plans and Test Results. This topic may warrant a separate SOP as outlined here, or a documentation section may be included in each appropriate SOP.

SOP on SOPs

This SOP defines the requirements for SOP format, content, structure, review, and control. It sets the framework for all other SOPs and provides the definitive standard for the contents, distribution, and approval of all other SOPs. It also provides a unifying appearance to all SOPs and ensures comprehensive, uniform contents.

Disaster Recovery

This SOP describes the plan for control and recovery in the event of a large-scale disaster. This plan must consider disasters such as fire, flood, prolonged power loss, and any other potential disasters which threaten a particular location or situation. Downtime operations must be addressed, and the disaster plan must be practiced at least annually.

The Disaster Plan is generally a detailed and substantial document. As a consequence, the Plan is usually only described and referenced in the Disaster SOP.

LOGS AND RELATED FORMS

Various data logs are referred to in previous sections in the context of the SOPs which generate them. These logs are critical elements in demonstrating control of the system. A properly designed and completed log serves not only as a record of data, but also as an indicator of management awareness and control, a link with future recall requirements, an insurance policy against disastrous loss, and a measurement of employee competence and skill. Consequently, logs are worth the extra effort to design carefully, monitor occasionally, and archive safely.

Some of the following logs, such as System Maintenance, are quite generic and should be applicable to most systems. Others, such as the Disaster Drill Log, will apply only to the system(s) affected by the corresponding Disaster Recovery Plan. Any style or format is acceptable as long as the basic contents described in the text are included.

System Backup Log

This log is used to document regular, incremental, or complete system backups performed in order to safeguard existing data and to minimize the future loss of data in the event of a system or application failure. This log is a part of the BACKUP AND RECOVERY SOP. It records the Serial or Code Number of the backup tape, the Date of the backup procedure, and the Initials of the backup technician. It may also record Incremental/Complete backup, Recycle date, and any notes pertinent to the backup or restoration procedure.

Routine Software Testing Log

Whenever a change is made to software, testing must be undertaken in proportion to the severity or scope of the change. Small changes may only require modular testing, while large-scale modifications may require a full acceptance retest. This log is a part of the SOFTWARE CHANGE CONTROL SOP or the INCLUSIVE CHANGE CONTROL SOP. It records the Software Change Identification Code or Date, the Tester's Initials, the Date and Scope of the testing, and a Pass/Fail judgement. It may also record the location of the Test Protocols and Test Results if the testing was extensive.

Software Change Control Log

This log is used to record all changes made to the system software. It is a part of the SOFTWARE CHANGE CONTROL SOP. It records the Work Request Code Number, the Change Request Date, the System experiencing the change, and a description of the change. It also records Testing status, Start and Close Dates, and the programmer's Initials. It may also record the requestor's initials and an impact statement for the system as a whole or for independent system functions.

User Problem Log

This log is used to record user-reported problems with the system or related software. These problems are generally related to specific applications. The log records the Date and Problem Description, the Repair Description, and the Initials of the Repair Technician or Tester. It may also record the Reporter's Initials and the Date of repair. This log is part of the PROBLEM REPORTS SOP.

System Maintenance Log

This log is used to record the preventative maintenance completed on particular hardware. The frequency and particulars of the maintenance schedule are generally provided in vendor-supplied documentation. This log records the Date of Maintenance, the Type of Maintenance performed, and the Initials of the maintenance person. It may also record the Date that maintenance was originally scheduled. This log is part of the HARDWARE MAINTENANCE SOP.

Training Log

This log is used to document all user training including orientation training for new users of existing systems, orientation training for individuals or groups of users of new systems or new versions of existing systems, and ongoing training for experienced users. This log records the Names and Departments of trained users, the Date of Completed Training, the Initials of the employee's supervisor, the Date of Testing or Skill Review, and the Testing Supervisor's Initials. It may also record the Training Synopsis or Syllabus and the name and affiliation of the Trainer. This log is part of the **TRAINING** SOP.

System Operator's Log

This log is used to record all activities related to the operation of a system. These system-level activities refer to all actions involving the actual operation of the system and include scheduled activities such as backups and periodic maintenance, and unexpected activities such as system malfunctions. This log records the Date of the activity, a Description of the procedure or function performed, and the Initials of the Operator. It may also record Problem Resolutions, System or Device Identification Numbers, and

SOPs or manuals referred to for specialized procedures or protocols. This log is part of the DOCUMENTATION SOP.

Security Log

This log is used to track and identify visitors, consultants, contractors, and other non-employees who are currently on the premises. This log records the Date and Name of the Visitor, the Visitor's Company, and the Times checked in and out. It may also record the employee being visited and the identifying Visitor's Badge Number. This log is part of the PHYSICAL SECURITY SOP.

Password Control Log

This log is used to track which users have access to the various clearance levels within the system and to monitor the passwords of all authorized users. Access to this log is generally restricted to the System Administrator or to the person responsible for system security. This log records the employee's Name and Password, the Date that password/security training was given, the Security Level or Clearance associated with the employee, the Date any entries or changes were made to the log, and the Security Supervisor's Initials. It may also record previous and revoked passwords and forbidden password selections. This log is part of the ACCESS SECURITY SOP.

Data Change Log

This log is used to record all changes made to data resident in the system. The record may be created and printed automatically as a part of the Audit Trail. This log records the Date and Time of the change, the system or file involved, the Data Values before and after change, the Reason for the data change, the Initials of the users making the change, and any required approval signatures. This log is part of the DATA CHANGE CONTROL SOP.

TRAINING DOCUMENTATION

This documentation includes a current summary of personnel training, experience, and job description, which should be available for all personnel involved in design or operation of an automated system. The purpose of this documentation is to provide evidence assuring that users of automated laboratory systems have the knowledge sufficient for their job requirements.

A computer system will perform best if its operators are familiar with its functioning. The comprehensive and complete training of all personnel interfacing with the automated data collection system must be delineated in a laboratory or manufacturing facility policy. The User Training SOP, discussed above, is an appropriate first step in documenting personnel training. Even in small firms or laboratories, the basic operational skills of the system users should be clearly defined.

A comprehensive employee training program must be established. Documentation must be available that identifies not only the quantity of training each laboratory employee receives, but also the quality of that training; training documentation should show that training meets its specified goals -- that system users have acquired the proper competence to meet job requirements. Additional documentation to provide this evidence can include reports of completed training, lists of basic operational skills required of system users for each performance function of the system, and performance evaluations which demonstrate proper levels of job knowledge and experience.

Training programs must fully document all phases of normal system function as they pertain to the particular user's responsibilities so that each user clearly understands the functions within their responsibility. It is equally important that system users understand enough about normal system function so that they can recognize any abnormal system function and report it to the appropriate responsible individual(s). All training procedures must undergo periodic review at least yearly, or whenever new or upgraded equipment or methodologies are installed.

The following is an example of an SOP on SOPs:

TITLE: SOP ON SOPs ______________________________

Related SOPs: :CODE:

:ISSUE DATE:

:SUPERSEDES:

SOURCE OF INFORMATION :PAGE 1 of X

DISTRIBUTION: :VERSION NUMBER: X.X

CONTENTS:

1. General Statement
2. Purpose
3. Scope
4. Definitions
5. Responsibilities
6. Execution
7. Appendices:
   7.1 Sample SOP Format
   7.2 Sample SOP Composition

		Signature:	Date:
AUTHOR	:	:	:
APPROVED BY	:	:	:
APPROVED BY	:	:	:
APPROVED BY	:	:	:
REVIEWED BY:	:	:	:
REVIEWED BY:	:	:	:

1. *General Statement:*

The SOP on SOPs is the governing document for all SOPs.
SOPs are the one vehicle for defining policy. SOPs must be reviewed, initialled, and dated at least every two years.

2. *Purpose:*

The purpose of the SOP on SOPs is to establish control policies for the way in which SOPs will be formatted, written, approved, distributed, changed, and archived.

3. *Scope:*

Identify the group whose SOPs are governed by this SOP on SOPs:

Corporate?
Divisional? (Reference corporate policy)
Departmental? (Reference corporate/divisional policy)

4. *Definitions:*

SOP

A statement of policy applying to a specific function or activity which follows clear rules of format, control, approval, distribution, and archiving.

SOP Format

See Appendix 7.1 Sample SOP Format

SOP Composition

See Appendix 7.2 Sample SOP Composition

5. *Responsibilities:*

5.1 Who is responsible for enforcing the SOPs and ensuring that all appropriate SOPs exist?
5.2 Who will evaluate the SOPs for:

Conflicts with other policies
Appropriate format
Appropriate definitions
Completeness

5.3 Who approves SOPs?
5.4 Who will circulate finalized drafts for approval?

5.4.1 Who will follow-up to ensure timely sign-off?

5.5 Who is responsible for distribution of approved SOPs?

5.5.1 Who will ensure that:

All authorized persons receive new SOPs and all SOP revisions?
All SOPs are collected or discarded when outdated or when an employee leaves the company?

5.6 Who will initiate review of SOPs every two years?
5.7 Who is responsible for maintaining the original versions of all current and superseded SOPs?

6. *Execution:*

Not Applicable.

7. *Appendix:*

7.1 Sample SOP Format
7.2 Sample SOP Composition

Appendix 7.1: Sample SOP Format

7.2.1 *Front Page:*

7.2.1.1 Department/Company Name
7.2.1.2 Title of SOP
7.2.1.3 Number of the SOP and revision indication
7.2.1.4 Number of superseded version
7.2.1.5 Numbers and names of all SOPs related to this policy
7.2.1.6 Source(s) (if any)
7.2.1.7 Number of pages of the policy
7.2.1.8 Table of Contents for the SOP

7.2.1.8.1 General Statement
7.2.1.8.2 Purpose
7.2.1.8.3 Scope
7.2.1.8.4 Definitions
7.2.1.8.5 Responsibilities
7.2.1.8.6 Execution
7.2.1.8.7 Appendices - listed by number and name or "NONE"

7.2.1.9 Approval and Review signatures section
7.2.1.10 Date of implementation

7.2.1.11 Review signature and date

7.2.2 *Following Pages:*

7.2.2.1 The header on each of the following pages must contain:

7.2.2.1.1 Department/Company name
7.2.2.1.2 SOP title
7.2.2.1.3 Number of SOP and revisions indication
7.2.2.1.4 Page number in sequence (e.g. Page 1 of 5)

Appendix 7.2: Sample SOP Composition

7.2.1 *General Statement*

The underlying policy, guideline, or objective which defines the SOP.

7.2.2 *Purpose*

This section describes the SOP and how the SOP helps to satisfy the policy defined in the General Statement section.

7.2.3 *Scope*

The scope section defines all areas in which the SOP applies. Examples of areas to be included in the scope include: departments, sections, and functional user groups.

7.2.4 *Definitions*

Common terms used in a special or specific context within the SOP, or terms not generally used, must be defined.

7.2.5 *Responsibilities*

This section establishes who is required to perform the duties necessary to ensure compliance with and adherence to the SOP.

7.2.6 *Execution*

Brief outline of a procedure or a complete procedure that is too short to require a separate procedure document.

7.2.7 *Appendix*

Appendixes should be used for additional and subordinate information only. Each appendix must be assigned an Appendix number and the same code number as the SOP to which it is attached.

Chapter 13

ELECTRONIC SIGNATURES

FDA TASK FORCE RECOGNIZES PROBLEM

On July 21, 1992, the FDA issued an advance notice of proposed rulemaking (ANPRM) in the *Federal Register* relating to the issue of electronic Identification/signatures. This notice, the result of eight months of work by an FDA task force and working group dealing with the issue, was a significant step on the part of the FDA and will have a great impact on how the FDA regulates a variety of industries. The ANPRM deals with an issue that is becoming increasingly important as industries move toward automated systems. A former PMA spokesman says that the acceptance of electronic identification is the "only significant issue left between industry and FDA."

The FDA Electronic Identification/Signature Task Force and Working Group were formed at the direction of the Commissioner in November 1991 to identify issues regarding acceptance of electronic identification in the pharmaceutical industry and recommend standards for such identification. The FDA took this step in response to issues raised in 1991 by the pharmaceutical industry regarding FDA's acceptance of electronic identification in terms of the Agency's Current Good Manufacturing Practices (CGMP) regulations.

At its first meeting, the Task Force quickly realized the complexity of the electronic identification issues. These issues potentially impact all of the agency's centers. A search of the agency's documents found 733 occurrences of the words "signature," "signatures," "sign," or "signed." Almost 400 of these occur in the Code of Federal Regulations, including 331 in 164 sections of Part 21 of the CFR. In this document search, however, no definitions of the word "signature" was found. The Task Force then appointed the FDA Electronic Identification/Signature Working Group to address the issues and make recommendations to the Task Force.

The Working Group is composed of representatives from the Center for Drug Evaluation & Research (CDER), Center for Biologics Evaluation & Research (CBER), Center for Devices and Radiological Health (CDRH), Center for Veterinary Medicine (CVM), Center for Food Safety and Applied Nutrition (CFSCAN), Office of Information

Resources Management, Office of General Counsel, Office of Enforcement, and Office of Regional Operations. The Working Group is chaired by Paul Motise of the CDER, who made the FDA's first public presentation on the subject to a joint PDA/FDA meeting in September 1992.

FDA DEFINES CONCERNS, TERMS, AND ISSUES

The working group submitted its first progress report on February 24, 1992, in which it identified three main areas of concern regarding the use of electronic identification/signatures:

- The use of alternative to handwritten signatures in industry records which are subject to FDA inspection (such as batch production records).
- The use of electronic alternatives in records submitted to the FDA for review and approval (such as Computerized New Drug Applications and Abbreviated New Drug Applications) where handwritten signatures have in the past provided certification of data or authentication of the records.
- The use of electronic alternatives to handwritten signatures by the FDA itself in internal administrative and regulatory documents (such as sample collection reports and electronic communications with regulated industries).

In what the ANPRM calls an attempt by FDA to consider the acceptance of various signature alternative technologies "according to their performance characteristics and the level of security they provide," the Working Group developed definitions for three terms: "Signature" (conventional, handwritten), "Electronic Signature" (e.g., retinal scan, voice print, hand print), "Signature Recorded Electronically" (handwritten signature captured on electronic media), and "Electronic Identification" (e.g., passwords, bar codes, personal identification codes). These would be accepted in descending levels, depending on the regulatory importance of the documents -- the most significant documents would need to have conventional signatures; less important documents could have Electronic Signatures or Electronic Identifications.

The Working Group report also identified seven major issues relating to the acceptance of electronic identification -- Legal Acceptance, Regulatory Acceptance, Enforcement Integrity, Validation, Security, Standards, and Freedom of Information. Six of these issues were discussed in the FDA's July 21 ANPRM:

Regulatory Acceptance

"Various FDA regulations are worded in ways that do not accept signature alternatives. For example, the CGMP regulations for human and veterinary drugs do not anticipate or allow electronic identification/signature substitutes for handwritten signatures; 21 CFR 211.186 explicitly requires full handwritten signatures. Various submissions to the agency regarding medical devices must contain signatures; 21 CFR 1005.25 requires signatures to be written 'in ink.' On the other hand, the agency has accepted encoded computerized endorsements where the low acid canned food regulations call for identification of individuals who perform certain actions. The agency believes it would be beneficial to take a uniform approach to accepting signature alternatives and is seeking comments on how such an approach may be codified and how various records required by the regulations might lend themselves to the use of electronic identification/signatures."

Enforcement Integrity

"Acceptance of signature alternatives must not hamper the agency's enforcement efforts. For example, FDA investigators must be able to obtain copies of electronic records which would be admissible evidence in regulatory actions to demonstrate individual responsibility. The agency conducted a review of relevant statutes and cases and found no court cases which clearly recognize the validity of signature alternatives. At the same time, no cases were found that would impede acceptance of signature alternatives. FDA's examination also disclosed several sections of Title 18 of the United States Code that FDA could use to pursue cases of electronic records falsification. (Details of the review are contained in the working group report.) The agency is particularly interested in comments on this aspect of adopting electronic identification/signatures."

Security

"The agency is concerned that electronic identification/signatures be secure from abuse and falsification. Although FDA recognizes that virtually any system can be corrupted and defeated by individuals who are intent on falsification, substitutes for handwritten signatures should nonetheless be at least as secure as conventional handwritten signatures. The agency requests comments on how electronic identification/signatures can be secured. This issue is critical because of the ease with which some electronic identification methods can be falsified without leaving an audit trail. The agency seeks comments on whether some types of signature alternatives may be too insecure to accept at all, and if the agency should establish a stratified system whereby the regulatory significance of a given record would determine the level of security that would be necessary for a signature alternative."

Validation

"The agency recognizes the importance of validation as a means of attaining confidence in the reliability of signature alternative technologies. Based on the agency's inspectional findings of inadequacies regarding computer systems validation in general, FDA is concerned that new technologies may be adopted before they are adequately validated. Therefore, the agency requests comments on key elements of validation of electronic identification/signature systems and on what type of guidance FDA should develop in this area."

Standards

"The agency's acceptance of signature alternatives would be facilitated if FDA could apply appropriate signature standards developed by other organizations. The agency is aware that a Digital Signature Standard has been proposed by the National Institute of Standards and Technolog[y] (NIST). A copy of the proposed NIST standards is included as an attachment to the working group report.... The agency seeks comments on the application of the proposed NIST standard to FDA matters, as well as comments on any other relevant standards."

Freedom of Information

"The agency has very limited experience in handling freedom of information (FOI) requests that: (1) are submitted in electronic form, (2) request that paper documents be supplied in electronic form, and (3) request that electronic documents be furnished in electronic form. However, FDA anticipates receiving such requests in the future as paperless systems are adopted. Comments are requested on the need to establish modified fees and procedures."

The seventh issue, which was discussed in the Working Group report but not included in the issues discussed in the *Federal Register*, was the *LEGAL ACCEPTANCE* issue: Will electronic identification be accepted by courts and have the same legal weight as handwritten signatures.

WORKING GROUP RECOMMENDATIONS

In its initial report, the Working Group made seven recommendations to the Task Force:

- The FDA should publish an advance notice of proposed rulemaking which would describe the main issues and invite comments and presentations from interested parties and developers of electronic identification/signature systems.
- If the comments received warranted such action, the FDA should codify, in one regulation, signature alternative acceptance provisions. Differences between the various FDA centers (and within the centers in their various regulations) could be reconciled by the proposed "multi-tiered approach" of accepting signatures, electronic signatures, and electronic identification.
- The FDA's Freedom of Information staff should consider how requests for electronic documents can be filled.
- The FDA should develop a program to train field investigators how to collect electronic records.
- Electronic identification systems, loaned by vendors, should be evaluated by the Winchester Engineering and Analytical Center to determine how the FDA should address weaknesses and positive features of such systems in agency regulatory and policy documents.
- The FDA should continue current signature requirements and interpretations until all of the issues addressed in the Working Group report have been resolved.

- The Working Group should continue to exist in order to address issues and develop regulatory and policy documents until a final regulation and guidance are in place.

COMMENTS ON THE ANPRM

The ANPRM published in the *Federal Register* in July allowed for a 90-day comment period which was to end October 19. The FDA received a request, dated October 7, 1992, from the Nonprescription Drug Manufacturers Association to extend the comment period for another 60 days. Accordingly, on October 21 the FDA published in the *Federal Register* an extension of the comment period to December 18, 1992, in order "to allow interested persons to submit meaningful comments on the advance notice of proposed rulemaking."

In the more than fifty comments from the pharmaceutical, medical device, and biologics industries responding to the ANPRM received by the FDA by December 18, 1992, there was little agreement with the issues raised by the FDA Working Group. A great deal of the difference of opinion revolved around the definition of "signature." The ANPRM stated that the FDA could not find "a codified definition of signature itself" when the agency Working Group searched agency regulations, and suggested that the term be defined as "the name of an individual, handwritten in script by that individual." Some industry representatives have suggested that this issue could be simply resolved, if the agency took the broader definition, based on the Uniform Commercial Code, found in Black's Law Dictionary:

> The act of putting one's name at the end of an instrument to attest its validity; the name thus written. A signature may be written by hand, printed, stamped, typewritten, engraved, photographed, or cut from one instrument and attached to another...; it being immaterial with what kind of instrument a signature is made. And whatever mark, symbol or device one may choose to employ as representative of himself is sufficient.

This definition would lend itself to use of electronic alternatives to handwritten signatures and/or initials.

Most comments took issue with the FDA's discussion of regulatory acceptance of electronic identification/signatures, pointing out that CGMP regulations should generally be interpreted as already allowing acceptance of electronic substitutes for handwritten signatures or initials. Comments from both Pfizer Inc., and the Du Pont Merck Pharmaceutical Company point out that the CGMP for finished pharmaceuticals in CFR 21.186(a), which requires a "full signature, handwritten," for master production and control records for drug products, is more of an exception in the CGMP, rather than an example of how FDA regulations do not allow electronic identification for handwritten signatures, since it is the only reference in the CGMP specifying "handwritten" signatures. The Pharmaceutical Manufacturers Association (PMA) recommended that the FDA "accept 'electronic identification' ... for purposes of meeting requirements for signatures', 'initials' or 'identification'" in FDA regulations, while "specific references to 'handwritten signatures' or 'initials' be removed from 21 CFR provisions and they be replaced with identification requirements that are non-specific with respect to the method employed to meet the requirement." The consensus of the comments seemed to be that a general policy statement from FDA accepting electronic identification was all that was necessary at this stage.

Another area of contention was the FDA's suggestion that a three-tiered approach be used for accepting electronic identification. Johnson & Johnson's response to the ANPRM called the tiered approach "an unnecessary complication for any proposed regulation," which seems to be the industry consensus. The industry comments in general point out that the FDA's system of definitions is more complex and confusing than necessary and that the FDA places too much reliance on the security, integrity, and legal acceptability of handwritten signatures as implied by the agency's proposed three-tiered approach to acceptance of electronic alternatives.

In addition to the comments sent in response to the ANPRM, there was further opportunity for industry comment. The November 1992 issue of *Pharmaceutical Technology* carried an article by Paul Motise which detailed the FDA's position. The same issue of *Pharmaceutical Technology* carried a counter article, written by a former PMA spokesman, which, in similar detail, discussed industry's position that it should be allowed to proceed with the application of properly compliant electronic identification technology.

THE VALIDATION ISSUE

The ANPRM sought comments on "key elements of validation of electronic identification/signature systems and on what type of guidance FDA should develop in this area." In its report, the Working Group stated that "validation would have to be a significant factor in acceptance of any signature alternatives." This conclusion is based on the FDA's experience that "inadequate or entirely lacking computer system validation has been a problem detected by our field investigators." Some of the issues regarding computer validation found by FDA investigators, and their relationship to electronic signatures, were discussed in an article by a former FDA Investigator in *Pharmaceutical Technology* (March 1992).

Some members of the Working Group provided additional comments on the various issues raised by the group, and these comments were appended to the group's initial report. In terms of validation, all agreed that reliability and authenticity of data, as well as security against unauthorized entry into and use of computer systems, would have to be ensured through validation before electronic alternatives to handwritten signatures could be accepted. Members of the Working Group also agreed that an agency-wide guidance policy should be developed for validation of computer systems that include electronic identifications.

Comments on Validation

All the comments submitted in response to the ANPRM agreed with the FDA that validation of any system including electronic identification is necessary to meet CGMP requirements. This is not surprising, as validation is probably the least controversial of the issues listed by the FDA. However, all the comments also agree that both the FDA and the industry have adequate validation guidelines and procedures already in place, and neither unique validation principles, standards, and policies nor new guidance, need to be developed to cover systems utilizing technologies for electronic identification. Electronic identifications do not need to be validated separately, but should be validated as part of the entire system employing the electronic identification technology.

The comments are consistent in the "key elements of validation of electronic identification/signature systems" which FDA requested be addressed. The major elements listed in the comments are established validation principles:

- Policies and plans for defining the firm's approach to system validation, and procedures for review and approval of the validation process.
- Proper system specifications delineating functional requirements, software requirements, the physical system and operating environment.
- Administrative control through policies, procedures and master plans defining a firm's philosophy and approach to system validation.
- Assurance of security.
- Verification of data entries.
- Installation, Operational, and Performance Qualifications for hardware and software.
- Assurance of implementation of proper change control procedures.
- Proper system maintenance.
- Personnel training.
- Well-organized documentation with defined meanings for approval signatures.
- Demonstration of performance through documented testing protocols and results.
- Record retention systems protect records and allow for accurate and efficient record retrieval.

PROPOSED RULE

On December 10, 1992, just before the end of the extended comment period, representatives of Burroughs Wellcome met with Paul Motise of CDER and addressed the issue of how the FDA could facilitate its reaction to the comments on the ANPRM. At that meeting, Paul Motise told the Burroughs Wellcome representatives that once the agency had summarized and analyzed the comments to the ANPRM, a proposed rule might be issued addressing the issue of all electronic records. At a meeting with representatives a firm which produced an electronic signature system, Motise was of the opinion that whenever FDA issued a proposed rule it would probably "codify general requirements as performance standards which any number of technologies could achieve." The FDA realizes that the issue needs to be addressed rapidly so as not to impede emerging technology for the industry. Although it was anticipated that a proposed rule might be issued as early as March or April 1993 to provide clarification of the electronic signatures required by the CGMP, it was not until more than a year later that a proposed rule was finalized.

On August 31, 1994, the FDA issued in the *Federal Register* proposed regulations on electronic signatures and electronic records. Given the comments on the agency's ANPRM, nothing in the proposed regulation is surprising. The FDA will emphasize systems controls and validation in accepting electronic records and electronic signatures in order to guarantee the integrity of electronic data.

The *Federal Register* relating to electronic signatures and electronic records as published by the FDA in 1992 and 1994 follow:

FEDERAL REGISTER, Vol. 57 No. 140 Tuesday, July 21, 1992, p. 32185 (Proposed Rule) 1/285

DEPARTMENT OF HEALTH AND HUMAN SERVICES

Food and Drug Administration

21 CFR Chapter I

[Docket No. 92N-0251]

57 FR 32185 -- ELECTRONIC IDENTIFICATION/SIGNATURES; ELECTRONIC RECORDS; REQUEST FOR INFORMATION AND COMMENTS

AGENCY: Food and Drug Administration, HHS.

ACTION: Advance notice of proposed rulemaking.

SUMMARY: The Food and Drug Administration (FDA) is announcing that it is considering whether the agency should propose regulations that would, under certain circumstances, accept electronic identification or electronic signatures in place of handwritten signatures where signatures are called for in Title 21 of the Code of Federal Regulations (CFR), and where the electronic form of the signature bearing record is allowable by the regulations. The decision on whether to propose such regulations will be based on information and comments submitted in response to this advance notice of proposed rulemaking, the recommendations and findings of the agency's Task Force on Electronic Identification/ Signatures, and the agency's experience with alternatives to conventional handwritten signatures.

SUPPLEMENTARY INFORMATION: The agency is aware that automated systems are being used more extensively in the various industries it regulates. Use of such systems is also expanding within the agency itself. An emerging objective of the use of automation is the implementation of paperless electronic records. Signatures are a key aspect of many records and the transition from paper records containing traditional handwritten signatures to paperless electronic records raises a set of issues relating to FDA acceptance of alternatives to handwritten signatures. Electronic records can contain human endorsements using various technologies. An alternative to the handwritten signature in such electronic records is termed an electronic identification/ signature, for purposes of this document.

The agency has met with members of the pharmaceutical industry who sought advice on how they could implement paperless records systems within regulations for 21 CFR Part 210-Current Good Manufacturing Practice in Manufacturing, Processing, Packing, or Holding of Drugs; General and 21 CFR Part 211-Current Good Manufacturing Practice Regulations for Finished Pharmaceuticals, i.e., CGMP's. Upon examining how the agency might accommodate such records and attendant endorsements by electronic identification/signature, FDA found the issues to be complex, affecting regulations beyond the CGMP area. For example, a review of Title 21 of the CFR found references to signatures in 132 different sections; a listing of these sections is contained in the report identified below. These regulations typically address signatures in manufacturing production records, clinical investigation records, and formal submissions to the agency. Absent, however, is a codified definition of signature itself. The same preliminary examination took note of the agency's own paperless electronic records which may contain electronic identifications/signatures.

To identify the issues and develop preliminary approaches on how FDA might accept signature alternatives in an agency-wide manner, the agency formed an FDA Task Force on Electronic Identification/Signatures. The task force created a subgroup, the Electronic Identification/Signature Working Group to address the issues in greater detail. A copy of the group's initial report dated February 24, 1992, has been placed on file and is on display at the Dockets Management Branch (address above), for public examination; copies may be obtained from NTIS (address above). The report recommended publication of an advance notice of proposed rulemaking to obtain wide public comment on the issues to determine whether the agency should promulgate regulations that would, under certain circumstances, accept electronic identification signatures in place of handwritten signatures where signatures are required by various FDA regulations.

The agency is considering the use of electronic identification/signatures within the general context of three categories of current and future paperless (electronic) records: (1) Records maintained by industry which are subject to FDA inspection (e.g., batch production records for drug products, low acid canned foods, or infant formulas); (2) records submitted to FDA for review and approval, usually as part of research or marketing applications (e.g., new drug applications and food additive petitions); and (3) FDA's own records (e.g., sample collection reports) and notifications to industry (e.g., electronic mail). The agency requests that comments address the issues within these three categories of records, and how these records may or may not lend themselves to the use of electronic identification signatures. The agency wishes to clarify, however, that at this time it does not seek to mandate the conversion of paper records to electronic form.

Although this notice pertains to records which are currently allowed by the regulations to be in electronic form, the issues related to electronic identification/signatures are germane to electronic records, in general. Therefore, the agency welcomes comments that identify any record now required by FDA that may be amenable to being in electronic form.

The agency is aware of a variety of signature alternative technologies and will consider their acceptance according to their performance characteristics and the level of security they provide. The level of performance and security would be commensurate with the regulatory significance of the electronic record. The agency is seeking comments on how this stratified acceptance might be accomplished. For example, FDA might establish definitions which incorporate technological and security distinctions for the terms signature, signatures recorded electronically, electronic signature, and electronic identification. In the agency's experience, a signature is generally the name of an individual, handwritten in script by that individual; the act of signing usually involves use of a writing or marking instrument and provides a unique and secure link to the individual. Where such a signature is captured on electronic media instead of on paper, the result might be termed "signatures recorded electronically."

The agency is aware of various technologies that dispense with the act of signing, but still apparently furnish an intrinsic biometric or behavioral link to an individual such that someone else cannot make use of that person's signature alternative (e.g., retinal scan systems, voice prints, hand prints). The agency is considering the term "electronic signatures" for such systems.

Another type of signature alternative appears to lack intrinsic biometric or behavioral links to the person being identified, but instead relies upon administrative controls to maintain the uniqueness and security of the identification method. Examples of such signature alternatives are systems which use (individually, or in combination) bar codes, passwords/identification codes, and personal identification devices (badges/cards). The agency is considering the term "electronic identification" for such systems.

The agency requests comments on how the above, or other classes of signature alternatives would fulfill three general purposes of the traditional handwritten signature: (1) To identify the actor and show his/her authority to act; (2) to document the action in a way that is legally binding and cannot be altered or repudiated; and (3) to create a record that would be admissible evidence in court.

To be acceptable to the agency, electronic identification signatures need to be as legally acceptable as conventional, handwritten signatures. The agency would have to conform to any acceptance criteria established by the courts. FDA expects electronic identification/signatures to carry the same commitment, legal weight, and significance as conventional handwritten signatures. Falsification of signature alternatives must be considered fraudulent to the same extent as is falsification of handwritten signatures.

The agency is seeking specific comments on each of the following issues which are addressed in greater detail in the above-referenced working group report:

1. Regulatory Acceptance. Various FDA regulations are worded in ways that do not accept signature alternatives. For example, the CGMP regulations for human and veterinary drugs do not anticipate or allow electronic identification/signature substitutes for handwritten signatures; 21 CFR 211.186 explicitly requires full handwritten signatures. Various submissions to the agency regarding medical devices must contain signatures; 21 CFR 1005.25 requires signatures to be written "in ink." On the other hand, the agency has accepted encoded computerized endorsements where the low acid canned food regulations call for identification of individuals who perform certain actions. The agency believes it would be beneficial to take a uniform approach to accepting signature alternatives and is seeking comments on how such an approach may be codified and how various records required by the regulations might lend themselves to the use of electronic identification/signatures.

2. Enforcement Integrity. Acceptance of signature alternatives must not hamper the agency's enforcement efforts. For example, FDA investigators must be able to obtain copies of electronic records which would be admissible evidence in regulatory actions to demonstrate individual responsibility. The agency conducted a review of relevant statutes and cases and found no court cases which clearly recognize the validity of signature alternatives. At the same time, no cases were found that would impede acceptance of signature alternatives. FDA's examination also disclosed several sections of Title 18 of the United States Code that FDA could use to pursue cases of electronic records falsification. (Details of the review are contained in the working group report.) The agency is particularly interested in comments on this aspect of adopting electronic identification/signatures.

3. Security. The agency is concerned that electronic identification/signatures be secure from abuse and falsification. Although FDA recognizes that virtually any system can be corrupted and defeated by individuals who are intent on falsification, substitutes for handwritten signatures should nonetheless be at least as secure as conventional handwritten signatures. The agency requests comments on how electronic identification/signatures can be secured. This issue is critical because of the ease with which some electronic identification methods

can be falsified without leaving an audit trail. The agency seeks comments on whether some types of signature alternatives may be too insecure to accept at all, and if the agency should establish a stratified system whereby the regulatory significance of a given record would determine the level of security that would be necessary for a signature alternative.

4. Validation. The agency recognizes the importance of validation as a means of attaining confidence in the reliability of signature alternative technologies. Based upon the agency's inspectional findings of inadequacies regarding computer systems validation in general, FDA is concerned that new technologies may be adopted before they are adequately validated. Therefore, the agency requests comments on key elements of validation of electronic identification/signature systems and on what type of guidance FDA should develop in this area.

5. Standards. The agency's acceptance of signature alternatives would be facilitated if FDA could apply appropriate signature standards developed by other organizations. The agency is aware that a Digital Signature Standard has been proposed by the National Institute of Standards and Technologies (NIST). A copy of the proposed NIST standard is included as an attachment to the working group report, on file at the Dockets Management Branch (address above). The agency seeks comments on the application of the proposed NIST standard to FDA matters, as well as comments on any other relevant standards.

6. Freedom of Information. The agency has very limited experience in handling freedom of information (FOI) requests that: (1) Are submitted in electronic form, (2) request that paper documents be supplied in electronic form, and (3) request that electronic documents be furnished in electronic form. However, FDA anticipates receiving such requests in the future as paperless systems are adopted. Comments are requested on the need to establish modified fees and procedures.

The agency also welcomes comments on any related issues. In addition, the agency would like to hear from developers of electronic identification/signaturesystems. Presentations before the task force are welcome and may be arranged through the contact person listed above. However, this request is for information only and does not undertake any commitment to purchase any system. In the event that any future acquisition results from this activity, such acquisitions will be conducted in accordance with the requirements for competition as set forth in the Federal Acquisition Regulation (Title 48 of the CFR). The agency emphasizes that the purpose of such presentations is for FDA to obtain information on emerging technologies and not to endorse or disapprove particular systems.

Interested persons may, on or before October 19, 1992, submit to the Dockets Management Branch (address above) written information and comments regarding this advance notice of proposed rulemaking. Two copies of the information and comments should be submitted, except that individuals may submit one copy. The information and comments are to be identified with the docket number found in brackets in the heading of this document. The information and received comments may be seen in the office above between 9 a.m. and 4 p.m., Monday through Friday.

FEDERAL REGISTER, Vol. 59 No. 168 Wednesday, August 31, 1994, p. 45160 (Proposed Rule) 1/285

DEPARTMENT OF HEALTH AND HUMAN SERVICES

Food and Drug Administration

21 CFR Part 11

Electronic Signatures; Electronic Records

[Docket No. 92N-0251]

AGENCY: Food and Drug Administration, HHS.

ACTION: Proposed rule.

SUMMARY: The Food and Drug Administration (FDA) is proposing regulations that would, under certain circumstances, permit the agency to accept electronic records, electronic signatures, and handwritten signatures executed to electronic records as generally equivalent to paper records and handwritten signatures executed on paper. These proposed regulations would apply to records when submitted in electronic form that are called for in Title 21 of the Code of Federal Regulations (CFR). The use of electronic forms of recordkeeping and submissions to FDA remains voluntary. This proposed rule is a followup to the agency's July 21, 1992, advance notice of proposed rulemaking (ANPRM). The intended effect of this proposed rule is to permit use of electronic technologies in a manner that is consistent with FDA's overall mission and that preserves the integrity of the agency's enforcement activities. This proposed rule is also intended to assist in achieving the objectives of the Vice President's National Performance Review.

SUPPLEMENTARY INFORMATION:

I. Background

In the *Federal Register* of July 21, 1992 (57 FR 32185), FDA published an ANPRM on whether the agency should propose regulations that would, under certain circumstances, permit the agency to accept electronic identification or electronic signatures in place of handwritten signatures where signatures are required in 21 CFR, and where the electronic form of the signature bearing record is allowable by the regulations. The ANPRM requested comments on current and future electronic records maintained by industry and subject to FDA inspections, submitted to FDA for review and approval, and FDA's own records and industry notifications. The ANPRM also identified and sought specific comment on the following issues: (1) Regulatory acceptance; (2) enforcement integrity; (3) security; (4) validation; (5) standards; and (6) freedom of information (FOI). In the *Federal Register* of October 21, 1992 (57 FR 48008), FDA published an extension of the comment period regarding the ANPRM. Interested persons were given until December 18, 1992, to comment on the ANPRM.

FDA received 53 comments from trade associations, pharmaceutical and medical device manufacturers, computer systems developers, private organizations, a Federal agency, a university, and consumers. The comments generally support the ANPRM's objectives. A number of the comments made suggestions. As appropriate, comments will be responded to in this document in the discussion of the proposed regulation set forth below.

II. Summary and Analysis of Comments to the ANPRM

A. Analysis of Comments

The agency received a total of 53 comments to the July 21, 1992, ANPRM. Comments came from a variety of sources including 6 trade associations, 27 pharmaceutical manufacturers, 2 medical device manufacturers, 1 contract laboratory, 8 computer systems developers, 1 law firm on behalf of a computer system developer, 1 law firm on behalf of a consortium of industrial research companies, 1 agency of the Federal Government, 1 drug sample distribution establishment, one medical center, 1 university food sciences unit, 1 express mail delivery service, and 2 individuals.

Comments generally supported the agency's efforts relative to electronic signatures and electronic records. One comment suggested that FDA's actions may provide a model for other Federal agencies. Several comments found the agency's electronic identification issues to be among the most important and immediate concerns currently facing the pharmaceutical industry.

One comment expressed concern that the ANPRM did not address medical devices and urged the agency to adopt uniform agencywide policies regarding electronic signatures.

In general, comments addressed the advantages of electronic records in enhancing product quality, control, production efficiency, and the conduct of nonclinical laboratory studies.

Comments urged the agency to follow a course of action that would not impede technological innovation. Comments also called for expedited resolution of the issues in order to facilitate industry's plans for implementing new technologies.

One comment commended the agency for making the February 24, 1992, progress report of the FDA Electronic Identification/Signatures Working Group available via e-mail and encouraged FDA to continue electronic distribution of agency documents. One comment submitted a 58-page paper which addressed legal considerations and a detailed stratification scheme based upon security risks.

Although the ANPRM stated that the scope of FDA's considerations extends to all articles that it regulates, and to all portions of 21 CFR under its jurisdiction, very few comments were received from sources outside the pharmaceutical industry. One medical device trade association mistakenly commented that medical devices were not covered. The agency emphasizes that all regulated articles are covered. The agency agrees that it is important to accommodate new technologies in a responsible manner. The agency also agrees with the comment that encouraged FDA to continue the electronic distribution of agency documents. FDA will be implementing this form of distribution increasingly in the future.

The decision to propose these rules is based upon: (1) The information and comments submitted in response to the July 21, 1992 ANPRM; (2) the recommendations and findings of the agency's Task Force on Electronic Identification/Signatures, which was reported in the progress report of FDA's Electronic Identification/Signature Working Group on February 24, 1992 (Ref. 1); and (3) the agency's experience with alternatives to conventional handwritten signatures and electronic records.

The agency is aware that automated systems are being used more extensively in the various industries that it regulates. Use of such systems is also expanding within the agency itself. Implementing paperless electronic records and attendant methods of "signing" such records is an emerging objective of the use of automation. Signatures are a key aspect of many records. The transition from paper records containing traditional handwritten signatures to paperless electronic records raises issues relating to FDA's acceptance of alternatives to handwritten signatures and their underlying trustworthiness.

FDA recognizes the importance of electronic records and their integration into a variety of automation efforts, such as manufacturing process controls, materials resources controls, laboratory information systems, clinical trial information systems, and electronic data interchange activities. The agency is aware that some new technologies and manufacturing methods require use of electronic records. For example, in certain highly controlled manufacturing environments, the presence of paper itself can pose a source of product contamination, and (for highly toxic compounds) paper can be a vehicle for exposing workers to dangerous compounds.

FDA is aware of the benefits of conducting official electronic communication with regulated industries and the public. However, the agency is also aware that legal, regulatory, and administrative concerns have delayed full use of electronic communication. FDA expects that promulgation of the regulations proposed in this document will begin to address the agency's concerns and facilitate the agency's modernization efforts.

Although most comments to the ANPRM addressed electronic records within the context of closed systems, where access is limited to people who are part of the organization that operates the system, the agency expects that near-term development and implementation of appropriate controls for open systems, where access extends to people outside of the operating organization, will facilitate secure, authoritative electronic communication between FDA and the regulated industries.

The Vice President's Report of the National Performance Review has as a stated objective the expanded use of new technologies and telecommunications to create an "electronic government." (September 7, 1993, Report of the Vice President's National Performance Review (pp. 113 through 117) (Ref. 2). This proposal would be a first step by FDA in implementing this objective, by, for example, allowing electronic filings of regulatory documents and expanded use of e-mail. This will result in significant benefits to the public, the regulated industry, and the agency. These benefits could include faster review and approval of new products, and rapid availability of a variety of agency documents around the clock.

FDA encourages the use of new technologies that will enhance the quality, safety, and efficacy of products it regulates, but is mindful of the need to maintain the ability to fulfill its consumer protection mandate. The agency believes that these proposed rules will accomplish both objectives.

B. Comments on Record Types

The ANPRM requested examples of records that: (1) Are maintained by industry and inspected by FDA, (2) are submitted to FDA, and (3) are created and maintained by FDA that may be amenable to electronic identification/signatures. Most respondents confined their comments to the first record type. However, a few comments provided the following examples of records in each category:

Records maintained by industry and inspected by FDA that may be in electronic form include:

1. Master and batch production and control records
2. Logs,
3. Standard operating procedures,
4. Laboratory notebooks,
5. Complaint records,
6. Validation protocols and data summaries,
7. Laboratory data summaries, and
8. Drug sample records under the Prescription Drug Marketing Act (the PDMA)(Pub. L. 102-353).

Although most comments addressed pharmaceutical records, the agency believes that it is necessary to recognize that records maintained by industry and inspected by FDA extend to other articles and include records such as:

1. Medical device history records, and medical device master records,
2. Master record files,
3. Blood bank donor records,
4. Thermally processed low-acid foods records, and
5. Hazard analysis critical control points.

Records submitted to FDA that may be in electronic form include:

1. New drug or new animal drug applications,
2. Product license applications,
3. Establishment license applications, and
4. Drug or veterinary drug master files.

Most comments focused on pharmaceutical documents. However, the agency recognizes that submissions for other FDA-regulated products would be applicable. Such records include, but are not limited to:

1. Medical device premarket approval applications,
2. Medical device premarket notifications,
3. Medicated feed applications,
4. Food additive petitions,
5. Color additive petitions,
6. Infant formula notifications,
7. Low acid canned food and acidified food firm, registration and scheduled process filing, and
8. Generally recognized as safe (GRAS) petitions.

One comment addressed records maintained by the agency and suggested that signatures recorded electronically (SRE's), as identified in the ANPRM, should be an acceptable alternative to signatures recorded on paper. The comment asserted that SRE's have sufficient uniqueness, are difficult to forge (especially when accompanied by the date and time the SRE was made), and would realize legal acceptance.

Two comments suggested that whatever policies are adopted for electronic records maintained by the industry, or records submitted to the agency, apply equally to FDA's own records.

Although the proposed rule focuses primarily on records maintained by industries inspected by FDA, and submissions to the agency, FDA will apply the principles in the new rule to its own electronic documents.

III. Definitions/Stratified Acceptance Approach

A. Definitions

One comment agreed with FDA's working definitions. The comment noted that electronic identification should suffice for all of the agency's applications and called for common codified definitions for the following words and phrases:

1. Signature

Several comments agreed with FDA's working definition of the term "signature." One categorized conventional signatures as "wet signatures" and one submission suggested renaming the term "handwritten signatures" for clarification.

2. Signatures Recorded Electronically

One comment suggested that the term "signatures recorded electronically" be defined as an electronically captured image of a handwritten signature on optical, magnetic or other electronic media. One comment agreed with the working definition.

3. Electronic Signature

Several comments called the working definition of the term "electronic signature" as acceptable and useful. However, some comments claimed that the term is imprecise and potentially confusing to the extent that the word "signature" also appears in other working definitions. Several comments suggested the alternative phrases: "Biometric/behavioral identification" and "biologically-based electronic identification."

One comment referred to its security code number assignment system as an electronic signature, used by physicians to phone in requests for additional drug samples previously reserved under the physicians' names. Telephone requests are followed up by confirmatory signed paper forms.

4. Electronic Identification

Many comments suggested that FDA define only two terms, "signatures"(meaning conventional handwritten signatures) and "electronic identification" (to encompass signatures recorded electronically, electronic signatures, and all other forms of electronic identification). Comments suggested that definitions should not imply superiority of one type of endorsement over another and offered the following definition of electronic identification: "any method for identifying an individual where the act of providing a personal mark (signing) is recognized and/or recorded electronically."

Comments asserted that secure, validated computer systems that use electronic identification provide better, or at least equivalent, authentication than systems using handwritten signatures.

One comment suggested that a more precise term would be "administratively controlled electronic identification." One comment said that its digital signature encryption technology, a system using encrypted "keys" and proprietary algorithms, would met the agency's working definition of electronic identification, but could be coupled with hardware and software that utilized biometric links to meet the definition of electronic signature.

5. Other Definitions

Two comments offered the following additional defined terms: "Signature Alternative" -- an electronically recorded mark from any type of electronic identification, not involving a signature recorded electronically, including electronic signature (biometric/behavioral identification) and, administratively controlled electronic identification.

"Signing" -- the act of providing a personal recorded mark that serves as identification. The mark can be, but is not necessarily, provided by handwriting. The mark may also be provided by a stamp, seal, or electronic device. The last example typically records the mark in magnetic or optical media rather than on paper.

The agency believes that the diversity of comments on definitions reflects the variety of signature technologies that are available, and the need for a simple codified definition of as few terms as possible The agency is persuaded by the general premise, expressed in many comments, that FDA should establish only two definitions based broadly on whether or not the "signature" is handwritten. Therefore, the agency is proposing to codify two definitions, one for "handwritten signature" and one for "electronic signature." Electronic signature would include electronic identification; handwritten signatures would include signatures recorded electronically.

FDA disagrees with the assertion that "electronic identification," rather than "electronic signature" should be one of the two broad terms, for several reasons. The agency believes the appearance of the word "signature" in both "electronic signature" and "handwritten signature" will not be confusing to the average person, especially where the codified definitions are clear.

More importantly, the agency believes that there are overriding advantages to maintaining the word "signature" in the term "electronic signature." The legal, regulatory, and psychological importance that the average person has come to associate with conventionally signing a paper document is more likely to be carried over and equally applied to technological alternatives if the word signature is preserved. On the other hand, substitution of the word "identification" for "signature" may, on its face, imply that the alternative is something quite different and perhaps less significant. Thus, terminology can help to establish the functional equivalency of different technologies.

In addition, the term "electronic identification" can be too limiting in scope because signatures do more than merely identify the person who signed something that could be done by a person who did not perform the action. However, retention of the word "signature" in the term "electronic signature" conveys by direct inference all of the purposes of a handwritten signature, including identification, authentication, and affirmation.

Accordingly, FDA is proposing in §11.3 to define "Handwritten signature" as the name of an individual, handwritten in script by that individual, executed or adopted with the present intention to authenticate a writing in a permanent form. The act of signing with a writing or marking instrument such as a pen, or stylus is preserved. However, the scripted name, while conventionally applied to paper, may also be applied to other devices which capture the written name.

"Electronic Signature" is defined in proposed §11.3 as the entry in the form of a magnetic impulse or other form of computer data compilation of any symbol or series of symbols, executed, adopted, or authorized by a person to be the legally binding equivalent of the person's handwritten signature.

B. Biometric/Behavioral Links as Part of the Electronic Signature

Systems which utilize biometric/behavioral links as part of the electronic signature verify a person's identity based on measurement of an individual's physical feature(s) or repeatable action.

One comment addressed the behavioral link incorporated in a software product designed for use in pen-based captures; it described how the system provides reliability and trustworthiness by calibrating and recognizing a set of characteristics attendant to the act of signing (pen strokes, speed, acceleration, etc.).

One comment provided a paper in support of a signature verification system that characterizes the act of signing to establish a behavioral link between the signer and the signature, noting the system's low error rate (0.19 percent false rejects and 0.56 percent false accepts), security, social acceptance, performance, low cost, and computer portability. The paper describes how the system could be used on networks or over phone lines, in conjunction with a microprocessor-based encryption card, to prevent transmission of a prerecorded (and possibly false) signature by requiring the generation of a signature for each endorsement.

One submission asserted that stable technologies exist to provide reliable and repeatable electronic verification of individuals based upon a biometric/behavioral link. The comment furnished a report summarizing testing on several such systems that use fingerprints, hand geometry, the act of signing, retinal scans and voiceprints; the comment cited access control as the primary type of application for such systems.

Several comments argued against technologies that incorporate biometric/behavioral links on the grounds of excessive cost; two comments said biometric based devices cost about $1,800 to $4,000 per unit and behavioral based devices cost $600 to $1,500 each.

Most comments argued against the premise that biometric/behavioral links are necessary or beneficial to electronic signatures. However, two comments asserted that appropriate application of an electronic signature requires a biometric or direct behavioral link to an individual, and one comment acknowledged that such links are less susceptible to procedural deviations than other authentication methods. One comment said biometric/behavioral links are appropriate to systems which control physical access to a facility.

Many comments urged FDA to refrain from requiring use of systems based on biometric/behavioral links (particularly where the drug current good manufacturing practice (CGMP) regulations require signatures) on the grounds that:

1. Such a requirement would be contrary to the objectives of the CGMP regulations;
2. Electronic signature systems are not routinely used in non-FDA regulated industry;
3. Electronic signature technology is relatively immature and unreliable;
4. The technology is relatively expensive; and
5. Electronic signature devices are impractical for pharmaceutical applications in which operators are garbed so as to obscure anatomical interaction with detection devices (e.g., hand or voiceprints would be difficult to manage where workers wear masks or gloves).

FDA believes it is important to allow firms to take advantage of a variety of new technologies. It is not the agency's intent to mandate use of systems that use biometric/behavioral links, although the agency recognizes the potential advantages of such systems and encourages their development and adoption. Comments generally indicate that biometric/behavioral link technologies have been developed, may have high levels of reliability, but have not yet been incorporated into manufacturing environments to any appreciable degree. Accordingly the agency's proposed regulations do not, at this time, specify the type of electronic signature technologies that are required.

However, because FDA recognizes the benefits of those electronic signatures which are inherently less vulnerable to falsification, and because the agency wishes to encourage the development of such technologies, the proposed regulations reflect the position that the robustness of biometric/behavioral based systems permits less stringent administrative controls to be used.

In addition, FDA considers that biometric/behavioral based systems may have greater application in open environments, which pose a greater challenge to signature integrity than closed environments.

C. *Purpose of Signatures*

One comment identified the following functions of a signature: To identify someone; to declare, to witness, to acknowledge or disclaim, to agree or disagree, and to exhibit responsibility or authorship, as a formalized personal act such that subsequent disavowal or disclaimer is highly unlikely. The comment added that good practice suggests that the signature be properly ascertained, clearly indicated, and appropriately exhibited in a prominent place, and that bilateral mechanisms can further this purpose, and focus the individual's attention on the gravity, solemnity, and formality of the event. The comment also noted that because the purpose of a signature is not always apparent, some documents include clarifying phrases such as "in witness thereof," or "agreed to by." The comment further stated that in the typical manufacturing environment custom governs the meaning of a signature (e.g., to acknowledge performance of a procedure, responsibility for proper performance of the procedure, or to whom that the person was merely present).

The agency believes the comment has identified an important aspect of a signed writing, namely the meaning ascribed to the signature. Accordingly the regulations proposed at §11.50(b) require the document being signed to clearly indicate the purpose of the electronic signature. FDA also agrees with the comment's view that bilateral mechanisms can help to establish the seriousness of the electronic endorsement, and the agency is proposing at §11.200(a)(1) to require certain electronic signatures to be composed of at least two elements.

Respondents also commented on how signature alternatives might fulfill the following traditional purposes of a signature:

1. To identify the actor and show his/her authority to act.

Many comments disagreed that presence of a signature shows the signer's authority to act, noting that such authority is generally determined by the individual's organization. However, several comments acknowledged that electronic identification systems can be programmed to confirm an individual's authority to act.

One comment said authority to act could be met by the use of identification codes/passwords for intra-establishment records and by public key encryption standards such as the Rivest-Shamir-Adleman (RSA) standard for inter-establishment records.

The agency agrees that the presence of a signature, per se, does not necessarily guarantee that the signer has the authority indicated. However, in general, the presence of the signature, in combination with the signer's title, is by custom a reasonable indication that the person does have the organization's authority to endorse the subject document. FDA believes that in most cases people will not sign a document if they lack the authority called for by the action of signing. In the kinds of electronic environments addressed by the comments, systems can check a cross-referenced authorization roster to see that an individual who attempts to sign a document has, in fact the requisite authority.

2. To document the action in a way that is legally binding and cannot be repudiated.

Comments generally asserted that properly validated and secure electronic identification systems would be legally binding.

The agency agrees with the comments regarding the importance of validation and security and the proposed rule places appropriate emphasis on these controls.

One comment suggested that documentation of the action, not the individual, should be of prime importance because FDA is concerned more with the actions of a company than with individuals within a company, and that concern with actions of individuals is the concern of the company itself. The comment added that the RSA encryption standard could be used in this area for inter-establishment electronic records.

FDA disagrees with the premise that FDA should be concerned more with corporate than individual actions. In FDA's enforcement activities, there is equal emphasis on the responsibility of both individuals and corporations. Furthermore, section 201(e) of the Federal Food, Drug, and Cosmetic Act (21 U.S.C. 321(e)) defines a person to include an individual, partnership, corporation, and association.

3. To create a record that would be admissible in court.

One comment suggested that a record should be admissible in court if it is shown that the record was generated by the responsible company, regardless of whether or not the record was signed; the RSA encryption standard was again cited as applicable for inter-establishment records. One submission said that electronic records would be admissible when authenticated by appropriate corporate officials under appropriate procedures relative to electronic identification.

The agency has found that court acceptance of records generally hinges on their reliability and trustworthiness. Although FDA agrees that a given unsigned record may be strictly admissible in a proceeding, establishing reliability and trustworthiness may well require that specific documents bear signatures of responsible individuals. In addition, as stated above, it is frequently important for FDA to establish individual, as well as corporate responsibility in pursuing regulatory actions, thus making it vital that evidentiary documents are signed by key individuals. The weight given to a piece of evidence may also depend upon the presence or absence of a verifiable signature.

D. Stratification

The ANPRM suggested that FDA might stratify acceptance of signature alternatives based upon the regulatory significance of the electronic record. Comments generally held that regulatory significance should not be the basis of stratification. Two comments argued against any regulatory stratification at all, one asserting that becuase conventional signatures are accepted in all situations, any alternative that provides security, identity, legibility and enforceability equal to or better than a handwritten signature should, likewise, be accepted for any applciation.

Two comments agreed with the concept of developing a stratified system whereby the regulatory significance of a record would determine the level of security needed for the signature alternative, but indicated that companies should individually define the various security categories and develop appropriate security procedures.

One comment said that electronic authorizations of high importance might require use of secondary passwords or codes to further augment security and verify data integrity.

Although most comments disagreed with the stratification approach suggested in the ANPRM, many comments suggested stratification along other lines, as follows:

1. Open Versus Closed Systems

Many comments suggested that stratification of signature alternatives be limited to security measures applied to inter versus intra company records. The distinction was stated in terms of "closed," versus "open" environments. Comments said that closed systems are typical in the pharmaceutical industry, and include administrative and physical controls to enhance reliability of the electronic endorsements.

Several comments described a typical CGMP closed system as: (1) Having controlled physical access; (2) having professionally written and approved procedures with employees and supervisors trained to follow them; (3) having records systems designed to facilitate quality assurance investigations when abnormalities may have occurred; and (4) being under legal obligation to the organization responsible for operating the system.

The following examples of documents in closed systems were given: CGMP records, GLP (good laboratory practice) and GCP (good clinical practice) records including clinical case reports, such submissions to FDA as new drug applications and adverse experience reports, and FDA internal records.

Comments generally characterized open systems as: (1) Having potentially greater exposure by outsiders; (2) entailing communication among multiple parties (e.g., communication by modem); and (3) extending system access to people who are not legally obligated to system managers.

Comments gave examples of open system documents including; Requests for drug samples, institutional review board (IRB) reviews of clinical protocols, GLP records, and Freedom of Information submissions to FDA.

2. Security Baseline Stratification for Open Systems

One comment presented a paper which addresses security stratification parameters based upon the risks of disclosure, where electronic messages are communicated in an "open" system. Stratification involves three security baselines, each of which considers the following message attributes: (1) Content sensitivity; (2) monetary value; (3) time sensitivity; (4) statutory security mandates; and (5) authentication certification requirements.

Message attributes, under the baseline system, determine the necessity and extent of the following security and reliability measures: (1) Noncryptographic identification and authentication; (2) systems controls to ensure authenticity, integrity, and availability; (3) audit trails; (4) message authentication codes (MAC's); (5) digital signatures/encryption; and (6) electronic notarization.

Message attributes combined with appropriate security and reliability measures then determine the electronic document's legal effect: The degree to which the documents are considered to be legal signed writings that are authentic and enforceable to the same extent as comparable documents prepared using conventional paper-based mechanisms.

The agency has carefully considered the divergent comments on acceptance stratification and is persuaded that the regulatory significance of a document need not be the basis of such stratification. However, the comments reflected a general premise that the nature and extent of security measures necessary to reasonably establish the reliability, authenticity, and confidentiality of an electronic signed writing will vary to the extent that the writings are vulnerable to unauthorized alteration or loss.

The agency agrees with comments that a fundamental two tier stratification based upon open and closed systems, as comments described, is warranted. FDA anticipates that most electronic documents which are maintained by industry and inspected by the agency would be considered as falling within "closed" systems. Electronic records that are submitted to the agency, however, as indicated by the comments, may be considered to be within either "closed" or "open" systems depending on how they are delivered (i.e., via "open" e-mail, or "closed" hand-delivery by submitters or postal services). Likewise, FDA's own electronic records may be stratified as existing in either open or closed systems depending on how they are originated and, for certain records, transmitted to correspondents.

The proposed regulations place primary emphasis on electronic records in closed systems, because that approach would cover most of the emerging electronic records and would respond to the most urgent of industry's needs in developing electronic record systems. FDA considers "open" systems to be nonetheless important because correspondence and regulatory submissions conveyed by public electronic networks are gaining wider implementation. Therefore, FDA may, in the future, propose more specific requirements relating to open systems, as the agency gains additional information and experience with open systems and the controls that may be necessary to maintain the integrity and authenticity of electronic documents in that environment.

IV. Legal Acceptance

Several comments said that electronic records would, in fact, be admissible in court, provided that there are controls in place to make the records reasonably reliable and trustworthy. One comment cited several recent court cases in support of this acceptability.

The agency notes that although the ANPRM did not specifically request comments on legal acceptability of electronic records and signatures, the gist of most of the comments is that legal acceptance will not be hindered, provided that the records are shown to be reliable and trustworthy. The case transcript cited by the comment included testimony from computer system operators which outlined key good computing practices that many of the comments also identified.

V. Regulatory Acceptance

A. General Considerations

One comment suggested that the disparity among FDA regulations regarding acceptance of signature alternatives was based upon definitions that are either too weak or restrictive, and called for common regulatory definitions.

The agency believes that any regulatory disparity derives from a number of factors, including the degree to which various regulations anticipate use of electronic records in place of paper records, and specific program needs of different FDA centers. FDA believes that differences can be dispelled by promulgation of these uniform broad based regulations on electronic records/signatures. The agency agrees that common definitions in such regulations would help to harmonize policy across different parts of FDA.

One comment recommended that FDA issue a broad policy statement or inspectional guideline that would broadly accept electronic identification/signaturesand that would at least establish criteria for the degree of security required for electronic identification/signature systems. The comment urged that no new regulations be issued.

The agency has determined that a policy statement, inspectional guide, or other guideline would be an inappropriate vehicle for accepting electronic signatures because such documents do not have the same legal significance as substantive regulations that require signatures. Guidance documents may be appropriate, however, to elaborate upon acceptance regulations.

B. Program Areas

1. Drug CGMP Regulations

Although the ANPRM applied to all FDA regulations in 21 CFR, most comments focused primarily on the CGMP regulations for drugs parts 210 and 211 (21 CFR parts 210 and 211). Some comments suggested that resolution of the issues in the CGMP context could be applied to resolve similar issues in the context of other FDA regulations.

Many comments argued that the existing CGMP regulations permit the use of electronic identification wherever documents are required to be signed, initialed, endorsed or approved, with the singular exception of §211.186 (master production and control records) which explicitly requires full handwritten signatures. Comments supported their assertions by citing preamble comment paragraphs 186, 282, and 447 in the final rule on CGMP's in the *Federal Register* of September 29, 1978 (43 FR 45014). FDA's Compliance Policy Guide (CPG) 7132a.08, and (unspecified) tacit acceptance by FDA field investigators who encounter electronic identification.

One comment identified several sections of the CGMP regulations as requiring signatures, including §211.188(b)(11) (batch production and control records), even though the word signatures, per se, does not appear ("Identification of the persons performing and directly supervising or checking each significant step in the operation").

Comments urged the agency to issue a policy statement (such as a CPG), in the near term, that would condone use of electronic identification for all applications of signatures in the regulations, except §211.186. Comments requested that in the long term, §211.186 be amended to delete reference to handwritten signatures and accept electronic identification.

The agency does not agree with the assertions that, except for §211.186, the CGMP regulations currently permit alternatives to handwritten signatures or initials. (See findings of the Electronic Identification/Signatures Working Group in its February 24, 1992, progress report.) The Center for Drug Evaluation and Research, in consultation with the Office of the General Counsel, considered and rejected as inappropriate the issuance of a CPG that would accept "electronic identification" or other signatures alternatives, even before the working group was formed.

The agency's conclusion regarding what the CGMP's allow was conveyed to the Pharmaceutical Manufacturers Association in a letter of December 5, 1991 (Ref. 3). Furthermore, the compliance policy guide cited by comments is not directly relevant because it addresses second check endorsements for operations executed by machine, rather than the form that human endorsements take. In addition, although comments cite several paragraphs of the 1978 *Federal Register* notice as supportive of their assertions, they overlook a key paragraph in which the agency clearly rejected substitution of employee numbers or codes for signatures or initials, on the basis of psychological differences from the act of signing and because of the ease of falsification (43 FR 45068, September 29, 1978 (comment 433)).

The agency advises that some sections of the CGMP regulations, while not using the words sign, signature, or initials, nonetheless implicitly requires endorsements to be in the form of handwritten signatures or initials. For example, the provisions of §211.188 require batch production and control records to contain the "[i]dentification of the persons performing and directly supervising or checking each significant step in the operation." FDA investigators have historically encountered and expect to find the identification to take the form of a signature. Some developers of automation systems also recognize that "identification" means "signature."

Accordingly, the agency is not issuing the suggested CPG, but is, instead, proposing these acceptance regulations that would cover records required by most FDA regulations, including the CGMP regulations. However, the agency may issue clarifying guidance documents, as needed, after such regulations are in effect.

2. Regulatory Submissions

Two comments said that regulations that require signatures on new drug applications necessitate substantial additional handling to furnish paper based signatures where the basic submissions are in electronic form. Comments suggested that the agency require submissions to contain, in lieu of the additional paper, a statement that signatures (handwritten or otherwise) are "on file." The comment added that FDA could verify those endorsements during its inspections. The comments observed further that when electronic submissions are copied or converted among various computer file formats, electronic endorsements might be omitted.

One comment stated that resolution of issues associated with electronic identification and the transfer or conversion of electronic data will be necessary if the benefits of electronic submissions are to be achieved.

The agency believes that codified acceptance of electronic signatures in lieu of handwritten signatures will address the issues relating to regulatory submissions. Acceptance of electronic signatures would, in most cases, obviate the need to have paper based handwritten signatures on file as a reference. However, the agency notes, from the comments, the importance of having the electronic records include the printed name of the signer so as to clearly identify the signer.

3. Prescription Drug Marketing Act

Several comments cited the signature requirements (for requesting and receiving samples of prescription drugs) in the PDMA provisions of the Federal Food, Drug, and Cosmetic Act, and based on the increasing use of computer technology to transact the handling of such requests, urged the agency to accept electronic identification in lieu of handwritten paper based signatures. Another comment echoed the same suggestion, recommending that biometric/behavioral links not be required, but noting also that physician requests for drug samples are generally made in "open" environments such that use of certain alternatives for full electronic or handwritten signatures needs review.

One comment requested that, for purposes of the PDMA, FDA accept SRE's based upon their uniqueness and reliability, and that such acceptance be codified in regulations. Another comment described its SRE pen-computer based system, emphasizing the nonalterability of signed electronic records to merit regulatory acceptance.

One comment assumed that the ANPRM did not pertain to the PDMA.

One comment asked that FDA issue implementing regulations under the PDMA that accept electronic signatures and that such issuance not be delayed pending the agency's broader consideration of electronic records and endorsements.

The proposed rule to implement certain parts of the PDMA and the Prescription Drug

Amendments of 1992 was published in the *Federal Register* of March 14, 1994 (59 FR 11842). That proposed rule would prohibit the imprinting or automatic reproduction of a signature by a device or machine such as a stamp, copier, or autopen at 21 CFR 203.61 (a). The agency recognizes that the PDMA proposal is not in total accord with this general proposed rule on electronic records and electronic signatures. As discussed in the preamble to the PDMA proposed rule (59 FR 11860), FDA will consider the comments concerning electronic signatures and other signature substitutes received in response to both proposed rules before final rules are published.

4. Good Laboratory Practice

One comment suggested that a uniform definition of electronic identification would facilitate application of computer based automated systems in the area of GLP's.

One comment cited the language of 21 CFR 58.130(e) (of the GLP regulations) as calling for handwritten signatures of paper based records, but allowing dated electronic identification for electronic systems.

FDA believes that, here again, broad acceptance regulations should resolve the issues related to GLP's.

VI. Acceptance Regulations

Several comments asserted that a general rule with a broad preamble and specific targeted subsection changes would be the most efficient means of accepting electronic signatures throughout the applicable regulations. Other comments also supported new regulations that would accept electronic identification/signatures throughout existing FDA regulations.

One comment suggested that FDA define the term electronic identification in the CFR in order to sanction use of those alternatives in place of handwritten signatures. Another comment said FDA's codified definition of signature should be clear yet general enough to allow industry the flexibility to use the most suitable technology. One comment said the agency should codify the terms signature, electronic signature, and electronic identification, provide examples of each term, and determine if there are substantive reasons for requiring handwritten signatures.

One comment suggested that to enhance the move from paper to electronic records, the agency should develop standards for the generation of portable electronic copies of records, copies that FDA may need in its enforcement activities. The comment also suggested that the agency require that systems be capable of generating such portable copies.

One comment suggested that regulations should consider an electronic record as "signed and final," once an operator endorses the record by entering a password.

One comment suggested that FDA's regulations would have to address both electronic integrity and administrative security.

One comment urged that FDA's final publication resolve several specific issues regarding: (1) Elimination of paper documents when they are converted to electronic form, and distinguishing originals from copies; (2) establishing the "legal original" between secure electronic copies of conventionally signed paper documents; and (3) whether or not an operation can be based upon a combination of electronic and paper records.

One comment suggested that, until legal and security issues are resolved, the agency should accept electronic submissions, encourage development of electronic records systems, but require supplementary or accompanying handwritten, paper based signatures. The comment added that such auxiliary endorsements would parallel the approach taken by the Internal Revenue Service regarding filing of electronic tax returns (based upon a conventionally signed paper form 8453) and would be relatively easy to implement. The same comment suggested that once electronic signatures are proven to be legally viable, FDA should not require them to be embodied in the electronic documents, but rather incorporated in supplementary documents so as to facilitate software modification. (As discussed in section VIII. of this document, one comment took the opposite view, stressing the importance of having the electronic signature securely bound to the signed document.)

One submission urged FDA to promulgate regulations regarding use of electronic signatures in the manufacture of blood components and subsequent testing and transfusion service laboratories.

FDA agrees with the comments that called for broad regulations that would clearly define the terms handwritten signature and electronic signature (and do so in a manner that affords industry the greatest latitude in adopting appropriate technologies), and set conditions under which the agency would accept alternatives to handwritten signatures. The proposed regulations apply to all FDA program areas, including blood components, which are regulated as either drugs or medical devices.

The agency does not believe it necessary to define the term "electronic identification" because the general meaning of the term, as suggested by comments, would be contained in the proposed definition of electronic signature.

The agency agrees that it is vital for FDA to be able to obtain copies of electronic documents and that systems should have the capability of generating such copies -- a provision that is in proposed §11.10(b). However, the agency does not, at this time, agree that FDA needs to develop specific performance standards for the "portability" suggested. FDA may develop appropriate guidelines in the future to address portability attributes.

Regarding the suggestion that FDA require parallel paper records to bear mandated signatures pending resolution of legal issues, the agency believes that such a provision need not be codified because there are no indications that legal acceptance of electronic records/signatures (per se) remains an issue, where the trustworthiness/reliability of such records/signatures has been established. The proposed acceptance regulations address measures to establish such trustworthiness and reliability. However, until the regulations are in effect, firms must supplement electronic records with paper documents for the purposes of having required signatures in conventional form.

The agency does not understand the basis for one comment's concern that electronic signatures not be required to be contained within the electronic records that are signed. The key factors in acceptability of electronic records/signatures have to do with establishing trustworthiness and reliability rather than facilitating software modification. Linking the electronic signature with the electronic document is an important attribute in establishing the authenticity of the endorsement, just as it is important to "affix" one's handwritten signature to a paper document. FDA believes that electronic signatures which are separate from their associated writings are less reliable and trustworthy than electronic signatures which are incorporated in their respective documents, to the extent that authors can more easily repudiate the authenticity of the separated signature.

VII. Enforcement Integrity

Most comments asserted that, based in part upon the provisions of Title 18 of the U.S. Code, use of signature alternatives should not adversely affect the agency's enforcement integrity. Comments asserted that laws against falsification of paper records apply equally to falsification of electronic records, and that FDA should have no difficulty in affixing individual responsibility when working with electronic records.

Comments also maintained that electronic record systems must, and can under current technology, be designed for reliable storage and retrieval, thus meeting industry and FDA audit needs. Comments added that electronic record systems can be validated and are at least as reliable, and more efficient than, paper-based records.

One comment asserted that copies of electronic records containing signature alternatives will be admissible evidence, in regulatory actions, to demonstrate individual responsibility when FDA informs the industry that signature alternatives are as binding as conventional signatures.

One comment asserted that within the context of the PDMA, electronic signatures would be admissible in court when combined with other system controls, such as phoned requests.

The agency recognizes that the ability to collect electronic records that are admissible as evidence, depends in large measure on whether or not the systems used to generate those records have been designed for reliable storage and retrieval. Accordingly, the proposed regulations, at proposed §11.10(c), require that systems that generate and maintain electronic records be designed so that the records can be reliably stored and retrieved. The storage/retrieval requirement should be coupled with the requirement that such systems be capable of generating accurate electronic copies that can readily be converted to human readable form. (See remarks on records "portability" in section VI. of this document.)

VIII. Security

Many comments contended that handwritten signatures are not intrinsically secure forms of identification because falsification can easily be executed unilaterally. Comments emphasized furthermore that properly validated and administered identification/password systems, which lack biometric links to individuals being identified, are more secure than handwritten signatures to the extent that falsification generally necessitates a bilateral action (i.e., two individuals must purposefully accomplish falsification). Comments asserted that security is fundamentally derived, not from the form of the identification, per se, but rather from the attendant system controls.

One comment argued against placing too high an emphasis on security and control measures for signature alternatives, noting that FDA has not instituted corresponding controls for conventional handwritten signatures on paper records. The comment elaborated that isolated forgeries are more apt to go unnoticed than repetitive forgeries of a manual signature, and that security of habitual signing derives more from the meaning attached to the signing process than the technical strength of the process itself. The comment concluded that the effectiveness of electronic signature alternatives should also derive less from technical security and more from the meaning attached to the signing process.

The agency finds merit in the comments' premise that the integrity of an electronic signature is derived more from the systems controls used to generate it than from the technology used to apply it. The emphasis on systems controls is justified and reflected in the provisions of the proposed regulations. However, FDA recognizes that electronic signatures based upon biometric/behavioral links can be more secure than others to the extent they are more difficult to falsify. Whereas the agency agrees that the meaning attached to the signing process is important (e.g., in establishing individual responsibility for an endorsed act such as approving a master production record), FDA does not agree that the meaning determines the security of the signing.

Regarding the comment that FDA has not instituted controls for the generation of handwritten signatures, the agency notes that specific FDA guidance on the matter has not been needed because conventional paper controls are well established in our culture and because falsification of paper documents can be readily investigated and documented by a long-standing body of forensic evidence (e.g., handwriting analysis, ink composition and dating, imprints on stacks of paper, erasure marks, etc.). On the other hand, a comparable body of evidence has yet to be established to pursue falsification of electronic documents and signatures.

The agency finds convincing the argument that electronic signatures based on user identification codes combined with passwords can be adequately secured in that the signature consists of multiple parts which require the collaborative efforts of two individuals to execute a falsification. FDA wishes to clarify, however, that contemporaneous use of both electronic signature elements must be executed for each signing. For example, if a person, having logged onto a system by entering both a password and a scanned employee badge containing an identification code, need only scan the badge to execute subsequent electronic signatures, then the safeguard of having multiple parts to the signature would be lost for those endorsements to the extent that another person could, unbeknownst to the badge owner, scan the badge and falsify the electronic signature. Should the owner carelessly leave the badge unattended, the required collaboration would be absent. On the other hand, if an "impersonator" needs to know the badge owners secret password in addition to physically possessing the badge in order to execute a signing, then collaborative efforts would be necessary to falsify the electronic signature; the badge owner would have to reveal the password to the would-be-imposter, as well as make the badge available. Accordingly, proposed §11.200(a)(1) requires electronic signatures that are not based on biometric/behavioral links to employ at least two distinct parts, all of which are contemporaneously executed at each signing. In addition, proposed §11.200(a)(3) requires that attempts at signature falsifications necessitate collaboration of at least two people.

The agency believes that the acceptance regulations need not require at least two distinct elements where the electronic signature employs a biometric/behaviorallink (e.g., retinal scan, voiceprint) to the signer. The bilateral security measure would not be necessary in such systems because only the genuine owner of the electronic signature would be capable of using it. The owner could not lose, lend, give away or otherwise transfer the signature in the first place.

One comment expressed the hope that security for alternatives to handwritten signatures will not result in lesser confidentiality.

FDA agrees that confidentiality of data in electronic records is as important as it is in paper records. Systems controls, for both paper and electronic documents, will determine the level of confidentiality.

One comment stated that signatures recorded electronically, if not somehow inalterably bound to the electronic document, are insecure to the extent the digitally recorded signature could be excised and superimposed upon other documents to falsify an endorsement. Another comment supported signatures recorded electronically when they are captured to inalterable media, such as optical disks, provided further, that access to such media is limited, thus reducing chances of alteration.

The agency agrees that binding an electronic signature to the signed electronic document is a vital systems control that helps to establish the authenticity of an electronically signed document. Accordingly, proposed §11.70 includes a "signature to document" binding provision. FDA notes that such a binding is usually inherent for handwritten signatures that are applied to paper documents.

As noted above regarding stratification, many comments made a distinction between the security needed for signature alternatives affixed to electronic documents contained within the administrative control of a given firm (closed system) and signature alternatives affixed to records (such as e-mail and submissions to FDA) that are transmitted from one establishment to another (open systems). Comments suggested that open systems require a higher level of security than closed systems, and that a combination of user identification codes and passwords, under suitable administrative controls, is sufficient for closed systems.

The agency agrees that because open systems are inherently more vulnerable to message compromise, additional security measures may be necessary to ensure electronic document integrity and authenticity. Such measures may include electronic document encryption and use of digital signatures. However, FDA believes that because such measures are still evolving, it would be premature to specifically require their use in documents submitted electronically to the agency. Instead, the proposed rule requires additional security measures, stated in general terms, that are designed to ensure document integrity, confidentiality, and authentication from point of creation to point of receipt.

One comment suggested that computer systems used within the CGMP and GLP regulations attain the security level of C2 within the Department of Defense Trusted

Computer System Evaluation Criteria (DoD 5200.28 -- STD), also known as the "Orange Book."

One comment concluded that, per the ANPRM working definitions, signatures recorded electronically (scripted signatures applied to devices other than paper) and conventional signatures applied to paper offer the greatest security.

FDA does not believe it necessary at this time to codify adherence to a specific security level that is stated in a standard. The agency believes that records under CGMP's and GLP's will have sufficient security when the provisions of the proposed rule are followed. However, should additional specific criteria be necessary to attain adequate levels of security, the agency may consider incorporating specific security standards such as the one suggested.

Many comments identified various administrative security controls attendant to the use of (what the ANPRM called) electronic identification (identification codes (ID)/passwords), and argued that appropriate use of such controls should make ID/password systems acceptable to FDA for use in closed systems. Comments generally emphasized the need to utilize such controls and not rely upon a single form of signature alternative in isolation. Suggested controls included the following:

1. Establish and follow employee policies which hold people accountable and liable for actions initiated under their (computer ID) accounts to deter forgery of electronic signatures. Comments suggested that employees who violate such policies would be subject to disciplinary action including termination.
2. Limit computer access to authorized individuals.
3. Execute carefully written and controlled operational procedures.
4. Train employees in the use of operational procedures.
5. Use fully documented production and control procedures
6. Validate systems.
7. Use identity checks; cross-checking to establish that machine readable codes on tokens and a personal identification number (PIN) are assigned to the same individual.
8. Use password checks, checking an independently entered password.
9. Change passwords periodically.
10. Use authority checks to determine if the identified individual has been authorized (or trained) to use the system, access, or operational device, or perform the operation at hand.
11. Use time stamped audit trails to document changes, record all write-to-file operations, and independently record the date and time of the operator's action or entry. Concerning audit trail integrity, comments emphasized the importance of creating back up files to re-create documentation and deter inappropriate records alterations.
12. Use operational checks to enforce permitted operational parameters such as functional sequencing or time.
13. Use records revision and change control procedures to maintain an electronic audit trail that documents time-sequenced development and modification of records.
14. Maintain control over the distribution, access, and usage of documentation required for various operations.
15. Encrypt records to provide secure, nonchangeable versions.
16. Use location (terminal) checks to determine that the physical source of the endorsement is valid.
17. Use intentions checks by providing confirming dialog that the signer understands precisely the intentions of a signature.
18. Use "time-outs" of under-utilized terminals to prevent their unauthorized use while unattended.
19. Use security against natural system failures.
20. Print the individual's name, along with time of "signing," on the electronic record to help reenforce the psychological link between the author and the endorsement.

The agency considers that most of the above systems controls have merit and they have been incorporated in the proposed regulations.

One comment identified the following steps to regulate and control the issuance of tokens, cards, PIN's, and other machine readable indicia of identity:

1. Chronological logging of each issuance;

2. Certifying the identity of each individual;

3. Noting and controlling the empowerment or authority of issuance;

4. Testing each token, card, or other indicia to make sure it works;

5. Keeping each issuance unique;

6. Assuring that issuances are periodically checked, recalled, or reissued;

7. Following loss management procedures to electronically deauthorize lost tokens, cards, etc. and to issue temporary or permanent replacements using suitable, rigorous controls for substitutes; and

8. Using reasonable transactional safeguards to prevent unauthorized use and detect and emergently report (with unmistakable notoriety) any unauthorized attempts.

The agency agrees that all of the above controls are reasonable and necessary measures to maintain password integrity. However, some of these controls may be more amenable to incorporation in guidelines rather than regulations, and therefore do not appear in the proposed rule.

In response to the ANPRM's request that comments identify any types of signature alternatives that would be too insecure to be acceptable, comments cited the use of unilateral methods, such as a user identification that is readily determined from a publication, or alternatives used in environments in which employees are motivated to falsify identifications. One comment stressed the importance of using bilateral systems, but urged the agency to permit industry to choose the exact methods (such as use of identification codes combined with passwords or tokens).

As explained above, the agency agrees that single entry signature alternatives that may be compromised ar not acceptable. Where bilateral signatures are used, both portions of the signature should be recorded contemporaneously with each "signing." Absent that duality, FDA would consider the signature to be unilateral and therefore, if capable of being compromised, unacceptable. The agency wishes to clarify, however, that single entity signatures based on biometric/behavioral links that cannot be implemented by people other than their genuine owners would be acceptable.

IX. Validation

Comments generally acknowledged the importance of validating signature alternative systems and said that there should be no difference between validation of signature alternatives and validation of other processes or systems. Most comments claimed that there already exists sufficient guidance, published by FDA and the industry, thus making it unnecessary for FDA to publish additional guidance on validation of signature alternatives.

Several comments acknowledged FDA's concerns about the adequacy of computer systems validation, but indicated that the primary issue concerns what constitutes adequate system specifications, a matter comments claimed is still developing.

Comments identified the following elements of signature alternative validation:

1. Correct specification;

2. Correct engineering;

3. Correct testing;

4. Correct operation;

5. System definition: functional requirements, software requirements, the physical system and its operating environment;

6. Assurance of software quality: structural and functional;

7. System documentation that is well organized and that includes policies, procedures and master plans defining the philosophy and approach to system validation, and defined meanings for approval signatures;

8. Security;

9. Verification of critical data entries;

10. Installation, operational, and performance qualification;

11. Change control and system maintenance;

12. Employee training;

13. A records retrieval system that protects records and enables their accurate and efficient retrieval throughout their retention period; and

14. Periodic system review and revalidation.

The agency is persuaded by the comments that although validation of electronic signature systems is important enough to be codified as a general requirement, publication of specifics as to what constitutes acceptable validation of such systems should be deferred at this time. Specific information on electronic signature validation may need to be provided in either future regulations and/or guidelines.

X. Standards

A. Standards in General

Several comments acknowledged the general utility of standards (e.g., for electronic signatures which use biometric/behavioral links), but suggested that the issue should be addressed separately on the basis that standards are not relevant to the forms of electronic identification anticipated for use in the pharmaceutical industry, and because they are seldom used in FDA-regulated industries generally.

Several comments said FDA should assess existing standards and provide input into development of new standards, but should not seek a lead role in their development. One comment suggested that FDA collaborate with industry in developing standards should they be warranted in the future.

Two comments argued that the absence of standards should not inhibit the agency from accepting electronic identification and that standards would not be necessary where there is an emphasis on validation, security, and well designed and enforced procedures.

One comment urged the agency to avoid adopting any single standard or technology for electronic signatures.

FDA recognizes the benefits of standards and their relevancy to legal and regulatory acceptance of electronic signatures. FDA regulations could be simplified by predicting acceptance of an electronic signature on adherence to one or more appropriate standards that have been derived from fair evaluation of public comments. Although industries regulated by FDA may not have participated in the development of the two emerging primary digital signature standards, i.e., the National Institute of Standards and Technology Digital Signature Standard (NIST DSS) or the RSA, either because (in the case of the RSA) the standard is proprietary, or because the industry did not anticipate their relevancy, the standards may nonetheless be valuable tools to ensure the authenticity and integrity of electronic records.

In general, the agency agrees with the premise that adherence to specific standards need not be codified at this time because adequate levels of security may be achieved by adherence to the controls contained in the proposed rule. However, the agency may need to address or adopt such standards in the future, as the industries become more familiar with them and their practical applications. The agency anticipates that its role will be that of a proactive participant in standards development. Absent the immediate application of such standards, the proposed rule emphasizes, as comments suggest, system security/integrity controls, and validation.

B. National Institute of Standards and Technology Digital Signature Standard

One comment suggested, without elaboration, that FDA obtain and consider three cited articles on digital signature standards.

Many comments cited the controversial nature, per published articles, of the NIST DSS and suggested that FDA not adopt the standard. Several comments inferred that FDA should favor the RSA over the NIST DSS on the basis that RSA is currently the de facto standard for commercial and some military applications.

One comment urged the agency to adopt a public, rather than proprietary standard, but noted the difficulty of modifying systems that are essentially completely developed to incorporate the NIST standard.

One comment encouraged FDA to adopt the NIST draft digital signature standard, on the grounds that the NIST DSS is a highly secure method of identification that will become mandatory for Federal agencies where a public-key based digital signature technique is needed and is to be the single standard for

Government communication with the private sector. The comment further supported the standard by noting its acceptance by the General Accounting Office as legal endorsement for Federal obligations. In addition, the comment asserted the nonrepudiation property of the NIST DSS. One comment acknowledged that the NIST standard offers the benefit, over handwritten signatures, of assuring that the document was not altered after being signed by the author.

The agency notes that subsequent to the working group's February 1992 progress report, several criticisms of the NIST DSS, specifically the absence of a "hash algorithm" and limited size of "keys," have been addressed. FDA has also become aware of several commercial products available to implement the standard, and the agency acknowledges that it may have direct applicability to FDA electronic communication with the agency's regulated industries. However, the standard is not yet finalized, and it has not yet achieved sufficiently wide utilization, in the agency's opinion, to merit mandatory use, at least in closed systems. The standard may have future applicability, though, in open systems, where documents are submitted to FDA via public electronic carriers, in which case adherence to a limited number of standards would be desirable to maintain practical communications. Accordingly, the agency is deferring a codified reference to the NIST DSS in particular. However, the agency is proposing in §11.30 to use established digital signature standards that are acceptable to FDA, as a system control that may be warranted to maintain record authenticity, integrity, and confidentiality in open systems.

XI. Freedom of Information

Several comments asserted that because matters relating to FOI are not relevant to the fundamental issues of electronic identification, such issues should be handled separately. However, comments expressed concern about the reliability of computer methods FDA might use to delete proprietary information from electronic records released under the FOI Act.

Two comments said that FDA should realize FOI processing cost savings when records are submitted electronically if the agency sets guidelines on such submissions.

Comments held diverse opinions about what form (electronic or otherwise) documents released under FOI should take. Several comments said FDA should establish standards to avoid having to copy and purge original records that exist in many different formats. Some comments said they would likely provide paper printouts of electronic records requested by FDA field investigators, and by so doing, the agency would not need to acquire specific software and hardware to handle proprietary formats. Likewise, two comments recommended that FDA respond to FOI requests by providing only paper copies of documents, regardless of the format requested. On the other hand, two comments encouraged the agency to develop systems whereby requesters could submit FOI requests by e-mail, or directly access an FDA data base to conduct on-line text searches. One of the comments suggested that resulting documents from such searches be mailed to requesters in a manner similar to the procedure used by the national Library of Medicine's Medline. The respondent suggested that modest connect time fees would be appropriate to such systems.

The agency disagrees with the assertion that FOI matters are irrelevant to electronic signature issues. When FOI requests are received electronically the agency must ensure that the requests are authoritative and genuine such that they may be processed and appropriate fees collected. In addition, as more firms implement electronic records, the agency will likely collect and store them electronically in the regular course of its investigational and inspectional activities. The consequent move from paper to electronic documents will necessitate use of appropriate purging technologies, as many of the comments have noted.

FDA finds the comment's suggestions that FOI records be handled strictly as paper documentation inconsistent with the implementation of electronic records systems. The agency believes the suggestion that FDA accept FOI requests by e-mail has merit, and it is exploring ways of implementing the suggestion within the context of electronic submissions in general. A data base of all available documents may not be practical at this time considering the scope of potential documents that may be in the data base. However, a publicly accessible on-line electronic data base of FOI-released documents may be in the public interest, and this suggestion may also be explored. The agency agrees that it should set technical standards for submission of electronic documents so as to allow the electronic handling of relevant FOI requests; this suggestion is also being explored within the context of electronic submissions in general.

XII. The Proposed Regulation for Electronic Signatures and Records

Proposed part 11 is made up of the following subparts: subpart A -- General provisions; subpart B -- Electronic records; and subpart C -- Electronic signatures:

A. General Provisions (Subpart A)

1. Scope (§11.1)

Although most of the comments to the ANPRM represented the pharmaceutical industry, the agency wishes to emphasize that the proposed rule applies to use of electronic records and signatures in the context of all FDA program areas and all industries regulated by FDA. Accordingly, proposed §11.1 states the extent of the regulation's scope to all parts of 21 CFR chapter I.

The agency recognizes, however, that in some instances records required by selected sections of chapter I may need to be retained in paper form and their associated conventional methods of signing may need to be preserved. In such instances, the agency would, by regulation, specify that electronic versions of those records would not be permitted. FDA does not anticipate many such situations, but is providing for them in proposed §11.1. The agency welcomes comments on any existing FDA regulations that address records where electronic versions of those records should not be permitted.

Under proposed §11.1, absent specific exemption by regulation, records required throughout chapter I could be created, modified, maintained, or transmitted in electronic form provided they meet the requirements of proposed part 11. Likewise, electronic signatures would be considered to be equivalent to full handwritten signatures, initials, and other general signings required throughout chapter I provided the electronic signatures and associated electronic records meet the requirements of the proposed part 11.

2. Implementation (§11.2)

The agency recognizes that the pace and extent of converting from paper to electronic records will vary significantly in industry and, in fact, within FDA itself. Adoption of electronic records technologies generally depends upon a number of factors, including systems availability, costs, integration into existing paper based records systems, and the need to train employees in developing and maintaining electronic systems. In order to implement the new rule in a fair and practical manner, the agency is dividing the types of records to be covered into two broad categories, namely records required by regulation to be maintained but not submitted to FDA (such as batch production records), and records submitted to FDA (such as food additive petitions and comments to proposed rules).

This approach is being taken for two reasons. First, the agency believes it is important to enable regulated industries to implement electronic records/signatures for records that are required by regulation to be maintained, but not submitted to the agency, as rapidly as possible. Some firms have already taken major steps toward implementing electronic production records and the agency does not wish to delay the appropriate adoption of new technologies.

Second, FDA is not yet prepared to accept and manage all submissions in electronic form. However, FDA believes it vital to enable those agency units that are prepared to receive and manage submissions in electronic form to do so as rapidly as practical. There are many different types of submissions to the agency. (A July 1991 FDA report entitled, "Basic Inventory of Submissions to the FDA," (Office of Planning and Evaluation) identified 87 different types of submissions (Ref. 4)). The agency is reviewing all of the various submissions to identify which documents it can accept and manage in electronic form (in whole or in part), and the corresponding capabilities of the receiving agency units. The agency is committed to accepting as many submissions in electronic form as possible, consistent with available resources, but realizes that the goal of accepting all submissions in electronic form will be achieved in phases over a period of time.

The agency intends to publish a public docket on electronic submissions. FDA proposes that this public docket will be established at the time that a final rule becomes effective. The docket would identify those submissions that may be made (in whole or in part) in electronic form, and the corresponding agency receiving units. Receiving units may also publish appropriate technical guidance documents on how submissions are to be made relative to the units' capabilities. In addition, FDA encourages submitters to work with the agency to develop appropriate pilot programs to implement electronic submissions that may be more complex in nature. The agency is committed to the goal of eventually accepting most submissions in electronic form because it recognizes the attendant benefits of using electronic records, benefits such as speedier document review times, cost savings in not having to store and manage paper, and the improved responsiveness to the general public and regulated industries that generally derives from electronic systems.

Therefore, proposed §11.2(a) enables persons to use electronic records/signatures in lieu of paper records/conventional signatures, in whole or in part, for records which are required by FDA regulation to be maintained, but not submitted to FDA. Proposed §11.2(b) enables persons to use electronic records/signatures in lieu of paper records/conventional signatures, in whole or in part, for records that are submitted to FDA, provided the type of submission has been identified in a public docket as one which FDA accepts in electronic form. The agency intends to announce changes to that public docket, on a periodic basis, by a variety of means. For example, a notice announcing changes may be published in the *Federal Register*.

FDA wishes to clarify that the requirements in proposed part 11 would apply to both types of electronic records (submissions FDA accepts in electronic form and records required by regulation to be maintained) unless, as stated above, a regulation specifically prohibits the record from being in electronic form.

3. Definitions (§11.3)

Proposed §11.3 sets forth definitions of key terms, including "biometric/behavioral links," "closed system," "open system," "electronic record," "electronic signature," and "handwritten signature."

A "biometric/behavioral link" (proposed §11.3(b)(3)) is a method of verifying a person's identity based on measurement of the person's physical feature(s) or repeatable action. The agency believes that biometric/behavioral links would be utilized in technologies that use, for example, voiceprints, handprints, and retinal scans to identify individuals. A system that characterizes the act of signing one's name, as a function of unique behavior (parameters of physical signing such as speed of stylus movement, pressure, pauses, etc.) is another example. A fundamental premise of biometric/behavioral link technologies is that the resulting electronic signatures are inherently unique to an individual and cannot, by ordinary means, be falsified.

A "closed system" (proposed §11.3(b)(4)) is an environment in which there is communication among multiple persons, where system access is restricted to people who are part of the organization that operates the system. FDA believes that electronic documents within a closed system are less likely to be compromised than those in an "open system" because they are not as vulnerable to disclosure to, and corruption by, unintended outsiders to the organization. Where a firm hand delivers to FDA a magnetic disk containing an electronic document, the agency would consider such communication to have been made in a closed system.

An "open system" (proposed §11.3(b)(8)) is an environment in which there is communication among multiple persons, where system access extends to people who are not part of the organization that operates the system. FDA believes electronic documents in open systems merit additional protection from unauthorized disclosure and corruption. Where a firm sends FDA an electronic document by electronic mail, the agency would consider such submission to have been made in an open system.

An "electronic record" (proposed §11.3(b)(5)) is a document or writing comprised of any combination of text, graphic representation, data, audio information, or video information, that is created, modified, maintained, or transmitted in digital form by a computer or related system. The agency is proposing a broadly based definition of this term in order to accommodate digital technologies that may incorporate pictures and sound, in addition to text and data.

Although, as discussed above, the ANPRM discussed four possible terms relating to different kinds of signatures, FDA is proposing two definitions based broadly on whether or not the "signature" is handwritten. Two definitions are proposed, one for "electronic signature" (proposed §11.3(b)(6)) and one for "handwritten signature" (proposed §11.3(b)(7)). The term electronic signature would include the meaning comments ascribed to electronic identification. Handwritten signatures would include signatures recorded electronically.

Proposed §11.3(b)(6) defines the term "electronic signature" as the entry in the form of a magnetic impulse or other form of computer data compilation of any symbol or series of symbols executed, adopted, or authorized by a person to be the legally binding equivalent of the person's handwritten signature. The fundamental premise is that an electronic signature is some combination of what a person possesses (such as an identification card), knows (such as a secret password), or is (the unique characteristic embodied in a biometric/behavioral link such as a voiceprint).

Proposed §11.3(b)(7) defines the term "handwritten signature" as the name of an individual, handwritten in script by that individual, executed or adopted with the present intention to authenticate a writing in a permanent form. An important aspect of a handwritten signature is that the act of signing with a writing or marking instrument such as a pen, or stylus is preserved. The agency is aware of electronic records systems which capture the image of a signature as a person applies a handwritten signature to a "screen" or sensing device. Because the traditional action of signing is preserved, the agency regards such a signature to be a handwritten signature even though it is written to an electronic document. The proposed definition includes wording to clarify this intent.

B. Electronic Records (Subpart B)

As discussed above, the agency has accepted the comments on the ANPRM that suggested that adequate system controls should be the basis for establishing the regulatory and legal acceptance of electronic records. The agency appreciates the extent of the suggested controls which are intended to ensure the authenticity, integrity, and confidentiality of electronic records and to ensure that signers cannot readily repudiate the electronic records as not genuine. FDA has incorporated most of the controls in the proposed regulations. Controls not adopted at this time may be incorporated in subsequent revisions to these regulations, or addressed in agency guidelines. In addition, FDA accepts the premise that some stratification of those controls should be codified based upon whether the electronic records are within closed or open systems. Therefore, this subpart includes separate controls for records in closed and open systems.

1. Controls for Closed Systems (§11.10)

Proposed §11.10 includes a general requirement that there be procedures and controls designed to ensure the authenticity, integrity, and confidentiality of electronic records, and to ensure that the signer cannot readily repudiate the signed record as not genuine. In addition, the agency is proposing 11 specific controls.

FDA wishes to emphasize that the proposed list of system controls is not intended to be all inclusive of what may be needed for a given electronic records system, and that some controls may not be necessary in all types of systems. The wording of the proposal is intended to clarify which controls are generally applicable and which are germane to certain types of systems depending upon their intended use. For example, operational checks to enforce permitted sequencing of events would not be appropriate to systems in which proper sequencing was not relevant to the events being recorded. Examples of system controls that would be applicable in all cases include validation and protection of records to ensure that records remain accurate and retrievable throughout their retention period.

Some of the proposed system controls (e.g., inspection and copying of records) are necessary to ensure that the agency can fulfill its enforcement responsibilities. The subject of enforcement integrity was extensively addressed in the ANPRM and by comments, most of whom asserted that properly validated and secured systems should not hamper the agency's enforcement activities.

As discussed above, many ANPRM comments asserted that enforcement integrity would not be hampered because, under Title 18 of the U.S. Code, falsification of electronic records would be equivalent to falsification of paper records.

The agency agrees that certain controls, such as system validation, are necessary to maintain the integrity of electronic documents it reviews and collects as part of its enforcement activities. It is also necessary for FDA to be able to review and copy electronic records in the same manner as paper records. Accordingly, the proposed rule contains several provisions designed to ensure that the agency's enforcement responsibilities are not impeded. For example, proposed §11.10(b), regarding the ability to generate true copies of electronic records that FDA can inspect, review, and copy, is intended to ensure that the agency will retain the ability to review electronic records on site and review copies of such records off site, in the same manner as is currently the case for paper records. Likewise, proposed §11.10(e), regarding time stamped audit trails to document record changes, is intended to ensure that changes to electronic records are evident and reviewable by the agency, to the same extent as paper records.

The agency encourages persons to consult with FDA prior to implementing electronic records systems if there are any questions regarding the ability of the agency to review and copy the electronic records. The proposed rule includes wording to that effect.

2. Controls for Open Systems (§11.30)

As discussed above, many comments to the ANPRM acknowledged that additional security measures, above and beyond those used for closed systems, may be needed to ensure the integrity, authenticity, and confidentiality of electronic records within open systems.

The agency agrees. FDA is aware that two kinds of additional systems controls can be effective in this regard -- use of document encryption, and use of digital signature standards. Digital signature standards use established mathematical algorithms and public and private signer numerical codes (called keys) to both authenticate an electronic record and establish its integrity. Several comments addressed these additional measures.

Accordingly, proposed §11.30 requires use of those controls identified in proposed §11.10 for closed systems (as appropriate to the nature of the records at issue) plus such additional measures as document encryption and use of digital signature standards acceptable to FDA as necessary to maintain record confidentiality and integrity under the circumstances. The agency intends to publish future guidance documents which identify acceptable digital signature standards.

3. Signature Manifestations (§11.50)

Proposed §11.50 requires several of the system controls suggested by comments to the ANPRM. This section requires electronically signed records to display the printed name of the signer and the date and time when the document was signed. The presence of the printed name, date, and time will assist the agency by clearly identifying the signing individual. In addition, the printed information will help firms to maintain an unambiguous method of readily and directly documenting the signer's identity and date of signing for as long as the electronic record is retained. Another benefit to having the name of the signer appear on the electronic document is to reinforce the solemnity and personal commitment associated with the act of signing.

Proposed §11.50 also requires that the meaning associated with the act of signing the electronic document be clearly indicated. As discussed in the ANPRM, the purpose of a signature can be varied (e.g., to affirm, review, approve, or indicate a person's presence or action). Many traditional paper records already contain statements that indicate the purpose of a signature, such as "material added by ***," "in witness thereof," and "approved by ***." The agency believes it is vital, for purposes of accurate documentation and establishment of individual responsibility, to include such statements in electronic records as well.

4. Signature/Record Binding (§11.70)

Signatures appearing on conventional paper documents cannot be readily excised, copied, or transferred to other documents so as to falsify another document. Attempts at such misdeeds can generally be revealed by available forensic methods. Such is not typically the case, however, with electronic signatures and handwritten signatures executed to electronic records (the image of the signature may be electronically "copied" from one location and "pasted" to another without evidence of the action.) In such cases, falsification of electronic documents would be relatively easy to achieve, yet difficult to detect. This problem could be solved by using available technologies to bind the signature to the electronic document in a secure manner analogous to the way conventional signatures are affixed to paper records.

As discussed above, two ANPRM comments specifically addressed signature to record binding. One comment stated that signatures recorded electronically, if not somehow inalterably bound to the electronic document, are insecure to the extent the digitally recorded signature could be excised and superimposed upon other documents to falsify an endorsement. Another comment supported signatures recorded electronically when they are captured to inalterable media, such as optical disks, provided, further, that access to such media is limited, thus reducing chances of alteration.

The agency agrees with the ANPRM comments and believes it is vital to verifiably bind a signed electronic record to its electronic or handwritten signature. Accordingly, proposed §11.70 includes a "signature to document" binding requirement to ensure that the signatures cannot be excised, copied or otherwise transferred so as to falsify another record. The agency believes that such binding is readily achievable under current technology. For example, the concept of such binding is part of digital signature standards to the extent that a message authentication operation will fail for a falsified document if the document's digital signature had been copied from a different document.

C. Electronic Signatures (Subpart C)

Proposed subpart C includes requirements for system controls that are relevant to electronic signatures. Here, as elsewhere throughout the proposed rule, the controls reflect suggestions made by the ANPRM comments.

In addition, the agency is including a requirement for providing certification to the agency that the electronic signature systems and, if necessary, specific electronic signatures are authentic, valid, and binding.

1. General Requirements (§11.100)

Proposed §11.100 requires each electronic signature to be unique to one individual and requires the issuing authority (for example, a systems security unit within a firm) to verify a person's identity before issuing an electronic signature. FDA considers these controls to be fundamental to the basic integrity of an electronic signature. Uniqueness is important because, if two or more people are assigned the same electronic signature (such as a combination of identification code and password) then the true identity of the signer could be in doubt and either of the two individuals could conceivably readily repudiate the recorded signature as not being his/her own. It is important for the assigning authority to verify a person's identity before issuing an electronic signature to prevent that person from wrongfully assuming someone else's identity and the privileges/authorizations that may be associated with that identity.

The agency is including a proposed requirement for providing certification to the agency that the electronic signature system guarantees the authenticity, validity, and binding of any electronic signature. Furthermore, upon agency request, additional certification or testimony that a specific electronic signature is authentic, valid, nd binding shall be provided. The certification should be submitted to the agency district office in which territory the electronic signature system is in use.

2. Identification Mechanisms and Controls (§11.200)

As noted above, electronic signatures are broadly based upon various combinations of what a person knows (such as a secret password), what a person possesses (such as an employee badge), and what a person is. The third element, what a person is, relates to what the agency is defining as a "biometric/behavioral link" to an individual -- a method of verifying a person's identity based on measurement of the person's physical feature(s) or repeatable actions. Examples of such features or actions include voiceprints, handprints, retinal scans, and the act of signing one's name in script. The most important attribute of an electronic signature that incorporates a biometric/behavioral link is that the measured feature or action is inherently unique to, and remains with, that individual. Unlike what a person knows or possesses, what a person "is" cannot be compromised by being lost, stolen, forgotten, loaned, re-assigned, or otherwise compromised by ordinary means.

Accordingly the agency is establishing two broad categories of electronic signatures, those based on biometric/behavioral links to individuals, and those that lack such links, as reflected in proposed §11.200.

Many of the ANPRM comments argued persuasively that FDA should not require biometric/behavioral links, but should accept electronic signatures that lack such links provided the electronic signatures are validated, secure, and administered under adequate system controls. Among those controls, comments emphasized the importance of maintaining electronic signatures that are made of multiple identification mechanisms (such as a combined identification code and password) and administrative measures to ensure that attempted use of an individual's electronic signature by anyone other than its genuine owner requires collaboration of two or more individuals. Such collaboration would prevent signature falsification by casual mishap -- a falsification that might result, for example, if someone acquired another person's unattended identification card or token. The provision would also help to impress people with the significance and solemnity of the electronic signature.

The agency agrees that biometric/behavioral links should not be a required feature of electronic signatures at this time. The agency also agrees that electronic signatures that lack biometric/behavior links should be acceptable when certain system controls are used. Accordingly, the agency has incorporated system controls for electronic signatures that lack such links, including multiple identification mechanisms and multiple part collaboration in proposed §11.200(a).

Although FDA is not, at this time, mandating use of biometric/behavioral links in electronic signatures, it is allowing for them and encourages their development and use. The premise behind the technology for electronic signatures based upon biometric/behavioral links is that the links are inherently secure such that a person's electronic signature could not be lost, stolen, loaned, or otherwise used by anyone other than the rightful owner. The agency is proposing to codify that premise at §11.200(b), to ensure that electronic signatures based on such links are designed so that they cannot be used by anyone other than their genuine owners.

3. Controls for Identification Codes/Passwords (§11.300)

The agency is aware that many electronic signatures are based upon combined identification codes and passwords. FDA believes that because of the relative ease with which such electronic signatures may be compromised, and because of their wide adoption, system controls to ensure their security and integrity merit specific coverage in these regulations.

Many of the ANPRM comments addressed specific administrative controls to ensure the security and integrity of electronic signatures that are based upon a combined identification code and password. One comment suggested eight controls specific to identification codes. The agency appreciates the various suggestions and agrees that five of them merit codification at this time. Proposed §11.300 includes those controls. Suggested controls that were not included in the proposed rule may be added in the future or addressed in future agency guidelines.

The agency wishes to emphasize that the controls listed in proposed §11.300 are not intended to be all inclusive of what may be needed to ensure the security and integrity of electronic signatures based on identification codes/passwords.

XIII. Analysis of Impacts [Omitted]

XIV. Paperwork Reduction Act of 1980 [Omitted]

XV. Environmental Impact [Omitted]

XVI. References

The following references have been placed on display in the Dockets Management Branch ... and may be seen by interested persons between 9 a.m. and 4 p.m., Monday through Friday.

1. FDA, Task Force on Electronic Identification/Signatures, Electronic Identification/Signature Working Group Progress Report, February 24, 1992.

2.National Performance Review, Report of the Vice President pp. 113-117, September 7, 1993.

3. FDA, Letter to Pharmaceutical Manufacturers Association, December 5, 1991.

4. FDA, Office of Planning and Evaluation. "Basic Inventory of Submissions to FDA," July 1991.

XVII. Comments [Omitted]

List of subjects in 21 CFR Part 11

Administrative practice and procedure, Electronic records, Electronic signatures, Reporting and recordkeeping requirements.

Therefore under the Federal Food, Drug, and Cosmetic Act, and under authority delegated to the Commissioner of Food and Drugs, it is proposed that 21 CFR part 11 be added to read as follows:

PART 11 -- ELECTRONIC RECORDS; ELECTRONIC SIGNATURES

Subpart A -- General Provisions

§11.1 Scope

(a) The regulations in this part set forth the criteria under which the Food and Drug Administration considers electronic records, electronic signatures, and handwritten signatures executed to electronic records, to be trustworthy, reliable, and generally equivalent to paper records and handwritten signatures executed on paper.

(b) These regulations apply to records in electronic form that are created, modified, maintained, or transmitted, pursuant to any records requirements set forth in chapter I of this title.

(c) Where electronic signatures and their associated electronic records meet the requirements of this part, the agency will consider the electronic signatures to be equivalent to full handwritten signatures, initials, and other general signings as required throughout this chapter, unless specifically exempted by regulation that is effective on or after the effective date of this part.

(d) Electronic records that meet the requirements of this part may be used in lieu of paper based records, in accordance with §11.2, unless paper based records are specifically required.

(e) Computer systems (including hardware and software), controls, and attendant documentation maintained pursuant to this part shall be readily available for, and subject to, FDA inspection.

§11.2 Implementation

(a) For records required by chapter I of this title to be maintained, but not submitted to the agency, persons may use electronic records/signatures in lieu of paper records/conventional signatures, in whole or in part, provided that the requirements of this part are met.

(b) For records submitted to the agency, persons may use electronic records/signatures in lieu of paper records/conventional signatures, in whole or in part, provided that:

(1) The requirements of this part are met; and

(2) The document or part(s) of a document to be submitted has/have been identified in a public docket as being the type of submission the agency accepts in electronic form. This docket will identify specifically what types of documents or parts of documents are acceptable for submission in electronic format without paper records and to which specific receiving unit(s) of the agency (e.g., specific center, office, division, branch) such submissions may be made. Documents to agency receiving unit(s) not specified in the public docket will not be considered as official if they are submitted in electronic form; paper forms of such documents will be considered as official and must accompany any electronic records. Persons should consult with the intended agency receiving unit for details on how and if to proceed with the electronic submission.

§11.3 Definitions

(a) The definitions and interpretations of terms contained in section 201 of the act apply to those terms when used in this part.

(b) The following definitions of terms also apply to this part:

(1) *Act* means the Federal Food, Drug, and Cosmetic Act (secs. 201-902, 52 Stat. 1040 *et seq.*, as amended (21 U.S.C. 301-392).

(2) *Agency* means the Food and Drug Administration.

(3) *Biometric/behavioral links* means a method of verifying a person's identity based on measurement of the person's physical feature(s) or repeatable action(s).

(4) *Closed system* means an environment in which there is communication among multiple persons, where system access is restricted to people who are part of the organization that operates the system.

(5) *Electronic record* means a document or writing comprised of any combination of text, graphic representation, data, audio information, or video information, that is created, modified, maintained, or transmitted in digital form by a computer or related system.

(6) *Electronic signature* means the entry in the form of a magnetic impulse or other form of computer data compilation of any symbol or series of symbols, executed, adopted or authorized by a person to be the legally binding equivalent of the person's handwritten signature.

(7) *Handwritten signature* means the name of an individual, handwritten in script by that

individual, executed or adopted with the present intention to authenticate a writing in a permanent form. The act of signing with a writing or marking instrument such as a pen, or stylus is preserved. However, the scripted name, while conventionally applied to paper, may also be applied to other devices which capture the written name.

(8) *Open system* means an environment in which there is electronic communication among multiple persons, where system access extends to people who are not part of the organization that operates the system.

Subpart B -- Electronic Records

§*11.10 Controls for closed systems.*

Closed systems used to create, modify, maintain, or transmit electronic records shall employ procedures and controls designed to ensure the authenticity, integrity, and confidentiality of electronic records, and to ensure that the signer cannot readily repudiate the signed record as not genuine. Such procedures and controls shall include the following:

(a) Validation of systems to ensure accuracy, reliability, consistent intended performance, and the ability to conclusively discern invalid or altered records.

(b) The ability to generate true copies of records in both human readable and electronic form suitable for inspection, review, and copying by the agency. Persons should contact the agency if there are any questions regarding the ability of the agency to perform such review and copying of the electronic records.

(c) Protection of records to enable their accurate and ready retrieval throughout the records retention period.

(d) Limiting system access to authorized individuals.

(e) Use of time stamped audit trails to document record changes, all write to file operations, and to independently record the date and time of operator entries and actions. Record changes shall not obscure previously recorded information. Such audit trail documentation shall be retained for a period at least as long as required for the subject electronic documents and shall be available for agency review and copying.

(f) Use of operational checks to enforce permitted sequencing of events, as appropriate.

(g) Use of authority checks to ensure that only those individuals who have been so authorized can use the system, electronically sign a record, access the operation or device, alter a record, or perform the operation at hand.

(h) Use of device (e.g., terminal) location checks to determine, as appropriate, the validity of the source of data input or operational instruction.

(i) Confirmation that persons who develop, maintain, or use electronic record/electronic signature systems have the education, training, and experience to perform their assigned tasks.

(j) The establishment of, and adherence to, written policies which hold individuals accountable and liable for actions initiated under their electronic signatures, so as to deter record and signature falsification.

(k) Use of appropriate systems documentation controls including:

(i) Adequate controls over the distribution, access to, and use of documentation for system operation and maintenance.

(ii) Records revision and change control procedures to maintain an electronic audit trail that documents time-sequenced development and modification of records.

§*11.30 Controls for open systems.*

Open systems used to create, modify, maintain, or transmit electronic records shall employ procedures and controls designed to ensure the authenticity, integrity and confidentiality of electronic records from the point of their creation to the point of their receipt. Such procedures and controls shall include those identified in §11.10, as appropriate, and such additional measures as document encryption and use of established digital signature standards acceptable to the agency, to ensure, as necessary under the circumstances, record authenticity, integrity, and confidentiality.

§11.50 Signature manifestations.

(a) Electronic records which are electronically signed shall display, in clear text, the printed name of the signer and the date and time when the electronic signature was executed.

(b) Electronic records shall clearly indicate the meaning (such as review, approval, responsibility, and authorship) associated with their attendant signatures.

§11.70 Signature/record binding.

Electronic signatures and handwritten signatures executed to electronic records shall be verifiably bound to their respective electronic records to ensure that the signatures cannot be excised, copied or otherwise transferred so as to falsify another electronic record.

Subpart C -- Electronic Signatures

§11.100 General requirements.

(a) Each electronic signature shall be unique to one individual and shall not be reused or reassigned to anyone else.

(b) Before an electronic signature is assigned to a person, the identity of the individual shall be verified by the assigning authority.

(c) Persons utilizing electronic signatures shall certify to the agency that their electronic signature system guarantees the authenticity, validity, and binding of any electronic signature. Persons utilizing electronic signatures shall, upon agency request, provide additional certification or testimony that a specific electronic signature is authentic, valid, and binding. The certification should be submitted to the agency district office in which territory the electronic signature system is in use.

§11.200 Identification mechanisms and controls.

(a) Electronic signatures which are not based upon biometric/behavioral links shall:

(1) Employ at least two distinct identification mechanisms (such as an identification code and password), each of which is contemporaneously executed at each signing.

(2) Be used only by their genuine owners; and

(3) Be administered and executed to ensure that attempted use of an individual's electronic signature by anyone other than its genuine owner requires collaboration of two or more individuals.

(b) Electronic signatures based upon biometric/behavioral links shall be designed to ensure that they cannot be used by anyone other than their genuine owners.

§11.300 Controls for identification codes/passwords

Electronic signatures based upon use of identification codes in combination with passwords shall employ controls to ensure their security and integrity. Such controls shall include:

(a) Maintaining the uniqueness of each issuance of identification code and password.

(b) Ensuring that identification code/password issuances are periodically checked, recalled, or revised.

(c) Following loss management procedures to electronically deauthorize lost tokens, cards, etc., and to issue temporary or permanent replacements using suitable, rigorous controls for substitutes.

(d) Use of transaction safeguards to prevent unauthorized use of passwords and/or identification codes, and detect and report in an emergent manner any attempts at their unauthorized use to the system security unit, and to organizational management.

(e) Initial and periodic testing of devices, such as tokens or cards, bearing the identifying information, for proper function.

SUMMARY

The issue of FDA acceptance of electronic substitutes for handwritten signatures is gaining more and more attention in industry and in the agency itself. On one hand, the FDA says it has not yet come across completely paperless production records because of requirements in the CGMP for handwritten signatures. On the other hand, the increase in automation in the pharmaceutical, medical device, and biologics industries makes paperless systems an ever nearer reality and heightens industry impatience for FDA acceptance of electronic identification/signatures so that manufacturers can take advantage of available and future technology to improve the manufacturing processes and quality of their products.

Until the FDA issues a final rule delineating its policy regarding electronic signatures, however, a former FDA investigator stated at the CANDA Conference '93 that electronic signatures will still be a volatile and problematic issue for field investigators. Those manufacturers who decide to use electronic identification for their records are likely to get a mixed response from FDA field investigators. Investigators will look at data integrity and security issues and will make their own decisions on accepting electronic identifications on a case-by-case basis, and some investigators may challenge an individual manufacturer's use of the new technology.

Chapter 14

DRUG INSPECTIONS

OVERVIEW

Although validation concepts had been used in analytical test methods, the real "validation story," as it has been called,[1] began in the mid-1970s when the FDA published its new Good Manufacturing Practice (GMP) regulations. Beginning with sterilization validation, the FDA later began looking at validation of aseptic manufacturing processes, and by the early 1980s non-aseptic processes had been included for FDA scrutiny.

In 1986, the FDA issued its "Guidelines on General Principles of Process Validation," still an important document for FDA-regulated industries to consider when applying validation principles and practices in manufacturing processes.

FDA began considering computer system validation as the pharmaceutical industry turned to automated production methods in the 1980s. Today, system validation is an increasingly important aspect of FDA inspections of pharmaceutical facilities, and new guidelines are being prepared by the agency to help the industry comply with newly-emerging regulations.

The documents which follow all relate to FDA's regulation and inspection of pharmaceutical products. These documents are:

1. FDA Investigations Operations Manual (IOM), Chapter 5 -- Establishment Inspections, Subchapter 540 -- DRUGS
2. CDER's "Guide to Inspection of Bulk Pharmaceutical Chemicals."
3. CDER's "Guide to Inspection of Pharmaceutical Quality Control Laboratories."
4. Compliance Program Guide 7346.832 -- PRE-APPROVAL INSPECTIONS/INVESTIGATIONS

---

[1]See Kenneth G. Chapman, "A History of Validation in the United States, Part I," *Pharmaceutical Technology*, Vol. 15 (October 1991): 82-96; Kenneth G. Chapman, "A History of Validation in the United States -- Part II, Validation of Computer-Related Systems," *Pharmaceutical Technology*, Vol. 15 (November 1991):54-68.

5. "Guideline on General Principles of Process Validation," prepared by CDER, CBER, and CDRH
6. FDA's "Guide to the Inspection of Foreign Pharmaceutical Plants"
7. CDER's "Guide to Inspection of Computerized Systems in Drug Processing."
8. Compliance Policy Guide 7132a.07 -- Computerized Drug Processing; Input/Output Checking.
9. Compliance Policy Guide 7132a.08 -- Computerized Drug Processing; Identification of "Persons" on Batch Production and Control Records.
10. Compliance Policy Guide 7132a.11 -- Computerized Drug Processing; CGMP Applicability to Hardware and Software.
11. Compliance Policy Guide 7132a.12 -- Computerized Drug Processing; Vendor Responsibility.
12. Compliance Policy Guide 7132a.15 -- Computerized Drug Processing; Source Code for Process Control Application Programs.

FDA INVESTIGATIONS OPERATIONS MANUAL
CHAPTER 5 -- ESTABLISHMENT INSPECTIONS
SUB CHAPTER 540 -- DRUGS

540 GENERAL

540.1 Authority

Section 704(a) of the FD&C Act provides general inspectional authority, and identifies the type of establishments and their records that may be examined during a prescription drug, new drug, or antibiotic drug inspection.

540.2 Objective

Section 501(a) (2) (B) of the FD&C Act states that a drug is adulterated if the methods used in, or the facilities or controls used for its manufacture, processing, packing, or holding do not conform to, or are not operated or administered in conformity with current good manufacturing practice, to assure that such drug meets the requirements of the Act as to safety and has the identity and strength, and meets the quality and purity characteristics, which it purports or is represented to possess. Therefore, the purpose of a drug inspection is:

a. to determine and evaluate a firm's adherence to the concepts of good manufacturing practice;
b. to ensure that production and control procedures include all reasonable precautions to ensure the identity, strength, quality, and purity of the finished products;
c. to identify deficiencies which could lead to the manufacturing and distribution of products in violation of the Act;
d. to obtain correction of those deficiencies; and,
e. to ensure that new drugs are manufactured by essentially the same procedures and the same formulations as the products used as the basis for approval.

540.3 Preparation and References

a. Preparation

1. Review the district files of the firm to be inspected including:

   Establishment Inspection Reports,
   District Profiles
   New Drug Applications,
   Sample results,
   Complaints,

Regulatory files,
Antibiotic Applications,

During this review identify products which:
Are difficult to manufacture;
Require special tests or assays, or they are not assayable;
Require special processes or equipment;
Are new drugs and/or potent low dosage drugs.

Review the factory jacket and remove all complaint forms (FDA-2516 and FDA-2516a), which are marked Follow-up Next Inspection. These complaints are to be investigated during the inspection and discussed with management. See IOM 516.4.

2. Become familiar with current programs relating to drugs.

3. Determine the nature of the assignment, i.e., a specific drug problem or a routine inspection, and if necessary, consult other district personnel such as chemists, microbiologists, etc.

   When making GMP inspections, discuss with your supervisor the advisability of using a microbiologist, analyst, engineer, or other technical personnel to aid in evaluating those areas of the firm germane to their expertise.

4. Review the FD&C Act, Chapter V, Drugs and Devices.

5. Review parts of 21 CFR applicable to the inspection involved and Bioavailability (21 CFR 320).

b. References

Review the current editions of the United States Pharmacopeia (USP), and Remington's Pharmaceutical Sciences for information on specific products or dosage forms.

540.4 Inspectional Approach

Because of the increased complexity associated with the production and control of drugs, in-depth inspection of all manufacturing and control operations is usually not feasible or practical. Therefore, an audit approach is recommended in which therapeutically significant drugs and those drugs which are difficult to manufacture are covered in greater detail.

The bioavailability regulations, 21 CFR 320.22, lists those drugs which are therapeutically significant and for which manufacturing changes can affect efficacy. Some manufacturers have difficulty in complying with the dissolution specifications established for many products. Also, significant problems are not uncommon with timed release or delayed release drugs requiring multiple dissolution (release) tests.

If reworked products are encountered, validation of their manufacturing procedures and justification for reworking should be reviewed. Written investigation reports which are required for any product failing to meet an established specification should also be reviewed and evaluated.

For those drug manufacturers marketing a number of bioavailability problem drugs, identify suspect products by:

a. Reviewing the firm's complaint files early in the inspection to determine relative numbers of complaints per product.
b. Inspecting the quarantine and/or rejected product storage area to identify rejected product.
c. Examining annual reviews performed under 21 CFR 211.180(e) to determine those products which have a high reject rate.
d. Reviewing summaries of laboratory data or laboratory workbooks.

Attempts should also be made to determine the attitude or philosophy of top management and how they react to problems, batch rejections, and investigate product failures.

541 DRUG REGISTRATION & LISTING

541.1 Authority

Section 510 of the FD&C Act and 21 CFR 207 & 607 delineate the requirements and exemptions relating to the registration of establishments engaged in the manufacture, preparation, compounding, or processing, of drugs including blood banks and other firms collecting, manufacturing, preparing, or processing human blood or blood products, and the listing of their products. Such establishments must register within five days after beginning operations and must also submit a listing of every drug product in commercial distribution at that time. Registration is accomplished by submitting an FDA-2656 (Registration of Drug Establishment) and the drug listing and subsequent June and December updating shall be on form FDA-2657 (Drug Product Listing). In lieu of an FDA-2657, tapes for computer inputs may be submitted if equivalent in all elements of information specified on the FDA-2657 after initial FDA review and approval of the formats.

Registration and Listing is required whether or not interstate commerce is involved.

541.2 Policy

Two or more companies occupying the same premises and having interlocking management are considered one establishment and will be assigned a single registration number. See IOM 501.14 - Multiple Occupancy Inspections for additional information.

Vitamin manufacturers are required to register unless their products are used solely in food supplements and do not become drugs or components of drugs. In most cases, bulk vitamin manufacturers should register unless they have knowledge of the use and labeling of each sale.

Independent laboratories providing analytical or other laboratory control services on commercially marketed drugs must register.

541.3 Responsibility

Your computer generated coversheet will indicate if the establishment is registered for the current year. If you determine that registration and listing is required, advise your supervisor. After checking for past registration, cancellation, etc., the district will provide the firm with the proper forms and instructions.

542 CURRENT GOOD MANUFACTURING PRACTICE - DRUGS

A detailed explanation of the manufacturing processes for each form of medication will not be undertaken in the IOM. However, IOM 542 & 543 outline areas which could be used as a guide for some of the areas to be investigated and fully developed if violations exist.

542.1 Building Facilities

Determine construction, size, and location of plant in relation to surroundings. Is there adequate lighting, ventilation, screening, proper physical barriers for operations requiring filtering, or dust, temperature, humidity, and bacteriological controls? Determine if there are adequate personnel washing, locker, and toilet facilities.

542.11 Plant Space

Determine whether adequate space is provided for the placement of equipment and materials to prevent mix-ups in the following operations:

Receiving, sampling, and storage of raw materials.
Manufacturing or processing.
Packaging and labeling.
Storage for containers, packaging materials, labeling, and finished products.
Control and Production laboratories.

542.12 Equipment

Determine the design, capacity, construction, and location of equipment used in the manufacturing, processing, packaging, labeling, and testing of drug products. Describe new or unusual equipment, including brief descriptions of operating principles. Consider the use of photographs and diagrams to supplement written descriptions.

Determine and report if the firm rents or leases equipment on a short term basis. If they do, report cleaning procedures utilized prior to using this equipment. Equipment of this type may have been used for pesticides, chemicals, drugs, etc. prior to being installed in the firm and could be a source of cross-contamination if not adequately cleaned and/or sanitized.

Watch for unorthodox or faulty applications of standard equipment.

To permit evaluation of manufacturing practices, determine if:

Mechanical parts which come in contact with drugs or drug components will react, add to, be absorptive, or adversely affect the identity, strength, quality, or purity of the drug.

The equipment is constructed in such a manner that lubricants or coolants may be used without becoming incorporated into the drug product.

The equipment is constructed or located to facilitate cleaning, adjustments, and maintenance, and to exclude possible contamination from previous as well as current manufacturing operations.

Equipment is thoroughly cleaned before reuse, and identified to indicate that it is ready for use.

Equipment is of suitable capacity and accuracy for use in measuring, weighing, or mixing operations. Most district laboratories have sets of official weights which may be used if necessary to check the accuracy of firm's weighing equipment.

Utensils and in-process containers are constructed to permit thorough cleaning.

Responsible supervisory or quality control personnel approve equipment before manufacturing and packaging operations begin, and are present while equipment is operating.

542.2 Personnel and Responsibilities

542.21 Responsible Individuals

Obtain the names, titles, and other information on key officers and production personnel as indicated in IOM 501.14 and IOM 525. Identify individuals who initiate and approve the master and batch production and control records and who investigates failures.

542.22 Employee Training

In the drug industry employee capabilities, education and training for the position they hold plays a significant part in the production of a quality product. Report whether the firm has any type of training program(s) for its employees. If so, report the type and length of training given, e.g., formal, on-the-job, professional, "buddy-system", practical, classroom, etc. Determine if instruction is general in nature and all employees take the same course, cover the same topic, or if the subjects are tailored to fit the operation and the employees involved in that operation. Also report who administers the training programs and their qualifications.

542.3 Components and Product Containers

542.31 Components

Check components or raw materials used in the manufacture or processing of drugs, regardless of whether they appear in the finished product. Report whether the following control procedures are utilized by the firm.

a. Receiving and Storage

Receiving records for components should be maintained, and contain:

1. name of component,
2. manufacturer,
3. supplier, if different from manufacturer,
4. carrier,
5. receiving date,
6. manufacturer's lot number,
7. quantity received, and
8. firm's assigned control number.

Check sanitary conditions in the storage area, stock rotation practices, and special storage conditions (protection from light, moisture, temperature, air, etc.). Examine all raw materials for new ingredients, proper labeling, etc.

Inspect glandular and botanical components for insect infestation.

Be alert for components that:

May be new drug substances. Appear to have no use in the plant. Appear to be from unknown suppliers.

Identify the name, number, and manufacturer of all colors used, and check against the Color Additives Status List, IOM Exhibit 530-C.

Determine:

If all components are received at one location in the plant and handled by one individual. If not, determine firm's receiving procedure and who is responsible for receiving components.

How components are coded or identified after receipt, and who has this responsibility.

If components are placed in quarantine areas or left in a general storage area.

The criteria used in removing components from quarantine. Determine the number and identity of individuals that have access to either the quarantine or storage area.

What records are maintained in the storage area to document movement of components to other areas.

How rejected components are handled.

b. Identification

Each component container should be identified by the firm to include a serial number or code which, by reference to receiving records, will identify the product, the supplier's name, quantity received, control number and date of receipt. The identification should be affixed to the container itself, rather than on a removable lid or in any fashion whereby it might become detached or defaced.

c. Testing

Determine the sampling and testing procedures for components, and the process by which approved materials are released for use.

d. Inventory

Determine the manner, validity, and accuracy of the firm's inventory system for drug components.

542.32 Product Containers

Evaluate the following to determine whether firm has shown that the containers and their component parts, are compatible with the product, will provide adequate protection for the drug against deterioration or contamination, are not additive or absorptive to a significant degree, and are otherwise suitable for use:

a. Specifications for containers, closures, and component parts.
b. What tests or checks are made (cracks, glass particles, durability of component parts, metal particles in ointment tubes, compliance with compendial specifications, etc.).
c. Cleaning procedures and how containers are stored.
d. Handling of labeled containers, e.g., preprinted ampules. Are these controlled as labeling, or as containers?

542.4 Control Records

542.41 Master Production and Control Records

The various master production and control records are of fundamental importance because all phases of production and control are governed by them. Master records, if erroneous, may adversely affect the product and should be prepared in accordance with, and embody the feature outlined in 21 CFR 211.186.

542.42 Batch Production and Control Records

The various batch production and control records should, reveal details of all significant steps in the manufacture, packaging, and control of specific batches of drugs. 21 CFR 211.188 provides the basic information for which the batch records should provide. A complete production and control record may consist of several separate records.

Routinely check batch record calculations against the master formula record and labeling. Give special attention to those products on which there have been complaints, and those where the formulae are hand-copied.

Be alert for errors in transcribing information from the master formula record to the batch record. Errors involving misinterpretation of symbols, abbreviations, decimal points, etc., arise from manual transcription.

It is important that batch production records be specific in terms of equipment (such as mixers) and processing times (such as mixing, granulating and drying times) especially for those products listed in 21 CFR 320.22. Since many firms conduct retrospective validation, it is necessary that the manufacturing processes for these products be standardized and controlled.

542.5 Production and Control Procedures

542.51 Critical Manufacturing Steps

Each critical step in the manufacturing process shall be performed by a competent, responsible individual and checked by a second competent, responsible individual. If such steps in the processing are controlled by automatic mechanical or electronic equipment, its performance should be routinely checked by one or more competent individuals.

These critical manufacturing steps include:

a. The selection, weighing, measuring and identifying of components.
b. The addition of ingredients during processing
c. Measures taken to prevent duplicate addition of ingredients.
d. The recording of any deviations from master formula on the batch record.
e. Adequate mixing time and testing of drug components during various stages of manufacture.
f. The determination of actual yield.

542.52 Batch Identification

All containers and equipment used in producing a batch of drugs should be clearly labeled at all times to identify fully and accurately their contents, batch number, and the stage of processing. The batch should also be stored and handled in a manner to prevent mixups or contamination with other drugs. Previous identification labels should be removed.

542.53 In-Line and Bulk Testing

To ensure the uniformity and integrity of products, there shall be adequate in-process controls, such as checking the weights and disintegration time of tablets, the fill of liquids, the adequacy of mixing, the homogeneity of suspensions, and the clarity of solutions.

Determine if in-process test equipment is available and that specified tests are made and properly recorded. Be alert for pre-recording test results such as tablet weight determinations.

Determine if finished bulk-drug items are held in quarantine until all tests are completed before release to the packaging and labeling department.

542.54 Actual Yield

Determine if competent and responsible personnel check the actual against the theoretical yield of each batch of drug. In the event of any significant unexplained discrepancies, determine if there is a procedure to prevent distribution of the bath in question, and other associated batches that may have been involved in a mix-up with it.

542.55 Personnel Habits

Observe the work habits of all plant personnel.

Determine:

Their attitudes and actions involving the jobs they perform. (Careless, lackadaisical, disgruntled, etc).

Their dress. (Clean dresses, coats, shirts and pants, head coverings, etc.).

Whether personnel in direct contact with drugs are given physical examinations or periodical health checks and how often.

If proper equipment is used for a given job or whether short cuts are taken. (i.e. use of hands and arms to mix or empty trays of drug components).

If there are significant language barriers between supervisors and employees, or between employees and written material.

542.56 Tablet and Capsule Products

Tablets and capsules are particularly susceptible to airborne contamination because of the handling of large quantities of ingredients, and the many manipulations, usually in the dry state.

The maintenance, cleaning, and location of equipment, and the storage of granulations and bulk dosage forms in the tableting area, all warrant close attention.

Determine the type of equipment and its location in the tableting operation. The equipment may include rotary or multiple layer rotary tableting machines, coating and polishing pans, punches and dies, etc. The equipment, as well as utensils and containers should be so constructed and located to facilitate maintenance and cleaning at the end of each batch or in the case of a continuous batch operation, at suitable intervals. Determine who is responsible for cleaning and the cleaning schedule.

Binders, disintegrators, bases, and lubricants, are significant components of tablets:

Identity and the amount of any binders added to the batch to keep the tablets together. Excess binder will make the tablet too hard for use.

Identity and amount of disintegrators used to facilitate break-up after administration.

Determine if the base, which should be an inert substance added to give tablet necessary size and weight, is compatible with active ingredients.

Identify the lubricant used. The lubricant assists in the flow of granulated material, prevents adhesion of the tablet material to the surface of punches and dies, and helps in ejection from the die cavity.

When necessary, mixing, granulation, drying or tableting operations should be segregated in enclosed areas to eliminate the possibility of cross-contamination. Determine what precautions are taken by the firm to prevent cross-contamination of drug products. See IOM 542.57.

Determine what temperature, humidity, and dust collecting controls are used by the firm in manufacturing operations.

Observe the actual operation of the equipment and determine:

a. Whether powders or granulations are mixed, ground or stored according to the firm's own specifications. This is especially important for potent low-dosage drugs.
b. Whether drying ovens are adequate and prevent cross-contamination of granulations in adjacent chambers. Does the firm record drying time and temperatures.
c. What in-line tests are conducted and whether these are performed by production or quality control personnel.
d. The disposition of in-process samples. Are they returned to the line, or retained for further testing?
e. Whether equipment is properly grounded.

Capsules may be either hard, soft, or pearl type. They may be hand or machine filled with powder or liquid. The manufacturing operation of powders for capsules should follow the same practice as for tablets. Determine manufacturing controls used, in-line testing and basis for evaluating test results, and individual directly responsible for filling operations.

A quick method for field examining granulations and powders for specks and off colors is:

For Specks:

Spread an ounce or so of fresh material in a thin layer on a piece of clear clean glass. (Microscope slides are OK).

Cover the material carefully with a clear clean glass.

Press gently, without sliding, to make full contact. Carefully examine. A magnifying glass helps.

For Off-color:

Make a small heap of each material side by side. Cover both with a clear glass and press so the two heaps run together.

Examine. A color difference will show a distinct dividing line.

542.57 Areas of Cross-Contamination

The potential for cross-contamination exists in every drug plant especially in those where different products are processed through common equipment.

Give special attention to the following:

a. Weighing, granulation, and tableting equipment and their respective areas. The granulation process may be especially vulnerable to cross-contamination or mix-ups since this area is often manned by relatively unskilled employees who may be unaware of the potential dangers of the potent drugs processed. Lack of supervision in the granulating room may lead to carelessness on the part of these workers.

   Weighing operations should be carefully observed. The use of common scoops, containers, or other such equipment should be carefully evaluated. Comminuting equipment and other mills should also be carefully checked.

   The Fitz mill, Pony mixer, tableting machines, and storage containers should be inspected for any accumulation of static material.

b. Determine the actual cleaning schedule for equipment and area to see if it is adequate to prevent cross-contamination.

c. Determine the type of equipment used to clean machinery and utensils and the ultimate disposition of the waste.

d. Determine what provisions are made to filter or remove drug particles from the air in those areas where drugs are processed.

e. Check whether component containers are brought into the manufacturing area, opened and exposed for any length of time while potent drug items are being processed.

Where cross-contamination of batches is suspected, production records should be reviewed to identify batches which could be contaminated with dangerous or potent ingredients. Unusual or questionable yield figures may also indicate addition of adulterating ingredients.

Where cross-contamination is suspected, collect in-plant Investigational Samples; e.g., tablets suspected of having become contaminated on an unclean tablet press, together with the uncontaminated granulation from which they were compressed. Each piece of equipment should have a log nearby which lists batches produced and documentation that the equipment was cleaned. If you suspect cross-contamination at a particular piece of equipment, the log may identify what other drugs were run. A copy of the log will assist the laboratory in knowing what specific contaminant to analyze for.

Follow-up with Official Samples representing the batch manufactured during the inspection and/or produced in close proximity to the inspection.

542.58 Sterile Products

Upon intravenous injection, filterable substances which cause an increase in body temperature lasting for several hours are called pyrogens. Since certain bacteria produce pyrogens, any condition that permits bacterial growth should be avoided. Pyrogens may develop in water standing in stills or other apparatus, distilled water standing overnight, or from surface contamination of ampules, vials, bottles or other equipment. Sterile parenterals may also contain chemical contaminants that will produce a pyretic response in humans or test animals although there are no pyrogens per se present.

Determine and evaluate the procedures used to minimize the hazard of contamination with micro-organisms and other particulate contaminants during the production of all drugs purporting to be sterile.

Remington's Pharmaceutical Sciences, 15th Edition, Chapter 84, is an excellent reference regarding pyrogens.

a. Personnel

Review the training program to ensure that personnel performing production and control procedures have experience and training commensurate with their intended duties. It is important that personnel be adequately trained in aseptic procedures.

b. Buildings

The non-sterile preparation areas for sterile drugs should be controlled. Refer to Subpart C of the proposed GMP's for LVP's; however, deviations from these proposed regulations are not necessarily deviations from the GMPR's.

Observe the formulation practices or procedures utilized in the preparation areas.

Determine how the firm minimizes traffic and unnecessary activity in the preparation area.

Determine if filling rooms and other aseptic areas are constructed to eliminate possible areas for microbiological contamination, such as, dust-collecting ledges, porous surfaces, etc.

Determine how aseptic areas are cleaned and rendered aseptic.

1. Air

Air supplied to the non-sterile preparation or formulation area for manufacturing solutions prior to sterilization should be filtered as necessary to control particulates. Air being supplied to product exposure areas where sterile bulk drugs are processed and handled should be High Efficiency Particulate Air (HEPA) filtered under positive pressure.

Review the firm's system for HEPA filters, determine if they are certified and/or Dioctyl Phthalate (DOP) tested and frequency of testing.

Check for the use of ultra-violet lights in the filling room. If U.V. lights are used, determine if the output is monitored by the firm and determine procedures, including records for assuring measurement are performed and weak bulbs replaced.

2. Environmental Controls

Specifications for viable and non-viable particulates should be established in conjunction with the validation of the sterilization and aseptic handling processes for sterile drugs. Specifications for viable particulates should include provisions for both air and surface sampling of aseptic processing areas and equipment.

Review the firm's environmental control program, specifications, and test data. Determine the firm's procedures if an out-of-limit test result is obtained. Also determine if review of environmental test data is included as a part of the firm's release procedures.

Note: In the preparation of media for environmental air and surface sampling, suitable inactivating agents should be added. For example, the addition of penicillinase to media used for monitoring sterile penicillin operations and cephalosporin products.

c. Equipment

1. Where deficient manufacturing procedures are encountered, determine and diagram the type, construction, and location of equipment used at any point in the manufacturing or processing of sterile products.
2. Determine the maintenance practices for this equipment.
3. Check the type of filters used. Determine the purpose of the filters, how they are cleaned, and how they are inspected for damage.
4. Determine the practice of sterilizing equipment after sterility has been broken.
5. Determine how equipment used in the filling room is sterilized.

d. Components, Containers, and Closures

1. Water for Injection

Water used in the production of sterile drugs should be controlled to assure that it meets USP specifications. Review the firm's Water for Injection production, storage, and delivery system and ensure that stills, filters, holding tanks, and plumbing are not likely to contribute to contamination. Evaluate the firm's procedures and specifications that assure the quality of the Water for Injection. Other water should be given coverage as outlined under subpart L of the proposed GMP's for LVP's.

2. Determine what controls, and special cleaning, sterilization, and depyrogenization procedures, are used in the handling and storage of containers and closures, prior to use.

e. Sterilization

1. Methods

Determine what method of sterilization is used. Review and evaluate validation data regardless of the method employed.

If steam under pressure is used, an essential control is a mercury thermometer and a recording thermometer installed in the exhaust line. The time required to heat the center of the largest container to the desired temperature must be known. Installation of a recording thermometer is desirable. Time and temperature, not pressure, are the governing factors in this type or method of sterilization. Steam must expel all air from the sterilizer chamber. The drain lines should be connected to the sewer by means of an air vent or open filter to prevent back siphoning. The use of paper layers or liners and other practices which might block the flow of steam should be avoided.

Charts of time and temperature of exposure should be filed for each sterilizer load.

If sterile filtration is employed, determine the firm's criteria for selecting the filter and the frequency of changing. Determine if the firm knows the bioburden and examine their procedures for integrity testing filters.

Filters might not be changed after each batch is sterilized. Determine if there is data to justify the integrity of the filters for the periods utilized and that "grow through" has not occurred.

Determine if the firm is utilizing asbestos filters. Asbestos filters are not generally permitted for liquid filtration (even if followed in line by membrane filters) in the preparation of human parenterals. See CFR 211.72 for limitations on the use of asbestos filters.

If ethylene oxide sterilization is employed, determine what tests are made for residues and degradation products.

Tyndallization or intermittent sterilization, frequently used in "sterilizing" parenterals is of questionable effectiveness. When used with a solution nutritive for bacteria, the holding period between heatings will permit the germination of heat-resistent spores of the heat labile vegetative forms. This will not occur in substances unable to support bacterial growth or in products containing bacteriostatic agents.

2. Indicators

Determine the type of indicators used to assure sterility, such as, lag thermometers, peak controls, Steam Klox, test cultures, Tell Tales, etc.

Caution: When spore test strips are used to test the effectiveness of ethylene oxide sterilization, determine if the spore test strips are moistened or refrigerated prior to use. Refrigeration may cause condensation on removal to room temperature. Moisture on the strips converts the spore to the more susceptible vegetative forms of the organism which may affect the reliability of the sterilization test.

If biological indicators are used, you should review the current USP on Sterilization and Biological Indicators. In some cases, testing biological indicators may become all or part of the sterility.

Biological indicators are of two forms, each of which incorporates a viable culture of a single species of microorganism. In one form, the culture is added to representative units of the lot to be sterilized or to a simulated product that offers no less resistance to sterilization than the product to be sterilized. The second form is used when the first form is not practical as in the case of solids. In the second form, the culture is added to disks or strips of filter paper, metal, glass, or plastic beads.

During EI's of firms relying on biological indicators, collect background data compiled by the firm to include:

(a) Surveys of the types and numbers of organisms in the product before sterilization.
(b) Data on the resistance of the organism to the specific sterilization process.
(c) Data used for selecting the most resistant organism and its form (spore or vegetative cell).

(d) Studies of the stability and resistance of the selected organism to the specific sterilization process.
(e) Studies on the recovery of the organism used to inoculate the product.
(f) If a simulated product or surface similar to the solid product is used, validation of the simulation or similarity. The simulated product or similar surface must not affect the recovery of the numbers of indicator organisms applied.
(g) Validation of the number of organisms used to inoculate the product, simulated product, or similar surface to include stability of the inoculum during the sterilization process.

Since qualified personnel are crucial to the selection and application of these indicators, review their qualifications. Education alone is not adequate but requires experience dealing with: the process, expected contaminants, testing of resistance of organisms, and other areas of esoteric information and technique.

Review the firm's instructions regarding use, control and testing of the biological indicator by product including a description of the method used to demonstrate presence or absence of viable indicator in or on the product.

Obtain the data used to support the use of the indicator each time it is used. Include:

Counts of the inoculum used.

Recovery data to control the method used to demonstrate the sterilization of the indicator organism.

Counts on unprocessed, inoculated material to indicate the stability of the inoculum for the process time.

Results of "Sterility" testing specifically designed to demonstrate the presence or absence of the indicator organism for each batch or filling operation.

In using indicators you must assure yourself that the organisms are properly handled. Procedures must positively safeguard the drug production. The USP cautions about contamination of the production and manufacturing areas with the indicator organisms. Special labeling of inoculated containers, tight closures, special attention by plant personnel, both manufacturing and Quality Control, adhering to well defined, written procedures are among the imperatives.

3. Filled Containers

Determine how the filled vials or ampules leave the filling room. Is the capping or sealing done in the sterile area? If not, where?

Determine what tests are made on finished vials, ampules, or other containers, to assure proper fill and seal.

Review examinations made for particulate contamination. You can quickly check for suspected particulate matter by using a Polariscope.

4. Personnel Practices

Check on how the equipment used in the filling room area is sterilized.

Observe filling room personnel practices (sterile gowns, masks, caps, shoe coverings, washing of hands, arms, etc.)

Determine if there is a dressing room adjacent to the filling area and how operators and supplies gain access to the sterile area.

Check on the practices after lunch and other absences. Is fresh sterile garb supplied, or are soiled garments reworn?

f. Laboratory Controls

1. Retesting for Sterility

Sterility retesting is acceptable provided the cause of the initial non-sterility is known, and thereby invalidates the original results. In the past, attempts have been made to justify retesting even though sterility controls were satisfactory. It cannot be assumed that the initial sterility test failure is a false positive. This conclusion must be justified by sufficient documented investigation. Additionally, spotty or low level contamination may not be identified by repeated sampling and testing.

Review sterility test failures and determine the incidence, procedures for handling, and final disposition of the batches involved.

2. Retesting for Pyrogens

As with sterility, pyrogen retesting can be performed provided it is known that the test system was compromised. It cannot be assumed that the failure is a false positive without documented justification.

Review any initial pyrogen test failures and determine the firm's justification for retesting.

3. Particulate Matter Testing

Cleanliness specifications or levels of non-viable particulate contamination should be established. Limits are usually based on the history of the process. The particulate matter test procedure and limits for LVP's in USP XXI can be used as a general guideline, but the levels of particulate contamination in sterile powders is generally greater than in LVP's. LVP solutions are filtered during the filling operation. However, sterile powders, except for powders lyophilized in vials, cannot include filtration as a part of the filling operation. Considerable particulate contamination is also present in sterile powders which are spray dried due to charring during the process.

Review the particulate matter test procedure and the release criteria. Review production and control records of any batches for which an excessive number of complaints of particulate matter have received.

g. Production Records

Production records should be similar to those for other dosage forms. Critical steps, such as integrity testing of filters, should be signed and dated by a second responsible person.

Review production records to ensure that directions for significant manufacturing steps are included and that completed records reflect process performed and reflect a complete history of production.

542.59 Ointments, Liquids, and Lotions

Major factors in the preparation of these drugs are the selection of raw materials, manufacturing practices, equipment, controls and laboratory testing.

Following the basic drug inspection fundamentals, fully evaluate the production procedures. In addition, obtain specific information regarding:

a. The selection and compatibility of ingredients.
b. Whether the drug is a homogeneous preparation free of extraneous matter.
c. The possibility of decomposition, separation, or crystallization of ingredients.
d. The adequacy of ultimate containers to hold and dispense contents.
e. Procedures to reduce metal particles in metal containers to established tolerance levels.
f. Maintenance of homogeneity during manufacturing and filling operations.

The most common problems associated with the production of these dosage forms is microbiological contamination resulting from faulty design and/or control of purified water systems. During inspections, test design and microbiological in-process and finished product testing results should be reviewed and evaluated.

Associated with microbiological contamination these products is the adequacy of preservative systems. Preservative effectiveness testing for these products should be reviewed.

Equipment employed for manufacturing topicals is sometimes difficult to clean. This is especially true for those which contain insoluble active ingredients, such as the sulfa drugs. The firm's cleaning procedures for those lines (equipment) employed for potent and/or insoluble drug products should be reviewed.

542.6 Finished Product Handling

542.61 Packaging and Labeling

Packaging and labeling operations must be adequately controlled to ensure that only those drugs that have met specifications established in the master formula records will be distributed, to prevent mixups between drugs during the packaging and labeling operations, and to ensure that correct labeling is employed.

Determine if packaging and labeling operations include:

a. Adequate physical separation of labeling and packaging operations from manufacturing process.
b. Review of:

    1. Label copy before delivery to printer.
    2. Printer's copy.
    3. Whether firm's representative inspects the printer.
    4. Whether or not gang printing is prohibited.
    5. Whether finished labels are checked before release to stock. Determine who is responsible for label review and whether label conforms to labeling specified in batch production records.

c. Separate storage of each label (including package inserts) to avoid mixups.
d. Inventory of label stocks. Determine if printer's count is accepted or if labels are counted upon receipt.
e. Designation of one individual to be responsible for storage and issuance of all labels.
f. Receipt by the packaging and labeling department of batch record, or similar type records, indicating amount of drug and number and size of units to be labeled. Determine if the batch record is retained by the packaging supervisor or accompanies the labels to the actual packaging and labeling line.

g. Adequate controls of the quantities of labeling issued. Determine if any excess labels are accounted for and if excess labels bearing specific control codes, and obsolete or changed labels are destroyed.
h. Inspection of the facilities prior to labeling to ensure that all previously used labeling and drugs have been removed.
i. Assurance that batch identification is maintained during packaging.
j. Control procedures to follow if a significant unexplained discrepancy occurs between quantity of drug finished and the quantity of labeling issued.
k. Segregated facilities for labeling one batch of drug at a time. If this is not practiced, determine what steps are taken to prevent mix-ups.
l. Methods for checking similar type labels of different drugs or potencies to prevent mixing.
m. Quarantine of finished packaged products to permit adequate examination or laboratory testing of a representative sample to safeguard against errors in the finishing operations and to prevent distribution of any batch until all specified tests have been met.
n. An individual who makes the final decision that the drug should go to the warehouse, or the shipping department.
o. Utilization of any outside firms, such as contract packagers, and what controls are exercised over such operations.

Review prescription drugs for which full disclosure information may be lacking. If such products are found, submit labels and other labeling as exhibits with the EIR. See 21 CFR 201.56 for the recommended sequence in which full disclosure information should be presented.

Review labels of OTC products for warnings required by 21 CFR 369.

Special attention should be devoted to firms using "rolls" of pressure sensitive labels. Investigators have found instances where:

a. Paper chips cut from label backing to facilitate running the labels through a coder interfered with the code printer causing digits in the lot number to be blocked out.
b. Some rolls contained spliced sections resulting in label changes in the roll.
c. Some labels shifted on the roll when the labels were printed resulting in omitting some of the required information.

Determine what controls or procedures that firm has to provide positive assurance that all labels are correct.

542.62 Control Code

A control code must be used to identify the finished product with a lot, or control number that permits determination of the complete history of the manufacture and control of the batch.

Determine:

a. The complete key to the code.
b. Whether the batch number is the same as the control number on the finished package. If not, determine how the finished package control number relates, and how it is used to determine the identity of the original batch.

542.63 Drug Warehousing. (See IOM 659)

Check the finished product storage and shipping areas for sanitary condition, stock rotation, and special storage conditions needed for specific drugs.

Whenever drugs are warehoused determine:

a. Storage conditions (temperature, humidity, refrigeration where necessary, etc.).
b. Rotation of stocks. Check for out-dated drugs or antibiotics, and other old or deteriorated drugs. Sample as indicated.
c. Presence of inadequately labeled or inadequately coded drugs, such as lack of warning statements, lack of inserts or wrong inserts, inadequate codes, lack of mandatory labeling, false claims, etc.
d. What products are being recalled and how this recall was brought to the attention of the dealer or warehouse.
e. The possible existence of new firms, as indicated by unknown or unfamiliar labels.
f. Unusual or new drug products marketed, but not covered by an approved NDA.

542.7 Laboratory Controls

542.71 Specifications and Test Procedures

Laboratory controls should include adequate specification and test procedures to assure that components, drug preparations in the course of processing, and finished products conform to appropriate standards of identity, strength, quality, and purity.

In order to permit proper evaluation of firm's laboratory controls, determine:

a. Whether firm has established a master file of specifications for all raw materials used in drug manufacture. This master file should include:

    1. Sampling procedures
    2. Sample Size
    3. Number of containers to be sampled.
    4. Manner in which samples will be identified.
    5. Tests to be performed.
    6. For components subject to deterioration, the provisions for periodic retesting.

b. The firm's policies relative to protocols of assay. These reports are often furnished by raw material suppliers; however, the manufacturer is responsible for verifying the validity of the protocols by periodically preforming their own complete testing and routinely conducting identity tests on all raw materials received.

c. Laboratory procedure for releasing raw materials, finished bulk drugs or packaged drugs from quarantine. Determine who is responsible for this decision. Raw material specifications should include approved suppliers. For NDA or ANDA drugs, the approved suppliers listed in their specifications should be the same as those approved in the NDA or ANDA.

d. If the laboratory is staffed and equipped to perform all raw material, in-process, and finished product testing that is claimed.

e. Whether drug preparations are tested during processing. If so, determine what type of tests are made and whether a representative sample is obtained from various stages of processing.

f. Specifications and description of laboratory testing procedures for finished products.

g. Procedures for checking the identity and strength of all active ingredients, and for pyrogen and sterility determinations where indicated.

   If the laboratory conducts pyrogen or safety tests, or bioassays, determine number of laboratory animals and if they are adequately fed and housed. Determine what care is exercised on weekends and holidays.

h. Sterility testing procedures

   Entries should be permanently recorded and show all results, both positive and negative. Examine representative samples being tested and their records.

   When checking the sterility testing procedures, determine:

   1. physical conditions of testing room.
   2. laboratory procedures for handling sterile sample.
   3. use of ultra-violet lights.
   4. use of sterilamps. They should not be used where they will shine on test material.
   5. number of units tested per batch.
   6. procedure for identifying test media with specific batches.
   7. test media's ability to support growth of organisms.
   8. length of incubation period.

9. procedure for diluting products to offset the effects of bacteriostatic agents.

i. Pyrogen testing procedures

Determine if animals involved in positive pyrogen tests are withdrawn from use for the required period.

j. If any tests are made by outside laboratories. Report the names of the laboratories and the tests they perform. Determine what precautions the firm takes to insure that the laboratories' work is bona fide.

k. Methods used to check the reliability, accuracy, and precision of laboratory test procedures and instrumentation.

l. How final acceptance or rejection of raw materials, intermediates, and finished products is determined. Review recent rejections and disposition of affected items.

m. The provisions for complete records of all data concerning laboratory tests performed, including dates and endorsements of individuals performing the tests, and provisions for relating the tests to the batch of component or product to which they apply.

n. The reserve sample program and procedures.

o. Whether stability tests are performed on:

   1. The drug product in the container in which it is marketed.
   2. Solutions prepared as directed in the labeling at the time of dispensing. Determine if expiration dates, based on appropriate stability studies, are placed on labels.

p. If penicillin and non-penicillin products are manufactured on the same premises, whether non-penicillin products are tested for of penicillin contamination.

Obtain copies of laboratory records, batch records, and any other documents that demonstrate errors or other deficiencies.

542.72 Returned Drugs

Returned drugs often serve as an indication that products may have decomposed during storage, are being recalled or discontinued.

Determine how returned drug items are handled, such as, quarantined, destroyed after credit, returned to storage, etc.

Note: Dumping salvage drug in the trash is not destruction and is a potentially dangerous practice. Advise management and caution them to properly dispose of or decharacterize drugs to preclude salvage.

If an abnormally large amount of a specific drug item is on hand, determine why. Check if returned drug items are examined in the laboratory, and, who make the ultimate decision as to the use of the returned drugs.

542.73 Complaint Files

21 CFR 211.198 requires that records of all written and oral complaints be maintained.

Although FDA has no authority to require a drug firm except for prescription drugs, to open its complaint files, attempt to review the firm's file on the premises. For prescription drugs the FD&C Act in Section 704(a) (1) (B) provides, "*** In the case of any factory, warehouse, establishments, or consulting laboratory in which prescription drugs or restricted devices are manufactured, processed, packed, or held, inspection shall extend to all things therein (including records, files, papers, processes, controls, and facilities) bearing on whether prescription drugs or restricted devices are adulterated or misbranded within the meaning of the Act, ***".

Complaints may not be assembled in a specific file but scattered throughout various files under the name of the product; customer, or injured or complaining party. Some firms may call these by other names, such as, "Adjustment Files", "Customer Relations", etc. Word your request broadly enough to include "strays", and all other allegations that challenge product integrity.

Where possible, copy the names and addresses of complainants and include a brief summary of the nature of the complaint in the body of the EIR. Submit copies of pertinent complaints.

Determine the policy of handling complaints. This should include such things as: how many must be received on a product or batch before action is taken; whether appropriate action (recall, withdrawal, etc.) is taken and what records are maintained.

Determine who reviews complaints and that person's qualifications. In some instances, this may be a medical advisor. Ascertain the criteria for determining the significance of complaints and how they are followed up. Determine if the control laboratory analyzes drugs involved in complaints. What records are maintained. Determine if records are kept of oral or telephone complaints.

During the inspection investigate any in-plant causes of complaints and determine if any further follow-up is necessary. If so, satisfy yourself that you have fully documented the complaint causes(s). See IOM 516.4 for discussion of complaints with management and IOM 592.31m for reporting in the EIR.

During the inspection follow-up on the complaints contained on the FDA-2516 and FDA-2516a forms you brought with you from the firm's district factory jacket.

Management may claim that all complaints are given to their insurance company. If so, determine the name of the company writing the product liability insurance, the policy number, and a list of the claims made against the firm.

Some firms may have all the complaints, regardless of which plant or product involved, sent to the home office or to the legal department. If this is the case, find out how the plant that actually made the product is advised and what followup is made to correct deficiencies.

542.8 Distribution and Promotion

542.81 Distribution

Complete distribution records should be maintained per 21 CFR 211.196.

Report the general distribution pattern of the firm to identify suspicious shipments of products subject to abuse or which have been targeted for high priority investigation by the Agency. These include steroids, counterfeits, diverted drugs (i.e.; physicians samples, clinical packs, etc.). Information concerning suspect shipments to legitimate or suspicious consignees should be forwarded to the home district for review and follow-up investigation. If no suspect products or consignees are found, obtain a list of the firm's larger consignees. Refer to IOM 537.11, when reporting specific shipments of suspect products.

Determine how and when firm checks on the authenticity of orders received. What references, if any, are used, e.g. current editions of the AMA Directory, Hays Directory, etc. Ask firm to demonstrate this procedure on selected large shipments or unusual consignees.

542.82 Promotion and Advertising

21 CFR 202.1 which pertains only to prescription drugs, covers advertisements in published journals, magazines, other periodicals, and newspapers, and advertisements broadcast through media such as radio, television, and telephone communication systems.

Determine what department or individual is responsible for promotion and advertising and how this responsibility is demonstrated. Ascertain what media (radio, television, newspapers, trade journals, etc.) are utilized to promote products.

Do not routinely collect examples of current advertising. Advertising should be collected only on assignment, or if, in your opinion, it is clearly in violation of Section 502(n) of the FD&C Act or 21 CFR 202.1.

542.83 Guarantees and Labeling Agreements

Determine the firm's policies relative to receiving guarantees for raw materials, and issuing guarantees on its own products. Also determine firm's practices relative to shipment of unlabeled drugs under labeling agreements. See IOM 526.

542.9 New Drugs, Antibiotics, Investigational Drugs

542.91 New Drugs

Check the current programs in your CPGM, Section 505 of the FD&C Act and 21 CFR 314.1 for required information. You may take the District's copy of the NDA into the plant as a reference during the inspection.

Document and report all deviations from representations in the NDA even though they may appear to be minor.

542.92 Antibiotics

Provide the same inspectional coverage as for other drugs. Refer to the approved antibiotic application to facilitate your evaluation of the firm's operation. In addition, refer to IOM 660.2 for inspectional guidance on antibiotic fermentation operations, antibiotic dosage form production and sterile bulk antibiotic operations.

542.93 Investigational Drugs

Follow the instructions in pertinent programs in your CPGM or as indicated in the specific assignment received.

542.94 Clinical Investigators/Clinical Pharmacologists

Inspections in this area will be on specific assignment which has been previously cleared by the Administration. Follow guidance in the CPGM or assignment.

543 INDEPENDENT COMMERCIAL DRUG CONTROL LABORATORIES

543.1 Objective

Determination of the capabilities of commercial drug control laboratories with respect to performance and qualifications of personnel, laboratory facilities, compliance with record keeping requirements, and reliability of reports rendered to customers. Performance and capabilities must be in conformance with current good manufacturing practice.

543.2 Equipment and Facilities

Check laboratory equipment and facilities to determine if they are consistent with the volume and type of service rendered, as indicated by the number of reports supplied to customers. Determine:

a. If necessary laboratory equipment is available, used, and properly maintained.
b. If the supply of animals for pyrogen tests and other biological tests is adequate and properly maintained.
c. Whether qualifications of key technical and professional personnel are consistent with types of services rendered.

543.3 Procedures

543.31 Methods

Determine if the analytical methods are properly selected and followed.

543.32 Record Keeping

Determine:

a. If analyst work data are retained and available, including necessary data on instrument calibration, checking of standards, and corrections necessary for blank determinations or other controls.
b. If there is documentary evidence that calculations are checked.

543.33 Pencilled Reports

Be alert for pencilled reports. Check selected work sheets for accuracy and, if practical, check reports against information available from laboratory reports.

Check for such unexplained inconsistences as:

a. Comparison of quantity of reserve sample with that received suggesting that insufficient material was used in conducting the analysis.
b. Reports of analysis too uniform or reported to accuracy beyond the capabilities of the method.
c. Analytical reports reflecting procedures beyond the capability of the laboratory equipment or facilities.
d. Total workload beyond the capacity of available personnel or facilities.

GUIDE TO INSPECTION OF BULK PHARMACEUTICAL CHEMICALS

Revised September 1991

Division of Field Investigations (HFC-130)
Office of Regional Operations
Office of Regulatory Affairs
AND
Division of Manufacturing and Product Quality (HFD-320)
Office of Compliance
Center for Drug Evaluation and Research
REFERENCE MATERIALS AND TRAINING AIDS FOR INVESTIGATORS

ACKNOWLEDGEMENTS

The original document was published in April 1984, and it was revised to include changes submitted by the Pharmaceutical Manufacturers Association (PMA) in February 1987. The current document incorporates revisions from a number of sources including the Field Drug Committee, ORA National Experts, field investigators, CDER personnel and the PMA. It also includes appropriate material contained in the March 1991 draft Mid-Atlantic Guide, which addresses the same topic.

The highlights of the current revision are: In Part I, two new sections have been added -- "Inspectional Approach" and "Relationship to Dosage Forms/Dosage Form Approval." In Part II, discussions have been added on: 1) aseptic/sterile processing under "Buildings and Facilities, 2) cleaning of product contact surfaces under "Equipment", and 3) validation of manufacturing processes under "Production and Process Controls". In addition, the Appendix has been added which includes references and a discussion of impurities. Also, there have been numerous other revisions throughout the document to increase emphasis on some information in existing sections or to streamline other sections.

We wish to express our appreciation to all who have contributed their time in the preparation of this revision, with special thanks to Ms. Jeanne Hutchinson (HFD-323) for her expert clerical assistance.

PART I -- GENERAL GUIDANCE

Introduction

This document is intended to aid agency personnel in determining whether the methods used in, and the facilities and manufacturing controls used for, the production of Bulk Pharmaceutical Chemicals (BPCs) are adequate to assure that they have the quality and purity which they purport or are represented to possess.

There are basic differences between the processes used for the production of BPCs and the processes used for the production of finished products. BPCs usually are made by chemical synthesis, by recombinant DNA technology, fermentation, enzymatic reactions, recovery from natural materials, or combinations of these processes. On the other hand, finished drug products are usually the result of a formulation from bulk materials whose quality can be measured against fixed specifications.

In almost every case in the production of BPCs, the starting materials, or derivatives of the starting materials, undergo some significant chemical change. Impurities, contaminants, carriers, vehicles, inerts, diluents, and/or unwanted crystalline or molecular forms which may be present in the raw materials are largely removed by various treatments in the production process. Purification is the ultimate objective and is effected by various chemical, physical, and/or biological processing steps. The effectiveness of these steps is in turn confirmed by various chemical, biological, and physical tests of the BPC.

In contrast, in finished drug product production, the quality of the drug ingredients (the components), and the care exercised in handling them, somewhat predetermines the purity of the finished drug product. Purification steps usually are not involved.

The use of precision automatic, mechanical, or electronic control and recording equipment and of automatic processing equipment is even more likely to be found in a BPC plant than in a finished drug product plant. Use of such equipment is appropriate when adequate inspection, calibration, and maintenance procedures are utilized.

Production equipment and operations will vary widely depending on the type of BPC in production, the scale of production, and the type of operation (batch vs. continuous). In general, the environmental conditions, equipment, and operational techniques employed are those associated with the chemical industry rather than the finished drug product industry. Chemical processes frequently are performed in closed systems, which tends to provide protection against contamination, even when the reaction vessels are not enclosed in buildings. However, this does not preclude the introduction of contaminants from equipment, materials used to protect equipment, corrosion, cleaning, and personnel.

In evaluating the adequacy of measures taken to preclude contamination of, or by, materials in the process, it is appropriate to consider the type of system (open or closed), form of the material (wet or dry), stage of processing and use of the equipment and/or area (multi-purpose or dedicated). "Closed" systems in chemical plants are often not closed when they are being charged and/or when the final product is being emptied. Also, the same reaction vessels are frequently used for different reactants.

Other factors that an investigator must consider in evaluating a BPC plant are:

(a) Degree of exposure of the material to adverse environmental conditions;
(b) Potential for cross-contamination from any source;
(c) Relative ease and thoroughness of clean-up;
(d) Sterile vs. non-sterile operations.

In the production of BPCs, the recycling of process liquors and recovery from waste streams which have been tested and meet appropriate standards often are necessary for quality, economic, and environmental reasons. In addition, the production of some BPCs involves processes in which chemical and biochemical mechanisms have not been fully understood and scientifically documented. Therefore, the methods and procedures for materials accountability will often differ from those applicable to the manufacture of dosage form drug products.

The producer of BPCs must recognize the need for appropriate evaluation, using appropriate standards and/or test procedures, of raw materials before their introduction into the process. In addition, as chemical processing proceeds, a chain of documentation should be established which at the minimum includes a written process and appropriate production records, records of raw materials used, records of initial and subsequent batch numbers, records of the critical processing steps accomplished, and intermediate test results with meaningful standards. It should be recognized that all intermediates need not be tested. A firm should, however, be able to identify critical or key points in the process where sampling and testing selective intermediates is necessary in order to monitor the performance of the process. As the end of the process is approached, the completeness of the records should increase, and the latter finishing steps should be thoroughly documented and conducted under appropriate conditions to avoid contamination and mixups.

Status of Bulk Pharmaceutical Chemicals

BPCs are components of drug products. The manufacture of BPCs should be carried out in accordance with concepts of good manufacturing practice (GMP) consistent with this guide whether or not the manufacturers are required to register under 21 CFR 207. The manufacturers of inactive ingredients may not be required to register with FDA, but they are not exempt from complying with GMP concepts, and they are not exempt from inspection. whether or not this type of firm will be inspected on a surveillance basis is generally discretionary. However, such a firm is always subject to "for cause" inspection.

The question of when an industrial chemical becomes a BPC can be complex, and there is no satisfactory answer. However, criteria such as the following can be used to identify a chemical as a BPC:

(a) When there is no recognized non-drug commercial use for the chemical.

(b) When it reaches the point in its isolation and purification where it is intended that the substances will be used in a drug product.

(c) When the manufacturer sells the product or offers it for sale to a pharmaceutical firm for use in a drug product.

Many elements and simple compounds that will ultimately comprise the molecule of BPC originate from botanicals, mines, oil wells, and sea water. It would be unrealistic to expect drug product GMP concepts to apply to the production of these progenitors. As a general rule, however, it is reasonable to expect GMP concepts to start to become applicable at that point where a starting material enters a biological or chemical synthesis or series of processing steps, where it is known that the end product will be a BPC.

Scope

This guide is applicable to all BPCs produced in the United States. It is also applicable to BPCs produced in foreign countries intended to be exported to the United States or to be delivered to a U.S. overseas base. This guide applies to: a) human drugs; b) veterinary drugs; and c) biologics.

The guide applies when the BPC is: a) a drug of animal, botanical, synthetic or biological origin, including those produced with rDNA technology; b) an inactive ingredient (although inspections will only be conducted by special assignment, or for cause); c) a component not appearing in the finished drug product; and d) a bulk intended for use in placebos.

Excluded from consideration are medical gases and bulk-packaged drug products (final dosage forms), which are subject to other requirements and full CGMPs.

General Guidance - Bulk GMPs

Although the GMP regulations under 21 CFR, Parts 210 and 211, apply only to finished dosage form drugs, Section 501(a) (2) (B) of the Federal Food, Drug, and Cosmetic Act requires that all drugs be manufactured, processed, packed, and held in accordance with current good manufacturing practice (CGMP). No distinction is made between BPCs and finished pharmaceuticals, and failure of either to comply with CGMP constitutes a failure to comply with the requirements of the Act. There are many cases where GMPs for dosage form drugs and BPCs are parallel. For this reason, the requirements under Part 211 will be used as guidelines for inspection of BPC manufacturers, as interpreted in this document. This document does not supersede the GMP regulations, rather it provides general guidance to inspectional personnel as to the extent and point of application of some of the concepts of Parts 210 and 211 to BPC production.

Although strict observance of GMPs, approaching or equaling those expected for finished drug products, may be expected in some types of bulk processes, in most others it is neither feasible nor required to apply rigid controls during the early processing steps. In all processes of this type, however, the requirements should be increasingly tightened according to some reasonable rationale. At some logical processing step, usually well before the final finishing operation, appropriate GMPs should be imposed and maintained throughout the remainder of the process.

Good judgement and a thorough knowledge of the process are required to permit sound evaluation of the processing step at which imposition of GMPs should take place. A detailed process flow diagram should be available for the processes used. This diagram should identify the unit operations, equipment used, stages at which various substances are added, key steps in the process, critical parameters (time, temperature, pressure, etc.) and monitoring points.

As briefly discussed in the introduction, the documentation system required for the early steps in the process must provide a chain of documentation but need not necessarily be as comprehensive as in the later parts of the process. Complete documentation should, at a minimum, be initiated where:

(a) The bulk pharmaceutical chemical can be identified and quantified for those processes where the molecule is produced during the course of the process (e.g., fermentation, synthesis, or recombinant DNA technology). In this regard, a theoretical yield should be established with appropriate limits, and there should be an investigation if the actual yield falls outside the limits.
(b) A contaminant, impurity, or other substance likely to adversely affect the purity, potency, or form of the molecule, is first identified and subsequent attempts are made to remove it (e.g., removal of crystalline occlusion, etc.).
(c) An attempt is initiated to separate a mixture of different forms of the same molecule and isolate a desired form of the molecule for pharmacological or other reasons (e.g., separation of racemic mixtures).

The complete documentation should be continued throughout the remainder of the process, including the application of full GMP concepts, for all significant processing steps until the BPC is packaged into a bulk container, or is transported without containerization to a location for subsequent manufacture into drug products.

Significant processing steps can involve a number of unit operations or unit processes. unit operations include those processing steps wherein the material is treated by physical means and/or the transfer and change of energy, but no chemical change of the molecule occurs; unit processes include those processing steps wherein the molecule undergoes a chemical change.

Significant processing steps can include: a) phase changes involving either the desired molecule or the solvent, inert carrier or vehicle, e.g., dissolution, crystallization, evaporation, sublimation, distillation or sorption; b) a phase separation such as filtration or centrifugation; c) any chemical change involving the desired molecule, e.g., removal or addition of water of hydration, acetylization, formation of the salt; d) an adjustment of the solution containing the molecule such as adjustment of pH or pO_2; e) a precision measurement of contained or added BPC components, in-process solutions, recycled materials is performed, i.e., weighing, volumetric measuring, optical rotation, spectrophotometric determinations, etc; and f) changes occur in surface area, particle size, or lot uniformity, e.g., milling, agglomeration, blending.

In order to promote uniformity in inspectional GMP coverage for a BPC, the following minimal criteria should be applied:

The lot of BPC to be released and/or certified is the essential element. A unique lot number should be assigned to this quantity of material. The firm should be prepared to demonstrate that this lot:

(a) Has been prepared under GMP conditions from the processing point as described above.
(b) Has a batch record (as described later in this document).
(c) Is homogenous.
(d) Is not intermingled with material from other lots for the purpose of hiding or diluting an adulterated substance while completing the processing through packaging.
(e) Has been sampled in accordance with a sampling plan which assures that the sample truly represents the lot.

(f) Has been analyzed using scientifically sound tests and methods designed to assure that the product meets established standards and specifications for quality, identity, and purity.
(g) Has stability data to support the intended period of use.

Inspectional Approach

The inspectional approach for coverage of a BPC operation should be the same whether or not that BPC is referenced as active ingredient in a pending application. The purpose, operational limitations and validation of the critical processing steps of a production process should be examined to determine that the firm adequately controls such steps to assure that the process works consistently. Overall, the inspection must determine the BPC manufacturer's capability to deliver a product that consistently meets the specifications of the bulk drug substance that the finished dosage form manufacturer listed in the application and/or the product needed for research purposes.

BPC manufacturing plants often produce laboratory scale or "pilot" batches. Scale-up to commercial full-scale (routine) production may involve several stages and data should be reviewed to demonstrate the adequacy of the scale-up process. Such scale-ups to commercial size production may produce significant problems in consistency among batches. Pilot batches serve as the basis for establishing in-process and finished product purity specifications.

Typically, manufacturers will generate reports that discuss the development and limitation of the manufacturing process. Summaries of such reports should be reviewed to determine if the plant is capable of producing adequately the bulk substance. The reports serve as the basis for the validation of the manufacturing and control process and the basic documentation that the process works consistently.

Drug Master Files (DMFs) are a valuable source of detailed information regarding the process and controls for BPCs. Although DMFs are not mandatory, most firms, particularly foreign manufacturers, have submitted them to FDA. A review of a process flow chart is helpful in understanding the various processing stages. Then, in conjunction with the review of the processing records, the critical stages should be identified, typically those where in-process samples are collected. The information expected from in-process testing should be determined along with the action to be taken by the firm should these specification limits be exceeded. For example, an in-process test result may show the presence of some unreacted material which may indicate that the process time should be extended.

A good starting point for the BPC inspection is a review of product failures evidenced by the rejection of a batch that did not meet specifications, return of a product by a customer, or recall of the product. The cause of the failure should have been determined by the manufacturer, a report of the investigation prepared, and subsequent corrective action initiated and documented. Such records and documents should be reviewed to ensure that such product failures are not the result of a process that has been poorly developed or one that does not perform consistently.

Complaint files should also be reviewed since customers may report some aspects of product attributes that are not entirely suitable for their use. These may be caused by impurities or inconsistencies in the BPC manufacturing process. Also, storage areas in the warehouse may hold rejected product. In addition, a review of change control logs, material review board documents, and master formula and batch production records showing frequent revisions may reveal problems in the BPC production process.

In the analytical laboratory, specifications for the presence of unreacted intermediates and solvent residues in the finished BPC should be reviewed. These ranges should be at or near irreducible levels.

An inspectional team consisting of investigators and engineers, laboratory analysts or computer experts should participate in the inspection, as appropriate, when resources permit.

Registration

Domestic manufacturers of bulk pharmaceutical chemicals are required to register (and list their products) in accordance with section 510 of the Act if they meet the definition of a "bulk drug substance" under 21 CFR 207.3 (a) (4), i.e., a substance that is represented as a drug and, when used, becomes an active ingredient or finished dosage form of such drug. Specifically excluded from registration are manufacturers of intermediates (21 CFR 207.3(a)(4)) and inactive ingredients which are excipients, colorings, etc. (21 CFR 207.10(e)).

Products of Foreign Origin

The results of inspections of foreign manufacturers of BPCs directly affect the status of these products when offered for entry into this country. BPC's may be sampled, detained, and/or refused entry into the United States if an inspection of the foreign manufacturer reveals that the firm is not complying with GMPs. This would also be the case if the products demonstrate actual adulteration or misbranding.

Although foreign firms are not required to register in accordance with section 510 of the Act, they are required to list all of their products (21 CFR 207.40(a)). Products not listed are subject to detention and/or refusal of entry.

Relationship to Dosage Forms and Dosage Form Approval

The finished product formulator is highly dependent on the BPC manufacturer to provide bulk substances uniform in chemical and physical characteristics. This is particularly important in the context of the product approval process where bioequivalency comparisons are made between clinical production or biobatches and commercial batches. The BPC used to manufacture commercial batches must not significantly differ from that used on these test batches to provide adequate assurance of product performance. Where significant differences occur, additional testing by the dose form manufacturer to establish the equivalence of the finished product may be required. This remains equally important post-approval for subsequent commercial batches to assure that marketed products are not adversely affected over time.

Manufacturers holding DMFs covering production of BPCs (21 CFR 314-420) must update such DMFs with any changes. The DMF holders must also notify each dose form manufacturer referencing the DMF of any such changes to the DMF.

In general, BPCs are used as purchased, with no further refining or purification taking place. Consequently, impurities present in the BPC will be present in the finished dosage form.

While dosage form manufacturers may have limited control over BPC quality (obtaining certificates of analysis and testing representative samples), the BPC manufacturer has ultimate control over physical characteristics, quality, and the presence of trace-level impurities in the BPC.

Many bulk substances are used in different types of dosage forms including oral, topical and parenteral products where physical characteristics, particularly particle size, may be important. While it is primarily the dosage form manufacturer's responsibility to identify the particular physical characteristics needed, it is the responsibility of the BPC manufacturer to adequately control processes to consistently provide BPCs complying with physical specifications.

The end use of the BPC should be identified and kept in mind during inspections of BPC manufacturers. A particularly important distinction involves whether or not the BPC will be used in the preparation of a sterile dosage form and whether or not it is represented as pyrogen free. The BPC manufacturer is responsible for ensuring that BPCs are pyrogen free if they make such a representation in specifications, labeling, or applications, including DMFs. In addition, any manipulation of sterile BPCs post-sterilization must be performed as a validated aseptic process. This is particularly important for those BPCs which are not further sterilized prior to packaging into final containers (e.g., bulk antibiotic powders).

In some instances, the USP monograph may specify that the BPCs not meeting parenteral grade standards be labeled as not suitable for use in the preparation of injectable products.

PART II -- SPECIFIC INTERPRETATIONS FOR BPC OPERATIONS

General

The following sections will discuss those specific points of the CGMPs which are clearly different in a BPC operation in contrast to a finished product operation. Points not separately discussed here should be viewed as appropriate to BPC manufacturing operations using finished product GMPs for guidance.

Buildings and Facilities

(a) Contamination/Cross Contamination

Cross contamination is not permitted under any circumstances. However, the fact that a BPC plant is, or can be, used for manufacturing multiple drugs, even simultaneously, is not in itself objectionable with only a few exceptions. There must be separate facilities and completely separate air handling systems for the production of penicillin as the CGMP regulations require for dosage form drug products. It is also encouraged that separate facilities and air handling systems be used for the production of certain steroids, alkaloids, cephalosporins, certain hazardous or toxic drugs, pesticides, chemicals, and/or starting materials.

NOTE: Containment via closed system is considered a separate facility. The intent is to require isolation of penicillin production operations from operations for non-penicillin products. Separation can be achieved in a facility, building, or plant by effectively isolating and sealing off from one another these two types of operations. Isolation of facilities does not necessarily mean separation by geographical distance or the placement of these operations in separate buildings. Effective means can almost certainly be developed to separate activities from one another to prevent cross-contamination problems within a single building. Containment in a fermentor would meet this criterion and they are applicable to both dry and liquid state penicillin production.

Even though penicillin production may take place in the same building as non-penicillin production, air handling systems must at all times be completely separate. This includes fermentation procedures. This is the only means by which cross-contamination can be prevented through air facilities.

The point at which the final BPC product is initially recovered (usually as a moist cake from a centrifuge or filter press) should be in a clean environment and not exposed to airborne contaminants such as dust from other drugs or industrial chemicals. Typically, the damp product will be unloaded into clean, covered containers and transported elsewhere for drying and other manipulations. These subsequent operations should be performed in separate areas because, once dry, the BPC is more likely to contaminate its environment; this in turn makes it likely that other products in the same area might become contaminated. The primary consideration is that the building and facilities should not contribute to an actual or potential contamination of the BPC.

Air handling systems for BPC plants should be designed to prevent cross-contamination. For economic reasons, it is a common practice to recycle a portion of the exhaust air back into the same area. For dedicated areas processing the same BPC, this is not objectionable. The adequacy of such a system of operation for multi-use areas, especially if several products are processed simultaneously, should be carefully analyzed. In multi-use areas where several products are completely confined in closed vessels and piping systems, the extent of filtration of the supply air (combined fresh make-up air and recycled air) is not a problem (although other regulatory agencies or company policy may impose restrictions) except when the closed system must be opened (charging). In those areas where the BPCs are in a damp or moistened form (such as filter or centrifuge cake) and may be exposed to the room air environment, filter efficiencies on the supply air system as low as 85% may be perfectly adequate. In those areas wherein one or more of the products is being processed in a dry form, even total filtration of the entire supply air flow with HEPA filters may not be adequate. In all cases, the firm should be able to demonstrate adequacy of their air handling system with data and (in case of doubt) the investigator should consider collection of product samples for analysis for cross-contamination.

Process wastes and unusable residues should be removed and disposed of in a manner that will insure that they do not interfere with subsequent steps of the process or adulterate the product.

Adequate sanitation of buildings and areas for BPCs requires considerable judgement. Many starting materials, particularly botanicals, may have some unavoidable contamination with rodent or other animal filth or be infested with insects. In such cases, it is not realistic to expect high standards in storage areas for starting materials and perhaps in the limited area of the plant wherein the initial steps of processing are conducted.

The control methods utilized by the firm to prevent an increase of such contamination or infestation in holding areas, or its spread to other areas of the plant, are of primary importance.

(b) Water Systems/Water Quality

Water used in the production of BPCs in many instances (e.g., fermentation of antibiotics) may be potable water obtained from wells or surface sources. This is acceptable provided that water quality standards are established that are consistent with compendial or other regulatory requirements for source drinking water. Although it is not expected that potable water be routinely tested as a component, sufficient data from periodic testing should be available to show compliance with standards from both chemical and microbiological standpoints, including freedom from pathogenic organisms. Where adequate data are available from municipal water authorities, it need not be generated by the manufacturer.

Purified water is widely used in the manufacture of BPCs. Because of the well recognized potential for microbial growth in deionizers and ultrafiltration (UF) or reverse osmosis (RO) systems used to produce purified water, such systems must be properly validated and controlled. Proper control methods include the establishment of water quality specifications and corresponding action levels, remedial action when microbial levels are exceeded, and adequate maintenance procedures such as regeneration and sanitation/sterilization. Appropriate specifications for chemical and microbial quality should be established and periodic testing conducted. Such specifications will vary depending on the process and the point in the process where the water is used. For example, if the water is used in later processing steps such as for a final wash of the filter cake, or if the BPC is crystallized from an aqueous system, the water quality standards should be higher than normally specified for purified water. This is particularly important where the BPC is intended for use in parenteral dosage forms. The frequency of microbial and chemical testing of purified water is dependent upon a variety of factors including the test results and the point in the process (e.g., final wash in centrifuge) at which such water is used.

The USP includes suggested microbial action guidelines for source drinking water and purified water in the General Chapter on Water for Pharmaceutical Purposes and includes standards for specific types of water in monographs (e.g. Purified Water, USP). If the firm specifies a water of compendial quality in an application, the water should meet the standards given in the compendium.

Similar principles to those discussed above for purified water apply to Water For Injection (WFI) utilized in sterile and pyrogen-free BPC processing. The WFI system must be monitored for microorganisms and the validation data and reports of monitoring should be reviewed as is required for the production of finished dosage forms.

Most purified and WFI water systems, including RO and UF systems, have the potential for the development of endotoxins. If the final BPC is purported to be pyrogen free or sterile, or will be used in preparing parenteral products, routine testing of the process water for endotoxins (preferably by the LAL method) is indicated. However, end point testing alone is not adequate and validation of the system to control endotoxin development should be conducted.

(c) Aseptic/Sterile Processing

One of the more difficult processes is the manufacture of a sterile BPC. The aseptic crystallization and subsequent processing (drying, milling, and blending) present unique challenges. Since the operators are the primary source of contamination in an aseptic operation, processes are being designed to eliminate direct operator contact. However, some aseptic bulk operations still utilize considerable operator involvement which requires adequate controls. Major potential problem areas include aseptic removal of the BPC from the centrifuge, manual transfer to drying trays and mills, and the inability to sterilize the dryer.

Unfortunately, not all equipment currently in use can be sterilized. The BPC manufacturer must have data to document the sanitizing of critical processing equipment such as centrifuges and dryers.

Sterilization by use of ethylene oxide is sometimes attempted for powders. In this operation, the powders are spread in a thin layer and exposed to the gas. Typically, however, ethylene oxide does not penetrate the BPC in this powdered form. The manufacturer should validate that the ethylene oxide exposure does, in fact, produce a sterile product.

The Sterile Drug Process Inspections Compliance Program (CP 7356.002A) provides detailed inspectional guidance for coverage of the manufacture of sterile BPCs. Also, the Aseptic Processing Guidelines, although intended for coverage of dosage forms, includes principles that are also applicable to aseptic processing of sterile bulks. Both documents should be reviewed in association with any inspections of the manufacture of sterile BPCs.

Equipment

(a) Multipurpose Equipment

As is the case with buildings, many BPCs are produced using multipurpose equipment. Fermentation tanks, reactors, centrifuges, and other pieces of equipment are readily used or adapted for a variety of products. With few exceptions, such multiple usage is satisfactory provided that the equipment is cleanable and is in fact cleaned according to written procedures. The cleaning program should take into consideration the need for different procedures depending on what product or intermediate was produced. Equipment that contains tarry or gummy residues that cannot be removed readily should be dedicated for use only with limited portions of a synthesis.

Where temperature control is important, temperature recording devices should be utilized, with recording charts retained as part of the batch record. For example, reactors may require narrow temperature ranges for consistent operation, and when recorders are absent, the manufacturer should justify their absence.

(b) Equipment Cleaning and Use Log

Where multipurpose equipment is in use, it is important to be able to determine previous usage as an aid in investigating cross-contamination or the possibility thereof.

An equipment cleaning and use log, while desirable and even preferable, is not the only method of determining prior use. Generally speaking, any documentation system that clearly identifies the previous batch and shows that the equipment was in fact cleaned is acceptable.

(c) Equipment Located Outdoors

Some fermentation tanks, reaction vessels, and other equipment are not situated within buildings; thus a considerable amount of processing occurs out-of-doors. Such processing is unobjectionable provided that it occurs in a closed system.

(d) Protected Environment

Isolation of intermediates or products may require the use of a protected environment to avoid microbial contamination or degradation caused by exposure to air or light. The degree of protection required may vary depending on the stage of the process. Equipment should be designed to minimize the possibility of contamination when used by the operator. Often, direct contact is involved in the unloading of centrifuge bags, transfer hoses (particularly those used to transfer powders), drying equipment and pumps.

Also, the sanitary design of transfer equipment such as pumps should be evaluated. Those with moving parts should be assessed in regard to the integrity of seals and other packing materials to avoid product contamination.

Processes requiring special environments to assure product quality (inert atmosphere, protection from light, etc.) should be carefully scrutinized for any lapses in the special environment. If any such lapses are found in the production process, adequate evidence and appropriate rationales must be shown that such lapses have not compromised the quality of the BPC. Such environmental concerns become more important after the purification of the BPC has been completed. The area where the BPC may be exposed, and especially those used to manufacture parenteral substances, should have environmental quality similar to that used for the manufacture of dosage forms. For example, controlled areas may need to be established along with appropriate air quality classifications. Such areas should be serviced by suitable air handling systems and there should be adequate environmental monitoring programs.

Any manipulation of sterile BPCs post-sterilization must be performed as an aseptic process, including the utilization of Class 100 air and other aseptic controls.

(e) Cleaning of Product Contact Surfaces

Cleaning of multiple use equipment is an area where validation must be carried out. The manufacturer should have determined the degree of effectiveness of the cleaning procedure for each BPC or intermediate used in that particular piece of equipment.

Validation data should verify that the cleaning process will remove residues to an acceptable level. However, it may not be possible to remove absolutely every trace of material, even with a reasonable number of cleaning cycles.

Specific inspectional coverage for cleaning should include:

1. Detailed Cleaning Procedure:

There should be a written equipment cleaning procedure that provides details of what should be done and materials to be utilized. Some manufacturers list the specific solvent for each BPC and intermediate.

For stationary vessels, often clean-in-place (CIP) apparatus may be encountered. For evaluation of these systems, diagrams will be necessary, along with identification of specific valves.

2. Sampling Plan:

After cleaning, there should be some periodic testing to assure that the surface has been cleaned to the validated level. One common method is the analysis of the final rinse water or solvent for the presence of the substance last used in that piece of equipment. There should always be a specific analytical determination for such a residual substance.

3. Analytical Method/Cleaning Limits:

Part of the answer to the question, "how clean is clean?", is, "how good is your analytical system?" The sensitivity of modern analytical apparatus has lowered some detection thresholds past parts per million, down to parts per billion.

The residue limits established for each piece of apparatus should be practical, achievable, and verifiable. When reviewing these limits, ascertain the rationale for establishment at that level. The manufacturer should be able to document, by means of data, that the residual level permitted is scientifically sound.

Another factor to consider is the possible non-uniformity of the residue. If residue is found, it may not necessarily be at the maximum detectable level due to the random sampling, such as taking a swab from a limited area on that piece of equipment.

Raw Materials

(a) Raw materials, especially those received in large quantities (hundreds of bags or in bulk), should not be physically moved from a quarantine area to a released area prior to quality control acceptance. However, such raw materials may remain in the quarantine area after release. The important consideration is that an unreleased material should not be used prior to quality control acceptance. Effective quarantine can be established with suitable identifying labels or signs, sound and valid documentation systems, etc. With increasing frequency, it is noted that such quarantine and documentation is widely being accomplished internally with a computer system in lieu of a physical stock control system. This is acceptable provided that system controls are adequate to prevent use of unreleased material.

(b) Film-wrapped pelletized bags may not be individually identified by information normally applied to every container in a lot. To insist otherwise would destroy many of the advantages of film wrapped pallets. This is acceptable provided the pallet load itself is adequately identified. If issued individually, bags should be identified with the necessary information at the time of issuance.

(c) Some raw materials are stored in silos or other large containers, making precise separation of lots difficult. Considering that such materials are usually nutrients or are inactive, such storage is acceptable. It should be possible, via inventory or other records, to show usage of such materials with reasonable accuracy.

(d) Solvents used in BPC production are frequently stored in large tanks. Often, fresh and recovered solvents are commingled so that precise lot identity is missing. This is satisfactory provided incoming solvents are identified and tested prior to being mixed with recovered solvents and if the latter are tested for contaminates from the process in which they were used previously. The quality of the solvent mixture must also be monitored at suitable intervals.

(e) Some raw materials are stored out-of-doors; e.g., acids, other corrosive substances, explosive materials, etc. Such storage conditions are satisfactory provided the containers give suitable protection to their contents, identifying labels remain legible, and containers are adequately cleaned prior to opening and use.

(f) Some raw materials may not be acceptance tested by the firm because of the hazards involved; e.g., phosphorus pentachloride, sodium azide, etc. This is acceptable where there is a reason based on safety or other valid considerations. In such a circumstance, assay certification from the vendor should be on file. There should always be some evidence of an attempt by the BPC manufacturer to establish identity even if it is only a visual examination of containers, examination of labels, and recording of lot numbers from the labels.

Containers, Closures, and Packaging Components

A system for BPC containers, closures, and packaging components should include the following features at a minimum:

(a) Suitable written specifications, examination or testing methods, and cleaning procedures where so indicated.

(b) Determination that the container-closure system is not reactive, additive, or absorptive so as to alter the quality of the BPC beyond its established specifications and that it provides adequate protection against deterioration and contamination.

(c) Storage and handling in a manner to protect containers and closures from contamination and deterioration and to avoid mixups (e.g., between containers that have different specifications but are similar in appearance).

(d) Use of bulk shipping containers in which bulk pharmaceutical components were received should be avoided for BPC storage or shipment unless a suitable polymer lining or inner bag is used.

Production and Process Controls

(a) Mother Liquors

Mother liquors containing recoverable amounts of BPCs are frequently re-used. Such re-use may consist of employing the mother liquor to dissolve the reactants in the next run of that step in the synthesis. Re-use may also consist of a separate reaction to obtain a "second crop" of final product. Finally, since crystallizations are sometimes slow, some second crops are obtained simply by allowing the second crystallization to continue for long periods after the first crop is removed. These secondary recovery procedures are acceptable providing the isolated BPC meets its original, or other suitable, specifications. The recovery procedures should be indicated in batch production records.

Similarly, mother liquors may contain unreacted starting materials or intermediates that are not recoverable. Secondary recovery procedures for these materials are acceptable provided that the materials meet suitable specifications.

(b) In Process Blending/Mixing

Deliberate in-process blending, or mixing, is that blending required in the process for a variety of reasons and is carried out with reasonable reproducibility from run to run during the process. Examples include: 1) Collection of multiple fermentation batches in a single holding tank (with a new batch number); 2) Recycling solution from one batch for further use in a succeeding batch; 3) Repeated crystallizations of the same mother liquor for better yield of crystals; and 4) Collecting several centrifuge loads in a single dryer/blender. Such in-process blending is acceptable provided it is adequately documented in batch production records.

Incidental carryover is another type of in-process mixing that occurs frequently. Examples include: 1) Residue adhering to the wall of a micronizer used for milling the finished BPC; 2) Residual layer of damp crystals remaining in a centrifuge bowl after discharge of the bulk of the crystals from a prior batch; and 3) Incomplete discharge of fluids or crystals from a processing vessel upon transfer of the material to the next step in the process. These practices are usually acceptable since we do not normally require complete cleanup between successive batches of the same drug during a production campaign. However, in the case of non-dedicated production units, complete cleaning procedures designed to prevent contamination that would alter the quality of the substance must be employed when changing from one BPC to another. The effectiveness of these cleaning procedures may require the use of analytical testing for the substances involved.

In contrast to in-process blending and incidental carryover discussed above, the process intent should be directed toward achieving homogeneity of the batch of finished BPC to the maximum extent feasible. Three areas in the processing of finished batches of BPCs should be examined carefully and critically. These are: 1) The final blending operation that will constitute the finished batch; 2) The point in the process at which the lot number is assigned; 3) The sampling procedure used to obtain the sample is intended to be representative of the batch.

Note: Blending of batches or lots that individually do not conform to specifications with other lots that do conform (to salvage adulterated material) is not acceptable practice.

(c) Validation of Process and Control Procedures

An important factor in the assurance of product quality includes the adequate design and control of the manufacturing process. Routine end product testing alone is not necessarily sufficient because of limited sensitivity of such testing to reveal all variations that may occur and affect the chemical, physical, and microbial characteristics of the product. Each step of the manufacturing process must be controlled to the extent necessary to assure that the product meets established specifications. The concept of process validation is a key element in assuring that these quality assurance goals are met.

Process validation is required in general and specific terms by the CGMP regulations for finished dosage forms (21 CFR Parts 210 and 211). More specific guidance on process validation is provided in guidelines (See References). Many of these concepts are applicable to BPCs to assure that such BPCs are manufacturered in accordance with CGMPs as required by the Act under Section 501(a)(2)(B).

BPC manufacturers are expected to adequately determine and document that significant manufacturing processes perform consistently. The type of BPC, the range of specifications and other factors determine the extent of the process development and documentation required. However, most bulk manufacturing processes and control procedures can be validated with less arduous procedures than would be required for finished dosage forms.

Many firms already possess the data necessary to prepare an evaluation of the process and demonstrate that it works consistently. For example, limitations of a reaction and/or purification steps are usually identified in the development phase. Impurities with acceptable levels and tests used to determine them are established at this phase. The report describing the process reactions and purifications, impurities, and key tests needed for process control provide the basis for validation. Thus, when the process is scaled up to production batch sizes, a comparison can be made with development batches. Scale-up and development reports, along with purity profiles would constitute such a validation report.

While validation can be applied to any process, greater emphasis should be placed on validation of the BPC production at the stage(s) in the synthesis and purification steps used for the bulk substance and/or the removal of impurities.

(d) Reprocessing

Where reprocessing occurs during the synthesis of a BPC, there should be written documentation covering the reason for the failure, the procedures involved in the reprocessing, and changes made to eliminate a recurrence of the problem. Merely relying on final testing of the reprocessed BPC as a means of demonstrating compliance with specifications, and neglecting the investigation and evaluation of the manufacturing process, is unacceptable.

Equivalence of the quality of reworked material to the original material must also be evaluated and documented to insure that the reprocessed batches will conform with all established standards, specifications, and characteristics. Obviously, if the product failure results from a human error, it will not reflect on the process, but may reflect on other aspects such as adequacy of training. However, there should be sufficient investigation, evaluation, and documentation to show that reprocessed product is at least equivalent to other acceptable product and that the failure did not result from an inadequate process.

(e) Process Change

Manufacturers should have a formal process change system in place with standard operating procedures covering such changes. Management of the change system should be assigned to an independent quality unit having responsibility and authority for final approval of process changes.

(f) Impurities

Characterization and control of impurities in a BPC are important because of the adverse effects that such impurities may have on dosage form stability, safety and efficacy. Consequently, it is important that manufacturers identify and set appropriate limits for impurities and adequately control manufacturing processes so that the impurities consistently meet established specifications.

The attached Appendix A (Impurities) includes a more detailed discussion of impurities and should be reviewed prior to conducting inspections.

In-Process Testing

BPCs are normally subjected to various in-process tests to show that a synthesis or fermentation is proceeding satisfactorily. Such tests are often performed by production personnel in production laboratory facilities. Approval to continue with the synthesis (process) is often issued within the production department. The important considerations are that specified tests are performed, recorded, and results are within specified limits. In addition, instruments should be calibrated at appropriate intervals.

It is important that a firm utilize a quality control unit independent from production that has the responsibility and authority to reject in-process materials not meeting specifications. Such responsibility and authority should also extend beyond testing to include overall quality assurance activities such as procedure approvals, investigation of product failures, process change approvals, and product record reviews.

Packaging and Labeling of Finished BPC

(a) Sound procedures must be employed to protect the quality and purity of the BPC when it is packaged and to assure that the correct label is applied to containers. A good system of packaging and labeling should have the following features at a minimum:

 (1) A file of master labels. A responsible individual reviews incoming labels against the appropriate master labels.
 (2) Storage of labels in separate containers, or compartments, to prevent mixups in storage.
 (3) Formal issuance by requisition or other document.
 (4) Issuance of an exact number of labels sufficient for the number of containers to be labeled, retention copies, and calculated excesses, if any.
 (5) The employment of a lot number from which the complete batch history can be determined.
 (6) Avoidance of labeling more than one batch at a time without adequate separation and controls.
 (7) Reconciliation of the number of labels issued with the number of units packaged, together with the destruction of excess labels bearing lot numbers.

(b) If returnable BPC containers are re-used, all previous labeling should be removed or defaced. If the containers are repetitively used soley for the same BPC, all previous lot numbers, or the entire label, should be removed or completely obliterated.
(c) Labeling for containers of BPCs is subject to all applicable provisions of 21 CFR, Parts 200 and 201. In case questionable labeling is encountered, collect samples of the labeling for submission to the appropriate Center(s) for review.

Expiration Dating or Re-evaluation Dating

(a) With few exceptions, expiration dates are not presently considered to be a general requirement for all BPCs. Thus the absence of an expiration date may not be objectionable. The chief exception is antibiotic BPCs where expiration dates are required by the antibiotics regulations.

(b) Where expiration or re-evaluation dates are used on BPCs either because of a regulatory requirement or voluntarily, they must be derived from appropriate stability testing.

(c) Where stability testing reveals a limited shelf life, e.g., less than two years, the label should declare a supportable expiration date or indicate the need for re-evaluation testing at an appropriate interval to assure quality at time of use.

Laboratory Controls

(a) Raw materials are usually subjected to an identity test and additional testing to determine if they meet appropriate specifications. Such specifications will vary in depth, sophistication, and the amount of testing required to show conformance. This in turn will depend on various factors such as the critical nature of the raw material, its function in the process, the stage of the synthesis, etc. Raw material specifications should be written documents, even if only minimal requirements are required/requested. The specifications should be organized to separate those tests that are routine from those that are performed infrequently or for new suppliers.

(b) Laboratory controls should include a comprehensive set of meaningful analytical procedures designed to substantiate that each batch of finished BPC meets established specifications for quality, purity, identity, and assay. Data derived from manufacturing processes and from in-process controls also provide some assurance that a batch may be acceptable.

(c) Many BPCs are extracted from, or purified by, the use of organic solvents in the later (final) stages of recovery. The solvents are normally removed by drying the moist BPC. In view of the varying (and sometimes unknown) toxicity of solvents, it is important that BPC specifications include tests and limits for residues of solvents and other reactants. Refer to the attached Appendix A for further information about impurities, including volatile organic impurities.

(d) Appropriate analytical methods should be validated.

Stability Testing

Most BPC manufacturers conduct stability testing programs for their products; however, such programs may be less comprehensive than the programs now required for finished pharmaceuticals.

Undetected changes in raw materials specifications, or subtle changes in manufacturing procedures, may affect the stability of BPCs. This, together with the generally widespread existence of stability testing programs, make it reasonable to require such programs for BPCs.

(a) A stability testing program for BPCs should contain the following features:

(1) The program should be formalized in writing.

(2) Stability samples should be stored in containers that approximate the market container. For example, where the product is marketed in polylined drums, it is acceptable to keep stability samples in the same container material/closure system within mini-fiber drums. Such samples may be stored in glass or other suitable containers only if there are data developed by the firm or others to show that results are comparable.
(3) The program should include samples from the first three commercial size batches.
(4) Thereafter, a minimum of one batch a year, if there is one, should be entered in the program.

NOTE: Lower levels of sampling may be acceptable if previous stability studies have shown the BPC to be stable for extended periods and the normal period between production and ultimate use of the BPC is relatively short.

(5) The samples should be stored under conditions specified on the label for the marketed product.
(6) It is recommended that additional samples be stored under stressful conditions (e.g., elevated temperature, light, humidity or freezing) if such conditions can be reasonably anticipated.
(7) Stability indicating methods should be used.

(b) Conducting a stability testing program does not usually lead to a requirement to employ expiration dates. If testing does not indicate a reasonable shelf life, e.g., two years or more, under anticipated storage conditions, then the BPC can be labeled with an expiration date or should be re-evaluated at appropriate intervals. If the need for special storage conditions exists, e.g., protection from light, such restrictions should be placed on the labeling.

Reserve Samples

Reserve samples of the released BPCs should be retained for one year after distribution is complete or for one year after expiration or re-evaluation date.

Batch Production Records

Documentation of the BPC manufacturing process should include a written description of the process and production records similar to those required for dosage form production. However, it is likely that computer systems will be associated with BPC production. There is increasing use of such computer systems to initiate, monitor, adjust, and otherwise control both fermentations and syntheses. These operations may be accompanied by recording charts that show key parameters (e.g., temperature) at suitable intervals, or even continuously throughout the process. In other cases, key measurements (e.g., pH) may be displayed on a television screen for that moment in time but are not available in hard copy.

In both cases, conventional hard-copy batch production records may be missing. In other words, records showing addition of ingredients, actual performance of operations by identifiable individuals, and other information usually seen in conventional records may be missing. As a practical matter, when computers and other sophisticated equipment are employed, the emphasis must change from conventional, hand-written records to:

(a) Systems and procedures that show the equipment is in fact performing as intended;
(b) Checking and calibration of the equipment at appropriate intervals;
(c) Retention of suitable backup systems such as copies of the program, duplicate tapes, or microfilm;
(d) Assurance that changes in the program are made only by authorized personnel and that they are clearly documented.

APPENDIX A - IMPURITIES

The United States Pharmacopeia (USP) defines an impurity as any component of a drug substance (excluding water) that is not the chemical entity defined as the drug substance.

It has been demonstrated that impurities in a finished drug product can cause degradation and lead to stability problems. Further, some adverse reactions in patients have been traced to impurities in the active ingredient. Therefore, the presence or absence of impurities at the time of clinical trial and stability testing is a very important element of drug testing and development, and the appearance of an impurity in scaled up product that was not present during test stages presents serious questions about the stability of the product and its impact on safety and efficacy.

We expect the manufacturer to establish an appropriate impurity profile for each BPC based on adequate consideration of the process and test results. Because different manufacturers synthesize drug substances by different processes and, therefore, will probably have different impurities, the USP has developed the Ordinary Impurities Test in an effort to establish some specification. Also, in order to protect proprietary information, tests for specific impurities and even solvents are typically not listed in the compendia.

The USP also notes that the impurity profile of a drug substance is a description of the impurities present in a typical lot of drug substance produced by a given manufacturing process. Such impurities should not only be detected and quanitated [quantitated], but should also be identified and characterized when this is possible with reasonable effort. Individual limits should be established for all major impurities.

During the inspection, compare the impurity profile for the pilot batch material to that of the commercial size BPC batches to determine if the profile has significant changes. In some cases, drug manufacturers have submitted purity profiles in filings. Yet, when covered in some detail in an inspection, it became apparent that additional impurity data obtained by other methods (gradient HPLC) had become available but not yet filed. Thus, manufacturers should be asked specifically for current complete purity profiles, and these profiles should include the levels of solvents normally found in the purified drug substance along with acceptable specifications. Determine if the current impurity profile is reported to dose form manufacturers, especially if it has changed. Also, determine if the DMF (or AADA for bulk antibiotics) is current.

The USP provides extensive coverage of impurities in the following three sections:

(a) USP Section 1086 -- Impurities In Official Articles

This section defines five different types of impurities, both known and unknown including foreign substances, toxic impurities, concomitant components (such as isomers or racemates), signal impurities (which are process related), and ordinary impurities. The USP notes that when a specific test and limit is specified for a known impurity, generally a reference standard for that impurity is required.

Two of the impurities are singled out for in-depth coverage, ordinary impurities and organic or volatile impurities.

(b) USP Section 466 -- Ordinary Impurities

These are generally specified for each BPC in the individual monograph. The method of detection involves comparison with a USP reference standard, on a thin layer chromatographic (TLC) plate, with a review for spots other than the principal spot. The ordinary impurity total should not exceed 2% as a general limit.

Be sure to review the extensive USP coverage of 8 factors that should be considered in setting limits for impurity levels.

Related substances are defined as those structurally related to a drug substance such as a degradation product or impurities arising from a manufacturing process or during storage of the BPC.

Process contaminants are substances including reagents, inorganics (e.g., heavy metals, chloride, or sulfate), raw materials, and solvents. The USP notes that these substances may be introduced during manufacturing or handling procedures.

The third and most recent USP section regarding impurities is one that appears in the USP-NF XXII third supplement:

(c) USP Section 467 -- Organic Volative Impurities

Several gas chromatography (GC) methods are given for the detection of specific toxic solvents and the determination involves use of a standard solution of solvents. There are limits for specified organic volatile impurities present in the BPC unless otherwise noted in the individual monograph.

As the USP notes, the setting of limits on impurities in a BPC for use in an approved new drug may be much lower than those levels encountered when the substance was initially synthesized.

Further, additional purity data may be obtained by other methods such as gradient high performance liquid chromatography (HPLC). Be sure to ask for complete impurity profiles.

In preparation for a BPC inspection, these sections of the USP should be given a detailed review.

APPENDIX B - REFERENCES

1. CP 7356.002A -- Sterile Drug Process Inspections.
2. CP 7356.002F -- Bulk Pharmaceutical Chemicals (BPCs).
3. Guideline on General Principles of Process Validation, May, 1987.
4. Guideline for Submitting Supporting Documentation in Drug Applications for the Manufacturer of Drug Substances, Feb. 1987.
5. Guideline on Sterile Drug Products Produced by Aseptic Processing, June 1987.
6. Code of Federal Regulations, Title 21
   Part 210 and 211, Drugs: Current Good Manufacturing Practice
   314.420 -- Drug Master Files
   201.122 -- Drugs for Processing, Repacking, or Manufacturing (bulk labeling requirements)
7. United States Pharmacopeia, Current Revision, and Supplements.

GUIDE TO INSPECTIONS OF PHARMACEUTICAL QUALITY CONTROL LABORATORIES

July, 1993

Division of Field Investigations
Office of Regional Operations
Office of Regulatory Affairs
U.S. Food & Drug Administration

1. INTRODUCTION

The pharmaceutical quality control laboratory serves one of the most important functions in pharmaceutical production and control. A significant portion of the CGMP regulations (21 CFR 211) pertain to the quality control laboratory and product testing. Similar concepts apply to bulk drugs.

This inspection guide supplements other inspectional information contained in other agency inspectional guidance documents. For example, Compliance Program 7346.832 requiring pre-approval NDA/ANDA inspections contains general instructions to conduct product specific NDA/ANDA inspection audits to measure compliance with the applications and CGMP requirements. This includes pharmaceutical laboratories used for in-process and finished product testing.

2. OBJECTIVE

The specific objective will be spelled out prior to the inspection. The laboratory inspection may be limited to specific issues, or the inspection may encompass a comprehensive evaluation of the laboratory's compliance with CGMP's. As a minimum, each pharmaceutical quality control laboratory should receive a comprehensive GMP evaluation each two years as part of the statutory inspection obligation.

In general these inspections may include

-- the specific methodology which will be used to test a new product
-- a complete assessment of laboratory's conformance with GMP's
-- a specific aspect of laboratory operations

3. INSPECTION PREPARATION

FDA Inspection Guides are based on the team inspection approach and our inspection of a laboratory is consistent with this concept. As part of our effort to achieve uniformity and consistency in laboratory inspections, we expect that complex, highly technical and specialized testing equipment, procedures and data manipulations, as well as scientific laboratory operations will be evaluated by an experienced laboratory analyst with specialized knowledge in such matters.

District management makes the final decision regarding the assignment of personnel to inspections. Nevertheless, we expect investigators, analysts and others to work as teams and to advise management when additional expertise is required to complete a meaningful inspection.

Team members participating in a pre-approval inspection must read and be familiar with Compliance Program 7346.832, Pre-Approval Inspections/Investigations. Relevant sections of the NDA or ANDA should be reviewed prior to the inspection; but if the application is not available from any other source, this review will have to be conducted using the company's copy of the application.

Team members should meet, if possible, prior to the inspection to discuss the approach to the inspection, to define the roles of the team members, and to establish goals for completion of the assignment. Responsibilities for development of all reports should also be established prior to the inspection. This includes the preparation of the FDA 483.

The Center for Drug Evaluation and Research (CDER) may have issued deficiency letters listing problems that the sponsor must correct prior to the approval of NDA/ANDA's and supplements. The inspection team is expected to review such letters on file at the district office, and they are expected to ask the plant for access to such letters. The team should evaluate the replies to these letters to assure that the data are accurate and authentic. Complete the inspection even though there has been no response to these letters or when the response is judged inadequate.

4. INSPECTION APPROACH

A. General

In addition to the general approach utilized in a drug CGMP inspection, the inspection of a laboratory requires the use of observations of the laboratory in operation and of the raw laboratory data to evaluate compliance with CGMP's and to specifically carry out the commitments in an application or DMF. When conducting a comprehensive inspection of a laboratory, all aspects of the laboratory operations will be evaluated.

Laboratory records and logs represent a vital source of information that allows a complete overview of the technical ability of the staff and of overall quality control procedures. SOPs should be complete and adequate and the operations of the laboratories should conform to the written procedures. Specifications and analytical procedures should be suitable and, as applicable, in conformance with application commitments and compendial requirements.

Evaluate raw laboratory data, laboratory procedures and methods, laboratory equipment, including maintenance and calibration, and methods validation data to determine the overall quality of the laboratory operation and the ability to comply with CGMP regulations.

Examine chromatograms and spectra for evidence of impurities, poor technique, or lack of instrument calibration.

Most manufacturers use systems that provide for the investigation of laboratory test failures. These are generally recorded in some type of log. Ask to see results of analyses for lots of product that have failed to meet specifications and review the analysis of lots that have been retested, rejected, or reworked. Evaluate the decision to release lots of product when the laboratory results indicate that the lot failed to meet specifications and determine who released them.

B. Pre-Approval

Documents relating to the formulation of the product, synthesis of the bulk drug substance, product specifications, analysis of the product, and others are examined during the review process in headquarters. However, these reviews and evaluations depend on accurate and authentic data that truly represents the product.

Pre-approval inspections are designed to determine if the data submitted in an application are authentic and accurate and if the procedures listed in the application were actually used to produce the data contained in the application. Additionally, they are designed to confirm that plants (including the quality control laboratory) are in compliance with CGMP regulations.

The analytical sections of drug applications usually contain only test results and the methods used to obtain them. Sponsors are not required to file all the test data because such action would require voluminous submissions and would often result in filing redundant information. Sponsors may deliberately or unintentionally select and report data showing that a drug is safe and effective and deserves to be approved. The inspection team must decide if there is valid and scientific justification for the failure to report data which demonstrates the product failed to meet its predetermined specifications.

Coordination between headquarters and the field is essential for a complete review of the application and the plant. Experienced investigators and analysts may contact the review chemist (with appropriate supervisory concurrence) when questions concerning specifications and standards arise.

Inspections should compare the results of analyses submitted with results of analysis of other batches that may have been produced. Evaluate the methods and note any exceptions to the procedures or equipment actually used from those listed in the application and confirm that it is the same method listed in the application. The analyst is expected to evaluate raw laboratory data for tests performed on the test batches (biobatches and clinical batches) and to compare this raw data to the data filed in the application.

5. FAILURE (OUT-OF-SPECIFICATION) LABORATORY RESULTS

Evaluate the company's system to investigate laboratory test failures. These investigations represent a key issue in deciding whether a product may be released or rejected and form the basis for retesting, and resampling.

In a recent court decision the judge used the term "out-of-specification" (OOS) laboratory result rather than the term "product failure" which is more common to FDA investigators and analysts. He ruled that an OOS result identified as a laboratory error by a failure investigation or an outlier test,[1] or overcome by retesting[2] is not a product failure. OOS results fall into three categories:

-- laboratory error
-- non-process related or operator error
-- process related or manufacturing process error

A. LABORATORY ERRORS

Laboratory errors occur when analysts make mistakes in following the method of analysis, use incorrect standards, and/or simply miscalculate the data. Laboratory errors must be determined through a failure investigation to identify the cause of the OOS. Once the nature of the OOS result has been identified it can be classified into one of the three categories above. The inquiry may vary with the object under investigation.

B. LABORATORY INVESTIGATIONS

The exact cause of analyst error or mistake can be difficult to determine specifically and it is unrealistic to expect that analyst error will always be determined and documented. Nevertheless, a laboratory investigation consists of more than a retest. The inability to identify an error's cause with confidence affects retesting procedures, not the investigation inquiry required for the initial OOS result.

The firm's analyst should follow a written procedure, checking off each step as it is completed during the analytical procedure. We expect laboratory test data to be recorded directly in notebooks; use of scrap paper and loose paper must be avoided. These common sense measures enhance the accuracy and integrity of data.

Review and evaluate the laboratory SOP for product failure investigations. Specific procedures must be followed when single and multiple OOS results are investigated. For the single OOS result the investigation should include the following steps and these inquiries must be conducted before there is a retest of the sample:

- the analyst conducting the test should report the OOS result to the supervisor
- the analyst and the supervisor should conduct an informal laboratory investigation which addresses the following areas:

    1. discuss the testing procedure
    2. discuss the calculation
    3. examine the instruments
    4. review the notebooks containing the OOS result

---

[1]The court provided explicit limitations on the use of outlier tests and these are discussed in a later segment of this document.

[2]The court ruled on the use of retesting which is covered in a later segment of this document.

An alternative means to invalidate an initial OOS result, provided the failure investigation proves inconclusive, is the "outlier" test. However, specific restrictions must be placed on the use of this test.

1. Firms cannot frequently reject results on this basis
2. The USP standards govern its use in specific cases only.
3. The test cannot be used for chemical testing results[3]
4. It is never appropriate to utilize outlier tests for a statistically based test, i.e. content uniformity and dissolution.

Determine if the firm uses an outlier test and evaluate the SOP.

Determine that a full scale inquiry has been made for multiple OOS results. This inquiry involves quality control and quality assurance personnel in addition to laboratory workers to identify exact process or non-process-related errors.

When the laboratory investigation is inconclusive (reason for the error is not identified) the firm:

1. Cannot conduct 2 retests and base release on average of three tests
2. Cannot use outlier test in chemical tests
3. Cannot use a re-sample to assume a sampling or preparation error
4. Can conduct a retest of different tablets from the same sample when a retest is considered appropriate (see criteria elsewhere)

C. FORMAL INVESTIGATIONS

Formal investigations extending beyond the laboratory must follow an outline with particular attention to corrective action. The company must:

1. State the reason for the investigation
2. Provide summation of the process sequences that may have caused the problem
3. Outline corrective actions necessary to save the batch and prevent similar recurrence
4. List other batches and products possibly affected, the results of investigation of these batches and products, and any corrective action. Specifically:

    - examine other batches of product made by the errant employee or machine
    - examine other products produced by the errant process or operation

[3]An initial content uniformity test was OOS followed by a passing retest. The initial OOS result was claimed the result of analyst error based on a statistical evaluation of the data. The court ruled that the use of an outlier test is inappropriate in this case.

5. Preserve the comments and signatures of all production and quality control personnel who conducted the investigation and approved any reprocessed material after additional testing

D. INVESTIGATION DOCUMENTATION

Analyst's mistakes, such as undetected calculation errors, should be specified with particularity and supported by evidence. Investigations along with conclusions reached must be preserved with written documentation that enumerates each step of the investigation. The evaluation, conclusion and corrective action, if any, should be preserved in an investigation or failure report and placed into a central file.

E. INVESTIGATION TIME FRAMES

All failure investigations should be performed within 20 business days of the problem's occurrence and recorded and written into a failure or investigation report.

6. PRODUCT FAILURES

An OOS laboratory result can be overcome (invalidated) when laboratory error has been documented. However, non-process and process related errors resulting from operators making mistakes, equipment (other than laboratory equipment) malfunctions, or a manufacturing process that is fundamentally deficient, such as an improper mixing time, represent product failures.

Examine the results of investigations using the guidance in section 5 above and evaluate the decision to release, retest, or rework products.

7. RETESTING

Evaluate the company's retesting SOP for compliance with scientifically sound and appropriate procedures. A very important ruling in one recent court decision sets forth a procedure to govern the retesting program. This district court ruling provides an excellent guide to use in evaluating some aspects of a pharmaceutical laboratory, but should not be considered as law, regulation or binding legal precedent. The court ruled that a firm should have a predetermined testing procedure and it should consider a point at which testing ends and the product is evaluated. If results are not satisfactory, the product is rejected.

Additionally, the company should consider all retest results in the context of the overall record of the product. This includes the history of the product,[4] type of test

[4]The court ordered a recall of one batch of product on the basis of an initial content uniformity failure and no basis to invalidate the test result and on a history of content uniformity problems with the product.

performed, and in-process test results. Failing assay results cannot be disregarded simply on the basis of acceptable content uniformity results.

The number of retests performed before a firm concludes that an unexplained OOS result is invalid or that a product is unacceptable is a matter of scientific judgment. The goal of retesting is to isolate OOS results but retesting cannot continue ad infinitum.

In the case of nonprocess and process-related errors, retesting is suspect. Because the initial tests are genuine, in these circumstances, additional testing alone cannot contribute to product quality. The court acknowledged that some retesting may precede a finding of nonprocess or process-based errors. Once this determination is made, however, additional retesting for purposes of testing a product into compliance is not acceptable.

For example, in the case of content uniformity testing designed to detect variability in the blend or tablets, failing and non-failing results are not inherently inconsistent and passing results on limited retesting do not rule out the possibility that the batch is not uniform. As part of the investigation firms should consider the record of previous batches, since similar or related failures on different batches would be a cause of concern.

Retesting following an OOS result is ruled appropriate only after the failure investigation is underway and the failure investigation determines in part whether retesting is appropriate. It is appropriate when analyst error is documented or the review of analyst's work is "inconclusive", but it is not appropriate for known and undisputed non-process or process related errors.

The court ruled that retesting:

- must be done on the same, not a different sample
- may be done on a second aliquot from the same portion of the sample that was the source of the first aliquot
- may be done on a portion of the same larger sample previously collected for laboratory purposes

8. RESAMPLING

Firms cannot rely on resampling[5] to release a product that has failed testing and retesting unless the failure investigation discloses evidence that the original sample is not representative or was improperly prepared.
Evaluate each resampling activity for compliance with this guidance.

9. AVERAGING RESULTS OF ANALYSIS

Averaging can be a rational and valid approach when the object under consideration is total product assay, but as a general rule this practice should be avoided[6] because

---

[5]The court ordered the recall of one batch of product after having concluded that a successful resample result alone cannot invalidate an initial OOS result.

[6]The court ruled that the firm must recall a batch that was released for content uniformity on the basis of averaged test results.

averages hide the variability among individual test results. This phenomenon is particularly troubling if testing generates both OOS and passing individual results which when averaged are within specification. Here, relying on the average figure without examining and explaining the individual OOS results is highly misleading and unacceptable.

Content uniformity and dissolution results never should be averaged to obtain a passing value.

In the case of microbiological turbidimetric and plate assays an average is preferred by the USP. In this case, it is good practice to include OOS results in the average unless an outlier test (microbiological assays) suggests the OOS is an anomaly.

10. BLEND SAMPLING AND TESTING

The laboratory serves a vital function in blend testing which is necessary to increase the likelihood of detecting inferior batches. Blend uniformity testing cannot be waived in favor of total reliance on finished product testing because finished product testing is limited.

One court has ruled that sample size influences ultimate blend test results and that the sample size should resemble the dosage size. Any other practice would blur differences in portions of the blend and defeat the object of the test. If a sample larger than the unit must be taken initially, aliquots which resemble the dosage size should be carefully removed for the test, retests, and reserve samples. Obviously, the initial larger sample should not be subjected to any additional mixing or manipulation prior to removing test aliquots as this may obscure non-homogeneity.

Multiple individual blend uniformity samples taken from different areas cannot be composited. However when variation testing is not the object of assay testing, compositing is permitted.

If firms sample product from sites other than the blender, they must demonstrate through validation that their sampling technique is representative of all portions and concentrations of the blend. This means that the samples must be representative of those sites that might be problems; e.g. weak or hot spots in the blend.

11. MICROBIOLOGICAL

The review of microbiological data on applicable dosage forms is best performed by the microbiologist (analyst). Data that should be reviewed include preservative effectiveness testing, bioburden data, and product specific microbiological testing and methods.

Review bioburden (before filtration and/or sterilization) from both an endotoxin and sterility perspective. For drug substance labs evaluate methods validation and raw data for sterility, endotoxin testing, environmental monitoring, and filter and filtration validation. Also, evaluate the methods used to test and establish bioburdens.

Refer to the Microbiological Inspection Guide for additional information concerning the inspection of microbiological laboratories.

12. SAMPLING

Samples will be collected on pre-approval inspections. Follow the sampling guidelines in CP 7346.832, Part III, pages 5 and 6.

13. LABORATORY RECORDS AND DOCUMENTATION

Review personal analytical notebooks kept by the analysts in the laboratory and compare them with the worksheets and general lab notebooks and records. Be prepared to examine all records and worksheets for accuracy and authenticity and to verify that raw data are retained to support the conclusions found in laboratory results.

Review laboratory logs for the sequence of analysis versus the sequence of manufacturing dates. Test dates should correspond to the dates when the sample should have been in the laboratory. If there is a computer data base, determine the protocols for making changes to the data. There should be an audit trail for changes to data.

We expect raw laboratory data to be maintained in bound, (not loose or scrap sheets of paper), books or on analytical sheets for which there is accountability, such as prenumbered sheets. For most of those manufacturers which had duplicate sets of records or "raw data", non-numbered loose sheets of paper were employed. Some companies use discs or tapes as raw data and for the storage of data. Such systems have also been accepted provided they have been defined (with raw data identified) and validated.

Carefully examine and evaluate laboratory logs, worksheets and other records containing the raw data such as weighings, dilutions, the condition of instruments, and calculations. Note whether raw data are missing, if records have been rewritten, or if correction fluid has been used to conceal errors. Results should not be changed without explanation. Cross reference the data that has been corrected to authenticate it. Products cannot be "tested into compliance" by arbitrarily labeling out-of-specification lab results as "laboratory errors" without an investigation resulting in scientifically valid criteria.

Test results should not have been transcribed without retention of the original records, nor should test results be recorded selectively. For example, investigations have uncovered the use of loose sheets of paper with subsequent selective transcriptions of good data to analyst worksheets and/or workbooks. Absorbance values and calculations have even been found on desk calendars.

Cut charts with injections missing, deletion of files in direct data entry systems, indirect data entry without verification, and changes to computerized programs to override program features should be carefully examined. These practices raise questions about the overall quality of data.

The firm should have a written explanation when injections, particularly from a series are missing from the official work-sheets or from files and are included among the raw data. Multiple injections recorded should be in consecutive files with consecutive injection times recorded. Expect to see written justification for the deletion of all files.

Determine the adequacy of the firm's procedures to ensure that all valid laboratory data are considered by the firm in their determination of acceptability of components, in-process, finished product, and retained stability samples. Laboratory logs and documents when cross referenced may show that data has been discarded by company officials who decided to release the product without a satisfactory explanation of the results showing the product fails to meet the specifications. Evaluate the justification for disregarding test results that show the product failed to meet specifications.

14. LABORATORY STANDARD SOLUTIONS

Ascertain that suitable standards are being used (i.e. in-date, stored properly). Check for the reuse of stock solutions without assuring their stability. Stock solutions are frequently stored in the laboratory refrigerator. Examine the laboratory refrigerators for these solutions and when found check for appropriate identification. Review records of standard solution preparation to assure complete and accurate documentation. It is highly unlikely that a firm can "accurately and consistently weigh" to the same microgram. Therefore data showing this level of standardization or pattern is suspect and should be carefully investigated.

15. METHODS VALIDATION

Information regarding the validation of methods should be carefully evaluated for completeness, accuracy and reliability. In particular, if a compendial method exists, but the firm chooses to use an alternate method instead, they must compare the two and demonstrate that the in-house method is equivalent or superior to the official procedure. For compendial methods firms must demonstrate that the method works under the actual conditions of use.

Methods can be validated in a number of ways. Methods appearing in the USP are considered validated and they are considered validated if part of an approved ANDA. Also a company can conduct a validation study on their method. System suitability data alone is insufficient for and does not constitute method validation.

In the review of method validation data, it is expected that data for repetitive testing be consistent and that the varying concentrations of test solutions provide linear results. Many assay and impurity tests are now HPLC, and it is expected that the precision of these assays be equal or less than the RSD's for system suitability testing. The analytical performance parameters listed in the USP XXII, <1225>, under the heading of Validation of Compendial Methods, can be used as a guide for determining the analytical parameters (e.g., accuracy, precision, linearity, ruggedness, etc.) needed to validate the method.

16. EQUIPMENT

Laboratory equipment usage, maintenance, calibration logs, repair records, and maintenance SOPs also should be examined. The existence of the equipment specified in the analytical methods should be confirmed and its condition noted. Verify that the equipment was present and in good working order at the time the batches were analyzed. Determine whether equipment is being used properly.

In addition, verify that the equipment in any application was in good working order when it was listed as used to produce clinical or biobatches. one would have to suspect the data that are generated from a piece of equipment that is known to be defective.

Therefore, continuing to use and release product on the basis of such equipment represents a serious violation of CGMP's.

17. RAW MATERIAL TESTING

Some inspections include the coverage of the manufacturer of the drug substance. The safety and efficacy of the finished dosage form is largely dependent on the purity and quality of the bulk active drug substance. Examine the raw data reflecting the analysis of the drug substance including purity tests, charts, etc.

Check the impurity profiles of the BPC used in the biobatch and clinical production batches to determine if it is the same as that being used to manufacture full scale production batches. Determine if the manufacturer has a program to audit the certificate of analysis of the BPC, and, if so, check the results of these tests. Report findings where there is substantial difference in impurity profiles and other test results.

Some older compendial methods may not be capable of detecting impurities as necessary to enable the control of the manufacturing process, and newer methods have been developed to test these products. Such methods must be validated to ensure that they are adequate for analytical purposes in the control and validation of the BPC manufacturing process. The drug substance manufacturer must have complete knowledge of the manufacturing process and the potential impurities that may appear in the drug substance. These impurities cannot be evaluated without a suitable method and one that has been validated.

Physical tests such as particle size for raw materials, adhesion tests for patches, and extrusion tests for syringes are essential tests to assure consistent operation of the production and control system and to assure quality and efficacy. Some of these tests are filed in applications and others may be established by the protocols used to manufacture the product. The validation of methods for such tests are as important as the test for chemical attributes.

Physical properties tests often require the use of unique equipment and protocols. These tests may not be reproducible in other laboratories, therefore, on site evaluation is essential.

18. IN PROCESS CONTROLS AND SPECIFICATIONS

Evaluate the test results from in-process tests performed in the production areas or laboratory for conformance with established sampling and testing protocols, analytical methods, and specifications. For example, evaluate the tests for weight variation, hardness, and friability. These tests may be performed every fifteen or thirty minutes during tableting or encapsulating procedures. All testing must comply with CGMP's.

The drug application may contain some of the in-process testing plan, including methods and specifications. The inspection must confirm that the in-process tests were done, as described in the plan, and ascertain that the results were within specifications. The laboratory work for the lengthier tests should also be reviewed.

The methods used for in-process testing may differ from those used for release testings. Usually, whether the methods are the same or different, the specifications may be tighter for the in-process tests. A product with a 90.0%-110.0% assay release specification may have a limit of 95.%-105.0% for the in-process blend. Some of the tests done may differ from those done at release. For example, a firm may perform disintegration testing as an in-process test but dissolution testing as a release test.

Expect to see consistent in-process test results within batches and between batches of the same formulation/process (including development or exhibit batches). If this is not the case, expect to see scientific data to justify the variation.

19. STABILITY

A stability-indicating method must be used to test the samples of the batch. If there is no stability-indicating assay additional assay procedures such as TLC should be used to supplement the general assay method. Evidence that the method is stability indicating must be presented, even for compendial methods. Manufacturers may be required to accelerate or force degradation of a product to demonstrate that the test is stability indicating. In some cases the sponsor of ANDA's may be able to search the literature and find background data for the specificity of a particular method. This information may also be obtained from the supplier of the drug substance. Validation would then be relatively straightforward, with the typical parameters listed in the USP in chapter <1225> on validation of compendial methods addressed as applicable.

Evaluate the manufacturer's validation report for their stability testing. Again, review the raw laboratory data and the results of testing at the various stations to determine if the data actually reported matches the data found in on site records.

Evaluate the raw data used to generate the data filed documenting that the method is stability indicating and the level of impurities.

20. COMPUTERIZED LABORATORY DATA ACQUISITION SYSTEMS

The use of computerized laboratory data acquisition systems is not new and is addressed in the following CGMP guidance documents:

- Compliance Policy Guide 7132a.07 Computerized Drug Processing: Input/Output Checking.
- Compliance Policy Guide 7132a.08 Computerized Drug Processing: Identification of "Persons" on Batch Production and Control Records.
- Compliance Policy Guide 7132a.11 Computerized Drug Processing: CGMP Applicability to Hardware and Software.
- Compliance Policy Guide 7132a.12 Computerized Drug Processing: Vendor Responsibility
- Compliance Policy Guide 7132a.15 Computerized Drug Processing: Source Code for Process Control Application Programs
- Guide to Inspection of Computerized Systems in Drug Processing.

It is important, for computerized and non computerized systems, to define the universe of data that will be collected, the procedures to collect it, and the means to verify its accuracy. Equally important are the procedure to audit data and programs and the process for correcting errors. Several issues must be addressed when evaluating computerized laboratory systems. These include data collection, processing, data integrity, and security.

Procedures should only be judged adequate when data are secure, raw data are not accidentally lost, and data cannot be tampered with. The system must assure that raw data are stored and actually processed.

The agency has provided some basic guidance on security and authenticity issues for computerized systems:

- Provision must be made so that only authorized individuals can make data entries.
- Data entries may not be deleted. Changes must be made in the form of amendments.
- The data base must be made as tamperproof as possible.
- The Standard Operating Procedures must describe the procedures for ensuring the validity of the data.

One basic aspect of validation of laboratory computerized data acquisition requires a comparison of data from the specific instrument with that same data electronically transmitted through the system and emanating on a printer. Periodic data comparisons would be sufficient only when such comparisons have been made over a sufficient period of time to assure that the computerized system produces consistent and valid results.

21. LABORATORY MANAGEMENT

Overall management of the laboratory work, its staff, and the evaluation of the results of analysis are important elements in the evaluation of a control laboratory. Span of supervisory control, personnel qualifications, turnover of analysts, and scope of the laboratory's responsibility are important issues to examine when determining the quality of overall management and supervision of work. Individually or collectively, these factors are the basis for an objection only when they are shown to result in inadequate performance of responsibilities required by the CGMPs.

Review laboratory logs for the sequence of analysis and the sequence of manufacturing dates. Examine laboratory records and logs for vital information about the technical competence of the staff and the quality control procedures used in the laboratory.

Observe analysts performing the operations described in the application. There is no substitute for actually seeing the work performed and noting whether good technique is used. You should not stand over the analysts, but watch from a distance and evaluate their actions.

Sometimes the company's employees have insufficient training or time to recognize situations that require further investigation and explanation. Instead they accept unexplained peaks in chromatograms with no effort to identify them. They may accept stability test results showing an apparent increase in the assay of the drug with the passage of time with no apparent question about the result. Also, diminishing reproducibility in HPLC chromatograms appearing several hours after system suitability is established is accepted without question. Good manufacturing practice regulations require an active training program and the documented evaluation of the training of analysts.

The authority to delete files and override computer systems should be thoroughly examined. Evaluate the history of changes to programs used for calculations. Certain changes may require management to re-examine the data for products already released.

FOOD AND DRUG ADMINISTRATION
COMPLIANCE PROGRAM GUIDANCE MANUAL
PROGRAM: 7346.832 -- PRE-APPROVAL INSPECTIONS/INVESTIGATIONS

DATE OF ISSUANCE: 6/1/92

FIELD REPORTING REQUIREMENTS

1. Inspectional

   A. The District Director or person acting in this position must submit a written response to the Division of Manufacturing and Product Quality within 10 calendar days. Their response should list their concurrence with the approval of each NDA/ANDA or state the reasons why approval should not be granted. This will be in response to a written CDER request. A copy of each response should be sent to MPQAS, HFC-120.
   B. Forward all violative Establishment Inspection Reports (EIRs) to HFD-300, Attention: Investigations and Compliance Evaluation Branch, with a recommendation to withhold NDA/ANDA approval and full documentation of non-compliance with CGMP regulations.
   C. *Submit a summary of findings and district recommendations, via EMS/FAX, to HFD-324 within 10 days after completion of the inspection*. Districts must not wait for completion of inspection reports before notifying the Center of the District's recommendation about the approval of an application. Therefore, use the FDA-483 and the judgement of highly trained employees as the basis for the District's recommendation. If the recommendation is to concur with approval no further documentation need be sent.*

      NOTE: EIRs for inspections that result in a District recommendation not to approve an application are to be submitted to CDER within 30 days of completion of the inspection.

      Districts should closely monitor corrections made by firms for those conditions deemed unacceptable. Reinspections of firms should be completed within 45 days of a firm's certification, indicating the problems have been corrected.

   D. Districts are required to update promptly the firm profile in accordance with Chapter 15, GWQAP Manual *when GMP inspections are conducted.*
   E. Special time frames apply in IND inspections. Assignments identified as "Treatment IND" or "Treatment Protocol" inspections must be completed within 10 calendar days of receipt.

Current Changes

F. Urokinase products; blood bags with anticoagulant; and plasma volume expander, are under the jurisdiction of the Center for Biologics Evaluation and Research. Consequently, all assignments covering these products will issue from CBER, Division of Inspections and Surveillance, Biological Product Inspections Branch (HFB-122) and all reporting will be forwarded to HFB-122.

2. Laboratory

A. Forward all laboratory reports on NDA and ANDA method validations to the CDER unit that requested the validation. Forward forensic examination results to the responsible headquarter unit (ODE or OGD).
B. Forward an EMS to the Division of Field Science (HFC-140) stating the District's recommendation concerning the validity of the proposed method and the date the validation report was sent to CDER.
C. Worksheets covering samples analyzed under this program are to be forwarded to the home district.
D. Report suspicious laboratory results, indicating possible fraudulent preapproval samples, to the home district and to HFD-324 as soon as possible.

PART I -- BACKGROUND

The Food, Drug, and Cosmetic Act provides that FDA may approve a New Drug Application (NDA) or an Abbreviated New Drug Application (ANDA) only if the methods used in, and the facilities and controls used for, the manufacture, processing, packing, and testing of the drug are found adequate to ensure and preserve its identity, strength, quality, and purity.

This program is designed to provide close inspectional and analytical attention to the *authenticity* of and accuracy of reporting of data in applications and provide information regarding facilities. Such coverage is necessary to assure applications are not approved if the applicant has not demonstrated an ability to operate with integrity and in compliance with all applicable regulations.

PART II -- IMPLEMENTATION

OBJECTIVE

The objective of this continuing compliance program is to assure that establishments involved in the manufacturing, testing, or other manipulation of new drug dosage forms

Current Changes

and new drug substances are investigated: 1.) through on-site inspections for compliance with CGMPs *2.) for conformance with application commitments 3.) to assure data is authentic and accurate,* and 4.) laboratory testing of products, including evaluations of the adequacy of analytical methodology.

Both foreign and domestic establishments are covered by this program. Such coverage is intended to be consistent to the extent possible.

This program provides guidance for inspections and for the laboratory evaluations of NDA and ANDA submissions. These activities are essential parts of the application *evaluation* and approval process. This program provides guidance for both headquarters and field initiated pre-approval inspections.

PROGRAM MANAGEMENT INSTRUCTIONS

Before any application[1] is approved by the Center for Drug Evaluation and Research, a determination will be made whether all establishments that will participate in the manufacture, packaging or testing of the finished dosage form or new drug substance are in compliance with CGMP and application commitments. This determination will be made by conducting pre-approval inspections and method validations. Post-approval inspections will monitor and enforce these requirements.

Pre-approval inspections will have as their objectives the following:

- Evaluation of the establishment's compliance with CGMP requirements, including coverage of the specific batches *upon which the application is based (e.g. pivotal clinical, bioavailability, bioequivalence, and stability).*
- Evaluation as to whether the establishment has adequate facilities, equipment, procedures, and controls to manufacture the product in conformance with application commitments.
- Audit of the *completeness and* accuracy of *preapproval batch* manufacturing and testing information submitted with the application.
- *Collection of forensic samples of the biobatch from the bioequivalence test laboratory and the applicant.
- Collection of application analytical method validation verification samples.*

The following strategy will be applied in assigning pre-approval inspections:

Current Changes

---

[1]"Application" as used in this program means NDA, ANDA, Antibiotic Drug Application or Abbreviated Antibiotic Drug Application and their supplements.

I. Pre-approval inspections* of the manufacturer* will be assigned by CDER in every instance for the following types of drugs:

   A. Narrow therapeutic range drugs (see list in Attachment B)
   B. New chemical entities
   C. Generic versions of the 200 most-prescribed drugs
   D. *Also, for drugs not represented in one of the above special categories, inspections will be assigned as follows: (1) when the plant and/or dosage form has not received a satisfactory inspection within the past two years (2) when the application is from a firm whose applications were the subject of a validity assessment (3) when the application is the initial one for the applicant, regardless of inspection history (4) when a headquarters review reveals discrepancies warranting investigation.

*II. CDER will also assign pre-approval GMP inspections of bulk pharmaceutical chemical (BPC) manufacturers, and ancillary firms based on whether there has been recent GMP coverage. This includes bulk drug substance manufacturers, sterilizers, labelers, packagers, testing laboratories, etc.

III. Where circumstances warrant*, districts may also, at their discretion, assign and conduct inspectional audits above and beyond those for which they receive specific headquarters assignments. Such inspections may be based on headquarters-supplied pending application lists or from receipt of district copies (third copies) of applications, when implemented.

For drugs not assigned inspections by headquarters, the districts will be consulted by CDER in writing prior to approving each application. The districts will respond within 10 calendar days to these requests with a recommendation relating to such approvals. The responses will be based on results of discretionary district inspections or other information that should be considered by CDER before approval.

IV. In addition to original applications, CDER may request that certain supplemental applications undergo a similar inspection/evaluation to determine the CGMP compliance status [see 21 CFR 314.70(b)(1)(v) and 2(vi)] and conformance with commitments made in the supplement. *Such supplements will include those providing for new facilities, major construction or new dosage form manufacturing processes normally requiring on-site GMP/process validation review. This will also include new raw material suppliers, new repackers and new outside control laboratories.

Current Changes

Supplements will generally be evaluated using the same criteria as for original applications; i.e., new dosage form manufacturing facilities would require product-specific inspections if the drug products fit the special inspection criteria, whereas raw material suppliers and other ancillary establishments would be evaluated based on their history where possible.*

Investigational Drugs

Inspections will be requested by the Center when:

1. The IND is a Treatment IND or Treatment Protocol (21 CFR 312.34 and 312.35).
2. The facility used to manufacture investigational drugs ("clinical supplies") is a commercial, contract manufacturer.
3. There is information available to the Center for Drug Evaluation and Research indicating that inspection is warranted to protect the health of patients.
4. The IND sponsor requests recovery of costs.

Special time frames will be established in the assignments from the Center. For Treatment INDs and Treatment Protocols, treatment may begin 30 days after submission unless placed on "clinical hold." In order to adequately protect patients' health, a comprehensive evaluation must be conducted within this 30-day period. Consequently, Districts must complete inspections within 10 calendar days after receipt of assignments.

Bioequivalence Testing Facilities (Biolabs)

Bioretention samples retained by biolabs will be collected routinely in accordance with the instructions in Part III Inspection -- Section G., under the subsection on "Biotest Facility Samples." However, under this program, inspections of facilities conducting bioequivalence/bioavailability studies will be initiated only on a for-cause basis as a follow-up to discrepancies discovered from FDA laboratory examination of such bioretention samples and/or from the District's pre-approval inspectional coverage of bio-batch manufacture and testing.

Specific guidance for conducting these for-cause inspections must be obtained through direct contact with the Division of Scientific Investigations, Clinical Investigations Branch, before initiating the inspection. These inspections are specialized and need careful direction to ensure that the inspectional efforts are maximized. The instructions under Compliance Program 7348.001 (In Vivo Bioequivalence) apply however, this coverage is distinct from that where inspections of facilities/studies are requested in conjunction with headquarters' reviews of biostudies. Those inspections are requested

Current Changes

by headquarters to validate bioequivalence claims, or where the facility is unknown to the agency, or where there are gross problems, fraud, or suspected conspiracy.

While CP7348.001 should be reviewed, resources should be charged to the appropriate pre-approval program assignment code.

District Responsibilities

The Act requires the Agency to review applications within a *certain* period, and prescribes the specific grounds for withholding approval. District pre-approval activities are an inherent part of the agency's *evaluation* process, and must be conducted according to these statutory requirements.* Therefore, Districts are responsible for timely responses to headquarters assignments, and District recommendations to delay, withhold or grant approval must provide adequate, documented grounds consistent with the Act, and must be fairly and equitably applied.

It is expected that any District conclusion or recommendation carries the approval of the District Director.

PART III -- INSPECTIONAL

Operations

A. Inspection Team

Inspection teams should consist of investigators, analysts, engineers, and/or computer experts, as appropriate. The team must be highly skilled in drug manufacturing and analytical technology. For example, these experts are typically the ones assigned to inspect new operations and to inspect highly complex systems that have produced products that failed specifications.

When possible, the analyst, who is involved in laboratory validation of the product under review, should participate in the inspection.

B. General Approach

Emphasis should be placed on an evaluation of the manufacturing process. As part of the process evaluation, correlation between the manufacturing processes for clinical batches and/or bioavailability studies should be reviewed. Conduct an in-depth review of the data, including research and development data, that the applicant/manufacturer uses to establish its proposed manufacturing processes and controls. The inspection will include *an audit* of the product and control data for the batch(es) used for clinical studies and/or batch(es) used for bioavailability studies.

Current Changes

A similar *audit* will be conducted for batches used to conduct stability studies since there can be problems if such studies used pilot batches that differ significantly from production batches. A comparison will then be made of production and control procedures submitted in the filing with procedures used for the manufacture of production batches.

*The inspection team will evaluate whether the applicant has scientific data to justify full scale production procedures and controls listed in the application. They will also evaluate the validation of pertinent manufacturing procedures that may have been performed, including equipment qualification. However, any significant problems found in the validation of production batches will not be grounds for recommending that application approval be withheld unless data is found to be of questionable validity.

Because of the volume of data associated with manufacturing and control processes, submission of all relevant data for complete evaluation at a headquarters level is neither practical nor possible. The inspection team should audit the data, focusing on accuracy and completeness. If that audit identifies questions regarding specifications or stability, the headquarters reviewer should be consulted.*

1. Manufacturing Process

a. Drug Product (Dosage Form)

In many cases, clinical production or trial runs of a new drug are produced in facilities other than the ones used for full scale production. The facilities and *production* controls used for the manufacture of the biobatch[2] must be reviewed *and audited*. Accurate documentation is essential so that the process can be defined and related to the batch(es) used for the biostudy.

FDA has published guidance for compliance with GMPs for the production of investigational drugs. Refer to this document for added guidance.

The application must be *referenced* as part of the inspection to assure that the proposed production process is the process that was used for the manufacture of the bio/stability batches. Some manufacturers have historically made small batches that were used for biostudies and stability studies and misrepresented

Current Changes

[2]Generic product biobatches are ANDA batches that are compared to the originator/reference product to establish their equivalence. NDA biobatches are NDA batches comparing the product planned for marketing with that studied during clinical trials to establish their equivalence.

them as larger batches in submissions. Documentation sometimes has included R&D notebooks and/or batch records. Inventory records and/or receiving records of drug substances have been found to be of value for the accountability of drug substances.

*b) Drug Substance (Bulk Drug Chemical)

The "Guide to Inspection of Bulk Pharmaceutical Chemical Manufacturing" and Compliance Program 7356.002F covering Bulk Pharmaceutical Chemicals should be used for guidance for inspections covering bulk drug chemical manufacturing processes. Refer to Part V for a discussion of validation requirements.*

2. Reprocessing

*The GMP regulations require reprocessing procedures to be written, and it is customary but not required that NDAs/ANDAs contain procedures covering foreseeable deviations from physical specifications (e.g., color capped tablets, deviations from hardness specifications, etc.).

If the NDA/ANDA contains a reprocess provision, the applicant must produce data to establish that the procedure will result in a product that is equivalent to the original product. Such reprocessing provisions listed in filings must be justified with scientific data.*

3. Laboratory

Laboratory equipment and procedures must be qualified and validated. Every NDA/ANDA inspection will include an evaluation of laboratory controls and procedures, and a review of some of the raw data used to generate results. This data may be located in research and development test logs. The authenticity and accuracy of data used in the development of a test method should be reviewed.

4. Components

The supplier and source of the active drug substance used in the manufacturing of the biobatch and/or clinical batch should be identified. The physical characteristics of the drug substance should be reviewed along with the physical specification for this substance. For solid oral dosage forms it is important to establish physical specifications both to assure dose uniformity and as a control for dissolution and absorption. Only in a few isolated cases have manufacturers been able to justify the absence of a physical specification for drug substances.

Current Changes

When a manufacturer changes suppliers of drug substance from that used for the manufacture of the bio-batch or clinical batches, then the application should include data demonstrating that the dosage forms produced from the drug substances from the two different suppliers are equivalent in terms of conformance with established specifications including those stated in the application.

5. Building and Facilities

The addition of any new drug to a production environment must be carefully evaluated as to its impact on other products already under production and changes that will be necessary to the building and facility. Construction of new walls, installation of new equipment, and other significant changes must be evaluated for their impact on the overall compliance with GMP requirements. For example, new products, such as cephalosporins, would require that the firm *demonstrate through appropriate separation and controls* that cross-contamination can not occur with regard to other products being made in the same facility. Also, facilities that may already be operating at full capacity may not have adequate space for additional products.

6. Equipment

New products, particularly potent drug products, can present cleaning problems in existing equipment. Manufacturers must validate their cleaning processes for the new drug/dosage form.

7. Packaging and Labeling Controls

If warranted, packaging and labeling control procedures are to be evaluated. Poor label control and accountability for other products may have an adverse impact on the firm's ability to assure that the new drug will always be properly labeled.

Review the label and packaging controls, taking into account past label mixups and recalls.

C. Inspection Scheduling

The scheduling of inspections is left to the discretion of the District within any assigned timeframes. This includes scheduling of inspections for which the district has received specific headquarters assignments. Every reasonable effort should be made, however, to conduct pre-approval inspections at the earliest possible opportunity, since unnecessary delays are not supportive of the Agency's responsibilities to provide timely review of NDAs and ANDAs.

Current Changes

*Such scheduling applies to biobatch facilities and production facilities, both of which must be covered with respect to conformance with CGMPs.

Districts may contact manufacturers to determine the readiness of facilities for inspections. In some cases facilities or the development of manufacturing processes may not have been completed. In addition, there may have been changes in the status of an application, e.g., major application deficiencies or the deletion of an ancillary facility, that will affect the need for an inspection.*

D. New Facility Reviews

For the inspection of major new facilities involving many applications, special coordination efforts are often beneficial. Field Management Directive No. 135, "Pre-Operational Reviews of Manufacturing Facilities," provides guidance in this area. Meetings or pre-operational inspections may be scheduled when such activities will contribute to the overall efficiency and effectiveness of the CGMP evaluation process.

E. Surveillance

Districts should be alert to the use of unapproved facilities or unapproved raw material suppliers during these inspections. If unapproved facilities are in use, they should be reported immediately to HFD-320. Inspections of these facilities are not required unless an assignment has been received from headquarters.

F. Application *Audit*

An audit of the drug application is essential to determine that commitments by the firm in the application are reflected in actual practice. Review of application information is also important in preparing for inspections of firms or processes with which the investigator is unfamiliar.

An additional copy of pertinent application information will be required of applicants and forwarded to Districts for use in making audits. The information required will include a copy of the manufacturing and controls section of the application, and information relating to batches used to conduct any bioequivalence, bioavailability and stability studies. Until the requirement to provide a District copy is promulgated, Districts must use the firm's copy. If Districts have any need to authenticate the copy obtained from the firm, they should consult with the appropriate CDER review unit to arrange access to the filed copy of the application.

Current Changes

G. Sample Collection

*This compliance program covers the collection of samples of the biobatches for forensic analysis. Such samples will be collected at both the applicant and the biotest facility in association with each drug application where an in-vivo or in-vitro study to demonstrate bioequivalence has been submitted to FDA.

The program also covers the collection of method validation samples for NDA products to confirm that applicants' proposed analytical methods are suitable for regulatory purposes. Similar coverage extends to ANDA products where non-compendial methods are proposed. For ANDA products for which compendial test methods are proposed, samples will be collected for method verification purposes to ensure that the product as compounded can be assayed satisfactorily with the compendial method. Such analytical verification is not intended to "validate" the compendial (USP) method.

When desirable, pre-approval inspections should be coordinated with the laboratory that is scheduled to perform the method validation so that they can participate in the inspection and in the collection of samples.

In most cases, the home district is responsible for pre-approval sample collections. Such samples are collected at facilities in its area or through assignments to other districts if samples are located elsewhere. However, headquarters is responsible for arranging sample collections in foreign countries. Foreign inspection staff will collect such samples in association with pre-approval inspections which they conduct. In countries with which FDA has bilateral inspection agreements (Canada, Sweden and Switzerland), the health authorities in those countries will collect or coordinate with HFD-320 for collection of all such samples.

The chart included as Attachment A summarizes pre-approval sampling operations, and the following narrative provides more detailed information.

METHOD VALIDATION SAMPLES

1. NDA -- These samples will be collected upon receipt by the district of the Method Validation forms FD 2871 and 2871A. These forms are issued by the reviewing chemist when satisfied that adequate analytical information has been submitted by the applicant and include information about the items and quantities to be collected and the assigned testing laboratories.

   Laboratory policies and procedures for validating analytical methods, submitted to CDER with NDA's, is available in Staff Manual Guide 4831.3 Validation of NDA Analytical Methods (undergoing clearance).

Current Changes

2. ANDA -- Sample for applications specifying noncompendial methods will be collected as for NDAs above. However, where methods are compendial (USP), samples for methods verification, as discussed below, are to be collected* *upon the initiative of the home district in association with assigned pre-approval inspections (or evaluations if an inspection is not assigned). Where samples are located at a site outside the district, the home district will assign their collection by the appropriate district. The following are the requested samples and they should be submitted to the designated field laboratory in Part IV -- Analytical unless otherwise assigned.

Products:		
	Tablets & Capsules --	300 Units
	Injections, Single --	100 Units
	Injections, Multi --	10 Units
	Oral Liquids --	36 Ounces
	Oral Powders for -- Reconstitution	10 Units

Documents: Copy/collect the portion of the application that describes the analytical methods and the report of the analysis performed by the firm on the product lot collected, including spectra and chromatograms.

Methods validation will be conducted where con-compendial analytical methods are included in the application. Methods verification will be conducted where the methods are compendial (USP). The latter will confirm whether the product as compounded can be analyzed satisfactorily with the USP method and is not intended to "validate" the USP analytical method.

Samples are to be collected for methods validation or verification for each application even if there is no requirement to establish bioequivalence with the innovator/reference product through in-vivo or in-vitro studies, (e.g., oral solutions, suspensions, injections). Where biostudies are required and have been submitted to the application, method validation/verification samples should represent the same batches used in such studies. However, if the biobatch is beyond expiration date, collect the next most recent pre-approval batch within expiration, if available. For adequate method validation or method verification, the samples should be within expiration date.*

*Each sample collected for method validation or verification under this program should be flagged, **"Method Validation/Verification NDA/ANDA__-_____"** **(application number of the product sampled).**

Current Changes

*FORENSIC SAMPLES

1. NDA -- Samples of the clinical production batch should be collected during pre-approval inspections.

    Finished Dosage Form -- 50 units*
    Inactive Ingredients -- 40 grams
    Capsule Shells -- 100 units
    Colors and Dyes -- 400 mg
    Active Ingredients -- 20 grams**

    *These samples should be from the batch(es) used to establish the bioavailability of the product. More than one batch should be sampled if the formulation/process for the batch used in pivotal clinical safety and efficacy studies is revised and the revised product is the subject of additional biostudies.

    **If the cost is prohibitive, or sample size is otherwise inappropriate, check with the designated laboratory for instructions.

    In addition, collect:

    -- a copy of the preapproval batch record;
    -- certificates of analysis (if applicable) received by the firm from suppliers for the specific batches of each active and inactive ingredient collected; and
    -- reports prepared by the firm of any tests performed by them on the batches of active ingredients collected. These reports should include methods, spectra, chromatograms and other charts.

    These samples and records should be forwarded to either the New York Regional Laboratory (NE and MA Regions) or the Division of Drug Analysis (other regions) according to the location of the applicant.

2. ANDA -- These samples should be collected during the pre-approval inspection when method validation or verification samples are collected. Collect the same lot number of the innovator and applicant product used in the biotesting as well as the same lot number of the ingredients used in the biobatch, if available.*

*Applicant Samples

Finished Dosage Form/Innovator -- 20 units*
Finished Dosage Form/Applicant -- 20 units**
Inactive Ingredients -- 40 grams

Current Changes

Capsule Shells -- 100 units
Colors and Dyes -- 400 mg***
Actives -- 20 grams***

*If the Innovator's product is not available at the applicant's facility, obtain the lot number of the product and include this information on the collection report.

**This sample should be collected from the retention sample of the applicant's product that was submitted to the bioequivalency testing laboratory.

***If the cost is prohibitive, or sample size is otherwise inappropriate, check with the designated laboratory for instructions.

In addition, collect:

- a copy of the preapproval batch record;
- a complete copy of approved testing method if other than USP (if USP, identify the edition and supplement, if applicable, for the test method used by the firm, e.g., USPXXII, First Supplement);
- certificates of analysis received by the firm from suppliers for the specific batches of each active and inactive ingredients collected;
- reports prepared by the firm of any tests performed by them on the batches of active and inactive ingredients collected. These reports should include methods, spectra, chromatograms, and other charts.
- analytical reports of the applicant's analysis of the lots of their product and the corresponding innovator product submitted to the private laboratory for bioequivalency testing; and
- a copy of reports received by the applicant from all bioequivalence laboratories for all in-vitro tests performed on the product covered by the ANDA.*

*In some cases, particularly for NDA products, clinical biobatches may be beyond expiration date at the time of sampling. This is associated with the sometimes lengthy development and application approval times and does not preclude sampling for forensic purposes (although such sampling would not be suitable for method validation/verification purposes).

These samples and records should be forwarded to either the New York Regional Laboratory (NE and MA Regions) or the Division of Drug Analysis (other regions) according to the location of the applicant.

Each sample collected for forensic analysis under this program should be flagged: **"Applicant Forensic-NDA/ANDA__-____" (application number of the product sampled).**

Current Changes

Biotest Facility Samples

This program directs the collection at the biotest facility, of innovator and applicant products used by the biotest facility in determining bioequivalence. The home district will collect such samples at biotest facilities within its own area. When the biotest facility is located outside the district, sampling assignments will be sent to the appropriate district.

Samples at biotest facilities will be collected for each application when in-vivo or in-vitro bioequivalence studies have been conducted. Although such samples should be collected promptly, they need not be collected individually in association with each pre-approval inspection or evaluation. Districts may "batch" such sample collections or combine sampling with other inspectional coverage of biotest facilities.*

*These samples should be forwarded to either the New York Regional Laboratory (NE and MA Regions) or the Division of Drug Analysis (other regions) according to the location of the applicant. The samples should consist of 20 Dosage Units each of the innovator and applicant products.

Each sample collected for forensic analysis under this program should be flagged: **"Biotest Forensic-ANDA_____________" (application number of the product sampled).***

H. Regional/District Recommendations

In those cases where CDER does not specifically assign a pre-approval inspection, districts will be requested to concur in approval, or to report information that would warrant delay of approval, such as the following:

1. An inspection of the subject establishment is underway, covering processes applicable to the product in question.

2. New information exists, such as an inspection, significant complaint or recall involving a health hazard, which casts doubt on the firm's ability to manufacture the product in question in compliance with GMP and NDA/ANDA requirements; or

3. Notwithstanding a lack of recent violations, the District believes that a special pre-approval inspection should be conducted to assure that the firm can manufacture products in compliance with GMP and NDA/ANDA requirements and to assure that commitments listed in the application have been met.

Current Changes

Districts will reply to CDER's notifications within 10 calendar days (with an information copy to MPQAS) with sufficient information to document the basis for delaying or withholding approval, if delay or withholding is *recommended.* A reply is required for each separate notification, and replies are expected to *reflect concurrence* by the District Director.

A CI/Potential OAI notification should be submitted to MPQAS, HFC-120, whenever an ongoing inspection finds CGMP deficiencies that may make the inspection OAI.

PART IV -- ANALYTICAL

*Two types of analyses are conducted under this program. One involves testing samples of the biobatch/clinical batch for forensic purposes, including fingerprinting. The other involves analysis of method validation/verification samples to ensure, for regulatory purposes, that products can be tested adequately with methods proposed in drug applications. See Part III, Section G, for further information.

All field laboratory pre-approval assay work will be conducted by the following designated laboratories:*

HOME DISTRICT OF APPLICANT	LAB
Atlanta, Nashville, New Orleans, Orlando	SRL
Boston, Buffalo	WEAC
New York	NYK-RL
Baltimore	BLT
Cincinnati	CIN
Newark, Philadelphia	PHI
San Juan	SJN
Chicago, Minneapolis, Detroit	DET
Dallas (ANDAs only), Kansas City (ANDAs only)	DAL
Dallas (NDAs only), and Kansas City (NDAs only)	DEN
Los Angeles, San Francisco, Seattle	SEA
*Foreign (Except Canada, Sweden, Switzerland)	NYK-RL* **

**Unless Otherwise Assigned

Headquarters' laboratory evaluation work will be handled by the Division of Drug Analysis (HFH-300) and the Division of Research and Testing (HFD-470).

Current Changes

Division of Drug Analysis will be the designated laboratory for all pre-approval samples from bilateral inspection agreement countries (Canada, Sweden, Switzerland). This includes both forensic and methods validation/verification work involving both dosage forms and ingredients.

*A. Forensic Analysis

Certain field laboratories and the Division of Drug Analysis will be responsible for biotest sample analyses under this program using methods such as FTIR to produce "fingerprints" for comparison with marketed products. The results of this analysis will also be utilized for comparison with the post-approval samples collected under the Post-Approval Audit Compliance Program (CP 7346.843).

The Division of Drug Analysis will maintain the reserve samples and a national database for all forensic samples analyzed under this program.*

B. Methods of Validation

1. When an application is ready for analytical methods validation, the reviewing chemist within the appropriate Division of the Office of Drug Evaluation (NDAs) or the Office of Generic Drugs (ANDAs) will send a letter to the firm and to the home district office notifying them of the need to conduct the methods validation and assay of the biobatch/clinical production batch. The Center will send the method validation package received with the initial submission, *along with a copy of the letter sent by the review chemist to the firm,* to the District laboratory and Headquarters laboratory, if applicable. The field investigator/chemist will collect samples of the batches for assay and methods validation testing *and verify that the methods used by the firm are the same as those filed in the application.* Copies of the letter to the firm will be sent to the ORA/ORO/Division of Field Science (HFC-140).
2. Upon receipt of the notification to conduct a methods validation, the district will promptly collect appropriate samples of the biobatch/clinical production batch *for method validation and forward to the home district's servicing laboratory.*
3. If the assigned field laboratory will be unable to complete the methods validation and other assays within 45 days, the laboratory should contact the Division of Field Science for assistance in assigning the work to another laboratory. The home district is responsible for collection of samples and for shipment of the samples to the testing laboratories.
4. Methods are to be run as described in the application. A 45 day period will normally be allowed for methods validation and the clock will start when the laboratories have received all the necessary samples, information, or equipment.

Current Changes

5. If a laboratory encounters problems in the methodology that require additional information from the applicant, the laboratory director will review and approve the need to contact the applicant for the information. The information will be requested in writing or confirmed in writing, if requested by telephone, and this written request will be included in the documentation submitted to the review chemist.
6. The laboratory will not advise the applicant of the final conclusion with respect to methods validation, but may tell them when a study is completed and submitted to headquarters.
7. While validation is in progress, validating laboratories are not to communicate with each other regarding test results or observations that may bias recommendations. However, purely technical or analytical problems may be discussed without revealing results. The review chemist and DFS are to be kept informed of all such communications. *Any such communication should be documented by memo with copies provided to the review chemist and DFS.*
8. When each laboratory has completed its portion of the methods validation, a final report should be compiled. Headquarters and field laboratory reports will be forwarded directly to the review chemist by the Laboratory Director *with a copy to the appropriate District Director.* Field laboratories will notify DFS by EMS of the District's recommendation concerning the validity of the proposed method and of the date the validation report was sent to CDER.
9. Each validation report will contain the following information:

(a) Identification of the application and test samples received and description of the product tested including confirmation that it complies with the product described in the application.

Use Attachment *C* to describe the product undergoing testing. Submit the completed Attachment *C* form with the worksheets to the review chemist. Send a copy of the Attachment *C*, collection report, and a copy of the Review Chemist's request memo to the Division of Drug Analysis, HFH-300.

(b) Original worksheets with calculations, results, comments by the analyst(s), associated spectra, chromatograms, etc. that include results of all tests performed and comparison of results obtained with the applicant's data and with the applicable specifications.

(c) An evaluation of each test provided by the firm and performed by the laboratory, accompanied by the original signed form FD-2871.

(d) A recommendation as to whether the methods are acceptable, acceptable after specified changes are made, or not acceptable.

Current Changes

Note: If the product is found to be compounded in such a way that it cannot accurately be analyzed by applicable USP methods, the product is not in compliance with the FD&C Act.

10. *The Office of Generic Drugs or the Office of New Drug Evaluation is responsible for all correspondence with the applicant including approval/disapproval of methods* The review chemist will forward a copy of the letter of approval accompanied by any approved NDA or ANDA method modifications to the laboratories that conducted the methods validation. Copies of all disapproval letters, subsequently approved changes, and similar correspondence with the firm will also be sent to the laboratories.
11. The laboratory responsible for the evaluation of methods in a given firm's NDA or ANDA application is required to maintain an orderly sample and data storage system that completely documents the materials received, as well as reserve samples, related to the application evaluation and the pre-approval inspection of the firm. This file will be up-dated with CDER approved changes after the initial approval. All related written material and samples will be available to CDER and the home district on short notice.

The reserve samples, associated documentation and copies of the FDA laboratory reports will be stored in an orderly and retrievable system for a minimum of three years.

PART V -- FOLLOW-UP: REGULATORY/ADMINISTRATIVE STRATEGY

1. GENERAL

The plant should be in substantial compliance with GMP regulations and should have the necessary facilities and equipment in pace to manufacture the specific product in the pending application. District Directors should recommend withholding approval until these conditions are met.

The district should recommend withholding approval when there are significant deviations from GMP regulations or other application commitments that may adversely impact on the product(s) covered by the pending applications. Some significant problems include, but are not limited to:

*Application misrepresents data or conditions relating to pre-approval batches

There are other inconsistencies and/or discrepancies raising significant questions about the validity of records

Current Changes

Pre-approval batches are not made in accordance with GMPs

There is a failure to report adverse findings or test data without adequate justification*

Districts are encouraged to discuss GMP and other problems having a direct bearing on the district's recommendation for NDA/ANDA approval and to obtain the firm's response to the discussion following preparation and issuance of the FD-483.

If applications are withheld because of significant CGMP non-compliance, and the GMP deficiencies also apply to commercially-marketed products, then action must be taken to assure that the deficiencies are corrected.

Refer to the Regulatory Procedures Manual for consideration of actions that may be indicated.

The district director is expected to send a letter advising the plant officials (or the sponsor when appropriate) that the district has recommended withholding approval of the application and shall state the reasons for this recommendation. This letter must be issued quickly and should be issued at the same time the district informs the Center of its recommendation withhold approval. This letter shall not be titled "Warning Letter" unless the documented conditions affect an approved and marketed product(s) and meets the requirements for a warning letter listed in the RPM.

No application should be recommended for approval if the applicant is found in a state of non-compliance with the CGMP regulations that may adversely impact on the product(s) covered by the pending applications until satisfactory correction is made. However, insignificant deviations should not result in a recommendation to withhold approval of applications.

2. *PROCESS VALIDATION

Districts must not recommend withholding approval of applications based on lack of complete full scale, multiple batch validation for sterile and non-sterile processes unless the data submitted in the filing is found to be of questionable validity or completeness.

Although the agency does not require the manufacturer to fully validate the manufacturing process and control procedures of the commercial batch production prior to approval, the Center reviewing offices will require that certain data be filed to demonstrate that a plant's sterilization and aseptic fill process have been qualified.

Current Changes

These filing issues are under the control of the Center's reviewing divisions. Manufacturers' questions about filings should be directed to the reviewing divisions and should not be discussed by the inspection team. However, the inspection team is expected to audit the data that has been submitted with the application to determine if the data is authentic, accurate, and complete.

Since complete process validation is not required prior to approval, the field is not required to audit complete process validation for sterile and non-sterile processes until the application has been approved. However, if *the plant has already validated the process prior to the pre-approval inspection, the validation should be evaluated during the pre-approval inspection. The inspection team should list deficiencies in the validation process on the FD 483 and advise the plant official that complete validation must be completed prior to shipment.

Do not recommend withholding approval based solely on absent/incomplete process validation, unless available data are of questionable validity or the firm disagrees with any significant problems found.

Sponsors must be able to justify filed specifications with scientific data. In other words, the sponsor should have conducted sufficient research on the test batches to establish specifications for the manufacturing and control procedures listed in the application. These data form the basis for the review and evaluation of the application and these specifications form the basis of the validation protocol which may be developed following the approval of the application. The final step in the product development process is validation that the process will perform consistently. Companies are expected to validate the process using the specifications listed in the filing.

Process validation requirements for the manufacture of bulk pharmaceutical chemicals (BPCs) differ somewhat from those involving dosage forms. The guide to Inspection of BPCs issued in 1991 states that BPC manufacturers are expected to adequately determine and document that significant manufacturing processes perform consistently. The type of BPC, the range of specifications and other factors determine the extent of the process development and documentation required. The documentation system required for early process steps must provide a chain of documentation, and while it need not be as comprehensive as in the later parts of the process, the manufacturer is required to identify and control the key steps in the process.

Many BPC manufacturers have recently initiated validation programs and we recognize that not all BPCs can be validated simultaneously. Therefore, we do not anticipate *taking legal action where a firm has an adequate program in place, including reasonable milestones. Regulatory action should be recommended where there is a lack of validation and evidence of a significant number of failed batches.

Current Changes

District offices are responsible for the implementation and management of a program to assure that manufacturing processes and laboratory control procedures have been validated prior to the shipment of recently approved dosage form drugs. No assignments will issue from CDER.*

PART VI -- REFERENCES, ATTACHMENTS, AND PROGRAM CONTACTS

References

A. Inspection Operations Manual, Subchapter 540
B. Code of Federal Regulations, Title 21
   Part 210 and 211, Drugs: Current Good Manufacturing Practice
   Part 310, New Drugs
   Part 314, New Drug Applications
   Part 429, Drugs Composed Wholly or Partly of Insulin
   Part 431, Certification of Antibiotics
C. Field Management Directive No. 135, "Pre-operational Reviews of Manufacturing Facilities."
D. CP 7356.002 -- Drug Process Inspections
E. United States Pharmacopeia, XXII Revision, 1990, and supplements.
F. *Guideline on Preparation of Investigational New Drug Products, March 1991.
G. Guide to Inspection of Bulk Pharmaceutical Chemical Manufacturers, September 1991.
H. Guideline on General Principles of Process Validation, May 1987. *
I. GWQAP Manual, Profile System, Chapter 15.
J. Regulatory Procedures Manual, Chapters 7 and 8.

ATTACHMENTS

Attachment A -- Preapproval samples
Attachment B -- List of Narrow Therapeutic-Range Drugs
Attachment C -- Characterization of Dosage Forms

Current Changes

GUIDELINE ON GENERAL PRINCIPLES OF PROCESS VALIDATION

MAY, 1987
(REPRINTED MAY 1990)

Prepared by: Center for Drug Evaluation and Research,
Center for Biologics Evaluation and Research, and
Center for Devices and Radiological Health
Food and Drug Administration

Maintained by: Division of Manufacturing and Product Quality (HFD-320)
Office of Compliance
Center for Drug Evaluation and Research

I. PURPOSE

This guideline outlines general principles that FDA considers to be acceptable elements of process validation for the preparation of human and animal drug products and medical devices.

II. SCOPE

This guideline is issued under Section 10.90 (21 CFR 10.90) and is applicable to the manufacture of pharmaceuticals and medical devices. It states principles and practices of general applicability that are not legal requirements but are acceptable to the FDA. A person may rely upon this guideline with the assurance of its acceptability to FDA, or may follow different procedures. When different procedures are used, a person may, but is not required to, discuss the matter in advance with FDA to prevent the expenditure of money and effort on activities that may later be determined to be unacceptable. In short, this guideline lists principles and practices which are acceptable to the FDA for the process validation of drug products and medical devices; it does not list the principles and practices that must, in all instances, be used to comply with law.

This guideline may be amended from time to time. Interested persons are invited to submit comments on this document and any subsequent revisions. Written comments should be submitted to the Dockets Management Branch (HFA-305), Food and Drug Administration, Room 4-62, 5600 Fishers Lane, Rockville, Maryland 20857. Received comments may be seen in that office between 9 a.m. and 4 p.m., Monday through Friday.

III. INTRODUCTION

Process validation is a requirement of the Current Good Manufacturing Practices Regulations for Finished Pharmaceuticals, 21 CFR Parts 210 and 211, and of the Good

Manufacturing Practice Regulations for Medical Devices, 21 CFR Part 820, and therefore, is applicable to the manufacture of pharmaceuticals and medical devices.

Several firms have asked FDA for specific guidance on what FDA expects firms to do to assure compliance with the requirements for process validation. This guideline discusses process validation elements and concepts that are considered by FDA as acceptable parts of a validation program. The constituents of validation presented in this document are not intended to be all-inclusive. FDA recognizes that, because of the great variety of medical products (drug products and medical devices), processes and manufacturing facilities, it is not possible to state in one document all of the specific validation elements that are applicable. Several broad concepts, however, have general applicability which manufacturers can use successfully as a guide in validating a manufacturing process. Although the particular requirements of process validation will vary according to such factors as the nature of the medical product (e.g., sterile vs. non-sterile) and the complexity of the process, the broad concepts stated in this document have general applicability and provide an acceptable framework for building a comprehensive approach to process validation.

Definitions

Installation qualification -- Establishing confidence that process equipment and ancillary systems are capable of consistently operating within established limits and tolerances.

Process performance qualification - Establishing confidence that the process is effective and reproducible.

Product performance qualification - Establishing confidence through appropriate testing that the finished product produced by a specified process meets all release requirements for functionality and safety.

Prospective validation - Validation conducted prior to the distribution of either a new product, or product made under a revised manufacturing process, where the revisions may affect the product's characteristics.

Retrospective validation - Validation of a process for a product already in distribution based upon accumulated production, testing and control data.

Validation - Establishing documented evidence which provides a high degree of assurance that a specific process will consistently produce a product meeting its pre-determined specifications and quality attributes.

Validation protocol - A written plan stating how validation will be conducted, including test parameters, product characteristics, production equipment, and decision points on what constitutes acceptable test results.

Worst case - A set of conditions encompassing upper and lower processing limits and circumstances, including those within standard operating procedures, which pose the greatest chance of process or product failure when compared to ideal conditions. Such conditions do not necessarily induce product or process failure.

IV. GENERAL CONCEPTS

Assurance of product quality is derived from careful attention to a number of factors including selection of quality parts and materials, adequate product and process design, control of the process, and in-process and end-product testing. Due to the complexity of today's medical products, routine end-product testing alone often is not sufficient to assure product quality for several reasons. Some end-product tests have limited sensitivity.[1] In some cases, destructive testing would be required to show that the manufacturing process was adequate, and in other situations end-product testing does not reveal all variations that may occur in the product that may impact on safety and effectiveness.[2]

The basic principles of quality assurance have as their goal the production of articles that are fit for their intended use. These principles may be stated as follows: (1) quality, safety, and effectiveness must be designed and built into the product; (2) quality cannot be inspected or tested into the finished product; and (3) each step of the manufacturing process must be controlled to maximize the probability that the finished product meets all quality and design specifications. Process validation is a key element in assuring that these quality assurance goals are met.

It is through careful design and validation of both the process and process controls that a manufacturer can establish a high degree of confidence that all manufactured units from successive lots will be acceptable. Successfully validating a process may reduce the dependence upon intensive in-process and finished product testing. It should be noted that in most all cases, end-product testing plays a major role in assuring that quality assurance goals are met; i.e., validation and end-product testing are not mutually exclusive.

The FDA defines process validation as follows:

---

[1]For example, USP XXI states: "No sampling plan for applying sterility tests to a specified proportion of discrete units selected from a sterilization load is capable of demonstrating with complete assurance that all of the untested units are in fact sterile."

[2]As an example, in one instance a visual inspection failed to detect a defective structural weld which resulted in the failure of an infant warmer. The defect could only have been detected by using destructive testing or expensive test equipment.

> Process validation is establishing documented evidence which provides a high degree of assurance that a specific process will consistently produce a product meeting its pre-determined specifications and quality characteristics.

It is important that the manufacturer prepare a written validation protocol which specifies the procedures (and tests) to be conducted and the data to be collected. The purpose for which data are collected must be clear, the data must reflect facts and be collected carefully and accurately. The protocol should specify a sufficient number of replicate process runs to demonstrate reproducibility and provide an accurate measure of variability among successive runs. The test conditions for these runs should encompass upper and lower processing limits and circumstances, including those within standard operating procedures, which pose the greatest chance of process or product failure compared to ideal conditions; such conditions have become widely known as "worst case" conditions. (They are sometimes called "most appropriate challenge" conditions.) Validation documentation should include evidence of the suitability of materials and the performance and reliability of equipment and systems.

Key process variables should be monitored and documented. Analysis of the data collected from monitoring will establish the variability of process parameters for individual runs and will establish whether or not the equipment and process controls are adequate to assure that product specifications are met.

Finished product and in-process test data can be of value in process validation, particularly in those situations where quality attributes and variabilities can be readily measured. Where finished (or in-process) testing cannot adequately measure certain attributes, process validation should be derived primarily from qualification of each system used in production and from consideration of the interaction of the various systems.

V. CGMP REGULATIONS FOR FINISHED PHARMACEUTICALS

Process validation is required, in both general and specific terms, by the Current Good Manufacturing Practice Regulations for Finished Pharmaceuticals, 21 CFR Parts 210 and 211. Examples of such requirements are listed below for informational purposes, and are not all-inclusive.

A requirement for process validation is set forth in general terms in section 211.100 -- Written procedures; deviations -- which states, in part:

> "There shall be written procedures for production and process control designed to assure that the drug products have the identity, strength, quality, and purity they purport or are represented to possess."

Several sections of the CGMP regulations state validation requirements in more specific terms. Excerpts from some of these sections are:

Section 211.110, Sampling and testing of in-process materials and drug products.

(a) "....control procedures shall be established to monitor the output and VALIDATE the performance of those manufacturing processes that may be responsible for causing variability in the characteristics of in-process material and the drug product." (emphasis added)

Section 211.113, Control of Microbiological Contamination.

(b) "Appropriate written procedures, designed to prevent microbiological contamination of drug products purporting to be sterile, shall be established and followed. Such procedures shall include VALIDATION of any sterilization process." (emphasis added)

VI. GMP REGULATION FOR MEDICAL DEVICES

Process validation is required by the medical device GMP Regulations, 21 CFR Part 820. Section 820.5 requires every finished device manufacturer to:

"...prepare and implement a quality assurance program that is appropriate to the specific device manufactured..."

Section 820.3(n) defines quality assurance as:

"...all activities necessary to verify confidence in the quality of the process used to manufacture a finished device."

When applicable to a specific process, process validation is an essential element in establishing confidence that a process will consistently produce a product meeting the designed quality characteristics.

A generally stated requirement for process validation is contained in section 820.100:

"Written manufacturing specifications and processing procedures shall be established, implemented, and controlled to assure that the device conforms to its original design or any approved changes in that design."

Validation is an essential element in the establishment and implementation of a process procedure, as well as in determining what process controls are required in order to assure conformance to specifications.

Section 820.100(a) (1) states:

"...control measures shall be established to assure that the design basis for the device, components and packaging is correctly translated into approved specifications."

Validation is an essential control for assuring that the specifications for the device and manufacturing process are adequate to produce a device that will conform to the approved design characteristics.

VII. PRELIMINARY CONSIDERATIONS

A manufacturer should evaluate all factors that affect product quality when designing and undertaking a process validation study. These factors may vary considerably among different products and manufacturing technologies and could include, for example, component specifications, air and water handling systems, environmental controls, equipment functions, and process control operations. No single approach to process validation will be appropriate and complete in all cases; however, the following quality activities should be undertaken in most situations.

During the research and development (R& D) phase, the desired product should be carefully defined in terms of its characteristics, such as physical, chemical, electrical and performance characteristics.[3] It is important to translate the product characteristics into specifications as a basis for description and control of the product.

Documentation of changes made during development provide traceability which can later be used to pinpoint solutions to future problems.

The product's end use should be a determining factor in the development of product (and component) characteristics and specifications. All pertinent aspects of the product which impact on safety and effectiveness should be considered. These aspects include performance, reliability and stability. Acceptable ranges or limits should be established for each characteristic to set up allowable variations.[4] These ranges should be expressed in readily measurable terms.

---

[3]For example, in the case of a compressed tablet, physical characteristics would include size, weight, hardness, and freedom from defects, such as capping and splitting. Chemical characteristics would include quantitative formulation/potency; performance characteristics may include bioavailability (reflected by disintegration and dissolution). In the case of blood tubing, physical attributes would include internal and external diameters, length and color. Chemical characteristics would include raw material formulation. Mechanical properties would include hardness and tensile strength; performance characteristics would include biocompatibility and durability.

[4]For example, in order to assure that an oral, ophthalmic, or parenteral solution has an acceptable pH, a specification may be established by which a lot is released only if it has been shown to have a pH within a narrow established range. For a device, a specification for the electrical resistance of a pacemaker lead would be established so that the lead would be acceptable only if the resistance was within a specified range.

The validity of acceptance specifications should be verified through testing and challenge of the product on a sound scientific basis during the initial development and production phase.

Once a specification is demonstrated as acceptable it is important that any changes to the specification be made in accordance with documented change control procedures.

VIII. ELEMENTS OF PROCESS VALIDATION

A. Prospective Validation

Prospective validation includes those considerations that should be made before an entirely new product is introduced by a firm or when there is a change in the manufacturing process which may affect the product's characteristics, such as uniformity and identity. The following are considered as key elements of prospective validation.

1. Equipment and Process

The equipment and process(es) should be designed and/or selected so that product specifications are consistently achieved. This should be done with the participation of all appropriate groups that are concerned with assuring a quality product, e.g., engineering design, production operations, and quality assurance personnel.

a. Equipment: Installation Qualification

Installation qualification studies establish confidence that the process equipment and ancillary systems are capable of consistently operating within established limits and tolerances. After process equipment is designed or selected, it should be evaluated and tested to verify that it is capable of operating satisfactorily within the operating limits required by the process.[5] This phase of validation includes examination of equipment design; determination of calibration, maintenance, and adjustment requirements; and identifying critical equipment features that could affect the process and product. Information obtained from these studies should be used to establish written procedures covering equipment calibration, maintenance, monitoring, and control.

In assessing the suitability of a given piece of equipment, it is usually insufficient to rely solely upon the representations of the equipment supplier, or upon

---

[5]Examples of equipment performance characteristics which may be measured include temperature and pressure of injection molding machines, uniformity of speed for mixers, temperature, speed and pressure for packaging machines, and temperature and pressure of sterilization chambers.

experience in producing some other product.[6] Sound theoretical and practical engineering principles and considerations are a first step in the assessment.

It is important that equipment qualification simulate actual production conditions, including those which are "worst case" situations.

Tests and challenges should be repeated a sufficient number of times to assure reliable and meaningful results. All acceptance criteria must be met during the test or challenge. If any test or challenge shows that the equipment does not perform within its specifications, an evaluation should be performed to identify the cause of the failure. Corrections should be made and additional test runs performed, as needed, to verify that the equipment performs within specifications. The observed variability of the equipment between and within runs can be used as a basis for determining the total number of trials selected for the subsequent performance qualification studies of the process.[7]

Once the equipment configuration and performance characteristics are established and qualified, they should be documented. The installation qualification should include a review of pertinent maintenance procedures, repair parts lists, and calibration methods for each piece of equipment. The objective is to assure that all repairs can be performed in such a way that will not affect the characteristics of material processed after the repair. In addition, special post-repair cleaning and calibration requirements should be developed to prevent inadvertent manufacture a of non-conforming product. Planning during the qualification phase can prevent confusion during emergency repairs which could lead to use of the wrong replacement part.

---

[6]The importance of assessing equipment suitability based upon how it will be used to attain desired product attributes is illustrated in the case of deionizers used to produce Purified Water, USP. In one case, a firm used such water to make a topical drug product solution which, in view of its intended use, should have been free from objectionable microorganisms. However, the product was found to be contaminated with a pathogenic microorganism. The apparent cause of the problem was failure to assess the performance of the deionizer from a microbiological standpoint. It is fairly well recognized that the deionizers are prone to build-up of microorganisms -- especially if the flow rates are low and the deionizers are not recharged and sanitized at suitable intervals. Therefore, these factors should have been considered. In this case, however, the firm relied upon the representations of the equipment itself, namely the "recharge" (i.e., conductivity) indicator, to signal the time for regeneration and cleaning. Considering the desired product characteristics, the firm should have determined the need for such procedures based upon pre-use testing, taking into account such factors as the length of time the equipment could produce deionized water of acceptable quality, flow rate, temperature, raw water quality, frequency of use, and surface area of deionizing resins.

[7]For example, the AAMI Guideline for Industrial Ethylene Oxide Sterilization of Medical Devices approved 2 December 1981, states: "The performance qualification should include a minimum of 3 successful, planned qualification runs, in which all of the acceptance criteria are met...." (5.3.1.2.)

b. Process: Performance Qualification

The purpose of performance qualification is to provide rigorous testing to demonstrate the effectiveness and reproducibility of the process. In entering the performance qualification phase of validation, it is understood that the process specifications have been established and essentially proven acceptable through laboratory or other trial methods and that the equipment has been judged acceptable on the basis of suitable installation studies.

Each process should be defined and described with sufficient specificity so that employees understand what is required. Parts of the process which may vary so as to affect important product quality should be challenged.[8]

In challenging a process to assess its adequacy, it is important that challenge conditions simulate those that will be encountered during actual production, including "worst case" conditions. The challenges should be repeated enough times to assure that the results are meaningful and consistent.

Each specific manufacturing process should be appropriately qualified and validated. There is an inherent danger in relying on what are perceived to be similarities between products, processes, and equipment without appropriate challenge.[9]

c. Product: Performance Qualification

For purposes of this guideline, product performance qualification activities apply only to medical devices. These steps should be viewed as pre-production quality assurance activities.

---

[8]For example, in electroplating the metal case of an implantable pacemaker, the significant process steps to define, describe, and challenge include establishment and control of current density and temperature values for assuring adequate composition of electrolyte and for assuring cleanliness of the metal to be plated. In the production of parenteral solutions by aseptic filling, the significant aseptic filling process steps to define and challenge should include the sterilization and depyrogenation of containers/closures, sterilization of solutions, filling equipment and product contact surfaces, and the filling and closing of containers.

[9]For example, in the production of a compressed tablet, a firm may switch from one type of granulation blender to another with the erroneous assumption that both types have similar performance characteristics, and, therefore, granulation mixing times and procedures need not be altered. However, if the blenders are substantially different, use of the new blender with procedures used for the previous blender may result in a granulation with poor content uniformity. This, in turn, may lead to tablets having significantly differing potencies. This situation may be averted if the quality assurance system detects the equipment change in the first place, challenges the blender performance, precipitates a revalidation of the process, and initiates appropriate changes. In this example, revalidation comprises installation qualification of the new equipment and performance qualification of the process intended for use in the new blender.

Before reaching the conclusion that a process has been successfully validated, it is necessary to demonstrate that the specified process has not adversely affected the finished product. Where possible, product performance qualification testing should include performance testing under conditions that simulate actual use. Product performance qualification testing should be conducted using product manufactured from the same type of production equipment, methods and procedures that will be used for routine production. Otherwise, the qualified product may not be representative of production units and cannot be used as evidence that the manufacturing process will produce a product that meets the pre-determined specifications and quality attributes.[10]

In another example, a manufacturer recalled insulin syringes because of complaints that the needles were clogged. Investigation revealed that the needles were clogged by silicone oil which was employed as a lubricant during manufacturing. Investigation further revealed that the method used to extract the silicone oil was only partially effective. Although visual inspection of the syringes seemed to support that the cleaning method was effective, actual use proved otherwise.

After actual production units have successfully passed product performance qualification, a formal technical review should be conducted and should include:

- o Comparison of the approved product specifications and the actual qualified product.
- o Determination of the validity of test methods used to determine compliance with the approved specifications.
- o Determination of the adequacy of the specification change control program.

2. System to Assure Timely Revalidation

There should be a quality assurance system in place which requires revalidation whenever there are changes in packaging, formulation, equipment, or processes which could impact on product effectiveness or product characteristics, and whenever there are changes in product characteristics. Furthermore, when a change is made in raw material supplier, the manufacturer should consider subtle, potentially adverse differences in the raw material characteristics. A determination of adverse differences in raw material indicates a need to revalidate the process.

---

[10]For example, a manufacturer of heart valves received complaints that the valve-support structure was fracturing under use. Investigation by the manufacturer revealed that all material and dimensional specifications had been met but the production machining process created microscopic scratches on the valve supporting wireform. These scratches caused metal fatigue and subsequent fracture. Comprehensive fatigue testing of production units under simulated use conditions could have detected the process deficiency.

One way of detecting the kind of changes that should initiate revalidation is the use of tests and methods of analysis which are capable of measuring characteristics which may vary. Such tests and methods usually yield specific results which go beyond the mere pass/fail basis, thereby detecting variations within product and process specifications and allowing determination of whether a process is slipping out of control.

The quality assurance procedures should establish the circumstances under which revalidation is required. These may be based upon equipment, process, and product performance observed during the initial validation challenge studies. It is desirable to designate individuals who have the responsibility to review product, process, equipment and personnel changes to determine if and when revalidation is warranted.

The extent of revalidation will depend upon the nature of the changes and how they impact upon different aspects of production that had previously been validated. It may not be necessary to revalidate a process from scratch merely because a given circumstance has changed. However, it is important to carefully assess the nature of the change to determine potential ripple effects and what needs to be considered as part of revalidation.

3. Documentation

It is essential that the validation program is documented and that the documentation is properly maintained. Approval and release of the process for use in routine manufacturing should be based upon a review of all the validation documentation, including data from the equipment qualification, process performance qualification, and product/package testing to ensure compatibility with the process.

For routine production, it is important to adequately record process details (e.g., time, temperature, equipment used) and to record any changes which have occurred. A maintenance log can be useful in performing failure investigations concerning a specific manufacturing lot. Validation data (along with specific test data) may also determine expected variance in product or equipment characteristics.

B. Retrospective Process Validation

In some cases a product may have been on the market without sufficient premarket process validation. In these cases, it may be possible to validate, in some measure, the adequacy of the process by examination of accumulated test data on the product and records of the manufacturing procedures used.

Retrospective validation can also be useful to augment initial premarket prospective validation for new products or changed processes. In such cases, preliminary prospective validation should have been sufficient to warrant product marketing. As additional data is gathered on production lots, such data can be used to build confidence in the adequacy of the process. Conversely, such data may indicate a declining confidence in the process and a commensurate need for corrective changes.

Test data may be useful only if the methods and results are adequately specific. As with prospective validation, it may be insufficient to assess the process solely on the basis of lot by lot conformance to specifications if test results are merely expressed in terms of pass/fail. Specific results, on the other hand, can be statistically analyzed and a determination can be made of what variance in data can be expected. It is important to maintain records which describe the operating characteristics of the process, e.g., time, temperature, humidity, and equipment settings.[11] Whenever test data are used to demonstrate conformance to specifications, it is important that the test methodology be qualified to assure that test results are objective and accurate.

IX. ACCEPTABILITY OF PRODUCT TESTING

In some cases, a drug product or medical device may be manufactured individually or on a one-time basis. The concept of prospective or retrospective validation as it relates to those situations may have limited applicability, and data obtained during the manufacturing and assembly process may be used in conjunction with product testing to demonstrate that the instant run yielded a finished product meeting all of its specifications and quality characteristics. Such evaluation of data and product testing would be expected to be much more extensive than the usual situation where more reliance would be placed on prospective validation.

---

[11]For example, sterilizer time and temperature data collected on recording equipment found to be accurate and precise could establish that process parameters had been reliably delivered to previously processed loads. A retrospective qualification of the equipment could be performed to demonstrate that the recorded data represented conditions that were uniform throughout the chamber and that product load configurations, personnel practices, initial temperature, and other variables had been adequately controlled during the earlier runs.

OFFICE OF REGIONAL OPERATIONS
INTERNATIONAL PROGRAMS AND
TECHNICAL SUPPORT BRANCH

GUIDE TO THE INSPECTION OF FOREIGN PHARMACEUTICAL MANUFACTURING PLANTS

September 14, 1992

BACKGROUND

There has been a significant increase in the number of foreign inspections of pharmaceutical manufacturing plants and especially the number of foreign pre-approval inspections. These increases have placed a significant burden on the foreign inspection program resources and along with Desert Storm produced an inspection backlog. Additionally, ORA managers are aware of the pharmaceutical industry's concerns about consistency and uniformity of inspection and enforcement between the domestic and foreign inspection programs.

To resolve these issues, ORA will not only increase the resources devoted to foreign inspection of pharmaceutical plants in FY'93, but will also implement significant changes in the administration and management of the foreign inspection program on 10-1-92. These changes are designed to reduce the inspection backlogs through strengthened teamwork and cooperation between the field and headquarters. We expect these new concepts to adjust consistency and uniformity where appropriate.

Consistency and uniformity of inspection and enforcement represent high priority goals in the ORA organization. These guidelines set forth clear instructions regarding the approach to the foreign inspection. Additionally only highly trained and experienced investigators and analysts with substantial experience in conducting pre-approval inspections will be assigned to conduct pre-approval inspections.

On September 14, 1992 ORA will provide training for all the members of the foreign inspection cadre to provide orientation to foreign travel and to explain the instructions contained in this document. We expect to monitor our compliance with the instructions in this document.

ORA needs to reduce the cost of travel to conduct on site foreign inspections. An assortment of documents are reviewed and evaluated during inspections of domestic and foreign establishments and ORA management has decided that some of these documents may be reviewed off site resulting in a significant saving of operational funds that would be spent if the inspection teams conducted the document review while in domestic or foreign travel status. On site audits of the raw data and the plant configurations and equipment will be required to verify the accuracy and authenticity of the data.

PURPOSE

The guidelines in this document are designed to reduce the inspection backlog, to establish greater inspection consistency and uniformity in both domestic and foreign inspections of drug manufacturers, to significantly reduce the cost of foreign travel, and to improve efficiency. Additionally, these guidelines will strengthen the foreign inspection program through greater use of the field resources at all levels of the program.

This document contains an outline of information and documentation that the inspection team is expected to review prior to the foreign inspection trip. As a minimum, the inspection team is expected to review the information contained in the documents listed in this guide. Since this information is requested from the plants prior to the inspection, part of the on site plant inspection will include verification of the authenticity, accuracy and completeness of the data provided. Different information will be requested and reviewed depending upon the type of manufacturer or ancillary firm (drug substance, oral solid dosage form, etc.) inspected.

These guidelines and instructions also explain some of the changes in program administration that will be used to manage this new foreign inspection program. They set forth specific information about the role of the inspection team and our staff in the Office of Regional Operations. They explain the communications process and the flow of work from the field to the Office of Regional Operations. They set forth guidance for inspection teams to communicate with officials of the regulated industry. And, there is information about the scheduling of inspection trips.

OVERALL PROGRAM ADMINISTRATION AND MANAGEMENT

The foreign inspection program continues to be under the management and direction of the International Programs and Technical Support Branch (IPTSB) of the Office of Regional Operations. Mr. Richard Klug heads this branch and many of the agency and industry officials are accustomed to dealing with him and with Peter Smith. These officers will continue to manage these programs however consideration is being given to increasing the staff in this unit. This branch provides overall management of the program, conducts liaison activities with CDER, industry, and field officials, and schedules inspection trips. Members of IPTSB reviewed and evaluated all inspection reports that were prepared from the foreign inspections; now the field will assist in this effort by evaluating and endorsing these reports before sending them to IPTSB. Additionally, the field is expected to draft a letter to the foreign plant when it recommends withholding approval of an application. This draft letter is to be sent to IPTSB for issuance to the foreign plant within the time frames specified in the ORA Management Plan.

Members of IPTSB will assure that consistent inspection approaches are being used and that decisions made by the field are consistent and uniform. They will resolve differences that may arise over inspection scheduling, availability of documents and other issues concerning the foreign inspection program.

The field foreign inspection cadre will be managed by the appropriate field managers. They must ensure that the assigned work is completed and that reports and decisions regarding necessary follow up are submitted to ORO/IPTSB on a timely basis. Field managers are accountable for the work that has been identified in the ORA field workplan covering foreign inspection responsibilities assigned to their districts and regions.

NEW PROCEDURES AND CONCEPTS

SUPPORT ACTIVITIES

Field managers will be accountable for typing and overall support of all the staff assigned to their regional foreign inspection cadre. This means that they will prepare expense accounts, review expense vouchers, and provide for other general support as may be required. They must also assure that each member of the foreign inspection cadre has a notebook computer assigned to them as well as other equipment that might be needed.

PRE-APPROVAL INSPECTIONS

IPTSB receives requests from the Center for pre-approval inspection clearance of specific NDA/ANDAs. Additionally, companies may have other applications pending before the agency and these applications may not have reached the point where they can be considered for approval. Nevertheless, the district NDA/ANDA program managers are expected to have background information on all pending applications filed by sponsors within their district. This information will be given to the inspection team and when time and priorities allow, these applications will be audited during the inspection. The inspection team leader is expected to advise a designated contact (now Peter Smith) by FAX of all the pending applications they expect to cover during the inspection. Upon return the report will clearly list all the applications that were audited during the inspection and the status of the application.

There will be cases where the inspection team discovers conditions that will impact on all pending applications or where a pending application is audited before CDER requests clearance. District NDA/ANDA program managers are expected to advise IPTSB of significant issues that must be resolved before an application can be approved even though CDER has not completed its review of the application. Such advance notice will be helpful to the reviewing office and to the sponsor.

REVIEWING DOCUMENTS PRIOR TO TRIP

Inspection teams are expected to review certain documents and records prior to the inspection trip. This will reduce the actual on site inspection time and reduce travel costs. Additionally it should increase efficiency and create greater uniformity.

Inspection teams are expected to carefully review the data in these documents and to select those issues that need careful attention during the on site inspection. As a minimum every inspection is to focus on an extensive evaluation of process failures, product failures, failures in laboratory tests, and process changes. Obviously, process failures are one of the most important points to cover during the inspection and these failures and process changes have a significant impact on the adequacy of process validation.

In the area of bulk drug substances, you are expected to carefully evaluate the company's compliance with the impurity profiles contained in the DMF or application.

The following documents represent the minimum which must be reviewed and evaluated prior to or during an inspection. The information contained in the following lists will be supplied by the pharmaceutical plant prior to the inspection trip. In some cases the plant may not have the information or the documents listed. When this occurs, the inspection team will have to develop the information contained in the documents and this will obviously have to be achieved on site. When this approach is required more time will be planned for the inspection to allow the inspection team to gain a complete assessment of the plant's ability to comply with GMPs and to audit the NDA/ANDA(s).

NOTE: Some of the information listed below may be found in the DMF or in the filed application. In such cases you are expected to review the information in the application rather than request the company to submit the same information again.

DRUG SUBSTANCE

Following is a list of the minimum information that must be reviewed and/or evaluated during an inspection of drug substance manufacturers. This information will be found in the filing or requested from the manufacturer or sponsor prior to the inspection trip as outlined in this document.

- DMF Copy and Review Chemist Comments (obtained from IPTSB or from the sponsor)
- DMF Facility (Type 1)
- List of Products Manufactured and Sites of Manufacture (list only products shipped to the U.S.)
- Manufacturing Record -- Final Purification Step (for an NDA product, drug substance used for biobatch)
- Validation Report for Drug Substance Manufacturing Process
  - Intermediate -- Source and Specifications
  - Processing Steps and Limitations
    - Rework, Reprocessing, Mother Liquors, Boildown, Second Crop and Recovered Solvents if utilized should be discussed
  - Operating Ranges and Equipments
  - Controls, Specifications and Tests
  - Purity Profile -- Mass Balance, List of all Impurities
  - Full Batch Size Consistency
  - Summary and Conclusion
- Annual Report -- History of Production
  - List of Batches Finished and all Test Results
  - List of Batches Through Reactor/Crystallizer

Note: Manufacturers may reprocess/recrystallize a batch and not assign a new number. Unless such a step is performed 100% of the time, it is a rework or reprocess and should be identified.

Reports of Failure Investigations of any Batches Requiring Rework/Reprocess within the last 100 batches manufactured.
Description/SOP Covering the Quality Assurance or Quality Control Dept.
  Failure Investigation
  Process Changes
  Monitoring Operations
  Review In-Process Testing
  Review Production and Control Records
Cleaning Validation SOP
Cleaning Validation Report for Specific Drug Substances
Process Change SOP
  Log/list of all processes changed (not just drug(s) for which pre-approval is being conducted)
List of all Complaints for the past year
Validation Report for Assay and Impurity Tests (unless report is filed)

Microbiological Data

NOTE: (For drug substances used for inhalation products and for parenterals, there are microbiological concerns). Microbiological data should be requested and should include:

Microbial Test Results for all batches
Environmental Test Results from the area where the drug substance is isolated, dried and packaged

STERILE BULK DRUG SUBSTANCE

In addition to the above:

List of all initial positive sterility test results and reports of investigation
List/log of all organisms isolated and source
SOP for Environmental Monitoring
Environmental Monitoring Monthly or Quarterly Reports and Investigations
SOP for Monitoring of WFI System
List/Summary of Daily Water for Injection Test Results (endotoxin/microbiological results)

ORAL AND TOPICAL DOSAGE FORMS

Biobatch Identity
Manufacturing and Controls Section of Filing
Development Report -- Data/documentation needed to support the filed process based on pilot/development batches
  Formulation
  Equipment

Manufacturing Procedures
In-Process Test Data
Final Dosage Form Test Results
Conclusion
Certification

Master Formula (if not filed in application)
Batch Record of First Full Scale Size Production Batch
Validation Protocol*
Validation Report*
Annual Report (History of Production)*
List of all batches and all test results
Reports of failure investigations
Description/SOP covering the Quality Assurance or Quality Control investigation of problems, defects, failures
Log/List of all Quality Assurance/Quality Control investigations
Process Change SOP
Cleaning Validation SOP
Cleaning Validation Report for specific drug product
List of all Applicable Complaints
SOP for Field Alerts
Validation Report for Stability Tests (Assay and Impurities)
Microbiological Data -- Microbiological data including test procedures and test results may be requested, particularly for topical and inhalation products.

STERILE DOSAGE FORM

Manufacturing and Controls Section of Filing
Development Report -- Data/documentation needed to support the filed process
Physical Factors (extrusion force, particulate levels)
Metal/light Sensitivity
Filter Compatibility and Integrity
Oxygen Sensitivity
Container Requirements
Preservative Effectiveness
Heat Sensitivity (autoclavability)
Drug Substance (source, chemical, microbial purity)
Manufacturing Procedures and Equipment
In-Process Test Data
Final Dosage Form Test Results
Conclusion
Certification
Master Formula (if not filed in application)
Batch Production Record of First Full Scale Production Batch

---

*Optional -- Dependent upon history of firm and post-approval inspection feasibility.

Validation Protocol*
Validation Report*
Annual Report (History of Production)*
  List of all batches and all test results
  Reports of failure investigations
Description/SOP covering the quality Assurance or Quality Control investigation of problems, defects, failures
Log/list of all Quality Assurance/Quality Control investigations
Process Change SOP
Cleaning Validation SOP
Cleaning Validation Report for Specific Drug Product*
List of all Applicable Complaints
SOP for Field Alerts
Validation Report for Stability Tests (Assay and Impurities)
List of all initial positive sterility test results and reports of investigations
Media fill results and reports of observations during the conduct of the fill
List/log of all organisms isolated and source
SOP for Environmental Monitoring
Environmental Monitoring Monthly or Quarterly Reports and Investigations
SOP for Monitoring of WFI System
List/Summary of Daily Water for Injection Test Results (endotoxin/microbiological results)

INSPECTION SCHEDULING

IPTSB will schedule inspections and to the extent possible assign the inspection to the district housing the sponsor who filed the application. For example, coverage of a foreign Merck plant would be assigned to the Mid Atlantic Region which allows the inspection team to visit the sponsor prior to the trip. In this case the inspection actually begins on the date that contact is made with the domestic sponsor of the application. An FD 482 should be issued to the domestic sponsor when visited in connection with a foreign plant inspection. It may be appropriate to issue an FD 483 to plant officials at the domestic sponsor location and at the foreign plant location depending on the person responsible for the issues listed on the FD 483.

Either a member of the IPTSB or the inspection team leader will contact the domestic sponsor of an application to obtain the information listed in the above sections to this document. The inspection team leader will be advised of the arrangements that have been made to obtain information. Arrangements for the collection of documents and data should be made so that the documents are sent directly to the inspection team leader and arrive within three weeks of the inspection team departure date. Include the time used to review documents prior to the inspection and time spent at the domestic sponsor as inspection time.

---

*Optional -- Dependent upon history of firm and post-approval inspection feasibility.

Documents and information will be collected and sent to the team leaders rather than the inspection team spending valuable time collecting documents from various headquarters offices. Where required, a briefing will be held prior to the trip, however briefings held to chiefly review documents in headquarters will be suspended unless the inspection issues are unusually complex.

COMMUNICATIONS

The inspection team is authorized to contact the domestic sponsor as part of the foreign inspection. You may contact them regarding answers to questions about the data that has been submitted, to arrange specific schedules to meet and discuss the data in the documents or to obtain additional information prior to the foreign inspection trip. When the inspection team needs to travel to the domestic sponsor, clear such travel through IPTSB.

Team members may contact the reviewing officers by telephone to discuss issues regarding specific applications. However, we expect the team to be supplied with written points of interest or issues that they will be expected to cover during the foreign inspection trip.

REVIEW AND ENDORSEMENT OF INSPECTION RESULTS

District office staff must review all inspection findings and make appropriate judgments (consistent with judgements made when similar issues are found in domestic plants), about the need for additional agency action. For example, districts are expected to make recommendations to withhold or to concur in approval of NDA/ANDAs using the same criteria they apply to domestic plant inspections. When the district decides to recommend that approval of an application be withheld, the district will draft a letter to the foreign plant and send the draft letter to IPTSB for issuance. Again consistency and uniformity is important.

Districts may concur in approval of applications following the implementation of corrections even though we do not conduct a follow up inspection to assess the implementation of corrective actions. On the other hand some corrections will require that a follow up on site evaluation be conducted prior to the approval of applications. Districts must use their own judgement and apply the same principles used for domestic plants. Obviously, the district will want to take into consideration such factors as the significance to the condition, the complexity of the technology and the compliance record of the plant.

Districts are expected to send decisions to concur or withhold approval to IPTSB. Send original inspection reports with the district's endorsement within the time frames specified in the ORA Management Plan for domestic inspections. Send the ten day notification to IPTSB with a copy to CDER as specified in the ORA Management Plan.

We expect companies to make corrections to the conditions reported by the inspection team. Documentation certifying that corrections have been made is sent to IPTSB, who will send it to the inspection team for review. District offices must review the documentation regarding corrective actions and submit a final recommendation to IPTSB along with copies of the documentation received from the foreign plant officials and from headquarters. (The Official Establishment File for all foreign plants is maintained in headquarters.)

VALIDATION POLICY

We expect that drug manufacturing processes for all dosage forms including bulk drug substances be validated prior to the pre-approval inspection. The agency will not likely be able to provide timely follow up evaluations of process validation because the costs associated with a follow up inspection are prohibitive at this time.

DOCUMENTATION OF VIOLATIONS OF GMPS

The same level of documentation for violations of GMPs and protocols is not required for foreign versus domestic inspections because the agency has authority under the FD&C act to restrict the importation of a product. Nevertheless, documentation should include copies of documents to prove that significant violations exist and to prove that restricting importation and/or non-approval of an application are warranted. Where fraud is discovered collect sufficient documentation to prove it.

CENTER FOR DRUG EVALUATION AND RESEARCH (CDER) GUIDELINES

GUIDE TO INSPECTION OF COMPUTERIZED SYSTEMS IN DRUG PROCESSING

FEBRUARY, 1983

National Center for Drugs and Biologics
and
Executive Director of Regional Operations
Division of Drug Quality Compliance (HFN-320)
Associate Director for Compliance
Office of Drugs
National Center for Drugs and Biologics
and
Division of Field Investigations (HFO-500)
Associate Director for Field Support
Executive Director of Regional Operations

REFERENCE MATERIALS AND TRAINING AIDS FOR INVESTIGATORS

U.S. DEPT. OF HEALTH AND HUMAN SERVICES
PUBLIC HEALTH SERVICE
FOOD AND DRUG ADMINISTRATION

I. *INTRODUCTION*

Computers are being used in increasing numbers in the pharmaceutical industry. As microprocessors become more powerful, reliable, and less expensive we can expect the proliferation of this technology, with increasing use by even very small pharmaceutical establishments. Computer systems are used in a wide variety of ways in a pharmaceutical establishment, such as, maintenance of quarantine systems for drug components, control of significant steps in manufacturing the dosage form, control of laboratory functions, management of warehousing and distribution activities. Computer systems may control one or more of these phases, either singly or as part of a highly automated integrated complex.

The purpose of this guide is to provide the field investigator with a framework upon which to build an inspection of drug establishments which utilize computer systems. This document is not intended to spell out how to conduct a CGMP drug inspection or set forth reporting requirements, but rather what aspects of computerized systems to address during such inspections and suggestions on how to address the systems.

This guide discusses some potential problem areas in application of computer systems, provides inspectional guidance, and includes a glossary of terms the investigator should be aware of prior to performing the inspection.

Questions and suggestions concerning this guide should be directed to the Manufacturing Standards and Industry Liaison Branch, Division of Drug Quality Compliance (HFN-323), 443-5307; or the Investigations and Engineering Branch, Division of Field Investigations (HFO-520), 443-3276.

II. *OVERVIEW*

When a computer system is first encountered in a drug establishment, it may be useful for inspectional purposes to begin with a broad overview of the system(s). Determine exactly what processes and functions are under computer control or monitoring and which are not. Computer involvement may be much more limited than it may initially appear. For example, computer application may be limited to control of a sterilization cycle in a single autoclave, or maintenance of distribution records.

For each drug process under computer control determine the general system loop (sensors, central processor, activator). For example, the general system loop for a steam autoclave under computer control could consist of temperature/pressure sensors connected to a microprocessor which transmits commands to steam/vacuum control valves.

The overview should enable the investigator to identify those computer controlled processes which are most critical to drug product quality. These are the systems which, of course, merit closer inspection.

III. *HARDWARE*

For each significant computerized system, it may be helpful to prepare or obtain a simplified schematic drawing of the attendant hardware. The drawing need only include major input devices, output devices, signal converters, central processing unit, distribution systems, significant peripheral devices and how they are linked.

Hardware Suppliers. During the inspection identify the manufacturers/suppliers of important computer hardware, including the make and model designations where possible. Hardware to identify this way includes CPUs, disk/tape devices, CRTs, printers, and signal converters. Proper identification of hardware will enable further follow-up at computer vendors should that be needed.

A. *Types*

1. *Input Devices*. Equipment which translates external information into electrical pulses which the computer can understand. Examples are thermocouples, flow meters, load cells, pH meters, pressure gauges, control panels, and operator keyboards. Examples of functions are:

a) Thermocouple provides temperature input for calculation of F value in a sterilizer.
b) Flow meter provides volume of liquid component going into a mixing tank.
c) Operator keyboard used to enter autoclave load pattern number.

2. *Output Devices*. Equipment which receives electrical pulses from the computer and either causes an action to occur, generally in controlling the manufacturing process, or passively records data. Examples are valves, switches, motors, solenoids, cathode ray tubes (CRTs), printers, and alarms. Examples of functions are:

a) Solenoid activates the impeller of a mixer.
b) Valve controls the amount of steam delivered to a sterilizer.
c) Printer records significant events during sterilization process.
d) Alarm (buzzer, bell, light, etc.) sounds when temperature in a holding tank drops below desired temperature.

Most active output devices will be in proximity to the drug processing equipment under control, but not necessarily close to the CPU. Passive output devices, however, may well be remote from the process or the CPU.

3. *Signal Converters*. Many input and output devices operate by issuing/receiving electrical signals which are in analog form. These analog signals must be converted to digital signals for use by the computer; conversely, digital signals from the computer must be converted into analog signals for use by analog devices. To accomplish this, signal converter devices are used.

4. *Central Processing Unit (CPU)*. This is the controller containing the logic circuitry of a computer system which conducts electronic switching. Logic circuits consist of three basic sections - memory, arithmetic, and control. The CPU receives electrical pulses from input devices and can send electrical pulses to output devices. It operates from input or memory instructions. Examples and functions are:

a) Programmable controllers can be used for relays, timers and counters.
b) Microprocessors can be used for controlling a steam valve, maintaining pH, etc. They consist of a single integrated circuit on a chip. This is the logic circuit of a microcomputer and microprocessors are often the same as a microcomputer.
c) Microcomputers and minicomputers can be used to control a sterilization cycle, keep records, run test programs, perform lab data analysis, etc.
d) Mainframe computers are usually used to coordinate an entire plant, such as environment, production, records, and inventory.

The distinction between CPUs is becoming less apparent with miniaturization of parts, CPUs are generally ranked by size from "large" mainframes to desk top microcomputers.

5. *Distribution System.* The interconnection of two or more computers. Also known as distributed processing. Generally, each computer is capable of independent operation but is connected to other computers in order to have a back-up system, to receive operating orders and to relay what is executed by other computers. Typical of such distribution systems is the linkage of smaller or less powerful units to larger or more powerful units. For example, a minicomputer may command and communicate with several microcomputers. A large CPU may also act as a "host" for one or more other CPU's. When such systems are encountered during an inspection, it is important to know the configuration of the system and exactly what commands and information can be relayed amongst the computers.

Networks are generally extensions of distributed processing. They typically consist of connections between complete computer systems which are geographically distant. Potentially, pharmaceutical companies could have international networks by using modems and satellites.

6. *Peripheral Devices.* All computer associated devices external to the CPU can be considered peripheral devices. This includes the previously discussed input and output devices. Many peripheral devices can be both input and output, they are commonly known as I/O devices. These include CRTs, printers, keyboards, disk drives, modems, and tape drives.

B. *Key Points*

1. *Location.* Three potential problems have been identified with location of CPUs and peripheral devices. These are:

a) *Hostile Environments.* Environmental extremes of temperature, humidity, static, dust, power feed line voltage fluctuations, and electromagnetic interference should be avoided. Such conditions may be common in certain pharmaceutical operations and the investigator should be alert to locating sensitive hardware in such areas. Environmental safeguards maybe necessary to ensure proper operation. There are numerous items on the market (such as line voltage monitors/controllers and anti-static floor mats) designated to obviate such problems.

Physical security is also a consideration in protecting computer hardware from damage; for example, books and bottles of reagents should not be stored on top of microprocessors. Likewise eating, drinking and smoking should be restricted in rooms housing mainframes.

b) *Excessive Distances between CPU and Peripheral Devices.* Excessively long low voltage electrical lines from input devices to the CPU are vulnerable to electromagnetic interference. This may result in inaccurate or distorted input data to the computer. Therefore, peripheral devices should be located as near to the CPU as practical and the lines should be shielded from such sources of electromagnetic interference as electrical power lines, motors, and fluorescent lighting fixtures. In a particularly "noisy" electronic environment this problem might be solved by the use of fiber optic lines to convey digital signals.

c) *Proximity of Input Devices to Drug Processing.* Input devices which are remote from (out of visual range of) the drug processing equipment are sometimes met with poor employee acceptance.

2. *Signal Conversion.* Proper analog/digital signal conversion is important if the computer system is to function accurately. Poor signal conversion can cause interface problems. For example, an input sensor may be feeding an accurate analog reading to a signal converter, but a faulty signal converter may be sending the CPU an inappropriate digital signal.

3. *I/O Device Operation.* The accuracy and performance of these devices are vital to the proper operation of the computer system. Improper inputs from thermocouples, pressure gauges, etc., can compromise the most sophisticated microprocessor controlled sterilizer. These sensors should be systematically calibrated and checked for accurate signal outputs.

4. *Command Over-rides.* In distributed systems it is important to know how errors and command over-rides at one computer are related to operations at another computer in the system. For example, if each of three interconnected microcomputers runs one of the three sterilizers, can a command entered at one unit inadvertently alter the sterilization cycle of a sterilizer under the control of a different microcomputer on the line? Can output data from one unit be incorrectly processed by another unit? The limits on information and command flow within a distributed system should be clearly established by the firm.

5. *Maintenance.* Computer systems usually require a minimum of complex maintenance. Electronic circuit boards, for example, are usually easily replaced and cleaning may be limited to dust removal. Diagnostic software is usually available from the vendor to check computer performance and isolate defective integrated circuits. Maintenance procedures should be stated in the firm's standard operating procedures. The availability of spare parts and access to qualified service personnel are important to the operation of the maintenance program.

C. *Validation of Hardware*

The suitability of computer hardware for the tasks assigned to pharmaceutical production must be demonstrated through appropriate tests and challenges. The depth and scope of hardware validation will depend upon the complexity of the system and its potential affect on drug quality.

The validation program need not be elaborate but should be sufficient to support a high degree of confidence that the system will consistently do what it is supposed to do. In considering hardware validation the following points should be addressed:

1. Does the capacity of the hardware match its assigned function? For example, in a firm using a computer system to maintain its labeling text, including foreign language labeling, do the CRT and printer have the capacity to write foreign language accent marks?

2. Have operational limits been identified and considered in establishing production procedures? For example, a computer's memory and connector input ports may limit the number of thermocouples a computer can monitor. These limits should be identified in the firm's standard operating procedures.

3. Have test conditions simulated "worst case" production conditions? A computer may function well under minimal production stress (as in a vendor's controlled environment) but falter under high stresses of equipment speed, data input overload or frequent or continuous multi-shift use (and a harsh environment). Therefore, it is insufficient to test computer hardware for proper operation during a one hour interval, when the system will be called upon in worst case conditions to run continuously for 14 days at a time. Some firms may test the circuits of a computer by "feeding" it electrical signals from a signal simulator. The simulator sends out voltages which are designed to correspond to voltages normally transmitted by input devices. When simulators are connected to the computer, the program should be executed as if the emulated input devices were actually connected. These signal simulators can be useful tools for validation; however, they may not pose worse case conditions and their accuracy in mimicking input device performance should be established. In addition, validation runs should be accomplished on line using actual input devices. Signal simulators can also be used to train employees on computer operations without actually using production equipment.

4. Have hardware tests been repeated enough times to assure a reasonable measure of reproducibility and consistency? In general, at least three test runs should be made to cover different operating conditions. If test results are widely divergent they may indicate an out of control state.

5. Has the validation program been thoroughly documented? Documentation should include a validation protocol and test results which are specific and meaningful in relation to the attribute being tested. For example, if a printer's reliability is being tested it would be insufficient to express the results merely as "passes," in the absence of other qualifying data such as printing speeds, duration of printing, and the number of input feeds to the printing devices.

6. Are systems in place to initiate revalidation when significant changes are made? Revalidation is indicated, for example, when a major piece of equipment such a circuit board or an entire CPU is replaced. In some instances identical hardware replacements may adequately be tested by the use of diagnostic programs available from the vendor. In other cases, as when different models of hardware are introduced, more extensive testing under worst case production conditions, is indicated.

Much of the hardware validation may be performed by the computer vendor. However, the ultimate responsibility for suitability of equipment used in drug processing rests with the pharmaceutical manufacturer. Hardware validation data and protocols should be kept at the drug manufacturer's facility. When validation information is produced by an outside firm, such as the computer vendor, the records maintained by the drug establishment need not be all inclusive of voluminous test data; however, such records should be reasonably complete (including general results and protocols) to allow the drug manufacturer to assess the adequacy of the validation. A mere certification of suitability from the vendor, for example, is inadequate.

IV. *SOFTWARE*

Software is the term used to describe the total set of programs used by a computer. These programs exist at different language levels, generally the higher the level, the closer the text is to human language. These levels are set forth below. During the inspection identify key computer programs used by the firm. Of particular importance are those programs which control and document dosage form production and laboratory testing. Usually a firm can readily list the names of such programs on a CRT display or in hard copy. Such a list is sometimes called a menu or main menu.

A. *Levels*

1. *Machine Language.* This language consists of coded instructions, represented by binary numbers, which are executed directly by the computer.

2. *Assembly Language.* Instructions are represented by alphanumeric abbreviations. These programs must be converted into machine language, sometimes called "object programs," before they can be executed. Programs which translate assembly programs to object programs are called assemblers. Different computers have different assembly languages. Computer manufacturers usually provide the assembler program.

3. *High Level Language.* This language is characterized by a vocabulary of English words and mathematical symbols. These are source programs which must be translated by a compiler or interpreter into an object program. High level languages generally operate the same on any computer which accepts the language although there may be different versions of the same language. Examples are FORTRAN, BASIC, and COBOL.

4. *Application Language.* This is generally based on a high level language but modified for a specific industry application and uses the vocabulary of that industry. Examples are AUTRAN (Control Data Corporation) and Foxboro Process Basic.

B. *Software Identification*

For the key computer programs used by a firm, the following items should be identified:

1. *Language.* High level or application name should be determined (or machine or assembly language).

2. *Name.* Programs are usually named with some relationship to what they do, i.e. Production Initiation, Batch History Transfer or Alarms.

3. *Function.* Determine what the purpose of the program is, i.e., start production, record and print alarms, or calculate F.

4. *Input.* Determine inputs, such as thermocouple signals, timer, or analytical test results.

5. *Output.* Determine what outputs the program generates. These may be a form of mechanical action (valve actuation) or recorded data (generation of batch records).

6. *Fixed Setpoint.* This is the desired value of a process variable which cannot be changed by the operator during execution. Determine major fixed setpoints, such as desired time/temperature curve, desired pH, etc. Time may also be used as a setpoint to stop the process to allow the operator to interact with the processing.

7. *Variable Setpoint.* This is the desired value of a process variable which may change from run to run and must usually be entered by the operator. For example, entering one of several sterilizer load patterns into a sterilization computer process.

8. *Edits.* A program may be written in such a manner as to reject or alter certain input or output information which does not conform to some pre-determined criterion or otherwise fall within certain pre-established limits. This is an edit and it can be a useful way of minimizing errors; for example, if a certain piece of input data must consist of a four character number, program edits can be used to reject erroneous entry of a five character number or four characters comprised of both numbers and letters. On the other hand, edits can also be used to falsify information and give the erroneous impression that a process is under control; for example, a program output edit may add a spurious "correction" factor to F values which fall outside of the pre-established limits, thus turning an unacceptable value into an "acceptable" value. It is, therefore, important to attempt to identify such significant program edits during the inspection, whenever possible. Sometimes such edits can manifest themselves in unusually consistent input/output information.

9. *Input manipulation.* Determine how a program is set up to handle input data. For example, determine what equations are used as the basis for calculations in a program. When a process is under computer control determine, in simplified form such as a

flow chart, how input is handled to accomplish the various steps in the process. This does not mean that a copy of the computer program itself needs to be reviewed. However, before computerized control can be applied to a pharmaceutical process there usually needs to be some source document, written in English, setting forth in logical steps what needs to be done; it would be useful to review such a document in evaluating the adequacy of conversion from manual to computerized processing.

10. *Program Over-rides* A program may be such that the sequence of program events or program edits can be over-ridden by the operator. For example, a process controlling program may cause a mixer to stop when the mixer's contents reach a predetermined temperature. The program may prevent the mixer from resuming activity until the temperature has dropped back to the established point. However, the same program may allow an operator to over-ride the stop and reactivate the mixer even at a temperature which exceeds the program limit. It is therefore important to know what over-rides are allowed, and if they conflict with the firm's SOP.

C. *Key Points*

1. *Software Development.* During the inspection determine if the computer programs used by the firm have been purchased as "canned" from outside vendors, developed within the firm, prepared on a customized basis by a software producer, or some combination of these sources. Some programs are highly specialized and may be licensed to pharmaceutical establishments. If the programs used by the firm are purchased or developed by outside vendors determine which firms prepared the programs.

In some cases "canned" or customized programs may contain segments (such as complex algorithms) which are proprietary to their authors and which cannot normally be readily retrieved in program code without executing complex code breaking schemes. In these cases the buyer must accept on faith that the software will perform properly. If the drug manufacturer is using such a program to control or monitor a significant process, determine what steps the firm has taken to assure itself that such program blind spots do not compromise the program performance.

Where drug firms develop their own application programs, review the firm's documentation of the approval process. This approval process should be addressed in the firm's SOP. It may be useful to review the firm's source (English) documents which formed the basis of the programs.

2. *Software Security.* Determine how the firm prevents unauthorized program changes and how data are secure from alteration, inadvertent erasures, or loss (21 CFR 211.68). Some computers can only be operated in a programming mode when two keys are used to unlock an appropriate device. When this security method is used, determine how use of keys is restricted. Another way of achieving program security is the use of ROM (read only memory), PROM (programmable read only memory), or EPROM (erasable programmable read only memory) modules within the

computer to "permanently" store programs. Usually, specialized equipment separate from the computer is needed to change an EPROM or establish a program in PROM so that changes would not be made by the operator. A program in EPROM is erased by exposure of the module (which has a quartz window) to ultraviolet light. In these cases a program is secure to the extent it can't be over-ridden by the operator.

Determine who in the firm has the ability and/or is authorized to write, alter or have access to programs. The firm's security procedures should be in writing. Security should also extend to devices used to store programs, such as tapes, disks and magnetic strip cards. Determine if accountability is maintained for these devices and if access to them is limited. For instance, magnetic strip cards containing a program to run a sterilization cycle may be kept in a locked cabinet and issued to operators on a charge-out basis with return of the card immediately after it is used.

D. *Validation of Software*

It is vital that a firm have assurance that computer programs, especially those that control manufacturing processing, will consistently perform as they are supposed to within pre-established operational limits. Determine who conducted software validation and how key programs were tested. In considering software validation the following points should be addressed:

1. Does the program match the assigned operational function? For example, if a program is assigned to generate batch records then it should account for the maximum number of different lots of each component that might be used in the formulation. Consider what might happen when three lots of a component are used with a program designed to record lot designations and quantities for up to two different lots of each component. The first lot may be accurately recorded, but the next two lots might be recorded as a single quantity having the second lot designation; the resultant computer generated record therefore would fail to show the use of three different lots and the quantities of each of the second and third lots going into the mixture.

2. Have test conditions simulated "worst case" production limits? A program should be tested, for example, under the most challenging conditions of process speed, data volume and frequency. Date field size should be considered in this aspect of validation. For example, the number of characters allowed for a lot number should be large enough to accommodate the longest lot number system that will be used.

3. Have tests been repeated enough times to assure consistent reliable results? Divergent results from replicate data entries may signify a program bug. In general, at least three separate runs should be made.

4. Has the software validation been thoroughly documented? Documentation should include a testing protocol and test results which are meaningful and specific to the attribute being tested; individuals who reviewed and approved the validation should be identified in the documentation.

5. Are systems in place to initiate revalidation when program changes are made? If process parameters such as time/temperature, sequence of program steps, or data editing/handling are changed then revalidation is indicated.

Although much of the software validation may be accomplished by outside firms, such as computer or software vendors, the ultimate responsibility for program suitability rests with the pharmaceutical manufacturer. Records of software validation should be maintained by the drug establishment, although when conducted by outside experts such records need not be voluminous but rather complete enough (including protocols and general results) to allow the drug manufacturer to assess the adequacy of the validation. Mere vendor certification of software suitability is inadequate. Signal simulators many be used in software validation. These are discussed in point No. 3 of *Validation of Hardware.*

V. *COMPUTERIZED OPERATIONS*

A. *Networks*

If the firm is on a computer network it is important to know: (1) what output, such as batch production records, is sent to other parts of the network; (2) what kinds of input (instructions, programs) are received; (3) the identity and location of establishments which interact with the firm; (4) the extent and nature of monitoring and controlling activities exercised by remote on-net establishments; and (5) what security measures are used to prevent unauthorized entry into the network and possible drug process sabotage.

It is possible under a computer network for manufacturing operations conducted in one part of the country to be documented in batch records on a real-time basis in some other part of the country. Such records must be immediately retrievable from the computer network at the establishment where the activity took place (21 CFR 211.180).

B. *Manual Back-up Systems*

Functions controlled by computer systems can generally also be controlled by parallel manual back-up systems. During the inspection determine what functions can be manually controlled and identify manual back-up devices. Process controls are particularly important. Determine the interaction of manual and computerized process controls and the degree to which manual intervention can over-ride or defeat the computerized process. The firm's SOP should describe what manual over-rides are allowed, who may execute them, how and under what circumstances. Determine if and how manual interventions are documented; a separate log may be kept of such interventions. The computer system may be such that it detects, reacts to and automatically records manual interventions and this should be addressed during the inspection.

C. *Input/Output Checks*

Section 211.68 of the CGMP regulations requires that input to and output from the computer system be checked for accuracy. While this does not mean that every bit of input and output need be checked it does mean that checking must be sufficient to provide a high degree of assurance that input and output are, in fact, accurate. In this regard there needs to be some reasonable judgment as to the extent and frequency of checking based upon a variety of factors such as the complexity of the computer systems. The right kinds of input edits, for example, could mitigate the need for extensive checks.

During the inspection determine the degree and nature of input/output checks and the use of edits and other built-in audits.

Input/output error handling has been a problem in computer systems. Determine the firm's error handling procedures including documentation, error verification, correction verification, and allowed error over-rides including documentation of over-rides.

As an illustration of inadequate input/output checks and error handling consider the situation of a firm which uses a computer system to maintain and revise labeling text. Master labeling is recorded on a disk and when a change is to be made the operator calls up a copy of the text from the master disk onto a CRT. The copy is then revised at the CRT, printed on paper and electronically printed onto another disk for storage until the paper copy is proofread and approved; once the paper copy is approved, the text on the temporary storage disk is transferred to the master disk replacing the previous text. As an example, the operator calls up a label to change the directions for use section, correctly makes the change but accidentally erases the quantity of content statement that read 100 ml. The operator "corrects" this error by re-entering what was believed to be the correct statement but what, in fact, was "150 ml." The proof-readers do not detect this error because their standard operating procedure is to proof only those portions of the labeling -- in this case directions for use -- which were supposed to be changed (a case of inadequate output check). In addition, the operator does not document the error or the "correction" and the "correction" is not verified. This would probably result in a misbranded product.

Section 211.68 of the CGMP regulations also requires maintenance of accurate back-up files of input data which are secure from alteration, loss or inadvertent erasure. These back-up files need not be on paper, however. They may, for instance, consist of duplicate tapes, disks or microfilm. During the inspection determine if the firm has such a back-up system, the form of such a system, and how it is protected. If a back up file is printed on thermal paper note if older files have faded. (It has been reported that the printing on thermal paper has a tendency to fade with time.)

D. *Process Documentation*

Most computer systems are capable of generating accurate and detailed documentation of the drug process under computer control. What is important is that records within the scope of the CGMP regulations, which happen to be in computerized form, do contain all of the information required. For example, if batch production records are generated by computer determine if they contain all of the information required to be in batch records.

E. *Monitoring of Computerized Operations*

Determine the degree to which the firm's personnel monitor computerized operations. Is such monitoring continuous or periodic? What functions are monitored? For example, a firm's computer system may be used to maintain the pH in a reaction vessel, but if the firm does not sufficiently monitor the system they may fail to detect a hardware problem which could allow the pH to be out of tolerance. During the inspection, where possible, spot-check computer operations such as:

1. Calculations; compare manual calculations of input data with the automated calculations or ask the firm to process a given set of input values and compare automated results against known results.

2. Input recording; compare sensor indications with what the computer indicates, for example. As mentioned previously, some analog signals may be incorrectly converted to digital signals and built-in programming edits may alter input data. For example, a thermocouple indicating 80°C may read out on a CRT as 100°C or any other temperature if the signal converter is malfunctioning.

3. Component quarantine control; for example, check the actual warehouse location of a particular lot against its location as reported by computer. If the computer indicates that a particular lot has passed a certain number of laboratory tests then the laboratory records may be checked to confirm the computer information.

4. Timekeeping; where computers are reporting events and controlling a process in real time, spot-check the time accuracy against a separate time piece; accurate timekeeping is especially important where time is a determinative or limiting factor in a process such as sterilization. It should be noted that some computer systems run on a 12 hour clock whereas others run on a 24 hour clock.

5. Automated cleaning in-place; determine the procedure used, how the firm assures adequacy of cleaning, and residue elimination.

6. Tailings accountability; where batches are produced back to back on a continuous basis under computer control are batch tailings accounted for in subsequent handling and formulation? For example, at the conclusion of a run the computer's memory may be downloaded and the controlling program reset. At an initial step the computer

may call for a programmed quantity of material to be added to a hopper; the amount to be added can be based upon the tare weight of an empty hopper. However, if the hopper is not, in fact, empty but contains tailings of a prior run the result may be a hopper with more material than called for in the batch formulation; thus, there may be errors of yield reconciliation or batch formulation. During the inspection determine what limits if any the firm places on tailings.

F. *Alarms*

A typical computer system will have a number of built-in alarms to alert personnel to some out-of-limits situation or malfunction. Determine what functions are linked to alarms. For example, alarms may be linked to power supply devices, feedback signals to confirm execution of commands, and pharmaceutical process conditions such as empty or overflowing tanks. Determine the alarm thresholds for critical process conditions and whether or not such thresholds can be changed by the operator. For example, if the temperature of water in a water for injection system is linked to an alarm which sounds when the temperature drops below 80°C, can the operator change the threshold to 75°C?

Determine how the firm responds when an alarm is activated. This should be covered in the firm's standard operating procedures.

Determine the types of alarms (lights, buzzers, whistles, etc.) and how the firm assures their proper performance. Are they tested periodically and equipped with in-line monitoring lights to show they are ready?

Because an activated alarm may signal a significant out of control situation it is important that such alarm activations are documented. Determine how alarm soundings are documented-in batch records, in separate logs or automatic electronic recording, for instance.

Can all alarm conditions be displayed simultaneously or must they be displayed and responded to consecutively? If an employee is monitoring a CRT display covering one phase of the operation will that display alert the employee to an alarm condition at a different phase? If so, how?

G. *Shutdown Recovery*

How a computer controlled process is handled in the event of computer shutdown (e.g. power failure) is significant and can pose a problem. Shutdown recovery procedures are not uniform in the industry. Some systems, for example, must be restarted from the initial step in the process sequence and memory of what has transpired is lost. Other systems have safeguards whereby memory is retained and the process is resumed at the point where it was halted. Determine the disposition of the computer's memory content (program and data) upon computer shutdown.

Determine the firm's shutdown recovery procedure and whether or not, in the event of computer failure, the process is brought into a "safe" condition to protect the product. Determine such safeguards and how they are implemented. Where is the point of restart in the cycle--at the initial step, a random step or the point of shutdown? Look for the inappropriate duplication of steps in the resumption of the process. The time it takes to resume a computerized process or switch to manual processing can be critical, especially where failure to maintain process conditions for a set time (e.g. pH control for antibiotic fermentation) compromises product integrity. Therefore, note recovery time for delay-sensitive processes and investigate instances where excessive delays compromise product quality or where established time limits (21 CFR 211.111) are exceeded.

Many systems have the ability to be run manually in the event of computer shutdown. It is important that such back-up manual systems provide adequate process control and documentation. Determine if back-up manual controls (valves, gates, etc.) are sufficient to operate the process and if employees are familiar with their operation. Records of manual operations may be less detailed, incomplete, and prone to error, compared to computerized documentation, especially when they are seldom exercised. Therefore, determine how manual operations are documented and if the information recorded manually conforms with CGMP requirements.

VI. *CGMP GUIDANCE*

A. *Hardware*

In general, the hardware of a computer system is considered to be equipment within the meaning of the CGMP regulations. Therefore those sections of the regulations which address equipment apply to hardware. For example, the following apply:

1. 21 CFR 211.63 requires that equipment be suitably located to facilitate operations for the equipment's intended use.

2. 21 CFR 211.67 requires a maintenance program for equipment.

3. 21 CFR 211.68(a) states that computers may be used and requires a calibration program.

B. *Software*

In general, software is regarded as records or standard operating procedures (instructions) within the meaning of the CGMP regulations and the corresponding sections of the CGMP regulations apply, for example:

1. *Record Controls*. 21 CFR 211.68(b) requires programs to ensure accuracy and security of computer inputs, outputs, and data.

2. *Record Access*. 21 CFR 211.180(c) states that records required by the regulations shall be available as part of an authorized inspection at the establishment for inspection and are subject to reproduction. Computer records retrievable from a remote location are acceptable.

In considering the copying of electronic records however, the act of copying must be reasonable, as the word reasonable is used in the FD& C Act to limit how we may conduct inspections. In some cases it may be reasonable to copy a disk or tape whereas in other cases it might not, particularly where we would have to physically remove the disk or tape from the establishment in order to copy it. (Consider the analogy of removing an entire file cabinet so that we can copy five pieces of paper.) We believe that, rather than copy an entire disk or tape ourselves, it is preferable to have the firm generate hard copies of only those portions of the disk or tape which we need to document.

3. *Record Medium*. 21 CFR 211.180(d) states that retained records may be originals or true copies and, when necessary, readers and photocopying equipment shall be available. This concept applies to magnetic tape and disks.

4. *Record Retention*. 21 CFR 211.180(a) states record retention requirements. They are the same for electronic media and paper.

5. *Computer Programs*. FD& C Act. Section 704(a), for prescription drug products, would allow inspectional access to computer programs if such inspection is performed within the constraint of being reasonable.

There are several factors which must be considered on a case by case basis in determining what is reasonable in accessing a firm's computer. For example, the effect on drug production is a factor; specifically, if the process of running a program disrupts drug production in an adverse manner then that would be unreasonable. Another factor is whether or not our manipulations give us access to unauthorized information; the data we may be searching with a program may contain some information we are not entitled to review such as financial data. Consider also that some computer programs are protected by copyright and carefully licensed to software users; thus, we would not be able to copy and use such programs without prior approval of their owners.

6. *Record Review*. 21 CFR 211.180(e) states that where appropriate records associated with every batch shall be reviewed as part of a periodic review of quality standards. It is acceptable for a firm to conduct part of the review by running a computer program which culls out analytical data from each batch and conducts trend analysis to determine the need to change product specifications, manufacturing methods, or control procedures. The data itself must be meaningful (i.e., specified and relevant to enable an evaluation to be performed). It is not necessary to review each and every bit of information on the batch record. However, the computerized trend analysis data would constitute only a portion of the data which must be reviewed. A review must also be made of records of complaints, recalls, returned or

salvaged products, and investigations of unexpected production discrepancies (e.g., yield reconciliations) and any failures of batches to meet their specifications. This information is usually separate from conventional batch records and so would not necessarily be reviewed by the trend analysis program.

7. *QC Record Review.* 21 CFR 211.192 requires the quality control unit to review and approve production and control records prior to batch release/distribution. If this record screening review (to check errors and anomalies) is computerized and is at least as comprehensive and accurate as a manual review, then it is acceptable for the QC unit to review a computer generated exception report as part of the batch release. The batch record information required by the regulation must still be retained. It is also important that the accuracy and reliability of the screening program be demonstrated. It is unlikely however, that all production and control records will be computerized; labeling, packaging, and analytical records may still be in manual form and would therefore be manually reviewed.

8. *Double Check on Computer.* 21 CFR 211.101(d) requires verification by a second person for components added to a batch. A single check automated system is acceptable if it provides at least the same assurance of freedom from errors as a double check. If it does provide the same assurance then we would gain nothing in applying a redundant second check which adds nothing to assuring product quality. The equivalency of an automated single check system to a manual double check must be shown, however, and this might not always be possible. For example, let's say 5 kilograms of a coarse white granular component must be added to a mixture. Two individuals checking the operation may check for the component's label accuracy, color and granularity, weight and finally the actual transfer of the material; if there were a mix-up prior to that transfer and a different component, say a white powder, was staged for addition to the batch it is probable that the double check screening would detect the error. On the other hand, a single check computer system might accurately check the component weight and physical transfer but not its granularity or other sign of identification. In this case the automated single check would not be as good as the manual double check.

9. *Documentation.* 21 CFR 211.188(b) (11) requires that batch production and control records include identification of each person who conducts, supervises or checks each significant step in the process. The intent is to assure that each step was, in fact, performed and that there is some record to show this, from which the history of the lot could be traced. It is quite possible that an automated system can achieve the same, or higher, level of assurance in which case it may not be necessary to have persons document the performance of each event in a series of unbranched automated events on the production line. For example, let us say an automated/computerized system is designed to perform steps A thru Z. If the program is such that every step must be executed properly before step Z is completed then an acceptable means of complying with the regulation would be all of the following: (1) documentation of the program; (2) validation that no step can be missed or poorly executed; and (3)

documentation of the initial and final steps. It would not be necessary in this example to document steps B thru Y.

10. *Reproduction Accuracy.* 21 CFR 211.188(a) requires the batch record to contain an accurate reproduction of the master record. The intent is to insure that the batch was, in fact, produced according to the approved formulation, manufacturing instructions and controls. The act of computer transcription can generate errors. The firm should check for such errors or otherwise assure that no errors can occur. During the inspection the investigator can ask to see the original approved, endorsed, master record and compare it to the batch record. The fact that the batch record is a second or third generation copy is not in and of itself objectionable provided it is accurate.

11. *NDA Considerations.* 21 CFR 314.8 requires that a supplement be submitted for changes in manufacturing/control processes or facilities from those stated in the approved NDA. If a firm has changed from a manual to a computerized system under an NDA, that change should be covered by a supplemental application. If the change gives increased assurance of product quality, then the change can be put into effect before the supplement has been approved. However, such supplements should state the anticipated implementation date (which should be sometime after the submission date) to allow reviewing chemists the lead time needed to determine if the type of change proposed is, in fact, of the kind which may be implemented prior to approval.

GLOSSARY

ADDRESS -- A switch pattern which identifies the location of a piece of data or a program step.

ALGORITHM -- A systematic procedure or equation designed to lead to the solution of a problem in a finite number of steps.

ALU -- Arithmetic Logic Unit; the circuitry within the CPU which performs all arithmetic functions.

ANALOG -- Continuous signal having a voltage which corresponds to the monitored value.

APPLICATIONS -- Term used to describe software written to perform tasks on a computer.

ASCII -- American Standard Code for Information Interchange; a system used to translate keyboard characters into bits.

ASSEMBLER -- Program which translates assembly code to executable machine code; e.g. assembly code ADD becomes machine code 04.

ASSEMBLY CODE -- Symbolics, a simple language; different computers have different assembly codes.

ASYNCHRONOUS -- Term used to describe the exchange of information piece by piece rather than in long segments.

AUXILIARY STORAGE -- Storage device other than main storage; disks and tapes.

BASIC -- Beginner's All Purpose Symbolic Instruction Code; a high level language.

BATCH PROCESSING -- Execution of programs serially with no interactive processing.

BAUD -- The rate at which data is received or transmitted in serial: one baud is one bit per second.

BINARY -- The base two number system. Permissible digits are 0 and 1.

BIT -- Binary Digit; the smallest unit of information in a computer, represented as 0 or 1, off or on for a switch.

BOOT -- An initialization program used to set up the computer when it is turned on.

BUFFER -- Part of memory used to temporarily hold data for further processing.

BUG -- A program error.

BUS -- Electrical pathway by which information flows to different devices.

BYTE -- A sequence of adjacent bits, usually eight, operated upon as a unit; the lowest addressable unit in a computer.

COMPILER -- Program which translates a computer language into executable machine code. A compiler translates an entire program before the program is run by the computer.

CP/M -- Control Program for Microcomputers; a registered trademark of Digital Research; an operating system.

CPU -- Central processing unit of a computer where the logic circuitry is located; the CPU controls the entire computer; it sends and receives data through input-output channels, retrieves data from memory and conducts all program processes.

CRT TERMINAL -- Cathode ray tube; an input/output device.

DATABASE -- Collection of data, at least one file, fundamental to a system.

DATA SET -- Term synonymous with file.

DIGITAL -- Relating to separate and discrete information.

DISK -- A circular rotating magnetic storage device. Disks come in different sizes and can be hard or flexible.

DISK DRIVE -- A device used to read from or write to a disk or diskette.

DISK OPERATING SYSTEM -- DOS, a program which operates a disk drive.

DISKETTE -- A floppy disk.

EPROM -- Erasable programmable read only memory: switch pattern in circuit can be erased by exposure to ultraviolet light.

FILE -- Set of related records treated as a unit, stored on tape or disk; synonymous with data set.

FIRMWARE -- A program permanently recorded, e.g., in ROM.

HARD COPY -- Output on paper.

HARDWARE -- Physical electronic circuitry and associated equipment.

HEXADECIMAL -- The base 16 number system. Digits are 0, 1, 2, 3, 4, 5, 6, 7, 8, 9, A, B, C, D, E, AND F. This is a convenient form in which to examine binary data because it collects 4 binary digits per hexadecimal digit. E.g. Decimal 15 is 1111 in binary and F in hexadecimal.

INTEGRATED CIRCUIT -- (IC) Small wafers of silicon etched or printed with extremely small electronic switching circuits; also called CHIPS.

INTERACTIVE PROCESSING -- An application in which each entry calls forth a response from a system or program, as in a ticket reservation system.

INTERFACE -- A device which permits two or more devices to communicate with each other.

INTERPRETER A program which translates a high level language into machine code one instruction at a time. Each instruction in the high level language is executed before the next instruction is interpreted.

I/O PORT -- Input/output connector.

JOB -- Set of data completely defining a unit of work for a computer.

K -- Symbol representing two to the tenth power, 1024, usually used to describe amounts of computer memory, and disk storage, in bytes.

LANGUAGE --Any symbolic communication media used to furnish information to a computer. Examples are PL/1, COBOL, BASIC, FORTRAN, AND ASSEMBLY.

LOADER -- A program which copies other programs from external to internal storage.

MACHINE CODE -- Numerical representations directly executable by a computer; sometimes called machine language.

MAIN STORAGE -- Term synonymous with MEMORY.

MAINFRAME -- Term used to describe a large computer.

MEGABYTE -- 1024K Bytes.

MEMORY -- A non-moving storage device utilizing one of a number of types of electronic circuitry to store information.

MENU -- A CRT display listing a number of options. The operator selects one of the options. Sometimes used to denote a list of programs.

MICROCOMPUTER -- A small computer (See MICROPROCESSOR).

MICROPROCESSOR -- Usually a single integrated circuit on a chip; logic circuitry of a microcomputer; frequently synonymous with a microcomputer. A microprocessor executes encoded instructions to perform arithmetic operations, internal data transfer, and communications with external devices.

MINICOMPUTER -- Medium sized computer.

MODEM -- Modulator - demodulation, a device which accepts data from a computer, and sends data to a computer, over telephone wires or cables. A half duplex MODEM can only receive or transmit data at one time. A full duplex MODEM can receive and transmit data at the same time.

MULTIPLEXER -- A device which takes information from any of several sources and places it on a single line.

NETWORK -- A system that ties together several remotely located computers via telecommunications.

OBJECT CODE -- Term synonymous with machine code.

OEM -- Original Equipment Manufacturer (i.e. maker of computer hardware).

OPERATING SYSTEM -- Set of machine language programs that run accessories, perform commands and interpret or translate high level language program (usually written into the ROM).

PARALLEL -- Term to describe transmission of data eight bits (one byte) at a time.

PARITY BIT -- An extra bit within a byte; used to verify the coded information in the byte itself. The extra bit is either a one or zero so as to make the total number of ones in a byte equal either an odd or even number (odd or even parity).

PERIPHERAL -- A general term used to describe an input or output device.

PROGRAM -- A collection of logically interrelated statements written in some computer language which, after translation into machine code, performs a predefined task when run on the computer.

PROM -- Programmable read only memory; once programmed the switch pattern on a PROM cannot be changed. Special equipment separate form the computer is usually used to "burn in" the switch pattern.

PROTOCOL -- Agreed upon set of standards which allow communication between computers, i.e. physical electrical links, message format, message priorities, etc.

RAM -- Random access memory; internal storage device containing volatile information which can be changed; read- write memory. When electrical power is cut off from a RAM IC its memory is lost.

RECORD -- Collection of related data treated as a unit.

ROM -- Read only memory; internal storage device in which information is permanent.

RS-232C An Electronic Industries Association (EIA) standard for connecting electronic equipment; data is transmitted and received in serial format. This is an interface standard that usually uses a 25 pin connector.

SERIAL -- Term to describe handling of data one bit at a time.

SOFTWARE -- Programs executable on a computer. Programs are written in any number of different languages.

SOURCE PROGRAM -- High level language program which the operator can read.

STORAGE DEVICE -- A unit into which data or programs can be placed, retained and retrieved.

SYNTAX -- Required grammar or structure of a language.

SYSTEM -- Term can refer to hardware or software. For hardware it is the collection of equipment that makes up the computer. For software it refers to an integrated number of computer programs to perform predefined tasks.

TAPE -- A liner magnetic storage device rolled onto a reel or cassette.

TELECOMMUNICATION SYSTEM -- The devices and functions relating to transmission of data between the central processing system and remotely located users.

TERMINAL -- A device, usually equipped with a CRT display and keyboard, used to send and receive information to and from a computer via a communication channel.

UTILITY PROGRAMS -- Special programs usually supplied by the producer of the operating system. They perform general functions such as making back up copies of diskettes and copying files from tape to disk.

VALIDATION -- The assurance, through testing, that hardware or software produces specified and predictable output for any given input.

WORD -- One or more adjacent bytes conveniently considered as an entity. A word is usually one to four bytes long, depending on make of computer.

FOOD AND DRUG ADMINISTRATION
COMPLIANCE POLICY GUIDE

GUIDE 7132a.07

ISSUING OFFICE: Office of Enforcement, Division of Compliance Policy

Issued: September 20, 1982
Reissued: September 4, 1987

SUBJECT: Computerized Drug Processing; Input/Output Checking

BACKGROUND

Section 211.68 (automatic, mechanical, and electronic equipment) of the Current Good Manufacturing Practice Regulations requires, in part, that input to and output from the computer or related system of formulas or other records or data be checked for accuracy. This requirement has generated questions as to the need for and extent of checking a computer's input and output.

The agency received several petitions to delete or modify the requirement on the grounds that a validated computer system need not have its input/output routinely checked. The request to delete or modify the requirement was denied because our experience has shown that input/output error can occur, even in validated systems. Printouts, for example, can contain errors as a result of faulty input, programming, or equipment malfunction. More significantly, there is the human element which can induce errors. At worst, input/output errors can result in serious production errors and distribution of adulterated or misbranded products. Several recalls have, in fact, been conducted because of insufficient input/output checks.

Despite the general need for input/output checks, not all input and output need be checked. The regulation is, in fact, deliberately silent on the required frequency and extent of data checking to afford firms the necessary flexibility. Also, the use of efficient input edits, for example, could mitigate the need for more detailed manual data checks.

POLICY

Input/Output checks of data for computer systems, as required by 21 CFR 211.68, are necessary to assure the quality of a drug product processed using such systems. The extent and frequency of input/output checking will be assessed on an individual basis, and should be determined based upon the complexity of the computer system and built in controls.

FOOD AND DRUG ADMINISTRATION
COMPLIANCE POLICY GUIDE

GUIDE 7132a.08

ISSUING OFFICE: Office of Enforcement, Division of Compliance Policy

Issued: November 2, 1982
Reissued: September 4, 1987

SUBJECT: Computerized Drug Processing; Identification of "Persons" on Batch Production and Control Records

BACKGROUND:

Section 211.188(b)(11) of the Current Good Manufacturing Practice Regulations requires that batch production and control records include documentation that each significant step in the manufacture, processing, packing, or holding of a batch was accomplished, including identification of the persons performing, directly supervising or checking each significant step in the operation.

Questions have been raised as to acceptable ways of complying with this requirement when the "person" performing, supervising or checking each step is, in fact, not a human being, but rather an automated piece of equipment, such as a computer system.

The intent of the regulation is to assure that each significant step in a process was, in fact, performed properly and that there is some record to show this. It is quite possible that a computerized system can achieve he same or higher degree of assurance. In this case it may not be necessary to specifically record the checks made on each of a series of steps in the production of the product.

POLICY:

When the significant steps in the manufacturing, processing, packing or holding of a batch are performed, supervised or checked by a computerized system an acceptable means of complying with the identification requirements of 21 CFR 211.188(b)(11) would consist of conformance to all of the following:

1. Documentation that the computer program controlling step execution contains adequate checks, and documentation of the performance of the program itself.
2. Validation of the performance of the computer program controlling the execution of the steps.
3. Recording specific checks in batch production and control records of the initial step, any branching steps and the final step

<u>NOTE:</u>

In assessing how well a computer system checks a process step it is necessary to demonstrate that the computer system examines the same conditions that a human being would look for, and that the degree of accuracy in the examination is at least equivalent.

FOOD AND DRUG ADMINISTRATION
COMPLIANCE POLICY GUIDE

GUIDE 7132a.11

ISSUING OFFICE: Office of Enforcement, Division of Compliance Policy

Material between asterisks is new or revised
Issued: October 19, 1984
Revised (minor edits): September 4, 1987

SUBJECT: *Computerized Drug Processing; CGMP Applicability To Hardware and Software*

BACKGROUND

The use of computers in the production and control of drug products is quickly increasing. Questions have been raised as to the applicability of various sections of the Current Good Manufacturing Practice Regulations to the physical devices (hardware) which constitute the computer systems and to the instructions (software) which make them function.

POLICY:

Where a computer system is performing a function covered by the CGMP regulations then, in general, hardware will be regarded as equipment and applications software[12] will be regarded as records. The kind of record (e.g., standard operating procedure, master production record) that the software constitutes and the kind of equipment (e.g., process controller, laboratory instrument) that the hardware constitutes will be governed by how the hardware and software are used in the manufacture, processing, packing, or holding of the drug product. Their exact use will then be used to determine and apply the appropriate sections of the regulations that address equipment and records.

[12]Applications software consists of programs written to specified user requirements for the purpose of performing a designated task such as process control, laboratory analyses, and acquisition/processing/storage of information required by the CGMP regulations.

FOOD AND DRUG ADMINISTRATION
COMPLIANCE POLICY GUIDE

GUIDE 7132a.12

ISSUING OFFICE: Office of Enforcement, Division of Compliance Policy

Material between asterisks is new or revised
Issued: January 18, 1985
Revised (minor edits): September 4, 1987

SUBJECT: Computerized Drug Processing; Vendor Responsibility

BACKGROUND

Computer systems used in the production and control of drug products can consist of various devices (hardware) and programs (software) supplied by different vendors, or in some cases by a single vendor. It is important that such computer systems perform accurately and reliably, *and* that they are suitable for their intended use.

Questions have arisen as to the vendor's responsibility in assuring computer systems performance and suitability. When an integrated system, composed of elements from several different vendors, fails, it can be especially difficult to attribute the cause of a problem to one particular vendor.

POLICY:

The end user is responsible for the suitability of computer systems (hardware and software) used in manufacture, processing or holding of a drug product.

The vendor may also be liable, under the FD&C Act, for causing the introduction of adulterated or misbranded drug products into interstate commerce, where the causative factors for the violation are attributable to intrinsic defects in the vendor's hardware or software. In addition vendors may incur liability for validation, as well as hardware/software maintenance performed on behalf of users.

FOOD AND DRUG ADMINISTRATION
COMPLIANCE POLICY GUIDE

GUIDE 7132a.15

ISSUING OFFICE: Office of Enforcement, Division of Compliance Policy

Issued: April 16, 1987

SUBJECT: Computerized Drug Processing; Source Code for Process Control Application Programs

BACKGROUND

An increasing number of pharmaceuticals are being manufactured under the control of computer systems. The manufacturing procedures, control, instructions, specifications and precautions to be followed within such automated systems are embodied in the computer program(s) which drive the computer. Depending on the complexity of the programs, they may also contain controlling data on product formulation, batch size, yields and automated in-process sampling/testing procedures. In a manual system such procedures, instructions, specifications, precautions and other controlling data would be embodied in master production records which must be reviewed and approved before implementation and which must be maintained, as required by the current good manufacturing practice regulations (CGMP's). Such manual records are, of course, prepared in human readable form.

In the case of computerized drug process control, certain information required by CGMP's to be in a master production record is contained in the source code for the application program. (An application program is software written to specified user requirements for the purpose of performing a designated task.) Source code is the human readable form of the program, written in its original (source) programming language. Source code must be compiled, assembled, or interpreted before it can be executed by a computer. Because the source code ultimately has a direct and significant bearing on drug product quality as manual master records, it is vital that source code and supporting documentation be reviewed and approved by the drug manufacturer prior to implementation, and be maintained as the CGMP's require for master production and control records. (E.g., see 21 CFR 211.100, 211.180, and 211.186.) Careful review of source code and its documentation is especially important for assuring that process specifications, conditions, sequencing, decision criteria, and formulas have been properly incorporated into the computer program; source code should also be reviewed to detect and remove dead code -- non-executable instructions which are usually artifacts of earlier versions of the program.

Supportive program documentation, such as flow diagrams and explanatory narratives, can be useful in understanding and reviewing source code. However, such documentation is not an acceptable substitute for source code itself.

POLICY:

We regard source code and its supporting documentation for application programs used in drug process control to be part of master production and control records, within the meaning of 21 CFR parts 210 and 211. Accordingly, those sections of the current good manufacturing practice regulations which pertain to master production and control records will be applied to source code.

Chapter 15

BIOLOGICS INSPECTIONS

OVERVIEW

Biologics and blood products are being increasingly scrutinized by the FDA, primarily because of the seriousness of potential contamination of these products. Validation of manufacturing processes, especially automated processes, therefore, has become a major issue in inspections of biologics and blood establishments, as can be seen in the number of Warning Letters issued to these establishments and other regulatory actions taken by the FDA.

The documents which follow all relate to FDA's regulation and inspection of biologics and blood products. These documents are:

1. FDA Investigations Operations Manual (IOM), Chapter 5 -- Establishment Inspections, Subchapter 550 -- BIOLOGICS
2. FDA Compliance Program Guide 7342.001 -- INSPECTION OF LICENSED AND UNLICENSED BLOOD BANKS
3. FDA Compliance Program Guide 7342.002 -- INSPECTION OF SOURCE PLASMA ESTABLISHMENTS
4. CBER "Draft Guideline for the Validation of Blood Establishment Computer Systems"
5. FDA's September 1994 "Guide to Inspections of Blood Banks"

FDA INVESTIGATIONS OPERATIONS MANUAL
CHAPTER 5 -- ESTABLISHMENT INSPECTIONS
SUB CHAPTER 560 -- BIOLOGICS

560 GENERAL

A "biological product" means any virus, therapeutic serum, toxin, antitoxin, vaccine, blood, blood component or derivative, allergenic product, or analogous product, or arsphenamine or its derivatives (or any trivalent organic arsenic compound), applicable to the prevention, treatment, or cure of diseases or injuries of man. Additional interpretation of the statutory language is found in 21 CFR 600.3.

Veterinary biologicals (use in non-human species) are subject to the animal Virus, Serum, and Toxin ACT which is enforced by USDA (21 U.S.C. 151-158).

560.1 Authority

Biologics are regulated under the authority of both the Public Health Service Act and as drugs or devices under the Food, Drug and Cosmetic Act. Blood and blood products for direct transfusion are prescription drugs under the FD&C Act. Recovered plasma and source plasma intended for manufacturing non-injectable products and in-vitro diagnostics are devices.

Section 704(a) of the FD&C Act provides general authority for drug and device inspections and identifies the type of establishments that may be inspected. Section 351 (c) of the PHS Act provides authority to inspect interstate biologics establishments.

Section 351 (d) of the PHS Act provides for licensure of biologic establishments and products and inspection of licensees. Most biological drugs are licensed rather than having a new drug or abbreviated new drug application (21 CFR 310.4.) Radioactive biologicals products require NDAs (21 CFR 505) unless they have an unrevoked and unsuspended license issued prior to August 25, 1975.

The investigational new drug application regulations (21 CFR 312) also apply to biological drugs subject to the licensing provisions of the PHS Act. However, investigations of blood grouping serum, reagent red blood cells, and anti-human globulin in-vitro diagnostic products may be exempted (21 CFR 31 2.2(b)).

Blood bank and plasmapheresis center inspections use the CGMPs for Blood and Blood Components (21 CFR 606) as well as the general requirements for biological products (part 600), the general biological standards (part 610), and the additional standards for human blood and blood products (part 640.) The umbrella drug GMPs (21 CFR 210/211) also apply to biologic drugs. In the event it is impossible to comply with both sets of regulations, the regulation specifically applicable to the product applies. This would generally be the 606s in a blood bank or plasma center.

Blood establishments are sensitive to maintaining confidentiality of donor names. The mere reluctance to provide records is not a refusal. However, FDA has the authority under both the PHS and the FD&C Act to make inspections and 21 CFR 600.22 (g)) provides for copying records during a blood establishment inspection. For prescription drugs, section 704 of the FD&C Act specifically identifies records, files, papers, processes, controls, and facilities as being subject to inspection.

On problems with access to records, explain FDA's authority to copy these records. IOM 514 should be followed if a refusal is encountered. When donor names or other identifiers are necessary, they may be copied, but the information must be protected from inappropriate release. IOM 130 provides guidance on disclosure of official information and records.

The CBER has load responsibility over certain medical devices per Intercenter Agreement (56 FR 58760 October 21, 1991.)

560.2 Objective

The inspectional objective is to ensure that biological products are safe, effective, and contain the quality and purity they purport to possess and that they are properly labeled. Facilities will be inspected for conformance with:

A. Provisions of the PHS Act and FD & C Act
B. Applicable GMPs in 21 CFR 210-211, 600-680, and 820.
C. FDA Policies

   1. Letters and memoranda (guidelines, recommendations) to the blood and blood products industry
   2. Compliance Policy Guides chapter 34

560.3 Preparation and References

A. Preparation

Review the district files of the facility to be inspected and familiarize yourself with its operation and compliance history. Review:

1. Appropriate Compliance Programs and related Compliance Policy Guides (CPG), Chapter 34. Note: CPG 7155d.01, MOU with the Department of Defense; 7155e.03, MOU with Health Care Financing Administration (HCFA) on transfusion services;
2. Correspondence from the firm depicting any changes since the last inspection;
3. Firm's registration and product listing information; The licensing summary.
4. The Plasmapheresis Inspection Checklist and Report, FDA 2722 &/or Blood Bank Inspection Checklist and Report, FDA-2609 (and instruction booklets);
5. Error and Accident Reports, Adverse Reaction Reports, complaints, and recalls;

6. Specific field guidance, e.g. inspectional strategies, including: Guidance on Blood and Blood Product Inspections, March 21, 1988 (Reporting to headquarters release of unsuitable blood units) Blood Bank and Plasma Establishment Inspectional Guidance, October 6, 1988

B. References

See exhibit 560-A for a list of references.

Recommendations, guidelines, information letters or memoranda are issued by the CBER to industry. Each district office has a compilation of these materials. CBER issues these as policy or as guidance is developed and a list is attached to Compliance Program 7342.001.

The OSHA regulation 29 CFR 1910.1030 December 6, 1991, intended to protect health care workers from bloodborne pathogens, including those involved in the collection and processing of blood products. The regulation defines expectations for the use of gloves, hand washing facilities, decontamination of work areas, waste containers, labeling and training of employees and exemptions for volunteer blood donor centers. FDA Investigators should adhere to these safety guidelines during inspections or related activities in establishments that process biologically hazardous materials.

Become familiar with these regulations and their applicability to 21 CFR 606.40(d)(1) & (2) which require the safe and sanitary disposal for trash, items used in the collection and processing of blood and for blood products not suitable for use.

Consult your district biologics monitor for copies of the above references. Additional copies may be obtained from ORO, Division of Field Investigations, HFC-132 (301) 443-3340.

560.4 Inspectional Approach

Use the compliance programs for inspectional guidance. In blood banks and plasma centers, the Plasmapheresis Inspection Checklist and Report (FDA-2722) and instruction booklet or the Blood Bank Inspection Checklist and Report (FDA-2609) and instruction booklet may be used. If an FDA 2609 or FDA 2722 is used, notes may be recorded directly on the checklist provided the checklist is filed as part of the report.

When a limited inspection is performed or for other reasons a complete establishment inspection is not performed, the EIR must clearly indicate areas covered. The report should include the summary page (page 1) of the appropriate checklist, the FDA-482, the FDA-483, if issued, and required cover sheets.

Particular attention should be given to error and accident reports indicative of problematic areas or processes. Follow up investigations to such reports should also be closely covered.

Complaints, in particular those involving criminal activity, must be promptly investigated and coordinated with other agency components as needed.

Inspections should cover infectious disease testing and control, including test procedures and interpretations of test results, decisions leading to the release of products, and donor deferrals. Investigators should also check on product returns which may indicate possible GMP deficiencies.

560.5 Regulations, Guidelines, Recommendations

The CBER issues memoranda, letters, recommendations and guidelines to registered blood establishments. The contents of most of these documents are incorporated into the establishment's SOPs and/or license applications or amendments.

The content of these various documents should not be referenced on an FDA-483. However, since these documents are often related to specific GMP requirements, in most cases deviations can be related back to the GMP. If a deviation is noted during an inspection and the investigator relates it to the regulations or law, then the item may be reported on the FDA-483. During the discussion with management, it should be clearly explained that the deviation relates to the regulation or law.

Copies of memoranda and recommendations are sent to all registered establishments when issued by CBER. If an establishment indicates they have not received any of these documents, provide copies and note this in the EIR.

Facilities are often given variances from FDA regulatory requirements and/or CBER guidelines and recommendations. If a firm indicates they have an approved variance, then they should be able to produce their written approval letter. If they cannot, then the variance may not have been approved. This can be verified by contacting the CBER.

561 REGISTRATION, LISTING AND LICENSING

561.1 Registration and listing

Section 510 of the FD&C Act and 21 CFR 607 delineate the requirements and exemptions relating to the registration of establishments engaged in the collection, preparation, or processing of blood products and the listing of these products. Such establishments must register within five days after beginning operations and must submit a list of blood products they distribute commercially.

Some transfusion services are exempt from registration under 21 CFR 607. This includes facilities approved for Medicare reimbursement and engaged in the compatibility testing and transfusion of blood and blood components, but which neither routinely collect nor process blood and blood components. Such facilities include establishments:

A. Collecting and or processing under documented emergency situations
B. Only performing therapeutic collection
C. Preparing recovered human plasma
D. Preparing red blood cells for transfusion

Although the Health Care Financing Administration (HCFA) was delegated routine inspectional responsibility for these establishments, the FDA still retains legal authority to inspect them if so warranted. If a routine inspection determines an establishment is a HCFA obligation, FDA's inspection should be terminated and reported as such. See CPG 7155e.03 - FDA/HCFA Memorandum of Understanding.

Laboratories that perform required testing for a source plasma or blood bank establishment are an FDA obligation and should be incorporated into the active OEI. Clinical laboratories are specifically exempt from registration per 21 CFR 607.65(g), but this does not exempt them from FDA inspections. Such laboratories should be encouraged to register voluntarily if performing any testing used by blood establishments to determine the suitability of donors or blood product quality. Inspections should focus on activities relevant to these testing operations.

Mobile sites are required to register if equipment is permanently fixed in place and requires quality control e.g. centrifuges, or if records, e.g. deferral or donor, are stored on the premises.

In the future, inspection of military blood banks will become a responsibility of the field. At the present time some joint inspections are being conducted by field and CBER personnel. These facilities are required to meet the same standards as other blood banks although military emergencies may require deviations from the standards. An individual license is held by each branch of the service; although each individual establishment may be licensed or unlicensed, all are required to register. (CPG 7155d.01, MOU with Department of Defense Regarding Licensure of Military Blood Banks.)

Registration and listing forms, FDA-2830, are provided by CBER to industry. The back side of these forms provide instructions and establishment and product definitions. Forms are available through the Office of Biological Product Review, Division of Product Certification, HFB-240, CBER. Registration and listing is required whether or not interstate commerce is involved.

Field Management Directive 92, Agency Establishment Registration and Control Procedures, details the registration process within the agency. Refer to Compliance Policy Guide (CPG) 7134.01 for additional information on registration.

Check and ensure that the firm's current registration forms reflect actual operations.

561.2 Establishment and Product license

Licensure is a requirement for manufacturers under section 351 of the Public Health Service Act only if products are shipped interstate.

Establishments apply for licensure directly to the CBER. An establishment license may cover multiple sites. For each and every product they ship in interstate commerce, firms must obtain a product license. For example, a firm may have an establishment license with product licenses for Red Blood Calls and cryoprecipitate and also manufacture additional products not shipped interstate for which they do not obtain a license.

Prior to granting a license, the CBER conducts a prelicense inspection of the firm for compliance with its license application and regulations. Through proper coordination, field personnel may accompany the headquarters CBER inspector on the inspection. Copies of CBER's prelicense inspection reports are forwarded to the districts and should be part of the firm's file.

Significant proposed changes in location, equipment, management and responsible personnel, or in manufacturing methods and labeling, of any product for which a license is in effect or for which an application for license is pending, shall be reported to CBER by the manufacturer, and in the case of an emergency, not less then 30 days in advance of the time such changes are intended to be made (21 CFR 601.12(a)).

562 RESPONSIBLE INDIVIDUALS

All blood establishments are required to have a designated qualified person (21 CFR 606.20, 600.10) responsible for compliance. For licensed establishments, this individual is designated the responsible head.

In licensed establishments the responsible head is approved by CBER and changes in the responsible head must be submitted to CBER for approval. If the responsible head is on the premises and available, the FDA-482 and FDA-483 should be issued to and discussed with this person. If a higher level management official is present, the FDA-483 may be issued to both individuals. If the FDA-483 is not issued to the responsible head, the name and address of the person should be reported for issuing a post inspection letter forwarding a copy of the inspectional observations. The name and address of the top management official should also be reported if this is different than the responsible head.

The designation as responsible head does not necessarily mean that individual is the most responsible for any non compliance of the firm. In licensed or unlicensed facilities, establish and document all individuals responsible for violations in addition to the responsible head. Determine who the responsible head reports to in the establishment and if these other individuals have more authority to make corrections.

563 TESTING LABORATORIES

Blood bank and plasmapheresis firms may use outside testing laboratories to perform required testing. Currently few of these testing laboratories operate under license and are routinely inspected by FDA. The remainder are not required to register and are inspected by HCFA. Voluntary registration is requested and these testing laboratories are added to the OEI and inspected for the required testing performed for blood banks.

During an inspection of a plasma center determine if outside testing laboratories were approved in the license application.

During an inspection of a blood bank which is using the services of an outside testing laboratory, determine if the firm is thoroughly knowledgeable of the procedures utilized by this outside laboratory; if there is a written contract which states specific test kits or reagents to be used, how test results will be reported, if training of the personnel occurred, and if the blood bank has audited the facility. If the blood bank is not knowledgeable of the procedures or a review of test results indicates problems, the reference laboratory must be scheduled for inspection. These inspections can be scheduled jointly through the Regional HCFA contact. If the reference laboratory is outside of the district refer to the appropriate district office.

564 BROKERS

Brokers are used by blood establishments to locate buyers for products such as recovered plasma or expired red blood cells which are used for further manufacture into products such as clinical chemistry controls and in-vitro diagnostic products not subject to licensure. Fractionators also use brokers to locate suppliers of plasma under short supply provisions (21 CFR 601.22.) During inspections, determine if the facility is selling products to any brokers. If brokers are used, determine if the brokered products are shipped to a facility operated by the broker or directly to the consignee.

Brokers who take physical possession of blood products, such as in storage, pooling, labeling, or distribution, are required to register and included in the OEI for routine inspection under the blood bank compliance program. If the broker is simply used to arrange the sale and all required records exist between the manufacturer and consignee, registration would not be required. If brokers supply labels, these labels should be checked to determine if they accurately show the proper intended use of the product.

565 TECHNICAL ASSISTANCE

Several regions and some districts have specialists in biologics who are available for technical assistance and consultation. Do not hesitate to avail yourself of their services.

The services of expert investigators in ORA/ORO/Division of Field Investigations (HFC-1 30) are available for telephone or on-site consultation and assistance in problem areas.

CBER/OC/Inspections and Surveillance (HFB-120) can provide technical assistance on blood banking principles, testing issues, and can coordinate assistance from other CBER offices.

FDA COMPLIANCE PROGRAM GUIDE 7342.001 -- INSPECTION OF LICENSED AND UNLICENSED BLOOD BANKS

PROGRAM: 7342.001
SUBJECT: Inspection of Licensed and Unlicensed Blood Banks (FY 92)
IMPLEMENTATION DATE: Upon Receipt
COMPLETION DATE: October 1, 1993
DATA REPORTING:
PRODUCT CODES: 57DY-[ ] [ ]
57YY-[ ] [ ]

PROGRAM/ASSIGNMENT CODES: 42001 - Unlicensed
42001A - Licensed
42001B - Donor Center
42001C - Reference Laboratories
42R825 - AIDS Related Activity

Note: For all joint prelicense inspections at blood or plasma establishments, use PAC 42832.

DATE OF ISSUANCE: AUG 1992

FIELD REPORTING REQUIREMENTS

Copies of all Coversheets and FDA-483's, regardless of final classification, are to be submitted to CBER, Surveillance and Policy Branch (HFB-124). The EIR, checklist, and exhibits do not need to be submitted, unless specifically requested.

If the District Office requests CBER review of the EIR regarding a particular issue that may need policy development or clarification, the ENDORSEMENT section of the computer generated (EIR) coversheet should be conspicuously flagged or a separate memorandum to HFB-120 should address the issue. Provide copies of the EIR, checklist, and exhibits for these issues, if applicable, in order to facilitate CBER evaluation.

PART I -- BACKGROUND

Blood and blood components which are biologics have been licensed and inspected since May, 1946, under the provisions of Section 351 of the Public Health Service (PHS) Act (42 U.S.C. 262). The control of biologic products was transferred to the FDA in 1972.

Most biologics also fall within the definition of a drug under the Federal Food, Drug, and Cosmetic (FD& C) Act, Section 201(g)(1) (21 U.S.C. 321). The Good Manufacturing Practices (GMP's) for Blood and Blood Components (21 CFR 606), are designed to assure the production of safe, pure, and effective blood and blood components by all unlicensed facilities (refer to Part III for discussion of 21 CFR 211). The Food and Drug Administration's authority to inspect and regulate unlicensed blood banks through the GMP's has assured more comprehensive, uniform, and efficient enforcement of the law.

Beginning in 1980, FDA and Health Care Financing Administration (HCFA) began coordinating visits to blood banks (unlicensed) according to a Memorandum of Understanding (see MOU 7155e.03, Chapter 55e in the CPG Manual). This agreement resulted in HCFA taking over the inspectional responsibility for a significant number of the blood banks (transfusion services) previously inspected by FDA. Currently, FDA inspects only "manufacturers" of blood and blood components (including those collected for autologous use). However, FDA may inspect transfusion services where there are indications of noncompliance with GMPs. These inspections are to be coordinated with the appropriate HCFA representative.

PART II -- IMPLEMENTATION

The current program represents a continuing compliance surveillance activity which began in 1972. The firms covered by the program are all registered with FDA pursuant to Section 510 of the FD& C Act and are inspected biennially to determine state of compliance with the applicable provisions of the regulations and their license application (if licensed). Parts 600, 601, 606, 607, 610, and 640 of Title 21, Code of Federal Regulations (21 CFR) apply to these inspections (refer to Part III for discussion of 21 CFR 211).

A. OBJECTIVE

To assure that blood and blood products are safe, effective, and adequately labeled by:

1. Providing guidance to investigators performing biennial inspections of licensed and unlicensed blood banks.
2. Providing regulatory/administrative guidance.
3. Encouraging voluntary compliance.
4. Obtaining information for evaluating the quality of blood and blood products.
5. Assessing systems used to prevent the release of unsuitable blood products.

B. PROGRAM MANAGEMENT INSTRUCTIONS

1. The firms covered under this program include:

   a. U.S. licensed blood banks.
   b. Donor centers.
   c. Unlicensed blood banks which "manufacture" blood and blood components.

   NOTE: Transfusion services which are a part of a facility approved for Medicare reimbursement and engaged in the compatibility testing and transfusion of blood and blood components, but which neither routinely collect nor process blood and blood components are exempt from registration. The collection and processing of blood and blood components in an emergency situation, therapeutic collection of blood or plasma, preparation of recovered plasma for further manufacturing use, or the preparation of red blood cells (from units of whole blood) are not acts requiring such transfusion services to register. However, if a facility prepares components such as Red Blood Cells, Washed: Red Blood Cells, Frozen; Red Blood Cells, Declycerolized; and/or collects blood, including autologous, it is considered to be a manufacturer and is subject to registration and FDA inspection.

   d. Clinical/reference laboratories which provide services to blood banks in such areas as testing for hepatitis B surface antigen, and antibody to HIV-1, and/or other viral markers for donor suitability evaluations or labeling statements relating to product quality. [See Part III(A)(1)(e)]
   e. Blood bank establishments and transfusion services which are not covered by HCFA, e.g., most Department of Veterans Affairs Hospitals, licensed military blood banks (see Part III,A,1,f) and Indian Health Service Hospitals. These establishments may function as transfusion services only. Since they are generally not Medicare reimbursed, they should be inspected by FDA, not HCFA.

2. The Division of Inspections and Surveillance (DIS), HFB-120, supplies a Blood Bank Checklist and Instruction Booklet, Form 2609 (5/91) for the inspection program. This checklist should be used for each inspection. It is intended to assure a uniform approach to facility evaluation and is designed as an investigational tool for the investigator.

3. Current agency policy is to inspect once each year.

   Each District is responsible for planning work to assure the annual inspection of each registered blood establishment in its inventory is completed. Scheduling priorities should take into consideration a firm's compliance history.

PART III -- INSPECTIONAL

A. OPERATIONS

Each inspection (except directed and follow-up inspections) will cover the manufacturing operation of the blood bank pursuant to the GMP's for blood and blood components (21 CFR 606). The GMPs have been written specifically for human blood and blood product collection, processing, compatibility testing, and storage facilities. In addition to the GMPs, specific regulations for certain blood products are contained in Part 640 - "Additional Standards for Blood and Blood Products" and in the general biologics regulations, Parts 600-610. These should be used in conjunction with the GMPs. As stated in Part I, biologics fall within the definition of a drug under the FD& C Act. The regulations in 21 CFR, Parts 211 through 226 are used to supplement, not supersede, the cGMP regulations in Part 600 (see 21 CFR 210.2 for the applicability of cGMP regulations). For example, 21 CFR 211.68 (automated equipment) is applicable to computers used in blood establishments. Through the quality assurance guidelines development process, the Agency is conducting a review of these regulations to determine which Part 211 cGMPs are appropriate to supplement the Part 606 cGMPs and to provide guidance as to how the new policy and expectations are to be interpreted to ensure uniform application and enforcement.

If the establishment employs mobile units to collect blood and blood components, an attempt should be made to inspect at least one operating mobile unit. This may be best accomplished by obtaining, during inspection of the central facility, a list of the scheduled mobile collections for the next 30 days. A convenient location can then be selected for an unannounced inspection prior to the completion of the EIR.

1. GENERAL

    a. This Compliance Program, Chapter 34 of the Compliance Policy Guides, CBER memoranda to blood establishments, and the checklist and instruction booklet, all contain instructions for inspecting blood banks. The Inspectional Operations Manual is being revised and will contain additional information for such inspections.
    b. The Blood Bank Inspection Checklist and Report, Form FDA-2609 (5/91), should be used to record observations. All "no" responses should be explained in the "comments" section of the checklist. CFR references are included for referring questions to specific sections of the regulations. Please note that IOM Chapter 592 states that if observations made during an inspection are recorded on a checklist, these same observations need not be recorded in the investigator's diary.
    c. Instructions for using Form FDA-2609, with explanations and interpretations of the regulations, are provided in the Instruction Booklet (Blood Bank Inspection Checklist and Report).
    d. Prelicense Inspections

CBER will conduct the prelicense inspections accompanied by a district investigator after the establishment and product license applications (ELA and PLA) have been reviewed by CBER and found to be satisfactory. The prelicense inspection is an in-depth review of the physical facilities, manufacturing methods, SOP manual, records, equipment, etc. This inspection is announced, and the establishment must be operational for at least 30 days. The inspection is intended to determine the firm's ability to meet commitments made in its license applications and to operate in compliance with applicable regulation. The Division of Product Certification (DPC), HFB-240, notifies Districts of pending applications which will be the subject of Center/District prelicense inspections. If, however, field investigators are advised at the beginning of an inspection that an unlicensed blood bank has submitted a license application, they should contact the DPC (301-295-8428) to verify the firm's status.

Note: If HFB-240 confirms that a license application has been received, inform the firm that your inspection is not the prelicense inspection but that your findings will be conveyed to the center to be included in its overall assessment of the license.

Arrangements for prelicense inspections are made as follows: 1) The CBER will contact the Director of Investigations Branch and provide information on the firm and the proposed inspection; 2) the District will select an investigator to accompany the CBER inspector; and 3) the District investigator will contact the CBER inspector and make final arrangements. The CBER inspector is responsible for completing the Establishment Inspection Report (EIR). CBER will retain the original report and send a copy to the district. CBER will issue post-inspectional correspondence and a copy of this will be included in the copy of the EIR sent to the district.

For other than prelicense inspections, the District investigators will perform the inspections, and the District will issue post-inspectional correspondence.

e. Clinical Laboratories

Blood bank establishments may use outside (off-site) clinical/reference laboratories to perform viral marker or other required testing. If these laboratories are Medicare reimbursed, such laboratories will be routinely inspected by HCFA. CLIA approved laboratories are currently exempt from the requirement to register with FDA; however, FDA does have the authority to conduct inspections of these laboratories when testing is for determination of product quality or donor suitability.

If a laboratory that is a HCFA obligation, i.e., does not perform testing for blood banks to determine product quality or donor suitability, should appear in the District's Official Establishment Inventory (OEI), a routine surveillance inspection by HCFA would be indicated for the laboratory. The district should remove clinical laboratories that do not perform testing for blood establishments from the OEI, and the FDA regional contact should notify the

HCFA regional contact to assure inspectional coverage by HCFA (see Part VI, the primary HCFA contact is the Associate Regional Administrator).

Laboratories performing testing for blood products or donor suitability should be added to the OEI for routine inspection.

NOTE: Laboratories performing routine viral marker screening tests for licensed blood banks must be licensed and are inspected by FDA. (See 21 CFR 640.2)

f. Military Blood Banks

Licensed military blood banks are currently inspected only by CBER personnel. FDA has a formal agreement (MOU, CPG 7155d.01) with the Department of Defense to inspect military blood banks. CBER is in the process of transferring the responsibility for inspecting a large percentage of licensed military blood banks to the district offices. Inspections of licensed military blood banks may be performed by District personnel either as a joint CBER/District inspection (CBER will notify the DIB) or after notification from Division of Field Investigations, Office of Regulatory Affairs, that the inspection of a particular military blood bank is the responsibility of the District.

Prior to initiating an inspection at a military blood bank (30 days, if possible), the district should notify the military liaison regarding the inspection. Refer to Part VI, page 9 for military contacts. It is not necessary to notify CBER of the inspection.

g. Department of Veterans Affairs (Veterans Administration) and Indian Health Service Hospitals

Veterans Administration and Indian Health Service Hospitals are inspected by the district as unlicensed blood banks unless the facility is licensed.

Note: Certain Veterans Administration hospitals have been issued U.S. licenses.

h. Foreign Blood Banks

Licensed foreign blood establishments, including military, are currently inspected by CBER inspectors.

i. Complaints

A complaint concerning a biological product must be entered into the consumer complaint system on a FDA-2516 (Complaint/Injury Report). For complaints involving blood establishments inspected by the districts, a copy of the FDA-2516 should be sent to the Division of Inspections and Surveillance, Surveillance & Policy Branch (HFB-124). For complaints involving licensed establishments manufacturing licensed biological products other than blood components (vaccines, etc.), that are routinely inspected by CBER, the FDA-2516 is to be sent to HFB-124 for follow-up.

On June 4, 1990, program responsibility for biostatistics and epidemiology support of CBER activities was transferred to CBER, Office of Biological Product Review, Biostatistics and Epidemiology Staff (HFB-250). This staff reviews and monitors biological product adverse reaction reports submitted on FDA Form FD-1639, and Vaccine Adverse Event Reporting System (VAERS) forms.

A complaint from an informant regarding GMP violations, serious heath hazards, or possible criminal activity should be promptly investigated by the District, and CBER, Division of Inspections and Surveillance (HFB-120) should be notified.

j. Donor Centers

Donor centers are registered facilities (fixed locations) where generally only collections occur. Inspections should be shorter in duration and less complex. When scheduling an inspection of a donor center, it may be necessary to call to verify the days and hours of operation.

k. Mobile Operations

Investigators may encounter two types of mobile systems used as blood collection centers:

1) self-contained mobile units which are motorized vehicles equipped for the collection and holding of blood and blood components (frequently referred to as blood mobiles). A licensed facility using blood mobiles should submit information regarding the use of blood mobiles to CBER, as a supplement to their ELA. Blood mobiles, since they move to various locations, are not registered establishments.
2) stationary mobile operations are temporary collection facilities such as schools, offices, churches, etc., where equipment, and personnel (mobile team) are brought to the facility for the collection of whole blood and removed after the blood collection is over. Stationary mobiles are not required to register, nor are they required to be submitted as information in the ELA.

Since mobile blood collections are usually sporadic and subject to change, the investigator should obtain schedules of operation from the main blood bank to facilitate inspection of at least one mobile unit as a part of the establishment inspection.

Field investigators may obtain information regarding donor centers, and self-contained mobile units for licensed establishments from the Division of Product Certification (DPC), HFB-240. Advise DPC of any fixed location or self-contained mobile unit which is not included in the pertinent licensing information.

3) Special attention should be directed to storage temperatures of units awaiting transfer to the main location. Units must be placed in temporary storage having the capacity to cool the blood toward 1-6 C unless platelets are to be prepared. Units of blood which have been collected for further manufacture into platelets may be stored between 20 and 24 C (room temperature) for no more than eight hours.

l. Automated (Mechanical) Component Preparation

Establishments have been licensed to collect Fresh Frozen Plasma (FFP); Platelets, Pheresis; and Source Plasma by automated procedures. If a blood bank is preparing platelets or other blood components by automated methods for transfusion, the following documents, in addition to the Blood Bank checklist and instruction booklet, can be used as references: Firm's SOP; Revised Guideline for the Collection of Platelets, Pheresis, October 7, 1988, or the Plasmapheresis Checklist and Instruction Booklet [Form 2722 (11/90)].

For establishments collecting Source Plasma, use the Plasmapheresis Checklist and Instruction Booklet and CP 7342.002, Inspection of Source Plasma Establishments, as inspectional guides.

Donors who donate at intervals greater than eight weeks need only meet the Whole Blood suitability criteria. For platelet donations at intervals less than eight weeks refer to the Revised Guideline for the Collection of Platelets, Pheresis. For FFP donations at intervals less than eight weeks refer to Plasmapheresis Checklist for donor suitability requirements.

m. Plasma Brokers

Refer to CPG 7134.22, Plasma Brokers, Registration and Compliance with GMPs, for more information. During routine inspections of licensed and unlicensed blood bank facilities:

1) Review disposition records for recovered plasma to identify plasma brokers.
2) Determine whether the identified plasma broker(s) is (are) registered. If the broker is in another district, forward the information via the EIR endorsement to that district for follow-up.
3) Inspections of the brokers that are not registered should be conducted to determine whether the broker performs functions that require registration, i.e., they take possession of the plasma. If the broker does not take possession of the plasma registration is not required; however, records must be maintained to trace the transfer of plasma products from the manufacturer (blood bank) to the consignee.
4) If the plasma broker takes possession of the plasma, evaluate the Broker's compliance with at least 21 CFR, Parts 606.100(b)(10) and (18), 606.121 (e) (5), 606.160(b) (3)(i) and (iii), 606.160(b)(2)(ii) and (v), and 606.165 regarding storage, records of receipt, pooling,

labeling, distribution, and use of the product. Refer to the Blood Bank Instruction Booklet for guidance regarding short supply agreements.

n. Autologous Blood Collections

Units of blood may be collected for autologous use from persons who would otherwise not be suitable to donate blood for homologous use.

It is required that blood and blood components from autologous units meet all donor requirements if intended to be "crossed over" for homologous use or shipped for further manufacturing (for both injectable and non-injectable use), unless otherwise approved by CBER. Standard operating procedures for donor suitability, e.g., medical history questions, donor examination, and AIDS educational information, may differ for autologous donors when the blood is intended for autologous use only. All autologous donations that do not meet the donor suitability requirements must be clearly labeled "For Autologous Use Only" and any unused and/or untested unit or component must be destroyed. An establishment may or may not test units for HBsAg and anti-HIV1, provided that units and components which are not tested are properly labeled and appropriate controls are established to preclude their inappropriate release and/or use. (See Reference 8, Part VI, Page 2). Establishments which collect autologous units should be evaluated for:

1) CGMP Compliance (see Blood Bank Inspection Checklist and Report). A number of establishments collecting autologous blood also prepare components, including recovered plasma. CBER has encountered several instances where recovered plasma, untested for anti-HIV-1 and/or HBsAg, has been shipped labeled as non-reactive for these tests. Such shipments represent significant violations.
2) The need to register with FDA if the firm collects autologous units or become licensed by FDA if autologous products are to be shipped in interstate commerce.

2. SPECIFIC INSTRUCTIONS

a. Documentation

To support administrative or regulatory action, deviations must be well documented and represent a continual pattern of operation rather than an isolated occurrence, unless an isolated deviation has the potential to jeopardize the safety of donors or products. When documenting deviations of a general nature, such as those dealing with the facilities or personnel, obtain specific detailed descriptions of the circumstances relative to the deviation.

To assist in the evaluation of the seriousness of deviations, lists of significant deviations are attached. (See Attachments A and B)

b. HBsAG and Anti-HIV-1 Units

Specimens testing initially reactive must be repeated in duplicate: if one or both of the duplicate retests are reactive, the unit may not be used for transfusion or further manufacturing (unless approved by CBER and/or in accordance with 21 CFR 610.40 or 610.45). There is no requirement that confirmatory or more specific testing, i.e., Western blot, neutralization, be conducted on the repeatedly reactive unit. Additional test(s) may be performed at the option of the establishment for the purpose of donor counseling, look-back, or re-entry.

For a more detailed explanation of the "donor re-entry" protocol, refer to the following memoranda to blood establishments issued by CBER: December 2, 1987, are Initially Reactive for Hepatitis B Surface Antigen (HBsAg)" and April 23, 1992, "Revised Recommendations for the Prevention of Human Immunodeficiency Virus (HIV) Transmission by Blood and Blood Products." In addition, the re-entry algorithms for anti-HIV-1 and HBsAg are included as Attachments F and G.

If the investigator encounters the following situations in which units of blood have been distributed and HBsAg or HIV antibody testing results reveal the units to be:

1. not tested (except for emergency situations and properly labeled [21 CFR 610.40 (b)(4)]) or tests not performed according to manufacturer's instructions; or
2. initially reactive with only one or no repeat testing; or
3. repeatedly reactive; or
4. tested into compliance, i.e., at least 2 reactive test results, and the firm releases the product on the basis of additional nonreactive results; or
5. HIV positive (repeatedly reactive by EIA and WB positive) or confirmed HBsAg reactive; or
6. nonreactive, but the donor had a previous repeatedly reactive test for HBsAg or HIV antibodies and has not been appropriately re-entered; or
7. repeatedly reactive units shipped for unapproved uses [see 21 CFR 610.40(d) and 610.45(c)].

Then the investigator should:

1. Document the disposition of the unit and components.
2. Discuss the observations in accordance with IOM 516.1.
3. Determine the firm's corrective actions.
4. Determine if the donor was appropriately deferred.
5. Determine if subsequent donations have occurred.
6. Inform the firm that an error/accident report is to be submitted to CBER (voluntary for unlicensed facilities).

Immediate action is imperative. HBsAg and anti-HIV-1 repeatedly reactive or positive units may represent significant health hazards to the recipients of transfusions or users of certain blood derivatives. They may also present a hazard to handlers or users of noninjectable products manufactured from them if not labeled as to the risk of HBsAg or anti-HIV being present.

The Office of Compliance, Division of Case Management (HFB-110) or Division of Inspections and Surveillance (HFB-120) and ORA/Division of Field Investigations (HFC-132) should be advised immediately by telephone. CBER will review such incidents on a case-by-case basis and determine whether the course of action the firm is taking is appropriate with regard to:

1. notifying consignees of the HBsAg or (*)anti-HIV test(*) results and mislabeling;
2. requesting return of unused products;
3. informing patient's physician if any of the products have been transfused.

In all the above situations, obtain a documentary sample to include, at least:

1. name and address of consignees;
2. the donor and/or unit number;
3. the collection date of the unit in question;
4. copies of all relevant HBsAg and/or anti-HIV test results including copies of the original results and repeat testing, WB or neutralization, (if applicable) and testing of subsequent donations, as indicated.

   In situations regarding HBsAg testing problems, collect test results for anti-HBc and ALT testing. These additional test results are used by CBER for health hazard evaluations;
5. copies of donor medical history records if the unit was collected from an individual who tested nonreactive for HBsAg or anti-HIV but had a history of hepatitis B or HBsAg reactivity, or signs and symptoms associated with AIDS;
6. for instances of no testing being performed, obtain copies of records indicating the collection date, unit number, and testing records with the adjacent numbers for the date the testing would have been performed;
7. for instances of improper testing being performed, obtain testing records for the donation date and unit number, a copy of the firm's SOP and the manufacturer's package insert. Also, obtain results of testing from prior and subsequent donations. In certain cases, a copy of the equipment operating instructions may also need to be collected;
8. obtain copies of invoices and shipping (distribution) records and copies of the labeling, preferably from units already labeled.

c. Educational Material for Donors Concerning AIDS and Donor History Questions Which Elicit a History of AIDS

All establishments should have established procedures for providing educational material for donors concerning AIDS, and donor medical history forms should be updated to include questions which will elicit a history of AIDS or behavior placing the donor in a high risk behavior category.

The educational materials and donor medical history forms should be reviewed to ensure that all of the high risk behavior categories and informed donor consents are consistent with current FDA recommendations.

Questions regarding educational materials and/or donor medical history forms should be directed to the Division of Transfusion Science (HFB-900) contact person listed in Part VI, CBER Program Contacts.

d. Errors and Accidents: Adverse Reactions: Transfusion Associated AIDS: Look-back

Title 21, Code of Federal Regulations, Part 606 contains specific requirements relating to standard operating procedures and records associated with errors and accidents in manufacturing, transfusion reactions, complaints, and adverse reactions.

21 CFR 606.160(b)(7)(iii) requires that records be maintained relating to errors and accidents. 21 CFR 606.160(b)(6) requires that records be maintained for transfusion reaction reports and complaints, including records of investigations and follow-up. 21 CFR 606.170 further requires that records be maintained of reports of adverse reactions, a thorough investigation of each adverse reaction be performed, and a written report of the investigation including conclusions and follow-up be prepared and maintained.

Records and files relating to errors and accidents in manufacturing, complaints, and adverse reaction reports are to be reviewed by FDA Investigators during routine inspections in order to assess the adequacy of the firm's operating procedures and recordkeeping systems for maintenance and evaluation of error & accident records. Investigators should also review procedures and recordkeeping systems for handling reports of transfusion-associated adverse reactions and for the review of manufacturing records. Investigators should review and evaluate the procedures followed and steps taken by blood establishments in their review of discrepancies found during manufacturing. The SOPs should be sufficiently detailed to provide for thorough investigations, documentation of the process followed, and review of the findings by appropriate and responsible personnel.

The following guidance should be used in assessing the completeness and effectiveness of the firm's system to ensure that recurring deficiencies are identified and appropriate corrective actions are taken.

ERRORS & ACCIDENTS: 21 CFR 600.14 requires that licensed manufacturers notify CBER of errors and accidents in the manufacture of biological products (including blood and blood components) that may affect the safety, purity, or potency of the product.

Examples of reportable errors and accidents in manufacturing which may affect product quality include, but are not limited to, the release of: (1) units repeatedly reactive to viral marker testing; (2) units from donors for whom test results were improperly interpreted due to testing errors related to improper use of equipment or failure to strictly follow the reagent manufacturer's directions for use; (3) units from donors who are (or should have been) either temporarily or permanently deferred due to medical history or a history of repeatedly reactive results in viral marker tests; (4) units prior to completion of all tests; (5) incorrectly labeled blood components [e.g., ABO, expiration date]; and (6) microbiologically contaminated blood components when the contamination is attributed to an error in manufacturing.

While 21 CFR 606.160(b)(7)(iii) requires all blood establishments to maintain records relating to errors and accidents, currently only licensed blood establishments are required to report errors and accidents to CBER. By memorandum dated March 20, 1991, CBER requested that unlicensed, registered blood establishments and transfusion services voluntarily report errors and accidents in manufacturing to CBER. Notification to CBER by licensed blood establishments under 21 CFR 600.14 is necessary when a finished unit is made available for distribution whether or not actual shipment has occurred. If an error is detected during product manufacturing or processing and prior to the finished unit being made available for release, notification is not required. Any deviation from established specifications must be investigated and resolved prior to release and distribution of blood and blood components.

Investigators should include the following deviations on the list of inspectional observations (Form FDA-483): (1) failure to maintain records of errors & accidents in manufacturing; (2) failure to establish and maintain written standard operating procedures relating to investigations of errors & accidents; (3) failure to document investigations and follow-up corrective actions in response to an error or accident; or (4) if licensed, failure to notify CBER of errors or accidents in manufacturing for blood products made available for release.

Investigators should not include the occurrence of an error or accident on the Form FDA-483 if the firm has discovered an error or accident, properly documented its investigation, and implemented corrective action(s) which prevent recurrence of the error or accident. The investigator should concentrate on the cause of the error and the corrective action taken, if applicable, to prevent its recurrence. Also, errors and accidents should be evaluated to determine if unsuitable products were shipped and if the firm notified consignees, if appropriate, according to 21 CFR, Part 7. Form FDA-483 observations resulting from a review of the error/accident files should focus on current deficiencies or violative practices, the firm's failure to implement effective corrective action, and/or shipment of violative product.

ADVERSE REACTIONS: Cases of post-transfusion infection with human immunodeficiency virus (HIV) or hepatitis virus are viewed as adverse reactions when associated with the transfusion of blood and blood components. Procedures to investigate adverse reactions must be established by the blood bank.

Reports to blood establishments from physicians, other competent health care providers, or legal counsel concerning blood recipients who have acquired AIDS, or who have developed antibodies to HIV following a transfusion, or who has been diagnosed as having contracted post-transfusion hepatitis should be recorded and evaluated as possible adverse reactions to a blood transfusion. These reports of adverse reactions must be investigated by all blood establishments (whether licensed or unlicensed) per 21 CFR 606.170, and records must be maintained as required by 21 CFR 606.160(b)-(1)(iii).

If a report of post-transfusion HIV infection indicates no high risk behavior on the part of the recipient, the blood establishment's investigation should include a review of donor and manufacturing records (i.e., donor deferral, processing, viral marker testing, and disposition records) to determine if an error or accident has occurred. If it is determined that the unit or component transfused was viral marker reactive or collected from a donor who did not meet donor suitability criteria, CBER should be notified of the error or accident concerning the release and distribution of the unsuitable unit(s) per 21 CFR 600.14. If the blood establishment's investigation reveals that no error or accident has occurred resulting in the release of an unsuitable unit, notification to CBER is not required.

Investigators should include the following deviations on the list of inspectional observations (Form FDA-483): (1) failure to maintain records of reported adverse reactions; (2) failure to establish and maintain written standard operating procedures relating to procedures to be followed in the investigation of reported adverse reactions; (3) failure to document the investigation, conclusions, and follow-up (if any); or (4) if licensed, failure to notify CBER of errors or accidents in manufacturing that were revealed during the blood establishment's investigation of the adverse reaction.

Transfusion Associated AIDS (TAA): Definitions of TAA vary according to the reasons for which they were developed. The Centers for Disease Control (CDC) has defined Transfusion Associated AIDS (TAA) as a case reported to meet its surveillance definition for AIDS with no apparent AIDS risk factors other than transfusion in the 5 years preceding onset of symptoms. The CDC surveillance definition for AIDS is contained in MMWR 1987; 36 [suppl. no. 1S].

The CDC definition of TAA may be more narrow and specific than what may be encountered during the review of SOP's for particular blood establishments. For example, an establishment may define a TAA case as when any person found to be anti-HIV-1 positive has received a blood transfusion at any time since 1977.

In certain instances, it may be found that cases of TAA which meet either the CDC, or any other definition for TAA, may have resulted from errors or accidents in the manufacture of blood and blood components (e.g., failure to properly identify, quarantine, and control products from donors placed on deferral registries due to medical history or a history of having tested reactive to viral markers; failure to properly review and interpret results of viral marker testing, etc.). For licensed facilities, any case of TAA, irrespective of the definition used, which is traceable to an error or accident in manufacturing that resulted in the release of an unsuitable blood component is reportable to CBER under 21 CFR 600.14. If the case cannot be traced to an error or accident in manufacturing, it is not required to be reported to CBER. However, the requirements of 21 CFR 606.100(c) and 21 CFR 606.170 as previously discussed, are applicable to all blood establishments.

Requirements for reporting communicable diseases are mandated by state laws or regulations, and the list of reportable diseases in each state varies. The Council of State and Territorial Epidemiologists has recommended that state health departments report cases of AIDS which meet the CDC surveillance definition to CDC's National Notifiable Diseases Surveillance System. Cases of AIDS which meet the CDC surveillance definition should be reported by blood establishments to the respective state health department.

"LOOK-BACK" See the recommended guidance for lookback as specified in the April 23, 1992, memorandum to blood establishments "Revised Recommendations for the Prevention of Human Immunodeficiency Virus (HIV) Transmission by Blood and Blood Products." In the near future, the Agency intends to propose requirements for written procedures, records, and notification to consignees concerning product "look-back".

If the firm's SOPs require that "look-back" be performed, it is expected that the procedure will be followed and that appropriate follow-up actions will be taken. Failure to perform "look-back" per an established SOP should be listed as an inspectional observation on Form FDA-483's. Failure to perform "look-back" in accordance with CBER recommendations should be discussed with management.

"Look-back" procedures were recognized by the blood banking industry as a desirable practice as early as June 7, 1985, when a joint statement was issued by the major blood organizations including the American Association of Blood Banks, Council of Community Blood Centers, and The American National Red Cross Blood Services.

If further clarification is necessary, contact the Division of Inspections and Surveillance (HFB-120), Phone: FTS 295-8191.

e. Other Tests -- Anti-HCV. Anti-HTLV-I. ALT and Anti-HBc

The current AABB/CCBC/ARC circular of information for the use of blood and blood components indicates that blood and blood components are tested and found negative for HBsAg, anti-HIV-I, anti-HCV, anti-HBc, anti HTLV-1 and within established limits for alanine aminotransferase (ALT). The circular is considered labeling for the blood product, and if the establishment has adopted the circular as labeling for the blood product(s), the tests listed in the circular must be performed. Failure to perform the tests listed in the labeling of the product, or failure to perform the tests according to manufacturer's test kit instructions may be considered misbranding if the labeling indicates the test result was negative. Refer to Part III, Section 2,

Specific Instructions, for documentation regarding the release of blood products.

Antibody to Human T-Lymphotropic Virus Type-I (Anti-HTLV-I): On November 29, 1988, FDA recommended testing of donations of whole blood and cellular components intended for transfusion for antibodies to HTLV-I. FDA also recommended that, with the exception of specific circumstances, units testing repeatedly reactive not be used for transfusion. The memorandum also gives criteria for donor deferral (see Attachment H). Western Blot and Radioimmunoprecipitation (RIPA) assays are used for concerning repeatedly reactive EIA results (neither of these tests are currently licensed).

Antibody to Hepatitis C Virus: The Food and Drug Administration (FDA) has approved enzyme immunoassay (EIA) test kits for detection of antibody to a antigens associated with a non-A, non-B hepatitis virus (hepatitis C virus). Refer to the April 23, 1992, memorandum to blood establishments "Revised Recommendation for Testing Whole Blood, Blood Components, Source Plasma and Source Leukocytes for Antibody to Hepatitis C Virus Encoded Antigen (Anti-HCV)". In general, instructions for interpreting the results obtained with the anti-HCV licensed tests are similar in concept to those for HBsAg and anti-HIV-1 testing. Tested units will fall into one of two categories; negative and repeatedly reactive. Only negative units are suitable for transfusion, with the exception of autologous donations under specified conditions. At the present time, the unavailability of a licensed confirmatory test precludes the development of a re-entry algorithm for repeatedly reactive donors.

Anti-HCV testing is not expected to eliminate all cases of post-transfusion hepatitis. Therefore, the FDA suggests that blood establishments continue all measures implemented previously to reduce the risk of post-transfusion hepatitis.

Antibody to Hepatitis B Core Antigen (Anti-HBc): The second marker to appear in hepatitis B infection is antibody to the hepatitis B core antigen. This screening procedure was implemented as a marker in the belief that exclusion of donors with anti-HBc would lead to a reduction of post transfusion hepatitis. This marker is also a criterion for re-entry of donors testing HBsAg repeatedly reactive. Only licensed anti-HBc test kits should be used by blood establishments. Manufacturers were granted licenses for anti-HBc kits in April, 1991. CBER issued a memorandum to blood establishments on September 10, 1991, recommending that all donations of blood and blood components intended for transfusion be screened by an FDA licensed test.

When conducting inspections of blood establishments and testing laboratories, determine if unlicensed or "Research Use Only" EIA test kits for anti-HTLV-I, anti-HCV, anti-HBc, anti-HIV, or HBsAg are used by these firms to test blood products or to re-enter donors. The appropriate use of unlicensed confirmatory tests for donor re-entry is discussed in the re-entry algorithms for anti-HIV and anti-HTLV. Currently, the use of other unlicensed confirmatory tests, i.e., anti-HCV, for donor re-entry is not appropriate. Obtain a copy of the label and package insert for the unlicensed ("Research Use Only") EIA test kit, and forward this information to the Division of Inspections and Surveillance (HFB-120). Note: Currently, there are no licensed confirmatory test kits for anti-HTLV-I, anti-HCV, or anti-HBc.

f. Standard Operating Procedures (SOP's)

21 CFR 606.100 lists (not all inclusive) the standard operating procedures (SOP's) required for blood establishments. In addition to those listed, firms must also have SOP's for the following procedures: viral testing; quarantine and disposition of repeatedly reactive and positive units; donor re-entry (if applicable); procedures to assure donor confidentiality; the maintenance of donor deferral registers; and, use of computer system(s) in blood bank operations (if applicable) Licensed establishments do not need to submit these additional SOP's to CBER for approval. Compliance with the regulatory requirements will be determined during inspections.

g. Laboratory Quality Assurance

An increasingly significant area on which blood establishments will focus in the future is laboratory quality assurance. In an effort to provide increased assurance of the quality of laboratory performance, a proposed rule, cGMP for Blood and Blood Components; Proficiency Testing Requirements, was published June 6, 1989, which is intended to supplement HCFA regulations regarding quality assurance.

The blood bank's proficiency test results should be reviewed during inspections to determine if the firm performed a follow-up investigation and corrective action of unacceptable or incorrect results. A rule is being prepared which would require successful participation in an approved proficiency testing program by all laboratories for testing blood and blood products. FDA's final rule has not been published.

h. Computerization

Computer systems are playing an increasingly important role in blood banking. Improper design of application software and inadequate validation of the systems can lead to significant problems, e.g, releasing of repeatedly reactive HBsAg, anti-HIV units, or subsequent donations from donors with previous repeatedly reactive results for HBsAg or anti-HIV. When a decision is made regarding donor suitability and/or product quality based upon the use of, and reliance on, the data maintained in the computer system, the computer system is performing a critical function in the manufacture of blood and blood products. This concept is discussed in April 6, 1988, and September 8, 1989, memoranda. Both of these memoranda are referenced in Part VI. Other agency references are also listed in Part VI.

The Blood Bank Inspection Checklist and Instruction Booklet, has been revised to include Part K, Computerization. This section of the checklist and instructions can be utilized to obtain detailed information on equipment, program, and user validation during inspections of computer systems. Part K of the checklist is used regardless if the blood bank has purchased the computer system from a vendor or developed an in-house system.

i. Misuse or Malfunction of Equipment or Supplies

During the course of an inspection, the investigator may observe or review documentation of instances where equipment or supplies are either being misused or not functioning as designed. Such instances may affect the safety of the product, operator, or donor. It is important, therefore, to examine the firm's overall use of equipment and supplies [most of which are medical devices] to assure that the equipment and supplies are being used according to directions, satisfactorily inspected, and operated properly by the establishment.

On May 10, 1989, a memorandum was issued by HFC-100 to field investigators regarding the need to review the use of key equipment and supplies to detect misuse and related problems. Please refer to that memorandum for additional detailed inspectional guidance of blood bank establishments relating to:

- Key Equipment
- Examples Of Problems Encountered/Observed
- Reasons Why Such Problems May Occur
- Items To Be Alert For During Inspections
- Steps To Take

If equipment or supply misuse or malfunction is observed, the following documentation will be important to include in the report.

1. Manufacturer and lot number.
2. Number of experiences (failures).
3. If misuse, a copy of the SOP and whether the personnel were following prescribed procedures.
4. Whether the firm has notified the manufacturer, supplier, or FDA of problems/defects, and if so, the type of response received.
5. Whether personnel have been adequately trained in the use of the equipment.
6. If observations concern misuse, possible defective equipment or supplies, or possible inadequate instructions for use, use your district contacts and the CBER contacts listed in Part VI Pages 5 and 6 as needed, to aid in your assessment of the observed condition or procedure.

Questions or guidance regarding this information should be directed to the Division of Inspections and Surveillance (HFB-120).

j. Sample Collection

Routine physical sample collection is not required under this program. If official, physical samples are required, they will be collected by the field on assignment from the Center (see Part IV).

If significant deviations are noted, a documentary (DOC) sample should be collected which indicates interstate shipment for sale, barter, or exchange of the product. To establish interstate jurisdiction, the blood product or any component, i.e., blood bag with anticoagulant, must have moved in interstate commerce. DOC samples should clearly distinguish whether the reported activity represents a violation of Section 301(a), introduction or delivery for introduction into interstate commerce of a blood product that is adulterated or misbranded, or Section 301(k), adulteration or misbranding of a blood product while the article is held for sale after shipment in interstate commerce, of the FD&C Act. [Samples may be collected using the policies/procedures developed for delayed documentation of interstate commerce.]

k. Imports

See RPM Chapter 9-77 for instructions concerning biological products. In addition, refer to the Import Alert for Blood and Blood Components, Including Human Plasma and Serum [N.57-01 Revised July 20, 1990].

3. REPORTING/SPECIAL INSTRUCTIONS

a. Submit copies of all coversheets and FDA-483's, regardless of final classification, to Surveillance and Policy Branch (HFB-124). The EIR, checklist, and exhibits do not need to be included (if necessary, HFB-124 will request copies of these).

b. The Form FDA-2609, Blood Bank Inspection Checklist and Report, should be used when performing each inspection. New or unusual situations should be described.

NOTE: This checklist is not intended to replace the inspection report, described in the IOM, Sections 591 and 592. However, NAI inspections generally do not require a narrative report. This will be at the discretion of the district.

c. Inspection reports shall consist of copies of the Forms FDA-2609, FDA-481 -- Computer Generated Cover Sheet, FDA-482, FDA-483, a narrative report (when necessary), and documentation of significant deviations.

PART IV -- ANALYTICAL

ANALYZING LABORATORIES

No field analyses are projected under this program.

Samples collected on assignment will be analyzed by CBER laboratories, e.g., Division of Transfusion Science (HFB-940) and/or the Division of Product Quality Control (HFB-210).

Prior to submission of physical samples to CBER laboratories, the Surveillance and Policy Branch (HFB-124) should be contacted for guidance and instructions.

Results of analysis will be forwarded to the appropriate District Compliance Branch. If there are any questions concerning analytical results, contact HFB-124.

Copies of collection reports for physical samples must be submitted to Division of Case Management (HFB-110).

PART V -- REGULATORY/ADMINISTRATIVE STRATEGY

Violative conditions must be evaluated promptly during and following an inspection to assure donor protection and product safety. In cases where an expeditious regulatory action recommendation (i.e, seizure, injunction, or suspension) appears appropriate, direct telephone contact with the Division of Case Management (HFB-110) is encouraged. While this Part emphasizes immediate remedies, consideration of other regulatory action is not precluded.

When allegations are received by the district office relating to the concealment of violative practices by means of inaccurate recordkeeping or other forms of cover-up, telephone contact with the Surveillance and Policy Branch (HFB-124) early in the investigation is encouraged.

The strategy and actions which are outlined below are those that are most appropriate and expeditious when donor protection or product quality are in question.

Attachments A and B list significant deviations which will be used to assist in making determinations requiring regulatory actions. Other deviations, which are not significant enough to require ceasing operations, but require correction, shall be brought to management's attention by appropriate compliance correspondence.

A. SUSPENSION/REVOCATION OF OPERATIONS UNDER LICENSE

When conditions in a licensed establishment are such that donor health or product safety is jeopardized (see Attachment A) and immediate correction must be effected, suspension of operations under license should be recommended. Telephone contact to the Office of Compliance Division of Case Management (HFB-110) should be initiated as soon as possible and before the conclusion of the establishment inspection.

Refer to Chapter 8-80 of the RPM for procedural instructions.

B. INJUNCTION

Injunction may be the action of choice for unlicensed and licensed establishments if there is a current and real health hazard; the establishment has a history of uncorrected deviations despite past warnings and the evidence indicates that serious deviations are likely to continue; or the suspension of the firm's license would result in an unacceptable shortage of products in the geographic area served by the blood bank.

Recommendations for injunction in a situation involving a health hazard should include a request for a temporary restraining order. All supporting documentation, e.g., inspection reports, previously issued notices or other pertinent correspondence, DOC samples, of Case Management (HFB-110) should be contacted as soon as possible.

C. PROSECUTIONS

1. It is agency policy to consider prosecution of individuals or firms when fraud, gross violations, health hazards, or continuing significant violations are encountered.
2. CBER has generally relied on the following statutes when pursuing prosecution:

   Title 18, United States Code (U.S.C.)
   The Federal Food, Drug, and Cosmetic (FD&C) Act (21 U.S.C. 201, et seq,).
   The Public Health Service (PHS) Act (42 U.S.C. 262, et seq.).

3. Types of Criminal Cases

   A. Felony cases

   Biologics cases have in the past involved fraud, i.e., intentional violation of the regulations governing the day-to-day processing of donors and/or product, and falsification of records to conceal the violations. Charges have been brought under title 18, U.S.C. Sections 1001, 371, 1505, and 2. The evidence supporting these charges has included the following:

   1. Statements from employees and/or donors (current and former) identifying individuals that knowingly and willfully created false records that concealed noncompliance with biologics regulations and standards. Falsified records have included those encountered in the areas of donor suitability determinations, (i.e., failure to ask donor history questions), whole blood weights recorded to conceal overbleeding, and entries recorded in quality control logs for quality control tests that had not been performed.
   2. For conspiracy charges, statements from employees and donors that employees agreed to falsify records. In general, statements from

various employees have provided evidence that management was involved in the falsification of records.

B. Misdemeanor Cases

1. FD& C Act Cases

a. Failure to comply with GMP's

CBER has addressed serious GMP noncompliance through the administrative sanctions of license suspension/revocation or, in the case of an unlicensed facility, by seeking injunctive relief. Few cases based solely on GMP deficiencies have been forwarded to CBER for prosecution. In evaluating the appropriateness of prosecution,* the district should consider: (1) whether an individual or the firm has demonstrated a careless or reckless disregard for the applicable requirements, (2) the seriousness of the violations, (3) the frequency of distribution of violative product and the amount of violative product distributed, and (4) whether the firm has had a continuous pattern of violative conduct involving repeated significant deviations. Violations should be indicative of more than isolated or sporadic errors or deficiencies; demonstrate a breakdown in a given system, such as viral marker testing, donor screening, or product distribution, which resulted in the release of unsuitable units; and, be directly capable of affecting the identity, safety, or quality of the product.

b. Misbranding

Violations relating to donor safety may also be considered for a misdemeanor case, particularly flagrant violations that represent a serious threat to donor health and safety or that have occurred in conjunction with significant GMP violations. Examples of deviations which may be indicative of a system-wide problem relating to donor safety include the failure to perform hemoglobin/hematocrit determinations, failure to perform blood pressure determinations, and gross, continued overbleeding of plasma donors.

Violations addressed in section B.1.a. above may also be addressed as misbranding violations. Blood and blood products must be prepared in accordance with the standards. Labels which bear the product name Whole Blood, for example, may be false and/or misleading if the product was not prepared in accordance

---

*Although this section focuses on criminal action guidance in the blood program, the criteria in this section are also applicable to misdemeanor violations in other program areas such as advertising and promotion.

with the regulations. In addition, licensed products must be manufactured in accordance with the provisions of the product and establishment licenses. The labels of products bearing the U.S. license number but not prepared in accordance with the provisions of the license may also be false and/or misleading.

2. PHS Act Cases

a. Shipping without a license

Shipping biological products for sale, barter, or exchange without a U.S. license is prohibited by the PHS Act. In cases where a firm has not charged for its product, exchange of the product for future business may meet the definition of exchange. In evaluating the appropriateness of prosecution, the district should consider whether interstate shipments are an ongoing practice and whether the firm has received prior warning of the violation. An isolated shipment of biological products in interstate commerce without a license or shipment in bona fide emergency situations would not warrant prosecution under the PHS Act. (CPG 7134.11)

b. Falsely labeling or marking

The PHS Act prohibits the false labeling or marking of a biological product. Examples of false marking or labeling include the labeling of units as nonreactive for a viral marker test although the unit is reactive and labeling a unit as Fresh Frozen Plasma although the unit was not manufactured in accordance with the regulations for Fresh Frozen Plasma.

c. Interference with an officer of FDA performing his/her duties

A person who interferes with an FDA officer performing an inspection is in violation of the PHS Act. In considering this cite the district must assure that the inspection was conducted during reasonable hours in a reasonable manner and that biological products are manufactured at the firm for sale, barter, or exchange in interstate commerce.

D. SEIZURES

Seizure is the action of choice for both licensed and unlicensed establishments to remove violative product from the market and must be acted on promptly by the District, CBER, Office of Enforcement, and Office of General Counsel. Violative products are those which are adulterated or misbranded in violation of the Federal Food, Drug, and Cosmetic Act. Because the shelf life of blood and blood products varies from three days to ten years, the shelf life of the product should be considered prior to recommending seizure. The district should submit a recommendation, draft letter to the U.S. Attorney, and draft complaint to CBER unless otherwise directed by the Regulatory Procedures Manual (RPM), Section 8-20. All supporting documents, e.g., inspection reports, pertinent correspondence, DOC samples, etc., should be submitted with the recommendation. CBER, Division of Case Management (HFB-110) should be contacted as soon as possible. Refer to the RPM for additional information.

E. WARNING LETTERS

1. All correspondence to licensed establishments must be addressed to the Responsible Head. For unlicensed establishments, correspondence should be addressed to the most responsible individual, e.g., blood bank director or hospital administrator.
2. The lists of significant deviations (Attachments A and B) serve as guides for determining the recommended course of action. Any significant deviation, if well documented and representative of a continual pattern of operation, as opposed to an occasional deviation, may warrant the issuance of a Warning Letter. Additionally, Warning Letters may be indicated when there are deviations which may not affect the donor's health or the integrity of the product, but because they continually occur, may reflect an overall inadequately operated establishment.
3. When the District Compliance Officer determines that a Warning Letter should be issued, follow Chapter 8-10 of the RPM. The following specific areas require center concurrence for the District Director to issue a Warning Letter, otherwise, the Warning Letters are issued directly by the district: viral marker testing violations; donor re-entry violations; violations in areas where specific guidance has not been provided; labeling violations (except areas where specific guidance has been provided, e.g., Compliance Programs, CPGs, Drug/Health Fraud Bulletins); computer application and software violations; product advertising violations.

   Schedule a follow-up inspection approximately 30 days after the response to the Warning Letter is received to determine the adequacy of the reported corrective actions. When corrective action has not been affected or the firm has failed to respond, the District should consider appropriate follow-up.
4. Submit copies of all Warning Letters and any correspondence between the firm and the District to Case Guidance Branch (HFB-114).

F. REFUSAL TO PERMIT INSPECTION

If an establishment refuses to allow an investigator entry for the purpose of conducting an inspection, follow the procedures outlined in Chapter 5 (Section 514) of the IOM.

If a licensed establishment refuses inspection [see 21 CFR 601.5, Revocation of License], contact Surveillance and Policy Branch (HFB-124) for additional guidance.

G. FAILURE TO REGISTER

Owners or operators of blood manufacturing establishments are required to register and list their products pursuant to 21 CFR, Part 607, Subparts A and B, and Section 510 of the Federal Food, Drug, and Cosmetic Act. Investigators and Compliance Officers should also refer to Field Management Directive (FMD) #92 and CPG #7134.01 for specific details, procedural instructions, and description of responsibilities involved in the registration process. For additional guidance or questions contact, Division of Product Certification (HFB-240).

H. FEDERAL/STATE RELATIONS

Currently the Agency has no formal cooperative program with State or local jurisdictions for the inspection or regulation of blood banks. Cooperation with these authorities is, however, encouraged; especially where it is known that a State or local jurisdiction has a regulatory program. Exchange of information should occur between all levels of State government whenever feasible.

I. VETERANS ADMINISTRATION/INDIAN HEALTH SERVICE HOSPITALS/ LICENSED MILITARY ESTABLISHMENTS

When deviations from the regulations are encountered during inspections of such blood banks, the District will provide a copy of the Form FDA-483 to the responsible individual(s) at the local level with copies to the headquarters of the Agency or the Responsible Head of the military establishment and to the Office of Compliance, Division of Inspections and Surveillance, Surveillance and Policy Branch (HFB-124). An example of the post inspection letter sent to the Responsible Heads of the licensed military blood establishments is provided as Attachment I.

If significant deviations are encountered, or if a reinspection demonstrates that adequate corrective actions have not been taken, notify the Division of Case Management (HFB-110), as soon as possible, to determine if the violations warrant issuance of a Warning Letter or other action.

Contacts for the Veterans Administration, Indian Health Service, and military will be found in Part VI of this program.

PART VI -- REFERENCES, PROGRAM CONTACTS, AND ATTACHMENTS

REFERENCES

1. Federal Food, Drug, and Cosmetic Act, as Amended, and Related Laws, Public Health Service Act, Biological Products.
2. Title 21, Code of Federal Regulations, Parts 211, 600, 601, 606, 607, 610, and 640.
3. Preamble to Good Manufacturing Practices for Human Blood and Blood Components, FR Vol. 40, No. 223, November 18, 1975.

4. Technical Manual, 10th Edition, American Association of Blood Banks, 1117 N. 19th Street, Arlington, VA. 22209.
5. Standards for Blood Banks and Transfusion Services, 14th Edition, American Association of Blood Banks, 1117 N. 19th Street, Arlington, VA. 22209.
6. Blood Bank Inspection Checklist and Report, Form FDA 2609 (5/91), and Instruction Booklet.
7. Regulatory Procedures Manual (RPM): Chapter 8-10 (Warning Letters); Chapter 8-80 (License Revocation and Suspension); Chapter 9-77 (Biological Products); Chapter 9-79 (Import Alert 57-01).
8. CBER Memoranda to Blood Establishments:

Exemptions to Permit Persons with a History of Viral Hepatitis Before the Age of Eleven Years to Serve as Donors of Whole Blood and Plasma: Alternative Procedures, 21 CFR 640.120, April 23, 1992.

Revised Recommendations for the Prevention of Human Immunodeficiency Virus (HIV) Transmission by Blood and Blood Products, April 23, 1992.

Revised Recommendation for Testing Whole Blood, Blood Components, Source Plasma and Source Leukocytes for Antibody to Hepatitis C Virus Encoded Antigen (Anti-HCV), April 23, 1992.

Use of Fluorognost HIV-1 Immunofluorescent Assay (IFA), April 23, 1992.

Clarification of FDA Recommendations for Donor Deferral and Product Distribution Based on the Results of Syphilis Testing, December 12, 1991.

FDA Recommendations Concerning Testing for Antibody to Hepatitis B Core Antigen (Anti-HBc), September 10, 1991.

Disposition of Blood Products Intended for Autologous Use that Test Repeatedly Reactive for Anti-HCV, September 10, 1991.

Responsibilities of Blood Establishments Related to Errors & Accidents in the Manufacture of Blood & Blood Components, March 20, 1991.

Deficiencies Relating to the Manufacture of Blood and Blood Components, March 20, 1991.

Revised Recommendations for the Prevention of Human Immunodeficiency Virus (HIV) Transmission by Blood and Blood Products - Section I, Parts A& B Only, Memorandum of December 5, 1990, to all registered blood establishments.

Testing for Antibody to Hepatitis C Virus Encoded Antigen (Anti-HCV), Memorandum of November 29, 1990, to all registered blood establishments.

Use of Genetic Systems HIV-2 EIA, Memorandum of June 21, 1990, to all registered blood establishments.

Autologous Blood Collection and Processing Procedures, Memorandum of February 12, 1990 to all registered blood establishments.

Recommendations for the Prevention of Human Immunodeficiency Virus (HIV) Transmission by Blood and Blood Products, Memorandum of February 5, 1990 to all registered blood establishments.

Guideline for Collection of Blood and Blood Products from Donors with Positive Tests for Infectious Disease Markers ("High Risk" Donors), Memorandum of October 26, 1989 to blood establishments.

Abbott Laboratories' HIVAG-1 test for HIV-1 antigen(s) not recommended for use as a donor screen, Memorandum of October 4, 1989, to all registered blood establishments.

Requirements for Computerization of Blood Establishments, Memorandum of Septem ber 8, 1989, to all registered blood establishments.

Use of the Recombigen HIV-1 Latex Agglutination (LA) Test, Memorandum of August 1, 1989, to all registered blood establishments.

HTLV-I Antibody Testing, Memorandum of July 6, 1989, to licensed source plasma manufacturers approved for immunization with Red Blood Cells.

Blood Bank and Plasma Establishment Inspectional Guidance (to check equipment and supplies for misuse or defects), Memorandum of May 10, 1989, to Regional Directors and District Directors.

Guidance for Autologous Blood and Blood Components, Memorandum of March 15, 1989, to all registered blood establishments.

Use of the Recombigen HIV-1 LA Test, Memorandum of February 1, 1989, to all registered blood establishments.

HTLV-I Antibody Testing, Memorandum of November 29, 1988, to all registered blood establishments.

Revised Guideline for the Collection of Platelets, Pheresis, Memorandum of 7 October 1988, to Facilities Preparing Platelets, Pheresis.

Control of Unsuitable Blood and Blood Components, Memorandum of April 6, 1988, to all registered blood establishments.

Recommendations for Implementation of Computerization in Blood Establishments, Memorandum of April 6, 1988, to all registered blood establishments.

Extension of Dating Period For Storage of Red Blood Cells, Frozen, Memorandum of December 4, 1987, to all registered blood establishments.

Recommendations for the Management of Donors and Units that are Initially Reactive for Hepatitis B Surface Antigen (HBsAg), Memorandum of December 2, 1987, to all registered blood establishments.

Deferral of Donors Who Have Received Human Pituitary-Derived Growth Hormone, Memorandum of November 25, 1987, to all registered blood establishment.

Reduction of the Maximum Platelet Storage Period to 5 Days in an Approved Container, Memorandum of June 2, 1986 to all registered blood establishments.

Collection and Shipment of HTLV-III Antibody-Positive Blood Products, Memoran dum of December 9, 1985, to all registered blood establishments.

Revised Definition of High-Risk Groups With Respect to Acquired Immunodeficiency Syndrome (AIDS) Transmission from Blood and Plasma Donors, Memorandum of September 3, 1985, to all registered blood establishments.

Testing for Antibodies to HTLV-III, Memorandum of May 7, 1985, to all registered blood establishments.

Implementation of Public Health Provisional Recommendations Concerning Testing Blood and Plasma for Antibodies to HTLV-III, Memorandum of February 19, 1985, to all registered blood establishments.

Plasma Derived from Therapeutic Plasma Exchange, Memorandum of December 14, 1984, to all source plasma establishments.

Equivalent Methods for Compatibility Testing, Memorandum of December 14, 1984 to all registered blood establishments.
Deferral of blood donors who have received the drug Accutane (isotretinoin/ Roche; 13-cis-retinoic acid), Memorandum of February 28, 1984, to all establishments collecting source plasma.
Compliance Policy Guide 7134.02, Human Blood and Blood Products as Drugs, updated 9/1/84.

9. Compliance Policy Guides:

7134.01:	Registration of Blood Banks and Other Firms Collecting, Manufacturing, Preparing or Processing Human Blood or Blood Products
7134.02:	Human Blood and Blood Products as Drugs
7134.05:	Licensing - Changes to be Reported per 21 CFR 601.12
7134.08:	Source Plasma (Human) - Guidelines for Informed Consent Forms
7134.10:	Quantitative Testing for Serum Proteins in Plasmapheresis Donors
7134.12:	Blood and Blood Products - Definitions
7134.11:	Interstate Shipment in Emergencies
7134.14:	Source Plasma (Human) - Regulatory Action Based on Overbleeding
7134.15:	Source Plasma (Human) - Use of units from Donors Subsequently Found to be Reactive To a Serologic Test for Syphilis
7134.17:	Adequate Space for Determination of Donor Suitability
7134.18:	Schedule of Physical Examinations for Donors Receiving Immunization Injections
7134.19:	Units of Single Donor Plasma (Human) Fresh Frozen and Single Donor Plasma (Human)
7134.22:	Plasma Brokers - Registration and Compliance with Good Manufacturing Practices
7134.23:	Plasmapheresis - 48-hour Period Between Plasmapheresis Procedures
7155d.01:	Memorandum of Understanding with the Department of Defense
7155e.03:	MOU with Health Care Financing Administration Concerning Blood Banking and Transfusion Programs

10. Computer Guidance Documents:

Software Development Activities, July 1987.
Guide to Inspection of Computerized Systems in Drug Processing, February 1983.
Guideline on General Principles of Process and Validation, May 1987.
Application of the Medical Device GMPs to Computerized Devices and Manufacturing Process.

PART VII -- CENTER FOR BIOLOGICS EVALUATION AND RESEARCH RESPONSIBILITIES

A. LICENSING

1. Applications for licenses are made on forms obtained from CBER, Division of Product Certification (DPC) (HFB-240). Initially, licenses for the establishment and at least one product must be requested and issued simultaneously; thereafter, additional product licenses may be issued as long as the applications are acceptable and the establishment license remains unsuspended and unrevoked.

   Once applications are reviewed by CBER personnel and all necessary data and information are found to be satisfactory, a prelicense inspection will be conducted. See Part III, (A)(1)(d) for discussion of the prelicense inspection. When the results of the prelicense inspection show the firm to be operating in a satisfactory manner, establishment and product(s) licenses will be issued.

   An establishment must designate a Responsible Head who will represent the manufacturer in all pertinent matters [21 CFR 600.10(a)]. The supporting documentation for the candidate designated as Responsible Head will be reviewed and approved by CBER, DPC (HFB-240).

   A manufacturer must file amendments to its product licenses to include the preparation of products at location(s) separate from the main establishment. After review of the amendments by CBER and a satisfactory prelicense inspection, the establishment license is reissued to include the additional location(s); separate licenses numbers for locations are not issued. Such locations are generally fully operating facilities, although they may not manufacture all of the products produced by the main establishment.

   Requests for inspections are issued to the field as assignments by the Surveillance and Policy Branch (HFB-124) for changes in licensed establishments, i.e., location changes, additional donor and distribution centers, etc.

B. PROGRAM EVALUATION

An evaluation of this Compliance Program, with conclusions and recommendations, will be coordinated and/or prepared, if appropriate, by CBER and will cover work completed in the preceding fiscal year. The ORO will receive a copy of any such evaluation for distribution to appropriate ORA units.

TABLE OF CONTENTS FOR ATTACHMENTS

Violations must be evaluated to determine if they warrant administrative or regulatory action or both. To assist in this determination, Attachments A and B are lists of significant deviations from the GMP regulations.

These two lists contain deviations which are sufficiently serious that if they are allowed to continue, might jeopardize donor health or might compromise product safety or integrity. The deviations on these lists apply to both unlicensed and licensed establishments. Any deviation listed has the potential to become a health hazard, and all deviations must be corrected as quickly as possible after responsible personnel have been made aware of them.

NOTE: Because all possible situations and variables cannot be anticipated, the lists are not all inclusive, nor are they mutually exclusive. The deviations are referenced to specific sections of the regulations.

C. Licensed Test Kits for the Detection of HBsAg, anti-HCV, anti-HBc
D. Licensed Test Kits for the Detection of Antibody to HIV and HTLV-I
E. Code of Federal Regulation Sections for Evaluating Laboratory Viral Marker Testing
F. Anti-HIV-1 Antibody Test Results Qualifying a Donor for Reentry
G. HBsAg Re-entry
H. Summary of Recommended Actions for HTLV-I Antibody Testing
I. Example of Post Inspection Letter to Military Blood Establishments

ATTACHMENT A

DEVIATIONS WHICH MAY WARRANT A RECOMMENDATION FOR IMMEDIATE ACTION. I.E.. SUSPENSION OF LICENSE OR INJUNCTION

1. ANY PRACTICE WHICH POSES AN IMMINENT HEALTH HAZARD

2. Inadequate donor suitability determinations.

Personnel incompetent or inadequately trained in the performance of their assigned functions, and who fail to ensure that the final product has the safety, purity, potency, identity, and effectiveness it purports or is represented to possess with a regard to donor suitability determination [606.20(b), 640.3(a), and 640.4(a)]. Examples are: a) unaware of or disregarding medical history questions to be asked of all donors; b) unfamiliar with and/or not performing hemoglobin, blood pressure, temperature determinations; c) any action by personnel that causes unsuitable donors to be accepted.

These deficiencies are best documented via direct observations and affidavits from employees which outline and describe the inadequacies in training that result in the final product not having the safety, purity, potency, identity and effectiveness it purports or is represented to possess. Donor affidavits documenting inadequate screening are often helpful, especially in establishing lack of understanding by the donors regarding AIDS educational information and informed consent.

3. Inadequate donor processing and testing (determination of blood groups, test for presence of HBsAg and anti-HIV-1).

   Personnel incompetent or inadequately trained in the performance of their assigned functions, and who fail to ensure that the final product has the safety, purity, potency, identity, and effectiveness it purports or is represented to possess with regard to donor processing and testing. Examples are:

   a. No attempt to determine ABO and Rh Blood groups. [640.5 (b) & (c)]

   b. Inadequate viral marker testing

      1. Failure to perform test or interpret results according to manufacturer's directions and specifications, e.g., use of outdated reagents or mixing of reagents from different master lots; failure to run the proper number of controls concurrently with the test; calculations incorrectly determined resulting in reactive results being interpreted as nonreactive; interpreting reactive results as nonreactive; failure to conduct necessary retests. [610.40, 610.45, and 606.65(e)]
      2. Anti-HIV-1 and/or HBsAg test not performed by approved (licensed) tests or as otherwise allowed in Section 610.40 and/or 610.45.
      3. Container, pilot, and laboratory samples not properly identified to relate them to the individual donor. [606.140(c), 606.160(c), and 640.4(e) & (g) (3)]
      4. Shipments of blood or components untested or reactive to anti-HIV or HBsAg and labeled negative or nonreactive.

4. Records of the performance of any significant step in the collection, processing, compatibility testing, storage and distribution of each unit of product are grossly deficient, nonexistent, or inaccurate to a degree which would constitute a danger to health. [606.160(a), (b), (c), (d) and (e)]

5. Changing or altering blood labels or records so as to falsify. [42 U.S.C. 262 (a), 21 U.S.C. 353 (a)]

6. Facilities grossly inadequate or filthy. [606.40]

7. Failure of licensed blood bank to notify CBER of any of the following: [601.12(a) & (b)]

   a. Changes in Responsible Head and/or responsible personnel where such personnel are determined to be inadequately performing their duties to an extent that a danger to health exists;

b. Changes in location, e.g., firm moved to inadequate facilities; or
c. Important changes in equipment, manufacturing methods, testing procedures, or labeling that are determined to be inadequate.

ATTACHMENT B

DEVIATIONS WHICH MAY WARRANT CONSIDERATION OF REGULATORY/ADMINISTRATIVE FOLLOW-UP

General

1. Changing or altering blood labels or records so as to falsify, but not constituting a danger to health or is not a current practice. [606.160].
2. Shipment of unlicensed blood or blood products in interstate commerce for sale, barter, or exchange. [42 U.S.C. 262(a)].
3. Inadequate or filthy facilities. [606.40; 640.4(b)].
4. Personnel inadequately trained or supervised in the operations they perform but not to an extent that constitutes a danger to health. [606.20(b); 640.3(a); 640.4(a)].
5. Records of the performance of any significant step in the collection, processing, compatibility testing, storage and distribution of each unit of product are significantly incomplete or inaccurate. [606.160(a)(b)(c) and (e)].
6. A history of similar or substantially similar deficiencies about which the manufacturer has been advised.

Specific

1. Donor Suitability:

   a. Specific deviations from the regulations or non-adherence to SOP's in: (1) providing donors with the AIDS educational material; (2) medical history questions to be asked of donors; (3) determining what makes a donor unacceptable for donation; and/or (4) not performing hemoglobin, blood pressure, or temperature determinations. [640.3; 606.20(b)]. Note: Autologous donations that will not be for homologous use or further manufacturing may differ from donor suitability determination applied to homologous donations, however, the personnel must follow the firm's SOP established for autologous donations.
   b. No written procedures for determining donor suitability, including the exclusion of high-risk donors in regard to AIDS. [606.100(b)(1) and (2); 640.3(b)(c)(e)(f)].
   c. Donor suitability records are incomplete or inaccurate. [606.160(a)(1) and (b)(1)].
   d. No records available from which unsuitable donors may be identified so that products from such individuals will not be distributed. [606.160(e)]

2. Blood Collection:

    a. Arm preparation technique is inadequate or deviates significantly from the firm's SOP. [640.4(f)].
    b. Container not permanently sealed in an acceptable manner. [640.2(c)].
    c. Container, pilot, and laboratory samples are incompletely identified so as to relate them to the individual donor. [606.140(c); 606.160(c); 640.4(e) and (g)(3)].
    d. Unapproved container used for collection of whole blood. [640.2(c); 640.4(c)].

3. Processing:

    a. Laboratory tests for the determination of ABO, Rh, HBsAg and other viral markers not performed on laboratory samples collected from the donor at the time of filling final container, except in unusual situations. [640.5; 610.40 (b)].
    b. Anti-HIV and/or HBsAg test(s) not performed with an approved (licensed) test kit or as otherwise allowed in Sections 610.40 and 610.45.
    c. Deviations in the performance of tests or interpretation of results according to manufacturer's directions and specifications, e.g., use of out-dated reagents; mixing reagents from different master lots; failure to run the proper number of controls concurrently with the test; failure to conduct necessary retests. [610.40; 610.45; 606.65(e)].
    d. Records of test results significantly incomplete or inaccurate. [606.160(b) (2)(i)].

NOTE: Refer to Attachment C for a list of licensed manufacturers of HBsAg test kits, and Attachment D for a list of licensed manufacturers of anti-HIV test kits.

4. Storage:

    a. Blood or blood components not stored at proper temperature. [606.100(b) (10)].
    b. Failure to maintain daily temperature records when blood and blood component are in storage. [606.160(b)(3)(iii)].
    c. Failure to have or follow a system which would prevent the issuance of any products not suitable for use. [606.40(a)(6)].

5. Components (Red Blood Cells, Plasma, Platelets, Cryoprecipitated AHF):

    a. Components prepared by a method which deviates significantly from the regulations or the firm's SOP. [640.16(b); 640.24(a); 640.30(a); 640.52(a)].
    b. Records of component preparation are significantly incomplete or inaccurate. [606.160(a)(1), (b)(2)(ii), (b)(2)(iii)].

6. Compatibility Testing:

   a. Records of test results are significantly incomplete or inaccurate. [606.160 (b)(4)(i)]

7. Adverse Reactions and Fatalities:

   a. Failure to investigate adverse reactions and maintain appropriate records. [606.170(a)]
   b. The Center for Biologics Evaluation and Research not advised of death(s) resulting from complications related to blood collection or transfusion. [606.170(b)].

8. Errors or Accidents:

   a. Errors or accidents in the manufacture of blood and/or blood products (licensed facilities) which might affect their safety, purity, and/or potency not reported to CBER. [600.14(a)]

9. Distribution and Receipt:

   a. Distribution and receipt procedures (system) of each unit can not be readily determined to facilitate its recall if necessary. [606.165(a)]

10. Computerization:

   a. Contact CBER (HFB-110 or HFB-120) for guidance.

ATTACHMENT E

CODE OF FEDERAL REGULATION SECTIONS FOR EVALUATING LABORATORY VIRAL MARKER TESTING

Sample Collection/Identification

21 CFR 606.140(c); 606.160(a)(1)

- o Record
- o Back-up
- o Storage
- o Labeling
- o SOP's -- 606.100(b)

Sample Integrity

- o In-house vs. Contract -- 606.140(c)
- o Test Kit Methodology -- 606.65(e)
- o Serum vs. Plasma
- o SOP's -- 606.100(b)

Reagents

21 CFR 606.65(e)

- o Licensed/unlicensed -- 610.45, 610.40(b)
- o Receipt Records -- 606.160(a)(2); (b)(7)(v)
- o Storage (conditions & records) -- 606.65(d) & (e)
- o Manufacturing Instructions -- 606.65(e)
- o Quality Control Testing -- 606.65(c) & (e)
- o Record -- 606.160(a)(1), (a)(2), (b)(5)(ii)
- o SOPs -- 606.100(b)(10) & (b)(14)

Equipment

21 CFR 606.60(a) & (b); 21 CFR 211.68

- o Intended use/design/compatibility -- if interfaced
- o User Manual -- 606.100(b)(7); 606.100(b)(15); 606.140(a) & (b)
- o Qualification/calibration -- 606.60(a) & (b)
- o Maintenance & Repair -- 606.60(a)(b), 606.160(b)(7)(iv)
- o Quality Control Records -- 606.160(b)(5)(i) & (ii)
- o SOP's -- 606.100(b)(15)

Personnel

- o Education & Experience -- 606.20(a) & (b), 600.10(a) & (b)
- o Training (initial & continued) -- 606.20(b), 600.10(b)
- o Records/SOP's -- 606.160(b)(7)(ii)

The Test

- o Sample ID -- 606.140(c)
- o Interpretation -- 606.140(a) & (b), 606.160(b)(2)(i), 606.100(b)(7)
- o Controls -- 606.140(a)
- o Invalid Tests -- 610.40, 610.45
- o Records/SOP's -- 606.100(b)(7), 606.160(b)(2)(i)

Records

21 CFR 610.40 & 610.45

- o All Tests/Retests & No Tests -- 606.65(e)
- o Written & Prior to Distribution -- 610.40(b)(4)
- o Review of Records -- 606.100(c)

Proficiency Testing

- o "Snap Shot in Time"
- o Registration Forms
- o Testing Procedures & Handling
- o Results
- o Follow-up

FDA COMPLIANCE PROGRAM GUIDE 7342.002 -- INSPECTION OF SOURCE PLASMA ESTABLISHMENTS
PROGRAM: 7342.002
SUBJECT: Inspection of Source Plasma Establishments (FY 91)
IMPLEMENTATION DATE: Upon Receipt
COMPLETION DATE: SEPTEMBER 30, 1991
DATA REPORTING:
PRODUCT CODES. For inspections
57DY-[ ] [ ]
57YY-[ ] [ ]
PROGRAM/ASSIGNMENT CODES: 42002
DATE OF ISSUANCE: MAY 1991

Field Reporting Requirements

Copies of all Inspection Reports (to include the EIR, coversheet and FDA-483, but NOT the exhibits) and Notice of Adverse Findings Letters and response letters are to be submitted to CBER, Surveillance and Policy Branch (HFB-124).

PART I -- BACKGROUND

Section 640.60, Title 21, Code of Federal Regulations, defines the product, Source Plasma, as the fluid portion of human blood collected by plasmapheresis and intended as source material for further manufacturing use. This definition excludes plasma intended for direct transfusion into patients.

Plasmapheresis is the procedure in which blood is removed from the donor, the plasma is separated from the formed elements and at least the red blood cells are returned to the donor. This procedure may be performed manually or through the use of automated equipment. *Regardless of the method used, (i.e., manual or automated), the collection procedures must be described in detail in the establishment's standard operating procedures manual.*

Source Plasma is a product subject to the licensure provisions of Section 351 of the Public Health Service Act (42 U.S.C. 262). For Source Plasma to be suitable for interstate (including international) shipment, an unsuspended and unrevoked U.S. license is required for both the establishment (i.e., collection facility) and the product (see 21 CFR 601.10 and 601.20). Before a product can be labeled as Source Plasma, it must first comply with the standards set forth in 21 CFR, Part 640, Subpart G, provided that the establishment also complies with the establishment standards prescribed in the regulations contained in 21 CFR 601.*

Current Changes

PART II -- IMPLEMENTATION

A. OBJECTIVE

To assure that Source Plasma is safe, effective, and adequately labeled by:

1. Determining if plasmapheresis establishments are operating in compliance with applicable regulations.
2. Assuring donor protection and a safe, pure, and potent product for use in further manufacturing.
3. Providing guidance to the field for performing annual inspections of all licensed establishments collecting Source Plasma.
4. Inducing voluntary compliance.

The firms covered under this program include all establishments engaged in the manufacturer of Source Plasma.

B. PROGRAM MANAGEMENT INSTRUCTIONS

Inspectional

1. Inspections of all establishments licensed to manufacture Source Plasma will be conducted at least once each fiscal year and will cover Parts 600, 601, 606, 607, 610.40, 610.41, 610.53, and 640.60 through 640.76 of Title 21, Code of Federal Regulations.
2. Each inspection under this Compliance Program will cover the total operation of the establishment, except that follow-up inspections to verify correction of previous violations need not cover all operations.

NOTE: If a licensed plasma center has a change in operations which requires a license amendment (for example, a change in location or responsible personnel) then CBER may request a directed inspection and limit coverage to the subject change.

*3. The *Center for Biologics Evaluation and Research (CBER) *supplies a checklist with accompanying instruction booklet for this program. The Plasmapheresis Inspection Checklist and Report (Form FDA-2722) is designed as an inspectional tool for the investigator and an information gathering document from which data can be compiled. Copies of the checklist and instruction booklet can be obtained from the *Biological Product Inspections Branch (HFB-122).*

Current Changes

PART III -- INSPECTIONAL

A. OPERATIONS

2. Prelicense Inspections

The *Center for Biologics Evaluation and Research (CBER)* and the District will conduct a joint prelicense inspection after the establishment and product license applications have been reviewed by CBER and found to be satisfactory. The prelicense inspection is an in-depth review of the physical facilities, manufacturing methods, equipment, etc. At this inspection, the establishment must be operational, records for at least 30 days should be available, and the establishment must demonstrate its ability to meet the commitments made in its license applications and to operate in compliance with applicable regulations in order for license to be issued. The Division of Product Certifications (DPC), *HFB-240*, notifies Districts of pending applications which will be subject of joint Center/District prelicense inspections.

NOTE: When a licensed plasmapheresis center decides to change from the manual to the automated method of collection, according to 21 CFR 601.12, the firm must, prior to the conversion, submit a product license amendment for collection of Source Plasma by automated procedures. This will usually prompt a prelicense inspection of the facility. In some cases, DPC will approve a firm's conversion to automated procedures on the basis of the manufacturer's training and internal audit process together with the district's compliance profile for the firm.

Arrangements for joint prelicense inspections are made as follows:

a) *CBER, Biological Product Inspections Branch (HFB-122*), will contact the District's Director of Investigations Branch and provide information on the firm and the proposed inspection date(s);

b) The district will select an investigator to accompany the *CBER inspector(s)*; and

c) The District investigator will contact the *CBER* inspector(s) and make final arrangements. The *CBER* inspector is responsible for completing the Establishment Inspection Report (EIR). The *CBER* will retain the original report and send a copy to the District. *CBER, Biological Product Inspections Branch, HFB-122,* will issue post-inspectional correspondence, and send a copy to the District office.

Current Changes

For other than prelicense inspections, the District Offices will perform the inspections and issue post-inspectional correspondence.

2. Activities Approved Under License

CBER, Division of Product Certification (DPC), HFB-240, provides establishment licensing information to District Offices on a quarterly basis. Attachment F is a list of typical licensing information supplied by HFB-240. The firm must have approval for changes. Unapproved changes should be documented.*

The Inspection Operations Manual (IOM) does not contain procedures for procedures for conducting inspections of plasmapheresis establishments. This Compliance Program, therefore, contains specific detailed instructions.

The most current version of the plasmapheresis checklist, Form FDA-2722, should be used when performing inspections of Source Plasma establishments. Instructions for using Form FDA-2722, together with explanations and interpretations of applicable regulations, are provided in the instruction booklet.

Investigators are empowered under the Food, Drug and Cosmetic Act and provisions of Section 351 of the Public Health Service Act to examine all parts of the manufacturing facility; manufacturing processes and controls; computer operations; and all pertinent records.

If the investigator is denied access to an establishment, for any reason, to conduct an inspection, or is denied access to records for review or copying during an inspection, follow the applicable instructions in Chapter 5 of the IOM, and contact the Surveillance and Policy Branch, HFB-124, at 301-295-8191. Subpart C of 21 CFR part 600 discusses inspections of biological product establishments and outlines the duties and powers of the inspector. Refusal to permit inspection of an establishment and refusal to permit access to or copying of required records are both bases for subsequent revocation of a source plasma establishment's license according to 21 CFR 601.5.

B. SAMPLING

Generally, documentary rather than physical samples are collected during plasmapheresis center inspections. Collection of physical samples for analysis should be coordinated through the Biological Product Inspections Branch (HFB-122) prior to collection.

Interstate Jurisdiction may be established by documenting the interstate movement of either the blood product (if applicable) or the anticoagulant in the blood collection container.

In general, documentary samples are almost always sufficient to support legal action. Refer to the Inspection Operations Manual for sample documentation. Regarding samples collected at plasmapheresis centers, it is often helpful to include the standard operating procedures applicable to the violation being documented. Documents which may be destruction/disposition and shipping records, and photographs of the violative conditions.

C. DOCUMENTATION

Investigators should contact their supervisors as soon as possible when violations are encountered which may warrant regulatory/administrative follow up.

To assist in determining whether administrative or regulatory action is warranted, two lists of significant deviations are provided as Attachments A and B.

To support regulatory action, deviations must be well documented and represent a continual pattern of operation rather than an isolated occurrence, unless an isolated deviation has jeopardized the safety of donors or products, or violated a mandatory licensing requirement.

Do not hesitate to take photographs to support verbal description of violative conditions in a plasmapheresis center. Investigators should be sure to collect sufficient evidence to document each of their FDA-483 observations. Toward that end, the following guidance is listed:

1. When documenting improper testing, obtain copies of the manufacturer's instructions, SOPs, appropriate test result records (*)that(*) contain the unit number, bleed date in question, and date testing.
2. When documenting lack of records, copy records before and after bleed date and unit number to show which are missing.
3. When documenting labeling deviations, make copies of the label.
4. For deviations concerning shipment of units prior to receiving written test results, obtain copies of invoices and shipping (distribution) records in addition to the *appropriate test result records.*

Current Changes

5. If any untested, or HBsAg *reactive* or anti-HIV reactive units have been labeled as nonreactive and shipped, the investigator should obtain a documentary sample, and immediately notify his/her supervisor. (See Attachment L, Shipments of Plasma Prior to Receipt of Written Test Results).

*6. If misuse or malfunction of equipment is observed the following documentation will be important to include in the report:

   a) Manufacturer and lot number.
   b) Number of experiences (failures).
   c) If misuse, a copy of the SOP, owner's manual and whether the personnel were following prescribed procedures.
   d) Whether the firm has notified the manufacturer, supplier, or FDA of problems/defects, and if so, the type of response received.
   e) Whether personnel had been adequately trained in the use of the equipment.

   NOTE: If observations noted with the use or performance of equipment or supplies represent potential design problems or possible inadequate instructions for use, alert CBER and the home district where the manufacturer is located.*

7. If the establishment is demonstrating a continual pattern of noncompliance with respect to donor safety, or conditions indicate the safety or integrity of the product is being compromised and a recommendation for suspension of operations under license is contemplated, an inventory of the plasma at the establishment on the last day of inspection should be obtained. This inventory may be determined by:

   a) subtracting the bleed number of the last unit shipped from the bleed number of the last unit collected; or
   b) the packing records (with unit numbers) which are often used as shipping records and list the numbers currently in storage. CBER will make a decision as to the disposition of the plasma in inventory at a suspended firm after review of all information.

8. Overbleeding

   a) When documenting overbleeding by volume, obtain copies of records that include the unite number, date, volume or weight of each unit of whole blood, the volume/weight of pooled plasma, and the weight of the donor on the day of donation. *Also consider collecting scale calibration and maintenance records.*

Current Changes

b) When documenting overbleeding by volume for automated plasmapheresis, obtain copies of records that include the unit number, date, total volume collected, and donor hematocrit, height and weight on the day of donation. Consider collecting donor scale calibration records, repair/maintenance records, and employee training records.

c) When documenting overbleeding by frequency, obtain copies of records that indicate which donors, unit numbers and bleeding dates are involved. Documenting the donor's name should be avoided unless absolutely essential. Documentary rather than physical samples are usually sufficient to document overbleeding.

NOTE: Overbleeding is further discussed in Attachment B, 5.

D. SPECIFIC INSTRUCTIONS

To supplement the information provided in the instruction booklet for the Plasmapheresis Inspection Checklist and Report, Form FDA-2722, specific instructions relative to the below listed topics are provided in Attachments G through X.

Attachment G:	HBsAg Reactive and Anti-HIV Repeatably Reactive Units or Untested Units
Attachment H:	Testing Blood and Plasma for Antibodies to HIV
Attachment I:	Off Site Plasma Storage
Attachment J:	Reference Laboratories and Other Outside Testing Laboratories
Attachment K:	Imported Biological Products
Attachment L:	Shipment of Plasma Prior To Receipt of Written Test Results
Attachment M:	Wrong Cell Infusion Incidents
Attachment N:	Infrequent Plasmapheresis
Attachment O:	Physician Substitutes
Attachment P:	Educational Material for Donors Concerning AIDs and Donor History Questions Which Will Elicit a History of AIDS and Other HIV Infections
Attachment Q:	Approved Shipments of Anti-HIV Repeatably Reactive or HBsAg Reactive Plasma
Attachment R:	Disposition of Anti-HIV Repeatably Reactive or HBsAg Reactive Units
Attachment S:	Automated Plasmapheresis
Attachment T:	Misuse or Malfunction of Equipment or Supplies
Attachment U:	Arm Preparation for Venipuncture
Attachment V:	Damaged Bags and Bottles of Source Plasma

Current Changes

Attachment W: Computerization
Attachment X: ALT and Anti-HBc Testing

E. Reporting

Copies of all Source Plasma establishment inspection reports, (to include the EIR, coversheet and FDA-483, but not the exhibits), regardless of classification, and copies of all Notice of Adverse Findings Letters and response letters should be sent to the attention of *Surveillance and Policy Branch, HFB-124*. Do not include the checklist (FDA-2722) with NAI and VAI-1 reports.

PART IV -- ANALYTICAL

No field analyses are projected under this Compliance Program.

Samples collected on assignment will be analyzed by *CBER/Office of Biological Product Review* laboratories.

Results of such analyses will be forwarded to the District Office Compliance Branch.

PART V -- REGULATORY/ADMINISTRATIVE STRATEGY

Violative conditions must be evaluated promptly during and following an inspection in order to assure donor protection and product safety. The strategy and actions which are outlined below are those we believe are most appropriate and expeditious when donor protection or product quality is in question. Consideration of other regulatory actions is not precluded. When allegations are received by the District Office relating to the routine concealment of violative practices by means of inaccurate record keeping or other forms of cover-up, telephone contact with the *Surveillance and Policy Branch, HFB-124,*, early in the investigation is encouraged. Prior to, or concurrent with, the initiation of the establishment inspection, visit the informant and obtain an affidavit documenting how the true records can be distinguished from the false records. During the establishment inspection, attempt to corroborate the information supplied by the informant.*

The lists of significant deviations which are included as Attachments A and B will assist in determining the appropriate course of action to take. Other deviations, which are significant and must be corrected, but may not require that a firm cease operations, shall be brought to management's attention by appropriate compliance correspondence.

A. SUSPENSION OF OPERATIONS UNDER LICENSE

If donor health and/or product *quality* is jeopardized, necessitating immediate corrective action, and the deviations are well documented, recommend suspension

Current Changes

of operations under license to the *CBER* Office of Compliance, *Division of Case Management, HFB-110*, as soon as possible and no later than 24 hours following the completion of the inspection. *Also, maintain contact with HFB-110 throughout the inspection and prior to issuance of the FDA-483.* Evaluate what effect the violative condition(s) have on product quality. The firm must make proper disposition of any product whose quality has been compromised.

Refer to the Regulatory Procedures Manual (RPM), Chapter 8-80, License Revocation or Suspension.

B. NOTICES OF ADVERSE FINDINGS/REGULATORY LETTERS

1. The lists of significant deviations serve as a guide for determining the recommended course of action (See Chapter 8-80 of the RPM for general guidance). Any one deviation on Attachment B, if well documented and representative of a continual pattern of operation, as opposed to an occasional deviation, may warrant the consideration of a Regulatory Letter recommendation to the OC. Additionally, there may be some continually occurring deviations which do not affect the donor's health or the integrity of the product, but reflect an overall inadequately operated establishment. In such cases, a Regulatory Letter recommendation may be warranted as opposed to a Notice of Adverse Findings.
2. When the District Compliance Officer determines that a Notice of Adverse Findings Letter or recommendation for a Regulatory Letter should issue, follow Chapter 8-10 (Notice of Adverse Findings and Regulatory Letters) of the RPM. *Copies of all Notice of Adverse Findings letters and response letters should be sent to the Surveillance and Policy Branch (HFB-124).*
3. *CBER* regulatory policies are stated in the Compliance Policy Guide Manual, Chapter 34 - Biologics. The current Compliance Policy Guides for biological products are listed in Part VI, Page 4.

C. FEDERAL/STATE RELATIONS

Currently, the Agency has no formal cooperative program with State or local jurisdictions for the inspection or regulation of Source Plasma manufacturers; however, cooperation with these authorities is encouraged, especially if a State or local jurisdiction has a regulatory program. Whenever feasible, exchanges of information should occur between all levels of government.

PART VI -- REFERENCES, PROGRAM CONTACTS AND ATTACHMENTS

REFERENCES

1. Federal Food, Drug, and Cosmetic Act, as Amended, and related Laws (Public Health Service Act, Biological Products).

2. Title 21, Code of Federal Regulations, Parts *211*, 600, 601, 606, 607, 610, and 640.

3. Preamble to Regulations for Additional Standards for Source Plasma (Human) Used in Preparation of Blood Derivatives Intended for Injection, FR, Vol. 38, No. 139, July 20, 1973.

4. Preamble to Regulations for Amendments to the Additional Standards for Source Plasma (Human), FR, Vol. 41, No. 50, March 12, 1976.

5. Title 42, Public Health Service Act, Part 72, Interstate Commerce, Subpart C, FR, Vol. 45, No. 141, Monday, July 21, 1980.

6. Technical Manual, Current Edition, American Association of Blood Banks, 1117 N. 19th Street, Arlington, VA 22209.

7. Current Plasmapheresis Inspection Checklist and Report, From FDA 2722, and Instruction Booklet.

8. Regulatory Procedures Manual (RPM): Chapter 8-10 (Notice of Adverse Findings and Regulatory Letter); Chapter 8-80 (License Revocation and Suspension); Chapter 9-77 (Biological Products); and Chapter 9-79 (Import Alert 57-01).

9. Revised Recommendations for the Prevention of Human Immunodeficiency Virus (HIV) Transmission by Blood and Blood Products - Section I, Parts A & B only, Memorandum of December 5, 1990 to all registered blood establishments.

10. Testing for Antibody to Hepatitis C Virus Encoded Antigen (Anti-HIV), Memorandum of November 29, 1990 to all registered blood establishments.

11. Use of Genetic Systems HIV-2 EIA, Memorandum of June 21, 1990 to all registered blood establishments.

12. Autologous Blood Collection and Processing Procedures, Memorandum of February 12, 1990 to all registered blood establishments.

13. Recommendations for the Prevention of Human Immunodeficiency Virus (HIV) Transmission by Blood and Blood Products, Memorandum of February 5, 1990 to all registered blood establishments.

14. Guideline for Collection of Blood and Blood Products from Donors with Positive Tests for Infectious Disease Markers ("High Risk" Donors), Memorandum of October 26, 1989 to blood establishments.

15. Abbott Laboratories' HIVAG-1 test for HIV-1 antigen(s) not recommended for use as a donor screen, Memorandum of October 4, 1989, to all registered blood establishments.

Current Changes

16. Requirements for Computerization of Blood Establishments, Memorandum of September 8, 1989, to all registered blood establishments.
17. Use of the Recombigen HIV-1 Latex Agglutination (LA) Test, Memorandum of August 1, 1989, to all registered blood establishments.
18. HTLV-I Antibody Testing, Memorandum of July 6, 1989, to licenses source plasma manufacturers approved for immunization with Red Blood Cells.
19. Blood Bank and Plasma Establishment Inspectional Guidance (to check equipment and supplies for misuse or defects), Memorandum of May 10, 1989, to Regional Directors and District Directors.
20. Guidance for Autologous Blood and Blood Components, Memorandum of March 15, 1989, to all registered blood establishments.
21. Use of the Recombigen HIV-1 LA Test, Memorandum of February 1, 1989, to all registered blood establishments.
22. HTLV-I Antibody testing, Memorandum of November 29, 1988, to all registered blood establishments.
23. Physician Substitutes, Memorandum of August 15, 1988, to all licensed manufacturers of source plasma.
24. Recommendations for Changeover from Use of Fresh Immunizing Red Blood Cells to Use of Frozen Immunizing Red Blood Cells Stored a Minimum of Six Months Prior to Use, Memorandum of October 7, 1988, to licensed source plasma establishments approved or applying to perform Red Blood Cell immunizations.
25. Revised Guideline for the Collection of Platelets, Pheresis, Memorandum of 7 October 1988, to Facilities Preparing Platelets, Pheresis.
26. Discontinuance of Pre-License Inspection of Immunization Using Licensed Tetanus Toxoid and Hepatitis B and Rabies Vaccine, Memorandum of July 7, 1988, to all licensed source plasma manufacturers.
27. Control of Unsuitable Blood and Blood Components, Memorandum of April 6, 1988, to all registered blood establishments.
28. Recommendations for Implementation of Computerization in Blood Establishments, Memorandum of April 6, 1988, to all registered blood establishments.
29. Recommendations Concerning Persons at Increased Risk of HIV-1 and HIV-2 Infection, Memorandum of April 6, 1988, to all registered blood establishments.
30. Extension of Dating Period For Storage of Red Blood Cells, Frozen, Memorandum of December 4, 1987, to all registered blood establishments.
31. Biosafety Responsibilities and Proposed Biosafety Workshop, Memorandum of December 4, 1987, to all registered blood establishments.

32. Recommendations for the Managements of Donors and Units that are Initially Reactive for Hepatitis B Surface Antigen (HBsAg), Memorandum of December 2, 1987, to all registered blood establishments.
33. Deferral of Donors Who Have Received Human Pituitary-Derived Growth Hormone, Memorandum of November 25, 1987, to all registered blood establishments.
34. Recommended Testing Protocol to Clarify Status of Donors With a Reactive Anti-HIV Screening Test, Memorandum of April 29, 1987, to all registered blood establishments.
35. Additional Recommendations for Reducing Further the Number of Units of Blood and Plasma Donated for Transfusion or for Further Manufacture by Persons at Increased Risk of HTLV-III/LAV Infection, Memorandum of October 30, 1986, to all registered blood establishments.
36. Reduction of the Maximum Platelet Storage Period to 5 Days in an Approved Container, Memorandum of June 2, 1986 to all registered blood establishments.
37. Collection and Shipment of HTLV-III Antibody-Positive Blood Products, Memorandum of December 9, 1985, to all registered blood establishments.
38. Revised Definition of High-Risk Groups With Respect to Acquired Immunodeficiency Syndrome (AIDS) Transmission from Blood and Plasma Donors, Memorandum of September 3, 1985, to all registered blood establishments.
39. Recommendations Concerning Blood or Plasma Previously Collected from Donors Currently HTLB-III Antibody Positive, Memorandum of July 22, 1985, to all registered blood establishments.
40. Testing for Antibodies to HTLV-III, Memorandum of May 1985, to all registered blood establishments.
41. Implementation of Public Health Service Provisional Recommendations Concerning Testing Blood and Plasma for Antibodies to HTLV-III, Memorandum of February 1985, to all registered blood establishments.
42. Revised Recommendations to Decrease the Risk of Transmitting Acquired Immunodeficiency Syndrome (AIDS) and Other Procedures Related to Blood Product Preparation, December 1984, to all registered blood establishments.
43. Plasma Derived from Therapeutic Plasma Exchange, Memorandum of December 14, 1984, to all Source Plasma establishments.
44. Physician Substitutes, Memorandum of December 14, 1984, to all manufacturers of source plasma.
45. Equivalent Methods for Compatibility Testing, Memorandum of December 14, 1984 to all registered blood establishments.

46. Deferral of Blood Donors Who Have Received the Drug Accutane (isotretinoin/Roche; 13-cis-retinoic acid), Memorandum of March 24, 1983, to all establishments collecting source plasma.

47. Recommendations to Decrease the Risk of Transmitting Acquired Immune Deficiency Syndrome (AIDS) from Plasma Donors, Memorandum of March 24, 1983, to all establishments collecting source plasma.

48. Source Material Used to Manufacture certain Plasma Derivatives, Memorandum of March 24, 1983, to all licensed manufacturers of plasma derivatives.

49. Requirements for Infrequent Plasmapheresis Donors, Memorandum of August 27, 1982, to all licensed manufacturers of source plasma.

50. Revised Guideline For Adding Heparin to Empty Containers for Collection of Heparinized Source Plasma (Human), Memorandum of August, 1981.

51. Guidelines for the Collection of Human Leukocytes For Further Manufacturing (Source Leukocytes). Memorandum of January, 1981, to all registered blood collecting facilities.

52. Guidelines For Immunization of Source Plasma (Human) Donors With Blood Substances, Revised June, 1980.

*53. Compliance Policy Guides:

7134.01:	Registration of Blood Banks and Other Firms Collecting, Manufacturing, Preparing or Processing Human Blood or Blood Products
7134.02:	Human Blood and Blood Products as Drugs
7134.05:	Licensing -- Changes to be Reported per 21 CFR 601.12
7134.08:	Source Plasma (Human) -- Guidelines for Informed Consent Forms
7134.10:	Quantitative Testing for Serum Proteins in Plasmapheresis Donors
7134.12:	Blood and Blood Products -- Definitions
7134.14:	Source Plasma (Human) -- Regulatory Action Based on Overbleeding
7314.15:	Source Plasma (Human) -- Use of Units from Donors Subsequently found to be Reactive To a Serologic Test for Syphilis
7314.17:	Adequate Space for Determination of Donor Suitability
7134.18:	Schedule of Physical Examinations for Donors Receiving Immunization Injections.
7134.22:	Plasma Brokers -- Registration and Compliance with Good Manufacturing Practices
7134.23:	Plasmapheresis - 48-hour Period Between Plasmapheresis Procedures
7155d.01:	Memorandum of Understanding with the Department of Defense

Current Changes

7155e.03: MOU with Health Care Financing Administration Concerning Blood Banking and Transfusion Programs*

ATTACHMENTS

LISTS OF SIGNIFICANT DEVIATIONS

Violations must be evaluated to determine if they warrant administrative or regulatory action or both. To assist in this determination, these lists of significant deviations from the GMP regulations were developed:

Attachment A: Deviations Which May Warrant Suspension Of Activities Under License.

Attachment B: Deviations Which May Warrant Regulatory/Administrative Follow-up.

NOTE: Because all possible situations and variables cannot be anticipated, the lists are not all inclusive, nor are they mutually exclusive. The deviations are referenced to specific sections of the regulations.

LIST OF LICENSED TEST KITS

Attachment C: Licensed Anti-HIV-1 Test Kits.

Attachment D: Licensed Test Kits for the Detection of HBsAg.

Attachment E: Licensed Test Kits for Detection of Anti-HIV-1 by Western blot.

SPECIFIC INSTRUCTIONS

The topics of Attachments G through X are listed in Part III, Page 6. The information provided in these attachments is designed to answer questions which are frequently asked by field investigators.

Current Changes

PART VII -- *CENTER FOR BIOLOGICS EVALUATION AND RESEARCH* RESPONSIBILITIES

A. LICENSING

1. Manufacturers of Source Plasma are required to submit establishment and product license applications to *CBER*. These applications must describe the manufacturing procedures in detail. The establishment must be operational at the time of the prelicensing inspection to demonstrate its ability to meet the specifications in the license applications, and to comply with the applicable regulations. *CBER* retains the responsibility for reviewing the license applications and performing prelicensing inspections.
2. *CBER/Division of Inspections and Surveillance will schedule the prelicense inspection and contact the District's Director of Investigations *Branch* (DIB) as far in advance as possible to request that an investigator be assigned to accompany the *CBER* inspector. The Field investigator will contact the *CBER* inspector to plan for the inspection. The *CBER* inspector will be the lead inspector and will be responsible for preparing the report and providing a copy to the District DIB.
3. The Division of Product Certification, *HFB-240*, will provide the Division of Inspections and Surveillance, Surveillance and Policy Branch (HFB-124) with quarterly revisions to establishment and product licenses. This information will be furnished to the appropriate district offices by HFB-124.
4. *CBER* requests all manufacturers to file, in the appropriate section of their Standard Operating Procedures Manual, approval for amendments so that the investigator can determine if a procedure has been approved for use.

B. PROGRAM REVIEW AND EVALUATION

An evaluation of this Compliance Program, with conclusions and recommendations, will be coordinated and/or prepared, if appropriate, by the Center and will cover work completed in the preceding fiscal year. The ORO would receive a copy of any such evaluations.

ATTACHMENT A

DEVIATIONS WHICH MAY WARRANT SUSPENSION OF ACTIVITIES UNDER LICENSE

If the inspection indicates a continuing pattern of deviations which constitute a danger to health, suspension of activities under license as prescribed by 21 CFR 601.6(a) may be warranted.

Current Changes

1. ANY PRACTICE WHICH POSES AN IMMINENT HEALTH HAZARD
* 2. Facilities grossly inadequate or filthy [606.40].
* 3. Failure to notify *CBER* of any of the following [601.12(a) and (b)] which may cause a danger to health:

   a. Changes in Responsible Head and/or responsible personnel.
   b. Change in location, which may be inadequate, i.e., of unsuitable size, construction, or location to facilitate adequate cleaning, maintenance, and proper operations.
   c. Important changes in equipment, manufacturing methods, testing procedures, or labeling.

* 4. Personnel:

   a) Inadequately trained or supervised in functions they perform [606.20 (b), 640.63(a), and 640.64(a)]
   *b) Unaware of or disregarding medical history questions to be asked of ALL donors; unaware of or disregarding the situations that make a donor unacceptable for donation; unfamiliar with or not performing hemoglobin, *total protein*, blood pressure, or temperature determinations. [640.3; 606.20(b); 640.63(C)].

* 5. Donor Suitability: Donors continued on plasmapheresis:

   *a) when total protein value is below 6 gm per 100 ml [640.63 (c) (5)].
   b) when hematocrit reading *is* below 38% or with hemoglobin values below 12.5 gms [640.63(c) (3)].
   c) when medical history questions were not asked;
   *d) Refractometer not adequately cleaned between total protein determinations, or instrument is such that [606.60] correct readings cannot be obtained.
   e) *when* responses to medical history questions indicated that the donor was unsuitable for plasmapheresis, *or
   f) when donors are determined to be unsuitable in any other respect.*

*6. Immunizing donors *by unacceptable procedure(s) and/or using unapproved antigens* [640.66].
*7. Collection of Whole Blood: Lack of proper identification procedures including a number on the container so as to relate unit of blood or red blood cells directly to donor (640.64(d) and 606.100(b)(4)].
*8. Storage and Shipment:

Current Changes

a. Failure to have an adequate quarantine system which would prevent issuance of *unsuitable units [606.40(a)(3)(4) and (6)] and [606.160 (e)]; and,

b. A demonstrated pattern of shipping Source Plasma labeled as nonreactive, which is found to be reactive to HBsAg and/or anti-HIV *[640.67(a) and (b)].*

NOTE: Some establishments have been approved to collect Source Plasma from donors who are reactive for HBsAg and/or anti-HIV repeatably reactive. Therefore, certain establishments will be making approved shipments of HBsAg reactive and/or anti-HIV repeatably reactive plasma which will not be in violation of the regulations. Refer to 42 CFR Part 72 for proper shipment of these products.

c. Shipping, in other than authorized situations, without written HBsAg [610.40(b)(*4*)] or Anti-HIV *[610.45(a)]* test results in possession.

*9. Records *grossly deficient or* non-existent for:

a) Determination of donor suitability [640.63(c) and 606.160(b)(1)]
b) Questions which would elicit a history of AIDS
c) Physician's review of laboratory and collection data in donor's chart [640.65(b)(2)].
d) Whole blood weights [640.65(b)(4),(5),(6) and 606.160(a)(1)]. *(Except for automated plasmapheresis.)*

NOTE: In situations where records are grossly deficient, the Investigator should establish through direct or indirect evidence, the underlying reasons for the lack of records.

10. Changing or altering product labels *which results in the release of unsuitable units.*

ATTACHMENT B

DEVIATIONS WHICH MAY WARRANT REGULATORY/ADMINISTRATIVE FOLLOW UP

Any of the listed deviations, if well-documented and representative of a continued pattern of operation, as opposed to isolated instances, could warrant regulatory/administrative follow-up.

Current Changes

1. Facilities: Inadequate or filthy [606.40].
2. Medical Supervision:

   a. Physical exam inadequately performed or no record of exam [640.63(b)].
   b. Hazards or risks not explained by physician; no signed informed consent form in donor's records [640.61].
   c. Physician not physically on premises when immunizations are performed [640.62].

3. Donor Suitability:

   a. Photographs, or some other equally assuring method, not available for positive identification of donors [640.65(b)(3)].
   b. Personnel are not following established procedures when determining donor suitability, including the exclusion of high risk donors in *high risk categories for* to AIDS [606.100(b)(1) and (2); 640.63(c)]:
   c. Donors plasmapheresed whose plasma protein composition was outside the normal range established by the testing laboratory [640.65(b)(2)(i)].

4. Blood Collection:

   a. Donor is not weighed prior to plasmapheresis [640.63(c)(6)].
   b. Improper container used for collection of whole blood [640.64(b)].

5. Plasmapheresis (Overbleeding):

   *a) Manual collection:

   A substantial degree of deviation from the requirements is significant. The critical factors in determining overbleeding are: 1) the weight (converted to volume) of the whole blood collected; and 2) the volume of plasma obtained from whole blood collections. While such factors as biological variability of the donors, human errors in blood collection, variability in bag weights and scale accuracy, and the amount and kind of anticoagulant must all be considered, the following general statements may be used as guides for determining substantial deviation:

      a. Whole blood volumes -- overbleeding is significant when: 1) the volume of whole blood removed routinely exceeds the maximum allowable limits in 21 CFR 640.65(b)(4), (5), and (6); and when there is no indication that overbleeding is recognized and/or corrected.

Current Changes

b. Plasma Volumes -- overbleeding is considered significant when units of Source Plasma routinely exceed the maximum allowable limit of 770 ml (790 gms) for donors weighing less than 175 pounds and 925 ml (949 gms) for those weighing 175 pounds or more. See CPG No. 7134.14. (Plasma collected in heparin solution will less due to the smaller volume of anticoagulant.)

NOTE: In determining a substantial degree of deviation, documentation must establish a pattern of conduct over a period of the time which illustrates a normal or customary practice of overbleeding.

b) Automated Collection

Overbleeding is considered significant if the firm, through neglect or intent routinely fails to perform those operations which are necessary to assure that the amount of plasma collected is in accordance with that listed in the approved nomogram. It would be necessary to document the apparent cause(s) of the overbleeding, for example: inadequate initial and periodic training of device internal weight monitors each day of use; or failure to use or improper use of the approved and/or gender-specific nomogram.*

6. Immunization:

*a) Inadequate records. For example, not recording, when appropriate: antigen administered; titers; lot number and manufacturer; dose; date; and route administration [606.160(a)(1) and (b)(1)(v)].

*b) Lack of written procedures for all tests related to red blood cell immunization, including antibody detection, identification, and titration [640.66].

7. Tests for Hepatitis B Surface Antigen (HBsAg):

a. Failure to perform test according to manufacturer's directions and specifications, e.g., use of out-dated reagents; failure to run the proper number of positive and negative controls concurrently with the test samples; failure to perform background count [640.67 and 606.65(e)].

b. Calculations for HBsAg test runs incorrectly determined [606.65(e)].

c. Test samples not adequately identified to relate them to donor [606.140(c)].

8. Tests to detect antibody to the Human Immunodeficiency Virus (anti-HIV) *[640.67 and 606.65(e)]*

a. Failure to perform test according to manufacturer's directions and specifications, e.g., using/mixing reagents from different lots; failure to run

Current Changes

proper number of positive and negative controls concurrently with the test samples *[606.65(e)].*

b. Calculations for anti-HIV test runs incorrectly determined *[606.65(e)].*

c. Test samples not adequately identified to relate them to the donor *[606.140(c)].*

9. Storage:

a) Failure to always store plasma intended for use in manufacturing injectable products at the proper temperature [640.69(b)].

b) Storage facilities not under supervision and control of the manufacturer [640.71(a)].

c) No daily storage temperature records maintained [606.160(b)(3)(iii)].

d) Final containers for Source Plasma are not stored or handled in a manner to prevent them from being damaged and results in exposure of the product to external factors that may cause deterioration or contamination [640.68(b).

10. A history of continuous violations of a similar nature about which the firm has been advised.

ATTACHMENT S

Automated Plasmapheresis

Many plasmapheresis facilities have recently received approval to collect Source Plasma using specific automated equipment such as the Haemonetics PCS and/or the Autopheresis C. The license approval letter will specify the equipment and model number.

Some areas in which investigators should direct specific attention are: personnel training (initial and ongoing); the number of automated devices each employee is assigned to monitor; individual device logs to document: operator problems, errors, testing and maintenance of each device, and donor adverse reactions.

The approved nomogram for the Haemonetics PCS and the Autopheresis C both use body weight as a variable in deriving the plasma collection volume. It is, therefore, very important for plasma centers using automated collection devices to have accurate donor weight scales. The donor weight scale(s) should be observed, standardized, and calibrated on a regularly scheduled basis as prescribed in the firm's SOP manual, and these scale(s) shall perform in the manner for which they were designed [21 CFR 606.60(a)].

Current Changes

The facility's approved SOP Manual and the approval letter from *CBER* are to be used as guides for inspecting these automated procedures.

ATTACHMENT T

Misuse or Malfunction of Equipment or Supplies

During the course of an inspection, the investigator may observe instances where equipment or supplies are either being misused or not functioning as designed. Such instances may adversely affect the product, operator/user, or donor. It is important, therefore, to examine the firm's overall use of equipment and supplies to be sure that they are: (a) used in accordance with manufacturer's instructions, (b) satisfactorily inspected and calibrated, and (c) in good repair.

On May 10, 1989, a memorandum issued from HFC-100 to field investigators discussing the need to review the use of key equipment and supplies for misuse and defect related problems. Please refer to that memorandum (listed in Part VI, References) for detailed inspectional guidance relating to:

-- Key Equipment
-- Examples Of Problems Encountered/Observed
-- Reasons Why Such Problems May Occur
-- Items To Be Alert For During Inspections
-- Steps To Take (appropriate follow-up)*

Current Changes

ATTACHMENT W

Computerization

Currently, there are only a limited number of Source Plasma establishments utilizing computer systems in control mode functions. However, we can expect over time to see more and more plasmapheresis centers employing computerization, just as we have seen many of them convert from manual to the automated means of plasmapheresis. When personnel depend on the computer system in making decisions regarding the suitability of donors and release of Source Plasma for further manufacture, the computer system is considered to have "control functions." This concept is discussed in the September 8, 1989 memorandum "Requirements for Computerization of Blood Establishments" from CBER to all registered blood establishments. The memorandum, also discusses validation of computer systems, and related operations. Another CBER memorandum on this topic was issued on April 6, 1988 (Recommendations for Implementation of Computerization in Blood Establishments). Both of these memoranda are referenced in Part VI.

The information to be determined regarding computerization during the inspection is included in Part K of the Current Blood Bank Inspection Checklist and Instruction Booklet.

DRAFT GUIDELINE FOR THE VALIDATION OF BLOOD ESTABLISHMENT COMPUTER SYSTEMS

[Docket No. 93N-0394]

Prepared September 28, 1993

For further information about this document, contact:

Center for Biologics Evaluation and Research
Suite 200N
1401 Rockville Pike
Rockville, MD 20852-1448
301-594-3074

Submit written comments on this document to:

Dockets Management Branch (HFA-305)
Food and Drug Administration
Rm. 1-23
12420 Parklawn Dr.
Rockville MD 20857

Submit requests for single copies of this document to:

Congressional and Consumer Affairs Branch (HFM-12)
Food and Drug Administration
1401 Rockville Pike
Rockville, MD 20852-1448
301-594-2000
FAX-301-594-1938 FAX ON DEMAND 301-594-1939
INTERNET "CBER_INFO@A1.CBER.FDA.GOV"

Comments and requests should be identified with the docket number found in brackets at the heading of this document.

I. PURPOSE

The purpose of this guideline is to assist manufacturers of blood and blood components, including blood banks and plasmapheresis centers, in developing a computerized system validation program consistent with recognized principles of system development methodology and quality assurance that are current good manufacturing practices. This is not intended to be a guideline for software manufacturers who may be subject to other guidance and requirements.

II. SCOPE

This guideline may be useful to blood establishments in developing and administering a computerized system validation program. Because the Food and Drug Administration (FDA) is in the process of revising 21 CFR 10.90(b), this document is not being issued under the authority of 21 CFR 10.90(b), and the document, although called a guideline, does not bind the agency and does not create or confer any rights, privileges, or benefits for or on any person. Blood establishments may follow the guideline or may choose to use alternative procedures not provided in the guideline. If a blood establishment chooses to use alternative procedures, the establishment may wish to discuss the matter further with the agency to prevent expenditure of resources on activities that may be unacceptable to the Food and Drug Administration (FDA).

Blood, blood components, or derivatives applicable to the prevention, treatment, or cure of human diseases or injuries are specifically identified as biological products subject to regulation pursuant to the Public Health Service (PHS) Act [42 U.S.C. 262, Section 351 (a)(1)]. Similarly, blood, blood components, or derivatives intended for use in the diagnosis, cure, mitigation, treatment, and prevention of diseases in humans are drugs as defined in Section 201 (g) of the Federal Food, Drug, and Cosmetic (FD&C) Act [21 U.S.C. 321 (g)]. Section 501 (a)(2)(B) of the FD&C Act states, in part, that a drug shall be deemed to be adulterated if "the methods used in, or the facilities or controls used for, its manufacture, processing, packing, and holding do not conform to, or are not operated or administered in conformity with, current good manufacturing practice to assure that such drug meets the requirements of the FD&C Act." Because blood and blood components are defined as drugs in the FD&C Act, the Current Good Manufacturing Practices (CGMP) in 21 CFR, Parts 210 and 211, are applicable.

This guideline is intended to be used in conjunction with the applicable federal standards in 21 CFR, Parts 600 through 680, and Parts 210 and 211, as they pertain to biological products for human use. Note that sections 210.2(a) and (b) provide that where it is impossible to comply with the applicable regulations in both Parts 600 through 680 and Parts 210 through 211, the regulation specifically applicable to the product shall apply and supersede the more general regulation. Applicable regulations may be found in Section XIV. A. and Section XVII. of this document.

The agency may amend this guideline after receiving comments submitted by interested persons, and as necessary thereafter.

III. INTRODUCTION

Blood establishments (e.g., blood banks, blood product testing laboratories, plasmapheresis centers, and transfusion services) have rapidly expanded the use of computers as tools in the manufacture of blood and blood components to assist in decision-making and in the management of critical data (e.g., donor registry, laboratory testing, storage of required records and data).

Proper performance and use of computerized systems in the manufacturing process is a critical part of final product quality. Blood establishments currently using and those considering use of computerized systems should clearly understand the need to establish and define the functions to be performed by the computerized system, the way in which it will interact with both manual and automated operations, and the importance of data integrity.[1]

This document provides general guidance on the process of validation of computerized systems used in blood establishments. The goal of validation is to perform and document systematic procedures that give assurance that the computerized system is functioning as specified and intended at each user site, including existing and new installations, and after every significant system change.

This document should also be useful for the hardware and software vendors that supply products used as key components of the blood establishment's computerized systems. By understanding the regulatory needs of the blood establishments, hardware and software vendors can better define the features, capabilities, and limitations of their products.

This document supersedes the information provided in the April 6, 1988, and September 8, 1989, memoranda regarding the use of computerized systems in registered blood establishments (see Section XIV. C. Memoranda).

Licensed blood establishments are reminded that they are required to report the proposed installation or major modification of a computerized system if it is to be used in the manufacture of blood and blood components to the Director, Center for Biologics Evaluation and Research (CBER), as a supplement to the firm's establishment license application (ELA) (see Section XIII. Reporting to CBER).[2]

IV. COMPUTERIZED SYSTEM DEFINITION

A blood establishment's computerized system includes: hardware, software, peripheral devices, personnel, and documentation [e.g., User's Manual, Standard Operating Procedure (SOP)]. Defining the computerized system logically begins by first documenting the user's needs and requirements. This also helps avoid misunderstanding between the user and the developer. The user should work with the system developer to define all steps in each function to be performed by the computerized system. The requirements should include a written description of the tasks to be performed including compliance with local, state, and federal regulations. If the computerized system is not

---

[1]See, e.g., 21 CFR 211.68, 211.100, 211.180, 211.188, 211.194, 600.10(a) and (b), 600.12(a) and (b), 606.20(a) and (b), 606.60, 606.100, 606.160.

[2]See, 21 CFR 601.12(a).

defined in terms of components and functions, a well designed validation program can not be established.

A. System Components

The development of a computerized system should include all the specifications for each of the system's components (e.g., parameters, limitations, and constraints). The user of the computerized system should account for operational needs and requirements, as well as establishing criteria for personnel training and quality assurance to ensure the consistent and reliable performance of the system in the working environment.[3]

Hardware, software, and peripheral device factors include:

- the compatibility between the hardware and software,
- the interfaces with peripheral devices,
- the physical constraints (e.g., space, power, air conditioning requirements, electromagnetic and radio frequency noise or interference, etc.), and
- maintenance and calibration procedures.

Personnel factors include:

- training,
- certification,
- unique identification,
- access limitations, and
- operating instructions (e.g., User's Manual, SOP).

The functions of the computerized system should be established with written descriptions outlining:

- inputs,
- processing,
- outputs,
- search criteria,
- data manipulation,
- data output format and content,
- structure of the database (e.g., definition of the fields and their contents), and
- communications with other equipment.[4]

---

[3]See, e.g., 21 CFR 600.10(a) and (b), 606.20(a) and (b), 606.60, 606.100, 606.160, 211.63, 211.67, 211.68, 211.100.

[4]See, e.g., 21 CFR 600.10(a) and (b), 606.20(a) and (b), 606.100, 606.160, 211.63, 211.67, 211.68, 211.100, 211.180.

B. System Documentation

The user should ensure that the documentation for the computerized system is current, accurate, and as detailed as necessary to ensure proper use and operation.[5] System documentation should include:

Vendor provided items, such as:

- operational specifications and requirements,
- instruction manuals [e.g., User's Manual, Operations Manual, Training Manual, (if the user will be maintaining the software, a program maintenance manual and other references are also needed)],
- diagrams, flow charts, and descriptions that define the interactions between modules, for software (including interfaces), and
- test databases.

User provided items, such as:

- user's requirements,
- Standard Operating Procedures (SOPs),
- the test database and a description of the test cases,
- records of operational validation of integrated hardware, software, and peripheral devices, and
- records of maintenance and changes made to system hardware, software, and peripheral devices.

V. SYSTEM VALIDATION PROTOCOLS

Validation is an essential component in the development and operation of a computerized system. A validation protocol is a written plan that is intended to be followed in testing a specific portion of the system or the total system in order to produce documented evidence that the system operates consistently and reliably. Validation protocols should be developed using system development methodology (SDM). SDM is a structured method used to plan, design, validate, and implement a system. SDM also incorporates the concepts and principles of quality control, quality assurance, and current good manufacturing practices (CGMP).

At a minimum, system validation protocols should include:

- a description of the function(s) to be tested,
- steps to be performed,
- equipment and methods to be used,
- inputs (from all sources), and
- expected outputs and evaluation criteria.

---

[5]See, e.g., 21 CFR 600.10(a) and (b), 606.20(a) and (b), 606.100, 606.160, 211.63, 211.67, 211.68, 211.100, 211.180.

Documentation of the performance of the validation protocols and an analysis of the results should be maintained by the blood establishment.[6] Blood establishments, as manufacturers of biological products, are responsible for ensuring the proper operation and validation for all procedures performed and supported by the computerized system relating to the manufacture and distribution of blood and blood components. Blood establishments also should ensure that SOPs include how the computerized system is used during manufacturing steps and that these procedures conform to current regulations (21 CFR, Parts 600 through 660 and Part 211).

Vendors or developers that commercially distribute software intended for use by blood establishments should be aware of the federal regulations relating to the manufacture of blood and blood components. In addition, because blood bank software products are devices, firms manufacturing and distributing such products must comply with statutory and regulatory requirements applicable to devices.

A. Software Development

Blood bank software products are medical devices. The user should ensure that vendors supply specifications, documentation, and validation for each feature and function of the software product, including:

- data inputs and outputs,
- data manipulation within each module,
- module to module interaction,
- interaction with peripheral devices,
- data storage and maintenance (including back-up procedures), and
- data search and retrieval algorithms.

The user should ensure that software vendor(s) provide information concerning product features and limitations for use in the development of computerized system validation protocols and ensure that:

- the software was developed and validated by qualified persons using SDM,
- all versions of the software are kept in readable form (including source code),
- information will be provided to an FDA Investigator upon request,
- software modifications have and will follow a systematic change control cycle,
- notification will be provided concerning known design defects or operational errors, and
- provisions have been made for providing information to the user in instances involving interruption or cessation of the vendor's business.

---

[6]See, e.g., 21 CFR 606.160, 211.68(b), 211.194.

B. User Responsibilities

Blood establishments as users should define the requirements for the system and establishing a written validation program that includes:

- system validation protocols,
- types of testing,
- detailed test cases,
- test database,
- documented results,
- documented evaluation and acceptance,
- schedule for periodic system evaluation, and
- revalidation as necessary.

When a new computerized system is being developed, close contact should be maintained between the blood establishment and the vendor(s) to gather information that may affect the system.

C. Blood Establishment As Software Developer

During inspections of blood establishments, three basic categories of software have been observed: 1) vendor supplied, 2) vendor supplied/"in-house" modified, and 3) "in-house" developed. Irrespective of the source of the software, the developer should follow accepted standards for software development (SDM) including, but not limited to, proper design, software validation and verification procedures, change control, and detailed documentation.

Blood establishments are responsible for the quality of the software and validation of its design and features if they modify vendor supplied software or develop software in-house. Due to the impact of any change on the proper operation and functioning of the software, modification and development should only be performed by qualified individuals. The software developer should be aware of the contents of the Technical Report on Software Development Activities (1987), Reviewer Guidance for Computer Controlled Medical Devices (1991) and other agency documents, which provide guidance concerning software product development and validation.

D. Prospective System Validation and Revalidation

The user, after appropriate consultation with the vendor(s), should design a plan for the prospective validation of the computerized system. Written procedures for production and process controls must be established and followed to ensure that products produced will meet specifications.[7] Validation includes defining all components of the system, testing each component of the system, and testing integrated components of the system to ensure that pre-determined specifications are met (see Guideline on General Principles of Process Validation). Prospective validation is performed by the user prior to

[7] See, 21 CFR 211.100, 211.110(a), 606.100.

implementation of the computerized system in manufacturing operations. Prospective validation is performed for all new systems and before changes, modifications, or enhancements to existing systems which may affect finished product characteristics.

The routine revalidation of existing computerized systems parallels the validation of new systems. Routine revalidation is based on the protocols that are used for the prospective validation and is performed regularly after the installation of the system.

E. Retrospective Validation and Revalidation

Retrospective validation is described in the FDA's "Guideline on General Principles of Process Validation" (see Section XIV. B. Food and Drug Administration Publications) It is extremely difficult to "retrospectively" validate a computerized system and is generally more costly and time consuming than prospective validation. If it must be attempted, significant information can be obtained through observation, documentation, and performance history. Generally, retrospective validation should be used only as a corrective measure in response to deficiencies noted concerning prior validation efforts.

For retrospective validation, the User's Manual, Operators Manual, Program Maintenance Manual, and the Training Manual provided by the vendor(s) will help to provide the general structure of the system. The blood establishment's SOPs should identify the tasks (including all steps) that are performed with the aid of the computerized system. Blood establishment personnel (or consultants) responsible for designing test cases can avoid making assumptions about the system functions if this documentation is complete. Missing or incomplete information may be reconstructed by researching:

- assembly specifications,
- maintenance logs, change logs, and records,
- project files,
- source code,
- software structure and data flow diagrams,
- computerized system test results (if any),
- previous versions of software, and
- failure reports.

F. Documentation

Computerized systems used in the processing and/or distribution of blood and blood components should have validation documentation that shows that the intended functions will consistently be performed accurately and reliably.[8] Documentation of validation includes:

- records demonstrating that the system met its predetermined specifications and requirements throughout development,
- records describing in detail the system's intended functions,

---

[8]See, e.g., 21 CFR 211.68, 211.110, 211.100(a) and (b), 606.160(b)(5)(i) and (ii), and Guideline on General Principles of Process Validation.

- records of testing (including all results) to demonstrate that the intended functions performed as designed during acceptance testing, and
- records documenting that hardware and software change controls are being followed as described.

VI. COMPUTERIZED SYSTEM TESTING

Evaluation of the system should be performed continuously to ensure that the hardware (including all peripherals) and system software (applications and the operating system) are operating as specified and that no unauthorized changes have been made. The user should ensure that the computerized system and the peripheral devices are being used in a manner that is consistent with their design. In addition, the user should ensure that the computerized system and the peripheral devices are calibrated in accordance with the equipment manufacturer's directions for use[9]. To evaluate the system, tests should be performed at the time of installation, after a change has been made, and on a routine basis according to schedules defined in the establishment's SOPs. Tests are an indispensable tool of validation.

Test case design should include normal inputs, boundary values, invalid inputs, stress conditions, and special cases. Properly analyzed test data are useful to identify additional testing needs, and the need to establish additional manufacturing control procedures (e.g., "work-arounds"). The test data will also be useful in establishing and documenting the repeatable and reliable performance of the computerized system.

CAUTION: Because the integrity of the data cannot be assured, system change/modification/enhancement testing should ONLY be performed using a TEST database.

Avoid the use of PASS/FAIL notation in the evaluation of the results of any type of testing. Actual test data, including inputs and test results, should be documented and available for review. PASS/FAIL notation is documentation of an interpretation which cannot be audited.

A. Testing Types

Software tests are intended to challenge the application software and other parts of the overall system functionally and structurally. Functional testing demonstrates only that the system outputs appear to be correct. It does not allow an assessment of whether the software is actually performing according to specifications and requirements. A complete functional test of every combination of inputs may not be feasible except for very small programs. Functional testing is essentially a subset of structural testing.

Structural testing should be designed to exercise all modules and branches of the software and their interrelationships with the hardware and peripheral devices. Structural testing should be performed to ensure that all relevant functions in the software perform as intended.

---

[9]See, 21 CFR 606.60 (a) and (b).

Each of the testing types described below should be conducted. Performing only one type of test will not prove that the system is working properly.

1. **Normal Testing** includes cases that test the functional and structural integrity of the computerized system. The input data for these test cases all fall within the range the user considers to be normal. Performing enough test cases can give a reasonable level of confidence that the system behaves as intended under normal conditions.

2. **Boundary Testing** is performed using values that force the system to discern whether the input is valid or invalid, or to make a decision as to which branch of the program to execute. Boundary test values are set at the edges (i.e., slightly below and above) of valid input ranges. Boundary testing does not mean making the computerized system "crash" or involuntarily stop.

3. **Invalid Case Testing** checks for input errors. The user should not assume that the system will detect invalid inputs. Invalid case testing demonstrates that, when the inputs are invalid values, the computerized system performs properly. Invalid inputs could be from the keyboard or from a peripheral device. Transposition or spelling errors are among the most common errors. Electrical interference, line noise, conversion of analog into digital data, or a software fault could cause the input data to be in error.

4. **Stress Testing** challenges the computerized system at its physical limits and documents its continued ability to perform correctly. Stress testing includes situations where the computerized system, peripheral devices, and operators are pushed to the operational limits of the system or worst case scenario.

5. **Special Case Testing**, also known as "exceptional case testing", documents the system's reactions to specific types of data or lack of data and is intended to ensure that the computerized system does not accept unsuitable data. These tests should be designed to document what happens when values that are not included in the ranges defined in the specifications are entered. Use of test cases with no data entry in a field will assist in establishing software system defaults. Special test cases should also challenge the system to determine if more than one person can add or edit important data at the same time, and if so, document how the system responds.

6. **Parallel Testing** is one of the most common types of tests performed by blood establishments. Parallel testing is performed by running two systems in parallel and comparing the outputs (e.g., two software application versions or software compared with a manual procedure). Modified software is often installed on the same physical drive as the production software, but in a separate region or partition. The parallel or test region system should emulate the production system except for authorized changes.

As commonly performed, parallel testing is not a completely effective form of functional testing because it focuses only on the cases presented during normal test runs; therefore, it checks only a narrow range of values. The design of parallel test cases does not always include invalid or boundary values. And, many times, the output is not properly analyzed. The comparison of the actual outputs to the predicted outputs is one analysis component frequently absent in the design of parallel testing plans.

> **NOTE: Due to the above, parallel testing cannot be relied upon as the sole criterion for validating a system. However, parallel testing can be a valuable tool when it is used in conjunction with other testing types for validation, or to train personnel to use the new computerized system.**

7. Vendor Simulations may be performed by the developer, vendor or a third party to test a prototype or a simulation [e.g., ALPHA (developer's site) or BETA (selected user site) testing]. Vendor simulations are not a substitute for validation by the user. Vendor simulations cannot completely emulate the user's environment or the people that will be using the system. Installing the computerized system in a different environment can cause changes in the system's operation. Because of the possibility of changes in the system's operation, testing at the time of installation should be performed by the user at each site, prior to approval and implementation.[10]

B. System Validation

System validation is based on test cases designed to verify and challenge the user's requirements, the system specifications, and the User's, Operations, and Maintenance Manuals. Analysis of the test data and an understanding of the system's structure should point out the need for other test cases required to complete testing. The user should determine and justify how many times each function should be tested. The criticality of the function will dictate the number and types of tests performed and the frequency of retesting and evaluation. Testing is complete when it has been demonstrated that the system repeatedly performs its intended functions and does not perform unintended functions.

The completion of validation, including an evaluation and summary of the test cases, the test data, the test database, and the results of testing should be approved through a formal process prior to implementation of the system or changes.[11]

C. Validation Documentation

Documentation of the validation process should reflect the tests and their results.[12] The SOPs should clearly define the protocol for each of the different types of systems tests. The documentation produced by good test plans will show that the system is running properly. At a minimum, validation protocols should document that:

---

[10]See, e.g., 21 CFR 211.63, 211.68.

[11]See, e.g., 21 CFR 211.22(c), 211.68, 211.100(a).

[12]See, e.g., 21 CFR 211.68(a), 606.160(b)(5)(i) and (ii).

- there are no unacceptable records,
- donor identification codes match the correct data,
- donor deferral codes are properly assigned and used,
- all units/components are properly identified,
- test results (including repeat tests) are properly entered and recorded,
- records of change (including audit trails) are correct,
- recorded and printed outputs are correct,
- manipulations and calculations are accurate and use the prescribed data,
- expiration dates are properly calculated, and
- video displays are correct.

VII. CHANGE CONTROL AND AUDIT TRAIL RECORDS

All changes that are made to the computerized system should be documented and formally authorized. Records of changes should include modifications and/or enhancements made to the hardware, software, and any other critical component of the system. The SOPs should clearly define the need for documentation of changes, an evaluation of the impact of a change, and an assessment concerning the necessity for revalidating portions or all of the system. It is recommended that changes to the computer hardware, software (including operating systems and interfaces), and peripheral devices be kept in separate logs.

Detailed records should be maintained for all hardware and software changes that are made to the system (e.g., corrections, modifications, upgrades, enhancements, etc.). At a minimum, these records should include:

- a description of the change,
- a description of the tasks performed to effect the change,
- date,
- person performing the task,
- equipment identification or modules that are affected,
- authorization signature,
- validation protocols,
- validation results, and
- documentation of approval and acceptance.

An audit trail documents changes made to the data. Audit trail records are part of the system's documentation and should only be accessible or reviewed by authorized persons (e.g., identified establishment personnel, FDA Investigators). Audit trails can be used to record access to the system. Each time an authorized or unauthorized user tries or gains access to the system there should be a log entry. At a minimum, an audit trail should include:

- the name of the person making the change,
- date,
- time,
- field name,

- previous data, and
- current data.

<u>Changes to data may not obscure the original data.</u>

VIII. MANUALS

A. Standard Operating Procedures (SOPs)

Standard Operating Procedures (SOPs) are procedural manuals that delineate the steps required to perform tasks. SOPs are the written "back-bone" of the blood establishment's operation. The SOPs should cross reference any vendor supplied documentation concerning instructions for use.

The SOPs should include the procedures for handling all operations.[13] These operations range from normal events to the total recovery from any type of disaster (including but not limited to system failure or crash, power outage, natural disaster, etc.). Particular attention should be paid to the procedures for archiving (backing up) the data. These procedures should include all administrative, routine, and emergency situations, such as:

- detailed descriptions of all validation protocols and records of their results,
- roles of the systems and maintenance personnel,
- detailed descriptions of all catastrophic validation routines and records of their results, and
- a complete disaster plan that describes the roles of all the facility's personnel.

Recovery from disasters should be a key concern. The methods and the location to be used for off-site storage of the archive should be carefully considered. Recovery procedures should define the responsibilities of all personnel during and while recovering from disasters. Disaster SOPs should include detailed descriptions, or cross-reference the locations, of:

- all validation procedures and records,
- the operator's role during and after the incident,
- interim operating instructions,
- access to off-site storage facilities,
- definitions of the user's, developer's, and vendor's roles when totally rebuilding the manual and automated systems within the facility, and
- training procedures.

The SOPs should contain detailed references to any "work arounds", or extra steps, that are necessary to perform the tasks that the computerized system does not or was not intended or designed to perform. Users should ensure that "work arounds" are included in their training program.

---

[13]<u>See, e.g.</u>, 21 CFR 606.100, 211.100.

B. User's Manual

The vendor supplied User's Manual should contain the instructions for use of the software. It is essential that the User's Manual be written in direct, easily understandable language wherever possible. Technical jargon should be used sparingly and defined when it is used. The User's Manual should clearly define normal system operations, back-up, and restoration methods. The vendor supplied User's Manual might not contain information on disaster recovery. Therefore, the blood establishment's SOPs for disaster recovery should be clearly defined.

1. Normal System Operations are tasks regularly performed by the user. Normal start-up procedures include logging on, menus and their uses, and automatic controls for user access to the functions and data [normally by password and user identification (USERID)]. Normal shut-down procedures include logging off and "house-keeping" procedures [diagnostics and archiving of data (including data back-up)] to be performed at pre-established intervals. Normal automatic functions that are performed by the computer include data edit checks, diagnostic routines, and automatic back-ups.

Normal input and output functions that are performed by the system's operators and by peripheral equipment include normal data input (e.g., keyboard entries for editing and deleting data), manipulating data that is input by peripheral equipment (e.g., test results), normal data/information output (e.g., screen displays, printed reports), and other tests to verify that the system is performing normally.

2. Back-up and Restoration Procedures should be established and followed regularly to help preserve and ensure the integrity of data. Data storage includes equipment descriptions and instructions for their use. "House-keeping" (e.g., daily, weekly, and monthly routines) include handling the physical media (e.g., mounting, labeling, and on and off site storage). SOPs should include actions to be taken when errors occur during this process (e.g., file not found, disk full, tape full, etc. [see Section IX. C. Data Storage)].

C. Training Manual

Vendor supplied Training Manuals, handouts, personnel proficiency tests, and training procedures are an integral part of the system's documentation. Vendor supplied materials may help provide guidance for the training of the facility's personnel, but are not a substitute for a Training Plan in the SOPs. The user's training material should be based on the concepts, principles, and instructions found in the SOPs and the vendor supplied User's Training and Maintenance Manuals. Different training plans may be designed for use by managers, supervisors, operations personnel, and other users.

D. Maintenance Manual

Vendor provided Maintenance Manuals address the technical details of the system's structure and operation. The Maintenance Manual should provide a clear understanding of the system, including detailed system descriptions that are cross referenced with the

User's and Training Manuals. Detailed descriptions of the system's hardware and software as well as their relationships within the facility should be in a user generated Maintenance Manual (including detailed descriptions of any known system and facility anomalies).

E. Document Maintenance

All SOPs and Manuals should be maintained to reflect current processes and procedures[14] that are accomplished through the use of the computerized system. The user should ensure that all SOPs and Manuals are cross referenced. Outdated SOPs and Manuals should be removed as appropriate and should be archived.[15] Proper document maintenance includes the ability to show when procedures were approved by management and implemented (i.e., effective date).

IX. MAINTENANCE

All systems require installation, validation, calibration, Quality Control (QC), and preventive maintenance, and should be included in the Quality Assurance (QA) program.[16] The user should ensure that the documentation is complete. The user has the responsibility for validation, proper use, and operation of the system.[17]

A. System Maintenance

Computerized system maintenance includes all parts of the system (e.g., hardware, software, and peripheral devices). System maintenance includes preventive and emergency maintenance as well as system evaluation audits (see Section XII. Audits). System maintenance is performed by the user in accordance with the SOPs and vendor recommendations.

1. Preventive Maintenance tests will help to confirm that the system performs correctly and will have a minimum of downtime. All vendors should provide protocols for maintenance and tests with documentation that recommend the intervals and any calibration needs. The user should integrate this documentation and these recommendations into the SOPs to ensure that the procedures are adequate for quality control and quality assurance.

2. Utilities or diagnostic software applications help to confirm that the system is working properly and assist in ensuring data integrity. These utilities should be included

---

[14]See, e.g., 21 CFR 606.100, 211.100(a).

[15]See, 21 CFR 606.100.

[16]See, e.g., 21 CFR 606.60, 606.140(b), 211.100, 211.110(a), 211.68(a).

[17]See, e.g., 21 CFR 600.10(a).

in testing and validation plans, and be clearly cross referenced in the SOPs. Blood establishments should document the use of and the results obtained from these utilities.

Diagnostics are available from hardware and software vendors. Lengthy diagnostics are normally run during non-peak hours ("house-keeping") to avoid interference with normal operations.

B. Data Maintenance

The user should develop procedures that ensure that the database is kept free of unwanted duplicate and discrepant data, and to help users recognize data entry errors. Users should be advised to develop "measures" that are realistic targets to be met to satisfy that adequate quality levels have been obtained.

1. Duplicate Data may be considered to be discrepant data. System design should minimize the occurrence of duplicate data. When duplicate data occur, these data should be removed from the database as soon as possible. The user's procedures for correcting undesired or duplicate data should be clearly defined in the SOPs.

2. Discrepant Data may have serious repercussions. The database should be routinely checked as part of the system's maintenance to ensure that data are correct and properly stored. Donor identification codes should match the other donor information (e.g., the address should match the donor, the donor's deferral status should be properly identified, the test results should correspond to the correct unit ID number, and the results of all tests and repeat tests should be properly entered).

C. Data Storage

Data storage and retrieval are critical parts of the system. A computer should be able to store and retrieve data accurately, on command, from both on-line and archival or back-up storage for as long as the record is required to be retained.[18] On-line and archival storage (e.g., magnetic media, microfiche, optical media, paper) should be validated to ensure that the system has correctly stored (created a true and permanent copy of) the data. Validation should also ensure that data is reliably retrieved for the duration of the retention requirement and that the expected life of the storage media is sufficient to meet the retention requirement.[19]

1. On-line Storage or routine storage should be handled by each system in accordance with its design. Software should contain a feature that recognizes system errors and warns the user that a problem has occurred. Some systems may not "mention" the fact that data will be lost or overwritten. The user should ensure that there is adequate protection against inadvertent data loss (e.g., when a storage device is "full" or "missing").

---

[18]See, 21 CFR 606.160(d), 211.180(a) and (b), 211.198.

[19]See, e.g., 21 CFR 211.68(b).

2. Archival Storage (long term back-up methods) should protect the user against catastrophic data loss. Archival storage uses long term storage media for data removed from the system. This storage media may be in the form of tapes, discs, or diskettes, and it should be stored off-site under appropriate storage conditions. Archival storage protocols should be clearly defined in the SOPs. The user should be cautioned to investigate lifetimes of backup media. For example it is recommended that tapes be "exercised" annually. Also, since tapes can be demagnetized with age, consideration should be given to procedures for copying data for longer storage.

X. SECURITY

All systems should have strict security controls. The security of the system and the integrity of the data should be strictly maintained. When the system or the data are changed, the changes should be clearly traceable through the use of audit trails. Systems that have connections to telephone communications through modems should have strict access controls and restrictions. Access restrictions on such systems should be designed to prevent unauthorized access or change.

Procedures that validate the hardware and software systems should rigidly test security. The use of USERIDs and passwords should be routinely tested. Validation protocols should ensure and document that sensitive donor information is not compromised, that critical data are valid, and that SOPs are correct.

A. Password and Electronic Identification

All persons authorized to have direct access to the software or data should be uniquely identified electronically. The security and access procedures selected should be strictly enforced to maintain the level of security established to ensure the integrity of the data. Records that identify system users should be maintained in accordance with requirements and recommendations. For example, the security of passwords should be ensured such that only selected persons are able to change the codes for deferred donors or quarantined units. Also, the use of both a USERID and password for electronic identification is rapidly becoming an industry standard for electronically signing a document. In state of the art computerized systems, both USERID and password are entered for each signing from LOGON to document endorsement.

Password security includes:

- periodically changing passwords, not to be reassigned or re-used,
- deleting the access authorization for persons that no longer have a need to work with the data,
- ensuring both unique identity of individuals and their access level,
- ensuring that users use passwords that are not recognizable as reflections of their personal life (i.e., birthday, license plate number, spouse's name),
- ensuring that passwords are not shared, and
- clearly defining SOPs for personnel guidance.

B. Confidentiality

Each establishment should develop and use clearly defined SOPs to provide for the maintenance of the confidentiality of sensitive donor information. These procedures should be designed to prevent inadvertent disclosure and to guard against unauthorized external and internal inquiries.

> **NOTE: In multipurpose systems, it is strongly recommended that sensitive files be protected by password and any other means deemed necessary.**

XI. TRAINING, SUPERVISION, AND PROFICIENCY TESTING

Computer systems should be managed like all other systems in blood establishments. Training manuals provided by the hardware or software vendors should be evaluated for accuracy and incorporated into training programs as appropriate. In addition, other written training procedures relevant to the system as a whole should be prepared and used appropriately. All standard supervisory controls should be applied, including, but not limited to, routine review of all critical procedures. There should be active monitoring of employee performance and formal proficiency testing programs.

XII. AUDITS

Regular audits should be performed as a routine part of the Quality Assurance (QA) program. QA consists of the actions, planned and taken, that provide confidence that all systems and elements that influence the quality of the product are working as expected. Quality Control (QC) is a component of QA that is performed to provide assurance that at the moment the test is run the procedure is working as expected.

The blood establishment should use vendor recommendations concerning procedures to develop a QA program.[20] At a minimum, these procedures should include:

- documentation,
- system testing,
- implementation, and
- maintenance.

The user should integrate and implement the principles of quality control and quality assurance in testing and validation.[21] In the SOPs, the user should define:

- the frequency of system testing,
- the manufacturing specifications,

---

[20] See, Center for Biologics Evaluation and Research (CBER), Draft Guideline for Quality Assurance in Blood Establishments, Docket No. 91N-0450, June 17, 1993.

[21] See, 21 CFR 211.22.

- the criteria for acceptability,
- the remedial action to be taken when the result of any test provides an unexpected result, and
- steps to be followed in investigating and correcting discrepancies that could lead to errors in manufacturing of blood and blood components.

A. Record Reviews

Routine reviews of the master database transaction logs, audit trail records and other logs should be performed by properly trained and authorized persons. The timing and scope of in-house record reviews should be clearly defined in the SOPs. Summaries of the review activities and corrective measures should be available to FDA Investigators upon request.

B. System QA Audits

System QA audits should be performed by specified personnel in accordance with plans that are clearly defined in the SOPs. Reviews of the computer system should be a part of each routine Quality Assurance audit. The audits of the computer system should evaluate those aspects of the system's use and performance necessary to ensure proper results. System QA audits should include, but not necessarily be limited to, review to determine that:

- all hardware and software has been subjected to all appropriate calibration, validation, and routine maintenance,
- appropriate documentation regarding the system's components is present,
- vendor upgrades, repairs, replacement of components have been evaluated for their impact on operations,
- appropriate training, supervision, and proficiency testing have been performed,
- system components are being used in accordance with applicable manufacturer's instructions, SOPs, and FDA requirements,
- procedures for change control, data maintenance, and integrity are followed, and
- problems that have been encountered with system malfunction have been appropriately evaluated and addressed.

XIII. REPORTING TO CBER

Licensed establishments must submit information describing the proposed computer system as a supplement to the Establishment License Application (ELA).[22] Requests for supplement approval should contain the following information:

1. Identification of the hardware, including the central and peripheral devices and network linkages.

---

[22]See, 21 CFR 601.12(a).

2. Identification of the source(s) of the application software used, this should include the name of the software vendor, the name and version number of the software package.
3. The location(s) where computers will be used. Include donor collection centers, self-contained mobiles, and the distribution centers.
4. A brief description of how the computer system will be used and in what areas of the facility.

An important proposed change to a computerized system is also reportable as a supplement to an ELA.[23] This includes any hardware or software modification or enhancement that may affect the way that the data are manipulated or interpreted prompting a revalidation of the computerized system (e.g., calculations for a decision making process such as donor suitability or unit release).

Supplements to the ELA should be sent to:

Food and Drug Administration
Center for Biologics Evaluation and Research
Document Control Center (HFM-99), Suite 200N
1401 Rockville Pike
Rockville, MD 20852-1448
Attn: Division of Blood Establishment and Product Applications, (HFM-370)

XIV. REFERENCES

A. Title 21, Code of Federal Regulations

A listing of the sections of Title 21, **Code of Federal Regulations** (21 CFR) that are related to computerized systems include:

1. **21 CFR, Part 200** -- Drugs: General
2. **21 CFR, Part 210** -- Current good manufacturing practice in manufacturing, processing, packing, or holding of drugs; general
3. **21 CFR, Part 211** -- Current good manufacturing practice for finished pharmaceuticals
4. **21 CFR, Part 600** -- Biological products: General
5. **21 CFR, Part 606** -- Current good manufacturing practice for blood and blood components
6. **21 CFR, Part 610** -- General biological products standards
7. **21 CFR, Part 640** -- Additional standards for human blood and blood products
8. **21 CFR, Part 800** -- Medical Devices: General
9. **21 CFR, Part 820** -- Good manufacturing practice for medical devices: general

---

[23]See, 21 CFR 601.12(a).

B. Food and Drug Administration Publications

The FDA documents that deal with computers and compliance issues, including reference materials and training aids for investigators, are:

Blood Bank Inspection Checklist and Report and accompanying instructions.

Form FDA 2609 (3/91), Section K, Computerization, contains questions that are asked by the Investigators during an inspection.

Guide to Inspection of Computerized Systems in Drug Processing

February, 1983. FDA, Division of Drug Quality Compliance and Division of Field Investigations. U.S. Government Printing Office 1983-381-166:2001

Technical Report on Software Development Activities

July, 1987. FDA, Office of Regulatory Affairs

Guideline on General Principles of Process Validation

May, 1987. FDA, Center for Drugs and Biologics and the Center for Devices and Radiological Health

Reviewer Guidance for Computer Controlled Medical Devices Undergoing 510(k) Review

August, 1991. FDA, Center for Devices and Radiological Health

Draft Guideline for Quality Assurance in Blood Establishments

Docket No. 91N-0450, June 17, 1993. FDA, Center for Biologics Evaluation and Research

C. Memoranda

Food and Drug Administration memoranda issued by the Center for Biologics Evaluation and Research (CBER) to All Registered Blood Establishments[24]:

---

[24]Copies of the Food and Drug Administration Memoranda issued by CBER to All Registered Blood Establishments are available through the Division of Congressional and Public Affairs, 1401 Rockville Pike Suite 200N, Rockville, MD 20852-1448. FAX (301) 594-1938. There may be a charge in accordance with Department of Health and Human Services regulations.

Subject: Responsibilities of Blood Establishments Related to Errors & Accidents in the Manufacture of Blood & Blood Components
Date: March 20, 1991

Subject: Deficiencies Relating to the Manufacture of Blood and Blood Components
Date: March 20, 1991

Subject: Requirements for Computerization of Blood Establishments
Date: September 8, 1989

Subject: Recommendations for Implementation of Computerization in Blood Establishments
Date: April 6, 1988

XV. GLOSSARY

ARCHIVE an indelible collection of computer system data or other records that are in long term storage. Backing-up the database is the first step in creating a computer generated archives. The steps in creating and safely maintaining an archive should be clearly defined in the SOPs.

CHANGE CONTROL the processes, authorities for, and procedures to be used for all changes that are made to the computerized system and/or the system's data. Change control is a vital subset of the Quality Assurance (QA) program within an establishment and should be clearly described in the establishment's SOPs.

COMPUTER SYSTEM an electronic device controlled by lists of instructions (programs). The computer system is used by its operators (personnel) to perform specific tasks. The "system" includes the resources used to complete a task (i.e., computer system [hardware, software], personnel, peripheral equipment [e.g., test equipment, printers, disk drives]). Validation confirms that the system is functioning according to specifications.

CONFIDENTIALITY OF DONOR INFORMATION broadly covers any means by which the unauthorized disclosure of sensitive data or information is prevented. Sensitive data are the personal details about the donor (i.e., test results, deferral status).

DATABASE (1) A set of data or part of a set of data consisting of at least one file that is sufficient for a given purpose or for a given data processing system. (2) A collection of data basic to a system. (3) A collection of data basic to an enterprise.

DATABASE SECURITY means by which the confidentiality and integrity of a database is maintained.

DEVELOPER a person, or group, that designs and/or builds and/or documents the hardware and/or software of computerized systems.

DEVELOPMENT METHODOLOGY a systematic approach to software creation that defines development phases and specifies the activities, products, verification procedures, and completion criteria for each phase.

DISASTER PLAN the part of an establishment's SOPs that contains detailed instructions on what to do in the event of a catastrophe. The disaster plan is an essential part of the establishment's SOPs.

DOCUMENTATION (1) A collection of documents on a given subject. (2) The management of documents, including the actions of identifying, acquiring, processing, storing and disseminating. (3) Any written or pictorial information describing, defining, specifying, reporting or certifying activities, requirements, procedures, or results. NOTE: As an indelible record, documentation is an indispensable tool of validation.

FIELD discrete location in a database that contains an unique piece of information. A field is a component of a record. A record is a component of a database.

FLOW CHART the graphical representation of the logical steps involved in a procedure or program. A flow chart is a part of the documentation of the structure of a computer program or system. Flow charts may be included in the Program Maintenance Manual.

HARDWARE the physical equipment that is part of the computer system (e.g., chassis, keyboard, monitor).

INSTALLATION includes installing a new computer system, new software or hardware, or otherwise modifying the current system.

LEVEL OF SECURITY determines what functions a user is authorized to perform. Authorization is determined by the level of security assigned to the USERID and password. The basic levels of security are READ and WRITE. There are many variations and combinations of security that can be achieved by limiting the user's ability to VIEW certain data. For example, a user could be permitted to READ a donor's address and deferral status but only WRITE or change the donor's address. Levels of security and access to the functions of the system are strictly controlled by the System Administrator.

MODEM an electronic device used to communicate between two or more computers using a communications protocol.

MODEM ACCESS using a MODEM to communicate between computers. MODEM access is often used between a remotely located computer and a computer that has a master database and applications software.

MULTIPURPOSE SYSTEMS computer systems that perform more than one primary function or task are considered to be multipurpose. In some situations the computer may be linked or networked with other computers that are used for administrative functions (e.g., accounting, word processing).

OPERATING SYSTEM software that controls the execution of programs, including such services as resource allocation and scheduling, input/output control, and data management. Although operating systems are predominately software, partial or complete hardware implementations are possible.

ORIGINAL EQUIPMENT MANUFACTURER (OEM) person or company that originally built a component part or system.

PASSWORD an alphanumeric word (string of characters) that is unique to each USERID. A password is used when logging on to a computer system. A password is used with a USERID to determine the levels of access that will be granted to a user.

PERIPHERAL DEVICE equipment that is connected to the Central Processing Unit (CPU). A peripheral device can be used to input data (e.g., laboratory test equipment) or to output data (e.g., printer, disk drive, video system, tape drive).

PRODUCTION DATABASE the computer file that contains the establishment's current production data.

RECORD OF CHANGE documentation of changes made to the system. A record of change can be a written document or a database. Normally there are two associated with a computer system (hardware and software). Changes made to the data are recorded in an audit trail.

REGION a clearly described area within the computer's storage that is logically and/or physically distinct from other regions. Regions are used to separate testing from production (normal use).

SEARCH and RETRIEVAL ALGORITHMS procedures that a computer uses to locate data within a database.

SITE SURVEY conducted to determine the characteristics of the site in which a system is to be installed. These surveys include the location, stability and availability of utilities (including air conditioning), and the physical layout of the site (including access to rooms and hallways).

SOFTWARE computer programs, procedures, rules, and associated documentation and data pertaining to the operation of a computer system.

SOURCE CODE human readable version of the list of instructions (program) that cause a computer to perform a task. A computer program that must be compiled, assembled, or interpreted before being executed by a computer.

SYSTEM ADMINISTRATOR the person that is charged by the Responsible Head with the overall design, administration, and operation of the organization's computerized systems. The System Administrator is normally an employee or a member of the establishment.

SYSTEM DEVELOPER may be the user, a contractor, a vender, or a software device manufacturer. See also developer.

SYSTEM DEVELOPMENT METHODOLOGY (SDM) or Life Cycle Methodology is a structured method used to plan, design, validate, and implement a system.

SYSTEM DOCUMENTATION the documentation of the requirements, design philosophy, design details, capabilities, limitations, and other characteristics of a system.

TEST DATABASE contains a mix of production and test data. Test data will have deliberately introduced errors. A test database is kept and used in a region (partition) that is segregated from the production software and database to prevent serious problems.

USER An individual or organization that normally supplies information for processing, or that receives, interprets, and uses the output or effects of such processing.

USERID an alphanumeric word (string of characters) that is unique to each user. The USERID is used when logging on to a computer system. The USERID is the primary key that determines the level of access that will be granted to the user.

VALIDATION the establishment of documented evidence (for example, data derived from rigorous testing) which provides a high degree of assurance that a specific process or system will consistently produce a product meeting its predetermined specifications and quality attributes.

VENDOR person or organization that provides, as a medical device manufacturer, software and/or hardware and/or firmware and/or documentation to the user for a fee or in exchange for services.

GUIDE TO INSPECTIONS OF BLOOD BANKS[25]

Division of Field Investigations
Office of Regional Operations
Office of Regulatory Affairs
U.S. Food & Drug Administration

September 1994

INTRODUCTION

The "Guide to Inspections of Blood Banks" is a consolidation of information previously provided in the Blood Bank Inspection Checklist and Report, and the Instruction Booklet for Blood Bank Inspection Checklist and Report, FDA-2609. This guide, which provides the most updated interpretation of certain regulations and guidelines, was prepared by the FDA, Office of Regulatory Affairs and the Center for Biologics Evaluation and Research.

The Blood Bank Inspection Checklist and Report, and the Instruction Booklet for Blood Bank Inspection Checklist and Report, FDA-2609, are no longer in effect. Consequently, field investigators are no longer required to fill out the checklist during establishment inspections nor submit it with inspectional reports.

The checklist and instruction booklet were last published in May 1991. Since May 1991, CBER issued a number of memoranda to industry, depicting new recommendations, and modifications to previous guidance, in addition to guidelines on quality assurance and validation of computer systems. Although the agency, at the time of this publication, is in the process of revising the regulations on blood and blood products, it will take some time before such revisions are in effect. The current guide provides interim inspectional direction.

This reference provides the most updated interpretation of certain regulations and guidelines. This reference is not intended to be a "How to ..." it is a technical reference and is intended to be used in conjunction with the Inspection Operations Manual (IOM), the Code of Federal Regulations, Title 21 (21 CFR), the Compliance Program for the Inspection of Licensed and Unlicensed Blood Banks (CP 7342.001) and the Compliance Policy Guides for biologics (CPG 7134).

In several instances the manual refers to memoranda published by CBER and sent to registered blood establishments. These memoranda should be available at the FDA District Offices, if not, a copy can be obtained from CBER, Division of Inspections and surveillance, Office of Compliance, (301) 594-1194.

---

[25]Note: This document is reference materials for investigators and other FDA personnel. The document does not bind FDA, and does not confer any rights, privileges, benefits, or immunities for or on any person(s).

The preparation of products for which there are no published Additional Standards must be described in the establishment's SOP manual and manufactured in accordance with the methods therein. The investigator should offer no advice or recommendation to the manufacturer regarding the preparation of such products. Questions concerning practices which may be hazardous should be addressed to the Division of Inspections and Surveillance (HFM-650) at (301) 594-1194.

GENERAL INFORMATION

Investigators should note the U.S. License number, if applicable (and location number) of facilities. This will serve to identify establishments in correspondence, applications and other forms of communications. Registration numbers, essentially central file numbers, are not the same as license numbers.

The establishment should have a validated pink copy of Form FDA-2830, Blood Establishment Registration and Product Listing, for the current calendar year or evidence of having submitted same. If the data on the registration form is not correct, list corrections to be made in comments, and instruct the establishment to submit in writing the updated information to the Division of Blood Applications, (HFM-370), 1401 Rockville Pike, Rockville, MD 20852-1448. If changes in the name and address, Medical Director, manufacturing procedures, or products have occurred, they should be reported to DBA in order to update the Form FDA-2830.

In unlicensed hospital blood banks, as a matter of courtesy, the hospital administrator should be notified that the blood bank is to be inspected.

OPERATIONS

Determine whether unlicensed products are shipped interstate for sale, barter, or exchange and, if so, thoroughly document such shipments. The requirements of 21 CFR 640, Subpart G apply to plasma exchange if the resulting plasma is sold.

Determine approximately how many units of whole blood are collected annually, how many are autologous, and how many are directed. Also, determine how many units of whole blood and red blood cells (RBCs) are received from outside sources each year, how many are autologous and how many are directed.

Determine which products are prepared and activities conducted. Licensed establishments are required to report changes in manufacturing to CBER. Some establishments submit SOPs in reporting these changes. Not all procedures are reviewed by CBER; therefore, if an investigator observes a procedure s/he considers unsafe for the donor or that will affect the safety, purity, or potency of the product contact the Division of Inspections and Surveillance at (301) 594-1191. Refer also to the July 21, 1992 memorandum, "Changes in Equipment for Processing Blood Donor Samples."

If the facility being inspected is licensed and pheresis products, i.e., Platelets, Pheresis and Fresh Frozen Plasma, are prepared by automated methods at other locations, the firm's license should be amended to permit these products collected at the other locations to be shipped in interstate commerce.

Leukocytes, granulocytes and monocytes collected by a pheresis are not licensed blood components. If these are shipped interstate, the license number must not be on the product label and the product is not to be sold, bartered, or exchanged.

RECORDS

In mass production of components, the on-site supervisor, or other responsible person, reviewing and approving the records may sign as the responsible person for each group of products prepared. For licensed blood banks, this procedure should receive approval from CBER.

Records must be maintained to prevent the distribution of subsequent units of blood drawn from unsuitable donors. Unsuitable donors include, but are not limited to, those who test repeatedly reactive for anti-HIV or HBsAg and have not been properly reentered or have a medical history which would preclude donation. The regulations do not prohibit firms from collecting blood from deferred donors. A firm may collect blood from deferred donors if it is not precluded by its SOP; however, the firm must have a system to prevent the distribution of these blood components.

Refer to FDA recommendations in current memoranda to registered blood establishments and/or the establishment's SOPs to determine that the appropriate disposition of blood components and deferral of donors occurs for units testing repeatedly reactive.

ERRORS, ACCIDENTS AND FATALITIES

Refer to CBER's memorandum to registered blood establishments, dated March 20, 1991, titled "Responsibilities of Blood Establishments Related to Errors and Accidents in the Manufacture of Blood and Blood Components."

Currently, there is no regulation that requires unlicensed, registered blood establishments or transfusion facilities to submit error and accident reports to the FDA. However, a thorough investigation and documentation of corrective action is required. A request has issued for unlicensed facilities and transfusion services to voluntarily report errors and accidents.

When a complication of blood collection or transfusion is confirmed to be fatal, the fatality must be reported to the Office of Compliance, 301-594-1194, by the collecting blood bank or by the facility that performed the compatibility test. This requirement is for any facility in which the fatality occurred.

In the event that the investigator becomes aware of a previously unreported fatal donor or recipient reaction which occurred since the last inspection, the Director, Division of Inspections and Surveillance, 301-594-1194, should be notified as soon as possible.

Lookback Policy

See the April 23, 1992 memorandum, "Revised Recommendations for the Prevention of Human Immunodeficiency Virus (HIV) Transmission by Blood and Blood Products" for additional guidance.

Post Donation Information Reports

Refer to CBER's memorandum to industry titled "Guidance Regarding Post Donation Information Reports" dated December 10, 1993.

FACILITIES, EQUIPMENT, PERSONNEL

Facilities

OSHA published the final rule for the "Occupational Exposure to Bloodborne Pathogens" in the December 6, 1991 Federal Register. Included in the rule are requirements for facilities to develop procedures to ensure the safety of employees with a potential for exposure to biohazardous materials and procedures for medical waste disposal. FDA requires hand-washing facilities for staff drawing and handling blood and the safe and sanitary disposal of trash. At some mobile sites where hand washing facilities may not be available, an alternate method to clean hands, i.e., bactericidal hand wipes, is acceptable.

The Centers for Disease Control (CDC) and the National Institutes of Health (NIH) published a booklet entitled Biosafety in Microbiological and Biomedical Laboratories which recommends precautions laboratory employees should follow. The booklet is available through the Department of Health and Human Services (DHHS). It is publication No. (CDC) 88-8395, 94-95, 2nd Edition, Washington, DC: US Government Printing Office, 1988.

Unauthorized persons may not wander through an area where blood is being drawn, therefore, the traffic flow of the facility, especially mobile sites, must be properly monitored and controlled.

The interview area has to offer the donor a degree of privacy so that the donor will be comfortable answering the questions without fear of being overheard.

Equipment

Refer to the CBER July 21, 1992 Memorandum to licensed establishments on "Changes in Equipment for Processing Blood Donor Samples." Discussion of changes in equipment for ABO/RH and antibody screening is presented under Part B Laboratory below.

If equipment is used in the establishment that is not listed in the CFR, performance checks and preventive maintenance should be performed by the firm according to the manufacturer's instructions and/or SOPs.

During the course of an inspection, the investigator may observe or review instances where equipment or supplies are either being misused or not functioning as designed. Because of misuse, or lack of adherence to SOPs and/or manufacturing instructions,the use of key equipment or supplies creates circumstances where donor, operator, or product safety is compromised. It is important to examine the firm's overall use of equipment and supplies to be certain that equipment and supplies are used according to directions, satisfactorily inspected, maintained and operating properly.

The standardization and calibration of the hematocrit centrifuge may be done with a commercially prepared control or by other methods, e.g., duplicate samples tested at multiple intervals.

Spectrophotometers used during viral testing should be checked periodically for linearity, drift and repeatability according to manufacturer's instructions.

Larger blood establishments may have purchased a central temperature monitoring system to monitor and record temperatures in blood storage units. Once the system is installed and its accuracy demonstrated and documented, a daily comparison of the internal thermometer to the recording chart/device is not required. However, periodic performance checks comparing calibrated thermometers to system printouts should be performed to assure the system is functioning accurately.

Procedures should provide for the calibration of the autoclave before initial use and after repairs. Calibration procedures should provide assurance that the autoclave functions as intended, i.e., sterilization of arm preparation supplies and/or decontamination of biohazardous material. Biologic indicators must be used periodically and a temperature control, such as heat sensitive tape, should be used with each run to verify that the materials are being sterilized. A minimum of 121.5°C (251°F) for 60 minutes by saturated steam at a pressure of 15 atmospheres is recommended for materials contaminated with blood; 20 minutes at the same temperature is required for arm preparation supplies.

Personnel

Blood bank personnel should be familiar with applicable regulations related to their respective tasks. Personnel should know the location of the SOP manual and be knowledgeable about those sections which pertain to their jobs.

Staff of mobile sites must have the same degree of training and supervision as for fixed donor sites located in the blood bank or in a donor center. Volunteers are permitted to assist in various areas and must be adequately trained. Training should be documented.

QUALITY ASSURANCE

Refer to the "Draft Guideline for Quality Assurance in Blood Establishments," dated June 17, 1993.

This publication assists manufacturers of blood and blood components on developing procedures and practices useful for administering a quality assurance program. Facilities may follow the guideline or choose to use alternative procedures not provided in the document. However, if an establishment chooses to use alternative procedures, the facility may wish to discuss such procedures with the FDA to prevent expenditures of resources on activities that may be unacceptable to the agency.

DISPOSAL OF INFECTIOUS WASTE

FDA advises that state and local laws should be followed. All blood contaminated waste should be autoclaved (121.5°C/251°F for 60 minutes) or incinerated. The firm's SOP should contain specific language for disposal of contaminated waste. Needles should be disposed of in a container designed to prevent accidental puncturing of personnel. If contaminated waste is disposed of by a contract waste disposal firm, a contracted agreement should be on file at the facility, and specify that Biohazardous material is disposed of appropriately according to EPA, state and/or local regulations. Inappropriate disposal practices should be referred to state authorities for follow-up.

PART A -- WHOLE BLOOD DONOR SUITABILITY

Donor Suitability

Refer to the following CBER memoranda for additional information:

1. "Donor Suitability Related to Laboratory Testing for Viral Hepatitis and a History of Viral Hepatitis," dated December 22, 1993.
2. "Revised Recommendations for Testing Whole Blood, Blood Components, Source Plasma and Source Leukocytes for Antibody to Hepatitis C Virus Encoded Antigen (ANT-HCV)," dated August 5, 1993.
3. "Deferral of Blood and Plasma Donors based on Medications," dated July 28, 1993.
4. "Revised Recommendations for Testing Whole Blood, Blood Components, Source Plasma and Source Leukocytes for Antibody to Hepatitis C Virus Encoded Antigen (Anti-HCV)," dated April 23, 1992.
5. "Revised Recommendations for the Prevention of HIV Transmission by Blood and Blood Products," dated April 23, 1992.
6. "Exemptions to Permit Persons with a History of Viral Hepatitis Before the Age of Eleven Years to Serve as Donors of Whole Blood and Plasma; Alternative Procedures," dated April 23, 1992.
7. "Clarification of FDA Recommendations for Donor Deferral and Product Distribution Based on the Results of Syphilis Testing," dated December 12, 1991.
8. "FDA Recommendations Concerning Testing for Antibody to Hepatitis B Core Antigen (Anti-HBc)," dated September 10, 1991.
9. "Recommendations for the Management of Donors and Units that are Initially Reactive for Hepatitis B Surface Antigen (HBsAg)," dated December 2, 1987.
10. "HTLV-I Antibody Testing," dated November 29, 1988.

All persons donating blood or blood components for transfusion or further manufacturing use should receive information about the safety of blood products in relation to AIDS epidemiology and the implications for donors who have engaged in certain high-risk activities. The information should be written in language that assures that the donor understands the definition of high-risk behaviors and the importance of self-exclusion. Procedures should also be available that permit communication of this information to visually impaired, illiterate, or non-English speaking persons if they are permitted to donate. The procedures applied should provide an opportunity at each visit for the donor to consider the information and to make an informed decision about whether to donate. 640.3

Donor History

Temperature, Blood Pressure: The suggested "normal" blood pressure value is 90-180 mm/50-100 mm. A low temperature is usually of no significance unless the donor has symptom of viral illness. Temperature conversion: °F = (°C x 9/5) = 32; °C = (°F -32 x 5/9. A temperature on the day of donation > = 99.6°F (37.5°C) would temporarily defer a donor. A temperature of 99°F (37.2°C) for more than 10 days should be evaluated before donation would be allowed.

Acute Respiratory Diseases: Symptoms of respiratory disease (colds, influenza, persistent cough, sore throat) or other manifestations of upper respiratory disease shall because for rejection until active symptoms have subsided. Such symptoms may be an early indication of a more serious illness.

Other Acute and Chronic Disqualifying Diseases: Convulsions, bleeding disorders, recent tooth extraction (within 72 hours), malaria, skin infection at the phlebotomy site, cancer, tuberculosis, diabetes, and heart disease should be cause for rejection.

Infectious Skin Diseases: Mild skin disorders such as acne, psoriasis, or the rash of poison ivy are not cause for deferral unless they affect the skin in the phlebotomy area. Donors with boils or other severe skin infections should be deferred until the phlebotomy area is free of infection.

Malaria: See the July 26, 1994 memorandum, "Recommendations for Deferral of Donors for Malaria Risk," for additional guidance. CDC provides a booklet, "Health Information for International Travelers," which should be used rather than the maps that were once used. This booklet is updated yearly and the firm should be using a recent edition.

Plasmapheresis donations that are also used for the preparation of Platelets are not exempted from the malaria restrictions.

Hepatitis: Donors who have had close contact with a patient with viral hepatitis should be deferred for at least twelve months. Hospital personnel working in areas where hepatitis is endemic, such as renal dialysis units, should be excluded for at least twelve months after employment in such areas. Prospective donors who have received HBIG following a hepatitis exposure should be deferred for at least 12 months.

Acupuncture patients may donate blood without a twelve month deferral if the needles used have been sterilized under proper conditions. If the needle sterilization procedure cannot be verified by the blood bank, the donor should be deferred for twelve months. The same principle can be applied to ear or other body piercing. If sterile procedures can be verified to have been used, the donor need not be deferred.

Donors who have been exposed to blood percutaneously, i.e., needle sticks or mucous membrane splashes, should be deferred from donating blood for at least twelve months.

Donors who have received a unit of blood, blood component, or derivative (other than clotting factor derivatives) which may be a possible source of infectious disease are deferred for twelve months after receipt of the product.

A memorandum notifying all blood establishments regarding deferrals for medications including Accutane, Tegison, Proscar and Pituitary Human Growth Hormone was issued July 28, 1993, "Deferral of Blood and Plasma Donors based on Medications." If a donor admits to taking medications, i.e., aspirin or antibiotics, there should be further questioning of the donor to determine the reason for taking the medication that might preclude donation (i.e., aspirin for fever or antibiotic for infection).

An acceptable reference for drug/medication deferral is the Drug Use and Blood Donor Acceptability Guide, Cutter Biological, 1983.

Directed Donors: should meet all suitability requirements and be tested as allogeneic donors. Occasionally, a directed donation may not meet all suitability and testing requirements, in which case, the patient's physician may make a medical decision to use the directed donation.

Screening Area Quality Control

The containers for the copper sulfate ($CuSO_4$) solution must be covered when not in use so that the solution will not evaporate. In order to prevent dilution of the copper sulfate, which might result in selection of donors with inadequate hemoglobin levels, containers must be thoroughly dry before use. Generally, 25 ml vials of copper sulfate solution are used. Because the blood:copper sulfate ratio is one drop of blood:one ml of copper sulfate, the copper sulfate should be discarded after 25 donors have been tested.

The copper sulfate solution used, which is equivalent to 12.5 gm/dl hemoglobin, has a specific gravity of 1.053; $CuSO_4$ equivalent to 13.5 gm/dl hemoglobin has a specific gravity of 1.055. The specific gravity should be checked periodically with a calibrated hydrometer.

If there is no clear distinction between acceptable and unacceptable donors, the solutions should be changed, or a different method of screening donors should be considered. Copper sulfate screening is not a quantitative procedure. Specific measurement of hemoglobin or hematocrit may indicate that a donor rejected by the copper sulfate procedure is, after all, acceptable.

There are newer laboratory instruments available to determine Hemoglobin or Hematocrit in blood donors (Hemocue by Leo Diagnostics AB, Helsinborg, Sweden and Hematastat by Separation Technology, Inc. are examples of such instruments). Quality control procedures should be in accordance with manufacturer's instructions and the firm's SOP.

Arm Preparation

Commercial arm preparation supplies are prepared by several manufacturers, e.g., Clinipad Corporation and Marion Scientific. These products are not marked sterile but may be used to prepare the donor's arm for phlebotomy.

Arm preparation supplies, such as gauze, cotton balls, and applicators may be prepared by the establishment. If the arm preparation supplies are sterilized by a central supply which is a part of a hospital's operation, it is not necessary to observe these preparation procedures or to review the records for these supplies.

Arm preparation should be done with care and for the full amount of time as stated in the SOP, and ideally with vigorous scrubbing.

There are several ways to do a satisfactory arm preparation. Sufficient duration and vigor of scrubbing are the key factors to removal of superficial microbes which is the goal of the arm preparation. The final step must be application of a bacteriostatic agent in a non-overlapping spiral beginning at the intended needle puncture site and extending outward.

Blood Collection

After the venipuncture area is prepared, the vein may be palpated above or below the prepared area; however the site of needle insertion should not be touched prior to venipuncture. It is not permissible to put iodine or sterile gauze on the finger to palpate the intended venipuncture site. Needles may be used for only one venipuncture. If a venipuncture is unsuccessful, a new needle must be used and the arm preparation repeated.

The completion of the donation may be signaled by a trip scale or vacuum-assist method based on mass or volume. Otherwise, the bag must be weighed (spring scales) and the flow stopped manually. The blood should be mixed gently (either manually or mechanically) during collection. The final unit should weigh approximately 425-525 grams plus the weight of the container and anticoagulant (approximately 90 grams). Low volume collections, e.g., pediatric autologous collections, are acceptable providing the establishment has an SOP. For low volume collections to be shipped in interstate commerce, the facility must have CBER approval.

Hermetic sealing of the blood bag tubing is accomplished by one of the following: dielectric sealing, metal clamp, or tying a tight "white knot" in the donor tubing. The contents of the tubing should be stripped into the bag, mixed well, and the anticoagulated blood allowed to refill the tubing. Following stripping of the tubing, an additional hermetic seal should be placed close to the bag so as to prevent tampering, thereby ensuring maintenance of a closed system.

Specific gravity of whole blood = 1.053 gm/mL for blood containing 12.5 gm/dL of hemoglobin. The following calculation is used to convert volume to weight:

$$1.053 \text{ gm/mL} \times 500 \text{ mL} = 526.5 \text{ gm}$$

There are firms with approval from CBER to collect FFP or other plasma byproducts using hemapheresis devices. Some firms are approved to aliquot the FFP (or other products) into smaller containers using a sterile tubing connecting device (STCD). An STCD is a device which seals two pieces of "like" (same size and composition) tubing together to make a sterile connection. Licensed establishments must have FDA approval for manufacturing components with an SCD. For additional guidance, see the July 29, 1994 memorandum, "Use of an FDA Cleared or Approved Sterile Connecting Device (STCD) in Blood Bank Practice."

Blood Collection Container

Names of blood container manufacturers may be found on the labels of the blood bags. The containers manufactured by Cutter Laboratories (Division of Miles), Fenwal Laboratories (Baxter Healthcare Corporation), and Terumo Corporation have been approved by CBER. Approved anticoagulants include anticoagulant citrate dextrose solution (ACD), anticoagulant citrate phosphate dextrose solution (CPD and CP2D), and anticoagulant citrate phosphate dextrose adenine solution (CPDA-1).

Collection sets must be stored in accordance with the manufacturer's instructions. Inappropriate storage may contribute to evaporation of the anticoagulant/preservative solution, or to mold growth on the container surface.

Laboratory Samples

Pilot (laboratory) samples containing blood for ABO, Rh, antibody screening and viral testing are collected at the time of donation. The tubes must be identified to accurately relate them to the unit. The method used for collection of the pilot sample should be one that precludes contamination of the donor unit, minimizes personnel exposure to blood, and maintains donor safety. See the AABB Technical Manual, for methods of pilot sample collection.

Dating Period

Expiration dates vary from 21 days for ACD, CPD, and CP2D, 35 days for CPDA-1 in whole blood and red blood cells, to 42 days for red blood cells to which additive solutions have been added, i.e., ADSOL^{R} (Fenwal), Nutricell^{R} (Cutter) and Optisol^{R} (Terumo).

Autologous Blood Collection

Autologous units are blood or blood products collected from a person for his/her own use at a later time. These predeposited blood products are handled and processed as similarly as possible to homologous units. CBER sent the following memoranda concerning autologous donations to all registered blood establishments: "Guidance for Autologous Blood and Blood Components" on March 15, 1989 and "Autologous Blood Collection and Processing Procedures" on February 12, 1990 and "Disposition of Blood Products Intended for Autologous Use that Test Repeatedly Reactive for Anti-HCV," dated September 11, 1991.

Therapeutic Phlebotomy

Therapeutic phlebotomies are considered to be a procedure intended to treat the patient for certain disease states, therefore, a written doctor's request should be available for the procedure. Units of blood that are collected for therapeutic phlebotomy must be fully tested and donors meet all suitability requirements if the units are to be transfused or if the plasma will be sold for further manufacture. Several recalls have occurred due to lack of testing of recovered plasma from therapeutic units prior to shipping for further manufacture.

Donor Reactions

A blood bank should be aware of the nature and frequency of donor reactions. Adequate records for each reaction should be kept and should include follow-up in order to insure adequate donor protection. The blood bank's SOP manual should list and describe donor reactions it considers to be adverse, as well as the procedures to be followed for handling and investigating these reactions. Severe donor reactions may include fainting, convulsions, severe hematomas (infiltration), or injury caused by falling. Mild donor reactions include feeling faint, nauseated, or dizzy.

PART B -- LABORATORY

ABO and RH Testing, and RBC Antibody Screening

Blood Grouping Reagents are used to determine the blood group. Anti-A will agglutinate or clump group A red cells. Anti-B will agglutinate group B red cells. Anti-D is used to determine Rho (D) factor.

Licensed antisera -- some establishments may "otherwise meet the requirements" for AGO and Rh licensed antibodies by producing their own antisera. If so, production records that are in compliance with the requirements for the manufacture of these products specified in 21 CFR 660, Subpart C must be kept. Licensed blood banks must have CBER approval to use such antisera.

Test methods used for ABO, Rh and antibody screening, which are different from the manufacturer's instructions, should not be cited as deviations if they are not prohibited by the manufacturer, have been demonstrated to be satisfactory, or have been approved for use by CBER. For example, the Microtiter™ plates and Groupamatic™ machines may be used for ABO, Rh, and antibody testing. The reagents must be tested and shown to perform adequately or satisfactorily when using these techniques.

Blood which tests Rho (D) negative must be confirmed by further testing (usually D^u) unless it is labeled in accordance with 21 CFR 640.5(c). Acceptable methods for further testing to confirm D negatives include use of the antiglobulin method, and use of a special channel on the Konron Groupamatic, Olympus PK700, or the Gamma STS-M automated blood groupers. Licensed blood establishments should have a letter from CBER approving their use of an automated blood grouper for D^u testing.

Not all anti-D reagents may be used for D^u testing; the package insert must include directions for D^u testing.

Procedures have been approved to use reagents with the Groupamatic™ beyond the dating period, provided a proper set of controls is used. Licensed establishments should have a letter from CBER on file indicating that a protocol has been submitted and approved for this procedure. Rare reagents, e.g., anti-Jk^b, anti-Le^b, etc., are sometimes used beyond the expiration date; this is acceptable only if adequate controls are used and the reactivity and specificity of the reagents are documented.

The reactivity and specificity of reagents are generally confirmed by testing at least one positive and one negative control sample. The negative controls are not essential for ABO reagents because the antithetical cell and serum results provide confirmation of test accuracy. The Anti-Human Globulin (Coombs) reagent must be tested each day of use with an IgG sensitized ("Coombs" control or "check") cell.

The manufacturer's instructions specify that the storage requirements for Anti-A, Anti-B, and Anti-D reagents is between 2-8°C; however, it is accepted practice for these reagents to stay at room temperature for the duration of the working day. This will usually not diminish the potency of the products throughout the normal period of use.

Automated Tests for ABO & RH Testing

Refer to the July 21, 1992 CBER memorandum to licensed establishments on "Changes in Equipment for Processing Blood Donor Samples." This document provides specific guidance on the documentation of changes in equipment, specifically calibration, validation, parallel testing, quality control, maintenance and emergency plans.

Laboratory records for automated testing should include the name of the person who prepared the reagents. If the system does not provide positive sample identification, a record must be made for the loading pattern and the record must include the name of the person(s) who loaded and unloaded the sampler; if results are visually interpreted, the record must include the name of the person(s) interpreting and transferring the results.

If the firm used automated methods for ABO and Rh typing at the time of licensure, a separate letter of approval from CBER for automated ABO and Rh testing will not be issued to the establishment.

Reagents used in microplate test systems should be recommended for this use by the manufacturer's package insert. If not originally licensed for the use of the microplate test system, a licensed establishment should have on file a letter from CBER approving its use. Gamma-Micro-U microtiter plats are approved for use only with the Gamma microtiter reagent unless the reagent has been evaluated and found acceptable for such use according to an established protocol.

Occasionally, an automated blood grouping instrument is unable to interpret an ABO or RH result. The firm should have an SOP to follow-up with further testing to obtain a result (usually by manual methods) and to up-date the testing record (data entry and verification).

Serological Test for Syphilis

All units of blood must be tested by an acceptable serological test for syphilis (STS). For further guidance, refer to CBER's December 12, 1991 memorandum to registered blood establishments on "Clarification of FDA Recommendations for Donor Deferral & Product Distribution Based on the Results of Syphilis Testing."

Disease Marker Testing

Each collection of whole blood and blood components must be tested for anti-HIV, syphilis, and HBsAg. In addition, they should be tested for anti-HCV, anti-HBc, and anti-HTLV-I per recommendations from FDA. Testing for alanine aminotransferase (ALT) was implemented as a surrogate marker for viral hepatitis. FDA has not made any recommendations as to whether or not blood banks should perform ALT testing; however, ALT testing has become an industry standard. If an establishment has implemented ALT testing, they should be following the manufacturer's instructions and their SOP for performing the test and interpreting the test results.

Each unit of blood must be tested for HBsAg by a licensed third generation test. Third generation tests include radioimmunoassay (RIA), reverse passive hemagglutination (RPHA), or enzyme-linked immunosorbent assay (ELISA or EIA).

Refer to CBER's December 22, 1993, memorandum to registered establishments entitled "Donor Suitability Related to Laboratory Testing for Viral Hepatitis and a History of Viral Hepatitis."

Each unit must be tested for antibody to HIV with a licensed test kit. There are three types of licensed kits based upon different manufacturing technologies:

> Whole viral lysates;
>
> Recombinant DNA technology; (See the February 1 and August 1,1989, memoranda "Use of the Recombigen HIV-1 LA Test" for further information.) and,
>
> Synthetic peptides,

Refer to Compliance Program 7342.001, Inspection of Licensed and Unlicensed Blood Banks, Attachments C and D, for a list of licensed test kits for HBsAg, anti-HIV-1, anti-HIV-2, anti-HIV-1/2, Anti-HCV, Anti-Hbc, Anti-HTLV-I, and Western blots.

<u>Testing Performed on Premises</u>

Refer to the following memoranda for further information:

1. "Recommendations for the Management of Donors and Units that are Initially Reactive for Hepatitis B Surface Antigen (HBsAg)," dated December 2, 1987
2. "HTLV-I Antibody Testing," dated November 29, 1987
3. "Revised Recommendations for the Prevention of HIV Transmission by Blood and Blood Products," dated April 23,1992
4. "Use of Fluorognost HIV-1 Immunofluorescent Assay (IFA)," dated April 23, 1992
5. "Revision to 26 October 1989 Guideline for Collection of Blood or Blood Products from Donors with Positive Tests for Infectious Disease Markers ('High Risk' Donors)," dated April 17, 1991
6. "FDA Recommendations Concerning Testing for Antibody to Hepatitis B Core Antigen (Anti-HBc)," dated September 10, 1991
7. "Revised Recommendations for Testing Whole Blood, Blood Components, Source Plasma and Source Leukocytes for Antibody to Hepatitis C Virus Encoded Antigen (Anti-HCV)" dated August 5, 1993
8. "Disposition of Blood Products Intended for Autologous Use That Test Repeatedly Reactive for Anti-HCV," dated September 11, 1991

If automated testing equipment is interfaced with a computer system see section on computerization, for further guidance.

In the past, the area used for HBsAg and anti-HIV testing would, by design, be in rooms separated from other blood bank activities. This is no longer considered to be important as all blood samples should be treated as capable of transmitting an infectious disease, and Biosafety Level 2 precautions should be applied in all areas where open samples are handled. However, if RIA procedures are used in the facility these areas still must be physically separated from other areas. Work areas, such as counter tops, should be constructed of non-porous materials and designed to permit thorough cleaning and disinfection. There should be policies to prevent excessive traffic of unauthorized personnel through viral testing areas.

Proficiency Testing

Most blood establishments will be participating in a proficiency testing program, either an in-house developed or an established program such as the College of American Pathologists (CAP), AABB or CDC.

A proposed rule was published in the June 6, 1989, Federal Register to require that each establishment or laboratory responsible for performing FDA required tests for HBsAg and anti-HIV participate in an approved program to demonstrate proficiency in performing these tests. The final regulation proposed by FDA has not been published, however, the final rule proposed by HCFA, which regulates laboratories receiving Medicare and Medicaid reimbursement, was published in the March 14, 1990, Federal Register. This final rule requires laboratories to have policies and procedures for an ongoing program to assure that employees are competent and maintain their competency to perform their duties.

Testing Performed by Outside Laboratories

The results and interpretations of all (initial and repeat) tests performed and an explanation of any symbols or phrases used in reporting results should be provided by the testing facility to the blood bank. The blood bank should have an SOP for the interpretation of the reports obtained from outside testing laboratories and written assurance that the outside testing laboratory interprets test results according to FDA requirements. The raw test data, i.e., absorbance readings from the spectrophotometer, need not be sent to the blood bank. In addition, if the blood establishment is re-entering donors with previously repeatedly reactive anti-HIV test results, the establishment must determine if the outside testing laboratory is performing Western blot assays with licensed test kits.

At this time, 21 CFR 607.65(g) exempts from the requirement to register clinical laboratories that are approved for Medicare reimbursement performing hepatitis and anti-HIV testing on donor blood for other registered facilities. If not registered, the testing laboratory should be asked to voluntarily register. Send the testing laboratory Form FDA-2830, Blood Establishment Registration and Product Listing, in accordance with the procedures described in Field Management Directive 92. Licensed blood establishments may have viral testing performed at testing facilities which are also licensed, and only with CBER approval.

Except for emergencies, no units should be issued until written hepatitis and HIV antibody test results are in the possession of the blood bank.

Use of Reactive Units

Refer to the following memoranda:

1. "Guideline for Collection of Blood or Blood Products from Donors with Positive Tests for Infectious Disease Markers ('High Risk' Donors)," dated October 26, 1989.
2. "Revision to 26 October 1989 Guideline for Collection of Blood or Blood Products from Donors with Positive Tests for Infectious Disease Markers ('High Risk' Donors)," dated April 17, 1991.
3. "Disposition of Blood Products Intended for Autologous Use That are Repeatedly Reactive for Anti-HCV," dated September 11, 1991.

Invalidation of Test Results

Refer to CBER's January 3, 1994 memorandum, titled "Recommendations for the Invalidation of Test Results When Using Licensed Viral Marker Assays to Screen Donors."

These recommendations clarify FDA's position on the invalidation of test results when screening donor blood using licensed viral marker assays, including the use of external control reagents.

PART C -- RED BLOOD CELLS

Open System

Blood components for transfusion are normally prepared in a closed, sterile system. Occasionally, however, the hermetic seal may be broken, thereby exposing the blood component to the outside environment. When this occurs the component should have an expiration date not to exceed 24 hours.

Additive Solutions

Manufacturers of certain blood collection systems have been approved for a room temperature eight hour hold period following the collection of whole blood prior to preparation of components. These systems are the ADSOL^R solution system (anticoagulant CPD), manufactured by Fenwal Laboratories, Division of Baxter Healthcare Corporation, the Nutricel^R system (anticoagulant CP2D), manufactured by Cutter Biological, Division of Miles Inc., and Optisol^R, manufactured by Terumo Corporation. In addition, Fenwal's collection system containing CPDA-1 is also approved for an eight hour hold prior to component preparation. Platelets, Fresh Frozen Plasma, Cryoprecipitated AHF and Recovered Plasma may be prepared from the whole blood within eight hours of collection.

If Fresh Frozen Plasma, Platelets, or Cryoprecipitated AHF are not prepared from the units of whole blood, the additive solution may be added to the red blood cells (RBCs) within three days of collection.

Red Blood Cells Frozen and Red Blood Cells Deglycerolized

The freezing of RBCs may be accomplished by two acceptable techniques, e.g., high concentration glycerol-slow freeze (storage at < = -65°C) and low concentration glycerol-rapid freeze (storage, usually in liquid N_2, at < = -120°C). The lot numbers of solutions and containers for glycerolization and deglycerolization must be recorded. Red Blood Cells, Frozen may be stored for ten years. If units in storage have not been tested for all currently required tests, there should be a procedure to prevent mislabeling when units are thawed.

Quality control testing should be performed and the firm should take corrective action when test results are out of the firm's established parameters as stated in the SOPs. Quality control testing may include periodic sterility testing, monitoring the removal of the glycerol and the amount of the free hemoglobin in the final product, and the RBC recovery. Certain blood centers have approval for an SOP that does not require sterility checks on Red Blood Cells, Deglycerolized, on a periodic basis if the facilities where the product is prepared are monitored for cleanliness and good housekeeping procedures, including proper maintenance of air filters in the heating, ventilation and air conditioning (HVAC) system.

An area of concern is that the firm has adequate procedures to assure that the final container is accurately identified to relate it to the donor. For example, if several units of blood are frozen and deglycerolized simultaneously, what controls does the firm utilize to assure against mix-ups? ABO and Rh checks should be performed after a unit is deglycerolized to verify the blood type.

Rejuvenating Solutions

Some establishments use rejuvenating solutions (Rejuvasol^R, Cytosol Labs, Red Blood Cell Processing Solution which contains pyruvate, inosine, phosphate and adenine) to restore normal characteristics of oxygen transport and delivery and improve post-transfusion survival of RBCs. These solutions should be used aseptically and according to the manufacturer's instructions. Rejuvasol^R may be used to rejuvenate RBCs which have been expired for up to three days. Once the rejuvenating solution is added to the red blood cells the unit may be washed and transfused as Red Blood Cells, Rejuvenated, or the red blood cells may be frozen and deglycerolized (labeled respectively as Red Blood Cells Frozen, Rejuvenated, and Red Blood Cells, Rejuvenated, Deglycerolized).

Red Blood Cells -- Leukocytes Removed

There are several methods for removing leukocytes from RBCs. These methods include centrifugation with or without saline washing, microaggregate blood filtration, freezing and deglycerolizing, washing (using either a manual or automated method) and filtration with filters designed and approved specifically for leukocyte removal.

For units labeled "Leukocytes Removed" quality control (QC) should be performed for all the methods listed above. This monthly QC is not necessary if: (1) the unit is not labeled "Leukocytes Removed"; or (2) the leukocytes are removed as the unit is being transfused using a filter that is connected to the transfusion set by the manufacturer or at the patient's bedside. The blood bank should have SOPs stating values of acceptance for leukocyte removal and corrective action to be taken if values are outside established limits. Filters specific for leukocyte removal are sometimes connected to the transfusion set at the patient's bedside and it would not be practical to monitor the post-filtration blood for leukocyte counts. The quality control for these filters was performed by the manufacturer as part of the approval process. Transfusion services that use filters that remove leukocytes during transfusion are not considered to be manufacturing a blood product.

Irradiated Blood

Experimental data have established that in certain immunodepressed patients transfusion of foreign immunocompetent cells (T lymphocytes) may lead to graft versus host disease (GVHD). GVHD occurs when donor T lymphocytes engraft, multiply and react against the tissues of the recipient. The radiosensitivity of these T lymphocytes is higher than that of other blood cells, so irradiation of cellular blood products before transfusion appears to be effective in preventing the transfusion induced form of this serious complication. The AABB has recently recommended that blood components prepared from directed blood donations from first degree family members (i.e., parents, children, siblings) be irradiated to decrease the risk of GVHD. The CBER issued guidance via a July 7, 1993, memorandum titled "Recommendations Regarding License Amendments and Procedures for Gamma Irradiation of Blood Products." This document addresses manufacturing and quality assurance procedures, labeling, other aspects of production and the use of irradiated blood and blood products. It also provides background on the effects of gamma radiation on product quality & stability, guidance on license amendments, record keeping and fatality reporting. If any questionable procedures for irradiating blood are encountered during an inspection notify the Division of Inspections and Surveillance, (301) 594-1191.

Red Blood Cells -- Records

Blood banks that prepare washed, frozen, deglycerolized and rejuvenated RBCs must record the lot numbers of solutions and/or containers. The lot numbers must be traceable to the unit number.

Red Blood Cells -- Immunization Programs

Refer to CBER's December 16, 1992 memorandum on the "Revision of October 7, 1992 Memorandum Concerning Red Blood Cell Immunization Programs." This document incorporates a 12 month deferral for donors and recipients of RBCs for immunization as part of Source Plasma programs. It also provides additional information on frozen storage, selection of safe donors and license amendments.

PART D -- PLASMA AND RECOVERED PLASMA

Each final container of plasma for transfusion prepared from a whole blood collection shall be in an integrally attached satellite bag at the time of collection; it must be transparent and hermetically sealed by a dielectric sealer, metal clamp, or tightly drawn "white knot;" and its label must be marked by number or other symbol so that it can be traced back to the donor.

Plasma Products for Transfusion

The final product should be stored in a manner which will show evidence of thawing. This may be accomplished in a variety of ways, e.g., by storing it upside down after freezing, or by placing a rubber band around the middle of the container and removing it when the unit has frozen.

Recovered Plasma

Recovered Plasma is an unlicensed source material intended for use in the manufacture of both licensed and unlicensed products. A license is not required to manufacture, distribute, or pool recovered plasma.

The short supply provision allows licensed manufacturers to use unlicensed or other licensed facilities, not a part of their own establishment, to perform the initial and partial manufacturing step of collecting blood or plasma. The Recovered Plasma is shipped solely to the licensee for further manufacture into licensed injectable or noninjectable products. Short supply agreements are between the licensed fractionator and the collection facility; not with brokers. The written agreements should be updated periodically, and a copy of this agreement should be on file at the collecting facility. Plasma brokers may be used as authorized agents and should be identified in the short supply agreement; refer the name of the broker taking possession of the Recovered Plasma or other blood components to the home district for follow-up, i.e., registration and inspection. Short supply agreements are also required between the registered collection facility and the licensed manufacturer for other blood components (i.e., RBCs and platelets) intended for further manufacture into licensed products.

Recovered Plasma does not have an expiration date, therefore, records are to be kept indefinitely.

PART E -- PLATELETS

Due to an increase in the number of post-transfusion sepsis reports, the seven day dating period for platelets reverted to five days, effective July 2, 1986.

Platelets may be pooled by the blood bank personnel, upon the request of a physician; however, this is done following designation of the platelets to a specific recipient, and the resultant pool is not considered a licensed product. The label of the pooled components should indicate the individual donor numbers comprising the pool or a pool number that relates it to individual donor numbers comprising the pool. Final containers must be transparent and hermetically sealed by a dielectric sealer, metal clamp, or tightly drawn "white knot." The expiration time for pooled platelets is limited to 4 hours. Until data are available indicating the effectiveness of the platelets is maintained for a longer period of time and CBER approval is obtained, the expiration time of platelets which are pooled with the aid of a sterile connecting device is also limited to four hours.

Quality Control

Quality control testing must be performed each month platelets are prepared, using one unit obtained from each of four different donors. Platelet counts (5.5 x 10^{10} in 75% of the units tested), pH determination (> = 6.0) and measurement of plasma volume should be made at the end of the storage/dating period.

If quality control testing is not under the supervision and control of the establishment, determine where the testing is performed and how test results are reviewed and handled by the establishment.

PART F -- CRYOPRECIPITATED AHF (ANTIHEMOPHILIC FACTOR)

There is no volume restriction for preparing Cryoprecipitated AHF. One unit of plasma may be used as a source of both Platelets and Cryoprecipitated AHF. Practical experience has demonstrated that using careful production techniques, an acceptable final product averaging no less than 80 IU can be manufactured from a single unit.

As with Platelets, Cryoprecipitated AHF may be pooled upon request of a physician. The label of pooled components should indicate the individual donor numbers comprising the pool or a pool number that relates the pooled AHF to the individual donor numbers comprising the pool. There are some firms manufacturing Cryoprecipitated AHF, Pooled as a licensed product.

Firms may have approval for variances under 21 CFR 640.120 from the regulations for manufacturing Cryoprecipitated AHF, i.e., preparing Cryoprecipitated AHF within 15 hours after phlebotomy. The firm should have written approval from CBER for any variance.

The final container should be transparent and hermetically sealed by a dielectric sealer, metal clamp, or a tight "white knot."

Quality Records

Quality control testing need be performed only in the months in which the product is prepared. Four units must be tested, but they may be pooled before the quality control assay is performed.

If quality control testing is not performed under the supervision of the establishment, see Compliance Policy Guide 7134.16.

PART G -- UNIFORM BLOOD LABELING

It is suggested that a label be collected for ready reference in checking the label items; however, it is important that the labels on units ready for issue are the labels in actual use and are properly completed. The Uniform Labeling Guidelines was published August 30, 1985, and the effective date was September 2, 1986. A revised Guideline for Uniform Blood Labeling is expected to be published in the near future. Specific label information should depict reference to viral marker testing results and other tests performed, e.g., ALT.

Circulars

The circulars are of great importance now that uniform or commonality labels are in use. The new labels have less information on them so that attention will be focused on the blood group (ABO and Rh). Information that has been deleted must be in the circular. The circular of information is an extension of the container labels and viewed as labeling containing statements of purported product quality. It should describe each component available for patient transfusion and give indications and contraindications for use. The blood supplier should have a plan for distributing the instruction circulars, assuring that the transfusion services have an adequate supply of the circulars, and the transfusion service should have a plan for distributing the circulars to the staff. Licensed blood banks will have submitted their circulars to CBER for approval. The firm's name and address should be on the circular.

PART H -- COMPATIBILITY TESTING AND TRANSFUSION REACTIONS

Compatibility Testing

Compatibility testing should be performed in an area sufficiently removed from other areas to eliminate distraction or the introduction of errors in testing.

Hospitals may elect not to crossmatch blood for certain surgical procedures that usually do not require the transfusion of blood. This procedure is referred to as "type and screen" and requires: 1) determination of the patient's blood group; 2) tests of patient's serum for unexpected antibodies; and 3) availability of units of blood in case the patient does need blood during the operation.

See the December 14, 1984, memorandum to blood establishments "Equivalent Methods for Compatibility Testing."

In addition, other methods for the compatibility testing may also be appropriate. These methods include, but, are not limited to, the use of patient specimens older than 48 hours and the use of plasma. Extended periods of time (e.g., more than three months) for holding test samples should be discussed with the Division of Inspections and Surveillance if encountered.

Antibody Testing

If an antibody screen (serum is tested with reagent red cells of known antigenic makeup to determine if there are antibodies in the serum) was performed on a unit by the supplier, the hospital or transfusion service need not repeat it, or perform a minor crossmatch.

Recipient Sample Identification

The recipient's blood sample should be identified by name and number to insure positive identification. Donor and recipient blood samples should be saved for at least seven days after transfusion in case there is a need for retesting.

Emergency Transfusions

SOPs should be available to expedite testing for transfusions in a life threatening emergency. Documentation should include signature of the requesting physician. If crossmatches are not completed sufficient documentation should be available.

Compatibility Test Records

Records should be kept of the receipt of recipient's sample and the ABO and Rh test results; lot numbers of reagents used for testing; and routine and emergency crossmatches and direct antiglobulin testing (if done). The vital signs of a recipient are not required to be on file at the blood bank.

Recipient Adverse Reactions

The blood bank's SOP manual should list and describe recipient reactions it considers to be adverse, as well as the procedures to be followed for handling and investigating these reactions. Recipient reactions usually not considered serious include low fever and chills of short duration, hives, or urticaria. Serious adverse recipient reactions usually include hemolysis, bacteremia, or septicemia.

If the blood bank acts as a transfusion service and receives blood from other sources, errors in the ABO and Rh grouping should be reported to the suppliers. Procedures should be established between the suppliers and users of blood and blood products for monitoring recipient adverse reactions which occur outside the supplying facility. If the supplier of mistyped blood is a licensed establishment, it is the responsibility of that establishment to report any errors to CBER, Office of Compliance. Refer to CBER's memorandum to industry dated March 20, 1991, titled "Responsibilities of Blood Establishments Related to Errors & Accidents in the Manufacture of Blood Components."

PART I -- STORAGE, DISTRIBUTION

Physical Storage

Blood products should be stored separately but not necessarily in a different refrigerator.

Quarantine procedures are a very important area in the control procedures to prevent the distribution of unsuitable units. Separate storage areas should be maintained for untested units, for units which are not suitable for use (units to be retested or repeatedly reactive), and for units which are suitable for distribution.

Units of blood intended for autologous use should be stored in an area separate from units for allogeneic use.

Storage Temperature and Records

All required temperatures should be maintained. Fluctuations outside storage temperature limits must be documented as to the possible reason, and any action required to maintain the blood or components at the proper storage temperature must also be documented.

The product storage temperature after drawing is 1-6°C unless room temperature platelets are to be prepared, in which case the blood should be held at 20-24°C. If the blood must be transported from a donor center to the processing laboratory, it must be placed in temporary storage having sufficient refrigeration capacity to cool the blood continuously toward a 1-6°C range, unless platelets are to be prepared. Temperatures below 20°C (68°F) and above 24°C (75.2°F) reduce platelet function and survival. Room temperature is usually not below 20°C.

Inspection

Blood should be visually inspected at the time of issue for any abnormality, such as hemoglobin in the plasma from red cell lysis, purple tinged red cells due to bacterial contamination, or blood clots.

Shipping

Facilities should have procedures to show that shipping containers maintain products at their appropriate temperature.

Reissue

Blood banks should have written criteria for reissuing blood that is returned to the blood bank. Studies have shown that the unit of blood sitting at room temperature usually maintains a temperature of ≤ 10°C for 30 minutes. Blood that has been issued for transfusion may be reissued if it is returned to the blood bank within 30 minutes, and was kept at room temperature or colder while out of the blood bank's control.

PART J -- PLATELETS, PHERESIS

Refer to the October, 1988, memorandum from CBER to all registered blood establishments" Revised Guideline for the Collection of Platelets, Pheresis."

PART K -- COMPUTERIZATION

This section will be used to evaluate USERS of computer systems, and it is not intended for use in the inspection of software developers.

Refer to the FDA "Draft Guideline for the Validation of Blood Establishment Computer Systems." This document focuses on computer system definitions, testing, manuals, maintenance, security, training, audits, FDA references and reportable activities to the FDA. The final document will supersede the April 6, 1988, memorandum from CBER to all registered blood establishments "Recommendations for Implementation of Computerization in Blood Establishments," and the September 8, 1989, memorandum "Requirements for Computerization of Blood Establishments."

The draft guideline should be used as the main guidance document on computerized systems for registered blood establishments. However, the document is subject to change, as public comments are being collected and evaluated for incorporation into the final guideline. Specific regulations which are currently applicable are 21 CFR 606.60 (Equipment), and 21 CFR 211.68 (Automatic, mechanical, and electrical equipment) for additional information.

If a blood establishment is developing and distributing software that was originally intended for in-house use, report the number and identity of the user sites. Blood establishments developing and distributing software for use in manufacturing blood and blood components should be advised that they area device manufacturer and be encouraged to register and list with the Center for Devices and Radiological Health.

If the software is vendor supplied, identify the developer and version. Software developers typically specify the software by a version number. When major changes have occurred in the software, a new number is assigned. If a version different from the original is in use, the facility should have procedures to test the software prior to implementation and documentation that the testing was performed.

A "shared" computer system is a system used by both the blood bank and other sections of a clinical laboratory or the entire hospital. Security procedures should have been established regarding limiting access to confidential blood bank information, e.g., donor deferral records. In addition, access to blood bank data should be limited so that inadvertent or unauthorized changes in data do not occur.

When a decision is made regarding donor suitability and/or product quality based upon the use of and reliance on the data maintained in the computer system, the computer system is performing a critical function in the manufacture of blood and blood products. An investigator should determine how the computer system is used and relied upon in critical manufacturing steps beginning with the donation and continuing through product release.

A short description with respect to each of the areas controlled by the computer should be reported, e.g., the firm depends on the computer to check for permanent and temporary deferrals; updates for the deferral list are maintained directly by the computer which evaluates all test results received directly from the test equipment; all component processing is performed using bar coded labels; quarantine is computer controlled and if reactive test results are noted, an automatic flag is put on the unit by the computer so that the unit (components included) cannot be released for distribution.

TERMS AND ABBREVIATIONS

Allogeneic Donor -- A donor who donates a unit of blood to be placed in the general blood supply. This donor must meet all suitability requirements and be fully tested.

Anti-A, Anti-B Blood Grouping Reagent -- Reagents used to determine blood group: Anti-A serum, from a group B individual, will agglutinate or clump group A red cells; Anti-B serum, from a group A individual, will agglutinate or clump group B red cells.

Anti-D Blood Grouping Reagent -- Reagents used to determine Rho (D) factor. Serum from an Rh negative individual who has been exposed to the D antigen by transfusion or pregnancy and has formed the antibody to the D antigen, or monoclonal antibodies which are prepared by hybridoma techniques.

Antibody Screen-- Donor or patient serum is tested with reagent red cells of known antigenic makeup; the purpose is to determine if the donor or patient has antibodies in his or her serum.

Autologous Donor -- A donor who donates a unit of blood (predeposit) for his/her own use.

Biosafety Level 2 -- Centers for Disease Control and National Institutes of Health, Department of Health and Human Services, published a booklet entitled Biosafety in Microbiological and Biomedical Laboratories which recommends precautions which laboratory employees should follow. The booklet is available through DHHS publication No. (CDC) 88-8395, 94-95, 2nd Edition, Washington DC: US Government Printing Office, 1988. A copy of the chapter concerning biosafety level 2 precautions can also be obtained from I&SS, (301) 295-8191. In addition, the AABB Technical Manual summarizes Biosafety Level 2 precautions.

Cryoprecipitated AHF (Antihemophilic Factor) -- The cold-insoluble portion of plasma remaining after FFP has been thawed between 1° and 6°C. Cryoprecipitated AHF (Cryo) is used to treat patients with hemophilia A, von Willibrand's disease and hypofibrinogenemia. Cryo contains fibrinogen and Factor VIII, a procoagulant present in normal plasma but deficient in the plasma of patients with hemophilia A.

Directed Donor -- A donor who donates a unit of blood for a specific patient. These donors should meet all suitability requirements and be tested as allogeneic donors. Occasionally, a directed donation may not meet all suitability and testing requirements, in which case, the patient's physician may make a medical decision to use the directed donation.

D^u -- A variant or weak form of the D antigen; no Anti-D^u reagent exists, but cells are tested for the variant by an indirect antiglobulin method, or the equivalent, such as a special channel on the Kontron Groupamatic, Olympus PK700, or Gamma STS-M automated blood groupers.

ELISA Screening Test -- ELISA (also referred to as EIA) is an acronym for enzyme-linked immunosorbent assay. This assay utilizes the principle of a solid phase, e.g., beads or microtiter plate wells, coated with antigen or antibody and an indicator reagent, antibody or antigen, respectively, to which an enzyme has been conjugated or "linked."

A typical ELISA test such as that used for the detection of antibody to HIV utilizes beads or microtiter wells coated with disrupted, inactivated HIV agents and goat anti-human Ig conjugated or "linked" to an enzyme, which on incubation with the appropriate substance, will produce a color.

When an unknown serum or plasma sample is tested for the presence of antibodies, it is placed in the antigen coated wells, the antibody in the sample links to the antigen on the solid-phase carrier and is detected by the anti-human antibodies conjugated to the enzyme. Possible results include:

Initially reactive -- Initial EIA test is reactive.

Repeatedly reactive -- One or both duplicate EIA retests is/are reactive.

Negative -- Initial EIA test is non-reactive or if reactive, both repeat duplicate EIAs are non-reactive.

Positive -- Repeatedly reactive EIA test. Western Blot (WB) positive.

WB indeterminate -- Western blot neither positive nor negative.

EIA indeterminate -- Repeatedly reactive EIA test, Western blot negative or indeterminate.

Hematocrit -- The percentage of red blood cells present in the whole blood volume.

Hemoglobin -- The main component of the red blood cell -- an iron containing protein which serves as the vehicle for the transportation of oxygen and carbon dioxide.

Major Crossmatch -- Patient's serum tested with donor's red cells; the purpose is to determine if the patient has an antibody to an antigen found on the donor's cells.

Minor Crossmatch -- Donor's serum tested with patient's red cells; the purpose is to determine if the donor has an antibody to an antigen found on the patient's cells.

Plasma -- Separated from red blood cells within 25 days after phlebotomy (within 40 days after phlebotomy when CPDA-1 solution is used as the anticoagulant). Stored at -18°C or colder within six hours after separation (see instruction booklet). Dating period is five years. Platelets and/or Cryoprecipitated AHF may be removed from product.

1. Liquid Plasma: Separated from red blood cells within 26 days after phlebotomy (within 40 days after phlebotomy when CPDA-1 solution is used as the anticoagulant). Stored at 1-6°C for a total of 26 days (40 days when CPDA-1 is used).

2. Fresh Frozen Plasma: Collected with minimal damage to tissues. Separated from red blood cells and frozen solid within six hours. Some manufacturers have received approval for their collection systems to allow room temperature storage before plasma separation to be extended to eight hours (see instruction booklet). Stored at -18°C or colder for up to one year.

3. Platelet Rich Plasma: Collected with minimal damage to tissues. Separated from red blood cells within six hours after phlebotomy (see instruction booklet). Procedure must produce a product with at least 250,000 platelets/microliter. Stored at 1-6°C for 72 hours, or at 20-24°C with gentle agitation for five days in approved containers.

4. Recovered Plasma: Obtained from single units of expired or unexpired whole blood, outdated plasma, or as a by-product in blood component preparation. For further manufacturing use only. Must be able to relate plasma unit to donor. Donors must meet donor suitability requirements and each unit must be tested for infectious agents, as required.

Plasmapheresis -- A process in which red blood cells are separated from the plasma of a blood donor and returned to the donor's circulatory system.

Plateletpheresis -- The removal of platelets from a donor, followed by the return of the red blood cells and sometimes the plasma to the donor.

Specific gravity of whole blood -- 1.053 gm/ml for blood containing 12.5 gm/dl of hemoglobin. The following calculation is used to convert volume to weight: 1.053 gm/ml X 500 ml = 526.5 gm.

SOP -- Standard operating procedures.

STS -- Serological test for syphilis, e.g., VDRL, RPR, and the treponema-based hemagglutination test on the Olympus automated blood grouping instrument.

Temperature conversion -- °F = (°C x 9/5) + 32; °C = (°F - 32) x 5/9.

Viral Marker Tests -- RIA -- radioimmunoassay; RPHA -- reverse passive hemagglutination; RPLA -- reverse passive latex agglutination; and ELISA -- enzyme-linked immunosorbent assay.

Western blot Technique -- The term "Western blot" (WB) refers to a technique for identifying antibodies to specific viral proteins based on molecular weight. The technique, therefore, allows the identification of antibodies to specific proteins associated with the HIV. Each manufacturer of WB kits has criteria for evaluating the results as negative, indeterminant or positive.

Chapter 16

MEDICAL DEVICE INSPECTIONS

OVERVIEW

In 1976, the Medical Device Amendments to the Federal Food, Drug and Cosmetic Act gave the FDA specific authority to regulate medical devices. The agency received additional authority under the Safe Medical Devices Act (SMDA) of 1990.

The level of regulation depends on the class in which a medical device is placed -- Class I, II, or III -- with Class I devices receiving the lowest level of regulation and Class III receiving the greatest amount of regulation.

Validation concerns are generally associated with FDA investigations under the Good Manufacturing Practice (GMP) regulations, which encompass the methods, facilities and controls used in pre-production design, manufacturing, storing, packaging, and installing medical devices.

The documents which follow all relate to FDA's regulation and inspection of medical devices. These documents are:

1. FDA Investigations Operations Manual (IOM), Chapter 5 -- Establishment Inspections, Subchapter 550 -- DEVICES
2. FDA Compliance Program Guide 7382.830 -- INSPECTION OF MEDICAL DEVICE MANUFACTURERS
3. CDRH, Office of Compliance and Surveillance, "Inspections of Foreign Device Manufacturers"
4. CDRH, Office of Compliance and Surveillance, "Application of the Medical Device GMPs to Computerized Devices and Manufacturing Processes: Medical Device GMP Guidance for FDA Investigators"

FDA INVESTIGATIONS MANUAL
CHAPTER 5 -- ESTABLISHMENT INSPECTIONS
SUB CHAPTER 550 -- DEVICES

550.1 Authority

Section 704(a) of the FD&C Act provides the general inspectional authority to inspect medical device manufacturers. The Medical Device Amendments of 1976, provided additional authority to inspect records, files, papers, processes, controls, and facilities to determine whether restricted devices are adulterated or misbranded. The Amendments also provide FDA authority under section 704(e) of the Act to inspect and copy records required under section 519 or 520(g), which includes records maintained in compliance with the Good Manufacturing Practice Regulations.

The Radiation Control for Health & Safety Act of 1968 (RCHSA) (Public Law 90-602) and the regulations (starting at 21 CFR 1000) for it's enforcement specifically cover electronic products that emit radiation. This Act amends Part F of Title II of the Public Health Service Act and thus is a part of the PHS Act. The prohibited acts are listed in Section 360B and "Enforcement" is contained in Section 360C. This Act contains authority (Sec. 263i(b)) allowing FDA to inspect all documents required by the Act. In addition, all radiation emitting medical devices are subject to all of the medical device requirements.

The Safe Medical Device Act of 1990 amends and provides authority to FDA for regulating device. The Act provide for such things as: user reporting of deaths and serious injuries; reclassification of certain devices; provisions for mandatory reporting of recalls initiated by manufacturers and additional authority to order the recall of devices; civil penalties; incorporation of the PCHSA into the FD&C Act; and, device design validation requirements. It also contains specific time frames for writing regulations to implement provisions of the Act. The provisions to order the recall of devices, to notify users, to temporarily suspend premarket approval of a device and to impose civil penalties became effective immediately.

550.2 Definitions

The term "device" is defined in Sec. 201(h) of the FD&C Act. In-vitro diagnostics (21 CFR 809) are devices, as defined in 201(h) of the Act, and may also be biological products subject to Section 351 of the PHS Act.

Electronic product radiation is defined as "any ionizing or nonionizing electromagnetic or particulate radiation and any sonic, infrasonic, or ultrasonic wave emitted from an electronic product as a result of the operation of an electronic circuit in such product" and is defined in 21 CFR 1000.

551 GENERAL

Inspections involving devices should be made only by those individuals qualified by experience and training in the device area.

Because of the specific nature of inspections and investigations involving radiation, only personnel who have special training in this field should be assigned such work. However, others may participate for training purposes. Specific Compliance Programs designate the type of individual and special training required for work in these areas.

Caution: Radiation-emitting devices and substances present a unique hazard and risk potential. Every effort should be taken to prevent any undue exposure or contamination. Monitoring devices must be used whenever radiation exposure is possible. Investigators should also be on the alert for, and avoid contact with, manufacturing materials and hazards associated with the manufacturing of many types of devices, which may present a threat to health, e.g., ethylene oxide, high voltage, pathogenic biomaterials, etc. See IOM 140 for additional safety information.

General device inspections will be conducted under various Compliance Programs found in the Compliance Program Guidance Manual.

552 DEVICE REGISTRATION AND LISTING

Section 510 of the FD&C Act and 21 CFR 807 describe the establishment registration, device listing, and premarket notification requirements and specify conditions under which firms are exempt from these requirements.

Manufacturers of finished devices (including contract manufacturers and device specification developers), repackers and relabelers, and initial distributors of imported devices are required to register their establishments by submitting an FDA-2891. After initial submission, annual registration is accomplished by use of the Center for Devices & Radiological Health (CDRH) computer generated FDA-2891(a). Component manufacturers are not required to register if the components are sold to registered device establishments for assembly into finished devices. Registration and listing is required, however, if the component is labeled for a health care purpose and sold to medical or clinical users. Optical laboratories, clinical laboratories, dental laboratories, x-ray assemblers, orthotic and prosthetic appliance assemblers, hearing aid dispensers and others who, using previously manufactured devices, perform a service function for physicians, dentists, other licensed practitioners or their patients, are exempt from establishment registration. An exemption from registration does not exempt a firm from inspection under Section 704 of the FD&C Act.

Each domestic establishment (except contract manufacturers & contract sterilizers) that is required to register, and all foreign device manufacturers who export their devices into the United States, must list their devices. Device listing by manufacturers and repackers/relabelers is accomplished using an FDA-2892; the same form is used to update listing information. Initial distributors satisfy listing requirements by providing CDRH with the names and addresses of foreign device manufacturers whose products they distribute in U.S. commerce.

Firms are required to register and list, even if interstate commerce is not involved.

See IOM EXHIBIT 550-A for types of medical device operations that require registration and listing.

552.1 Premarket Notification - Section 510(k)

The Medical Device Amendments of 1976 require device manufacturers to notify the CDRH at least 90 days before commercially distributing a device. This is known as a "Premarket Notification" or a "510(k)" submission. "Commercial distribution" for practical purposes means that the device is held or offered for sale.

A manufacturer must submit a Premarket Notification to FDA in any of the following situations:

a. Introducing a device into commercial distribution for the first time.
b. Introducing a new device or product line for the first time which may already be marketed by another firm.
c. Introducing or reintroducing a device which is significantly changed or modified which could affect the safety or effectiveness of the device. Such changes or modifications could relate to design, material, chemical composition, energy source, manufacturing method, or intended use.

The above requirements do not apply to "custom devices." A "custom device" is a device made exclusively for and to meet the special needs of an individual physician or health professional or for use by an individual patient named in the order of a physician or dentist (such as specially designed orthopedic footwear); is not generally available in finished form for purchase; and is not offered through labeling or advertising for commercial distribution.

Refer to IOM EXHIBIT 550-A for types of medical devices that require 510(k) submissions.

The investigator should document for CDRH review failures to submit required 510(k)s.

552.2 Premarket Approval

Devices classified into Class III are required to undergo premarket approval in accordance with the provisions of Section 515 of the Act. Premarket approval for a device is initiated with the submission of an application to FDA. Prior to approval of a premarket approval application (PMA) or a supplemental PMA, FDA has the authority to inspect the applicant's facilities and those records pertinent to the PMA.

C.P. 7383.001 contains specific guidance on performing PMA pre-approval and post-approval inspections.

Inspections of manufacturing facilities are usually required prior to approval of a Premarket Approval Application. A full GMP inspection may not be necessary if there has been a recent satisfactory inspection covering a device similar to the PMA device.

Requests for PMA inspections issue from HFZ-331. The assignments will request that the firm be inspected for compliance with the GMP regulations, and with their commitments in the PMA.

553 CLASSIFICATION OF DEVICES

All medical devices subject to the FD&C Act will be classified into one of the following:

a. Class I.

This group includes devices for which general controls (i.e., the controls in Section 501, 502, 510, 516, 518, 519 and 520 of the FD&C Act) provide reasonable assurance of safety and effectiveness.

b. Class II.

This group, requiring Performance Standards, consists of devices for which the general controls alone are insufficient to provide reasonable assurance of safety and effectiveness, and for which there is sufficient information to establish performance standards that will provide such assurance. Such standards may be established under Section 514 of the FD&C Act. Until a standard is published for a Class II device it remains subject only to the general controls described for Class I devices.

c. Class III.

This group requires premarket approval. It includes devices which:

1. Cannot be placed into Class I because insufficient information exists to determine that the general controls alone are sufficient to provide assurance of safety and effectiveness, and cannot be placed into Class II because insufficient information exists for establishing performance standards, and
2. Are purported or represented to be for use in supporting or sustaining human life or in preventing impairment of human health, or otherwise present a potential unreasonable risk of illness or injury.

Unless they are determined substantially equivalent to devices distributed prior to the 1976 Medical Device Amendments, devices proposed for marketing after May 28, 1976 fall automatically into Class III. Class III medical devices marketed before May 28, 1976, and the substantially equivalent devices marketed after that date remain subject to the premarket notification requirements until required to have an approved PMA by a Final Order under Section 515(b) of the Act. Petitioners can request to have such devices reclassified into Class I or II. Transitional devices, those regulated as new drugs before May 28, 1976, are automatically assigned to Class III

Manufacturers who have questions regarding the classification of a device can write CDRH under Section 513(g) of the FD&C Act and request an opinion as to the status of the device.

554 INVESTIGATIONAL DEVICE EXEMPTION (IDE) REGULATION

The IDE regulation in 21 CFR 812 contains requirements for Sponsors, Institutional Review Boards (IRBs) and Clinical Investigators. Additional requirements are found in 21 CFR 50, Informed Consent, and 21 CFR 56, IRB's. All Sponsors of device clinical investigations must have an approved IDE unless specifically exempted by the regulation. Sponsors who have an approved IDE are exempt from requirements on labeling, registration and listing, premarket notification, performance standards, premarket approval, GMPs, banning of devices, restricted devices, and color additives.

Provisions for obtaining an IDE and the sections of the regulations with which sponsors, investigators, and IRBs must comply differ according to the risks posed by the device. Sponsors of nonsignificant risk devices must obtain IRB approval and are subject to a limited number of provisions; sponsors of significant risk (See 21 CFR 812.3(m).) investigations are subject to the entire regulation.

There are investigations, described in 21 CFR 812.2(c), that are exempt from the IDE regulation. Exempted investigations apply to devices and diagnostics which meet the criteria in the regulation. These devices are, however, still subject to other regulatory requirements of the Act, such as labeling, premarket approval of Class III devices, and GMPs (as stated in the preamble to the IDE regulation).

A Sponsor who knows a new device is not "substantially equivalent" to a preamendments device or who is not sure if a device is "substantially equivalent" without conducting a clinical investigation must obtain an approved IDE to conduct the clinical investigation. After collecting clinical data, a sponsor who desires to market a device must either submit a premarket notification (510k) or premarket approval application to FDA. A premarket notification may be submitted if the sponsor believes the data supports a finding of substantial equivalence.

Certain radiation-emitting electronic devices that are investigational are also subject to regulations enforced by FDA.

Transitional devices, must have an approved IDE in order to be investigated.

Sponsors, Monitors, IRBs, Investigators, and Non-Clinical Toxicological Laboratories will be covered under the Bioresearch Monitoring Program. FDA has the authority to inspect and copy records relating to investigations. Records identifying patients by name will be copied only if there is reason to believe that adequate informed consent was not obtained or that investigator records are incomplete, false, or misleading.

554.1 Devices Based on Unorthodox Principles

The utility of some devices may be based on new information or unexplainable principles, not available to the established scientific community. These devices may be encountered during routine surveillance or while doing a specific assignment. In these cases obtain the following in as much detail as possible:

a. The mode of operation of the device and the theory of it's effect upon the body, including any references made to, and (if available) copies of published literature.
b. Copies of the device labeling.

554.2 Banned Devices

Section 516 of the FD&C Act provides under General Controls that a device for human use may be banned by regulation if it presents substantial deception or an unreasonable and substantial risk of illness or injury. The banning normally becomes effective when a final regulation is published. However, in cases where the deception or risk of illness or injury presents an unreasonable, direct, and substantial danger to the health of individuals, the banning can be effective on the date of the publication of a proposed regulation and remain in force until the FDA makes a final decision on the matter. If labeling can be devised to correct the problem, the device will not be banned, unless the labeling is not corrected within the time specified by the FDA. See 21 CFR 895.

When the investigator determines that banned devices are being distributed, the distribution, manufacture, etc. should be documented as for any other violative product.

554.3 Detention

Section 304(g) of the FD&C Act provides for detention of devices as described in 21 CFR 800, Subpart C.

Procedures to be followed in the detention of devices are outlined in IOM 950.

555 MEDICAL DEVICE GOOD MANUFACTURING PRACTICES REGULATION (GMP)

Section 520(f) of the Act provides the Agency with authority to prescribe regulations requiring that the methods used in, and the facilities and controls used for, the manufacture, packing, storage, and installation of medical devices conform to good manufacturing practices. The medical device Good Manufacturing Practices Regulation (21 CFR 820) became effective on December 18, 1978.

21 CFR 820 is established and promulgated under the authority of Sections 501, 502, 518, 519, 520(f), and 701(a) of the FD&C Act (21 U.S.C. 351, 352, 360h, 360i, 360j(f), and 371(a)). Failure to comply with the provisions of 21 CFR 820 renders a device adulterated under Section 501(h) of the Act.

The regulations promulgated under 21 CFR 820 establish minimum requirements applicable to finished devices, as defined in 820.3(j). This regulation is not intended to apply to manufacturers of components or parts of finished devices, but is recommended to them as a guide. In some special cases, components have been classified as finished devices (dental resins, alloys, etc.) and are subject to the GMP. Manufacturers of human blood and blood components are not subject to this part, but are subject to 21 CFR 606.

The medical device GMP is an umbrella GMP that specifies general objectives rather than methods. It is left to the manufacturer to develop the best methods to meet these objectives. You must use good judgement in determining compliance with the GMP, keeping in mind that it is an umbrella GMP and all requirements may not apply or be necessary. The purpose of the GMP is to assure conformance to specifications; and, to ensure that all requirements that will contribute to assuring the finished device meets specifications are implemented. You should not insist that a manufacturer meet nonapplicable requirements. Refer to IOM EXHIBIT 550-A for types of establishments that are required to comply with the GMPs.

555.1 Critical Devices

The GMP utilizes a two-tier approach, which creates an individual category for devices that are surgically implanted or intended to support or sustain human life and whose failure, when used in accordance with instructions provided in the labeling, could reasonably be expected to result in significant injury or illness. This group of devices is designated as critical. All other devices are described as noncritical. General requirements not specifically identified as critical requirements apply to all devices. Critical devices must meet both the general requirements and those identified as specific to critical devices.

When identifying critical devices, FDA uses recommendations received from the Device GMP Advisory Committee and the device classification panels. The selection of critical devices is independent of the classification of devices into Class I, II, III. Only inspect as critical devices those identified by CDRH as such. An updated list appears as an attachment to the Compliance Program - Inspection of Medical Device Manufacturers.

Should you have a question regarding the critical status of a specific device, (i.e., you believe the device is critical and the firm is taking the position that it is not), and you do not believe the issue can be resolved by phone, a written inquiry should be submitted to the Division of Compliance Programs (HFZ-330), CDRH. Pertinent sections of the EIR and a sample of the device labeling should accompany the inquiry.

Guidance is also available from the Division of Compliance Programs in situations where a firm acknowledges that it manufactures a critical device but the critical process and/or components have not been identified.

When an assignment is made to conduct a GMP inspection of a new PMA device, CDRH will specify whether it is to be inspected as a critical or noncritical device.

555.2 Quality Assurance Program

The term "quality assurance" as specified in the GMP encompasses all activities necessary to assure the finished device meets its design specifications. This includes assuring that manufacturing processes are controlled and adequate for their intended use, documentation is controlled and maintained, equipment is calibrated, inspected, tested, etc. Some manufacturers may use the term "quality control" or "GMP Control" instead of quality assurance. It doesn't matter what term is used as long as the quality assurance concept is understood and implemented. Historically, "quality control" has meant inspection and test which, although the primary mechanisms for detecting defects, only sets aside nonconforming product and does not prevent the deficiency which caused the defect. Quality assurance activities are intended to prevent the production of non-conforming products and includes quality control activities. The GMP is based on quality assurance principles and was designed to prevent the production of nonconforming product. A manufacturer's implementation of the GMP is implementation of a quality assurance program.

All manufacturers of medical devices are expected to establish and implement a quality assurance program tailored to the device manufactured. Each manufacturer must prepare and implement all activities, including but not necessarily limited to the applicable requirements of the GMP, that are necessary to assure the finished device, the manufacturing process, and all related procedures conform to approved specifications. The quality assurance activities needed to achieve this may be more or less than those required by the GMP.

One aspect of a quality assurance program is that it identify, recommend or provide solutions for quality assurance problems and verify their implementation, as stated in 21 CFR 820.20(a)(3).

Trend analysis is a method of complying with this QA requirement. Process and product accept/reject data collected by the firm through their documented systems, along with the complaint system, can be used in identifying conditions or situations which might not be apparent, or may be dismissed as isolated incidents. Once identified, measures can then be implemented to control or eliminate their reoccurrence.

Investigators should not make general FDA 483 observations that a manufacturer does not have a quality assurance system. If an adequate response is expected from the manufacturer the charge must be more specific and point out the controls that are missing or believed inadequate.

555.3 Quality Audit

Medical device manufacturers subject to the GMP must conduct periodic quality audits. Quality audits should consist of a formal, planned check of all elements in the quality assurance program. They are not product audits. Such audits must be conducted using adequate detailed written procedures by appropriately trained individuals. If conducted properly, a quality audit can detect program defects and, through isolation of unsatisfactory trends and correction of factors that cause defective products, prevent the production of unsafe or nonconforming devices. Without an effective quality audit function the quality assurance program is incomplete and there is no assurance that the manufacturer is consistently in a state-of-control.

The inspectional approach for identifying inadequate auditing of a quality assurance program is limited by the agency's policy which prohibits access to audit results. The policy is stated in CPG 7151.02. Evidence of inadequate auditing may exist without gaining access to the written audit reports.

This evidence may be obtained by relating the audit program to deficiencies observed in complaint files, change control and/or calibration, or other problems in quality assurance systems. If significant quality assurance problems have existed both before and after the firm's last self audit, then you should critically review the written audit procedures. The audit procedures should cover each quality assurance area within the plant, and should be specific enough to enable the person conducting the audit to perform an adequate audit. The auditors must be adequately trained. If it is possible to interview an auditor, ask how the audits are performed; what documents are examined; how long audits take; etc.

Audits should be conducted by individuals not having direct responsibility for matters being audited. In one-person and very small firms, where hiring an outside auditor to meet this requirement would be impractical or overly burdensome, self-audit may be acceptable and the auditor need not be independent. Consult with CDRH, Manufacturing Quality Assurance Branch at (301) 427-1128 when necessary. If there are significant FDA-483 observations, and independent audits are being performed but deficiencies are apparently not being identified by the auditor, then an FDA-483 should contain an observation indicating a lack of adequate audits.

If possible, attempt to determine whether corrective action by upper management is being taken. Auditors may be asked if they observed any of the ongoing GMP deficiencies during their prior audits. (Ongoing GMP deficiencies may also be identified by reviewing prior FDA-483's). If the answer is yes, check the written audit schedule to determine if follow up audit is scheduled for the deficient areas. Also, check the written audit procedure for instructions for review of audits by upper management, and re-audit of deficient areas. A failure to implement follow-up corrective actions may be listed as a GMP deficiency on the FDA-483.

The GMP requires a manufacturer to certify in writing that audits have been conducted whenever requested to do so by an investigator. Investigators through their supervisors should consult with the CDRH (HFZ-332), at (301) 427-1128 prior to requesting such certification.

555.4 Control of Components (Critical and Non-Critical)

Every manufacturer must have a written procedure or protocol for accepting components which specifies how they are to be received, handled, held and examined for acceptance and rejection. One or more designated individual must be assigned to accept and reject components and a record of this activity must be maintained.

The device GMP contains no mandatory reserve sampling requirements, as in the drug GMP. A discretionary reserve sampling requirement is provided for critical components.

It is permissible for component suppliers to establish and implement the controls and activities required to meet the GMP component requirements. However, the finished device manufacturer has the ultimate responsibility for assuring components are acceptable for use. When device manufacturers rely on suppliers for component QA, the device manufacturers must have procedures that will assure contracts are properly reviewed, component requirements are clearly defined and the supplier has the capability to meet contractual requirements. Supplier capability may be assessed by the manufacturer through site visits, historical data, etc. Suppliers should be audited periodically as part of the finished device manufacturers quality audit. The inspections and tests to be performed and those responsible must be defined and documented.

555.5 Critical Components

Critical device manufacturers must evaluate their device and identify the components that meet the critical component definition, as provided in 820.3(e). Only a critical device may have critical components 820.3(e).

When an investigator chooses to challenge a manufacturer's selection or nonselection of critical components, documentation for CDRH evaluation must be collected to support this challenge.

The GMP requires that the percentage of critical component rejects per lot be recorded, as well as the percentage of lots rejected. If the manufacturer has the raw accept/reject data in a form that can be evaluated to determine supplier performance, and is periodically doing this evaluation, this would be acceptable in lieu of recorded percentages.

555.6 Production and Process Controls

a. Manufacturing Specifications and Processes

Verify that specifications and documented work instructions are provided for all processes in which variations could result in failure of the finished device to meet specifications. Typical process examples are molding, heat treatment, welding, sterilization, blending, package sealing, solvent bonding, etc. Drawings may often be used for assembly, fabrication, etc. and in some cases training and workmanship standards may suffice in lieu of written procedures e.g., hand soldering.

Verify that specification and procedure changes, and new specifications and procedures are reviewed and approved using a formal process. Changes in work instructions, drawings and other instructional procedures should also be made according to formal controls. Approval is often documented by using a standard form such as a Engineering Change Notice (ECN), Drawing Change Request (DCR), or other approval documents.

When written procedures are absent, determine if they are necessary before citing the absence as a deficiency. The need for written instructional procedures should be based on factors such as the complexity of the activity and operator training. Failure to carry out operations properly or consistently may be an indication that written instructions are necessary.

b. Manufacturing Specifications and Processes for Critical Devices

Determine if the critical device manufacturer has identified critical operations in the manufacturing sequence. The definition of a critical operation is provided in 820.3(g). If you believe any operations are critical, and the manufacturer has not identified them as such, collect documentation to support your position for CDRH evaluation.

Verify that critical operations are noted in the device history record with appropriate verification signatures.

c. Process Validation

While the GMP does not specifically require process validation the requirement is interpreted to be contained in Sections 820.100, 820.100(a)(1), 820.5, and 820.3(n). The FDA Guideline on General Principles of Process Validation, dated May 1987, offers guidance.

Generally, process validation is a preproduction activity. Prospective validation includes considerations made before a new product is introduced, or when there is a manufacturing process change which may affect the product's characteristics. The validation program must be planned and documented, and the validation results must be documented and maintained. Retrospective process validation may be used, if adequate, for products which may have been on the market without sufficient preproduction process validation. Extensive review of manufacturing and assembly process data, along with product testing, may be used as a type of validation for devices manufactured individually or on a one time basis.

Qualification/validation should be performed (as applicable) for processes such as sterilization processes (steam, dry heat, ETO, radiation, filtration, aseptic fill), manufacturing processes (lyophilization, molding, soldering, machining, blending/mixing, water purification systems), environmental control systems (clean rooms and laminar air flow units), test methodology, and packaging/labeling processes.

555.7 Reprocessing

Determine if devices or components are reprocessed or reworked during manufacture. Generally, routine reconditioning or repair prior to distribution is not considered reprocessing unless the activity would adversely affect the reliability, safety, or effectiveness of the device.

Routine replacement of defective parts would normally not be considered reprocessing, unless the structure of the supporting materials such as adhesives, epoxies, solders, etc., must be changed or modified. Replacing a defective plug-in circuit board would not normally be considered reprocessing. Replacement of integrated circuit chips where supporting material must be removed, replaced or modified would be considered reprocessing. Patching, regrinding, remelting, structure-strengthening, resterilization, reheating, etc., would also be considered reprocessing. If, in a reconditioning or repair process, the device is modified so that it does not conform to its original specifications, the activity then becomes manufacturing.

Where reprocessing is not a routine part of the process, a high reprocessing rate may indicate production problems such as inadequate training or procedures which should be evaluated and corrected.

555.71 Reprocessing Critical Devices

Verify that written reprocessing procedures are provided for guidance when critical devices are reprocessed. In many cases the reprocessing procedure may be the same as those used for routine processing.

In cases of routine reprocessing, determine if the manufacturer has assessed the effect of reprocessing on the finished device or component and if the evaluation is documented.

555.8 Records

FDA has clear authority under section 704(e) of the Act to inspect and copy records required under section 519 or 520(g). Investigators should only collect copies of documents as necessary to support observations or to satisfy assignments.

Manufacturers who have petitioned for and obtained exemption from the GMP are not exempted from FDA authority to review and copy complaints and records associated with investigation of device failures and complaints.

You may advise manufacturers that they may mark as confidential those records they deem proprietary to aid FDA in determining which information may be disclosed under Freedom of Information.

Records must be maintained for as long as necessary to facilitate evaluation of any report of adverse performance, but not less than two years from the date the device is released for distribution. Records required by the Radiation Control for Health and Safety Act must be maintained for five years.

It is permissible to retain records in photostatic copy form, providing the copies are true and accurate reproductions.

a. Device master record

A device master record may exist in many forms. For example;

1. One or more files or volumes, or,
2. A list referring to the location of all documentation required by the master record, reflecting the latest revisions, and signed and dated as having been checked for accuracy and approved, or,
3. Any combination thereof.

Ensure that all documentation required by the GMP is included or listed in the device master record and that there is a formal method for approving and making changes to this documentation, including procedures.

b. Critical device master record

In addition to the requirements for noncritical devices, critical device master records must contain labeling procedures, copies of all approved labeling, and information on critical components and their suppliers.

c. Device history record

Verify that history records representing individual devices or lots of devices exist for all finished devices manufactured. The history record should reflect that all operations, processes, etc., described in the master record have been accomplished. In addition, the history record is specifically required to contain:

1. date(s) of manufacturing

2. quantity manufactured
3. quantity released for distribution
4. control number(s) if required
5. a record of examination of labeling materials issued for devices, including the date and person performing the examination.

The GMP does not require control numbers or traceability for noncritical devices. Control numbers/traceability are, however, required for in vitro diagnostics (IVD's) subject to the labeling requirements of 21 CFR 809.10 and devices subject to the Radiation Control for Health and Safety Act. The GMP does require control numbers/traceability for critical devices. Verify that history records contain evidence that labeling was examined prior to actual use.

The GMP does not require that noncritical device history records be maintained at one location. For example, the history record for an operation may be maintained at that operation location in logbook form. However, history records for critical devices should be maintained in a form where they can be readily reviewed and signed prior to device distribution as required by 21 CFR 820.161. Section 820.20(a)(1) requires that production records be reviewed; however, it does not specify when. In most cases this would require that history records be maintained as one or more documents or files in a single location.

555.9 Complaint Files

Complaints are written or oral expressions of dissatisfaction with finished device identity, quality, durability, reliability, safety, effectiveness or performance. Routine requests for service would not normally be considered complaints. However, service requests should be reviewed to detect complaints, and as part of the trend analysis system, and to comply with 820.20(a)(3).

FDA has the authority to require a device firm to open its complaint files, and review and copy documents from the file.

Provisions in the FD&C Act pertaining to FDA review of records are:

a. For restricted devices the FD&C Act in Section 704(a) (2) provides, "***In the case of any factory, warehouse, establishment, or consulting laboratory in which prescription drugs or restricted devices are manufactured, processed, packed, or held, inspection shall extend to all things therein (including records, files, papers, processes, controls and facilities), bearing on whether prescription drugs or restricted devices are adulterated or misbranded within the meaning of this Act***." See FD&C Act Sec. 704 for a full explanation and for a list of the items, e.g., financial data, which are exempt from disclosure to FDA.
b. For all devices, including restricted devices, refer to Section 704(e) of the FD&C Act, which states "Every person required under Section 519 or 520(g) to maintain records and every person who is in charge or custody of such records, shall, upon request of an officer *** permit such officer *** to have access to and to copy and verify, such records."

c. Section 519 of the FD&C Act requires manufacturers, importers, or distributors of devices intended for human use to maintain such records, and provide information as the Secretary may by Regulation reasonably require.
d. Section 520(g) of the FD&C Act covers the establishment of exemptions for devices for investigational use and the records which must be maintained and open for inspection.

GMP requirements for complaint files are found in 21 CFR 820.198 and became effective on December 18, 1978. Additionally, the GMP regulation contains a provision that records maintained in compliance with the GMP must be available for review and copying by FDA (21 CFR 820.180). Complaint files are GMP required records, therefore, the manufacturer must make all complaints received on or after December 18, 1978 and the records of their investigation available for FDA review and copying.

Look for patterns or trends of similar complaints and complaints that appear related to possible GMP deficiencies. Investigate these complaints during the inspection to determine if deficiencies actually exist in the manufacturing process, or have been corrected. Where problems have been corrected determine if appropriate remedial action was successfully accomplished.

Complaints of injury, death or any hazard to safety must be maintained in a separate portion of the complaint file and must be investigated by the firm.

The GMP does not require manufacturers to maintain written complaint handling procedures. However, such procedures are usually necessary for assuring consistency in processing complaints. Where available, you should attempt to review complaint handling procedures.

When a firm determines that complaint handling will be conducted at a place other than the manufacturing site, copies of the record of investigation of complaints must be forwarded to and maintained at the actual manufacturing site.

Determine the identity and qualifications of those who review complaints. Ascertain the basis for determining significance of complaints and how follow-up is conducted. Determine if oral or telephone complaints are documented.

Some firms file complaints under other names, such as "Trade Inquiries," "Customer Contacts," "Service Requests" etc., while others make a distinction between physical/mechanical and medical complaints. Make sure all complaints are adequately covered and reported.

When a manufacturer claims to have received no complaints, determine if provisions have been made for the review and investigation of complaints when received. Determine who has been assigned the responsibility to evaluate them when and if they occur.

Specific instructions for investigating medical device complaints can be found in the Compliance Program - Medical Device Problem Reporting.

555.91 Medical Device Reporting

The Medical Device Reporting (MDR) regulation is one part of the Center's health protection program. It is a mandatory information reporting system that requires manufacturers and importers to report to FDA certain adverse experiences caused or contributed to by their devices. This program is administered by the Product Monitoring Branch, Division of Product Surveillance, of the Center's Office of Compliance and Surveillance. The regulation requires that a report be submitted to FDA whenever a manufacturer or an importer becomes aware of information that its device:

may have caused or contributed to a death or serious injury,

or

has malfunctioned and, if the malfunction recurs, is likely to cause or contribute to a death or serious injury.

The Office of Compliance guidance, "Medical Device Reporting Questions and Answers" should be referenced for further guidance about the MDR regulation. Inspections for compliance with the MDR regulation are conducted following the guidance contained in the MDR Compliance Program. When reviewing the manufacturers complaint files, look for complaints that are reportable and have not been reported by the manufacturer.

556 GMP EXEMPTION AND VARIANCES

Section 520f(2) (A) of the FD&C Act makes provisions whereby manufacturers, trade organizations, or other interested persons may petition for exemption or variance from all or part of the GMP which in their judgment are not appropriate to their manufacturing operations. Filing a petition does not defer compliance with the GMP requirements and petitions will not be processed while an investigation is ongoing, or while regulatory action is pending or ongoing.

Some Class I devices have been exempted from the GMP through the classification process. Each classification panel was required to consider the Class I devices reviewed by that panel and recommend if they should be exempt from the GMP. Devices exempted from the GMP by the classification process are published in classification regulations in the Federal Register.

Devices labeled or otherwise represented as sterile are not eligible for exemption from the GMP regulation. A sterile device is subject to all GMP requirements pertinent to assuring that sterility specifications are met and that the sterilization process is properly performed and adequately controlled.

No exemptions will be granted from 820.198 - Complaint Files, which requires the device manufacturer to have an adequate system for complaint investigation and follow-up, or from 820.180 - General Requirements, which gives authorized FDA employees access to complaint files, device related injury reports, and failure analysis records for review and copying. When FDA has granted a manufacturer an exemption from one or more GMP requirements, the manufacturer still has the responsibility to implement appropriate quality control measures to assure that the finished device has the quality it purports to possess, as stated in Section 501(c) of the FD&C Act. A manufacturer who has been granted a GMP exemption is subject to inspection under Section 704(a) of the FD&C Act, and may be subject to regulatory action if devices are adulterated or misbranded.

Whenever a question arises as to whether a device is exempt from the GMP, contact the Manufacturing Quality Assurance Branch at (301) 427-1128 for clarification.

556.1 Sterile Devices

Inspections of sterile device manufacturers are conducted under a specific Compliance Program.

556.2 Labeling

Specific labeling requirements for in vitro diagnostics (IVDs) are contained in 21 CFR 809.10. Part 809.10(a) contains specific labeling requirements for the individual IVD containers, and for the outer package labeling and/or kit labeling. Part 809.10(b) contains specific labeling requirements for the product insert, which must be included with all IVD products. These two sections also contain the requirements for: lot numbers, allowing traceability to components (for reagents) or subassemblies (for IVD instruments); stability studies for all forms of the product; an expiration date, or other indication to assure the product meets appropriate standards; and, the requirements for establishing accuracy, precision, specificity and sensitivity (as applicable).

Part 809.10(c) lists the labeling statements required for IVD's which are being sold for investigational and research use. Determine whether the firm is limiting the sale of IVD's, labeled as such, only to investigators or researchers. Document any questionable products, and submit to CDRH for review.

Warning and caution statements recommended for certain devices, along with certain restrictions for use, are described in 21 CFR 801. This same section also contains the general labeling regulations which apply to all medical devices.

557 SMALL BUSINESS REPRESENTATIVE AND DIVISION OF SMALL MANUFACTURERS ASSISTANCE

Small Business Representatives (SBRs) are located in various Regions. Their primary function is to assist small manufacturers in the interpretation of regulations and other regulatory problems. CDRH has a Division of Small Manufacturers Assistance (DSMA) which operates as a nationwide resource, as mandated by Section 10 of the Medical Device Amendments of 1976. DMSA's function is to provide technical and other nonfinancial assistance to small manufacturers of medical devices to help them comply with the FD&C Act.

When establishments have questions regarding the Act, Amendments of 1976, or a specific device and it is determined they are in a state-of-control and no regulatory action is pending, an investigator should refer these firms to DSMA, and/or to SBRs in regions that have them. DSMA can be reached at (301) 443-6597. A toll free number is available and is 1-800-638-2041. However, where a regulatory action is pending, or an inspection indicates an establishment may be in violation, no such comments should be made until you have discussed them with your supervisor.

557.1 Governments-Wide Quality Assurance Program (GWQAP) Inspections

Inspections done under the GWQAP are done upon request by CDRH (HFZ-331) through ORO (HFC-120). Each assignment is specific and may involve more than one compliance program. These inspections must be completed no later than 6 days from the date of the request from HFC-120. Specific questions arising during or a result of these inspections should be directed to HFC-120 who will relay the request to HFZ-331.

557.2 Contract Facilities

Many device manufacturers may employ the services of outside laboratories, sterilization facilities, or other manufacturers (i.e., injection molders, packagers, etc). In such cases, the finished device manufacturer is responsible for assuring that these contractors comply with the GMP and that the product or service provided is adequate. These contractors may also be subject to FDA inspection and the GMP regulation.

Determine how a manufacturer assures the capability and performance of contractors. In many cases contractors should be audited as part of the finished device manufacturer's quality audit program.

557.3 Small Manufacturers

When inspecting one-person or very small manufacturers for compliance with the GMP master record and written procedure requirements, the investigator should realize that written assembly, process, and other instructional procedures required for larger firms may not be needed. In a small firm, division of work is at a minimum, with one person often assembling and testing the finished device. In many cases, blueprints or engineering drawings would be adequate in lieu of procedures. In assessing the need for written procedures, the investigator should make judgments based on training and experience of the individuals doing the work and the complexity of the manufacturing process. However, this does not mean that small manufacturers have any less responsibility for assuring safe and effective devices are produced. Small manufacturers usually do not need the complex documentation programs typically maintained by larger firms to assure uniformity because fewer people are involved in production.

558 TECHNICAL ASSISTANCE

Each region and some districts have engineers and radiological health personnel who are available for technical assistance and consultation. Do not hesitate to avail yourself of their services.

The services of engineers, quality assurance specialists, and expert investigators in ORA/ORO/Division of Field Investigations (HFC-130) are available for telephone or on-site consultation and assistance in problem areas.

The CDRH, Manufacturing Quality Assurance Branch (MQAB), is available at (301) 427-1128 to answer questions regarding GMP interpretation and application, and on-site consultation and assistance as requested.

WEAC has various personnel (biomedical, sterility, electronic, materials, mechanical, nuclear and plastics engineers) available for telephone consultation and on-site assistance. They can be reached at FTS 8-839-8700 or commercially at (617) 729-5700.

559 SAMPLE COLLECTION

Because of the limited funds available for samples and the relatively high cost of device samples, it is essential that you consider, in consultation with your supervisor, the following factors before collecting a physical sample of a device:

a. If follow-up to a GMP deviation, will sampling demonstrate the deviation and/or a defective product. Documentary Samples may be more suitable for GMP purposes.
b. Likelihood of the analysis showing that the device is unfit for its intended use.
c. Samples costing over $250.00.
d. Laboratory capability to analyze the sample. See IOM 454.21e for sample routing information.

If you are still uncertain, discuss with your supervisor and contact Lillian Gill at the CDRH Laboratory at FTS 443-2444 or (301) 443-2444. In an effort to consolidate resources, certain samples, especially those of a complex nature, will be analyzed by the Center. Contact CDRH for assistance as follows:

a. In-vitro Diagnostic Devices.

   Office of Science & Technology
   Beth Green (HFZ-113) (301) 443-4052
   Lillian Gill (HFZ-100) (301) 443-2444

b. General Medical Devices.

   Office of Science & Technology (HFZ-100)
   Lillian Gill (HFZ-100) (301) 443-2444

NOTE: Device samples do not require 702(b) portions. Include in the FDA-525 and with the C/R if destined for different locations, a copy of the firm's finished device specifications, test methods, and acceptance and/or rejection criteria.

FDA COMPLIANCE POLICY GUIDE 7382.830 -- INSPECTION OF MEDICAL DEVICE MANUFACTURERS

PROGRAM: 7382.830
SUBJECT: Inspection of Medical Device Manufacturers
IMPLEMENTATION DATE: Upon Receipt
COMPLETION DATE: September 30, 1995
DATA REPORTING:
PRODUCT CODES: 73-91
PROGRAM/ASSIGNMENT CODES: 82830L 42830L -- All Routine Inspections
82830C 42830C -- All F/U Inspections
82011 -- Report Time Spent on Assessment of Firm's MDR Practices
DATE OF ISSUANCE: 5/20/94

Field Reporting Requirements

Send only an EIR to CDRH if the inspection resulted in the issuance of a direct reference Warning Letter to the firm, or is part of a recommendation for other administrative/regulatory action, unless the district wishes to obtain comment from CDRH. This policy does not relieve the district from COMSTAT reporting requirements.

All EIRs and administrative/regulatory action recommendations shall be sent to HFZ-300.

As soon as the district becomes aware of any information which might effect the agency's new product approval decision with respect to a firm, the district should immediately notify HFC-240, Division of Medical Products Quality Assurance, via EMS or FAX. They will convey the information by FAX or equivalent expeditious means to the appropriate Center regulatory units.

PART I -- BACKGROUND

- **This compliance program provides guidance to the FDA field staff for the enforcement of the requirements of the GMP regulation (21 CFR 820), the MDR regulation (21 CFR 803) and the Medical Device Tracking regulation (21 CFR Part 821).**

A. THE GMP REGULATION

This FY 1994 compliance program provides guidance for a new GMP inspectional strategy designed to compensate, in part, for the increasing strain on field resources while, at the same time, providing for rapid regulatory action when required. The strategy also places additional emphasis on manufacturers' responsibility to monitor their compliance with the requirements of the GMP regulation, and to make appropriate and timely corrections of problems in their manufacturing and quality assurance systems.

B. THE MDR REGULATION

The Medical Device Reporting (MDR) regulation became effective on December 13, 1984. This rule requires manufacturers and importers (initial distributors) of medical devices, including in-vitro diagnostic devices, to report to FDA whenever the manufacturer or importer receives or otherwise becomes aware of information that reasonably suggests that one of its marketed devices:

1) may have caused or contributed to a death or serous injury or,
2) has malfunctioned and that the device or any other device marketed by the manufacturer or importer would be likely to cause or contribute to a death or serous injury if the malfunction were to recur.

Every time a GMP inspection is conducted under this Compliance Program, an MDR inspection will also be conducted.

C. THE MEDICAL DEVICE TRACKING REGULATION

Congress recently amended the Federal Food, Drug and Cosmetic (FD&C) Act to add new requirements and provisions concerning the regulation of medical devices. These requirements are described in the Safe Medical Devices Act (SMDA) of 1990, signed into law by President Bush on November 28, 1990. One of those provisions was the Medical Device Tracking Requirements promulgated as 21 CFR 821.

The Tracking regulation requires manufacturers to implement a method of tracking permanently implanted or life sustaining/supporting devices used outside a device user facility, the failure of which would be reasonably likely to have serious adverse health consequences.

This regulation is intended to ensure that tracked devices can be traced from the device manufacturing facility to the patient.

PART II -- IMPLEMENTATION

A. OBJECTIVE

GMP's

1. To identify domestic and foreign manufacturers who are not operating in a state-of-control. To bring such manufacturers into a state-of-control through voluntary, administrative or regulatory means, as appropriate.

MDR

2. To identify manufacturers and importers who are not in compliance with the Medical Device Reporting regulation. To bring such firms into compliance through voluntary, administrative or regulatory means, as appropriate.

Medical Device Tracking

3. To determine if manufacturers are making good faith efforts to comply with the Medical Device Tracking regulation. To assure industry awareness of this requirement and encourage compliance as a prerequisite to a phased enforcement policy assuring compliance.

B. PROGRAM MANAGEMENT INSTRUCTIONS

1. The following guidelines are suggested for implementing this compliance program:

   a. This compliance program is to be used to conduct Compliance Status Information System (COMSTAT) inspections (formerly GWQAP inspections) of devices when directed by HFC-240. This is in accordance with the current COMSTAT Manual and to obtain data for COMSTAT profiles and/or updates during regularly scheduled GMP inspections.
   b. If the device is labeled as sterile, also use Compliance Program Circular 7382.830A to inspect the sterilization process.
   c. If the establishment is a contract sterilizer (see 7382.830B for the definition of a contract sterilizer), it is subject to applicable requirements of the GMP regulation and should be covered using this compliance program as well as 7382.830B.
   d. Some manufacturers produce their own devices labeled as sterile and act as a contract sterilizer for other manufacturers. Such manufacturers should be covered under 7382.830A and 7382.830B as well as this compliance program.
      NOTE: A device which is subject to a process designed to reduce its microbial load, but which is NOT labeled as sterile, is to be covered under 7382.830, not under 7382.830A/B.
   e. Medical Devices related to blood banking and/or human blood processing will be inspected under this Compliance Program. For guidance, see: Working Relationships Agreement Among the Bureaus of Medical Devices (BMD), Radiological Health (BRH), and Biologics (BOB), April 1, 1982.

2. Intensified Review of the Complaint File

   Part III and **Attachments B** and E of this Compliance Program contain special guidance for reviewing manufacturers' product experience reports to determine compliance with the GMP/MDR requirements for handling complaints.

When a location other than the manufacturing facility or importer is responsible for investigating complaints and submitting MDR reports, the home district should forward a copy of the EIR to the district office where the complaints are handled, with a request for additional follow-up. That district should send an "FYI" copy of its complaint file review to the home district. (Both districts should also follow the reporting requirements shown on the cover page of this program).

3. <u>Scheduling Biennial Inspections of Medical Device Manufacturers</u>

   a. <u>Priorities for GMP Inspections</u>

   In order to assure the best use of resources, and to assure that manufacturers of devices which present a greater risk to the public are inspected before those which pose a lesser risk, the following manufacturers should, for scheduling purposes, be given top priority:

   (1) Manufacturers whose last GMP inspection was violative and there is no compelling evidence of correction.
   (2) Manufacturers who received a 510(k) approval decision for a critical/significant risk device within the past year, and have not been inspected within the last two years for processes similar to those used to manufacture the 510(k) device.
   (3) Manufacturers of all other Class II or III devices that have never received a GMP inspection.
   (4) Manufacturers of all other critical/significant risk devices. Within this group assign the highest inspectional priority to those establishments which have gone the longest without a GMP inspection.

Note: Inspections of manufacturers of devices with a pending PMA approval will be assigned under the PMA Compliance Program (7383.001).

Inspections of Manufacturers that have submitted 510(k)s for preamendment Class III devices will be assigned under Compliance Program 7383.003

All other manufacturers should be inspected as resources permit. The primary goal of emphasizing inspection of the above device manufacturers is to change the scheduling of inspections from one that is purely based on the interval since the last inspection to one that also considers the health hazard significance of the device. Conducting the inspection shortly after a 510(k) has received approval will also allow an evaluation of manufacturers of significant devices at the most critical stage of production. Because most manufacturing and design problems develop or become apparent within the first year of the device's life cycle, inspecting at this time should provide a better opportunity for identifying manufacturing and design problems. GMP inspectional coverage will be focused on that segment of the industry that is actively bringing devices to market and thus presenting the most risk to the public. Those firms that may not receive a biennial inspection should be those producing lower risk products.

b. GMP Pre-Clearance Inspection Program for Class III 510(k) Pre-Amendments Devices (CP 7383.003)

When top-priority inspectional assignments that support this program cover all profile classes (except those associated exclusively with certain Class I devices) the district may count the inspection as a qualifying GMP inspection.

c. Initial Inspections

Newly registered and listed firms should receive a directed inspection per Section I of Attachment A as soon as possible after manufacturing operations commence. Generally, firms that manufacture Class III devices and critical/significant risk devices should be inspected within 6 months and firms that manufacture all other Class II devices within 12 months. If the device(s) classification is not known in advance and cannot be determined otherwise, i.e., phone contact, catalog review, etc., schedule the inspection and determine the appropriate inspectional approach after identifying the device(s). For guidance in determining if an establishment should be subject to the GMP regulation refer to page 50 of Medical Device GMP Guidance for FDA Investigators.

If it cannot be determined that at least one device is Class II, III, or Non-GMP exempt Class I, as discussed in section B.3.f. below, review the firm's complaint handling practices; then terminate the inspection. Charge the time to PAC 82R800 (District Initiated Assignment).

d. Routine Reinspections

All manufacturers of Class II or III devices should receive a directed GMP/MDR inspection with 24 months of the previous "qualifying inspection". See part II, B.3.a [Priorities for GMP Inspections.]

e. Statutory Coverage List (formerly the Alert List)

Any registered firm that manufactures Class II or III devices and has not had a "qualifying inspection" during the 24 months since they registered will appear on the district's Statutory Coverage List (formerly the alert list).

- The Statutory Coverage List (formerly the Alert List) will be based on the date of the last "qualifying inspection" (i.e., the last GMP/MDR inspection under PAC's 82830 C, L, or F, 83001, 83003, or 42830 C, L, or F.

f. Class I Device Manufacturers

NOTE: All Class I devices, including those exempted from most of the GMP requirements, must comply with the complaint file requirements of the GMP regulation as well as the reporting requirements of the MDR regulation. Class I manufacturers should receive lowest inspectional priority unless addressed by a special assignment or a health hazard is apparent. See Attachment C for those Class I devices which are exempt from most GMP requirements.

g. Follow-up Inspections

Part III of this program instructs investigators to discontinue the inspection when they encounter conditions that meet the criteria for Situation I in Part V.A.1. The Warning Letter to the manufacturer warns the manufacturer of its responsibility for reviewing all manufacturing and quality assurance systems. Because other problems may have existed which the manufacturer should have identified and corrected, the follow-up inspection should be a comprehensive inspection.

Follow-up inspections conducted to determine if deviations have been corrected may be counted as qualifying inspections and should be reported against either PAC 8283OC. All other follow-up inspections, including washouts, are to be reported against PAC 82R800.

4. Scheduling Inspections of Importers (Initial Distributors)

MDR Compliance inspections of importers (initial distributors) will be under separate assignment.

5. Resource Instructions

When possible, Electro-Optical Specialists should be used for inspections of laser devices.

Experienced and knowledgeable investigators should conduct inspections of establishments that are manufacturing critical devices. Contact DFI/International and Technical Operations Branch (HFC-133) should the need for expertise, not available in the Region, become apparent (Refer to FMD No. 142).

PART III -- INSPECTIONAL

BACKGROUND

This program includes guidance for determining compliance with the Good Manufacturing Practices (GMP), Medical Device Reporting (MDR) and Medical Device Tracking regulations.

A. OPERATIONS

Inspectional Strategy

A "qualifying inspection" is a GMP inspection conducted under this program as per the inspectional strategy presented in Attachment A.

With this FY'94 compliance program, CDRH is initiating a major change in inspectional strategy. Investigators will conduct a comprehensive inspection only when conducting a follow-up inspection following enforcement action. All other inspections will be directed inspections as per Section I in Attachment A.

When conducting all routine GMP inspections you are required to start the inspection with a review of: (1) complaints and MDR reports (see Attachment A, Section I(B)), (2) changes which the manufacturer has made in the design or manufacturing process, and (3) records of production lots which failed in-process or finished device testing. Any indications of problems that your review identifies will provide a focus for your inspection. If you do not find indications of problems after reviewing the above records, complete the inspection as directed in Section I in Attachment A and issue a FDA 483, listing any objectionable conditions that you have observed.

WHEN THE INSPECTION IDENTIFIES SYSTEM WIDE DEFICIENCIES WHICH, IN TOTAL, MEET THE CRITERIA FOR SITUATION I IN PART V.A.1. OF THIS PROGRAM, DOCUMENT THE CONDITIONS THAT CONTRIBUTED TO THE PROBLEM(S), AND CLOSEOUT THE INSPECTION.

The FDA 483 should contain the following statement:

THE OBSERVATIONS NOTED IN THIS FDA 483 ARE NOT AN EXHAUSTIVE LISTING OF OBJECTIONABLE CONDITIONS. UNDER THE LAW, YOUR FIRM IS RESPONSIBLE FOR CONDUCTING INTERNAL SELF-AUDITS TO IDENTIFY AND CORRECT ANY AND ALL VIOLATIONS OF THE GMP REGULATION.

1. Initial Inspections

   Follow the instructions in Section I of Attachment A only. Consider the guidance in A.3 through A.8 below.

2. Routine Inspections

   You will be required to tailor your inspection to the particular situation presented by each manufacturer. For all routine inspections follow only the instructions in Section I of Attachment A. It will also be necessary to consider the guidance in A.3 through A.8 below.

3. Important Points to Remember for an Inspection

a. Cover all of the profile classes (except those associated exclusively with certain GMP exempt Class I devices -- see Part II, B.3.f.). If your district has not determined all of the major profile classes the establishment is involved in, consult with the COMSTAT coordinator to obtain this information.

b. In determining whether a firm's systems are adequate and appropriate, consider applicable factors such as:

-- size of manufacturer;
-- the devices and their intended use; and
-- whether the total manufacturing and quality assurance program fulfills the intent of the GMP regulation.

c. Device manufacturers subject to existing FDA performance standards such as 21 CFR 1020, 21 CFR 1040, 21 CFR 1050 should include in their device master and history records those procedures and records demonstrating compliance with the applicable standard.

d. Special Instructions for Sterilization Process

A device subjected to a process designed to reduce its microbial load, but which is NOT labeled as sterile, is to be covered only under this program.
If the device is labeled as sterile, inspectional coverage indicated in 7382.830A and B, where appropriate, is to occur.

e. Special Instructions for Inspecting Small Manufacturers

Refer to Section I(I) of Attachment A, or Page 12 of the book entitled **"A Pocket Guide to Device GMP Inspections."**

f. Inspection of Radiation Emitting Devices

When conducting GMP inspections of radiation emitting devices, also cover the requirements of the applicable standard promulgated under PL90-602.
If the device is a sunlamp product as defined in 21 CFR 1040.20, also use Compliance Program Circular 7382.830X.

g. Recalls

Under the provisions of the Safe Medical Device Act manufacturers must now report all recalls to FDA. Confirm that all recalls conducted by the firm since the last inspection have, in fact, been reported to the district office.

4. Selection of Device(s) for Inspection

See Part II, B.3.a. for information on GMP Inspection Priorities.

The selection of the device or devices to be covered should be based first on evidence of defective and/or nonconforming devices identified by your review of the complaint files, change control records, and finished device testing records. The selection also will depend on the total number of appropriate profile classes used in the establishment. This can best be accomplished at the beginning of your inspection by reviewing current product catalogs and/or product listings. Cover all profile classes (except those associated exclusively with GMP exempt Class I devices (see Part II, B.3.f.) by examining the manufacturing of as few device lines representative of those classes as possible.

5. Critical Devices

Investigators should refer to Attachment A or the booklet entitled, **"A Pocket Guide to Device GMP Inspections"** regarding questions on the proper application of the critical device requirements. Questions relevant to the critical status of a device should be directed to the contact person designated in Part VI of this program.

6. FDA Compliance Status Information System (COMSTAT) (formerly known as GWQAP)

   a. When selecting specific devices to represent profile classes, give preference to critical devices and devices which have had problems. Where possible, select those devices which represent multiple, mutually exclusive profile classes. A list of the device related profile classes appears in the current FDA COMSTAT Manual (formerly the GWQAP Manual).

   b. Inspections conducted under a COMSTAT assignment should include:

      (1) coverage of the device(s) specified in the assignment, or devices representing all the same profile classes as the assigned device; and,
      (2) other devices as required to provide coverage of any remaining profile classes, except GMP exempt Class I devices (See Part II, B.3.f.)

7. Complaint File Review

   a. **GMP Complaint Handling Requirements**

   FDA surveys of firms' complaint files have shown that some firms were deficient in their complaint handling practices. These deficiencies were caused by a firm's failure to:

   (1) follow its own procedures for processing product experience reports; or,

(2) develop procedures which meet the requirements of 21 CFR 820.198, .162, .20 and .5, and
(3) firms identify problems but do not take timely corrective actions.

The GMP regulation requires that all complaints be reviewed, evaluated and maintained by a formally designated unit. This unit could be one appropriately trained individual, or a department which is staffed with appropriately trained individuals. This unit must decide whether an investigation of the complaint needs to be performed. There is NO requirement in the GMP regulation that all complaints must be maintained in one file, or that there be a written procedure for processing complaints. Typically, manufacturers will keep complaints in a customer sales file, product returns/credits file, warranty file, medical file, or legal file. The inspection should ascertain what files are maintained that meet the definition of a complaint, as found in 21 CFR 820.198(a).
[Note: If GMP defined complaints are not maintained by the formally designated unit, it should be noted on the FDA 483.]

By placing these complaints in different files, manufacturers have not noted instances of repeated component/device failure with a common cause. Ask the firm if it trends complaints. If no trending or problem identification is done, then the inspection should begin with the investigator conducting a trend of the complaints.

Evaluate the firm's complaint handling practices for the devices you are inspecting since the last GMP/MDR inspection. Refer to Attachments A, B, and E for guidance.
Note: The actual complaints or deficiencies in complaint handling practices may provide leads in identifying product defects which have not been adequately corrected by the firm, either by recall, and/or changes in the device and/or manufacturing process.

Reference #4 in Part VI explains how the GMP complaint files relate to reports required under MDR.

b. **Operational Procedure for Determining Compliance with the MDR Regulation.**

The guidance formerly included in C.P. 7382.011, Enforcement of the Medical Device Reporting (MDR) Regulation, is now part of this compliance program. See Attachments A and E for detailed directions.

8. Medical Device Tracking

Determine if the firm makes any medical device(s) subject to the Medical Device Tracking regulation. See **Attachment F** for a list of subject devices. If so, follow the guidance in Attachment A, Section I G.3 to determine if the firm has device tracking procedures in place and, if it does, to assess their adequacy.

If there is a complete absence of a tracking system, you should cite this violation on the FDA 483. Other violations should be addressed verbally and reported in the EIR.

9. Sample Collection

For GMP or MDR violations, documentary samples will be collected as necessary.

Physical samples are not required to support GMP violations, and should not be routinely collected for GMP cases. If the district should reference violative documentary or physical samples as evidence to support GMP deviations, the condition of the samples should be tied to the GMP deviation to show a cause/effect relationship.

If you are uncertain as to whether a sample should be collected, the district should consult with the CDRH Headquarters Laboratory Liaisons or the division of field science in ORA on the laboratory capability to conduct the analysis (See Part VI, C. Program Contacts).

10. **Imports**

No import wharf examinations or sample collections are scheduled under this program.

11. **Exports**

Devices that are manufactured in the U.S., but not marketed in the U.S., are not subject to the GMP requirements, provided that the manufacturer has documented proof that its devices are offered for sale only in foreign countries.

Devices that are manufactured in the U.S., but not marketed in the U.S. are subject to MDR including those exported under the provisions of Section 801(e) of the Act. This includes devices exported for investigational use.

12. Follow-up Inspections

Unless the previous violative inspection was a comprehensive inspection, the Situation I violations that were identified may have been part of more widespread systems problems. After receiving the Warning Letter the manufacturer should have investigated its manufacturing and quality assurance systems and initiated appropriate corrections. To assure that the manufacturer has fulfilled its responsibility, the follow-up inspection should be conducted as a comprehensive inspection using both sections I and II of Attachment A. If problems similar to those originally identified, or new problems which meet the criteria for Situation I are identified, complete the comprehensive inspection and document all observations.

13. <u>Foreign Inspections</u>

Use FDA/CDRH document entitled, <u>"Inspections of Foreign Device Manufacturers"</u> for guidance in conducting foreign device inspections.

B. <u>REPORTING</u>

1. General Reporting requirements are listed on the cover page. As a general rule the time used for preparing the EIR should not exceed the time spent conducting the inspection.
2. <u>MDR Observations</u> -- If the establishment failed to comply with any of the MDR requirements, include those observations on the FDA 483.
3. <u>GMP Observations</u> -- If you observe any violations of the GMP requirements, you should place them on the FDA 483. The most serious violations (e.g., those that could potentially result in production of defective devices, or identification of production/design problems) should be noted on the FDA 483 first. Your FDA 483 comment, however, should differentiate between problems which are indicative of a <u>system</u> failure and rare isolated situations.
4. <u>Critical Devices</u> -- The EIR must indicate whether the device being covered is a critical device or non-critical device. Please place this information in the Summary of Findings.
5. <u>510(k) Observations</u> -- If the firm failed to have a 510(k), or made significant changes which requires a new 510(k), <u>do not place them on the FDA 483</u> unless you obtain concurrence from CDRH/OC.
6. <u>Medical Device Tracking Requirements</u> -- The EIR must indicate whether or not the establishment inspected:

   a. makes any devices subject to the "Medical Device Tracking Requirements" and, if so, whether the firm is meeting its tracking obligations. These new requirements have an implementation date of 8/29/93. See **Attachment F** for questions and answers about the Medical Device Device Tracking regulation. See the April 1994 Edition of 21 CFR, Part 821 for a corrected copy of the medical device tracking regulation.

   If there is a complete absence of a tracking system, you should cite this violation on the FDA 483. Other violations should be addressed verbally and reported in the EIR.

<u>PART IV -- ANALYTICAL</u>

A. <u>ANALYZING LABORATORIES</u>

The district will make all the necessary arrangements for proper handling of samples with the following designated testing facilities:

TYPES OF DEVICES	ANALYZING LABORATORIES
All General Medical Devices	WEAC
Radioimmunoassay Devices	PHI
All other In Vitro Diagnostic Devices	BLT
Sterility Analysis, Bioburden Bioindicators, and Endotoxins	Refer to CP 7382.830A

See PART VI regarding those persons designated as contacts for designated laboratories and specific products.

B. ANALYSES TO BE CONDUCTED

Sample collection and analysis will be determined on a case-by-case basis through consideration of inspectional findings, compliance and scientific capabilities and expertise. Full collaboration between investigations and analytical personnel is essential at this phase. See Part III A.9 for additional information.

PART V -- REGULATORY/ADMINISTRATIVE FOLLOW-UP

A. REGULATORY/ADMINISTRATIVE FOLLOW-UP

1. SITUATION 1

The district has evidence indicating that the manufacturing process is producing nonconforming and/or defective finished devices. Such evidence would include information from sample analyses, complain files, the firm's failure analyses, MDR/PRP reports, or GMP observations made during the inspection.

OR

The inspection documents GMP deviations of a significant type or quantity to conclude that there is a reasonable probability -- in light of the relationship between GMP deviations observed, and the particular product and manufacturing process involved -- that the firm will likely produce nonconforming and/or defective finished devices. Such deviations include one or more of the following:

-- failure to establish and document a formal quality assurance program;
-- failure to validate significant manufacturing processes and quality assurance tests (refer to document titled "Guidelines on the General Principles of Process Validation" for the validation requirements);

-- failure to document, review, approve, implement and validate changes to components, finished devices, labeling, packaging or manufacturing process specifications;
-- failure to establish and implement adequate recordkeeping procedures (e.g., device history record, device master record);
-- failure to establish and implement an adequate complaint handling program;
-- failure to establish and implement an adequate failure investigation program; and,
-- failure to ensure that finished devices meet all specifications prior to distribution.

If any of these deviations exist, and the significance of the deviation and the device warrants it, the district should consider administrative and/or regulatory action, e.g., Warning Letter, injunction, administrative detention, seizure, civil penalty, and/or prosecution. The district is expected to classify the EIR as OAI.

If any of these deviations exist for foreign manufacturers, and the device warrants it, regulatory action should be considered, i.e., Warning Letter and/or Warning Letter with automatic detention.

If a serious health hazard is identified, FDA initiated recall and/or administrative detention/seizure should be considered as the initial action to bring the situation under prompt control.

2. SITUATION 2

The inspection documents GMP deviations of a quantity or type to conclude that there is minimal probability -- in light of the relationship between GMP deviations observed and the particular product and manufacturing processes involved -- that the firm will produce nonconforming and/or defective finished devices. The Form FDA 483, Inspectional Observations, will serve to inform the firm of any objectional findings.

Firms receiving an FDA 483 under Situation 2 should receive follow-up coverage as district resources permit. However, the continued presence of GMP deviations which have a low probability of leading to an unsafe or ineffective device will not usually warrant recommendation of an administrative and/or regulatory action if no new or more serious deviations are found.

3. POLICY

Agency policy requires that products sold to the federal government should be treated in the same manner as products sold to commercial accounts. Consequently, when FDA recommends against acceptance of a device because that device, or its manufacturer, is in violation of the FD&C act, FDA must also include appropriate regulatory/administrative action against the same or similar device sold to commercial accounts. Compliance Policy Guide 7132c.02, and 7153.04 should be followed in making an evaluation of the violations and the action of choice.

If a firm has shipped a violative product to a Government agency, regulatory action consistent with the nature of the violation(s) may be taken even though there have been no shipments to commercial customers. Formal regulatory action in connection with a violative shipment may not be necessary in some cases. For example, the firm promptly corrects the violative condition, and existing Agency policy would not require further action if the matter involved a product shipped to a non-government customer. However, where corrections are not or can not be made promptly, the main concern is preventing the subsequent shipment of the product to another customer. When the product has been shipped solely to a Government agency and is under the control of that agency and there is no threat to the public, DMPQA should ascertain the intention of the agency holding the goods (e.g., will they return or destroy the goods; will they request FDA to initiate seizure, etc). If the procuring agency requests FDA action, DMPQA will refer the matter to the home district for their consideration of an appropriate recommendation.

Action which may be considered are FDA-requested recall, Warning Letter, seizure, injunction, and prosecution.

All voluntary corrective action proposals submitted to the district in response to the issuance of an FDA 483 or Warning Letter should be carefully reviewed and **a response issued to the manufacturer.** Corrective action proposals should be submitted by a responsible official of the firm in writing, detailing the action(s) to be taken to bring the violative process or product into compliance within a specified time frame. Voluntary correction does not preclude the initiation of administrative and/or regulatory action.

In determining whether GMP deviations are sufficient to support legal action, consideration should be given to the significance of the device, the firm's quality history, and whether the problem is widespread or continuing.

When CDRH does not agree with a district's recommendation for a regulatory action, the district will be notified of the reasons for disapproval in writing.

a. Warning Letter

Issuance of a Warning Letter should be in accordance with Chapters 8-10 of the Regulatory Procedures Manual. See also **Attachment G** for model Warning Letters.

If the District determines that issuance of the Warning Letter has resulted in corrective action by the firm, the district shall, within ten (10) days after confirmation, update the firm's Quality Assurance profile and inform CDRH, Field Programs Branch of the firm's return to compliance.

All firms which have been sent a Warning Letter must be inspected no later than 6 months after the violative EI.

b. Violative Follow-Up Inspections

Investigators are, with the issuance of this FY'94 compliance program, being instructed to close out an inspection as soon as they have documented conditions which meet the criteria of Situation 1. The new model Warning Letter (**Attachment G**) advises manufacturers that the conditions identified by the investigator may be symptomatic of systems problems, and that the manufacturer is responsible for investigating, identifying, and correcting systems problems. The model Warning Letter further directs the firm to discuss in its response how it will address the systems problems related to the conditions identified by the investigator.

To assure that the manufacturer did, in fact, review all manufacturing and quality assurance systems, investigators are instructed to conduct a comprehensive follow-up inspection. When investigators identify the same or additional conditions that meet the criteria for Situation 1, the district should consider seizure or injunction as the regulatory action of choice.

c. Enforcement Strategy For Firms With Repeated Violative Inspections

(1) Some firms have a high rate of recidivism. They have adopted a pattern of correcting violative conditions in response to Warning Letters or other regulatory actions, and usually maintain those corrections long enough to pass the follow-up inspection. When FDA inspects the firm several years later, however, the investigator identifies similar conditions that again meet the criteria for Situation 1. This tendency toward recidivism is often due to the failure of the firm to have in place basic manufacturing and quality assurance systems which meet the requirements of the GMP regulation.

(2) When dealing with another violative inspection for such a firm the District should consider using the following strategy, which may be used either alone or in conjunction with a seizure when the severity of the conditions do not justify mass seizure:

(a) Issue a Warning Letter that follows the model Warning Letter in Attachment G-2. This Warning Letter requests the manufacturer to submit to the District (for up to 3-years) an annual certification by an outside expert consultant stating that it has conducted a complete audit of the firm's manufacturing and quality assurance systems relative to the requirements of the device GMP regulation. The firm must submit a copy of the consultant's report, and certification by the firm's CEO that he or she personally has received and reviewed the consultant's report and that the firm has made all corrections called for in the report.

(b) You have the option of limiting your review of the certifications only to the extent necessary to confirm that the consultant and the firm have met the requirements set forth in the Warning Letter. You may also request a technical evaluation of the consultant's report by the appropriate branch within the Office of Compliance. You have no obligations, however, to send to the firm comments regarding the adequacy of the consultant's report or the firm's corrections.

(c) It will not be necessary to schedule a follow-up inspection for at least 6-months after the firm certifies that it has completed all corrections. The District may remove the firm from the COMSTAT list as soon as the firm has certified that it has **completed** all corrections recommended by the consultant.

(d) If the follow-up inspection indicates that the corrections are satisfactory, notify the firm that you have no objections to the corrections. Remind the firm, however, that it must continue to submit to the District, on the schedule specified in the Warning Letter, certification by an outside expert consultant that it has conducted an update audit and certification by the firm's CEO that any corrections noted to be needed by the consultant have been made, and that it remains in compliance with the requirements of the GMP regulation. The firm should also continue to submit copies of the audit results.

(3) If conditions identified by the follow-up inspection meet the criteria for Situation I, initiate action per 3b above.
(4) If the evidence indicates that the consultant's or firm's certifications are fraudulent, the District may wish to request participation by the Office of Criminal Investigations. When there is clear evidence that the firm falsified its status reports to the District, initiate appropriate charges under USC 18, Section 1001.

d. Recall

If the district believes that prompt removal of a violative product from channels of commerce is necessary, it shall proceed in accordance with established recall procedures in the RPM and 21 CFR, Part 7 (Enforcement Policy), Subpart C (Recalls). In the event there exist serious adverse health consequences or a death, CDRH may request discontinuation of distribution and recall a device to the user level in accordance with Section 518(e) of the Act.

e. Administrative Detention/Seizure

Prior to approving an administrative detention, the District Director must have reason to believe the device is misbranded or adulterated and the firm holding the device is likely to quickly distribute or otherwise dispose of the device, or detention is necessary to prevent use of the device by the public until appropriate regulatory action may be taken by the Agency. District Directors must consult with the Center for Devices and Radiological Health by telephone. Contact the appropriate Div./Branch in OC for the subject device by consulting the CDRH/OC organization chart in PART VI, C. Program Contacts. Concurrence must be given by the Director, OC, CDRH, based on a recommendation by his/her staff.

The district must immediately recommend a seizure.

Of course, a seizure action can be recommended without administrative detention to remove violative devices from commercial distribution, either at the manufacturer, distributor, repacker or at a device user location.

f. Injunction

If a firm has a continuing pattern of significant deviations in spite of past warning, or its operations constitute a serious health hazard, injunction shall be the recommended action of choice. If a serious health hazard exists, the recommendation should include a request for a temporary restraining order (TRO) to prevent the distribution of devices which have been manufactured under the violative conditions documented in the inspection report (See RPM 8 60 30C). The recommendation shall be accompanied by copies of all necessary documentation, e.g., complete inspection reports, Warning Letters issued, sample analysis reports, firm's response to Warning Letters and/or FDA 483s. In the absence of samples, the inspectional evidence must clearly show that the establishment has substantially deviated from GMPs. These deviations must be well documented.

g. Citation

A citation shall be recommended if appropriate as stated in RPM 8-30.

h. Prosecution

The criteria stated in RPM-8-50 shall be the criteria for consideration of prosecution of individuals in violative GMP cases.

i. PMA Disapproval/Withdrawal

Refer to Compliance Program 7383.001, Part V.

j. Automatic Detention

Automatic detention should be recommended whenever there is clear documented evidence to suggest that the foreign manufacturer is producing nonconforming and/or defective devices.

k. Civil Penalties

Section 303(f)(1)(B)(i) of the Act states that civil penalties shall not apply to GMP violations "unless such violation constitutes (I) a significant or knowing departure from such requirements, or (II) a risk to public health." Section 303(f)(1)(B)(iii) further stipulates that civil penalties shall not apply to "section 501(a)(2)(A) which involve one or more devices which are not defective." However, if it can be shown that the firm has received appropriate warning (e.g., FDA 483, Warning Letter, etc.) and that the same GMP violations are subsequently encountered, civil penalties should be considered. Recommendations for civil penalties should be made to address past violations as opposed to ongoing violations.

1. <u>Facilitating Review of Regulatory Recommendations</u>

(1) The district is <u>expected</u> to consult with OC both prior to, <u>and</u> especially during the inspection, once it is determined that regulatory action is being considered. Contact the appropriate Div./Branch in OC for the subject device by consulting the CDRH/OC organization chart in PART VI, C. Program Contacts.

(2) When the district knows a regulatory action will be forthcoming as a result of the inspection, FAX a copy of the <u>issued</u> FDA 483 to the appropriate division in the Office of Compliance. A copy of the FDA 483 annotated with exhibit numbers, and EIR page numbers, helps the reviewers. Therefore, the review process can begin within CDRH while the EIR and recommendation are being written by the district.

(3) It is the responsibility of district management to ensure that the documentation and evidence presented with each legal action recommendation is sufficient to justify <u>each charge</u>. The volume of material submitted must be minimized, and should include only the basic documentation needed to support each GMP charge/example.

(4) All necessary samples and other supporting documentation <u>must</u> be tabbed and their location cross referenced in the recommendation in order to assist in a timely review. It is <u>highly recommended</u> that you provide a table which cross references the violation with the FDA 483 item number, the inspection report page number and the exhibit.
Photographs submitted with recommendation for 502(a) and/or 502(f) charges must be <u>prints</u> rather than reproductions, such as copies. Photos will be returned upon request.

(5) It is essential that all significant questions, problems, or other weaknesses in the evidence regarding the recommended action be stated, along with pertinent district comments. Otherwise, reviewers may miss a problem entirely until litigation is commenced.

(6) The recommendation must begin with the most serious violation of the regulations <u>with reference to the EIR pages, exhibits and sample results</u> which document the violation. <u>Each charge must be parenthetically referenced in the Recommendation Memorandum and the page location of the supporting evidence given</u>. Violations must be listed in decreasing order of importance. Each violation should be related to its effect on device quality in light of overall controls, and should be separated according to the type of manufacturing activity.

(7) Physical samples are <u>not</u> required to support GMP violations, and should <u>not</u> be routinely collected for GMP cases. If the district should reference violative documentary or physical samples as evidence to support GMP deviations, <u>the condition of the samples should be tied to the GMP deviation to show a cause/effect relationship</u>.

(8) Evidence of previous warning and other regulatory actions should be referenced along with a description of corrective actions. If the

recommendation or current EIR references a previous report, either copy the cited EIR pages, or summarize the information.

B. MDR REGULATORY/ADMINISTRATIVE FOLLOW-UP

1. **GENERAL INFORMATION**

MDR violations must be fully documented in any of the situations described below. If there are no complaint or MDR records due to the firm's failure to maintain such records, the EIR narrative should document those findings. When review of the firm's complaint files and service/repair/warranty records has identified MDR reportable events involving deaths, serious injuries, or malfunctions that were not reported to the FDA, failure to report should be listed on the FDA 483.

CDRH requests that copies of Warning Letters and information regarding Direct Reference Seizures be sent to HFZ-300 and HFZ-530. If an alternate means is selected to advise the firm that they are in violation of the MDR regulation, please inform HFZ-300 and HFZ-530 accordingly, by forwarding copies of the FDA 483, memorandum of meeting, or any other pertinent information as appropriate.

2. **SITUATION 1**

NATURE OF VIOLATION	ACTION
o The firm has received prior notice, but continues to not submit MDR's for deaths or serious injuries (other than those which are "unanticipated temporary injuries), and has not established procedures (GMP) for complaint handling and investigation.	Direct Reference Seizure and/or Civil Penalties
o Reinspection determines that a firm is a chronic or flagrant violator which continues to fail to report MDR's, particularly those involving deaths or serious injuries.	Injunction

Consideration should be given to the significance of the unreported incident(s) (the apparent risk to health or level of injury involved), the scope of the problem (whether the problem is widespread or continuing), and the firm's history.

o Inspection reveals the firm has failed to obtain the data needed to file a complete MDR report as described in Part 803.24(c)	Direct Reference Warning Letter

This failure is the result of an incomplete investigation or the failure to conduct any investigation into the event. The failure to conduct a follow-up investigation may be a violation of the GMP requirements for failure investigation and Complaint Files (Parts 820.162 and 820.198) and a GMP charge should be considered. The MDR regulation itself does not require that an investigation be performed.

o Failure to submit MDR's for deaths or serious injuries (other than those which are "unanticipated temporary injuries") of which the firm is aware.	Direct Reference Warning Letter

Authority does not extend to citing the failure to report medical intervention to relieve unanticipated temporary injuries as defined in 21 CFR 803.3(h)(3)(ii), or the failure to report malfunctions as defined in 803.3(c). CDRH must be consulted prior to citing failures to report serious injuries that necessitate medical or surgical intervention, by a health care professional, to relieve unanticipated temporary impairment of a body function or unanticipated temporary damage to a body structure, or malfunctions that would be likely to cause or contribute to a death or serious injury.

o Failure to submit MDR's for unanticipated temporary injuries or malfunctions.	Warning Letter (Requires CDRH Concurrence)

PART VI REFERENCES, ATTACHMENTS AND PROGRAM CONTACTS

A. APPLICABLE REFERENCES OR AIDS

1. Code of Federal Regulations, Title 21, Part 820 Good Manufacturing Practice for Medical Devices: Regulations Establishing Good Manufacturing Practices for the Manufacture, Packing, Storage, and Installation of Medical Devices.
2. Federal Food, Drug, and Cosmetic Act, As Amended.
3. Investigations Operations Manual -- Chapter 5, Subchapter 550.
4. GMP Complaint Files: How They Relate to Reports Required Under MDR, Originally published in Medical Device & Diagnostic Industry, Volume 7, Number 5, May 1985, Revised version (4/10/85) distributed to all district offices.
5. Medical Device Reporting -- Questions and Answers, HHS Publication (FDA) 88-4226, February 1988.

6. An Overview of the MEDICAL DEVICE REPORTING Regulation, DHHS publication No. (FDA) 85/4194, December 1984.
7. MDR -- Federal Register, Vol. 49, No. 180, Friday, September 14, 1984, Rules and Regulations, pages 36326 through 36351, Medical Device Reporting; Final Rule. 21 CFR PARTS 600, 803, 1002, and 1003.
8. STERILIZATION -- QUESTIONS AND ANSWERS, MARCH 1985.

9. DEVICE GOOD MANUFACTURING PRACTICES MANUAL. HHS Pub. No. (FDA) 91-4179, Fifth Edition August 1991.
10. NBS Special Publication 250 -- May 1984 (or update) Calibration and Related Measurement Services U.S. Dept. of Commerce NBS Washington, D.C. 20234.
11. Medical Device GMP Guidance for FDA Investigators, DHHS Publication No. (FDA) 84-4191, April 1984.
12. Sterile Medical Devices. A GMP Workshop Manual. Fifth Edition, March 1984. Prepared by Division of Small Manufacturers Assistance, Office of Training and Assistance, HHS Publication FDA 84-4174.
13. Guideline on General Principles of Process Validation: Notice of Availability published in the Federal Register on May 1987.
14. Working Relationships Agreement Among The Bureaus of Medical Devices (BMD), Radiological Health (BRH), Biologics (BOB), Published April, 1982. Refer to District Reference File for copies.
15. Device GMP -- Applications and Interpretations for the Field Investigator, November 1980.
16. Plastic Medical Devices: A Study of Quality in The Making. September 1980. A copy of this film has been supplied to each FDA District office. This film is intended for use with the Medical Device Reference Files on plastics (see reference #17 below).
17. Medical Device Reference Files on syringes, catheters, tubes and airways, IOL and contact lenses, IUDs and filters. September 1980. One hard copy and one microfiche copy of each of these reference files has been supplied to each FDA District office.
18. GMP Questions and Answers, November, 1979.
19. Device Good Manufacturing Practices. A quality Audit Program for Industry. September, 1979.
20. Quality Control Handbook, Juran, J.M., 3rd edition, McGraw-Hill, 1974.
21. ANSI Z80.1-1972 Revision of X80.1-1964 American National Standard Requirements of First-Quality Prescription Ophthalmic Lenses.
22. MIL-STD-105E, Sampling Procedures and Tables for Inspection by Attribute.
23. GWQAP Manual.
24. Classification Names for Medical Devices and In Vitro Diagnostic Products, HHS Publication No. (FDA) 91-42246, August 1991. This directory is organized by "keywords" in alphabetical order. The classification number (5 digit product code), class, and CFR regulation number is given for each entry listed.
25. Code of Federal Regulations, Title 21, Part 809.10, Labeling for In Vitro Diagnostic products.

26. Advisory List of Critical Devices -- 1988; Notice Published in the Federal Register on March 17, 1988.
27. Overview of Metallic Orthopedic Implants; Technical report, reference material and training aid for investigators prepared by the Division of Field Investigations (HFC-130), Office of Regional Operations, Office of Regulatory Affairs, HHS, Public Health Service, FDA, June, 1988.
28. AQL Inspector's Rule and Manual. This special purpose plastic slide rule that rigidly adheres to MIL-STD-105E can be obtained from Infor. Inc., P.O. Box 606, Ayer, MA. 01432. Phone (509) 772-0713. Cost is approximately $20 each excluding shipping and packaging.
29. A Pocket Guide to Device GMP Inspections, DHHS Publication No. (FDA) 92-4248, November 1991. NOTE: THE POCKET GUIDE DOES NOT INCLUDE GUIDANCE REGARDING COMPLAINT FILES AND THE TRACKING REGULATION NOW CONTAINED IN ATTACHMENT A.
30. Inspections of Foreign Device Manufacturers, February, 1993. This eight page document was prepared by CDRH, Office of Compliance.
31. Code of Federal Regulations, Title 21, Part 821, Medical Device Tracking Requirements.
32. Medical Device Tracking -- Questions and Answers Based on the Final Rule, HHS Publication No. (FDA) 93-4259, August 26, 1993.

Copies of CDRH GMP publications are available from the Division of Small Manufacturers Assistance (DSMA), Telephone: 800-638-2041 or FAX 301-443-8818.

B. ATTACHMENTS

ATTACHMENT A --	Device GMP Inspections.
ATTACHMENT B --	Instructions for Reviewing Records Kept in a Manufacturer's Device Complaint File(s).
ATTACHMENT C --	Class I Devices Exempt from Most of the GMP Requirements by Classification Regulations.
ATTACHMENT D --	Advisory List of Critical/Significant Risk Devices.
ATTACHMENT E --	Inspectional Guidance for Performing a Review of the Firm's Compliance with the MDR Regulation.
ATTACHMENT F --	Medical Device Tracking, Questions and Answers Based on the Final Rule, August 26, 1993.
ATTACHMENT G --	Model Warning Letters.

C. PROGRAM CONTACTS

1. ORA Contacts

a. Questions regarding inspectional requirements:

Investigations Branch (HFC-132)
Telephone: (301) 443-3340

b. For technical assistance or advice while conducting an inspection:

International Programs and Technical Support Branch (HFC-133)
Telephone: (301) 443-1855

c. Questions about accessing or connecting to the Parklawn Computer Center and Model 204

Leo Bauman
Division of Information Systems, ORA
(301) 443-1314

An easy method for Field Users to access the system is to log on to the regional VAX, then type:

TELNET PCCSNA.FDA.GOV <return>

Field Users should set up their communication program to emulate a VT100 or other of the options before logging on to the Regional VAX.

d. Questions regarding sampling of devices and laboratory capabilities:

Division of Field Science (DFS), HFC-140,
Telephone: (301) 443-3320

e. The WEAC contact point for testing medical devices is:

Director
WEAC Engineering Branch, HFR-NE480
Telephone: (617) 729-5700

2. CDRH Contacts

See the New CDRH/OC Organizational Chart on Page 7 of Part VI to identify which unit within OC is responsible for answering your question or giving you guidance, depending on the type of device.

a. Questions regarding the interpretation and applicability of the MDR regulation:

   Bryan H. Benesch
   Reporting Systems Monitoring Branch, HFZ--533
   Telephone: (301) 594-2735
   Electronic Mail (BHB)

b. MDR Report and Data Summaries:

   Dale Burke
   Division of Surveillance Systems, OSB, CDRH
   Information and Analysis Branch, HFZ-531
   Telephone: (301) 594-2731

c. Industry MDR Report: (301) 427-7500. Do not call this phone number to make inquiries.

d. Questions regarding sampling and/or testing of **general medical** devices.

   Edward Mueller or Donald Marlowe
   Division of Mechanics and Material Sciences, HFZ-150
   Telephone: (301) 443-7003

e. Questions about retrieving information from the MDR database and using MDRAPSY.

   Dale Burke
   Division of Surveillance Systems, OSB, CDRH
   Information and Analysis Branch, HFZ-531
   Telephone: (301) 594-2731

f. Express Mail Address for All Regulatory Action Recommendations:

   OC Document Control, HFZ-300
   Office of Compliance
   Center for Devices and Radiological Health
   2098 Gaither Road
   Rockville, Maryland 20850

g. Questions regarding the interpretation and applicability of the device GMP regulation and GMP exemptions:

   Contact the appropriate Division/Branch in OC for the subject device by consulting the CDRH/OC organization chart in Part VI, page 7. Program Contacts.

h. Questions regarding GMP critical device status:

   Sharon M. Kalokerinos
   Division of Enforcement II
   Telephone: (301) 594-4613

i. Questions regarding this Compliance Program:

   Bill Keene
   Field Programs Branch
   Telephone: (301) 594-4695 ext. 111

PART VII -- CENTER RESPONSIBILITIES

A. Program Evaluation for GMP Inspections

   1. Unit Assigned: Field Programs Branch
      Associate Director for Program Operations
      Office of Compliance
   2. An evaluation of this program will be completed every two years within six months of the close of the fiscal year.

B. MDR Report Summaries

   1. Unit Assigned

      Office of Surveillance and Biometrics
      Information Management and Analysis Branch, HFZ-531

   2. Summary of Work

      Upon request, OSB will forward to an investigator, reports containing summaries of MDR reports submitted to CDRH by a particular firm.

C. Selection and Assignment of Importers and Manufacturers Requiring Special Inspection.

1. Unit Assigned

   Office of Surveillance and Biometrics
   Device Administration Experience and Monitoring Branch

2. Summary of Work

   On a fiscal year as well as an ad hoc basis, firms will be selected and assigned for inspection.

ATTACHMENT A

DEVICE GMP INSPECTIONS

Inspections of
Medical Device Manufacturers
and
GMP Regulation Requirements

GMP INSPECTIONAL STRATEGY

With this FY'94 compliance program CDRH is initiating a new strategy for conducting GMP inspections. The traditional "Two-Track GMP Strategy" required, at a minimum, that every manufacturer receive a comprehensive GMP quality audit once every four years, and a limited GMP quality audit at the intervening two-year interval. If manufacturers experienced problems, however, they were subject to comprehensive quality audits more frequently.

The new strategy requires that only a follow-up inspection must be a comprehensive quality audit. All other inspections, including initial inspections, will be directed inspections per Section I of this Attachment.

All inspections must be started with a review of the complaint and MDR files, recalls, manufacturing change control records, and finished device testing records. When these records indicate that the manufacturer has experienced device failures and/or produced nonconforming devices you should focus on those particular devices and manufacturing processes. WITH THE EXCEPTION OF COMPREHENSIVE FOLLOW-UP INSPECTIONS, AS SOON AS YOU HAVE IDENTIFIED AND DOCUMENTED SYSTEM WIDE DEFICIENCIES WHICH, IN TOTAL, MEET THE CRITERIA FOR SITUATION I IN PART V.A.1., DISCONTINUE THE INSPECTION AND ISSUE THE FDA 483

When you identify objectionable conditions that, in total, **do not** meet the criteria for Situation I in Part V.A.1., complete the inspection as directed per Section I of Attachment A and issue the FDA 483.

This Attachment includes the following information:

-- Section I of this Attachment is designed to help you identify and obtain evidence of nonconforming product and also contains general inspectional guidance. This information should be used for both initial inspections and reinspections.
-- Section II, which should be used in addition to Section I for comprehensive follow-up inspections, contains the GMP Regulation requirements presented as questions, with important references cited where needed. The sequence in which the GMP requirements are listed follows the sequence of the narrative guidance in Section I.

The inspectional guidance that follows requires not only a review of procedures and records, but also familiarity with the actual manufacturing process and the conditions in the manufacturing area. Problems in the manufacturing process or quality assurance system, which you identify, should be documented. When these problems are related to any nonconforming devices uncovered during the inspection, refer to Part V of C.P. 7382.830 for guidance on documenting potential regulatory actions.

SECTION I

Inspections of Medical Device Manufacturers

A. General GMP Application Notes
B. Pre-Inspectional Activity
C. 510(k) Device Inspections
D. PMA Device Inspections
E. Device Selection for Inspection
F. Critical Devices
G. All Device Inspections
   1. Complaint Handling System
   2. MDR Compliance
   3. Medical Device Tracking
   4. Failure Investigation
   5. In-Process & Finished Device Rejects & Rework
   6. Evaluation of Procedures for Change Control
   7. Validation
   8. Components
   9. Audits
   10. PMA Devices
H. Sterile Devices
I. The Small Manufacturer
J. Written Procedures

SECTION I
INSPECTIONS OF MEDICAL DEVICE MANUFACTURERS

By following this approach, investigators will be able to focus on actual/potential problem areas during the Establishment Inspection (EI). Briefly summarized below are some of the key points that should be kept in mind during GMP inspections.

A. GENERAL GMP APPLICATION NOTES

The Center for Devices and Radiological Health has determined that certain types of establishments are either wholly or partially exempt from the GMP. Refer to pages 50-53 of the guideline booklet entitled "Medical Device GMP Guidance for FDA Investigators" for information regarding the application of device GMPs in the following subject areas:

1. Determine when a **product** is exempt from the regulation
2. Applicability of the GMP to the **component manufacturers**
3. Applicability of the GMP to **service firms**
4. The GMP and **custom device manufacturers**
5. Responsibilities of **contract manufacturers and specifications developers** under the GMP
6. Application to **Contract Test Laboratories**

B. PRE-INSPECTIONAL ACTIVITY

A review of the MDR/PRP data base should be the starting point of any inspection. A trend analysis of the data can be obtained from DMI, Information and Analysis Branch, HFZ-531, by calling Dale Burke at 301-594-2731. Please make requests at least five days in advance of the date the inspection will begin.

C. 510(k) DEVICE INSPECTIONS

A majority of inspections will cover devices that were marketed after May 28, 1976, via a 510(k) premarket notification. The inspection should determine whether all the devices being covered have a 510(k) substantial equivalence (SE) ruling. For those devices selected for inspection, it should be determined if the company is complying with any product and/or process specifications listed in the 510(k). Also, if significant product and/or process changes have occurred, it should be determined if a new 510(k) has been submitted and found to be substantially equivalent.

D. PMA DEVICE INSPECTIONS

Refer to C.P. 7383.001 for guidance regarding PMA inspections.

E. DEVICE SELECTION FOR INSPECTION

All profile classes (manufacturing technologies) used by the manufacturer must be covered by selecting as few device lines as possible. If a device is selected because it appears to be a nonconforming product, and that device does not involve all of the profile classes used by the manufacturer, select an additional device(s) which requires the missing profile class(es).

F. CRITICAL DEVICES

If the firm manufactures any critical devices, usually one or more of these should be covered. Determine whether the critical device sections of the GMPs have been applied to the critical device. In particular: (a) Which components are critical? What mechanism was used to define critical components, and what types of tests are performed on incoming critical components prior to acceptance? (b) Which operations are considered critical? What mechanism was used to define critical operations, and how are critical operations documented?

G. ALL DEVICE INSPECTIONS

1. Complaint Handling System -- 820.198

This should be the beginning point of every inspection to determine whether the firm has received complaints of possible (or potentially) defective devices.

If the firm states that because it has never received complaints it does not need a complaint file system, it should be cited on the 483 for failure to have a complaint file system.

-- Does the firm repair and service devices it produces?

If so, the firm must have an adequate system in place for screening repair and service requests to assure whether any of these meet the definition of a complaint. Report in the EIR any refusals to provide access to service or repair records.

-- Also, does the firm screen repair and service requests and conduct trend analyses to determine whether there are premature failures within the warranty period, and to detect problems with particular components, subassemblies, or design?
-- Does the firm review complaints, service records, or repair records for MDR reportability?
-- Who has responsibility for MDR related event review and reporting?

Remember when reviewing complaints to check for MDR reportable events.

-- Does the firm perform trending of components that are replaced? If so, attempt to review the data. Report refusals.

2. MDR Compliance -- 803.24

-- Has the firm submitted MDR reports to FDA for the types of events considered to be reportable?

Prior to an inspection, obtain copies of MDR reports submitted to CDRH in the last 12 months for the firm to be inspected. When writing the EIR, be sure to state which 12 month time period was covered. If a firm has submitted no MDR report, you have the option to ask the D.P.U. to search through the MDR data base for another firm that manufactures or imports similar types of devices and familiarize yourself with those incidents. Look at examples of each type of event reported in each MDR category (death, serious injury, malfunction). These can be discussed with the firm to illustrate the types of problems that should be reported or used by the investigator as a guide in determining if the firm has information that should have been reported under the MDR regulation.

Examine records at the firm according to instructions in Attachment E. THIS ACTIVITY IS NOT OPTIONAL.

3. Medical Device Tracking

-- Does the firm manufacture any of the devices listed in Attachment F, (page 6)?

If so, determine if the firm has device tracking procedures in place and, if it does, assess their adequacy.

If there is a complete absence of a tracking system, you should cite this violation on the FDA 483. Other violations should be addressed verbally and reported in the EIR.

4. Failure Investigation -- 820.162

-- Does the firm have procedures in place for a formally documented failure analysis program?

Written procedures are not required but must be followed if established by the firm. Review the firm's procedures to determine if they are adequate.

It should be remembered that the complaint section of the GMP (21 CFR 820.198) refers to possible complaints, and performing investigations to determine whether the complaint can be confirmed. Once the complaint is confirmed as an actual failure of the device, the failure investigation section of the GMP (21 CFR 820.162) takes effect.

-- Review records for indications of defects.
-- Determine whether investigations have been performed and ascertain if there is a common failure trend (e.g., a particular component(s), subassembly(ies), manufacturing error(s), or employee training).

These common failure trends would provide clues on areas or products to focus on during the inspection.

5. **In-Process and Finished Device Rejects and Rework** -- 820.20(a)(2), .100 & .101; .115 & .116; 820.160 & .161

-- Does an examination of the device history records show that any lots, or portions of lots, have been rejected during either in-process or finished device inspection for failing to meet any or all of the product's specifications?

Instances of release and distribution of lots that failed to meet any specifications should be reported and documented. Records should be examined to determine whether any lots that have failed specifications were reworked/reprocessed, and whether this reprocessing is adequate to assure that specifications will be met without affecting the safety or performance of the device.

A high rate of rework/reprocessing would also provide strong evidence that the manufacturing process is not operating in a state-of-control.

Are all sampling plans for inspection and rework based on an acceptable statistical rationale (i.e., MIL STD 105E)?

The sampling plan used should be examined to determine whether adequate samples are obtained for inspection, and should be described completely in the EIR.

6. **Evaluation of Procedures for Change Control** -- 820.100(a)(2), 820.100(b)(3), and 820.181 (DMR).

-- Have any changes been made to the device and/or process?

These may have occurred as a result of complaints, services/repairs, failure investigations, in-process/finished device inspection, or rework/reinspection, and can be determined by examining the change control documents. If changes have been made to either the device design or the device manufacturing process, determine whether they have been adequately validated to assure that:

- -- The change corrected the problem
- -- The change did not adversely affect any other component/subassembly in the device

Avoid getting bogged down in cosmetic changes. In addition, changes for economic reasons (e.g., to cut costs) may sometimes lead to performance problems with the device.

-- Can you find a series of change orders (found by reviewing the DMR and various engineering drawings) for the same device which are intended to sequentially correct the same or similar problems?

If yes, this may indicate that the original change was not effective and that subsequent changes have been or are being made to correct the problem.

-- Did the firm receive complaints on lots manufactured after the design change?

All these are potential indications that the change was not effective and was not adequately validated.

-- Review the change control system.
-- Who must authorize changes?
-- What are the provisions for validating the changes?
-- When is a change actually effective?
-- Are the changes adequately documented?
-- Were personnel affected by any changes adequately notified?

7. Validation -- 820.100(a)(1)

Process Validation is the act of establishing documented evidence that provides a high degree of assurance that a specific process will consistently produce a product meeting its pre-determined specifications and quality attributes.

Prospective process validation is validation conducted prior to the distribution of either a new product, or a product made under a revised manufacturing process, where the revisions may affect the product's characteristics.

Retrospective process validation is validation of a process for a product already in distribution based upon accumulated production, testing and control data.

The inspection should determine whether adequate prospective or retrospective validation of the manufacturing process has been performed. Validation should ensure that the quality of the product will be maintained if the process is controlled within established parameters and that the validation, either prospective or retrospective, has addressed the limits of these parameters. Validation, depending on the scope of the operation, can cover all aspects from the selection of components to various manufacturing processes to end-product testing.

-- What are the process controls and were they validated?
-- The firm should be able to document that the product will meet specifications when the process is operated at the optimal process parameters.
-- The firm should be able to document that it can control the process within limits.
-- Pay attention to the process parameters; i.e., temperature, humidity, tensile strength, viscosity. Did the manufacturer include all the necessary parameters in the processing procedures?

8. Components -- 820.20(a)(2), 820.80 and .81

The selection and acceptance of components is critical to the confidence that will be built into a finished device.

-- Does the manufacturer have component specifications?
-- If yes, do the methods used to evaluate components assure that received components meet these specifications?

Particular attention should be paid to critical components and those components that would require special handling or storage to maintain their integrity. Be wary of numerous changes of suppliers without having adequate change control. Also, be wary of situations where a manufacturer has relaxed specifications because the supplier could not meet the original specifications. Check device history records for affected lots and preceding lots to determine if finished device reject rates have changed. Also check complaint files to determine if the changes have resulted in increased complaints.

Vendor audits, at this time, are not required and failure to perform them should not be placed on the FDA 483, unless it is required by the firm's written procedures.

Where a manufacturer uses Just In Time (JIT) component acceptance, and relies on component vendor audits in place of incoming component inspection, review the manufacturer's audit schedule and audits. Pay special attention to the vendor's rejection rate.

9. Audits -- 820.20(b)

-- Does this finished device manufacturer conduct planned and periodic audits of its quality assurance program as required?

The quality audit is the foundation of the quality assurance program. Determine if the manufacturer has a written procedure for conducting quality audits and how often these audits are conducted. It is recommended that the time between audits not exceed a 12-month period. More frequent audits may be recommended if the firm has a serious GMP problem.

Does the written audit procedure include the following:

-- Objectives
-- Responsibilities
-- Scope
-- Evaluation criteria
-- Scheduling

Consult Subchapter 550 of the IOM for updated coverage of Audits (Revised 2-9-90).

10. PMA Devices -- 814.39(a)(4)

-- If any PMA devices are manufactured at this facility, is this manufacturing site approved in either the original PMA application or a PMA Supplement, even if only partially manufactured here?

H. STERILE DEVICES

1. If the firm is manufacturing sterile devices, the criteria for Track I and II inspections in C.P. 7382.830 A, Attachment A, must be followed.
2. There should be records to show how the sterilization process was validated. Obtain appropriate records to document any deficiencies.
3. Determine whether the firm may be producing nonsterile devices. Some of the ways of determining this are:

-- Have any lots had positive sterility test results and/or positive BI results? Were these lots adequately reprocessed, or were they just released? Was an adequate investigation performed prior to discrediting the initial test results?
-- Are reprocessing (resterilization) procedures adequate, and do they assure that device performance or package seals will not be adversely affected?
-- Are test methods based on accepted test methodologies?

  -- Are they performed properly so as to be able to detect positive sterility/BI test results?
  -- If an in-house test method is used, has it been validated to show it is capable of detecting nonsterile units?

Note: In evaluating in-house sterility test methods, the sample size is an important factor. (For example, the USP test of 40 samples is unable to detect 15% of nonsterile units with 95% confidence.)

-- Have any lots been released even though cycle specifications have not been met?
-- Does the cycle have an adequate Sterility Assurance Level (SAL) for the device being sterilized?

The SAL is the probability of a unit being nonsterile after exposure to a valid sterilization process. The SAL varies according to the intended use of the device. Sterilized articles not intended to contact compromised tissues are generally thought to be safe for use with an SAL of 10^{-3}; that is, a probability of one nonsterile unit in a thousand. Invasive and implantable devices should have an SAL of 10^{-6}; that is, no more than one nonsterile unit in a million. In practice, many firms use overkill cycles which assure an even lower probability that a device will be nonsterile.

-- What is the bioburden on the device, and is the sterilization cycle able to destroy the bioburden to an appropriate SAL?

I. THE SMALL MANUFACTURER

At the present time there is no official definition of a small manufacturer, although 10 or fewer employees is often used as the definition. An investigator should realize, however, that a small firm usually does not need the same degree of documentation necessary for a large firm to achieve a state of control and that **many of the GMP**

written procedure requirements are not necessary for a small manufacturer, unless the firm is producing nonconforming devices.

At a minimum, the small manufacturer must have the following:

-- Space for components, finished devices, and manufacturing processes
-- A DMR containing these specifications
-- Testing/release according to specifications
-- A DHR containing the test results
-- Training/qualifications of employees
-- Audit procedures

The flexibility of the GMP requires greater effort from the investigator in determining compliance with the GMP than a GMP where all requirements are more specific. An investigator who understands the purpose and meaning of the various GMP requirements will not merely follow the words verbatim, but will apply the GMP properly to any operation to which the GMPs apply. If the guidelines in the booklet entitled "Medical Device GMP Guidance for FDA Investigators" are followed, an investigator can apply the GMP to any situation, irrespective of the size of the firm or the complexity of its operations.

J. WRITTEN PROCEDURES

The purpose of written procedures is to provide guidance, to assure uniformity and completeness, and for communicating and managing operations.

In large manufacturing operations involving many unskilled people and operations, **written procedures** are usually necessary. In a **small firm**, communication lines are usually short, few people are involved, and management is readily available to provide guidance so that the need for written procedures is usually considerably less.

Often, training and work experience are valid substitutes for written procedures. For example, machinists are typically skilled personnel who fabricate components and finished devices using dimensioned drawings for guidance instead of written procedures. The investigator must evaluate each situation and determine the need for a **written procedure** on the basis of training and the knowledge possessed by the operator and the control needed. Typically, a written procedure may not be necessary when:

-- The activity is very simple
-- Straightforward quantitative rather than qualitative standards determine acceptability
-- The operation is performed by skilled personnel

In a **small firm**, the investigator may conclude that a GMP requirement for a written procedure is not needed. Such a decision, however, must be supported by observation that sufficient control is present to meet the intent of the written procedure, and the fact that nonconforming devices are not being produced.

When the GMP requires a procedure, but does not specify a "**written**" procedure (e.g., 820.115, 820.100(a)) and the manufacturer does not have the applicable procedure,

the investigator must evaluate the controls in place and determine if they are adequate without written procedures. Such decisions can be made taking into consideration factors such as the number of mistakes, rework, rejects, complaints, etc., that can be related to the operation for which the procedures are intended.

An investigator should not insist that a manufacturer meet GMP requirements that do not contribute to assuring conformance to specifications simply because they are part of the GMP. Section 519(a)(1) of the Act prohibits recordkeeping requirements that are unduly burdensome to a device manufacturer.

SECTION II

820.198 Complaint Files
820.162 Failure Investigation
820.100 Manufacturing Specifications and Processes
820.101 Critical Devices, Manufacturing Specifications and Processes
820.115 Reprocessing of Devices or Components
820.116 Critical Devices, Reprocessing of Devices or Components
820.160 Finished Device Inspection
820.161 Critical Devices, Finished Device Inspection
820.180 General Requirements
820.181 Device Master Record
820.182 Critical Devices, DMR
820.80 Components
820.81 Critical Devices, Components
820.20 Organization
820.25 Personnel
820.40 Buildings
820.46 Environmental Controls
820.56 Cleaning and Sanitation
820.60 Equipment
820.61 Measurement Equipment
820.184 Device History Record
820.185 Critical Device, DHR
820.195 Critical Devices, Automated Data Processing
820.120 Device Labeling
820.121 Critical Devices, Labeling
820.130 Device Packaging
820.150 Distribution
820.151 Critical Devices, Distribution Records
820.152 Installation

Throughout Section II, **boldface** type is used to designate that one of the following is necessary:

1. Record required
2. Written procedure necessary
3. Documentation of procedure indicated

- - - - - - - - - - - - - - - - - -

820.198 COMPLAINT FILES

-- Are **written** and oral complaints related to the identity, quality, durability, reliability, safety, effectiveness, or performance of a device reviewed, evaluated, and maintained by a formally designated unit?
-- Does this unit determine whether or not an investigation is necessary?
-- When no investigation is made, does the unit **maintain a record** that includes the reason and the name of the individual responsible for the decision not to investigate?
-- Are all complaints involving the possible failure of a device to meet any of its performance specifications reviewed, evaluated, and investigated?
-- Are all complaints pertaining to injury, death, or any hazard to safety immediately reviewed, evaluated, and investigated by a designated individual(s), and are they maintained in a separate portion of the complaint file?
-- When an investigation is made, is a **written record** of each investigation maintained by the unit that is designated to process complaints?
-- Does the record of investigation include the following?
    -- The name of the device
    -- Any control number used
    -- Name of complainant
    -- Nature of complaint (i.e., specific details of the complaint, as obtained from the complainant)
    -- Reply to complainant
-- Where the formally designated unit is located at a site separate from the actual manufacturing establishment, is there a duplicate copy of the record of investigation of any complaint transmitted to and maintained at the manufacturing establishment where the device is produced in a file designated for device complaints?

820.162 FAILURE INVESTIGATION

-- After a device has been released for distribution, is any failure of that device or any of its components to meet performance specifications investigated?
-- Is a **written record** of such investigation, including conclusions and followup, made?

820.100 MANUFACTURING SPECIFICATIONS AND PROCESSES

(a) Specification controls

-- Are **written** manufacturing specifications and processing procedures **(including sterilization)** established, implemented, and controlled to assure the device conforms to its original design or any approved changes in that design?

-- Are there procedures for the development of manufacturing specifications for the device?
-- Are there procedures for making **changes** in the manufacturing specifications of the device? Are such changes approved and documented by a designated individual(s) and do they include the approval date and the date the change becomes effective?

(b) Processing controls

-- Are there **written** procedures describing any processing controls necessary to assure conformance to specifications?
-- Is any approved **change** in the manufacturing process communicated to appropriate personnel in a timely manner?

820.101 CRITICAL DEVICES, MANUFACTURING SPECIFICATIONS AND PROCESSES

(a) Critical operation performance

-- Are all critical operations performed by a designated individual(s) on suitable equipment and are they verified?

(b) Record of critical operation

-- Does the device history record contain or refer to the location of each critical operation identifying the date performed, the designated individual(s) performing the operation and, when appropriate, the major equipment used?

820.115 REPROCESSING OF DEVICES OR COMPONENTS

-- Are reprocessing procedures established, implemented, and controlled to assure that the reprocessed device or component meets the original, or subsequently modified and approved, specifications?

820.116 CRITICAL DEVICES, REPROCESSING OF DEVICES OR COMPONENTS

(a) Reprocessing procedures

-- Are there **written** procedures for any reprocessing associated with the production of a critical device or component?
-- Do these procedures describe the reprocessing equipment and include any special quality assurance methods or tests?
-- Is the device or component to be reprocessed clearly identified and separated from like devices or components not to be reprocessed?

-- When there is a constant reprocessing of a device or component, is a determination of the effect of the reprocessing upon the device or component made and **documented**?

(b) Reprocessing control

-- Is any prior quality assurance check repeated on the reprocessed device or components if reprocessing could adversely affect any performance characteristic previously inspected?

820.160 FINISHED DEVICE INSPECTION

-- Are there **written** procedures and acceptance criteria for the finished device inspection to assure that device specifications are met?
-- Prior to release for distribution, is each production run, lot, or batch CHECKED for conformance with device specifications?
-- Where necessary, is each production run, lot, or batch TESTED for conformance with device specifications?
-- Where practical, is a device selected from a production run, lot, or batch, and tested under simulated use conditions?
-- In your narrative report, describe the sampling plan for checking, testing, and release of the device. CDRH has applied this requirement for noncritical devices using 820.20 as justification.
-- Are finished devices held in quarantine or otherwise adequately controlled until released?
-- Are devices that fail finished device inspection, and are not to be reprocessed, disposed of in a manner which precludes their release as finished goods?

820.161 CRITICAL DEVICES, FINISHED DEVICE INSPECTION

-- Is a critical device or component that does not meet its performance specifications investigated?
-- Is a **written record** of the investigation, including conclusions and followup, made?
-- Does a critical device leave the control of the manufacturer for distribution only after all acceptance records and test results have been checked by a designated individual(s)?
-- Does a designated individual(s) assure that all records and documentation required for the device history record are present, complete, and show that release of the device was consistent with the release criteria?
-- Does a designated individual(s) authorize, by signature, the release of the device for distribution?

820.180 GENERAL REQUIREMENTS

-- Are the following **records** maintained at the manufacturing establishment or other reasonably accessible location and are they available for review and copying?

-- Device master record
-- Device history record
-- Complaint files
-- Are all **records** required for this device retained for a period of at least 2 years from the date of release for commercial distribution by the manufacturer?

820.181 DEVICE MASTER RECORD

-- Is the device master **record** prepared, dated, and signed by a designated individual(s)?

Records and Change Control Procedures

-- Are any **CHANGES** in the device master record authorized in **writing** by the signature of a designated individual(s)?
-- Do such **CHANGES** include the approval date and the date the changes become effective?
-- Are any forms used to approve **CHANGES** in the device master record part of the device master record?

Does the device master record for this device include, or refer to the location of, the following information?

-- Device specifications including appropriate drawings, composition, formulation, and component specifications
-- Production process specifications including appropriate equipment specifications, production methods, production procedures, and production environment specifications
-- Quality assurance procedures and specifications including quality assurance checks used and the quality assurance apparatus used
-- Packaging and labeling specifications including appropriate methods and processes used

820.182 CRITICAL DEVICES, DEVICE MASTER RECORD

In addition to the requirements of 820.181, does the device master record for a critical device include or refer to the location of the following?

-- Full information concerning critical components and critical component suppliers, including the complete specifications of all critical components, the sources where they may be obtained, and **written** copies of any agreements made with suppliers
-- Complete labeling procedures for the individual device and copies of all approved labels and other labeling.

820.80 COMPONENTS

-- Are components used in manufacturing received, stored, and handled in a manner designed to prevent damage, mixup, contamination, and other adverse effects?

-- Are components quarantined prior to acceptance or clearly identified as not yet accepted?

(a) Acceptance of components

-- Are there **written** acceptance criteria and procedures for the acceptance of components?
-- Does a designated individual(s) accept or reject components?
-- Is a **record** maintained of component acceptance and rejection?
-- Where deviations from component specifications could result in the device being unfit for its intended use, are components inspected, sampled, and tested for conformance to specifications?

(b) Storage and handling of components

-- Are components whose fitness for use deteriorates over time stored in a manner to facilitate proper stock rotation?
-- Are component control numbers or other identifications easily viewable?
-- Are all obsolete, rejected, or deteriorated components clearly identified and segregated from accepted components?
-- Are **records** maintained of the disposition of all obsolete, rejected, or deteriorated components?

820.81 CRITICAL DEVICES, COMPONENTS

(a) Acceptance of critical components

-- Are there **written procedures** for sampling, testing, and inspecting all lots of critical components to assure that critical components conform to specifications?
-- You may want to collect or record the firm's procedures for your EIR regarding the statistical rationale for sampling critical components.
-- Is each lot of critical components identified with a control number(s) upon receipt?
-- Are the percentage of defective critical components in each sample and the percentage of lots rejected **recorded** and identified by supplier name?

(b) Critical component supplier agreement

-- Where possible, has the manufacturer secured from the critical component supplier a **written** agreement whereby the supplier agrees to notify the manufacturer of any proposed change in a critical component?

820.20 ORGANIZATION

-- Where possible, does a designated individual(s), not having direct responsibility for the performance of a manufacturing operation, take responsibility for the quality assurance program?

(a) Quality assurance program requirements

-- Does the QA program assure that production records are reviewed?
-- Are there acceptance criteria for the receipt of all components, manufacturing materials, in-process materials, packaging materials, labeling, and finished devices?
-- Does the quality assurance program consist of procedures that assure the approval or rejection of devices manufactured, processed, packaged, or held under contract by another company?
-- Does the QA program assure the function of identifying, recommending, or providing solutions for quality assurance problems and verifying that the implementation of such solutions is performed?
-- Is the QA program assuring that all quality assurance checks are appropriate and adequate for their purpose and are performed correctly?

(b) Audit procedures

-- Are there **written audit procedures**? (Note: The investigator should not seek access to the firm's internal audit reports.)
-- Are there planned and periodic audits of the quality assurance program implemented to verify compliance with the quality assurance program?
-- Are audits performed in accordance with **written procedures** by appropriately trained individuals not having direct responsibilities for the matters being audited?
-- Are audit results documented in **written** audit reports that are reviewed by management having responsibility for the matters audited?
-- Is followup corrective action, including re-audit of deficient matters, taken when indicated?

The following additional requirements should be covered for Track II inspections.

820.25 PERSONNEL

(a) Personnel training

-- Where training programs are necessary to assure that personnel have a thorough understanding of their jobs, are such programs conducted and **documented**?
-- Are employees made aware of device defects which may occur from the improper performance of their specific jobs?
-- Are quality assurance personnel made aware of defects and errors likely to be encountered as part of their quality assurance functions?
-- Are personnel in contact with the subject device or its environment: clean, healthy, and suitable attired where lack of cleanliness, good health, or suitable attire could adversely affect the device?
-- Are any personnel who, by medical examination or supervisory observation, appear to have a condition which could adversely affect the device excluded from affected operations until the condition is corrected?
-- Are personnel with such conditions instructed to report them to their supervisors?

-- Look for examples where personnel failed to perform, or inadequately performed, a task because of training; e.g., bonding, molding, assembly.

820.40 BUILDINGS

-- Do the facilities provide ADEQUATE SPACE DESIGNED to PREVENT mixups and to assure orderly handling of the following?

- Incoming components
- Rejected or obsolete components
- In-process components
- Finished devices
- Labeling
- Devices that have been reprocessed, reworked, or repaired
- Equipment
- Molds, patterns, tools, records, drawings, blueprints
- Testing and laboratory operations
- Quarantined products

820.46 ENVIRONMENTAL CONTROLS

-- Where environmental conditions at the manufacturing site could have an adverse effect on a device's fitness for use, are these environmental conditions controlled to prevent contamination of the device? See C2.0 of Medical Device GMP Guidance for FDA Investigators.

-- Are there periodic **documented** inspections of these environmental controls to verify that the environmental control system is properly functioning?

820.56 CLEANING AND SANITATION

-- Are **written** cleaning procedures and schedules adequate to meet manufacturing process specifications?

(a) Personnel sanitation

-- Are washing and toilet facilities clean and adequate?

-- Where special clothing requirements are necessary to ensure that a device is fit for its intended use, are clean dressing rooms provided for personnel?

(b) Contamination control

-- Are there procedures designed to prevent contamination of equipment, components, or finished devices by rodenticides, insecticides, fungicides, fumigants, hazardous substances, and other cleaning and sanitizing substances?

-- Is adherence to such procedures **documented**?

(c) Personnel practices

-- Where eating, drinking, and smoking by personnel could have an adverse effect on a device's fitness for use, are such practices limited to designated areas?

(d) Sewage and refuse disposal

-- Are sewage, trash, byproducts, chemical effluents, and other refuse disposed of in a timely, safe, and sanitary manner?

820.60 EQUIPMENT

(a) Maintenance schedule

-- Is equipment used in the manufacturing process appropriately designed, constructed, placed, and installed to facilitate maintenance, adjustment, and cleaning?
-- Where maintenance of equipment is necessary to assure that manufacturing specifications are met, is a **written record** maintained **documenting** when schedule maintenance activities are performed?
-- Is such schedule visibly posted on or near each piece of equipment, or is the schedule readily available to personnel performing maintenance activities?

(b) Inspection

-- Are periodic **documented** inspections made to assure adherence to applicable equipment maintenance schedules?

(c) Adjustment

-- Are any inherent limitations or allowable tolerances visibly posted on or near equipment requiring periodic adjustments, or are they readily available to personnel performing these adjustments?

(d) Manufacturing material

-- Is manufacturing material subsequently removed from the device or limited to a specified amount that does not adversely affect the device's fitness for use?
-- Are there **written procedures** for the use and removal of manufacturing material?
-- Is the removal of manufacturing material **documented**?

820.61 MEASUREMENT EQUIPMENT

-- Is all measurement equipment suitable, and capable of making the required measurement?
-- Is all QA and production measurement equipment routinely checked, calibrated, and inspected according to **written procedures**?

-- Are records documenting these activities maintained?
-- When computers are used as part of an automated production or quality assurance system, are all program changes made by a designated individual(s) through a formal approval procedure?

(a) Calibration

Do calibration procedures include specific directions and limits for accuracy and precision?
-- Are there provisions for remedial action when accuracy and precision limits are not met?

(b) Calibration standards

-- Does all production and quality assurance measurement equipment have calibration standards?

(c) Calibration records

-- Are the calibration date, the calibrator, and the next calibration date **recorded** and displayed, or are **records** containing such information readily available for each piece of equipment requiring calibration?
-- Does a designated individual(s) maintain a **record** of calibration dates and also of the individual performing the calibration?

820.184 DEVICE HISTORY RECORD

-- Is a DHR maintained to **demonstrate that the device is manufactured in accordance with the device master record**?
-- In addition, does the DHR include or refer to the location of: (1) the dates of manufacture, (2) the quantity manufactured, (3) the quantity released for distribution, and (4) any control number used?

820.185 CRITICAL DEVICES, DEVICE HISTORY RECORD

-- Is there a critical device history **record** for each control number which identifies the specific label, labeling, and control number used for each production unit, and is this **record** readily accessible and maintained by a designated individual(s)?

(a) Component documentation

-- Does the DHR contain the control number designating each critical component or lot of critical components used in the manufacture of the device?
-- Does the DHR contain the acceptance **record** of the critical component, including acceptance date and signature of the recipient?

(b) Inspection checks

-- Does the DHR contain or refer to the location of **records** containing: the inspection checks performed, the methods and equipment used, results, the date, and signature of the inspecting individual(s)? (See 820.101 Record of critical operation.)

820.195 CRITICAL DEVICES, AUTOMATED DATA PROCESSING

-- When automated data processing is used for manufacturing or quality assurance purposes, are there **documented** procedures designed to prevent inaccurate data input, output, and programming errors?

820.120 DEVICE LABELING

(a) Label integrity

-- Are labels designed, printed, and applied so as to remain legible during the customary conditions of processing, storage, handling, distribution, and use?
-- Are labels and other labeling not released to inventory until a designated individual has proofread samples of the labeling for accuracy?

(b) Separation of operations

-- Is each labeling or packaging operation separated physically or spatially in a manner designed to prevent mixups?

(c) Area inspection

-- Prior to the implementation of any labeling and packaging operation, is there an inspection of the area where the operation is to occur by a designated individual to assure that devices and labeling materials from prior operations do not remain in the labeling or packaging area?
-- Are any such items found destroyed, disposed of, or returned to storage prior to the onset of a new or different labeling or packaging operation?

(d) Storage

-- Are labels and labeling stored and maintained in a manner that provides proper identification and that is designed to prevent mixups?

(e) Labeling materials

-- Are labeling materials issued for devices examined for identity and accuracy?
-- Is a **record** of such examination, including the date and person performing the examination, maintained in the device history record?

820.121 CRITICAL DEVICES, DEVICE LABELING

(a) Control number

-- Do labels issued for critical devices contain a control number?

(b) Labeling check

-- Is the signature of the individual who proofreads the label and other labeling **recorded**, and is the date of the proofreading **recorded**?

(c) Access restriction

-- Is access to labels and labeling restricted to authorized personnel?

820.130 DEVICE PACKAGING

-- Are the device package and any shipping container for a device designed and constructed to protect the device from alteration or damage during the customary conditions of processing, storage, handling, and distribution?

820.150 DISTRIBUTION

-- Are there **written** procedures and release criteria to assure that only those finished devices approved for release are distributed?
-- Where a device's fitness for use or quality deteriorates over time, is there a system to assure that the oldest approved devices are distributed first?

820.151 CRITICAL DEVICES, DISTRIBUTION RECORDS

-- Do the distribution **records** for critical devices include or make reference to the location of: the name and address of the consignee, the name and quantity of devices, the date shipped, and the control number used?

820.152 INSTALLATION

-- Where a device can be installed by someone other than the manufacturer or his authorized representatives, does the manufacturer provide instructions and procedures for proper installation?

INSPECTIONS OF FOREIGN DEVICE MANUFACTURERS

Center for Devices and Radiological Health
Office of Compliance and Surveillance
Division of Compliance Programs
Manufacturing Quality Assurance Branch
February 1993

Each inspection of a foreign device manufacturer should be a thorough GMP inspection in accordance with CP 7382.830, with emphasis on the following key points:

A. PRE-INSPECTIONAL ACTIVITY:

The MDR/PRP data base should be reviewed prior to starting the inspection. A trend analysis can be performed of the data to determine whether there are any potential product and/or process problems. These can then be used to provide focus during the inspection. Printouts of MDR/PRP reports will be provided as part of the preinspection package from IPTSB. If a printout is not received, contact Deborah Paulsgrove, IPTSB (301) 443-1855.

B. 510(k) DEVICE INSPECTIONS:

A majority of inspections will cover devices which are marketed via a 510(k) premarket notification. The inspection should determine whether all devices being shipped to the U.S. have a 510(k) which has been found substantially equivalent. For these devices, it should be determined if the company is complying with any product and/or process specifications listed in the 510(k). Also, if significant product and/or process changes have occurred, it should be determined if a new 510(k) has been submitted.

In some cases, the initial importer submits the 510(k) and the foreign manufacturer may not see the submission. If so, determine who are the initial importers of the device.

In addition, some inspections may be driven by the 510(k) pre-clearance inspection program. In these instances, the investigator must FAX a short summary of inspectional findings, along with the FDA 483 (if issued) to headquarters immediately after the inspection.

C. PMA DEVICE INSPECTIONS:

Some inspections will be for premarket approval (PMA) devices. The inspection may be a pre-approval type for either an original PMA or a PMA supplement or it may be a postmarket inspection. In either instance, the inspection must be conducted in accordance with Compliance Program 7383.001. The investigator will be provided with the appropriate manufacturing sections of the PMA.

Some routine inspections will be conducted for PMA devices which have previously received clearance, and are not subject to the post-approval inspection criteria of the Compliance Program. These inspections should determine if the company is manufacturing, and shipping to the U.S., any variations of the device which do not have PMA clearance. Also, it should be determined if there are any significant changes in either the device's design or in the manufacturing process, and whether PMA clearance has been obtained for these, or whether the manufacturing site has changed.

D. ELECTRONIC PRODUCT RADIATION PRODUCING DEVICES:

During the inspection, determine whether the firm manufactures any devices subject to Sub Chapter C -- Electronic Product Radiation Control, formerly the Radiation Control Health and Safety Act (RCHSA). If so, determine whether the firm has submitted initial reports. Are any of the devices subject to performance standards? If so, are they meeting the performance standards? Determine whether the firm reported all Accidental Radiation Occurrences (AROs) to FDA. Determine whether the firm performed any Corrective Action Plans (CAPs) without notifying FDA.

E. CRITICAL DEVICES:

1. If any critical devices are being manufactured for export to the U.S., these should be given priority coverage. Determine whether the critical device sections of the GMPs have been applied to the critical device. In particular:

   a. which components are critical, what mechanism (i.e., FMEA) was used to define critical components, and what types of tests are performed on incoming critical components prior to acceptance.
   b. which operations are considered critical, what mechanism (i.e., FMEA) was used to define critical operations, and how are critical operations documented.

F. ALL DEVICE INSPECTIONS:

1. Complaint Handling System -- This should be the beginning point of every inspection to determine whether the firm has received complaints of <u>possible</u> (or <u>potentially</u>) defective devices. The review of complaints should not extend to just products or lots shipped to the U.S., but to <u>all</u> lots shipped by the firm. Any complaints received can be used to focus on a potential problem area during the inspection. Collect hard copies of enough complaints to provide a representation of actual complaints and a print-out or copy of the complaint database if there are numerous complaints.

   Even if the firm states they have never received complaints, or have no complaint file, determine whether a complaint handling unit has been formally designated. As long as the item is placed on the FDA 483, the next inspection of the firm can use the newly-established complaint file to focus in on problem areas. If the firm has received complaints from non-U.S. customers, but none form their U.S. customers, try to determine whether U.S. distributors are forwarding complaints to the firm.

If the firm repairs and services devices they produce, they must have an adequate system in place for screening repair and service requests to assure whether any of these meet the definition of a complaint. Also, repair and service requests, shipments of spare parts/assemblies, and warranty replacements to distributors should be screened to determine whether there are infant mortality problems, or failures within the warranty period, and to detect problems with particular components, subassemblies or design. Repairs/replacements are often performed by the distributor or subsidiary to save shipping costs to/from the manufacturer. In such cases, review shipments of spare parts to determine if product failures are occurring.

Complaints, service or repair records for devices shipped to the U.S. should be reviewed for MDR reporting. Also, it should be determined whether the foreign manufacturer or the initial importer has responsibility for MDR related event review and reporting.

2. Failure Investigation and Analysis System -- Determine whether there is a formal failure analysis program in place that includes written records of the actual failure investigations. It should be remembered that the complaint section of the GMP (§21 CFR 820.198) refers to possible device failures, and performing investigations to determine whether the complaint can be confirmed. Once the complaint is confirmed as an actual failure of the device, the failure investigation section of the GMP (§21 CFR 820.162) takes effect. Keep in mind that this section requires that any failures of released devices must be investigated. If the foreign firm has a formally documented system, a review of failure investigation records from the date of the last inspection or two years, whichever is longer, should indicate the types of problems the firm has experienced. These may provide clues on areas or products to focus on during the inspection.
3. In-Process and Finished Device Rejects and Rework -- Records should be examined to determine whether there has been excessive lots, or portions of lots, rejected during either in-process or finished device inspection for failing to meet any or all of the product's specifications. Excessive rejects may be an indication of poor process control. Instances of release and distribution of lots which failed to meet any or all specifications should be reported and documented. Records should be examined to determine whether any lots which have failed specifications were reworked/reprocessed, and whether this reprocessing is adequate to assure that specifications will be met without affecting the safety or performance of the device.

   All sampling plans for inspection and rework should be examined to determine whether they are based on an acceptable statistical rationale (i.e., MIL STD 105E). The sampling plan used should be examined to determine whether adequate samples are obtained for inspection, and should be described in the EIR.
4. Change Control Procedures Evaluation -- The inspection should determine if any changes have been made to the device and/or process. These may have occurred as a result of complaints, failure investigations, in-process/finished device

inspection, or rework/reinspection, and can be determined by examining the change control documents.

If changes have been made to either the device design or the device manufacturing process, determine whether they have been adequately validated (820.100 (a)(2) and 820.100(b)(3)) to assure that: the change does indeed correct the problem; and, the change does not adversely affect any other component/subassembly in the device. Avoid getting bogged down in cosmetic changes. However, changes for economic reasons (i.e., to cut costs) may sometimes lead to performance problems with the device.

One way to determine if the change has not been adequately validated is if a series of change orders (found by reviewing the DMR and various Engineering Drawings) are found for the same device which are intended to sequentially correct the same or similar problems. This may indicate that the original change was not effective, and subsequent changes have been or are being made to correct the problem. Or, complaints may be received on lots manufactured after the design change.

5. Validation -- The inspection should determine whether adequate validation of significant manufacturing processes has been performed (820.100(a)(1)). Validation should ensure that the product will consistently meet specifications if the process is controlled within established parameters; and, that the validation, either prospective or retrospective, has addressed the limits of these parameters. Validation, depending on the scope of the operation, can cover all aspects from the selection of components, to various manufacturing processes, to end-product testing. Note that retrospective validation requires extensive device history records and a protocol which includes the limits of the operation and historical data. Refer to the Guideline on General Principles of Process Validation.
6. Components -- The proper selection and assurance of acceptance of components is crucial. It should be determined whether a manufacturer has component specifications and the methods used to assure that received components meet these specifications. Particular attention should be paid to critical components and those components which would require special handling or storage to maintain their integrity. Be wary of numerous changes of suppliers, and situations where a manufacturer has relaxed component specifications because the supplier could not meet the original specifications.

   Vendor audits, at this time, are not required and failure to perform them should not be placed on the FDA 483, unless it is required by the firm's own written procedures.
7. Audits -- Finished device manufacturers are required to conduct planned and periodic audits of their quality assurance programs. The quality audit is the foundation of the quality assurance program. You should determine if the manufacturer has a written procedure for conducting quality audits and how often these audits are conducted. It is recommended that the time between audits not exceed a 12 month period. Audits more frequently than 12 months may be needed if the firm has GMP problems.
8. Initial Distributors -- The inspection should determine who are the initial distributors (U.S.) for each device. The EIR must include the complete name and

address of the initial distributors and the specific device(s) distributed by each must be identified. This information is vital for determining MDR reporting requirements. Foreign firms are not subject to MDR because they are not required to register. However, in cases where there is joint ownership and control, knowledge of a reportable event by the foreign manufacturer is imputed to its U.S. subsidiary, i.e., the initial distributor. In all cases, the initial distributor is subject to reporting when it becomes aware that its foreign manufacturer has information that meets the threshold requirements of MDR.

Foreign firms that are owned or jointly controlled by a U.S. firm are subject to MDR. If a U.S. firm provides the foreign firm with device specifications then this is considered to be joint control. If the U.S. firm owns part of the foreign firm and has the right to dictate policy, orders, etc., then this is considered to be joint ownership.

During the inspection, the initial distributor should be identified for all MDR reportable complaints found. Also, determine whether the foreign firm has received any complaints from initial distributors.

9. Product Labeling. -- Collect specimens of representative labels and labeling including product brochures and catalogues, and other promotional material. Collect three copies of labels and labeling for all violative devices, e.g., recall, lack of 510(k), etc.
10. Recalls -- Collect hard copies of any and all records indicating the possible recall of devices. This includes complaints, change orders, and quality messages to customers.

G. STERILE DEVICES:

1. If the firm is manufacturing sterile devices, the criteria for Track I and II inspections in CP 7382.830A, Attachment A must be followed.
2. There should be documentation showing how the sterilization process has been validated. If there appear to be deficiencies, appropriate records must be obtained to document the deficiencies.
3. Whether or not a sterilization process has been validated, determine whether the firm is producing non-sterile devices. Some of the ways of determining this are:

- Have any lots had positive sterility test results and/or positive biological indicator (BI) results.
- Were any lots reprocessed and then released.
- Are reprocessing procedures adequate, and do they assure device performance or package seals will not be adversely affected.
- Are test methods based on an accepted method and are they performed properly to be able to detect positive sterility/BI test results. If an in-house test method is used, has it been validated to show it is capable of detecting non-sterile units. (Remember that the USP test of 40 samples is unable to detect 15% of the non-sterile units with 95% confidence.)
- Have any lots been released even though cycle specifications have not been met.

- Does the cycle have a sterility assurance level (SAL), or an adequate SAL, for the device being sterilized.
- What is the bioburden on the device, and is the sterilization cycle able to destroy the bioburden to an appropriate SAL level.

H. DOCUMENTATION:

1. FDA 483 -- Each listed observation should be clear and concise. The facts should be clearly stated so that, if the company chooses, they can respond appropriately to the pertinent observations.
2. EIR -- Make certain that each item on the FDA 483 is fully discussed in the report. Where possible, documentation should be collected to support each observation (even if the documentation is in a foreign language).
3. All EIRs must include the FAX number of the foreign manufacturers, if available. Written correspondence, e.g., Warning Letters, will be mailed and FAXed to foreign manufacturers simultaneously. FAXing the correspondence is necessary to assure that the foreign manufacturer receives the correspondence before it is available under FOI.

I. DISCUSSIONS WITH MANAGEMENT:

During the discussion, do not under any circumstances indicate any inspectional conclusions.

All discussions regarding FDA 483 observations should be made in accordance with IOM 516.1.

If no FDA 483 is issued, inform the firm that your establishment report is subject to further review and they may or may not hear from the agency. Keep in mind that FDA 483 observations may be determined to be non-supportable upon further review, and in instances where no FDA 483 is issued, further review of the EIR may disclose significant deviations which could result in a regulatory action.

If the firm indicates that a response to FDA 483 inspectional observations will be sent to the agency, inform the firm that all responses must (A) include the documentation supporting and or showing corrections (e.g., records, data) and (B) all records and documents must be translated into English.

J. REINSPECTION OF AUTOMATIC DETENTION FIRMS:

1. If the reinspection has clearly determined that the foreign manufacturer has made corrections and no FDA 483 is issued, then FAX a short note to HFC-134 (Debbie Paulsgrove), stating the same and that the firm should be removed from automatic detention. In addition, identify exactly what devices should be removed from automatic detention if there are devices which should remain on automatic detention. After receipt of the FAXed information, HFC-134 will inform Manufacturing Quality Assurance Branch (MQAB) HFZ-332, of the inspectional findings. MQAB will in turn inform Regulatory Guidance Branch,

HFZ-323, who is responsible for notifying Import Operations, HFC-131, that the automatic detention should be removed.

A foreign manufacturer may be considered for removal from automatic detention when a FDA 483 is issued if the FDA 483 observations are new, insufficient, and/or minor deviations and the reinspection has determined that adequate corrections have been made regarding GMP deviations which resulted in the Warning Letter and Automatic Detention.

2. Foreign manufacturers should remain on automatic detention if the reinspection results in the issuance of a FDA 483 which contains significant continuing, and/or new GMP deviations. If this is the case, then a short note should be FAXed to HFC-134 (Debbie Paulsgrove) stating the same and specifying whether any device should be added or removed from the ongoing automatic detention.

APPLICATION OF THE MEDICAL DEVICE GMPS TO COMPUTERIZED DEVICES AND MANUFACTURING PROCESSES: MEDICAL DEVICE GMP GUIDANCE FOR FDA INVESTIGATORS

Office of Compliance and Surveillance
Division of Compliance Programs
FIRST DRAFT: November 1990

FOREWORD

In October 1982, the Food and Drug Administration established the Center for Devices and Radiological Health (CDRH) by merging the Bureau of Medical Devices and the Bureau of Radiological Health.

The Center develops and implements national programs to protect the public health in the fields of medical devices and radiological health. These programs are intended to assure the safety, effectiveness, and proper labeling of medical devices, to control unnecessary human exposure to potentially hazardous ionizing and nonionizing radiation, and to ensure the safe, efficacious use of such radiation.

The Center publishes the results of its work in scientific journals and in its own technical reports. These reports provide a mechanism for disseminating results of CDRH and contractor projects. They are sold by the Government Printing Office and/or the National Technical Information Service.

We welcome your comments and requests for further information.

Walter E. Gundaker
Acting Director
Center for Devices
and Radiological Health

PREFACE

This document contains general information for applying the Device GMPs to computerized manufacturing processes and to the manufacture of computerized medical devices. It is intended to provide guidance to FDA investigators and supplement document FDA 84-4191, Medical Device GMP Guidance for FDA Investigators.

This document is not intended to suggest that the procedures and practices discussed are the only ones that the Food and Drug Administration finds acceptable. Computer technology, as well as the procedures for duplication and evaluation of software, is changing at a fast pace. Therefore, it is the responsibility of the manufacturer to establish procedures and controls adequate to assure compliance with applicable GMP requirements. The adequacy of each manufacturer's practices will be evaluated on a case by case basis.

This document is also designed to supplement FDA issued compliance policy statements and reference manuals, such as Compliance Policy Guides, [(1)] and FDA's Technical Reference on Software Development Activities.[(2)] These materials should be reviewed in conjunction with this document.

[[(1),(2)] See Appendix B for references.]

Application of Medical Device GMPs to Computerized Devices and Manufacturing Processes

1.0 PURPOSE

This document outlines GMP requirements as applied to the manufacture of computerized devices and the control of computerized manufacturing and quality assurance systems. It is intended to provide guidance to FDA investigators and to supplement FDA document 84-4191, Medical Device GMP Guidance for FDA Investigators.[(1)] This document is also designed to supplement FDA issued compliance policy statements and references on Software Development Activities Policy Guides [(2)] and FDA's Technical Reference on Software Development Activities. [(3)]

2.0 SCOPE

This document applies to manufacturers who utilize automated systems for manufacturing, quality assurance, and/or recordkeeping. It also applies to manufacturers of medical devices that are driven or controlled by software.

3.0 INTRODUCTION

The GMP contains requirements which assure that specifications are established for the devices, components, labeling, and packaging and that these specifications are met. The GMP is written in general terms in order that it may apply to a broad diversity of medical devices and manufacturing processes found in the medical device industry. Because of this, FDA investigators sometimes have difficulty in applying the GMP to certain aspects of the industry. Automation is one area where investigators have expressed difficulty in applying the GMP, whether it is automation of individual devices or automation of a manufacturing system.

This document is intended to assist investigators in properly interpreting and applying the GMP to this industry. However, investigators should understand that while the procedures and controls described in this document are acceptable to FDA, they may not be the only procedures and controls acceptable to FDA. Manufacturers are free to use other approaches as long as they can provide assurance that they are adequate in meeting the applicable GMP requirements.

4.0 APPLICATION OF THE GMP

4.1 GENERAL

In order to assure that only safe and effective devices are distributed, devices must be designed and manufactured under adequate quality assurance controls. The following is a section by section discussion of the GMP as it applies to computers and describes the types of controls that would typically be expected. The actual controls utilized by a manufacturer may differ from those described. When they do, investigators should obtain justification from the manufacturer.

The validity of the manufacturer's approach should be evaluated in terms of the manufacturer's demonstrated degree of success in applying the approach to the manufacture and distribution of only safe and effective devices.

4.2 ORGANIZATION (820.20)

The GMPs require all manufacturers of medical devices to establish and implement an organizational structure that includes a formal quality assurance program and sufficient personnel to assure that all devices are manufactured in accordance with the GMPs. The program that a manufacturer establishes to implement the GMP requirements effectively becomes the firm's quality assurance program.

In order to comply with the GMP requirements, manufacturers organize themselves in such a way that there is adequate and continuous control over all activities affecting quality. Technical, administrative and human factors affecting quality of the products produced are properly controlled. Such controls are oriented towards the reduction, elimination and prevention of quality deficiencies.

The responsibility, authority and the interrelation of all personnel who manage, perform and verify work affecting quality is defined. The program emphasizes the identification of actual or potential quality problems and the initiation of remedial or preventive measures.

All the elements, requirements and provisions adopted by a company for its quality assurance program are documented in a systematic and orderly manner in the form of written policies and procedures. Such documentation (e.g. quality plans, manuals, records, etc.) ensures a common understanding of quality policies and procedures.

Management assigns an individual the responsibility and authority for ensuring that the requirements of the GMP are implemented and maintained.

Part 820.20(a) contains some specific responsibilities of the QA program.

4.2.1 Quality Assurance Program Requirements (820.20(a))

Part 820.20(a)(1) of the GMP mandates that all production records must be reviewed. This requirement applies equally to manual and computerized records.

Per Part 820.20(a) (2) each manufacturer is responsible for assuring the acceptability of components and labeling, as well as the finished device, regardless of whether they are manufactured in house or provided under contract by another company (vendor supplied). Therefore, a manufacturer's quality assurance program includes procedures for assuring approval or rejection of contract supplied software.

To assure that only acceptable software is received, manufacturers who purchase software from vendors establish a program for assuring that the vendor has demonstrated a capability to produce quality software. The program provides assurance that the requirements for the software are clearly defined, communicated and completely understood by the vendor. This may require written procedures for the preparation of requirements and purchase orders, vendor conferences prior to contract release and other appropriate methods. In order to assure understanding, manufacturers establish a close working relationship and feedback system with the vendor. In this way a program of continual quality improvements can be maintained and quality disputes avoided or settled quickly.

Acceptance procedures for contract supplied software may vary. For example, they may include third party certification. The finished device manufacturer, however, has the primary responsibility for assuring the software is adequate for its intended use. When third party certification is used, the certification package includes adequate documented evidence that the software complies with specified requirements. Examples of such evidence include documentation of the review, including procedures used to evaluate the software, the results of the evaluation and evidence of the decision making process used by the manufacturer to conclude that the software will fulfill its requirements. When the contract supplied software includes more functions than are utilized, those portions of the program which will be used are evaluated for their application. Also, the software is evaluated to assure the unused portions do not interfere with proper performance. Specific requirements which apply to these activities are covered under 820.80(a), 820.160, and 820.161 and are discussed later in this document.

Part 820.20(a)(3) requires manufacturers to identify quality assurance problems and to verify the implementation of solutions to those problems. Thus, quality data collected by a firm through its various documented process and control systems, such as work operations, processes, quality records, service reports and customer complaints, are evaluated by appropriate methods (e.g., trend analysis) to determine if there are trends or recurring problems which warrant corrective action.

These reviews are an important element of an effective quality assurance program and are important for identifying conditions or situations (e.g., device design problems or problems associated with the manufacturing process) which might not otherwise be apparent or might be dismissed as isolated incidents. The results of investigations and corrective actions are, of course, documented.

Per 820.20(a)(4) all quality assurance checks must be appropriate and adequate for their purpose and must be performed correctly. QA software checks may be both quantitative or qualitative; testing is not restricted to quantitative measurements. Testing of software involves evaluation of conformance to specifications and ability to perform as intended. Therefore, test results may vary in expression from a numerical value which is the check sum for the program or the result of a complex mathematical calculation to a qualitative determination, such as the functional adequacy of the illumination of a light or a display.

QA checks of the original program before it is released to manufacturing include review of documentation to assure that the program conforms to its design specifications, which are covered by 820.181(a), as well as an evaluation to assure it performs as intended. It is common for the program to be evaluated as segments or modules first, then as an integrated unit, and finally a system. The documented test results are evidence of the evaluation. When software is involved in manufacturing and quality assurance, evaluation is covered by 820.100(a)(1) and 820.61. This evaluation is performed when software is developed in house, and, when it has been supplied by a vendor.

After the program has been accepted, and released to production, it is evaluated to assure that it is accurately transferred/copied from storage medium to storage medium (e.g., magnetic disks to integrated electronic circuitry chips). The evaluation also assures that the working master copy remains an exact duplicate of the program that was approved and released to production and that the software has not undergone unapproved revision or modification.

Section 820.20(b) requires manufacturers to conduct planned and periodic audits of their quality assurance program. Every quality audit includes a review of procedures and activities to assure standard operating procedures are adequate and are being followed and that all elements of the quality system are effective in achieving stated quality objectives. As applied to software, this audit includes evaluation of procedures used to assure that hardware and software are adequate for their intended use, and that SOPs remain adequate. The audits extend to all phases of software design, development, testing, design transfer, implementation, and maintenance activities related to computerized processes and devices.

Since in many cases manufacturers rely on suppliers to assure quality of software, the quality audit includes the supplier. On site assessments are made of the supplier's capability to produce quality software and other components.

Audits are conducted by individuals qualified to perform the task. Evidence is available to show that individuals involved in software QA review and evaluation have been adequately trained. As with any other device manufacturing process, these individuals have a working knowledge of how the device is made, and they should also have a working knowledge of developing and documenting software.

The results of the audits are documented and brought to the attention of the personnel responsible for the areas audited and upper management. Timely corrective action is carried out and verified as necessary.

4.3 PERSONNEL TRAINING (820.25(A))

All personnel must have adequate training to perform their assigned responsibilities. This means that individuals responsible for producing and evaluating software have the necessary education, training, and experience to assure that the software is properly prepared and maintained. These individuals know how to develop the software and have an understanding of how to properly document and test the program to minimize, with an adequate degree of confidence, the effect of latent faults.

Also of concern are the training, experience, and knowledge of employees responsible for duplicating software and handling magnetic storage media (e.g., floppy disks, tapes, PROM chips, etc.). Training is conducted to assure these individuals are fully aware of their responsibilities, particularly of the controls and procedures they must follow to assure that software incorporated into the final medical device is not adversely effected and performs as intended. Appropriate records of training/experience are maintained.

4.4 ENVIRONMENTAL CONTROL (820.46)

Where environmental conditions could have an adverse impact on a device's fitness for use, the conditions must be controlled. In general, this applies to the manufacturing environment and areas used for storage of components, and the finished device.

Computers and software storage media may be sensitive to the environment. Mainframe (and some mini) computers generally call for stringent temperature and humidity controls, and all computers are subject to some degree of environmental limitations.

Overheating, whether from an external source or from the computer's own electronic circuits, can have an adverse effect on a computer's ability to operate properly. Failures caused by system overheating may range from total failure or shutdown of the system to intermittent errors. The maximum temperature at which a microprocessor or central processing unit (CPU) can operate is usually stated in the processor/CPU specifications established by the system's manufacturer.

Humidity may also adversely affect a computer system. Because the computer system is an electronic unit, excessive humidity can have a detrimental effect on electrical contacts and circuitry within the system. Conversely, a dry environment will increase the possibility of static discharges that can damage electrical circuits, software storage components (e.g., chips, and other static sensitive components) and, in turn, have an adverse effect on the software.

The degree of environmental control required is determined by the manufacturer of the finished device. Specifications are developed for the environment and maintained in the device master record. A control system is implemented to assure that the environmental specifications are not exceeded. This environmental control system is periodically inspected for proper functioning and the inspections are documented.

Some electronic hardware components such as computer chips which house the software assembled in the device, are sensitive to electrostatic discharge (ESD). When ESD is a concern, only personnel at properly ESD controlled work stations handle blank static sensitive chips, preprogrammed chips, and the circuit boards containing these chips. ESD controls include grounding, humidity control, negative ion generation, etc. The firm's system for controlling ESD is periodically inspected to assure it is exercising adequate control: Are work stations properly grounded? Are employees, working with ESD sensitive components grounded to work stations? Is humidity monitored? Routine inspections are part of the equipment maintenance procedures.

Other forms of preprogrammed media such as disks (hard and floppy) and magnetic tapes are also handled only in environmentally controlled areas. In areas where these are used, the ability to retrieve data may also be adversely affected by exposure to dust and dirt; therefore, dust and dirt is controlled in addition to ESD.

Components and other media are protected from sources of magnetic interference which can result in the potential accidental erasure of the software by a magnetic field from a permanent magnet or electromagnet. If the product is electromagnetic interference (EMI) sensitive, then efforts to control and/or test for EMI are documented.

4.5 EQUIPMENT (820.60)

Section 820.60 of the GMP mandates periodic maintenance of equipment used in the manufacturing process, when applicable. When applied to the software used in production, working master copies of software are periodically challenged and compared against the archived master as a means of assuring that the working copy of the released version is a true copy of the master. Unauthorized changes may compromise the accuracy and reliability of the process. Comparison of two or more computer programs may be accomplished by a number of different procedures. One common method uses a software utility program which compares two programs and prints differences found between them. A comparison of disk directories between the master and working copies as well as the use of some comparative programs can assist in identifying the differences. The differences may be as simple as one copy containing additional utility programs while the others do not. Another procedure involves comparing the checksums of the preprogrammed chips. The checksum is the value which results from the addition of the values stored in each address on the chip. The values from each chip of the working copy are then compared with the checksums of the archived master. Any difference between the two reflects a discrepancy in the programs and indicates a change in either of the two copies, but it does not identify the location of the difference(s). This is accomplished separately.

Only the current version of software that has been approved and released for use by the device manufacturer is available in the manufacturing/quality control area. When software revisions have been made and released for use, obsolete versions of the program are removed from use. Appropriate corresponding documentation (e.g., written manufacturing procedures and/or design specifications) is also updated and distributed in a timely manner.

A written maintenance procedure is recommended as a dependable means for assuring that all aspects of equipment maintenance are covered.

4.6 MEASUREMENT EQUIPMENT (820.61)

All computerized production and quality assurance measurement equipment must be suitable for its intended use and capable of producing valid results. To establish confidence in the adequacy of computerized equipment, the hardware (sensors, transmitters, etc.) is calibrated and the software is challenged and validated to assure fitness for its intended use. Calibration is done in accordance with written calibration procedures and schedules. The frequency of calibration may be dependent upon the purpose of the measurement taken, stability, and how often the equipment is used.

Calibration of computer hardware is similar to calibration of any other electromechanical system. The sensor's measurement of temperature, voltage, resistance, etc., is compared against the measurement of a known standard traceable to the National Institute of Standards and Technology (formally the National Bureau of Standards) or other acceptable standard. An important part of the calibration activity is to assure that measurements are properly transmitted across computer communication lines and properly interpreted by the computer system.

Verification of properly transmitted measurements is accomplished by comparing the measured value that has been input into the computer system with the value of the traceable standard.

The PROM programmer, a piece of manufacturing equipment, is used for programming integrated circuits (ICs). To assure that integrated circuits are adequately programmed, equipment maintenance and calibration needs for programmers should be considered, including proper voltage, current, and pulse shape.

Modification of the hardware and/or configuration of the system may require system recalibration. Procedures for handling modifications are addressed in standard operating procedures.

When automated production or QA systems are used, the software programs are validated. Validation of system software is a complex activity which must be carefully planned and performed, before use of the software package, and after significant revision of the system occurs. Validation may also be required after any revision of the operating system software. Software verification and validation activities are discussed in greater detail in FDA's reference manual, Software Development Activities.[2]

4.7 COMPONENTS (820.80)

The GMP regulation requires that manufacturers establish adequate procedures for acceptance and storage of components to assure that only those components that are acceptable for use are released to manufacturing.

4.7.1 Acceptance of Components (820.80(a))

The components of a software driven device typically consist of circuit boards, resistors, transistors, and other discrete items commonly found in electrical devices. However, there are two additional components of special concern in a software driven device: the actual software that controls functions of the device and the hardware on which the software is stored or mounted.

Software inspection and testing are normally accomplished in a manner different from that performed on the discrete components of the device. The component specifications for software are usually referred to as the software requirements. These include user or device requirements (e.g., the device will respond in a specific manner to a specific input), and they also cover system requirements which are functions associated with the internal workings of, or handling of, data by the software. System requirements may include functions such as error checking, polling, fault tolerance, etc. Ability of the software to meet both user and system requirements is crucial to proper operation of the device.

Preparation and use of an adequate test plan, based on knowledge of software logic and the hardware environment in which the software will run, will assure that software is adequately tested or evaluated, and thereby establish confidence that it does meet specifications. In most cases, evaluation of the software requires not only testing separately from the device by simulation testing (which may use a database with known inputs) but also by testing it in the environment in which it will be used (i.e., the finished device). For example, a database which includes signals of ventricular fibrillation may be used to evaluate one function of a cardiac monitor. Because of the complexity of both software and hardware, this testing may be performed as part of the manufacturer's software development and software quality assurance activities. These tests are routinely referenced as software verification and validation. Final versions of approved test procedures constitute written component acceptance procedures for the software and results of the final tests document that acceptance criteria have been met.

After it is determined that the software is acceptable for use, consideration is given to the need for periodic retests. Retests are usually necessary when the software (operating system or application program) is revised or a software failure is encountered.

When preprogrammed storage media such as chips, disks, etc., are received as components, acceptance procedures assure that software contained in these components is the current version and that it has been adequately duplicated. Acceptance evaluation can be accomplished in a number of ways. One method is a bit by bit comparison of the software program in the incoming component against a known correct master copy of the program. Another method consists of determining the checksum of the software in the incoming component and comparing it against the known checksum for the current version of the program. (This test method has been previously discussed in the "Equipment" section of this document, however, the method is also applicable to acceptance of components.) These tests only assure accuracy of the reproduction efforts; they do not reflect the quality of the software program, which can only be determined through the verification and validation test efforts previously discussed.

Incoming acceptance procedures for unprogrammed (blank) ICs vary. They may consist of electrical tests or only a visual examination, depending upon whether history has demonstrated that the supplier can consistently provide a quality product.

Some medical device manufacturers may purchase OEM (Original Equipment Manufacture) products such as CRTs, computers, etc., and combine these products into a medical device system. These may be considered components rather than finished devices. In such cases, it is the medical device manufacturer's responsibility to assure the OEM products are acceptable for use. This may include testing the products individually and as part of the finished system to assure they conform to specifications.

4.7.2 Storage and Handling of Components (820.80(b))

As with all finished device components, software must be adequately identified to prevent mix up and adequately stored to prevent damage.

Software contained on media such as disks, etc., is identified by providing name or title and version or revision level of the software. This serves to prevent use of obsolete versions of the program.

Programmed media can be damaged by the environment. For example, it is possible that software may be accidentally altered if the hardware which contains the software program is exposed to electrostatic discharge (ESD), or to ultraviolet radiation. Therefore, manufacturers exercise care in the handling and storage of magnetic media and programmed chips. (ESD control has already been covered in this document under 4.4 Environmental Control (820.46)).

Employees engaged in handling ESD sensitive components are properly trained and made aware of the results of improper performance or poor ESD control practices.

Electrostatic sensitive chips are stored in ESD protective carriers before they are assembled into circuit boards. Under some conditions, materials promoted for ESD control can actually contribute to ESD. Therefore, materials used are qualified for their use to assure adequacy. Circuit boards containing these components are also protected against ESD damage. It is also important that preprogrammed chips and circuit boards that contain preprogrammed chips be handled only by properly trained personnel at properly ESD controlled work stations.

Some hardware components, whose software can be erased by ultraviolet light, require a protective covering over the erasing "window" of the component. If the window is left uncovered, the program contained on the component may possibly "fade" in time through exposure to fluorescent light, sunlight, or other sources of UV radiation. Protective covers may include a special plastic cap or a piece of light resistant tape placed over the window.

As previously described under 4.4 Environmental Control (820.46), exposure to dust or dirt may affect the ability of preprogrammed magnetic media, such as disks or tape, to record and read data. Contents of the disk or tape may also be altered if stored in the vicinity of a strong magnetic field. Therefore, these media are protected from rough handling and temperature extremes, as well as magnetic fields and electromagnetic radiation.

4.8 CRITICAL DEVICES, COMPONENTS (820.81)

In addition to the requirements of 820.80 as described above, additional controls are established and implemented for handling critical components of a critical device. Section 820.81(a) requires that specific controls be in place for the acceptance of critical components. Computer components such as integrated circuits (ICs) may be identified as critical components when they are used in a critical device. The complexity of these components can make it difficult for the device manufacturer to adequately test these components for acceptance. In this situation, the device manufacturer may have to rely on the component manufacturer to certify in writing that the required specifications have been met, and require the component manufacturer to provide actual test data. A vendor QA program (as discussed on page 3) is also established to assure confidence in the data.

GMP section 820.81(b) requires that, where possible, the finished device manufacturer must obtain a written agreement from the supplier of critical components which states that the device manufacturer will be notified of any proposed change in a critical component. In relation to computerized devices, this section applies to both hardware and software components that are critical. Hardware may include custom designed components such as gate arrays, programmable logic arrays, ROMs, and analog arrays which may have been made specifically to the finished device manufacturer's specifications. Critical component software may include programs which perform and control critical functions of a device. Whether the components are customized hardware or are a software program, it is important that the finished device manufacturer know when the component supplier makes any changes because a change to a component may adversely impact the finished device.

4.9 MANUFACTURING SPECIFICATIONS AND PROCESSES (820.100)

Specifications and procedures for manufacturing a device must be established, implemented and controlled to assure that the device conforms to its original design or to any approved changes in that design (820.100).

Section 820.100(a)(1) requires all manufacturers to assure that design requirements are properly translated into device and component specifications which are used in production. When applied to computerized operations, this means that manufacturers are prepared to provide evidence that the software used for duplicating the device software and the software used in automated manufacturing or quality assurance meets the software design specifications.

This section is interpreted by FDA to include process validation. When a manufacturing process is automated, the computerized system is validated to assure it performs as intended. In validating computerized equipment, parameters that the system is designed to measure, record, and/or control are evaluated by an independent method until it is demonstrated that the computer system will function properly in its intended environment.

When a manufacturing process is controlled by computer, functional evaluation of the control system may include, but is not limited to, the following activities:

- o equipment (peripherals, etc.) and sensor checks using known inputs, which may consist of processing test or simulated data;
- o alarm checks at, within, and beyond their operational limits; and,
- o evaluation of operator override mechanisms for how they are used by operators and how they are documented.

In case of system failure, evaluations would include:

- o how data is updated when in manual operation;
- o what happens to data "in process" when the system shuts down;
- o what procedures are in place to handle system shutdown; and,
- o how product or information handled by the computerized process is affected.

Process validation is conducted to evaluate the effectiveness and repeatability of the process and its impact on the device during both expected operation and worst case situations. When software is involved, this activity may in many cases have to be accomplished in two steps: first, the software is integrated into the system and the system is evaluated independently of the system it is to control; second, the software is integrated into the system and the system is evaluated.

Section 820.100(a)(2) requires that changes to specifications of a device, which includes software specifications, must be subject to controls as stringent as those applied to the original software program. Usually, this means validation that includes an evaluation of how the change impacts on the rest of the software. For example, if the addition of a subroutine or function is determined to have little effect on the device or process, only a limited number of modules may require retesting and revalidation. On the other hand, changes such as updating the operating system software could have an impact on the entire application software, thereby requiring more intensive evaluation. In any event, all changes are evaluated to assure that they are appropriate (that they achieve their intended purpose) and that they do not adversely affect the unchanged software.

Revisions to software follow established change control procedures to assure that the history of the changes are maintained and that each change is properly reviewed, approved and dated before implementation.

In order to control and maintain the software and to know its configuration at any time, documented evidence is needed to demonstrate why each change was made, that each change is adequate, and that it has been approved for use. As with any device, this information is essential for investigating device defects.

Also, if the change significantly extends the indication for use, or affects the safety or effectiveness of the device, a new 510(k) or PMA supplement may need to be submitted to FDA. If the change is made to correct a problem with respect to safety, effectiveness or performance, a recall may be needed.

4.9.1 Processing Controls (820.100(b))

When the possibility exists for the device to deviate from its design specifications as a result of an inadequately controlled manufacturing process, written manufacturing procedures must be established. As applied to software, this GMP provision includes the process of duplicating the currently released, approved "master" software program onto other storage media, generally for assembly into the device. To assure that this process is adequate and produces consistent results, manufacturers have established written procedures.

Standard operating procedures (SOPs) for software handling and duplication are controlled documents. Any changes or revisions made in these documents are subjected to formal review and approval by designated individual(s) before implementation. Once approved, the revised procedures are conveyed to appropriate personnel in a timely manner.

Process controls also include computer security and may involve limiting physical access to the computer on which the software is written and/or tested and also may include limiting access to the software itself to prevent unauthorized changes. Software security may include the use of passwords, passkeys, etc. Assignment and use of these security measures should also be controlled.

4.10 FINISHED DEVICE INSPECTION (820.160)

Adequate procedures must be in place and implemented to assure that the finished device meets its design specifications. Testing should verify that the software functions utilized perform as intended and that unused functions do not adversely affect performance. For software driven devices, it is sometimes impossible to fully qualify the computer program through performance of function tests. Because of the computer program's logic and branching capabilities, a specific task performed by the device may be accomplished in one manner, one time, and depending on the logic of the program and the data entered, in a totally different manner another time. Therefore, independent testing of the software itself is conducted if the true capabilities and limitations of the device and software are to be known. This was discussed earlier in the "Components" section of this document. Rarely can the full functional capabilities of the software be demonstrated by testing only the finished device.

Therefore, once the software has been accepted as a component for use, and adequate control of the duplication process during manufacturing has been established through validation and process control, it is usually not necessary to re-verify performance of software in each unit, batch, or lot of devices manufactured. Instead, assurance is established that the correct version of the software program is included with the device. One way to do this is to access the program and call up its current revision or version identification either on a visual display or a printout. This method, however, is not always possible. A second method consists of verifying that the labels on the program chips or magnetic media reflect the proper software revision level identified in the device master record (DMR).

Finished product inspection of a software driven medical device also includes tests normally associated with an electromechanical device. Although these tests may not fully challenge the software, they help to assure that the device has been properly assembled.

4.11 FAILURE INVESTIGATION (820.162)

When failure occurs in a distributed software driven device or in a distributed device which consists solely of software, an adequate investigation must be conducted to identify the cause. For example, in a software driven device, the failure may be related to the device design, the manufacturing process, or the quality assurance equipment used in the evaluation of the device. In a device that consists only of software, the cause of the failure may be related to the software design or the process used for duplicating the software. When the failure of a finished device is attributable to the software used in manufacturing or quality assurance, identification of the cause of a failure may require review of the software program's logic and of the test procedures and results, as well as a retesting of the program. Further reviews may be required of the duplication process and of environmental control records for those areas where ESD sensitive components were handled and assembled.

If the software error is in the device, similar investigative activities are conducted. In either situation, the investigation extends to determining effects on other products, and results in a written record of the investigation and any follow up action and corrective action taken.

4.12 RECORDS, GENERAL REQUIREMENTS (820.180)

Recordkeeping requirements that apply to nonautomated devices also apply to software controlled devices. Records must be available for review and copying by FDA employees, including those records which have been computerized and placed on computer storage media such as magnetic tape, disks, etc.

All records maintained in accordance with 21 CFR 820 are required to be retained for a period of time equivalent to the design and expected life of the device, but in no case less than two years from the date of release of the device for commercial distribution.

4.13 DEVICE MASTER RECORD (820.181)

The device master record (DMR) consists of diagrams, descriptions, schematics, etc., that constitute the specifications for the medical device product, the manufacturing process, and QA program. In addition to items detailing specifications for the device hardware, the device master record for a software driven product also includes detailed specifications for the device software. Detailed specifications are also required when the device consists of only software.

All records and documents contained in the device master record are controlled documents, including documentation related to software. Any revision or change of the software program or its supporting documentation are made in accordance with formal change control procedures and authorized by signature of the designated individual(s). Magnetically coded badges and other electronic identifiers may be used in lieu of signatures if adequate controls are in place to prevent their misuse.

4.13.1 Specifications (820.181(a))

The device master record must include specifications for the device. When software is part of the device, specifications include or refer to:

- o the final, complete, approved software design requirements, which describe in narrative and/or pictorial form, such as a flow chart, what the software is intended to do (e.g., to control or monitor something) and how it will accomplish these tasks. Also included is a description of how the software will interact with the hardware to accomplish various functions of the device's design. The specifications may also include a checksum for the program. The description is in a form that can be understood by all individuals who work on and/or will maintain the program during its life. Note that the description does not include documentation of the working drafts (or in process steps) of the software design; it only includes the final approved specifications. The procedures for evaluation of the software to assure specifications are met are covered by 820. 181(c).
- o a description of the device's computer hardware system specifications, such as interfaces, connections, and media for storage of the program in the device.
- o the computer source code as either hard copy or on magnetic medium. It usually is necessary for the finished device manufacturer to have the source code. This documentation is indispensable for adequately maintaining the program and evaluating the impact of any change on the rest of the program.

It is important that the device manufacturer collaborate with the software vendor in the initial stages when software specifications are being developed and when any changes are introduced in order to assure that the intent of the design is adequately translated into software code. In these situations, the device manufacturer and software vendor establish a contract that delineates responsibilities relating to the development and maintenance of the software.

The program source code typically includes or refers to adequate documentation which describes the subroutines or modules for the language used. Additional documentation that describes the design of the program is maintained. The intent is to assure that individuals maintaining the program have sufficient documentation to fully understand the purpose of the software design. Depth and detail of the documentation are proportionate to the complexity of the systems involved.

4.13.2 Production Process Procedures (820.181 (b))

The DMR must contain production process specifications. When applied to software controlled processes the DMR includes procedures for environmental control and specifications where applicable; procedures for duplication of software for assembly into the finished device; specifications for use of any automated or computerized manufacturing equipment or processes; and specifications for any computerized packaging and labeling operations. The DMR also includes procedures for computer/software security, if implemented.

To assure consistency of results, the DMR includes written change control procedures and any change in software that is part of the device or that is used in manufacturing or in quality assurance.

4.13.3 Quality Assurance Procedures and Specifications (820.181(c))

The DMR must also include all documentation used to determine quality of conformance to established specifications for the components, device, packaging, labeling and manufacturing processes. For software, this includes, but is not limited to, identification of any automated test equipment, as well as test procedures and criteria used to evaluate the current device software program for acceptance for use in manufacturing (820.80)(a) and for acceptance of hardware components used to store the software in the device (also 820.80(a)). For computerized manufacturing processes, this also includes any tests which are performed to evaluate the adequacy of the process, such as evaluating the integrity of package seals and verifying that the correct label was applied.

4.13.4 Labeling (820.181(d))

The final element required by the device master record concerns labeling for the finished medical device. Because of the possible complexity of a software driven device, extensive labeling may be required for adequate user instructions. This labeling may take the form of user manuals or it may be embedded directly into the software for the device, appearing on screen as instructions and menus.

User manuals or directions are written in clearly understood terminology and consist of operating instructions that explain how the system works and the procedures to be followed. Manuals include an explanation of all advisories, alarm and error messages, as well as corrective actions to be taken when these situations occur.

4.14 DEVICE HISTORY RECORD (820.184)

The device history record (DHR) demonstrates that the device is manufactured in accordance with the specifications in the device master record. This agreement is shown by documented evidence that manufacturing and test procedures have been followed and that the results meet acceptance criteria. When software is part of the device, this documentation includes a record of the version of the software which was assembled into the device, results from evaluating the device software (e.g., performance), in addition to all documentation needed to show that the software was adequately reproduced during manufacturing.

Adequate production records are in place to properly document all significant activities. For example, software that is part of a device may be copied into components, such as PROMs (Programmable Read Only Memory Chips), which are then assembled into the device. Production records for this activity document the results of the duplication process. For example, when checksums are used to identify the revision of the software which is duplicated into components, the production record documents the checksum and the number of components which were copied as well as the date the activity was performed. All production records are included in, or referred to in, the device history record.

4.15 CRITICAL DEVICES, AUTOMATED DATA PROCESSING (820.195)

Section 820.195 applies only to manufacturing or quality assurance activities associated with critical devices. Automated data processing is the means used to gather and analyze information on some characteristic of the device manufacturing process or QA program without direct use of an operator to control the activity or verify the results. Automated data processing systems provide an effective method for performing routine, repetitive tasks. Although generally more reliable than manual equivalents, such systems demand adequate controls for equipment setup and programming. The GMP regulation requires a manufacturer to implement controls that will assure the correctness and appropriateness of these programs, program changes, equipment and data input and output.

4.16 COMPLAINT FILES (820.198)

Firms must prepare and implement adequate complaint handling systems including the review, investigation, and evaluation of both hardware and software failures of distributed devices. A notation in the complaint file that a system has failed as a result of a software error is supported with data or evidence to justify that conclusion. When a software failure is encountered, an investigation is conducted to determine the cause of the error and its impact on the capabilities of the device and similar devices.

Many manufacturers use computers for recording and tracking complaint information contained in paper documents, such as letters from complainants or laboratory reports. The complete information may be copied into the computer system in lieu of maintaining the original documents. If the documents are retained, however, the computerized complaint record makes reference to corresponding paperwork.

Complaints are an excellent source of information about device design and the manufacturing process by which the device was produced. When complaint files are computerized, a software program that provides a means of determining the existence of any similar recurring problems with the device, similar devices, or with the manufacturing process, which might indicate a need for possible corrective action, is invaluable.

APPENDIX A

DEFINITIONS

Archived Master (copy of software)	A software library which contains formally approved and released versions of software and documentation from which copies are made.
Checksum	The value from adding the individual values at each address of the hardware component which contains the software program. This value may also be used to indicate software versions.
Chips	An electronic hardware component consisting of integrated microcircuits which perform a significant number of functions.
Disk Directories	Index of the file names on the disk. It may also include file size, date of creation, and date last altered.
Electrostatic Discharge (ESD)	A discharge of the potential energy that electric charges possess by virtue of their positions relative to each other. This discharge may adversely affect hardware sensitive to potential differences.
Error Checking	A means of determining if recording of data, its input into a computer system, and its transfer within the system, including transmission, is correct.
Fault Tolerance	Systems that continue to operate satisfactorily in the presence of faults (i.e., hardware failures).
Integrated Circuits (IC)	Complex electronic circuits etched on small semiconductor chips.
Master Copies of Software	The approved versions of the software from which copies are made for use and reproduction in the manufacturing environment.
Polling	In a data communications system, a line control method in which the computer asks each terminal on the system, in turn, if it has a message to send.
Programmable Read Only Memory	Memory chips of which the contents can be read but not altered during program execution. However, the contents of the (PROMS) memory can be altered before it is assembled in the computer system.

PROM Programmer	Electronic equipment which is used to transfer a software program into a PROM.
Third-Party Certification	The procedure and action, by a duly authorized independent body, of confirming that a system, software subsystem, or computer program is capable of satisfying its specified requirements in an operational environment. Certification usually takes place in the field under actual or simulated operational conditions, and is used to evaluate the software itself and the specifications to which the software was designed. Certification activities take place under a written, approved (by the manufacturer) protocol.
Validation	Establishing documented evidence which provides a high degree of assurance that a specific process will consistently produce a product meeting its predetermined specifications and quality attributes.
Validation Testing	Testing that commences after the completion of the development testing and includes module and subsystem level testing. These tests can be considered to be "rehearsals;" they are basically gross tests of the coding against specifications.
Verification	The process of reviewing, inspecting, testing, checking, auditing, or otherwise establishing and documenting whether or not items, processes, services, or documents conform to specified requirements.
Verification Testing	An acceptance test of software. These tests are rigorous and detailed and will result in the software quality certification that the coding is in complete agreement with the specifications, design, and test documentation.
Worst Case	A set of conditions encompassing upper and lower processing limits and circumstances, including those within standard operating procedures, which pose the greatest chance of process or product failure when compared to ideal conditions. Such conditions do not necessarily induce product or process failure.

APPENDIX B

REFERENCES

1. FDA 84-4191: Medical Device GMP Guidance for FDA Investigators (April 1984).
2. FDA Compliance Policy Guides

Office of Enforcement, Compliance Policy Guide 7132A.07: Computerized Drug Processing: Input/Output Checking (October 1,1982).
Office of Enforcement, Compliance Policy Guide 7132A.08: Computerized Drug Processing; Identification of "Persons" on Batch Production and Control Records, (December 1, 1982).
Office of Enforcement, Compliance Policy Guide 7132A.11: Computerized Drug Processing., CGMP Applicability to Hardware and Software, (December 1, 1984).
Office of Enforcement, Compliance Policy Guide 7132A.12: Computerized Drug Processing., Vendor Responsibility, (January 18, 1985).
Office of Enforcement, Compliance Policy Guide 7132A.15: Computerized Drug Processing; Source Code for Process Control Application Programs, (April 16, 1987).

3. FDA 84-4191: Software Development Activities, Reference Materials and Training Aids for Investigators (July 1987).
4. FDA Field Computer Specialists and their location: Martin Browning, DFI/ORA, Rockville, MD; John Kunkel, Minneapolis District; Philip Piasecki, Boston District; Sam Clark, Atlanta District; Paul Figarole, Baltimore District; Dwight Herd, San Juan District.
5. References for Definitions:

Capron, H.L. and Williams, Brian K., Computers and Data Processing, 1982, The Benjamin/Cummings Publishing Company, Inc.
Parker, Sybil P., Editor in Chief, McGraw-Hill Dictionary of Scientific and Technical Terms, 1984, McGraw-Hill Book Company.
Ralston, Anthony, Editor, Encyclopedia of Computer Science and Engineering, 1983, Van Nostrand Reinhold Company.
Foster, Richard A., Introduction to Software Quality Assurance, 1975, R. A. Foster.
Dersey, Roger M., Digital Circuits and Devices, 1985, John Wiley and Sons, Inc.
Fraf, Rudolf F., Modern Dictionary of Electronics, 1977, Howard W. Sams and Company, Inc.
Jay, Frank, Editor in Chief, IEEE Standard Dictionary of Electrical and Electronics Terms, 1984, The Institute of Electrical and Electronics Engineers, Inc.

6. Recommended References:

FDA 87-4179: CDRH, Device Good Manufacturing Practices Manual, 4th Edition, Division of Small Manufacturers Assistance, OTA (November 1987).

FDA Compliance Program Guidance Manual, Compliance Program 7382.830, Inspection of Medical Device Manufacturers (October 1985).

Center for Drugs and Biologics and Center for Devices and Radiological Health, Guideline on General Principles of Process Validation (May 1987).

FDA 90-4236: CDRH, Preproduction Quality Assurance Planning; Recommendations for Medical Device Manufacturers, Office of Compliance and Surveillance, Division of Compliance Programs (September 1989).

Chapter 17

HAZARD ANALYSIS

OVERVIEW

Recently, reinterpretation and expansion of the FDA's 510(k) requirements have led to growing concern about system validation issues related to computer-related medical devices, and the FDA continues to provide guidance concerning whether or not validation evidence is relevant to a 510(k) submission.

SYSTEM VALIDATION AND THE 510(k)

System validation concerns may enter consideration in the 510(k) preparation and submission process at two stages: Outcome Probability Analysis, and Causal/Preventive Analysis. Both stages are related to issues of Hazard Analysis.

Under current FDA procedures, a medical device that relies upon, is controlled by, has a critical performance factor, or otherwise is substantially dependent upon a software (or firmware)-driven component is subject to a special Device Safety Issues (Hazard) Analysis. This analysis is conducted by a computer engineer using a table construction tool to examine and assess the following factors:

- Potential (or reported) Device Error or Failure
- Patient Effect
- Outcome Potential
- Criticality of Outcome (death, serious injury, non-serious injury, or no injury)
- Qualitative Probability (rare, seldom, occasional, frequent)

These factors are utilized to assign an appropriate "Issue Category" that, in effect, determines the degree of scrutiny and system testing that a particular device may merit.

LEVELS OF CONCERN

The FDA has identified three "levels of concern" based on the risk to patients from potential device failures. These levels are:

Major Concern

"The level of concern is major if operation of the device or software function directly affects the patient, so that the failures or latent design flaws could result in death or serious injury to the patient, or if it indirectly affects the patient (e.g., through the action of a care provider) such that incorrect or delayed information could result in serious injury or death of the patient."

Moderate Concern

"The level of concern is moderate if the operation of the device or software function directly affects the patient so that failures or latent design flaws could result in minor to moderate injury to the patient, or if it indirectly affects the patient (e.g., through the action of a care provider) where incorrect or delayed information could result in injury of the patient."

Minor Concern

"The level of concern is minor if failures or latent design flaws would not be expected to result in death or injury to the patient. This level is assigned to a software component that the manufacturer can show to be totally independent of other software or hardware that may be involved in a potential hazard and would not directly or indirectly lead to failure of the device that could cause a hazardous condition to occur."

Five factors are considered in the FDA review of a software-controlled device to determine the level of concern for particular devices:

1. The nature of a device; i.e, its intended use.
2. The severity of potential harm that a device failure could inflict or permit.
3. The benefits a patient might receive from using the device.
4. The degree of the device's involvement in therapy or diagnosis.
5. the degree of the software involvement in the function of the device.

Criticality of Outcome Analysis

Potential hazards should be identified. Hazards relating to the malfunction of a computerized medical device can generally be described in four categories: death, serious injury, non-serious injury, or no injury.

Outcome Probability Analysis

In the assessment process, the determination of Qualitative Probability is based upon two factors:

1. The product history, including pre-approval testing; and
2. Evidence that the controlling system (presumably software) meets design and control requirements sufficient to prevent or correct emergent problems (validation).

Should the history of the product contain evidence of potential errors or failures (which presumably have been corrected in the final versions of the product upon which the 510(k) is based), or should the product be dependent upon the software or firmware component to insure safe performance to prevent or avoid a potential error or failure, or should the product be of a type attracting particular regulatory attention, the investment in an independent and expert validation study will provide potentially cost-effective evidence in favor of low probability of system failure.

The outcome probabliity analysis results in a criticality rating of each hazard -- rare, seldom, occasional, or frequent.

Combining the criticality of outcome and outcome proability analysis can result in a chart depicting the level of concern for each device:

	Rare	Seldom	Occasional	Frequent
Death	Moderate	Major	Major	Major
Serious Injury	Minor	Moderate	Major	Major
Non-Serious Injury	Minor	Minor	Moderate	Moderate
Non-Injury	Minor	Minor	Minor	Minor

Causal/Preventive Analysis

If the result of the Device Safety Issues (Hazard) Analysis places the product in category three or four (Death or serious injury) independent and expert validation evidence will likely speed the 510(k) approval process, and it will likely insure that the testing and scrutiny to which the device is subject is appropriate and technically accurate.

A system validation study is recommended for all devices characterized as producing or potentially producing a device error or failure resulting in a category three or four effect.

THE SYSTEM VALIDATION STUDY

The purpose of the system validation study as it relates to hazard analysis is to provide appropriate evidence that the software or firmware components of a device are designed and functioning to: (a) minimize the probability of a potential outcome; and (b) demonstrate appropriate controls to prevent or minimize the negative effects of such a failure. To effectively succeed in this role, a validation study must pass two tests:

1. Independence of the validation: evidence gathered, analyzed, and presented by person or persons independent of the 510(k) submitter.
2. Expertise: decision supported by and testimonial report authored by highly credible and expert person or persons.

The actual design of the study will vary significantly, but should generally include evidence of:

- A thorough source code review examining for conformity of standards of documentation, design, change control, audit capability, and control.
- A performance test review, in which the validating expert analyzes the available product test results to determine whether reported problems were related to system design or performance.
- A system audit review, in which the validating expert audits all system logs, notes, developmental flows, and other evidence of design and testing.

FDA GUIDELINES

Three FDA guidelines on 510(k) reviews are attached. The first is the July 1986 Guidance on the Premarket Notification Review Program. Next is the August 1991 Reviewer Guidance for Computer Controlled Medical Devices Undergoing 510(k) Review. The third guideline is the April 8, 1994, draft document on "Deciding When to Submit a 510(k) for Change to an Existing Device."

GUIDANCE ON THE CENTER FOR DEVICES AND RADIOLOGICAL HEALTH'S PREMARKET NOTIFICATION REVIEW PROGRAM (June 30, 1986)

PURPOSE OF THE GUIDANCE

This guidance document outlines requirements for premarket notification review, and explains some points of the Center for Devices and Radiological Health (CDRH) will consider when making a determination that a device is, or is not, "substantially equivalent." A consistent application of the principles set forth in the guidance should be useful to Center staff in conducting complete and expeditious review of 510(k) submissions.

PURPOSE OF THE PROGRAM

In brief, the premarket notification program is meant to:

- identify new devices that must be placed automatically into class III and undergo premarket approval or reclassification before they are marketed.
- classify new devices; a not substantially equivalent (NSE) new device is in class III, and a substantially equivalent (SE) new device is in the same regulatory class as the device to which it is found equivalent; and
- achieve marketing equity by allowing manufacturers of new devices that are SE to pre-Amendments devices to market their devices without facing any greater regulatory burdens than faced by manufacturers of the pre-Amendments devices.

BACKGROUND

THE STATUTE: the Medical Device Amendments were enacted on May 28, 1976. They direct FDA to issue regulations that classify all devices that were in commercial distribution at that time into one of three regulatory control categories, class I, II, or III, depending upon the degree of regulation necessary to provide reasonable assurance of their safety and effectiveness. The class into which a device is placed determines the requirements that must be met before a manufacturer may distribute the device in interstate commerce.

Class I devices are subject to a comprehensive set of regulatory authorities applicable to all classes of devices, e.g., premarket notification, registration and listing, prohibitions against adulteration and misbranding, and rules for good manufacturing practices. Class II devices also need performance standards, and class III devices need premarket approval.

The Act also specifies how a new device, i.e., a post-Amendments device, is to be classified. A post-Amendments device is automatically in class III and must undergo premarket approval or reclassification before it can be marketed, unless it is within a type of device that was in commercial distribution prior to May 28, 1976 and is SE to another device within such type; or, it is within a type of device introduced after May 28, 1976 that has been reclassified into class I or II and is SE to another device within such type. A SE device is in the same class, and is subject to the same requirements, as the device to which it is SE.

Section 510(k) of the Act requires a person who wishes to introduce a device into commerce to notify the Center at least ninety days in advance. This premarket notification is referred to as a "510(k)." The Agency uses 510(k)s to determine if new devices are, or are not, SE to either a pre-Amendments device or a reclassified post-Amendments device.

CONGRESSIONAL GUIDANCE: The statutory objectives for the 510(k) program are clear. Also, the Report by the Committee on Interstate and Foreign Commerce on the Medical Device Amendments of 1976 (House Report) offers some additional guidance on program implementation, describing the term "substantial equivalence" as follows:

> The term "substantially equivalent" is not intended to be so narrow as to refer only to devices that are identical to marketed devices nor so broad as to refer to devices which are intended to be used for the same purposes as marketed products. The committee believes that the term should be construed narrowly where necessary to assure the safety and effectiveness of a device but not narrowly where differences between a new device and a marketed device do not relate to safety and effectiveness. Thus, differences between "new" and marketed devices in materials, design, or energy source, for example, would have a bearing on the adequacy of information as to a new device's safety and effectiveness, and such devices should be automatically classified into class III. On the other hand, copies of devices marketed prior to enactment, or devices whose variations are immaterial to safety and effectiveness would not necessarily fall under the automatic classification scheme.[1]

If substantial equivalence were judged too narrowly, the marketing of devices that would benefit the public would be delayed; the device industry would be unnecessarily exposed to the greater burdens of premarket approval; new devices would not be properly classified; and new manufacturers of pre-Amendments type devices would not have marketing equity. If substantial equivalence were judged too broadly, the statutory purpose may not be served, i.e., devices with new uses or those presenting new or different risks would be marketed without adequate regulatory control.

THE 510(K) REGULATION: On August 23, 1977, FDA issued a final regulation for Establishment Registration and Device Listing for Manufacturers of Devices (21 CFR 807). Subpart E describes premarket notification procedures. The 510(k) rules were

---

[1]H.R. Rep. No. 94-853 pp. 36-37. The Amendments, as enacted, are comprised primarily of the bill H.R. 11124 as drafted by the Subcommittee on Health and the Environment of the Committee on Interstate and Foreign Commerce and adopted by the House of Representatives. This report provides the only Congressional guidance on the criteria for substantial equivalence.

incorporated into the registration and listing regulation because the 510(k) requirements apply only to those persons required to register.

The regulation describes the circumstances under which a 510(k) must be submitted, the information required, and the FDA rules for maintaining confidentiality of submitted information. Essentially, the regulation provides that a person required to register must submit a 510(k) at least ninety days before marketing a device that: (1) is being introduced into distribution for the first time by that person, or (2) is in distribution but is being significantly modified in design, or use, etc.[2]

A 510(k) must contain, among other things:

- proposed labeling sufficient to describe the device's intended use;
- a description of how the device is similar to or different from other devices of comparable type, or information about what consequences a proposed device modification may have on the device's safety and effectiveness; and
- any other information the Center needs to determine whether the device is SE.

The statute provides that a new device can be SE to one of two types of devices (from here on referred to as a "predicate device"). A predicate device is one that was in distribution prior to enactment of the Amendments, or it is a post-Amendments device that was subject to reclassification from class III to class I or II. The Center does not routinely require that manufacturers perform research to determine what specific predicate devices were available in 1976, or were available at the time a post-Amendments device was reclassified from class III to class I or II; nor does the Center routinely require that all 510(k)s initially provide information on a predicate device. Instead, the Center requires submitters to provide information that compares the new device to a marketed device of a similar type, regardless of whether this marketed device was marketed before or after enactment of the Amendments, or before or after a type of post-Amendments device was reclassified.

This means that the manufacturer can submit a 510(k) comparing a new device to a device that has been found to be SE. It does not mean, however, that the Center finds new devices SE to devices that have been found SE. It only reflects the agency's position that the similarity of a new device to a marketed device is evidence that can be considered in determining that a new device is, as is the marketed device to which it is compared, SE to a predicate device. Nevertheless, the ultimate burden of demonstrating the substantial equivalence of a new device to a predicate device remains with the 510(k) submitter, and in those occasions when the Center is unfamiliar with certain aspects of the predicate device, the submitter will be required to provide information that substantiates a claim of substantial equivalence.

The guidance that follows assumes that 510(k)s provide comparisons with marketed devices.

---

[2]This document does not provide guidance on what changes in marketed devices are significant enough to require the submission of a 510(k), but the Center plans to issue such guidance in the near future.

GENERAL POINTS CONSIDERED BY CDRH DURING REVIEW OF 510(K) SUBMISSIONS

CDRH normally approaches the review of 510(k)s by considering the following points:

> Does the new device have the same intended use as a predicate device? Does the new device have the same technological characteristics, i.e., same materials, design, energy source, etc.? If it has new technological characteristics, could they affect safety or effectiveness?

If the new device has a new intended use (what constitutes a new intended use is discussed below), it is considered NSE. If the new device has the same intended use as a predicate device and the same technological characteristics related to safety and effectiveness, the new device is considered SE. If the new device has the same intended use as a predicate device, and it has new technological features that could affect safety or effectiveness, additional issues are considered:

> Do the new technological features pose the same type of questions about safety or effectiveness as are posed by the predicate device with the same intended use? Are there accepted scientific methods for evaluating whether safety or effectiveness has been adversely affected as a result of the use of new technological characteristics? Is there information to demonstrate that the new technological features have not diminished safety and effectiveness?

If the answers to any of the above questions are negative, the device is generally considered NSE.

Thus, as a matter of practice, CDRH generally considers a device to be SE to a predicate device if, in comparison to the predicate device:

- the new device has the same intended use (as discussed below); and,
- the new device has the same technological characteristics, (i.e., same materials, design, energy source, etc.); or, it has new technological characteristics that could not affect safety or effectiveness; or,
- it has new technological characteristics that could affect safety or effectiveness, and

  -- it generates the same type of questions about safety or effectiveness;
  -- there are accepted scientific methods for evaluating whether safety or effectiveness has been adversely affected as a result of the use of new technological characteristics; and
  -- there are data to demonstrate that the new technological features have not diminished safety or effectiveness.

Premarket notification submissions should be designed to answer these questions. The guidance below describes how the Center draws conclusions about the questions. The guidance covers the meaning of "intended use," technological changes that can be reviewed under the 510(k) process, data that are required in a 510(k), and what "combination" devices can be 510(k)'ed. Also, flow charts on Attachments I and II illustrate the decision making process.

THE MEANING OF INTENDED USE

While a new device must have the same intended use as a predicate device in order to be SE, the Center does not require that a new device be labeled with precise therapeutic or diagnostic statements identical to those that appear on the predicate device labeling IN order for the new device to have the same intended use. Label statements may vary. Certain elements of a predicate device's labeled indication may not be critical to its intended therapeutic, diagnostic, prosthetic, surgical, etc., use. The Center's scientific expertise enables it to exercise considerable discretion in construing intended uses in the labeling and promotional materials for predicate and new devices.[3] Thus, a new device with the same intended use as a predicate device may have different specific indication statements, and, as long as these label indications do not introduce questions about safety or effectiveness different from those that were posed by the predicate device's intended use, the new device may be found SE.

For the purposes of determining whether or not the new device has the same intended use as a predicate device, the Center assesses any differences in label indications in terms of the safety and effectiveness questions they may raise. The Center considers such points as physiological purpose (e.g. removes water from blood, transports blood, cuts tissue), conditions or disease to be treated or diagnosed, professional or lay use, parts of the body or types of tissue involved, frequency of use, etc. If a new device is determined to have the same intended use, the Center may then proceed to determine whether or not it is substantially equivalent. (Devices which do not have the same intended use cannot be substantially equivalent.) The following examples illustrate these concepts.

Nonimplanted Blood Access Device: This type of pre-Amendments class II device is intended to provide access to a patient's blood so that blood can, for example, be subjected to hemodialysis. Pre-1976 devices are labeled for insertion into the femoral vein. New devices that are labeled for insertion into the subclavian vein may have the same intended use to the extent that labeling for the new and predicate devices indicates that both are intended to provide access to a patient's blood for similar purposes. The differences in labeling described above do not undercut the fact that the new and predicate devices have the same intended use. CDRH was further able to conclude that, because the differences in labeling related only to a method of use not relevant to the effect that the insertion is to achieve (and therefore did not pose new safety or effectiveness questions) and because there were not other significant changes (in technology, design, etc.), the new device was substantially equivalent.

Conventional Dialyzer: This type of pre-Amendments device is in class II. The pre-Amendments devices are labeled for use as part of artificial kidney systems for patients with renal failure. The principal purpose of the device is to remove excess water from

---

[3]Ordinarily, intended use is determined by reference to "labeling" or promotional claims; only in rare cases might it be necessary to infer intended use from other types of information.

the vascular system. Some new devices that have been found SE are labeled for use as part of a heart-lung machine to remove excess water from the vascular system at the end of surgery. Again, the Center concluded that this is not a different intended use. Differences in the labeling relate only to a nonessential condition that does not bear materially on the safe and effective use of the device, and moreover, there are not other significant changes (in technology, design, etc.); therefore the devices are substantially equivalent.

Blood Tubing Set: This pre-Amendment class II device consists of tubing that enables blood to flow from a patient to a dialyzer and back to the patient. Some new devices that have been found SE are labeled for use in carrying blood to and from a plasmapheresis device. The Center concluded that this is not a different use for the two tubings.

The labeling differences relating to the subclavian catheter, the conventional dialyzer for use with the heart-lung machine, and the blood tubing set for plasmapheresis, are not significant enough to require a finding that the devices are for different intended uses. Moreover, the specific uses associated with the labeling modifications do not present issues of safety and effectiveness different from those posed by the use of their predicate devices, and therefore, the devices can be found SE in terms of intended use. On the other hand, for some new devices, modifications in label indications will not be found to represent the same intended use as a predicate device, even though the intended effect of the new device is very similar to that of the predicate device. This is because slight modifications in intended use can be significant to the claimed effect or purpose of the predicate device. If a device has a different intended use, there is no reason to proceed further to decide whether the devices are substantially equivalent. (Obviously, however, some of the same issues relevant to substantial equivalence determinations may arise in assessing whether or not devices are for the same intended use, i.e., differences in labeling are judged in terms of any safety and effectiveness questions they may raise.) Two examples follow:

Long-term Percutaneous Intravascular Catheter: This pre-Amendments, unclassified device consists of a slender tube labeled for insertion into the vascular system for extended periods in order to sample blood, monitor blood pressure, or administer drugs. A similarly designed device has been found NSE because it is labeled for use as a spinal canal access catheter. The new device was intended to deliver drugs over an extended period of time to the spinal canal. While both devices deliver drugs into the human body, each device relates to a different body system. The concern raised by this difference led to the conclusion that the intended uses of the devices are not the same. Inserting the device into, and maintaining it in, the spinal canal raises significantly different safety issues compared to the issues posed by intravascular insertion; namely, the new device potentially posed risks to the spinal cord, whereas the predicate device did not.

Power Aspirators: These pre-Amendments devices, currently proposed for class II, were labeled to remove loose bone chips, blood, or tissue from the body during surgery. Similarly designed new devices have been found NSE when labeled for use in "suction lipectomy." In this new surgical procedure, the aspirating device is inserted under the skin to remove body fat, sometimes very large amounts of body fat, for cosmetic purposes. In suction lipectomy, the aspiration process performed by the device becomes the surgical procedure. It is no longer an aspirator merely used as an aid in performing another surgical procedure that is not aimed solely at the removal of fat. For these reasons, the new devices, as labeled, do not have the same intended uses as the predicate device. The difference in intended use is underscored by the fact that removal of large amounts of body fat through aspiration raises questions of safety and effectiveness not posed by the labeled use of pre-Amendments aspirators, e.g., possible metabolic changes, and permanent bagging of the skin resulting when the fat removed from the area exceeds the ability of the skin to contract. Thus, the intended uses are not the same.

As the examples above illustrate, for a device to be found SE, it is important that a 510(k) for the device demonstrate that the intended use of the device does not differ from the intended use of marketed devices within the same type and that any changes in labeling are immaterial in terms of safety and effectiveness issues.

DEVICES WITH NEW TECHNOLOGICAL FEATURES

If a new device has the same intended use as a predicate device, and there are no technological differences between the new and a predicate device, the new device is SE. If the device has the same intended use and technological differences, but the technological differences could not affect safety or effectiveness, it is SE. If the device has the same intended use and technological differences that could affect safety or effectiveness, the new device may not be SE. Technological differences may include modifications in design, materials, or energy sources; for example, changes in the power levels of electrical surgical instruments, the use of new reagents in in vitro diagnostic devices, the use of new materials in orthopedic implants, and the use of new battery designs in implanted pacemakers. The Center finds devices with new technological features to be NSE when the new feature could adversely affect safety or effectiveness in a way that is consequential under the conditions of intended use.

In taking this approach, the Center focuses on the technological differences that are medically and scientifically significant and avoids differences that would arise from a mechanistic application of rigid formal criteria to the wide variety of substantial equivalence questions posed by new devices proposed for marketing under a 510(k). Substantial equivalence determinations of necessity require the Center to exercise reasonable scientific judgment. For example, the strict application of a rule that would make devices NSE if they have a new material would be inappropriate for such devices as bed pans. For other devices, a "materials rule" may seem more appropriate, e.g., for implants. Even for implants, however, such a rule would be too encompassing if applied to substitute materials known to be, or easily shown to be, equivalent or superior; for example, if applied to a new post-Amendments hip made of titanium, which is generally known to be, and can easily be shown to be, stronger and less corrosive than some stainless steel that was used in pre-Amendments hips.

Thus, from a scientific perspective, to determine which technological changes are consequential, the Center considers whether:

- the new device poses the same type of questions about safety or effectiveness as a predicate device;
- there are accepted scientific methods for evaluating whether safety or effectiveness has been adversely affected as a result of the use of new technological characteristics; and
- there are data to demonstrate that new technological characteristics have not diminished safety or effectiveness.

The examples below should help clarify these concepts.

Electrocardiographs: This device is in class II. Pre-Amendments electrocardiographs produce an analog visual display of the electrical signals produced by the heart. Some post-Amendments devices that have been found SE display these signals in a digital, rather than analog, form. This technological innovation could affect the effectiveness of the device, i.e., the electronic components used to produce the digital display might not achieve the accuracy of an analog display. The Center found the new devices SE because the modification presented the same questions of effectiveness as posed by the predicate device, i.e., does the display accurately represent the electrical activity of the heart? Also, there are accepted bench test procedures to demonstrate comparability of accuracy between the new and predicate devices, and the necessary comparability information was presented in the 510(k).

Inflatable Penile Implant: This device is in class III. The pre-Amendments implants consist of inflatable cylinders implanted in the penis, connected to a reservoir filled with fluid and implanted in the abdomen, and a subcutaneous manual pump implanted in the scrotum. When the cylinders are inflated, they provide rigidity to the penis. The connections between the various parts of this system sometimes leaked. Manufacturers of post-Amendments versions of these devices modified the connector designs. These new devices have been found SE even though such a change could affect safety and effectiveness. This is because the new connectors presented the same questions of safety and effectiveness as posed by the predicate device, i.e., are the connector materials nontoxic and do the connectors leak? Since the materials were the same, questions about the comparability of toxicity were obviated, and bench testing and clinical evidence easily and clearly demonstrated that the modified connectors were not more prone to leaks than the connectors in the predicate devices.

Perimeter: This device is proposed for classification in class I or II, depending on whether it is manually or electrically operated. The pre-Amendments devices are AC-powered or manual devices that project light on various points of a curved surface, and the patient manually indicates whether he or she sees the light. The devices are designed to determine the extent of a patient's peripheral vision. Manufacturers of post-Amendments devices have incorporated software into the device. The software automatically lights various points on the curved surface, and keeps track of whether the patient indicates that the light has been seen by recording that the patient has or has not pressed a button. The new device posed the same safety and effectiveness questions as the predicate devices, i.e., are the light-points in the appropriate places, is there risk of shock, do the points light up when they are meant to, and can the patient's reaction be accurately recorded? Because the questions are the same, and obviously accepted testing procedures can ensure comparability of performance between the new and predicate device, and because clinical data showed comparability, the new devices with a software package were found to be SE.

Hemodialyzers: This is a class II device that is used as an artificial kidney system. The device allows the removal of water and waste from the patient's blood through a semipermeable membrane. Pre-Amendments devices used cellulose based semipermeable membranes. Manufacturers of post-Amendments devices have used various other types of membranes in the filters and these devices have been found SE. This is because the new membranes raised the same questions of safety and effectiveness as the cellulose membrane, e.g., is the degree of permeability acceptable, is the material toxic, and is the degree of particulate contamination acceptable? Accepted preclinical tests, with clinical confirmation, demonstrated that the new membrane and the cellulose membranes were comparable.

On the other hand, there will be many technological changes that will make a device NSE. For example:

Heart Valves: This pre-Amendments type device is in class III. It is intended to perform the function of any of the heart's natural valves. It is constructed of prosthetic or biological materials, or both. The Center has found many new valves to be NSE even though the new valves, in some cases, incorporated only very slight design modifications in relation to predicate valves. This is not because the device is a high risk device. Rather, it is because of the nature of the device. A minor change in design can raise questions about safety and effectiveness that are inappropriate for evaluation under 510(k). Specifically, different valve designs can lead to a different balance in the performance factors related to valves. Minor changes in a valve may make it more susceptible to fractures than its predicate, but less susceptible to occlusion. In this case, it is difficult to compare the performance of one valve to another because they present different risk/benefit equations. As a result, there are not accepted methods for assessing comparability. Consequently, these devices are typically found NSE.

Electrohydraulic Lithotripter: This pre-Amendments device is in class III. The pre-Amendments device was used to fragment urinary bladder stones. It consists of a high voltage cable connected to a bipolar electrode introduced into the bladder through a cystoscope. The Center found an electrohydraulic lithotripter inserted into the ureter through a cystoscope for use on ureteral stones to be SE. The principle of operation for both devices are identical. The stone is fragmented by the discharge of electricity between the two electrical poles. Another post-Amendments device, known as the Extracorporeal Shock Wave Lithotripter was NSE. This device breaks kidney stones through the use of a succession of externally generated shock waves focused on the stone. The shock waves are transmitted through water, in which the patient is sitting, and then transmitted through body tissue to the stone. The device raised questions about safety and effectiveness not present in the previous devices, e.g,, the shock waves, which enter large portions of the body, could affect cardiopulmonary functions or could have had unknown effects on much of the tissue and organs in the abdomen. An evaluation of these questions was not appropriate for a 510(k) because they presented issues outside the questions of safety and effectiveness raised by the electrohydraulic lithotripter.

Femoral Stem: This lower portion of a total artificial hip replacement is a pre-Amendments device proposed for class II. Some of these devices are uncemented and are fixed by impaction of the stem into the top of the femur. The pre-Amendments impact/fit stems had smooth, straight (uncontoured) surfaces on the stem that fit into the femur. The Center found SE some post-Amendments versions of the stem that have surface variations, i.e., grooves, irregular or regular contours, etc., that allow the bone to remodel over time. The Center found to be NSE post-Amendments stems with porous coatings (for uncemented use) that require bone ingrowth into the device for fixation. The contoured stem relied on impact placement and bone remodeling to an essentially smooth surface. The health of the bone would still depend upon normal bone vascularization systems even if the bone conformed to some of the slightly wavey contours. The questions of safety and effectiveness were the same as those for the traditional impact fit stem. On the other hand, for bone to grow into, and remain viable in, a porous material, the bone would need to depend upon an intricate vascularization system not normally occurring in bone that may not, in fact, develop or be sustainable. Thus, the porous coated device was found NSE because it presented new scientific questions about long term fixation.

As the examples above illustrate, for a device to be found SE, it is important that a 510(k) for the device either demonstrate that the technology of the device does not differ from the technology of marketed devices within the same type; or, to the extent that the technology does differ: explain why the device does not present new questions about safety or effectiveness, describe what accepted methods exist to test comparability of performance, and present data, generated from such tests, that demonstrate comparability. (See the 510(k) data discussion below. 510(k)s for new devices that are different from predicate devices, but are not different from a marketed device within the same type, may not need to provide testing information).

DATA REQUIREMENTS FOR 510(k)s

The requirements of section 510(k) of the act are intended not only to notify FDA that a device is about to be marketed but primarily to enable FDA to determine whether the device is SE to one already in commercial distribution. To fulfill this responsibility, the Center requires that a 510(k) include descriptive data needed to understand a new device's intended use, physical composition, method of operation, specifications, performance claims, etc. Similar information about the device to which the new device is being compared may also be required. In addition, under certain circumstances, the Center requires performance testing information, i.e., data from bench, animal, or clinical tests, in order to determine that a device performs according to its description.

While the Center has concluded that it should sometimes require performance testing data in order to confirm that a new device is SE, the 510(k) review process is not a substitute for premarket approval, and the Center does not attempt to address all of the issues that would be answered in a PMA in its review of 510(k)s. Data in a 510(k) should show comparability of a new device to a predicate device, whereas demonstration, in an absolute sense, of a device's safety and effectiveness, is reserved for PMAs.

Although there are no hard and fast rules governing all the possible data collection issues that can arise in 510(k)s, the following guidance will govern many situations.

The Center will always require descriptive data necessary to understand a new device's intended use, physical composition, method of operation, specifications, performance claims, etc. Also, the Center will always require any descriptive data necessary to understand the same characteristics of the device to which the new device is being compared.

The Center normally will not require performance testing data to substantiate equivalence if a new device does not have an important descriptive difference in comparison to marketed devices within its type, i.e., it does not have a new material, method of operation, etc., and it has descriptive characteristics that are precise enough to ensure that comparability will be achieved if the new device is manufactured according to its description. For example, the Center will accept the descriptive statement that a new surgical instrument will be made of surgical grade stainless steel, as are other marketed surgical instruments, as substantiation that the new instrument is equivalent in material/composition to a predicate device. (See below for an explanation of how a comparison to any marketed device can lead to a conclusion about the substantial equivalence of a new device to a predicate device.)

The Center normally will require performance testing data to substantiate equivalence if a new device has an important descriptive difference in comparison to marketed devices within its type, and it is not clear from an initial review that the device has an intended use or technological change that makes it NSE; or, the new device has descriptive characteristics that are too imprecise to guarantee that comparability in performance will be achieved even if the new device is produced as described. Following is an example of when descriptive similarities are inadequate to ensure that comparability of performance will be achieved when a new device has the same descriptive characteristics as a marketed device. The fact that a new pacemaker lead and a marketed pacemaker lead are both in the polyurethane "family" does not ensure that both devices have the same chemical formulations, and differences in the chemical formulations of polyurethanes could affect safety or effectiveness. Differences in design, extrusion, and assembly of polyurethanes with the same chemical formulation could also affect safety or effectiveness. Thus, the Center would generally require performance testing information to confirm that the new lead is SE to the predicate lead.

If it is clear from an initial review that a new device has an intended use or technological feature that makes it NSE, the Center will not review or require performance information in the 510(k). Instead, the applicant will be notified that the device is NSE, and any performance data will be reviewed in a PMA or reclassification petition.

Typically, a 510(k) provides descriptive and testing data that compares the new device to another marketed device within the type, but does not necessarily compare the new device directly to a predicate device. In these cases, the Center can rely, as necessary, on performance data appearing in previously reviewed 510(k) files, in Center classification files, or in the literature, to determine that the device is not only comparable to another marketed device within its type, but is also SE to a predicate device.

Nevertheless, the ultimate burden of demonstrating the substantial equivalence of a new device to a predicate device remains with the 510(k) submitter, and in circumstances when the Center is unfamiliar with certain aspects of the predicate device, the submitter will be required to provide information that substantiates a claim of substantial equivalence.

NEW DEVICES THAT ARE COMBINATIONS OF OLD DEVICES

A new device is a so-called "combination" device when it claims to have the same intended uses as two or more different types of predicate devices. Normally, this is achieved by combining two ore more predicate devices into a device that is sold as a unit, e.g., a urinary catheter may incorporate a temperature measuring device; or a cardiac monitor, electrocardiograph, and blood pressure computer that were sold separately prior to May 28, 1976, might be combined into one electronic monitoring device.

When a new device combines types of devices from different classes, questions have arisen about what classification the new device will have.

In its review of the 510(k), the Center will subject the combination device to the same sorts of questions and documentation requirements that are applied to a single device. When such a device is found to be SE, it combines devices from different classes and is classified in the highest of the predicate device classifications unless the combined devices are regulatable as separate articles, e.g., they are detachable. In that case, the separately regulatable articles will be regulated in separate classes.

REVIEWER GUIDANCE FOR COMPUTER-CONTROLLED MEDICAL DEVICES UNDERGOING 510(k) REVIEW

August 29, 1991

SCOPE: This guidance applies to the software aspects of premarket notification [510(k)] submissions for medical devices. It should be noted that "software" encompasses programs and/or data that pertain to the operation of a computer-controlled system, electronic system, data base, or expert system, whether they are contained on floppy disks, hard disks, magnetic tapes, "laser" disks, or embedded in the hardware of a device, usually referred to as "firmware."

1.0 INTRODUCTION

The FDA is receiving an increasing number of applications to market medical devices that depend on software. Most electrically powered medical device now employ some form of computer control, and because computer-controlled systems are complex and difficult to test adequately, the FDA examines evidence from every phase of system development including the preproduction phases of computer program development. The FDA is focusing attention on the software development process to assure that potential hazardous failures have been addressed, effective performance has been defined, and means of verifying both safe and effective performance have been planned, carried out, and properly reviewed. FDA believes that, in addition to testing, device manufacturers should conduct appropriate analyses and reviews in order to avoid errors that may affect operational safety.

This guidance presents an overview of:

1) the kind of information on software FDA reviewers may expect to be included in 510(k) submissions, and
2) the approach FDA reviewers should take in reviewing computer-controlled devices.

The nature and depth of the software documentation should depend on: (1) the intended use and function of the device, (2) the effect on and risk to the patient of potential device faults, (3) the role of software in the device, and (4) the regulatory level of control assigned to the device by the classification panels. These factors are expressed later in Section 2 of this document as the "level of concern."

This guidance IS NOT a tutorial on software development, quality assurance, or testing, and IS NOT a standard for such activities. Numerous texts on these subjects have been written, and there are many established consensus standards (see Bibliography Attachment B.)

This guidance IS intended for professionals who understand current software development theory and practice; thus, the concepts and special terminology should also be familiar. However, to avoid misunderstanding unfamiliar terms, refer to the Glossary (Attachment A), which defines and explains software terms as they are used in this document.

Section 3.2 discusses the types of information that reviewers may expect in a submission, and Section 4 discusses the types of questions that may need to be addressed during 510(k) review. Labeling expectations are discussed in Section 5.

For convenient reference, Section 3.4 contains a matrix which outlines in general terms the nature and depth of documentation to be expected in 510(k)s for each level of concern. Note that this table only illustrates some reference points on a continuous scale of "concern." Reviewers should be guided not only by the table but also by the detailed discussion in Sections 2.0, 3.2, 3.3, and 4.0 dealing with the level of concern, applicable review items, and possible review questions. They should also be guided by their own experience and judgment, as well as any guidance offered by division management in each review case.

Reviewers should expect the submitter of a 510(k) for a computer-controlled medical device to have considered the relevance and importance of each element in the matrix, and either to submit information addressing the element or to present a sound argument as to why the information is not needed for the review. Further guidance concerning a specific device should be obtained from the reviewer's division management.

2.0 LEVELS OF CONCERN

Risk to patients from potential failures or possible design flaws is a significant concern in all medical device reviews but may not be the only concern. The nature of the device and its intended use are additional factors. The 'level of concern' for a particular device is a function of the nature of the device, its indications for use, and the severity of harm that the device could inflict (or permit) as a result of failures. The severity of harm that the device could inflict, or permit, as a result of failures could be unsafe or ineffective delivery of energy, unsafe or ineffective administration of drugs, unsafe or ineffective control of life sustaining functions or incorrect information that could lead to a misdiagnosis or the wrong treatment or therapy. This section suggests evaluation criteria that could be applied generally and should be used to establish the level of concern and hence the depth of review for computer-controlled devices.

The depth of review appropriate for the software of the device is properly dictated both by the level of concern of the device and the role that the software plays in the functioning of the device, including the presence or lack of safeguards that may be incorporated in the device.

In establishing a level of concern, the following factors should be considered:

1. The risk or danger to the patient of using (and not using) the device. Examples of questions that reflect this concern are:

    -- Could the device immediately threaten the patient's life under plausible conditions?
    -- Could the device directly cause irreversible illness or permanent injury under plausible conditions?

2. The degree of the device's influence on therapy and diagnosis or use in production of a licensed biological product. Examples of questions that reflect this concern are:

   -- Does the device directly control (without operator intervention) delivery of energy, administration of parenteral drugs, or life-sustaining functions?
   -- Does the device provide a diagnosis?
   -- Does the device provide information/data only? Is the device an information system that is part of a regulated classified device or an accessory to it?
   -- Would competent health professionals be reasonably expected to exercise judgment in the use of the diagnosis or information?
   -- Does the device provide alarms for life threatening conditions?
   -- Does the device consolidate or obscure information or data that is available to the user in the predicate device or similar devices:
   -- Could software errors cause diagnostic or monitoring information to be missed or inaccurate? Could the errors lead to the release of an unsuitable licensed biologic product?
   -- Could software errors cause alarms provided for life threatening conditions to be inoperative or missed?

3. Any guidance concerning a specific device obtained from the reviewer's division management.

Assessing the level of concern for a medical device is a two-stage process. First, it is necessary to look at the intended use, the degree of the device's influence on therapy and diagnosis, and the potential hazards created by the device, regardless of the method of control for a particular function or the method of incorporating a safety feature into the design of the device. Remember, any guidance obtained from the reviewer's division management should also be considered.

The second stage in evaluating the level of concern of the device for software review is to consider how the software is directly or indirectly involved in the factors previously mentioned, i.e., the role the software plays in the functioning of the device, and to assign a level of concern for each component (or function) of the software. Are some components of the software independent of others and/or do they have little or no adverse affects while others play a more active role in controlling a delivery, diagnostic, or monitoring function? Could some units of the software indirectly control or affect a delivery, diagnostic, or monitoring function of the device even though the unit does not directly control those aspects of the device? Some aspects of the software may have a higher level of concern than others and should be considered when reviewing software documentation.

In order to make an assessment of the level of concern, the 510(k) should include the labeling for the device, information about the intended use, system and software requirements, and hazards associated with the device, and a software hazard analysis. The hazard analysis at the "device" level should include the hazardous event, the cause, the method of control, and corrective action, as well as an indication of the level of concern. The software hazard analysis should include similar information for each software component/function: the hazardous event, the cause, corrective action, and the level of concern. Corrective action may include but is not limited to fail-safe mechanisms, redundant controls, error-handling routines, fault-tolerance, alarms, testing activities, etc. Software safety requirements implemented to eliminate, minimize, or warn of a specific hazard should be traceable to the hazard analysis.

Three levels of concern are discussed below.

> MAJOR: The level of concern is major if operation of the device or software function directly affects the patient so that failures or latent design flaws could result in death or serious injury to the patient, or if it indirectly affects the patient (e.g., through the action of a care provider) such that incorrect or delayed information could result in death or serious injury of the patient.
>
> MODERATE: The level of concern is moderate if the operation of the device or software function directly affects the patient so that failures or latent design flaws could result in minor to moderate injury to the patient, or if it indirectly affects the patient (e.g., through the action of a care provider) where incorrect or delayed information could result in injury of the patient.
>
> MINOR: The level of concern is minor if failures or latent design flaws would not be expected to result in death or injury to the patient. This level is assigned to a software component that the manufacturer can show to be totally independent of other software or hardware that may be involved in a potential hazard and would not directly or indirectly lead to a failure of the device that could cause a hazardous condition to occur.

When a level of concern is assigned for each functioning component of the software, the highest level of concern generated is that assigned to the software aspect of the device. This level of concern assignment for the software aspect is to be documented in your review; i.e., state what the level of concern for the software aspect of the device is, and why. Aspects of the software for a device that are of a lower level of concern do not always require the same level of documentation as those of a higher level of concern for the device. For software components/functions of a lower level of concern, the 510(k) should include an explanation that discusses how the lower level of concern items are independent and do not affect any of the higher level of concern issues.

In general, reviewers should expect reasonably complete evidence to document software development for software components with a high level of concern and correspondingly less detail for software components with a lower concern in the 510(k).

3.0 SOFTWARE DEVELOPMENT, DOCUMENTATION, AND REVIEW

3.1.1 Development

For software development documentation, reviewers may expect manufacturers to provide information that relates to the five phases of software development: specifications/requirements, design, implementation, verification and validation, and maintenance.

After "high-level" system requirements for the device are established, they are elaborated and made more specific in order to characterize the requirements of the software aspect of the device. After the software requirements are established and formally reviewed for completeness, the software design process may begin.

It has been said that the "design process involves developing a conceptual view of the system, establishing system structure, identifying data streams and data stores, decomposing high level functions into subfunctions, establishing relationships and interconnections among components, developing concrete data representations, and specifying algorithmic details." (Fairley) Reviewers may expect manufacturers to provide documentation regarding the structure and processing details, modules (components) and modularization criteria, design verification activities, and the objectives of testing and the test plan criteria for evaluating the software design.

Translating the design specifications into source code is the bulk of the implementation phase. "The primary goal of implementation is to write source code and internal documentation so that conformance of the code to its specifications can be easily verified, and so that debugging, testing, and modification are eased." (Fairley)

Verification and validation activities include testing and activities performed at the unit, integration, and system level, but verification activities are also used to assess the quality of the software throughout the development process. Test plan generation usually begins early in the software development process to assure consistency between requirements, design, implementation, and testing.

Maintenance activities are those activities associated with software changes, which may include making enhancements or correcting problems. The firm should include configuration management procedures that allow for tracking of changes throughout the software development process. Changes, updates, and enhancements may occur at any stage during the software development process, so controls should be in place that allow for updating requirements, design, code, and testing, with verification and assurance activities performed at each level.

Reviewers may also expect that computer-controlled medical devices have been analyzed for potential hazards throughout the software development cycle, beginning with the requirements phase and integrated into the development and testing phases. The reviewer should expect the 510(k) to show that the results of these analyses to be applied in the design to eliminate or minimize the hazards are identified. Reviewers may look for evidence of up-to-date tools and practices appropriate to software design and development. Some examples of such practices are: requirements analysis; specification reviews; design reviews; code walk-throughs; control flow analysis; software complexity metrics; data flow analysis; fault tree analysis; and failure mode, effects, and criticality analysis (FMECA).

3.1.3 Testing

Although complete detailed review of software testing is impractical at times, in some circumstances reviewers may expect and seek evidence for the testing that has been done. In devices with large programs this may include evidence of intermediate testing of modules, functions, subroutines, and so forth. Evidence for analysis and testing during integration of these program components and the device may be sought. Reviewers may look for test procedures and data that stress the programs in a system, for instance, by applying extremely rapid streams of input and output and/or large data loads. Reviewers should particularly seek complete evidence for tests that exercise safety related functions of the software and the device. Testing requirements should be traceable to the system/software requirements and design.

3.1.3 Documentation

Manufacturers should properly document software policies, procedures, and activities associated with the software in medical devices. Documentation in a 510(k) submission is expected to be a subset of this documented information which may include documentation regarding development of system level and software requirements; software development; software design; assurance activities; planning software verification, validation, and testing; conducting verification, validation, and testing (results of testing and analyses); and certification of readiness for production and distribution. Section 3.2 contains advice on the expected content of documentation applicable to these tasks. Reviewers may expect that system and software requirements can be traced through the documents applicable to each of these software development tasks.

3.1.4 Software Changes

If a device has undergone software changes, documentation should be provided demonstrating that these changes have been performed and analyzed according to a manufacturer's current software quality assurance procedures. This may include an updated description of the software requirements, documentation showing an analysis of software components that are directly or indirectly affected by the changes, update software design documentation, an update hazard analysis, testing and analyses performed to evaluate the impact of the changes on the device at each phase of development, and testing that demonstrates that current software requirements are met.

3.2 SUMMARY OF APPLICABLE REVIEW ITEMS

In general, Section 510(k) premarket notifications for computer-controlled medical devices should contain documentation as mentioned in Section 3.1.3 above, which is presented in this section as an outline. The amount and detail should be consistent with the type of application (in this case, 510(k)) and commensurate with the Level of Concern discussed in Section 2. Section 3.4 includes a matrix outlining the typical documentation for a 510(k), but this section (3.2), Chapter 2, Section 3.3, and Chapter 4 should be read in order to get the "big picture" of the types of documentation to expect in a 510(k). Reviewers in some divisions of the Office of Device Evaluation may have more detailed guidance information for specific devices. Please refer to the glossary for definitions of unfamiliar terms.

In addition to the hazard analysis for the device and software that was discussed in Chapter 2, the following sections summarize other relevant information that may be needed for review.

3.2.1 FUNCTIONAL REQUIREMENTS AND SYSTEM SPECIFICATIONS

Relevant information may include but is not limited to the following:

- system and software requirements;
- safety critical functions implemented into the design as a result of an on going hazard analysis (software safety requirements, redundant controls, fail-safe mechanisms, error handling, etc.);
- applicable algorithms; and
- device limitations due to the software.

3.2.2 SOFTWARE DESIGN

Relevant information may include but is not limited to the following:

- traceability between modularization criteria and software requirements;
- traceability between software safety requirements, software units, and hazardous events discussed in the hazard analysis;
- software architecture; and
- software structure chart depicting the partitioning of the system into functional units, the hierarchy and organization of those units, and the communication interfaces, including a written description.

3.2.3 SOFTWARE DEVELOPMENT

Relevant information may include, but is not limited to the following:

Software engineering activities performed for each phase of software development for the device under review, including assurance activities performed at each of the following phases:

- requirements;
- design;
- implementation;
- verification and validation; and
- maintenance.

This may include any SOP for software development.

3.2.4 SOFTWARE VERIFICATION, VALIDATION AND TESTING

Relevant information may include but is not limited to the following:

- techniques and methods used at module, integration, and system level (the verification and validation plan);

- requirements to be verified;
- how identified hazards and safety functions have been tested and analyzed;
- test strategies and methodologies (may include functional, performance, stress, and structure tests);
- environmental testing; and
- test acceptance and completion criteria.

3.2.5 RESULTS OF SOFTWARE VALIDATION, VERIFICATION AND TESTING

Relevant information may include results, reports, data, and summaries that provide evidence that requirements and goals were met at the various stages of software development. These may be in summary form or complete with data, as necessary, and commensurate with the Level of Concern.

3.2.6 CERTIFICATION

In this guide, certification means documentation affirming that the software development process was followed, that quality assurance procedures were adhered to, and that testing demonstrates that the requirements were met and the software and system requirements were fulfilled.

3.3 REVIEW TASK

Reviewers will seek to determine whether the device, if it performs as intended, would be substantially equivalent to a predicate device, following the established ODE "510(k) Substantial Equivalence Decision-Making Process."

The reviewer is reminded that, as discussed in Section 2, the level of required software documentation and the depth of review should be consistent with the level of concern of the device and the role that the software plays in the functioning of the device (including the presence or lack of safeguards that may be incorporated).

3.3.1 INTENDED USE

Does the new device have the same indication statements?

The reviewer should first consider the intended use and the system functional requirements as stated in the manufacturer's application and labeling. If the computer-control aspect changes the intended uses and/or indications, the device may be found not substantially equivalent through the appropriate decision-making process.

3.3.2 TECHNOLOGICAL CHARACTERISTICS

Does the new device have the same technological characteristics, e.g., design, materials, etc.?

Whether the answer to this question is yes or no, the reviewer should establish a "Level of Concern," as discussed in Section 2. The medical device manufacturer should first provide a hazard analysis that identifies the potential hazards associated with the device, the method of control, the corrective action (including design considerations employed to reduce or eliminate the occurrence of the hazardous conditions), and the level of concern. The labeling of the device should provide the user with adequate warnings or cautions about such occurrences, and instruction about the device operation if one of the hazardous conditions should occur.

In addition to the hazard analysis for the entire device, the medical device manufacturer should provide documentation that identifies the various software components that comprise the software, the undesirable events (hazards) that may occur if a failure should occur, the risk to the patient, corrective actions, and the level of concern as defined in Section 2 of this document. This information can be provided in tabular form. This should also be supplemented with information that would discuss how the hazards were identified, or how components identified as a "minor" level of concern were shown to be totally independent of other software/hardware that may be involved in a potential hazard and would not directly or indirectly lead to a failure of the device that would cause a hazardous condition to occur. If all software is considered to be a "minor" level of concern, the manufacturer should provide a discussion as to how this determination was made.

After establishing the Level of Concern, the next step is to decide on the extent of the review to be undertaken. The reviewer must decide what documentation will be required to reasonably determine substantial equivalence and must decide how that documentation should be studied. (Section 3.2 outlines what documentation is expected.)

3.3.2.1 NEW TECHNOLOGICAL CHARACTERISTICS

If the device has new technological characteristics such as microprocessor control versus electro-mechanical control or other technological changes, the reviewer should follow the appropriate path of the 510(k) substantial equivalence decision-making process and ask appropriate questions in determining substantial equivalence.

-- Could the new characteristics affect safety or effectiveness? For a computer-controlled device, the answer would be 'yes' since the way the device performs or the way the device provides information may affect the safety and effectiveness issues regarding a device.

-- Do the new characteristics raise new types of safety or effectiveness questions? The fact that the device is software-controlled does not necessarily mean that there are new types of safety and effectiveness questions?

• Do the software requirements for the device reveal that new functions are performed or new algorithms/techniques are used that may raise new types of safety and effectiveness questions. If so, the device may be determined to be not substantially equivalent. If not, the reviewer should proceed to the next question in the review process.

-- Do accepted scientific methods exist for assessing effects of the new characteristics?

- Is there a sound factual basis on which the system design is based and does it satisfy the functional requirements and systems specifications (see Sections 4.1.1, 4.2.1 or 4.3.1 for more information and questions), and is this design consistent with predicate devices in terms of device performance (safety and effectiveness)?
- Were hazards identified in the hazard analysis, have results been incorporated into the design for safety considerations, and do these make the device as safe as the predicate device (see Sections 4.1.2, 4.2.2, or 4.3.2 for more information and questions)?
- Does the firm's software development include quality assurance activities and verification, validation, and other testing methods and/or procedures (see Sections 4.1.2, 4.1.3, 4.2.2, 4.2.3, 4.3.2, or 4.3.3 for more information and questions) to demonstrate that specifications for the device will be met and that proper safety measures have been incorporated in order to show that the device is as safe and as effective as the predicate device? Has each software component and hazard associated with the software been addressed?

-- Are performance data available to assess effects of new characteristics?

- Do quality assurance activities and verification, validation and the results of other testing (see Sections 4.1.3, 4.1.4, 4.2.3, 4.2.4, 4.3.3, or 4.3.4 for more information and questions) adequately demonstrate (or assure) that the device is as safe and effective as the predicate device, and was certification provided (see Sections 4.1.5, 4.2.5, or 4.3.5 for more information and questions)? Has each software component and hazard associated with the software been addressed?

3.3.2.2 SAME TECHNOLOGICAL CHARACTERISTICS

If the device uses the same technology as the predicate device, the reviewer should follow the appropriate path of the 510(k) substantial equivalence decision-making process and ask appropriate questions in determining equivalence.

-- Are the descriptive characteristics precise enough to ensure equivalence? The fact that the device is software-controlled does not necessarily mean that there are descriptive characteristics precise enough to ensure equivalence. Software for new devices has generally been developed independently and, therefore, is unique.

- Is there a sound factual basis on which the system design is based, does is satisfy the system and software requirements, and is this design consistent with predicate devices in terms of device performance (safety and effectiveness)?
- Were hazards identified in the hazard analysis, have results of the analysis been incorporated into the design for safety considerations, and do these make the device as safe as the predicate device?
- Does the firm's software development include quality assurance activities, and verification, validation, and other testing methods and/or procedures to demonstrate that specifications for the device will be met and that proper safety measures have been incorporated in order to show that the device is as safe and

as effective as the predicate device? Is there sufficient documentation that traces back to the level of concern for each component of the software?

-- Are performance data available to assess equivalence?

- Do verification, validation, and the results of other testing and quality assurance activities adequately demonstrate that the device is as safe and as effective as the predicate device and was certification provided? Is there sufficient documentation that traces back to the level of concern for each component of the software?

3.4 SUMMARY OF REVIEW TASK AND DOCUMENTATION MATRIX

In summary, the information in the submission should enable the reviewer to assess the following questions based on the Level of Concern of the device and the concomitant level of detail the reviewer feels is necessary to adequately review the software of the device:

-- How thoroughly has the manufacturer analyzed the safety of critical functions of the system and software? If testing has not been completed, how thoroughly will these functions be analyzed?

-- How well has the manufacturer established the appropriateness of the function(s), algorithm(s), and/or knowledge on which the system and software are based?

-- How carefully has the manufacturer implemented the safety and performance requirements in the system and the software? If testing has not been completed, will the test plan adequately demonstrate that the manufacturer implemented the safety and performance requirements?

-- Is there sufficient documentation that traces back to the level of concern for each component of the software?

-- Based on this information about the software component of a medical device and review of the other aspects of the device, is the device substantially equivalent to the predicate device?

When completing a 510(k) review, the highest level of concern for a software component should be stated as the "level of concern" for the software review.

A matrix table provided on the following page outlines the documentation regarding 510(k) submissions. Appropriate documentation is listed under Minor, Moderate, and Major Level of Concern for each of the five software development tasks mentioned earlier. This table is a "summary guide" that can only be used adequately when information from Chapters 1, 2, Sections 3.2 and 3.3, and Chapter 4 are applied to the items in the table. For more detailed questions regarding 510(k) software review, refer to section 4.

4. 510(k) REVIEW QUESTIONS

Has the manufacturer carefully established and formally documented a basis for determining whether the device is substantially equivalent to a predicate device: (see Section 3.3) The questions presented in this section for a minor, moderate, or major level of concern may be more detailed and/or rigorous than necessary for a specific review of a given device. Therefore, the reviewer should use discretion in reviewing the software of a software-controlled medical device and should follow any guidance provided by division management and/or guidelines that may exist for specific devices that incorporate any software documentation requirements.

Remember, the submission should include a hazard analysis that identifies the potential hazards associated with the device, the method of control (hardware/software), safeguards incorporated, and the identified level of concern. The level of concern for each component of the software should be identified as well as the undesirable event if a failure were to occur and the risk to the patient. The documentation regarding development, verification and validation, testing and analyses of results, and handling of software changes should be traceable to the identified software components and consistent with the level of concern of the software.

4.1 Minor Concern

4.1 Minor Concern -- The level of concern is minor if failures or latent design flaws would not be expected to result in death or physical injury to the patient.

4.1.1 System and Software Requirements and Design

This documentation should summarize the overall performance requirements of the device and the functional requirements of the software/computer system necessary to achieve them.

Some key questions in reviewing this type of device:

- -- Has a discussion of the safety features implemented as a result of the hazard analysis been provided?
- -- Are the requirements consistent with the intended use and claims of the device under review and of predicate devices?
- -- Do the requirements establish a basis for deciding whether the device will be effective? If achieved, will the device be equivalent to other legally marketed devices?
- -- Are there conflicting requirements between the overall performance objectives and the software requirements?
- -- If interpretation and/or calculations are made by the device, is there a basis in the application and/or scientific literature to support the algorithms used?
- -- Do safety measures address both environmental influences and random component failure?
- -- Has a software structure chart depicting the partitioning of the system into functional units, the hierarchy and organization of those units, and the communication interfaces, including a written description, been provided?

510(k) SOFTWARE DOCUMENTATION MATRIX

The 510(k) should include a hazard analysis for the device and software, and a discussion of the level of concern.

**Before using this matrix, read Sections 2, 3.1, 3.2, 3.3, and 4 of this document.

DATA GROUP	MINOR CONCERN	MODERATE CONCERN	MAJOR CONCERN
System and Software Requirements & Design	A written description of the system and software requirements (a structure chart depicting the partitioning of the system into functional units should be provided for all levels of concern)	A written description of the system and software requirements, including a listing of functional units/modules and a description of how each fulfills the requirements (traceability between safety requirements and hazards should be included)	System specifications and software requirements document, including a listing of the functional units/modules and a description of how each fulfills the requirements (traceability between safety requirements and identified hazards should be included)
Software Development	A summary description of how the system and software was developed and a description of how changes will be handled	A description of the software development activities and SW quality assurance procedures over the software life cycle for the device	Same as for moderate, except the software development SOPs should be provided
Verification & Validation	A description of the verification and validation activities	A description of the verification and validation activities at unit, integration, and system level, including pass/fail and test completion criteria, and the system level functional test plan (traceability between hazards, safety functions, and testing should be demonstrated)	Same as for moderate, except test plans should be provided for unit, integration, and system level testing

Test Results and Analyses	A summary of the verification and validation results	A summary of the verification and validation results in sufficient detail to demonstrate requirements were met at various levels of testing, and the results of system level testing	A verification and validation test report and summary, and results of testing at each level (unit, integration, system)
Certification	Written affirmation stating that software development was followed, that good Quality Assurance procedures were adhered to, and that test results demonstrate that the system specifications and the functional requirements were met. If the software development process has not yet been completed, and you and your division management have agreed that an equivalence determination can be made prior to completion, all elements of this table as discussed in this document should be provided: the hazard analyses, the system and software requirements, the documentation of the software development process, all test plans and test completion criteria, all expected results and outcomes, a report of what testing should demonstrate, and certification indicating that the described process will be completed with the indicated results and device specifications.		

4.1.2 Software Development

This documentation should summarize the software development process of the device under review.

Some key questions in reviewing this type of documentation:

- -- Are there appropriate company procedures for software development, quality assurance, and documentation of computer code?
- -- Do procedures encourage dividing code into functional segments (modular design)?
- -- Are design reviews, assurance activities, verification, validation, and testing performed prior to formal release of software?
- -- Are there adequate company procedures for documentation, structure, and acceptance of computer code developed for the manufacturer by outside contractors?
- -- Is there a description and explanation of the responsibilities and accountability of personnel involved in the software development process? Has the manufacturer indicated what personnel are responsible for approving software changes, documentation, and testing during the development process?

4.1.3 Verification, Validation, and Testing

This documentation should summarize activities performed to assess and improve the quality of the software throughout development and to determine conformance to specifications. It should include a description of how requirements were verified at various levels of the software. (See Section 4.3.)

Some key questions in reviewing this type of device:

- -- Has the manufacturer conscientiously tried to assure that its documented software requirements and specifications are complete, consistent, and met by the software?
- -- Are software tests traceable back to system and software requirements?
- -- Were the tests developed from system and software requirements?
- -- Were appropriate techniques used to assess and improve software quality, such as walk-throughs, code inspections, debugging and unit testing, integration testing, acceptance testing, or other quality assurance procedures?

4.1.4 Results of Testing

This documentation should include a system level test protocol with data, and a summary of the results.

Some key questions in reviewing this documentation:

- -- Does the manufacturer discuss how the results and analyses demonstrate conformance to requirements established for the device?

4.1.5 Certification

This documentation should consist of a statement from an authorized employee of the device manufacturer stating that the stated software development process was followed, that quality assurance procedures were adhered to, and that testing demonstrates that the functional requirements were met and that system specifications were fulfilled.

4.2 Moderate Concern -- The level of concern is moderate if the operation of the device or software function directly affects the patient so that failures or latent design flaws could result in minor to moderate injury to the patient, or if it indirectly affects the patient in that incorrect or delayed information could result in injury of the patient.

4.2.1 System and Software Requirements and Design

In addition to the documentation for a minor concern device, the documentation should be more complete in discussing the requirements and design, including safety requirements incorporated into the device.

The following are, in addition to the questions addressed for a minor level of concern device, key questions to be answered for this part of the review:

- -- Are the requirements clear and specific enough to be the basis for detailed design specifications and functional test cases?
- -- Are safety requirements consistent with the hazard analysis and the predicate device?
- -- Has the manufacturer identified the portions of controlling software that affect device safety?
- -- If warranted, is there redundancy between electro-mechanical controls and software controls?
- -- Do safety measures address both environmental influences and random component failures?
- -- Are there measures for detecting errors in finished software at installation time or during operation?
- -- If applicable, are there measures to prevent unauthorized access and changes to programs?
- -- Are potentially hazardous functions of the device monitored?
- -- Are safe and unsafe operating states of the device identified?
- -- Are potentially hazardous functions of the device limited by the software? Is an explanation provided to describe these in sufficient detail?

4.2.2 Software Development

The following are, in addition to the questions addressed for a minor concern device, key questions to be answered for this part of the review:

- -- Do procedures encourage dividing code into functional segments (modular design)?

-- Has the manufacturer identified the modular breakdown design of the program and the functions associated with each (modularization criteria)?
-- How is the hazard analysis performed and how are the results used in the design and testing?
-- Does the development plan include hazard analysis procedures and the identification of SAFETY REQUIREMENTS for the software and system?
-- Are verification and testing of safety requirements addressed?
-- Does the firm's quality assurance procedures or configuration management policy assure that software revisions and testing requirements are updated and are traceable to corresponding software or program requirements when changes are made? Is appropriate regression testing performed?
-- Are the assurance activities performed at each phase of software development for the device under review discussed?
-- Are design reviews discussed?

4.2.3 Verification, Validation, and Testing

This documentation should describe activities performed to assess and improve the quality of the software throughout development by determining the conformance to system and software requirements and by assuring that safeguards against potential hazards function properly. This information may include functional test protocols, and a discussion of test strategies and test completion criteria.

The following are, in addition to the questions addressed for a minor concern device, key questions to be answered for this part of the review:

-- Is the software evaluated at the module level prior to integration?
-- Are tests performed to demonstrate that the integrated subunits interact properly under nominal conditions?
-- Are tests performed to verify software response times under varying loads, including safety functions?
-- Are tests performed to stress the overall system?
-- Are the test cases used adequate to evaluate the design of the software and the device? Does the manufacturer include a discussion of how the test coverage of all operating conditions, input/output parameters, calculations, decisions, safety requirements, or any other requirement or specification is performed or assured through testing activities?
-- Are safety functions tested and analyzed?
-- Are revisions adequately evaluated to assure that they do not adversely affect other aspects of the software requirements or design?

4.2.4 Results of Testing

This documentation should summarize the results of verification, validation, and testing in sufficient detail so that the reviewer can assess their adequacy. Functional testing results should be provided, and a summary addressing how the results demonstrate conformance to requirements, with emphasis on safety requirements of the identified hazards.

Some key questions in reviewing this documentation are:

-- Do the results of tests and analyses demonstrate conformance to requirements established for the device?
-- Have the safety functions of software been rigorously tested and analyzed?
-- Did the system and software tests try marginal data, incorrect commands and data, complete testing of conditional sequences, momentary power failures, and malfunctioning data sensors?
-- Has a summary been provided that discusses the adequacy of the results and how they demonstrate that requirements were met?

4.2.5 Certification

This documentation should consist of a statement from an authorized employee of the device manufacturer stating that the software development process was followed, that quality assurance procedures were adhered to, that testing demonstrates that the functional requirements were met, and that system specifications were fulfilled.

4.3 Major Concern -- The level of concern is major if an operation of the device or software function directly affects the patient so that failures or latent design flaws could result in death or serious injury to the patient, or if it indirectly affects the patient such that incorrect or delayed information could result in death or serous injury of the patient.

4.3.1 Requirements and Specifications

In addition to the documentation required for a device of moderate concern, documentation for a major concern device should also include a more detailed description of device functional performance requirements and software/computer specifications and a more detailed description of potential software system hazards and software functions implemented as a result of such potential hazards. Particular attention should be paid to the thoroughness of the hazards analysis and the safeguards designed into the device, including the software.

The following questions are, in addition to the questions addressed for a moderate concern device, key questions to be answered for this part of the review:

-- How are hazardous processes monitored directly?
-- What redundant controls exist between electro-mechanical and software controls to reduce/prevent the occurrence of potential hazards that could lead to serious injury or death? What steps have been taken to limit the impact of these hazards should they occur?
-- What potentially hazardous functions of the device are limited by the software and how is this done?

4.3.2 Software Development

Documentation for a major concern device should include a written description of the design and development methods used, a thorough discussion of the hazard analysis performed, the results of the analysis, a description of how those results were incorporated into the design, and a description of quality assurance procedures.

The following are, in addition to the questions asked for a moderate concern device, key questions to be answered for this part of the review:

-- How is a software failure analysis conducted? Were failures derived by tracing "backward" from safety system faults to software (and components) where the fault could originate (top down) or by tracing fault propagation from a postulated failure origin to all possible outcomes, including system faults, classifying the criticality of each (bottom up)? (Failure Modes and Effects Criticality Analysis (FMECA))

-- Do the results of the hazard analysis make sense in comparison to predicate devices and the requirements of the device under review?

-- Has the software been structured into functional segments that make sense?

-- Does the development plan call for rigorous testing of the built-in safety features?

4.3.3 Verification, Validation, and Testing

This documentation should describe activities performed to assess and improve the quality of the software throughout development and to determine conformance to functional specifications. This includes the verification and validation protocol or other quality assurance testing performed and should include test strategies, test completion criteria, and the input and performance range over which the testing and analysis was performed.

The following are, in addition to the questions asked for a moderate concern device, key questions to be answered for this part of the review:

-- Are the test strategies and acceptance criteria consistent with requirements and are they adequate to evaluate the design of the device?

-- How are the safety features of the device tested and analyzed?

-- Are test strategies sufficient to determine that a device is as safe and as effective as a predicate device.

4.3.4 Results of Testing

This documentation should present the verification and validation and other quality assurance testing results and analysis so that the reviewer can assess and determine that the specifications were met at each appropriate level of the software architecture.

The following are, in addition to the questions asked for a moderate concern device, key questions to be answered for this part of the review:

-- Do the results of tests and analysis demonstrate conformance to requirements established at each level (module/integration/system) of the software architecture and the overall system?
-- Do results show that safety related functions were tested adequately and perform reliably?
-- Was the planned approach to designing, coding, and testing followed?

4.3.5 Certification

This documentation should consist of a statement from an authorized employee of the device manufacturer stating that the specified software development process was followed, that quality assurance procedures were adhered to, that testing demonstrates that the functional requirements were met, and that system specifications were fulfilled.

5.0 Labeling

When considering the labeling of a software/software-controlled device, the device labeling should adequately address special requirements and considerations of the software. This may include special equipment or software required for use, instructions for operator response to a fault or failure condition, likely causes of faults or failures, and how this information is presented to the operator and its accessibility.

Labeling is also relevant when software is a separate or separable part of the system and requires customer installation (usually when software is distributed on media such as magnetic tape, magnetic disk or plug-in modules, or via telecommunication). When third parties commercially introduce noncustom software for the original manufacturer's equipment, labeling is especially important.

The following questions should be considered when reviewing labeling for a software/software-controlled device:

1. Can it be determined from the labeling, software, and functional requirements whether the software is appropriate for a given task or intended use?
2. Will the software operate a given system or work with available equipment? Is particular equipment or additional software required for operation?
3. Can the purchaser use the software in a system by consulting the instruction manual, or is training necessary?

    a. Does the labeling provide the training requirements of the operator?
    b. If training is necessary, what labeling is available for training purposes? Is it adequate?

4. Does the manufacturer provide installation or is the software user-installed? If it is user-installed, are directions adequate?

    a. Is an installation qualification/checkout performed regardless of who performs the installation? Is this qualification/checkout adequate?

5. Does the manufacturer list probable sources/causes of likely software and hardware faults?

   a. Do the operating instructions include, when appropriate, a description of possible faults, the hazards presented to the patient if the faults should occur, the device configuration for each fault, and reasonable guidance for recommended operator response?
   b. Is the information about potential system faults readily accessible to the user (i.e., visual or audible alarms, displays, operator's manual or index guide that is attached to or placed with device)?
   c. If hazardous operating procedures are possible (cannot be obviated by design), are they identified and proscribed in warnings?

Labeling for software/software-controlled devices should be identified in appropriate manuals, instructions, and promotional material.

Attachment A

GLOSSARY

ALGORITHM: roughly means step-by-step instructions or rules for accomplishing a specific task or goal. By this broad interpretation an algorithm may be an equation, a recipe, an instruction manual, a part of a computer program or an entire computer program. An algorithm can be written in a natural language, like English, but not always very precisely.

ARCHITECTURE: Program architecture design involves identifying the software components, decoupling and decomposing them into processing modules and conceptual data structures, and specifying the interconnection among components. An architectural design specification may include data flow diagrams for the software, conceptual layout of data structures and data bases, name and functional description of each module, interface specifications for each module, interconnections among modules and data structures, timing constraints, exception conditions, and names, dimensional units, and other attributes of data objects.

CHECKSUM: A method for detecting errors in a series of bytes. The checksum is a one-byte number that can be computed fairly simply and appended to the series much as a parity bit is appended to a single byte. Errors are detected by computing a new checksum and comparing it to the checksum that came with the data. Checksums can detect some errors but there are many error combinations that leave the checksum unaltered. It is safest to assume that the checksum is useful for detecting single-byte errors and not useful for error correction.

COMPUTER-CONTROLLED: A computer-controlled device is one whose operation is influenced or directly controlled by a program-controlled mechanism that retrieves, decodes, and executes instruction to perform its designed tasks, or whose operation is controlled by the program itself.

CONFIGURATION CONTROL: The process of evaluating, coordinating, approving, disapproving, and implementing changes to identified CONFIGURATION ITEMS. (See CONFIGURATION ITEM LIST.)

CONFIGURATION ITEM LIST: A list that identifies all the collections of hardware and software elements that are treated as units (CONFIGURATION ITEMS) for the purpose of CONFIGURATION CONTROL of a specific product or system.

CYCLIC REDUNDANCY: A method for detecting errors in a series of bytes. Two extra bytes are computed and appended to the series. The computation is complex and there are fewer undetectable combinations of errors than for CHECKSUM or PARITY. Error detection is the primary goal of this method.

DESIGN ANALYSIS: The evaluation of a design to determine correctness with respect to stated requirements, conformance to design standards, and other criteria.

DESIGN ANALYSIS REPORT: A written report of the proceedings and recommendations of a DESIGN ANALYSIS.

DESIGN REQUIREMENTS SPECIFICATION: A document that prescribes completely and precisely the design of a system, a system component, or software. Typical contents may include, but are not limited to, algorithms, control logic, data structures, input/output formats, interface descriptions, physical requirements, structural requirements, and functional requirements.

{Applicable algorithms may or may not be part of software design requirements. Whether or not they are included depends on the particular data structures involved. Numeric structures may require specific equations that closely specify algorithms, but lists and strings may occupy different data structures depending on the performance required. Search and match within X seconds (as above) is a requirement; element-by-element search of a matrix table would specify a general algorithm and search a matrix table starting with element 0,0 followed by 0,1 followed by ... 0,n followed by 1,0 followed by ... m, would be a very tight algorithm for a search.}

DESIGN REVIEW: The formal review of an existing or proposed design to detect and remedy deficiencies that could affect fitness for use and environmental aspects of the product, process, or service, and to identify potential improvements in performance, safety, or economy.

DESIGN REVIEW REPORT: A written report of the proceedings and recommenda tions of a DESIGN REVIEW.

DIGITAL: means that quantities are represented by discrete whole numbers, nearly always base 2 (binary), as opposed to continuously variable, quantitative analogs such as electrical voltage or current. Applied to early computers, analog means operating on an electrical signal that changes continuously in proportion to a real quantity (such as velocity or pressure). To over-simplify, digital computers generate "stepped" data curves, while analog computers generate smooth data curves.

EMBEDDED: Embedded software usually refers to software in a Read Only Memory; that is, computer memory that cannot be changed by the computer. Embedded data usually means data that appears as part of a program.

ERROR DETECTION: Error detection schemes are described in computer science literature. Among them are PARITY, CHECKSUM, CYCLIC REDUNDANCY and LONGITUDINAL REDUNDANCY.

FUNCTION: The specific purpose or characteristic action of an entity, such as a system, component, or program.

FUNCTIONAL TESTING: Functional tests execute typical operating conditions of the software and system, with typical input values and results. These tests should test boundary conditions, system defaults, ordering, and special values.

INITIAL HAZARD ASSESSMENT: The initial classification and estimation of an identified system hazard. It includes the hazard description, an estimate of its criticality, and an estimate of probability that it will occur. The initial probability estimate may be qualitative or quantitative. INITIAL HAZARD ASSESSMENT applies to hazards identified any time during the LIFE CYCLE of a product or system. PRELIMINARY HAZARD ANALYSIS is an important part of INITIAL HAZARD ASSESSMENT.

INITIAL HAZARD ASSESSMENT REPORT: A complete list of potential system hazards identified throughout product design and development, a qualitative estimate of criticality (cost) of each, a qualitative (or quantitative if data are available) estimate of the probability of each, and recommendations for resolving any items that seem likely to pose problems.

INSPECTION: A formal evaluation technique wherein a person or group other than the author examines software requirements, design, or code in detail to detect faults, violations of development standards and other problems.

LIFE CYCLE: A period of time that starts when a product is conceived (as an idea) and ends when it is no longer available for use. The LIFE CYCLE has several phases, typically designated in order as: requirements phase, design phase, implementation phase, test phase, installation and checkout phase, operation and maintenance phase,and, sometimes, retirement phase. Readers should note that the preceding phase designations represent only one LIFE CYCLE model; other models may have more or fewer phases, and tasks that characterize one phase may overlap the tasks of one or more subsequent phases.

LONGITUDINAL REDUNDANCY: A method for detecting errors in a series of bytes. Error detection is done as in the CHECKSUM method, but the computation differs slightly. The method can detect single-bit errors, and when used with PARITY, it can localize single-bit errors for correction. There are, however, many combinations of errors that can defeat this method for both correction and detection.

MANUFACTURER: The terms "manufacturer" or "firm" used in this document refer to a manufacturer seeking market clearance of a medical device, a distributor seeking market clearance of a medical device, or an applicant representing a manufacturer or distributor who is seeking market clearance of a medical device.

MODEL: A model is like an algorithm in one sense. An algorithm can model a physical process (an equation) or it can model some aspect of human reasoning (a so-called expert system computer program). In another sense, a MODEL (data) can be a collection of numbers that represent the properties of an object. A model (computer program) can operate on a model (data) to estimate the effects of a real process on a real object.

MODULARIZATION: Modularization allows the designer to decompose a system into functional units, to impose hierarchical ordering on function usage, to implement data abstractions, and to develop independently useful subsystems. It can be used to isolate machine dependencies, to improve the performance of a software product, or to ease debugging, testing, integration, tuning, and modification of the system.

MODULARIZATION CRITERIA: This includes the conventional criterion in which each module and its submodules correspond to a step in the execution sequence, the information hiding criterion in which each module hides a difficult or changeable design decision from other modules, the data abstraction criteria in which each module hides the representation details of a major data structure behind functions that access and modify the data structure, levels of abstraction in which modules and collections of modules provide a hierarchial set of increasingly complex services, coupling-cohesion, in which a system is structured to maximize the cohesion of elements in each module and to minimize the coupling between modules, and problem modeling in which the modular structure of the modular structure of the system matches the structure of the problem being solved.

OPERATING SYSTEM: Essentially the software that enables users to interact with a computer. Operating systems generally handle interactions with PERIPHERAL EQUIPMENT, programming languages, and so forth.

PARITY: An error detection method for small units of data (usually 8 bits, a BYTE) wherein an extra bit is appended to the data unit. It is useful for detecting odd numbers of incorrect bits in a single byte. Even numbers of incorrect bits leave parity unchanged. By itself, PARITY is not useful for error correction.

PERFORMANCE TESTS: Performance tests are usually designed to verify system response times, system throughput, memory utilization, system execution, and traffic rates on data channels and communication links.

PERIPHERAL EQUIPMENT: Peripheral equipment is all computer system equipment other than the computer processor circuits and memory circuits; for instance, disk drives, tape drives and terminals.

PRELIMINARY HAZARD ANALYSIS: An INITIAL HAZARD ASSESSMENT of system hazards undertaken early in the product LIFE CYCLE. It includes describing the hazard, estimating its criticality, and estimating the probability that it will occur. The initial probability estimate may be qualitative or quantitative.

PROCEDURE: A set of steps to be followed to accomplish a task each time the task is to be done.

PRODUCT REQUIREMENTS SPECIFICATION: See SYSTEM REQUIREMENTS SPECIFICATION.

PROGRAM PATH: The order in which a computer executes its programmed instructions under given internal and environmental conditions.

PROGRAM PATH ANALYSIS: The identification of data and conditions that will cause a computer to execute a given PROGRAM PATH.

PROJECT PLAN: A management document that describes the approach to be taken for a project; for instance, developing a system. The plan typically describes the work to be done, resources needed, methods to be used, CONFIGURATION MANAGEMENT and QUALITY ASSURANCE procedures to be followed, schedules to be met, project organization, and so forth.

REQUIREMENT: A condition or capability that a system or component must meet or posses to satisfy a contract, standard, specification or other formally imposed document. Subsequent development of a system or component is based on the set of all requirements for that system or component.

REQUIREMENTS INSPECTION REPORT: A report of the proceedings and recommendations of an inspection review of a REQUIREMENTS SPECIFICATION.

REQUIREMENTS TRACE MATRIX: A document that records or traces the links or correspondence between REQUIREMENTS identified at each stage of product or system development. A matrix should at least document individual system requirements, elements of the design intended to fulfill a requirement, and test procedures and data needed to verify meeting the requirement.

SAFETY-CRITICAL SOFTWARE: Computer programs that control functions that affect system safety. Such programs include those causing the host computer to directly command accessories that affect safety or may communicate with (provide data for) other safety-critical software or systems.

SOFTWARE COMPLEXITY METRICS: A measure of the degree of complication of a system or system component: number of branches, nesting, data structures, lines of code, program size, etc.

SOFTWARE COMPONENT: A part of the computer program that implements some but not all of the functions or tasks of the complete program, viz., a subroutine, a code module, a function, or a procedure. Subroutine components may contain several more primitive subroutines integrated to perform a certain task.

SOFTWARE CONFIGURATION CONTROL PROCEDURES: A document that prescribes the procedures to be followed in evaluating, coordinating, approving, and disapproving changes to software.

SOFTWARE DEVELOPMENT PLAN: See PROJECT PLAN. A SOFTWARE DEVELOPMENT PLAN should include requisite inspections, reviews, audits, and tests for VERIFICATION and VALIDATION of software.

SOFTWARE HAZARD ANALYSIS: The identification of SAFETY-CRITICAL SOFTWARE, the classification and estimation of potential hazards, and identification of PROGRAM PATH ANALYSIS to identify hazardous combinations of internal and environmental program conditions.

SOFTWARE HAZARD ANALYSIS REPORT: A document that records the proceedings and recommendations of a SOFTWARE HAZARD ANALYSIS.

SOFTWARE REQUIREMENTS SPECIFICATION: A document that prescribes completely and precisely the conditions and/or capabilities that system or product software must possess or meet to satisfy the System or Product Requirements Specification. All typical inclusions of Requirements Specifications apply here. If a product or system is comprised of software alone, then the System or Product Requirements Specification replaces the Software Requirements Specification.

STRESS TESTS: Stress tests are designed to overload a system, such as input out of range data, rapid data entry, over processing, disconnecting interfaces, etc.

STRUCTURE TESTS: Structure tests examine the internal processing logic of the software. The goal is to achieve an optimum level of test coverage, both branching and path-traversing.

SYSTEM DEVELOPMENT PLAN: See PROJECT PLAN.

SYSTEM HAZARD ANALYSIS: The identification of specific internal and environmental conditions that may cause hazards in a system. The analysis should identify components and subsystems in which faults or failures would cause hazardous conditions or actions, should describe the fault or failure, and should estimate the probability and severity of the hazard.

SYSTEM HAZARD ANALYSIS REPORT: A document that records the proceedings and recommendations of a SYSTEM HAZARD ANALYSIS.

SYSTEM REQUIREMENTS SPECIFICATION: A document that prescribes completely and precisely the conditions and/or capabilities necessary for the system or product to fulfill ins intended use and/or to satisfy any additional contracts, standards, specifications, or other formal agreements with the firm's customers. SYSTEM REQUIREMENTS SPECIFICATIONS typically should include safety requirements, functional requirements, performance requirements, interface requirements, design requirements, and development standards. Synonymous with PRODUCT REQUIREMENTS SPECIFICATION.

TEST CASE: One or more sets of test data and associated PROCEDURES developed to exercise a particular FUNCTION or PROGRAM PATH, or to verify compliance with a specific REQUIREMENT.

TEST PLAN: A document prescribing the approach to be taken for intended testing activities, typically including, items to be tested, tests to be performed, schedules, personnel requirements, evaluation criteria, risks, and contingency plans.

TEST PROCEDURES DOCUMENT: A document that contains detailed instructions (TEST PROCEDURES) to set up and conduct one or more TEST CASES and to evaluate the results.

TEST REPORT: A document that records the conduct and results of TEST PROCE DURES carried out for a system or system component. (See TEST PROCEDURES DOCUMENT).

TEST REQUIREMENTS SPECIFICATION: A document that prescribes completely and precisely the TEST CASES, including correct results, necessary to verify that the system, product, or software possesses, meets, or fulfills the appropriate Requirements Specification. A Test Requirements Specification may stand alone or may be part of a TEST PROCEDURES DOCUMENT.

VALIDATION: The process of evaluating software at the end of the software development process to ensure compliance with software requirements.

VERIFICATION: The act of reviewing, inspecting, testing, checking, auditing, or otherwise establishing and documenting whether or not items, processes, services, or documents conform to specified requirements.

Attachment B

BIBLIOGRAPHY

Leveson, N.G. "Software Safety: Why, What and How," Computing Surveys, V18, #2, pp. 126-163, June, 1986.

Quirk, Verification and Validation of Real Time Software, Springer-Verlag, New York, 1985.

Sommerville, Software Engineering, Addison-Wesley, Reading, MA. 1982.

Fairley, Software Engineering Concepts, McGraw-Hill, New York, 1985.

Kernihan and Plauger, The Elements of Programming Style, McGraw-Hill, New York, 1978.

Beizer, Software System Testing and Quality Assurance, Van Nostrand Reinhold, New York, 1984.

Houston, M.F. "What Do the Simple Folk Do? Software Safety in the Cottage Industry," Supplemental Proc. Compass '87, 2nd Annual Conf. on Computer Assurance, June 1987, pp. S20-S24.

ANSI/IEEE Std 729-1983, Glossary of Software Engineering Terminology, Institute of Electrical and Electronic Engineers, New York, 1983.

Alexandridis, Microprocessor System Design Concepts, Computer Science Press, 1984.

Deciding When to Submit a 510(k) for Change to an Existing Device
Draft, April 8, 1994

Introduction

Almost from the establishment of the Medical Device Amendments to the Food, Drug and Cosmetic Act in 1976, FDA staff have attempted to define better when a change in a medical device would trigger a need for manufacturers to submit a new premarket notification (510(k)) to the Agency. The regulatory criteria contained in §807.81(a)(3) state that a premarket notification must be submitted when:

> *(3) The device ... is about to be significantly changed or modified in design, components, method of manufacture, or intended use. The following constitute significant changes or modifications that require a premarket notification:*
>
> > *(i) A change or modification in the device that could significantly affect the safety or effectiveness of the device, e.g., a significant change or modification in design, material, chemical composition, energy source, or manufacturing process.*
> > *(ii) A major change or modification in the intended use of the device.*

The key issue here is that the phrase "could significantly affect the safety or effectiveness of the device" and the use of adjectives such as "major" and "significant" lead to subjective interpretations.

Previous attempts to provide guidance on interpreting the regulations have focused generally on defining broad issues or principles that should be used in making the decision of when to submit a 510(k). A complicating factor has been that the variety of device types currently marketed as well as the myriad changes that occur as technology evolves are so diverse that one or two unifying principles cannot possibly account for all decision-making opportunities.

An alternative approach would be to enumerate all device types and all potential types of changes and then match each combination of device and change with a decision. Given the thousands of individual device types and possible tens or hundreds of enumerable changes, this would be an extremely difficult task. Furthermore, the resulting guidance could fill volumes and not be very easy to use.

Between the two extremes of broad principles and detailed enumeration is the area where models can be developed to assist in the decision making. If done correctly, such a model could provide the definitive answer regarding whether a 510(k) is necessary in the large majority of circumstances. This document proposes such a model to be used by manufacturers in their decision making. In the model, we attempt to address device types and changes to devices at a detailed enough level that application of the broad principles espoused by the agency would assure a very high percentage of correct decisions. The goal of the model is to provide definitive answers to a manufacturer's questions on whether a 510(k) is necessary for a particular type of change and to minimize the number of situations where the answer would be uncertain. If this goal is achievable, then the model could profitably be applied even in the anticipated user fee environment.

The 510(k) Process and Good Manufacturing Practices

Any guidance on 510(k)s for changes to a marketed device must consider the role of Good Manufacturing Practices (GMP) regulation in changes in device design. For some types of changes to a device, the Agency continues to find that a 510(k) is not necessary and that reliance on existing GMP requirements may reasonably assure safety and effectiveness of the changed device.

It is important to note that the current GMP does not directly address the original design of a device. It was the recognition of the need for this type of control for many types of devices that led to the inclusion of pre-production design controls in the Safe Medical Devices Act of 1990. The recently proposed new GMP regulation will implement the new authority and directly address design controls for new devices.

The current GMP, however, is not silent on device design. It requires manufacturers to document in the device master record (§820.181) any changes (and approval to changes) to device design and any associated testing. It also requires process validation to assure that products meeting the designed quality characteristics will consistently be produced (§820.5 and §820.100). Finally, manufacturers must have a formal approval procedure for any change in the manufacturing process of a device (including those dictated by design changes) (§820.100(b)(3)).

The net effect of the GMP is to require that, when manufacturers make a change in the design of a device, they must have a process in place to demonstrate that the manufactured product meets the change in design specifications (or the original specifications, if no change was intended). They must keep records, and these records must be made available to FDA. Thus, although the current GMP does not address the initial design of a device (like the proposed new GMP), it does exert significant influence when changes to a device are implemented.

Scope of the Proposed Model

The decision model outlined in this document has been developed to provide guidance to the manufacturer or distributor of Class I, Class II or pre-enactment Class III devices who are intending to modify their device and are in the process of deciding whether the modification requires the submission of a new 510(k). The decision model guidance is intended to supplement the general guidance on submitting 510(k)s, contained in ODE Bluebook Memorandum No. 86-3 (June 30, 1986), "Premarket Notification Review Program."

The types of modifications that are addressed in this model include both modifications to design, as well as modifications to device labeling, such as modified indications for use, claims for safety or effectiveness, etc. Furthermore, the model is also intended to apply to those situations when a device is recalled, modified or retrofitted as a result of problems with the device. In this instance, the changes need to be evaluated, using the model, to determine whether a 510(k) is necessary.

The model is not intended to apply to manufacturers of device kits or combination products, e.g., device/device, device/drug, or device/biologic combinations. Nor is the model intended to address the need for submitting 510(k)s by refurbishers or remanufacturers of devices. CDRH intends to develop additional guidance specific to these situations.

The model attempts to incorporate as much as possible existing guidance and policy regarding when 510(k)s are necessary for modifications to a device. In some cases, the existing guidance derives from advice given on only a few devices, and this guidance has been generalized to apply to a broader range of products. However, special cases exist where both manufacturers and FDA have worked to establish definitive guidance for modifications to specific devices, e.g., "Premarket Notification Document (510(k)) for Daily Wear Contact Lenses." This guidance is not intended to supply such existing device-specific guidance.

Principles for Building the Model

In developing this model for aiding in the decision on when to submit a 510(k), several principles had to be established. Some derive from existing 510(k) policy and are widely known; others are necessary for the model to work and need to be recognized in applying the model. To understand and use the model, the following principles need to be borne in mind:

- First time manufacturers of a device will always have to submit a new 510(k) or PMA.
- The model uses as a basis the change proposed by the manufacturer and not any unforeseen results of implementing that change.
- The basis for comparison of any changed device is the device described by the most recently cleared 510(k) (or a legally marketed pre-1976 device). That is, manufacturers may make changes without having to submit a 510(k), but each time they make a change, the device they must compare to is their most recently cleared device (or their pre-1976 device), not the currently marketed device. Thus, manufacturers may make a number of incremental changes in their device without having to submit a 510(k). Only when the sum of the incremental changes exceeds any of the thresholds or criteria in the model would they need to submit a new 510(k).
- Each type of change must be assessed separately, and, when any one change requires the submission of a 510(k), then a 510(k) incorporating all the changes, including any prior incremental changes, must be submitted. This new 510(k), once cleared, would form the basis of comparison for any future changes. The new 510(k) should incorporate the old 510(k) and identify and discuss all the changes made to date.
- When manufacturers change their device, they must comply with the Good Manufacturing Practice regulation. This regulation requires that specification changes be subject to controls as stringent as those applied to the original design specifications of the device, and that such changes be approved and documented by a designated individual(s), which includes the approval date and the date the change becomes effective (21 CFR 820.100(a)(2)). This means that when a change is made to the device, there is verification through testing that the change does not adversely

affect the device's safety or effectiveness. Only then can manufacturers assure an accurate assessment of the change(s) in the device in applying the model. They must maintain records of the verification testing under the current GMP. It is this validation of the changes and the documentation in the device master record that will obviate the need for a 510(k) in certain instances.

- To derive maximum benefit from this guidance, manufacturers should have in place a formal mechanism for evaluating whether a proposed device change meets the regulatory threshold for a new 510(k). This mechanism should document consideration of each decision point in this model and the basis for which all conclusions were made. Where such a decision mechanism is objective, neutral, prudent and well documented, it will serve to create a strong case favoring the validity of the manufacturer's conclusion.
- The model will not be capable of addressing <u>every</u> type of change to <u>every</u> type of device. Decisions in the "gray areas" will have to be made by FDA on a case-by-case basis. Manufacturers are invited to discuss these decisions with the Agency's Offices of Compliance and Device Evaluation. Once the gray areas are better defined and understood (i.e., sufficient numbers of case-by-case decisions have been made), then the model should be refined to encompass the new areas. It will be FDA's responsibility to document such decisions and to use them in revising the model at a future date.
- Any change to a device may be sufficiently significant that a 510(k) would not receive clearance (would be found "not substantially equivalent") and premarket approval would thus be required.

Proposed Model

The model that is proposed is a matrix model. This means nothing more than that we have attempted to define a small number of device types and a small number of change types, and have arrayed them as the columns and rows, respectively, of a matrix. The elements of the matrix contain the decision of whether to submit a 510(k) or merely to document their decision-making in the device master record as required under the current Good Manufacturing Practices regulation (21 CFR 820) for changes in device design, components, packaging or labeling. The rows of the matrix are identified by the type of device for which the change is being considered. The major categories of device types are defined to be consistent with (not the same as) the categories of devices currently used by the ISO and rapidly becoming standard in the European Community, if not the world. Thus, major categories of device types are identified as follows:

Implant
- Active
- Passive

Non-Implant
- Invasive
- Body Contact
- Non-Body Contact

In Vitro

These major categories are further subdivided in to device types according to the clinical use of the device, similarly to the organization of devices in FDA's Classification Regulations:

Diagnostic
Prosthetic
Monitoring
Surgical
Therapeutic

Originally, the columns of the matrix were defined by enumerating the major categories of changes that could be made to a device.

Labeling Changes
Technology Changes
Performance Changes

These categories were further subdivided by attempting to enumerate reasonably broad (but definitive) types of changes within these three major categories, e.g., a type of labeling change would be a new indication for use, a type of technological change would be a change in material formulation, and a type of performance change would be a change in performance specifications. (Definitions for most of the types of changes are contained in the attached definitions list.)

Once the rows and columns of the matrix are identified, one can, in theory, pick a type of change and look down that column to identify which device types would need a 510(k) if that type of change were made. In some instances, it was found that certain types of changes would almost always lead to a decision to submit a 510(k), while others would almost always lead to a decision to keep records. In order to simplify the matrix, columns representing such changes were removed from the matrix, and a flow chart was developed which contained all of the questions for which there were definitive answers independent of device type as well as those questions for which the matrix would have to be used. The attached flow chart thus asks a series of questions and directs the reader either to submit a 510(k), to maintain records or to go to the attached matrix to determine which should be done.

The matrix now contains only six columns (we started with about 20!), each of which is associated with a type of change that may or may not require a new 510(k). The rows of the matrix have similarly been condensed since some types of devices, as defined above, do not exist. For example, there are currently no surgical or prosthetic active implants. Thus, the matrix contains only 21 rows.

To use the model, one must answer the questions posed in the flow chart until either directed to submit a 510(k), document the decision-making, or go to the matrix. The matrix allows only two options: submit a 510(k) or document the decision-making. When contemplating changes to a device, manufacturers need to apply the flow chart and matrix for each individual type of proposed change, e.g., performance specification change, manufacturing process change, etc. If any one of the changes results in the need to submit a 510(k), then the 510(k) should be submitted and should incorporate all of the intended changes, as well as any previous changes to the device that did not result in the need to submit a 510(k). If all proposed changes result in the advice to document decision-making, manufacturers should document the application of the model in the device master record along with the necessary records of their validation of changes to their product.

For those circumstances where the proposed change is not addressed in the flow chart or in the matrix or in a device-specific guidance document, manufacturers should contact the Office of Device Evaluation in CDRH in writing to find out whether or not a 510(k) should be submitted. CDRH will document all such inquiries and its responses to them and will use this information in future revisions to this guidance. Manufacturers should file the FDA response with all other 510(k) decision-making documentation in the device master record.

510(k) Matrix Definitions

The following definitions are intended to clarify the meaning of all terms used in the flow chart and matrix used to determine when a 510(k) is necessary for a change to an existing device. Wherever possible, existing definitions from the Food, Drug and Cosmetic Act, the medical device regulations or ODE Bluebook memoranda have been used. In some cases, we have relied on strict dictionary definitions of terms. In a few instances, we have adapted existing definitions or have created new ones to fit the needs of this model.

1. Indications for Use: For what patient conditions the product is used, e.g., in cardiovascular surgery, in adult patient populations, etc. In some instances, a change in the Indications for Use may also imply a new intended use for the device, in which case a PMA is usually necessary. Included in Indications for Use would be a proposed change from a prescription device to over-the-counter (OTC) sales. In addition, a change in the locale where the device is to be used (e.g., hospital setting to the home) would also be considered a change in Indications for Use.
2. Labelling: *The term "labelling" means all labels and other written, printed, or graphic matter (1) upon any article or its containers or wrappers, or (2) accompanying such article.* (Section 201(m) of the Food, Drug and Cosmetic Act). This can include, among other things, any user or maintenance manuals, as well as and all advertising and promotional literature.
3. Warnings, precautions and contraindications:

    - Warnings describe serious adverse reactions and potential safety hazards that can occur in the proper use or misuse of a device, along with consequent limitations in use and mitigating steps to take if they occur.
    - Precautions describe any special care to be exercised by a practitioner or patient for the safe and effective use of a device.
    - Contraindications describe situations in which the device should not be used because the risk of use clearly outweighs any reasonably foreseeable benefits.

4. Claim: A claim may be any representation of the intended use through descriptive terms or pictorials that describe the use, capabilities, limitations or attributes of a

device when found in the labeling, promotional materials, advertising or verbal statements.

5. Documentation: Recording the results of applying the model to proposed changes in a device. Consideration of each decision point should be recorded as well as the final conclusions reached. A copy of this documentation should be maintained in the device master record.
6. Control Mechanism: The manner by which the actions of a device are directed. An example of a change in control mechanism would be the replacement of an electro mechanical control with a microprocessor control.
7. Operating Principle: The mode of operation or mechanism of action through which a device fulfills (or achieves) its intended use. An example of a change in operating principle would be using a new algorithm to compress images in a picture archiving and communications system.
8. Energy Type or Character: The type of power input to or output from the device. Examples of a change in energy type or character would be a change from AC to battery power (input) or a change from ionizing radiation to ultrasound to measure a property of the body (output).
9. Environmental Specifications: The (range of) acceptable levels of environmental parameters or operating conditions under which the device will perform safely and effectively. Examples of changes in environmental specifications would be expanding the acceptable temperature range in which the device will operate properly or hardening the device to significantly higher levels of electromagnetic interference.
10. Performance Specifications: The performance characteristics of a device as listed in device labeling or in finished product release specifications. Some examples of performance specifications are measurement accuracy, output accuracy, energy output level, stability criteria, etc. Generally speaking, changes in performance specifications such that additional capabilities are added within a previously cleared range are not considered significant. Providing kilovoltage settings in a diagnostic x-ray system every 5 kVp rather than every 10 kVp, within a previously cleared range of 60 to 120 kVp, is a good example. However, the addition of 130, 140 and 150 kilovoltage settings on the same diagnostic x-ray system or allowing operation only at 60, 90 and 120 kVp could be considered significant performance specification changes. Generally speaking also, when performance is specified by a measured parameter with an associated accuracy, changes in a device which do not affect that measured parameter (within the stated accuracy) are not considered performance changes. Changes outside the labeled accuracy, on the other hand are of concern because they could imply a new clinical effect.

Thus, for the purpose of this guidance, changes in performance specifications occur when the new performance is specified outside of a previously cleared range, when previously cleared performance specifications are eliminated, or when measured parameters are outside of the accuracy specified in the labeling for the device.

11. Ergonomic Change: A change in the way in which the device and the user are intended to interact. Examples of this would be the addition of an audible alarm to

an indicator light alarm, a change in the layout of a control panel, or a change from analog to digital presentation of information.

12. Operating Instructions: Directions for use. These are the directions provided to the practitioner or other user of the device, as appropriate, such that the device can be used safely and effectively.
13. Packaging: Any wrapping, containers, etc., used to protect, to preserve the sterility of, or to group medical devices.
14. Method of Sterilization: The physical or chemical mechanism used to achieve sterility or to achieve a specific sterility assurance level (SAL). Traditional methods of sterilization are steam, dry heat, ethylene oxide or gamma radiation sterilization methods.
15. Software Change: A change in the software program or in the firmware program which is used to control the actions or output of a device or which provides the input to a device. This includes any changes to a software program which is by itself a medical device.
16. Material Supplier: The firm supplying the raw material to a finished device manufacturer.
17. Reuse/reprocessing: Actions necessary for using the device with a new patient or for using a device more than once on a single patient. This may include instructions for assembly/disassembly, on-site sterilization or disinfection, etc. This does not include the refurbishing or repair of a device for resale.
18. Material Type: The generic name of the material from which the device is manufactured. (Use the generic name in the biomaterial compendium.) An example of a material type change would be the change from natural latex rubber to synthetic rubber.
19. Material Formulation: The base polymer formulation or the alloy, additives, colors, etc., used to establish a property or the stability of the material. This does not include processing aids, mold release agents, residual contaminants, or other manufacturing aids that are not intended to be a part of the material. An example of a change in material formulation would be a change from a series 300 stainless steel to a series 400 stainless steel.
20. Manufacturing Process: The chemical, physical, biological or other procedures used in the preparation, compounding, processing or packaging of a device. An example of a manufacturing process change would be a change in the way plastic joints or seams are permanently joined.
21. Dimensional Specifications: The physical size and shape of the device. Such specifications may include the length, width, thickness or diameter of a device, as well as changes in location or a part or component of the device. Generally speaking, changes in dimensional specifications are significant only when the properties of the new device cannot be predicted based upon the properties of the previously cleared device(s). Thus, changes in dimensional specifications that provide a new size outside of the already cleared range are almost always considered significant, e.g., providing a catheter of size French 21 when only sizes of 15 and 19 were previously cleared would be significant, but providing a size 17 catheter would not. On the other hand, providing intermediate sizes of some orthopedic implants could be significant, especially when diligent testing is necessary to

measure how the performance characteristics scale with the size change. For the purpose of this guidance, where a device line is offered in multiple sizes, changes in dimensional specifications occur only when the properties of the new device cannot be predicted based upon the properties of the previously cleared device(s) or when the size of the device is outside a previously cleared range.

22. Implant: A device which is intended to reside within a surgically or naturally formed cavity of the human body for a period of 30 days or more.
23. Active Implant: An implanted device that imparts energy to the body to achieve its intended purpose, e.g., an implanted cardiac pacemaker or an implanted muscle stimulator, or which requires energy to achieve its intended use, e.g., an implanted insulin pump.
24. Inactive Implant: An implanted device that does not impart energy to the body or that does not require outside energy to achieve its intended purpose, e.g., a total knee prosthesis or a porcine heart valve.
25. Non-implant: Any device not defined to be an implant or an in vitro device.
26. Invasive Non-Implant: Any device that communicates with body fluids or tissues (excluding skin) either through a natural or surgically produced opening. This includes any device that would otherwise be called an implant if it is intended to reside within the body for a period of less than thirty days.
27. Body Contact Non-implant: Any device that must contact the external surface of the body to achieve its intended use.
28. Non-body Contact Non-implant: Any device that need not be in contact with the body to achieve its intended use.
29. In Vitro Device: *Those reagents, instruments, and systems intended for use in the diagnosis of disease or other conditions, including a determination of the state of health, in order to cure, mitigate, treat, or prevent disease or its sequelae. Such products are intended for use in the collection, preparation, and examination of specimens taken from the human body.* (21 CFR 809.3(a))
30. Diagnostic Device: Any device used in the act or process of deciding the nature of a diseased condition or other state of health.
31. Prosthetic Device: Any device used to replace a missing, defective, or diseased part of the body.
32. Monitoring Device: Any device used to assess the state of a known disease or other state of health.
33. Surgical Device: Any device used in the treatment of disease or other conditions of health by manual or instrumental procedures, such as the removal of diseased tissue by cutting.
34. Therapeutic Device: Any device (other than a surgical device) used in the cure or mitigation of disease or other conditions.
35. Change: As used in the model, this means a proposed change and not the impact of a proposed change. Important impacts of a proposed change are identified on the flow chart. For example, a manufacturer may propose a change in method of sterilization. This change could impact on performance specifications because of potential chemical or physical damage to the device. The proposed change (in method of sterilization) is the change that must be used in the model.

Chapter 18

FDA ADVERSE FINDINGS

OVERVIEW

FDA-regulated industries are inspected by investigators from the agency's district offices. Following the inspection of a firm, FDA investigators submit to the firm's management a Form FDA 483 ("Inspectional Observations"), which lists violative findings.

FDA can take further regulatory actions if violations noted on the FDA 483 are not corrected, including disapproval or withdrawal of applications, the issuance of Warning Letters, FDA-requested recall of a product, product detention and seizures, injunctions, prosecutions, and civil penalties.

Warning Letters

The most common regulatory action following the issuance of an FDA 483 is the Warning Letter. Recently, warning letters regarding violations of GMP regulations issued to the pharmaceutical, medical device and biologics industries have been emphasizing validation-related problems.

A 1990 FDA memorandum relating to computerized systems in nonclinical laboratories noted that there were 84 computer-related observations cited on FDA-483s beetween 1979 and 1990. This was only 3% of the total reported FDA-483 observations. Most of the computer-related deficiencies were in the areas of general validation, inadequate SOPs, and inaccurate data.[1]

Recently, the FDA is citing more violations relating to computer system. Of the 167 Warning Letter citations reviewed for the last two years (1992-1994), the majority (21%) deal with general validation violations. The next largest amounts (15% and 14%, respectively) relate to lack of adequate systems documentation or lack of specifications. Eleven percent of citations relate to inadequate testing. Smaller numbers of violations were found relating to change congrol procedures (9%), SOPs and protocols (6%), systems security problems (5%), inadequate personnel training (4%), quality

[1]Memorandum, Ty Fugiwara, Nonclinical Laboratory Studies Branch (HFD-345), to Dr. Paul Lepore, HFC-230), March 29, 1990.

assurance/quality control (4%), maintenance (4%), 510(k) submissions (3%), and vendors (2%).

Recalls

Another regulatory action is product recall. Some recalls can be initiated by the FDA. Often, when problems are discovered the product-manufacturing firm will initiate a recall on its own accord.

In 1992, the FDA prepared a report on software-related recalls of medical devices for the period between 1983 and 1992. The FDA prepared this report to "provide manufacturers with insights on how and where to increase their quality assurance efforts in software development, quality assurance and production, or to minimize certain problems that may lead to software related problems resulting in device failure." Of the 2,792 quality problems analyzed during the report period which resulted in recalls, 165 (6%), were quality problems attributable to software failures. This report is attached to this chapter following the section on Warning Letters.

WARNING LETTER CITATIONS

The following listing indicates how widespread those problems are in the FDA-regulated industries.

General Validation

Medical Device Manufacturer, New Jersey. Letter issued by Atlanta District Office, 12/8/92: "... your firm has not validated the software used on ... test machines."

Medical Device Manufacturer, Michigan. Letter issued by Detroit District Office, 2/18/93: "A computer used in ... finished product testing has not been validated prior to its implementation."

Medical Device Manufacturer, Florida. Letter issued by Orlando District Office, 3/5/93: "Failure to validate all computer software."

Medical Device Manufacturer and Analytical Laboratory, Illinois. Letter issued by Chicago District Office, 5/4/93: "The *** test software system has not been validated to assure proper performance."

Blood Bank, Illinois. Letter issued by Chicago District Office, 5/10/93: "Failure to validate the computer system which is used in the control of testing and the analysis of data."

Drug Manufacturer, Texas. Letter issued by Dallas District Office, 6/23/93: "Failure to establish validation of computer and related systems used in ... testing."

Medical Device Manufacturer, California. Letter issued by Los Angeles District Office, 7/15/93: "Failure to validate the performance of software used for evaluating finished **** devices."

Methadone Treatment Center, Michigan. Letter issued by Detroit District Office, 7/23/93: "The computer system in use at your facility in a control function has not been validated prior to being place 'on-line'."

Medical Device Manufacturer, Texas. 483 observations, 7/29/93: "There is no evidence that the computer software used to control the ETO sterilizer cycles has been validated/verified by either the firm or the manufacturer to determine whether the software program performs as intended."

Medical Device Manufacturer, Texas. Letter issued by Dallas District Office, 8/26/93: "Your firm in conjunction with your contract sterilizer, continue to show deficiencies in validation procedures for the ETO sterilization monitoring process in relation to software and hardware configurations."

Hospital Blood Bank, Tennessee. Letter issued by Nashville District Office, 8/30/93: "The inspection revealed deviations from the requirements for controls of your computer system ... in that: ... the computer system was not properly validated; and the validation that was reportedly performed was not documented."

Medical Device Manufacturer, Alabama. Letter issued by Nashville District Office, 10/5/93: "The inspection revealed deficiencies related to the computer systems/programs in use for processing/controls. There were no records of validation after installation."

Medical Device Manufacturer, California. Letter issued by Los Angeles District Office, 10/5/93: "The computer system used for complaints documentation and the computer system utilized for label printing have not been validated."

Medical Device Manufacturer, Indiana. Letter issued by Detroit District Office, 10/27/93: "The software used to operate the chemistry analyzer was not validated prior to use. In addition, you have not followed your SOPs for Software Development in that there is no functional specifications, testing plan, and no documentation of testing performed for the O/S 9 version of software for the chemistry analyzer."

Drug Repacking Facility, Tennessee. Letter issued by Nashville District Office, 9/21/93: "During this inspection it was found that there appears to be ... inadequate controls for the computer system used in the processing/control of your repacking operation including no validation, no written specifications, and lack of documentation, instructions for the system."

Systems Manufacturer (Medical Device), Florida. Letter issued by Orlando District Office, 9/24/93: "Failure ... to validate all computer software."

Medical Device Manufacturer, Germany. Letter issued by CDRH Office of Compliance, 10/12/93: "Failure to validate computer software programs when computers are used as part of an automated production or quality assurance system.... For example, there were not written procedures nor any written records of validations performed for: a. The software that tracks the calibration of all test and production equipment; b. The software program which tracks all critical components by test lot number; and c. The software program for component acceptance tests to include all tests to be performed, including sampling levels/sizes."

Medical Device Manufacturer, Japan. Letter issued by CDRH Office of Compliance, 12/10/93: "Failure to validate the computer software programs by adequate and documented testing when the computers are used as part of an automated production or quality assurance system.... For example, the computer programs used to operate the sterilizers have not been validated."

Blood Bank, Missouri. Letter issued by CBER, 12/27/93: "Failure to assure that input to and output from a computer has been checked for accuracy ..., in that: a.

The entire computer system has not been validated to ensure that it consistently performs within pre-established limits, as required."

Medical Device Manufacturer, North Carolina. Letter issued by Atlanta District Office, 12/13/93: "Your firm had failed to adequately validate the software utilized by the ***."

Drug Manufacturer, Tennessee. Letter issued by Nashville District Office, 1/19/94: "Failure to assure that the ***Software program used for potency calculations has been validated."

Drug Manufacturer, New York. Letter issued by New York District Office, 1/20/94: "Failure to maintain appropriate validation data for computer software used to formulate varied batch sizes."

Medical Device Manufacturer, Colorado. Letter issued by Denver District Office, 3/2/94: "Device history records revealed instances of inadequate controls over manufacturing, testing and final Quality Assurance criteria.... e. Complaints indicate that defective software was distributed, this was verified by your firm." Also: "In-process manufacturing, testing procedures and controls are inadequate.... d. Final verification of software is not indicated in the DHR for two *** units."

Medical Device Manufacturer, South Carolina. Letter issued by Atlanta District Office, 3/14/94: "Your firm failed to appropriately validate changes to the software utilized in the *** monitoring equipment you manufacture and distribute. Our review of the *** machines found the software design, testing, and maintenance to be inadequate. Deficiencies documented in the software development included a lack of procedures for software design, specification review, structural testing, or functional testing of a finished software package; failure to properly document the design phase as required by your own Product Design Standards procedure; and insufficient documentation of the software code testing."

Medical Device Manufacturer, Colorado. Letter issued by Los Angeles District Office, 3/18/94: "Process validation studies for the firm's *** Lyophilizer were not performed prior to manufacturing of finished products after a computer unit was installed."

Medical Device Manufacturer, California. Letter issued by Los Angeles District Office, 4/6/94: "Failure to adequately validate manufacturing and quality assurance processes, where variation in process or operation parameters could adversely affect the product, or result in release of a product that does not conform to specifications. For example: ... b. There was no documented evidence to indicate that the software for the *** had been validated for its intended use in product testing."

Medical device Manufacturer, Germany. Letter issued by CDRH Office of Compliance, 4/18/94: "Failure to validate software programs by adequate and documented testing when computers are used as part of an automated production or quality assurance system.... For example, the software for the computer controlled lathe system used in processing the handle components has not been validated in order to establish confidence that the process will consistently produce components meeting their design specifications."

Medical Device Manufacturer, Tennessee. Letter issued by Nashville District Office, 5/2/94: "Deviations included ... lack of validation of the computer software program used as a part of controls."

Medical Device Manufacturer, Italy. Letter issued by CDRH Office of Compliance, 6/7/94: "Failure to validate software programs by adequate and documented testing, when computers are used as part of an automated production or quality assurance system.... For example, there is no validation data for the software programs used to control the automated lathes and milling machines. There are no version numbers for these programs nor are they signed and dated."

Medical Device Manufacturer, New York. FDA form 483, dated 2/16/94: "Protocol ... (*** Retrospective Validation ...) does not provide for the verification of raw data (for the most part handwritten) to the data entered into the computerized database (utilized in the statistical analysis of these data)."

Medical Device Manufacturer, Texas. Letter issued by Dallas District Office, 7/1/94: "Failure to validate software programs by adequate and documented testing, when computers are used as part of an automated production or quality assurance system."

Drug Manufacturer, New Jersey. Letter issued by Newark District Office, 7/7/94: "The computer system used [to] control sample receipt, sample number assignment, sample tracking, test results, storage and reporting, and stability is not validated."

Medical Device Manufacturer, Canada. Letter issued by CDRH Office of Compliance, 7/7/94: "Failure to ... validate computer software programs used in production.... For example ... the software for the automated injection molders has not been validated."

Medical Device Manufacturer, Texas. Letter issued by Dallas District Office, 7/27/94: "Failure to have adequate checks designed and implemented for automated data processing used for manufacturing or quality assurance purposes to prevent inaccurate data output, input, and programming errors."

Medical Device Manufacturer, North Carolina. Letter issued by Atlanta District Office, 8/1/94: "Failure to qualify/validate the microprocessor for on ETO sterilizer controller."

Security

Drug Manufacturer, New Jersey. Letter issued by Newark District Office, 11/5/92: "Your firm lacks a computer system validation report addressing areas such as data security."

Blood Bank, California. Letter issued by San Francisco District Office, 1/13/93: "We note with regard to your computer system that employees do not use confidential passwords to prevent entry into their user accounts, and a two-way modem allows an outside service vendor access to viral marker test data without prior authorization from a responsible individual in the blood bank."

Medical Device Manufacturer, New York. Letter issued by New York District Office, 2/18/93: (4) "Failure to establish adequate processing controls. Security for the **** Machine consisted of a key-lock mechanism, to limit access to the keyboard. Employees were using the keyboard without a key."

Medical Device Manufacturer and Analytical Laboratory, Illinois. Letter issued by Chicago District Office, 5/4/93: "The *** test software system ... lacks a program to assure system security and to create back-ups of assay methods."

Medical Device Manufacturer, New Mexico. Letter issued by Denver District Office, 5/7/93: "Your firm has ... failed to periodically review and validate computer security change codes."

Blood Bank, California. Letter issued by Los Angeles District Office, 9/28/93: "Failure to establish/implement adequate computer security to assure data integrity ... in that written security procedures fail to: define which job functions are authorized to have access to programs within the *** system and who may modify, add, or delete data; include standards for the selection and use of passwords; and require that passwords be changed on a regular basis and deleted when users leave the employment of your establishment."

Blood Bank, Florida. Letter issued by Orlando District Office, 9/29/93: "Inspection revealed: written procedures for manual operations when the computer system is down are not established ... and the procedure in use for volume verification on the automated system used for viral marker testing is not in accordance with the manufacturer's recommended procedure."

Drug Manufacturer, New Jersey. Letter issued by Newark District Office, 7/7/94: "Other than the 'on/off' switch, there is no system security."

Change Control Procedures

Drug Manufacturer, New Jersey. Letter issued by Newark District Office, 11/5/92: "Your firm lacks a computer system validation report addressing areas such as ... change procedures."

Blood Bank, Indiana. Letter issued by Detroit District Office, 11/12/92: "... you do not routinely revalidate the Plasmapheresis instruments when software changes are implemented."

Medical Device Manufacturer, California. Letter issued by Boston District Office, 12/2/92: "Failure to have specification changes approved and documented by a designated individual(s). There is no procedure for signing off or formally approving changes that are made to the SAIL software. Also, the changes that are made to the source code of your software regarding individual revisions or code changes are not documented to show that the appropriate change was made."

Medical Device Manufacturer, Texas. Letter issued by Los Angeles District Office, 12/4/92: "Failure to validate each significant change in the computer software used in continuous passive motion devices."

Blood Bank, Florida. Letter issued by Orlando District Office, 9/29/93: "Inspection revealed the release and subsequent transfusion of a unit of HBc positive platelets. Your investigation revealed the unit was inadvertently released because of a modification in a software program of your computer system. No written validation protocol is available and there is no documentation that the modification in the software that caused the release of the HBc positive platelets was validated."

Medical Device Manufacturer, California. Letter issued by Los Angeles District Office, 11/18/93: "There is no written procedure designating the person or persons who are allowed to make or alter software programs or changes used on device production equipment."

Blood Bank, Missouri. Letter issued by CBER, 12/27/93: "Failure to maintain and follow adequate written standard operating procedures (SOPs) for the collection, processing, compatibility testing, storage and distribution of blood and blood components.... For example: ... There are no written procedures for computer system validation." Also: "Failure to assure that input to and output from a computer has been checked for accuracy ..., in that: a. The entire computer system has not been validated to ensure that it consistently performs within pre-established limits, as required. b. Software program modifications that affected decisions in donor or unit acceptability were not validated. c. The validation of changes to the labeling program to print on-demand ABO/Rh labels was inadequate. For example: i. There is no documentation that describes the changes made to the various programs. ii. The validation protocol was written after the validation was completed. iii. There is no documentation that describes what was done to verify the software after it was modified."

Medical Device Manufacturer, Michigan. Letter issued by Detroit District Office, 1/24/94: "Device history records (DHRs) are incomplete in that computer software revisions are not implemented by an engineering change procedure."

Medical Device Manufacturer, New Jersey. Letter issued by Newark District Office, 3/11/94: "There was no software validation results for the *** on file at the time of the inspection. For example: A. *** modified the software program so that when the unit is in 'run mode' and a 'low battery' condition is detected, the unit is placed into standby and is not allowed to operate in a 'low battery' condition. B. Modification of the software program to go into leak alarm at 1 cc per 1/2 hour instead of 5 cc per 1/2 hour." Also: "Your firm failed to adequately validate modifications to the ***. For example: ... C. Modification of software program so that when the unit is in 'run mode' and 'low battery' the condition is detected. D. Modify the software program to go into a leak alarm at *** instead of ***." The FDA Warning Letter also commented on the firm's response to the 483 issued after the inspection: "Your response ... did not contain validation protocols or results for the non-software modifications to the ***. With respect to ... validation of software changes, there appears to be no standard procedures for evaluating future proposed software changes to the system."

Medical Device Manufacturer, Illinois. Letter issued by Chicago District Office, 3/28/94: "Failure to control changes to equipment used in the manufacture of diagnostic kits, in that unauthorized and undocumented changes were made to software of the feeder track sensor of the *** Line in the lyophilization room."

Medical Device Manufacturer, New York. Letter issued by New York District Office, 4/6/94: "Failure to have specification changes approved and documented by a designated individual(s), including an approval date and the date the changes become effective.... For example: ... Changes in the software are not performed through a formal change order approval system."

Blood Bank Software Manufacturer, New York. Letter issued by New York District Office, 4/13/94: "The records did not always identify the individual who approved changes and revisions in the software, nor the approval date and effective date."

Blood Bank, Washington. Letter issued by Seattle District Office, 4/15/94: "Failure to maintain manual or computer-generated records of corrections made to inaccurate data"; "Failure to have an adequate SOP covering change control."

Medical Device Manufacturer, New York. FDA form 483, dated 2/16/94: "Failure to have maintained a log which would have documented each time the Validation study *** database (which included direct entry raw data) had been entered and updated or otherwise changed and would have identified the person responsible for making those entries."

Blood Bank Computer Software Manufacturer, Washington. Letter issued by Seattle District Office, 6/28/94: "Program Temporary Fix Reports (PTFs) lacks documentation as to who approved the decision not to complete the validation protocol subsequent to changes made to programs of *** software and the basis for that decision.... PTFs lack documentation of the change approval date and the date the change became effective.... There are no written procedures for the development of specifications, validation requirements, and change control protocols for the laboratory transfer software and the mobile software which are developed and maintained by a consultant to ***. Additionally, review of SRF ... suggest inadequate validation of the mobile software distributed to users by your firm in March 1993. The software allowed the registration of donors who had been previously deferred when the donor record was accessed by social security number or donor account number."

Documentation

Drug Manufacturer, New Jersey. Letter issued by Newark District Office, 11/5/92: "Your firm lacks a computer system validation report addressing areas such as ... record retention for HPLC computerized systems at your ... Laboratory."

Drug Manufacturer, Tennessee. Letter issued by Nashville District Office, 4/2/93: Repackaging operations failed "to have written computer program specifications."

Medical Device Manufacturer, Ohio. Letter issued by Atlanta District Office, 4/16/93: "Your firm had failed to maintain the level of documentation required in regards to the development/testing/validation of the software component of the *****. Written specifications were inadequate in that they failed to provide measurable, objective criteria which could be tested for conformance with the component requirements. The testing documentation for this software was incomplete in that it did not always define the test cases to be used, did not clearly define the expected results of the test, and often did not include actual test results." Also: "There was no documentation available that the checksum procedure used to guard against microprocessor malfunction would detect multibit changes in the MPU RAM. A multibit error has been documented in this MPU unit...."

Medical Device Manufacturer, Pennsylvania. Letter issued by Philadelphia District Office, 4/23/93: "There is no documentation to assure that the manufacturing procedures for programming any programmable device are followed."

Medical Device Manufacturer and Analytical Laboratory, Illinois. Letter issued by Chicago District Office, 5/4/93: "The *** test software system ... does not provide hard copy to allow verification of instrument parameters actually used."

Medical Device Manufacturer, Ohio. Letter issued by Cincinnati District Office, 5/6/93: "The Device Master Record for the *** is incomplete in that it does not include documentation of validation of the software that is part of the device; it does not include the computer source code as either hard copy or on magnetic medium; it does not include changes (upgrades) made to the software that is part of the device; and it does not include the test procedures and criteria that were used to evaluate the current device software program for acceptance for use." Also: "There is no documentation that your firm conducts an evaluation of the assembled and installed *** devices to assure that they will perform as intended."

Medical Device Manufacturer, Sweden. Letter issued by CDRH Office of Compliance and Surveillance, 5/14/93: "Failure to collect all information necessary to investigate a complaint, and to determine whether an injury, death, or hazard to safety exists.... For example, information regarding the programmer model type and software revision level ... are not documented."

Medical Device Manufacturer, Colorado. Letter issued by Denver District Office, 5/17/93: "No documentation to show that software is developed, tested, validated, reviewed, and accepted under a formally approved or adequately controlled system."

Drug Manufacturer, Texas. Letter issued by Dallas District Office, 6/23/93: "Failure to establish a written procedure designed to assure proper performance of mechanical, automatic, and electronic equipment."

Medical Device Manufacturer, Germany. Letter issued by CDRH Office of Compliance, 7/23/93: "Validation records for the software utilized in the sterilization chamber's **** are not maintained."

Medical Device Manufacturer, Illinois. Letter issued by Chicago District Office, 10/19/93: "Failure to maintain adequate written processing procedure in that the *** Filler Inspection System does not contain a written procedure describing the function of the *** system, nor does it define all defect patterns that are loaded into the automated inspection system."

Blood Bank, New Jersey. Letter issued by Mid-Atlantic Regional Office, 8/17/93: "Failure to maintain accurate records from which unsuitable donors may be identified so that products from such individuals will not be distributed in that: ... Comparison of the donor deferral files with the computer-generated deferral list found several discrepancies."

Blood Bank, Pennsylvania. Letter issued by Philadelphia District Office, 9/3/93: "Failure to have and follow an adequate written program designed to assure proper performance of the computer system, and to maintain appropriate validation data."

Medical Device Manufacturer, North Carolina. Letter issued by Atlanta District Office, 12/13/93: "The facility lacked any formal documentation which described the requirements and specifications for the software."

Medical Device Manufacturer, Michigan. Letter issued by Detroit District Office, 1/24/94: "Device history records (DHRs) are incomplete in that computer software revisions are not implemented by an engineering change procedure; and documentation for the manufacture of *** software lacks several elements, such as number manufactured, programming results, and individual performing testing. In addition, there is not documentation to show what modifications to the device hardware and software have been properly validated prior to implementation...."

Blood Bank, New York. Letter issued by New York District Office, 2/3/94: "Written procedures were not available for the *** Computer System describing the following: a. Maintenance and change control procedures for system hardware and software. b. Validation protocols to independently test critical operation functions. c. Procedures for the active review of system function and the timely correction of errors. d. Procedures for the initial and continued training of personnel."

Medical Device Manufacturer, Ireland. Letter issued by CDRH Office of Compliance, 2/1/94; "Failure to have adequate checks designed and implemented for automated data processing used for manufacturing or quality assurance purposes to prevent inaccurate data output, input, and programming errors.... For example: All device history records (DHR) are generated for an *** which is not validated, the *** not approved, and an approved hard copy of the DHR does not exist."

Medical Device Manufacturer, New Jersey. Letter issued by Newark District Office, 3/11/94: "Your Device Master Record documents for the *** are not dated or signed by a designated individual(s). For example: A. SOP for Programming *** microcomputer chips."

Medical Device Manufacturer, New York. Letter issued by New York District Office, 4/6/94: "Failure to review, evaluate, and maintain by a formally designated unit all records of written and oral complaints relative to the identity, quality, reliability, safety, effectiveness, or performance of the device.... For example: ... Defects (bugs) identified in software through complaints are not handled and investigated as complaints. The complaint file does not contain any complaints concerning software problems even though a number of bugs were identified through service reporting system and user complaints that would qualify them as complaints."

Blood Bank Software Manufacturer, New York. Letter issued by New York District Office, 4/13/94: "Failure to establish and implement an adequate complaint handling program ... in that written records of complaints and investigations for the 1.6x revisions, including conclusions and follow-up, were not prepared contemporaneously with the investigations. When such records were prepared ... they were dated to reflect the date of the original investigation which occurred up to five months previously."

Blood Bank, California. Letter issued by Los Angeles District Office, 6/27/94: "The documentation to assure that the computer system in use is performing in the manner for which it is intended and that any changes are evaluated, is deficient. For example, there was no approved and dated validation protocol. Detailed written procedures for computer use have not been prepared for all functions. There is a lack of documentation that a complete validation has been performed."

Blood Bank Computer Software Manufacturer, Washington. Letter issued by Seattle District Office, 6/28/94: "Failure to establish a quality assurance program which identifies, recommends, or provides solutions for quality assurance problems and verifies the implementation of such solutions ... in that: "... 2. Documentation accompanying the tape distributions of important program corrections does not adequately indicate the significance of the software program corrections or the need of the user to install the software changes and review/correct related records. 3. Documentation supplied to users describing the functions of *** software is incorrect or inadequate; limitations of the software are not sufficiently explained." Also:

Service Request Forms "lack documentation of all information necessary to conduct a detailed investigation of a failure. Reports do not specify which PTF level of a software program was being utilized or the configuration table settings of the user's system at the time failure occurred.... SRF's lack consistent documentation of the conclusion of the investigation, follow-up, and response to the complainant."

Drug Manufacturer, New Jersey. Letter issued by Newark District Office, 7/7/94: "There is no documentation of the system or its operation, only menu presentations on the computer screen."

Medical Device Manufacturer/Distributor, Oregon. Letter issued by Seattle District Office, 7/15/94: "... information required by the Medical Device Reporting (MDR) regulations as specified in 21 CFR 803 to be furnished to the Food and Drug Administration was not submitted as follows: ... Complaint ... reported the return of a *** ... due to programming/telemetry difficulties."

Drug Manufacturer, Michigan. Letter issued by Detroit District Office, 7/22/94: "Your laboratory records do not document the validity of the computer program used to calculate analytical test results."

Medical Device Manufacturer, Germany. Letter issued by CDRH Office of Compliance, 8/8/94: "Failure to validate software programs by adequate and documented testing.... For example, no validation documentation (validation protocol, data, or conclusions) could be provided for the *** (software controlled) machines used for manufacturing."

Training

Blood Bank, Indiana. Letter issued by Detroit District Office, 11/12/92: "Documentation of employee training for operation of the ... instruments is incomplete, and there is no provision for annual update training of employees as required. Also, a designated trainer has not been identified to train employees in the operation of your automated instruments."

Medical Device Manufacturer and Analytical Laboratory, Illinois. Letter issued by Chicago District Office, 5/4/93: "There is no established training program to assure that laboratory personnel are familiar with laboratory control procedures and current good manufacturing practice."

Blood Bank, Illinois. Letter issued by Chicago District Office, 5/10/93: "Failure to document the initial or update training of personnel who use the computer system."

Medical Device Manufacturer, Alabama. Letter issued by Nashville District Office, 10/5/93: "The inspection revealed deficiencies related to the computer systems/programs in use for processing/controls. There were no ... computer system user procedures/instructions for the computer systems."

Medical Device Manufacturer, Utah. Letter issued by Denver District Office, 9/7/93: "Failure to Document Adequate Training of Personnel. Your firm lacks written procedures to document that all personnel have the necessary training to use the automated equipment."

Medical Device Manufacturer, France. Letter issued by CDRH Office of Compliance, 8/23/94: "Failure to provide personnel with training necessary to perform their assigned responsibilities adequately.... For example, no formal training program or documentation exists to assure that the operator assigned to the reconfiguring of the RAM is adequately trained or qualified to operate this equipment."

Maintenance/Calibration

Blood Bank, Indiana. Letter issued by Detroit District Office, 11/12/92: "You do not routinely maintain calibration and maintenance records for equipment in use at this facility...."

Medical Device Manufacturer, New York. Letter issued by New York District Office, 12/21/92: "Failure to establish written procedures for routinely calibrating, inspecting and checking all production and quality assurance measurement equipment, such as mechanical, automated or electronic equipment, so that they are suitable for their intended purposes and are capable of producing valid results."

Medical Device Manufacturer, New York. Letter issued by New York District Office, 2/18/93: (1) "Failure to establish a written schedule for the maintenance, adjustment and cleaning of equipment. For example ... there were no cleaning procedures or schedule for the automated **** Machine." (2) "Failure to assure production and quality assurance measurement equipment was being routinely calibrated, inspected and checked. Records documenting these activities have not been maintained. For example... there were no validation protocols and studies for automated manufacturing, using the **** Machine and the operating software."

Medical Device Manufacturer, California. Letter issued by Los Angeles District Office, 2/23/93: "You have failed to assure that all production and quality assurance measurement equipment are routinely calibrated.... For example, there is no record documenting the calibration of the **** EPROM programmer. Also, there are not provisions to document remedial action taken when accuracy and precision limits are not met. Your written reply to the FDA-483 ... stated that the EPROM programmer manufacturer was contacted and that they acknowledged the need for routine calibration of the programmer and therefore it was returned to the vendor for calibration." In addition, this Warning Letter noted that this firm had had several product recalls, including one for "incompatibility between software and the flow control valves."

Contract Sterilization Facility, France. Letter issued by CDRH Office of Compliance and Surveillance, 6/24/93: "Failure to adequately calibrate, inspect, and check measurement equipment according to written procedures.... For example, ... there is no documentation of recalibration of the *** following a repair which involved replacement of a logic circuit and related subassemblies."

Medical Device Manufacturer, New York. Letter issued by New York District Office, 7/26/93: "Failure to routinely calibrate quality assurance measurement equipment according to written procedures and to maintain complete calibration records, such as: ... There was no documentation that analyzers A and B were calibrated. According to the service manager, PC boards inside the computers require quarterly calibration to assure proper gain."

Medical device Manufacturer, Germany. Letter issued by CDRH Office of Compliance, 4/18/94: "Failure to develop and adhere to a written schedule for maintenance, adjustment, and cleaning of equipment to assure that manufacturing specifications are met.... For example, you have no written procedures or maintenance schedule for your ... computerized lathe...."

SOPs/Protocols

Blood Bank, Indiana. Letter issued by Detroit District Office, 11/12/92: "Your standard operating procedures (SOPs) are incomplete and/or contain incorrect information...."

Medical Device Manufacturer, New York. Letter issued by New York District Office, 2/18/93: "Records documenting these activities have not been maintained. For example... there were no validation protocols and studies for automated manufacturing, using the **** Machine and the operating software."

Blood Bank, Illinois. Letter issued by Chicago District Office, 5/10/93: "SOPs are incomplete in that they ... do not provide for the documentation of overrides and changes which are made to computer system data."

Blood Bank, California. Letter issued by Los Angeles District Office, 5/13/93: "Your facility still has no written validation protocols for the computer system in use and demonstrated validation of that system in accordance with written protocols." Also: "Incomplete written standard operating procedures for the computer system in use."

Pathology Laboratory, Kansas. Letter issued by Kansas City District Office, 6/16/93: "Failure to maintain and/or follow written standard operating procedures ... in that: .... Your firm lacks a written validation protocol for the *** Software Program which your firm is currently validating."

Blood Bank, Pennsylvania. Letter issued by Philadelphia District Office, 9/3/93: "Failure to maintain and/or follow written standard operating procedures (SOP's) that include all steps to be followed in the collection, processing, storage, and distribution of blood and blood products ... in that: ... An SOP does not exist that adequately incorporates use of the computer into daily operations and establishes procedures for the active review of computer system functioning."

Blood Bank, Missouri. Letter issued by CBER, 12/27/93: "Failure to maintain and follow adequate written standard operating procedures (SOPs) for the collection, processing, compatibility testing, storage and distribution of blood and blood components.... For example: ... There are no written procedures for computer system validation."

Drug Manufacturer, New Jersey. Letter issued by Newark District Office, 2/10/94: "Quality Control action limits established by your firm can be printed from a computer file, but there are no written SOPs requiring documentation of sign-off & approval of these specifications.

Medical Device Manufacturer, New Jersey. Letter issued by Newark District Office, 4/12/94: "Your firm's validation for the *** software lacked the following: A. A protocol identifying the objectives, critical parameters and critical ranges. B. A report/conclusion to support expected outputs."

Blood Bank, Washington. Letter issued by Seattle District Office, 4/15/94: "Failure to have adequate written SOPs for the verification of data transfer when the computer system is updated via these tapes provided by the software vendor." "Failure to have an adequate SOP covering change control."

Specifications

Medical Device Manufacturer, California. Letter issued by Boston District Office, 12/2/92: "Failure to conduct processing control operations in a manner to assure that the device conforms to applicable specifications. Validation for the N.3 Software as well as finished device inspection of the ... System in which this software is used, failed to detect a defect in the software. This defect led to a recall of this product...."

Medical Device Manufacturer, New York. Letter issued by New York District Office, 2/18/93: "Failure to establish, implement and control manufacturing specifications and processes. For example ... there were no procedures for operation of the automated **** Machine," and "there was no documented monitoring of the automated **** Machine's manufacturing specifications, such as processing time, temperature and pressure." "Failure to establish adequate processing controls." **Medical Device Manufacturer, Wisconsin.** Letter issued by Minneapolis District Office, 2/22/93: "Your Good Manufacturing Practice control system lacks an adequate engineering change control procedure in that changes to software and device specifications are not controlled and validated before being used in the finished device."

Medical Device Manufacturer, Sweden. Letter issued by CDRH Office of Compliance and Surveillance, 5/14/93: "Failure to subject specification changes to controls as stringent as those applied to the original design specifications of the device, and to have a designated individual(s) document the approval of such changes with approval date and the date the change becomes effective.... For example, the software used in the computer test systems have been revised *** times in the past *** years, and the revisions have not been evaluated to determine if the software changes require revalidation."

Software Manufacturer (Software used in Blood Banks), Missouri. Letter issued by Kansas City District Office, 5/14/93: "Failure to establish, implement, and control manufacturing specifications in a manner to assure that the device conforms to its original design or any approved changes in that design."

Medical Device Manufacturer, New Jersey. Letter issued by Newark District Office, 6/18/93: "Procedures have not been established to assure that the design basis for the device is correctly translated into approved specifications. For example: There is no documented evidence that software revisions result in a device meeting specified quality attributes." Also: "Specification changes are not approved and documented by a designated individual and do not include the approval date of the change. For example: Software upgrade Revision 1.00 to 1.01."

Medical Device Manufacturer, Oregon. Letter issued by Seattle District Office, 7/7/93: "Failure to subject specification changes to controls as stringent as those applied to the original design specifications of the device, and to include change approval and documentation."

Medical Device Manufacturer, California. Letter issued by Los Angeles District Office, 7/15/93: "Failure to establish complete specifications for device software and

failure to adequately validate the performance of software used in the **** analyzers."
Medical Device Manufacturer, New York. Letter issued by New York District Office, 8/20/93: "Failure to document/validate software changes to assure the **** conforms to the original approved design specifications."

Medical Device Manufacturer, California. Letter issued by Los Angeles District Office, 8/23/93, after CDRH review of firm's response to 483 observations: "You addressed the specific observations and provided reasonable explanations. However, in so doing you identified another problem for which you provided no resolution. You state that hybrid testing by the manufacturer was not reliable because of spurious ATI results. You indicated the problem was caused by the vendor's use of software of a version different than used by your firm. You have not provided information regarding steps taken to assure that vendors provide components which adhere to your specifications."

Software Manufacturer, North Carolina. Manufactures software to calculate radiation dosages for teletherapy and brachytherapy systems. Letter issued by Atlanta District Office, 8/10/93:

"2. Failure of the device master record to include device specifications and quality assurance procedures and specifications.... For example:

* "The device master record form indicates the record is comprised primarily of procedures for the duplicating and customizing disks from the master software program. The device master record does not contain device specifications and quality assurance procedures and specifications.
* "During the inspection, the investigator observed the absence of a requirement/specification document for the *** software program. As a response to this observation, you submitted a 'requirement/specification document and a program design document'.... Your 'requirement/ specification document and a program design document' is a general design outline of the *** software program and not a detailed specification document suitable for the development of the software program."

Systems Manufacturer (Medical Device), Florida. Letter issued by Orlando District Office, 9/24/93: "Failure ... to ensure changes to the Device Master Record (DMR) are authorized in writing by a responsible individual and to include written specifications or procedures used to manufacture, test, and to deliver the device."

Blood Bank, California. Letter issued by Los Angeles District Office, 9/28/93: "Failure to assure that equipment (the computer system) performs in the manner for which it was designed so as to assure compliance with the official requirements prescribed for blood and blood products.... For example, failure to maintain original specifications or develop a current set of specifications for the *** computer software."

Medical Device Manufacturer, Alabama. Letter issued by Nashville District Office, 10/5/93: "The inspection revealed deficiencies related to the computer systems/programs in use for processing/controls. There were ... no systems/programs specifications or source codes available during this inspection."

Medical Device Manufacturer, Germany. Letter issued by CDRH Office of Compliance, 10/12/93: "Failure to control specification changes as stringently as those applied to the original design specifications of the device.... For example: a.

Software design documents for computerized devices are not adequate in that they contain only upper level information; and, b. No final test/validation procedures or report was done when there was a change from the VAX to the UNIX based system to assure no effect on the application software."

Medical Device Distributor, New Jersey. Letter issued by Newark District Office, 10/21/93: "Software used in your ATAC systems has not been validated to make sure your specifications for the device and process are consistently met."

Medical Device Manufacturer, New Mexico. Letter issued by Denver District Office, 11/5/93: "No written manufacturing specifications and processing procedures, including validation, have been established, implemented, and controlled for software, to assure that the device conforms to its original design or any approved changes in that design."

Medical Device Manufacturer, Ireland. Letter issued by CDRH Office of Compliance, 12/3/93: "Failure to establish and implement specification control measures to assure that the design basis for the device and packaging is correctly translated into approved specifications.... For example, the validation of software used to control the *** did not contain all process setting parameters with associated tolerances and/or ranges, verification of actual operating parameters, any challenge to out of bounds parameters and/or invalid parameters, or verification of current temperature pressure and speed calibration sensors."

Medical Device Manufacturer, Illinois. Letter issued by Chicago District Office, 12/16/93: "Failure to subject specification changes to controls as stringent as those applied to the original design specifications of the device, and to have such changes approved by a designated individual(s) including the approval date.... For example: a. There was no Engineering Change Notice issued for modification made to computer software which was changed to accommodate a two button instead of a three button mA and kVp console.... b. You have not validated (by documented testing) the current computer software, version 2.02.... You stated in your response that detailed documentation is available and the software was validated; however, you have not provided documentation of this validation, either in your response or during the inspection."

Blood Bank Software Manufacturer, New York. Letter issued by New York District Office, 4/13/94: "1. Failure to assure that specification changes are subjected to controls as stringent as those applied to the original design specifications of the device ... in that: a. There were no records of the software specifications, production/process specifications, quality assurance specifications, and labeling specifications of versions 1.60 through 1.65 of the device."

Medical Device Manufacturer, Massachusetts. Letter issued by Boston District Office, 6/22/94: "Failure to follow ***'s established 'Engineering Change Notice Procedure' when the software change was implemented on the ***. The response from your firm, claimed that the ECN procedure was followed, however, these records were not available to the investigator during the inspection and the software engineer had not completed an ECN describing the change. We acknowledge receipt of the new ECN's covering these changes, however, changes must be reviewed and approved before they are released to production.."

Blood Bank Computer Software Manufacturer, Washington. Letter issued by Seattle District Office, 6/28/94: "Failure to maintain an adequate device master record for *** software which includes device specifications, production process specifications and procedures, quality assurance procedures and specifications, and packaging and labeling specifications.... Failure to maintain adequate device history records which demonstrate that *** software is manufactured in accordance with its corresponding device master record...." Also: "Failure to maintain and/or follow adequate written procedures for finished device inspection to assure that *** specifications are met prior to release for distribution ... in that software distributed in March 1993 was issued to one *** user prior to the completion of all program corrections. The user subsequently identified a failure in the software directly related to one of the uncorrected programs." Also: "Failure to establish, implement, and control written manufacturing specifications and processing procedures to assure that the device conforms to its original design or any approved changes in that design ... in that:

"1. There are no written procedures for documentation and approval of changes in specifications for the programs of the *** software.

"2. Review of program changes issued to users noted many were completed and distributed without documentation regarding validation of the changes.

"3. Software functions including *** (Donor Merge), *** (laboratory Results Evaluation), and the soundex and alpha search were released on multiple occasions after in-house testing failed to reveal that the functions did not meet stated specifications.

"4. Review of Service Request Forms (SRFs) submitted by users of *** software indicates that numerous software defects have been identified by users. For example:

"a. SRFs ... indicate that deferral statuses for donors were not properly updated; and SRF ... indicates that the next donation date was not accurately calculated by the system, when donor records were merged using the donor merge function.

"b. SRF ... indicates that the donor deferral status and date of next acceptable donation was not appropriately transferred with the blood unit record to the correct donor account number; and SRFs ... indicate that users have encountered record locking problems, while utilizing the post phlebotomy correction function.

"c. SRFs ... indicate that the system had inadvertently assigned the same blood unit number to two different donors.

"d. SRFs indicate that users were allowed to register donors using social security numbers which were already in the system, thereby creating duplicate records."

Medical Device Manufacturer, Massachusetts. Letter issued by Boston District Office, 7/1/94: "Failure to assure that specification changes shall be subject to controls as stringent as those applied to the original device.... For example, changes made to the *** software, versions 4.13 through 4.60 lack validation."

Medical Device Manufacturer, Colorado. Letter issued by Denver District Office, 7/12/94: "Your firm failed to assure that the software used in the *** will function as intended. There are no written specifications and the testing protocol for the software was not signed and dated as approved for use."

Medical Device Manufacturer, Denmark. Letter issued by CDRH Office of Compliance, 8/24/94: "Failure to assure that the data processing used for manufacturing or quality assurance purposes are adequately designed and implemented.... For example, there was no software requirement specification document for the *** to support the test plan."

510(k)

Medical Device Manufacturer, Nebraska. Letter issued by Kansas City District Office, 12/2/92: "Computerization and software capability changes ... constitute significant changes that could affect the safety and effectiveness of the devices."

Medical Device Manufacturer, Oregon. Letter issued by Seattle District Office, 7/7/93: Lack of proper 510(k) submission "because the technology used is significantly different from the technology used in the 1960 ... version of these devices. Examples of these significant changes include ... the addition of computers."

Medical Device Manufacturer, Washington. Letter issued by CDRH Office of Compliance, 12/13/93: Firm had distributed medical device "which contains a neurometric analysis program (EEG software)" without an approved application for premarket approval.

Medical Device Manufacturer, New York. Letter issued by New York District Office, 1/6/94: "The ... inspection revealed that the **** is misbranded in that a notice or other information respecting the new intended uses of the device and the additional software capabilities was not provided to the Food and Drug Administration.... These new intended uses ... and additional software capabilities, such as attention deficit disorder, pattern recognition program, and topographic brain mapping are major changes in the intended use of the device and constitute a change in the design of the device that could affect safety or effectiveness."

Medical Device Manufacturer, Sweden. Letter issued by CDRH Office of Compliance, 12/23/93: Device requires 510(k) submission, because "based on the information we have reviewed, we believe that the described changes in the software for the wheelchairs appear to constitute a significant change, as described in 21 CFR 807.81."

Testing

Blood Bank, Pennsylvania. Letter issued by Philadelphia District Office, 12/24/92: "Failure to assure that equipment performs in the manner for which it was designed ... in that the computer system was placed in use before all functions had been tested."

Medical Device Manufacturer, New York. Letter issued by New York District Office, 1/25/93: "There is no documented validation testing of the software program for the **** to demonstrate that it is suitable for its intended purpose and capable of producing valid results."

Medical Device Manufacturer, Texas. Letter issued by Dallas District Office, 3/10/93: "Failure to assure that all quality assurance checks are appropriate and adequate for their purpose and are performed correctly.... For example: ... The *** is used as a finished device test, but it is not able to detect multiple ROM address failures contained in the integrated circuit.... **** passed its final device electrical test even though the test lead wires were incorrectly installed."

Medical Device Manufacturer, Sweden. Letter issued by CDRH Office of Compliance and Surveillance, 5/4/93: "Failure to adequately test for conformance to specifications each production run, lot or batch prior to release for distribution.... For example: ... Final device testing for the *** fails to include a check of all microprocessor alarm functions and messages."

Software Manufacturer (Software used in Blood Banks), Missouri. Letter issued by Kansas City District Office, 5/14/93: "Failure to employ testing procedures adequate to assure device specifications are met."

Medical Device Manufacturer, Ireland. Letter issued by CDRH Office of Compliance and Surveillance, 6/9/93: "Failure to assure that the equipment used in the manufacturing process is appropriately designed, constructed, and installed.... For example, Installation Qualification Records for the Vacudyne Programmable Air Wash Controller do not include a description/reference to what tests were performed nor the actual results of these tests."

Contract Sterilization Facility, France. Letter issued by CDRH Office of Compliance and Surveillance, 6/24/93: "Failure to validate computer software programs by adequate and documented testing, when computers are used as part of an automated production system.... For example, the software used in the computer controlled sterilization process has not been validated."

Medical Device Manufacturer, Japan. Letter issued by CDRH Office of Compliance, 7/13/93: "The software for the **** has not been adequately validated.... Software update versions ****, ****, and **** have not even received functional input/output testing.... Software validation documentation ... does not include test results showing the software has been fully tested over the expected performance range, other than functional and integrated testing.... Software development standards do not require documentation of structural software testing when changes are made.... The system for confirming, testing, and approval of software 'bug' corrections does not require an effective date for releasing the changes after approval of the change." Also, "failure to routinely calibrate, inspect and check according to written procedures, all production and Quality Assurance measurement equipment, and to validate by adequate and documented testing computer software programs when used as part of an automated production or quality assurance system."

Hospital Blood Bank, Tennessee. Letter issued by Nashville District Office, 8/30/93: "The inspection revealed deviations from the requirements for controls of your computer system ... in that: there was no written validation plan providing for testing under normal, stress, exceptional, boundary, and invalid conditions to assure that the system operates properly in the intended environment."

Blood Bank, South Carolina. Letter issued by Nashville District Office, 8/30/93: "Please note that any software used to determine donor suitability, such as a program that keeps a record of unsuitable donors, must be functionally tested and challenged under user conditions prior to its actual use."

Software Manufacturer, North Carolina. Manufactures software to calculate radiation dosages for teletherapy and brachytherapy systems. Letter issued by Atlanta District Office, 8/10/93:

"1. Failure to have specification changes approved and documented by a designated individual(s), including an approval date and the date the change becomes effective.... For example:

* "There is no documentation of a plan, protocol, nor test strategies to determine conformance to specifications and requirements at any level of software development (modular, integrated, system).
* "Test cases or protocols for the evaluation of the *** software design have not been developed. Consequently, there are no test data or results to demonstrate conformance to specifications/requirements.
* "Suitable test cases to evaluate the design of the software have not been developed. Tests should cover all operating conditions, input/output parameters, calculations, decisions, safety and other requirements.
* "Comparing calculated results of algorithms in different software programs to demonstrate software programming accuracy is not sufficient to validate the design of the software. The fact that the software programs are written in different programming languages for different operating systems and hardware makes such a comparison even less credible.
* "The testing of the original Basic HP program is not well documented. You indicate 30 tests were initially planned to validate the Basic HP program and that 25 tests were completed. Only 7 of the tests were documented. The tests are related to the portion of the software dedicated to the radiation dose calculations and not to parts of the program involving interface with hardware and user.

"In your June 14, 1993 letter in which you respond to the investigator's observations, you claim that test cases for the main-software program (dose calculations) do exist. You indicate that existing tests ... will be performed on the current C++ version of the software within 90 days.... You also state that documentation for non-technical portions of the software (user and hardware interfaces) will be developed.

"The test cases referred to in your response are validation tests for the original HP platform. They have not been performed on the current C++ language, PC version to date. Tests for user and hardware interfaces have not been developed. Your response is not adequate because the testing plan is not sufficient to validate the design of the *** software package."

Blood Bank, California. Letter issued by Los Angeles District Office, 9/28/93: "Failure to draft/implement a software analysis procedure which defines the critical operations of the *** System to be used for development and testing. Failure during the software evaluation/testing to always compare the test inputs against undesired/invalid data to ensure that only valid data is input or handled correctly by the *** System (i.e. errors or invalid entries)."

Medical Device Manufacturer, Ireland. Letter issued by CDRH Office of Compliance, 2/1/94: "Failure to validate software programs by adequate and documented testing, when computers are used as part of an automated production or quality assurance system.... For example: ... Finished product labels are printed from an ***. There is no validation of this ... program."

Medical Device Manufacturer, South Carolina. Letter issued by Atlanta District Office, 3/14/94: "No formalized written procedures have been established for the testing of software after changes are made in the program. No such procedures have been established to address software maintenance. Your firm could provide no documentation of any testing performed after software changes were made to existing modules. The software in use has been changed from version 5.00 to version 11.00. At least three *** units were shipped with software version 10.00 before the version had been formally approved for distribution. Your firm could provide no justification for their failure to test or correct units in the field with software later found to be flawed."

Blood Bank Software Manufacturer, New York. Letter issued by New York District Office, 4/13/94: "In-process test protocols required to evaluate the software and its revisions, test data, and test results, and records of failures encountered were not maintained, nor were records of in-process testing and debugging maintained. c. There was no record of the validation test protocol for version 1.63." "Failure to adequately test the finished device for conformance with device specifications ... in that 14 user-reported problems requiring revisions and/or programming error changes in version 2.00 were not found through the use of your in-house validation protocol."

Blood Center, Colorado. Letter issued by Denver District Office, 5/10/94: "Written validation test plans for software programs are not always followed. For example, validation records of the software *** programs, used for the shipment of research units, indicate that units with a status of consigned or infectious risk are not included in testing as required by your validation test plan described in the functional specifications. No documentation was found to support the exclusion of these units from the testing."

Medical Device Manufacturer, North Carolina. Letter issued by Atlanta District Office, 6/21/94: "Failure to document and/or validate the testing of software used in the ***." "Failure to have written procedures for software development and testing and EPROM verification checks."

Medical Device Manufacturer, North Carolina. Letter issued by Atlanta District Office, 6/21/94: "Failure to document and/or validate the testing of software used in the ***." "Failure to have written procedures for software development and testing and EPROM verification checks."

Medical Device Manufacturer, Thailand. Letter issued by CDRH Office of Compliance, 7/29/94: "Failure to assure that all quality assurance checks are appropriate and adequate for their purposes and are performed correctly.... For example, the investigator noted variations in the test outcomes on the electronic testing machines. These variations were described by representatives of your firm as being due to improper operator techniques. Your firm must assure that this testing is being performed correctly, and that reproducible test results are being obtained." "Failure to establish that quality assurance measurement equipment is capable of producing valid results.... For example, the electronic testing machines used by your firm have not been validated."

Medical Device Manufacturer, France. Letter issued by CDRH Office of Compliance, 8/23/94: "Failure to assure that components are inspected, sampled, and tested for conformance to specifications, and a record maintained of acceptance or rejection.... For example, no certificate of analysis nor documentation exists that microprocessor wafers are reviewed upon receipt. ***'s response dated July 12, 1994, to the FDA 483 is inadequate because ... either the certificate of analysis needs more detail or a copy of the test specifications must be provided for the microprocessor."

Audit Trail

Blood Bank, California. Letter issued by San Francisco District Office, 1/13/93: "Failure to maintain and/or follow written standard operating procedures to include all steps to be followed in the collection, processing, storage, and distribution of blood and blood components ... in that: ... There is no system to track computer problems or to contact the computer service vendor to insure that problems are investigated and resolved."

Drug Manufacturer, New Jersey. Letter issued by Newark District Office, 7/7/94: "There is no audit trail in the system to document changes to sample results or other data in the system."

Acceptance Criteria

Medical Device Manufacturer, New York. Letter issued by New York District Office, 2/11/93: "Failure to have written acceptance criteria and procedures for the control component used on microprocessor controlled sterilizers. Also, there is no quality control system in place to assure that this hardware/software system was ever validated or that it functions as per specifications."

Quality Assurance/Quality Control

Medical Device Manufacturer, New York. Letter issued by New York District Office, 2/11/93: "There is no quality control system in place to assure that this hardware/software system was ever validated or that it functions as per specifications."

Medical Device Manufacturer, Wisconsin. Letter issued by Minneapolis District Office, 4/14/93: "Failed to identify, recommend, or provide solutions for quality assurance problems and verify the implementation of such solutions" following a computer search for device failures.

Medical Device Manufacturer, California. Letter issued by Los Angeles District Office, 7/15/93: "Failure of the quality assurance system to identify and implement solutions to quality assurance problems. For example: ... The quality assurance system failed to identify problems associated with the development and configuration of device software."

Systems Manufacturer (Medical Device), Florida. Letter issued by Orlando District Office, 9/24/93: "Failure to establish and document a Quality Assurance program;

failure to establish, document, and conduct periodic audits of the manufacturing operations"

Medical Device Manufacturer, California. Letter issued by Los Angeles District Office, 3/18/94: "Failure to establish an adequate procedure for periodic audits of the quality assurance system. For example, the audit procedure does not specifically include a requirement for auditing software development and configuration activities."

Blood Bank Software Manufacturer, New York. Letter issued by New York District Office, 4/13/94: "Failure to have a quality assurance program or written procedures for conducting planned and periodic quality assurance audits ... to assure that production records are reviewed and all quality assurance checks are appropriate and adequate for their purpose and are performed correctly."

Medical Device Manufacturer, Italy. Letter issued by CDRH Office of Compliance, 6/7/94: "Failure to perform planned and periodic audits in accordance with written procedures by appropriately trained individuals not having direct responsibility for the matters being audited, and to have written procedures for performing planned and periodic audits of the quality assurance program.... For example, audits are being performed by individuals having direct responsibility over those matters being audited, and there are no written audit procedures."

Blood Bank Computer Software Manufacturer, Washington. Letter issued by Seattle District Office, 6/28/94: "Failure to establish a quality assurance program which identifies, recommends, or provides solutions for quality assurance problems and verifies the implementation of such solutions ... in that: 1. Inadequate records are maintained to identify consignees receiving software updates." Also: "Failure to perform periodic audits of a quality assurance program in accordance with written audit procedures performed by appropriately trained individuals not having direct responsibility for the matters being audited."

Vendors

Systems Manufacturer (Medical Device), Florida. Letter issued by Orlando District Office, 9/24/93: "Failure ... to have written specifications and procedures to test incoming vendor supplied components prior to use."

Medical Device Manufacturer, Alabama. Letter issued by Nashville District Office, 10/5/93: "The inspection revealed deficiencies related to the computer systems/programs in use for processing/controls.... In addition, the inspection documented a failure to audit the computer system vendor for an appropriate Quality Assurance program."

Medical Device Manufacturer, North Carolina. Letter issued by Atlanta District Office, 12/13/93: "No formal agreement existed between your firm and the software vendor which addressed issues such as development standards, access to source code, and responsibilities for testing, maintenance, and error notification. The software developer had not been audited by *** and there was no formal procedure addressing such audits. The software validation test plan lacked any structural testing requirements or results."

Blood Bank, Washington. Letter issued by Seattle District Office, 4/15/94: "Failure to have written procedures for resolving software errors reported to the vendor and for review and assessment of 'work arounds' for the errors."

EVALUATION OF SOFTWARE RELATED RECALLS FOR FISCAL YEARS 1983-91

Prepared by: Division of Compliance Programs
Office of Compliance and surveillance
Center for Devices and Radiological Health

May 1992

INTRODUCTION

In July 1984, the Division of Compliance Programs (DCP), Office of Compliance and Surveillance, CDRH initiated a program to identify the types of problems that result in recalls and to determine the causes of these problems. This report evaluates recall data associated with Software/firmware failures of computerized devices for Fiscal Years 1983 through 1991, inclusive.

"Software" encompasses programs and/or data that pertain to the operation of a computer controlled system, whether they are stand alone systems, contained on floppy disks, hard disks, magnetic tapes, "laser" disks, or embedded in the hardware of a device, usually referred to as "firmware."

The information presented in this report may provide manufacturers with insights on how and where to increase their quality assurance efforts in software development, quality assurance and production, or to minimize certain problems that may lead to software related problems resulting in device failure. This information is especially important considering the increasing number and variety of software-controlled medical devices used by the healthcare industry.

METHOD OF ANALYSIS

This report examines device problems in terms of their causes which are referred to as "quality problems." A single quality problem may result in one or more "recalls." A recall is a firm's removal or correction of a marketed product that the Agency considers to be in violation of the laws it administers and against which the Agency would initiate legal action, (e.g., seizure.) A correction is the repair, modification, adjustment, relabeling, destruction or inspection (including patient monitoring) of a product without its physical removal to some other location. A device firm experiencing a single problem with a specific step in a manufacturing operation may find that the problem has adversely affected many different types of devices processed through that operation. For example, a single process problem in the manufacture of a catheter resulted in a defect in a wide variety of catheters. This resulted in over 100 recalls conducted by the manufacturer in order to properly identify and account for all defective devices. Thus, counting individual recalls, instead of quality problems, could give an inflated view of the actual magnitude of the problems existing at a manufacturer's facility. For this reason, this report will address quality problems, not individual recalls.

For purposes of analysis, software quality problems are divided into the following specific categories:

Software Design-Device	A defect or error was designed into the software during the device software development phase.
Software Design-Manufacturing Process	A defect or error was designed into the software during the software development phase.
Software Change Control	A defect or error was introduced into the device software as a result of a change to the software.

FINDINGS

From Fiscal Year 1983 through 1991, there were two thousand, seven hundred and ninety-two (2,792) quality problems which resulted in recalls of medical devices. As shown by Figure 1, one hundred and sixty-five (165), or six percent (6%), of all quality problems resulting in recalls during this period were attributed to software failures. Of the one hundred and sixty five (165) software related quality problems, one hundred and thirty-three (133), or eighty-one percent (81%), were due to inadequate design of the device's software and/or inadequate design of the software controlling the manufacturing process. The other thirty-two (32) quality problems or nineteen percent (19%), were incorporated into the device when changes were made after production began.

The Appendix of this report contains specific information about the recalls falling into the above three categories of software quality problems.

The following table shows the number of software quality problems and percent of the total quality problems by fiscal year. These rates do not demonstrate any significant trends.

SOFTWARE QUALITY PROBLEMS BY FISCAL YEAR

Fiscal Year	Number	Percent
83	7	4%
84	7	4%
85	17	6%
86	19	6%
87	16	5%
88	24	8%
89	26	9%
90	27	7%
91	22	5%

Software quality problems, as they relate to classification panels, are shown in Figure 2 below. (These panels are groups into which all medical devices are divided depending on their function and/or purpose. For example, pacemakers are in the cardiovascular panel and ventilators are in the anesthesiology panel.) The largest single area of software quality problems involved cardiovascular panel devices (23%) followed by radiological devices (19%) and in vitro diagnostic devices (18%).

CONCLUSIONS/RECOMMENDATIONS

The above data suggests that manufacturers should place increased emphasis on software development quality assurance programs. Manufacturers of software that is either identified as a medical device or used to operate medical devices, have quality assurance obligation under the GMP and established premarketing requirements. For example, 21 CFR 820.61 Measurement Equipment, requires all medical device manufacturers to validate software used in production and quality assurance measurement equipment. All changes to software used in the device, quality assurance program or manufacturing process are subject to 21 CFR 820.100(a)(2) and (b)(3) and must be validated before implementation. When software is used for manufacturing or quality assurance it is subject to 21 CFR 820.195, which requires that adequate checks be designed and implemented to prevent programming errors. In addition, finished software and software driven devices must be tested periodically under simulated use conditions (21 CFR 820.160). Manufacturers who are uncertain about how the GMP applies to their manufacturing processes should refer to the May 1992 guidance for FDA investigators, "Computerized Devices/Processes Guidance. Application of the Medical Device GMP to Computerized Devices and Manufacturing Processes.

All software and software driven medical device manufacturers subject to premarket clearance must provide the software development information described in "Reviewer Guidance for Computer Controlled Medical Devices Undergoing 510(k) Review," dated August 29, 1991. These publications are available from the Division of Small Manufacturers Assistance, Office of Training and Assistance, Center for Devices and Radiological Health, Food and Drug Administration, 5600 Fishers Lane, Rockville MD 20857.

APPENDIX

Software Related Device Quality Problems

The data used in compiling this report were taken from actual recall files maintained by FDA. This Appendix contains summaries of the recall data divided into the following specific categories: software design-device, software design-manufacturing process, and software change control. Each recall record is identified by a control number beginning with "U" or "Z," and contains the product name, the problem, and a statement about the cause, i.e., what led to the problem. (The amount of information found in the "cause" section of the recall summary varies with the detail sought by the individual investigator and/or the information provided by the manufacturer). Within each of the following sections, the quality problems are sorted by product frequency.

Section 1 -- Software Design-Device

Section 2 -- Software Design-Manufacturing Process

Section 3 -- Software Change Control

Section I

Software Design-Device

Recall: U2033
Product: Insulin Infusion Pump
Problem: Excessive or insufficient insulin delivery if anomalous entry appears on display panel and manual dose, supplemental dose, or basal rate is allowed to continue.
Cause: Software error.

Recall: Z0015
Product: Volumetric Infusion Pump
Problem: When the solution container is empty, unit does not alarm. This could result in the pumping of air into the patient.
Cause: Software error.

Recall: Z4215
Product: Infusion Pump
Problem: Delivers fluid at rate other than intended when specific circumstances are met. If volume remains manually set to 0, flow rate set to less than 1 ml/jr, pump will deliver fluids at the rate previously set which could be up to 699 ml/hr.
Cause: A software logic error in the master EPROM that has existed during the entire product history, allows pump to deliver fluid at a rate faster than set on display.

Recall: Z2757
Product: Infusion Pump
Problem: Over infusion.
Cause: Software error.

Recall: Z0108
Product: Infusion Pump
Problem: Infuser fails to accurately deliver drugs within claimed accuracy, 2%.
Cause: Error in software program.

Recall: Z8049
Product: Infusion Pump
Problem: Firm discovered that if a bolus delivery is made while pumps are operating in the body weight mode, the middle LCD fails to display a continual update in mg or mcg per kg.
Cause: Firm's investigation indicated a software error, that was undetected during validation, causes middle LCD window to exhibit 000kg when body weight mode and other conditions exist.

Recall: Z5229
Product: Infusion Pump
Problem: If the one hour limit value is set between 26 ml & 39 ml, then each time the pump is stopped & restarted 25.6 ml is added to the set one hour limit.
Cause: Software defect, EPROM, directly affected the one hour limit feature of the device.

Recall: Z2030
Product: Infusion Pump
Problem: If device primary check fails, secondary back-up does not activate and goes to a third back-up system. The drug therefore will be infused at a one ml rate greater than the infusion rate.
Cause: Defect discovered in the software during quality engineering tests of the pump. Defect exists in back-up delivery check system.

Recall: Z5040
Product: Infusion Pump
Problem: The pump inaccurately reports the dose given to the patient as only one-half that actually delivered. This occurs only at the 4mg/ml concentration.
Cause: A one byte error in a software table of values causes the device to inaccurately report the drug dose given.

Recall: Z7980
Product: Infusion Pump
Problem: Word "rate" instead of "total" is displayed when the tamper alarm feature is activated by the unintentional change of the "hundredths" total volume digital switch.
Cause: Software error.

Recall: Z8041
Product: Infusion Pump
Problem: Infusion rate may be higher than programmed.

Cause: Engineering analysis confirmed software error that occurs in two of the program modes, the percentage mode and time per second between pulses.

Recall: Z1355
Product: ECG & Vital Signs Monitor/Central Station Monitors
Problem: Audio alarm not activated when cardiac output signal is interrupted.
Cause: Circuitry and programming design wouldn't allow activation of audio alarm when cardiac output was interrupted due to lead fault

Recall: Z1188
Product: Patient Monitor
Problem: The device fails to alarm under certain conditions.
Cause: Radio frequency interference picked up by some units caused by manner of hook up with central station necessitated software reprogramming.

Recall: Z0069
Product: Cardiac Monitor
Problem: Detection of asystole in paced patients is greatly complicated by the presence of pacemaker artifacts.
Cause: The software has a defect in the arrhythmia algorithm.

Recall: Z7739
Product: Pt. Monitor Software System ECG Monitor
Problem: Dose rate display on titration chart may be inaccurate when calculation contains significantly different rates of measure.
Cause: Problem in the software program leading to errors in the display when the calculation contains significantly different units of measure.

Recall: Z9709
Product: Computer Software program
Problem: Under certain display situation conditions pressing the pharmaset or dataset keys will turn off the high systolic limit alarm.
Cause: Software problems exist under certain circumstances.

Recall: Z4400
Product: Cardiovascular Monitor, Central Station
Problem: If ECG alarm condition occurs and the alarm has been silenced and patient has not returned to within programmed alarm limits, a second ECG alarm from that patient would not be indicated with an alarm tone or documented on strip.
Cause: Software errors.

Recall: Z4970
Product: Cardiac Monitor

Problem: The waveforms of one may superimpose on the waveforms of another bedside unit in remote view of a central station cluster screen. This may cause repeat display of waveforms.
Cause: Software errors.

Recall: Z1491
Product: Cardiac Monitor
Problem: Device may fail to detect apnea under certain conditions of use.
Cause: Software errors.

Recall: Z2216
Product: Ventilator, Continuous (Respirator)
Problem: False positive alarm reports were received by firm.
Cause: Software error.

Recall: Z3036
Product: Ventilator, Continuous (Respirator)
Problem: The 02 concentration being delivered to the patient can go to 21% when device returns to operation from certain settings or from self check.
Cause: Software error.

Recall: Z3557
Product: Ventilator, Continuous (Respirator)
Problem: The high peak pressure limits may be exceeded under certain conditions.
Cause: Software errors.

Recall: Z5047
Product: Microprocessor Ventilator
Problem: Device did not cycle as expected and the patient was not ventilated. Event would occur only if disconnect ventilation is declared simultaneous with patient initiated breath in modes other than CMV.
Cause: Software error.

Recall: Z3508
Product: Ventilator Continuous
Problem: Under certain conditions the ventilator may switch into disconnect ventilatory mode inappropriately. The alarm reset key may not return the ventilator to original settings.
Cause: Software error.

Recall: Z5878
Product: Ventilator with (and without) Pressure Support
Problem: Ventilator delivered pressures greater than set pressure without alarm.
Cause: Malfunction due to failure in P.O. chip to which software did not respond appropriately. P.O. chip had two different simultaneous failures.

Recall: Z3235
Product: Diagnostic Ultrasound System
Problem: Output of doppler blood flow measurement can be misinterpreted when the operator changes certain settings while strip chart is running.
Cause: Software that controls printer allows printer lag when machine settings are changed. This results in either late printing or loss of data.

Recall: Z5166
Product: Ultrasonic Diagnostic Imaging System
Problem: Inaccurate flow values.
Cause: Software programming error in EPROM.

Recall: Z6118
Product: System Imaging, Doppler Pulsed, Ultrasonic
Problem: Power emitted from scanheads when used in conjunction with certain systems in doppler mode, exceeds 510(k) filings.
Cause: Error in software that controls the power output of scanheads.

Recall: Z4639
Product: Ultrasound Imaging System
Problem: Blood velocity may be incorrectly calibrated.
Cause: Software defect was found in the PW doppler cursor angle function on the inverted 2-D sector display. Revision of the software will include changes that correct the software defect.

Recall: Z7210
Product: System, Imaging, Doppler Pulsed Ultrasonic
Problem: Scaling errors.
Cause: Equipment software did not prevent use of data in specific operational modes of the cardiac analysis package as intended.

Recall: Z7201
Product: Accelerator, Linear

Problem: Computer caused x-radiation system to provide radiation in excess of that intended by the operator.
Cause: The computer prompt requested units of radiation activity of MGM units instead of the required computation of MCI dosage units resulting in greater than expected dose.

Recall: Z3827
Product: Accelerator, Linear
Problem: There is a potential for massive radiation overdoses.
Cause: Two software defects.

Recall: Z5239
Product: Accelerator, Linear
Problem: Noncompliant with the RCHS Act -- accidental radiation could occur.
Cause: Software problem allowed a noise to be interpreted as a command to return the gantry back to the original start angle.

Recall: Z1071
Product: Accelerator, Linear, Medical, Accy.
Problem: In circumstances where wedge is wider than 6cm, will not extrapolate out as stated in the operator's manual.
Cause: Software program contains a calculation error.

Recall: U1653
Product: Patient Monitor
Problem: There is distortion of the delayed channel outputs.
Cause: Software timing of output signals caused one channel's wave to distort the other channel's output. Software has been modified.

Recall: Z2834
Product: Patient Monitor
Problem: Unstable blood pressure indications at low pressure readings.
Cause: Analog pressure waveform may be replaced by low frequency sine waves at low pressures. Pressure processing algorithm collapses to only mean pressure value below 10-20 mm hg.

Recall: Z0608
Product Monitor, Bed, Patient
Problem: Device causes under or overinfusion.
Cause: The rounding function of the software was not precise enough, resulting in overinfusion.

Recall: Z0304
Product: Chemistry Analyzer
Problem: When the instrument is operated in the pediatric mode, program revision 5.6.2 may cause the module immediately to the left of the total bilirubin module to recalibrate improperly.
Cause: Error in software causes problem when certain variables are present.

Recall: Z4117
Product: Chemistry Analyzer
Problem: Sequencing errors and analyzer calculation errors.
Cause: Error in software results in calculation-misplaced decimal.

Recall: Z5390
Product: Chemistry Analyzer

Problem: A potential sample transport "home position" problem.
Cause: Software error.

Recall: Z6256
Product: Pacemaker Software, Programmer Cartridge
Problem: Software contains six programming errors, that concern indicators for telemetry message and display.
Cause: Software programming error.

Recall: Z7256
Product: Pacemaker Programmer
Problem: May incorrectly confirm a pacemaker programming event.
Cause: Software errors.

Recall: Z1307
Product: Electronic Memory Cartridge for Pacemaker
Problem: A "high" lead impedance may be displayed instead of the actual measured lead impedance.
Cause: For lead impedances over 600 ohms the pulse generators displayed a high lead impedance instead of the actual measured value. Software design problem existed.

Recall: Z5707
Product: Automated Sample Handler
Problem: Samples may be skipped.
Cause: Sample handler skips one or more sample tubes and misidentifies test results. Cause has not yet been determined.

Recall: Z0048
Product: Automated Cell Counter
Problem: There is the possibility that units may be skipped.
Cause: Some level of operator misuse, but firm has made software change to eliminate possibility of using device while bar code reader is disabled.

Recall: Z7230
Product: Automated differential Cell Counter
Problem: The auto-reporter under certain conditions of use may print results of the prior sample instead of the sample being run. Data printed out on report is inaccurate.
Cause: Software errors.

Recall: U1893
Product: Pacemaker Programmer
Problem: Anomalous pacemaker behavior
Cause: Programmer allowed pacemakers to be programmed into a configuration that allowed for anomalous behavior.

Recall: Z9669
Product: Generator, Pulse, Pacemaker, Implantable
Problem: Pulse generator may have a reduction in output voltage if programmed to certain modes, i.e., less than 2.5 volts in the nonsensing mode.
Cause: Inadequate design specification for upper and lower values of the integrated chip.

Recall: Z1204
Product: Pulsed Echo Imaging Ultrasound System
Problem: The fetal weight formula used in the software for the "auto calc" option results in an overstatement of the fetal weight of 3-15% and an error in volume calculation.
Cause: When designing the "auto calc," the firm used a formula for determining fetal weight obtained from the literature. Investigation discovered a transposition error in the formula in the literature.

Recall: Z4659
Product: Pulsed Echo Imaging Ultrasound System
Problem: System may produce incorrectly calibrated N-mode strip chart.
Cause: Software defect permits 77500A/B/C strip chart recorder to produce incorrectly calibrated recordings. The error occurs with a particular sequence of control changes involving the expanded N-mode function.

Recall: Z3364
Product: Ultrasound Scanning System
Problem: The 280 SL trace function calculates an image greater than the actual object.
Cause: Error in the software.

Recall: Z1296
Product: Imager, Ultrasonic Obstetric/Gynecologic
Problem: Device was too sensitive to the operators input which caused the circumference measurements to be too large, 20 + 5% over actual.
Cause: Software error.

Recall: Z2605
Product: Accessory Kit for Cardiac Output Monitor
Problem: Alarm may be disabled without alerting the central station.
Cause: As a result of errors in the software, the alarm could be disabled when control buttons were pushed in a specific sequence.

Recall: Z1867
Product: Monitor, Cardiac Output, Trend
Problem: Device may lock up.
Cause: Software design error. Possible to lock up with or without error message being displayed.

Recall: Z5475
Product: Electrosurgical Generator
Problem: When the generator is used with a patient having an impedance of 20-25 ohms., a false alarm may result. Activating this alarm disables the generator output.
Cause: Software error.

Recall: Z2266
Product: Electrosurgical Device
Problem: Hand switch accessory keying causes other accessory outlets to be energized.
Cause: Defect resulted from a programming error.

Recall: Z1346
Product: Programmable Cardiac Pacer
Problem: Improper programming can change pacer pulse rate from 70 to 35 bpm and sensitivity from 0.8 to off, or rate from 50 to 100 or as high as 120 bpm.
Cause: Software and hardware defect.

Recall: Z2196
Product: Pacemaker, External Pulse Generator
Problem: Under certain programming parameters, the EP stimulators can cause a patient's heart to fibrillate.
Cause: It was determined that high rate of pacing can occur under a limited set of parameter settings.

Recall: Z6346
Product: Surgical Laser
Problem: Software control panel had no way of monitoring or displaying to the operator the actual laser output power in comparison with requested power.
Cause: Software error.

Recall: Z7670
Product: Instrument, Surgical, Powered, Laser
Problem: The device is adulterated in that there is potential for low-power instability in the device in the super power pulse mode settings below 5 watts.
Cause: The problem originated in the PC circuit board and the CPU, but the exact cause has not been determined.

Recall: Z5886
Product: Spirometer, Diagnostic
Problem: The polgar predicted values for pediatric FVC were off by approximately 30%
Cause: Programming error in the pediatric polgar.

Recall: Z9989
Product: Spirometer, Diagnostic

Problem: Inaccurate values
Cause: Software errors in the extrapolation method.

Recall: Z4667
Product: Operating Software for Chemistry Analyzer, Photometric, Discrete
Problem: Specimens with extremely elevated SCOT or SGPT levels may yield an inaccurate result and could be reported as normal values.
Cause: Software problem. Operating tests for only SGOT or SGPT were inaccurate. Error "flag" samples above linear range did not cover all possible values.

Recall: Z0118
Product: Automated Cell Counter
Problem: Analyzer may skip samples, associates results with incorrect patient
Cause: Software errors.

Recall: Z2148
Product: Gas Machine For Anesthesia
Problem: During clinical cases both CRT screens went blank, other gauges could be read to monitor machine operation.
Cause: The alarm stack pointer allows 20 different alarms to be accumulated, however, the software improperly causes a loss of sync pulse when the 18th alarm is invoked.

Recall: Z2040
Product: Anesthesia Machine
Problem: A defect in the 700 ventilator EPROM software may cause the ventilator to cease operation.
Cause: EPROM software in the ventilator could fail when the altitude adjustment feature was set on "0" meters and the total flow volume was set at a delivery rate of less than 2.2 liters per minute.

Recall: Z4848
Product: Medical Computers and Software
Problem: Calculation error in "counts same" program.
Cause: When performing an area of interest study using the program "counts save" in the 256 X 256 mode, the DP53300 or DPS33000 incorrectly calculates the radiation level. Software design problem.

Recall: Z5678
Product: Medical Computers and Software
Problem: Point dose calculation was a maximum of 5% lower than the isodose distribution data, varied depending on whether input was clockwise or the reverse.
Cause: Linear source program contained errors in the point dose calculation section.

Recall: Z6058
Product: Blood Gas Analyzer, Blood Phase, Accessory
Problem: Potential incorrect blood gas results.
Cause: Problem caused by software of 201 printer, error occurs when combined mode is selected.

Recall: Z6068
Product: Blood Gas Analyzer Accessory
Problem: There is a patient operation sequence associated with the automatic "assign patient" function of the device.
Cause: Potential software problem operation sequence associated with the automatic "assign patient" function. May obtain incorrect patient report.

Recall: Z6268
Product: Electrocardiograph (Bed Monitor)
Problem: Display, clock function problems.
Cause: Software error.

Recall: Z3869
Product: Electrocardiograph
Problem: Incomplete electrocardiograph tracing.
Cause: Software defect signals having amplitudes of plus or minus 3 or more millivolts would be truncated when printed on paper. Only occurs when the gain set in the X1/2 position and the amplitude was 3 millivolts or greater.

Recall: Z2670
Product: Defibrillator, Low Energy
Problem: Removing the memory module may cause unexpected operational difficulties.
Cause: Software error. Test data shows that when memory module removed, operational difficulties occur between 7-8% of the time.

Recall: Z2381
Product: Defibrillator, Low Energy
Problem: Immediately following a defibrillation discharge, the ECG monitor shows a flat trace for 10 seconds or longer when monitoring through a three-lead patient cable.
Cause: Firm found that the "flat line" condition related to two software modules and hardware dependent thresholds.

Recall: U0353
Product: Computer, Diagnostic, Programmable
Problem: Inaccurate answers.
Cause: Error in computer program which calculated shunt fraction: (QS/QT) was replacing arterial CO_2 pressure for arterial O2 pressure and rounding off numbers incorrectly. Unit reprogrammed.

Recall: U0173
Product: Automated Cell Counter (Particle)
Problem: False readings when white blood count exceeds 100,000.
Cause: Error in software allowed counter to reset to "0" when 100,000 value was reached, value actually was 101,000, etc.

Recall: U0523
Product: Insulin Infusion System
Problem: System infuses continuous amount of insulin when value of "0" is entered as value for constant maximum dextrose infused.
Cause: Software error.

Recall: U2103
Product: Programmable Enzyme Calculator
Problem: Calculates erroneous results.
Cause: An error in computer algorithm.

Recall: Z0424
Product: Data Processing Module
Problem: Program matched name of one patient with data from another.
Cause: Error in software.

Recall: Z3394
Product: Blood Pressure Monitors
Problem: Unit screen will not display "error" as it should when it doesn't get a good reading.
Cause: Software error existed in the ROM circuit, which under certain circumstances would prevent the unit from displaying an "error" message.

Recall: Z0825
Product: Glucose Oxidase, Glucose
Problem: Erroneous reading.
Cause: Defect in the arithmetic routine that causes a failure to carry forward 256 to the output routine.

Recall: Z2505
Product: Cardiac Computer (Output)
Problem: Inflated cardiac output determinations resulted with the thermodilution catheter at any injectable volumes.
Cause: There was an error factor in the software for some injectable volumes and catheter sizes.

Recall: Z3245
Product: Electromyograph

Problem: False positive or negative responses when the device was in the automatic mode for testing the eight electrodes on the patient.
Cause: Machine could not be manipulated into accepting bad electrode reading or rejecting good readings because of errors in software. Software requested the user to perform electrode check before use.

Recall: Z2875
Product: Microplate Reader
Problem: An incorrect reading of 0.0000 may appear between the readings of .999 and 1.000.
Cause: Error in software.

Recall: Z3605
Product: Analyzer, 02, Partial Pressure
Problem: Inaccurate test results occurred if device was used when the batteries were low or depleting.
Cause: The low battery warning scheme in the software did not provide sufficient warning to the user of battery depletion.

Recall: Z5315
Product: Program Tape for Analyzer, Chemistry Micro
Problem: Inaccurate results.
Cause: Tape was communicating transposed numbers under certain circumstances due to software error.

Recall: Z5195
Product: Chemistry Analyzer with CO_2, Enzymatic
Problem: Customer complaint of high carbon dioxide recoveries with analyzer. The microliter diluent option does not consistently perform with acceptable precision tolerances.
Cause: Error in the software that controls the performance of the CO_2 recovery test.

Recall: Z5285
Product: Electrocardiograph Monitor
Problem: Ventricular fibrillation bedside alarm would not trigger a central station alarm if the bedside heart rate limit controls are set at the extreme high and low rate alarm points such as 0 bpm and 250 bpm.
Cause: The coded memory was found to be inadequate.

Recall: Z5435
Product: Whole Body Scanner
Problem: CT computer failed to restrict uncommanded movement of the table top.
Cause: Software error.

Recall: Z5406
Product: Ophthalmic Laser

Problem: Out of compliance with the performance standard.
Cause: Software error

Recall: Z2496
Product: Software Cartridges, Spinal Cord Stimulator
Problem: No output obtained when certain programming sequence was used.
Cause: Software error.

Recall: Z6296
Product: Patient Monitors/Arrhythmia Detector and Alarm
Problem: Potential for all patient alarms to be indefinitely suspended.
Cause: Software error.

Recall: Z6876
Product: CO_2 Analyzer
Problem: Upper limit CO_2 alarm can be manually set above upper limit without alarm sounding.
Cause: Software design allowed movement above the upper limits without alarm sounding.

Recall: Z6436
Product: Gamma-Ray Scintillation Camera
Problem: The detector head could rotate in a path to contact the patient or table if the detector head is programmed to collect data at a >180 degrees angle in the noncircular ECT studies.
Cause: Caused by problem with the noncircular ECT software.

Recall: Z6896
Product: Noninvasive Blood Monitor
Problem: Over sensitivity of monitor to artifacts leading to problems in obtaining blood pressure readings.
Cause: Design deficiency in software allowed for over sensitivity of the monitor to artifacts.

Recall: Z7336
Product: Monitor, Carbon Dioxide, Cutaneous
Problem: The high CO_2 alarm limit can be disabled under one circumstance without user being aware of the fault, namely, depressing the high CO_2 alarm increment arrow for a short period of time after it reaches maximum value.
Cause: EPROM programming error.

Recall: Z4446
Product: Phototherapy Device (PUVA)
Problem: Failure of timing device.
Cause: Failure caused by defective circuit boards, firmware timing.

Recall: Z2287
Product: Telemetry arrhythmia Central Station Monitoring System
Problem: Alarm and arrhythmia detection problems.
Cause: Software errors.

Recall: Z2727
Product: Diagnostic Ultrasound System
Problem: Excessive down time due to lock up and image quality problems.
Cause: Software errors.

Recall: Z3377
Product: Electrosurgical Unit and Accessories
Problem: The generator microprocessor may lock up while in the spray mode.
Cause: A missing time delay in the software program.

Recall: Z4567
Product: Ventilator, Infant, Computerized
Problem: Digital display for mean airway pressure may be inaccurate when rate control is adjusted to 50 or 100 breaths per minute.
Cause: Digital display calculation of mean airway pressure inaccuracy is due to one EPROM being programmed for these settings.

Recall: Z5127
Product: Monitor/Treadmill Controller Equipment
Problem: Loss of treadmill control and unexpected speed and grade increase.
Cause: Malfunction in the monitor/controller due to software problem in control boards.

Recall: Z5167
Product: X-Ray, Generator, High-Voltage, Therapeutic
Problem: Errors in radiation dose calculation when times exceeding 20 half lives are keyed in, at 20 half lives the calculated dose is reportedly 12% low.
Cause: Software error.

Recall: Z1318
Product: Electrosurgical Cutting Device
Problem: Unintended activation of electrode poses a potential burn hazard to patient and surgical personnel.
Cause: When standard microbipolar output is activated, power simultaneously delivered to the monopolar handswitch jack causes unintended activation of the handswitch in this location due to error in software logic.

Recall: Z1408
Product: Tourniquet System, Pneumatic
Problem: Incomplete deflation and erroneous failure alarm.
Cause: Software error.

Recall: Z0858
Product: Monitor, Ultrasonic, Fetal
Problem: In dual mode display, a calculation error occurs resulting in 15% over estimation of fetal heart rate.
Cause: Software error, and new software was created.

Recall: Z0998
Product: Calibration, Flow Meter, Gas
Problem: Under certain conditions, the RT-200 with Software versions 1.7 and 1.8 could provide data that is not within its performance specs. When a test lung is used, triggering error can result in volume measurements that are out of specifications.
Cause: Programming error.

Recall: Z3888
Product: Infusion Syringe Pump
Problem: Failures of product to meet the occlusion pressure alarm activation.
Cause: The investigation indicated the pump has two software-related faults: one of the motor encoder and a second on the liner grip which is used to check the liner rate of travel of the syringe drive.

Recall: Z3908
Product: Digital Image Storage Devices
Problem: The mean pressure gradient was displayed incorrectly.
Cause: Calculation error in software.

Recall: Z4858
Product: Air Gas Monitor
Problem: Potential oxygen sensor problem. Sensor will incorrectly read oxygen after exposure to high oxygen levels for several minutes.
Cause: Software error.

Recall: Z5658
Product: Angiographic Injector
Problem: Device injects prematurely which may lead to repeat injection(s) when interfaced with related equipment or electrical shorts occur within the system.
Cause: Microprocessor malfunction due to false signal. Problem appears to be with the initial design and programming of the device. The "short" also appears to be related to extensive usage.

Recall: Z0719
Product: System, X-Ray, Fluoroscopic, Nonimage Intensified
Problem: The system can produce radiation without an image appearing on the monitor during transfer of video from RAM to disk storage.
Cause: Firmware defect on controller PCB.

Recall: Z2939
Product: System X-Ray, Fluoroscopic
Problem: Device continues to produce x-rays after release of the foot switch.
Cause: Software problem, replacement of EPROMs indicated as solution.

Recall: Z3309
Product: Microtiter Diluting/Dispensing Device
Problem: Test results are inaccurate when device is in horizontal plate position. The micro plate wells did not fill and read "nonreactive."
Cause: Software error.

Recall: Z4489
Product: Blood Pressure Monitor, Non-Invasive
Problem: Will not measure low end blood pressure readings to 40 mm as claimed in the literature.
Cause: Monitor had not been tested at that range, since software instructs to abort at 40 mm and the monitor will read "No Data." Software has been corrected. New software does not abort until cuff pressure deflates below 20 mm.

Recall: Z4709
Product: Coagulation System, Automated
Problem: "Rollover" values I.N.R. show not fault message, neither are they otherwise flagged.
Cause: Software error, version 6.0 will be exchanged with software version 6.1.

Recall: Z6329
Product: Holograph, Fetal, Acoustical
Problem: Software prints incorrectly.
Cause: Software error.

Recall: Z6429
Product: Nephelometer for Clinical Use
Problem: Inaccurate test results, test values may not be obtained due to software and reagent incompatibility.
Cause: Software error.

Recall: Z8139
Product: System, Imaging, X Ray
Problem: X-ray tube current may exceed specification.
Cause: Software problem at some combinations of techniques may affect accuracy of biased focal spot, tubes will exceed stated limits.

Recall: Z0150
Product: Intra-Aortic Balloon Pump

Problem: While in battery operation, stops pumping due to processor lock-up, freezing screen and keyboard. Also shuts off power to unit prematurely without warning message or after displaying less than 20 minute battery warning message.
Cause: Software error.

Recall: Z0630
Product: Vestibular Analysis Apparatus
Problem: Software that operates the balance evaluation system contains a calculation error that may affect composite score by less than 5% in the worst case.
Cause: A software design error resulting in miscalculation of test results when height is entered as inches.

Recall: Z4220
Product: Programmer, Pacemaker, Module
Problem: Pulse amplitude of pacer is inadvertently changed to a lower value during mode reprogramming from an unregulated to a regulated pulse amplitude.
Cause: The software was mistakenly written so that when there was reprogramming, there are also changes in the programmed output amplitude.

Recall: Z2060
Product: Stimulator, Spinal Cord, Implanted
Problem: If patient uses the programmer in an amplitude drop sequence, may receive unpleasant stimulation which could cause involuntary patient reaction.
Cause: Firm states that problem is caused by a software defect. The programmer has a "Clock" that rotates through all 105 volt amplitudes. Software causes rotation up to highest amplitude before going to desired settings which are sent to patient.

Recall: Z5430
Product: Autotransfusion Apparatus
Problem: Error in the autowash function which could cause the wash bag to overfill and burst.
Cause: A software error in the auto cycle option function of the configuration option of the program.

Recall: Z6240
Product: Blood, Separator, Automated
Problem: Problems with plasma line, pump runaways, failure alarms and alarm tests.
Cause: Software error.

Recall: Z8930
Product: System X-Ray, Tomography, Computed
Problem: May provide erroneous diagnostic and anatomic information during certain modes of operation.
Cause: Defective computer software allowed display of previous patient's image.

Recall: Z1711
Product: Nuclear Magnetic Resonance Imaging System
Problem: As a result of a technician entering incorrect date of study during an imaging session, entire system shut down.
Cause: The original software program had a start date of 1863; this caused changes in other areas, including tomographic reconstructions, series manipulations, and heart studies.

Recall: Z5471
Product: Generator, High Volt X-Ray, Diagnostic
Problem: Display fails to indicate the true technique factors. The display is not updated when the cast modifiers are deselected.
Cause: Software not designed and validated properly for the necessary function to operate all displays.

Recall: Z5531
Product: Colorimeter Photometer Spectro, Clinical Use
Problem: Software yielded erroneous laboratory results.
Cause: The manufacturer failed to validate the software and to audit its supplier to assure that the software was validated.

Recall: Z6071
Product: Hexokinase Glucose
Problem: Nonlinear Readings
Cause: To correct deficiency the firm is reducing linearity claim of the HK glucose on the software from 750 mg/dl to 600 mg/dl and the high absorbance limit from 2.4A to 2.0A so that the software is in line with the labeling claims.

Section II

Software Design-Manufacturing Process

Recall: Z1567
Product: Prescription Daily and Extended Wear Contact Lenses
Problem: Lenses were labeled with the incorrect expiration dates.
Cause: Software error

Recall: Z0050
Product: Guide Wire, Angiographic Catheter
Problem: Mislabeled product is labeled as a straight tip guide wire but actually is a "J" tip end.
Cause: The new printing system in the computer printed the words "straight tip" instead of "J tip." This was a software error.

Recall: Z2621
Product: Lens, Intraocular
Problem: The lenses were labeled with optic diameter of 6.0 mm but their optic diameter of 6.5 mm.
Cause: Problem caused by misprogramming computer software which was configured to capture the optic size printing on the internal label that came from the test file instead of from the file that controls all information on the labels.

Section III

Software Change Control

Recall: Z9551
Product: Pulse Generator, Pacemaker, Implant
Problem: Pacemaker programmer could cause temporary loss of pacemaker function.
Cause: Revised software contained errors that occur when pulse width is set to decrement and user releases program button within 60 millisecs of the command.

Recall: Z10661
Product: Pulse Generator, Pacer, Implant
Problem: Unexpected rate scans can be initiated during the rate programming of Dialog II and Sensilog III.
Cause: Computer software defects, noncompatability with modified software.

Recall: Z5218
Product: Ventilator, Continuous
Problem: Device delivers unscheduled breath due to autocycling by ventilator.
Cause: Failure investigation found that at the "release 9," software was determining monitored peep at an inappropriate point in the software which could cause the device to autocycle. "Release 9" version had been in use since 1987.

Recall: Z2449
Product: Ventilator, Continuous
Problem: The ventilator may cease ventilation.
Cause: The watch dog timer circuit (WTC) may fail and the device will cease ventilating without audible or visual alarm. This failure resulted during one of the numerous software updates since 1983.

Recall: Z2429
Product: Pulse Oximeter

Problem: Unstable readings and inaccurate.
Cause: Software revision no. 6 contains an error which involves constant writing into the calibration memory which destroys some memory positions causing an EPROM failure at 32xxx message.

Recall: Z7590
Product: Oximeter
Problem: Device may lock up and display test pattern values for oximeter parameters when initially turned on.
Cause: During a software revision, circuit board may not receive command to leave calibration at end of self-check process and initialization.

Recall: Z8560
Product: Chemistry Analyzer (Photometric), Discrete
Problem: Results produce sample mix-ups.
Cause: Firm's failure analysis concluded that an error condition in the 3.2 software version can cause the sample identification number to be transferred to another patient in the instrumentation load list.

Recall: Z5171
Product: Chemistry Analyzer (Photometric) for Clinical Use
Problem: There is the potential for device to report incorrect ratio results to the host computer.
Cause: The problem, caused by an earlier software revision, 5.1, provided a 41-second time frame for ration calculations.

Recall: Z0731
Product: Computer Software, Linear Accelerator
Problem: Inherent error involving a coefficient and mirror beam calculations.
Cause: Problem occurred with an unvalidated version of 2.65 software. If coefficient set to value other than zero, could cause high or low off axis calculation.

Recall: Z4511
Product: Accelerator Linear Medical
Problem: System may cause a release of all brakes on the table.
Cause: The last updated software, version 1.3.1, was defective, causing system to detect unauthorized movement and react by releasing all brakes on the table.

Recall: Z1504
Product: Digital X-Ray System
Problem: The projected X-ray field may be larger than the diameter of the image reception when used in the fluoroscopic mode.
Cause: At the introduction of the current version, No. 9, a typographical error occurred in the software algorithm, that when a certain mode selection sequence is followed, projected an out of spec. x-ray field.

Recall: Z2285
Product: Calculation Data, Proc Module for Clinical Use
Problem: Incorrect analytical results for triglycerides, cholesterol, and/or uric acid may occur when certain conditions exist.
Cause: An error existed in the revised software.

Recall: Z5035
Product: Pressure, Blood, Computer
Problem: Inaccuracy can occur in the computed mean pressure readings.
Cause: Erroneous algorithm in "A" revision software creating an approximate 5-15% low mean reading depending on waveform analyzed.

Recall: Z0508
Product: Incubator, Neonatal
Problem: System lock-up.
Cause: Lock-up condition results in the LCD temperature to remain frozen while incubator is not being controlled. Software revision to eliminate "lock up" caused erroneous analog-digital conversion.

Recall: Z0299
Product: Susceptibility Test Panels, Antimicrobial
Problem: Inaccurate results.
Cause: Error due to an incorrect incorporation of the National Committee for Clinical Laboratory standards breakpoints into the software.

Recall: Z4549
Product: Ventilator, Frequency Monitor
Problem: Less than accurate read out of patient's CO_2 and NO_2 levels.
Cause: Error was introduced when a software change was made in an attempt to enhance the versatility of the device by covering the possibility of a pressure transducer failure, the software revision was not validated for original label claims.

Recall: Z4809
Product: Ultrasonic Pulse Echo Imaging system
Problem: Problem affects color flow image, blood flow appears slightly displaced from the anatomy.
Cause: Software defect was introduced in a recent software release, affecting color flow image.

Recall: Z5189
Product: Bone Densitometer

Problem: Software releases through 3.6 contain error making some patient files inaccessible
Cause: If the "delete" key is used, an error message will display instead of requested patient data. Under certain sequence of user interaction with the data files, software can lose track of which patient is associated with which data.

Recall: Z5659
Product: Auto Urinalysis System
Problem: The unit will read "trace" protein level as "negative."
Cause: The wrong master software program was used in making a software revision.

Recall: Z1430
Product: System, X-Ray, Tomography, Computed
Problem: in certain modes the system reverses the video image with "R" showing the left side and "L" showing the right side.
Cause: Error in software and hardware changes to the system caused the reversal.

Recall: Z4060
Product: Automated Pipette
Problem: The sample management center 10 ml teflon syringe leaked during use.
Cause: Software design in previous version, placed the syringe in such a position as to allow the barrel to dry. This eventually caused the syringe to leak. In the new software, the rest position for the syringe in a moist location.

Recall: Z5560
Product: Blood Gas Monitor, On-Line
Problem: Miscalculation in software resulted in erroneous oxygen saturation value.
Cause: Error was instituted into software during previous revision. A parenthesis in source code was located in the wrong space.

Recall: Z5730
Product: Radioimmunoassay, carbamazepine
Problem: Instrument could not be calibrated with the kit, it rejected the second code for 5.2 activation.
Cause: Investigation concluded that the original code for the software, set 2 was incorrect.

Recall: Z8910
Product: Imager, Ultrasonic, Ob-Gyn
Problem: Does not allow proper programming of the device.
Cause: A programming error in version C:3.1 prevents access to proper menu selections when one attempts to enter certain parameters.

Recall: Z6480
Product: Infusion Pump

Problem: Inappropriate storing of fluid delivery without display of information to the user.
Cause: Hospitals installed the device prior to software being tested, and there had been continual software revision due to customer complaints.

Recall: Z1011
Product: Automated Differential Cell Counter
Problem: Device has the potential to dilute patient samples and lose pre-set calibration factors under certain conditions.
Cause: Software errors.

Recall: Z2931
Product: Camera, Scintillation (Gamma)
Problem: Tomo comparison images are displayed for side-by-side comparison as stress/stress images instead of stress/redistribution images.
Cause: The problem, attributed to the 3.1 software version, occurs when stress data image is selected first, followed by the redistribution image.

Recall: Z4491
Product: Transmitter and Receiver, ECT, Telephone
Problem: ECGs can be matched with wrong patient I.D. at time of data transmission between two sites.
Cause: Problem was caused by software revision 3.16, that instead of deleting old waveforms, deleted the new one.

Recall: Z5361
Product: Nuclear Magnetic Resonance Imaging System
Problem: Image Processing equipment for diagnostic nuclear medicine systems gave inaccurate curve statistics.
Cause: The curve statistics command in the release 5 software was found to be inaccurate, leading to incorrect results.

Recall: Z5751
Product: Spirometer, Diagnostic
Problem: Inaccurate readings.
Cause: Software errors regarding the accuracy of the values were programmed into the printer/charger.

Recall: Z6701
Product: System, Radiation Therapy, Radionuclide
Problem: Incorrect calculation of the delay factor.
Cause: Software versions V14-B, etc. included errors that caused inaccurate delay factor for isotope units, ignores finish month and reverses dates.

Recall: Z3511
Product: Electrocardiograph

Problem: Device does not have capability of noting that an electronic pacer is in place.
Cause: A defect in the revised software (V.1017) causes the statement "electronic pacemaker no further interpretation possible."

Chapter 19

CASE STUDY EXPERIENCE: SIXTEEN THESES[1]

INTRODUCTION

The importance and related frequency of system validation analyses have exploded over the past five years proportionate with increased pressures from United States and European regulatory agencies.[2] The demand for evidence of control of the development, installation, and utilization of computer systems has produced an appropriate response in the pharmaceutical, medical device, and biologics communities. System validation, the collection and analysis of that evidence of control,[3] has emerged as a high priority throughout the regulated industries as the USFDA, USEPA, UK Department of Health, and other EC equivalent agencies properly perform their agenda-setting functions.

Between 1987 and 1991 Weinberg, Spelton & Sax, Inc., audited 314 regulated computer systems. These 314 cases represent a wide diversity of system purpose, platform, design, application, and control. In composite they are summarized in more than 100,000 pages of reports, test results, and notes representing a reasonable cross section of the field of system validation.

This monograph is the result of an analytical summary of those 314 cases. Each case includes an initial audit including on site investigation, functional test, code review, and extensive interviews with system developers, users, and operators. In most cases (288), that initial audit was followed within six months by a second study, which generally (in 237 cases) resulted in the validation (collection of persuasive evidence of control) of the system. Using these audits, studies, and evidences as a database, this monograph addresses the fundamental question, "What have we learned?" The answer

[1]This chapter is based on an article entitled "Case Studies in System Validation: Sixteen Theses," that appeared in the *Drug Information Journal*, Volume 27, No. 2 (April-June 1993), pp. 253-262. Permission to publish it here has been granted by the *Drug Information Journal*.

[2]A.F. Hirsch, *Good Laboratory Practice Regulations* (New York: Marcel Dekker, 1989).

[3]Dr. Sandy Weinberg, *System Validation Standards* (Dubuque, Iowa: Kendall/Hunt Publishing Co., 1990).

is contained in 16 emergent theses, which collectively represent a proposed summary of the state-of-understanding of system validation in the health-related regulated industries.

These 16 theses have been organized for convenience into three categories, intended to focus upon the segment of the community likely to be most interested in or able to profit from the conclusions:

a) regulatory theses,
b) technical theses, and
c) business theses.

No doubt, interest and applicability will overlap, but this organizational scheme does seem to lend itself most appropriately to efficient field consideration of the conclusions.

THE CASE STUDY FILE

The 314 cases included in this analysis can be characterized according to several dimensions. All are drawn from the health-related regulated industries: pharmaceutical, medical device, biologic, and environmental health laboratory. Since many systems overlap several of these environments (for example, a chromatography system might be characterized as a medical device utilized to study biologics in an environmental laboratory), no clear breakdown into subcategories is possible beyond the business categorization of the companies commissioning the system validation audits. Thirty-one of the systems included are utilized by companies describing themselves as "medical device (or diagnostics) companies;" 48 come from the biologics industries; six studies involved environmental health laboratories; 49 of the systems included are drawn from the pharmaceutical industry.

The vast majority of these systems were utilized (or, in the case of medical devices, manufactured) in North America; this conclusion is not surprising, since most of the system validation pressure has originated with the United States Food and Drug Administration. Two hundred twelve systems are North American (the vast majority of those from the United States); four are Asian (largely utilized in Japan); 98 are European (Germany, Great Britain, Switzerland, Sweden, Netherlands, Finland, France, Italy, Ireland).

Categorizing the systems by regulatory area can be confusing. All systems probably are (or ought to be) subject to Good Manufacturing Practices (GMP) for the production of those systems. More meaningful, though, is the classification of systems into the regulatory area in which they are used (or are intended to be used). Under this scheme a laboratory system might fall under Good Laboratory Practices (GLP) if used in research facilities, under GMP guidelines if used in the final stability check of a manufactured product, under Good Automated Laboratory Practices[4] (GALP) if used in an EPA setting, or under the emergent Good Clinical Practices (GCP) if utilized in a clinical setting. Within this limitation, the case file can be described as falling within: GLP

---

[4]U.S., Environmental Protection Agency, *Good Automated Laboratory Practices (GALPs)* (Research Triangle Park, N.C.: U.S. EPA, December 28, 1990).

guidelines, 82 systems; GMP guidelines, 182 systems; GALP guidelines, 3 systems;[5] GCP guidelines, 47 systems.

Finally, the 314 systems constituting the case file can be characterized according to two meaningful technical dimensions. One hundred, ninety-nine of the systems utilized a mini-computer (or super micro serving in a host-slave relationship); 59 systems were supported by a mainframe platform; 34 systems utilized networked microcomputers; and the remaining 22 systems operated on a freestanding microcomputer, PLC, or microchip.

Six of the systems were programmed (application code) in proprietary languages unique to the manufacturer. Nine of the systems used a manufacturer-variation on ladder logic coding. Seventy-four systems were programmed in versions of C, 19 in BASIC, 52 in COBOL, 82 in FORTRAN, 16 in PASCAL, 11 in versions of Assembler, and 45 in a variety of database sublanguages (including DBIII, DBIV, FoxBase, Oracle, and mainframe databases).

Together, the case file represents systems in use by more than 200 clients in 33 states and 16 different countries, all struggling to develop and provide evidence of control of their diverse computer systems.

REGULATORY THESES

These five theses are of largely regulatory impact, related most closely to the nature of the evidence that is utilized to demonstrate control of the system.

1. Most systems are absent development documentation. Since system development has often preceded regulatory interest, most users find themselves having invested time, energy, and dollars in systems that may be effective but for which necessary design and development specifications have never been documented. Of the 314 systems in the case file, only 73 systems were supported with appropriate and adequate design specifications.

The problem is significant in terms of system support as well as regulatory focus. Sam Clark, former computer investigator for the USFDA, has quoted a series of informal studies estimating that 60% of system problems are related to a lack of specifications.[6] While it may be impossible to truly qualify the extent of the relationship, the degree of regulatory concern over specifications is clear, and the difficulty in evaluating and managing a system without specifications has long been understood.[7]

Retrospective recreation of system specifications is possible and, in absent situations, appropriate. The expense and difficulty of that process, however, is sufficiently daunting to urge consideration of replacement of borderline systems, potentially resulting in a minor boom in the support software industry.

---

[5]GALPs still in draft; 1991 studies present pre-GALP development studies.

[6]Society of Quality Assurance Meeting, Atlanta, GA, 1992.

[7]E. Yourdon, *Decline and Fall of the American Programmer* (New York: Yourdon Press, 1992).

2. Most systems are absent effective audit trails. The audit trail concept, as popularized by the financial industries, focuses upon bottom line reconciliation. The regulatory concern for audit trails in the health industries utilizes the concept in a very different manner: the trail is intended as a method of fixing responsibility. Data points may be changed for a variety of reasons, some acceptable, some questionable. Regulators have a legitimate concern in seeking assurance that any data change is accounted for and can be identified by the nature of the change, its date, and the identity of the responsible person.

Only 81 of the systems in the case file were initially equipped with effective audit trails. Management responsible for an additional 16 systems had designed supplemental manual audit paper trails. In more than two-thirds of the cases, no methods of reliably tracking data changes were included in the system design or operation.

3. System Version Control is largely lacking in software-based systems. Firmware systems are generally characterized by effective version control, as one generation of chip, PLC, or EPROM is replaced in its entirety by another clearly designated generation. In complex systems using resident software, formal version control is often neglected in favor of undocumented patches, "work arounds," and quick fixes. Two hundred seven of the cases in the file involved systems in which significant variations were found to exist between the official version record and the actual version in use. Most commonly, these variations were the result of system maintenance personnel responding to emergency change or debug requests without updating the supporting documentation appropriately.

While this thesis does represent a regulatory control problem, the practical implications and impact of this problem are debatable. The inevitable rush to fix immediate problems is likely to continuously result in a documentation lag regardless of enforcement pressure. At the same time, the danger of such fixes is often minimal, as many changes are insignificant in nature. The appropriate response may be a new regulatory flexibility, combining a classification of changes into two categories ("critical" and "non-critical") and imposing the immediate version/change control requirements only on critical changes. Non-critical fixes could be documented through a "batch" rather than "on-line" change control procedure, with a periodic (annual?) reversioning of software to reflect all changes made in that interim period.

4. Most systems are appropriately characterized as medical devices. Recently the USFDA Center for Biologics Evaluation and Research has begun to treat some decision making systems[8] as medical devices, subject to section 510(k) of the Safe Medical Devices Act. The definition of medical devices incorporated and implied in that act seems to include all software systems that potentially affect the public health and safety. With the exclusion of financial, business, and personnel systems, that broad definition seems to encompass the vast majority of systems used in the health-related industries.

Of the 314 systems represented by the case file, 302 can be classified as medical devices under the current, widely-accepted definition promulgated by the Health Industry Manufacturers Association and other related organizations. Perhaps moving responsibili-

---

[8]For example, donor deferral file of blood processing organizations.

ty for system validation to the Center for Medical Devices and Radiological Health of the USFDA (and the European and Japanese equivalent agency divisions) represents an effective end to regulatory confusion and an appropriate inclusion of vendors and developers within the regulatory umbrella.

5. Most system Standard Operating Procedures do not accurately describe actual system management practices. This gap is often found in environments in which SOPs are unwritten, unclear, or inappropriate. Two hundred sixty-five of the systems included in the case file had negative SOP comments included in the initial audit report. These comments most often referred to practices out of coordination with documented procedures, or with ambiguously or unclearly written SOPs.

The problem, however, may well extend beyond system managers or SOP writers. One of the fundamental characteristics of a complex system environment is the speed of change; in fact, ability to change programming is the defining difference between a calculator and a computer. Flexibility to change is the definitive characteristic of a computer system. The gap between practices and SOPs may reflect institutional restrictions on the ability to change formal Standard Operating Procedures as rapidly as new technologies and applications obsolete old SOPs. Many organizations report a six month review and sign-off process for new Standard Operating Procedure approval. Yet, in recent years, technological innovations for software data storage, for example, have emerged an average of every 60 days![9] While not many of these innovations are relevant to a particular setting, the perceived discrepancy may be demoralizing and may potentially result in a belief that SOPs are never the useful directive tools they are intended to represent.

The solution to this problem may lie in a combination of general SOPs, representing topics of system management, and a secondary "working procedures" system of more detailed practices. For example, an SOP may call for periodic back-up of datafiles and refer to a detailed working procedure that may summarize current definitions of frequency, process, and storage location.

TECHNICAL THESES

These eight technical theses represent observations concerning software code development, system organization, and testing.

6. Language selection is directly related to system effectiveness. Although it is possible that increased developer sophistication over time will mitigate the accuracy and significance of this thesis, current data argues for a significant relationship between the language selected for a system project and the effectiveness and control of that final system. Eighteen of the systems included in the case file, for example, involved projects utilizing prototyping languages. In each of these cases, the ease of prototype development resulted in a seemingly related lack of care in documentation of the development process. The premium placed on user flexibility resulted in a disincentive to reduce that

---

[9]John Diebold, *The Innovators* (E.P. Dutton, 1990).

flexibility with a carefully pre-defined set of specifications and subsequent modification trail of documented changes.

Similarly, the selection of database languages apparently resulted in forcing database solutions to emergent problems that might more effectively have been handled through reconsideration of analytical formula rather than restructure of data files. Contrapositively, formula driven languages, such as FORTRAN and COBOL, superimposed algorithm solutions to database problems.

These observations are not intended to argue for or against any particular approach or language but, rather, to suggest that selection of such a language is a fundamental design issue and should not be left to the chance of programmer preference. An inappropriate selection can result in convoluted solutions and complex problems over the lifespan of a project.

7. De-Centralized Systems generally represent superior control. Of the 298 systems to which centralization had appropriately been considered,[10] approximately two-thirds (203) were de-centralized, defined as having a user-oriented focus of control. In the remaining centralized cases, institutional distance from the end-product application and related GLP, GMP, GALP, and GCP regulations resulted in near universal problems of communication, documentation, and change control. In one situation in which management attempted to limit these problems by providing systems operations personnel with GMP training and a firm grounding in the implications of the (pharmaceutical) product under development, the different orientation of central systems professional still resulted in serious problems of versioning and change control maintenance documentation.

While there is no theory supporting evidence to prove the inevitability of increased regulatory control problems in centralized environments, the preponderance of cases makes strong inductive argument for adopting a de-centralized strategy in most cases. Exceptions might be reasonably suggested in very small operations, in situations in which applications personnel have been integrated into centralized systems departments, and in organizations relying on very strong quality assurance controls in lieu of quality integration throughout the company.

8. "Ownership" of a system is a critical determinant of control. The factor that most commonly serves as a predictor of serious system validation problems is the inability of an organization to clearly and consistently identify the person or department responsible for a system. That lack of ownership inevitably leads to confusion of processes, documentation, and performance.

Some organizations have successfully managed to control "multiply owned" systems. Seven such cases are included in the case file. In all of these situations a mainframe (or, in one case, a large mini-computer) contained two or more distinct software systems used by different departments. In each case, the distinct system was "owned" by a single department, while the overall system was shared (in two cases, owned by one of the sharing departments, in the other five cases, "owned" by a consortium of the sharing

[10]The remaining systems represent direct impact devices, such as PLCs, for which centralization of control is not a relevant dimension.

groups). This arrangement requires some careful organizational designs, but it can obviously be successful. The key to multiple ownerships seems to rest in each department feeling an overlapping control, rather than a structure in which no department seems to "own" the system.

9. Functional (performance) Testing is critical even in "live" systems. The argument is often advanced that "live" systems need not be separately tested, in that normal operation and quality assurance review represents a continuous test scenario. The case file experience suggests that this position, while logically persuasive, is undermined by conflicting results.

In functional tests of the 219 "live" systems contained within the case files, problems unreported by the system users were found in 44 cases (almost 25% of the population). Nine of those problems were deemed "critical," requiring major redesign, modification, correction, or replacement.

Two factors are apparently at work here. Technically, normal operation of a system may never access existent but rarely used pathways within a complex software code. Well designed testing, examining a sample of all pathways, may discover hidden problems that lay dormant in normal operation but which may impact significantly in usual but important situations.

Second, there is a human tendency to accept unskeptically complex computer generated results. In one case, for example, blood absorption results produced by a (faulty) software graphics system were inappropriate and, upon even casual examination, impossible. But system users, sensitized to accept such diagnostics without checking, had been using the flawed analyses in several important studies. The objective and fresh look implicit in functional testing identified the problem readily.

10. Erroneous Data Testing often uncovers hidden problems. Functional testing tracks system performance with "real" or potentially real data. An Erroneous Data Test examines a system's ability to identify and reject inappropriate data. For example, a blood processing system may be designed to read inventory codes of five meaningful digits. The Erroneous Data Test challenges the system with a missing digit in the fourth position, with an alphabetic rather than numeric character, with a number beyond defined limits, and with similar "boundary stretching" exceptions.

In 69 cases the Erroneous Data Test identified system problems not found through any other review. Most tests and reviews appropriately look for functionality according to specifications. The Source Code Review, for example, examines algorithms and formula for appropriateness. The Erroneous Data Test, however, operates in the crisp atmospheric level free from assumptive clouds. Every operator knows the system won't accept inappropriate data; few have evidence of that impossibility, however, despite experience with misprinted bar codes and miswired keyboards. All too often, in gathering that evidence, it becomes apparent that the unlikely event has not been anticipated and that the potential results of a dual key strike or blotted magnetic ink character is a significant system misinterpretation.

11. "Expert" or "Artificially Intelligent" software tools are generally inappropriate in a working, regulated environment. These advanced languages, sometimes called "fourth (or fifth) generation languages," in effect allow a user to specify relationship or function. The computer system then generates (or selects) the code to accomplish that directed task. In development, these tools can prove effective with appropriate controls (see Thesis Six), but in an operating environment they represent a seemingly unsurmountable regulatory dilemma.

If a user can create a new code with every use, how can the control of that code be assured? How can the new function be tested, the new source code be reviewed, the appropriate documentation assured? Without these controls the user is effectively unchecked, an insupportable situation in a regulated environment.

There are, of course, reasonable exceptions. Fifteen cases included statistical analysis processes in which users routinely wrote original "PROCS" to direct analysis of specific data segments. With some guidance, departmental cross-checking of these procedures coupled with careful documentation provided the appropriate level of control. Similarly, in one case a highly sophisticated expert system was accessible to users but was restricted to report generation functions. That is, a user could request a report on a unique combination of circumstances but could not restructure files, change algorithms, or permanently affect code.

These latest generation languages, and the future developments they portend, do represent a significant challenge for regulators and should raise careful concerns in all technical users.

12. Code Organization is directly related to maintenance capability. Convoluted "spaghetti" code is generally the result of a patchwork program, constructed from bits and pieces of other programs and modified over time. Generally, the program jumps back and forth between code sections without rationale (hence the spaghetti path of a flow trace), following no apparent scheme or logic. Such a program can be fully functional and is often a heritage of "pre-regulated" software days. One hundred three of the systems in the case file could be accurately characterized as "spaghetti."

In confirmation of regulatory concerns, such spaghetti systems are generally characterized by maintenance difficulties and increased maintenance costs. Ninety of the 103 spaghetti systems in the case files were classified as having serious maintenance problems.

While it may not be practical to replace functioning, older spaghetti systems, long term plans will probably find such replacement ultimately cost effective. In the short term, increased documentation and careful source code review should be budgeted to avoid support problems.

13. There is no adequate substitute for a Source Code Review. While a review may appropriately focus upon a sample of all code paths and routines, all programs in a system should be reviewed for accuracy and appropriateness of algorithm and formulae; clarity of organization; internal documentation; and change control. Other substitutes, including increased testing, error log analysis, and manual controls, have consistently failed to find serious problems revealed in source code reviews.

Five cases stand out from the case file. In one system a case review found an erroneous "t-test" formula not identified in all other testing and analysis. A review of a second system found a condition flaw in a PLC pathway normally activated only in an emergency. Two other system reviews found flaws in security/auditing files. A fifth case involved an erroneous decision condition that could have potentially resulted in the mislabeling of a critical biologic product. While five cases out of 314 may not represent a statistically imposing situation, the kinds of problems identified, coupled with the reinforcing effect a source code review has in defining other testing, more than justifies the regulatory requirement.

BUSINESS THESES

Three theses have business or financial implications, and will be of greatest interest to managers balancing budgetary pressures against the drive to validate systems.

14. In the development of new systems, the costs of validations are offset by savings in the design and implementation process. The expenses involved in "building in" validation to the development of a new system are limited and easily cost-justified. While some consulting assistance may be necessary in planning the validation requirements for a system, even these fees are quickly offset by process savings: the actual changes are generally the "free" results of quality, and often non-incremental alternate paths of development.

System validation guidelines are simply summaries of good system development practices. Specifications ought to be synonymous with good contracting. Appropriately, that contract includes specific final performance demonstrations; these demonstrations can effectively serve as the functional test protocols. Code review is as appropriate a part of purchasing software as looking under the hood is appropriate (and critical) in purchasing a new car (and, incidentally, serves the same "future maintenance" function). The increased documentation requirements of a regulated system are easily woven into the development process, becoming a matter of retention rather than creation. Without these appropriate practices, and the consulting that can assure they are designed-in, a system is likely to prove more expensive to develop and to maintain.

While it is not possible to meaningfully compare the cases that involved development in accordance with validation principles with those that did not (since the systems in each category, nine in compliance and eleven in non-compliance, varied widely in purpose and complexity), analysis of all twenty of these developmental projects suggests that building-in validation could or did result in significant cost (and time) savings.

15. In retrospective environments, the investment in system validation compliance is generally offset with savings in system maintenance costs within four years. The actual average (mode) offset was 3.75 years for the 283 systems tested in place.[11] This cost estimate includes the fees associated with the auditing and consulting process and the person days, equipment, and other expenses involved in changes required for compliance. In no case did the (user reported) expenses involved in compliance result in expenses exceeding 10% of the annual operating budget.

---

[11]The unaccounted-for 11 systems were commercial products "accepted" as part of the validation process.

To provide additional information concerning the potential costs of validation, however, it is important to note that a small percentage (2%) of the systems audited were never brought into compliance, perhaps for financial reasons. In other cases (an additional 3.42%) in which the initial audit did not result in final validation, the system was replaced rather than modified.

These expected costs seem to be reasonable, particularly when viewed in the light of regulatory pressure. The cost of a New Drug Application delay has been estimated at $100,000 per day,[12] for example; failure to provide validation evidence is one of many stated reasons for delay.

16. System Validation is generally most cost effective with outside assistance. While this thesis can be criticized as self-serving, it is nonetheless data supported. Regulatory requirements call for independence in the validation process. That independence can be provided in house through the quality assurance structure, or externally through an independent testing and validation organization (not involved in system sales or development). Because of the relatively unusual nature of a validation study, the technical expertise required, the cost of developing tools and test protocols, and the relative credibility or prestige of external certification versus internal review, the external group is more likely to be cost effective.

The 314 cases in the file, of course, all involved external help (a criterion de facto for inclusion) at an average client cost of $62,000 (with a wide range from $16,000 to $270,000 for a single system, dependent largely on complexity of system). Ninety-three cases, however, involved situations in which clients had previously attempted self-validation. In each of those 93, the final cost of external validation was lower than the previous (unsuccessful) investment in self-validation. Particularly in situations in which system validation is time critical, is an unusual organizational activity, or in which a high degree of regulatory scrutiny is likely, external validation is likely to prove a cost savings alternative.

CONCLUSIONS

The 16 theses summarized here are intended to reflect observations and conclusions culled from an analysis of five years of system validation projects. Over that five-year time frame the frequency of projects per year has increased five fold, though the kinds of problems (and solutions) have proven remarkably consistent.

Consistent, too, have been the recommendations, guidelines, and field determinations of regulatory authorities. Though the field is certainly not mature, with new questions arising regularly and new technology-inspired issues emerging on a seeming daily basis, the principles underlying system validation have remained clear and consistent. Further, agreement on those principles has been international in scope and largely consistent across varying national agencies (including, to large a extent, agreement between FDA and EPA in the United States).

---

[12]Discussion with System Validation Committee, Pharmaceutical Manufacturers Association, 1991.

Over the next few years, additional experience in validation will allow for further testing and refinement of these 16 theses and, no doubt, the development of additional guidelines that can assist in developing a theoretical and applicational body of knowledge for the emergent field of system validation.

Chapter 20

FUTURE DEVELOPMENTS

Predicting the future is always a dangerous business; stating those predictions in a published book is a doubly so. Not only are guesses preserved in undeniable print, but the lead time of publishing makes a book the least immediate of all modern media. In a sense, every printed statement is a prediction, anticipating by six to eighteen months the copyright date of the publication and by three to five years the "read date" of the final ingestion. Predicting three to five years beyond that time, into the next decade and, in this case, the next century and next millennium, is particularly frightening.

To preserve a sense of calm, and to allow some assurance of accuracy, prediction is best based upon projection rather than fantasy. Consider, for example, the differences between prediction based upon a crystal ball, and prognostication based upon the examination of animals.

A crystal ball allows the clever predictor to see not the future, but rather the hopes of his audience. Surrounded by meaningless symbols and exotic chants, the false prophet looks for feedback from hopeful listeners, projecting their perceived expectations into his predictions. The result is satisfying in the short term, but it provides little real information and no accurate views of the future.

At first glance, all other ancient methods of prophesy seem to suffer from the same charlatanism. But while astrology, I Ching, and runes may do little more than provide vague generalities for interpretation by susceptible believers, there is an exception.

Consider the ancient Greek practice of examining the entrails of a sacrificial animal. While such a barbaric approach may seem to be worthless mysticism, it in fact has a valuable methodology hidden within. For a society based on husbandry, selecting one animal at random and examining it for disease is a fairly effective way of predicting the future health of the tribe. While the occult trappings that surround the ceremony are simply illusion, examining the sacrificial animal is, in reality, a method of predicting by looking at the present and projecting that reality into the knowable future.

Without the necessity of sacrificing any animals, that projective technique can provide a meaningful look at validation over the next five years. These predictions are based upon the projection of known or expected events, rather than upon a combination of wishful thinking and inspired dreaming.

Twelve independent, predictive hypotheses have been generated by extending current trends, projecting seeded predispositions, and extrapolating from expected events. While these hypotheses carry no guarantees, they do paint a composite picture that ought to accurately depict the extent, influence, importance, and approaches to validation over the next five years.

1. All software used in the production, testing, tracking, and development of pharmaceutical and biologics products will be classified as "medical devices."
2. The European Union will have evolved a three-tier model of regulation, with validation serving as the defining characteristics of the three tiers.
3. In politically stable portions of eastern Europe, pharmaceutical development technology will overtake Western development, with greater internal validation controls.
4. Standardized clinical database formats will allow more effective CANDA submissions, comparison of study results, and sharing of data across borders. These data transfers and comparisons will significantly magnify the importance of system validation.
5. The U.S. Environmental Protection Agency will consolidate its databases to allow cross-data comparisons, requiring more extensive validation.
6. Technological advances in telecommunications will require additional controls for protecting data and communication systems.
7. Compromises concerning the acceptance of electronic signatures will lead to limited use of electronic files as primary source documents.
8. The Food and Drug Aadministration will increasingly utilize the "SEC model" for increasing regulatory efficiency while controlling governmental costs.
9. An expansion of the use and extent of the COSTART (adverse reactions reporting system) will expand validation issues into hospital and clinical systems.
10. Interest in and requirements for validation will expand in three dimensions: laterally, through other U.S. agencies; horizontally, through other national regulatory bodies; and vertically, as additional kinds of tests and systems emerge.
11. Internal software verification tools will make continuous validation testing a reality.
12. Increased regulatory sophistication will shift attention from primary focus upon documentation to more detailed considerations of testing and design.

ALL SOFTWARE USED IN THE PRODUCTION, TESTING, TRACKING, AND DEVELOPMENT OF PHARMACEUTICAL AND BIOLOGICS PRODUCTS WILL BE CLASSIFIED AS "MEDICAL DEVICES."

This trend has already begun, first in March 1994, with the classification of systems used in the blood processing industry (inventory control and donor tracking), and over the subsequent years throughout the biologics industries. This classification makes good sense, since it passes a significant portion of the validation burden on to software vendors, where that responsibility is realized most cost-effectively. Validation evidence, submitted as part of medical device filings (510(k) or PMA), can then be incorporated into the final validation studies conducted by all users of that software product.

Two other trends are acting to propel this classification process. First, there is a natural tendency for regulators to expand their range of interest. Field investigators are tacitly encouraged to continue their visitations until problems are found (and beyond). As an increasingly sophisticated industry does an continually better job of meeting Good Manufacturing Practices (and other regulations), investigators will find themselves looking to the fringes of those requirements. While validation requirements for software are central to the operation of a facility, they are on the periphery in terms of the normal core of investigation.

Much as clinical researchers began concentrating on contract laboratories once in-house laboratories were found in compliance, and process validation reviews began looking at vendors once user equipment was well controlled, so the energies and concerns of investigators have increasingly been directed to the developers of software. Those software shops, where many real research and manufacturing decisions are *de facto* being made, are the new focus of investigations once investigators find themselves with excess capacity, time, and energy.

Complimenting this natural, expanding focus from regulatory agencies is a positive pressure, which is eliciting enthusiastic cooperation from knowledgeable vendors. While the word has yet to permeate the software-development industry fully, many vendors are now aware of the decision of the First Circuit Court of Appeals (and other courts) providing significant restrictions on the liability of medical-device vendors whose products are FDA-registered. Except in cases of extreme negligence, registration provides a primary defense against the high costs of negligence litigation. That umbrella of protection is highly valuable to many vendors who would otherwise be hesitant to sell their inventory, MRP, or database software in such a lawsuit-prone enterprise as the pharmaceutical industry. When protection from future suits is offered in exchange, the cost of validation and registration shrinks to insignificance.

There are two kinds of industries in the United States: Those that are regulated, and wish they weren't; and those that are not regulated, and wish they were. In all the grousing about the FDA and other federal regulators, the pharmaceutical industry often forgets that the alternative to regulation is, most likely, costly litigation and loses sight of the protective advantage regulation brings.

As a result of these two trends, and of the dissemination of interpretations of the Safe Medical Device Act of 1990, a number of software companies have voluntarily complied with registration requirements. This core of voluntary compilers will burgeon in the near future, in effect defining registration as an industry standard. Once that standard is accepted, all non-compliant vendors will be forced to conform by a combination of regulatory, market, and litigation pressures.

THE EUROPEAN UNION WILL HAVE EVOLVED A THREE-TIER MODEL OF REGULATION, WITH VALIDATION SERVING AS THE DEFINING CHARACTERISTICS OF THE THREE TIERS.

The tiers are already defined: contrary to media hoopla, and consistent with general opinion, national entities have proven unwilling to surrender sovereignty in a variety of areas, despite the allure of a European Union. In the drug regulation area, that unwillingness has been clear. While the requirements for drug approval may be uniform throughout Europe, the perceived level of enforcement of those requirements is not. The British express private fears about weak inspections in Spain; the Spanish worry about Italy; the Italians about Ireland; the Irish about Luxembourg, ad infinitum. Joint inspection, cross inspection, and central inspection schemes have been proposed, tried, and largely abandoned. Using national labeling requirements as a rationalization, most drugs sold in the EU require both EU endorsement and a tacit or explicit endorsement by the country of consumption.

The third tier comes from the United States and Japan. Both countries are exporters of pharmaceutical products (Japan having increased its share of the medical device marketplace sharply in the latter half of the decade). But the United States and Japan also represent the two largest markets for pharmaceuticals, medical devices, and biologics. This consumer clout has led to the increased use of regulation for the protection of home industries, imposing a higher standard for indirect quality than that imposed in the rest of the world.

In order to comply, the harmonized U.S. and Japanese standard has evolved as a third tier of regulation. Products may be approved by the EU, by specific countries within the EU (particularly by the UK Department of Health, the German BGA, and French Ministry of Health), and by the U.S. FDA and/or Japanese Ministry of Health and Welfare. Because the U.S./Japanese requirements increase product cost without a (European-perceived) increase in quality, differential labeling is used, along with specialized production facilities.

This trend has already emerged, sinking its roots as early as 1992. Over the remainder of the decade it will continue, with those specialized production facilities shifting to Ireland, Brazil, Mexico, and Malaysia. More importantly, though, the criteria for differentiation will solidify.

Approval from tier one to tier two will differ, largely based upon level of validation. Uniform regulation (in tier one) will be reinforced through uniform enforcement of that regulation (in tier two) through the insistence upon validation of all processes, systems, and products. As validation represents a high degree of quality control of existing processes and procedures, that secondary inspection will provide skeptical nations with the assurances that countries whose enforcement is perceived to be less rigorous are enforcing common standards.

Tier three approval will rest upon secondary validation, the review by third parties of the systems controlling through validated processes. In effect, tier one approval will assure that the final product meets standards. Tier two will assure that the method of producing that product is appropriate, assuring consistency, safety, and quality. Tier three will assure that the systems controlling those methods are appropriate and under reasonable controls, adding cost-efficient production, product reliability and recall, and legalistic assurances to the tier two quality assurances.

Since some countries of the world may be unwilling or unable to pay for the performance improvements represented by tier two and tier three, all three levels will be utilized as reasonable market approaches. But open marketing, aimed toward all potential consumers, will require all three levels. In effect, the U.S. FDA (deferred to by the Japanese) will become the ultimate approver of products. For products also regulated elsewhere, the primary qualification for add-on FDA approval will be the validation of the development, testing, trials, production, and tracking systems.

IN POLITICALLY STABLE PORTIONS OF EASTERN EUROPE, PHARMACEUTICAL DEVELOPMENT TECHNOLOGY WILL OVERTAKE WESTERN DEVELOPMENT, WITH GREATER INTERNAL VALIDATION CONTROLS.

Those portions of the old Soviet Union and its client states that avoided significant political upheaval have quickly realized that biologics and pharmaceutical industries represent a near-perfect cure for the need for Western currency, the need to upgrade home standards of living, and the need to avoid an intellectual brain drain. While the development of new products is an expensive process, the production of established biologics, in particular, is not capital intensive. The output of biologics facilities also produces both immediate domestic value and high export potential. Israel first realized this opportunity in the early 1990's, giving the development of a biologics (and medical devices) industry high priority as political circumstances allowed a partial shift from a military economy. As early as 1993, similar policies emerged in Russia, the Czech Republic, Poland, and Hungary.

In the process of developing these production capabilities, technological advances in chromatographic separation and gas chromatography have allowed biologics facilities to sidestep more primitive separation methods. Equally important, facilities built post-1992 do not need to be re-engineered to meet current validation philosophies. Computerized production facilities and computerized testing and tracking systems are designed to meet even the most stringent validation requirements. The result is a shift of production sites, both for domestic and export production of established biologics products.

This shift toward increased biologics and pharmaceutical development in Eastern Europe will also serve to accelerate the general remix of products, favoring biologic agents over chemical products. Biologics are generally more expensive to develop, but less expensive to produce. While a replacement of chemical compounds cannot be expected, an increasing number of treatments and cures will rely upon DNA-manipulated and related biological components.

STANDARDIZED CLINICAL DATABASE FORMATS WILL ALLOW MORE EFFECTIVE CANDA SUBMISSIONS, COMPARISON OF STUDY RESULTS, AND SHARING OF DATA ACROSS BORDERS. THESE DATA TRANSFERS AND COMPARISONS WILL SIGNIFICANTLY MAGNIFY THE IMPORTANCE OF SYSTEM VALIDATION.

The U.S. FDA's "SMART" (Submission Management and Review Tracking) Project, intended to define standards for clinical data, is already far behind original timeframe projections, and much modified in scope and concept. But SMART is progressing on target in the one arena that is most critical: the teams have achieved consensus on the value and importance of the conceptual framework. Arriving at a widely-acceptable format for data input, and developing effective software to control that data, may still be years away. Yet the battle to persuade the industry of the advantages of that struggle has been won. The debate now centers upon the final details of "how," rather than on "whether" or "why."

By standardizing the format (and hence the requirements) for clinical data, a single system is possible for FDA receipt of Computer Assisted New Drug Applications (CANDAs). That single system speeds the user process, since retraining in every individual company's system is no longer necessary, and since multiple experience with the standardized system can potentially result in improved, simplified, and user-friendly approaches.

With agreement upon a single system, tacit agreement on the data to be collected is achieved. Tagging every study with a subject's age and weight, but not with other non-critical clinical characteristics, speeds the study design process and eliminates the most common need for study reanalysis or replication. The format defines the acceptable reporting parameters, eliminating much of the guesswork that has previously characterized clinical research design.

While the processes of review and design are accelerated, the real benefits of the clinical database system derive from its data sharing capability. With a standardized format, it is possible to compare the results of two treatments developed by different companies, even if those companies conduct, for example, one set of studies in Germany for the BGA and a second set in the United States for the FDA. Control groups can be compared, too, to determine the degree to which they are truly representative and to discover data anomalies that may prove to be significant pointers. In some cases, clinic patients may be spared by using a single control group for a composite study of competing compounds or treatments.

These kinds of comparison examinations ought to save costs and maximize safety. Cross-comparison improves study controls without increasing time or costs. The comparison of study results, again focusing largely on control groups, can serve as a warning of study error, fraud, or misinterpretation.

Finally, a uniform (and internationally accepted) clinical database format can speed the approval process for drugs licensed in one country and applying for licensure elsewhere. That speed will lower developmental costs in itself, and will reduce the need for duplicating studies that are expensive and time consuming. In a competitive market place, where every year of regulatory delay reduces the value of a patent, time delays contribute significantly to the bottom-line cost of product.

All of these benefits depend, first, upon cooperative agreement, which seems to be slowly reaching consensus, and, second upon the technology for implementation. That technology is available in terms of bytes and code, but it will require a special emphasis upon protection of data. If a variety of researchers beyond the control of the submitting company have access to data, it is all the more critical to have effective write restrictions in place, to build data security precautions into hardware and software, to establish audit trails, to design disaster-recovery precautions that will protect against inevitable problems, and to design and implement controls that will assure data integrity.

In short, data sharing will increase industry reliance upon system validation, which in turn is the warrantee that these protections have been developed. An unauthorized user, unprotected virus, unintentional erasure, or inappropriate access can have profound results on an international database. Validation of the central system, and of the individual systems that will be entering and accessing that data, represents protection against a nightmare scenario of data corrupted, data changed, data lost, or data frozen.

THE U.S. ENVIRONMENTAL PROTECTION AGENCY WILL CONSOLIDATE ITS DATABASES TO ALLOW CROSS-DATA COMPARISONS, REQUIRING MORE EXTENSIVE VALIDATION.

The EPA has made a strong commitment to validation, particularly in its Good Automated Laboratory Practice (GALP) guidelines that control contract laboratory computer systems. Much as the FDA is struggling under SMART to allow cross comparison of data through a standardized database format, the EPA has launched a project of its own that will eventually allow comparison and composite of data from its regional laboratories, contract laboratories, and industry laboratories. However, there has not yet been an effort to assure compatibility of FDA and EPA data, a development that would make great scientific sense.

The EPA data compatibility effort will prove useful in combining information from different programs and concerns. A good portion of the difficulty in controlling acid rain, for example, comes from the cross-program nature of the problem: Acid rain is caused by air pollution emissions which fall into the water supply. Acid rain, therefore is classified under both the EPA's air quality and water quality programs. When issues focus on the effect of acid rain on forest timber, debate tends to focus on the origin of the problem -- trees absorbing chemicals through inspiration, or through root ground water -- rather than solutions to the problem.

As logical as data comparison and combination may be, they are dependent upon the same two factors facing the FDA -- the development of a standardized format and the validation of all systems interacting with that centralized database. Standardization allows communication; validation controls corruption of the data once communication is established.

Progress on data standardization will reach fruition in the next three years. As validation issues are identified and addressed, that database will be designed and protected with the appropriate validation controls to assure the quality of the standardized database contents.

TECHNOLOGICAL ADVANCES IN TELECOMMUNICATIONS WILL REQUIRE ADDITIONAL CONTROLS FOR PROTECTING DATA AND COMMUNICATION SYSTEMS.

As local area networks (LANS) are increasingly replaced with wireless configurations (the so-called "LAWNs" -- local area wireless networks), a new concern has emerged. Data integrity and command-stream integrity depend upon the proper message packet arriving at the appropriate receiver point. In the first half of the 1990's, concern focused upon the integrity of the message packet, with validation controls imposed and confirmed to assure that data was not changed by poorly written or followed Standard Operating Procedures, software errors, or hardware problems.

While data protection issues still exist, new wireless technologies increase the dangers once restricted to intercontinental satellite transmission. In wireless telecommunication systems, there is a high potentiality for misreceived data. Interference, misaddressing, partial reception, and improper packeting all can distort or lose information.

One of the key components in a validation study is data integrity testing. As wireless transmissions become increasingly common, that data integrity testing is supplemented by confirmation of design features intended to verify the accurate receipt of information. These features generally include bisyncronous communication strategies, the use of check-sums or confirmatory codes, human confirmation of data sets or randomly selected data points, and mode/deviation analyses of data sets. All of these strategies provide increased validation evidence of control.

In non-data systems, such as programmable logic controllers and transmitting (or receiving) medical devices, the problems are potentially greater and the validation issues proportionately more challenging. Consider, for example, the difference between a portable electrocardiogram (EKG) recorder that stores an electronic record of heart readings, compared to a similar device that broadcasts (or narrowcasts) the same information to a central receiving point for storage and analysis at that remote location. Both situations require the same control of measurement, translation, and recording. But the remote system adds additional complications to the digitation, transmission, reception, and digital interpretation of the results.

COMPROMISES CONCERNING THE ACCEPTANCE OF ELECTRONIC SIGNATURES WILL LEAD TO LIMITED USE OF ELECTRONIC FILES AS PRIMARY SOURCE DOCUMENTS.

In 1993, the FDA and the Pharmaceutical Manufacturers Association (PMA) reached an agreement about the acceptance of "electronic signatures" on required documents. Under the agreement, a so-called electronic signature had to pass three evidentiary tests: Identification of the unique author (or approver) of the document, identification of the fixed date and time of the approval, and identification of the exact text at the time of the approval. In practicality, these three tests required the use of some sort of a password protection system, some text stamping system (with date and time), and some sort of check code to confirm that the text was unchanged from time of initial approval.

In 1994, the FDA promised a formal announcement of that new policy, and informally de-emphasized adverse finding reports for use of electronic documents. As the fledgling policy now stands, most documents in most situations can be legally designated as "reviewed and approved" with the use of carefully controlled electronic "signatures."

Potentially, this new policy can have tremendous advantages in controlling Standard Operating Procedures. By placing SOPs in "read only" files on computers throughout a laboratory or manufacturing facility, management can assure widespread availability, consistency, and instant correction as new versions are developed.

Unfortunately, the reliability of document management systems -- especially exaggerated fears about the relative lack of reliability of those systems -- have hampered their widespread use. What would a laboratory do if the document system were no longer available to technicians? Must the entire operation be halted, since the reference SOPs are (temporarily) not available?

In answer, many facilities have chosen to print out a copy of the relevant SOPs. But in such a situation, is not the printed copy (used only in emergencies) the true copy? And, if so, can the electronic copy be considered a legal replacement? The dilemmas are serious, and they have led many users back to paper.

Two developments are increasing the trend toward electronic documentation. First, more and more facilities are installing integrated systems, in which both documents and operations reside. If that system is non-functional, all production halts, making the documentation issue moot. Second, increases in the actual and perceived reliability of word processing systems allows a greater reliance on electronic systems, without back-up reliance on paper undermining the credibility of the primary system.

In one sense, electronic documentation has developed, and has been narrowly accepted by regulators, in advance of need or technology. But as complex CIM and laboratory operations create that need, and as increased reliability improves technology, electronic signatures -- and the requisite reliance upon the validation of those systems -- will meet the emergent demand.

THE FDA WILL INCREASINGLY UTILIZE THE "SEC MODEL" FOR INCREASING REGULATORY EFFICIENCY WHILE CONTROLLING GOVERNMENTAL COSTS.

The Securities and Exchange Commission is responsible for (among other things) auditing specified financial data from public corporations. Rather than attempt to perform that huge array of audits itself, however, the SEC has evolved a system of deferred responsibility. Public companies are required to hire an independent firm of auditors (a Certified Public Accountant firm). The CPA audits the corporate books and prepares a report to stockholders and the regulatory agency. The SEC, of course, reserves the right to conduct its own investigation if circumstances warrant.

This model has several advantages. It provides management with guidance as well as audit results (a government auditor cannot suggest improvements, while an outside consulting auditor often can). It creates a "user fee" system in which public companies pay directly (through audit fees) for the assistance and controls they utilize. It reserves government intervention for areas of real need, rather than routine investigation, allowing the SEC to spend more of its energy in setting policy than in enforcement. And, perhaps most importantly, it provides much more expert augmentation to the SEC than a government agency can generally afford to hire.

Over the past four years, the FDA has been slowly evolving the same pattern, particularly in the area of system validation. Investigators tend to defer to the opinions of known auditors, reviewing their reports rather than conducting redundant investigations. Client companies gain the advisory as well as auditing benefits of outside consultants, while getting a review of much greater depth and expertise than the FDA can afford. And the FDA can effectively regulate a much wider area of interest without increasing government fees; regulated companies pay for their own independent audits.

Two additional elements are emerging that will cement this trend. First, a not-for-profit association known as the International Validation Forum, Inc., has emerged, providing a consensus of validation standards and an method of accrediting validations (the "Certified Validation Professional" program). The Forum includes government, industry, and auditor representatives, and it works much like the CPA accrediting board.

Second, increasing pressures on the FDA have reinforced the "user fee" concept, both in the areas of new drug application reviews and, indirectly, in investigations. As FDA expands its responsibility in an era of government fiscal control, some form of indirect regulation is inevitable, and the auditor concept is receiving tacit endorsement.

AN EXPANSION OF THE USE AND EXTENT OF THE COSTART (ADVERSE REACTIONS REPORTING SYSTEM) WILL EXPAND VALIDATION ISSUES INTO HOSPITAL AND CLINICAL SYSTEMS.

As physicians, hospitals, clinics, and the general public encounter side effects or adverse reactions that may be a result of treatments, they notify the vendors of the medical devices, pharmaceuticals, or biologics responsible. Those vendors investigate and, where appropriate, notify the FDA (and other international agencies) through adverse reactions reporting systems such as COSTART.

COSTART defines the kinds of adverse reactions which are reportable and summarizes for comparison and trend analysis the kinds of reactions observed. The system is generally reliable, both technically and as a policy, and fulfills an important data collection and information dissemination need.

The limitation to COSTART is its entry point. Adverse reactions that are found to be isolated -- not directly related to a specific drug or product, or not categorized by a COSTART reaction -- are often lost from the database. The system, in short, is reactive and not proactive; it does not trace treatments to determine which combinations of therapies may prove adverse or, more positively, clinically successful.

The potential for COSTART to be used or modified in this kind of proactive manner is powerful and positive. Technically, only two major modifications would be necessary to add a significant tool to health-care research: connection to clinical input, and a regression model tracking results against treatments. Such a COSTART application, in mirror to the SMART system (a post-clinical reporting database) described above is under design. It will proliferate as regulatory agencies increasingly turn their attention to hospitals and clinics.

As with SMART, such an expanded COSTART system will require an expanded emphasis upon validation. All patient data systems interacting with a COSTART database will require validation to avoid the potential of data corruption and to maintain controls on system use and data security. With these validation controls, patient rights can be protected, databases can be isolated and controlled, and researchers will have a powerful new tool for comparing therapies.

In an era of increasing financial pressures forcing health-care reform, it will be difficult not to use post-clinical data as a valuative and planning tool. Under pressure to evaluate the effectiveness of varying therapies, COSTART -- or a parallel system -- coupled with validation controls on the input systems that interact with it, will prove a necessary and valuable contribution.

INTEREST IN, AND REQUIREMENTS FOR, VALIDATION WILL EXPAND IN THREE DIMENSIONS: LATERALLY, THROUGH OTHER U.S. AGENCIES; HORIZONTALLY, THROUGH OTHER NATIONAL REGULATORY BODIES; AND VERTICALLY, AS ADDITIONAL KINDS OF TESTS AND SYSTEMS EMERGE.

One of the enduring truisms about validation is its longevity. In marketing parlance, regulatory concerns about computer systems have more legs than a centipede.

FDA interest in the control of computer systems started in the pharmaceutical divisions, spread rapidly through biologics, and now dominates medical devices. Food products have come under similar scrutiny, particularly in the label/recall scenario made necessary by publicity concerning false reports of tampering with soft drinks and other products. Cosmetics, long regulated as a cross between "foods" and OTC drugs, have similarly received new focus of validation investigations. The efficacy of dentifrice additives, particularly, is based upon studies summarized with a LIMS and analyzed using a statistical software package. While veterinary products have never driven much regulation, recent concerns about the link between antibiotics in cattle feed and inadvertent food additives in beef have led to extension of validation issues, along with other regulatory concerns, in the veterinary arena.

That veterinary interest has led to validation requirements in the Department of Agriculture. Simultaneously, FDA computer validation guidelines for laboratories led to the Good Automated Laboratory Practices for EPA laboratories, and hence to laboratory requirements for validation in Department of Energy facilities.

In the next three years, look for validation requirements in the nuclear industry (where reliance on computer control gauges is already high), in aeronautics (particularly with publicity surrounding the overhaul of, and the discovery of inadequacies in, the air controller system), and throughout the defense industries, where NASA-style quality assurance through redundancy has proven to be ineffective.

Outside the United States, regulatory attention is turning toward validation requirements in China (where quality assurance is slowly shifting from a reliance on human checks to computer controls), and in Brazil (where new, automated CIM facilities are now on line or under construction). Over the next five years, expect a similar emergence of validation interest in India, South Africa, and Russia as their pharmaceutical industries mature and their export opportunities expand.

Vertical shifts are now leading to increased focus upon networks and communication systems, as well as on analog-to-digital conversion controllers. These patterns of use require more extensive data integrity testing and more elaborate design scenarios, including built-in software verifications.

INTERNAL SOFTWARE VERIFICATION TOOLS WILL MAKE CONTINUOUS VALIDATION TESTING A REALITY.

Validation testing has always suffered from the same flaw. Regardless of how frequently it is repeated, each test itself is fundamentally a snapshot of a system at a given moment, from which continuous functionality is projected. The possibility that the snapshot represents anomaly rather than typicality may be small, but it is always real.

Of course, computer systems do not spontaneously mutate. A snapshot that represents a current reality ought to approximate the functional future. But ranges of data may tap extremes that involve otherwise untested pathways, streams, or file boundaries. As new "artificially intelligent" programs and programming devices are utilized, that possibility grows in proportion. As users generate new software though inquiry and use, a program does change over time. As those changes accumulate, the snapshot decreases in value.

The alternative is some sort of continuously self-testing mechanism. In 1993, for example, Weinberg, Spelton & Sax, Inc. began recommending the inclusion of mode analyzers in bloodbank software. Simplified, these analyzers are code sections that compare software labeling results with known norms -- percentage of the population with Type O blood, for example -- to signal variations that may be caused by computer (or other) problems. Over the past few years, other internal checking systems, including Sign Tests, standard deviation analyzers, data comparison functions, and self-valuative algorithms, have been designed into software products to provide a "moving picture" rather than snapshot of functional activities.

While these continuous testing approaches will not replace validation, they will significantly strengthen the testing component of the validation process. As more software is designed specifically for the health-care industries, and as the use of intelligent or expert systems increases, the importance of these internal testers will grow significantly. By the end of the decade, all critical software ought to be designed with the inclusion of appropriate, internal-testing algorithms to provide an early warning of anomalies that might potentially signal software problems.

INCREASED REGULATORY SOPHISTICATION WILL SHIFT ATTENTION FROM PRIMARY FOCUS UPON DOCUMENTATION TO MORE DETAILED CONSIDERATIONS OF TESTING AND DESIGN.

The final hypothesis for consideration involves the regulators themselves. In recent years, the FDA has been plagued by staff freezes and salary erosions that have undermined efforts to increase education levels. Many of the best computer experts within the agency have been lured into the private sector. Other talented experts have been promoted out of direct field contact and into administrative and management positions.

The problem has been compounded by the relative age of the agency. As mandatory federal retirement frames have been extended, the average age of employees has increased. In an agency that experienced its greatest period of growth at the end of the Second World War, in the late 1940's, this aging effect is particularly dramatic. And while older employees are generally more productive than less experienced and focused younger employees, the FDA has found itself with a work force that largely received its formal education prior to the computer explosion, and that has never been recruited for, or oriented to, the high-technology, computer-related disciplines.

In recent years, three phenomena have synthesized to greatly increase the technical sophistication of the FDA. First, retirements have led to recruitment of employees whose skill sets and orientations generally include some basic level of computer expertise. This level has provided a base upon which in-house training programs have successfully built, and we have found a gratifying increase in the expertise of entry-level FDA personnel over the past five years.

Second, field experience has proven an effective educator, no doubt at some cost to inspected facilities. As a field investigator observes both weak and effective validation practices, the depth of understanding of what to look for and how to evaluate those practices increases dramatically. The situation in which a field investigator is involved in his or her first system validation review is now unusual, while it was a standard event even three years ago.

Finally, industry awareness of the vulnerability of computer systems and subsequent interest in validation has led to a long and valuable series of discussions, presentations, and seminars at national meetings of all kinds. The FDA has successfully performed its agenda-setting function. Validation has become important, and with that recognition have come a number of educational programs for investigators as well as industry personnel.

As a result, there has been a general increase in the level of refinement of understanding of validation's underlying principles. The need to examine source code algorithms has supplanted the inappropriate generality of analyzing all lines of code. Testing is now generally divided into different approaches that are appropriate for different situations. The concept of hazard analysis, tying depth of evidence to vulnerability, has gained widespread acceptance.

As a result of this sophistication process, focus has shifted to software design issues, including the concepts of conservative defaulting, redundancy, and ergonomics. These theories and concepts have provided developers with new strategies to increase controls and improve design configurations.

At the same time, new testing approaches, including the internal testing mechanisms described above, have improved the functional picture of even complex software. Testing methods of data transfer, file structure, and stress operations have finally honed confidence levels.

Sophistication has also shifted attention away from less technical concerns. While documentation is no less important, it represents a smaller portion of the regulatory pie as design and testing issues have increased in significance. While SOPs are still critical, less investigator energy is dissipated in mundane reviews of wording and style.

The result, of course, is positive for all concerned. Regulators are increasingly confident in their processes and procedures. Vendors are focusing on the critical issues of software performance. Users have greater confidence in, and evidence of, performance. The consumer public finds itself with better-controlled research and production processes without a significant increase in costs.

SUMMARY

While every prognosticator must fear unexpected events that may accelerate, trivialize, bypass, or negate some predictions, the hypotheses summarized above seem to posses a margin of safety. They represent extensions of existing trends, and they are firmly based in today's reality rather than tomorrow's dreams.

There is no doubt that validation issues continue to dominate regulatory concerns; validation is, ultimately, the operationalization of quality control. Over time, validation requirements will be clarified, tools will emerge, strategies will solidify, and tactics will strengthen. It is difficult to imagine a research laboratory or production facility that will not increasingly rely upon computer systems, and hence the place of computer validation in the pantheon of quality concerns is secure and expanding.

Change, no doubt, is a constant, and many unanticipated concerns and issues will emerge over the final years of the millennium to provide a rich context in which validation will be placed. But regardless of that context, and regardless of the relative counterpoints that may influence it, the growing importance of control of computers will be a major theme.

Appendix A

GLOSSARY

A to D (See Analog-to-Digital Converter)

ABNORMAL TERMINATION or ABEND Any unexpected interruption of processing. A key factor in the validation process is the ability to determine the cause of such events. The system should not simply freeze and require a restart. If a problem develops somewhere in the system, there should be a flag indicator and a way for the user to bypass or correct the problem, including tracing the fault. If the system simply freezes, then data in the process loop may be corrupted. Input not fully defined should not cause an abend. An abend is considered to be an Adverse Event (AE). Proper response to an abend includes diagnosis of the cause (with appropriate response) as well as correction of the problem and retesting prior to on-line production use.

ACCELERATED VALIDATION PROGRAM A program of Hewlett-Packard Co. in cooperation with Weinberg, Spelton & Sax, Inc., in which the vendor of HP LIMS makes available to purchasers a validation package and a Weinberg, Spelton & Sax, Inc. certificate. These tools provide an advanced starting point for users in developing the systems validation of their integrated laboratory. Users are then responsible for on-site validation.

ACCEPTANCE TEST Formal test conducted to determine whether a system satisfies its acceptance criteria and to enable the customer to determine whether to accept the system. (ANSI/IEEE Std. 729-1983)

ACCEPTANCE VALIDATION Testing, analysis, and review of a system incorporated into the formal initial approval process for live use of that system to prove control of that system. Acceptance Validation is the "present tense," compared to Developmental Validation (applying to future systems), and Retrospective Validation (of existing or past systems). A Validation Standard Operating Procedure (SOP) should require an Acceptance Validation of all new systems prior to or as part of the user acceptance process. Validation is a process of testing the testers and of making sure that the criterion set by the end users is appropriate. Given the current state of regulation, there is no easily defensible rationale for accepting a new or replacement system without including an Acceptance Validation in that process.

ACCESS, AUTHORIZED The parameters of system resource utilization given to end users. In many systems, various levels of access are used to restrict certain users.

These restrictions, and procedures for documenting the access granted, should be described in detail in an appropriate SOP.

ACCESS CONTROL Setting parameters for end user utilization of a system. Automated teller machines which require both an artifact (card) and a personal code are examples of clear and effective access control devices. For some systems, access is restricted physically by housing a limited system in a locked or controlled room. The key determining factor in access control is accountability; the unique identities of all users should be clearly determinable.

ACCESS PROTECTION The devices and/or procedures designed to control all levels of utilization of system resources. The most common access protection is a differential level password control system, often appropriately coupled with limitations on physical access and restrictions on remote (modem) access.

ACCESS, UNAUTHORIZED/UNINTENTIONAL Inappropriate use of a system or a portion of system resources. Depending upon the critical or hazardous nature of the system, that unauthorized access (hacking) may be of life-threatening significance. In a validated system, the users at each level should not be able to review their own work. For example, a system manager should not be able to overwrite or alter the numerical output of a process without appropriate authorization. Limited access to system locations, restrictions on modems, and general environmental security precautions are appropriate. Physical access to a system should be restricted to authorized individuals who are trained in proper operation. Security precautions should be designed that are appropriate to the vulnerability of the system to prevent unintentional system access. Unauthorized persons should be physically restricted from system access.

ACCURACY In order to be validated, the system as configured must provide appropriate and unerring processing of entered data and activated programs when operated within established standard procedures. Operating instructions should be prepared to reflect current and appropriate action sequences to assure safe and effective use of the system and its components.

ADC (See ANALOG-TO-DIGITAL CONVERTER)

ADDRESS A character or group of characters that identifies a register, a particular part of storage, or some other data source or destination.

ADVERSE DRUG REACTION Any undesirable effect, or interaction with human physiology, that is caused by a drug. This includes abuse, dependency, and interaction with drugs and/or foods.

ADVERSE EVENT (AE) 1) In the validation context, any unexpected interruption of processing anywhere in the system may be considered to be an AE. Systems are particularly vulnerable to environmental factors including power surges, temperature, humidity, dust, and radiation levels which fail to meet hardware manufacturers' requirements. To control AEs, a log should be maintained which indicates the date and time of the occurrence of each Adverse Event, the diagnosis, and the actions taken. Evidence that such a log is regularly reviewed and analyzed to identify trends or hidden problems is important to the validation process. 2) In the context of FDA guidelines, an adverse experience/adverse event is any undesirable experience occurring to a subject, whether or not it is considered to be related to the drug(s) being tested.

AFTERIMAGE Magnetic media may retain a ghost image, or afterimage, even after they have been erased. There are two solutions to this potential source of contaminated data in a high-hazard environment: either constantly use new media or constantly fill all fields with some null code. For example, consider a computer-controlled medical device which releases a chemical when three conditions are met and/or the chemical passes three tests. In this environment, start with x-filled fields rather than blanks or empty ones to prevent the computer from reading previous results.

AI (See ARTIFICIAL INTELLIGENCE)

ALGORITHM 1) A finite set of well-defined rules for the solution of a problem in a finite number of steps; for example, a complete specification of a sequence of arithmetic operations for evaluating sin x to a given precision. 2) An algorithm may be an equation, a recipe, an instruction manual, a part of a computer program, or an entire computer program. An algorithm can be written in a natural language like English, though relatively imprecisely. The source code review includes an analysis of the algorithms utilized in a program to assure conformity to the user's intended approach or method. For the purpose of validation, formulae and decision algorithms employed by the system must be accurate and appropriate. Users cannot assume that the test or decision criteria are correct; those formulae must be inspected and verified.

ALU (See ARITHMETIC LOGIC UNIT)

AMERICAN STANDARD CODE FOR INFORMATION INTERCHANGE (ASCII) A byte-oriented coding system based upon the use of seven-bit codes and used as a standard coding format for data communications.

ANALOG 1) Representation by means of continuously variable physical quantities; for example: a dial speedometer or tachometer. 2) Continuous signal having a voltage which corresponds to the monitored value.

ANALOG-TO-DIGITAL CONVERTER (ADC or A to D) Digital computers perform operations in binary (or equivalent) code. Information that is to be input into a digital system must be converted to binary form to be processed by the digital circuits. Outputs of a digital system are in some type of a binary code and often must be converted to a different form.

Many devices are used on either side of a digital computer in order to communicate with people and/or other computer systems. Input or output devices like keyboards, temperature probes, flow-meters, or pressure sensors provide the means for a digital computer to monitor and control a physical process. On the input side, measurements of process parameters that are analog in nature are usually transduced (changed to a proportional electronic voltage or current) and sent to an analog-to-digital converter, ADC, which then changes the analog quantity into a corresponding digital code. Some process parameters are already digital in nature, but must be changed to the appropriate form for a particular computer system.

Analog-to-digital converters (ADCs) and digital-to-analog converters (DACs) function as devices for communicating between a completely digital system or device, such as a computer, and people. This function is becoming increasingly more important as microcomputers move into areas of process control where computer control was previously not feasible. The conversion of a signal from A to D (or back) creates an area of high-error vulnerability. Careful testing of the converter and the conversion is recommended in all high-hazard environments. (See also Digital-to-Analog Converter.)

ANALYSIS PHASE 1) The stage in the validation process during which existing operations are observed and evaluated. Particular attention is paid to the identification of any differences between Standard Operating Procedures (SOPs), what actually happens, and what the system is supposed to do. 2) In the System Development Life Cycle, the process of determining the expected or intended performance parameters of the system.

APPLICATION (or application program) Software that does relatively complex tasks and that lets the user create and modify data. Some common types of applications are word processors, databases, graphics programs, spreadsheets, and statistical analysis packages. In high-hazard environments, critical applications should be validated.

APPLICATION INDUSTRY Regulated environment or environments in which the system under review will be utilized.

APPLICATION PLATFORM Refers to hardware necessary to support a system. An Application Platform may incorporate firmware. Platforms are generally designated by the name of a specific mainframe or minicomputer manufacturer.

APPLICATION SOFTWARE Programs specifically produced for the functional use of a computer system; for example: software for inventory, data analysis, temperature monitoring, general ledger, etc.

ARCHITECTURE Program architecture design involves identifying the software components, decoupling and decomposing them into processing modules and conceptual data structures, and specifying the interconnection among components. An architectural design specification may include data flow diagrams for the software; conceptual layout of data structures and data bases; name and functional description of each module; interface specifications for constraints; exception conditions; and names, dimensional units, and other attributes of data objects.

ARCHIVE Central repository for all information and records concerning a validated system. The archive should also include testing documentation, change control documentation, and validation records for all phases of SDLC. The archive is dynamic, since it contains current and previous copies of system design specifications and SOPs. The archive is both a record and a template. Regulatory requirements for the retention of an archive vary somewhat: in some circumstances, it may be necessary to retain data seven, ten, or even more years. Archives may be retained in any of a number of media, including paper, microfiche, and magnetic media.

ARCHIVE COPY A historical record of system components maintained for proof of regulatory compliance. Such a record may include (in addition to current version master copy) previous versions of software data-files, SOPs, etc., but does not need to include actual hardware. The validation concern is that the regulated site must be able to access information in a timely fashion. To facilitate such access in light of changes in hardware and software, a reliable retrieval method should be used; for example: historical systems may be saved on microfilm or dumped into ASCII files.

ARCHIVIST The original Good Laboratory Practice (GLP) standards and adapted Good Automated Laboratory Practice (GALP) standards require the archiving of a variety of records for periods of time ranging up to ten years. The archivist is responsible for the safe storage and retrieval of these records. [In the GALP Standards section

5.0, the code letter "A" is used to identify all specific and direct responsibilities of the Archivist.]

ARITHMETIC LOGIC UNIT (ALU) A computational subsystem of the central processing unit (CPU) that performs the mathematical and logical operation of a digital system.

ARTIFICIAL INTELLIGENCE (AI) Refers to programs which emulate human thought processes. Such programs have the potential to make decisions and take initiatives. AI is built into control programs and decision matrices. Relatively primitive AI programs are currently used in a few regulated areas, including molecular modeling in both pharmaceutical and biotech R&D efforts. (See Fifth-Generation Languages.) AI components of systems in high-hazard environments should be validated.

ASCII (See AMERICAN STANDARD CODE FOR INFORMATION INTERCHANGE)

ASSEMBLE To translate a program expressed in an assembly language into a machine language, and perhaps to link subroutines. Assembling is usually accomplished by substituting machine language operation codes for assembly language operation codes and by substituting absolute addresses, immediate addresses, relocatable addresses, or virtual addresses for symbolic addresses. (ANSI/IEEE)

ASSEMBLY LANGUAGE A computer-oriented language whose instructions are written in a one-to-one correspondence with computer instructions and that may provide facilities such as the use of macro instructions.

ASSISTANCE REQUESTS The developer, vendor, or supporter of the system has developed a mechanism for responding to user requests for additional training or assistance.

ASYNCHRONOUS COMMUNICATION The term used to describe the irregular, serial transmission of data in bursts or segments. During asynchronous transmission, at the point that the burst is sent, there is vulnerability due to the possibility that the transmission could be terminated or lost before it is confirmed by the receiver. While it is true that the error detection routine used determines the data integrity, and hence the reliability, of the transmission, it is also true that synchronous transmissions, using a bisync protocol, are theoretically more reliable. Since checking can be both continuous and simultaneous, they are preferable in high-hazard environments. (See Bisync.)

ATTEMPTED ACCESS Many systems will disconnect after three improper access attempts; some will log the improper attempts for later review. These features may be particularly valuable in systems of high criticality. A validated system should have a means of logging and documenting any unusual and/or unsuccessful resource utilization for regular review by the systems administrator.

AUDIT A review of the system validation plan and study results, conducted by a quality assurance group or an outside, independent agency or a regulator. The audit confirms conformity to validation standards and the appropriateness of those standards.

AUDITABILITY The successful inclusion within a system of automated or enforced manual mechanisms to allow tracking of the utilization and performance of that system. Common mechanisms include error trapping routines, console logs, audit trails, activity files, and similar records of functionality. A system may require manual auditing, or may create its own electronic audit trail. (See Audit Trail.)

Validation is a tool for identifying the existence of automatic or manual audit capabilities or the need to devise audit trails.

AUDIT TEST Independent check on the output and/or processing of a system, usually based on comparing controlled data input, with expected results, to the actual output. The audit test is a subset of functionality testing. (See Function Test.)

AUDIT TRAIL Sequential tracking of changes to data records or files that allows an oversight function to determine what the changes were, when they were made, and who made them. An audit trail is a key factor used to determine data integrity. The trail may also indicate unauthorized access or unauthorized modification of data or file contents.

AUDIT TRAIL REVIEW The automatic log of the audit trail is reviewed regularly by management personnel and annually by validators or regulators.

AUDIT TRAIL TEST The audit trail generated by the system has been tested in a "black box" (expected outcomes) test and conforms to the following guidelines:
a) Testing is conducted in a special test library or other environment independent of the actual system's real database; in other words, "off-line."
b) A minimum of 20 test changes have been included in the testing methodology.
c) A perfect conformity of the audit trail to the 20 test changes has been observed.
d) The test is repeated annually, whenever a new version or update is installed, whenever a new hardware component is installed, and immediately upon discovery of any audit trail discrepancy.

AUTHORIZATION A system control feature that requires specific approval before processing and/or increased access can take place. The authorization feature may create an audit trail, or indicate the electronic code of the authorizing agent.

AUTOMATIC AUDIT TRAIL The system shall have a mechanism in place for generating an audit trail of system changes or modifications.

AUTOMATIC CODE GENERATOR The ACG may be used to create source code, particularly in inventory and material requirements planning (MRP) systems. It is not necessary to validate the ACG if the actual source code produced by the process is subject to thorough review. (See Fourth- and Fifth-Generation Languages).

AUXILIARY STORAGE A device for holding data or programs, other than main memory/RAM (e.g., disks and tapes). All auxiliary storage devices have a specific shelf life, after which time data stored on them should not be considered reliable (see archive). The manufacturer-specified parameters for storage life assume location in an appropriately controlled environment. In any case, stored data should be transferred to fresh media well within those parameters.

AVAILABILITY (OF DATA) For regulatory purposes, clinical raw data must be readily available to regulators for review of system(s). Data, including the tests used to test the system, must be available to regulators. Availability of Data should be defined within a site's SOPs. Although no consistent time parameters have been applied to the "availability issue," in most regulatory situations the time frame of 48 hours has emerged as the consensus minimum limit on "ready access" or "availability."

BACKGROUND The electronic and radiation energy environment. In certain parts of the world, naturally occurring radon and other radioactive gases may be present in sufficient concentrations to distort sensitive system-data storage-media. In addition, artificial radiation from nearby high-voltage equipment and power lines and certain other apparatus may have the potential to cause data distortion. In such cases, the vulnerable parts of a system may have to be hardened to the radiation in order for the system to be validated.

BACK-OUT The procedures used to remove software input changes, program versions, and/or input data (files) if serious problems are encountered during program execution and/or system processing. The validation concern is that data integrity be maintained. The system user can demonstrate that the process is still under control if he/she can provide documentation of a Back-Out process that got the system back to a point before the processing problems began.

BACK-UP 1) Alternative systems, procedures, methods, and data files that can be used to restore service in the event of system malfunction or the loss of data resources. Frequently, large operations with many sites rely on back-up sites in the event that the main system goes down. These back-up sites need to be tested and validated as the system changes. There should be back-ups for data, software, critical operating supplies, and hardware spares. Both the location of, and the media chosen for, storage must be appropriate and compatible. The process of backing up should be habitually or automatically cycled every 15 to 30 minutes to avoid the loss of data in the event of power modulation. Only data entered prior to the last back-up can be considered to be pure after an adverse event. 2) Full: a copy of the current version of all files on disk. 3) Incremental: a copy of the current version of all files on disk that have been changed since the previous backup.

BAR CODE Series of solid bars of different widths and intermittent spaces used to encode data for reading by light-emitting scanning devices. As long as bar codes are uniquely and positively affixed to the sample or data point, the use of bar coding can eliminate or mitigate concerns of confusion of identity. Bar codes are also utilized to protect confidentiality (particularly in blood processing systems) and to provide double entry for sensitive data.

BASIC Beginner's All-Purpose Symbolic Instruction Code: A high-level programming language.

BASIC INPUT OUTPUT SYSTEM (BIOS) Firmware that is not modifiable by systems people. Different brands of otherwise compatible personal computers ("clones") may utilize differing BIOS, causing rare and generally harmless compatibility problems. These problems are generally related to transfer speed and to "overflow" conditions of large data sets. In particularly sensitive and hazardous environments, BIOS differences may prove significant and problematic.

BATCH 1) A group of data records processed together as a unit. 2) A specified quantity of a drug or other material that is intended to have uniform characteristics and quality within specified parameters and is produced according to a single manufacturing order during a particular manufacturing cycle.

BATCH INTERFACE A link for the transfer of data between systems in large file segments. In batch transfer, internal control codes may be used to assure that the batch sent and the batch received are identical. (See checksum.)

BATCH JOB 1) A group of programs that are run at controlled intervals. These may be controlled by an operating system and be run based on the availability of system resources. Also refers to programs run at regular intervals, traditionally overnight. Example: Updating data files. 2) Execution of programs serially with no interactive processing. Batch Job processing allows for easier auditing, because each batch can be checked before processing as well as after. In the event of an interrupted batch job, great care must be exercised upon restart to avoid loss or unintended duplication of data.

BAUD A unit for expressing the speed of transmitting data over distances, usually in terms of bits per second. One baud is one bit per second.

BENCHMARK Standard acceptable measure of performance of hardware and software. Independent reviews published in reputable journals and magazines may provide reliable "benchmark" information about a system or software.

BENCHMARKING A process whereby a company compares its strategies, operations, processes, and procedures to those of companies generally regarded as best, in the same class, and uses the results of those comparisons to improve its own performance.

BETA TESTING PHASE In this phase, in the Software Development Life Cycle (SDLC), the Beta Test Plan is executed by the user of the system. The testing is performed on an actual operational system, but not using the system as a live system (online). The results of the testing and any problems encountered with the system or the test plan are reported and documented.

A review of the output from Beta Testing is part of this phase. Documentation of the results of the Beta Testing must be analyzed, and every system or test plan inadequacy must be addressed. Also, suggested system enhancements need to be reviewed and considered as possible additions to the system for inclusion in a later revision.

BETA TEST PLAN DEVELOPMENT PHASE This phase in the Software Develop ment Life Cycle (SDLC) defines instructions for the user to follow to verify the system is functioning as expected. The Beta Test Plan is written from approved functional specifications and the system test plan.

A review of the Beta Test Plan, to ensure that the plan defines all the necessary steps required to test all system functionalities as perceived by the user, is part of this phase. If applicable, the plan must address system configurations. The plan must be written specifically so that a user, not a programmer or tester, can execute the test plan.

BINARY 1) Refers to the base-two number system. Permissible digits are 0 and 1. 2) A characteristic or property involving a choice or condition in which there are two possibilities.

BIOS (See Basic Input Output System)

BISYNC (Binary Synchronous Communication) Bisync protocol is a product of IBM Corporation. It is character-oriented, meaning that each character has a specified boundary. Since synchronous transmission is a continuous process, whereas asynchronous data transmission is discontinuous, the use of a Bisync protocol is preferable in high-hazard environments. (See Asynchronous Communication.)

BLACK BOX TEST A test of a system which compares input data with known and predictable outcome, with actual output produced by the system. In accounting, for example, manually checking the balances of an electronic spreadsheet, without considering the methods which the spreadsheet program used to arrive at those balances, is an example of black box testing. In the crudest sense, a system that has passed the black box test can be said to be functional. The black box test, however, does not carry any implied or explicit assumption that the system will continue to work reliably under varying real-world conditions.

BLOCK DIAGRAM A diagram that shows the operation, interrelationships, and interdependencies of components in a system. Boxes, or blocks (hence the name), represent the components; connecting lines between the blocks represent interfaces. There are two types of block diagrams: a functional block diagram, which shows a system's subsystems and lower-level products and their interrelationships, interfaces, and interdependencies with other systems; and a reliability block diagram, which is similar to the functional block diagram except that it is modified to emphasize those aspects which influence reliability.

BMD A statistical analysis package of programs (Biomedical Data) originally used in the medical field. BMD is similar to SAS in purpose and function, though the package has been criticized as overly "user friendly." The fact that it is difficult for a user to tell when statistical assumptions and accepted methodological conventions have been violated suggests that, in high-hazard environments, its use should be validated.

BOOTSTRAP A short computer program that is permanently resident or easily loaded into a computer, whose execution brings another, larger program, such as an operating system or its loader into memory. (ANSI/IEEE)

BRANCH An instruction which causes program execution to jump to a new point in the program sequence rather than execute the next instruction. In the event of a program modification, tracking functionality back to the last branch will allow determination of whether or not a change has affected other parts of the program.

BRANCH ADDRESS Refers to a specific location in a decision tree within a computer program. If a problem is discovered in part of the program, tracking up to the branch address will allow isolation of that problem and the program parts that may have been affected.

BRANCH COVERAGE Area affected by branch. If a problem is discovered in a part of the program, tracking down through the branch coverage will allow identification of aspects of functionality that may have been affected.

BUFFER An area in memory or auxiliary storage used to temporarily hold data for further processing. Buffers used in high-hazard environments should be secure.

BUG 1) An accidental condition that causes a functional unit to fail to perform its required function. 2) A manifestation of an error in software (ANSI/IEEE). In the validation process, the proper response to a discovered bug is a three-step sequence of: a) identification, b) correction, and c) retest.

BUS Electrical pathway by which information flows to different devices.

BYTE A binary character string operated upon as a unit and usually shorter than a computer word.

CAD (See Computer-Assisted Design)

CALIBRATION The purpose of calibration is to make sure that a device is measuring reality accurately. If a device is calibrated, then its measurements agree with some accepted standard or benchmark. The issue in the context of the validation process is that there must be documentation of appropriate calibration of devices that are part of the system being validated. During the validation process, measurements may be taken from the devices themselves.

CANDA Computer-Assisted New Drug Application.

CAPLA Computer-Assisted Product Licensing Application.

CCRO Clinical Contract Research Organization.

CE MARK The European Community (EC) mark of conformity, indicating that a product is considered to meet the essential requirements of EC Council Directives. Once the mark is given, a product can be marketed throughout the European Union.

CEN European Committee for Standardization. An organization based in Brussels that has established industrial standards for the European Community.

CENELEC European Committee for ElectroTechnical Standardization. An organization based in Brussels that has established electrical standards for the European Community.

CERTIFICATION In the context of system validation, certification means documentation affirming that the software development process was followed, that quality assurance procedures were adhered to, and that testing demonstrates that the requirements were met and the software and system requirements were fulfilled.

CHANGE CONTROL Change control exists when all programming changes and modifications show evidence of consistent oversight, including documentation of the following: effective date of change, the responsible programmer, and retesting certification. Evidence of change control may include an audit trail of patches. The ability to assign responsibility for modification of data or software or any element of a system is also essential.

CHANGE CONTROL COORDINATOR A person who analyzes each change request to ensure that no conflict exists with other requests, coordinates with DP personnel to develop a scheduled date for completion of each change task, monitors all change reports to the change committee, maintains the change files, and assures that the change libraries are maintained. This employee helps support the activities of the project administrator, project manager or phase/project leader, and quality assurance personnel.

CHANGE CONTROL LOG Usage of this form will document all changes made to software. Refer to SOP of software change control.

CHANGE IMPACT STATEMENT Use this document to compile information relative to the impact of a proposed software change. Refer to SOP of software change control.

CHANGE REQUEST Formalized method of notification to system support organization for any modification of system functionality. In a validated system the process will be defined within change control SOPs.

CHECKLIST 1) A series of steps which are to be followed in sequence and crossed off as they are completed. 2) A special tool developed by Weinberg Seplton & Sax, Inc. to provide comprehensive documentation. This report is based upon information provided by the professional and other personnel of the site being validated. It includes observations and field and laboratory tests conducted by Weinberg, Spelton & Sax, Inc., in cooperation with the site managers and the vendor of the system used at the site (if applicable). The standards upon which this report is based have been developed in cooperation with the Institute of Biomedical Science and Engineering of Drexel University; they conform to validation standards of the Institute of Pharmaceutical Sciences of the Center for Professional Advancement and industry standards of system validation as developed by the American Institute of Industrial Engineers and the International Institute of Industrial Engineers. All information, observations, tests, and conclusions are effective as of the date of the final report signature.

CHECKSUM A method for detecting errors in a series of bytes. The checksum is a one-byte number that can be computed fairly simply and appended to the series much as a parity bit is appended to a single byte. Errors are detected by computing a new checksum and comparing it to the checksum that came with the data. Checksums can detect some errors but there are many error combinations that leave the checksum unaltered. It is safest to assume that the checksum is useful for detecting single byte errors and not useful for error correction.

CIOMS (See Council for International Organizations of Medical Science)

CIS Clinical Information System.

CLARKE'S LAW Any sufficiently advanced technology is indistinguishable from magic.

"CLEAN COMPILE" A "clean compile" is a computer-generated output free of error messages or syntax errors. All final printouts should be "clean compile."

CLINICAL TRIAL Clinical trials in the context of FDA guidelines mean systematic studies in humans in order to discover or verify the effects of and/or adverse reactions to drugs (pharmacodynamics) and studies of the absorption, distribution, metabolism and/or excretion of drugs (pharmacokinetics).

CODING AND MODULE TEST PLAN DEVELOPMENT PHASE In this phase in the Software Development Life Cycle (SDLC), the source code is developed and the program/module test plan is defined. The test plan contains detailed instructions for testing the internal logic of each program/module. The test plan must ensure that each statement in the program/ module will be executed at least once, and that eachdecision in the program/module will achieve each possible outcome at least once (although not necessarily in every combination).

This phase includes a review of the source code and program/module test plan to verify their completeness and consistency with the applicable standard operating procedure for coding and testing software.

CODING PHASE The step in system design when computer-language instructions are written to meet the specifications required.

COMMUNICATION INTERFACE The methodology, software, and equipment chosen to allow users to communicate with a system and vice versa.

COMPILE To translate a higher-order language program into its relocatable or absolute machine code equivalent.

COMPILER A computer program that produces machine language from a source program that is usually written in a higher-level language by a computer user. The compiler is capable of replacing single source-program statements with a series of machine language instructions or with a subroutine.

COMPREHENSIVE SYSTEM PERFORMANCE TEST As an initial validation, Weinberg, Spelton & Sax, Inc. accelerates and enhances the ongoing system performance test by following a statistically valid sample of cases (up to 200) from entry into a processing center through labeling and shipping, checking for system accuracy, reliability, and consistency. This procedure has been used effectively in blood processing environments.

COMPUTER-ASSISTED DESIGN (CAD) When CAD programs are used to design or test designs of medical devices, the Computer-Assisted Design software should be tested and validated.

COMPUTER-CONTROLLED A computer-controlled device is one whose operation is influenced or directly controlled by a program-controlled mechanism that retrieves, decodes, and executes instruction to perform its designed tasks, or whose operation is controlled by the program itself.

CONDITION TEST Data processing term signifying checking for a yes/no, armed/disarmed condition.

CONFIDENTIALITY An agreement for the protection of proprietary and/or person al information.

CONFIGURATION CONTROL The process of evaluating, coordinating, approving, disapproving, and implementing changes to identified configuration items. (See configuration item list.)

CONFIGURATION ITEM LIST A list that identifies all the collections of hardware and software elements that are treated as units (configuration items) for the purpose of configuration control of a specific product or system.

CONGRUOUS ENVIRONMENTS The standards of validation apply to the specific user environment, independent of the system supplier, manufacturer, or vendor. These standards apply to each individual application site within a given user organization, regardless of whether or not multiple setting applications are operating under a uniform FDA license or corporate structure. Multiple systems operating in the same laboratory or congruous environment may be validated simultaneously, but a given inspection or investigation is expected to apply only to those systems specifically examined on the occasion of that inspection.

CONNECTIVITY Refers to the ability of various components in a system to communicate with each other. It may also refer to the ability of systems to talk with each other. The validation concern is the maintenance of accuracy, reliability, and integrity through each point of connection.

CONSERVATIVE DEFAULT A regulatory concept which says one should always err on the safe side. Always minimize the chance of a more serious problem. The FDA would rather not release a medication that will have side effects, even if it might cure someone. Consider a substance poisonous until you prove that it is not, rather than safe until someone gets sick from the substance. The concept of conservative default may be summarized by the phrase: when in doubt, delay.

CONTENT ERROR Refers to wrong data; for example a wrong phone number. The common kinds of content errors include: truncation, concatenation (incorrect combination), transposition (order change), and substitution. The way to control most content errors is to list the data fields separately and cross reference them.

CONTINUED ACCURACY AND RELIABILITY Refers to an essential characteristic of a validated system. If a system has continued accuracy and reliability, then the system as configured will continue to provide appropriate and unerring processing of entered data and activated programs when operated within established standard procedures and will monitor and provide warning of deviation from appropriate performance.

CONTRACT RESEARCH ORGANIZATION (CRO) Regulators within certain countries may allow specific tests and/or regulatory requirements to be carried out and/or verified by an organization other than the manufacturer who is seeking regulatory approval. A sponsor may transfer responsibility for any or all of its obligations to a contract research organization. Any such transfer must be in writing and must describe each of the obligations assumed by the CRO.

CONTROL Refers to a characteristic of systems, processes, and organizations that is a minimal requirement of regulatory agencies. Control requires the ability to ensure reliability, to calibrate appropriately, and to recognize abnormal events within the contexts of regulated systems, processes and/or organizations. The utilization of a system without understanding its function and/or without concern for its appropriate use reflects a lack of control. Users should retain conceptual checks upon results and should utilize systems as effective tools.

CONTROL OF ENVIRONMENT The physical environment of a validated system must conform to hardware, firmware, and equipment manufacturers' requirements concerning such things as temperature, power surges, dust, radiation, and humidity.

COSTART A system for reporting to the FDA on adverse reactions as they are documented by manufacturers and distributors.

COUNCIL FOR INTERNATIONAL ORGANIZATIONS OF MEDICAL SCIENCE (CIOMS) Provides a forum for regulators and manufacturers to meet and discuss the development of standard procedures. CIOMS is under the auspices of the World Health Organization (WHO) and UNESCO.

CPR Component Production Rules.

CPU Refers to the Central Processing Unit of a computer where the logic circuitry is located. The CPU controls the entire computer; it sends and receives data through input-output channels, retrieves data from memory, and conducts all program processes.

CRC (See Cyclic Redundancy Check)

CRO 1) Chief Regulatory Officer. 2) See Contract Research Organization.

CRT Cathode Ray Tube; an output device.

CRITICAL SOFTWARE Software whose failure could have an impact on safety, or could cause large financial or social loss.

CUMULATIVE DOCUMENTATION TRAIL Refers to the mass of evidence (written procedures, test data, electronic records) that support the certification of a system as validated.

CURRENT ACCURACY AND RELIABILITY A system has the necessary characteristic of current accuracy and reliability if, at the time of review, the system as configured provides appropriate and unerring processing of entered data and activated programs when operated within established standard procedures.

CURRENT GOOD MANUFACTURING PRACTICES (CGMP) These are the most recent modifications of Good Manufacturing Practices or a plan for total quality control in the manufacture of pharmaceuticals. (See Good Manufacturing Practice.)

CYCLIC REDUNDANCY CHECK (CRC) A method for detecting errors in a series of bytes. Two extra bytes are computed and appended to the series. The computation is complex, and there are fewer undetectable combinations of errors than for checksum or parity. The Cyclic Redundancy Check technique does not burden the transmitter with the continuous sending of parity bits, and thus offers more efficient data transmission, combined with a powerful error-detection technique.

DAC (See DIGITAL-TO-ANALOG CONVERTER)

DATA ACCESS Refers to the ability to retrieve, view, analyze and/or modify data. In a validated system, access must be controlled (i.e., monitored, reviewed, and approved) to assure accuracy, integrity and the ability to audit. (See Data Integrity.)

DATA ACCESS DIAGRAM Refers to a graphic representation of the organization of and access paths between data files.

DATA ANALYSIS Periodic testing review.

DATABASE 1) A set of data, part or the whole of another set of data, and consist ing of at least one file that is sufficient for a given purpose or for a given data processing system. 2) A collection of data fundamental to a system. (ANSI.)

DATABASE DESIGN PHASE This phase in the Software Development Life Cycle (SDLC) defines the structure of the database and the fields. The database design includes a graphical representation of its structure in conjunction with documentation describing each field within the database. This design must be consistent with the global system design defined in the system architecture phase and the system features defined in the functional specifications phase.

A review of the database design and fields to ensure that they meet the database requirements defined above is included in this phase.

DATA CHANGE LOG Whenever data is changed, an entry must be made into a log indicating the data change, reason for the change, and the various approvals for the change. Refer to SOP of system data changes.

DATA DICTIONARY A definition list of all fields, elements, and individual component elements. It must include a description of what they mean, their purpose, the range of appropriate values, and each element's location.

DATA FIDELITY The data is accurate; that is, there is true correlation between the data and reality. Data fidelity should be checked independently of data integrity.

DATA-FLOW DIAGRAM Refers to a graphic tool which represents a system with emphasis on data movement, processing functions (transformations), and the files that are used to support processing.

DATA INTEGRITY The ability to view and/or utilize data should be controlled, i.e. monitored, reviewed and approved. The system should receive, store, transmit, operate upon, process, recall, sort, and otherwise manipulate user-provided or generated data without system-imposed inappropriate modification, distortion, addition, or deletion.

DDR Donor Deferral Registry.

DEBUG The process of locating, analyzing, and correcting suspected faults (bugs). (ANSI/IEEE.) In validated systems, debugging activities should be fully documented.

DESIGN A system design should be elegant -- the carefully executed result of a planned approach, rather than a patchwork reflecting trial-and-error-without-method approach. Design control implies: organization; a selection of components to reflect function, efficiency, and reliability; and a construction to minimize emergent problems.

DESIGN ANALYSIS The evaluation of a design to determine correctness with respect to stated requirements, conformance to design standards, and other criteria.

DESIGN ANALYSIS REPORT A written report of the proceedings and recommendations of a design analysis.

DESIGN GOALS PHASE This phase in the Software Development Life Cycle (SDLC) defines the overall design objectives of the product. This phase primarily addresses the needs of the user and the system. The design goals document generated should include: the description of the system environment, overall system functions, identification of any other system(s) that will interface, input/output data requirements, any known constraints of the system, and a description of the target user.

DESIGN PHASE The period of time in the Software Development Life Cycle (SDLC) during which designs for architecture, software components, interfaces, and data are created, documented, and verified to satisfy requirements (See ANSI/IEEE Std 729-1983)

DESIGN REQUIREMENTS SPECIFICATION A document that prescribes completely and precisely the design of a system, a system component, or software. Typical contents may include, but are not limited to: algorithms, control logic, data structures, input/output formats, interface descriptions, physical requirements, structural requirements, and functional requirements. {Applicable algorithms may or may not be part of software design requirements. Whether they are included depends on the particular data structures involved. Numeric structures may require specific equations that closely specify algorithms, but lists and strings may occupy different data structures depending on the performance required. Search and match within X seconds, is a requirement; element-by-element search of a matrix table would specify a general algorithm and search a matrix table, starting with element 0,0 followed by 0,1 followed by ...0,n followed by...m,n would be a very tight algorithm specification for a search.}

DESIGN REVIEW Refers to a formal evaluation of a pre-coding plan, measuring the proposed steps against the system design specification and/or the software requirement specification.

DESIGN REVIEW REPORT A written report of the proceedings and recommenda tions of a design review.

DETAILED DESIGN SPECIFICATIONS PHASE This phase in the Software Development Life Cycle (SDLC) defines the overall design, behavior, and data flow of a program or module. The Detailed Design Specifications include the name, functional description, and variables used within the program/module/data file. For each applicable program/module, the interactive relationship(s) with other programs/modules, data input/output, database usage, and file access are defined. For each applicable data file, the format and program/module access is defined.

This phase also includes a review of the Detailed Design Specifications to verify their completeness and consistency with the applicable standard operating procedure for software development documentation.

DEVELOPMENT PHASE This phase in the Software Development Life Cycle (SDLC) develops a system test plan which defines all the necessary steps required to test the system's functionality as defined in the functional specifications. The system test plan must include setting up the system's environment, must provide valid and invalid input data, and must specify the appropriate output. The system test plan must be specific, so that a tester, not a programmer, can execute the test plan.

DEVELOPMENTAL TOOLS Usually refers to 4GL based programming technology; e.g., screen generators.

DEVELOPMENTAL VALIDATION Refers to a system review which is integrated into the process of customizing, configuring, adapting, or inventing a unique system or system component. In the case of new systems being designed or under construction, appropriate features and functions can be "built in" so as to appropriately satisfy defined acceptance criteria.

DEVICE HISTORY RECORD This is the production record which documents that the DMR specs were properly followed in the manufacturing process. It is the key document when a recall is ordered for tracing problems.

DEVICE MASTER RECORD This is the total design history, including the current and prior versions, required for approval of a computerized medical device.

DHR (See Device History Record.)

DIAGNOSTIC 1) Hardware-related self-check equipment or routines. 2) A message generated by a computer program pertaining to the detection and the isolation of errors, faults or failures.

DIGITAL In a digital system or device, quantities are represented by discrete whole numbers, nearly always base 2 (binary), as opposed to continuously variable numbers. To oversimplify, digital computers generate "stepped" data curves, while analog computers generate smooth data curves.

DIGITAL-TO-ANALOG CONVERTER (DAC) Process-related output devices translate the computer's outputs to appropriate actuating signals needed to control the process. These actuating signals might be simply opening and closing of switch contacts, or pulses to a stepping motor. Many times, however, the required actuating signal has to be analog in nature, such as a voltage for controlling the speed of a dc motor. In these cases, a digital-to- analog converter (DAC) is needed to convert the digital system output to the required analog form. (See also Analog-to-Digital Converter.)

DIRECT CONNECT MODEM Communication device allowing an immediate computer-to-computer communication. These raise control concerns with sensitive systems. Ideally, modem access should be through a callback system.

DIRECT DATA ENTRY Refers to the on-line submission of forms, clinical data, etc. into the system by field personnel without the use of a paper form submitted to a data entry function. While economical in time, direct data entry raises control concerns by removing one of the traditional documentation steps from the system.

DISASTER An incident that causes loss of data processing services. Possible disaster causes include fire, explosion, lightning, power disruption/surge, flood, sabotage, and wind damage.

DISASTER RECOVERY PLAN The goal of a Disaster Recovery Plan is to identify probable disasters such as flood, fire, civil unrest, etc., and to detail corrective action. Since the purpose of corrective action is to eliminate confusion and recover, key people and their back-ups must know what the plan is and where to find it.

DISK A circular rotating electromagnetic storage device. Disks come in different sizes and can be hard or flexible. Because of likely degeneration of the substrate, data and applications stored on disk should be replicated every five years.

DISK DRIVE A device used to read from or write to a disk or diskette. Since disk drives are mechanical and therefore relatively vulnerable, disaster recovery SOPs should account for crashes.

DISKETTE A flexible or semi-flexible plastic disk storage medium, also called a floppy disk. Data and applications stored on diskettes should be replicated every three years. (See Disk.)

DISK OPERATING SYSTEM (DOS) The system that controls the input and output to and from a disk.

DMR (See Device Master Record.)

DOCUMENTATION UPDATE Any and all written materials that explain the use, function, and/or evolution of a system. These include, but are not limited to: design specifications, authorizations, procedures, training materials, work aids, and various historical logs, disks, or systems.

DOWNLOAD Passing information (usually data or documents) from a host system to another computer, usually from a mainframe or mini to a workstation or PC.

DRIVER PROGRAM Triggers or enables other programs in a system. Specialized software designed to allow the CPU to deal with peripherals, such as printers.

DUAL MATRIX PROCESS OF VALIDATION The process of validation requires the identification of potential problem areas and the design of two methods of checking each area. For example, to check audit-trail accuracy, the validation process would include both a functionality test and a source code review. The Dual Matrix Methodology was developed to assure maximal practical quality assurance to control, for example, the release of AIDS-contaminated (or other impure) blood products for human transfusion. The methodology is based upon theoretical principles incorporated in the standards developed for the DEP auditing field (Halper, et al., 1985), now considered to be "commonly acceptable accounting practices," and upon hardware and software testing approaches evolving from the system design and development industries. Four key principles form the foundation of the approach: 1) Potential Vulnerability Analysis; 2) Test Design Redundancy; 3) White, Grey, and Black-Box Test Matrix Mix; and 4) Independence of Validation Procedure.

The Dual Matrix Methodology begins with an environment-specific, configuration-specific, version-specific, and implementation-specific analysis of the areas of potential vulnerability for the system to be tested. Following established auditing practices (Baird, 1974), researchers began with a comprehensive system analysis of the test site, including the development of a data-flow analysis, hardware inventory analysis, and structured needs analysis (Semprevivo, 1982). This analysis includes an examination of the computer system hardware (all components), software (operating and application), personnel (training, background, and standard operating procedures), and user requirements (inquiries, reports, and outputs). Inputs to the analysis include direct inventory of components and software; observed processes; interviews with users, management, DP management, DP personnel, and supervisory personnel; examination of operating logs; examination of output reports; and a review of supporting documents, including error logs, repair records, regulatory reports, and internal analyses. The analysis of the data collected focuses upon potential vulnerability of the data entered, stored, and manipulated by the system; of the programmed decision matrix acting upon that data; and upon the system-generated disposition decision constituting the significant output of the system. The areas of potential vulnerability are further analyzed through a review of the actual problems encountered during the previous 24 months of operation and are expanded as new vulnerabilities emerge as secondary results of the tests conducted (below).

For each of the identified areas of potential vulnerability, a total of two distinct tests are designed and implemented, representing two of the three general testing categories. These tests are situationally modified adaptations of the standardized tests designed to assure system accuracy, including source-code analysis tests (Stark, 1978), throughput analysis (Willingham and Carmichael, 1975), stress analysis (Goren, 1989), error-log analysis (Figarole, 1985), and continuous testing (Garwood, 1984).

The redundancy concept described and implemented provides more than just the quality-assurance levels discussed above and required as a result of the sensitivity of data in this environment (Bronstein, 1986). The use of redundant tests also adds a new level of managerial control, providing a divergent approach that, in effect, can recheck the work of in-house teams and provide a control on the efforts of outside validators. Particularly in light of the personal and organizational liability in a compliance-regulated industry, this level of oversight provides a significant "added value."

As generally used in the industry, "White-Box" tests refer to specific line-by-line, or signal-by-signal, traces through a system; "Black-Box" tests examine the input and output of a system, or expected-versus-actual output, without examining the actual system itself; and the compromise "Grey Box" tests blend the two, building internal "flags" or "checkpoints" into a system that provide intermediate Black Box points. Together, the three-pronged approach constitutes the industry norm level of testing recommended for system acceptance (Swanson, 1988).

In the environment described herein, the FDA is concerned that management tends to rely primarily upon the "black box" result. All too often, managerial focus begins and ends with the question of whether products labeled as safe do indeed meet established and appropriate safety standards. But the consistency of that confidence can only be based upon the internal expert procedures and standards used by the system to make that labeling determination. A mixture of White, Black, and Grey-Box tests is appropriate to assure that consistency of confidence.

In accordance with the redundancy value described above, the Dual Matrix methodology's effectiveness is maximized when the duality provides for the use of two tests representing two of the three testing categories, to check for the designed corrections and protections against the identified vulnerabilities. For each identified area of potential weakness, two tests are implemented: either a White and Black-Box test, a White and Grey, or a Grey and Black-Box test.

Standard managerial data-processing controls call for an acceptance test of all new systems, and a constant "debugging" check on all developed and functioning software. These tests are generally built into the responsibilities of the internal personnel responsible for the implementation of systems and the development and maintenance of software.

The testing procedures described herein, of necessity, are designed, implemented, and interpreted by personnel outside of the normal process, in much the same way that independent auditors annually certify a company's financial records. This independence protects against the carelessness of assumptive acceptance; the shortcutting encouraged by performance pressures; the perceptual difficulties of self-proofing; as well as the purposeful problems of sloth, fraud or misplaced loyalty (Gilhooley, 1985).

The Dual Matrix Methodology calls for the use of a validation team of individuals not permanently employed by the organization to be validated; reporting directly to executive or senior managerial-level personnel; prepared to testify under oath concerning procedures and findings; and operating on a pre-approved, unconditional, fixed-fee, grant or contract. These safeguards assure complete independence of validation staff and findings.

EBCDIC (Extended Binary Coded Decimal Interchange Code) Eight-bit code used to represent data in most computers.

ECHO Display of characters on a terminal output device as each character is keyed.

ECHO CHECK Refers to a means of verifying the accuracy of a data transfer op eration by transmitting received data back to the source and comparing it with the original data.

EFTA European Free Trade Association. As part of its program, the EFTA runs the Pharmaceutical Inspection Convention (PIC), an agreement among member countries for mutual recognition of national Good Manufacturing Practice inspections.

ELECTRONIC DATA INTERCHANGE (EDI) Direct, paperless transmission of requirements from system to system. For example, an MRP system places a JIT order with a vendor system, which responds with an electronic bill of lading and invoice, receiving an electronic funds transfer authorized by the customer's accounts payable system. These paperless systems raise control issues that need to be addressed with more rigorous auditing procedures.

ELECTRONIC IDENTIFICATION Indicates the identity of the individual who created the document. When electronic identification is used, a system thinks it knows who is utilizing its resources.

ELECTRONIC SIGNATURE Refers to a method of imprinting an I.D. code to a specific document within the system, which provides evidence that an individual has reviewed and approved a document that has not been subsequently modified.

EMBEDDED A term that usually refers to software in a read-only memory chip (ROM); that is, computer memory that cannot be changed by the computer. Embedded data usually means data that appears as part of a program.

EMERGENT CONTROLS Acceptable documentation and training for a validated system includes the establishment of procedures to deal with emerging situations, including updates and version changes, hardware recalls, user-discovered problems, and assistance requests.

ENCRYPTION A process by which information once entered into a system is then unrecoverable to all system users without the machine-language encryption key. For all practical purposes, only the computer can recognize and translate encrypted data. Encryption can be used to protect confidentiality; for example, the identification of potential blood donors whose samples test positive for HIV or some other sexually-transmitted disease. Encryption provides the highest level of security.

END USER A person who has the direct use of an automated system to perform their work function (client is a synonym). In a validated system, documentation of end-user training and adherence to SOPs are critical.

ENVIRONMENTAL CONDITIONS The range of temperature, power surges, dust, humidity, and radiation which can have an impact on the reliability of a system. The validation concern is the maintenance of control of such conditions.

ERASABLE PROGRAMMABLE READ-ONLY MEMORY (EPROM) A firm ware chip that can be programmed a single time, then is burned in as firmware. The switch pattern in circuit can be erased and reprogrammed using an EPROM programmer.

ERRONEOUS DATA TEST The "Grey-Box," flagged operations test designed to confirm the system's ability to appropriately reject bad data. The test uses a special test file to avoid corruption of the live data base in the event that inaccurate input is erroneously accepted. Erroneous data testing involves the input of data violating protocols of field size, composition, range, complexity, and coherence, as well as testing of major logical flows and decisions within a representative sub-program. A comparison of the actual results to expected results, discrepancy reports, and a summation of critical and non-critical issues identified during the test are all provided.

ERROR An error is a reported malfunction generated automatically by hardware or system software.

ERROR DETECTION Error detection schemes are described in computer science literature. Among them are parity, checksum, cyclic redundancy, and longitudinal redundancy. A validated system usually employs several of these schemes throughout the system.

In a validated system, the source code supporter shows evidence of appropriate system maintenance procedures, including error detection and correction, such as: internal syntax error detection tools; internal logic error detection tools; error reporting system; non-critical error correction procedure through revisions and updates; critical error correction procedure through immediate user notification; and on-line, hands on, or telephone user assistance procedure, with generated error log.

ERROR LOGS Official documentation of problems (malfunctions, deficiencies, failures) encountered in the operation of a system.

ERROR RECOVERY ROUTINES System software programs for handling problems and resuming work flow. In validated systems, the concern is that the error recovery software includes the ability to trace the problem and that the problem does not threaten data integrity.

ETSI European Telecommunication Standards Institute.

EVIDENCE OF VALIDATION The collection of documentation, test results, observations, expert opinion, or other supporting materials demonstrating that an independent review has determined conformity to regulatory guidelines and/or a convincing demonstration of a system's current accuracy and reliability, expected continued accuracy and reliability, auditability, managerial control, and maintenance of data integrity.

EXECUTABLE CODE Successfully compiled instructions that the CPU can follow to perform a task ("execute a program").

EXPERT SYSTEM An Expert System generates code every time it operates; therefore, it is impossible to review source code prior to use. Application source code does not exist prior to use. The solution in the validation context is to validate the expert system rather than the end code of the expert system and to make sure that the end code is regularly audited within SOPs. Example 1: a system that sorts and delivers ingredients. The manager requires different sorts at different times. Therefore, the sorting system must be validated rather than individual user sorts. There must be an audit trail that is regularly reviewed by management. The validator must have evidence of that. Example 2: Statisticians using SAS are constantly writing code which generates different statistical tests every time they use the system. Therefore, the validator must determine that SAS does what it is supposed to do, and that the statistician is following appropriate protocols. Most of the real use of Expert Systems is in the development of systems. Applications have fixed code. Examples of Expert Systems include SAS, SPSS, and BMD. (See also Artificial Intelligence.)

FAILSOFT Refers to a programming concept or practice that allows a system to continue to operate even when there has been a failure, error, or problem. From a validation point of view, the existence of failsoft path within a system is extremely negative. In any high-hazard system, one always wants some notification of a failure, problem, or error. For example, cardiac pacemakers have double batteries so that if one fails, the pacemaker continues to operate. However, if the user has no indication that one battery is gone, then there is no backup (a failsoft condition). When the second one dies, so does the user.

FAILURE MODE ANALYSIS Refers to a procedure that determines which mal function symptoms appear immediately before and/or after a failure of a critical parameter in a system. After all the possible causes are listed for each symptom, the product is designed in such a fashion that the problems are eliminated.

FAILURE MODE EFFECTS ANALYSIS A procedure in which each potential failure mode in every sub-item of an item is analyzed to determine its effect on other sub-items and on the required function of the item.

FAILURE MODE EFFECTS AND CRITICALITY ANALYSIS (FMECA) A procedure that is performed after a failure mode effects analysis to classify each potential failure according to its severity and probability of occurrence. The determination of how a software-failure analysis was conducted. Were failures derived by tracing "backward" from safety system faults to software (and components) where the fault could originate (top down), or by tracing fault propagation from a postulated failure origin to all possible outcomes (including system faults), classifying the criticality of each (bottom up)?

FAST TRACK VALIDATION This is a program developed by Pharmacia Bioprocess in cooperation with Weinberg, Spelton & Sax, Inc., and other consultants. Under a Fast Track system, clients are provided with assistance in any and all phases of validating biotech processes, including controlling and analyzing computer systems.

FDA-483 An indication of an adverse finding by the Food and Drug Administration. A 483 is often used in conjunction with a regulatory letter which contains a detailed description of the finding and the action proposed.

FEATHERSTONE'S STEPS IN SYSTEM DEVELOPMENT

1) wild enthusiasm
2) disillusionment
3) total confusion
4) search for the guilty
5) punishment of the innocent

6) promotion of non-participants

FIBER DISTRIBUTED-DATA INTERFACE (FDDI) Refers to a standard for 100 Mbps token ring LAN, based on a fiber-optic medium. The validation concern is that data be transferred without being corrupted.

5GL (See Fifth-Generation Language)

FIFTH-GENERATION LANGUAGES (5GLs) A fifth-generation language allows the end user to interact with a computer system in conversational English. It may incorporate artificial intelligence and may seem, to the user, to be forgiving of mistakes. A system using a 5GL expects anomalies and corrects according to set parameters by using interactive querying. (See 4GLs and Generation of Languages.) Because of the difficulty in reviewing source code in a situation where users, in effect, create new code every time they utilize the system, many regulated environments have opted for a viable alternative: they concentrate upon the language mechanism for creating the appropriate code from user commands. The review of that "internal language code", coupled with careful documentation of the user commands issued and appropriate review of the accuracy of those commands, has generally met regulatory acceptability guidelines.

FILE ERROR A file error is a combined set of fields that make up a record. A file is reading the wrong entry for a particular record. For example, reading a telephone book and dialing the number above or below the correct name. The correct data is there; you simply assigned it wrongly.

FINAL COMPILATION All final printout reports or data-sets designed for archive, decision point, regulatory report, or review meet criteria including "clean compile," vendor and version identification, and completeness.

FINAL INSPECTION CERTIFICATE After a system has been completely validated, a certificate is issued which says so.

FINGERPRINT A time/date stamp which proves that a document has not been changed since the time of the stamp. In some jurisdictions, a Fingerprint is acceptable evidence that a document has not been modified. It may be accepted as an electronic signature.

FIRMWARE 1) Machine instructions burned into a chip. The instructions are then part of a machine and is not user modifiable (e.g., BIOS chips). 2) Computer programs and data loaded in a class of memory that cannot be dynamically modified by the computer during processing. (ANSI/IEEE) 3) Hardware that contains a computer program and data that cannot be changed in its user environment. (ANSI/IEEE)

FIRST LAW OF HARDWARE Plug it in.

FIRST VALIDATION The initial attempt to validate a system. This usually requires greater resources than subsequent attempts. It is often much more efficient to use external expertise for this first attempt.

510(k) A section of the Food, Drug and Cosmetic Act requiring medical device manufacturers to submit a premarket notification to the FDA to allow the FDA to determine whether the device they intend to market is equivalent to a device already legally marketed. The 510(k) submission is a thorough description of the medical device, including computer hardware and software components. The 510(k) indicates whether the device is new, unique, bypasses human intervention, is potentially hazardous, and/or meets certain other technical measures that classify the (amount of) supporting materials that may be required. It is recommended that evidence of validation be included in the 510(k) for all computer- controlled or computer-produced medical devices and for all hazardous medical devices. Since some system and application software is now classified as medical devices, many vendors are appropriately filing 510(k) submissions for their software. Obtaining a copy of this 510(k) filing (either through the Freedom of Information Act or from the vendor) may help in the process of selecting a system and may provide important data to be included in the validation package for appropriate systems.

FLAG An indicator that signals the occurrence of an error, state, or other specified condition. (ANSI/IEEE)

FLOWCHART A graphical representation of the definition, analysis, or solution of a problem, in which symbols are used to represent operations, data flow, and equipment. (ISO)

FMECA (see Failure Mode Effects and Criticality Analysis)

FOOTPRINT Refers to the physical space taken up by a piece of equipment. The term is sometimes used to indicate a trail or signs followed in an audit trail.

FORMWARE Refers to firmware with data entry gaps. BIOS-controlled password system in which firmware requires correct user response before advancing.

4GL (See Fourth-Generation Language)

FOURTH-GENERATION LANGUAGE (4GL) Refers to a code generator which produces an application language code such as COBOL. The application code is then compiled to executable or object code for the computer. The purpose of fourth-and fifth-generation languages is to enable an end user who is not a programmer to utilize the resources of a system. A 4GL like NATURAL, FOCUS, or SAS tends to trade efficient use of system resources for "user friendliness."

FRACTIONATION The process of separating blood or plasma into its component parts.

FUNCTION 1) General application or purpose of the system. 2) A process that is performed on a number or character string. 3) A precoded routine.

FUNCTIONAL DESIGN The specification of the working relations between the parts of a system in terms of their characteristic actions.

FUNCTIONAL ENHANCEMENTS Upgrades or improvements in software to provide new activities or operations. The validation concern revolves around how any new or improved functions interact with other parts of the system. Changes made to a validated system require a thorough testing of the changed module(s), as well as of the modules that interface with it (them).

FUNCTIONALITY Refers to the ability of a system to consistently provide the quality results required in a specific environment.

FUNCTIONAL SPECIFICATION PHASE This phase in the Software Development Life Cycle (SDLC) defines the system features necessary to fulfill the needs of users. Specifications must be consistent with the design goals and must define the system function, the system usage and operations, the input domain, and procedures for generating output. These specifications must also serve to generate the user's manual, continue software definition and development, and be reviewable and understandable by the project and user personnel. A review of the functional specifications to ensure they are consistent with the design goals is a part of this phase.

FUNCTION TEST The primary system "Black-Box" functionality test, which is designed to determine the degree of accuracy with which the system correlates to the reality it is intended to describe. The function test compares known outputs with obtained outputs from a set of selected inputs. Function tests are usually designed to verify system response times, system throughput, memory utilization, system execution, and traffic rates on data channels and communication links. Functional tests execute typical operating conditions of the software and system with typical input values and results. These tests should test boundary conditions, system defaults, ordering, and special values.

FUZZ Refers to the range of acceptability in specifications. The tighter the FUZZ the more rigorous the requirements for the validation process must be. The range of acceptability in specifications would be greater in a system that controlled the manufacture of bandages as compared to one which controlled the manufacture of cardiac pacemakers. In the hazard analysis procedure, establishing the probability of a hazard is a way of determining FUZZ.

GALEF'S DEDUCTION Fallible people design fallible computers.

GALP (see Good Automated Laboratory Practice)

GCP (see Good Clinical Practice)

GENERATION OF LANGUAGES The first-generation language that activates a computer system consists of positive and negative electrical charges. The second-generation language contains the zeroes and ones of Assembly. Third-generation (3GLs) are higher-level, more user-friendly languages like BASIC, COBOL, and FORTRAN. Each third-generation language has a code with its own unique syntax and logic. The computer takes the source code from a programmer in a 3GL and generates object code to the machine. Fourth-Generation Languages (4GLs) are more like the conversational native language of the user. A word in a 4GL represents a string of 3GL code which is then compiled to the object code that the machine uses. Fifth-Generation Languages (5GLs) are interactive with or query the user in a conversational manner. There is usually some level of artificial intelligence (AI) active with 5GLs.

GLP (see Good Laboratory Practice)

GMP (see Good Manufacturing Practice)

GOOD AUTOMATED LABORATORY PRACTICE (GALP) Since laboratories are relying increasingly upon computer systems to collect, analyze, summarize, and report data, a wide range of situations has arisen for which the GLPs provide little or ambiguous guidance. To avoid the confusion and potential problems that such uncertain situations can create, and to deal with the newly arising problems of possible data corruption, loss, or inappropriate modification that computerized systems may inadvertently generate, the Environmental Protection Agency has developed a series of supplemental guidelines: the Good Automated Laboratory Practice (GALP). These supplemental guidelines are intended to provide clarification of expectations of performance and control for laboratories electing to utilize computer systems for data collection, storage, manipulation, or reporting. The GALPs, representing an addition to, but not a replacement of, GLPs, have been distributed to the industry and represent current EPA compliance policy.

The ultimate issue in the extension of the GLP concepts to an automated laboratory through the implementation of GALPs is that of control. Effective management and operations of an automated laboratory cannot be assured unless use and design of that system are consistent with standards intended to assure system control.

The GALP standards are based upon a foundation of six concepts, detailed below. Together, these concepts define the necessary control issue that underlies the GALPs. Recognizing that wide variations in computer system designs, technologies, laboratory purposes, and applications are likely to create situations in which appropriate and successful alternate control strategies may evolve that were not anticipated in the GALP standards, these six principles are provided as guidelines for evaluating proposed equivalencies. They are intended to serve as guideposts in the interpretation of the standards and as evaluation criteria in the consideration of proposed equivalencies.

1. DATA: The system must provide a method of assuring the integrity of all entered data. Communication, transfer, manipulation, and the storage/recall process all offer potentiality for data corruption. The demonstration of control necessitates the collection of evidence to prove that the system provides reasonable protections against data corruption.

2. FORMULAE: The formulae and decision algorithms employed by the system must be accurate and appropriate. Users cannot assume that the test or decision criteria are correct; those formulae must be inspected and verified.

3. AUDIT: An audit trail, allowing the tracking of data entry and modification to the responsible individual, is a critical element in the control process. The trail generally utilizes a password system or equivalent to identify the person or persons entering a data point and generates a protected file logging all unusual events.

4. CHANGE: A consistent and appropriate change-control procedure capable of tracking the system operation and application software is a critical element in the control process. All software changes should follow carefully planned procedures, including a pre-install test protocol and appropriate documentation update.

5. SOP: Control of even the most carefully designed and implemented system will be thwarted if appropriate user procedures are not followed. This principle implies the development of clear directions and Standard Operating Procedures, the training of all users, and the availability of appropriate user support documentation.

6. DISASTER: Consistent control of a system requires the development of alternate plans for system failure, disaster recovery, and unauthorized access. The principle of control must extend to planning for reasonable unusual events and system stresses.

GOOD CLINICAL PRACTICE (GCP) Good Clinical Practice (GCP) was made law in Japan and went into effect on October 1, 1990. All clinical studies for Product License (PL) application now have to comply with this GCP. The four main issues in Japanese GCP are: Institutional Review Board (IRB), contract between sponsor and research center, informed consent, and the maintenance of records.

GOOD LABORATORY PRACTICE (GLP) The Environmental Protection Agency's performance and operational guidelines, summarized as the Good Laboratory

Practice (GLP) have been widely accepted and implemented by agency laboratories, independent laboratories contracting with the EPA, and by industry laboratories submitting findings and results to the Environmental Protection Agency. These standards have weathered the practical tests of field experience and continue to represent the expectations and requirements for laboratory management and performance.

GOOD MANUFACTURING PRACTICE (GMP) The purpose of the Good Manufacturing Practice regulations of the FDA (21 CFR) is to assure that all pharmaceutical products and biologic products and medical devices meet the requirements of the Federal Food, Drug and Cosmetic Act as to safety and efficacy and have the identity and strength to meet the quality and purity characteristics which they purport to have, as required by section 501(a)(2)(b) of the Act. Nonconformity with GMP, therefore, establishes adulteration prohibited by section 301.

The Food and Drug Administration strives to ensure that the regulated industries comply with a total quality control concept through participation in voluntary CGMP compliance seminars and workshops sponsored jointly with the industries or with educational institutions. The concept of a total quality control system is neither limited in scope to the analytical methods of assay, control charts, product inspections made during the manufacturing processes and prior to finished dosage form distribution, nor to the statistical techniques utilized in these discrete operations.

GRACEFUL DEGRADATION Refers to the characteristics built into a computer system to assure that system failures default conservatively; that is, system characteristics that assure that failures do not produce critical problems.

GREY BOX TEST Testing which builds flags or interim-result points into a system and compares expected to obtained results at those predetermined control points. The "grey-box" test is, in effect, a compromise between the more comprehensive but cumbersome "white box" and the efficient but less complete "black box."

GROUP TEST The process of determining if something fits within the Fuzzy Set.

GUIDELINE Regulatory agency requirements that are framed as suggestions without the force of law. Failure to respond effectively to them may trigger negative responses from their source agency.

HARDWARE Physical equipment used in data processing, as opposed to computer programs, procedures, rules, and associated documentation.

HARDWARE CONSTRAINTS Limitations imposed upon operations by equipment itself. The most common limitations are capacity and speed. If a data set is too large for a machine, then a malfunction could occur and some data could be lost.

HARDWARE INTERFACE Refers to the physical equipment point where peripherals link into the hardware platform. In a validation effort, the concern is for standard interfaces that can readily be tested, maintained and/or replaced.

HARDWARE REPLACEMENTS All replacements of interacting hardware components with non-identical components, differing in BIOS, in chip, or in configuration, as in non-PROM ROM, should result in an immediate revalidation study of the system or sub-system isolate.

HASH TOTAL Refers to a method of determining whether a transmission has been completed. Taking a total of ASCII values and confirming them. (See Checksum.)

HAZARD ANALYSIS A determination of the potential degree of human harm represented by failure of a device.

HIGH-LEVEL LANGUAGE A programming language that usually includes language features such as nested expressions, user-defined data types, and parameter passing not normally found in lower-order languages; that does not reflect the structure of any given computer or class of computers; and that can be used to write machine independent source programs. (ANSI/IEEE)

HIGH-LEVEL TESTING Managerial or executive testing which is a more cursory review of results and/or general performance test. A regulatory analysis is generally a high-level review which is looking for, among other elements, evidence of low level-testing. Validation of a system, process, or product incorporates high-level testing.

HOROWITZ'S RULE A computer makes as many errors in two seconds as 20 men working 20 years make.

ID CARD Identification card; required for access to any secure facility.

IMPLEMENTATION PHASE The period of bringing a system into actual use; it includes end-user training and has traditionally led to the discovery of the need for patches, requests for functional enhancements, or for both.

INFO-GLUT An excess of information that has the effect of slowing down a decision process. Any time input speed exceeds analysis speed.

INITIAL HAZARD ASSESSMENT The initial classification and estimation of an identified system hazard. It includes the hazard description, an estimate of its criticality, and an estimate of probability that it will occur. The initial probability estimate may be qualitative or quantitative. Initial Hazard Assessment applies to hazards identified any time during the life cycle of a product or system. Preliminary hazard analysis is an important part of Initial Hazard Assessment.

INITIAL HAZARD ASSESSMENT REPORT A complete list of potential system hazards identified throughout product design and development, a qualitative estimate of criticality (cost) of each, a qualitative (or quantitative, if data are available) estimate of the probability of each, and recommendations for resolving any items that seem likely to pose problems.

INPUT PARAMETER The range of values that will be accepted by a system when data is entered.

INSPECTION A formal evaluation technique wherein a person or group other than the author examines software requirements, design, or code in detail to detect faults, violations of development standards, and other problems.

INSTALLATION QUALIFICATION (IQ) The process of checking the system against the vendor/developer standards of operating environment, physical connection, safety parameters, and functional parameters prior to the initial utilization of the system in production. An IQ test generally includes functional testing of system outputs, analog devices, and connections.

INSTRUCTION (1) A program statement that causes a computer to perform a particular operation or set of operations. (ANSI/IEEE) (2) In a programming language, a meaningful expression that specifies one operation and identifies its operands, if any.

INSTRUMENT CONTROL Refers to supervision of a process through a data collection device that can also initiate corrective action without human intervention being necessary. Processes controlled by instruments, in high-hazard environments, should be validated.

INTEGRATED SOLUTION A system linking several software (and possibly hardware) platforms to solve a business problem in a way that makes them transparent to the end user. Example: A document control system that provides word processing, publishing, version control, electronic distribution, and archiving within the same shell.

INTEGRATION TESTING An orderly progression of testing in which software elements, hardware elements, or both, are combined and testeduntil the entire system has been integrated (ANSI/IEEE Std. 729-1983).

INTERACTIVE PROCESSING An application in which each entry calls forth a response from a system or program, as in a ticket reservation system.

INTERFACE A shared boundary. An interface might be a hardware or software component to link two devices, or it might be a portion of a storage or registers accessed by two or more computer programs. (ANSI)

INTERNATIONAL ORGANIZATION FOR STANDARDIZATION (See ISO)

INTERPRETER (1) Software, hardware, or firmware used to interpret computer programs. (ANSI/IEEE) (2) A computer program used to interpret. (ISO) (3) A program which translates a high-level language into machine code, one instruction at a time. Each instruction in the high-level language is executed before the next instruction is interpreted.

INTERRUPT 1) Refers to a signal which causes the hardware to transfer program control to some specific location in internal storage, thereby breaking the normal flow of the program being executed. After the interrupt has been processed, program control is again returned to the interrupted program. 2) A signal that breaks or halts a program. These can be initiated by the user, the program, or elsewhere in the system. Uncontrolled interrupts are AEs.

INVENTORY CONTROL Procedures for tracking materials required for a commercial activity, especially in a production environment. These activities can be automated and are subject to validation as they relate to materials supplied for a product that can be subject to recall. They also track where a released product subject to recall is sent.

I/O PORT Input/Output connection port.

ISO (International Organization for Standardization) A worldwide federation of national standards bodies (ISO member bodies). The work of preparing international standards is normally carried out through ISO technical committees. Each member body interested in a subject for which a technical committee has been established has the right to be represented on that committee. International organizations, governmental and non-governmental, in liaison with ISO, also take part in the work. ISO collaborates closely with the International Electrotechnical Commission (IEC) on all matters of electrotechnical standardization.

ISO 9000 SERIES STANDARDS A series of international standards involved with internal quality management and external quality assurance. ISO 9000 series standards are compatible with, but do not replace, the obligation to validate a system. For vendors developing computerized medical devices, including software products and diagnostic systems, ISO standards serve as an excellent definition of appropriate operating protocols likely to lead to a validated end product.

ITERATION 1) Refers to automatic repetition, under program control, of the same series of processing steps until a predetermined stop or branch condition is reached. 2) Refers to the current stage of development within the gestation period of policies, procedures, documentation, or systems, prior to approval of a formal version.

JOB 1) Set of data completely defining a unit of work for a computer. 2) A collection of specified tasks constituting a unit of work for a computer; e.g., a program or related group of programs used as a unit.

JOB DESCRIPTION Identification of the tasks associated with a job and the skills required to perform those tasks satisfactorily.

JOINT EUROPEAN STANDARDS INSTITUTE (JESI) which consists of CEN, CENELEC, and ETSI.

JUST-IN-TIME MANUFACTURING (JIT) Refers to an optimal material require ment planning system for a manufacturing process, in which there is little or no manufacturing material inventory on hand at the manufacturing site and little or no incoming inspection. The lack of incoming inspection raises control concerns in high-hazard sites. At a manufacturing site where a JIT system is in place, the computer system is the major source of component information that would be critical in the event of a recall.

KERBEROS A network authentication system developed at the Michigan Institute of Technology for its Project Athena. It is an emerging industry standard for security in client-server computing environments. KERBEROS provides a means of identifying individuals on a network through the use of secret encryption key technology.

KERNEL A portion of a program remaining resident in the main storage (RAM) of a computer, which enables that program to be called when needed. This resident portion can cause unforeseen conflicts with applications unless there is stringent prior testing to insure compatibility.

KOSEISHO The Koseisho is the government Ministry of Health and Welfare responsible for drug registration in Japan. It is similar to the Department of Health (DOH) in the UK, the Bundesgesundheitsamt (BGA) in Germany, and the Food and Drug Administration (FDA) in the U.S. The minister of the Koseisho is a member of parliament and may be changed when the cabinet is reorganized by the prime minister. There are nine bureaus and a secretariat in the central office. The pharmaceutical affairs bureau is responsible for the various aspects pertaining to drugs. All clinical studies for a product license (PL) application to the Koseisho have to comply with Japanese Good Clinical Practice (GCP). (See Good Clinical Practice.)

LABORATORY INFORMATION MANAGEMENT SYSTEM (LIMS) A computerized system through which data can be retrieved, manipulated, analyzed and reported on; e.g., toxicology, reproductive, and pathology studies.

LABORATORY MANAGEMENT In compliance with statutes and/or regulations, the laboratory management has ultimate responsibility for compliance with all GALP standards. Laboratory management may designate responsible person(s) to meet the individual requirements and unique set of circumstances inherent in specific operating environment(s) and system(s). However, laboratory management is both accountable and responsible for compliance with statutes and/or regulations. (See Responsible Person.) [In the GALP Standards, Section 5.0, the code letter "M" is used to identify all specific and direct responsibilities of Laboratory Management.]

LANDAU'S PARADOX When system programmers declare that a system works, the system has worked and will work again some day.

LANGUAGE Refers to any symbolic communication media used to furnish information to a computer. Examples are PL/1, COBOL, BASIC, FORTRAN, and ASSEMBLY.

LAUNEGAYER'S OBSERVATION Asking dumb questions is easier than correcting dumb mistakes.

LIFE CYCLE Refers to a period of time that starts when a product is conceived (as an idea) and ends when it is no longer available for use. The life cycle has several phases, typically designated in order as: requirements phase, design phase, implementation phase, test phase, installation and checkout phase, operation and maintenance phase, and, sometimes, retirement phase. Readers should note that the preceding phase designations represent only one Life Cycle model; other models may have more or fewer phases, and tasks that characterize one phase may overlap the tasks of one or more subsequent phases.

LIMITED ACCESS The specific locations of systems equipment should be carefully restricted through the use of limited access doors, restricted hallways, restricted buildings, or other appropriate limitations.

Specifically:

All data processing systems rooms should be locked, with access limited to currently authorized personnel.

Rooms housing minicomputers, mainframes, and their data storage peripherals (disk and tape drives) are included.

All laboratories in which computerized systems are actively in use but not constantly monitored by authorized personnel should have locked-door limited access, locked keyboard restrictions, or both.

LIMS (see LABORATORY INFORMATION MANAGEMENT SYSTEM)

LIVE ENVIRONMENT ERROR LOG REVIEW Over the course of one or more weeks, a real-time audit log of all programs is generated. For validation purposes, review that log to trace the source and disposition of all encountered errors and to identify any unusual programs (not normally accessed) that may require special consideration.

LOADER A routine that reads an object program into main storage prior to its execution. (ANSI/IEEE)

LOG A record documenting all system activity. These must be complete, signed, available, reviewed, audited, and, where appropriate, include corrective actions taken.

LOGGING The documentation of detected errors and problems, so that review and audit trails are created and maintained in hard copy.

LONGFELLOW'S OBSERVATION It takes less time to do something right than it takes to explain why you did it wrong.

LONGITUDINAL REDUNDANCY A method for detecting errors in a series of bytes. Error detection is done as in the checksum method, but the computation differs slightly. The method can detect single-bit errors, and when used with parity it can localize single-bit errors for correction. There are, however, many combinations of errors that can defeat this method for both correction and detection.

LOW-LEVEL TESTING More intensive, detailed testing, looking "down" into a system through comprehensive performance or stress testing. Includes verification of instruments, periodic recalibration of sensing and measurement devices, and complete user acceptance testing of new systems. Development and performance of IQ, OQ, and/or PQ is low-level testing.

MACHINE LANGUAGE A representation of instructions and data that is directly executable by a computer. (ANSI/IEEE)

MAINFRAME A large computer.

MAIN MEMORY A non-moving storage device utilizing one of a number of types of electronic circuitry to store information.

MAIN STORAGE SIZE The amount of memory available to the CPU, usually described in number of bytes. Originally done in thousands of bytes (K or kilobytes), the more current number now is in millions (megabytes), rapidly approaching the billion (gigabytes) range.

MAINTAINABILITY A characteristic of a computer system that refers to its ability to remain functional over extended time.

MAINTENANCE The task of updating software (and hardware) to keep it current. Based upon vendor recommendation and/or common operational parameters, clear maintenance standards and schedules must be established. There is a qualitative difference between controlled maintenance, which is essential to a validated system, and having a programmer simply provide a patch immediately.

MAINTENANCE CONTRACT An agreement with any vendor, especially a software provider, to keep code functionally current and to provide enhancements for an agreed-upon time and fee. Inability to maintain a system will preclude its use in any business environment, not just a regulated one.

MAINTENANCE LOG Refers to documentation of all routine, extraordinary, and recommended maintenance procedures. A system-maintenance log must be designed and established in order to demonstrate control.

MAINTENANCE ORDER A written request which provides a record of maintenance work requested and performed. These orders, properly logged, provide a part of the evidence trail of control.

MAINTENANCE QUALIFICATION A review of the maintenance schedule, log, and repair list. Analysis confirms that the schedule is met according to the log and that the repair list is accurate according to log entries. Attach the maintenance schedule, repair list, and sample of log.

MAINTENANCE SCHEDULE For each main component part in the system, a maintenance schedule has been established. This schedule identifies the frequency of regular and routine maintenance, identifies trigger points (chronometric, performance, or alternate) for additional maintenance, and provides trouble-shooting guidelines identifying special maintenance procedures.

MANAGEMENT AWARENESS AND CONTROL The organization and/or system management maintains appropriate control mechanisms to assure proper operations, use, maintenance, and procedural supervision of the system. Management awareness and control is generally maintained and demonstrated through the development, distribution, and enforcement of clearly designed standard operating procedures (SOPs).

MANAGER OF SOFTWARE ENGINEERING The Manager of Software Engineering is responsible for appointing and supervising the product manager, approving the project plan, and ensuring that all standard operating procedures (SOPs) are being followed. The manager also completes the product approval phase.

MANUFACTURER In the context of regulated environments, the terms "manufacturer" or "firm" refer to a manufacturer seeking market clearance of a product, a distributor seeking market clearance of a product, or an applicant representing a manufacturer or distributor who is seeking market clearance of a product.

MARK TWAIN'S POSTULATE Always do right. This will gratify some people, and astonish the rest.

MATERIAL REQUIREMENTS PLANNING (MRP) Refers to a computerized system which provides for manufacturing, planning, and control for all activities from acquisition of raw materials to delivery of completed products and is designed to support key management interfaces in various functions of the corporation. The validation concerns come from the fact that the MRP would be utilized for potential product recalls.

McFADDEN'S LAW The trouble with experience as a teacher is that the test comes first, and the lesson afterward.

MEDICAL CLASS I, II, III / IV Refers to three FDA-defined levels of medical devices. Class I requires practically no regulatory action. It includes diapers, bandages, etc. Class II includes some testing kits (glucose, pregnancy tests) and some devices that require testing of reliability (condoms) or sterility. Class III / IV are the most highly regulated. These devices directly impact human health, either with or without human intervention. Prior to being combined, Class III included incubators and monitors, while Class IV included pacemakers, infusion pumps, etc. There is some indication that the FDA will be including laboratory and manufacturing devices as Class II or III / IV devices in the near future.

MEDICAL DEVICE - GENERAL REGULATORY ASSESSMENT Medical device GMP regulations are found in 21 CFR Part 820. The fact that process validation is mandated under this regulation is clearly noted.

Section 820.5 requires the medical device manufacturer to "prepare and implement a quality assurance program that is appropriate to the specific device manufactured."

Relevant questions then, focus on (or should reveal) whether a quality assurance program (those activities required to assure confidence in the quality of the process used to manufacture a finished medical device) has been established that is either specific for a particular medical device or general for all medical devices from the company.

A generalized QA program must identify specific issues related to specific medical devices.

Section 820.100 states that process validation itself is an essential element in establishing confidence that a process will consistently produce a product meeting the designated quality characteristics, specifications and goals.

Written procedures are in place that specify methodologies of control that assure that the medical device conforms to its original design.

Written procedures are in place that assure that the original basis for the medical device design falls within current approved specifications (and required regulatory guidelines).

Written procedures are in place that specify methodologies of control that assure that the medical device conforms to any design changes.

Written procedures are in place that specify methodologies of control that assure that any design changes are fully tracked and show evidence of supervisory approval at all stages of development including: design and assessment stage, prototype production and testing stage, finished product and testing stage, finished product validation stage, and the final review stage.

Written procedures are in place that call for the description of the medical device by physical, electrical, performance, and safety characteristics.

MEGABYTE 1024K bytes.

MENU A computer display listing a number of options. The operator selects one of the options. Sometimes used to denote a list of programs.

MERSKEY'S RULE Do a silly test and you get a silly answer.

MICROFICHE Microfilm record, usually 3" by 5," which can be used for permanent storage of text and/or data archives.

MICROPROCESSOR A microprocessor executes encoded instructions to perform arithmetic operations, internal data transfer, and communications with external devices. It is usually a single integrated circuit on a chip. Logic circuitry of a microcomputer; frequently synonymous with a microcomputer.

MILESTONES Most complex projects are divided into measurable sub goals. These subdivisions of a large system or project can usually be utilized independently. Such independent subdivisions are called Milestones.

MILITARY SAMPLING A method of sampling used in quality control in which a predetermined table (usually randomly generated) is used to select a specified number of representative individual cases.

MINIMAX A theory of efficiency that says minimize effort in order to maximize output.

MIRACLE ERROR An error in data and/or processing for which no known cause has been identified. Validation does not allow for continued utilization of a system that has been found to have a Miracle Error.

MNEMONIC A code or device designed to aid someone's memory. A positive characteristic of well-designed source code is the use of meaningful mnemonics as variable names and internal labels.

MOBIL'S MAXIM Bad regulation begets worse regulation.

MODEL A Model is like an algorithm in one sense. An algorithm can Model a physical process (an equation), or it can Model some aspect of human reasoning (a so-called expert-system computer program). In another sense, a Model (data) can be a collection of numbers that represent the properties of an object. A Model (computer program) can operate on a Model (data) to estimate the effects of a real process on a real object.

MODEMS (modulator-demodulator units) Permit outside access to systems via telephone lines. Careful modem security is vital to the restriction of unauthorized access. Specific security procedures may vary according to the environment, and may include any or all of the following:

Removal of modem at times other than those necessary to authorized use.

Use of "bit check" or "bisync" protocols.

Use of a modem printer to track all access attempts.

Use of a "call back" system.

Use of dedicated or restricted phone lines.

Use of originating phone number verification equipment.

Use of other emerging restrictive technology.

MODULARIZATION The process of modularization allows the designer to decompose a system into functional units, to impose hierarchical ordering on function usage, to implement data abstractions, and to develop independently useful subsystems. It can be used to isolate machine dependencies; to improve the performance of a software product; or to ease debugging, testing, integration, tuning, and modification of the system.

MODULARIZATION CRITERIA These include the conventional criterion in which each module and its sub modules correspond to a step in the execution sequence; the information-hiding criterion, in which each module hides a difficult or changeable design decision from other modules; the data abstraction criterion in which each module hides the representation details of a major data structure behind functions that access and modify the data structure; levels of abstraction, in which modules and collections of modules provide a hierarchical set of increasingly complex services; coupling-cohesion, in which a system is structured to maximize the cohesion of elements in each module and to minimize the coupling between modules; and problem modeling, in which the modular structure of the system matches the structure of the problem being solved.

MODULE TESTING PHASE This phase in the Software Development Life Cycle (SDLC) performs the module test plan and generates all supporting documentation.

A review of the output, to ensure the results are as expected, is a part of this phase. Any errors found are reported in the supporting documentation.

MONITORING The ongoing survey of the functioning of system components.

MRP (See MATERIAL REQUIREMENTS PLANNING)

MULTIPLEXER A device which takes information from any of several sources and places it on a single line.

NASSI-SCHNEIDERMANN CHART An alternate method of flowcharting.

NATURAL LANGUAGE 1) Computer code that mimics conversational human speech in its commands. 2) A 4GL developed for use with ADABASE.

NETWORK A system that ties together several remotely located computers via telecommunications.

NETWORK SECURITY Refers to the protection of network resources against unauthorized disclosure, modification, utilization, or destruction.

NODE In a series or network of interacting computers, each computer represents a node or communications link in the network.

OBJECT CODE Compiled, assembly-language code. A fully compiled or assembled program that is ready to be loaded into the computer. (ANSI/IEEE)

OBJECTIVITY AND INDEPENDENCE Traditionally, regulated industries have relied upon an in-house quality assurance or quality control division with a separate line of reporting authority to provide the necessary objectivity and independence. The same industries follow a different set of standards in situations of financial reporting to the SEC, relying upon outside CPA firms despite the high level of in-house expertise, recognizing the added value of the independence and objectivity this approach provides.

The area of system validation more closely parallels the SEC model since, in effect, it is reviewing the appropriate role of quality assurance, as well as data processing and information systems, and because an unusually high level of expertise is required to follow the intricacies of various source-code and hardware components as well as the regulatory issues involved.

Based upon these requirements, an underlying assumption of appropriate system validation standards is the total independence, high level of expertise, and complete objectivity of the signatory validators. Should the validator have a reporting relationship within the organizational structure, or a real or perceived direct or indirect benefit from the outcomes of the validation, the validation process and results could be severely compromised.

OEM Original Equipment Manufacturer (i.e. maker of computer hardware).

OFF-LINE A term describing equipment, devices, or persons not in direct communi cation with the central processing unit of a computer. Function testing of systems should be done Off-Line.

ONGOING SYSTEM PERFORMANCE TEST Refers to system tests conducted by in-house staff on an ongoing basis, providing immediate indication of problems and an annual base of at least 50 cases.

An example of an ongoing system performance test within an environment that distributes blood products might include a daily comparison of whole blood samples to computer-generated switch status, labeling, and attribute matrix labeling protocol.

ON-LINE 1) Direct accessing of a computer system. 2) A device in current communication with the CPU, versus off-line. Function or stress testing of systems in high-hazard environments should not be done on-line. (See Off-Line.)

OPEN SYSTEMS INTERCONNECTION A model developed by the ISO as a means of simplifying information exchange. Designing communication hardware and software in accordance with OSI standards allows users to customize a system by selecting the combination of products which offers the exact features required. The ISO has divided communications into seven sub-processes called layers and has written OSI specifications for each one. OSI specifications are functional only; the tasks to be performed in each layer are defined exactly, but the methods used to accomplish these tasks are left to the designer. Any method is acceptable as long as the same method is used at both ends of the communications link. The International Organization for Standardization (ISO) defines the model, in 1979, as a framework for defining network protocols.

OPERATING PERSONNEL Refers to staff who use the system. They should have GMP, GLP and/or GALP training.

OPERATING QUALIFICATION (OQ) Refers to a check of process, equipment, and control-system prior to use in a live production environment. The OQ is often combined with the PQ (Performance Qualification). In effect, the OQ is the "present" low level test, in direct parallel to the acceptance validation high-level review.

Decision Criteria: Because of the sensitive nature of an OQ test, the decision criterion is generally one of ACCEPTANCE = 100% conformity to established norms. Any deviation from 100% constitutes a failure. If any means of determining acceptance is met by an equivalent means, or additional comments are needed, a separate document should be attached with an explanation of reasoning or results.

OPERATING SYSTEM (OS) Essentially, the software that enables users to interact with a computer. Operating systems generally handle interactions with peripheral equipment, programming languages, and so forth. Operating systems generally include software that controls the execution of programs. An operating system may provide services such as resource allocation, scheduling, input/output control, and data management. Although operating systems are predominantly software, partial or complete hardware implementations are possible.

OPERATIONAL LIMITS Prescribed ranges within which a system is expected to function reliably.

OPERATOR 1) The individual responsible for the physical use of a mainframe com puter system. 2) A mathematical and algorithmic term for the subject of an operation. For example, if c=axb then a and b are the operators of the multiplicative function.

PARALLEL Term to describe transmission of data eight bits (one byte) at a time.

PARITY An error-detection method for small units of data (usually 8 bits, a byte), wherein an extra bit is appended to the data unit. It is useful in detecting odd numbers of incorrect bits in a single byte. Even numbers of incorrect bits leave parity unchanged. By itself, Parity is not useful for error correction.

PARITY BIT An extra bit within a byte, used to verify the coded information in the byte itself. The extra bit is either a one or zero, so as to make the total number of ones in a byte equal either an odd or even number (odd or even parity).

PASSWORD A unique string of alphanumeric characters used to verify an end-user's identity. To maintain proper control, these must be secret and changed according to a controlled cycle.

PASSWORD CONTROL LOG The system administrator or individual in charge of system security must keep track of which users have access to various secure database levels. This document will help organize such information, as well as keep track of the passwords of the various users. Refer to SOP on system security and passwords.

PASSWORD FILE A password or user file should restrict unauthorized users and log access attempts. The password file should be characterized by:

1. Maintenance procedures to delete passwords of individuals no longer authorized, and to change passwords periodically.
2. An active log generated automatically by every unusual event.
3. Periodic managerial review of the password log.

PASSWORD SYSTEM A password system should be included in all appropriate devices to uniquely identify the system user. The password system should include the following characteristics:

- Secret Passwords (random preferred over user-selected),
- Passwords of not fewer than four digits.
- Passwords changed at least annually.
- Password system with automatic cut off and log generation after three erroneous

sign-on attempts.

Password systems are not necessary in environments in which a single authorized user has physical access to a device, or with devices that do not allow for user input, modification, or analysis.

PATCH A change to the code of a program that is usually done quickly and for a very specific purpose, most often to correct an immediate problem. In a high-hazard environment, Patches should be validated.

PATH 1) A logical chain of instructions to accomplish a single program action with in a module. 2) A link between modules or programs.

PERFORMANCE Refers to intended specific results or objectives of a system.

PERFORMANCE QUALIFICATIONS Refers to a process during which all automatic systems logs, error messages, variable data, process attributes, and product characteristics (QC and QA) will be examined for validity and accuracy. The purpose of performance qualifications is to ensure that the system will operate and produce expected production-quality results under load conditions. This will give a representation of long-term performance, verification of vendor guarantees for availability and failure rates, and a better debugged and more reliable system.

PERFORMANCE TEST (See Function Test).

PERIPHERAL EQUIPMENT All computer system equipment other than the computer processor circuits and memory circuits; for instance: disk drives, tape drives, and terminals.

PHANTOM PHASE Refers to a tongue-in-cheek label for the retrospective failure to perform a system analysis prior to designing a system. The process of defining the system's needs and objectives may be a phantom or non-existent phase in the SDLC.

PHARMACEUTICAL RESEARCH AND MANUFACTURERS OF AMERICA (PhRMA) Formerly the Pharmaceutical Manufacturers Association (PMA). A national association representing more than 135 major manufacturers. It keeps close contact with the industry and technological developments, specializing in the guidelines to follow in the manufacturing of pharmaceuticals. PhRMA puts out more than 70 publications in the field and a complete listing is available free of charge. The publications cover such topics as regulatory issues of the FDA, de-ionized water for the water industry and water treatment, clinical guidelines and manufacturing guidelines, and the proceedings of seminars. Some of the publications are in a report format of four or five pages; others are 200 pages in length.

PHASE DEFINITION FORMAT The software methodology documentation is based on a standard format used to define all of the software development life cycle. The phases within the development cycle include the following (A definition of each phase is listed alphabetically in this glossary):

1) DESIGN GOALS PHASE
2) FUNCTIONAL SPECIFICATION PHASE
3) PROJECT PLAN PHASE
4) SYSTEM ARCHITECTURE PHASE

5) USER'S MANUAL AND DOCUMENTATION PHASE
6) DEVELOPMENT PHASE
7) DATABASE DESIGN PHASE
8) BETA TEST PLAN DEVELOPMENT PHASE
9) DETAILED DESIGN SPECIFICATIONS PHASE
10) CODING AND MODULE TEST PLAN DEVELOPMENT PHASE,
11) TRAINING SUPPORT PERSONNEL PHASE
12) MODULE TESTING PHASE
13) SYSTEM INTEGRATION TESTING PHASE
14) SYSTEM TESTING PHASE
15) BETA TESTING PHASE
16) PRODUCT APPROVAL PHASE
17) RELEASE PHASE

PHYSICAL INSPECTION Observation and review of system installation and physical site for factors relating to safety, continual operation, and disaster recovery capability.

PORTABILITY Ability of an application code to run under different operation systems and/or on different hardware platforms.

PRELIMINARY HAZARD ANALYSIS (PHA) Refers to an initial assessment of system hazards undertaken early in the product life cycle. It includes describing the hazard, estimating its criticality, and estimating the probability of its occurrence. The estimate may be qualitative or quantitative.

PRELIMINARY HAZARD ANALYSIS REVIEW When the preliminary hazard analysis (PHA) has been performed for the medical device and is available for review, it will meet the following criteria which demonstrates at least four areas of safety criticality[1]:

Class 1 = Catastrophic (may cause death; may cause total system or medical device loss)

Class 2 = Critical (may cause severe injury or occupational illness; may cause severe damage to the system or medical device)

Class 3 = Minor (may cause minor injury or occupational illness; may cause minor damage to the system or medical device)

Class 4 = Negligible (a problem that does not result in injury, occupational illness; problems that do not damage system or medical device)

PRE-MARKETING APPLICATION (PMA) A submission to the FDA required for Class III Medical Devices.

PRIVACY 1) Anonymity guaranteed to the personal details of a clinical subject. 2) A right which is potentially compromised by a company under the Freedom of Information Act once a regulatory filing is made.

PROBLEMS A reported malfunction of a computer component observed by end users of an application.

PROCEDURE A set of steps to be followed to accomplish a task each time the task

---

[1]M. Frank Houston, "Designing Safter, More Reliablee Software Systems," U.S. Food and Drug Administration, Center for Devices and Radiological Health, 1989.

is to be done.

PROCESS CONTROL SYSTEM Automated management of an industrial production system. Even though a system may be used in other areas, such as petroleum and chemicals, validation is required when it is adapted for use in the pharmaceutical industry.

PROCESS PERFORMANCE CHARACTERISTICS Those characteristics which identify components that need constant re-calibration, are easily breakable, and are otherwise essential for proper process performance.

PROCESS PROCEDURE The step in a production process that needs to be documented within a standard operating procedure (SOP).

PROCESS QUALIFICATION (PQ) Refers to a check of the ongoing process equipment and control system as it is utilized in the live production environment. A PQ generally includes testing a known sample or known data against the system performance, and it may include stress testing through a non-live library replica or analysis of Error File stress reactions.

PROCESS VALIDATION The accumulation of evidence in support of expert testimony demonstrating the appropriate functionality of process equipment utilized in the manufacture of a biologic or pharmaceutical product. This evidence includes the development and performance of an installation qualification, a process qualification, and an operating qualification. When the process equipment under review is computer-controlled, computer-regulated, or computer-activated, or when said equipment reports to or through a computer or receives data from or through a computer, a system validation is incorporated into the Process Validation. Because process validation is a current requirement of the FDA's "Current Good Manufacturing Practice Regulations for Finished Pharmaceuticals and Medical Devices," it becomes necessary and applicable to the manufacture of pharmaceuticals and medical devices.

The FDA requires that each of the following three principle be followed:

1. Quality, safety and effectiveness must be designed and built into the product itself.
2. Quality cannot be inspected or tested into the finished product (i.e., quality must begin with manufacturing).
3. Each step of the manufacturing process must be controlled to maximize the probability that the finished product meets all quality and design specifications.

The apparent key to meeting these three criteria is Process Validation.

<u>The FDA Definition:</u>

"Process validation is establishing documented evidence which provides a high degree of assurance that a specific process will consistently produce a product meeting its predetermined specifications and quality characteristics."

It is therefore of the utmost importance that the manufacturer (vendor) composes a validation protocol which clearly states how system validation will be conducted. This must include testing parameters, product characteristics, inventories, acceptance testing of production equipment, and a checklist of acceptance criteria.

In order to maintain quality assurance of a particular product, the manufacturer must be able to track the production stages and current configuration of the product itself. A version-control plan is equally important.

PRODUCT AND PROCESS TESTING CONSIDERATIONS Prior to executing the initial OQ (Operational Qualification Acceptance Test) and rigorous PQ (Product and Performance Qualification Test), product characteristics and specifications must be established. All products intending to be manufactured must have the following characteristics defined: physical, chemical, electromagnetic, and product performance.

PRODUCT APPROVAL PHASE This phase in the Software Development Life Cycle (SDLC) determines whether or not the product can be released. The review committee reviews portions of documentation from each phase. The product is approved if all phases leading to this phase have been performed following established standard operating procedures.

PRODUCT DATABASE A product database contains information maintained on a regulated product that tracks released products for adverse events and possible recall.

PRODUCT MANAGER The Product Manager is responsible for the planning and coordinating of all aspects of a specific product. The product manager defines the work responsibilities for the project and allocates the resources required for completion. The product manager also ensures that all the requirements for product release are met.

PROGRAM 1) A computer program. 2) A schedule or plan that specifies actions to be taken. 3) To design, write, and test computer programs.

PROGRAM CHANGE REQUEST FORM This form is used to initiate the program change process. Refer to SOP of software change control.

PROGRAM LOGIC SOURCE CODE REVIEW An examination of the internal logic of all operating programs, based upon a detailed trace of source code and a review of program documentation.

PROGRAM PATH Refers to the order in which a computer executes its programmed instructions under given internal and environmental conditions.

PROGRAM PATH ANALYSIS The identification of data and conditions that will cause a computer to execute a given Program Path.

PROGRAM REQUEST PROCESSING CHECKLIST This form will help track the progress of a requested programming change. Refer to SOP of software change control.

PROJECT ADMINISTRATOR Refers to a person who reviews the initial change request form from the requestor. He/she records the request in log (or begins log page). He/she participates in ultimate approval/rejection of the request. This person also enters/checks logbook for future updates/enhancements.

PROJECT LEADER The Project Leader is responsible for the completion of all development and testing aspects of a specific software project. The Project Leader assigns the tasks of the project to the appropriate individuals and coordinates the testing of the product. Also, the Project Leader is responsible for the technical quality of the project. All steps related to the project are reported to the Project Leader.

PROJECT MANAGER OR PHASE/PROJECT LEADER May share several responsibilities. They assess required action, prepare impact statement, participate in approval/rejection decision, obtain approval of change from end user (if applicable). and place the actual change into work.

PROJECT PLAN Refers to a management document that describes the approach to be taken for a project (e.g., developing a system). The plan typically describes the work to be done, resources needed, methods to be used, configuration management and quality assurance procedures to be followed, schedules to be met, project organization, and so forth.

PROJECT PLAN PHASE This phase in the Software Development Life Cycle (SDLC) develops the Project Plan for the system. The plan may include the assignment of roles defined in this methodology, resource requirements, hardware requirements, scheduling of milestones, and a definition of budgetary considerations. Any assumptions made about the project are also listed in this document.

PROM Programmable read-only memory; once programmed, the switch pattern on a PROM cannot be changed. Special equipment, separate from the computer, is usually used to "burn in" the switch pattern.

PROTECTED STORAGE Refers to storage locations which are protected from heat, dust, humidity, and electromagnetic interference. Their intended use is usually for long-term archiving.

PROTOCOL 1) A set of conventions or rules that governs the interactions of processes or applications within a computer system or network. (ANSI/IEEE) 2) A set of rules that governs the operation of functional units to achieve communication. 3) Within FDA guidelines, protocol refers to a document which provides the background, rationale, and objectives of the clinical trial and describes its design and organization and the conditions under which it is to be performed and managed.

PSEUDO CODE Refers to user language which describes the operation of a program by having each native language statement equivalent to one computer statement. Example:

1. Get 3 numbers.
2. Add 3 numbers together.
3. Divide the total by 3.
4. Print out total.

The preceding is a pseudo code program for averaging three numbers. One could write that program in any computer language.

PURGE The permanent removal of data from a file or archive, whether physical or electronic. Unlike "delete," the purge function wipes a magnetic record clean, rather than simply erasing the directory listing.

QUALITY ASSURANCE (QA) 1) In the context of FDA guidelines, Quality Assurance refers to the implementation of appropriate planned and systematic actions necessary to provide adequate confidence that the required quality will be obtained. 2) In the context of EPA guidelines, the inclusion of a data and procedural "double check" through a Quality Assurance Unit or individual is an established and widespread GLP principle, extended here to automated lab systems. The legitimacy and credibility of that checking function, of course, rests with the independence of QA, assured through a separate reporting relationship. While it is possible that QA may have additional responsibilities in the organization, those responsibilities should not compromise that independence: the QA should not be and should not report to or through the RP. Specific QA responsibilities include review of system SOPs, inspection and audit of the system, review of final reports for data integrity, and review of record archives. [In the GALPs Standards section 5.0, the code letter "Q" is used to identify all specific and direct responsibilities of QA.]

QUALITY ASSURANCE PERSONNEL Should be involved in all phases of configuration management control, but the involvement should be restricted to that of consultant in the early stages (prior to revision release), and active participant thereafter. QA should periodically review and update the change control logbook. QA also must act as liaison in those cases of change initially requested by end users.

QUALITY ASSURANCE SYSTEM Refers to a set of principles followed to allow effective control of quality throughout a process.

QUALITY CHARACTERISTIC 1) Refers to those characteristics of a product or service that impact on its ability to satisfy stated or implied needs. 2) A particular, measurable characteristic examined to test product quality. 3) A product or service free of deficiencies.

QUALITY CONTROL 1) The operational techniques and procedures that are used to fulfill requirements for quality. 2) May refer to any technique which evaluates the quality of a product being processed by comparing it to a predetermined standard and taking the proper corrective action if the quality falls below that standard.

QUALITY CONTROL UNIT Refers to a separate, independent organization, including staff (with a reporting responsibility), equipment, and physical location which has the task of ensuring that the required second check is carried out prior to release of product.

QUALITY PROOF Evidence that the entity (system, process, or product) meets preestablished standards. A product, process, or system is not considered to be of adequate quality until it has been proven to be so.

QUICK AND DIRTY A phrase commonly used to describe 1) A one-time ad hoc report, where statistical rigor is sacrificed to the interest of speed. The likelihood that this report will be used to support long-range strategic decisions is inversely proportionate to the likelihood of ever replicating its results. 2) Programmers' favored approach to producing a patch, where documentation and convention are also sacrificed to the god of speed.

RAM Random Access Memory; internal storage device containing volatile information which can be changed; read-write memory. When electrical power is cut off from a RAM integrated circuit (IC), its memory is lost.

RAW DATA 1) Data that cannot be easily derived or recalculated from other information (US-EPA, GALP-7.11, DEC 1990) and may include:

Handwritten lab worksheets/records/notes of original observation.

Exact copies or transcripts of the above.
Photographs, microfilm, microfiche.
Computer printouts.
Magnetic tapes of dictated observations.
Database files and reports stored on floppy diskettes or tape.
Version-specific notes, source code for developed software.

The specific definition is still subject to some regulatory controversy as of this writing. (see 21 CFR 58.3(k) 2) In the context of FDA guidelines, Raw Data means all records of original observations or activities in a clinical trial necessary for the construction and evaluation of the trial. Such material includes laboratory notes, memoranda, calculations, documents, recorded data from automated instruments etc., or exact verified copies in the form of photocopies, microfiches, etc. thereof. The term can also include photographic negatives, microfilm or magnetic media. In the case of the latter, dated and verified written transcriptions may replace the originals.

REAGENT A substance involved in a chemical reaction used to test a blood or plasma product.

RECALIBRATION Refers to the process of making a comparison of system outputs to predetermined output standards and subsequently adjusting that system to conform to the expected, normalized output.

RECALL OF INFORMATION The ability to retrieve data from archives. Recall is a mandatory ability for demonstrating control.

RECORD Collection of related data or words treated as a unit.

RECOVERY Restoration of an automated system (hardware, software, and/or data) to the point immediately before an AE or a system failure.

RECOVERY CAPABILITY Since regulatory guidelines require the retention of records for a period of time, systems generating, storing, or analyzing that data must be capable of maintenance and continued functionality despite the ravages of reasonably anticipated natural and man-made disasters.

REGULATORY QUALITY Refers to the need for an independent, second check that must be configured to meet specialized regulatory requirements which go beyond corporate and/or industry standards.

RELEASE PHASE This phase in the Software Development Life Cycle (SDLC) releases the system to a limited number of users as a live system. The number and identification of users the system is released to is determined by the product manager.

Review the limited release to ensure the goals from the design goal phase are complete, that the system meets all functional specifications, and analyze all problem reports. If there are no existing problem reports that are considered to be related to a hazardous situation, then the product can be released to all other customers. Otherwise, the problems need to be addressed by identifying and fixing the faulty phase and following through the methodology.

RELEASE PROTOCOL Standardized procedures required to be followed before any GMP regulated product is release for market use.

REPAIR PARTS LIST A list of all the new parts added or replaced in the course of maintenance must be available and current. The Repair Parts List may, in effect, be the current inventory list, identifying major system components.

REQUIREMENTS DEFINITION Those conditions or capabilities that a system or component must meet or possess to satisfy a contract, standard, specification, or other formally imposed document. Subsequent development of a system or component is based on the set of all requirements for that system or component.

REQUIREMENTS INSPECTION REPORT A report of the proceedings and recommendations of an inspection review of a (product, software, or system) Requirements Specification.

REQUIREMENTS TRACE MATRIX Refers to a document that records or traces the links or correspondence between requirements identified at each stage of product or system development. A matrix should at least document individual system requirements, elements of the design intended to fulfill a requirement, and test procedures and data needed to verify meeting the requirement.

REQUISITE STANDARD OPERATING PROCEDURES AND LOGS Studies have shown that nearly 70% of all 483s written by FDA inspectors are related to SOP or log-maintenance issues. Good system and medical device documentation and record keeping is a requirement of a regulatory audit. Our research has identified 35 particulars which must appear as individual, or a part of, corporate or departmental standard operating procedures.[2]

RESILIENCE (See SOP Resilience).

RESISTANT TO STRESS Refers to design features within the system that take into account the reality of conditions exceeding input parameters and operational limits.

RESPONSE TIME The interval elapsing before a computer accepts an input. Failure to allow for elongated response times can lead to unintentional double entry of data or use of inappropriate functions during an otherwise normal job processing.

RESPONSIBLE PERSON (RP) Laboratory management is ultimately responsible for the operations conducted by computerized equipment, laboratory and information systems, and associated operating environments. As such, laboratory management may directly supervise and/or assign/delegate its authority to meet the individual requirements and unique set of circumstances inherent in specific operating environment(s) and system(s). By permitting laboratory management to delegate its authority and responsibility, appropriate personnel (i.e., Responsible Persons) will have the opportunity to put into place those policies, plans, and practices that will provide the required evidence of control.

The RP is generally a professional, with some computer background, in a position of authority related to the control and operation of the automated data system. The RP's responsibilities include training of users, implementing appropriate security measures, developing or reviewing SOPs for system use, enforcing change-control procedures, and responding to emergent problems. [In the GALP Standards section, 5.0, the code letter "R" is used to identify all specific and direct responsibilities of the RP.]

RESTART Refers to the process of reestablishing the execution of a routine.

RETROSPECTIVE VALIDATION System review which determines the reliability, accuracy, and completeness of a system in current or historical use.

REUSABILITY The ability to recycle routines, modules and even entire programs in

---

[2]For more details on items to include in these SOPs, see S. Weinberg, System Validation Standards (Kendall Hunt, 1990).

the development of new software systems.

REVALIDATION Recheck of the reliability of a system which is undertaken at regular intervals. It is recommended that these intervals be no longer than every two years. It is also performed after any major system enhancements or failures.

REVIEWER CREDENTIALS The credentials held by a system reviewer are of critical importance, because the judgments upon which conclusions are based should pass expert-witness credibility review for technical and theoretical expertise, appropriate experience, and complete independence.

ROBOTICS Use of automated equipment to replace human labor and/or operators in repetitive industrial functions.

ROM Read only memory; internal storage device in which information is permanent.

ROUTINE TESTING OF SOFTWARE LOG Any time a change is made to the software, at least module testing must be performed. Such information must be documented. Refer to SOP on retesting/revalidation of software.

RP (See Responsible Person)

RS 232 INTERFACE 1) An Electronic Industries Association (EIA) standard for connecting electronic equipment; data is transmitted and received in serial format. 2) Standard communications connection for most personal computers in use today.

SAFETY ANALYSIS AS RELATED TO COMPUTER SOFTWARE OR SYSTEM FIRMWARE When computer software (or EPROM firmware) is used in the control of a medical device system, its inherent safety becomes a regulatory concern. The concern itself emanates from the links between the software and safety-critical functions. Therefore, as soon as the software functionality is identified, a Software Safety Analysis (SSA) should begin.

SAFETY-CRITICAL SOFTWARE Computer programs that control functions that affect system safety. Such programs include those causing the host computer to directly command accessories that affect safety, or that may communicate with (provide data for) other safety-critical software systems.

SAMPLE The Sample Size Issue

In many ways the performance testing of computer systems parallels the QA testing common to the pharmaceutical industry. The goal--identifying and diagnosing either error sources or results--is consistent. The importance of random selection of test access is congruous. Also, obviously, the criticality of documenting procedures, results, and analyses is also identical. In a very important way, though, the testing of systems differs from the testing of pills on an assembly line or laboratory animals injected with a new drug. System testing uses different rules for defining sample size, largely because it uses a different conceptualization of "sample."

Both the "Military Standard" and the "Power Formula" methods of selecting sample size are inappropriate for system testing, potentially resulting in samples either significantly too small, or unnecessarily and redundantly too large. These statistical approaches are designed to assure that sufficient cases are included to provide confidence that "outlyers" -- unusual cases and abnormal reactions -- will, in effect, be canceled out by mirror-image outlyers on the opposite end of the scale. The results, a smoothed normal (or other unbiased sampling distribution) is then suitable for complex and sophisticated statistical analysis.

If we want to judge average heights in a population, then, we use a sizing rule that will provide reasonable confidence that all the basketball players will be "canceled out" by the jockeys, providing a sample unskewed by the unusual extremes. To create a sample of 20, for example, we might randomly select 20 people from the population and measure each individual's height. It would, of course, be inappropriate to select only two people, write down each of their heights 10 times, and pretend that we were sampling 20 different individuals. Yet that procedure is, in effect, the process of testing a computer system with a random population of 20 test cases: the computer may actually be exercising the same two programming pathways 10 times each. The difference is simple: in product testing (or in most social science and scientific research) a sample is used to test the results on a data population, utilizing a data sample. But in testing computer systems, we are instead testing not a sample of the data, but a processing sample; the sample is designed to be representative of the system pathways, not representative of the data set upon which the system operates. Ideally, statistical sampling will, in appropriate sizings, generate a high level of assurance that most pathways are explored. But the possibility that some pathways are neglected, creating the potential that the system is inaccurate or inoperative (not 1% inoperative, but 100% unreliable) is very real. At the same time, testing multiple cases through the same pathway is an unnecessary time and resource expenditure. Program testing must define its sample through careful stratification of the data set, instead of relying merely on the rules of large numbers to carry the day.

What, then, is the guideline for sample size in system performance testing? A sample is constructed by defining the relevant system pathways and constructing one data case to follow each of those paths. The sample size is determined not by the population of the data to be manipulated nor by arbitrary rules limiting "enough is enough." Instead, the sample size is defined by an extremely simple but sophisticated formula: $Y = X$, where Y is the sample size, and X represents the number of paths to be tested.

The sample size should never be a number smaller than X, and gains nothing in reliability, credibility, or value by being any larger. In system performance testing, the test sample size should be equal to, and constructed to reflect, the relevant system pathways to be tested.

SAS (Statistical Analysis Software) Refers to widely used analytic software support ed by the SAS Institute. though never formally validated, it is still used by the FDA for analytic work and is widely used in the industry.

SAVE The process of making a backup copy of text or data as a precaution against the next system failure and/or to maintain an historical record.

SCENARIO ANALYSIS Application users are responsible for the identification and description of likely, probable, or potential disaster scenarios and should have developed a reasoned response plan to assure data integrity in the event of these scenarios. Possible scenarios include, but are not limited to: power interruption, flood, fire, and sabotage.

SCHEMA Specific and preestablished procedure. In effect, a standard for activity or action connected to a particular environment.

SCHROEDER'S CAT A philosophical conundrum that posits a cat being both alive and dead simultaneously. The Hawthorne Effect is a more formal reference to the phenomena. The validation concern addressed is the fact that the process of measuring something may change it. This problem necessitates unobtrusive observation on the part of validators.

SDLC (SOFTWARE DEVELOPMENT LIFE CYCLE) The process of producing a system from design to implementation and maintenance. The FDA has used the 8-Step or waterfall model in its training, but the key is to have an SDLC that works for you, and to follow it.

SECURITY LOG Used by security personnel, this document indicates who may currently be found on the premises. Refer to System Validation Standards Checklist.

SEGAL'S LAW A person with one watch knows the time; a person with two watches is never sure.

SELF-CHECKING CODE Computer code that automatically reviews itself for errors against certain internal rules and may make corrections to compensate. These correction procedures must be understood and confirmed to be sure they generate acceptable results.

SELF EXPLANATORY A documentation principle, especially applied to source code, requiring that a properly trained professional be able to easily and quickly understand the activity explained in the text. This is a key principle in maintaining a compliant system.

SERIAL Term to describe handling of data one bit at a time.

SHARED LOCK Refers to a security principle which requires that two "keys" be used simultaneously to open a "lock." The "keys" and "lock" may be physical, software related, or a combination of the two.

GEORGE BERNARD SHAW'S PRINCIPLE Build a system that even a fool can use, and only a fool will want to use it.

SITUATION-UNIQUE TESTS Special tests of security, of override access potential, of log accuracy, etc., are designed and conducted as needs arise during the context of other testing. These situational tests are developed, conducted, and interpreted by Weinberg Associates, Inc., or by in-house staff, as appropriate.

SIX SIGMA (6Σ) An alternate notation for expression of a quality-level goal of 99.999% accuracy, the general research standard of the pharmaceutical industry. A 6Σ level of quality represents a minimum of 4 defects per million items. In comparison, the current estimate of software quality in the United States is 3Σ, about three to four defects per thousand lines of code. The airline industry utilizes a standard of 6.4Σ, or approximately three errors per 2.4 million items (or incidents).

SKUNK WORKS A project or activity that uses a trial-and-major-error approach. One does a very rough model and keeps correcting until it works. A skunk works is outrageously difficult to validate.

SMART MODEM 1) A communication device which manages most of the connection process and provides error checking transparent to the user. 2) A modem with a CPU and programmable memory.

SOFTWARE Programs, procedures, rules, and any associated documentation pertaining to the operation of a computer system.

SOFTWARE COMPLEXITY METRICS A measure of the degree of complication of a system or system component: number of branches, nesting, data structures, lines of code, program size, etc.

SOFTWARE CONFIGURATION CONTROL PROCEDURES A document that prescribes the procedures to be followed in evaluating, coordinating, approving, and disapproving changes to software.

SOFTWARE DEVELOPER Refers to an individual or organization that provides software for commercial purposes. Also referred to as vendors. Until recently, these were unregulated by the FDA. A new interpretation has classified certain pharmaceutical industry software as medical devices, now requiring them to follow validation procedures.

SOFTWARE DEVELOPMENT PLAN A software development plan should include requisite inspections, reviews, audits, and tests for verification and validation of software. (See Project Plan.)

SOFTWARE ENGINEER The Software Engineer is responsible for translating the design from documentation into code. The code generated is documented in such a way that it can be audited for conformance. This individual also writes module test plans and performs module and integration test plans.

SOFTWARE ERROR LOG Usage of this document is found in cases of errors reported and corrected in a particular piece of software at a specific site.

SOFTWARE HAZARD ANALYSIS The identification of safety-critical software, the classification and estimation of potential hazards, and the creation of a program path analysis to identify hazardous combinations of internal and environmental program conditions.

SOFTWARE HAZARD ANALYSIS REPORT A document that records the proceedings and recommendations of a Software Hazard Analysis.

SOFTWARE INTERFACE Refers to code written specifically to allow otherwise incompatible programs or systems to pass or share data. Example: Programming to allow the QA LIMS to pass its release decision to the MRP or order management systems, bypassing manual input of a paper report.

SOFTWARE REQUIREMENTS SPECIFICATION A document that prescribes completely and precisely the conditions and/or capabilities that system or product software must possess or meet to satisfy the system or product requirements specification. All typical inclusions of requirements specifications apply here. If a product or system is comprised of software alone, then the System or Product Requirements Specification replaces the Software Requirements Specification.

SOFTWARE, REUSABLE Refers to programs that are able to be recycled for re lated purposes. This is a key premise for the current interest in CASE --Computer-Aided Software Engineering. (Also see Reusability.)

SOPs (See Standard Operating Procedures)

SOP ACCURACY All identified SOPs conform to operating requirements, recom mendations, and procedures developed and distributed by system suppliers, designers, vendors, or supporters.

SOP ADHERENCE Exists when current SOPs conform to and accurately reflect actual system use and function, as confirmed through unobtrusive observation, managerial report, and user personnel report.

SOP RESILIENCE The characteristics of a standard operating procedure that permit it to apply over a period of time despite minor situational changes, as opposed to the non-resilience of "working procedures," guides, or other less formal and more technically specific procedures.

SOURCE CODE Software language instructions and logic that are produced by programmers. These can be read and reviewed relatively easily by other programmers and auditors. This code must be passed through a compiler to produce executable code or object code.

SOURCE-CODE REVIEW Examination of software programs to determine their availability, internal documentation, display of evidence of change control, accuracy of algorithms and formulae, logical construction, error detection and correction, internal organization, and consistency of terminology.

SOURCE-CODE TEST RECORD This is a checklist that will aid a programmer in evaluating developed source code.

SOURCE PROGRAM A computer program that must be compiled, assembled, or interpreted before being executed by a computer. (ANSI/IEEE)

SPAGHETTI CODE Refers to highly unreadable and hence virtually unmaintainable source code produced by programmers to guarantee job security. (See Weinberg's Law.) There are many systems currently operating which contain varying amounts of Spaghetti Code. Some level of control of such systems, or parts of such systems, may be demonstrated by historical documentation of their reliability. In high-hazard environments, Spaghetti Code should be replaced as soon as possible.

SPSS Statistical Package for the Social Sciences

STANDARD OPERATING PROCEDURES (SOPs) 1) In the context of FDA guidelines, SOPs consist of written guidelines for dealing with any defined situation occurring during the conduct of clinical trials. 2) Formalized documentation describing and prescribing steps to be taken in normal and defined conditions. Standard Operating Practices refer to the actual activities with which procedures ought to correlate. Effectively controlled systems are managed through a series of SOPs that define security and disaster recovery procedures; data archiving, storage, and retrieval; software versioning and change control; user training; safety; and related operations.

STRESS TEST (See Erroneous Data Test).

STRUCTURE TESTS An examination of the internal processing logic of the soft ware. The goal is to achieve an optimum level of test coverage, both branching and path-traversing.

STUBS Temporary modules placed in a computer program until the real module has been written. In high-hazard environments, stubs should be validated.

SUPPORT The existence of support control implies a built-in clarity of training materials, manuals, and diagrams; the accessibility of experts and assistance; and a concern for finding the cause of problems rather than merely patching holes as they occur. Both system support and user support are intended to solve to solve problems as they emerge.

SUPPORT PERSONNEL The Support Personnel are responsible for providing custo mer support and informing the product manager and other appropriate personnel of specific problems with the product in the field.

SUPPORT PERSONNEL TRAINER The Support Personnel Trainer is responsible for developing a training manual and training the support personnel.

SYNTAX Rules of grammar and structure that must be followed to compile a program in a particular programming language.

SYNTAX REVIEW A second check of the adherence to rules by a programmer. This is facilitated by the fact that virtually all programming languages have a syntax check built in that will point out errors which must be corrected. The major concern in system validation is dealing with vendor proprietary languages that may not have checks built in.

SYSTEM 1) Totality of resources (personal, material, procedural) required to meet an objective. 2) Hardware, firmware, and software components combined to accomplish a variable task, such as data analysis, control, manipulation, transmission, storage, and retrieval.

The significant, defining factor of a computer system is the task performed by the system. This factor may lead to the definition of a system as a smaller, self-contained set of hardware, software and components. In more complex cases, it may lead to the definition of system as a group of hardware, software, and component sets functioning together to perform a given task.

SYSTEM ARCHITECT The System Architect is responsible for interpreting the functional specifications and translating them into a system design or architecture. This System Architecture is the global design of the system. The System Architect also ensures consistency among the modules as they are developed. This individual identifies the parts of the system that can utilize development tools, communicates system interdependencies to the system integrator and makes recommendations to the project manager.

SYSTEM ARCHITECTURE PHASE This phase in the Software Development Life Cycle (SDLC) defines the global system design. The global design must be consistent with the functional specifications and the project plan. The System Architect defines which modules need to be designed, how they interact, and how they interface. The System architect also identifies the need for development tools. The detailed design protocol is defined so any subsequent detailed design that follows the protocol is deemed correct. A review of the System Architecture Document to ensure the concepts in the functional specifications and project plan are addressed is a part of this phase.

SYSTEM BACKUP LOG Used to document system backups as a regularly performed activity done to safeguard existing data, as well as to minimize loss of data in the event of a system and/or application failure. Refer to SOP on system backup and restoration.

SYSTEM DEVELOPMENT PLAN (See Project Plan).

SYSTEM ERROR Failure by the automated system to perform according to expectations. This failure constitutes an AE and must be entered in an error log for review and possible maintenance.

SYSTEM HAZARD ANALYSIS The identification of specific internal and environmental conditions that may cause hazards in a system. The analysis should identify components and subsystems in which faults or failures would cause hazardous conditions or actions, should describe the fault or failure, and should estimate the probability and severity of the hazard.

SYSTEM HAZARD ANALYSIS CROSS CHECK The System Hazard Analysis Cross Check (SHACC) is complied later in the development process, usually during the design phase. It is used to identify which hazards occur during which particular medical device functions. A matrix format is most commonly used, listing those hazards identified previously (in the PHA) vertically, and the medical device functions horizontally. A check mark is placed in the cell that correctly matches a potential hazard with a relevant function.

SYSTEM HAZARD ANALYSIS REPORT A document that records the proceedings and recommendations of a system hazard analysis.

SYSTEM INTEGRATION TESTING PHASE This phase in the Software Develop ment Life Cycle (SDLC) defines the system integration test plan. The test plan is performed to ensure that the modules are functioning together.

A review of the testing to ensure that the interface logic of the units, rather than the internal logic of the units, is sound is part of this phase. The internal logic of the units is tested in the implementation and module test plan development phase. Any errors found are reported with the supporting documents.

SYSTEM INTEGRATOR The System Integrator is responsible for ensuring that all the interdependencies between modules are consistent. The System Integrator writes a test plan to ensure that any data passed between programs is consistent. The test plan is performed on a system that is representative of a real-life system.

SYSTEM MAINTENANCE LOG This document may be used to demonstrate preventive maintenance of a particular piece of equipment at a certain time by qualified individuals. Refer to SOP on general physical safety of the system.

SYSTEM NETWORK ARCHITECTURE (SNA) A software protocol designed by IBM to allow networking of the various generations of IBM computers. Most major computer manufacturers have developed a network architecture protocol to support their own computer equipment, and many of these designs include a means to include computers from other manufacturers so as to increase the versatility of the network.

SYSTEM OPERATOR'S LOG This form is used for logging all activities related to the operation of a system at a particular site. These system-level activities refer to all actions involving the actual operation of the system and include regularly scheduled activities (such as backups and periodic maintenance) and unexpected activities (such as malfunctions).

SYSTEM PROBLEM LOG This document is used to chronicle all problems (and related fixes) to general system operations. Refer to SOP on system log -- procedures and review.

SYSTEM REPAIR LOG Used to document all repairs (and related sign-offs) made to hardware related to a particular system. Refer to SOP on general physical safety of the system.

SYSTEM REQUIREMENTS SPECIFICATION (DOCUMENT) A document that prescribes completely and precisely the conditions and/or capabilities necessary for the system or product to fulfill its intended use and/or to satisfy any additional contracts, standards, specifications, or other formal agreements with the firm's customers. System Requirements Specifications typically should include safety requirements, functional requirements, and development standards. Synonymous with product requirements specification.

SYSTEM SECURITY Sensitive equipment may be damaged or erroneous results reported if the system is purposefully or accidentally subject to tampering. System security standards are designed to permit tracing of a product from raw materials to final product, and from final product back to raw materials, in the event of a recall or serious problem; to track all data input or modification back to the individual responsible for that input or modification; and to restrict unauthorized remote or direct access to confidential data and programs.

SYSTEM SPECIFIER The System Specifier is responsible for writing the functional specifications with input from the user advocate. The System Specifier is also responsible for reviewing the system test plan and the user's manual.

SYSTEM TESTER The System Tester is responsible for performing the system test plan and for documenting each step taken during testing. Any errors found are reported with supporting documentation. The System Tester is also responsible for verifying the technical accuracy of the user's manual.

SYSTEM TESTING PHASE In this phase of the Software Development Life Cycle (SDLC), the system tester performs the system test plan. The output documentation is generated to verify the system performs as described in the system test plan. A review of the output documentation to ensure it meets the expected output described in the system test plan is part of this phase.

SYSTEM TEST PLAN Refers to a test plan oriented to the system's environment, providing valid and invalid input data, specifying appropriate output, and writing detailed test instructions. The code is reviewed in a "black box" format, in which the specifics of the module functionality are ignored.

SYSTEM VALIDATION The accumulation of evidence in support of expert testimony demonstrating present and continuing control of a computer system utilized in manufacturing, research, quality assurance, or the function of a medical device. This evidence includes testing of the overall performance of the system, a review of the source code controlling the system applications, an analysis of the support documentation and standard operating procedures, and a review of the system security and disaster recovery policies.

TAKING OUT OF PRODUCTION Removing an automated system from the work flow, or a specific product or recipe/design from the manufacturing process.

TECHNICAL OPERATING PROCEDURES (TOPs) These procedures support SOPs. However, they need not have the same level of resilience as SOPs because they are more likely to change, and the process for their approval is less stringent. TOPs might be, for example, the instructions in an operating manual of some specific piece of equipment, or the reference guide of application software.

TECHNICAL WRITER The Technical Writer is responsible for writing and formatting the user's manual for the system. The Technical Writer obtains information from the functional specifications, test plans, on-line training, and interviews with the software engineers. The Technical Writer continually edits and updates the user's manual to ensure its validity during the product development cycle. This individual is also responsible for reviewing the manual for consistency, readability, and ease of use.

TEST CASE One or more sets of test data and associated procedures developed to exercise a particular function or program path, or to verify compliance with a specific requirement.

TEST DATA Refers to data-sets created to enable a judgement of system performance. It should include normal, unusual, and abnormal values for possible range of entries.

TEST-FILE STRESS ANALYSIS Using a test file in which programs can be examined without affecting the system, each significant program is tested with expected and unexpected stress input. Programs of special vulnerability, and programs not normally accessed in other tests, receive special consideration.

TESTING Designing, conducting, analyzing, and interpreting a series of white, black, and grey box tests to assure the current accuracy and reliability of computer hardware and software system components. These tests generally include a comprehensive source code review, performance test, stress test, and other tests as appropriate.

TESTING COMPONENTS As part of the overall validation process, a series of tests to assure the current accuracy and reliability of computer hardware and software system components. In this study, specific testing includes:
Ongoing System Performance Testing
Comprehensive (50 Case) System Performance Testing
Test File Stress Analysis
Program Logic Source Code Review
Live-Environment Error Log Review

TEST PLAN A document prescribing the approach to be taken for intended testing activities, typically including: items to be tested, tests to be performed, schedules, personnel requirements, evaluation criteria, risks, and contingency plans.

TEST PROCEDURES DOCUMENT A document that contains detailed instructions (Test Procedures) to set up and conduct one or more test cases and to evaluate the results.

TEST REPORT A document that records the conduct and results of test procedures carried out for a system or system component. (See Test Procedures Document).

TEST REQUIREMENTS SPECIFICATION A document that prescribes completely and precisely the test cases, including correct results, necessary to verify that the system, product, or software possesses, meets, or fulfills the appropriate requirements specification. A Test Requirements Specification may stand alone or may be part of a test procedures document.

THROUGHPUT That which is processed. The status of data before it is output. In effect the vulnerability of data to things like truncation or erroneous algorithms during throughput needs to be checked.

TICKIT Refers to a specific set of software quality-assurance guidelines required in the United Kingdom and similar to the ISO 9000 series.

TOOLS Refers to automated aids for the development of systems.

TQM Total Quality Management

TRACKING Inventory, case, subject, condition, and code identifiers should permit dual-directional tracking in the event that recall necessity emerges or source questions arise.

TRAINING The system developer, supporter, or vendor has designed and implemented a system of user training designed to provide basic operations and interpretation background. The training program includes identification of sub-skills or lessons; identification of appropriate user background; and identification of appropriate user authorization, approval, testing or certification examinations.

TRAINING LOG All user training must be documented. The training supervisor should track employee training and follow-up review of the users' skills.

TRAINING SOPs A user organization (in-house) developed system of training and authorizing procedures for user access, which includes the following elements:

1) Identification, outline, or listing of specific training modules, skills, parameters, or tasks.

2) Log of training activities identifying delivery of above listed training elements, specific by user.

3) System of testing, checking, or approving user mastery of the training elements listed above.

4) System of periodic review, recheck, or reinforcement of user skills.

TRAINING SUPPORT PERSONNEL PHASE This phase in the Software Develop ment Life Cycle (SDLC) defines a training program and trains the support personnel on the new system. Written documentation must be provided to support personnel to aid them in understanding and supporting the system. This phase includes a review of the support training documentation to ensure its consistency with the developed product.

UNOBTRUSIVE OBSERVATION Refers to the process of observing a system or personnel without creating subconscious demands; observation without modifying the entity being observed.

UPDATES AND VERSION CHANGES All new versions and updates of software utilized in or in connection with the system under review is subject to careful retesting prior to implementation, meeting appropriate verification standards.

USER The User is responsible for performing the limited release phase and for reporting any problems encountered with supporting documentation. The User is the person who uses the final product on a day-to-day basis. Responsibility for familiarity with and conformity to SOPs is shared by all system users. Similarly, while responsibility for the enforcement of security controls and for the delivery of adequate training may be vested elsewhere, all users are expected to cooperatively comply with and support management policies. [In the GALP Standards section 5.0 the code letter "U" is used to identify all specific and direct responsibilities of the system users.]

USER ADVOCATE The User Advocate, with the user(s) input, is responsible for proposing the system design goals. The User Advocate reviews the functional specifications and the user's manual to ensure compatibility with the system design goals.

USER APPLICATION DOCUMENTATION All appropriate system documents should be carefully logged in an accessible location. These documents should include current and historical user manuals; source code or source code access information, current and historical technical manuals; validation and testing studies; maintenance records; component inventories; and appropriate regulatory, managerial, or quality control reviews.

USER IDENTIFICATION The system should permit ready identification of the unique user responsible for data input, modification, interpretation, or storage. Such identification is dependent upon a reliable password system and upon the maintenance of a user log and access file.

USER PROBLEM LOG This form is used in the logging of user-reported problems related to the operation of a system and associated applications software, located at a specific site. The User Problem Log differs from the system log in that The User Log pertains to problems related to specific applications. The system log entries are more related to the computer system as a whole. Refer to SOP on user-reported problem log.

USER'S MANUAL AND DOCUMENTATION PHASE This phase in the Software Development Life Cycle (SDLC) develops the user documentation for the system. This documentation includes detailed instructions of the system features and operations and may also contain user validation test plans. The documentation is generated from the functional specifications, test plans, on-line training, and interviews with software engineers. The manual must be technically accurate. The format, readability, ease of use, and consistency must be preserved between different revisions. The user documentation must be continually reviewed and edited to reflect changes to the system.

USER VALIDATION A high-level review conducted to determine the conformity of a system, process, or product to internally established standards relating to its use in a specific (user site) environment; and the appropriateness of those standards to that specific site.

VALIDATION A high-level review to determine the conformity of a system, process, or product to internally established standards and the appropriateness of those standards. Validation is a second "quality assurance" examination by an independent party designed to provide a high level of confidence of control of the system or process for managerial and regulatory personnel.

VALIDATION CERTIFICATE Written document certifying that a system has successfully been through the Validation process. This is the first proof that an inspector will expect to see if s/he asks if a system is validated. FDA field investigators have stated that the signature on the certificate may be as far as they go in assessing the validity of the system.

VALIDATION COMPONENTS Systems validation is the comprehensive process of determining the reliability, accuracy, and completeness of a computer system and the ability of that system to track, recall, preserve, and manipulate data free of system-induced variance. The validation process includes an examination of: standard operating procedures (SOPs), system testing procedures; vendor related verification studies; system technical and user documentation, security protocols, and the planning and procedures for recovery in the event of disaster.

VALIDATION SOPs Standard operating procedures outlining a corporate-wide or organization-wide validation policy should be included and available for review.

VENDOR The organization or individual that designs, codes, supports, licenses, and/or distributes automated systems has some responsibilities under the GALP standards. These responsibilities generally impose design, support, and notification requirements. If the vendor is an outside source, providing an awareness of these guidelines may be a necessary role of the laboratory (via its management). In an in-house developed system, these controls may fall under the general purview of the RP. [In the GALP standards section 5.0 the code letter "V" is used to identify all specific and direct responsibilities of the system vendor.]

VENDOR VALIDATION A high-level review conducted to determine the conformity of a system, process, or product to those internally established standards in use by the developer, distributor, or seller and the appropriateness of those standards. The vendor package, supplied to a user, becomes the core of the user validation.

VERIFICATION A review of the vendor supplied documentation and support for a new system, designed to determine the degree of conformity of the vendor's procedures with the standards established in the user's validation policy. The verification may include spot checks on the accuracy of the data, as well as an examination of the procedures used collecting that test data.

VERIFIER Person who re-keys information to ensure its accuracy.

VERSION, UPDATE

PROPOSED POLICY: All functional user and developmental areas (working with the specific software system) shall endorse and adopt a uniform and consistent methodology for defining software versions. This methodology shall be based upon a three-part designation: **X Y Z** with the following denotations:

1. The *X* designation shall represent the primary version. All software in development (prior to initial release) shall carry an *X* designation of "0;" the initial release of that software shall carry the *X* designation of "1." The *X* designation shall increase (to "2," "3," etc.) in the following circumstances:

a. Software is converted to run on a new hardware platform, within or between competing vendor brands;

b. Software is converted to run under a new operating system, including a new release of the existing operating system if that release carries an increased *X* designation itself;

c. Software modifications of any extent or degree directly affecting fundamental safety issues surrounding the function of the computer software system or its final product, including but not limited to modifications prompted by: newly instituted regulatory policies, when those policies interactively determine the suitability of the end product for human use; in-house specified new release criteria or restrictions; FDA-mandated changes under formal 483 notification; and modifications resulting

from discovered major functional errors in previous versions;
d. Software modifications rendering the current (replacement) version incompatible with the new version;
e. Software modifications redefining or restructuring files, tables, or case definitions.
2. The *Y* designation shall represent the secondary version. The *Y* designation increases according to the following rules within each *X* number; a new primary version resets the *Y* designation at "0". Initial release of new software systems carries the designation "1.0.0," regardless of the *Y* designation under the development version. The *Y* designation increases incrementally under the following conditions:
a. A release of a correction version in which the errors modified did NOT affect the fundamental safety of the final product processing operation;
b. Software is subject to minor modifications to run on a new compatible hardware platform;
c. Software is subject to minor modifications to run under a modified operating system not designated by a new *X* version number;
d. Software is enhanced with a modification including any or all of the following:
-- expansion or expanded definition of file, decision table, or case structure;
-- a new feature or capacity not fundamental to or critical to the basic operation of the system, nor to conformity with FDA (or other regulatory) mandated changes;
-- a new feature intended as a user aid, help tool, or other support attribute;
3. The *Z* designation shall represent the tertiary version. The *Z* designation increases incremental whenever one of the listed modifications is released, reverting to "0" whenever the *Y* designation is changed. Changes affecting only the *Z* designation include:
a. Enhancements in internal ("comment") documentation;
b. Software modifications to screen layout not affecting the input data or variable assignment;
c. The addition of compatible, independently-validated or accepted, non-customized additional software programs. For example, adding a word processor package to a system residing within the software shell, but not interacting with the files, tables, cases, or other applications programs in that system, would result in an increased *Z* designation.
4. In all "borderline," ambiguous, or questionable cases, judgement shall default in a <u>conservative</u> direction, including:
a. Changes that may represent either *Z* or *Y* modifications shall be considered *Y* modifications, resulting in an incremental increase in the *Y* version designation;
b. Changes that may represent either *Y* or *X* modifications shall be considered *X* modifications, resulting in an incremental increase in the *X* version designation.

WALKTHROUGH 1) Physical tour of an installation or system to review its compliance to regulations, SOPs, etc. 2) Logical tour of software where a coder explains to one or more peers what has been done and why.

WEINBERG'S LAW If builders built buildings the way programmers program pro-

WORLD HEALTH ORGANIZATION (WHO) Based in Geneva, Switzerland, the WHO assists countries researching their health needs. Many publications are available, including over 50 on pharmacology. A free publications directory is available upon request.

WRITE-ACCESS TO LIBRARIES Ability to manipulate/create/modify programs in a given system section. Developmental personnel should be confined to the test libraries and should be unable to move programs into production or access the programs once they are in production.

X.25 Communication protocol for packet switching.

X.400 Communication protocol for E-mail.

YURNAUKAS' OBSERVATION To err is human; to really foul things up takes a computer.

SYSTEM VALIDATION TERMS ENGLISH-TO-GERMAN

ABNORMAL PROGRAM TERMINATION, OR ABEND -- Programmabbruch
ACCELERATED VALIDATION PROGRAM -- Beschleunigtes Validierungsprogramm
ACCEPTANCE TEST -- Akzeptanztest
ACCEPTANCE VALIDATION -- Akzeptanzvalidierung
ACCESS, AUTHORIZED -- Zugriff, berechtigter
ACCESS CONTROL -- Zugriffsbeschränkung
ACCESS PROTECTION -- Zugriffsschutz
ACCESS, UNAUTHORIZED -- Zugriff, unberechtigter
ACCESS, UNINTENTIONAL -- Zugriff, unbeabsichtigter
ACCURACY -- Fehlerfreiheit
ADDRESS -- Adresse
AFTERIMAGE -- Nachabbild
ARTIFICIAL INTELLIGENCE (AI) -- Künstliche Intelligenz
ALGORITHM -- Algorithmus
ANALOG -- Analog
ANALOG-TO-DIGITAL CONVERTER (ADC) -- Analog-Digital-Umsetzer
ANALYSIS PHASE -- Systemplannungsphase
APPLICATION -- Anwendung
APPLICATION INDUSTRY -- Anwendungsindustrie
APPLICATION PLATFORM -- Benutzeroberflache
APPLICATION SOFTWARE -- Benutzersoftware
ARCHITECTURE -- Architektur
ARCHIVE -- Archiv
ARCHIVE COPY -- Hinterlegungskopie
ARCHIVIST -- Archivar
ARITHMETIC LOGIC UNIT (ALU) -- Arithmetischer Logikbaustein
ASSEMBLE -- Übersetzen
ASSISTANCE REQUEST -- Unterstützungsantrag
ASYNCHRONOUS -- Asynchron
ATTEMPTED ACCESS -- Zugriffsversuch
AUDIT -- Audit
AUDITABILITY -- Auditfähigkeit
AUDIT TEST -- Audittest
AUDIT TRAIL -- Protokoll sämtlicher Eingaben und Änderungen
AUDIT TRAIL REVIEW -- Kontrolle des Audit Trails
AUDIT TRAIL TEST -- Test des Audit Trails
AUTHORIZATION -- Autorisierung
AUTOMATIC AUDIT TRAIL -- Automatischer Audit Trail
AUTOMATIC CODE GENERATOR -- Automatischer Codeerzeuger
AUXILIARY STORAGE -- Hilfsspeicher
AVAILABILITY (OF DATA) -- Verfügbarkeit von Daten

BACKGROUND -- Hintergrund
BACK-UP -- Batensicherung
BAR CODE -- Streifencode
BATCH -- Batch
BATCH INTERFACE -- Batchverbindung
BATCH JOB -- Batchjob
BENCHMARK -- Bezugsgröße
BETA TESTING PHASE -- Beta-Testphase
BETA TEST PLAN DEVELOPMENT -- Entwicklungsphase des Beta-Testplans
BINARY -- Binär
BLACK BOX TEST -- Black-Box Test
BOOTSTRAP -- Urladeprogramm
BRANCH -- Zweig
BRANCH ADDRESS -- Sprungadresse
BRANCH COVERAGE -- Zweigabdeckung
BUFFER -- Puffer
BUG -- Programmfehler
BUS -- Bus
BYTE -- Byte

CAD -- Rechnerunterstüzter Entwurf
CALIBRATION -- Kalibrierung
CENTRAL PROCESSING UNIT (CPU) -- Zentraleinheit
CERTIFICATION -- Zertifizierung
CHANGE -- Änderung
CHANGE CONTROL -- Änderungslenkung
CHANGE CONTROL COORDINATOR -- Koordinator für Änderungslenkung
CHANGE CONTROL LOG -- Protokoll der Änderungslenkung
CHANGE IMPACT STATEMENT -- Protokoll der Änderungsauswirkungen
CHANGE REQUEST -- Änderungsantrag
CHECK CONDITION -- Prüfbedingung
CHECKLIST, CHECKED OFF -- Checkliste, abgearbeitete
CHECKSUM -- Prüfsumme
"CLEAN COMPILE" -- Formfehlerfreie Programmübersetzung
CODING AND MODULE TEST PLAN DEVELOPMENT PHASE -- Entwicklungsphase des Codier und Modultestplans
CODING PHASE -- Codierphase
COMMUNICATION INTERFACE -- Kommunikationsschnittstelle
COMPILE -- Umwandeln
COMPILER -- Compiler
COMPREHENSIVE SYSTEM PERFORMANCE TEST -- Umfassender Systemsleistungstest
COMPUTER-CONTROLLED -- Computergesteuert
CONDITION TEST -- Bedingungstest
CONFIDENTIALITY -- Vertraulichkeit
CONFIGURATION CONTROL -- Umgebungskontrolle

CONFIGURATION ITEM LIST -- Konfigurationsliste
CONGRUOUS ENVIRONMENTS -- Kongruentes Umfeld
CONNECTIVITY -- Vernetzung
CONTENT ERROR -- Inhaltsfehler
CONTINUED ACCURACY AND RELIABILITY -- Stetige Fehlerfreiheit und Zuverlässigkeit
CONTROL OF ENVIRONMENT -- Umfeldüberwachtung
CUMULATIVE DOCUMENTATION TRAIL -- Durchgängige, kummulierende Dokumentation
CURRENT ACCURACY AND RELIABILITY -- Aktuelle Fehlerfreiheit und Zuverlässigkeit
CYCLIC REDUNDANCY -- Zyklische Redundanz

DATA -- Daten
DATA ACCESS -- Datenzugriff
DATA ANALYSIS -- Datenanalyse
DATABASE -- Datenbank
DATABASE DESIGN PHASE -- Datenbank-Entwurfsphase
DATA CHANGE LOG -- Datenänderungsprotokoll
DATA DICTIONARY -- Data Dictionary
DATA FIDELITY -- Datentreue
DATA FLOW DIAGRAM -- Datenflußdiagramm
DATA INTEGRITY -- Datenintegrität
DATA PROCESSING COORDINATOR -- Datenverarbeitungskoordinator
DEBUG -- Programmfehlersuche
DEPARTMENT OF HEALTH (DOH) -- Gesundheitsministerium
DESIGN -- Entwurf
DESIGN ANALYSIS -- Entwurfsanalyse
DESIGN ANALYSIS REPORT -- Entwurfsanalyseprotokoll
DESIGN PHASE -- entwurfphase
DESIGN REQUIREMENTS SPECIFICATION -- Anforderungsspezifikation für den Entwurf
DESIGN REVIEW -- Entwurfskontrolle
DETAILED DESIGN SPECIFICATIONS PHASE -- Entwurfsphase der Detail-spezifikationen
DEVELOPMENTAL TOOLS -- Entwicklungswerkzeuge
DEVELOPMENTAL VALIDATION -- Entwicklungsvalidierung
DEVICE HISTORY RECORD -- Gerätehistorie
DEVICE MASTER RECORD (DMR) -- Gerätebestandsdatei
DIAGNOSTIC -- Diagnostik
DIGITAL -- Digital
DIGITAL-TO-ANALOG CONVERTER -- Digital-Analog Umsetzer
DIRECT CONNECT MODEM -- Direktanschlußmodem
DIRECT DATA ENTRY -- Direkte Dateneingabe
DISASTER -- Notfall
DISASTER RECOVERY PLAN -- Wiederherstellung im Notfall

DISK -- Platte
DISK DRIVE -- Plattenlaufwerk
DISKETTE -- Diskette
DISK OPERATING SYSTEM -- Plattenbetriebssystem
DOCUMENTATION PHASE -- Dokumentationsphase
DOCUMENTATION UPDATE -- Nachführung der Dokumentation
DOWNLOAD -- Download
DRIVER PROGRAM -- Programmrahmen
DRUG UTILIZATION REVIEW (DUR) -- Drogenbenutzungskontrolle
DUAL MATRIX PROCESS OF VALIDATION -- Doppelmatrix-Validierungsverfahren
DYNAMIC PROTOCOLS -- Dynamisches Protokoll

ELECTRONIC DATA INTERCHANGE (EDI) -- Elektronischer Datenaustausch
ELECTRONIC IDENTIFICATION -- Elektronische Identifizierung
ELECTRONIC SIGNATURE -- Elektronische Unterscrift
EMBEDDED -- Eingebettet
EMERGENT CONTROLS -- Entstehende Kontrollen
ENCRYPTION -- Verschlüsselung
END USER -- Endbenutzer
ENVIRONMENTAL CONDITIONS -- Umfeldbedingungen
ERASABLE PROGRAMMABLE READ-ONLY MEMORY (EPROM) -- Löschbarer und programmierbarer Festspeicher
ERRONEOUS DATA TEST -- Test Fehlerhafter Daten
ERROR DETECTION -- Fehlererkennung
ERROR LOGS -- Fehlerkontrollaufzeichnungen
ERROR RECOVERY ROUTINES -- Fehlerbeseitigungsroutinen
EVIDENCE OF VALIDATION -- Validierungsnachweis
EXECUTABLE CODE -- Ablauffähiger Code
EXPERT SYSTEM -- Expertensystem

FAILURE MODE EFFECTS AND CRITICALITY ANALYSIS -- Schadenanalyse von Ausfallsarten und deren Folgen
FAST TRACK VALIDATION -- Beschleunigtes Validierungsverfahren
FIFTH-GENERATION LANGUAGE (5GL) -- Sprache der fünften Generation
FILE ERROR -- Dateifehler
FINAL COMPILATION -- Abschlußübersetzung
FINAL INSPECTION CERTIFICATE -- Ausgangsprotokoll
FINAL PROTCOL OF VALIDATION -- Validierungsabschlußprotokoll
FINGERPRINT -- Fingerabdruck
FIRMWARE -- Firmware
FIRST LAW OF HARDWARE -- Erstes Hardware-Gesetz
FIRST VALIDATION -- Erstvalidierung
FLAG -- Anzeiger
FLOWCHART -- Flußdiagramm
FOOTPRINT -- Grundriß

FORMULAE -- Formeln
FOURTH-GENERATION LANGUAGE (4GL) -- Sprache der vierten Generation
FRACTIONATION -- Fraktion
FUNCTION -- Funktion
FUNCTIONAL ENHANCEMENTS -- Funktionelle Erweiterungen
FUNCTIONALITY -- Funktionalität
FUNCTIONAL SPECIFICATION PHASE -- Erstellungsphase des Pflichtenhefts
FUNCTION TEST -- Funktionstest

GENERATION OF LANGUAGES -- Sprachgenerationen
GOOD AUTOMATED LABORATORY PRACTICES (GALPs) -- Gute Automatisierte Laborpraktiken
GOOD LABORATORY PRACTICES (GLPs) -- Gute Laborpraktiken
GOOD MANUFACTURING PRACTICES (GMPs) -- Gute Herstellungspraktiken
GREY-BOX TEST -- Grey-Box Test
GROUP TEST -- Fachabteilungstest
GUIDELINE -- Richtlinie

HARDWARE -- Hardware
HARDWARE CONSTRAINTS -- Hardware-Beschrankungen
HARDWARE INTERFACE -- Hardware-Schnittsellen
HARDWARE REPLACEMENTS -- Hardware-Nachrüstung
HASH TOTAL -- Mischsumme
HAZARD ANALYSIS -- Gefahrenanalyse
HIGH-LEVEL LANGUAGE -- Höhere Programmiersprache
HIGH-LEVEL TESTING -- Testen auf höherer Ebene
HYPERCARD -- Hyperkarte
HYPERTEXT -- Hypertext

ID CARD -- Identitätskarte
IMPLEMENTATION PHASE -- Implementierungsphase
INFO-GLUT -- Informationsschwemme
INITIAL HAZARD ASSESSMENT -- Erstmalige Gefahreneinschätzung
INITIAL HAZARD ASSESSMENT REPORT -- Erstmaliges Gefahreneinschätzungsprotokoll
INPUT PARAMETER -- Eingabeparameter
INSPECTION -- Prüfung
INSTALLATION QUALIFICATION (IQ) -- Installationsqualifizierung
INSTRUCTION -- Befehl
INSTRUMENT CONTROL -- Instrumentenkontrolle
INTEGRATED SOLUTION -- Integrierte Lösung
INTERACTIVE PROCESSING -- Dialogverarbeitung
INTERFACE -- Schnittstelle
INTERPRETER -- Interpretierer
INTERRUPT -- Unterbrechung
INVENTORY CONTROL -- Bestandsführung

I/O PORT -- Ein-und Ausgabe Anschluß
ITERATION -- Iteration

JOB -- Aufgabe
JUST-IN-TIME (JIT) -- Just-in-time

KERNEL -- Nukleus

LABORATORY INFORMATION MANAGEMENT SYSTEM (LIMS) -- Laborinformationssystem
LABORATORY MANAGEMENT -- Laborführung
LANGUAGE -- Sprache
LIFE CYCLE -- Lebenszyklus
LIMITED ACCESS -- Beschränkter Zugang
LIVE ENVIRONMENT ERROR LOG REVIEW -- Kontrolle der Fehleraufzeichnungen im aktiven Umfeld
LOADER -- Ladeprogramm
LOG -- Kontrollaufzeichnungen
LONGITUDINAL REDUNDANCY -- Longitudinalredundanz
LOW-LEVEL TESTING -- Test auf niederer Ebene

MACHINE LANGUAGE -- Maschinensprache
MAINFRAME -- Großrechner
MAIN MEMORY -- Hauptspeicher
MAIN STORAGE SIZE -- Hauptspeichergröße
MAINTAINABILITY -- Wartbarkeit
MAINTENANCE -- Wartung
MAINTENANCE CONTRACT -- Wartungsvertrag
MAINTENANCE LOG -- Wartungsprotokoll
MAINTENANCE ORDER -- Wartungsauftrag
MAINTENANCE QUALIFICATION -- Wartungsqualifizierung
MAINTENANCE SCHEDULE -- Wartungsplan
MANAGEMENT AWARENESS AND CONTROL -- Managementbewußtsein und Aufsicht
MANAGER OF SOFTWARE ENGINEERING -- Leiter der Software-Entwicklung
MANUFACTURER -- Hersteller
MATERIAL REQUIREMENTS PLANNING -- Materialbedarfsplannung
MEDICAL CLASS I, II, III/IV -- Medizinische Klasse I, II, III, IV
MEDICAL DEVICE -- GENERAL REGULATORY ASSESSMENT -- Medizinische Geräte -- allgemeine Verordnungsbegutachtung
MENU -- Menü
MICROFICHE -- Mikrofiche
MICROPROCESSOR -- Mikroprozessor
MILESTONES -- Meilensteine
MILITARY SAMPLING -- Militärische Probenentnahme
MNEMONIC -- Mnemonisch

MODEL -- Modell
MODEMS -- Modems
MODULARIZATION -- Modularisierung
MODULARIZATION CRITERIA -- Modularisierungskriterien
MODULE TESTING PHASE -- Modultestphase
MULTIPLEXER -- Multiplexkanal

NATURAL LANGUAGE -- Natürliche Sprache
NETWORK -- Netzwerk
NODE -- Knoten

OBJECT CODE -- Phase
OBJECTIVITY AND INDEPENDENCE -- Objektivität und Unabhängigkeit
ONGOING SYSTEM PERFORMANCE TEST -- Andauernder Systemleistungstest
ON-LINE -- Online
OPEN SYSTEMS INTERCONNECTION -- Offenes Kommunikationssystem
OPERATING PERSONNEL -- Bedienungspersonal
OPERATING QUALIFICATION (OQ) -- Betriebsqualifizierung
OPERATING SYSTEM -- Betriebssystem
OPERATIONAL LIMITS -- Grenzen der Anwendbarkeit
OPERATOR -- Bediener
ORIGINAL EQUIPMENT MANUFACTURER (OEM) -- Ursprünglicher Gerätehersteller

PARALLEL -- Parallel
PARITY -- Parität
PARITY BIT -- Paritätsbit
PASSWORD -- Paßwort
PASSWORD CONTROL LOG -- Paßwortkontrollaufzeichnungen
PASSWORD FILE -- Paßwortdatei
PASSWORD SYSTEM -- Paßwortsystem
PATCH -- Direktkorrektur
PATH -- Pfad
PERFORMANCE -- Leistung
PERFORMANCE QUALIFICATION -- Leistungseignung
PERFORMANCE TEST -- Leistungstest
PERIPHERAL EQUIPMENT -- Peripheres Gerät
PHANTOM PHASE -- Phantomphase
PHASE DEFINITION FORMAT -- Format der Phasendefinition
PHYSICAL INSPECTION -- Physische Kontrolle
PORTABILITY -- Portabilitat
PRELIMINARY HAZARD ANALYSIS -- Vorläufige Gefahrenanalyse
PRELIMINARY HAZARD ANALYSIS REVIEW -- Vorläufige Gefahrenanalysenkontrolle
PREMARKETING APPLICATION -- Antrag bevor Kommerzialisierung
PRIVACY -- Intimsphäre

PROCEDURE -- Prozedur
PROCESS CONTROL SYSTEM -- Prozeßsteuerungssystem
PROCESS PERFORMANCE CHARACTERISTICS -- Prozeßleistungsmerkmale
PROCESS PROCEDURE -- Prozeßablauf
PROCESS QUALIFICATION (PQ) -- Prozeßqualifizierung
PROCESS VALIDATION -- Prozeßvalidierung
PRODUCT AND PROCESS TESTING CONSIDERATIONS -- Produkt-und Prozeßtesterwägungen
PRODUCT APPROVAL PHASE -- Produktzulassungsphase
PRODUCT DATABASE -- Produkt-Datenbank
PRODUCT MANAGER -- Produktmanager
PRODUCT REQUIREMENTS SPECIFICATION -- Spezifizierung der Produktanforderungen
PROGRAM -- Programm
PROGRAM CHANGE REQUEST FORM -- Antragsformular für Programmänderungen
PROGRAM LOGIC SOURCE CODE REVIEW -- Kontrolle des Quellencodes des Programmablaufs
PROGRAM PATH -- Programmpfad
PROGRAM PATH ANALYSIS -- Analyse des Programmpfades
PROGRAM REQUEST PROCESSING CHECKLIST -- Bearbeitungscheckliste für Programmanforderungen
PROGRAMMABLE READ-ONLY MEMORY (PROM) -- Programmierbare Festspeicher
PROJECT ADMINISTRATOR -- Projektverwalter
PROJECT LEADER -- Projektleiter
PROJECT MANAGER OR PHASE/PROJECT LEADER -- Projektmanager oder Phasen/Projectleiter
PROJECT PLAN -- Projektplan
PROJECT PLAN PHASE -- Projektplannungphase
PROTECTION -- Sicherheit
PROTOCOL -- Protokoll
PSEUDO CODE -- Pseudocode
PURGE -- Löschung

QUALITY ASSURANCE (QA) -- Qualitätssicherung
QUALITY ASSURANCE PERSONNEL -- Qualitätssicherungspersonal
QUALITY ASSURANCE SYSTEM -- Qualitätssicherungssystem
QUALITY CHARACTERISTIC -- Qualitätsmerkmale
QUALITY CONTROL -- Qualitätskontrolle
QUALITY CONTROL UNIT -- Qualitätslenkungsgruppe
QUALITY PROOF -- Qualitätsnachweis
QUICK AND DIRTY -- Quick and dirty

RANDOM ACCESS MEMORY (RAM) -- Direktzugriffsspeicher
RAW DATA -- Ausgangsdaten

REAGENT -- Reagens
RECALIBRATION -- Rekalibrierung
RECALL OF INFORMATION -- Rückruf von Information
RECORD -- Datensatz
RECOVERY -- Recovery
RECOVERY CAPABILITY -- Wiederherstellbarkeit
REGULATORY QUALITY -- Qualität der Verordnungsmäßigkeit
RELEASE PHASE -- Freigabephase
RELEASE PROTOCOL -- Freigabeprotokoll
REPAIR PARTS LIST -- Ersatzteilliste
REQUIREMENTS DEFINITION -- Anforderungsbestimmung
REQUIREMENTS INSPECTION REPORT -- Anforderungskontrollprotokoll
REQUIREMENTS TRACE MATRIX -- Anforderungsverfolgungsmatrix
REQUISITE STANDARD OPERATING PROCEDURES AND LOGS -- Erforderliche Standardarbeitsvorschriften und Protokolle
RESISTANT TO STRESS -- Robustheit
RESPONSE TIME -- Antwortzeit
RESPONSIBLE PERSON (RP) -- Verantwortliche Person
RESTART -- Wiederanlauf
RETROSPECTIVE VALIDATION -- Retrospektive Validierung
REUSABILITY -- Wiederverwendbarkeit
REVALIDATION -- Revalidierung
REVIEWER CREDENTIALS -- Qualifizierung des Kontrolleurs
ROBOTICS -- Robotik
ROM -- Festspeicher
ROUTINE TESTING OF SOFTWARE LOG -- Routinetest des Software-Protokolls

SAFETY ANALYSIS AS RELATED TO COMPUTER SOFTWARE OR SYSTEM FIRMWARE -- Sicherheitsanalyse relativ zur Software oder zur System-Firmware
SAFETY-CRITICAL SOFTWARE -- Sicherheitskritische Software
SAMPLE -- Probe
SAVE -- Abspeichern
SCENARIO ANALYSIS -- Szenarioanalyse
SCHEMA -- Datenbankbeschreibung
SECURITY LOG -- Sicherheitsprotokoll
SEGAL'S LAW -- Segals Gesetz
SELF-CHECKING CODE -- Selbstprüfender Code
SELF-EXPLANATORY -- Selbsterklärend
SERIAL -- Scriell
SHARED LOCK -- Gemeinsame Sperre
SITUATIN-UNIQUE TEST -- Einmaligkeitstest der Situation
SMART MODEM -- Intelligentes Modem
SOFTWARE -- Software
SOFTWARE COMPLEXITY METRIX -- Software-Komplexitätsmetrik
SOFTWARE CONFIGURATION CONTROL PROCEDURES -- Kontrollprozeduren der Software-Konfiguration

SOFTWARE DEVELOPER -- Software-Entwickler
SOFTWARE DEVELOPMENT LIFE CYCLE (SDLC) -- Lebenszyklus der Software-Entwicklung
SOFTWARE DEVELOPMENT PLAN -- Software-Entwicklungsplan
SOFTWARE ENGINEER -- Software-Enwickler
SOFTWARE ERROR LOG -- Software-Fehlerprotokoll
SOFTWARE HAZARD ANALYSIS -- Software-Gefahrenanalyse
SOFTWARE HAZARD ANALYSIS REPORT -- Software-Gefahrenanalyseprotokoll
SOFTWARE INTERFACE -- Software-Schnittstellen
SOFTWARE REQUIREMENTS SPECIFICATION -- Software-Pflichtenheft
SOFTWARE, REUSABLE -- Software, wiederverwendbar
SOP -- SAV
SOP ACCURACY -- Genauigkeit der SAVs
SOP ADHERENCE -- Befolgung der SAVs
SOP RESILIENCE -- Anpassungsvermögen der SAVs
SOURCE CODE -- Quelcode
SOURCE CODE REVIEW -- Quellcodekontrlle
SOURCE CODE TEST RECORD -- Quellcodedatensatz
SOURCE PROGRAM -- Quellenprogramm
SPAGHETTI CODE -- Spaghetticode
STANDARD OPERATING PROCEDURES (SOPs) -- Standardarbeitsvorschriften (SAVs)
STATIC PROTOCOLS -- Statisches Protokoll
STRESS TEST (SEE ERRONEOUS DATA TEST) -- Streß-Test (vgl. Test fehlerhafter Daten)
SUPPORT -- Unterstützung
SUPPORT PERSONNEL -- Unterstützungspersonal
SUPPORT PERSNNEL TRAINER -- Ausbilder für Unterstützungspersonal
SYNTAX -- Syntax
SYNTAX REVIEW -- Syntaxkontrolle
SYSTEM -- System
SYSTEM ARCHITECT -- Systemarchitekt
SYSTEM ARCHITECTURE PHASE -- Systemarchitekturphase
SYSTEM BACKUP LOG -- Protokoll der Systemdatensicherung
SYSTEM DEVELOPMENT PLAN -- Systementwicklungsplan
SYSTEM ERROR -- Systemfehler
SYSTEM HAZARD ANALYSIS -- Systemgefahrenanalyse
SYSTEM HAZARD ANALYSIS CROSS CHECK -- Kreuzsicherung der Systemgefahrenanalyse
SYSTEM HAZARD ANALYSIS REPORT -- Protokoll der Systemgefahrenanalyse
SYSTEM INTEGRATION TESTING PHASE -- Testphase der Systemintegration
SYSTEM INTEGRATOR -- Systemintegrator
SYSTEM MAINTENANCE LOG -- Protokoll der Systemwartung
SYSTEM NETWORK ARCHITECTURE -- Architektur des Systemnetzwerkes
SYSTEM OPERATOR'S LOG -- Protokoll der Systembenutzer
SYSTEM PROBLEM LOG -- Protokoll der Systemprobleme

SYSTEM REPAIR LOG -- Protokoll der Systemreparaturen
SYSTEM REQUIREMENTS SPECIFICATION -- Spezifizierung des Systempflichtenhefts
SYSTEM REQUIREMENTS SPECIFICATION DOCUMENT -- Systempflichtenheft
SYSTEM SECURITY -- Systemsicherheit
SYSTEM SPECIFIER -- Systemspezifizierer
SYSTEM TESTER -- Systemtester
SYSTEM TESTING PHASE -- Systemtestphase
SYSTEM TEST PLAN WRITER -- Verfasser des Systemtestplanes
SYSTEM VALIDATION -- Systemvalidierung

TAKING OUT OF PRODUCTION -- Außereinsatznahme
TECHNICAL WRITER -- Technischer Redakteur
TEST CASE -- Testfall
TEST DATA -- Testdaten
TEST FILE STRESS ANALYSIS -- Streßanalyse der Testdatei
TESTING -- Test
TESTING COMPONENTS -- Testkomponenten
TEST PLAN -- Testplan
TEST PROCEDURES DOCUMENT -- Testablaufdokument
TEST REPORT -- Testprotokoll
TEST REQUIREMENTS SPECIFICATIONS -- Spezifikationen der Testanfrderungen
THROUGHPUT -- Durchsatz
TOOLS -- Werkzeuge
TRACKING -- Nachführen
TRAINING -- Schulung
TRAINING LOG -- Schulungsprotokoll
TRAINING SOPs -- Schulungs SAVs
TRAINING SUPPORT PERSONNEL -- Schulung des Unterstützungspersonal

UNOBTRUSIVE OBSERVATION -- Unauffällige Beobachtung
UPDATES AND VERSION CHANGES -- Versionsänderungen und Aktualisierung
USER -- Benutzer
USER ADVOCATE -- Benutzeranwalt
USER APPLICATION DOCUMENTATION -- Betreiberdokumentation
USER IDENTIFICATION -- Benutzeridentifizierung
USER NUMBER -- Benutzernummer
USER PROBLEM LOG -- Benutzerproblemprotokoll
USER VALIDATION -- Benutzervalidierung

VALIDATION -- Validierung
VALIDATION CERTIFICATE -- Validierungszertifikat
VALIDATION COMPONENTS -- Validierungskomponente
VALIDATION SOPs -- SAVs für Validierung
VENDOR -- Lieferant
VENDOR VALIDATION -- Lieferantenvalidierung

VERIFICATION -- Verifikation
VERIFIER -- Überprufer
VERSION, UPDATE -- Version, Nachführung der

WALKTHROUGH -- Schreibtischtest
WHITE BOX TEST -- White-Box Test
WORLD HEALTH ORGANIZATION -- Weltgesundheitsorganisation
WRITE-ACCESS TO LIBRARIES -- Schreibzugriff auf Bibliotheken

Appendix B

BIBLIOGRAPHY

Accomando, William P. CANDA Overview: What Should a CANDA Look Like? *Drug Information Journal*, Vol. 27 (April-June 1993).

Ackoff, Russell L. Management Misinformation Systems. *Management Science*, December 1967, pp. 147-156.

Adamson, J.R. An Approach to Validation. *Pharmaceutical Engineering*, Vol. 12 (May 1992).

Advanced Micro Devices. The AM7990 Family Ethernet Node. Sunnyvale, CA: Advanced Micro Devices, 1983.

Agalloco, James. Validation of Existing Computer Systems. *Pharmaceutical Technology*, Vol. 11 (January 1987).

Agalloco, James. The Validation Life Cycle. *Journal of Parenteral Science & Technology*, Vol. 47 (May-June 1993).

AICPA. Audit Approaches for a Computerized Inventory System. New York: American Institute of Certified Public Accountants, 1980.

AICPA. Audit Considerations in Electronic Fund Transfer Systems. New York: American Institute of Certified Public Accountants, 1979.

AICPA. Auditing Standards and Procedures. Committee on Auditing Procedure Statement No. 33. New York: American Institute of Certified Public Accountants, 1963.

AICPA. Guidelines to Assess Computerized General Ledger and Financial Reporting Systems for Use in CPA Firms. New York: American Institute of Certified Public Accountants, 1979.

AICPA. Management, Control and Audit of Advanced EDP Systems. New York: American Institute of Certified Public Accountants, 1977.

AICPA. Statement on Auditing Standards No. 3, The Effects of EDP on the Auditor's Study and Evaluation of Internal Control. New York: American Institute of Certified Public Accountants, 1974.

AICPA. The Auditor's Study and Evaluation of Internal Control in EDP Systems. New York: American Institute of Certified Public Accountants, 1977.

Aitman, Mark. Validation of High Integrity Systems. *Drug Information Journal*, Vol. 27 (April-June 1993).

Almgren, Raymond. Software Development Strategies for PC-Based Testing. *Evaluation Engineering*, January 1991.

Allocca, John A. and Allen Stuart. Transducers Theory and Applications. Reston, VA: Reston Publishing Company, Inc., 1984.

American Chemical Society Committee. Guidelines for Data Acquisition and Data Quality Evaluation in Environmental Chemistry. *Analytical Chemistry*, Vol. 52, 1980.

Alter, Steven. Decision Support Systems: Current Practice and Continuing Challenges. Reading, MA: Addison-Wesley Publishing Company, 1980.

Alwitt, Josh and John Kinney. The Impact of Document Image Management on Clinical Data Processing. *Drug Information Journal*, Vol. 27, (October-December 1993).

AMF Potter & Brumfield Catalog. Princeton, IN: Potter & Brumfield, 1984.

Analog Devices, Inc. Analog Devices 3B Industrial Control Series Data Sheet. Norwood, MA, 1981.

Anon. Products-Information Management. *Laboratory Practice*, Vol. 38 (May 1989).

Andersen, Arthur. A Guide for Studying and Evaluating Internal Controls. Chicago: Arthur Andersen and Co., 1978.

Andersen, Niels Erik, *et al.* Professional Systems Development. New York: Prentice-Hall International, 1990.

Anderson, C.A. Approach to Data Processing Auditing. *The Interpreter*, Insurance Accounting and Statistical Association, Durham, NC, April 1975, pp. 23-26.

Anderson, John. Object-Oriented Software Development: Method Assessment. George Mason University, Spring 1990.

Anisfeld, Michael H. Validation -- How much Can the World Afford: Are We Getting Value for Money? *Journal of Pharmaceutical Science and Technology*, Vol. 48 (January/February 1994).

Ariv, Gad and Michael J. Ginzberg. DDS Design: A Systematic View of Decision Support. *Communications of the ACM*, Vol. 28 (October 1985), pp. 1045-1052.

Arthus, L.J. Measuring Programmer Productivity and Software Quality. New York: John Wiley & Sons, 1985.

Aron, Joel. The Program Development Process, Part II: The Programming Team. Reading, MA: Addison-Wesley, 1983.

Auerbach Staff. What Every Auditor Should Know About DP. Data Processing Management Service, Auerbach Publishing Co., Portfolio 3-09-03, 1975.

Auslander, David M. and Paul Sagues. Microprocessors for Measurement and Control. Berkeley: Osborne/McGraw-Hill, 1981.

Awad, Elias M. Introduction to Computers in Business. Englewood Cliffs, NJ: Prentice-Hall, Inc., 1977.

Axner, David H. and Fonnie H. Regan. Alphanumeric Display Terminal Survey. *Datamation*, Vol. 24, (June 1978), pp. 183-219.

Baber, Robert L. Error-Free Software: Know-How and Know-Why of Program Correctness. John Wiley and Sons, Ltd., 1991.

Baer, Robert M. The Digital Villain. Reading, MA: Addison-Wesley Publishing Co., Inc., 1972.

Baird, Lindsay L. Identifying Computer Vulnerability. *Data Management*, June 1974, pp. 14-17.

Bar-Hava, Ne. Training EDP Auditors. *Information and Management*, No. 4, 1981, pp. 30-42.

Barnard, Chester I. The Functions of the Executive. Cambridge: Harvard University Press, 1968.

Barr, David B. FDA's Aseptic Processing: Proposed Regulation. *Journal of Parenteral Science & Technology*, Vol. 47 (March-April 1993).

Barrett, M.J. Education for Internal Auditors. *EDP Auditor*, Spring 1981, pp. 11-20.

Basalla, George. The Evolution of Technology. Cambridge: Cambridge University Press, 1988.

Basden, A. and E.M. Clark. Data Integrity in a General Practice Computer System (CLINICS), *International Journal of Bio-Medical Computing*, Vol. 11 (1980), pp. 511-519.

Bauer, Nancy, Bruce Binkowitz, and Wanda Bidlack. The Progression and Integration of the Statistical CANDA at Merck. *Drug Information Journal*, Vol. 27 (January-March, 1993).

Beckman Instruments. Hall Effect Manual. 2nd ed. Fullerton, CA: Helipot Division of Beckman Instruments, Inc., 1964.

Beckman Instruments. Interfacing Liquid Crystal Displays in Digital Systems, Application Note AN-B. Beckman Instruments, Inc., Scottsdale, AZ, 1980.

Beizer, Boris. The Frozen Keyboard. Blue Ridge Summit, PA: PAB Books, 1988.

Beizer, Boris. Software System Testing and Quality Assurance. New York: Van Nostrand Reinhold, 1984.

Bellin, David and Gary Chapman, eds. Computers in Battle. New York: Harcourt, Brace, Javanovich, 1987.

Bennett, John L., ed. Building Decision Support Systems. Reading, MA: Addison Wesley Publishing Company, 1983.

Bentley, Jon. More Programming Pearls: Confessions of a Coder. Reading, MA: Addison-Wesley, Publishing Company, 1988.

Benton, F. Warren. EXECUCOMP: Maximum Management with the New Computers. New York: John Wiley & Sons, 1983.

Bequai, August. How to Prevent Computer Crime. New York: John Wiley & Sons, 1984.

Berkeley, Peter E. Computer Operations Training: A Strategy for Change. New York: Van Nostrand Reinhold Company, 1984.

Berkland, Richard D. Validatable Software Design. *Pharmaceutical Technology*, Vol. 15 (September 1991).

Biggerstaff, Ted. J. and Alan J. Perlis, eds. Software Reusability. Vol. 2, Applications and Experience. New York: ACM Press/Addison-Wesley, 1989.

Birnbarum, N., J. Hains, J. Farkas, and P. Cruz. A LIMS Spectrophotometer Interface. *American Laboratory*, March 1990.

Black, Henry C. Black's Law Dictionary. Revised Fourth Edition. St. Paul: West Publishing Co., 1968.

Blake, Chuck (American Red Cross Vice President for Computer and Information Systems), private interview, March 1989.

Block, Robert. The Politics of Projects. Englewood Cliffs, NJ: Yourdon Press/Prentice-Hall, 1983.

Boar, Bernard. Application Prototyping. New York: John Wiley & Sons, 1984.

Board of Governors of the Federal Reserve System. Electronic Fund Transfers, Regulation E (12 CFR 205), Effective March 30, 1979 (as amended effective May 10, 1980).

Boehm, B.W. Software Engineering Economics. Englewood Cliffs, NJ: Prentice-Hall, 1982.

Bolt, Richard A. The Human Interface. London: Wadsworth, Inc., 1984.

Bonczek, Robert H., Clyde W. Holsapple, and Andrew B. Whinston. Foundations of Decision Support Systems. New York: Academic Press, Inc., 1981.

Boskin, Michael I. Bashing the Myths About LIMS. *Plastics Technology*, October 1989.

Boskin, Michael I. and Rick A. Ziebell. LIMS: An SPC Tool. *Quality*, May 1990.

Bottenberg, R.A. and R.E. Christal. An Iterative Technique for Clustering Criteria Which Retains Optimum Predictive Efficiency. Lackland Air Force Base, Texas: Personnel Laboratory, Wright Air Development Division, March 1961 (WADD-IN-61-3, ASTIA Document AS-261 615).

Bouldin, Barbara. Agents of Change. Englewood Cliffs, NJ: Prentice-Hall, 1989.

Braff, Terry. A Test Program -- Actual Costs and Benefits. 5th International Conference on Testing Computer Software, Measuring Testing Success, 1988.

Breitenberg, Maureen. Questions and Answers on Quality, the ISO 9000 Standard Series, Quality System Registration, and Related Issues. NISTIR 4721. Gaithersburg, MD: U.S. Department of Commerce, National Institute of Standards and Technology, Standards Code and Information Program, April 1993.

Bretzin, M.J. and M.L. Hess. Quality Assurance of the Computer Validation Process in the Control Laboratory. *Pharmaceutical Engineering*, Vol. 12 (May 1992).

Bronstein, Robert J. The Concept of a Validation Plan. *Drug Information Journal*, Vol. 20 (1986), pp. 37-42.

Brooks, Fred. The Mythical Man-Month. Reading, MA: Addison-Wesley, 1975.

Brooks, F.P. No Silver Bullet: Essence and Accidents of Software Engineering. *Computer*, Vol. 20 (April 1987).

Brown, Elizabeth H. Procedures and Their Documentation for a LIMS in a Regulated Environment. In R.D. McDowall, ed. Laboratory Information Management Systems. Wilmslow, U.K.: Sigma Press, 1987, pp. 346-358.

Bruckheimer, Michael. FDA's Inspection of Clinical Investigators. *Drug Information Journal*, Vol. 27 (January-March 1993).

Buxton, J.M., P. Naur, and B. Randell. Software Engineering, Concepts and Techniques. New York: Petrocelli/Charter, 1976.

Buyse, Marc. Regulatory Versus Public Health Requirements in Clinical Trials. *Drug Information Journal*, Vol. 27 (October-December 1993).

Callahan, John J. Needed: Professional Management in Data Processing. Englewood Cliffs, NJ: Prentice-Hall, 1983.

Campbell, D.T. and J.C. Stanley. Experimental and Quasi-Experimental Designs for Research. Chicago: Rand McNally, 1963.

Campbell, S.K. Flaws and Fallacies in Statistical Thinking. Englewood Cliffs, NJ: Prentice-Hall, 1974.

Campo, Jose A. Design Transfer. *Medical Device & Diagnostic Industry*, Vol. 16 (February 1994).

Canadian Institute of Chartered Accountants Committee. Competence and Professional Development in EDP for the CA. *CA Magazine*, September 1974, pp. 26-70.

Canning, Richard G. COBOL Aid Packages. *EDP Analyzer*, January 1972.

Canning, Richard G. Computer Security: Backup and Recovery Methods. *EDP Analyzer*, January 1972.

Canning, Richard G. That Maintenance "Iceberg." *EDP Analyzer*, October 1972.

Canning, Richard G. Project Management Systems. *EDP Analyzer*, September 1976.

Caputo, C.A. Managing the EDP Audit Function. *COM-SAC, Computer Security, Auditing and Controls*, Vol. 8, January 1981, pp. A-1 to A-8.

Card, David N. with Robert L. Glass. Measuring Software: Design Quality. Englewood Cliffs, NJ: Prentice-Hall, 1990.

Cardarelli, Joseph S., Sami A. Halaby, Chris Gopal, and Art Mlodozeniec. Initiating CIM: Manufacturing vs. Pilot-Plant Operations. *Pharmaceutical Technology*, Vol. 15 May 1991).

Carr, Jim, Michael Eddy, and Albert Rego. How to Implement an Electronic Document Storage and Retrieval System. *Medical Device & Diagnostic Industry*, Vol. 15 (February 1993).

Cash, James I., Jr., F. Warren McFarlan, and James L. McKenney. Corporate Information Systems Management: Text and Cases. Homewood, IL: Richard D. Irwin, 1983.

Casti, John L. Paradigms Lost. New York: William Morrow, 1989.

Chamberlain, Richard. Computer Systems Validation for the Pharmaceutical and Medical Device Industries. 2nd Edition. Buffalo Grove, IL: Interpharm Press, 1994.

Chapman, Kenneth G. A History of Validation in the United States: Part I. *Pharmaceutical Technology*, Vol. 15 (October 1991).

Chapman, Kenneth G. A History of Validation in the United States: Part II, Validation of Computer-Related Systems. *Pharmaceutical Technology*, Vol. 15 (November 1991).

Chapman, Kenneth G. and Paul F. Winter. Electronic Identification and Signatures: A Response. *Pharmaceutical Technology*, Vol. 16 (November 1992).

Chapman, M. Audit and Control in Data Base/IMS Environment. *COM-SAC, Computer Security, Auditing and Controls*, Vol. 6, July 1979, pp. A-9 to A-14.

Charette, Robert N. Software Engineering Risk Analysis and Management. New York: McGraw-Hill, 1989.

Cheney, P.H. Educating the Computer Audit Specialist. *EDP Auditor*, Fall 1980, pp. 9-15.

Chew, Nancy. Validation: When Too Much is Not Enough. *BioPharm*, Vol. 6 (September 1993).

Christ, Oliver P. Software Validation: Requirements for Programmable Electronic Medical Systems, Part I, How to Comply with the Essential Requirements of the EC Directives. *Medical Device Technology*, March 1994.

Christian, Kaare. The UNIX Operating System. New York: John Wiley & Sons, 1983.

Chudzicki, Mark J. LIMS: Family Designs for Growth, Flexibility, and Corporate Productivity. *Research & Development*, November 1989.

CICA. Computer Audit Guidelines. Canadian Institute of Chartered Accountants, Toronto, Ontario, Canada, 1975.

CICA. Computer Control Guidelines. Canadian Institute of Chartered Accountants, Toronto, Ontario, Canada, 1971.

Circuit Court of Tennessee. Integrated Medical Systems, Inc. vs. Methodist Health-System, Inc., case number 16316-2. Circuit Court of Tennessee, Thirteenth Judicial Circuit at Memphis, July 1986.

Clark, A. Samuel. Computer Systems Validation: An Investigator's View. *Pharmaceutical Technology*, Vol. 12 (January 1988).

Clark, C.T. and L.L. Schkade. Statistical Methods for Business Decisions. Cincinnati: South-Western Publishing, 1969.

Clark, Robin and Paul Redstone. Establishing the Functions to be Validated. *Drug Information Journal*, Vol. 27 (April-June 1993).

Clary, J.B. and R.A. Sacane. Self-Testing Computers. IEEE Computer, October 1979, pp. 45-59.

Clinical Laboratory Improvement Act of 1967 (P.L. 90-174, December 5, 1967).

Clinical Laboratory Improvement Amendments of 1988 (P.L. 100-578, October 31, 1988).

Clocksin, W.F. and C.S. Mellish. Programming in PROLOG. New York: Springer-Verlag, 1981.

Cochran, W.G. Sampling Techniques. New York: John Wiley, 1963.

Cochran, W.G. and G.M. Cox. Experimental Designs. 2nd ed. New York: John Wiley, 1957.

College of American Pathologists. Standards for Laboratory Accreditation. Skokie, IL: Commission on Laboratory Accreditation, College of American Pathologists, 1988.

Cole, Jon, Tyler W. Mainquist, and James R. Cichoracki. Computerization and Computer Validation: Managing the Risks. *Applied Clinical Trials*, Vol. 2 (December 1993).

Collins, Mary, ed. CANDA, 1995: An International Regulatory and Strategy Report. Waltham, MA: Parexel International Corp., 1994.

Connell, John L. and Linda Brice Shafer. Structured Rapid Prototyping. Englewood Cliffs, NJ: Prentice-Hall, 1989.

Connolly, Thomas M. EPA Investigations of Laboratory Fraud in the CLP and GLP Programs. Presentation at the Society of Quality Assurance 1993 Annual Meeting, San Francisco, October 6, 1993.

Connor, J.E. and B.H. De Vos. Guide to Accounting Controls -- Establishing, Evaluating and Monitoring Control Systems. Boston: Warren, Gorham & Lamont, Inc., 1980.

Conover, W.J. Practical Nonparametric Statistics. New York: John Wiley, 1971.

Contrac Corporation. Raster Graphics Handbook. Covina, CA: Conrac Corporation, 1980.

Cooke, John E. and Donald H. Drury. Management Planning and Control of Information Systems. Hamilton, Ontario: Society of Management Accountants of Canada, April 1980.

Cummings, Stuart W. Distributed Databases for Clinical Data Processing. *Drug Information Journal*, Vol. 27 (October-December 1993).

Crosby, P.B. Quality is Free. New York: McGraw-Hill, 1979.

Cutaia, Al. Technology Projection Modeling of Future Computer Systems. Englewood Cliffs, NJ: Prentice-Hall, 1990.

Dahl, O.J., Edsger Dijkstra, and C.A.R. Hoare. Structured Programming. Englewood Cliffs, NJ: Prentice-Hall, 1972.

Daniel, W.W. Applied Nonparametric Statistics. Boston: Houghton Mifflin, 1978.

Daniels, Michael W., Douglas O. Fegenbush, and Edward F. Frees. An Alphanumeric and Graphic LIMS for Routine Physical Chemistry Analyses. *American Laboratory*, July 1992.

Data Acquisition Telecommunications. Local Area Networks 1983 Data Book, Sunnyvale, CA: Advanced Micro Devices, Inc., 1983.

Datapro Research. Datapro Reports on Information Security. Delran NJ: McGraw-Hill, Inc., 1989.

Date, Chris. An Introduction to Database Systems. Vol. 1, 4th ed. Reading, MA: Addison-Wesley, 1986.

Davey, Judith. Electronic Data Transfer for Clinical Trials. *Drug Information Journal*, Vol. 28 (April-June 1994).

David, F.N. Games, Gods, and Gambling. London: Charles Griffin and Company, 1962.

Davies, O.L., ed. The Design and Analysis of Industrial Experiments. New York: Hafner Publishing, 1956.

Davis, G. Ensuring On-Line System Integrity Using Parallel Simulation on a Continuing Basis. *COM-SAC, Computer Security, Auditing and Controls*, Vol. 7 (July 1978), pp. A-9 to A-11.

Davis, Gordon W. Early Development of the Clinical Information System of a Start-Up Pharmaceutical Company. *Applied Clinical Trials*, Vol. 3 (June 1994).

Davis, Keith. Human Behavior at Work. New York: McGraw-Hill, 1972.

Davis, Randal, Bruce Buchanan, and Edward Shortliffe. Production Rules as a Representation for a Knowledge-Based Consultation Program. *Artificial Intelligence*, Vol. 8 (1977).

Davis, Stanley M. Future Perfect. Reading, MA: Addison-Wesley, 1987.

Dean, R.N. A Comparison of Decision Tables Against Conventional COBOL as a Programming Tool for Commercial Applications. *Software World*, Spring 1971.

Dearden, John. MIS is a Mirage. *Harvard Business Review*, January-February, 1972, pp. 90-99.

De Feu, Marc. ISO 9000 and its Relation to System Validation. *Drug Information Journal*, Vol. 27 (April-June 1993).

DeMarco, Thomas. Controlling Software Projects. New York: Yourdon Press, 1982.
DeMarco, Thomas. Structured Analysis and System Specification. Englewood Cliffs, NJ: Yourdon Press/Prentice-Hall, 1978.
DeMarco, Thomas and Tim Lister. Peopleware. New York: Dorset House, 1987.
DeMarco, Thomas and Tim Lister. Software State-of-the-Art: Selected Papers. New York: Dorset House, 1991.
Dent, Nigel J., ed. Implementing International Good Practices: GAPs, GCPs, GLPs, and GMPs. Buffalo Grove, IL: Interpharm Press, 1993.
DeSain, Carol. Drug, Device and Diagnostic Manufacturing: The Ultimate Resource Handbook. 2nd Edition. Buffalo Grove, IL: Interpharm Press, 1992.
DeSain, Carol and Charmaine Vercimak. Documentation Basics that Support Good Clinical Practices: The Master Plan. *Applied Clinical Trials*, Vol. 2 (June 1993).
DeSain, Carol and Charmaine Vercimak. Documentation Basics that Support Good Clinical Practices: Part II. *Applied Clinical Trials*, Vol. 2 (July 1993).
DeSain, Carol and Charmaine Vercimak. Documentation Basics that Support Good Clinical Practices: Part III, Protocols and SOPs. *Applied Clinical Trials*, Vol. 2 (September 1993).
DeSain, Carol and Charmaine Vercimak. Documentation Basics that Support Good Clinical Practice: Part IV, Clinical Data Collection Documents. *Applied Clinical Trials*, Vol. 2 (October 1993).
DeSain, Carol and Charmaine Vercimak. Documentation Basics that Support Good Clinical Practice: Part V, Clinical Systems of Accountability and Traceability. *Applied Clinical Trials*, Vol. 2 (November 1993).
DeSain, Carol and Charmaine Vercimak. Implementing International Drug, Device, and Diagnostic GMPs. Buffalo Grove, IL: Interpharm Press, 1994.
Dessy, Raymond E. The Electronic Laboratory. Washington, D.C.: American Chemical Society, 1985.
Dewoskin, Robert S. and Stephanie M. Taulbee. International GLPs. Buffalo Grove, IL: Interpharm Press, 1993.
Dice, Barry, Operations Manager, Sovran Financial Corp., Telephone Interview, April 25, 1990 (Hyattsville, MD).
Dietz, David L. and Christie J. Herald. Reconciling a Software Development Methodology with the PMA Validation Life Cycle. *Pharmaceutical Technology*, Vol. 16 (June 1992).
Dijkstra, Edsger. A Discipline of Programming. Englewood Cliffs, NJ: Prentice-Hall, 1976.
Dixon, W.J. and F.J. Massey, Jr. Introduction to Statistical Analysis. 3rd ed. New York: McGraw-Hill, 1969.
Dolotta, T.A., ***et al.*** Data Processing in 1980-1985: A Study of Potential Limitations to Progress. New York: John Wiley & Sons, Inc., 1976.
Dorf, Richard C. Robotics and Automated Manufacturing. Reston, VA: Reston Publishing Company, 1983.
Dorfmann, Joan. Data Center Housekeeping. *Infosystems*, July 1984.
Dorricott, K.O. Organizing a Computer Audit Specialist. *CA Magazine*, May 1979, pp. 66-68.

Double, Mary Ellen and Maryann McKendry. Computer Validation Compliance: A Quality Assurance Perspective. Buffalo Grove, IL: Interpharm Press, 1993.

Draper, N. and H. Smith. Applied Regression Analysis. New York: John Wiley, 1966.

Drug Information Association. Computerized Data Systems for Nonclinical Safety Assessment: Current Concepts and Quality Assurance. Maple Glen, PA: Drug Information Association, 1988.

Dunn, O.J. and V.A. Clark. Applied Statistics: Analysis of Variance and Regression. New York: John Wiley, 1974.

Dunn, Robert. Software Defect Removal. New York: McGraw-Hill, 1984.

Dunn, Robert and Richard Ullman. Quality Assurance for Computer Software. New York: McGraw-Hill, 1982.

EDP Auditors Foundation. Control Objectives -- 1980. Streamwood, IL: EDP Auditors Foundation, 1980.

Edwards, Perry and Bruce Broadwell. Data Processing. Belmont, CA: Wadsworth Publishing Co., Inc., 1979.

Egol, Len. Putting PCS to Work in the Lab. *Chemical Engineering*, August 1990.

Electronic Industries Association. EIA Standard RS-422, Electrical Characteristics of Balanced Voltage Digital Interface Circuits. Washington, D.C.: EIA, Engineering Department, 1975.

Electronic Industries Association. EIA Standard RS-232-C, Interface Between Data Terminal Equipment and Data Communication Equipment Employing Serial Binary Data Interchange. Washington, D.C.: EIA, Engineering Department, 1969.

Electronic Industries Association. Equipment Employing Serial Binary Data Interchange. Washington, D.C.: EIA, Engineering Department, 1969.

Electronic Fund Transfer Act. 15 USC (1979), sec. 1693 et seq.

Emens, K.L. A Survey of Internal EDP Audit Activity Among EDPAC Member Companies. *EDP Auditor*, Fall 1976, pp. 11-17.

Enger, Norman L. and Howerton, Paul W. Computer Security: A Management Audit Approach. New York: AMACOM, 1981.

English, John T. ISO 9000 and System Validation. Presentation at Pharmaceutical and Medical Device Manufacturing Seminars, San Juan, Puerto Rico, March 1-19, 1994.

Epstein, R.A. The Theory of Gambling and Statistical Logic. New York: Academic Press, 1967.

Fagan, M.E. Design and Code Inspections to Reduce Errors in Program Development. *IBM Systems Journal*, Vol 15, (No. 3, 1976).

Feigenbaum, Edward, Pamela McCorduck, and H. Penny Nii. The Rise of the Expert Company. New York: Times Books, 1988.

Figarole, Paul L. Computer Software Validation Techniques. DIA Conference on Computer Validation, January 21-23, 1985.

Fike, John L. and George E. Friend. Understanding Telephone Electronics. Dallas: Texas Instruments, 1984.

Firebaugh, Morris, *et al.* A Feast of Microcomputers. *Personal Computing*, Vol. 2 (November 1978), pp. 60-81.

Fischer, Renate and Karl Schick. The Role of Quality Assurance in Good Clinical Practice. *Drug Information Journal*, Vol. 27 (July-Sept. 1993).

Fisher, F.E. Fundamental Statistical Concepts. New York: Canfield Press, 1973.

Fisher, Royal P. Information Systems Security. Englewood Cliffs, NJ: Prentice-Hall, Inc., 1984.

Flavin, Matt. Fundamental Concepts of Information Modelling. New York: Yourdon Press, 1981.

Forsyth, A. An Approach to Audit an On-Line System. *COM-SAC, Computer Security, Auditing and Controls*, Vol. 7, January 1980, pp. A-1 to A-4.

Frank, E. Integrating Reviews of EDP Systems with Regular Audit Project. *EDP Auditor*, Summer 1974, pp. 10-11.

Freedman, Daniel and Gerald Weinberg. Handbook of Walkthroughs, Inspections and Technical Reviews. Boston: Little, Brown, 1982.

Freedman, David H. Is the Micro-Mainframe Link Connecting with MIS? *Infosystems*, February 1984.

Freedman, David H. Stress and the MIS Manager. *Infosystems*, March 1984.

Freeman, Peter. Software Perspectives: The System Is the Message. Reading, MA: Addison-Wesley, 1987.

Fribush, Howard M., Gary Robertson and Joseph Vagaggini. CLP Quick-Turn-Around Method. *Environmental Testing & Analysis*, March/April 1994.

Friedson, Eliot. The Social Construction of Illness. In Profession of Medicine. New York: Harper and Row, 1970, pp. 203-332.

Friend, George E., *et al.* Understanding Data Communications. Dallas: Texas Instruments, 1984.

Friis, M. William. Putting Information Systems Theory into Practice. *Infosystems*, February 1984.

Gagnon, Steven R. and Jesse Rodriguez. Applying Quality Assurance Concepts to Ensure the Performance of Pharmaceutical High-Purity Water Treatment Systems. *Microcontamination*, January 1994.

Gallegos, Frederick and Doug Bieber. What Every Auditor Should Know About Computer Information Systems. Available as Accession Number 130454 from the General Accounting Office (GAO) and reprinted from pp. 1-11 in EDP Auditing, Auerbach Publishers, 1986.

Gane, Chris and Trish Sarson. Structured Systems Analysis: Tools and Techniques. New York: Improved System Technologies, 1977.

Gardine, Thomas. Automation of Import Procedures. *RAPS News*, October 1993.

Gardner, Elizabeth. Computer Dilemma: Clinical Access vs. Confidentiality. *Modern Healthcare*, November 3, 1989, pp. 32-42.

Gardner, Elizabeth. Secure Passwords and Audit Trails. *Modern Healthcare*, November 3, 1989, p. 33 (sidebar).

Gardner, Elizabeth. System Assigns Passwords, Beeps at Security Breaches. *Modern Healthcare*, November 3, 1989, p. 34 (sidebar).

Gardner, Elizabeth. System Opens to Physicians, Restricts it to Others. *Modern Healthcare*, November 3, 1989, p. 38 (sidebar).

Gardner, Elizabeth. "Borrowed" Passwords Borrow Trouble. *Modern Healthcare*, November 3, 1989, p. 42 (sidebar).

Gardner, Elizabeth. Recording Results of AIDS Tests Can be a Balancing Act. *Modern Healthcare*, November 3, 1989, p. 40 (sidebar).

Garwood, R.M. FDA's Viewpoint on Inspection of Computer Systems. Course Notes from "Computers in Process Control: FDA Course," March 19-23, 1984.

Gaul, Gilbert M. A Threat to the Safety of Red Cross' Blood Supply. *The Philadelphia Inquirer*, March 21, 1989, page 1.

Gauthier, Richard. Using the UNIX System. Reston, VA: Reston Publishing Company, Inc., 1981.

GE-Intersil. Hot Ideas in CMOS Data Book. Cupertino, CA: GE-Intersil, 1984.

Gerrity, T.P., Jr. The Design of Man-Machine Decision Systems: An Application to Portfolio Management. *Sloan Management Review*, Vol. 112 (Winter 1971), pp. 59-75.

Gibbons, J.D. Nonparametric Methods for Quantitative Analysis. New York: Holt, Rinehart and Winston, 1976.

Gild, Tom. Principles of Software Engineering Management. Reading MA: Addison-Wesley, 1988.

Gilhooley, Ian. Improving the Relationship Between Internal and External Auditors. *Journal of Accounting and EDP*, Vol. 1 (Spring 1985), pp. 4-9.

Ginsbury, Karen and Gil Bismuth. Compliance Auditing for Pharmaceutical Manufacturers: A Practical Guide to In-Depth Systems Auditing. Buffalo Grove, IL: Interpharm Press, 1994.

Glass, R.L. Real-Time: The "Lost World" of Software Debugging and Testing. *Communications of ACM*, Vol. 23, (May 1980), pp. 264-271.

Glass, R.L. Software Reliability Guidebook. Englewood Cliffs, NJ: Prentice-Hall, 1979.

Glover, Donald E., Robert G. Hall, Arthur W. Coston, and Richard J. Trilling. Validation of Data Obtained During Exposure of Human Volunteers to Air Pollutants. *Computers and Biomedical Research*, Vol. 15 (Number 3, 1982), pp. 240-249.

Goodman, Daniel B. Standardized and Centralized Electrocardiographic Data for Clinical Trials. *Applied Clinical Trials*, Vol. 2 (June 1993).

Goodwin, Connie. Considerate, Carefully Planned Office Environments Reduce Employee Anxiety. *Data Management*, June 1985.

Goren, Leonard J. Computer System Validation, Part II. *BioPharm*, Vol. 2 (February 1989), pp. 38-42.

Grady, Robert and Deborah Caswell. Software Metrics: Establishing a Company-Wide Program. Englewood Cliffs, NJ: Prentice-Hall, 1987.

Grimes, J. and E.A. Gentile. Maintaining International Integrity of On Line Data Bases. *EDPACS Newsletter*, February 1977, pp. 1-14.

Guenther, William C. Analysis of Variance. Englewood Cliffs, NJ: Prentice-Hall, 1964.

Guilford, J.P. Fundamental Statistics in Psychology and Education. New York: McGraw-Hill, 1965.

Guynes, Steve. EFTS Impact on Computer Security. *Computers & Security*, Vol. 2 (1983), pp. 73-77.

Haider, Trong and Kjell Skaug. Forming an Understanding of GCP and a Validation Strategy at Nycomed Imaging. *Drug Information Journal*, Vol. 27 (April-June 1993).

Hall, Douglas V. Microprocessors and Digital Systems 2nd ed., New York: McGraw-Hill, Inc., 1983.

Halper, Stanley D., Glen C. Davis, P. Jarlath O'Neil-Dunne, and Pamela R. Pfau. Handbook of EDP Auditing. Chapter 28 of Warren, Gorham, & Lamont, Testing Techniques for Computer-Based Systems, 1985, pp. 28-1 to 28-26.

Halstead, Maurice. Elements of Software Science. New York: Elsevier, 1977.

Hamrell, Michael R. The Use of Computers in Multinational Drug Development. *Regulatory Affairs*, Vol. 5 (Winter 1993).

Harnett, D.L. and J.L. Murphy. Introductory Statistical Analysis. Reading MA: Addison-Wesley, 1975.

Hartmann, David C. Software Safety in Medical Devices. McLean, VA: The MITRE Corporation, 1993.

Hartwig, Frederick and Brian E. Dearing. Explanatory Data Analysis. Beverly Hills: Sage Publications, 1979.

Harz, Nan S. Telecommuting -- There's No Place Like Home. *Data Management*, June 1985.

Hatley, Derek and Imtiaz, Pirbhai. Strategies for Real-Time System Specification. New York: Dorset House, 1987.

Havard, J., J. Weaver, B. Gray, and P. Granade. An Automated In-Plant Analytical System. *American Laboratory*, October 1990.

Hayes Microcomputer Products, Inc. Smartmodem 1200 Hardware Reference Manual. Norcross, GA: Hayes Microcomputer Products, Inc., 1983.

Hayes, John R. The Complete Problem Solver. Philadelphia: The Franklin Institute Press, 1981.

Hayes-Roth, Frederick, Donald A. Waterman, and Douglas B. Lenat. Building Expert Systems. Reading, MA: Addison-Wesley Publishing Company, 1983.

Hedman, Gregory A. Increases in Laboratory Productivity Through LIMS Migration: Part I. *American Laboratory*, October 1992.

Hedman, Gregory A. Increases in Laboratory Productivity Through LIMS Migration: Part II. *American Laboratory*, January 1993.

Heidorn, G.E. Automatic Programming through Natural Language Dialogue: A Survey. *IBM Journal of Research and Development*, Vol. 20 (July 1976), pp. 302-313.

Henke, Cliff. International Electronics Standards: Compatibility or Interference? *Medical Device & Diagnostic Industry*, Vol. 15 (February 1993).

Herrick, Steven S. and Peter Berthrong. Validation of Computer Systems. *Scientific Computing and Automation*, September 1989.

Hershey, P. and K.H. Blanchard. Management of Organization Behavior. Englewood Cliffs, NJ: Prentice-Hall, 1972.

Hetzel, William. The Complete Guide to Software Testing. 2nd ed., Wellesley, MA: QED Information Sciences, 1988.

Hewlett-Packard. Improving Measurements in Engineering and Manufacturing -- HP-1B. Palo Alto, CA: Hewlett-Packard, 1976.

Hewlett-Packard. Optoelectronics Designer's Catalog. Palo Alto, CA: Hewlett-Packard, 1985.

Highland, Harold Joseph. Protecting Your Computer System. New York: John Wiley & Sons.

Hinton, M.D. ACS-LIMS: Manual of Operation. Mobile: Applied Computer Systems,1994.

Hinton, M.D., M.E. Sawyer, and D. Davis. LIMS and the Quality Assurance Laboratory. *American Laboratory*, October 1992.

Hirsch, Allen F. Good Laboratory Practice Regulations. New York: Marcel Dekker, Inc., 1989.

Hoaglin, David C., Frederick Mosteller, and John W. Tukey. Understanding Robust and Exploratory Data Analysis. New York: John Wiley & Sons, 1983.

Hollingworth, Dennis. Minicomputers: A Review of Current Technology, Systems, and Applications. Rand Report R-1279. Santa Monica, CA: The Rand Corporation, July 1973.

Holscher H. and J. Rader. Microcomputers in Safety Technique. Bayern, Germany: Verlag, 1984.

Honan, Patrick. Telecommuting: Will It Work for You? *Computer Decisions*, June 15, 1984.

Hopper, E.L. Staffing of EDP Auditors on the Internal Audit Staff. *The Interpreter*, August 1975, pp. 24-26.

Hordeski, Michael. Computer Integrated Manufacturing. Tab Professional and Reference Books, 1988.

Horwitz, G. Needed: A Computer Audit Philosophy. *Journal of Accountancy*, April 1976, pp. 69-72.

House, William C., ed. Data Base Management. New York: Petrocelli Books, 1974. Part V, pp. 330-97.

Houston, M. Frank. Designing Safer, More Reliable Software Systems. U.S. Food and Drug Administration, Center for Devices and Radiological Health, 1989.

Houston, M. Frank. What Do the Simple Folk Do? Software Safety in the Cottage Industry. *Supplemental Proceedings, Compass '87*, 2nd Annual Conference on Computer Assurance (June 1987).

Huang, Linda G. *et al.* The Guide to Biomedical Standards. 18th Edition. Brea, CA: Quest Publishing Company, 1992.

Hubbert, J. Data Base Concepts. *EDP Auditor*, Spring 1980.

Huff, D. How to Lie with Statistics. New York: W.W. Norton, 1954.

Hulme, K. and M.E. Aiken. The Normative Approach to Internal Control Evaluation of On-Line/Real-Time Systems. *The Chartered Accountant in Australia*, July 1976, pp. 7-16.

Humphrey, Watts S. Managing the Software Process. Reading, MA: Addison-Wesley, 1989.

Hunter, Ronald P. Automated Process Control Systems Concepts and Hardware. Englewood Cliffs, NJ: Prentice-Hall, 1978.

Huntsberger, D.V. Elements of Statistical Inference. 3rd ed. Boston: Allyn and Bacon, 1971.

Hussain, Donna and K.M. Hussain. Information Resource Management. Homewood, IL: Richard D. Irwin, 1984.

IEEE. Fiber Optic Applications in Electrical Substations. IEEE 83TH0104 PWR, IEEE Service Center, Piscataway, NJ, 1983.

IIA. Establishing the Internal Audit Function. Altamonte Springs, FL: Institute of Internal Auditors, 1974.

IIA. Hatching the EDP Audit Function. Altamonte Springs, FL: Institute of Internal Auditors, 1975.

IIA. How to Acquire and Use Generalized Audit Software. Altamonte Springs, FL: Institute of Internal Auditors, 1979.

IIA. Systems Auditability and Control -- Audit Practices. Altamonte Springs, FL: Institute of Internal Auditors, 1977.

IIA. Systems Auditability and Control -- Control Practices. Altamonte Springs, FL: Institute of Internal Auditors, 1977.

IIA. Systems Auditability and Control -- Executive Report. Altamonte Springs, FL: Institute of Internal Auditors, 1977.

Ince, Darrel, ed. Software Quality and Reliability. London: Chapman and Hall, 1991.

Inhorn, Stanley L., ed. Quality Assurance Practices for Health Laboratories. Washington, D.C.: American Public Health Association, 1978.

Inlander, Charles B., Lowell S. Levin, and Ed Weiner. Medicine on Trial, Prentice-Hall, 1988; "Blood, Sweat, and Fears," pp. 92-111.

Intel Corporation. A History of Microprocessor Development at Intel. Intel Article Reprint/AR-173. Intel Corporation, Santa Clara, CA.

Intel Corporation. An Intelligent Data Base System Using the 8272. Application Note AP-116. Intel Corporation, Santa Clara, CA, 1981.

Intel Corporation. ASM286 Assembly Language Reference Manual. Intel Corporation, Santa Clara, CA, 1983.

Intel Corporation. 8086/8087/8088 Macro Assembly Language Reference Manual for 8086-Based Development Systems. Intel Corporation, Santa Clara, CA, 1980.

Intel Corporation. Error Detecting and Correcting Codes, Application Note AP-46. Santa Clara, CA: Intel Corporation, 1979.

Intel Corporation. Getting Started With the Numeric Data Processor, Application Note AP-113. Intel Corporation, Santa Clara, CA, 1981.

Intel Corporation. iAPX 86/88, 186/188 User's Manual Programmer's Reference, Intel Corporation, Santa Clara, CA. 1983.

Intel Corporation. iAPX 286, Hardware Reference Manual, Intel Corporation, Santa Clara, CA. 1983.

Intel Corporation. iAPX 286, Operating Systems Writer's Guide, **Intel Corporation.** Santa Clara, CA. 1983.

Intel Corporation. iAPX 286, Programmer's Reference Manual, Intel Corporation, Santa Clara, CA. 1983.

Intel Corporation. iAPX 386, High Performance 32-bit Microcomputer Product Preview, Advance Information, Intel Corporation, Santa Clara, CA.

Intel Corporation. iRMX 86, Introduction and Operator's Reference Manual. Intel Corporation, Santa Clara, CA. 1984.

Intel Corporation. Intel SBC 202 Double Density Diskette Controller Hardware Reference Manual. Intel Corporation, Santa Clara, CA, 1977.

Intel Corporation. Introduction to the 80186 Microprocessor Application Note AP-186, Intel Corporation, Santa Clara, CA, 1983.

Intel Corporation. Introduction to the iAPX 286. Intel Corporation, Santa Clara, CA, 1985.

Intel Corporation. Microsystem Components Handbook Microprocessors and Peripherals, vols. 1 and 2. Intel Corporation, Santa Clara, CA, 1985.

Intel Corporation. Multibus Handbook. Intel Corporation, Santa Clara, CA, 1983.

Intel Corporation. Multibus Interfacing. Application Note AP-28A, Intel Corporation, Santa Clara, CA, 1979.

Intel Corporation. Multiprogramming with the iAPX 88 and iAPX 86 Microsystems. Application Note AP-106, Intel Corporation, Santa Clara, CA, 1980.

Intel Corporation. SDK-86 MCS-86 System Design Kit User's Guide. Intel Corporation, Santa Clara, CA 1981.

Intel Corporation. Using the 8273 SDLC/HDLC Protocol Controller, Application Note AP-36. Intel Corporation, Santa Clara, CA 1978.

Intel Corporation. Using the 8292 GPIB Controller, Application Note AP-66, Intel Corporation, Santa Clara, CA, 1980.

Jackson, Barbara Bund. Computer Models in Management. Homewood: Richard D. Irwin, Inc., 1979.

Jackson, Michael. System Development. Englewood Cliffs, NJ: Prentice-Hall, 1983.

Jackson, Michael. Principles of Program Design. New York: Academic Press, 1975.

Jancura, E.G. Developing Concepts of Technical Proficiency in EDP Auditing. *The Ohio CPA*, Spring 1979.

Japanese Ministry of Health and Welfare. Guide to Medical Device Registration in Japan. 5th Edition. Tokyo: Yakuji Nippo, LTD., 1994.

Jenkins, A. Milton. MIS Design Variables and Decision Making Performance. Ann Arbor: UMI Research Press, 1983.

Johnson, Curtis D. Process Control Instrumentation Technology. 2nd ed., New York: John Wiley & Sons, 1982.

Johnson, R.R. Elementary Statistics. N. Scituate, MA: Duxbury Press, 1973.

Jones, T. Capers. Programming Productivity. New York: McGraw-Hill, 1986.

Justice, Richard M., Jr., Jole O. Rodriguez, and William J. Chiasson. Ten Steps to Ensure a Successful PreNDA Approval Inspection. *Journal of Parenteral Science & Technology*, Vol. 47 (March-April 1993).

Kahan, Jonathan S. Medical Device Reclassification: An Opportunity Ignored. *Medical Device & Diagnostic Industry*, Vol. 15 (January 1993).

Kaisler, Stephen H. The Design of Operating Systems for Small Computer Systems. New York: John Wiley & Sons, 1983.

Kaitin, Kenneth I. and Helen L. Walsh. Are Initiatives to Speed the New Drug Approval Process Working? *Drug Information Journal*, Vol. 26, 1992.

Kay, Robert L. The Systems Engineering Approach to Quality Software. *Medical Device & Diagnostic Industry*, Vol. 16 (June 1994).

Keane, John Austin. Computers and Quality Assurance in a Regulated Industry. *Medical Device & Diagnostic Industry*, Vol. 3 (October 1981).

Kearns, J. Are We Ready for Continuous Process Auditing. *CA Magazine*, September 1980, pp. 68-71.

Keeler, R. Manage Your Data Better With a LIMS System. *R&D Magazine*, April 1991.

Keen, Peter G.W. "Interactive" Computer Systems for Managers: A Modest Proposal. *Sloan Management Review*, Fall 1976, pp. 1-17.

Keen, Peter G.W. and Michael S. Scott Morton. Decision Support Systems: An Organizational Perspective. Reading, MA: Addison-Wesley, 1978.

Kelly, F.J., D.L. Beggs, K.A. McNeil, T. Eichelberger, and J. Lyon. Multiple Regression Approach. Carbondale, IL: Southern Illinois Press, 1969.

Kendrick, John. Implementation of a Software Quality Assurance Program. Final Project, Fall 1989.

Kenney, Donald P. Minicomputers. New York: AMACOM, 1973.

Kent, Phyllis. Internal Clinical Audits of Computerized Data for NDA Submissions. *Clinical Research Practices & Drug Regulatory Affairs*, Vol. 22 (No. 3 1984).

Kerlinger, F.N. Foundations of Behavioral Research. New York: Holt, Rinehart and Winston. 1965.

Kernighan, B.W. and P.J. Plauger. The Elements of Programming Style. New York: McGraw-Hill, 1978.

Kernighan, B.W. and P.J. Plauger. Software Tools. Reading, MA: Addison-Wesley, 1976.

Kimbrell, John Y. CANDA as a Strategic Initiative for the Pharmaceutical Industry. *Pharmaceutical Technology*, Vol. 16 (October 1992).

King, J.P. A Vendor's Contribution to Computer System Software Validation -- Change Control. *Pharmaceutical Engineering*, Vol. 12 (May 1992).

Kish, L. Survey Sampling. New York: John Wiley, 1965.

Kleijnen, Jack P.C. Computer and Profits: Quantifying Financial Benefits of Information. Reading, MA: Addison-Wesley Publishing Company, 1980.

Koller, Alexander J. and Gerard W. Liesegang. Information Management Architecture Within the Analytical Laboratory: Part I. *American Laboratory*, September 1992.

Korteweg, Marijke. Computer Validation in Clinical Research: Regulatory Require ments in the European Community. *Drug Information Journal*, Vol. 27 (April-June 1993).

Korteweg, Marijke. GCP Audit Certificates and Compliance Statements in the European Community. *Applied Clinical Trials*, Vol. 2 (June 1993).

Korteweg, Marijke. General Concepts with Regard to Quality Control/Quality Assurance in Data Management. *Drug Information Journal*, Vol. 27 (October-December 1993).

Kosmala, Rupert M. The Relationship Between Functional Requirements and Software Design. *Pharmaceutical Technology*, Vol. 14 (November 1990).

Kraak, Frans J. and Karel J. de Neef. In-Process Data Quality Control: A Practical Implementation. *Drug Information Journal*, Vol. 27 (October-December 1993).

Kuhn, Thomas. The Structure of Scientific Revolutions. 2nd ed. Chicago: University of Chicago Press, 1970.

Kull, David. Demystifying Ergonomics. Computer Decisions, September 1984.

Kuong, J.F. A Framework for EDP Auditing. *COM-SAC, Computer Security, Auditing and Controls*, Vol. 3 (1976), pp. A-1 to A-8.

Kuong, J.F. Advanced Tools and Techniques for Systems Auditing. Management Advisory Publication, 1978.

Kuong, J.F. Approaches to Justifying EDP Controls and Auditability Provisions. *COM-SAC, Computer Security, Auditing and Controls*, Vol. 7, July 1980, pp. A-1 to A-8.

Kuong, J.F., ed. Audit and Control of Advanced/On-Line Systems. Management Advisory Publications (MAP-7), 1983.

Kuong, J.F. Audit and Control of Computerized Systems. Management Advisory Publications (MAP 6), 1979.

Kuong, J.F. Auditor Involvement in System Development and the Need to Develop Effective, Efficient, Secure, Auditable, and Controllable Systems. Keynote Speech at the First Regional EDP Auditors Conference, Tel-Aviv, Israel, June 3, 1982.

Kuong, J.F. Checklists and Guidelines for Reviewing Computer Security and Installations. Management Advisory Publications (MAP-4), 1976.

Kuong, J.F. Computer Auditing and Security: Manual-Operations and System Audits. Management Advisory Publications (MAP-5), 1976.

Kuong, J.F. Computer Security, Auditing, and Controls: Text and Readings. Management Advisory Publications, 1974.

Kuong, J.F. Controls for Advanced/On-Line/Data-Base Systems, vols. 1 and 2. Management Advisory Publications, 1985.

Kuong, J.F. Managing the EDP Audit Function. Paper Presented at the First Regional EDP Auditors Conference, Tel-Aviv, Israel, May 29-June 3, 1982.

Kuong, J.F. Organizing, Managing, and Controlling the EDP Auditing Function. Seminar Text, Management Advisory Publications, 1980.

Kuong, J.F. Organizing and Staffing for EDP Auditing. *COM-SAC, Computer Security, Auditing and Controls*, Vol. 2 (1975).

Lagan, J.J. The Development of an EDP Audit Function in a Complex Environment. *Employment Research Management*, 680, 1978.

Lainhart, J.W. and B.R. Snyder. A Simultaneous-Parallel Approach to Testing Computerized Systems. *EDPACS Newsletter*, October 1977, pp. 7-12.

Lamprecht, James L. Implementing the ISO 9000 Series. New York: Marcell Dekker, Inc., 1993.

La Niece, Chris and Dai Bedford. Best Practice in the Validation of Business Critical Systems. *Drug Information Journal*, Vol. 27 (April-June 1993).

Langmead, J.M. and R.V. Boos. How Do You Train EDP Auditors? *Management Focus*, September-October 1978, pp. 6-11.

Laposki, Edward J. Planning and Managing the Computer Automation Project. *Pharmaceutical Manufacturing*, August 1985.

Laurel, Brenda, ed. The Art of Human-Computer Interface Design. Reading, MA: Addison-Wesley, 1990.

Lawrence, Gill. On-Line to Central Database Processing at Searle. *Drug Information Journal*, Vol. 27 (October-December 1993.

Lee, John Y. Documentation Requirements for Preapproval Inspections. *Pharmaceutical Technology*, Vol. 17 (March 1993).

Lesea, Austin and Rodnay Zaks. Microprocessor Interfacing Techniques. 2d ed., Berkeley: Sybex Inc., 1977.

Leveson, Nancy G. Software Safety: Why, What and How. *ACM Computing Surveys*, Vol. 18 (June 1986).

Leveson, Nancy G. Software Safety in Embedded Computer Systems. *Communications of the ACM*, Vol. 34 (February 1991).

Leveson, Nancy G. and J.L. Stolzy. Safety Analysis Using Petri Nets. *IEEE Transactions on Software Engineering*, Vol. 13 (March 1987).

Leveson, Nancy G. and P.R. Harvey. Analyzing Software Safety. *IEEE Transactions on Software Engineering*, Vol. 9 (September 1983).

Levey, Brian and William Leonard. LIMS Operational Qualification. *American Laboratory*, March 1993.

Lientz, B.P. and E.B. Swanson. Software Maintenance Management. Reading, MA: Addison-Wesley, 1980.

Lindauer, Richard F. An Official, Electronically Searchable Version of USP-NF. *Pharmaceutical Technology*, Vol. 17 (March 1993).

Linger, R.C., H.D. Mills and B.L. Witt. Structured Programming: Theory and Practices. Reading, MA: Addison-Wesley, 1979.

Liscouski, Joseph G. Issues and Directions in Laboratory Automation. *Analytical Chemistry*, Vol. 60 (January 15, 1988).

Litecki, C.R. and J.E. McEnroe. EDP Audit Job Definitions: How Does Your Staff Compare? *The Internal Auditor*, April 1981, pp. 57-61.

Littlewood, Bev and Lorenzo Strigini. The Risks of Software. *Scientific American*, November 1992.

Lumato, Ann Nicols and Nico Lumato. A UNIX Primer. Englewood Cliffs, NJ: Prentice-Hall, 1983.

Lyles, Richard I. Practical Management Problem Solving and Decision Making. New York: Van Nostrand Reinhold, 1982.

Macchiaverna, P.R. Auditing Corporate Data Processing Activities. New York: The Conference Board, Inc., 1980.

Mackintosh, Douglas R. Building Quality Assurance Into Clinical Trials. *Applied Clinical Trials*, Vol. 2 (April 1993).

Mackintosh, Douglas R. Quality Assurance of Data Management and Statistical Analysis. *Applied Clinical Trials*, Vol. 3 (June 1994).

Mader, Chris and Robert Hagin. Information Systems: Technology, Economics and Applications. Chicago: Science Research Associates, Inc., 1974, pp. 291-309.

Maddox, M., C. Higgins, and D. Mathews. Client-Server LIMS Solutions: Part I. *American Laboratory*, March 1993.

Maggon, Krishan K. and Daniel Brandt. Standard Operating Procedures and the Conduct of Clinical Trials. *Applied Clinical Trials*, Vol. 3 (July 1994).

Magnani, Harry H. and Guus A Sturk. Centralized Laboratory Testing in Europe. *Applied Clinical Trials*, Vol. 3 (March 1994).

Mair, W.C., K. Davis, and D. Wood. Computer Control and Audit. Altamonte Springs, FL: Institute of Internal Auditors, 1976.

Mallory, Steven R. Building Quality into Medical Product Software Design. *Biomedical Instrumentation & Technology*, Vol. 27 (March/April 1993).

Manell, Paul. CANDAs: A New Standard for Electronic Regulatory Submissions in Europe. *Applied Clinical Trials*, Vol. 2 (April 1993).

Manell, Per. Computer Validation Applied in the Regulatory Environment. *Drug Information Journal*, Vol. 27 (April-June 1993).

Mannion, James C., Anthony Abruzzini, George Nichoalds, Siegrid Hessenthaler, and Christine Uhlinger. Predictable Challengers in the Preparation of the Clinical Sections of NDAs and PLAs. *Applied Clinical Trials*, Vol. 3 (March 1994).

Marks, R.C. Performance Appraisal of EDP Auditors. Speech Given at the 11th Conference on Computer Auditing, Security, and Control, ATC/IIA, New York, May 4-8, 1981.

Marks, William E. Evaluating the Information Systems Staff. *Information Systems News*, December 24, 1984.

Martin, J. Accuracy, Security and Privacy in Computer Systems. Englewood Cliffs, NJ: Prentice-Hall, 1973.

Martin, J. Principles of Data Base Management. Englewood Cliffs, NJ: Prentice-Hall, Inc., 1976.

Martin, James. Strategic Data Planning Methodologies. Englewood Cliffs, NJ: Prentice-Hall, 1982.

Martin, James. The Wired Society. Englewood Cliffs, NJ: Prentice-Hall, Inc., 1978.

Martin, James and Carma McClure. Software Maintenance: The Problem and its Solutions. Englewood Cliffs, NJ: Prentice-Hall, 1983.

Martin, James and Carma McClure. Structured Techniques: The Basis for CASE. Englewood Cliffs, NJ: Prentice-Hall, 1987.

Mason, Richard O. and E. Burton Swanson. Measurement for Management Decision. Reading, MA: Addison-Wesley, 1981.

Masters, George and Paul Figarole. Validation Principles for Computer Systems -- FDA's Perspective. *Pharmaceutical Technology*, Vol. 10 (November 1986).

Matsuda, Takehiko. Guideline on Control of Computerized Systems in Drug Manufacturing. *Journal of Pharmaceutical Science and Technology*. January/February 1994.

Mattes, D.C. LIMS and Good Laboratory Practice. In R.D. McDowall, ed., Laboratory Information Management Systems. Wilmslow, U.K.: Sigma Press, 1987, pp. 332-345.

Maynard, David W. Validation Master Planning. *Journal of Parenteral Science & Technology*, Vol. 47 (March-April 1993).

McClure, Carma. CASE is Software Automation. Englewood Cliffs, NJ: Prentice-Hall, 1989.

McClure, Carma. Managing Software Development and Maintenance. New York: Van Nostrand Reinhold, 1981.

McCorduck, Pamela. Machines Who Think. San Francisco: W.H. Freeman, 1979.

McDonald, Thomas F. TechnoAcceptance Cycle Soothes Automation Blind Spots. *Data Management*, May 1985.

McDowall, R.D., ed. Laboratory Information Management Systems. Wilmslow, U.K.: Sigma Press, 1987.

McFarlan, F. Warren and James L. McKenney. Corporate Information Systems Management: The Issues Facing Senior Executives. Homewood, IL: Richard D. Irwin, 1983.

McGuire, P.T. EDP Auditing -- Why? How? What? *The Internal Auditor*, June 1977, pp. 28-34.

McKibbin, Wendy Lea. A Solution to PC Management? *Infosystems*, February 1984.

McMenamin, Steve and John Palmer. Essential Systems Analysis. Englewood Cliffs, NJ: Yourdon Press/Prentice-Hall, 1984.

McNamara, John E. Technical Aspects of Data Communication. Digital Equipment Corporation, Maynard, MA, 1977.

Megargle, Robert. Laboratory Information Management Systems. *Analytical Chemistry*, Vol. 61 (May 1989), pp. 612/a-621A.

Mendenhall, W. An Introduction to Linear Models and the Design and Analysis of Experiments. Belmont, CA: Wadsworth, 1967.

Mendenhall, W., L. Ott, and R.L. Scheaffer. Elementary Survey Sampling. Belmont, CA: Wadsworth, 1971.

Mendenhall, W. and James E. Reinmuth. Statistics for Management and Economics, 3rd ed. N. Scituatc, MA: Duxbury Press, 1978.

Merrer, Robert J. and Peter G. Berthrong. Academic LIMS: Concept and Practice. *American Laboratory*, Vol. 21 (March 1989), pp. 36-45.

Merrett, T.H. Relational Information Systems. Reston, VA: Reston Publishing Co., 1984.

Metzger, Philip. Managing a Programming Project. 2nd ed. Englewood Cliffs, NJ: Prentice-Hall, 1983.

Michie, Donald, ed. Introductory Readings in Expert Systems. New York: Gordon and Breach Science Publishers, Inc., 1982.

Mick, John and Jim Brick. Bit-Slice Microprocessor Design. New York: McGraw-Hill, Inc., 1980.

Miller, T.L. EEDP -- A Matter of Definition. *The Internal Auditor*, July-August 1975, pp. 31-38.

Mills, Harlan, Richard Linger, and Alan Hevner. Principles of Information Systems Analysis and Design. New York: Academic Press, 1986.

Minsky, Marvin. The Society of Mind. New York: Simon & Schuster, 1986.

Mitchell, Tony. Computer-Assisted Clinical Research. *Drug Information Journal*, Vol. 27 (April-June 1993).

Moehle, Frederick L. W. A Knowledge Base Model for CANDAs, Part I. *Applied Clinical Trials*, Vol. 2 (June 1993).

Moore, Richard W. Introduction to the Use of Computer Packages for Statistical Analyses. Englewood Cliffs, NJ: Prentice-Hall, 1978.

Morel, Guillemette. Validation: From Theory to Practice, A Quality Assurance Perspective. *Drug Information Journal*, Vol. 27 (April-June 1993).

Morozoff, Paul E. Specifying Software Validation and Verification for a Biomedical Application. *Biomedical Instrumentation & Technology*, Vol. 28 (May/June 1994).

Morris, R., III. The Internal Auditors and Data Processing. *The Internal Auditor*, August 1978.

Motise, Paul. FDA Considerations on Electronic Identification and Signatures. *Pharmaceutical Technology*, Vol. 16 (November 1992).

Motise, Paul. What to Expect When FDA Audits Computer-Controlled Processes. *Pharmaceutical Manufacturing*, July 1984.

Motorola Semiconductor Products, Inc. Optoelectronics Device Data Book, DL118R1. Phoenix: Motorola Semiconductor Products Inc., 1983.

Motorola Semiconductor Products, Inc. Power Mosfet Selector Guide and Cross Reference SG56R4. Phoenix: Motorola Semiconductor Products, Inc., 1983.

Murty, Ram and Sreemantula Satyanarayana. Evaluation of an Automated Washer for Processing Parenteral Closures. *Pharmaceutical Technology*, Vol. 17 (March 1993).

Musa, John, Anthony Iannino, and Kazuhira Okumoto. Software Reliability: Measurement, Prediction, Application. New York: McGraw-Hill, 1987.

Myers, Glenford. The Art of Software Testing. New York: Wiley-Interscience, 1979.

Myers, Glenford. Digital System Design with LSI Bit Slice Logic. New York: John Wiley and Sons, 1980.

Myers, Glenford. Software Reliability. New York: John Wiley & Sons, 1976.

Myers, Glenford. Reliable Software Through Composite Design. New York: Petrocelli/Charter, 1975.

Myers, Ware. Build Defect-Free Software, Fagan Urges. *IEEE Computer*, August 1990.

Naisbitt, John and Patricia Aburdene. Megatrends 2000. New York: William Morrow, 1990.

Nally, J. and R. Kieffer. The Future of Validation: From QC/QA to TQ. *Pharmaceutical Technology*, Vol. 17 (October 1993).

Nelson, D.C., V. Dauciunas, R. Lysakowski, and M. Duff. The Standard for Chromatography Data Communication from the Analytical Instruments Association. *American Laboratory*, September 1990.

Neter, J. and W. Wasserman. Applied Linear Statistical Models. Homewood, IL: Richard D. Irwin, 1974.

Neuman, Anita. Get a Head Start: Perform Process Validation in Pilot Production. *Medical Device & Diagnostic Industry*, Vol. 16 (January 1994).

Neumann, Kurt. Some Aspects of Total Quality Management in the Computer System

Validation Process in Clinical Research. *Drug Information Journal*, Vol. 27 (April-June 1993).

Niehoff, Ken. Using Rovira Diagrams to Specify the User Interface. *Medical Device & Diagnostic Industry*, Vol. 16 (January 1994).

Noether, G.E. Introduction to Statistics: A Nonparametric Approach. 2nd ed. Boston: Houghton Mifflin, 1976.

Nolan, Richard L. Managing the Crises in Data Processing. *Harvard Business Review* (March-April 1979), pp. 115-126.

Nolan, Richard L. Managing the Data Processing Function. 2nd ed. St. Paul, MN: West Publishing, 1982.

Norman, Donald A. The Design of Everyday Things. New York: Doubleday/Cur rency, 1990.

Norris, P.M. EDP Audit and Control -- A Practitioner's Viewpoint. *EDP Auditor*, Winter 1976, pp. 8-14.

Nowak, Horst. A Systematic Approach for Handling Adverse Events. *Drug Information Journal*, Vol. 27 (October-December 1993).

O'Brien, James A. Computers in Business Management, An Introduction. Homewood, IL: Richard D. Irwin, Inc., 1979, pp. 150-74.

O'Donnell, Peter. Computers and Numbers. *Applied Clinical Trials*, Vol. 3 (July 1994).

O'Donnell, Peter. GCP and Computers: Mixing People with Processes. *Applied Clinical Trials*, Vol. 3 (March 1994).

O'Shea, Tim and Marc Eisenstadt, eds. Artificial Intelligence. New York: Harper and Row, Publishers, 1984.

Odiorne, George S. Management Decisions by Objectives. Englewood Cliffs, NJ: Prentice-Hall, Inc., 1969.

Olivier, Daniel P., Michael D. Konrad, and Markus Weber. Software Development: Comparing Software Assessment Standards. *Medical Device and Diagnostic Industry*, Vol. 16 (June 1994).

Orr, Ken. Structured Requirements Definition. Topeka, KS: Ken Orr & Associates, 1981.

Orr, Ken. Structured Systems Development. Englewood Cliffs, NJ: Yourdin Press/Prentice-Hall, 1977.

Outterson, Dave. Color Visual Inspection in Pharmaceutical Manufacturing. *Pharmaceutical Technology*, Vol. 18 (January 1994).

Pabst, C.A. What's All the Fuss About EDP? *California CPA Quarterly*, June 1975, pp. 10-16.

Page-Jones, Meilir. The Practical Guide to Structured Systems Design. 2nd ed. Englewood Cliffs, NJ: Yourdon Press/Prentice-Hall, 1988.

Parikh, Girish. The Politics of Software Maintenance. Infosystems, August 1984.

Parker, Marilyn and Robert J. Benson. Information Economics. Englewood Cliffs, NJ: Prentice-Hall, 1988.

Parnas, David Lorge and Paul C. Clements. A Rational Design Process: How and

Why to Fake It. *IEEE Transactions on Software Engineering*, Vol. SE-12 (February 1986).

Patrick, R.L. Performance Assurance and Data Integrity Practices. Washington, D.C.: National Bureau of Standards, 1978.

Paulson, Daryl S. and Dennis G. Vogel. Preparatory Software Documentation for Validation of Computer-Controlled Manufacturing Operations. *Pharmaceutical Manufacturing*, May 1984.

Pava, Calvin H.P. Managing New Office Technology; An Organizational Strategy. New York: Free Press, 1983.

Peach, Robert W., ed. The ISO 9000 Handbook. 2nd Edition. Fairfax, VA: CEEM Information Services, 1994.

Pearson, E.S. and M.G. Kendall. Studies in the History of Statistics and Probability. Darien, CT Hafner Publishing Company, 1970.

Peine, Ira C. Quality Assurance Compliance: Procedures for Pharmaceutical and Biotechnology Manufacturers. Buffalo Grove, IL: Interpharm Press, 1994.

Pency, Laura, ed. Interpharm Pharmaceutical Manager's Compliance Guide. Buffalo Grove, IL: Interpharm Press, 1994.

Perry, W.E. Adding a Computer Programmer to the Audit Staff. *The Internal Auditor*, July 1974, pp. 1-7.

Perry, W.E. Auditing Computer Systems. Melville, NY: FAIM Technical Products, Inc., 1977.

Perry, W.E. Career Advancement for the EDP Auditor. *EDPACS Newsletter*, February 1975, pp. 1-6.

Perry, W.E. Ensuring Data Base Integrity. New York: John Wiley & Sons.

Perry, W.E. Internal Control. Melville, NY: FAIM Technical Products, Inc., 1980.

Perry, W.E. The Making of a Computer Auditor. *The Internal Auditor*, November-December 1974.

Perry, W.E. Snapshot -- A Technique for Tagging and Tracing Transactions. *EDPACS Newsletter*, March 1974, pp. 1-7.

Perry, W.E. Trends in EDP Auditing. *EDPACS Newsletter*, December 1976, pp. 1-6.

Perry, W.E. Using SMF as an Audit Tool -- Accounting Information. *EDPACS Newsletter*, February 1975.

Perry, W.E. and Donald L. Adams. SMF -- An Untapped Audit Resource. *EDPACS Newsletter*, September 1974.

Perry, W.E. and J.F. Kuong. Developing an Integrated Test Facility for Testing Computerized Systems. Management Advisory Publications (MAP-12), 1979.

Perry, W.E. and J.F. Kuong. EDP Risk Analysis and Controls Justification. Management Advisory Publications, 1981.

Perry, W.E. and J.F. Kuong. Effective Computer Audit Practices Manual (ECAP). Management Advisory Publications.

Perry, W.E. and J.F. Kuong. Generalized Computer Audit Software-Selection and Application. Management Advisory Publications (MAP-14) 1980.

Pharmaceutical Manufacturers Association. PMA Computer Systems Validation Committee Report. *Pharmaceutical Technology*, Vol. 10 (May 1986), pp. 24-26.

Phipps, Gail. Practical Application of Software Testing. Computer Sciences Corporation, presented October 8, 1986.

PhRMA, Computer Systems Validation Committee. Computer System Validation: Auditing Computer Systems for Quality. *BioPharm*, Vol. 7 (September 1994).

Pieterse, Herman, *et al.* Source Documents: Definitions, Verification Procedures, and Archiving. *Applied Clinical Trials*, Vol. 3 (September 1994).

Pirsig, Robert. Zen and the Art of Motorcycle Maintenance. New York: Bantam Books, 1975.

Pinkus, Karen V. Financial Auditing and Fraud Detection: Implications for Scientific Data Audit. *Accountability in Research*, Vol. 1 (1989), pp. 53-70.

PMA, Clinical Data Management Group Task Force. Electronic Exchange of Laboratory Data: Recommendations for Improving Quality and Reducing the Hidden Costs. *Drug Information Journal*, Vol. 27 (July-Sept. 1993).

PMA, Computer System Validation Committee. Computer System Validation -- Staying Current: Installation Qualification. *Pharmaceutical Technology*, Vol. 14 (September 1990).

PMA, Computer System Validation Committee. Computer System Validation -- Staying Current: Introduction. *Pharmaceutical Technology*, Vol. 13 (May 1989).

PMA, Computer System Validation Committee. Computer System Validation -- Staying Current: Security in Computerized Systems. *Pharmaceutical Technology*, Vol. 17 (May, 1993).

PMA, Computer System Validation Committee. Computer System Validation -- Staying Current: Software Development Testing Strategies. *Pharmaceutical Technology*, Vol. 13 (September 1989).

PMA, Computer System Validation Committee. Computer System Validation -- Staying Current: Vendor-User Relationships. *Pharmaceutical Technology*, Vol. 17 (September 1993).

PMA, Computer System Validation Committee. Validation Concepts for Computer Systems Used in the Manufacture of Drug Products. *Pharmaceutical Technology*, Vol. 10 (May 1986).

Polanis, M.F. Choosing an EDP Auditor. *Bank Administration*, January 1973, pp. 52-53.

Polya, G. How to Solve It. Garden City: Doubleday and Company, 1957.

Pratt, Terrence W. Programming Languages: Design and Implementation. Englewood Cliffs, NJ: Prentice-Hall, 1975.

Pressman, Roger S. Making Software Engineering Happen. Englewood Cliffs, NJ: Prentice-Hall, 1988.

Prutch, Shirley F. In Praise of Operations Managers. *Datamation*, June 15, 1984.

Quinn, Thomas. Benefits-Driven Phased Implementations for EBR/MES Projects, Part II. *Pharmaceutical Technology*, Vol. 17 (November 1993).

Rabbitt, John T. and Peter A. Bergh. The ISO 9000 Book: A Global Competitor's Guide to Compliance & Certification. White Plains, NY: Quality Resources, 1993.

Raiffa, Howard. Decision Analysis. Reading, MA: Addison-Wesley Publishing Company, 1970.

Ramsey, James B. and Gerald L. Musgrave. APL-STAT: A Do-It-Yourself Guide to Computational Statistics. Belmont, CA: Lifetime Learning Publications, 1981.

Ravden, Susannah and Graham Johnson. Evaluating Usability of Human-Computer Interfaces. New York: John Wiley & Sons, 1989.

Rawling, Alan K. Computer-Aided Report Auditing. *Drug Information Journal*, January-March 1994.

Ray, Curtis. An Advanced Chromatography Data System. *American Laboratory*, September 1990.

Reber, S. Laboratory Information Management Systems. *American Laboratory*, February 1985.

Reilly, R.F. and J.A. Lee. Developing an EDP Audit Function: A Case Study. *EDPACS Newsletter*, May 1981, pp. 1-10.

Reimann, Bernard C. and Allen D. Warren. User-Oriented Criteria for the Selection of DDS Software. *Communications of the ACM*, Vol. 28 (February 1985), pp. 166-79.

Rhodes, Wayne L., Jr. The "Mythical" Datacom Manager. *Infosystems*, March 1984.

Rockart, John F. Chief Executives Define Their Own Data Needs. *Harvard Business Review* (March-April 1979), pp. 81-93.

Roland, John. The Microelectronic Revolution. *The Futurist*, Vol. 13 (April 1979), pp. 81-90.

Romano, Carol A. Privacy, Confidentiality, and Security of Computerized Systems: The Nursing Responsibility. *Computers in Nursing*, May/June 1987, pp. 99-104.

Rothman, Lawrence J. Sharing the Informational Wealth. *Pharmaceutical Technology*, Vol. 14 (November 1990).

Rubinstein, Moshe F. Patterns of Problem Solving. Englewood Cliffs, NJ: Prentice-Hall, 1975.

Ruff, P. SQL*LIMS -- Providing Information Management Solutions for the Pharmaceutical Industry. *Laboratory Practice*, June 1990.

Rugg, Tom. LANtastic. Berkeley, CA: Osborne McGraw-Hill, 1992.

Ryan, James. LIMS Aid Environmental Labs. *R&D Magazine*, July 1991.

Ryan, James. Data Quality with LIMS. *Quality*, May 1988.

Sackman, Harold. Mass Information Utilities and Social Excellence. Princeton: Auerbach Publishers, 1971.

Salazar, Joanne M., Chris Gopal, and Arthur Mlodozeniec. Computer Migration and Validation: A Vendor's Perspective. *Pharmaceutical Technology*, Vol. 15 (June 1991).

Sammet, Jean E. The Use of English as a Programming Language. *Communications of the ACM*, vol. 9, (March 1966), pp. 228-30.

Sandberg, Jim. Design Output. *Medical Device & Diagnostic Industry*, Vol. 15 (December 1993).

Sandhu, H.S. Hands-On-Introduction to ROBOTICS -- The Manual for the XR-Series Robots. Champaign, IL: Rhino Robots, 1983.

Sandowski, C. and G. Lawler. A Relational Data Base Management System for LIMS. *American Laboratory*, Vol. 21 (March 1989), pp. 70-79.

Savich, R.S. The Care and Feeding of an EDP Auditor. *EDP Auditor*, Summer 1974, pp. 12-13.

Sayre, James E., Jr. Good Clinical Practices (GCP) Quality Audit Manual. 2nd Edition. Buffalo Grove IL: Interpharm Press, 1994.

Sax, Philip E. Compatibility of Environmental Quality Requirements and ISO 9000 Recommendations. Presentation at 1st Annual International Conference on ISO 9000, Lake Buena Vista, FL, November 1-2, 1993.

Sax, Philip E. Product and Process Inspections Using ISO 9001. Proceedings of the HIMA Conference on Medical Device Software Good Manufacturing Practices. HIMA Publication 93-6, September 1993.

Schatt, Stan. Understanding Local Area Networks. Third Edition. Carmel, IN: SAMS.

Scheffe, Henry. The Analysis of Variance. New York: John Wiley, 1959.

Schindler, Max. Computer-Aided Software Design. New York: John Wiley & Sons, 1990.

Schmidt, Ian. Client/Server ODBMSs. *Software Development*, October 1993.

Schneidermann, B. Improving the Human Factors Aspect of Database Interactions. *ACM Transactions on Database Systems*, Vol. 3, (December 1978), pp. 417-39.

Schneidermann, B. Designing the User Interface: Strategies for Effective Human-Computer Interaction. Reading, MA: Addison-Wesley, 1987.

Schneidman, A. A Need for Auditors' Computer Education. *The CPA Journal*, June 1979, pp. 29-35.

Schroeder, Frederick J. Developments in Consumer Electronic Fund Transfers. *Federal Reserve Bulletin*, Vol. 69 (1983), pp. 395-403.

Schuber, Stefan. Product, Not Paper: A Manufacturing Execution System Installation at SmithKline Beecham. *Pharmaceutical Technology*, Vol. 17 (November 1993).

Schuetze, Guenter. Computerized Production of Artwork for Labeling. *Drug Information Journal*, Vol. 27 (April-June 1993).

Schulmeyer, G. Gordon. Zero Defect Software. New York: McGraw-Hill, Inc., 1990.

Schuyler, Michael. Now What? How to Get Your Computer Up & Keep it Running. New York: Neal-Schuman Publishers, 1988.

Scientific American. Microelectronics. San Francisco: W.H. Freeman and Co., 1977.

Scoma, I., Jr. The EDP Auditor. *Data Management*, May 1977, pp. 14-17.

Sculley, John. Odyssey. New York: Harper & Row, 1987.

Seippel, Robert G. Transducers, Sensors, and Detectors. Reston, VA: Reston Publishing Company, Inc., 1983.

Sellers, G.R. Elementary Statistics. Philadelphia: W.B. Saunders Co., 1977.

Semprevivo, Philip C. Systems Analysis, Science Research Associates, 1982.

Sharp, John. European Pharmaceutical Technical and Regulatory Compendium. Buffalo Grove, IL: Interpharm Press, 1994.

Sheingold, Daniel H., ed. Transducer Interfacing Handbook -- A Guide to Analog Signal Conditioning. Norwood, MA: Analog Devices, Inc., 1981.

Sheridan, Robert. How to Manage Regulatory Submissions for Device Modifications. *Medical Device & Diagnostic Industry*, Vol. 15 (October 1993).

Shlaer, Sally and Steve Mellor. Object-Oriented Systems Analysis: Modeling the World in Data. Englewood Cliffs, NJ: Yourdon Press/Prentice-Hall, 1988.

Shore, John. The Sachertorte Algorithm. New York: Penguin Books, 1986.

Siegel, S. Nonparametric Statistics for the Behavioral Sciences. New York: McGraw-Hill, 1956.

Silva, Marianne A. A Personal Experience with Validation and Regulation. Presentation at the 1st Annual Validation Forum, Williamsburg, VA, May 3, 1994; published in *Weinberg Validation Review*, Vol. 3 (No. 4, 1994).

Simon, H.A. The New Science of Management. New York: Harper and Row, Publishers, 1960.

Simon, H.A. The Science of the Artificial. Cambridge: Massachusetts Institute of Technology Press, 1981.

Simon, H.A. What Computers Mean for Man and Society. *Science*, Vol. 195, (March 18, 1977), pp. 1186-91.

Singer, Miller, and Witt. Structured Programming, Theory and Practice. Reading, MA: Addison-Wesley Publishing Co., 1979.

Smith, Martin R. Commonsense Computer Security. London: McGraw-Hill Book Company, 1989.

Smith, K. and J. Farrie. Data Management in a New Generation Chromatography System. *American Laboratory*, March 1990.

Snyder, Dean E., ed. FDA-Speak: The Interpharm Glossary of Acronyms and Regulatory Terms. Buffalo Grove, IL: Interpharm Press, 1992.

Spelton, Janie. Validation Requirements for CANDA Systems. Presentation at Drug Information Association Workshop, "Clinical Data Management," Brussels, Belgium, November 8-10, 1993.

Sprague, Ralph H., Jr. and Eric D. Carlson. Building Effective Decision Support Systems. Englewood Cliffs, NJ: Prentice-Hall, Inc., 1982.

Spriet, Alain and Therese Dupin-Spriet. Imperfect Data Analysis. *Drug Information Journal*, Vol. 27 (October-December 1993).

Stallings, William. Local Networks -- An Introduction. New York: Macmillan, 1984.

Stark, Peter A. Computer Programming Handbook. Tab Professional and Reference Books, 1978, pp. 229-232.

Stearns, E.W. The Information Center: The Best Answer? *Infosystems*, June 1984.

Steifel, M. Floppy Disk System. *Mini-Micro Systems*, Vol. 11 (November 1978), pp. 37-51.

Stein, Gary C. A Comparison of Validation Aspects of Statistical Data Analysis Systems. Presentation at Drug Information Association Workshop, "Statistical Issues in the Pharmaceutical Industry," Hilton Head, SC, March 30, 1993.

Stein, Gary C. Document Architecture in a Global Document Management System. Presentation at Drug Information Association, 29th Annual Meeting, Chicago, IL, July 11-15, 1993.

Stein, Gary C. Ensuring Computer System Validation: The Regulatory Imperative. Presentation at IBC Computer Assisted Product License Applications Workshop, Washington, DC, March 24-25, 1994.

Stein, Gary C. The Good Automated Laboratory Practices (GALPs) as a Guide to Laboratory Regulatory Compliance. Presentation at Analytical Laboratory Exposition and Conference, San Francisco, CA, October 5-7, 1993.

Stein, Gary C. Issues in the Regulated Laboratory and the Demand for Validation of Analytical Systems. Presentation at Hewlett-Packard Analytical Forum, Chepstow, UK, October 26-28, 1993.

Stein, Gary C. Validation of Computer Systems: What FDA Needs and Expects. Presentation at Drug Information Association Workshop, "FDA Pre-NDA Approvals in Europe," London, UK, April 28-29, 1993.

Stevens, J. Lawrence. Design Verification. *Medical Device & Diagnostic Industry*, Vol. 16 (January 1994).

Stevenson, Robert. HPLC '93: Environmental Applications of HPLC. *American Environmental Laboratory*, August 1993.

Stokes, Theresa E. Clinical Computing and GCP Impact Study. *Applied Clinical Trials*, Vol. 2 (April 1993).

Stokes, Theresa E. Hardware and Software System Validation Issues Under EC Regulations: Vendor Support for Validation. *Drug Information Journal*, Vol. 27 (April-June 1993).

Stokes, Theresa E., *et al.* Good Computer Validation Practices. Buffalo Grove, IL: Interpharm Press, 1994.

Sturm, Robert M. Technology Update: Introduction to a Global Document Manage ment System. *Applied Clinical Trials*, Vol. 3 (February 1994).

Sudman, S. Applied Sampling. New York: Academic Press, 1976.

Sulcas, P. Planning Timing of Computer Auditing. *The South African Chartered Accountant*, July 1975, p. 232.

Svenson, Debra and Michael McLean. Computer-Integrated Electronic Batch Records and Pharmaceutical Documents: A First Step Toward Paperless Factories. *Pharmaceutical Technology*, Vol. 14 (November 1990).

Swanson, E. Burton. Information System Implementation. Irwin, 1988, "Evaluation: Taking the Measure of Performance." pp. 74-83.

Swanson, E. Burton and Cynthia Mathis Beath. Maintaining Information Systems in Organizations. New York: John Wiley & Sons, 1989.

Sweeney, Fergus. Practical Application of ISO 9000 and EC Good Clinical Practice in Validating Clinical Data Management Systems. *Drug Information Journal*, Vol. 27 (April-June 1993).

Taylor, John K. Validation of Analytical Methods. *Analytical Chemistry*, Vol. 55 (May 1983).

Teagarden, C.J. A Stepwise Approach to Software Validation. *Pharmaceutical Technology*, Vol. 13 (September 1989).

Tektronix Inc. Essentials of Data Communications. Beaverton, OR: Tektronix Inc., 1978.

Tetzlaff, Ronald F. GMP Documentation Requirements for Automated Systems. Part I. *Pharmaceutical Technology*, Vol. 16 (March 1992).

Tetzlaff, Ronald F. GMP Documentation Requirements for Automated Systems. Part

Tetzlaff, Ronald F. GMP Documentation Requirements for Automated Systems. Part III: FDA Inspections of Computerized Laboratory Systems. *Pharmaceutical Technology*, Vol. 16 (May 1992).

Tetzlaff, Ronald F. Validation Issues for New Drug Development: Part I, Review of Current FDA Policies. *Pharmaceutical Technology*, Vol. 16 (September 1992).

Tetzlaff, Ronald F. Validation Issues for New Drug Development: Part II, Systematic Assessment Strategies. *Pharmaceutical Technology*, Vol. 16 (October 1992).

Tetzlaff, Ronald F. Validation Issues for New Drug Development: Part III, Systematic Audit Techniques. *Pharmaceutical Technology*, Vol. 17 (January 1993).

Tetzlaff, Ronald F., Richard E. Shepherd, and Armand J. LeBlanc. The Validation Story: Perspectives on the GMP Inspection Approach and Validation Development. *Pharmaceutical Technology*, Vol. 17 (March 1993).

Thiel, Carol Tomme. Managing the Highly Skilled. *Infosystems*, March 1984.

Thierauf, Robert J. Decision Support Systems for Effective Planning and Control. Englewood Cliffs, NJ: Prentice-Hall, Inc., 1982.

Thompson, P. William. Advances in the Automation of Bioprocessing. *Pharmaceutical Technology*, Vol. 14 (November 1990).

Tomlinson, Edward A. Automating a Manufacturing Operation: Selecting the Right Software. *Pharmaceutical Technology*, Vol. 12 (February 1988).

Tsokos, C.P. Probability Distributions: An Introduction to Probability Theory with Applications. Bellmont, CA: Duxbury Press, 1972.

Tucker, Jane and Iain Menzies Smith. Electronic Data Transfer for Clinical Trials. *Drug Information Journal*, Vol. 28 (April-June 1994).

Turn, Klein. Computers in the 1980s. New York: Columbia University Press, 1977.

Tussing, R.T. and G.L. Helms. Training Computer Audit Specialists. *Journal of Accountancy*, July 1980, pp. 71-74.

Ungar, Wendy. Good Clinical Practices: Principles and Application of the European Community Note for Guidance. *Drug Information Journal*, Vol. 28 (January-March 1994).

United Kingdom, GLP Compliance Programme, Department of Health. Good Laboratory Practice, Advisory Leaflet Number 1: The Application of GLP Principles to Computer Systems. London: Department of Health, 1989.

UK Pharmaceutical Industry Computer Systems Validation Forum. Pharmaceutical Industry Supplier Guidance: Validation of Automated Systems in Pharmaceutical Manufacture. First Draft (Consultation Document), February 1994.

U.S., Department of Commerce, National Bureau of Standards. Glossary for Computer Systems Security. FIPS Publication 39.

U.S., Department of Commerce, National Bureau of Standards. Guidelines for Automatic Data Processing Risk Analysis. FIPS Publication 65, 1979.

U.S., Department of Defense. Defense System Software Development. DOD-STD-2167A. February 29, 1988.

U.S., Department of Defense. Defense System Software Development Data Item. Description for the Software Design Document. DI-MCCR-80012A. Washington, D.C.: Government Printing Office, February 29, 1988.

U.S., Department of Defense, National Computer Security Center. Glossary of Computer Security. NCSC-TG-004-88, Version 1, 1988.

U.S., Department of Health and Human Services. Mandatory Guidelines for Federal Workplace Drug Testing Programs; Final Guidelines. Federal Register, Vol. 53, No. 69, April 11, 1988, pp. 11969-11989.

U.S., Department of Health and Human Services. Medicare, Medicaid and CLIA Programs; Final Rule with Comment Period. Federal Register, Vol. 55, No. 50, March 14, 1990, pp. 9537-9610.

U.S., Department of Health and Human Services, Food and Drug Administration. Compliance Program Guidance Manual, Chapter 28, "Human Drugs and Biologics: Bioresearch Monitoring," Document 7348.808/TN-FY 87-2, July 29, 1986.

U.S., Department of Health and Human Services, Food and Drug Administration. Compliance Review Team (for American Red Cross), March 1989.

U.S., Department of Health and Human Services, Food and Drug Administration. Report of the FDA Task Force on International Harmonization. December 1992.

U.S., Department of Transportation. Procedures for Transportation Workplace Drug Testing Programs; Final Rule. Federal Register, Vol. 54, No. 230, December 1, 1989, pp. 49854-49884.

U.S., Environmental Protection Agency, Office of Compliance Monitoring, Pesticide Enforcement Branch. Enforcement Response Policy for the Federal Insecticide, Fungicide, and Rodenticide Good Laboratory Practice (GLP). n.d.

U.S., Environmental Protection Agency, Office of Information Resources, Management. Automated Laboratory Standards: Current Automated Laboratory Data Management Practices (Final, June 1990).

U.S., Environmental Protection Agency, Office of Information Resources, Management. Automated Laboratory Standards: Evaluation of the Standards and Procedures Used in Automated Clinical Laboratories (Draft, May 1990).

U.S., Environmental Protection Agency, Office of Information Resources, Management. Automated Laboratory Standards: Evaluation of the Use of Automated Financial System Procedures (Final, June 1990).

U.S., Environmental Protection Agency, Office of Information Resources, Management. Automated Laboratory Standards: Good Laboratory Practices for EPA Programs (Draft, June 1990).

U.S., Environmental Protection Agency, Office of Information Resources, Management. Automated Laboratory Standards: Survey of Current Automated Technology (Final, June 1990).

U.S., Environmental Protection Agency, Office of Information Resources, Management. EPA LIMS Functional Specifications (March 1988).

U.S., Environmental Protection Agency, Office of Information Resources, Management. EPA System Design and Development Guidance, Vols. A, B, and C (1989).

U.S., Environmental Protection Agency, Office of Information Resources, Management. Good Automated Laboratory Practices: EPA's Recommendations for Ensuring Data Integrity in Automated Laboratory Operations, With Implementation Guidance (Draft, December 28, 1990).

U.S., Environmental Protection Agency, Office of Information Resources, Management. Survey of Laboratory Automated Data Management Practices (1989).

U.S., Environmental Protection Agency, Office of Prevention, Pesticides, and Toxic Substances. Questions and Answers: Enforcement Action Against Bio-Tek Industries and Microbac Labs for Violations of Good Laboratory Practice Standards. For Your Information Bulletin, June 26, 1992.

U.S., General Accounting Office. Bibliography of GAO Documents, ADP, IRM, & Telecommunications, 1986. Washington, D.C.: GAO, 1987.

U.S., General Accounting Office. Evaluating the Acquisition and Operation of Information Systems. Washington, D.C.: GAO, 1986.

U.S., General Accounting Office. Improvements Needed in Managing Automated Decision-making by Computers Throughout the Federal Government. The Comptroller General of the United States, Report FGMSDD-76-5, 1976.

U.S., Office of Management and Budget. Guidance for Preparation and Submission of Security Plans for Federal Computer Systems Containing Sensitive Information. OMB Bulletin No. 88-16, July 6, 1988.

Van Bruwaene, Ben, Luc Dekeyser, and Marleen Stevens. Integrated Study Automation: Computerization at the Site of Investigation. *Applied Clinical Trials*, Vol. 2 (October 1993).

Van Manen, Robbert P. and Erik Houthoff. A Central Database with Distributed Off-Line Data Entry and Correction: The Best of Both Worlds? *Drug Information Journal*, Vol. 27 (October-December 1993).

Van Manen, Robbert P. and Karel J. de Neef. Validating a Clinical Data System: Technical, Organizational, and Managerial Aspects. *Drug Information Journal*, Vol. 27 (April-June 1993).

Vasarhelyi, M.A., C.A. Pabst, and I. Daley. Organizational and Career Aspects of the EDP Audit Function. *EDP Auditor*, 1980, pp. 35-43.

Votrax. SC-01 Speech Synthesizer Data Sheet. Troy, MI: Votrax, 1980.

Walsh, James E. Handbook of Nonparametric Statistics. 3 vols. New York: Holt, Rinehart and Winston, 1953.

Warburton, David. Establishing an Effective Calibration and preventive Maintenance Program. *Medical Device & Diagnostic Industry*, Vol. 15 (January 1993).

Ward, J.H., Jr. and E. Jennings. Introduction to Linear Models. Englewood Cliffs, NJ: Prentice-Hall, 1973.

Ward, Paul and Steve Mellor. Structured Development for Real-Time Systems, Vols. 1-3. Englewood Cliffs, NJ: Yourdon Press/Prentice-Hall, 1986.

Warnier, Jean-Dominique. The Logical Construction of Programs. New York: Van Nostrand Reinhold, 1976.

Webster, Edward. Figuring the Economics of the New Page Printers. *Datamation*, Vol. 24 (May 1978), pp. 171-77.

Wegner, Peter. Programming Languages-the First 25 Years. *IEEE Transactions on Computers*, Vol. C-25 (December 1976), pp. 1207-25.

Weinberg, Gerald. An Introduction to General Systems Theory. New York: John Wiley & Sons, 1975.

Weinberg, Gerald. An Introduction to General Systems Theory. New York: John Wiley & Sons, 1975.

Weinberg, Gerald. Rethinking Systems Analysis and Design. Boston: Little, Brown, 1982.

Weinberg, Gerald. The Psychology of Computer Programming. New York: Van Nostrand Reinhold, 1971.

Weinberg, Sanford B. Case Studies in System Validation: Sixteen Theses. *Drug Information Journal*, Vol. 27 (April-June 1993).

Weinberg, Sanford B. Development of a Central Database for Pharmaceutical and Environmental Results: Technical Considerations. Presented at ACHEMA Conference, Frankfurt, Germany, June 7, 1994; published in *Weinberg Validation Review*, Vol. 3 (No.3, 1994).

Weinberg, Sanford B. FDA Issues in System Validation of the QC Laboratory. Address to the Investigations Branch, Compliance Branch, IBr Steering Committee, U.S. Food and Drug Administration, December 10, 1993; published in *Weinberg Validation Review*, Vol. 3 (No. 1, 1994).

Weinberg, Sanford B. Good Automated Laboratory Practices. *Quality Assurance: Good Practice, Regulation, and Law*, Vol. 2 (March-June 1993).

Weinberg, Sanford B. Impact of GLP and GALP Requirements on Customers and Laboratory Users of Instruments. Presentation at Analytical Instrument Association, Spring Meeting, April 25-27, 1993.

Weinberg, Sanford B. Methodology for the Validation of a Laboratory Information System. Presented at Pittcon '94, Chicago, IL, February 27-March 4, 1994.

Weinberg, Sanford B. 1993 International LIMS Conference. *American Environmental Laboratory*, August 1993.

Weinberg, Sanford B. System Validation: The Regulatory Environment. Presentation at the Neu Ulm Conference on the Evolution to Paperless Clinical Trials, Neu Ulm, Germany, March 10-11, 1993.

Weinberg, Sanford B. System Validation Standards. Dubuque: Kendall/Hunt Publishing Company, 1990.

Weinberg, Sanford B. Toward Harmonization of Raw Data. *American Environmental Laboratory*, April 1993.

Weinberg, Sanford B. Validation of Computerized Biotechnology Systems. Presentation at Drug Information Association Workshop on "Biotechnology: Meeting the Challenges of the 1990s," May 19-21, 1993.

Weinberg, Sanford B. Why the GALPs? *Environmental Lab*, February/March 1993.

Weinberg, Sanford B., Ronald M. Romoff, and Gary C. Stein. Handbook of System Validation. Philadelphia: Weinberg, Spelton & Sax, Inc., 1993.

Weinberg, Sanford B., *et al.* System Validation Checklist. Copyrighted Monograph, 1988.

Weinberg, Sanford B., *et al.* Testing Protocols for the Blood Processing Industry. Copyrighted Monograph, 1989.

Weinberg, Spelton & Sax, Inc. GALP Regulatory Handbook. Boca Raton, FL: Lewis Publishers, 1994.

Weizenbaum, Joseph. Computer Power and Human Reason. San Francisco: W.H. Freeman and Company, 1976.

Wells, Heather J., Terry M. Price, and Caroline A. Evans. Design and Construction of an Automated Validation Procedure for a Clinical Data Management System. *Drug Information Journal*, Vol. 27 (April-June 1993).

Wesley, Roy L. and John A. Wanat. A Guide to Internal Loss Prevention. Stoneham, MA: Butterworth Publishers.

Wickelgren, Wayne A. How to Solve Problems. San Francisco: W.H. Freeman and Company, 1974.

Wiederhold, Gio. Database Design. New York: McGraw-Hill, 1983.

Wilkins, B.J. The Internal Auditor's Information Security Handbook. Altamonte Springs, FL: Institute of Internal Auditors, 1979.

Williams, Michael H. Good Computer Validation Practice is Good Business Practice. *Drug Information Journal*, Vol. 27 (April-June 1993).

Willingham, J. and D.R. Carmichael. Auditing Concepts and Methods. New York: McGraw-Hill, 1975.

Winer, B.J. Statistical Principles in Experimental Design. 2nd ed. New York: McGraw-Hill, 1991.

Wirth, Niklaus. Algorithms + Data = Programs. Englewood Cliffs, NJ: Prentice-Hall, 1976.

Wirth, Niklaus. Systematic Programming. Englewood Cliffs, NJ: Prentice-Hall, 1983.

Wood, Bill J. and Julia W. Ermes. Applying Hazard Analysis to Medical Devices; Part II: Detailed Hazard Analysis. *Medical Device & Diagnostic Industry*, Vol. 15 (March 1993).

Wood, Bill J. and Julia W. Ermes. Applying Hazard Analysis to Medical Devices; Part I: Initial Hazard Analysis. *Medical Device & Diagnostic Industry*, Vol. 15 (January 1993).

Wood, Janie and Denise Silver. Joint Application Design. New York: John Wiley & Sons, 1989.

Wonnacott, T.H. and R.J. Wonnacott. Introductory Statistics for Business and Economics. 2nd edition. New York: John Wiley and Sons, 1977.

Written, Ian H. Communicating with Microcomputers. New York: Academic Press, Inc., 1980.

Yourdon, E. Structured Walkthroughs. New York: Yourdon Press, 1982.

Zaffarano, Joan. Calculators: The Programmables and Prepos. *Administrative Management*, August 1973, pp. 40-48.

Zeisel, H. Say It with Figures. New York: Harper & Row, 1957.

Zimmerli, Wesley D., Gale P. DeHaven and Frank R. Funderburk. Computerized Behavioral Testing: Implications for Quality-of-Life and Outcomes Assessment. *Applied Clinical Trials*, Vol. 2 (July 1993).

Zuboff, Shoshana. In the Age of the Smart Machine. New York: Basic Books, 1988.

INDEX

Printed and bound by CPI Group (UK) Ltd, Croydon, CR0 4YY
17/10/2024
01775713-0001